T3-BGM-215

Concurrent Engineering: Tools and Technologies for Mechanical System Design

NATO ASI Series

Advanced Science Institutes Series

A series presenting the results of activities sponsored by the NATO Science Committee, which aims at the dissemination of advanced scientific and technological knowledge, with a view to strengthening links between scientific communities.

The Series is published by an international board of publishers in conjunction with the NATO Scientific Affairs Division

A Life Sciences B Physics	Plenum Publishing Corporation London and New York
C Mathematical and Physical Sciences D Behavioural and Social Sciences E Applied Sciences	Kluwer Academic Publishers Dordrecht, Boston and London
F Computer and Systems Sciences G Ecological Sciences H Cell Biology I Global Environmental Change	Springer-Verlag Berlin Heidelberg New York London Paris Tokyo Hong Kong Barcelona Budapest

NATO-PCO DATABASE

The electronic index to the NATO ASI Series provides full bibliographical references (with keywords and/or abstracts) to more than 30 000 contributions from international scientists published in all sections of the NATO ASI Series. Access to the NATO-PCO DATABASE compiled by the NATO Publication Coordination Office is possible in two ways:

- via online FILE 128 (NATO-PCO DATABASE) hosted by ESRIN, Via Galileo Galilei, I-00044 Frascati, Italy.
- via CD-ROM "NATO Science & Technology Disk" with user-friendly retrieval software in English, French and German (© WTV GmbH and DATAWARE Technologies Inc. 1992).

The CD-ROM can be ordered through any member of the Board of Publishers or through NATO-PCO, Overijse, Belgium.

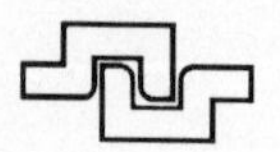

Series F: Computer and Systems Sciences Vol. 108

Concurrent Engineering: Tools and Technologies for Mechanical System Design

Edited by

Edward J. Haug

The University of Iowa, Center for Computer-Aided Design
208 Engineering Research Facility
Iowa City, IA 52242-1000, USA

Springer-Verlag
Berlin Heidelberg New York London Paris Tokyo
Hong Kong Barcelona Budapest
Published in cooperation with NATO Scientific Affairs Division

Proceedings of the NATO Advanced Study Institute on Concurrent Engineering Tools and Technologies for Mechanical System Design, held in Iowa City, Iowa, May 25–June 5, 1992

CR Subject Classification (1991): J.6

ISBN 3-540-56532-9 Springer-Verlag Berlin Heidelberg New York
ISBN 0-387-56532-9 Springer-Verlag New York Berlin Heidelberg

Library of Congress Cataloging-in-Publication Data. Concurrent engineering: tools and technologies for mechanical system design/edited by Edward J. Haug. p. cm. – (NATO ASI series. Series F, Computer and system sciences; vol. 108). Includes bibliographical references and index. ISBN 0-387-56532-9
1. Concurrent engineering–Congresses. 2. Engineering design–Congresses. 3. Mechanical engineering–Congresses. I. Haug, Edward J. II. Series: NATO ASI series. Series F, Computer and system sciences; v. 108. TA174.C587 1993 658.5–dc20 93-1624

Printed in Germany

Typesetting: Camera ready by authors
45/3140 - 5 4 3 2 1 0 - Printed on acid-free paper

Preface

These proceedings contain lectures presented at the NATO Advanced Study Institute on Concurrent Engineering Tools and Technologies for Mechanical System Design held in Iowa City, Iowa, 25 May - 5 June, 1992. Lectures were presented by leaders from Europe and North America in disciplines contributing to the emerging international focus on Concurrent Engineering of mechanical systems. Participants in the Institute were specialists from throughout NATO in disciplines constituting Concurrent Engineering, many of whom presented contributed papers during the Institute and all of whom participated actively in discussions on technical aspects of the subject.

The proceedings are organized into the following five parts:

Part 1	Basic Concepts and Methods
Part 2	Application Sectors
Part 3	Manufacturing
Part 4	Design Sensitivity Analysis and Optimization
Part 5	Virtual Prototyping and Human Factors

Each of the parts is comprised of papers that present state-of-the-art concepts and methods in fields contributing to Concurrent Engineering of mechanical systems. The lead-off papers in each part are based on invited lectures, followed by papers based on contributed presentations made by participants in the Institute.

The basic concepts and methods presented in Part 1 provide an overview of Concurrent Engineering concepts and technical approaches to integrating tools and technologies for multidisciplinary Concurrent Engineering of mechanical systems. While it is not possible to be comprehensive in treatment of the extraordinarily broad field of Concurrent Engineering of mechanical systems, these papers provide a balanced introduction to and development of underlying methods that support the integration of a wide variety of tools and technologies that now constitute the scope of Concurrent Engineering of mechanical systems and will continue to evolve during the decade.

In order to be more concrete regarding implementation and use of tools and technologies for multidisciplinary Concurrent Engineering, specific application sectors are highlighted in Part 2. Even though technical aspects of the various mechanical system sectors addressed are quite different, a central theme of tool integration to support a broad range of discipline-specific applications, all deriving information from a central database and returning results to the central database, is clearly evident. Much as in Part 1, the scope of applications addressed is only a modest sampling of the breadth of Concurrent Engineering applications that are currently under development and will continue to evolve in mechanical system design.

Of special importance in Concurrent Engineering of mechanical systems are manufacturing considerations presented in Part 3. The sampling of manufacturing approaches presented highlights the importance of trade-offs that exist between design of mechanical systems and manufacturing processes to be used in their fabrication. As indicated in the first two papers of Part 3, complex and diverse trade-offs exists between product design and the quality of the manufactured product, as well as approaches for optimizing designs for manufacturability and control of manufacturing processes.

In view of the importance of trade-offs associated with Concurrent Engineering, design sensitivity analysis and optimization methods suitable for this purpose are presented in Part 4. Design sensitivity analysis methods that have emerged during the past decade are summarized and their use in design optimization involving a broad spectrum of disciplines is illustrated using selected applications. While work continues in developing and implementing design sensitivity analysis and optimization methods in specialized disciplines, a trend toward multidisciplinary trade-off analysis and design optimization is apparent.

The emerging field of virtual prototyping and human factors associated with simulation-based design of mechanical systems is addressed in Part 5. High-speed dynamic simulation methods developed in the late 1980s are shown to provide the foundation for revolutionary new tools for real-time simulation of mechanical systems, at a design level of detail and fidelity. This new capability will permit operator-in-the-loop simulation for tuning the design of mechanical systems to the capability of the intended population of human operators. Realization of this new capability is shown to be dependent upon fundamental human factors analysis methods that involve both engineering and psychology specialists. This emerging field represents the essence of Concurrent Engineering, bringing both mechanical and human performance into an integrated environment where trade-off analysis and design optimization of operator-machine systems is possible.

The extent and variety of the lectures and contributed papers presented in these proceedings illustrate the contribution of numerous individuals in preparation and conduct of the Institute. The Institute Director wishes to thank all contributors to these proceedings and participants in the Institute, who refused to be passive listeners and participated actively in discussions and contributed presentations. Special thanks go to M. Bender, L. Handsaker, R. Huff, M. Laverman, and D. Dawes for their efforts in planning and support for the Institute. Finally, without the financial support* of the NATO Office of Scientific Affairs, the US Army Tank-Automotive Command, and the NASA Goddard Space Flight Center, the Institute and these proceedings would not have been possible. Their support is gratefully acknowledged by all concerned with the Institute.

January 1993 E.J. Haug

* The views, opinions, and/or findings contained in these proceedings are those of the authors and should not be construed as an official position, policy, or decision of the sponsors, unless so designated by other documentation.

Contents

NATO Advanced Study Institute

Concurrent Engineering Tools And Technologies For Mechanical System Design

Iowa City, Iowa USA
25 May - 5 June, 1992

Sponsors

NATO : North Atlantic Treaty Organization
TACOM : The US Army Tank-Automotive Command
NASA : The US National Aeronautics and Space Administration

Director : E. J. Haug, The University of Iowa, USA

Organizing Committee

N. Olhoff, Aalborg University, Denmark
W. Schiehlen, University of Stuttgart, Germany
C. Soares, Technical University of Lisbon, Portugal

Lecturers

J. Ashton, FMC Naval Systems Division, USA
N. Badler, University of Pennsylvania, USA
R. Beck, US Army Tank-Automotive Command, USA
K. Choi, The University of Iowa, USA
D. Clausing, Massachusetts Institute of Technology, USA
J. Cleetus, West Virginia University, USA
H. Eschenauer, University of Siegen, Germany
H. Frisch, NASA Goddard Space Flight Center, USA
P. Hancock, University of Minnesota, USA
E. Haug, The University of Iowa, USA
E. Mettala, Defense Advanced Research Projects Agency, USA
N. Olhoff, Aalborg University, Denmark
P. Pedersen, The Technical University of Denmark, Denmark
W. Schiehlen, University of Stuttgart, Germany
R. Vos, Boeing Aerospace and Electronics Company, USA
S. Wu, University of Michigan, USA

Participants

M. Acar, Loughborough University of Technology, United Kingdom
M. Akkurt, Instanbul Technical University, Turkey
C. Alessandri, University of Florence, Italy
J. Ambrosio, Technical University of Lisbon, Portugal
A. Andre, NASA Ames Research Center, USA
J. Baatrup, The Technical University of Denmark, Denmark
J. Bals, DLR Institute for Flight Systems Dynamics, Germany
N. Bau-Madsen, Aalborg University, Denmark
D. Bestle, University of Stuttgart, Germany
H. Bordett, The University of Texas at Arlington, USA
M. Bossak, Warsaw University of Technology, Poland
M. Botz, Technische Hochschule Darmstadt, Germany
J. Caird, University of Minnesota, USA
K. Ciarelli, US Army Tank-Automotive Command, USA
J. Cyklis, Cracow University of Technology, Poland
P. De Castro, Universidade do Porto, Portugal
P. Dehombreux, Faculte Polytechnique de Mons, Belgium
R. DeVries, Ford Motor Company, USA
J. Downie, Brighton Polytechnic, United Kingdom
N. Ertugrul, Dokuz Eylül University, Turkey
J. Flach, Wright State University, USA
K. Grabowiecki, Industrial Institute of Construction Machinery, Poland
R. Grandhi, Wright State University, USA
P. Green, University of Michigan, USA
J. Hansen, The Technical University of Denmark, Denmark
C. Inan, Dokuz Eylül University, Turkey
S. Jendo, Polish Academy of Sciences, Poland
O. Kaynak, Bogazici University, Turkey
A. Keil, Institute of Mechatronics, Germany
P. Kiriazov, Bulgarian Academy of Sciences, Bulgaria
L. Krog, Aalborg University, Denmark
E. Kurpinar, Ege Universitesi, Turkey
T. Lekszycki, Polish Academy of Sciences, Poland
P. Level, University of Valenciennes, France
S. Ligaro, University of Pisa, Italy
E. Lund, Aalborg University, Denmark
N. Maia, Technical University of Lisbon, Portugal
L. Markov, Bulgarian Academy of Sciences, Bulgaria
J. McPhee, University of Waterloo, Canada
M. Otter, DLR Institute for Flight Systems Dynamics, Germany
T. Ozel, Dokuz Eylül University, Turkey

M. Pereira, Technical University of Lisbon, Portugal
G. Pratten, ICL, United Kingdom
R. Riesenfeld, University of Utah, USA
P. Rosko, Slovak Technical University, Czechoslovakia
J. Santos, Technical University of Lisbon, Portugal
O. Sigmund, Essen University, Germany
A. Stensson, Lulea University of Technology, Sweden
E. Stephens, Georgia Institute of Technology, USA
S. Strzelecki, Institute of Machine Design 1-6, Poland
K. Svendsen, Technical University of Denmark, Denmark
M. Tekelioglu, Dokuz Eylül University, Turkey
O. Thomsen, Aalborg University, Denmark
D. Tortorelli, University of Illinois at Urbana-Champaign, USA
S. Twu, Cummins Technical Center, USA
F. Uldum, Technical University of Denmark, Denmark
G. Ulusoy, Bogazici University, Turkey
C. Vibet, University of Paris, France
J. Wargo, Defense Advanced Research Projects Agency, USA
C. Wilmers, Technical University of Hamburg-Harburg, Germany
R. Zobel, University of Manchester, United Kingdom

Part 1

Basic Concepts and Methods

World-Class Concurrent Engineering

Don P. Clausing

Bernard M. Gordon Adjunct Professor of Engineering Innovation & Practice
Massachusetts Institute of Technology, Cambridge, MA 02139 USA

Abstract: World-class concurrent engineering is the modern way to develop new products. It features basic concurrent engineering (improved process and teamwork), enhanced quality function deployment (enhanced QFD), and quality engineering using robust design (Taguchi methods). This replaces the many dysfunctions, some of which are elegant, of traditional practice. The result is much better functionality, lower costs, and shorter development time.

Keywords: concurrent engineering / product development / teams / QFD / robust design / Taguchi / optimization / quality

1 Better Products

World-class concurrent engineering is the improved *total development* process for the development of new products that are competitive in the global economy. It combines the best engineering methods, the best management approaches, and the best teamwork to greatly reduce development time and all costs, improve quality, and increase product variety, all of which greatly improve customer satisfaction. But first, let us begin at the beginning.

1.1 Basic Engineering—The Foundation

The essential core of all development work is concept creation. We create concepts by combining our knowledge of engineering science, physics and materials, with technological insights to achieve a new idea, Figure 1, a creative rearrangement of physics and materials, occasionally brilliant. We cannot rearrange the physics and chemistry and materials unless we understand them. A sound grounding in the engineering sciences is required. Likewise, we cannot invent a new mechanism if we have never seen a mechanism; technological insights provide the springboard to new technology. These are the fundamentals that engineers learn in engineering school and during the first few years of industrial practice; they enable concepts to be created. However, concept creation is not enough for success.

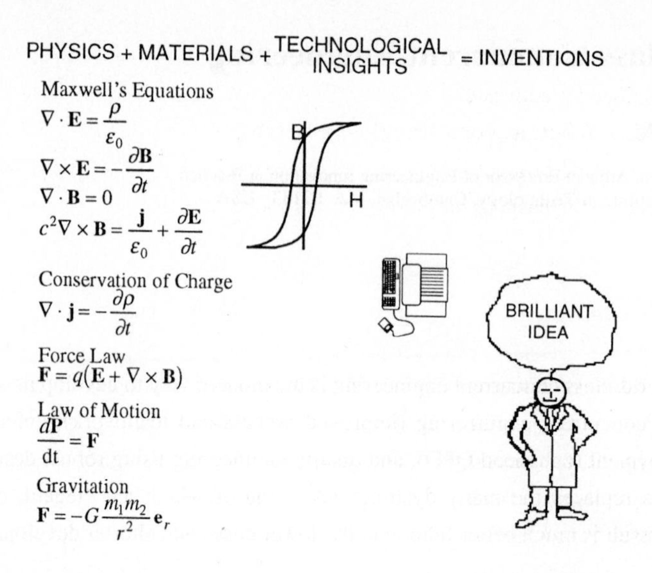

Figure 1. Concept Creation Process

Will the concept work? Not necessarily; many concepts are not feasible. When design is in conflict with physics, then physics always wins. This is the overwhelming case for the engineering sciences as the required foundation for all product development. If the inertial forces, Poynting vectors, and entropy changes are not carefully planned in concert with nature, then the product is doomed to failure. Perpetual motion machines are the classical examples.

The undergraduate engineering curriculum typically includes one or two design courses. These concentrate on creative concepts and feasibility, the assurance of a first-order compatibility with the laws of nature. This limited introduction is partial design.

The basic engineering curriculum is the foundation for success in product development. Newtonian mechanics, electromagnetic theory (Maxwell's equations), thermodynamics, and the mathematics to succinctly summarize and apply this collective experience must be at the core of all device developments. The specialization of these to mechanisms, circuits, and energy converters, and an introduction to partial design, enables simple devices to be developed so that they function well.

Getting the engineering fundamentals right is the foundation for a good product. However, two products can both have the mechanisms and circuits very competently done, yet differ greatly in their success. They can differ in important ways in their responsiveness to customer needs, the viability of the core concepts, the producibility of the design, the robustness of its functional quality, the economical precision of its production, the success of integration, the effective use of standardization, and strategic impact.

It is the intent of this article to go beyond the engineering fundamentals to communicate the structured, disciplined practices that build on the fundamentals to make the critical difference between success and failure in product development. These practices are essentially independent of the specific type of product, applying to mechanisms, circuits, energy converters, and structures. These are the basic principles for product success.

1.2 Beyond The Basics

Beyond the basics, engineering science and partial design, success with products is achieved by *total development.* The improved total development process builds on the engineering sciences and partial design with the successful approach that addresses customer needs, concept selection, functional robustness, integration into the total system, beneficial standardization, producibility and maintainability, and strategic coherence for corporate success.

Success with products requires the improved total development process for bringing new products into being, starting with customer needs and ending with the product in production. Total development is the link between strategy and partial design. This relationship is displayed in Figure 2.

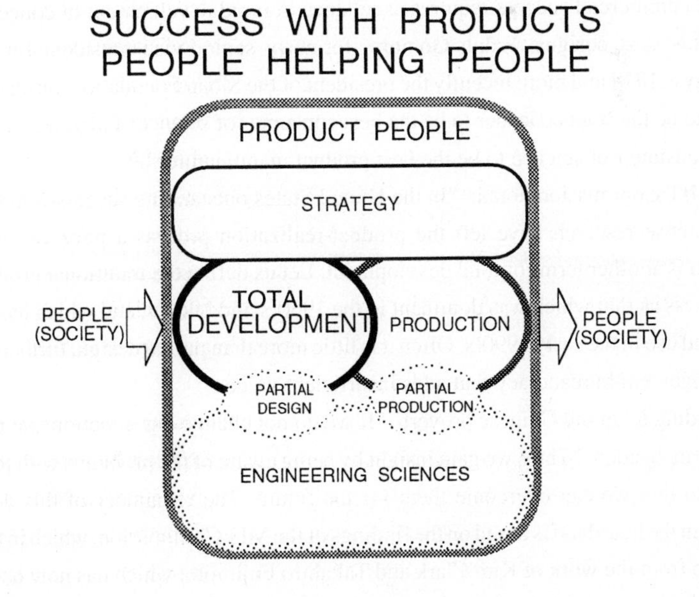

Figure 2. Roles of Product People

Total development and production are the linked activities in which product people help all people (society). The needs of society are received within the total development activity (left side of Figure 2), and returned to society as new products (right side of Figure 2). The improved total development process incorporates the application of the engineering sciences and partial design, and successfully address customer needs, concept selection, robust functionality, integration, standardization, producibility, and strategy. First, let us examine the frequent problems.

1.3 Problems In The Heartland

The young engineer with a few years of design experience in industry has mastered partial design. Unfortunately, many product development activities have moved only a short distance beyond partial design. The product development process has not been the subject of much study; there is much opportunity to improve it.

The MIT Commission on Industrial Productivity, in its report Made in America[1], found that six weaknesses hamper American manufacturing industries. The two weaknesses that are most relevant to product development are (1) technological weaknesses in development and production, and (2) failures of cooperation. Technological weakness seems paradoxical, as the United States is widely believed to be the technological leader of the world. However, the strength of the United States is in research and advanced development, creating new technological concepts. Sustained commercial success requires excellence in rapid development of concepts into high quality, low cost designs. Ralph Gomory, formerly senior vice president for science and technology at IBM and more recently the president of the Sloan Foundation, observed: "You do not have to be the science leader to be the best consumer of science; and you do not have to be the best consumer of science to be the best product manufacturer."[2]

The MIT Commission noted: "In the United States outstanding successes in basic science and in defense research have left the product-realization process a poor cousin." Product realization is another term for total development. Let us define the traditional product development process as that which was dominant in the 1950's and 1960's, and which has remained in widespread use in the early 1990's. Often it is little more than partial design, further encumbered by a management bureaucracy that adds insufficient value.

According to an old Chinese proverb, "If we do not change our direction, we might end up where we are headed." Thus, we gain insight by being aware of the problems with the traditional process, so that we can overcome them for the future. The remainder of this description of problems in the heartland is based on the findings of the MIT Commission, which in turn received much help from the work of Kim Clark and Takahiro Fujimoto, which has now been published as a book.[3]

Many problems with the traditional processes are cited: difficulty in design for simplicity and reliability, failure to pay enough attention at the design stage to the likely quality of the manufactured product, excessive development times, weak design for producibility, inadequate attention to customers, weak links with suppliers, and neglect of continuous improvement. The process and management suffer from a lightweight program manager, fuzzy objectives and roles at the outset, and serial development. Lightweight program managers have insufficient authority relative to the functional "silos" in the organization, with the result that there is inadequate concentration on the specific products, and decisions are slow in being made. Inadequate consensus at the outset leads to considerable divergence and chaos as the work progresses. Serial development has been caricaturized as the "throw it over the wall" style. A group does work and makes decisions, and throws them over the wall to the next group. This creates great problems in wasted time, weak understanding, and inadequate commitment to earlier decisions.

The MIT Commission observed:

> "The key to major additional reductions in product-development time will be to make further progress in what manufacturing specialists term "design for quality" techniques. The challenge here is for product-development teams to arrive at a product design that has been systematically optimized to meet customers' needs as early as possible in the development project. The design must be robust enough to ensure that the product will provide customer satisfaction even when subject to the real conditions of the factory and customer use. The more problems prevented early on through careful design, the fewer problems that have to be corrected later through a time-consuming and often confusing process of prototype iterations."

Inadequate attention to customers' needs and weak development of robust quality are major shortcomings of the traditional process.

The traditional approach emphasizes specialized excellence, all too often at the expense of overall success. Cloistered groups of technical specialists looking inward within their own specialty can be overwhelmingly competent at a specialized conference, but the result in a product development program usually lacks integration and often focuses on suboptimal objectives. Brilliant results are of little value if within the receiving group they are little noted nor long remembered. Also, the brilliant results are often answers to the wrong question. High-bay, automated inventory storage and retrieval systems were a brilliant solution to the objective of the materials management specialty, the improved storage and retrieval of inventory. A much better approach was to redefine the objective: minimization of inventory; which led to JIT production. Although this example is from production operations, the same problem is prevalent in product development.

The problems can be summarized as two problems: (1) failure of process, and (2) failures of cooperation. During the same time that the MIT Commission was doing its work, the CEO of a Japanese company said, "In the United States you have all-star teams, but you keep losing all of the games." Why do all-star teams lose games? If a championship team plays an all-star team, for example, if the Super Bowl winner were to play the NFL all stars, the championship team will normally win because they have a better *game plan* and *better team work*. The findings of the MIT Commission and the sports metaphor give the same message; the traditional approach suffers from a weak development process and weak teamwork.

1.4 Improved Total Development Process—An Overview

The improved total development process (ITDP) has three major elements: basic improvements in clarity and unity, enhanced quality function deployment (EQFD), and quality engineering using robust design. There is little common usage about nomenclature. The term concurrent engineering is widely used, although some prefer the synonym simultaneous engineering, or other names such as integrated product and process development. As Shakespeare said, "A rose by any other name smells just as sweet." In this article the names improved total development process (ITDP) and world-class concurrent engineering will be used as synonyms. Then the basic improvements in clarity and unity can be known as basic concurrent engineering, which has been implemented by many American companies during the 1980's, and has become the subject of many conferences and seminars. Most of the companies who have implemented basic concurrent engineering are not nearly as far along in their practice of enhanced quality function deployment (EQFD) and quality engineering using robust design. Therefore, their results have been both very promising and somewhat disappointing. The entire improved total development process is needed for success in the global economy.

Basic concurrent engineering consists of two elements: (1) improved process (better game plan), which provides greater clarity to the activities, and (2) closer cooperation (better teamwork), which creates greater unity within the team that does the work. Thus, basic concurrent engineering addresses the problems that were found by the MIT Commission on Industrial Productivity, and which were supported by the sports metaphor; all-star teams losing games. Basic concurrent engineering is essentially basic improvements in clarity and unity. The improved process (better game plan) that provides improved clarity has four primary features: (1) concurrent process, (2) focus on quality, cost and delivery (QCD), (3) emphasis on customer satisfaction, and (4) emphasis on competitive benchmarking.

In the concurrent process, key activities are done together at the same time—a big improvement over the traditional sequential, throw-it-over-the-wall style. As an example, process engineering to prepare for production is done concurrently (simultaneously) with the design of the product.

The closer cooperation (better teamwork) that improves unity consists of: (1) integrated organization (multifunctional teams), (2) employee involvement (participative management), and (3) strategic relations with suppliers.

Beyond basic concurrent engineering we need a strengthened approach to the satisfaction of customer needs and to the consistency of the product performance. Enhanced quality function deployment (EQFD) and quality engineering using robust design provide dramatic enhancement within the improved total development process by providing these capabilities. Together they provide strong responsiveness to the voice of the customer, the viability of the core concepts, the robustness of functional quality (consistency of product performance), the economical precision of production, the success of integration, and the effective use of standardization.

2 Basic Concurrent Engineering

The heart of basic concurrent engineering is the concurrent process that is carried out by a multifunctional product development team (PDT). Product design, production-process engineering, field-support development, and all other elements of product success are addressed from the beginning as an integrated set of activities and objectives. The ideal is simple: one team working on one system in one total development activity, all focused on the benefit of the customer. The system is the product, the production capability, and the field-support capability. The design parameters, production parameters, and field-support parameters all integrated together define the unified system. It is the responsibility of the PDT to define and quantify all of the parameters in one total development activity.

The major benefits of the basic concurrent engineering stem from a few principles: (1) start all tasks as early as possible, (2) utilize all relevant information as early as possible, (3) everyone participates in defining the objectives of their work, (4) operational understanding is achieved for all relevant information, (5) a strong commitment is made to adhere to decisions and utilize all previous relevant work, (6) decisions are made in a single trade-off space, (7) decisions are robust, overcoming a natural tendency to resort to quick, novel decisions, (8) trust among teammates, (9) the team strives for consensus, and (10) the team uses a visible concurrent process. All of these principles seem to be unexceptionable. However, the experienced reader will recognize that in the traditional process they have been frequently violated.

The essence of the concurrent process and the multifunctional PDT is both simple and subtle. The mind cannot literally work on two tasks concurrently (at least not consciously). In the concurrent process, frequent information exchanges occur at the level of the small tasks. In the traditional process, the work blocks were huge before information transfer occurred. At the micro level the tasks in the concurrent process are sequential or iterative, but at the macro level the effect is the concurrent process. The difference between two people being members of the PDT and the alternative in which they are assigned to the same project, but remain in separate organizations, may seem small, but it is critical. If they are not members of the same PDT, the probability is greatly increased that some of the ten principles above will be violated, with a very detrimental effect on the development program. When an individual's primary allegiance is to a cloistered group of functional specialists, the specialty will be performed elegantly, but usually with inadequate benefit to the overall development program. When an individual is a member of the PDT, he or she is much more likely to participate in defining the objectives of their work, and thus the objectives are likely to be both more relevant to the program and better understood by the participants. Often the definition of the objectives is the difficult part of any task. Membership in the PDT greatly improves this. Membership in the PDT also greatly improves the exchange of information. Otherwise the communication is apt to be couched in terms that are dear to the functional specialty, but not fully understood by other people on the program. Membership in the

PDT greatly increases the probability that the individuals and small groups will help others to effectively utilize the output from their work, going beyond a perfunctory communication. This greatly improves understanding and commitment to the decisions that are made, which are vital success factors.

2.1 Concurrent Process

In the best form of concurrent engineering, the design of the production system and the field-support system is started early, concurrently with the design of the product, Figure 3. This has five major benefits: (1) the development of the production and field-support systems has an early start, (2) trade-offs are made among design, production, and logistics concurrently, as one system, (3) good design for manufacturability and field supportability are facilitated, (4) the production and field-support people gain a clear understanding of the design, and are committed to its success, and (5) prototype iterations are reduced, because the design is more mature before the first full-system prototypes are built.

An additional facet of concurrent engineering occurs even earlier in the product development cycle, during completion of the system concept. In the past it was common for market research to determine customer/user needs, throw them over the wall to planning, who outlined the requirements for the product and then threw them over the wall to product design engineering. This sequential development of the requirements and the system concept made it unlikely that the needs of the customer/user were adequately considered in choosing the system. These activities are now in best practice combined together and carried out by one multifunctional team. This is another application of the basic principles of concurrent engineering.

The best concurrent process, envisioned in Figure 3, treats development as one activity that incorporates product, production system, and field-support system. There are no upstream and downstream activities in the traditional sense. Of course, in the natural flow of the work some things are done before others. Concepts are selected before detailed design, and production tools are designed before they are built, for example. However, the best concurrent process avoids the unnatural separation of work into "upstream" and "downstream" in accordance with organizational rigidities.

In the traditional process, tasks were clearly labeled product-development, production-capability-development, or field-support-development, and the first type (upstream) were completed before the other two (downstream) were started. In the best concurrent process the tasks are not defined in this divisive style. All tasks now incorporate the product view, the production view, and the field support view. Sub tasks may still remain "pure." For example, the finite element analysis (FEA) is still usually done by a specialist. However, the utilization of the results is now much more beneficial. The FEA specialist works closely with a subsystem design team, for the frame for example. The FEA is combined with producibility and used to minimize the deflection and cost of the frame. The FEA specialist works closely with the subsystem team to define the objectives of the FEA and then presents the results to the team in a format that can

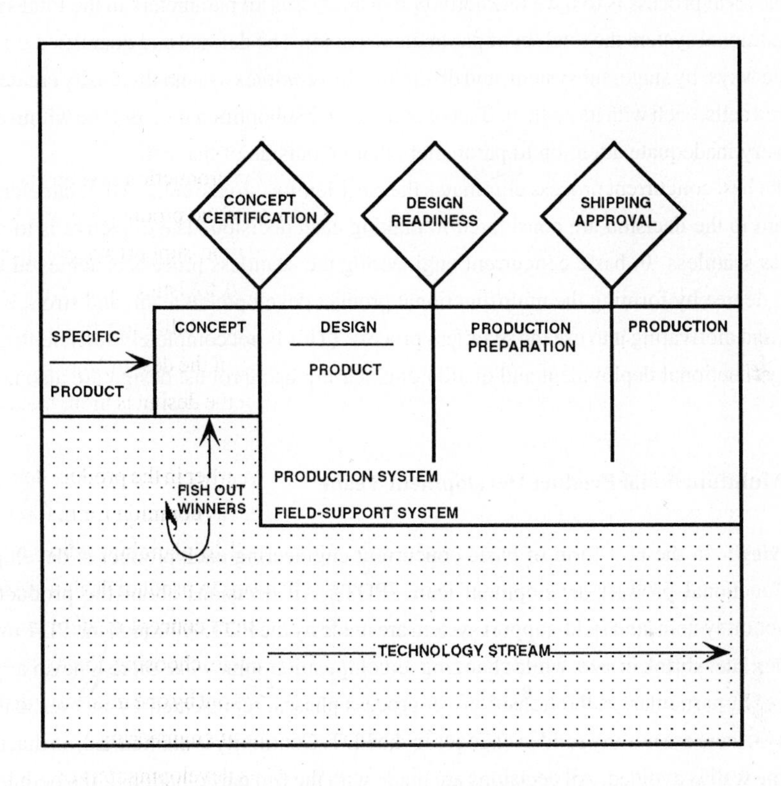

Figure 3. Concurrent Process

be easily understood and utilized by the team. The FEA specialist works with the team to help in the application of the results, probably as electronic data, in optimizing the frame design to achieve minimum cost and deflection—producibility and functionality optimized together. For the duration of this task the FEA specialist is effectively a member of the team. Contrast this with the dysfunctions of the traditional process. The FEA specialist received drawings of the preliminary frame design, did the FEA, and tossed the printout "over the wall" to the frame design engineer—no team, so producibility was not considered. Commonly the design engineer put the FEA results on the shelf to collect dust, occasionally pulling them out to be used as a talisman to ward off "evil" managers and other status seekers (people whose main job was to roam about seeking the status of all activities). Even when the design engineer wanted to use the FEA, it was difficult because of an inconvenient format. If this barrier were overcome, the FEA was typically used to assure that some deflection specification was not exceeded—not optimization, and producibility (downstream activity) was not considered. Multiply this vignette by a thousand, and the superiority of the concurrent process is apparent.

The ideal process is to have one activity that addresses all parameters in the total system. In the traditional system the total set of parameters that must be defined and quantified is separated in three ways: by stage, subsystem, and discipline. In a complex system this easily creates several hundred cells, each with its dedicated set of parameters. Suboptimization is done within each cell, with very inadequate attention to parameters that lie outside of that cell.

The best concurrent process eliminates the partitioning into the cells. All parameters that are relevant to the decision are considered in making each decision. The objective is to make the process seamless. In basic concurrent engineering the seamless process is achieved to a very useful degree by forming the multifunctional product development team, and strongly encouraging and motivating it to use the seamless process. (This is not completely sufficient; enhanced quality functional deployment and quality engineering using robust design are also needed).

2.2 Multifunctional Product Development Team

Overview: In the best form of basic concurrent engineering each product is developed by a multifunctional product development team (PDT). All decisions about the product design, production system, and field-support system are made by the PDT. Although the PDT must grow and then later shrink in size, while changing its composition somewhat, there is never any sudden change. In particular, at the transitions in process phases, represented by the vertical lines in Figure 3, there are not any sudden changes in the PDT. Continuity is maintained, so that throwing over the wall is avoided. All decisions are made with the full participation of the people with all of the relevant functional knowledge.

Basic concurrent engineering is best carried out by a multifunctional product development team (PDT) led by a strong product manager. All functions of the corporations should participate. People who are doing significant work for the specific product development program should be part of the PDT while they are doing this work. There is a vast psychological difference between doing a task within a support group and doing it as a member of the PDT. As a PDT team player the contributor will (1) understand the specific requirements, (2) have the necessary close communications with other members of the PDT, and (3) be dedicated to the utilization of the task results to make design decisions. All three of these have a much lower probability of happening if the contributor remains outside of the PDT.

Specialization: The effectiveness of the PDT is strongly influenced by the generalist-specialist spectrum of capability and style. Prior to 1940 most of product development was done by generalists, Figure 4, and the problems of segmentalism were not usually severe. This approach

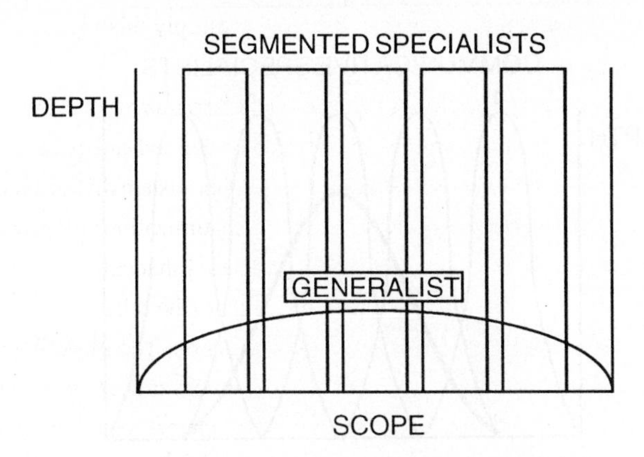

Figure 4. Dysfunctional Specialization

sufficed for products that were not high in technical sophistication. However, during the period of 1940 to 1960 the shortcomings of this approach became obvious, and the emphasis was shifted to technical depth and sophistication. However, this led to segmentalism, Figure 4, cloistered groups of technical specialists looking inward within their own specialty. This caused tremendous problems that concurrent engineering is now overcoming.

The successful PDT utilizes a balanced modulation of specialization, Figure 5. Even the most specialized people broaden themselves sufficiently so that they can communicate effectively with the internal customers for their work. The core of the PDT consists of people who are not narrow specialists, but combine a good combination of breadth and depth, curve 2 in Figure 5. Thus, the traditional product design engineers become quite knowledgeable about production, and traditional production-process engineers become knowledgeable about customer needs and product function. This enables them to function effectively as a team; to work on the complete set of parameters as one system to be developed in one activity.

The FEA specialist who was mentioned earlier is a good example of the new model in specialization. In the days of dysfunctional specialization, the FEA specialist was an example of the segmented specialist in Figure 4. The FEA specialist did not understand design and production, and the design and production people did not understand FEA. Therefore, they often did not reach a common definition of objectives, and the FEA results were thrown over the wall in a format that the design and production people did not interact with effectively. Now in the new model PDT they have all broadened sufficiently to reach common objectives and utilize the FEA to quickly improve cost and quality early in the development process.

The example of FEA can be taken another step. Should sophisticated design tasks be done by specialists, or should they be moved into the work domain of the core PDT people? Should

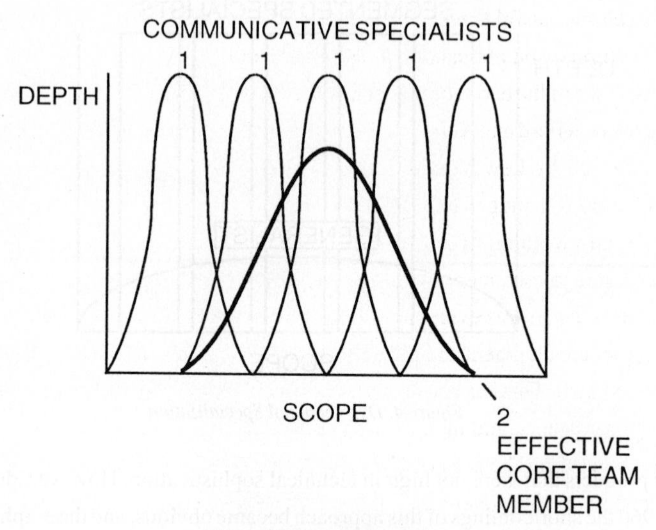

Figure 5. Successful Breadth

the design engineer do the FEA or go to a specialist? If the design engineer can do the FEA, that is preferred because it avoids human-interaction process loss. As computers become more user friendly, the design engineer can incorporate more and more specialized tasks into his or her portfolio of capabilities. Specialized knowledge is utilized both by bringing specialists into the PDT and by making the knowledge available to the core PDT people via user-friendly computers. The best balance between the two is constantly evolving by a process of continuous improvement. The same principle of broadened perspective to enable effective cooperative work applies here also. The specialists and the core PDT people must cooperate to produce computer systems that are effective in the PDT environment.

The Team Is A Start: The formation of the multifunctional PDT is a good start, but teams can go wrong with disastrous results. Hosking and Morley[4], based strongly on the work of Janis, have analyzed the nature of the problem (1) "stress generates strong need for affiliation within the group." "People who have misgivings keep silent and increasingly give the benefit of the doubt to the emerging group consensus." (2) The team members seek to "avoid the stress of actively open minded thinking." They tend to focus on the popular option, and use "non-vigilant information processing" to downplay the risks that later become all too obvious.

Overcoming the possible dysfunctions of teams is straightforward. Teams use vigilant information processing, which is the improved total development process. Also, teams can help themselves by simply being on guard against the problem; teams are not a mystical panacea. The improved total development process (ITDP) helps the team to be vigilant in processing information. The successful team runs down a clear path between facile consensus on the one hand and egocentric disputatiousness on the other hand.

Applications: Two of the early, outstanding successes with basic concurrent engineering in the United States were at Ford and Xerox. At Ford the development of the new Ford Taurus was done by Team Taurus led by Lew Veraldi during 1980 to 1985. This was a heavyweight product manager mode, and was judged to be far more successful than the previous lightweight product manager mode. Lew Veraldi: "Teamwork was a major factor in the success of Taurus and Sable. Early and dedicated involvement by all members of the team was key."[5]

At Xerox the change was even more radical. Implemented by Frank Pipp when he became manager of copier development and production, the change in 1982 was to the independent PDT (no functional homes). This has been highly successful in completing the development of the 10 series (Marathon) copiers, and in developing the more recent 50 series.

No matter what form the organization takes, there will still be boundaries that must not be allowed to create a throw-it-over-the-wall style. The modes that focus on the product have been the most successful. The product-oriented PDT using the concurrent process has been found to be much more successful than functional groups using a sequential process. The switch to the multifunctional PDT using the concurrent process can be made in less than a year with strong leaderships. Examples are Lew Veraldi and Team Taurus and Frank Pipp and the Xerox PDT's, both in the early 1980's.

2.3 Other Enablers

In addition to the multifunctional PDT practicing the concurrent process, there are three other improvements for a better game plan, and two for better team work. The process improvements that reinforce the concurrent process are: (1) focus on quality, cost, and delivery (QCD), (2) emphasis on customers satisfaction, and (3) emphasis on competitive benchmarking. The two improvements that reinforce the multifunctional PDT are: (1) employee involvement, also known as participative management, and (2) strategic relationships with suppliers.

One aspect of the better game plan is the focus of all activities on the quality, cost, and delivery (development schedule) of the new product. This overcomes many fragmented bits of game plans with other local objectives, which often have adverse effects on quality, cost, and delivery (QCD). In the past much work that appeared very elegant by some functional criteria eventually was found to add little to the QCD of the product, and in many cases was actually dysfunctional. The focus on QCD is part of the general approach of utilizing all relevant information to make decisions that satisfy all of the relevant objectives.

Another aspect of the better game plan is the emphasis on customer satisfaction. Internal corporate metrics are de-emphasized, and replaced by responses from customers. The emphasis on customer satisfaction extends throughout all of product development, and all other corporate activities. All objectives are put to the test of the effect upon the customer. Much effort is devoted to learning and understanding the opinions of customers. This extends from the customers' needs at the start of a new development to the reactions of the users of the finished product.

The fourth aspect of the better game plan is the emphasis on competitive benchmarking (CBM). Not only are the products benchmarked against the best of the competition, but also all processes are subject to being benchmarked. Concurrent engineering itself is to a considerable extent the result of competitive benchmarking, which is applied to many detailed sub-processes within concurrent engineering, and is important for continuous improvement. The MIT Commission found that parochialism was a major weakness of American companies. It is very unlikely that a large percentage of all improvements around the globe will occur within one organization. Therefore, an important element of success is vigilance in finding, understanding, bringing in, and implementing major improvements.

The second aspect of closer cooperation, reinforcing the multifunctional PDT, is employee involvement and participative management. The full talents of all people are utilized, and responsibility is decentralized to local areas of expertise and actionability.

The third aspect of closer cooperation is strategic relationships with suppliers. In addition to the corporation bringing its own production-capability and field-support-capability people upstream to work concurrently with the product design engineers, it is also an essential element of concurrent engineering to bring in suppliers to play a major role in the design of the new product.

2.4 Summary

Basic concurrent engineering is a significant improvement in game plan and teamwork relative to the traditional functional, serial process. The better game plan is the concurrent process, reinforced by emphasis on quality, cost, and delivery, customer satisfaction, and competitive benchmarking. Better teamwork is achieved by the multifunctional product development team (PDT), reinforced by participative management and strategic relationships with suppliers.

Basic concurrent engineering makes large improvements; Ford and Xerox made major comebacks in the early 1980's by switching to it. However, even more vigilant information processing is needed for global competitiveness. Enhanced quality function deployment and quality engineering using robust design combine to elevate basic concurrent engineering to world-class concurrent engineering. These are introduced in the next section.

3 Better Decisions

Global competitiveness requires vigilant information processing that goes beyond basic concurrent engineering. The development program starts with broad goals which are then focused by customer needs, and a complex product requires millions of decisions to carry it into production and into the marketplace. Inevitably, most of these decisions can be and must be made by individuals, based on experience. Utilizing experience includes the use of analyses, the most concise records of experience. It also includes the use of handbooks, computerized records, and other repositories of experience.

In developing a complex product, there may be 10 million decisions to be made. Most of them are within the realm of individual experience. However, the roughly ten thousand most critical decisions require more attention, and most of them do not lie entirely within the experience of any individual. However, collective experience properly concentrated is sufficient. The right multifunctional team using a disciplined approach can make good decisions. The primary approach for these decisions is Quality Function Deployment (QFD)[6], now extended into Enhanced QFD (EQFD).[7]

Still fewer decisions, perhaps roughly one thousand, are very critical and cannot be made successfully on the basis of even the collective team experience. These decisions must be arrived at by systematic optimization. The process that has been found most broadly useful in total development work is Dr. Taguchi's System of Quality Engineering using Robust Design.[8,9] This also uses teams, and EQFD can be effectively used to lead into Dr. Taguchi's optimization system.

In addition to the benefits of collective team experience and systematic optimization, the improved total development process is also superior to the traditional process in having a strong focus on the customer, in emphasizing problem prevention, and in using visual, connective methods to further strengthen team performance. In the traditional process there has been insufficient attention to the customer. Instead the work has often focused on satisfying the professional standards of cloistered groups of technical specialists, to the detriment of customer satisfaction. Basic concurrent engineering greatly improves this by emphasizing customer satisfaction, and QFD further improves the effectiveness of meeting customer needs.

In basic concurrent engineering decision making by teams is a key feature, and simple team problem solving methods are used. In EQFD this is taken much farther by using visual, connective methods. These methods use large visual displays, usually on a wall, to help the team to concentrate on the decision and the relevant information. Usually these displays are on paper, and vary from a few feet in both dimensions to as large as five feet by fifteen feet. Many of them also provide connectivity; the output from one activity becomes the input to the next.

The improved total development process (world-class concurrent engineering) focuses on the customer with a problem-prevention style that uses team experience to make decisions with the help of visual, connective methods, which also lead into systematic optimization for the most difficult and critical decisions. This is summarized in table below.

	TRADITIONAL	BASIC CONCURRENT ENGINEERING	E.Q.F.D.	TAGUCHI SYSTEM
BENEFIT	Specialist	Customer	Customer	Customer
STYLE	Reaction	Prevention	Prevention	Prevention
WHO	Individual	Team	Team	Team
COMMUNICATION	Over the wall	Verbal	Visual, Connected	Uses EQFD
SOURCE	Speciality	Experience	Experience	Optimization

Basic concurrent engineering by itself overcomes the problems of the traditional process to some degree. EQFD and Dr. Taguchi's System of Quality Engineering greatly add to the vigilance of the information processing by the use of visual connected team methods and systematic optimization. EQFD provides a good entrance into optimization, and the two are most effective when integrated together. In summary, most decisions are made by individuals; this is partial design, which is a key foundation element within the improved total development process. To provide a much stronger framework for the partial design tasks, EQFD is practiced to effectively utilize team experience, and Dr. Taguchi's System is used when experience alone is insufficient. This is summarized in Figure 6. EQFD and Dr. Taguchi's System are critical to success with products.

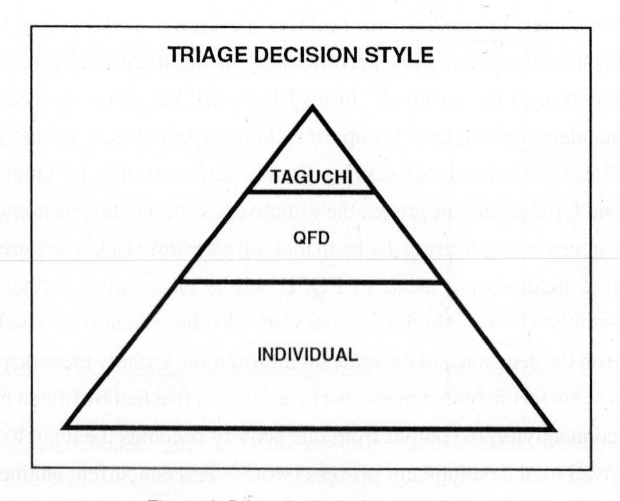

Figure 6. Select the Right Decision Style

3.1 Enhanced Quality Function Deployment

Quality Function Deployment (QFD) is a visual, connective process that helps teams to focus on the needs of customers throughout total development. Enhanced QFD (EQFD) builds on QFD to make it more relevant to complex products that are conceptually dynamic.

QFD is a systematic process to help identify customer desires and deploy them throughout all functions of the corporation, remaining faithful to the voice of the customer. QFD has been evolved by development people in response to major problems in the traditional process. These are disregard for the voice of the customer, disregard of competition, consideration of specifications in isolation, low expectations, little input from design and production in developing the product planning specifications, divergent interpretations of the specifications, lack of structure, lost information, and weak commitment to previous decisions.

Another view that helps to further understand the nature of the problem is to observe the strong vertical orientation of the traditional organization, Figure 7. The analogy with vertical fibers in a fabric has been emphasized by the late Professor Ishikawa, one of the founders of total quality management. A fabric with strong fibers in only one direction will be very weak. The strong vertical fibers have also been spoken of as silos or chimneys. Each vertical fiber, or chimney or silo, contains cloistered groups of specialists, looking inward within their specialty. The vertical cloisters of specialists are good for deploying the voice of the executives vertically, for enabling fast trackers to rise quickly, and for polishing specialties to a brilliant luster. Unfortunately, new products do not move through the organization vertically. They must move through the organization horizontally, starting with the customers and returning to the customers. QFD is a strong weaver of horizontal threads to provide a strong organizational fabric, overcoming the traditional weaknesses of the vertical orientation Figure 8.

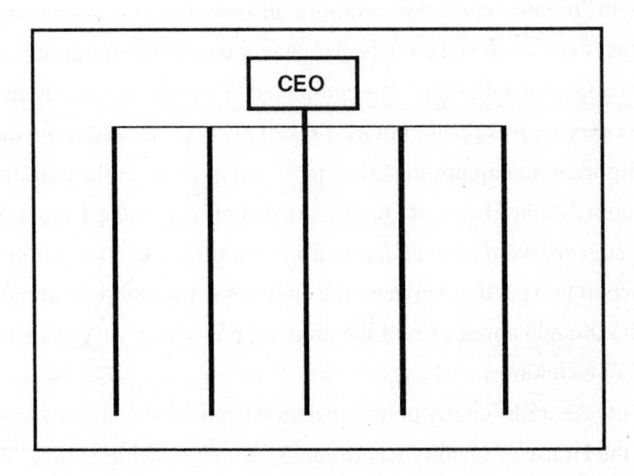

Figure 7. Vertical Orientation

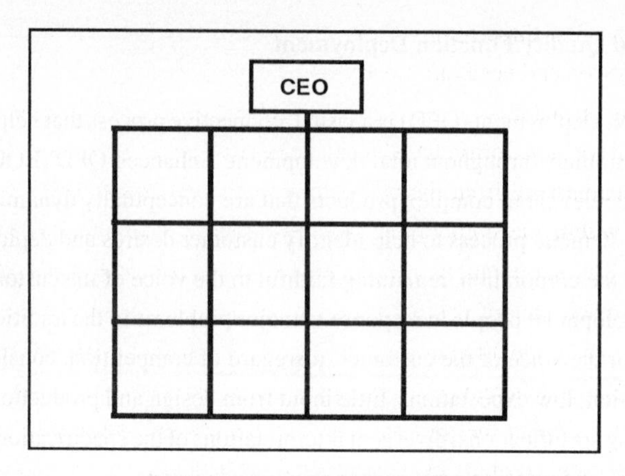

Figure 8. Strong Organizational Fabric

QFD helps the multifunctional team to deploy the voice of the customer out into production operations on the factory floor. To start this process QFD uses the House of Quality to do the planning for the new product.

House of Quality: The House of Quality is a large visual format that the multifunctional team uses to plan the new product. By eliminating much rework that has traditionally been done later in the development process. The House of Quality greatly reduces the development time, in addition to providing greater customer satisfaction as a result of the much sharper concentration on the voice of the customer. In the House of Quality the multifunctional team deploys from the voice of the customer to the corporate expectations for the new product. The House of Quality helps the team to maintain a high fidelity to the voice of the customer.

The House of Quality format is shown schematically in Figure 9. The standard House of Quality has eight "rooms," each representing a different facet of product planning. Room 1 is the voice of the customer, in the customers' language; the needs and desires of the customers. An example, from a famous QFD that was completed at Toyota Autobody in 1979, is *the van will not rust while carrying fresh fruit*, nor *as a result of being washed in an automatic car wash*. It is critically important to capture the customers' perspective in the corporate language. This is recorded in Room 2 of the House of Quality. An example from the Toyota Autobody QFD is *will not rust at the edges of the sheet metal*, a well-known site for severe rusting to start. Note the very great difference in perspective between the engineers, who thought about technology, and the customers, who thought about uses of the product. It is one of the great strengths of QFD that it helps to connect technology and uses.

To overcome the traditionally poor translation from the voice of the customer into corporate expectations, the House of Quality has Room 3, the relationship matrix. This helps the team to verify and improve the fidelity of the translation. Once the multifunctional team has reached consensus on the first three rooms, an excellent start has been made on improved product planning. The customers' needs are clearly stated in corporate terms that are well understood in the same way by the team.

To help achieve superiority, the House of Quality has the benchmarking Rooms 4 and 5. Room 4 is benchmarking from the perspective of the customer, while Room 5 is technical benchmarking within the corporation. The two sets of benchmarkings are compared for consistency, and further iterations are performed until consistency is achieved.

In Room 6 each pair of specifications is considered for interference or reinforcement (synergy). This early identification of interactions among the specifications enables early planning to overcome inherent conflicts, much better than rework.

Room 7 helps to plan the project. Typically it contains at least the importance of each corporate expectation (column) and the expected difficulty in achieving the expectation. If two expectations are both very important, both very difficult, and Room 6 shows that they are conflicting, then the strongest development effort should be devoted to them.

Room 8 is the final objective for the entire House of Quality activity, the quantification of the corporate expectations for the new product. These numbers are strongly based on Room 5, the technical benchmarking, which provides strong guidance on the values that will characterize a superior product. The multifunctional team reaches consensus about the values that will bring success. Even more importantly they have a consistent understanding of the definition of each corporate expectation, and have a common commitment to achieving the goals for success that they have worked so hard to define in the House of Quality.

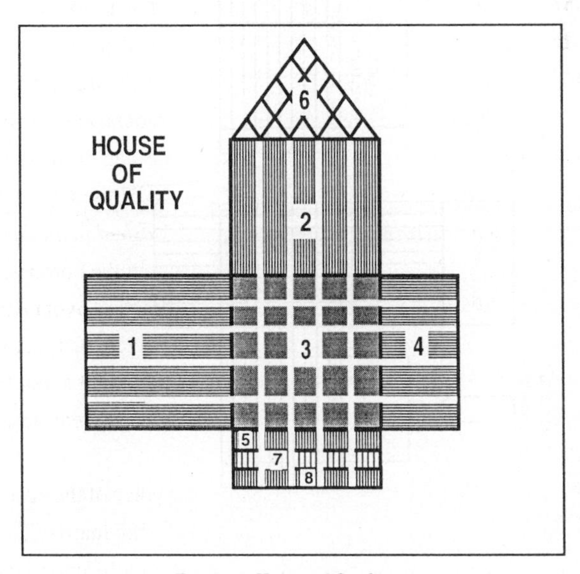

Figure 9. House of Quality

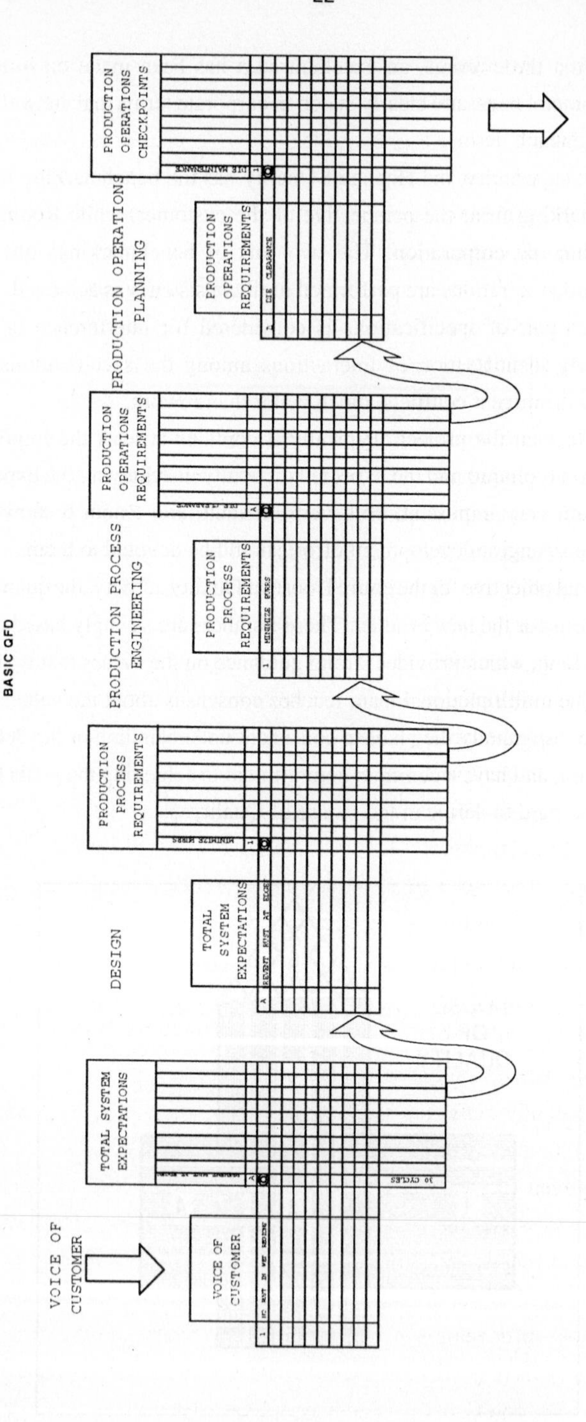

Figure 10. Structure of Basic QFD

To The Factory Floor: After the completion of the House of Quality, the multifunctional team uses QFD to deploy into design and production-process engineering, and thus into production operations planning, which defines the operations on the factory floor. The output from each matrix is in the columns, which then become the input in the rows of the next matrix, Figure 10.

The matrices of QFD provide connectivity. Production operations can be traced back to the customers' needs that are driving them, along the way finding the design and process engineering decisions that responded to the needs and led to the specific production operations.

QFD has the advantages of visual, connective methods: (1) relevant information is found and used, (2) the large, visual formats focus the team discussion on the objectives of the work session, (3) the large displays allow the eye to rove over the relevant information, and logical connections are observed that otherwise would remain hidden, (4) the different types of decisions are connected; for example, product design and process engineering decisions are consistent and compatible, (5) the team members develop a common understanding about the decisions, their rationale, and their implications, and (6) the team members become committed to the common enterprise of implementing the decisions. In addition, QFD provides the added specific advantage of focusing all decisions on the voice of the customer. For these reasons QFD provides a very large increase in the vigilance of the information processing. When holistically integrated into the improved total development process, QFD provides very large benefits. Once the team becomes adept at capturing the spirit of QFD, it becomes a tremendous organizational capability, which provides strong competitive advantage.

QFD provides strong capability beyond basic concurrent engineering. However in its simplest form the total-system expectations are directly deployed into piece-part characteristics. This is straightforward to do for simple products that are conceptually static. However, for conceptually dynamic products the total system concept must be selected before piece parts can be considered. Also, for complex products the total-system expectations must be deployed down to the subsystem level and then to the piece part level. Enhanced QFD (EQFD) has been developed to extend the applicability of QFD to dynamic, complex products. It also contains three enhancements to strengthen product planning.

The most important of the enhancements is the Pugh concept selection process.[10] Although originally developed completely independently from QFD, it too is a visual, connective process, and therefore has proven very easy to integrate into QFD. As shown in Figure 11, the Pugh concept selection process can be used to develop the total system architecture, subsystem concepts, piece-part concepts, and production concepts. Although referred to as a selection process, it almost always leads to the generation of new concepts that did not previously exist. The criteria for the concept development and selection are based on the House of Quality, so that the total-system concept is responsive to the needs of the customers. The criteria are entered as the row headings, A, B, and C in the schematic of the concept selection matrix, Figure 12. Typically fifteen to twenty criteria are used. The concepts are posted as the column headings,

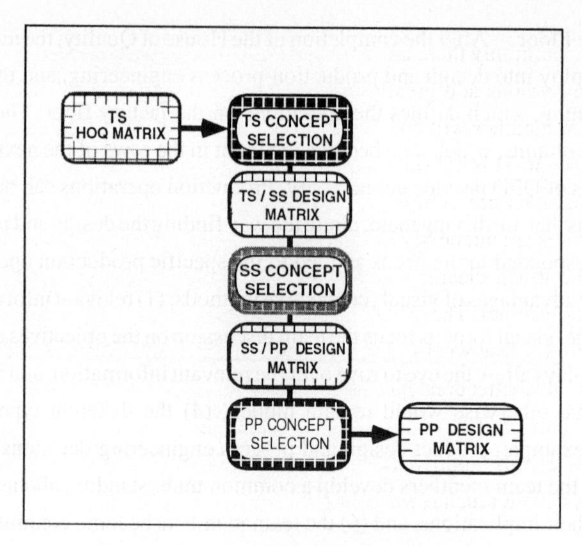

Figure 11. Pugh Concept Selection Process

often more than fifteen. The team chooses one concept as the initial datum to which all of the other concepts will be compared. If there is an existing dominant concept in the marketplace, then it is usually used as the initial datum. Otherwise, the team makes a quick judgment to pick a concept that seems to be very good, and it becomes the datum. Then the team judges each concept relative to the datum in its ability to satisfy criterion A. If it is clearly superior, then it is given a plus mark. If clearly inferior, it is given a minus mark. If it is not clearly superior nor inferior, then it is assigned S for "the same." The team judges all concepts on the basis of criterion A, and then proceeds to criterion B. As this evaluation is being done, much new insight is developed by the

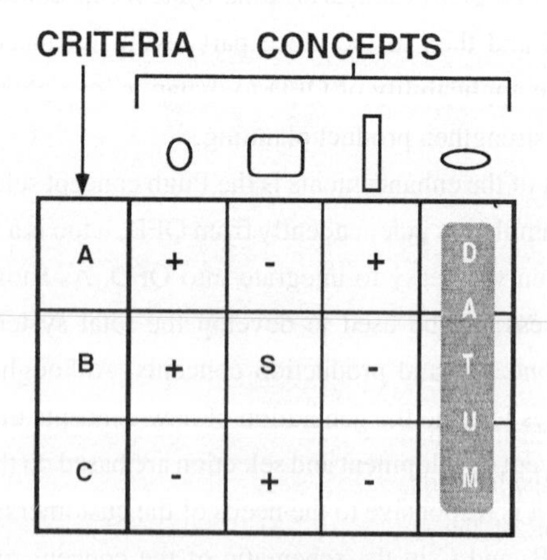

Figure 12. Concept Selection Matrix

team members. Commonly there are different initial perceptions about the evaluations, and the ensuing brief discussions add greatly to clear understanding of the concepts and the criteria. Frequently a team member will observe that if concepts 2 and 13 are combined, the result will be a superior concept. This hybrid is added to the matrix as a new concept. After several hours the matrix evaluations are completed, and the total scores are summed for each concept. However, the scores are not nearly as important as the insight that has been gained by the team. They now have a much clearer understanding of the type of concept features that will be responsive to the customer. The team agrees on additional work, and comes back together to run the matrix again in a few weeks. Typically roughly half of the original concepts have been dropped, but new and better concepts have been generated and are added to the matrix. The team iterates in this fashion until they converge on the dominant concept, Figure 13. The Pugh concept selection process greatly improves the capability of the team to move forward with a strong concept, and thus avoid much rework later in the development process.

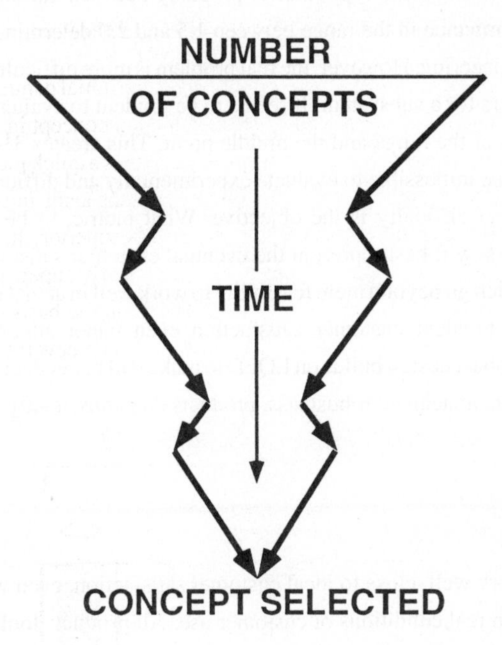

Figure 13. Iterative Convergence to Dominant Cancept

Deployment down through the levels of the product, Figure 11, is done in matrices that are essentially the same as the basic design and process engineering matrices. Instead of deploying forward in the direction of the factory, the total system/subsystem (TS/SS) matrix and the subsystem/piece part (SS/PP) matrix deploy downward from total system level to the piece part level. Again the columns from one matrix become the rows of the next matrix.

In summary, EQFD is an extensive corporate capability that integrates the corporation holistically with a concentrated focus on customer satisfaction. The visual, connective process greatly improves team decision making that is based on the collective experience.

3.2 Robust Quality—When Experience Is Not Enough

The disciplined practice of EQFD goes far beyond basic concurrent engineering in integrating the organization together in a holistic pursuit of customer satisfaction. However, some of the most critical decisions must go beyond the limitations of experience, and systematically optimize to select the values for the critical design parameters that will most consistently achieve customer satisfaction. EQFD marshals the team's experience to identify the most critical parameters, and to judge the range that will be best. For example, a friction coefficient can be identified as critical, and experience teaches that the best value is probably between 1.5 and 2.0. Then systematic evaluations of performance in the range between 1.5 and 2.0 determine the best value. This is traditional good engineering. However, the real problem is more difficult. The team may identify 13 critical parameters for a subsystem. It seems to be prudent to evaluate three values for each parameter, the ends of the range and the middle point. This creates 3^{13}, 1,594,323, options, a number that would be impossible to evaluate experimentally and difficult to evaluate for many analyses. The second difficulty is the objective. What metric, to be evaluated early in the development process, will best represent the eventual customer satisfaction? Customers want robust products, which go beyond mere feasibility to work well in actual conditions, maintaining performance close to ideal customer satisfaction even under adverse conditions. Quality engineering using robust design builds on EQFD to make still better decisions by optimizing sets of critical parameters to achieve robustness, products that consistently satisfy the customers.

3.3 Robustness

Robust products work well, close to ideal customer satisfaction, even when produced in real factories and used in real conditions of customer use. All products look good when they are precisely made in a model shop and are tested under carefully controlled laboratory conditions. Only robust products provide consistent customer satisfaction. Robustness also greatly shortens the development time by eliminating much of the rework that is known as build, test and fix.

Robustness is small variation. For example, Sam and John go to the target range, and each shoot an initial round of ten shots, Figure 14. Sam has his shots in a tight cluster, which lies outside of the bulls-eye. John actually has one shot in the bulls-eye, but his pattern is widely dispersed. In this initial round John has one more bulls-eye, but Sam is the robust shooter. By a simple

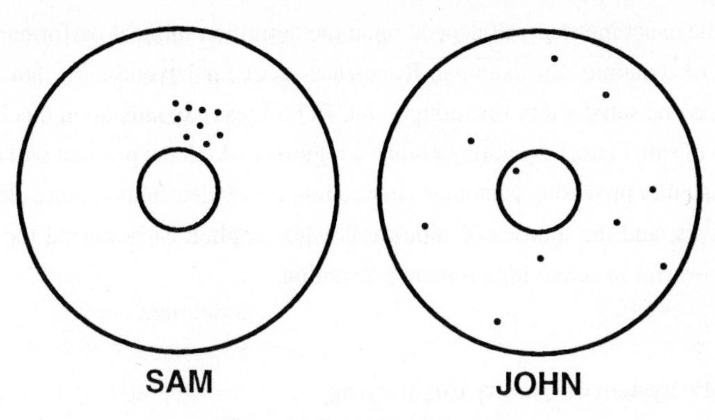

Figure 14. Example of Robustness

adjustment of his sights, Sam will move his tight cluster into the bulls-eye for the next round. John faces a much more difficult task. He must systematically optimize his arm position, the tension of his sling, and the other critical parameters that control the robustness of his shooting. This example reveals several important facts about robustness: (1) the application of the ultimate performance metrics to initial performance is often misleading; Sam had no bulls-eyes even though he is an excellent marksman. (2) Adjustment to the target is usually a simple secondary step. (3) Reduction of variation is the difficult step. (4) A metric is needed that recognizes that Sam is a good marksman, that measures his expected performance after he adjusts his sights to the target.

Automobiles give further insight about robustness. Customers want a car that is not a lemon, that is robust against production variations. A lemon is a car that has production variations that cause great customer dissatisfaction. To overcome this the production processes have to be more robust so that they produce less variation, and the car design has to be more robust so that its performance is less sensitive to production variations. The customers also want a car that will start readily in northern Canada in the winter, and will not overheat in southern Arizona during the summer; a car that is robust with respect to the variations of the conditions of customer use. Customers also would prefer cars that are as good at 50,000 miles as when new; that are robust against time and wear. This example reveals the three sources of undesirable variation in products: (1) production variations, (2) variations in conditions of use, and (3) deterioration. These three types of noises inevitably cause some degradation of performance, some deviation away from ideal customer satisfaction. The optimization of robustness minimizes these deviations, and keeps performance economically close to ideal customer satisfaction. Once a product has been designed as far as possible by experience, then the systematic optimization of robustness quickly determines the best values for the critical parameters; the friction coefficient should be 1.7, when the collective team experience had narrowed the range to 1.5 to 2.0. Robustness is the objective; for efficient optimization it is essential to define the best metrics for robustness. The exact metrics will depend upon the definition of ideal performance, analog or digital, static or dynamic, for example. Robustness goes far beyond mere feasibility, which simply requires one satisfactory operating point. Robustness expands upon this by developing a large region of satisfactory operating points, a region in which the product will nearly always lie, thus consistently providing customer satisfaction. Better decisions require clear statements of the objectives, and the metrics of robustness when applied early during the development process are powerful in achieving problem prevention.

3.4 Taguchi's System of Quality Engineering

Although there have been many independent approaches to robustness, by far the most complete and powerful has been developed by Dr. Genichi Taguchi starting in the late 1940's. His system of quality engineering using robust design is summarized in Figure 15. Lack of robustness causes the performance variations in the output from the product to be excessive, which produces large quality loss, the financial loss after the product is in the hands of the customer. The performance variations are caused by the three types of noises that afflict the product. There are four activities to reduce the total cost, primarily quality loss and manufacturing cost: (1) product parameter design, the systematic optimization of the robustness of the product design, (2) tolerance design, to select the economical precision levels around the nominal (target) values, (3) process parameter design, the systematic optimization of the most important production processes so that they will inherently produce more consistent products, and (4) on-line quality control, prudent

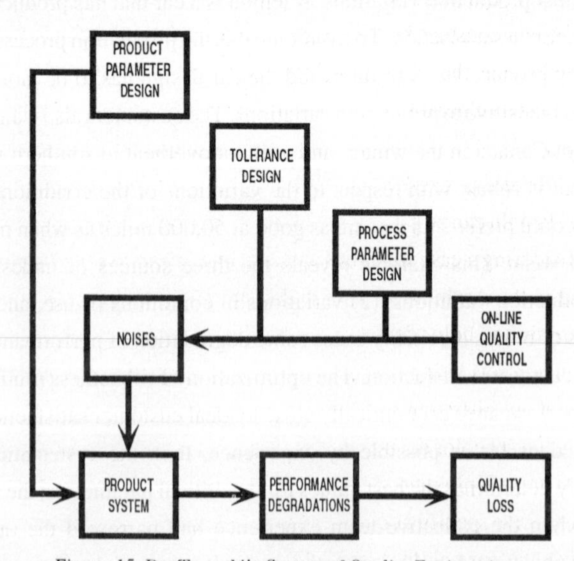

Figure 15. Dr. Taguchi's System of Quality Engineering

intervention on the factory floor to further improve production consistency. The first activity, product parameter design, is the most powerful because it provides protection against all three kinds of noises. The remaining three activities all economically reduce production variations, one of the three types of noise that afflict the product. Although they are very valuable, these activities cannot help overcome the problems that are caused by the variations in the conditions of customer use nor the problems that stem from wear and other forms of deterioration. Therefore, the rapid, early optimization of the robustness of the product system is the most important activity in Dr. Taguchi's system. Now we will look in more detail at each of the eight elements of this system as displayed in Figure 15.

Quality Loss is the financial loss after the product is in the hands of the customer. It is caused by the deviations from ideal performance. Of course, customer dissatisfaction can also be caused because the ideal performance is not what the customer wanted; the product planning did not follow the voice of the customer. In working on robustness we do not reconsider the product planning; our objective now is to avoid excessive quality loss that is the result of deviations from ideal performance as defined during product planning. A car with air conditioning may be said to be higher in quality as a result, but it is not more robust. Obviously there is more than one dimension to the total image of quality. Here we are considering one dimension, the discrepancy between actual performance and ideal performance.

Two products, both of which meet the corporate specifications, can impart very different quality losses. Therefore, meeting specifications is a poor measure of quality. This is counter to the tradition that zero defects is sufficient. Zero defects is simply an interesting and sometimes useful, rather arbitrary, milepost on the road to quality improvement. Ideal performance is unobtainable, but it is a constant beacon that shows the direction for quality improvement. If lesser quality is taken as the objective, then achieving that quality will tend to stop all improvement, leaving us vulnerable to further improvement by the competition. Robust products reduce quality loss.

Performance Variations cause quality loss; the customers are completely satisfied only by ideal performance. Insight can be gained into the relationship between performance variation and quality loss by considering a case study that is based on a report about Sony color television sets that was published in the Asahi newspaper in the late 1970's. The top half of Figure 16 shows the production variation in color density from the Sony factories in San Diego and Tokyo, the design of the television sets being the same. The curves that are plotted are probability density functions; smoothed out histograms of the actual sets that were shipped.

T represents the target color density that provides ideal customer satisfaction. The corporate specification limits are at T plus and minus five. All of the sets that were shipped from San Diego were within the corporate specification limits, zero defects. The probability density function for the Tokyo factory is the normal (Gaussian) distribution with six standard deviations within the specification range. From the probability tables we can read that the small tails that lie outside

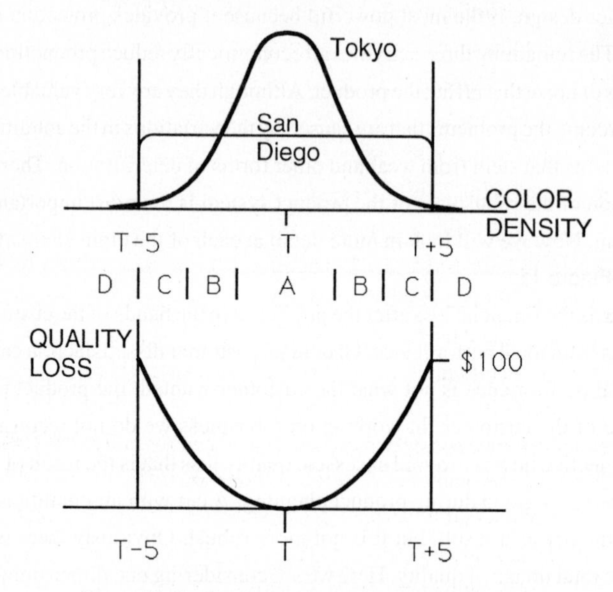

Figure 16. Quality Loss—The Modern Quality Paradigm

the corporate specification limits contain 0.3% of the total sets; three out of every thousand sets that were shipped were "defective", that is they did not satisfy the corporation's definition of acceptability. Thus, San Diego was shipping zero defects, while Tokyo was shipping 0.3% defective; on the basis of per cent defects San Diego was doing better.

However, grades are also placed on Figure 16; A grade being within one standard deviation of the target, B grade being the second standard deviation, and C grade being the third standard deviation. We see that Tokyo had about twice as many A sets as San Diego, and many fewer C sets. The customers really wanted A sets, so the Tokyo sets were pleasing twice as many customers. So which should we prefer, zero defects or twice as many pleased customers?

To help us further analyze the comparison. Dr. Taguchi developed the quality loss function, displayed in the bottom half of Figure 16. The abscissa is color density, as in the top half of the figure. The ordinate is the average quality loss per set. If we ship 1000 sets with color densities between 1.95 and 2.05 and their total quality loss is $16,000, then the average quality loss per set of $16 would be plotted at 2. Now there are two key observations that provide us with great insight.

First we must associate a dollar value with the specification limit. The corporation has defined this to be the limit of acceptability, so the expected quality loss must be unacceptably high. The dollar value can be estimated as the cost of the countermeasure. At the specification limit many customers will demand service. We know fairly accurately the average cost of such service, say $100, and enter this as the ordinate at T+5. We also expect this to be symmetrical, and therefore also enter $100 at T-5. Of course at the customer-satisfaction target T we expect the quality loss to be zero, so now we have three points on the quality loss curve. Now what shape do we expect the curve to have between zero and the specification limits?

The second key observation is to observe two sets near a specification limit. Suppose we pick two sets from the assembly line, and measure their color densities to be T+4.99 and 5.01, one just inside the corporate specification limit , and the other just outside the corporate specification limit. Now let us position the sets in adjacent rooms, and walk back and forth between the two sets. Can we tell any difference in the two pictures? Of course not! The corporate specification limit is completely unnoticed by the customers. This means that the expected quality loss does not have a jump at the specification limit, but is continuous. The average quality loss at 4.99 is only slightly less that at 5.01. This is different from the concept of zero defects, which lumps all products within the corporate specification limits together as non-defective, from which it might be inferred that the expected quality loss is zero and then jumps to a large value at the specification limits. It seems reasonable that the expected quality loss will increase slowly at first, anything in the A range will be considered good, and there will be little quality loss. Then the curve of expected quality loss will swing upward with increasing slope to the value of $100 at T+5.

After considering these simple common sense observations, Dr. Taguchi proposed that the simplest approximation that conforms to reality is the quadratic quality loss curve that is shown in the bottom half of Figure 16. Actual quality loss curves undoubtedly differ somewhat, but the quadratic curve powerfully changes our thinking, and is sufficiently accurate for nearly all purposes. Furthermore, it is relatively easy to determine, while the actual curve would be very difficult to determine.

Now let us apply Dr. Taguchi's quality loss function to the Sony production. We see that the Tokyo factory had about two thirds of its shipments in the A range, where the quality loss per set is very small. In contrast, the San Diego factory had nearly one third of its production in the C range where the quality loss per set is very high. If we simplify by assuming that the San Diego straight line extends all the way to the specification limits, multiply the probability density functions by the quadratic quality loss function, and then integrate over all of the sets that were shipped, the result is that the expected quality loss from the Tokyo sets was one third of that from the sets that were shipped from San Diego.

A major conclusion is that it more important to keep performance close to the ideal-customer-satisfaction target than it is to avoid a few defects. The fact that the factories were in San Diego and Tokyo is incidental, and even the normal distribution curve is not the important message. The objective must be to reduce the variance from the ideal-customer-satisfaction target. The quality loss, averaged over the entire population of products that are shipped, is proportional to the variance, expressed as:

$$L_a = kV$$

This assumes that the mean value for all of the sets is equal to the target value. More completely expressed as:

$$L_a = k[(m-T)^2 + V]$$

Good quality is a mean value that is near to the target and a small variance. The value of k is directly determined from the value of L at the specification limit. Thus, $100 at five gives a k value of four.

One of the great advantages of this understanding of quality is that it always provides a goal for continuous improvement. The value of the variance will never become zero, so we can always evaluate it, and make improvements if the cost is justified. One of the problems with intermediate quality goals, such as zero defects, is that they can be achieved, and then there is not any motivation for further improvement. However, a competitor may then greatly further reduce the variance, and gain competitive advantage. By concentrating on the variance, and the deviation of the mean from the target, we always know our exact position, and can determine the most productive areas for our quality improvement activities.

A little reflection about the source of the shape of the probability density function for the San Diego factory leads to the conclusion that they were inspecting the sets after the final assembly, and only shipping the sets that were within the corporate specification limits. Sets that were outside of the limits had to be reworked. The inspection and rework cost money, increasing the unit manufacturing cost (UMC). Tokyo was obviously not inspecting; their inherent quality was good enough so that they did not have to inspect, thus reducing their manufacturing cost. This is another lesson; robustness reduces the manufacturing cost.

This learning about performance is a key element of better decisions. The reduction of performance variance is an essential goal.

Noises are the causes of the performance variations. When all conditions are nominal, then the performance is on target. Three type of noises afflict products. The most important noise is the variations in the conditions of use by customers. The product will be used at high temperatures and low temperatures, high humidity and in dry conditions, at high voltage and low voltage, and in many other varying conditions that challenge the product. We product people have very little control over these conditions. Sometimes in exasperation we may exclaim that they should not have used the product that way, but customers are going to use the products as they see fit. Competitive advantage comes to the robust products that can best withstand the varying conditions of use. The second type of noise is production variations. The Sony case is an example. The third type of noise is wear and other forms of deterioration. This is similar to production variations; the critical part characteristics deviate from their nominal values. Thus, the second and third type of noises are really the same, except one is the initial condition, while the other is time dependent. This time-dependent component of quality is often referred to as reliability. We do have control over the production variations, and during design we have some control over the rate of wear and other deteriorations.

The Product System is afflicted by the noises, which cause its output performance to vary. We want the product to be robust, to dampen the effects of the noises, not passing them through to the output. Once the system concept is selected, robustness is achieved by finding and operating on the flat part of the response curve. However, in real systems there are many critical design parameters, and plotting response curves is not feasible. We need a direct engineering approach to find the flat part of the curve for the many-dimensional curves that represent realistic systems.

Product Parameter Design is the optimization of the nominal values of the critical design parameters to achieve robustness in a short time. The process of robustness optimization is:

1. Define objective; best metric
2. Define feasible alternative design values
3. Select some alternatives for evaluation
4. Impose noises
5. Evaluate performance of selected alternatives
6. Select best design values
7. Confirm robust performance.

The most important guideline of all is to do the robustness optimization early so that problems are prevented.

BEST METRICS concisely measure robustness, the performance variation after the mean value has been adjusted to the target. The best metrics have been given the name signal-to-noise ratio. Signal refers to the performance that we want, and noise refers to the variation of performance away from the ideal-customer-satisfaction value, which we do not want. Thus, we want a high signal-to-noise ratio. The terms signal and noise are used for both inputs and outputs. Earlier the three types of input noises were described. Now the noise in the performance output has been introduced as part of the signal-to-noise ratio. Usually it is clear from the context which is intended.

For a television set the picture from the station is the input signal, and lightning and other electromagnetic disturbances are the principal input noises. The picture on the screen is the performance signal, and snow and other forms of picture degradation are the noises in the performance. The signal-to-noise ratio measures the ratio of good performance to undesirable performance for fixed values of the input signal and noises. Thus, in comparing different designs it is important to keep the input signal and noises fixed, or else the comparison of the performance signal-to-noise ratios will not be valid.

FEASIBLE ALTERNATIVE VALUES for the critical design parameters are chosen primarily on the basis of engineering judgment. The multifunctional team uses EQFD, especially fault trees, to select the most critical parameters. For a subsystem the team may prioritize the 20 most critical parameters, and then decide to use the top 13 in the initial optimization. The best judgment about the nominal value, based on experience including analyses, has already been exercised in the normal course of design. Now it seems prudent to also try a larger value and a smaller value, seeking improvement. Thus, commonly we use three feasible alternative values for each critical design parameter. For 13 critical design parameters we have 3^{13} or 1,594,323 combinations. It is easy to comprehend that the traditional poke and hope approaches are unlikely to find the needle in such a big haystack.

SELECT ALTERNATIVES for evaluation so that a combination that is close to the optimum will be quickly found. It is better to have 95% of optimum performance in two months rather than obtain 98% of optimum performance in one year. When each critical design parameter has three levels, there are two increments of variation, a total of 26 for the 13 parameters. In addition to variation we need to evaluate the mean value of performance, so a minimum of 27 evaluations are necessary for 13 critical design parameters at three levels each. It seems simple prudence to evaluate each level of each critical parameter nine times; we do not have any basis for preferring one level over another. Likewise, in considering pairs of critical design parameters, it seems prudent to select each combination an equal number of times. For three levels there are nine combinations of pairs, 1-1, 1-2,..., 3-3, so for 27 evaluations each will be selected three times. Finding 27 combinations that obey these common-sense guidelines, seems as though it might be quite difficult. Fortunately it has been done for us and tabulated in ready references. Combinations that obey our simple guidelines are called orthogonal arrays. For three levels of 13 parameters the orthogonal array is referred to as the L_{27} orthogonal array. The L stands for Latin, which is rooted in the history of such arrays. The 27 refers to the number of evaluations that are defined by the array.

	1	2	3	4	5	6	7	8	9	10	11	12	13
1	1	1	1	1	1	1	1	1	1	1	1	1	1
2	1	1	1	1	2	2	2	2	2	2	2	2	2
3	1	1	1	1	3	3	3	3	3	3	3	3	3
4	1	2	2	2	1	1	1	2	2	2	3	3	3
5	1	2	2	2	2	2	2	3	3	3	1	1	1
6	1	2	2	2	3	3	3	1	1	1	2	2	2
7	1	3	3	3	1	1	1	3	3	3	2	2	2
8	1	3	3	3	2	2	2	1	1	1	3	3	3
9	1	3	3	3	3	3	3	2	2	2	1	1	1
10	2	1	2	3	1	2	3	1	2	3	1	2	3
11	2	1	2	3	2	3	1	2	3	1	2	3	1
12	2	1	2	3	3	1	2	3	1	2	3	1	2
13	2	2	3	1	1	2	3	2	3	1	3	1	2
14	2	2	3	1	2	3	1	3	1	2	1	2	3
15	2	2	3	1	3	1	2	1	2	3	2	3	1
16	2	3	1	2	1	2	3	3	1	2	2	3	1
17	2	3	1	2	2	3	1	1	2	3	3	1	2
18	2	3	1	2	3	1	2	2	3	1	1	2	3
19	3	1	3	2	1	3	2	1	3	2	1	3	2
20	3	1	3	2	2	1	3	2	1	3	2	1	3
21	3	1	3	2	3	2	1	3	2	1	3	2	1
22	3	2	1	3	1	3	2	2	1	3	3	2	1
23	3	2	1	3	2	1	3	3	2	1	1	3	2
24	3	2	1	3	3	2	1	1	3	2	2	1	3
25	3	3	2	1	1	3	2	3	2	1	2	1	3
26	3	3	2	1	2	1	3	1	3	2	3	2	1
27	3	3	2	1	3	2	1	2	1	3	1	3	2

The L_{27} array that is above obeys our prudent guidelines. The numbers 1,2, and 3 in the table refer to the three levels of each critical design parameter. Thus, in the first evaluation each of the critical design parameters has its level one. Each row defines a different design combination for evaluation. The performance of each of the 27 design combinations is evaluated. If a good computer simulation, FEA for example, is available, then it can be used to evaluate the performance. Otherwise experimental evaluation is used.

IMPOSE NOISES on the product during the evaluations to simulate real conditions so that the comparisons of the 27 options will be realistic, not merely reflecting laboratory comparisons under ideal conditions. If the winning combination is chosen under ideal operating conditions, it is not likely to be successful in the real operating conditions of the market place. Therefore, the multifunctional team carefully identifies the most important noises in the actual use of the product. EQFD including fault trees is again very helpful here. Then each option, each row of the orthogonal array, is subjected to the same array of noises, and the performance is evaluated for each noise combination. In the simplest approach two combinations of noise are used. If high temperature, high humidity, and low input voltage are known to stress the performance in one direction, then these noises are taken to reasonable extremes, based on customers' conditions, and then the performance is evaluated. Then the noises are taken to their opposite extremes, and then the performance is again evaluated. This gives two performance evaluations for each design point, just enough to test its sensitivity to noises. The exact amount of stress in the values of the noises is not too important; we are not going to use these performance values to predict the field performance, but simply to compare the 27 options. As long as the values of the noises that we select are roughly the same as in the field, then the comparison will be valid, and the combination that we select will be robust in the marketplace.

Alternatively, if each performance evaluation is quick and simple, then more noise combinations are typically used. For example, we might identify the seven most important noises, and decide to test each of them at their two extreme values. This gives 2^7, or 128 combinations. We might decide that 128 is too many to evaluate, and use an orthogonal array to select a representative sample. Typically we would use an L_8 to select eight combinations of noise for actual evaluation. Then each of the design points, 27 in the above example, are subjected to the eight different noise combinations, a total of 216 performance evaluations. When computer simulations that take only a very short time to run are available, then it is common to use more performance evaluations. It is common to use the L_{36} orthogonal array to select 36 design points, and then also use the L_{36} orthogonal array to select 36 noise combinations, a total of 1296 performance evaluations.

Thus we see that the number of noise combinations that we select is strongly influenced by the ease and time to perform the evaluations. For experimental evaluations it is common to take two noise combinations, and rare to use more than eight.

EVALUATE PERFORMANCE of the selected design and noise combinations; this is straightforward engineering work, either experimental or computer determination. For each design point, each row of the design orthogonal array, the evaluation of the performance at several noise conditions enables the effect of noises to be evaluated in the signal-to-noise ratio. Thus, a signal-to-noise ratio is calculated for each design point. If the L_{27} orthogonal array was used to select the design options for comparison, then 27 signal-to-noise ratios are calculated.

SELECT BEST VALUE By comparing these values the best of the 27 options is quickly identified. However, there were a total of 1,594,323 options, so it is highly improbable that the best overall option was in the 27 that were selected for initial evaluation. Therefore, we must interpolate from our data to select the best point from among the 1,594,323 options. Taking column one, page 27, as an example, the first nine rows were tested with this critical design parameter at its first level. Therefore, the average of these nine signal-to-noise ratios characterizes the performance that is associated with level one of the critical design parameter that we had assigned to the first column. Likewise, the average of the second set of nine signal-to-noise ratios characterizes the second level of the critical design parameter, and the average of the third set of nine signal-to-noise ratios characterizes the third level. Then the level that has the highest average signal-to-noise ratio is chosen as the best. The same type of analysis is done for the other critical design parameters, and the best level is chosen for each. A little reflection will lead to the insight that this analysis opens the door to each of the options, any of the 1,594,323 could be the winner. If level one was best for critical parameter one (first column of the design array), level three was best for the second critical parameter, level two was best for the third parameter etc., then the best combination can be written as 1323312113223, each number giving the best level for that critical parameter. It is our prediction that this combination will give the best performance, the best signal-to-noise ratio, of all of the 1,594,323 design options. Next we need to verify the performance of our selected winner.

CONFIRM ROBUST PERFORMANCE The performance of the selected winner is evaluated, and its signal-to-noise ratio is calculated. Usually it will be significantly better than the best of the 27 design options that were originally evaluated, and much better than the original design point. The robustness of our product or subsystem has been optimized, with tremendous benefit.

Of course, using 27 data points to estimate the most likely best design point out of 1,594,323 options cannot be a precise process. The design point that is selected by the simple process that has been described will usually be close to the best, but not actually the very best. To find the very best would require 1,594,323 evaluations, and even then we could not be absolutely certain that the best had been found. Therefore, our goal is simply to get close to the best in a short time.

There are also two other types of uncertainties in our optimization. First, the 13 design parameters that we selected may not have been the most critical. Second, the best value of signal-to-noise ratio for each parameter probably did not lie exactly on one of the three values that we initially selected. Therefore, one or two more iterations are usually in order. Initially we selected 20 critical design parameters in prioritized order, based on the collective experience of the team. In the first iteration of evaluations we evaluated the top 13 on our list. Typically some of these are found to actually have little effect on the signal-to-noise ratio, and can be dropped from the second iteration of evaluations. They are replaced by some of the design parameters that were lower on our original list, in the 14 to 20 positions. To overcome the second uncertainty the signal-to-noise values from the first iteration are plotted against the three levels of each design parameter to enable interpolation. For example, if the curve appears to have a peak between the second and third levels, then in the second iteration the three values are selected in this range, thus providing a fine tuning of the optimization. Commonly a third iteration is judged to have captured most of the potential optimization in a short time, and the optimization of product robustness is complete. The benefits in improved customer satisfaction, reduced costs, and shorter development time are tremendous.

COMPARISON OF TAGUCHI OPTIMIZATION AND DESIGN OPTIMIZATION Although generally similar, there are significant differences between Dr. Taguchi's optimization (just described) and Design Optimization (NLP). It is an interesting fact that Dr. Taguchi and coworkers have developed many different specific objective functions, while the Design Optimization workers have developed many different search methods. The many different SN ratios have been developed in response to the needs of many fields, and enable Dr. Taguchi's methods to be very widely applied. There are a total of more than fifty types of SN ratios. This is indicative of the rich variety of applications for Dr. Taguchi's methods. The Design Optimization workers have emphasized the development of fast and robust search methods. Some of the newer programs are much improved, so the emphasis on search methods has been rewarded. The major distinctions between Dr. Taguchi's methods and Design Optimization are summarized below:

TAGUCHI	DESIGN OPTIMIZATION (NON-LINEAR PROGRAMMING)
Emphasis on objectives	Emphasis on search methods
Analytical and experimental	Analytical only
Minimization of Quality Loss is emphasized	Quality Loss is implicitly assumed to be zero (inequality constraint)

Tolerance Design After the nominal values of the critical design parameters have been selected, the economical precision levels are selected. The nominal values are the target for production, whether by our factory or our suppliers. However, production will always vary, and during product development we have some choice about the amount of precision that will be achieved. Product parameter design is often all benefit, but tolerance design always involves a trade-off. If we want more precision, we must pay for it.

The quadratic quality loss function can be used during tolerance design to estimate the expected quality loss in the field, and a quantitative trade-off can be made. During product development the primary precision decision to be made is the selection of the production process. For example, a shaft can be turned or it can be turned and ground. The second option provides much better precision, but also costs much more. Is it worth it? If the expected quality loss is reduced enough, then the extra manufacturing cost will be justified.

Process Parameter Design is done to optimize the robustness of the production processes, which reduces the production variations, the second type of noise to the product. Production variations are caused by factory noises such as the variations in the properties of raw materials, variations in the ambient temperature and humidity, vibrations, variations in the values of the critical process parameters, variations in the maintenance of the production equipment, and variations in the practices of the workers. Optimization of the robustness of the processes makes the production more uniform despite the presence of these noises. This improves both customer-perceived quality and the manufacturing cost. The process for optimizing the robustness of the production processes is essentially the same as for the product. In process parameter design we optimize the values of the critical production-process parameters, such as feeds, speeds, and depth of cut. This has often been applied to existing processes, frequently with very large improvements that have brought substantial cost reductions.

On-Line Quality Control further reduces variations during actual production. No matter how robust the production processes are, if there is not any intervention during production the values will deviate substantially from the target. If we never intervene, the quality loss will be very much excessive. If we intervene too frequently, the reduction in quality loss will not compensate for the immediate cost of the excessive frequency of intervention. This is another trade-off, similar to tolerance design. We add the cost of intervention (measurement cost plus adjustment cost) to the expected quality loss because of the deviation from the target, and choose the frequency of intervention that minimizes the total cost.

3.5 Summary

Dr. Taguchi's system of quality engineering using robust design has four improvement activities. The last three of these, tolerance design, process parameter design, and on-line QC, all reduce production variations, the second type of noise to afflict the product. This is very beneficial, but these activities do not provide any protection against the customer-use noises, and very little protection against the deterioration noises. Product parameter design is the most important improvement activity, as it provides protection against all three types of noises. This keeps *performance* close to the ideal customer-satisfaction value, which is our real objective.

Traditionally quality was a factory activity to make parts conform to corporate specification limits. In the new quality paradigm quality is primarily a design activity to make performance adhere closely to the ideal customer-satisfaction value. This paradigm shift is a very essential element of the improved total development process (world-class concurrent engineering).

The primary features of Dr. Taguchi's system of quality engineering are:

Focus on robust performance
Early optimization of robustness
Best quality metrics (signal-to-noise ratios)
Integrated system of quality engineering
Systematic design changes for efficient optimization.

These are arranged in descending order of importance. The systematic design changes, through the use of the orthogonal arrays, is an important feature, but it is the least important. There has been a misunderstanding by some upon their initial introduction to this system that it is only a slight wrinkle in the design of experiments, but that is actually its least important feature. At the Xerox Corporation during the 1970's, for example, great success was achieved although only the first two of the above features was utilized.

Quality engineering using robust design greatly improves customer satisfaction, reduces costs, and most importantly of all greatly shortens the development time. The early optimization of robustness eliminates great amounts of rework later in the development process, and thus the product is gotten to market much sooner with much less confusion during production start up.

3.6 The Right Decision Style

Most decisions can be made by individuals, with only informal interactions with their colleagues. However, the most critical decisions require intensified vigilant information processing to achieve world-class concurrent engineering. This is provided by Enhanced QFD and quality engineering using robust design. QFD is a visual, connected methodology that helps teams to reach the best decisions when individual decision making is not adequate. When the team finds that even their collective experience is insufficient to complete the decision, they then branch out of QFD into the systematic optimization of robust design.

The most critical decisions are those that most strongly influence customer satisfaction, including costs. Starting with the voice of the customer, the application of EQFD and quality engineering using robust design leads the team to identify the critical parameters and make the best decisions. It is important that these vigilant information processing methodologies be tightly integrated into world-class concurrent engineering.

4 The Benefits

The improved total development process (ITDP) enables success in the global economy. Anything substantially short of the ITDP that is described in this article will lead inexorably to a bedraggled economic condition; corporate demise or increasing dependence on an ever-declining standard of living as the primary basis for marginal competitiveness. The ITDP provides a buoyancy that separates the swimmers from those who must resort to cries to politicians to save them from drowning.

The benefits of the improved total development process are enhanced customer satisfaction, reduced costs, and shorter time to market. This enables more product cycles, which provide greater product variety and increased corporate learning. The lead rapidly grows over the companies that are still stuck in the rut of the traditional process.

The improved total development process is needed in order to be competitive in civilian industries and in defense industries, in mature industries and in dynamic industries, in industries with complex products and in industries with simple products, and in high-volume industries as well as low-volume products. Companies that do not practice the improved total development process (also known as world-class Concurrent Engineering) will continue to have technological weaknesses in development. The ITDP is the only modern way to develop products and processes.

References

1. Dertouzos, M.L., Lester, R.K., Solow, R.M., and the MIT Commission on Industrial Productivity: *Made in America*. Cambridge, MA: The MIT Press 1989
2. Gomory, R.: A dialogue on competitiveness. *In: Issues in Science and Technology*, Summer 1988
3. Clark, K.B., and Fujimoto, T.: Product Development Performance. Boston: Harvard Business School Press 1991
4. Hosking, D.M., and Morley, I.E.: *A Social Psychology of Organizing*. New York: Harvester Wheatsheaf 1991
5. Veraldi, L.: "New Program Management at Ford" At: The First Dartmouth Conference on Next Generation Management and Engineering, (Hanover, NH) September 12, 1988
6. Hauser, J.R. and Clausing, D.P.: The House of Quality. In: *Harvard Business School Review*, pp. 63-73, May-June 1988
7. Clausing, D.P. and Pugh, S.: Enhanced Quality Function Deployment. Proceedings of the Design Productivity International Conference, Vol. 1, pp. 15-25, (Honolulu) 1991
8. Taguchi, G. and Clausing D.P.: Robust Quality. In: *Harvard Business School Review*, pp. 65-75, January-February 1990
9. Phadke, M.S.: *Quality Engineering Using Robust Design*. Englewood Cliffs, NJ: Prentice Hall 1989
10. Pugh, S.: Concept Selection--A method that works. Proceedings of the International Conference on Engineering Design (ICED), (Rome) 1981

Virtual Team Framework and Support Technology

K.J. Cleetus

Concurrent Engineering Research Center, West Virginia University, 886 Chestnut Ridge Road, Morgantown, WV 26506 USA

Abstract: This paper first discusses the motivation for the concept of a Virtual Team and outlines its essential features. The merits of implementing such a concept on the computer network to support Concurrent Engineering between distributed team members are then highlighted. In the second part of the paper the efforts in the DICE (DARPA Initiative in Concurrent Engineering) project to realize virtual teams through a set of generic software services are reviewed.

Keywords: concurrent engineering / simultaneous engineering / groupware / enterprise integration / collaboration / virtual team / team / coordination / design notebook / integration framework / framework / desktop conference / wrapper / CAD / communication protocol / multimedia / distributed computing / group decision support / workgroup application / workflow

1 The Virtual Team Framework

1.1 Problems of Industry

The M.I.T. Commission on Industrial Productivity named six broad weaknesses afflicting American companies [1]. Two of them are worth stating in the context of this paper.

1) Failure of cooperation.
2) Technological weaknesses in development and production.

Elaborating on the first, the Commission cited the widespread failure of cooperation within and among companies. Partly as a result of the cultural conditioning to compete, a once hallowed source of advantage to the American economy -- competition -- is now seen to be detrimental to success in the emerging world in certain contexts. "Communication and coordination," said the Commission, "is often inhibited [in many U.S. firms] by steep hierarchical ladders and organizational walls." This constitutes a confirmation of the urgent need to develop and deploy forceful mechanisms of communication and coordination and those are the themes of this paper.

1.2 Competitive Problems in Industry

We should note that the resolution of these problems will not be achieved solely through the use of computer networks and advanced software for group work. There are organizational and cultural imperatives too which have to be dealt with. Technology is the facilitator, and in some cases by making a certain style of working very easy to use for employees, it may even catalyze a more close-knit mode of employee participation. But where the legacy of the past causes delays and alienation of employees from their work, the first and necessary change has to be cultural, a kind of *glasnost* before the technological *perestroika.*

The second weakness as described in the Commission's report is more an effect than a cause, an effect resulting from the barriers to concurrent approaches to problems on all fronts from the inception. This deficiency is underscored in the following words from the Red Book of Specifications of the DICE Architecture [2]: "Not treating life cycle issues early in the design process guarantees the late discovery of problems and very long and costly design iterations." From this viewpoint the development of a new product is not a problem of design alone, but design in the context of customer requirements, engineering, manufacture, vendor supply, standards, customer support, compatibility, cost, and many other issues such as the environment and the problems of product disposal at the end of the life cycle.

The Department of Defense (DoD) came to the same conclusions in a different way. Because it is primarily a customer of products, not a producer, the unique and incisive insights of the customer were available to the DoD. The symptom is that products "take too long to develop, cost too much to produce, and often do not perform as promised or expected" [3]. The root cause identified was that the design of the product is isolated from the design of the manufacturing processes employed later. The two functions are separated in time, and performed by quite different persons with little interaction -- perhaps in geographically dispersed departments. The resultant loss of information and intent, and the lack of exploitation of production knowledge and manufacturing constraints early in the development project, led to many malfunctions and cycles of changes to fix fundamental problems discovered late in the project.

For some time now there has been a realization that product development lags because there is no process of constant evaluation and exploitation of ideas whether originating in the research departments of individual companies or in the global research reported in journals. There is too little motivation on the part of research departments to communicate downstream and inspire development workers; and conversely, development groups neglect to involve research departments closely even in the initial launch of a project. The unfortunate separation between research and product development and how it may be alleviated is the subject of an interesting article [4].

1.3 The Solution is Concurrent Engineering

The solution proposed to the lack of cooperation among departments of a company on the one hand, and the lack of involvement of all relevant decision makers on the other hand, is Concurrent Engineering (CE). A definition quite suggestive of what is actually involved in practicing CE is as follows [5]:

CE is a **systematic approach**
to integrated product development that emphasizes

response to customer expectations
and embodies

team values of cooperation, trust and sharing
in such a manner that

decision making proceeds
with large intervals of **parallel working** by all life-cycle perspectives early in the process,

synchronized by comparatively **brief exchanges**
to produce **consensus**.

The definition of Winner et al. [3] emphasized the consideration of all elements of the product life cycle from the outset. The new definition places team values as the centerpiece and recognizes that CE, for all the top management commitment that might back it up, will fail if the organization, at all levels, does not embrace the norms of team working and does not learn to live by them each day. The desired sharing of early information, for example, has no chance of taking root institutionally, unless there is implicit trust that the information will be used without prejudice to its donor.

The next important value that is embedded in the definition is the focus on the customer. In the wonderful phrase of a Xerox Corporation advertisement: "Learn to look at your product through the customer's eyes, even if it is painful." The definition places the goal of the team's work squarely in terms of customer satisfaction, rather than in achieving some internal standard of the firm, which often may be once or twice removed from the metrics applied by the customer.

A further clause defines the activity in CE as being one of decision-making. The term CE is something of a misnomer. There is no value of CE, or methodology or technology to support CE, that is not applicable, with a simple transposition of terms, to any business decision. Indeed, any decision at all that requires the collective intelligence of several roles is approachable through CE. This highlights the importance of propagating CE across the enterprise in all functions, since the precise role that might have a bearing on a particular decision cannot be judged in advance. It is a mere extension, from stipulating the role of the customer as part of the product development team with an important voice, to think of vendors, shareholders, and society itself as having unique roles in product development decisions.

CE is a systematic approach. It entails new systems, and needs to be integrated into the organization as procedures and policies that will replace the less efficient ones of yesterday which caused information to be unavailable to those who had an implicit dependence on it, or caused it to travel by labyrinthine paths before reaching the role that needed to exploit it. CE replaces those tardy systems. If the "steep hierarchical ladders and organizational walls" that the MIT Commission on Industrial Productivity referred to are not to inhibit a firm, then new information transport systems must be developed to support the coordinated working of persons involved in making a decision, no matter where they are located.

The next important characteristic of CE is that it is a process, and one which entails parallel working, synchronized from time to time. The efficiency of CE is seriously handicapped if the work is sequential among the roles. A good team leader tries at every stage to partition the work so that many tasks can advance in parallel. This is a separate aspect of hastening the group progress, and the word *concurrent* may be used to describe it too, though the essential concurrency that CE encourages is that of applying all the decision-making perspectives in each phase of the project so that the judgement of several minds is brought to bear on every issue.

Parallel working has the inevitable penalty of inconsistency after some time, because the partitioning of tasks along roles is approximate in the best of situations; an overlap often remains. This leads to conflicts among the individual decisions as the parallel tasks progress in time. The resolution of these conflicts has to be planned, and this is done is by announcing a point of synchronization every so often so that the emerging alternatives and the details of the decisions may surface to the entire group's view. The team decides how to trade off among alternatives and render the decisions consistent again before the next regime of parallel tasks is set in motion.

The thesis is that these cycles of convergence may be more numerous in the life of a project than the find-and-fix cycles of the past, but the time required to achieve consensus is less, since the minds meet more often over smaller divergences of viewpoint. Coordinated working among team members demands frequent exchange of information so that the impact of decisions made at one stage is not allowed to lie unexamined till late in the project by those whom it might affect.

The need for this parallel working early in the project is emphasized through an important observation: fuller consideration has to be given to the large-grained decisions made initially because they set lower bounds on cost, time, complexity, and risk, as well as establishing upper bounds on achievable product reliability and customer satisfaction. The wisdom of spending a longer time in the design, rather than in the engineering stage, is the equivalent principle.

Finally, the definition returns to the importance of consensus, which is hard to achieve but nevertheless important if the team is to remain a team, governed by openness but also aware of the imperative of making decisions at every stage. All decisions include a rich world of options, but also exclude an even richer world of rejected options. Unless decisions are taken consciously with the knowledge of the group, rather than an individual, the faith in the basic team approach will be dashed. Consensus is seen as the process of bringing about harmony among conflicting requirements, and it ends in the sense of group achievement which overcomes the individual's loss from having had to yield ground somewhere along the way.

CE is not a single idea but a whole cluster of concerns during product development. It would be strange if an organization did not have a single CE approach implemented in some way. CE can be introduced in a straightforward way by constituting teams who have responsibility for the life of the product and interact to plan all aspects of the product from design to logistics and disposal. Through inter-disciplinary teams for the product development project the organization puts the members in close proximity, empowers them to take decisions and acquire resources, provides enthusiastic leaders, and then lets it all be worked out within the teams. The argument goes that the development time will be reduced because the communication from one team member to another has no barrier of distance or hierarchical access protocols to impede information flow and work requests. The product quality stands to gain also because the traditionally upstream designers are collocated with the traditionally downstream and neglected manufacturing or maintenance engineers. Thereby the downstream is kept aware of the design as it unfolds from the beginning, and can give the benefit of critique and suggestions, and convey the constraints of the downstream perspectives to the upstream designers. Time is also saved because one can plan well in time for changes and new acquisitions or developments in downstream facilities to accommodate decisions upstream. Cost is reduced indirectly by associating a money value with the reduced time to market.

Another approach directly addresses the quality issue by employing methods that go by the name of Quality Function Deployment [6] and Taguchi methods [7]. The first is sharply focused on defining precisely what the customer wants, and then evaluating every alternative developed in brainstorming sessions from the standpoint of the customers. The method is recursively applied to the product in a top-down decomposition along product assembly structure lines. It is traditionally done by humans with paper and pencil charts using the house of quality of Juran. Taguchi methods are probably the most celebrated of quality attainment methods, because they have been associated with the reputation enjoyed by well accepted consumer durables and industrial products from Japan. The central idea is that you can reduce variability in a product without using costly high quality components if you exploit non-linearities in the influence of design parameters on the product quality. Finding the best values of the design parameters is the subject of an elegant and economical method, first discovered by Rao, and widely applied and propagated by Genichi Taguchi and his disciples. It is a direct cost-reduction and quality-improvement method. As with the previous organizational approach it is apparent that *implementing CE does not necessarily call for extensive use of information technology and knowledge systems*.

While it may be comforting that there need be no specific reliance on computer-based information technology for CE, it would yet be a wasted opportunity if the great potential of computers is not realized in achieving the CE vision, at least for those organizations that have already traveled far along the road to CE by adopting the aforementioned approaches. One may start by instituting new direct information channels, and enforcing the percolation of information at early stages in the design to all the downstream areas of responsibility, and agreeing on the content and format of data that will be exchanged among the various perspectives. These things and the notion of the empowered team that can decide on its own how conflicting requirements are to be resolved, form a sufficient basis for CE introduction. But if the exchange of data among team members is largely *ad hoc* and across the table, if their means of coordination are largely by face-to-face meetings, and if available mechanisms like telephone, fax and paper memos are the basic media of communication -- then there is still scope for advancing the CE vision with additional technology.

1.4 Right the First Time

The aim of CE is often succinctly stated as getting things right the first time for all perspectives.

How does this affect *quality*? By getting it right the first time for all perspectives by being focused on customer needs and propagating the influence of decisions to all affected perspectives.

How does this affect *time*? By getting it right the first time for all perspectives so that the number of iterations through the whole loop and through each major stage of product development is just once. All perspectives should have participated and contributed modifications to achieve mutual consistency.

How does this affect *cost*? By getting it right the first time for all perspectives so that re-work is avoided, and by having considered costs over the total life-cycle, rather than delivery cost alone.

With this discursus on CE, we can proceed to the virtual team concept. Computers are becoming ubiquitous appliances for information processing and communication. They do not support teams and group working now, but that is being remedied by the growth of the new field called *groupware*. Besides much of the important data that needs to be shared among design teams is already in machine readable form. If communications and computers are married the separation of people over long periods of time and large geographical distances can be ameliorated. Imagine then the formation and operation of a team of distributed participants on an integrated product development project who are virtually collocated -- hence the name Virtual Team. They share data, communicate, and coordinate their work using a computer network and information technology as the essential means of accomplishing the close team work necessary for CE.

1.5 Barriers to Implementing CE

The idea of exploiting computer networks for team working is no longer a new one. Tools like electronic mail have been in existence, but this is far from sufficient to foster the structure of the team and the sharing of data that are necessary to coordinate the work of team members in the product development process. Computers and communication networks form the substrate of technology to create a new kind of cooperative environment. However there are many advances in computer communications that have been spurred by the requirements of virtual team working -- transparency in access to programs and people on the network is an important theme. The early attempts at using networks to improve the communication among team members utilized special equipment and the facilities installed in a separate room. Advances in communications and computer hardware have now made it possible to satisfy the special requirements of multimedia team communication without leaving the accustomed workplace of the employee. Today it is possible to think of a team being brought together on the network and sharing distributed data and having group transactions like meetings, notifications, scheduling of work, propagation of results, and so forth, being conducted entirely via the network with the aid of some CE services. This constitutes the Virtual Team. For CE the Virtual Team consists of team members in different perspectives of the product development, wielding powerful CAD tools in their private workspace, but publishing their results to a shared information base. Besides, an over-arching coordination service exists to schedule work, report progress, notify persons, generate work authorizations, and carry out entire suites of coordinated processes among multiple perspectives, without the benefit of face-to-face meetings.

Teams owe their high potential for problem-solving to an assortment of factors. The most important is the direct, unimpeded flow of information in many different forms, employing every one of the senses. The communication incurs no time delays; it is multi-way and interactive. Ideas can form and one expert can influence another's thinking. Solutions can therefore emerge more rapidly in those cases where the solution depends intrinsically on placing together two or more pieces of the puzzle which are held by different people. Moreover creative synergy can only appear when active minds inspire thoughts that would not perhaps arise in solitude.

Not only is it necessary to see the problem from many different points of view to comprehend its multi-dimensional character, but the sharing of the vision is a pre-requisite for the team to work at the highest level of achievement. More limited sharing and tighter control on the information flow will inevitably result in a less optimal solution -- one that does not attempt to keep in mind all the requirements from the beginning. If it is not possible by constant interaction to create a common vision of where the team is going and what are the criteria by which success in the project will be appraised, and the kind of product features and characteristics which will excite the potential customer, it is more than likely that team members will be working at cross purposes, and perhaps at variance with the customer's interest.

When team members interact there is always a phase when there is disorder, when ideas are being generated and there is no solution in sight. This period of divergence and confusion is a necessary part of the problem solution process. The argument here is that close team interaction, as occurs in a collocated team, accelerates this process and gets the team rapidly through the initial mind-expanding phase of the project. Closeness, rapidity of interaction, and the sense that ideas are falling on fertile ground and critical listeners who share the same goal are tuned in, is important. It encourages the free flow of ideas and corresponds to a Socratic role for team members to become midwives to the birth of new ideas. The ideas must not be lost at this stage, and even if the time for a full consideration is not available, at the very minimum everything must be captured for the future.

In a later phase, when the team starts subjecting the ideas to communal critique, each perspective of the problem surfaces. This is a critical point. New ideas and new points of view are being urged and the greatest openness is needed on the part of team members to be willing to accept unfamiliar notions, but not to shy away from asking all the pertinent and rigorous questions from the unique vantage point of each perspective. The combination of openness to "foreign" ideas and rigor in applying the professional criteria of a discipline and the wisdom gathered bv experience, is what characterizes team interaction at its best.

1.6 The Computer Network -- a Powerful Resource

It is not strange that computers should be able to play a central role in overcoming the barriers of information flow across distances. Information management is the chief strength of computers, and when this strength is exploited in concert with the potential to reach out by telecommunications across distances, there is at hand a powerful tool to build just the kind of bridges by which human thoughts and decisions can be managed in a distributed team over the life of a project.

There is much work that is routine (e.g., reporting) and much that is peripheral to the skills of a designer or decision-maker (e.g., data management). At the very least in a computer supported environment for team-work, the labor entailed by such activity could be reduced greatly. But there is more that can be accomplished if computer systems are designed to support a framework of policies and procedures that accompany decision-making. With such systems a participant in the process is walked through the process by task fragments that appear in the person's work list and a set of packaged instructions.

The various forms of interactions between team members can be structured to provide great clarity in the interchange and eliminate incomplete or omitted information. Most of the day-to-day interactions are routine requests, messages, data feeding and reporting. This can be aided by structured forms-oriented data entry in the CE environment for carrying out these various individual submissions.

The access to tools of work can be managed from such an environment in a way that goes beyond mere icon double-clicking. Tools may be on the network. Can they still be thus accessed? Tools may need to be connected to data files which are on the network. It would also be desirable to access the capabilities of one tool from another remote tool. Making this possible with ease for the end-user and not much work for the software support group is one of the many things a better group working environment is expected to deliver to the CE-inspired organization.

The concept of virtual collocation will be developed further later on. The essence is that by providing transparency at a foundation level, as part of the network services, for treating users and programs in the same way, whether remote or local, the path is made easier for the development of CE services and end-user applications to exploit a unitary view of the team on the network. Virtual collocation then becomes an efficient replacement of the erstwhile physically collocated team, a replacement which saves on wait time and travel time.

1.7 Compunications Challenges

The marriage of computers and communications has been referred to as "compunications." The network can compensate for physical separation, and its associated penalty of time, by making the entire team and its work and the computer-based tools present to every team member on his screen. These are challenges for the technology. Since the exchange can involve large engineering data files of millions of characters, and images that can be even larger, the bandwidth of the network is a matter of concern. Today's networks are fast only in local groupings within one office or plant complex. They become progressively degraded in speed as the lines reach out across states and countries. High speed networks over wide areas are not common today.

The computer's ability to deal with human input is also severely limited. Media such as voice input, video cameras, writing tablets, and so on are not standard equipment. What is standard is a color screen and a keyboard, and this places a limit on using the network as a comprehensive replacement for telephone, tele-conferences, fax, and other common means of conquering distance and time. Fortunately, workstation computers and personal computers are becoming available with multimedia support as standard, or as inexpensive add-on features.

Using all of these capabilities is going to pose another challenge for computer hardware and software vendors: keeping things simple. Not to distract the user intent on his work by having a plethora of manual set-ups and switches of work context to exercise the CE environment through a rich set of multimedia transactions, is a requirement. All must be natural and simple. And herein lies a great challenge for those would exploit the workstation of the future to create the ideal groupwork environment for CE. It will be characterized by transparency among the network resources, simplicity of organization of CE services, and a completely natural adaptation to the mode of working of the user.

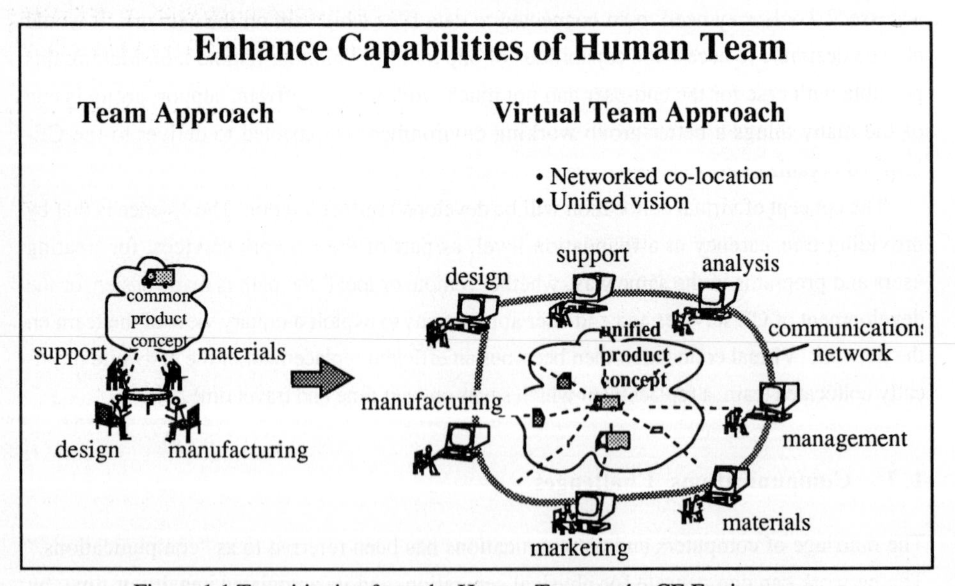

Figure 1.1 Transitioning from the Human Team to the CE Environment

1.8 The Virtual Team Concept

The notion of the Virtual Team is at the heart of efforts at the Concurrent Engineering Research Center (CERC). The well-known format of the tiger team is the base metaphor for getting things done in concert by a group of experts who each bring a unique viewpoint to the project. They have their roles well established and arrive at a shared view of their goal as a team. They are also empowered to do the needful without undue interference and usurpation of responsibility; but additional input may always be given or brought in by the team members themselves.

The team is thrown together for long sessions of intense exchange and thinking. The expectation is that out of this process will come a well-judged result which none can disavow, since it is the group's solution not an individual's solution. And a solution which, furthermore, does indeed contain a response to multi-faceted objectives. It will take time for a harmonious adjustment to often contradictory demands to come about. But early feedback on ideas and constant critique bring about great appositeness of thought during the problem solving process. The central thesis is that individual expertise is never enough to uncover the waterfront of requirements, and one may be led to totally wrong solutions, and most certainly to less efficient ones, by failing to represent in the team all the "perspectives" from which the consensus decision will be judged when played out in the real world.

Human tiger-teams are episodic in their work. They form and dissolve as soon as the mission is accomplished. Is it possible that tiger-teams can be a way of doing work day in and day out? Can one maintain the same energy in the grind of daily routine as that which workers, blue- and white-collar alike, experience when working in tiger-teams? That is the goal which many organizations have espoused, from factory floors to product engineering.

Seen thus, CE is a people-oriented solution to planning and executing work. Its mandate is to employ representative teams, and empower them to act as the owners of a project. Many companies have made their first hesitant steps to learning the new paradigm of team-working in just such a way. The teams have to be chosen carefully. If there is to be constant interaction you must respect those you interact with and be able to share ideas frankly without raising the human fear of being challenged or criticized. Training in the process of interaction so that negative comments can be given in a reasoned way is part of the learning. And the best experience suggests that this is not learnt in a day, but needs to be constantly reinforced by other team members when they observe disruptive forms of interaction.

The notion of the virtual team is fraught with difficulties if one is going to rely on the existing communication services and existing means of sharing data. These amount to little more than sending electronic mail and transmitting files. There is far less plenitude in this exchange as compared to the exchange among persons in a room, and too many veils of inaccessibility to penetrate before the data actually becomes visible. Therefore these primitive services need to be enhanced before they can serve as surrogates for the face-to-face teams they are expected to replace or supplement. This provides wonderful opportunities to do, not merely as well, but better than face-to-face teams. Computers are begging to be exploited for group work, a fact which has let loose creative minds on a range of problems. The ideas that have emerged promise to change the team working process in organizations, and endow such capabilities on the team, that the organization of the future will depend almost entirely on the new CE services to conduct their work in the team format.

This is a radical change in the application of computers, from the development of ever more powerful tools in the service of the individual decision maker or engineer or office worker, to inventing tools that serve a group and are exercised in a group and allow a group to go through all the stages of decision making.

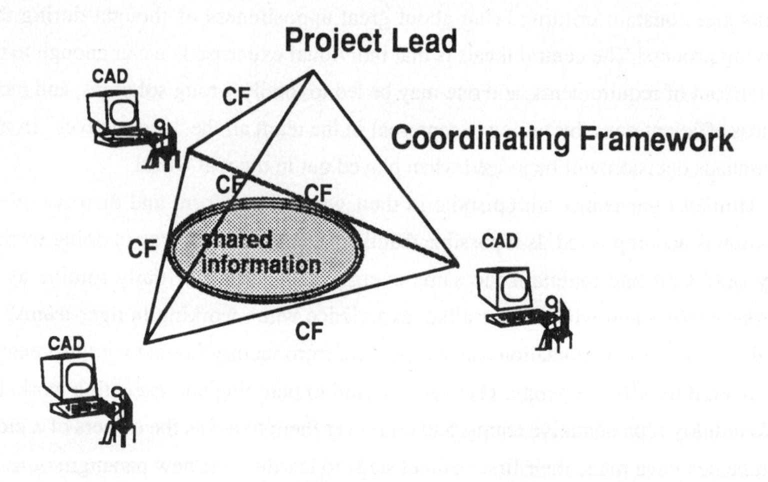

Figure 1.2 Cooperative Decision Support

1.9 Virtual Team Model

Apart from the underlying communication facilities, there are two basic capabilities needed for teams to perform over a network. A central repository of shared data must exist so that whenever information is to be shared it is placed there, and automatically becomes accessible to the entire team. This scheme is more general and less burdensome than having a series of point-to-point data flows to share data.

The second important requirement is a coordinating framework to enable a project leader to communicate tasks, goals, and deadlines; keep the team abreast of work-in-progress by team members; and cause the individual decisions to propagate their influence throughout the team. Such an over-arching coordination will force the team to render consistency to their decisions from time to time, and thus achieve consensus, so that parallel working does not result in a permanent and irreconcilable divergence of viewpoints.

The virtual team defined in this way is a good model for decision-making in the corporate realm. There is nothing that is specific in this model to engineering or product development. All manner of decisions that require multiple perspectives and the creative synthesis of a solution from disparate individual contributions, are amenable to such a treatment. The process described and modeled by concurrent engineering is really a decision-making paradigm, not solely a product development paradigm, and is therefore worthy of adoption across the enterprise.

'Product data model' is the correct term to use if product development is the aim of the project, as it is in the original notion of concurrent engineering. If it is any other type of decision, one still needs to have a data model of the decision, to represent what it is that is being decided, and to make trade-off analyses in terms of these 'what's'. The computer representation of the 'what' is the data model, and this brings a crispness and definition to the debate before the decision is taken. Of course, the presentation of alternative decisions may actually take place under different data models, because the 'what's' of one alternative are not those of another alternative. This shows that *data models* for a decision need to have versions, just as the *data* of one data model will have versions, as the decision evolves.

What has been sketched is not an automated design system. It is very much a system to provide extensive data sharing and coordination and communication aids to the human experts who need to work together. The expert is still in command, and if he has any automated design tool it would be automation resident in the CAx tool used, and not in this system for collaborative decision-making.

1.10 Workflow Management

If the shared information base holds the 'what' of the decision, the how is even more important, because the productive efficiency in an organization resides in its ability to repeatedly use the same processes to arrive at different decisions from the data characterizing the different situations. The 'how' is what we call the workflow. Modeling it and then causing the fragments of work to be executed by the networked participants in a project is the concern of Workflow Management. If CE is above all a process by which things are done, then Workflow Management is at the heart of CE, because it enforces the coordinated parallel or sequential tasks that implement the detailed CE process over a network.

Admittedly, there is a lack of strict determinacy in the work on a product development project. Everything can not be foreseen and recorded in a PERT network soon after the project starts. CE is not an orchestrated action among team members who are marching to a score written in advance. At best the project leader has a detailed plan of work for the near future, and a knowledge of various milestones to be achieved in the more remote future; though with no assured plan that can mechanically and routinely compass those milestones. Workflow management will cause the near term tasks to be carried out by the team, and gather the results so that they can be disseminated among the members.

The workflow manager can be enhanced to draw upon a library of processes that are frequently used to effect a piece of work. For example, if a mechanical engineering task to analyze the natural modes of oscillation and evaluate the energy fed to those modes from a continuum of driven frequencies is a recurring task, then it can be represented as a task fragment, described in terms of the manual and computer operations and laboratory investigations it entails, and kept in a library. In that way every time the task needs to be done it can be retrieved and instantiated to fit the current problem. In a sense expertise to do a skilled job has been captured as a recipe that can be applied again and again. But workflow is not necessarily linear. There could be decision points to decide between two alternative workflows based on data measured in the process.

The basic parallelism desired in CE is realized by a judicious task breakdown. There are no set rules by which this is done. The experience of the project leader will determine how successful is the partitioning of work among perspectives and what is done in parallel. With CE there is a great premium upon doing work in parallel, even if it means some temporary inconsistency. The project leader must risk the inconsistency in the confidence that soon enough there is a point at which team members will be required to synchronize and remove conflicts, if any such have surfaced.

Seen in this light the CE process differs from the classical sequential process chiefly in the replacement of fewer long find-and-fix cycles by more frequent but shorter cycles between synchronization points early in the product development process. The product development process progresses in stages. Each stage comprises the activities shown schematically in Figure 1.3. The steps indicated in the shaded blocks are conducted using the CE services, and the transactions they make possible on the network, without the product developer ever needing to leave the workplace.

Virtual Teams are today's answer to bring about intimate collaboration, speed of interaction, sharpness of communication and clarity of roles. Virtual teams leverage the model of tiger team interactions and implement it in the modern context where it is all but impossible to have the contributing expertise assembled in one location. It pays heed to the promised

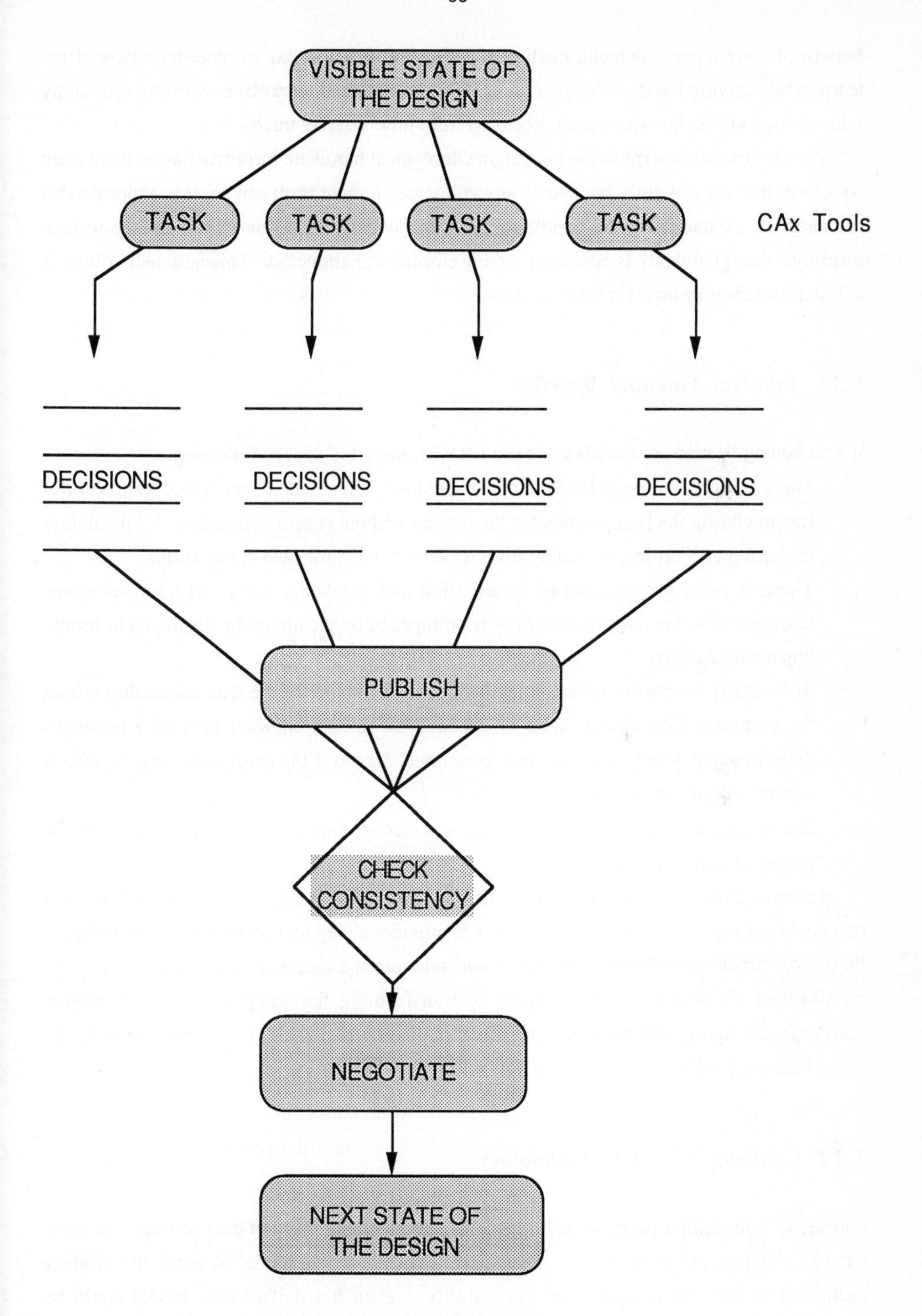

Figure 1.3 The Basic CE Process

benefit of computers which can mediate the vital information link to other members of the team, while leaving the knowledge worker in his accustomed workplace, working efficiently with all the tools on his workbench, liberated from time-wasting travel.

Virtual Teams can exploit the existing technological hardware infrastructure to form team structures that are as tightly bonded as any collocated human team can be. It is expected that the need for exchange can be handled most often without travel and without face-to-face meetings, though nobody is advocating their elimination altogether. This new technology is still in its infancy, untried for the most part.

1.11 Inherent Potential Benefits

If a reckoning be made of the advantages a list will surely include the following:

- There is no limitation in location or composition of team members. The organization is free to choose the best persons for the project without regard to location. All the delays in starting up a project stemming from relocation are eliminated at one stroke.
- There is rapid propagation of information and decisions across all team members electronically. The information flow is unimpeded by the hierarchy displayed in formal organization charts.
- Role clarity across disciplines is embedded in the notion of the data ownership within the common data model. Who is expected to decide on what part of a product's definition and who has a voice in it is defined. So too is the team awareness of who is currently doing some task.
- There is great flexibility in constituting and reconstituting the team as the needs of the project change over time.

Response time in the marketplace is critical. Companies that have fallen behind are ones that could not exploit opportunities in time. CE provides a way for corporations to reconfigure their most precious resource -- people -- and reinvent the corporation in a new form. The possibilities are limitless. The promise of transforming the company into one, whole, constantly changing, and adapting, organism is thereby realizable. It is a resolution of the Chief Executive Officer's constant anxiety.

1.12 Challenges for CE Technology

Sharing of information poses several obstacles: the incompatibility of data formats and tools used in different perspectives, the absence of a common data model to unify the product definition in multiple perspectives, the inability -- even if a unified data model could be prescribed -- for the data to be accessed from the target CAD file or repository and made available locally at the user's workstation, the lack of transparency in data access when the actual data are lying in many different and distributed data bases and CAD files, the lack of standards for data interchange to overcome the incompatibility of basic hardware and software.

In the *communication* arena the outstanding difficulties to be overcome are: the difficulty of accessing programs which are distributed, the lack of uniform ways of communicating among programs, the lack of transparency in communications which implies that you need to know the location and capabilities of each program to send a message to it, the lack of network directory and registration services for programs, people and resources, the inability to treat programs as community resources that anyone can access from anywhere if authorized, the lack of computing features to manage a stream of jobs to be run in some serial/parallel sequence on different computers, the lack of communication applications that work for a group and permit exchange using multiple media, the inability to archive exchanged information for easy retrieval by indexes, the lack of features to run a structured negotiation session over the network.

In *team coordination* today these flaws are often apparent: person and function roles are not clear, there is no responsible team or leader, all life-cycle concerns are not represented, requirements are not formalized and clear, no tools exist to monitor product performance criteria, there is a failure to partition work, there is no tool to manage the flow of work dynamically over the communication network, tasks under way in other perspectives are not visible, the influence on one discipline of decisions taken elsewhere is not realized till late, consistency is not fully assured at the time of signing off, tools to evaluate trade-offs among multiple criteria belonging to different perspectives are lacking, the current product data are not in public view, a strong sense of the team presence is difficult to maintain over wires.

These barriers call for the development of a new generation of computer software (and hardware) that will be the foundation on which team operation on the network for the engineering domain is built. You might say that the provision of these services, in as integrated a fashion as possible, and conforming to reasonably widely adopted standards, will ensure a far better distributed environment for the practice of CE through Virtual Teams. It will form the technological support for practicing CE, and like any technology it has the potential to accelerate and institutionalize the practice -- beyond the power of recalcitrant non-believers in CE to sabotage. Once the wires are strung all through the enterprise and these means of disseminating data and coordinating work and communicating intent exist, the formation and empowerment of teams is all that is necessary to spark radically new modes of collaboration.

In the future vision of CE technology the present-day collocated teams with face-to-face communication occasionally assisted by telephone, fax, and email will give way to mechanisms of communication that are more powerful, more swift and more rich in presentation capability. These technologies of multimedia and shared workspaces spanning multiple machines will effectively subsume the current weaker forms of communication and effortlessly render the team (virtually) collocated from the workplace. One need not get up and go to a meeting; one is never out of a meeting or out of reach to the team members.

The engineering tools of customary use which accommodate the solo user today will become 'collaboration aware' and work off a unified data model and even enjoy a common representation for limited subsets of the data. Tools will become integrated and pass data seamlessly across the network, both through the systems integration efforts of manufacturing companies and through adoption of standards by tool vendors. The interfaces to interact with the engineering tools will become more intuitive and natural; they will employ 3-dimensional interaction and visualization technologies and allow a better touch and feel and manipulation of the artifacts modelled and simulated in the computers.

And from being hierarchical organizations with sluggish and rigid information flows across disciplines, companies will become alert self-managed teams which propagate their decisions swiftly across boundaries of functions and coordinate their work by tracking hundreds of constraints and requirements automatically over the life of the project.

Add to this the great reduction in travel, and the savings of time previously invested in data chasing and reporting which can be channeled into the important design and trade-off activities, and we can envision not only a faster, but a "greener", route to better products.

2 The Virtual Team Support Technology

2.1 The DICE Technology

Put simply, the DICE technology goal is to enable members of a Virtual Team connected by a high speed computer network to communicate and to coordinate product information and workflows so as to promote cooperation and achieve rapid consensus. DICE advocates using a set of computer-based services to enable a team, cooperating over a network, to transcend the barriers of distance, platform and tool heterogeneity, and insular viewpoints. The domain experts, of course, retain their specialized computer tools as the primary means of design, analysis, and simulation, but these are no longer wielded in a private workspace for single perspectives. There is sharing at every level.

DICE is not concerned with developing techniques for automating the design and development process. These functions are well-served by computer-aided design and engineering software already on the market. Rather, the underlying motivation is to confer on large teams composed of many perspectives and geographically distributed information resources, the same benefits that small tiger teams enjoy when working in close proximity. This is accomplished by creating an open system that allows the deployment of certain generic CE services that can be accessed from heterogeneous platforms. It is based on the client-server model shown in Figure 2.1.

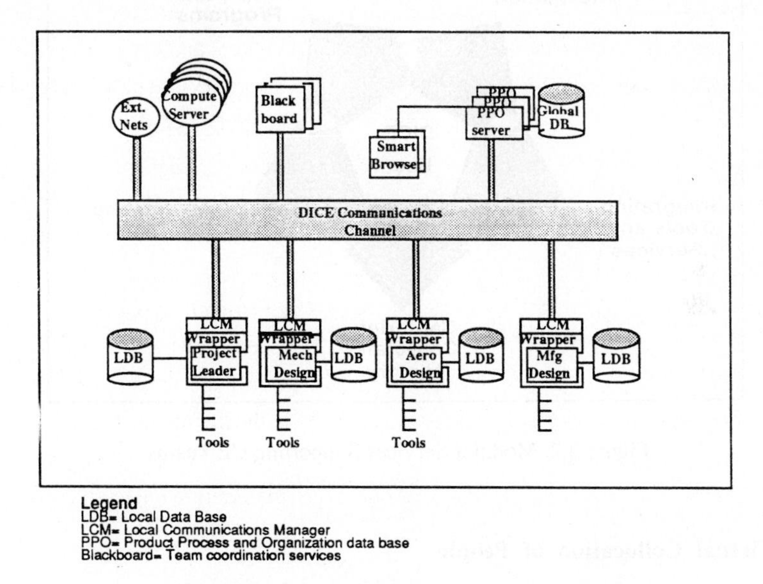

Figure 2.1 The Information Bus Architecture

The DICE services are based on such existing standards as X-Windows and TCP/IP data communication protocols, and on such commercial products as spreadsheets and hypermedia software packages. These services are still evolving and have many gaps (for example, workflow control). Nevertheless they exist, as do documents that explain how to incorporate and use the DICE services. And they have been demonstrated successfully in an environment of multiple, geographically-dispersed perspectives with many teams, and they cover a large part of the life cycle of a product. Several organizations are in the process of integrating them into their own development environments.

Users are presumed to have workstations from which they invoke computer assistance for their tasks. The data resides in a company-wide information repository called the Product, Process and Organization data base, composed of all the data regarding past products, process models, and resource characteristics. The engineering applications are "wrapped" to access data from the common engineering data base. In addition, local data can be used. The system provides a set of generic services to the virtual team, and these are shown on the upper part of the channel through which they communicate. The DICE generic services fall into five categories, as shown in Figure 2.2. They are described at length in several papers in the Second and Third Symposia in Concurrent Engineering [8, 9]. These categories are set forth below with brief descriptions.

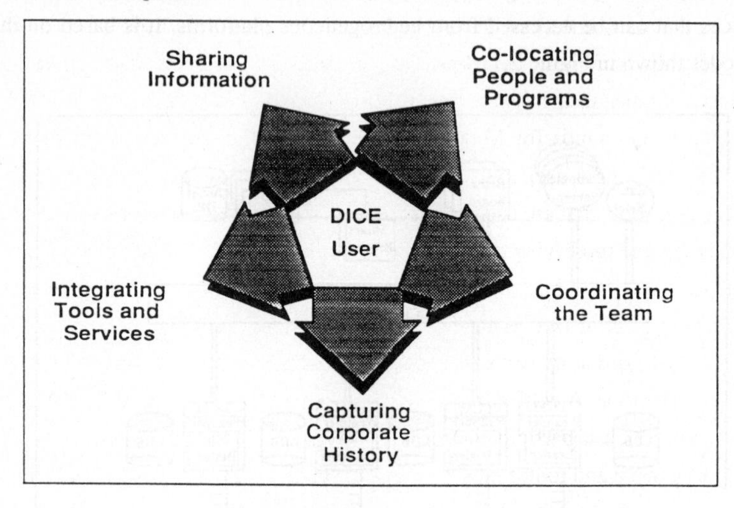

Figure 2.2 Modular Services Supporting CE Teams

2.2 Virtual Collocation of People

This service envisages enhanced communications among team members by providing them with the following capabilities:

- Conferencing among multiple remote, networked participants, supported by facilities for archiving and management of the proceedings.
- Applications and display sharing, based on the "client-server" model that allows all team members to interact synchronously with a shared application by using the same interface, including graphical outputs.
- Multimedia capabilities for creating, editing, and saving compound documents containing multimedia information by using such X-window-based document editors as FrameMaker. Multimedia communication capabilities should be enhanced with real-time audio transmission, voice mail, and real-time video capture and transmission.

MONET is a real-time multimedia conferencing system to facilitate virtual collocation of people, computers and information resources. It is an example of a desktop conferencing system, developed in the DICE project to bring the immediacy of human face-to-face meetings to computer-mediated meetings on the network. The potential benefit is that team members are able to confer without leaving their accustomed workplace, thus saving time and reducing the considerable energy consumption in going to and from meetings (a form of "green engineering", one might say). Besides, for the sacrifice of some body warmth, the significant advantage accrues that one may respond faster and advance decision-making with greater rapidity because many questions that come up and otherwise would remain unanswered till the next meeting, can now be responded to with the data bases, engineering tools, spreadsheets and other such software accessible while the meeting is in progress.

MONET, which stands for Meeting on the Network, also enables the sharing of the output displays of single-user tools, provided that they are X-Windows compatible. Thus many single-user tools that are used to visualize data can now be simultaneously shared and used for viewing and modifying detailed product information in the public view of a team of concurrent participants. While these changes are being made, voice and text exchanges can take place, and even be affixed as an annotation to the product data.

MONET uses a directory server to locate and ascertain the availability of a participant in the network at any time. A conference server executes in every host machine that is currently the locus from which a participant is meeting; it manages all conferences in which the participant is engaged and routes messages (text, graphics, audio, and still video) that come to that host machine from other hosts, to the appropriate conference window. Therefore messages of many kinds may be exchanged over the data network: spoken voice, graphic screens cut from a CAD tool running in one window and pasted into the conference window, and typed text. If there is provision for a video camera and corresponding hardware boards in the workstation, it is easy to capture a frame and display it on all other participants' screens. However, live motion video is inhibited by the severe inadequacy of bandwidth on most networks, even local area networks, in spite of the data compression and differencing schemes employed in some advanced protocols. The underlying communication is provided by TCP/IP with the BSD-Unix socket entities. Each conference server waits for messages by listening on a pre-assigned port. Functions are provided to start a conference, leave a conference, call participants, add participants, close a conference and some additional functions for housekeeping. The typical appearance of the conference windows is shown in Figure 2.3 [10]

Participants needing to interactively share an X-Window based application start up an additional server called the Cooperative Multiuser Interface to X (COMIX). This sharing is transparent to the application and thus collaboration via that application can be effected without changing a single line of source code -- this is the great advantage. One copy of the single-user application is shared among multiple users, and naturally this necessitates the serializing of inputs reaching the application server. One of the implemented serialization mechanisms is "chalk passing", whereby each participant may request the chalk, and when it is obtained, the participant can interact and modify the data at will (all the while changes are seen simultaneously by others in the conference -- this is referred to as WYSIWIS or

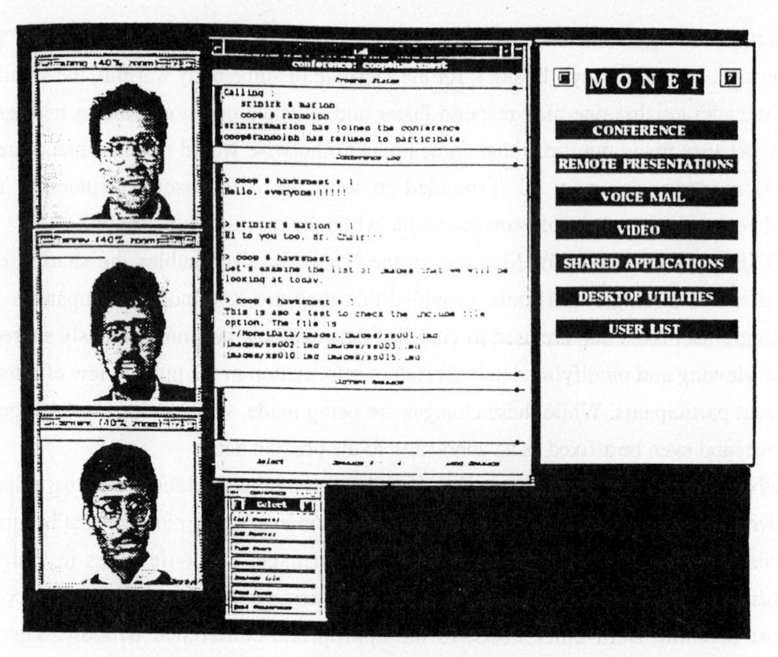

Figure 2.3 Participant Icons and the Conference Window

what-you-see-is-what-I-see) until the chalk is voluntarily released. Pre-emption by a moderator is permitted. The underlying architecture is one in which the COMIX sharing server intervenes between the multiple X servers and the single client application, intercepting and multiplexing the mouse/keyboard inputs and the display outputs. In effect, COMIX simulates the scenario of multiple participants gathered around a chalkboard and manipulating the information on it in full view of all the participants, who can listen, see, and contemplate everything in a team context.

MONET has a simple archiving capability, permitting the whole meeting to be saved as a file, and played back later for review. It still lacks the indexing mechanisms that can turn it into a corporate memory, at least as far as the proceedings of meetings are concerned. It has been proposed that before a meeting is adjourned a basic dialog box should be used to capture the conference chairperson's summary: the agenda discussed, the decisions arrived at, the names of parts or subsystems that figured in the meeting, any particular constraints that weighed in the discussion, assumptions, and so forth. These will be used to provide keywords by which the meeting would be indexed. Some indexing elements would be automatic such as the perspectives involved, date, and project. With this minimal post-meeting summary, it would be possible to answer many questions even years in the future; for instance, when was the material of this part decided? What were the assumptions at that time? Who authorized this modification? Indeed, MONET-like meeting facilities could provide archives that constitute one facet of the corporate memory (see below at Section 2.7).

2.3 Virtual Collocation of Computer Programs

Facilities to exchange messages through synchronous meetings and asynchronous (but structured) electronic mail will be strong pillars of the virtual team support technology. It is just as important to improve the ease of communication among distributed programs, and the ease of sharing common programs. Enhanced communication among *applications* in heterogeneous networks will allow a greater degree of cooperation than now is generally possible. The following capabilities are indicative:

- A distributed yellow pages of resources allowing access to people, software, and machines available on the network, without prior knowledge of their physical location, as well as transport of data between computer programs by using uniform interfaces.
- Application and task management supporting *transparent execution* of remote applications which reside anywhere in the network, including job control capabilities over the network similar to those commonly available in operating systems for a single computer.

The Communications Manager (CM) suite of software in DICE is a collection of hierarchical modules that provide important services for collaboration by messaging among programs to bring about interaction and integration. It is based on BSD-Unix sockets. A picture showing the layering of functions is given in Figure 2.4 [11]. The bottom layer, the CS (Communication Services), has two components, one that provides a connection-oriented service whereby the two processes are synchronized during the entire time of communication; and one which is connection-less allowing messages to be conveyed when the process at the receiving end is not actually executing at the same time as the sending process.

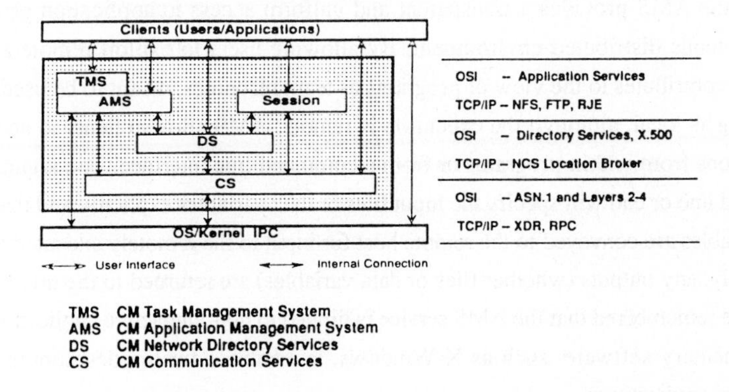

Figure 2.4 Layered Services of the Communications Manager

In the connection-oriented scheme, the client process (which wishes to establish a connection) and the server (which wishes to accept connection requests) acquire communication end points on their host systems and bind it with addresses they wish to be named by. Then the server indicates that it is prepared to accept connection requests from the potential clients and listens for any calls. The client makes the connection request to be connected to the server with a known address. Thereafter the two processes can engage in passing messages to one another and a variety of data types is supported.

The connection-less protocol is implemented with a library of callable C functions used by the programs, and two daemons are set up in the sender and receiver. The sender specifies the receiver's address, the message contents, and any special flags. The receiver can specify that it will receive from a specific sender or from any sender. The messages received are stored for future retrieval, and once again a variety of data types is supported.

To enhance transparency as to the physical address of the sender and receiver, a dynamic directory service is provided. A directory server daemon runs on at least one machine in the network. With this facility whenever a communicating server or client comes up, a database is updated by the directory server indicating on which host machine the named server or client is currently executing. Send requests need only identify a symbolic name of the server and the messages will be routed automatically to the appropriate host, and the CM daemon running within it will deliver the message to the right server application in that host. This is very valuable in distributed systems because application programs should be developed without relying on physical addresses which never remain fixed in a large network. It also simplifies the migration of distributed applications from one network to another.

The Application Management System (AMS) is a higher level component and it serves efficiently to keep track of applications which are registered for invocation by any team member on the network. The user does not need to know the protocol for executing a remote application, or discover in which host it is meant to execute. Combined with the directory service, the AMS provides a transparent and uniform access to application programs in a heterogeneous distributed environment. By allowing users to exploit remote applications easily, it contributes to the view of programs as team resources, meant to be used by all, and appearing as local, although the execution is remote. AMS allows clients to access remote applications from within programs or from a command line interface. The arguments of the command line or call will specify the input files or data variables explicitly and these files and data variables are conveyed to the remote host for input to the remotely executed application; conversely, any outputs (whether files or data variables) are returned to the invocation point. It must be remembered that the AMS service is limited to non-interactive applications, because complementary software, such as X-Windows, is available for the developing interactive distributed applications.

The Task Management System (TMS) is designed to satisfy the need to execute an entire suite of applications in some sequence (parallel/serial) and transport data variables and data files from the output of one application to the input of subsequent applications. This is often the case for performing engineering (and even business) analyses. The TMS is, in effect, the conventional job control facility of an operating system, extended to heterogeneous distributed systems, without requiring any changes to the native operating systems. The key element is the task files through which the user needs to specify once the stream of tasks and their execution dependence on other tasks. The entire suite can then be invoked as a single macro command repeatedly.

2.4 Team Coordination

The following activities in a distributed team would contribute to enhancing the coordination when multi-functional concerns in an enterprise have to be taken into account:

- Team formation, structuring, and management of the personnel deployed throughout the organization in multiple teams, working together on several projects at once in different roles.
- Planning, scheduling, and managing of project related team activities, based on activity models that use a shared information data base. This capability needs to be provided through a "Workflow Manager" module that manages the resources, analyzes the project for schedule optimization, generates reports, and disseminates task assignments via the generic communication services.
- Sharing of common views of the evolving product design to allow prompt notification of any changes and decisions to all the team members. This capability can be offered through a blackboard scheme that provides common visibility of the product model, at different levels of granularity, through a graphical interface.
- Directing the flow of information among the team members and supporting negotiations for conflict resolution.
- Managing product requirements and constraints in a persistent manner throughout the product development process. This activity is supported by a "Constraint Manager" that will capture and maintain a constraint network for continuous evaluation of the product/process parameters as they are "posted" on the blackboard.
- Monitoring of the product development cycle through continuous tracking and assessment of pre-selected "figures of merit" that reflect the quality of the design and the decision-making process. This activity is supported by a "Design Assessment Tool" that will apply quality evaluation models to an attribute of the the product, and evaluate and display performance metrics in the form of graphical and/or tabular reports.

The Project Coordination Board (PCB) [12] is a DICE coordination service for teams of users who are linked together in the performance of work over a long period of time and build up the product's data definition gradually over that time. It works on the basis of a common workspace in which firstly, the tasks that are being worked on by team members at any time are visible to all, and, secondly, the actual product data decisions that result from these tasks in multiple perspectives are asserted onto slots in a structured data model and made visible to team members. These commonly visible tasks and product data are the fundamental bases of the coordination provided. In addition to this there are notification mechanisms, attached to the product data model, so that changes to specific product data are notified by messages to those perspectives that have expressed a dependence on them.

A Constraints Management System (CMS) is also attached to the product data, expressing the known customer, manufacturing, or other constraints that tie the various product data attributes into dependent relationships. In this way the project and the organization can, at once, enforce customer focus and take into account the physical or manufacturing constraints that are needed to achieve consistent, manufacturable products, satisfying or exceeding requirements. Conflicts are noted at the earliest time and notified to the project leader and team members for action.

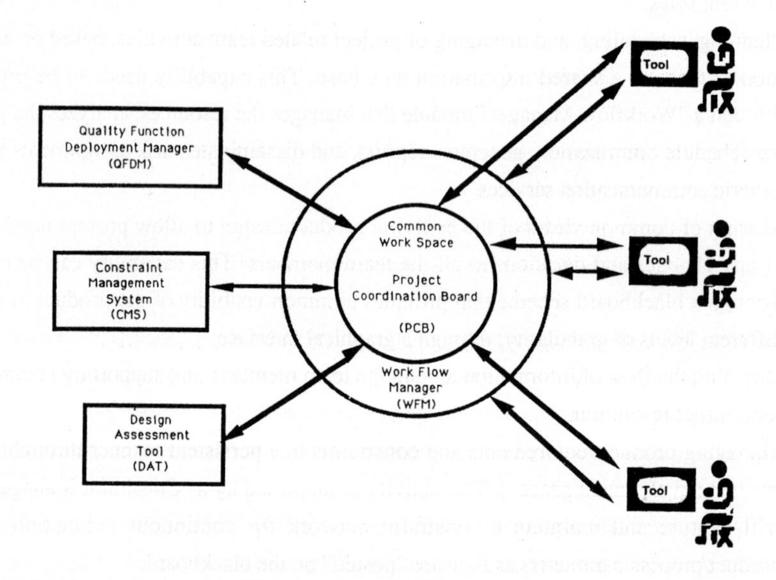

Figure 2.5 Coordinating the Team

Reporting and assessing of product performance is also available as an automated service once the the project leader specifies which product data attributes are to be directly tracked and what performance measures are to be computed by specified programs based on the product data held or pointed to in the PCB. It must be stated that product data attributes may be explicit data attributes, or pointers to CAD files, that contain data attributes. The performance monitoring of the product is done by a module called the Design Assessment Tool (DAT), a module that maintains the format and content of a typical assessment procedure that has been specified. The DAT executes whenever invoked (or at periodic intervals) and displays the results graphically and in the form of a report. In this way the project leader can obtain a "goodness" measure of the current state of the design based on multiple criteria, presumably reflecting the customer priorities.

The notion of the team itself as something that is configured into a data base of team members with specific roles during a project is reinforced by the PCB. A team member may belong to several projects, each with its own set of team members, tasks, and product data model; whenever users in the organization sign on to a project via the PCB from their workplace, they immediately see tasks waiting for them, other tasks being done by other members of that project team, and the current state of the product data. Consensus building is emphasized by team members being notified of changes and having to respond to them. The PCB, with its significant capabilities for sharing and disseminating vital project information, is visualized as a key element for coordinated team work when the participants are distributed.

2.5 Information Sharing

Shared information services are central to the notion of working rapidly and consistently in a group. Some standards are inevitable to enable information sharing, and the following specific capabilities are needed for engineering information services:

- Information modeling and representation in conformance with the emerging STEP standard of the PDES (Product Data Exchange using STEP) consortium.
- Storage, retrieval, and exchange of product data by using an underlying Object-Oriented Data Base (OODB), in combination with uniform methods to access legacy data bases.
- Configuration control of multiple versions of technical data, engineering releases, parts, and documentation, tailored to the rules and procedures of specific organizations.

The PCB requires a common data model that evolves over time and specifies which data in various perspectives, organized as a part-of hierarchy (the product structure tree of engineering data management), suffice to define the product comprehensively. The PCB is dynamic, in the sense that the product data model is not specified ahead of time, but evolves even as the project work advances.

On the other hand, there is the need to access the completely defined data of past products, whose data model can be specified with finality. However, the usual case is that the totality of data for a single product lies in several different target data bases, and even in the form of multimedia files, CAD files, and mere paper. How then to provide a uniform way of accessing and presenting such variegated data from several sources, heterogeneous in format, different as to access mechanisms, and distributed over the whole network? This very large and challenging problem is not going to be solved in the near future, at least without considerable restrictions that may make the solution irrelevant for organizations engaged in developing complex engineering products.

The goal of one of the DICE efforts is to provide a unified way of accessing representative information repositories, and to develop methodologies for implementing it systematically.

One component of the information sharing system is called the MIND (Model-based Information Directory) [13]. This model defines what data exist regarding the product and indicates where that data ultimately resides. There are back-end interfaces to representative data repositories -- a commercial relational data base, a multimedia document repository, an object oriented data base, and a configuration management system. A picture of the information sharing architecture is shown in Figure 2.6.

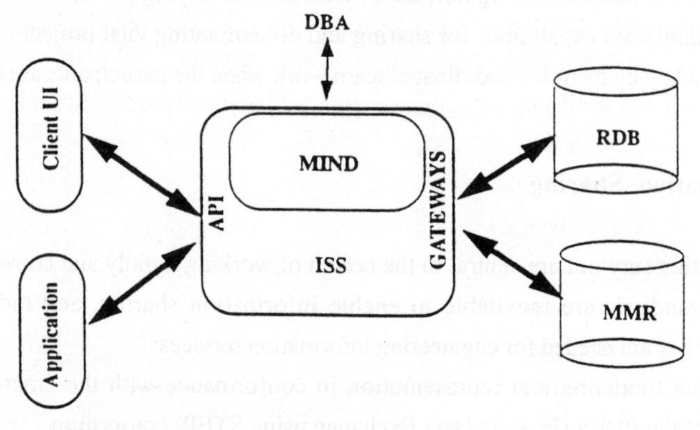

ISS: Information Sharing System
API:Application Programmers Interface
MIND:Model-based INformation Directory
DBA:Database Administrator
RDB:Relational Database
MMR:Multi-Media Repository

Figure 2.6 Overview of the Information Sharing System

As a result of the current work there will be available the capability to develop a product model directory based on the Express language, translated to C++ by a translator and maintained in a process that uses a commercial object data base. The repositories from which data can be supplied will include a relational data base system and a multimedia repository. Besides there will be an application programming interface for the information sharing system to allow applications to access data over the network. Modifications to data or the model of data are not envisaged since this system is meant to serve legacy product data. Later, a user interface could be built to access the information visually. It should be emphasized that the information sharing server here envisaged has a substantial CE flavor in that it supports a virtually unified data model; though the data itself resides in disparate data bases, the MIND sub-system brings it all together in a single model of the data and provides read access to the data elements rather transparently.

2.6 Integration of Engineering Tools with the Architecture

Engineering organizations have long had to contend with the problem of integrating different engineering tools whose internal representations or external data formats did not coincide. The following considerations now apply, in addition:

- Integration of DICE and commercial software into usable systems. This includes modeling data and processes, building wrappers for methods and services, implementing communications between modules, and customizing applications.
- Creation of a CE shell through which team members can conduct their group transactions with relative ease.

From one viewpoint, that of the end-user employing tools, the essence of the problem in working with other perspectives is the difficulty of exchanging the data of proprietary tools. This makes it necessary to employ translators -- which are often unavailable, and partial at best in their efficacy. Incompatibilities in operating systems, hardware platforms, and communication protocols aggravate the problem. A wide variety of tools is typically used in organizations and they have overlapping functions. It is difficult even to decide which tool should be used for what purpose. The sheer variety of interfaces also makes it difficult for the user to learn their interaction protocol.

The integration methodologies that have been invented to get over this problem are nowhere near being systematic and comprehensive, yet. But it is worth delineating a few approaches that have achieved their purpose in a limited context. Firstly, it is possible to distinguish between tight and loose integration methods. The latter approach makes no effort to modify the tool itself, on the assumption that such modifications are not feasible. What is left is to see whether the tool's data can be extracted and transmitted to another program. If it is rather open, the tool itself will permit operations like exporting (and importing) a subset of the data residing in the tool. Then the other tool may import the data directly or after a translator has been applied to transform the format to be consistent with that of the target tool. If a complete translator can be written this process of extracting data from one tool, applying a translator to the resultant file, and importing it into another tool can be built as a single TMS job (see Section 2.3 above). When tools are not open such data exchange becomes problematic, but can still be accomplished if the detailed structure of the data file is at least made available by the vendor of the CAD tool. The user organization can do the bespoke programming to transform the data into what can be accepted by another tool. Unfortunately, today there are very meager portions of the complete data that have any widely accepted and usable exchange standards directly and fully supported by CAD vendors.

In the case of tight integration, the approach is to modify existing applications so that the process of data extraction, translation and data importation (repeated any number of times serially among several applications or back and forth between two programs iteratively) is accomplished directly by one program calling another and connecting over the network by some communication protocol (for instance, the Sun Remote Procedure Call) that is built at the two end points. To construct such an interaction one must either have the hooks embedded in the CAD application, or open it up and put it in oneself. Obviously, this is labor intensive. But the thesis of this approach is that most users prefer to work from a single application's interface, if possible, and the application will have lasting use. These applications can be extended, using what are called wrappers, to host external data, tools, and CE services (see Figure 2.7). The wrapper manifests itself in one way as an extension to the tool's interface to incorporate the menus and dialog boxes by which other applications and their necessary data can be specified. The wrapper also has to perform the back-end work of extracting data from the tool's internal data structures and writing it out as a message or a file. In the latter case, the wrapper should translate the file format and then execute the other program that will use the data. Examples of such research prototypes include ASPRIN [14, 15], a system developed at GE Corporate Research and Development, to illustrate the use of a spreadsheet for integrating a collection of engineering tools. Data from the engineering tools are brought into a spreadsheet, and other analysis tools are invoked using the data present in the spreadsheet cells. The AMS (see Section 2.3 above) can also be used to perform the remote application invocation. The success of this approach depends critically on the tool selected to be the host-environment (or the integrating framework). The tools should be extensible, as regards the user interface and the function call interface; in other words, they must be fairly "open" tools.

2.7 Capture of Design Intent in a Corporate Memory

The need for continuing access to the knowledge acquired in past projects, and the preservation of experience gained and lessons learned, has been the motivation for yet another generic service. It is hoped that by preserving contextual information along with the raw data, the past can be a guide in future projects, and at the same time, the pitfalls of mere imitation can be averted. Since intelligent re-use is the key, the following requirement arises:

- Archiving of design, history, intent, and rationale used in the decision-making process. This function should be provided by an indexed electronic notebook that will be interfaced with other CE generic services and the engineering tools used by the team.

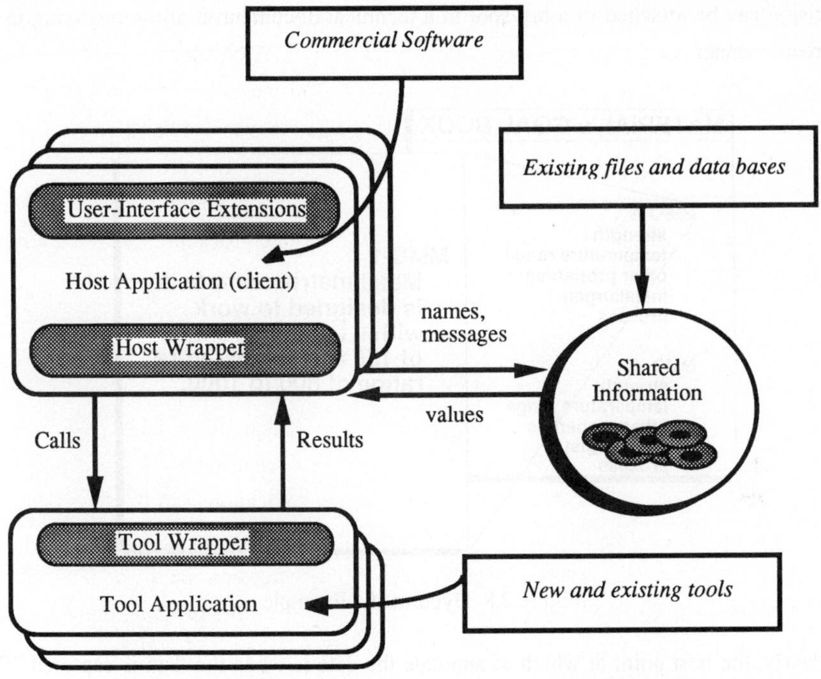

Figure 2.7 Sharing Tools and Services in a Host Application

The corporate memory of a product can be defined as all the meta-data, i.e. the why and wherefore concerning the product data which indicates why certain decisions were taken during the life-cycle. The effort to record the data itself is so taxing that little energy is left to narrate the reasoning behind the data, and what other data configurations were tried and what trade-offs have been made, based on what customer requirements or external constraints. One approach used in the Electronic Design Notebook (EDN) [16] of GE Corporate Research and Development is to provide services to create textual descriptions of the why and link them to the data itself. Such an EDN can be based on any word processing tool, or even better, a tool that allows sketching, drawing and organization of data, as in a spreadsheet. At one extreme, the EDN can be used as a private diary of the team member. But it can be formalized with some standard data authorization forms (for example, engineering change notices) to obtain a team accessible repository of the justification for certain product definition data.

The EDN used at one site has the capability to hot link an icon in the document to another application, or to another document. Thus the annotation of the result in one document can be found by clicking on the icon. Such icons may also be attached to a CAD file, linking another document to it which contains the reasoning of how the data came to be. The annotations can also be attached as a voice fragment, if the Multimedia editor developed at CERC (as an extension of FrameMaker™) is employed [10]. Figure 2.8 shows how the reasoning behind a decision can be attached to a hot spot in a technical document to allow browsing in the hypercard manner.

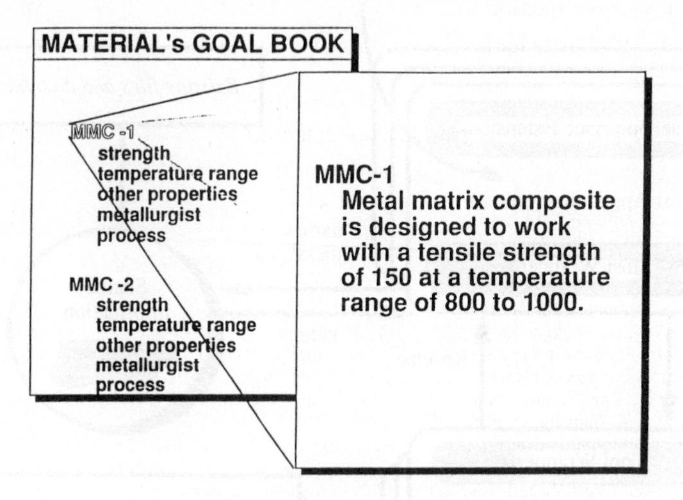

Figure 2.8 Hypermedia Example

Clearly, the best point at which to annotate the data is when the data is captured. This could be in the CAD file itself, provided the CAD tool is open enough to allow an extension of its interface, followed by a remote procedure call to invoke some documentation program like FrameMaker which will bring up the relevant notebook in which the annotation should go. The annotation could also be attached to a slot in the virtually unified data model held by the PCB, or by the MIND sub-system, for those attributes that require a raison d'etre to be recorded. The indexing of the notebook is important. Later, when one wishes to recover the reasoning and constraints that led to a decision, it should be accessible by date, by part number, by person who made the decision, by perspective involved, by engineering change notice number, etc.. This has yet to be done comprehensively, but a sort of index is present in the EDN in the form of a collection of Design Study Summary forms used to annotate all major decisions.

Acknowledgements

It is a pleasure to acknowledge the wealth of discussions and wide cooperation among many researchers at CERC and the General Electric Corporate Research and Development in the DICE project. I have benefited greatly from their generosity and openness. I especially thank the Director of CERC, Dr. Ramana Reddy, for his infectious enthusiasm. This paper is a partial record of what I have learned by participating in the project, and I hope I have given due credit to all the contributors in the references. Further references to related work done in other projects will be found there.

This work has been sponsored by the Defense Advanced Research Projects Agency under Grant MDA-972-91-J-1022 for the DARPA Initiative in Concurrent Engineering (DICE).

References

1. Berger, S., Dertouzos, M.L., Lester, R.K., Solow, R.M., Thurow, L.C. *Toward a New Industrial America.* Scientific American. June 1989.
2. Cleetus, K.J., Uejio, W. eds. *Red Book of Functional Specifications for the DICE Architecture.* Concurrent Engineering Research Center, West Virginia University, Morgantown, WV. Feb 1989.
3. Winner, R.I., Pennell, J.P., Bertrand, H.E., Slusarzuk, Marko M.G. *The Role of Concurrent Engineering in Weapon Systems Acquisition.* Institute of Defense Analyses Report R-338. December 1988.
4. Reich, R.B. *The Quiet Path to Technological Preeminence.* Scientific American . Oct 1989.
5. Cleetus, K.J. *Definition of Concurrent Engineering.* Technical Report CERC-TR-RN-92-003. Concurrent Engineering Research Center, West Virginia University, Morgantown, WV. Apr 1992.
6. Hauser, J. R. and Clausing, D. *The House of Quality.* Harvard Business Review. May - June 1988.
7. Dehnad, K. (ed.). *Quality Control, Robust Design, and the Taguchi Method.* Wadsworth & Brooks. 1989.
8. *Proceedings of the Second National Symposium on Concurrent Engineering.* Concurrent Engineering Research Center, West Virginia University, Morgantown, WV. Feb 1990.
9. *Proceedings of the Third National Symposium on Concurrent Engineering.* Society for Computer Aided Engineering, Rockford, IL, and Concurrent Engineering Research Center, West Virginia University, Morgantown, WV. June 1991.
10. Srinivas, K., Reddy, R., Babadi, A., Kamana, S., Dai, Z. *MONET: A Multimedia System for Conferencing and Application Sharing in Distributed Systems.* Proceedings of the First Workshop on Enabling Technologies for Concurrent Engineering. Concurrent Engineering Research Center, West Virginia University, Morgantown, WV. April 1992.
11. Kannan, R., Cleetus, K.J., Reddy, Y.V. *The Local Concurrency Manager in Distributed Computing.* Proceedings of the Second National Symposium on Concurrent Engineering. Concurrent Engineering Research Center, West Virginia University, Morgantown, WV. Feb 1990.
12. Londono, F., Cleetus, K.J., Nichols, D.M., Iyer, S., Karandikar, H.M., Reddy, S.M., Potnis, S.M., Massey, B., Reddy, A.L.N., Ganti, V. *Managing Chaos: Coordinating a Virtual Team.* Proceedings of the First Workshop on Enabling Technologies for Concurrent Engineering. Concurrent Engineering Research Center, West Virginia University, Morgantown, WV. April 1992.
13. Karinthi, R., Jagannathan, V., Montan, V., Petro, J., Raman, R., Trapp, G. *Integrating Heterogeneous Information Repositories in a Concurrent Engineering Environment.* Proceedings of the First Workshop on Enabling Technologies for Concurrent Engineering. Concurrent Engineering Research Center, West Virginia University, Morgantown, WV. April 1992.
14. Lewis, J.W., and the DICE Team. *Wrappers: Integration Utilities and Services for the DICE Architecture.* Proceedings of the Third National Symposium on Concurrent Engineering. Society for Computer Aided Engineering, Rockford, IL, and Concurrent Engineering Research Center, West Virginia University, Morgantown, WV. June 1991.
15. Lewis, J.W., Fogg, H.W., Uejio, W.H., Sum, R.N., Sarachan, B.D., Kenny, K.B., Czechowski, J.W. *A CE Toolkit for DICE.* Proceedings of the First Workshop on Enabling Technologies for Concurrent Engineering. Concurrent Engineering Research Center, West Virginia University, Morgantown, WV. April 1992.
16. Uejio, W.H., Carmody, S., Ross, B. *An Electronic Project Notebook from the Electronic Design Notebook (EDN).* Proceedings of the Third National Symposium on Concurrent Engineering. Society for Computer Aided Engineering, Rockford, IL, and Concurrent Engineering Research Center, West Virginia University, Morgantown, WV. June 1991.

Integrated Tools and Technologies for Concurrent Engineering of Mechanical Systems

Edward J. Haug

Center for Computer Aided Design, The University of Iowa, Iowa City, Iowa 52242-1000 USA

Abstract: Computer Aided Engineering tools and technologies that hold the potential for creating a simulation based Concurrent Engineering environment in the near-term and a design optimization based capability for the future are analyzed. Technical challenges and opportunities associated with integrating these tools into a software environment that can support multidisciplinary engineering teams are defined and illustrated. A road map for evolutionary creation of a simulation based Concurrent Engineering design environment in the near-term and a design optimization environment in the longer term is presented. Projects underway to create the capability advocated by the road map are presented, to illustrate technical considerations peculiar to Concurrent Engineering of mechanical systems and challenges associated with network based multidisciplinary CAE system integration for Concurrent Engineering.

Keywords: simulation based design / tool integration for Concurrent Engineering / Concurrent Engineering road map

1 Introduction

Current mechanical system development approaches tend to revolve around improving characteristics of a product such as Reliability, Maintainability, and Producibility (RM&P), using prototype hardware test and analysis techniques that are applied after the active design phase, called the "post design phase." Little integration between designers and RM&P specialists currently exists [1]. This creates a competitive situation among engineering disciplines that inhibits effective communication and cooperation. RM&P analysis must thus compete for both resources and schedule in the post design phase with system integration and operational testing, significantly compromising any benefits that might be achieved [1].

Advances and growth in Computer Aided Engineering (CAE) technology create new possibilities for meaningful integration of RM&P considerations early in the design process, when a significant impact can be made. The proliferation of engineering workstations, computer networks, and advanced computer simulation methods offers the potential for both design engineers and RM&P specialists to share data and analysis tools throughout the design process, in order to:

(1) evaluate accessibility and other related maintainability characteristics as the design layout is created in a solid modeling Computer Aided Design (CAD) environment
(2) identify failure modes and carry out failure analysis to predict reliability using computer simulation, prior to prototype fabrication and test; i.e., "soft prototyping"
(3) simulate manufacturing processes and the effect of design changes on the cost, effectiveness, and robustness of manufacturing processes
(4) capture and apply rules to help the design engineer arrive at a design that optimizes tradeoffs between RM&P and other product characteristics.

This paper explores the potential impact that state-of-the-art CAE tools can have on Concurrent Engineering of mechanical systems and uses a road map, presented first in Reference 1, that may be followed for their orderly infusion into the design process. Enabling technologies for Concurrent Engineering of mechanical systems are outlined and illustrated via examples. Technical considerations associated with software integration to accomplish the intended goals are outlined here and addressed in greater detail elsewhere in this book [2-5]. Finally, two projects that are underway to integrate tools for specific mechanical system Concurrent Engineering applications are described, to indicate the scope and approach being used in software integration.

While this paper addresses issues that are common to all disciplines contributing to Concurrent Engineering of mechanical systems, the application focus is on system performance, reliability, and maintainability. Consideration of manufacturability [6-8], design optimization [9-15], and human factors [16-18] are treated elsewhere in this book.

2 The Impact of CAE in Concurrent Engineering

Computer Aided Engineering is having a profound impact on the way in which engineering, manufacturing, and support departments in some companies conduct their activities. Design cultures cut across the academic world, the source of engineers, into industrial and government activities that collectively develop, produce, and procure complex products. In order for the full potential of CAE to be realized in improving RM&P characteristics of mechanical systems, even greater cultural changes will be required in the design community. All personnel that influence the RM&P features of a product, including designers and RM&P specialists, must become "on-line" in an integrated design process [1,4,19]. This means that RM&P specialists must become familiar with the CAE product development process and designers must develop relationships between specific design features and RM&P characteristics of end items. Both must perform their functions in a highly interactive environment of shared responsibility and authority.

Past emphasis on and organizational isolation of research, design, and RM&P specialties has caused communities comprised of these specialties to be isolated, both physically and organizationally. They have tended to become autonomous with respect to mainstream design functions. Consequently, the priorities of these isolated functions have become independent, or have diverged from the principles of sound engineering. Functional isolation has promoted sequential design practices, with complete designs being "thrown over the wall" [1,4,19] by design engineers for review and negotiation of changes by "ilities" specialists. Using the traditional design process, it becomes very expensive, in both cost and schedule, to change the design at this stage to achieve desirable RM&P characteristics that could have been "designed in" if the opportunity had been seized early in the design process. Furthermore, incentives for reduced weight, increased power, and improved performance have been easily identified. Thus, they have overridden the less tangible incentives for improved RM&P characteristics.

Many product development and manufacturing sectors, both private and public, are now applying heavy emphasis to bringing the "ilities" back into the mainstream of the design process. This requires a major change in management viewpoint, toward a CAE based Concurrent Engineering approach in which appropriate RM&P trade-offs are considered equally with performance, throughout the design process. While Concurrent Engineering concepts are still being refined [1,4,19], the basic characteristics of a Concurrent Engineering design process include the following:

(1) concurrent design of products and their manufacturing and support processes
(2) establishment of multi-function systems engineering and design teams
(3) practical engineering optimization of product and process characteristics, to create process and environmentally tolerant (robust), efficient, and cost effective designs
(4) computer simulation (soft prototyping) before prototype hardware testing (hard prototyping)
(5) laboratory experiments to confirm predictions of process and product characteristics.

One of the effects of applying a Concurrent Engineering philosophy in the mechanical system design environment is a marked change in how development testing is used to achieve good RM&P attributes of the product. Historically, development testing has concentrated on a Test, Analyze, and Fix (TAAF) approach [1]. Instead of excessive reliance on TAAF after completion of the detailed design to "test in" RM&P, Concurrent Engineering encompasses a broad scope of computer simulation and development testing, directed to early verification of product characteristics and processes that are used to create and support the product. The long-term goal is for simulation and development testing to replace the TAAF approach and serve two primary functions within the CE environment;

(1) continually validate and refine simulation tools and databases that are used in support of product and process definition

(2) discover RM&P shortfalls that could not be predicted within the scope of the available suite of CAE analysis tools.

To a significant extent, RM&P research during the past decade has focused on electronic items and systems, due to their expense and contribution to life cycle cost for complex systems. The defense and commercial industrial communities have a reasonable understanding of top level electrical engineering design practices and a fairly detailed understanding of the current CAD, CAE, and Computer Aided Manufacturing (CAM) tools that are used for design and supportability functions. Design practices and design related capabilities that are applicable to mechanical systems, however, are not as well understood. For these reasons, contrasts are drawn in Reference 1 between CAE methods that support RM&P design of electronic and mechanical systems. The abbreviated treatment of this topic here is to provide a frame of reference for considering emerging issues in mechanical design for RM&P.

Electronic RM&P engineering (especially diagnostics) has had significant success, due partly to the fact that the majority of electronic failure characteristics can be represented as independent failure modes. Mechanical systems tend to be much more complex, with each component having diverse failure mechanisms, interdependent failure modes, and dissimilar failure distributions; e.g., normal distribution due to wear and fatigue, versus exponential distribution due to independent failure rates. There is also a substantial difference in sources of maintenance actions, due to differences in accounting for failure modes between mechanical and electronic systems. In most cases, maintenance actions in electronics are induced by the occurrence of a failure that is accounted for in maintenance workload calculations, as well as mean times between unscheduled maintenance actions, the primary end user's measure of system reliability. On the other hand, wear mechanisms in mechanical systems do not count against Mean Time Between Maintenance (MTBM) figures, although they add to total maintenance burden if they result in replacement at scheduled maintenance intervals. This implies much greater complexity in making design trade-off decisions that impact mechanical system RM&P.

Prior to discussing specific mechanical CAE tools that hold potential for improved design for RM&P, it is of value to review some of the challenges facing mechanical RM&P design, in comparison to more advanced tools and guidelines for electronic system RM&P design. A number of important and broadly valid points are concisely made in an Army Management

Engineering College textbook [20] on the topic "General Guidelines For Reliability Prediction of Mechanical Equipment":

(1) In comparison to the analysis of electronic systems, confidence levels of reliability predictions for mechanical systems are generally low, because a standard and widely accepted approach for mechanical reliability prediction does not exist. Electronic equipments are far easier to model, because they are usually composed of standardized parts that perform a single function and fail at a constant and predictable rate. In comparison, mechanical systems are subject to a variety of dynamic stresses that are determined by the design configuration.

(2) The difficulty of establishing standard procedures for mechanical reliability prediction is due, in part, to the complexity of developing realistic models of mechanical systems and the lack of accurate failure data on most mechanical equipment.

(3) In order to be effective, reliability prediction techniques must relate reliability criteria and data to design engineering parameters.

(4) The model of a mechanical system must realistically describe all failure modes and their effects on functionality of the system.

(5) Utilization of electronics oriented procedures for reliability prediction of mechanical systems usually limits the accuracy and usefulness of the quantitative results.

(6) A major obstacle to creating accurate models of mechanical systems is the number of engineering variables (and their complex interrelationships) that must be systematically accounted for.

This assessment of the potential for reliability prediction of mechanical systems is, on the surface, extraordinarily pessimistic. While it appears to be an accurate assessment of the current state-of-the-art of tools used by reliability specialists in mechanical system reliability analysis, it does not reflect the potential that is offered by adapting well established and broadly used engineering simulation and CAD tools to create an environment in which these tools can become an integral part of the process of mechanical design for RM&P. Advanced simulation and CAD tools for geometric modeling, structural analysis, dynamic analysis, thermal analysis, design sensitivity analysis, etc., are used in virtually every engineering organization that does mechanical system design. These tools, however, tend to be isolated in the analysis and design departments of development organizations and have not yet found their way into formal procedures for RM&P analysis and design. The challenge, is to adapt, integrate, and apply established CAE tools by both the design and RM&P communities, in a form that serves their needs.

3 Enabling Technologies For Simulation Based Concurrent Engineering

The ability to create and modify a design several times in the time that was previously required for a single design iteration has led to a phenomenal growth in the use of CAE technology in some industrial sectors; e.g., the aerospace community. While this process has been evolving over the past twenty years, the recent emergence of powerful CAE and CAD workstations that integrate design capture with analysis, in a graphical context at the engineer's desk, has rapidly accelerated this revolution. In 1985, it was estimated that 10% of American designs were accomplished using CAE methods. By the year 2000 this figure is expected to exceed 80% [1]. This is causing profound changes in the design process, making it much faster and far more integrated than ever before.

Simultaneous advances in computer hardware and software performance, improved understanding of the behavior of complex mechanical systems in diverse environments, and markedly improved control and visibility of manufacturing processes hold the potential for bringing the CAE revolution to bear on mechanical system design for RM&P. Key supportability analysis procedures such as accessibility analysis, dynamic load analysis, stress analysis, and failure analysis are becoming available as computer tools, to replace previously error prone and time consuming manual methods. Initial steps are being taken to integrate these tools into the designer's CAE environment. Integration of engineering data in a common database has the potential for enhanced timeliness, while decreasing the opportunity for errors, thus encouraging increased interaction in the design process by RM&P specialists.

New CAE tools provide the following potential for improvement of mechanical system design for RM&P [1]:

(1) allow the designer to accomplish RM&P analyses and iterate trade-offs
(2) permit simulation-based TAAF techniques to begin while the design is still fluid
(3) increase the capability to manage data and data flow, improve timeliness of feedback, and ensure automatic update of the integrated database
(4) permit RM&P specialists to proactively influence the creative design process, both directly and by installation of design rules and design algorithms for use by the designer
(5) allow significant improvements in communication and data transmission between the design engineer and the RM&P community.

This section identifies a few CAD and CAE tools that can be adapted from the engineering design and analysis community to directly support design for mechanical RM&P. The immediate potential for CAE tool application in design for RM&P and the immense potential in the future for further advances depend on exploiting the following emerging technologies and engineering tools:

(1) powerful low and moderate cost graphics based workstations that permit user interaction with simulation and CAD tools
(2) mini supercomputers and super workstations that have the power of last generation supercomputers, at a very small fraction of their cost
(3) networked computer hardware and software that permits integrated use of workstations and compute servers, as well as access to central databases, on a timely basis
(4) transition of the CAD industry from producing electronic drafting systems to creating geometric modeling systems that characterize designs and support design analysis and improvement, with dimensioned drawings being a byproduct of the design process.

The simulation and CAD functions defined below outline the role that engineering simulation and design tools can play in support of the RM&P design process. The approach recommended uses CAD and CAE tools, integrated into the process of design for RM&P.

Geometric Modeling defines the geometry of components and subsystems, in wire frame, surface, or solid form, for support of design, analysis, and production planning. Geometric models provide a geometry database that is essential for component stress analysis, thermal analysis, fluid flow analysis, kinematic and dynamic analysis, accessibility analysis, and specification of manufacturing processes. Geometric modeling is becoming an integral part of modern CAD systems, which provide access to geometry data to support engineering design, RM&P analysis, and production planning. Surface modeling is commonly used in CAD systems. Solid modeling is evolving and is, or will soon be available in most CAD systems and CAE software. CAD systems marketed and supported by companies such as Intergraph, Catia, CADAM, Computervision, Unigraphics, etc.; software such as PATRAN [21]; and other specialized modelers are currently used in engineering design and analysis and can support design for RM&P.

The use of geometric modeling in support of design for RM&P requires reassessment of geometric databases to assure utility for the RM&P community, rather than developing a new stand alone capability. It also requires the development of RM&P specific graphics capability to utilize the data; e.g., anthropomorphic models to lift and place parts, assemblies, and tools [17,22,23]. Graphics communications standards are emerging that will permit the entire engineering design and RM&P community to access a unified geometry database and carry out Concurrent Engineering in a timely and practical way [4]. As noted earlier, solid modeling reduces or eliminates the need for physical mock ups and supports design for maintenance of mechanical packaging configurations. It also supports a broad range of simulation and design steps that are required for analysis of failures and their consequences; e.g., dynamic, structural, and thermal analysis and design for producibility.

Dynamic Simulation provides a method for carrying out computer simulation to predict system dynamic performance and loads and stresses that act on components of mechanical systems [24-28]. Interfaces with CAD geometry files permit the engineer to generate model data, establish the simulation environment, carry out computer simulations, and display high quality animations of system motion with a degree of realism that approaches that of an actual field test [25,26]. Interfaces with structural finite element models are required to define compliance of flexible system components, determine stress and strain time histories in components of mechanical systems, and predict component failures and component life [27,28]. System dynamic simulation tools such as CONTOPS [29], DADS [30], and DISCOS [31] provide a capability for rigid and flexible dynamic system simulation, using current computer technology. Major advancements are on the horizon that will provide high speed dynamic simulation of large scale mechanical systems, with a high degree of realism. Real-time interactive simulation and computer graphics animation are becoming feasible and create a "virtual prototyping" capability that will come into common use in the near future [32-34].

Recent emphasis on creating workstation and computer graphics interfaces for system dynamic modeling and analysis tools is making the technology available to both development and RM&P journeyman engineers [25]. Considering the importance of loads and stresses in assessing reliability of a mechanical system, system dynamic simulation tools will become imperative in speeding the design process, identifying reliability problems early in design, and correcting difficulties prior to fabrication and test. Use of dynamic simulation tools in test planning will also be of significant value, permitting test engineers to carry out numerous simulations, observe realistic animated dynamic performance predictions, and identify critical modes of operation that require evaluation during test. The result will be much better planned tests, both for creating critical environments that may lead to failure and for minimizing the amount of costly testing that needs to be carried out in operations that are of little threat to the system.

Structural Analysis provides the capability for using structural geometric and material data and loads that are predicted by dynamic simulation or measured by experiment to predict stresses, strains, displacements, natural frequencies, and related structural performance factors. This capability permits prediction of failure modes for structural components and subsystems and provides data needed to evaluate durability and fatigue life. Finite element structural analysis computer codes such as NASTRAN [35] and ANSYS [36] are broadly used in design departments for analysis of structural components and subsystems of mechanical systems. Graphics based post-processors [21] that are available with finite element codes and as part of CAE and CAD software systems permit the engineer to create finite element structural models, using CAD geometric data, and to carry out analysis of structural performance.

Dynamic Stress and Life Prediction methods utilize finite element structural analysis tools that are well established in aerospace and automotive industries, for support of component design [27,28]. Application of finite element structural analysis by the RM&P community for failure prediction, in system environments for which component loads are known [24-26], will be of great value. More important, coupling finite element structural analysis codes into an integrated software system that includes system dynamics will permit prediction of component loads and stresses due to component interaction more accurately than has been possible in the past. This capability, coupled with historical data on component failure rates and fundamental data on material failure due to loads and stresses, will permit creation of new tools for reliability prediction that can be applied in a timely way, as soon as a design or design modification is defined.

To be more concrete regarding simulation tools that are used in dynamic stress and life prediction, consider the off-road High Mobility Multipurpose Wheeled Vehicle (HMMWV) shown in Figure 1. This vehicle is intended for use in severe off-road environments that subject the vehicle to extreme loads and structural fatigue failure modes. In order to determine loads acting on components within the vehicle, a dynamic simulation model is created for dynamic response and load analysis. Figure 2 shows a schematic representation of the HMMWV vehicle suspension and steering subsystems.

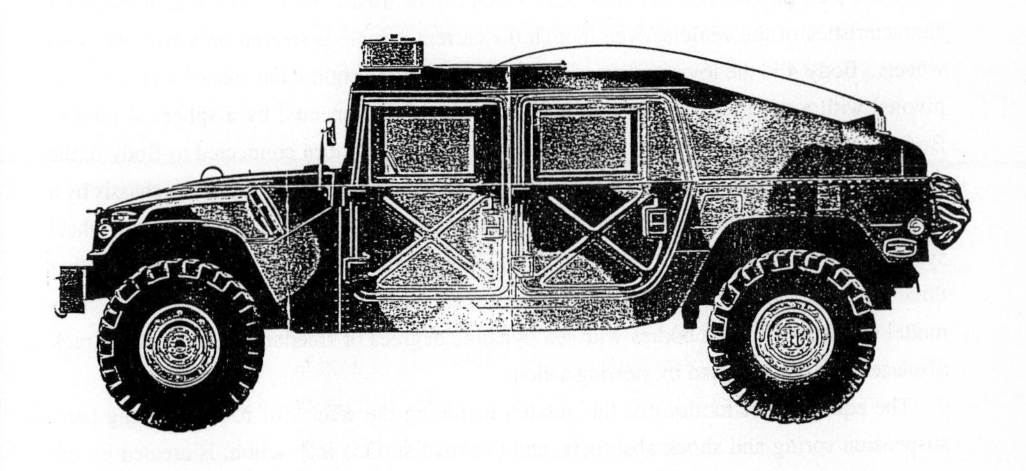

Figure 1. High Mobility Multipurpose Wheeled Vehicle (HMMWV)

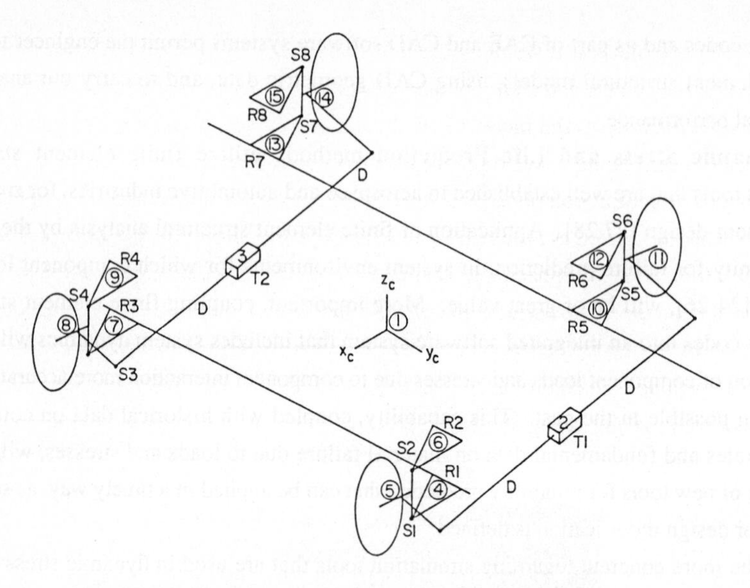

Figure 2. Schematic Representation of a HMMWV Vehicle

Numbers inside circles represent bodies that can move relative to one another during motion of the vehicle. In this model, each of the bodies is modeled as being rigid, for the purposes of dynamic simulation. Body 1 is the chassis of the vehicle, which includes the basic vehicle structure, engine, and transmission. Body 2, which translates relative to the chassis, is the front steering rack, whose motion is controlled by turning the steering wheel. Body 3 is a conceptual rear steering rack, which can be used to investigate four-wheel steer characteristics of the vehicle, even though the current vehicle is steered only with the front wheels. Body 4 is the lower suspension control arm of the right front suspension, which is pivoted with a revolute joint relative to the chassis and connected by a spherical joint to Body 5, which is the kingpen-wheel assembly. Body 5 is in turn connected to Body 6, the upper suspension control arm, by a spherical joint and Body 6 is connected to the chassis by a revolute joint. This closed chain kinematic subsystem permits vertical motion of the wheel system relative to the chassis (jounce) and steering motion of the wheel. Each of the other three suspension subsystems has identical form to that described at the right front. This model consists of fifteen bodies with ten dynamic degrees of freedom, presuming the rack displacements are specified by steering action.

The equations of motion for this model, including the effects of roll stabilizing bars, suspension spring and shock absorbers, and tire-road surface interaction, is created by the DADS [30] dynamic simulation software system. Road profiles, off-road terrain profiles, and extreme bump courses are modeled geometrically and the vehicle is driven over a series of such test courses, much as it would be in actual field tests. Results of the simulation include the motion of the vehicle and all its components, as well as forces and torques that act on each of the bodies making up the model shown in Figure 2. The component of the vehicle of immediate concern is the lower right suspension control arm shown in Figure 3. Note that this y-shaped structural component has its left segment constructed as a weldment and the right segment constructed as a stamped component that mounts the spring and shock absorber assembly through bolted connections. This component transmits extreme loads created by tire interaction with the road surface and obstacles. Under extreme conditions, the shock absorber bottoms-out; i.e., encounters a hard bump-stop, which imposes extreme loads on the right segment of the control arm. High stress zones are anticipated in the vicinity of the attachment points of the spring-shock absorber mounting assembly to the control arm.

Figure 3. HMMWV Lower Suspension Control Arm

The 16,000 degree of freedom (DOF) finite element model; i.e., 16,000 equations in 16,000 variables, of the control arm shown in Figure 4 was constructed [28] to analyze stresses in the control arms due to applied loads. Extreme loads predicted from dynamic simulation were applied to the finite element model and stress distributions were studied, to determine locations and orientations of critical stresses. While the finite element model shown in Figure 4 is adequate for predicting distribution of load and stress patterns, it is not adequate for accurate prediction of localized stresses in critical zones. For this purpose, finite element submodels shown in Figure 5 were constructed in the vicinity of critically stressed zones, to provide higher resolution, more accurate localized stress prediction. These submodels are coupled at their boundary nodes with displacement results obtained from the finite element model shown in Figure 4, to transmit global finite element deflection data to the local region for highly accurate stress prediction.

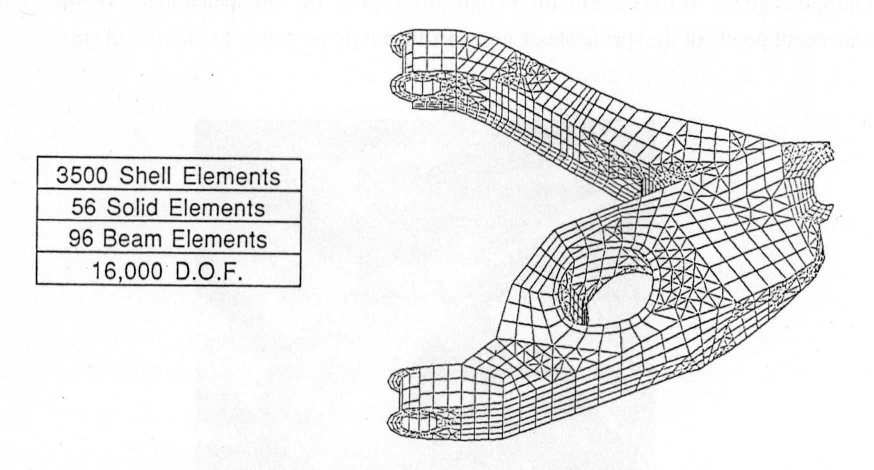

3500 Shell Elements
56 Solid Elements
96 Beam Elements
16,000 D.O.F.

Figure 4. Finite Element Model of HMMWV Lower Control Arm

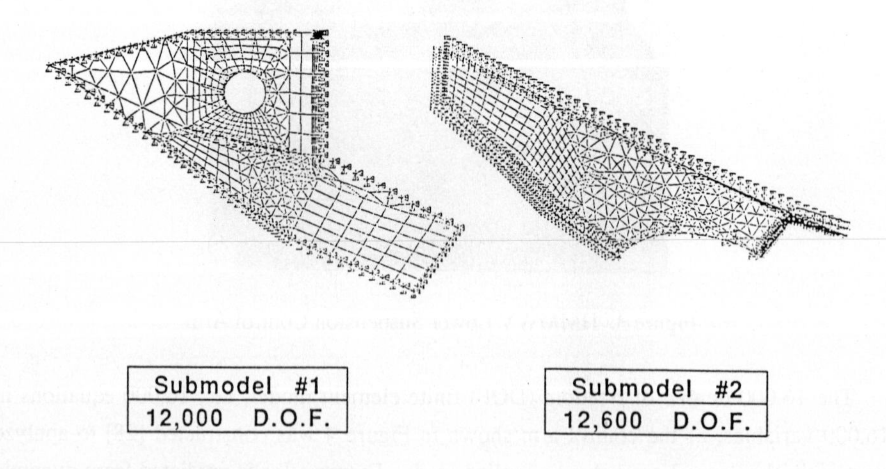

Figure 5. Global-Local Finite Element Submodels

Using stress influence coefficients [27,28] calculated from finite element models shown in Figures 4 and 5 and dynamic load histories using the model of Figure 2, stress time histories in critical zones modeled in Figure 5 were calculated, much as if strain gauges had been applied to these zones and data collected from actual tests. These stress time histories were then communicated to a stress amplitude-cycle counting computer program that applies Minor's rule to estimate contribution to the formation and propagation of cracks in the structural material [28]. This accumulated damage prediction can then be extrapolated to estimate the number of repetitions of the load history required to create a crack of critical size, in this application a crack of 2 mm length. This methodology allows failure prediction and reliability estimation using simulation, early in design before fabrication and test of full components and certainly the full vehicle system.

In order to confirm predictions made, three control arms were instrumented with strain gauges and tested in the laboratory dynamic loading environment shown in Figure 6. Static loads were first applied and stresses measured to verify that the finite element model was adequate. Once confidence was gained in the accuracy of the finite element model, load time histories obtained from vehicle dynamic simulation using the model of Figure 2 were applied using this computer controlled test equipment, replicating the load history associated with one encounter of the series of road, terrain, and obstacle maneuvers to be tested via simulation. These load histories were repetitively applied until 2 mm cracks were observed, at which time the experiment was stopped.

Figure 6. Life Prediction Experimental Verification

Results of this analysis and experiment are shown in Figure 7. As noted in the first line of the tabulated results, static stresses from analysis and experiment differed by at most 6%, with most differences in the 1% range. This level of accuracy was judged to be more than adequate for purposes of life prediction. As shown in the second line of the tabulated results in Figure 7, the analytical prediction of life was that the control arm would survive 813 repetitions of the load history before 2 mm cracks formed. The results of experimentation for three control arms, shown under experimental results in Figure 7, indicate that one experiment essentially precisely confirmed the analytical prediction and the worst disagreement was a 50% over estimate of the life of the component. Since this type of life prediction estimate is understood to be accurate only to within an order of magnitude, these results are exceptionally good and suggest that life prediction and reliability-based design can indeed be carried out using simulation early in the design process, prior to fabrication and test of component hardware.

	Analysis	Experiment	Difference
Static Stresses	Finite Element Analysis	Strain Gage	Max. 5.8% Min. 1%
(Fatigue Life Number of Load Histories)	813	1218 1043 850 Average 1037	21/6%

$$\text{Difference} = \left|\frac{\text{Experiment} - \text{Analysis}}{\text{Experiment}}\right| \times 100\%$$

Figure 7. Comparisons of Prediction and Experiment

To illustrate the complexity of analysis involved in this application, the life prediction/reliability based design runstream that involves numerous computer programs and data sets employed in this analysis is shown in Figure 8. Design data created on a CAD system are extracted and passed to a structural modeler to create the finite element models shown in Figures 4 and 5. This extensive finite element model data set is then passed to a finite element structural analysis computer code to create mass and stiffness matrices, lumped mass distribution data, deformation modes (both vibration and static correction), modal mass and stiffness matrices, and stress influence coefficients [28,37]. This extensive data set is then passed to the flexible dynamics interface computer program that generates flexible component and control data required for vehicle dynamic simulation. At the lower right of Figure 8 is shown the operations analysis function that defines the field test data to be used in simulation, scenarios in which the simulation is to be performed, and the frequency of encounter of various road, terrain, and obstacle features. These data are then passed, with results from the flexible dynamics interface, to the system dynamic simulation program that carries out the actual vehicle simulation and predicts component loads that are combined with stress influence coefficients to create stress and strain time histories at critical zones within the component. These data are then passed to the accumulated component fatigue prediction computer program that estimates accumulated damage for critical sections in welds, due to one encounter of the load history associated with the tests being simulated. The number of repetitions of the load history to create critical crack size is then estimated. Finally, these data are passed to the system life prediction software to estimate life of critical components, and failure rates. Shown at the bottom of Figure 8 is the option to carry out failure mode effects and criticality analysis [38], and ultimately estimate system reliability.

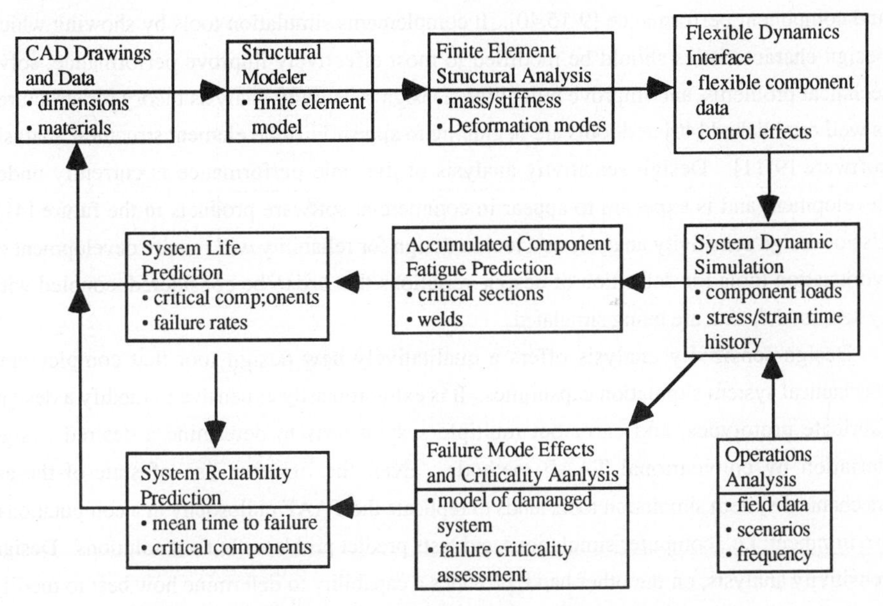

Figure 8. Life Prediction/Reliability Based Design Runstream

This extensive computational environment and runstream requires numerous computer programs, computers, and data storage and retrieval mechanisms to carry out the runstream. Methods and technologies that enable this major application are the topic of later sections of this paper.

Failure Analysis provides a modeling environment that permits definition of component or subsystem failure and prediction of degraded component and system performance, as the consequence of a failure [39]. Dynamic and structural models of the mechanical system can account for a failure and carry out simulations to define the consequences of a failure, predict degradation in system performance due to a failure, and determine changes in loads on all components in the system due to a failure. Failure analysis is feasible, but is not currently available in a CAD environment. Failure analysis relies on dynamic and structural analysis tools that are traditionally used in the design process for simulating the undamaged system. Use of these tools for failure analysis requires implementation of a workstation modeling environment that permits definition of failures that are to be considered, modification of the model of the undamaged system that has been created earlier, definition of the environment in which the failure is to be investigated, and reporting simulation results in a form that is suitable for analysis of failure effects and their criticality [38].

Design Sensitivity Analysis defines the influence of variations in design on system and component performance [9-15,40]. It complements simulation tools by showing which design characteristics should be modified to most effectively improve performance, solve technical problems, and improve reliability. Design sensitivity analysis theory for structures is well established [40] and tools are beginning to appear in finite element structural analysis software [9-11]. Design sensitivity analysis of dynamic performance is currently under development and is expected to appear in commercial software products in the future [41]. Use of design sensitivity analysis as a tool in design for reliability will require development of workstation tools for definition of design variations that are to be considered, coupled with system models that are being simulated.

Design sensitivity analysis offers a qualitatively new design tool that complements mechanical system simulation capabilities. It is extraordinarily expensive to modify a design, fabricate prototypes, and carry out multiple system tests to determine a desired design variation by conventional TAAF methods. Even the iterative use of state-of-the-art mechanical system simulation tools tends to replicate the TAAF philosophy in a computational environment; i.e., computer simulations and tests predict problems but not solutions. Design sensitivity analysis, on the other hand, provides a capability to determine how best to modify a design to improve performance and correct problems. The additional computational cost of obtaining design sensitivity information for structural systems is only about 10% of the cost of the basic structural analysis [40]. Coupling simulation methods with design sensitivity analysis tools will permit timely and cost effective prediction of problems and determination of design fixes, prior to building and testing hardware.

Human Factors analysis may be used to analyze designs for ease of maintenance and servicing, using a human geometric (anthropomorphic) model to simulate performance of remove, replace, and service tasks in an operational use envelope [17,22,23]. Operator-in-the-loop simulation is used for evaluating and modifying designs to enhance performance of the human-machine system [16,18,32-34]. Maintainability analysis tools exist and are being improved. Current capabilities are useful [22] and enhancements that are under development [17] will improve the utility of the tools for accessibility analysis, analysis of maintenance considerations, and in evaluating human capacity to carry out required maintenance functions. The technology for real-time operator-in-the-loop simulation to support investigation of the operator's ability to control equipment, operator cumulative trauma, and ease of training personnel to operate equipment is under development and can be exploited in the near future.

Qualitatively new human factors tools will permit human factors specialists to interact with the design team, throughout the system design cycle [16]. This will permit design of optimal interfaces between the operator and the machine, allowing users to employ machines as natural extensions of their minds and bodies. In the near term, benefits are primarily derived from more easily maintainable systems in which accessibility and strength requirements are addressed during design.

Environmental Sensitivity Analysis quantifies the sensitivity of equipment reliability and performance to variations in anticipated operational environments; e.g., component reliability changes due to variations in temperature, humidity, dust, and vibration. Based on this sensitivity, the designer can provide the proper environmental control; e.g., component placement, cooling allocation, material selection, sealing conditions, and damping, in order to maximize reliability, subject to other system design constraints. Sensitivity analysis tools are available but are used after the fact in design analysis. Computerized tools to optimize the effect of environmental sensitivity on reliability are within the reach of current technology. However, the effects of environments on equipment reliability need to be quantified. Further work to integrate environmental analysis tools with constrained optimization programs is necessary to improve the analysis and reduce cost.

4 A Concurrent Engineering Tool Development Road Map

While the definition of a detailed development plan for Concurrent Engineering tool development was beyond the scope of the report of Reference 1, a timeline describing the phases in which tools to support Concurrent Engineering for Reliability and Maintainability (R&M) could be brought to maturity was presented. Creating the desired evolution from the present post-design review consideration of R&M to an engineering design mode of operation in which R&M characteristics of mechanical systems are given heavy weighting requires developments in three areas. First, certain design functions that must be carried out in order to improve R&M must be defined, independent of the design methodology that is employed or the tools that are available to the designer. Second, there must be an evolution in design methods that yields a design synthesis approach in which R&M properties are designed into mechanical systems. Third, intertwined between the design functions and evolving design methods is the development of CAE tools to support the process of design of mechanical systems for R&M. Achieving the desired evolution of design methods for R&M not only requires that individual design tools be improved, but also that design tools and computer technology be integrated into a CAE based interdisciplinary Concurrent Engineering design environment.

Evolution of Design Methods for R&M that is needed to support the goal of design for mechanical system R&M was defined in Reference 1, in the form of a matrix that relates design functions for R&M to a projected evolution in design methods due to advancing CAE technology. Some of the major design functions that exploit tools identified in the previous section of this paper, which directly support enhanced R&M properties of mechanical systems, are tabulated along the vertical axis of the matrix of Figure 9. These functions are an essential subset of the myriad of design functions that make up the tool kit of Concurrent Engineering. A summary of these functions is provided, for convenient reference, in the following subsection. More detail on these functions and the CAE tools used in carrying them out may be found in Reference 1.

The evolution of design methodology that will yield the desired improvements in R&M of mechanical systems is displayed along the top of the figure. This evolution begins with the present **post design review** method, makes the transition to **simulation based design** that employs engineering simulation tools in advance of prototype or system testing, and proceeds to a **design optimization** approach that uses simulation models and computer optimization methods to yield the best R&M properties possible, within technological and cost constraints. In describing the evolution of tools that support the various functions in Figure 9, it is understood that any tool used under a given methodology is available for use in all succeeding methodologies. A description of each of the three design methodologies envisioned is given in a subsequent subsection.

Tools that must evolve or be created to support the seven R&M design functions shown along with left of Figure 9, for each of the progressively advancing design methodologies, are described in the ext of Figure 9. The evolution from post design review to simulation based design can be supported in the near term by exploiting and extending existing design simulation tools to meet R&M goals. Realization of the design optimization methodology is a longer term goal that builds on simulation tools, advancements in optimization methods, and integration of design software with CAD and CAE methods and databases. The evolution of methodologies represented in Figure 9 thus embodies significant advancements that can be initially realized in two to three years and culminate in revolutionary developments that will occur only in about a decade.

	POST DESIGN REVIEW	SIMULATION BASED DESIGN	DESIGN OPTIMIZATION
GEOMETRIC MODELING	Line oriented CAD drafting systems or manual drafting used to generate orthogonal view drawings and system layouts. Manual extraction of data required to support each independent analysis program. Component data developed manually.	Solid modeling based CAE and CAD systems used to create electronic spatial definition of part, subsystem, and system geometry. Inertial, structural, and related data extracted automatically and transmitted to appropriate analysis program. Analysis output enhanced with realistic graphics.	Parameterized designs permit treatment of geometry, inertial properties, etc. as attributes that are automatically updated and extracted during iterative design. Supports parametric component and subsystem design optimization for performance, R&M, and producibility.
FAILURE ANALYSIS	Failure data on existing systems and comparability analysis based on field data for related systems used to make qualitative and rough quantitative estimates of failures of a proposed design. Qualitative lessons learned employed to suggest problems and design improvements.	Dynamics system simulation and structural finite element analysis carried out to predict loads and subsystem failures, using fundamental material failure criteria and failure rate data for existing components. Qualitative lessons learned data used with predicted loads and stresses to identify problems and design improvements.	Simulation models extended to include constraints on design to avoid predicted failures. Design optimization methods used to fine-tune design of components and subsystems to avoid failures, consistent with constraints on performance, R&M, and producibility.
STRESS/LIFE ANALYSIS	Failure rates observed in previous systems used with comparability analysis for estimation of loads on the present system to estimate failure rates and life.	Dynamic and structural duty cycle analysis performed to generate component and subsystem load, stress, and strain data for a variety of operational scenarios. Component and subsystem life predicted, based on fundamental material failure characteristics, available component failure rate information, and frequency of operational use information.	Stress and life models extended to include constraints on design, to achieve acceptable component life. Optimization methods used to fine-tune design of components and subsystems, in the presence of stress and life constraints.
FAILURE MODE ANALYSIS	Failure modes defined by qualitative analysis and comparison with existing systems. Failure mode interaction and propagation identified subjectively and from analysis of comparable existing systems, to estimate criticality.	Failure modes and criteria for occurrence of failures defined by simulation used to establish models of system behavior in the presence of a variety of failure modes. System dynamic and structural analysis carried out in the presence of failures, to evaluate propagation of failures due to load redistribution. Quantitative FMECA generated.	Composite failure mode models extended to include constraints on system level performance, R&M, and producibility. Optimization methods used to fine-tune design to avoid catastrophic effects of individual and interacting failure modes.

Figure 9. Evolution of Mechanical Design for R&M

ACCESSIBILITY ANALYSIS	Qualitative judgments and wooden mockups used to evaluate accessibility of parts and subsystems, to evaluate the ability of maintenance personnel to remove and replace components.	Solid models used, with anthropomorphic data and computer graphics that describe operator characteristics and capabilities, to evaluate accessibility for removal and replacement functions, prior to fabrication of prototypes or mockups.	Geometric constraints that assure component and subsystem removal and replacement capability by the required percentile of maintenance operators formulated as requirements for design optimization. Provision made for simultaneous optimization of system performance, R&M, and producibility to include requirements on accessibility. Optimization used to fine-tune design of panels, placement of fasteners, and component geometry for tool access.
DESIGN SENSITIVITY ANALYSIS	The influence of variations in design characteristics on performance and R&M properties estimated, based on experience and comparability analysis with existing systems.	Quantitative sensitivity of performance and R&M characteristics to variations in design parameters predicted by both dynamic and structural simulation tools. Design sensitivity information displayed using computer graphics to guide design improvement, prior to fabrication and test.	Simulation based design sensitivity analysis integrated with component, subsystem, and system design optimization formulations, to improve system performance and R&M constraints.
ENVIRONMENTAL SENSITIVITY ANALYSIS	Sensitivity of system performance and R&M properties to environmental variations estimated, based on comparability analysis with existing systems and limited component and system test data.	Simulation methods used to carry out sensitivity analysis with respect to variations in system and component environments. Performance sensitivity used with R&M prediction tools and design sensitivity analysis results to identify design modifications that enhance system performance and R&M insensitivity to variations in environment.	Simulation based sensitivity of performance and R&M characteristics to environmental variations integrated with design sensitivity analysis and R&M prediction tools into a system design optimization formulation. Component and subsystem designs optimized to meet performance and R&M goals with bounds on the sensitivity of R&M characteristics to variations in environment.

Figure 9. Evolution of Mechanical Design for R&M (Continued)

A note on the terse statements in Figure 9 and the descriptions of design functions and methods in succeeding subsections may assist the reader in associating them with his or her specific class of system, subsystem, or component. Since the scope of Reference 1 includes disparate classes of defense mechanical systems; e.g., nuclear submarines, tanks, helicopters, fighter aircraft, and a myriad of mechanical components, the discussion was necessarily relatively general. The concepts embodied in the matrix of Figure 9 can be sharpened for specific classes of systems, subsystems, and components to serve as road maps for different sectors of the mechanical system design community.

Functions in Design for R&M that must be carried out are a subset of those employed in the broader context of Concurrent Engineering. The specific functions identified in Figure 9 that must be carried out in design for R&M, using tools identified in Section 3 of this paper, are summarized here as follows:

(1) **Geometric Modeling** involves definition of part, subsystem, and system geometry and associated nongeometric data that determine inertial and structural properties. Parameterized geometry supports future design optimization and feature-based design methods.

(2) **Failure Analysis** involves identification and quantitative prediction of failures that may occur in a mechanical system; e.g., a vehicle suspension spring breaks, a jet engine looses 30% of its thrust capacity, or a seal in a submarine periscope leaks. Failure analysis includes both the qualitative definition of the nature of failures and the quantitative prediction of the onset of a failure, as a function of loads and stresses on components and subsystems.

(3) **Stress/Life Analysis** involves determination of loads and stresses on subsystems and components; e.g., engine torque, stress in a component, or pressure acting on a seal, and prediction of component life due to the loads and stresses encountered in an operational environment.

(4) **Failure Mode Analysis** involves analysis of failure modes to determine their impact on performance and redistribution of loads and stresses on other components of the system; e.g., suspension spring breakage in a vehicle leads to large wheel travel and fracture of a suspension bump stop, or leakage in a submarine seal leads to intake of fluid that damages optical components.

(5) **Accessibility Analysis** involves analysis of the ability of maintenance personnel to remove and replace parts of a mechanical system; e.g., the ability to remove and replace a vehicle suspension spring without disassembling the suspension subsystem, or the ability to remove and replace a seal in a submarine periscope.

(6) **Design Sensitivity Analysis** involves determination of the influence of changing a design parameter on failure characteristics; e.g., increasing the diameter of wire in a coil spring of a vehicle suspension system by one percent yields a three percent increase in the ultimate failure load of the spring, or increasing the bulk modulus of rubber seal material by one percent yields a ten percent increase in pressure that can be supported by the seal.

(7) **Environmental Sensitivity Analysis** involves determination of the influence of variations in environments that are defined by operational scenarios on R&M

characteristics; e.g., a ten degree increase in the extreme temperature encountered by the tire of a vehicle reduces its load capacity by five percent, or a five degree reduction in the temperature of a seal decreases its pressure sealing capability by twelve percent.

Evolution of Design Methodology required to carry out the design functions described above and reflected at the left of the matrix of Figure 9 involves a significant change in design methods that will bring concurrent consideration of system performance and RM&P characteristics, of a system into the design process, for broad classes of mechanical systems. While the evolution in design methods is relatively independent of design function, it is highly dependent on the evolution of CAE and simulation tools. In order to make the evolution of design methods projected in Figure 9 more meaningful, two stages of evolution are projected beyond the current post design review process, as follows:

(1) **Post Design Review** is the current serial process of reviewing completed designs for R&M properties, after the engineering design community has essentially completed its design activity. This methodology severely limits the impact that R&M specialists can have on design, since they inherit the design after it is complete and can only suggest modest modifications to improve R&M properties. The post design review methodology is based on either paper or electronic drawing characterization of designs and is heavily dependent on experience based judgements and on analysis of test data for mechanical R&M evaluation and improvement.

(2) **Simulation Based Design** is the process of using modern workstation based CAE and graphics capability for system performance, reliability, and maintainability simulation, by both engineering design and R&M personnel, throughout the design process. Fundamental advancements in simulation technology and powerful engineering simulation software permit quantitative prediction of system R&M properties, based on conceptual component and subsystem structure and fundamental component data, long before system, component, or subsystem prototypes are fabricated and tested. Availability of validated simulation technology provides design teams with the capability to predict performance and R&M properties of systems and to determine the effect of design variations on these properties. The result is a set of design tools that permit concurrent design for performance and R&M characteristics.

(3) **Design Optimization** is the use of automated computer trade-off and optimization methods in a computer aided engineering environment that exploits simulation tools as models of mechanical systems being designed, to support system design optimization for both performance and R&M properties. Implementation of formal computer optimization methods will permit fine tuning the design of individual components, as well as creating a system level design optimization capability that will

permit enforcing constraints on failures, failure mode interaction, accessibility, and environmental sensitivity. Using design sensitivity analysis tools, systematic trade-offs and design optimization can be carried out at both the system and subsystem levels.

The evolution from post design review to simulation based design can be supported in the near term by exploiting and extending existing design simulation tools to meet R&M goals. Realization of the design optimization methodology is a longer term goal that builds on simulation tools, advancements in optimization methods, and integration of design software with CAD and CAE methods and databases. The evolution of design methods represented in Figure 9 embodies significant advancements that can be initially realized in two to three years and culminates in revolutionary developments that will occur only in a decade.

No attempt is made in this paper to discuss each of the twenty-one summary entries in the matrix of Figure 9. As noted above, these very brief descriptions must be sharpened to provide a concrete road map that is applicable to guide the development of design methods for a specific class of mechanical system. It is interesting to note that many of the papers contained in this book address specific applications and tools summarized in Figure 9. Rather than commenting further on specifics of tools that contribute to the evolution in design methodology, the remainder of this paper is devoted to a brief discussion of emerging CAE tools and their integration into environments that can support Concurrent Engineering of mechanical systems for improved R&M characteristics.

5 Software Integration to Support Design for RM&P

As a concrete example of CAE tool integration to support mechanical system design in a Concurrent Engineering environment, the software environment shown schematically in Figure 10 may be considered. Development and implementation of this software environment is a goal of the NSF-Army-NASA Industry/University Cooperative Research Center for Simulation and Design Optimization at The University of Iowa. The goal of this development is integration of state-of-the-art advanced engineering simulation tools discussed in Section 3 of this paper with a database and command processing system that permits interdisciplinary engineering teams to carry out system simulation and design evaluation throughout the design process.

The heart of the software environment shown in Figure 10 is a database and command processor [2-4] that is a repository of design, environmental, and performance data that characterizes the mechanical system under development. These data, stored in a variety of formats on a variety of computers, are accessible to the broad range of engineering simulation and design analysis tools shown on the periphery of the diagram in Figure 10. Most tools

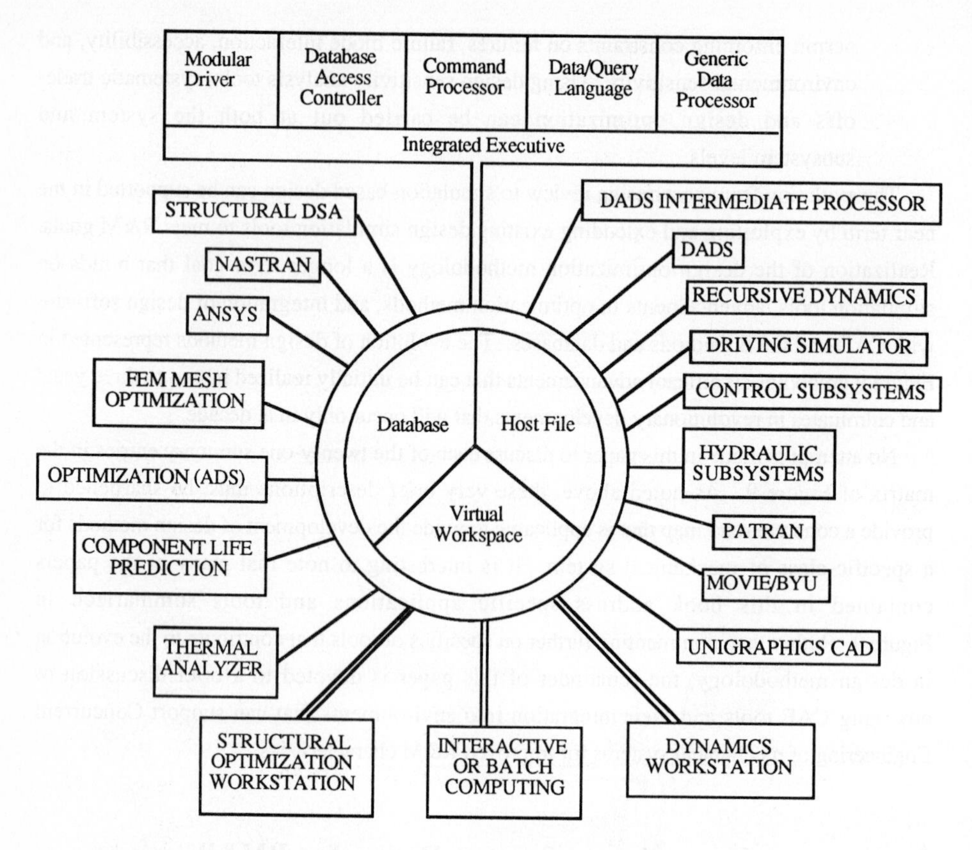

Figure 10. A Software Environment for Simulation Based Concurrent Engineering

contained in blocks in Figure 10 are commercially available software packages that are in common use in industrial design organizations. The spokes that connect these tools to the database are schematic representations of data passing mechanisms that permit the engineering team to pass data required for a variety of simulations from geometry files, CAD definition of the design, and simulation environment data to the appropriate simulation tool for analysis and return of appropriate data to the central database for performance, reliability, and maintainability evaluation.

As noted in Figure 10, numerous discipline specific tools; e.g., structural analysis, thermal analysis, dynamic analysis, control analysis, and hydraulic analysis software, must be used in the mechanical system design process. This implies that contributions are required from specialists in numerous engineering disciplines to support development of a practical environment that can be used by an engineering team. This leads to the requirement for graphics based computer workstation environments that will support design definition, engineering design, data generation, and simulation test definition. This diverse collection of activities requires use of a variety of computing subsystems, ranging from CAD workstations to mainframe computers. A schematic diagram of the computing environment on which this software system is being implemented in the Center for Simulation and Design Optimization at The University of Iowa is shown in Figure 11. Four types of engineering workstations are used to support the variety of modeling and simulation tools shown in Figure 10, as indicated at the top of Figure 11. Two CAD systems, shown at the upper right of Figure 11, are used to generate and capture design data and create modeling information that is passed to the central database to support a variety of simulation tools. While modeling is typically done on a CAE or CAD workstation, many of the analyses required involve extensive data sets and massive amounts of computation. For this reason, a separate data control computer (VAX) is used to manage the database, and mini supercomputers (Alliants) are used to support number crunching. In order to support the multidisciplinary team, however, details of this underlying computing environment must be transparent to the user.

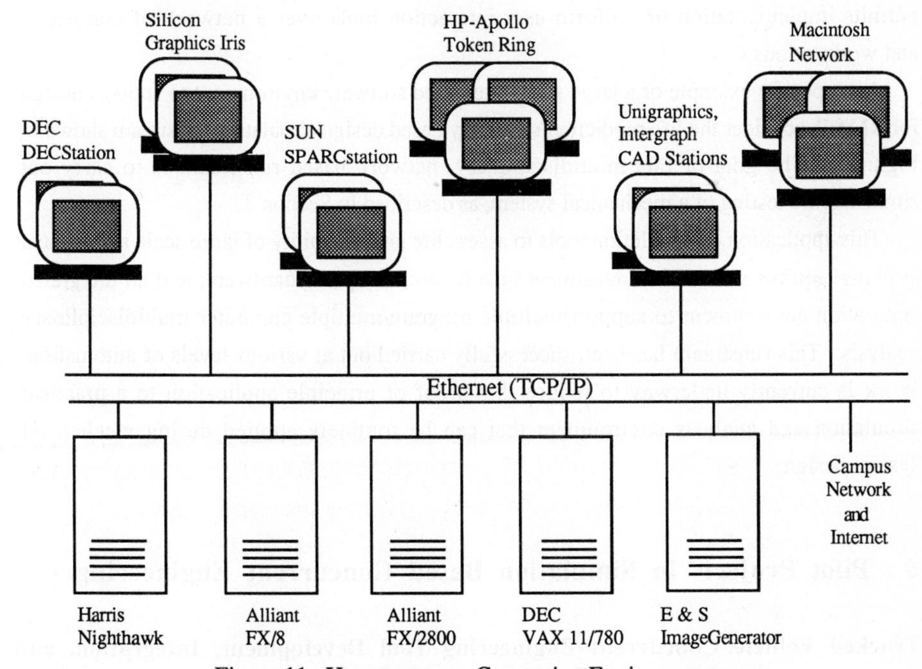

Figure 11. Heterogeneous Computing Environment

In order to support the wide variety of simulation and design applications shown in Figure 10 on the heterogeneous computing environment shown in Figure 11, network computing tools are employed to control the simulation process. The Network File System is used to locate data files on the various computers in the network and to pass data required from sources in the network to the network node on which modeling or analysis is to occur. The Network Computing System is employed to automate the process of data passing and for assignment of specific computational tasks to computer nodes on the network shown in Figure 11. For example, an engineer carrying out structural analysis may employ PATRAN on a HP-Apollo workstation to create a structural model and call for a stress analysis. If the stress analysis is to be carried out using a large scale finite element code such as ANSYS that resides on an Alliant mini supercomputer, logic embedded in the network controls the process of passing data required for the analysis specified from the HP-Apollo to the Alliant, launching the ANSYS finite element run and returning data to the HP-Apollo workstation for analysis by the engineer. With these networking tools, the engineer need not know which machine on the network has actually carried out the computational support tasks. Finally, using X-window network windowing standards, information may be displayed in multiple windows on the engineer's workstation or CAD system, under the control of another computer on the network, without overt involvement by the engineer. Use of such standards permits implementation of uniform user interaction tools over a network of computers and workstations.

As a specific example of a large scale integrated software environment to support design for RM&P, consider the life prediction/reliability based design simulation runstream shown in Figure 8. The goal of this multidisciplinary, network based runstream is to carry out simulated life testing of a mechanical system, as described in Section 3.

This application of simulation tools to assess life and reliability of large scale mechanical systems requires substantial investment in software, computer hardware, and an integrated simulation environment to support multiple program/multiple computer multidisciplinary analysis. This runstream has been successfully carried out at various levels of automation. Work is currently underway to reduce this proof of principle application to a practical simulation and analysis environment that can be routinely applied during mechanical system design.

6 Pilot Projects in Simulation Based Concurrent Engineering

Tracked Vehicle Concurrent Engineering Tool Development, Integration, and Validation. The purpose of this section is to outline developments in two substantial projects that are developing concrete capabilities, using the software integration model of Section 5. In October of 1991, the Center for Simulation and Design Optimization was awarded a two-year project in tracked vehicle Concurrent Engineering. This project exploits recent technical developments and investments by DARPA, the Army, the National Science

Foundation, and US industry to develop and validate a Concurrent Engineering capability for tracked combat vehicles, with the military vehicle development industry. Simulation based design technologies that have not yet been exploited by this industrial sector are being integrated using the DARPA DICE framework described in Reference 4 on a network of state-of-the-art workstations, parallel computers, and vehicle driving simulator facilities to support Concurrent Engineering of tracked combat vehicles. A partnership with three military vehicle industrial firms and support from West Virginia University and the Army Tank-Automotive Command leverages the results of ongoing research and development programs to create a qualitatively new tracked combat vehicle Concurrent Engineering capability. The purpose of this subsection is to explain the approach being used in this development, to create a practical, usable, and validated Concurrent Engineering environment for a specific class of mechanical systems.

Integral to this project is the continuous use, refinement, and validation of the Concurrent Engineering tools by the industrial partners, using computer and support facilities provided by the Center and facilities at each of their sites. Computer network connections are being provided to give the industrial partners access to Center facilities. The tools developed are being implemented and tested at Iowa, using network connection to the industrial sites. Each industrial partner will apply and tools developed on a generic tracked vehicle application, to guide the development of an effective Concurrent Engineering environment. Each industrial partner will then carry out a substantial application project using the tools and environment developed. This mode of development is recommended to create a proven Concurrent Engineering environment that will be functional in operating industrial environments.

A suite of CAE and operator-in-the-loop simulation tools is being integrated into the DARPA DICE environment [4], with assistance from West Virginia University and industrial and government partners in the project. Tools to be integrated include tracked combat vehicle mechanical system performance analysis, geometric modeling, load analysis, dynamic analysis, structural analysis, dynamic stress computation, life prediction, structural design sensitivity analysis, optimization, and operator-in-the-loop driving simulation, as shown in Figure 12.

Each of the software systems to be integrated provides design and analysis capabilities for a single discipline; tracked vehicle design [26], fatigue life prediction [27,28], design sensitivity analysis and optimization [9-11], and driving simulation [32]. A Concurrent Engineering environment that allows engineers using these tools to share information and coordinate their activities in tracked vehicle design is being built using the basic services and tools provided by the DICE architecture [4]. Two fundamental requirements for creating such a DICE based tracked vehicle CE environment are (1) the software must be integrated to allow engineers to exchange and share engineering data, and (2) a means must be provided in the

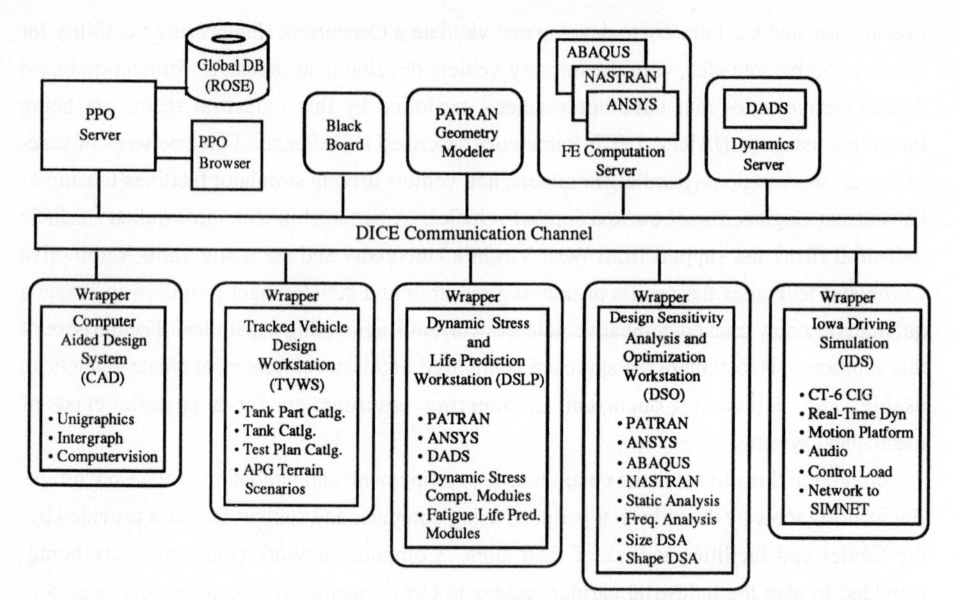

Figure 12. Tracked Vehicle Concurrent Engineering Environment Being Developed

environment to manage and coordinate activities for maximum concurrency as the design evolves.

With respect to the first requirement, each software package is being integrated into the DICE architecture by means of a DICE data wrapper [4], so that it can communicate with the global Product, Process, and Organization (PPO) server for shared information and the DICE Blackboard for coordination with other wrapped Center software. The DICE Concurrency Manager will be used for remote process communication in the DICE architecture.

With respect to the second requirement, a PPO model is being created for tracked vehicle system Concurrent Engineering. This PPO model will contain a global database schema for tracked vehicle systems (the Product), specifications of design and analysis activities for tracked vehicle systems (the Process), and information on personnel and resources (the Organization). The PPO model will be implemented in the ROSE database system and the DICE PPO server will provide services within the tracked vehicle design environment to allow the PPO model to be accessed, inspected, and shared.

The global database schema for tracked vehicle system information will emphasize engineering data sharing. For example, the database will contain pointers to engineering models (Finite Element, Geometry) that are created by engineer A on workstation B, so that a different engineer may obtain these pointers and read the same engineering information as A. This sharing of engineering model information assists in maintaining model consistency across different engineering discipline software and provides for information distribution.

The Process part of the PPO will contain specifications of engineering activities, pre- and post-conditions, and data requirements for each activity. Engineers and the project leader will use the Process part of the PPO model and use the DICE Blackboard and DICE Concurrency software to coordinate their activities, in a heterogeneous computing environment.

The DICE architecture tools and software modules to be used in building this tracked vehicle CE environment are as follows:

(1) DICE wrapper tools, which facilitate development of software interfaces to DICE architecture services, tools, and other wrapped DICE applications

(2) DICE Blackboard, which provides a software forum for engineers to cooperate and coordinate their activities

(3) DICE Concurrency manager, which provides software mechanisms for network transparent communication between workstations and design and analysis run-stream authoring, scheduling, and execution services

(4) ROSE and PPO information management tools.

Using the above Concurrent Engineering environment, the typical scenario shown in Figure 13 for tracked vehicle design is described below, where activities 2 and 3 will be concurrently carried out, once activity 1 is completed. Likewise, activities 4 and 5 will be carried out concurrently, once activities 2 and 3 are completed.

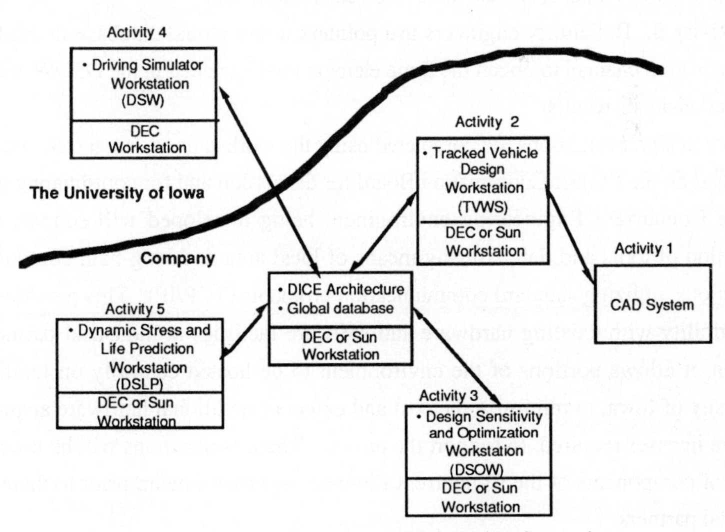

Figure 13. A Typical Concurrent Engineering Scenario

Activity 1. The CAD system is used to specify the design of vehicle parts and to assemble parts into part assembles. Assembly information and mass property data for each part; e.g., mass, moments of inertia, and center of gravity, are transferred from the CAD database to the Tracked Vehicle Design Workstation (TVWS) [26] database. In the global database, pointers are structured in the same way that vehicle part information is structured in the TVWS to store pointers to part characteristics obtained from all CAE tools.

Activity 2. The TVWS is used to generate tracked vehicle dynamic analysis models and to launch standard dynamic simulation tests using the Dynamic Analysis and Design System (DADS) [30]. Pointers to dynamic simulation models and dynamic analysis results that are need for other CAE workstations will be stored in the global database, through the PPO and communication channel.

Activity 3. Design engineers use part information pointers available in the global database to review the geometric characteristics of parts and build finite element structural models using PATRAN, through the Design Sensitivity Analysis and Optimization Workstation (DSOW) [9-11]. The structural design engineer uses parameterization capabilities provided in the DSOW to define design parameters and launch structural analysis. Pointers to these structural models and analysis results, which are to be used for dynamic stress computation, are stored in the global database through the PPO and communication channel. The design engineer can then carry out structural design sensitivity analysis.

Activity 4. The soldier-in-the-loop Dynamics Workstation (DSW) will use the dynamic simulation model generated by the TVWS to control soldier-in-the-loop simulations on the IDS that provide soldier feedback about the vehicle designed.

Activity 5. Reliability engineers use pointers in the global database through the DICE communication channel to obtain the finite element model created in the DSOW and associated structural analysis results.

Once design evaluations are completed using the workstations, suggested design changes are posted on the Project Coordination Board for discussion and for coordinating teamwork.

The Concurrent Engineering environment being developed will support transparent integration of tools and databases by means of local area and long-haul computer network connections, utilizing standard communication protocols (TCP/IP). This provides maximum compatibility with existing hardware and software facilities at industrial partner sites. In addition, it allows portions of the environment to be hosted centrally on facilities at The University of Iowa, to minimize the cost and extent of additional hardware acquisitions and software licenses required to support the project. These connections will be used for remote testing of components of the Concurrent Engineering environment, prior to their delivery to industrial partners.

Realistic prototype examples are being used as the foundation for guiding development of the Concurrent Engineering environment, for testing and evaluating the effectiveness of the tools implemented. To form a common foundation for joint design, test, and evaluation of the environment, a generic MI tank test application is being used by all participants in the project.

Design definition and rigid-body off-line simulation will be carried out with the TVWS at each site. Life prediction and structural design sensitivity analysis of a road arm is being used to provide experience in use of the DSLP workstation and the DSOW. Communication of load information from the TVWS, detailed structural design parameterization from the DSOW, and deformation mode and dynamic stress computation for the DSLP workstation is communicated through the DICE tool sets. Project design information is stored in the global database, with interactive project information posted on the blackboard. Driving simulator applications will be carried out using an initially fixed base to demonstrate the capability to support soldier-in-the-loop simulation, in parallel with more traditional design activity.

Simulation Based Design for Military System Supportability and Human Factors. An extended simulation based Concurrent Engineering environment is being developed in a related project recently awarded by the Defense Modeling and Simulation Office (DMSO) to bring realistic consideration of military system supportability and human factors into the early phases of the design process.

The specific objectives of this project are as follows:

(1) Broaden the scope of applicability of simulation based design to ground tactical vehicles, material handling equipment, construction equipment, and maintenance equipment.

(2) Incorporate tools for maintainability evaluation and design for maintainability into a Concurrent Engineering environment, using advanced anthropomorphic modeling methods and computer graphics that permit consideration of protective clothing, restricted vision, and special tools [17].

(3) Enhance the ability of military personnel-in-the-loop simulators under development to create realistic duty cycle information needed in design for durability and reliability early in the design process.

(4) Use simulator generated duty cycle information, which has heretofore been available only after hardware has been developed and tested, early in the design process when design latitude remains to optimize military equipment for durability and reliability.

Since the project outlined earlier in this section is focused on tracked combat vehicles, this pilot project addresses extensions of the methods and software being developed to support wheeled tactical vehicles, material handling equipment, construction equipment, and other mechanical systems. To meet this objective, the tracked vehicle Concurrent Engineering environment shown in Figure 12 will be broadened, as shown in Figure 14, to support a broad range of applications. As indicated at the lower left of Figure 14, the CAD environment will include three major CAD systems that are used by participants in this project, providing the capability to define components, structures, and other design characteristics of military equipment. Parts catalogs and databases associated with design for supportability and human factors will be defined, to form the foundation for intercommunication among the various workspaces shown at the bottom of Figure 14. Wrappers will be developed for the CAD systems, to permit access to design data from each of the four functional workspaces shown at the lower right of Figure 14, as transparently as possible to application oriented engineering teams.

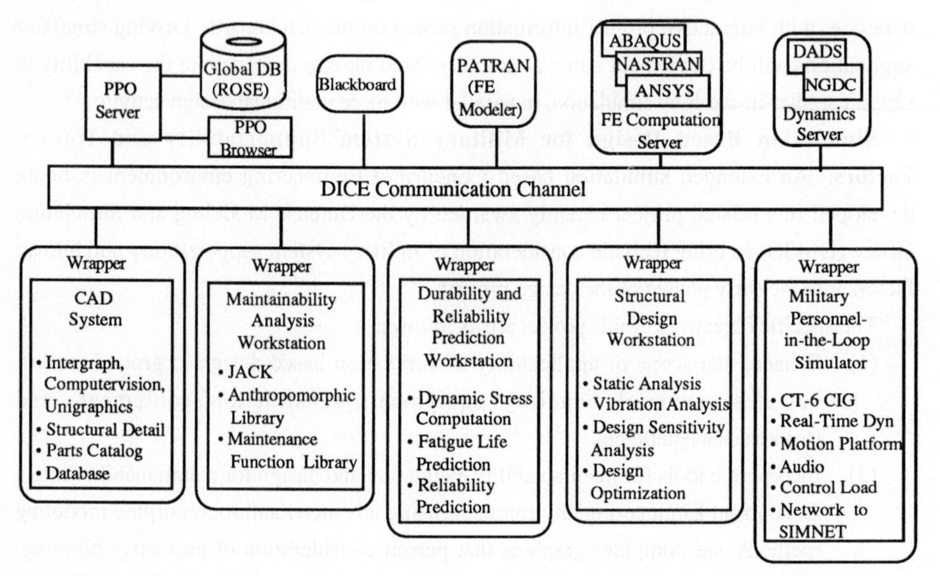

Figure 14. Simulation Aided Design System for Supportability and Human Factors

The CAD interfaces, wrappers, and data structures [5] developed in this project will be documented to support each of the four application workspaces. Initial application experience with the participating organizations will be used to refine the software environment, the database structure, and communication interfaces to ease the burden on specialists in the application workspaces.

Wrapper software and communication channels will be developed to support the Maintainability Analysis Workstation shown in Figure 14, based on the JACK anthropomorphic modeling system [17]. Geometry that characterizes design of the environment in which maintenance operations are to be investigated will be made accessible to the maintenance analysis workstation from the CAD environment, permitting identification, extraction, and transmission of geometric information needed for remove, manipulate, and replace functions encountered in maintenance activities, with personnel wearing protective clothing and carrying tools.

The dynamic stress and life prediction workstation that is the foundation for durability and reliability analysis of tracked combat vehicles shown in Figure 12 will be extended to be more broadly applicable and incorporated in the environment proposed for Durability and Reliability Prediction Workstation shown in Figure 14. Wrapper software and interfaces will be extended to provide communication with each of the four other workspaces/workstations, to provide data flow needed for durability and reliability assessment of broad classes of mechanical subsystems and components.

The Structural Design Workstation shown in Figure 14 will include a capability for static and vibration analysis, as well as design sensitivity analysis and optimization of structural components. Interfaces between the Structural Design Workstation and the CAD environment will be strengthened over currently existing capabilities, to enhance the ability of structural specialists to access structural detail from the CAD system, carry out structural analysis of component failure, evaluate design sensitivities, and contribute deformation, mass, and stiffness information to the Durability and Reliability Prediction Workstation.

Interfaces between the Military Personnel-in-the-Loop Simulator shown at the lower right of Figure 14 and other workspaces will be created to implement design for supportability and human factors. The software wrappers and communication channels required to communicate information associated with duty cycles, loads, and human factors information to other workspaces of the proposed simulation aided design environment will be developed and implemented into the DICE architecture. Since the advanced military personnel-in-the-loop simulator will be located only at The University of Iowa, communication channels and interfaces will be designed to be remotely accessible by participants in this project and future users of the capability developed under this project.

8 Conclusions

Basic CAE tools and technologies are in place to support an immediate evolution from the current state of affairs to a simulation based Concurrent Engineering environment that can effectively bring consideration of downstream disciplines such as reliability, maintainability, and producibility into the early stages of the design process. These tools are reviewed in this paper, illustrated in much more detail in other papers in this book, and shown to provide realistic design trade-off and analysis and optimization tools that can be used as a "soft prototyping" capability throughout the design process. Their integration into a practical and effective Concurrent Engineering environment, however, requires substantial effort and should be carried out using a systematic development, test, and evaluation mode, much as hardware systems are developed and tested. The road map presented and analyzed in the paper can serve as an effective guide to evolutionary development of major new Concurrent

Engineering capabilities that exploit a broad existing base of CAE software, CAD workstations, compute servers, databases, and network computing capability that largely exist and only need to be integrated. If systematically developed in an evolutionary way, the result can be a revolutionary new Concurrent Engineering capability for design of mechanical systems.

Acknowledgment: Research supported by NSF-Army-NASA Industry/University Cooperative Research Center for Simulation and Design Optimization of Mechanical Systems.

References

1. "Application of Concurrent Engineering to Mechanical Systems Design." CALS Technical Report 002, Suite 300, 1025 Connecticut Avenue NW, Washington, DC 20036, 1989.
2. Vos, R.G., "Implementation and Applications of Multidisciplinary Concurrent Engineering," *Concurrent Engineering Tools and Technologies for Mechanical System Design* (E. J. Haug ed.), Springer-Verlag, Heidelberg, 1993.
3. Vos, R.G., "The Emerging Basis for Multidisciplinary Concurrent Engineering," *Concurrent Engineering Tools and Technologies for Mechanical System Design* (E. J. Haug ed.), Springer-Verlag, Heidelberg, 1993.
4. Cleetus,K. J., "Virtual Team Framework and Support Technology," *Concurrent Engineering Tools and Technologies for Mechanical System Design* (E. J. Haug ed.), Springer-Verlag, Heidelberg, 1993.
5. Wu, J.K., "Data and Process Models for Concurrent Engineering of Mechanical Systems," *Concurrent Engineering Tools and Technologies for Mechanical System Design* (E. J. Haug ed.), Springer-Verlag, Heidelberg, 1993.
6. Wu, S.M., and Hu, S.J., "Defect Preventive Quality Control in Manufacturing," *Concurrent Engineering Tools and Technologies for Mechanical System Design* (E. J. Haug ed.), Springer-Verlag, Heidelberg, 1993.
7. Ashton, J.E., "Relationship Between Design for Manufacturing, a Responsive Manufacturing Approach, and Continuous Improvement," *Concurrent Engineering Tools and Technologies for Mechanical System Design* (E. J. Haug ed.), Springer-Verlag, Heidelberg, 1993.
8. Liu, T.H., and Fischer, G.W., "An Approach for PDES/STEP Compatible Concurrent Engineering Applications," *Concurrent Engineering Tools and Technologies for Mechanical System Design* (E. J. Haug ed.), Springer-Verlag, Heidelberg, 1993.
9. Choi, K.K., and Chang, K.H., "Design Sensitivity Analysis and Optimization Tool for Concurrent Engineering," *Concurrent Engineering Tools and Technologies for Mechanical System Design* (E. J. Haug ed.), Springer-Verlag, Heidelberg, 1993.
10. Twu, S.L., and Choi, K.K., "Confirmation Design Sensitivity Analysis for Design Optimization," *Concurrent Engineering Tools and Technologies for Mechanical System Design* (E. J. Haug ed.), Springer-Verlag, Heidelberg, 1993.
11. Chang, K.H., and Choi, K.K., "Shape Design Sensitivity Analysis and What-if for 3-D Design Applications," *Concurrent Engineering Tools and Technologies for Mechanical System Design* (E. J. Haug ed.), Springer-Verlag, Heidelberg, 1993.
12. Olhoff, N, Lund, E., and Rasmussen, J., "Concurrent Engineering Design Optimization in a CAD Environment," *Concurrent Engineering Tools and Technologies for Mechanical System Design* (E. J. Haug ed.), Springer-Verlag, Heidelberg, 1993.
13. Pedersen, P., "Concurrent Engineering Design with and of Advanced Materials," *Concurrent Engineering Tools and Technologies for Mechanical System Design* (E. J. Haug ed.), Springer-Verlag, Heidelberg, 1993.
14. Tortorelli, D.A., "Design Sensitivity Analysis for Coupled Systems and their Application for Concurrent Engineering," *Concurrent Engineering Tools and Technologies for Mechanical System Design* (E. J. Haug ed.), Springer-Verlag, Heidelberg, 1993.

15. Grandhi, R.V. and Srinivasan, R., "Concurrent Engineering Tools for Forging Die and Process Design," *Concurrent Engineering Tools and Technologies for Mechanical System Design* (E. J. Haug ed.), Springer-Verlag, Heidelberg, 1993.
16. Hancock, P.A., Andre, A.D., Caird, J.K., Flach, J.M., and Green, P., "Human Factors in Vehicle Driving Simulation," *Concurrent Engineering Tools and Technologies for Mechanical System Design* (E. J. Haug ed.), Springer-Verlag, Heidelberg, 1993.
17. Badler, N.I., "Simulated Humans, Graphical Behaviors, and Animated Agents," *Concurrent Engineering Tools and Technologies for Mechanical System Design* (E. J. Haug ed.), Springer-Verlag, Heidelberg, 1993.
18. Frisch, H.P., "Man/Machine Interaction Dynamics and Performance Analysis," *Concurrent Engineering Tools and Technologies for Mechanical System Design* (E. J. Haug ed.), Springer-Verlag, Heidelberg, 1993.
19. Clausing, D.P., "World-Class Concurrent Engineering," *Concurrent Engineering Tools and Technologies for Mechanical System Design* (E. J. Haug ed.), Springer-Verlag, Heidelberg, 1993.
20. *Basic Reliability Design.* Army Management Engineering College, Rock Island Arsenal, IL.
21. *PATRAN PLUS User's Manual, Vols. 1 and 2.* Software Products Division, PDA Engineering, 1560 Brookhollow Drive, Santa Ana, CA, 1989.
22. Easterly, J., and Ianni-Wright, J.D., "Crew Chief: A Model of a Maintenance Technician," *Concurrent Engineering of Mechanical Systems, Vol. 1* (E. J. Haug ed.), Center for Computer Aided Design, The University of Iowa, Iowa City, Iowa, pp. 199-204, 1989.
23. Badler, N., Lee, P., Phillips, C., and Otani, R.M., "The *JACK* Interactive Human Model," *Concurrent Engineering of Mechanical Systems, Vol. 1* (E. J. Haug ed.), Center for Computer Aided Design, The University of Iowa, Iowa City, Iowa, pp. 179-198, 1989.
24. Beck, R.R., "Simulation Based Design of Off-Road Vehicles," *Concurrent Engineering Tools and Technologies for Mechanical System Design* (E. J. Haug ed.), Springer-Verlag, Heidelberg, 1993.
25. Wu, J.K., M.A. Fogle, J.Y. Wang, and J.K. Lu., "A Dynamic Workstation," *Proceedings, ASME 15th Design Automation Conference,* Montreal, Canada, Sept. 17-21, 1989.
26. Ciarelli, K., "Integrated CAE System for Military Vehicle Applications," *Proceedings of the First Annual Symposium on Mechanical System Design in a Concurrent Engineering Environment*, Iowa City, Iowa, pp. 301-318, October 24-25, 1989.
27. Dopker, B., Yim, H.J., and Haug, E.J., "Computational Methods for Stress Analysis of Mechanical Components in Dynamic Systems," *Concurrent Engineering of Mechanical Systems, Vol. 1* (E. J. Haug ed.), Center for Computer Aided Design, The University of Iowa, Iowa City, Iowa, pp. 217-238, 1989.
28. Baek, W.K., and Stephens, R.I., "Computational Life Prediction Methodology for Mechanical Systems Using Dynamic Simulation, Finite Element Analysis, and Fatigue Life Prediction Methods," Technical Report R-71, Center for Computer Aided Design, The University of Iowa, Iowa City, IA, May 1990.
29. Singh, R.P., and Likins, P.W., "Singular Value Decomposition for Constrained Dynamical Systems," *J. of Applied Mechanics*, Vol 52, No. 4, pp. 943-948, 1985.
30. *DADS User's Manual Rev. 5.0.* Computer Aided Design Software, Inc., Oakdale, IA, 1988.
31. Bodley, C., Devers, A., Park, A., and Frisch H., *A Digital Computer Program for Dynamic Interaction Simulation of Controls and Structures (DISCOS)*, NASA Technical Paper 1219, 1978.
32. Haug, E.J., Kuhl, J.G., and Tsai, F.F., "Virtual Prototyping for Mechanical System Concurrent Engineering," *Concurrent Engineering Tools and Technologies for Mechanical System Design* (E. J. Haug ed.), Springer-Verlag, Heidelberg, 1993.
33. Kuhl, J.G., Papelis, Y.E., Romano, R.A., "An Open Software Architecture for Operator-in-the-Loop Simulator Design and Integration," *Concurrent Engineering Tools and Technologies for Mechanical System Design* (E. J. Haug ed.), Springer-Verlag, Heidelberg, 1993.
34. Yae, K.H., "Teleoperation of Redundant Manipulator," *Concurrent Engineering Tools and Technologies for Mechanical System Design* (E. J. Haug ed.), Springer-Verlag, Heidelberg, 1993.
35. *MSC/NASTRAN User's Manual, Vols. I and II.* The Macneal Schwendler Corporation, 815 Colorado Boulevard, Los Angeles, CA, 1988.
36. DeSalvo, G.J., and Swanson, J.A., *ANSYS Engineering Analysis System User's Manual, Vols. 1 and 2.* Swanson Analysis Systems, Inc., P.O. Box 65, Houston, PA, 1987.
37. Yim, H.J., and Haug, E.J., "Computational Methods for Dynamic Stress Analysis of Mechanical Systems," Technical Report R-84, Center for Computer Aided Design, The University of Iowa, Iowa City, IA, August 1990.
38. Chiang, A.C., and Wu, J.K., "Automated Failure Mode Effects and Criticality Analyses for Reliability Prediction of Multibody Mechanical Systems," Technical Report R-139, Center for Computer Aided Design, The University of Iowa, Iowa City, IA, August 1992.

39. Haug, E.J., and Yeh, H.F., "Dynamic Simulation for Failure Analysis," *Concurrent Engineering of Mechanical Systems, Vol. 1* (E. J. Haug ed.), Center for Computer Aided Design, The University of Iowa, Iowa City, Iowa, pp. 107-126, 1989.
40. Haug, E.J., Choi, K.K., and Komkov, V., *Design Sensitivity Analysis of Structural Systems*, Academic Press, New York, 1986.
41. Bestle, D., "Optimization of Automotive Systems," *Concurrent Engineering Tools and Technologies for Mechanical System Design* (E. J. Haug ed.), Springer-Verlag, Heidelberg, 1993.

The Emerging Basis for Multidisciplinary Concurrent Engineering

Robert G. Vos

Boeing Aerospace and Electronics Company, Seattle, WA 98124, USA

Abstract: Some industry perspectives on Concurrent Engineering are examined as a background for discussing the technology basis and level of maturity. Integrated multidisciplinary analysis tools are presented as a major contributor to concurrency. Key islands of integrated technology are identified and the potential for connecting these islands is discussed.

Keywords: quality management / integrated analysis / concurrent engineering / multidisciplinary analysis / optimization / computer software / database / user interface

1 Introduction

In the early 1970s, when optimization methods were beginning to emerge for specific technologies in the engineering environment, there were those who lamented that "Optimization is the buzz-word of a dying technology." Their implication was, of course, that as a technology becomes relatively mature and its growth begins to level off, a natural tendency of the affected technologists is to "optimize" applications. In this case, it turned out during the 1980s that optimization methods finally began to come to fruition, supported by faster computing hardware and better algorithms. In this writer's opinion, optimization methodology will expand even faster during the 1990s, and will become effective in treating multidiscipline as well as single discipline problems.

Today, Concurrent Engineering is becoming an issue of intense interest throughout government and industry. Major established companies are using the terminology in sales presentations and seminars. The government CALS (Computer aided Acquisition and Logistics Support) initiative is attempting to achieve greater concurrency in the procurement and support process. A new bimonthly journal called Concurrent Engineering has just been announced by Auerbach Publishers. At the same time there seems to be an undercurrent of opinion, mostly by those not involved in the process, that Concurrent Engineering technology is at best lacking in substance, and at worst may be "the buzz-word of a dying industry." The coming decade will undoubtedly prove or disprove the worth of Concurrent Engineering. Already, however, there is evidence that the Concurrent Engineering concept is poised for rapid growth, even while many other technologies are reaching plateaus or slowing in their level of development activity.

Some quantitative insight into this issue can be realized by observing the recent levels of reported research in Concurrent Engineering (CE) and related fields. Figure 1 is a tabular summary of research abstracts, covering the 1980s decade and taken from a computerized version of the Engineering Index. The literature search used 13 keyword topics related to CE. Included are the 13-topic and yearly sums, as well as the total number of abstracts on all topics compiled in the Index. Note that the 1990 values probably do not yet represent complete counts. From the data shown, one can draw several interesting conclusions:

1. General research activity (measured in terms of Index abstracted publications) seems to have peaked about 1985, and it forms a very rough bell shaped curve for the decade.
2. Development in the automation area, at least under that title, has slowed significantly from its peak.
3. Research activities showing definite growth are PDES, quality management, CALS, and CE. These growth activities are also the smallest, i.e., they are the most immature.
4. Quality Management (QM) and Concurrent Engineering (CE) have the highest growth rates, again measured by number of abstracted publications.

Before examining some industry perspectives, the writer would like to offer a few initial comments. Although it might seem risky to extrapolate technology growth rates based on the still relatively small numbers of published works, the trends for CE and QM are so pronounced that highly positive extrapolations are justified. Reviewing the publications themselves, or attending a few of the many ongoing related conferences and seminars, reinforces the conclusion that these research areas are still immature. There does not yet exist a complete consensus on the "what" and "why" of QM/CE, although there is agreement on the fundamental characteristics. CE is basically seen as a way to get products to market "faster, cheaper and better." QM is viewed as a team oriented approach to "doing it right the first time, avoiding waste, and surviving competition."

Some of the published work in CE has legitimately resulted from new concepts and applications of concurrency. A perceptive observer will also notice that many of the current publications, perhaps even the mainstream works, really represent efforts to bring previously established technologies under the CE banner. These technologies may provide useful enabling support for concurrency, but on their own they have received less recognition and support. Whatever the motivations, such efforts should be welcomed and judged on their merits, because the success of the broad CE objectives will require the coalescing of many supporting and interacting technologies. In fact, many of these technologies are CE pioneers

Keywords	1981	1982	1983	1984	1985	1986	1987	1988	1989	1990	Σ
Integration/Integrated	4669	5573	7354	9692	11501	11725	9737	8236	10141	4686	83314
Automation/Automated	1664	2323	3248	4316	5008	5187	3707	2975	2945	1202	32575
Database/Data Management	2236	2410	2838	3315	3362	3308	2939	2314	2748	1241	26711
Standards	1682	1955	2639	3234	3233	3377	2587	2179	2117	1038	24041
Computer Software	798	965	1690	2644	3240	3325	2923	3039	3620	1595	23839
Computer Hardware	58	77	116	220	435	516	462	374	415	179	2852
User Interface	93	143	192	295	324	395	313	317	381	135	2588
Networking	0	0	0	0	0	0	0	176	178	71	425
IGES	2	4	12	21	25	11	15	12	12	4	118
PDES	2	1	5	6	16	11	17	12	26	11	107
Quality Management	0	0	1	3	4	2	3	5	22	43	83
CALS	3	0	2	0	0	2	5	10	12	2	36
Concurrent Engineering	0	0	0	1	0	2	1	2	11	14	31
Σ	11207	13451	18097	23747	27148	27861	22709	19651	22628	10221	196720
Total Abstracts	150142	174393	200674	217811	205255	196383	166653	134219	147947	74690	1668167

Figure 1. Research Abstract Counts (Engineering Index)

in that they recognized the importance and essential ingredients of CE long before the terminology even became popular.

A final comment regarding QM and CE since these two topics are difficult to treat separately; QM has probably been the primary creative force behind the CE concept, because the basic goals of QM can not be achieved using only sequential engineering approaches. QM and CE technologies are tending to closely track one another, from their common presentation forums and champions, to their common high growth rate and terminology. Although this lecture is primarily devoted to the emerging basis for CE, we will need to view CE as a subset of, and an essential ingredient for, QM.

2 What is CE/QM?

The question is very relevant to ask for any technology which is immature and in a stage of rapid expansion. There have been various government and industry perspectives on the answer, related to organizational objectives, methods for implementation, and the anticipated payoff.

Deming's "fourteen point" philosophy on management for quality [1] is well known, and is being covered by others in this lecture series. The philosophy is generally viewed as a major factor in helping Japan to gain a competitive position in the post World War II era. It can be summarized as a total and personal commitment to building quality into the production process. CE, although not explicitly stated, is implicit in the recommendations to eliminate barriers between groups and involve everyone together in the process.

Grunenwald [2] defines Total Quality Management (TQM) as "a search for opportunities, or things to fix." We can be sure this is intended to apply to the production process, not the resulting product! He credits A. Fiegenbaum in the early 1950s with origination of the Total Quality term.

A recent TQM article in Aviation Week and Space Technology [3] summarizes the Defense Department's TQM implementation guide concept as "Doing the right thing, right the first time, on time, all the time; always striving for improvement, and always satisfying the customer." Interestingly, producer/consumer relationships are sometimes reversed from those normally considered. The Air Force Systems Command is quoted as considering the contractor community as its customer for written proposal solicitations (RFPs) it sends out, and is trying to improve that product through concurrent dialogue with industry.

Jones, in a Computer-Aided Design article [4] provides an informative summary of the DoD CALS effort. The aim is to "improve the productivity and quality of the defense industry." The emphasis is on standards and information transfer, and considering aspects of future weapons systems long term support concurrently with the design phase.

Turning to industry, a dramatic reorganization of Douglas Aircraft in February 1989 was accomplished to support the company's concept of QM and to reduce operating losses. A major effect was to flatten the management hierarchy from nine levels to five, cutting management positions by half.

The best overall presentation of CE directions, goals and concerns is available in the report "Industrial Insights on the DoD Concurrent Engineering Program" [5]. This report resulted from a meeting of industrial officials, gathered by DoD to provide feedback on their CE program. There were four major recommendations from this forum, providing considerable insight into industry thoughts on CE and QM:

1. Develop methods to encourage, but not mandate, CE implementation within industry.
2. Acknowledge CE as a principal means for achieving TQM.
3. Establish a government sponsored CE initiative to provide funding for education and research.
4. Create a procurement process which allows flexibility in on-going trade-offs of system requirements, including cost/schedule/performance, emphasis on end user needs and involvement, and training of acquisition personnel.

The report stated that "the current educational emphasis on specialized disciplines needs to change to an emphasis on integrated interdisciplinary processes," and that "concurrent engineering requires large numbers of disciplines to work much more closely and interactively together."

It is a commonly held perception that the goals of CE are basically to achieve the "ilities" - producibility, interoperability, reliability, maintainability, and traceability. To these we should like to at least add testability and analyzability. A successful CE process functions by getting the product intents and requirements downstream, and moving the constraints and problems upstream. If well implemented, it forces more interaction between all affected organizations and technologies.

We should guard against an oversimplification of the basic character of CE. There is a valuable adage, known in the software development field as Brook's Law: "Adding manpower to a late software project makes it later." CE is not merely adding resources to a critical stage of a project, whether early or late. CE does mean a more effective integration and distribution of resources. Concurrency must be accomplished in at least three directions: across the involved organizations, through integration of the required technologies, and over the complete life cycle of the product.

3 Impetus for Development

The growth and use of QM/CE is obviously motivated by the industry crucial factors of cost, profit and competition. In the last decade we have seen these factors influenced strongly by the foreign environment; by the availability of skilled technical manpower; and by some plateaus we perceive being reached in technology, the economy, and the environment. To these influences we must add the increasing level of technology interaction and sophistication required to obtain improved performance, and the pressure exerted on industry by government mandates and shrinking budgets.

Many of these motivations have been discussed in-depth in the literature or are present in other parts of this lecture series. A factor we should like to discuss here is the influence of higher performance requirements and the resulting need for greater technology sophistication and integration. Figure 2 summarizes a past/present/future scenario of product lines crucial to the aerospace industry over the last 20-30 years. In the days of the Saturn V booster and the large ballistic missile, the mechanical systems were characterized as single body problems with basically low-frequency attitude control. There were a few significant technology interactions, such as fuel slosh and pogo dynamics, but the major technology challenges were single-discipline performance and reliability. In the current space shuttle and space station era, our aerospace systems involve several interconnected, but still relatively stiff, bodies; environmental and on-board thermal effects management; nonlinear actuator mechanisms; and vibration isolation. This era, not surprisingly, has fostered much of the development of multidiscipline analysis tools (thermal, structures, controls, multibody interactions), as well as large problem data management systems and an emphasis on consistent user/software interfaces. As we plan for the future, aerospace products will involve many flexible bodies; broad band control/structure interaction; requirements for vibration isolation and vibration suppression; more nonlinearities in both structures and control; severe thermal loads and more potential thermal interactions; and precision optical pointing and shape control.

The CE concepts are certainly no stranger to the process of multidiscipline design, simulation and analysis. In fact it should be recognized that CE concepts have been an integral part of this process for the last decade, because they were required simply in order to address the engineering interactions and data flow needs, and to obtain the right answers.

There are interesting insights to be gained by considering two CE related developments during the 1980s. The first was the evolution of computer hardware and the role of parallel processing. During the mid 1980s, and prior to the introduction of today's RISC architecture and faster chips, there was a period when affordable computing speed seemed to be reaching a plateau. As many computationally intense applications were frustrated, there was an increased spurt of interest in parallel processing. Parallel processing remains an important technology, although some of the interest has been drained away by faster serial and vector processing,

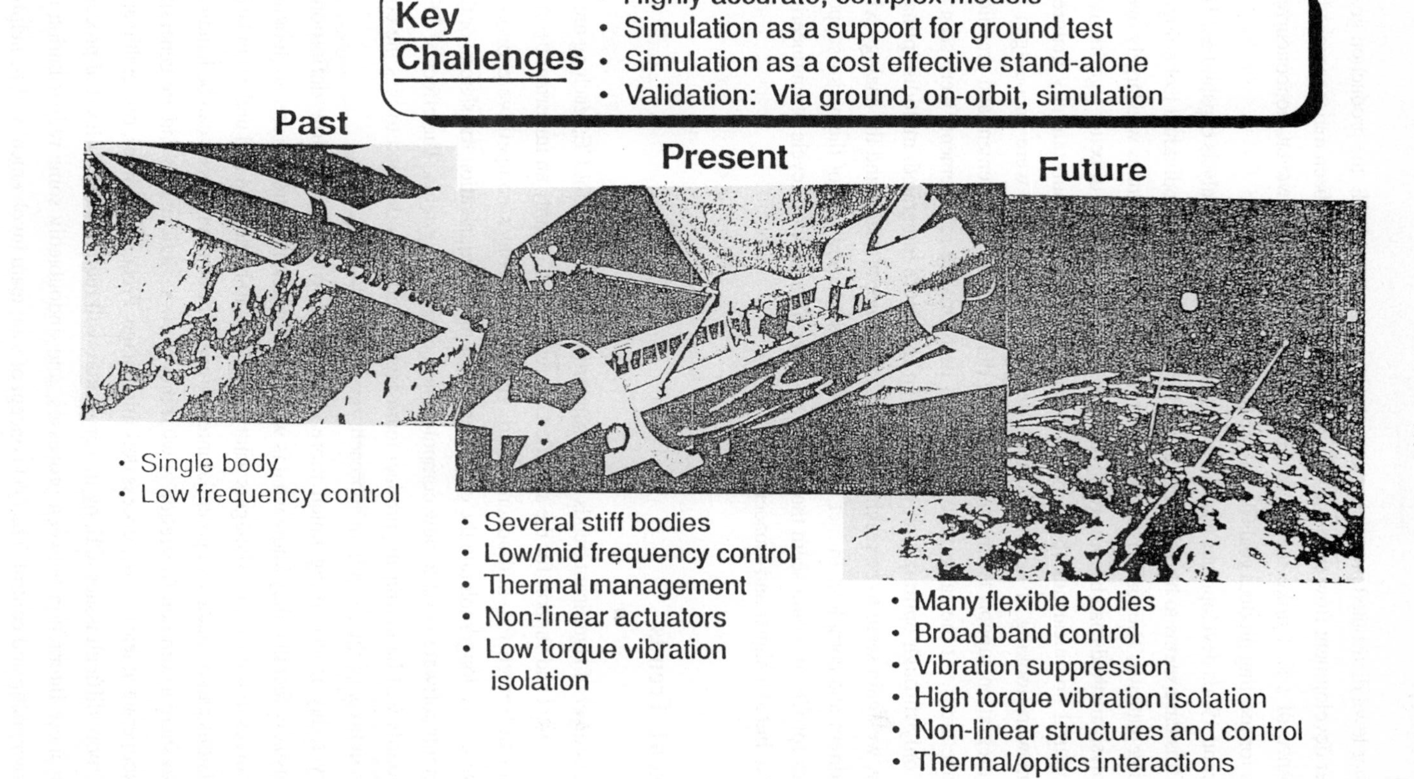
Key Challenges
• Highly accurate, complex models
• Simulation as a support for ground test
• Simulation as a cost effective stand-alone
• Validation: Via ground, on-orbit, simulation
Past
Present
Future
• Single body
• Low frequency control
• Several stiff bodies
• Low/mid frequency control
• Thermal management
• Non-linear actuators
• Low torque vibration isolation
• Many flexible bodies
• Broad band control
• Vibration suppression
• High torque vibration isolation
• Non-linear structures and control
• Thermal/optics interactions

Figure 2. Product Performance Impacts

and by higher level distributed processing. Drawing an analogy with CE, the production goals for a shorter development flow time and a faster time-to market have been major reasons for the emergence of CE. Unlike the computer hardware analogy, however, nonconcurrent alternatives for meeting product goals do not seem to be forthcoming.

The second 1980s development was the emergence of integrated analysis capabilities. The impetus in this case seems to have been twofold. First, several of the individual technologies were reaching plateaus. The rush to develop new finite element formulations was largely over, thermal analysis methods had become relatively stable, and the creation of better algorithms to handle structural nonlinearities (both material and geometric) had substantially matured. Researchers were looking at new directions for their energies, and were encouraged by emerging integration support technologies, e.g., in the database management and graphical user interface areas. At the same time, increasing performance requirements were seen as achievable only through optimization of the design in a more integrated, multidisciplinary sense. Now, we again seem to be reaching some goals, if not plateaus, and the analogy with CE seems direct and complete. In fact, integrated analysis is now one of the major support technologies for CE. It would seem that we are re-entering a previous cycle of engineering development, but at a higher and broader level.

4 Current Trends

There are a number of significant trends in the supporting technologies for CE, reaching much beyond the interest and support for the specific CE area itself. Certainly an important trend is the enthusiastic search for, and adherence to, standards. Today's competitive economic environment can no longer tolerate the cost of unnecessary communication translators, or the cost of difficult software ports to new computers and operating systems. The effect of CE's use of standards will be to knit the product team more tightly together and to bias the cost trades more in favor of the concurrent approach.

Industry today is becoming much more conversant with process flow definitions. Engineers usually feel that they know what is wrong with a technical process and are looking for the resources to improve it. Managers want the process to be defined and understandable so that decisions can be made and potential interactions with other processes can be handled. Both tend to share a common frustration at what they view as deficiencies and are generally willing to cooperate at least in the process definition stage. Process resources, ownership and control are more difficult issues. CE, by its very nature will require a higher level of process integration; more interaction between processes; and undoubtedly some redistribution of resources, ownership and control. The QM concept of total team involvement will be helpful in addressing the combined human and technical issues.

Data management concepts and terminology are no longer unfamiliar to the field of engineering. The interest in standardization has certainly been a strong encouragement for the trend toward integrated engineering databases. As more people and modules become involved in an organization or project, especially for a concurrent environment, the integrated database becomes essential. It is also true that data today is being viewed as a more valuable resource, and any data cataloging/query/manipulation tools which make this resource more readily accessed, protected and maintained over time, is viewed in a very positive manner. Finally, data today is becoming much more voluminous, and requirements for exchanging large amounts of data between organizations are increasing. The database system is becoming a necessity to avoid being overwhelmed by data storage and management demands. All of these factors are crucial to the CE philosophy. CE will not only benefit from the trend toward improved data management capabilities, but will itself be a major accelerating factor.

Common geometry and graphics are really separate issues, but share some common technology and needs. The trend to common geometry was fostered by early CAD requirements. It was a necessary initial step in both the design process and the analysis model building process. Common geometry really means a standard, automated way of representing the product physical definition, and of interfacing to the design and analysis requirements. "Common" did not really mean common, initially, because a diverse set of CAD packages and geometric representations were being developed. Today there is a stronger trend toward IGES/PDES commonality and a more ready exchange of geometric data.

The trend toward more available and standard graphics has two major aspects--graphical user interfaces and data visualization. In a rather brief time span (the last 2-3 years) interfaces for most of the workstation based software packages have shifted to a graphics oriented approach. The most obvious reason for this is the emergence of graphical toolkits and standards. A competent and recognized interface toolkit greatly reduces the development time and costs. Perhaps more important is that it encourages the standardization which then establishes a basis for trust in the portability and long term viability of the software. Interfacing standards are also important to the user, in establishing a consistent look and feel, and in reducing the required user familiarization time. The Motif interface toolkit supported by the Open Software Foundation (OSF) is becoming the most prominent choice, but the SUN developed Open Look toolkit also has a significant following.

Data visualization tools are less standardized than those in the user interface area. Reasons for this include the more diverse nature of the applications, the greater complexity of the associated data, and the 3-dimensional nature of much of the required graphics. Many vendor-specific visualization support packages are available, while at the same time recognized 3-D standards (Phigs and PEX) are beginning to emerge. More and faster 3-D graphics hardware options are also becoming available. Visualization requires considerably more development to reach the desired level of analysis concurrency.

The trend toward increased use of multidiscipline packages is visible and wide reaching. On the one hand, CAD and FEM model building packages are now used to support a variety of disciplines throughout the broad mechanical engineering field. Likewise, FEM analysis software, initially developed for structural stress and dynamics applications within the aerospace industry, is continuing its spread into diverse applications such as electromagnetics and computational fluid dynamics. Another type of multidiscipline package is that which takes several existing, often mature technologies, and combines them with appropriate interfaces into an integrated capability. The interfaces may involve data flow between different modules, data formatting between modules and a common database, a common graphical user interface look and feel. The trend is toward greater use of such integrated packages, as well as their expansion to incorporate more technologies and modules.

A readily visible trend during the last 5-10 years has been continued software commercialization and competitive shakeout. A number of years ago, it seemed that anyone who developed a clever algorithm or process, and was able to implement it in software, could expect recognition and use at least within a specialized application. Today, the software environment is keyed more to "survival of only the fittest," and a key ingredient to survival is often being large and well known. Individual "desk drawer" software is no longer viewed with much enthusiasm. Significant concerns are standardization, user learning curves, documentation, validation, and maintenance--all concerns which are becoming increasingly important to the more sophisticated and interactive working environment of CE. A bottom line is perhaps that potential software users no longer have the time to deal with dozens of individual packages or different systems. They want more encompassing software that can be trusted to deliver all of the important characteristics mentioned above.

A final trend worthy of note is the tendency, at least in some areas, toward greater reliance on simulation. A prime example in the aerospace industry is the use of computational fluid dynamics as a supplement or alternative to wind tunnel testing. Costs for testing have not decreased at as great a rate as those for simulation. Buoyed by rapid computing hardware improvements and more sophisticated software, simulation is also more conducive to rapid up-front trade studies. Good engineers will recognize that some systems require testing, and may not even be tractable to analysis by simulation. Simulation will of course not identify mistakes in manufacturing. But where simulation can be used, it is a substantial benefit to product flow time and also provides the designer/analyst with early insight into the product design limitations and potential improvements. Aside from cost, time, and insight, a major issue may simply be technical feasibility. A space structure, for example, operates in a zero-G environment. In fact, it may not be able in its deployed configuration to survive, or to perform as designed, within a one-G environment. Since ground test may therefore not be feasible, and since on-orbit testing is costly and risky at best, a greater reliance on simulation is an attractive alternative.

5 Existing Islands of Concurrent Support

As we share thoughts on where we are and where we can go in the CE field, we obviously need to identify the existing building blocks which are supportive of the CE process. We use the term "islands" here to describe these building blocks because the support for CE has grown in a largely bottom-up fashion. Individual islands seem to be constantly enlarging, bumping into one another, and/or forming bridges between themselves. We will briefly examine the evolution of these islands and bridges during the 1970-1990 time frame, as a basis for looking at current barriers and potential future directions.

Although it is impossible to define a beginning time point for the CE technology, it was the 1970s when several of the key technology islands were formed, and the 1980s when developments accelerated and some of the integrating bridges began to evolve. Figure 3 depicts significant areas which were part of the 1970s and 1980s history.

FEM analysis methods had of course already developed to a significant degree during the 1960s, but the next decade saw them extended, formalized and applied to broader classes of problems, including structural nonlinearities (material and geometric). As this technology matured during the 1970s, it freed important engineering energies and resources to be directed at other pressing problems. The groundwork was laid, and incentives were provided, to develop later links to CAD geometry and modeling. Matrix computational utilities were also created to support the FEM technology, and later began to evolve to support more general simulation requirements. It is difficult to overestimate the significance of FEM technology in underpinning the early CE process.

Mainframe computers, available before this time period, became more accessible to the engineering work force during the 1970s. Their use evolved first through remote card readers, then through teletype terminals, and finally to the early interactive editing and graphics terminals. Interactive terminals represented a major step forward toward achieving more rapid modeling and analysis. It also accelerated software development, which in turn supported other crucial CE building blocks. The evolution was not really effective, however, until the more interactive VAX type of minicomputer arrived on the scene in the late 1970s. The joining of these resources provided the computer hardware environment for launching the activities of the next decade.

Relational database concepts began making their appearance at this time. Initially developed and oriented more for the commercial business applications, relational concepts began to be viewed with interest in engineering for two reasons. First, relations promised to provide a very general data structure for representing any kind of tabular data. Second, relational query seemed to offer users a powerful way to extract general information from the data, without being overwhelmed by the large data volumes. Relational capabilities were

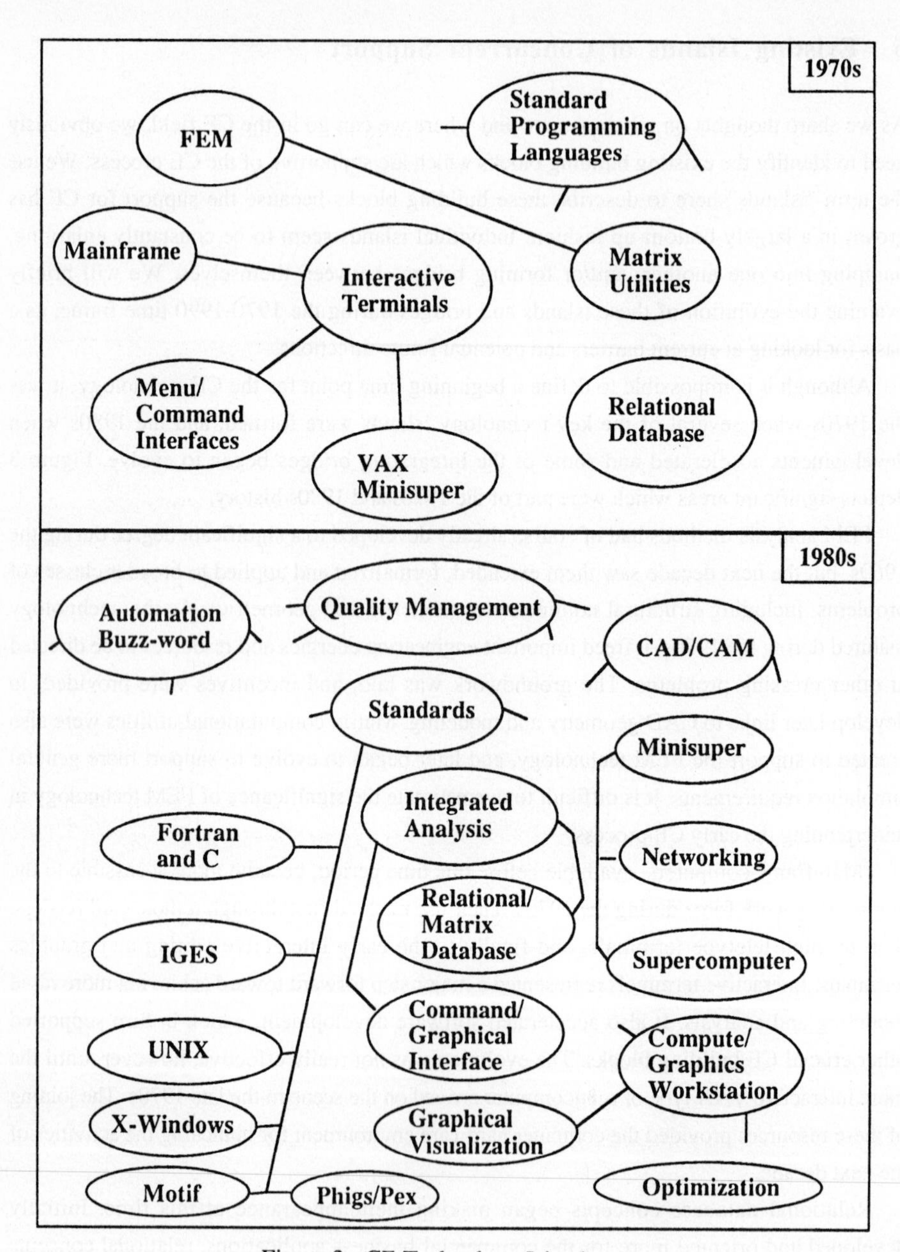

Figure 3. CE Technology Islands

implemented in some pilot studies, but would have to await the 1980s to evaluate their utility and overcome some significant deficiencies in the engineering environment.

Programming languages began to exhibit some degree of standardization, at least to the point of providing development teams a way of communicating and combining software modules. Fortran finally came of age with the introduction of the Fortran 77 standard. This breathed new life into what had become a fairly mature and universal, but limited, engineering language. Other new generation language concepts were typified by Algol and Pascal, which still threatened to supersede Fortran in the minds of many enthusiastic users. Fortran, however, demonstrated its staying power, as some of the other languages gradually retreated into obsolescence.

Finally, in the user interface area, menu concepts became popular as a simple and more friendly approach, and were an essential complement to the interactive terminal. Command languages were also introduced as a more sophisticated and powerful altemative, with the additional advantage that they could serve both interactive and batch interface needs. Batch support remained important, because of computationally intense tasks and because it was the only effective way to automate and interconnect the pieces of engineering analysis into repeatable runstreams.

As the 1980s accelerated the CE process, one of the most prevalent buzz-words (and probably the most significant emphasis), was on all aspects of automation. Automation was viewed as the primary hope for maintaining a competitive edge in the industrial production process. One of Webster's definitions for "automated" implies a "process which occurs without conscious thought or act of the will." This concept obviously leaves something to be desired in supporting the engineering goal of greater insight and optimization. Many engineers displayed considerable negativism toward automation efforts, and toward the hypothesis that many engineering tasks could be performed more effectively by the computer. Of course, engineers and their productivity have derived considerable benefits from the automation of non-creative labor intensive tasks.

The CAD/CAM island developed significantly during this time period, into practical capabilities for defining geometric configurations, building analysis modules, and automating production machining methods. On the path to gaining widespread acceptance, CAD/CAM needed to emphasize standard data formats for sharing between systems and companies. It was also necessary to reduce the number of CAD/CAM systems, through a process of competitive obsolescence and mergers. Key linkages were formed between CAD/CAM and most of the other technologies shown in Figure 3.

Integrated analysis capabilities were developed to address different technology combinations, applications and levels of sophistication. Capabilities include government-sponsored systems such as ASTROS, for aircraft structures/aerodynamic analysis and optimization; the NASA/LaRC IDEAS package, emphasizing spacecraft preliminary design;

and the NASA/GSFC and Air Force sponsored IAC/ISM capability, for thermal/structural/controls/optics simulation. Large commercial software firms also view themselves as providing integrated analysis/modeling packages, e.g., SDRC's IDEAS[2] package and PDA's Patran based system. All these capabilities have, or are evolving to have, significant database management packages and command/graphical interfaces.

Engineering database management capabilities now are integrated with many of the other CE supportive technology islands. With few exceptions, however, the two critical features (matrix computational support and relational query) have demonstrated little true integration. Matrices are sometimes stored as elements of a relation, but data access and manipulation for such strategies are quite limited. The general engineering effects of database management, however, have been far reaching. The common database often provides greater portability, and better dataflow bridges to other technology islands.

Graphical interfaces to many of the technology areas have been a good idea waiting to happen. After some limited and costly attempts do develop such interfaces from the ground up, Motif and other graphics development toolkits are now readily available. The bridges here are only starting to be developed, but offer great potential. The toolkits are being used to build capabilities from a set of graphics "widgets," such as pull-down or pop-up menus, buttons, dials, text boxes, and scrollable list boxes. One of the most important advantages is the ability to create higher level "dialogue boxes," where the user can define, and see at a single glance, all of the options and parameters for a particular application. This is a tremendous advantage over the old style cascaded menus, where non-pertinent data remained to confuse the user and pertinent data was often not visible because of screen space limitations.

Combined compute/graphics workstations have been around for most of the decade, but have gained widespread popularity only in the last several years. As graphics software support has improved, and hardware speed and capabilities have increased, these workstations are now the platform interface of choice for most engineering applications.

Emerging standards have been one of the most significant support islands for CE during the 1980s. Without them, there would have been little willingness to invest the resources required for software development. Standards for operating systems, programming languages, data transfer mechanisms, windows, networking, and graphical user interfaces are now fairly mature, and are penetrating most of the other islands. Standards for 3-D graphics visualization (Phigs/PEX) are somewhat less mature, and are yet to be exploited.

Finally, in the networking area, there is widespread use of communication protocols and data flow/translation software, while remote procedure call (RPC) capabilities are also in place. Much advantage has been taken of these, although the RPC support for distributed processing and synchronized task control is still underutilized.

6 Barriers to Greater Concurrency

As the CE technology matures further, more barriers will be recognized, understood and perhaps overcome. Present experiences indicate several areas of concern, ranging from technical, to cost/manpower constraints, to competitive and human concerns. We examine a few aspects here, and more will be identified in other lectures.

The basic technical problem with CE comes immediately to mind. Concurrency means more interconnectivities in the process, and each task then becomes more affected by the others. This is a primary goal of CE implementation. Since each task is of finite length and tends to be iterative, it will have associated errors and uncertainties prior to completion. This will negatively impact the time and resources required for the interacting tasks. Hopefully, this will be counterbalanced by the positive mutual optimization benefits.

The specialist/multidisciplinarian dichotomy seems to be another difficulty. As we strive to achieve ever greater performance in our systems, we must rely on the expertise of the specialist to meet the design goals. The specialist, however, usually works best in a sequential mode because he/she has less understanding of the interactions with other disciplines. Since resources are limited, we must either be satisfied with a tradeoff, or have a very good approach to people communication/management.

Many disciplines require a significant time for simulation model building or change. This tends to be true for thermal, structures and controls modeling. Not only may resources be wasted if models are built based on incomplete or erroneous information, and then redone, but the flow time involved may not allow full interaction with other disciplines (aerodynamics, electronics, packaging) within the required development schedule. If the interactions are not critical, it may be decided to design separately, simulate concurrently, and then redesign the fixes. Future improvement can gradually be expected as model building technology is enhanced.

It may be true that manpower constraints favor a sequential (non-optimized) product development approach over a concurrent one. Getting all the technologists involved on a design team at the same time may not be practical. The choice then is to rely on more of the non-specialist multidisciplinarian support personnel, and hope that increased integration will compensate for less technical depth.

A lack of resources or long-term project commitment may lead to a decision to select only the most significant technical discriminator for design study. If the project proceeds, interactions may be treated at a later stage. In fact, sometimes the scope of CE must be expanded to include status reports and political or public opinion feedback necessary to obtain a consensus. CE, of course, could be of benefit in completing a project before the consensus changes.

One must always be aware of the potential barriers to full, cooperative human interaction. An excellent and candid description of some practical experiences in this regard is given by E. Midkiff [6] based on his practical experiences with TQC efforts. In a sequential process, resistance from one person can slow the product development. In a concurrent process, the same resistance can stop it completely. The TQM concept of total team involvement and consensus building should not be forgotten.

Several other concerns with CE are worth mentioning. In a CE design/build team environment, the team members perform both the design and analysis, so that the historical check and balance system is not present. A true CE process must be far reaching, involving contractors, subcontractors and the customer (e.g., government). Data sharing is a primary means for CE, and industry has a very valid concern about protection of company proprietary data. If the customer is to be certain of maintenance and spares during the product life cycle, it is necessary for all aspects of the product to be recorded and traceable. This also means, however, that the original producer may lose the downstream competitive advantage and might therefore be less than enthusiastic about providing such data.

The list of barriers seems significant and lengthy. It is not intended to be discouraging. If we are to benefit from CE, we must employ CE where appropriate, be aware of and overcome as many barriers as possible, and supplement CE with sequential engineering methods where necessary.

7 Future Directions

We should expect substantial evolution and advancement of CE in the next few years, as the existing technology base is integrated and expanded. A pictorial summary of key areas expected in the 1990s is shown in Figure 4.

Quality management concepts seem to be here to stay, and will go hand in hand with CE. Standards will become increasingly important, extending to 3-D graphics, distributed processing database management components, and complete product descriptions via PDES. CAD will continue to mature in the geometry and automated modeling areas; and CAM will make further great strides, in areas from computer chip manufacturing, to robotics, to superplastic metal forming. Design/build teams will become more popular in an effort to cut costs and reduce product development time. We can expect automated voice interfaces to become everyday practical tools. Vector/parallel processor support software will become more mature, leading to increased practicality of such machines, and massively parallel processors will be used for specific applications. More emphasis on visualization will support greater insight and reduced human resources, and greater reliance on networking and distributed processing will help to avoid short-term plateaus in computer hardware capabilities.

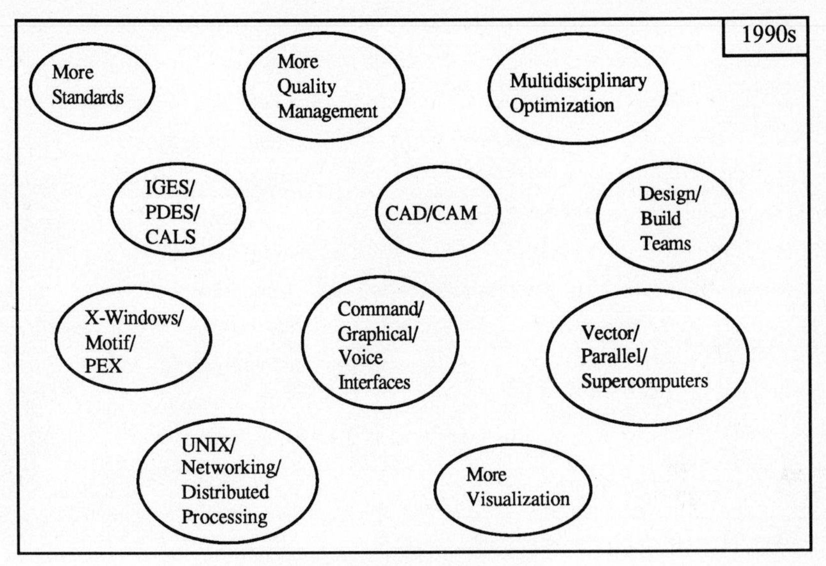

Figure 4. CE Future Directions

References

1. W.E. Deming, Out of the Crises, MIT Press, Boston, MA, 1986.
2. W.J. Grunenwald, "Cost of Quality as a Baseline for Total Quality Management (TQM) Implementation," Transactions of the IEEE, 1989, p. 1611-1613.
3. W.B. Scott, "TQM Expected to Boost Productivity, Ensure Survival of U.S. Industry," Aviation Week and Space Technology, December 4,1989.
4. P.F. Jones, "The CALS Initiative," Computer Aided Design, Volume 22, p. 388-392, 1990.
5. The Pymatuning Group, Inc., "Industrial Insights on the DoD Concurrent Engineering Program," Report under Contract MDA972-88-C-OO10, sponsored by DoD (DARPA) and DoT (TSC), October 1988.
6. E.B. Midkiff, "The Unknown Factor in the Human Equation," Transactions of the IEEE, 1989, p. 1982-1985.

Presentation and Evaluation of New Design Formulations for Concurrent Engineering

Eric R. Stephens

Georgia Institute of Technology, Atlanta, GA 30332-0140 USA

Abstract: Design is an important activity which drives many other disciplines. To insure future competitiveness in a world wide market place, reliable design processes and products must be understood. To develop this understanding, design must be studied so that in the future it can become a reliable science. To begin this study, as with other sciences, design formulations must be developed and laboratories constructed. To begin a Ph.D. research program, a survey of design directed activities was conducted, a new formulation for design conceived, and a design laboratory proposed. This paper summaries the initial phase of this research.

Keywords: design formulation / object oriented / multidisciplinary / EXPRESS / PDES / QFD

1 Introduction

There are many issues which can be addressed to improve design. These issues include the creation of a more complete design representation, the coordination of diverse design resources, and the advancement of designer interaction. Many of the issues arise from the desire to perform multidisciplinary design. To develop an informed strategy to cope with these issues, a survey of current trends in information modeling, design methodology, and concurrent engineering was conducted.

To respond to these challenges in design, a formulation for design has been proposed. Stated concisely, it is proposed that **designs can be represented through a centralized definition which encapsulates a distributed instantiation.** To evaluate this design formulation, LEGEND, a Laboratory Environment for the Generation, Evaluation, and Navigation of Design has been designed and is being constructed. A key criterion for this formulation is that it should enable multidisciplinary design.

In the following section the design formulation is explained. Important concepts and terms are introduced by means of examples. As much as possible, the examples flow from a ground vehicle design discussion. However, some of the examples will be from an aircraft design problem. In general, no discipline-specific knowledge is required to understand the ideas.

At this point no detailed information is presented with respect to implementation. The skills required to create LEGEND are themselves a subject of research. At present the basic structure of LEGEND has been prototyped and is running on the workstation network within the CAE/CAD Lab at the Georgia Institute of Technology.

The design of LEGEND will be described in terms of the following parts: design representations, encapsulated central definitions, and distributed instantiations. In each case the basic idea is presented and an example is included. For a more detailed explanation and a review of the related current thoughts, please examine the LEGEND proposal [1].

2 Legend

2.1 Design Representation

Design has come to mean both noun and verb. Design used as a noun refers to a specification of the effect a system should have on its environment and the specification of the means which will be used to achieve the desired effect [2]. Design used as a verb refers to the process of converting information that characterizes the needs and requirements for a product into knowledge about a product [3]. To represent design, both its use as a noun and as a verb must be accommodated. Design specification will be accomplished by design entities. The design process will be described by a design-decision strategy.

2.1.1 Design Entities

To represent a design specification four types of design entities will be used: form, function, process, and engineering model:

- **Form** is defined as what something is.
- **Function** is defined as what a form does or has done to it.
- **Process** is defined as how some form does some function.
- **Engineering Model** is defined as a idealized representation of a process.

These design entities can be structured so that very complex designs can be decomposed and then described at different levels of detail which can expand as the design process progresses.

Forms and functions can be structured hierarchically. Figure 1 illustrates the progression from general to specific as being from top to bottom. This type of architecture facilitates the capture and navigation of both known and needed information. It also treats part and assembly knowledge.

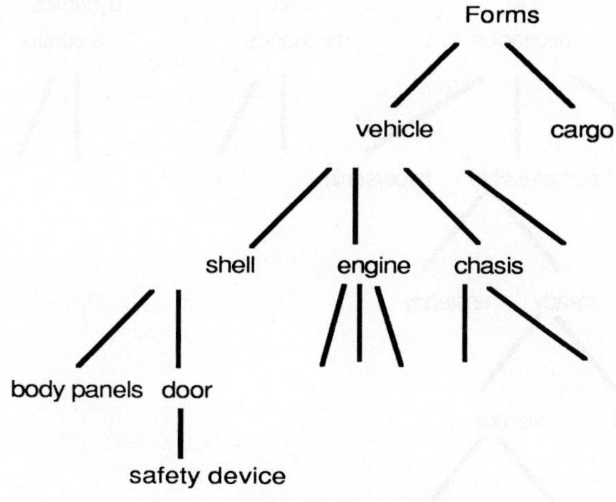

Figure 1. Form Tree

To support analysis within design it is important to indicate where it is needed and to what extend it is developed. Forms performing their functions imply processes. A safety-device form performs the function protect through a solid mechanics process. Processes are used to indicate the need for engineering models.

Engineering models can be related hierarchically as well. As shown in Figure 2 multidisciplinary models can be organized efficiently in a tree. Engineering models can also vary from elementary to detailed, e.g. simple beam theory to nonlinear dynamic deformations. These models can typically be directly linked to computation resources.

Using these entities a design specification can be described. What a design is can be viewed in forms. What a design does can be viewed in its functions. How a design performs those functions with those forms can be seen with engineering models linked the indicated processes.

2.1.2 Decision Strategy

Design used as a verb refers to the process of designing. A very basic representation of this process is as follows. Given a design specification, state a decision to be made, evaluate that decision, use the results of the decisions to effect a change in the specification. This process repeats as changes in the specification trigger decisions. With this description of the

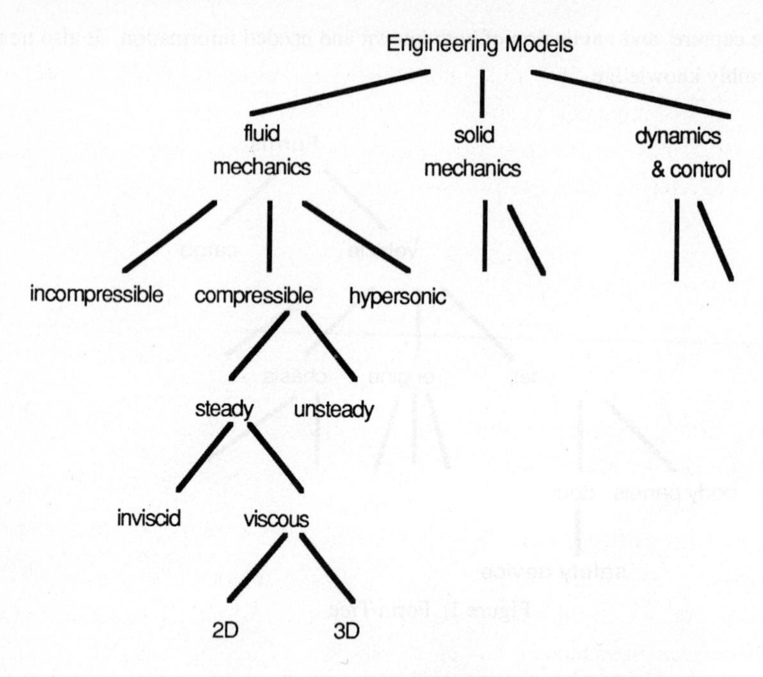

Figure 2. Hierarchy for Engineering Models

designing process it becomes possible to look at what types of decisions can be stated and how can they be evaluated.

Decisions are mixtures of selection and compromise [4]. Selection involves making a choice between a number of possibilities. This choice should take into account a measure of merit. As an example, one would select either wheels, tractor-treads, or legs as a mechanism for delivering land-based forward momentum. The second category of decisions, compromise, does not exclude like selection. Compromise indicates the determination of the "right" values of design attributes, such that, the system being designed will be feasible and optimal [5]. As an example, one could compromise over the radius and thickness of a wheel based on optimum cost and weight.

To state interrelated sets of decisions and to store them with the design, two techniques are being combined. An Enhanced Quality Function Deployment interface to the design is being used to enable the designer to associate decision with explicit form-function interactions. This use of EQFD has already been shown help insure that the Voice of the Customer remains in focus throughout the designing process [6]. To capture complex decision strategies, either interactively or prepared off-line, Decision Support Problems are being constructed. [Marinopolus]

To represent the design, both its specification and its process, a number techniques are being combined. Design entities can be linked through hierarchies of forms, functions, processes, and engineering models. Decision strategies can be viewed through Enhanced Quality Function Deployment to insure VoC clarity, while decisions can be described as Decision Support Problems. This new synthesis approach represents the design.

2.2 Encapsulated Centralized Definition

In response to the growing concern for information management in design, a scalable but transportable implementation has been sought. This implementation must support growth as the design evolves and should not be limited by a confined architecture. This implementation must also allow exchange of designs between co-workers, industry, and academia.

To accomplish this a PDES/EXPRESS schema is being used to capture the design. Both the design's specification and decision strategy are developed in EXPRESS. This enables a centralized definition which utilizes encapsulated objects. EXPRESS is an emerging language to describe an internationally accepted Product Data Exchange using STEP. STEP is the STandard for Exchange of Product data models [7].

> "EXPRESS is not a programming language. It consists of language elements which allow unambiguous object definition with the constraints on the data clearly and concisely stated. The EXPRESS language does not contain language elements which allow input/output, information processing or exception handling." [8]

Research in the use of EXPRESS to enable PDES will not only permit the exchange of product models, but will also include the development of EXPRESS to transport process oriented engineering models. It is believed that all of the design entities and decision strategies can be centrally defined and encapsulated with EXPRESS.

2.3 Distributed Instantiation

Instantiation is a computer science term. It refers to an explicit enactment or an instance. A definition or more exactly a schema provides a template for describing a set of similar instances. Without becoming too engrossed in jargon, a schema could describe cars in terms of a set of yet unspecified attributes such as model, number of doors, color, cylinders, etc. An instance would be a red V8 Corvette.

The need for a distributed instantiation comes from the way design engineers work together using different types of tools for different tasks. Given a central definition of a vehicle, one would expect the shell to be mainly worked on by stylists using CAD system A, the engine to be worked on by another team using another CAD system B and a variety of analysis tools, and so forth. In this way, a design's instantiation is distributed.

With an understanding of design representation in LEGEND, EXPRESS was introduced to capture both specification and decision strategies. EXPRESS can only describe the schemas; it is not intended to be used to create distributed instantiations. What other technologies must be utilized to enable this capability?

To construct distributed instantiations and manage user interactions in LEGEND, a number of emerging techniques and developing resources are being adapted. Both relational and object oriented databases are being used. Established engineering tools, such as CAD environments and analysis codes, will be used as design agents. Decision support tools will be available within LEGEND. And to document and browse design histories, a hypermedia capability is being constructed.

To enable distributed instantiations requires a mature treatment of the well-known application integration problem faced by virtually all design organizations. Existing tools are being *wrapped* and act as design agents. The wrapping process varies from tool to tool, but in each case, it enables data interchange between a design instance and a tool. This strategy is described in greater detail the LEGEND proposal [1] and is an extension of the DICE wrapping technique for integration.

At this point is important to remember that the research oriented goals are aimed at the verification of the design formulation. It is not intended that LEGEND be a production-grade integrated engineering environment. It is, however, intended that these issues be investigated and the resulting environment serve as an effective learning tool which could be extended to production.

At this time is possible to describe the components which comprise LEGEND. The above-mentioned tools have already been acquired and some level of experience has been gained with each of them. Figure 3 illustrates the proposed environment.

2.4 LEGEND User Interface

LEGEND itself exists independently of the agents used in designing. As a software-based laboratory environment, agents can be added or removed like equipment. The selection of active agents is based on the what design or designing process is being studied. To support a flexible well established user interface, an X Windows virtual display is used to provide access to all agents from a common location. Processes will be distributed in order to support the need for large scale computation and to use agents in their native installations.

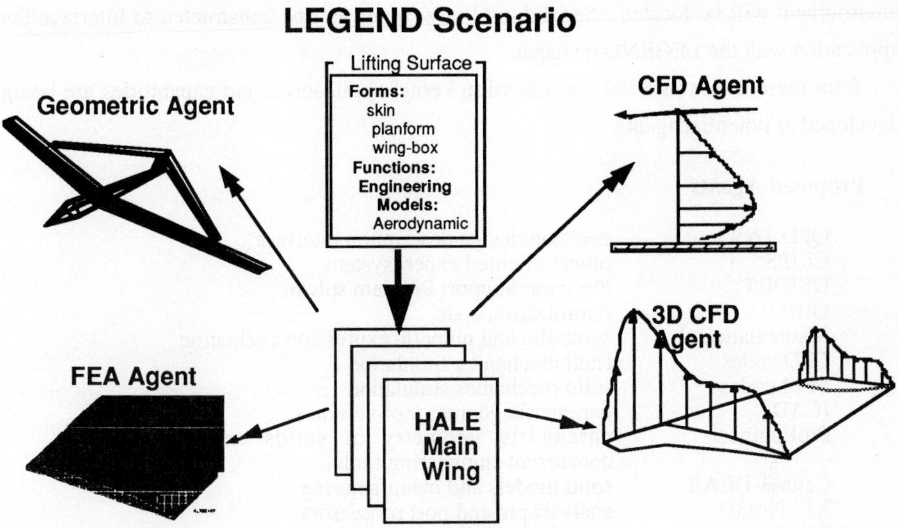

LEGEND Scenario

Lifting Surface
Forms:
skin
planform
wing-box
Functions:
Engineering
Models:
Aerodynamic

Geometric Agent

CFD Agent

3D CFD Agent

FEA Agent

HALE Main Wing

Figure 3. LEGEND Scenario

At present several workstations are being employed, but it is anticipated that ten to twenty workstations will be coordinated with LEGEND.

2.5 Information Management

Using *STEP Tools*, a set of translators, and *ET++*, a C++ programming environment, the design's description can be converted from EXPRESS into an evaluatable form. Once a description has been conventionally coded, the design can be instantiated.

To manage instantiated designs within LEGEND, at least two types of databases are being used. To provide rapid search of large sets of existing information, a relational database such as ORACLE is used. A ROSE object-oriented database contains all active design components. When a design is active, its description changes dynamically in response to agent activity. An object-oriented instantiation is best suited to manage this type of manipulation.

2.6 Design Agents

Agents suggest and complete all decisions. The designer is the principle agent. A LEGEND manager administers all agent interaction. To evaluate decisions, agents work in series or in parallel. To create agents with these capabilities, a two phase process has been conceived. First, an existing kernel application which has a capability to be added to the environment will be located. Second, a C++ wrapper will be constructed to interface the application with the LEGEND manager.

With these issues in mind the following kernel applications and capabilities are being developed as potential agents.

Proposed Agents

QFD Tool	decision/design description interface
CLIPS	object oriented expert system
DSIDES	Decision Support Problem solver
OPT	optimization code
Mathematica	symbolic and numeric expression evaluation
CFD codes	fluid mechanics simulation
FEM codes	solid mechanics simulation
ICAD	parametric geometry of volumes
ProEngineer	parametric geometry of solids and links to concurrent engineering tools
Catia, I-DEAS	solid models and manufacturing
XL, Plot3D	analysis pre and post processors
WingZ	spreadsheet interface to designing history

These types of agents are already at use in many designing environments. The difference in LEGEND is the integration, decision strategy, and the use of a central design model.

3 Concluding Remarks

A survey was conducted to locate pertinent design issues and technologies. From this intensive review, a proposed formulation for design emerged and a laboratory in which to test it has been planned.

In phase two LEGEND is being built to test the resulting formulation. The usefulness of representing designs through centralized definitions encapsulating distributed instantiations is being studied. Also, LEGEND's construction process provides much needed experience in designing and building integrated computing environments to support life-cycle engineering design.

Specifically, one test in LEGEND will be to complete a high altitude long endurance (HALE) aircraft design. Using LEGEND the design will be expanded to include the necessary analysis to design for passive controlled lifting surfaces.

The Voice of the Customer HALE design will set the stage for an increased understanding of lifting surface technology and its effects on total aircraft design. Although the emphasis will be on the determination of trends, the improved analysis methods and descriptions will also benefit the complete aircraft designing process.

References

1. Stephens E. R.: "LEGEND: Laboratory Environment for the Generation, Evaluation, and Navigation of Design", Ph.D. Proposal: Georgia Institute of Technology, School of Aerospace Engineering, May 1992
2. Fortenberry N. L.: Analysis and Synthesis of Structured Systematic Design Theories," MIT Mechanical Engineering Sc. D. Thesis, January 1991
3. Karandikar HM, Rao J, and Mistree F: "Sequential Vs. Concurrent Formulations for the Synthesis of Engineering Design," ASME:Advances in Design Automation, DE-Vol.32-2, December 1991
4. Bascaran E, Karandikar H. M., and Mistree F: "A Decision Based Approach For Modeling Hierarchy in Design:A Conceptual Exposition," ASME Design Theory and Methodology Conference (submitted), July 1992
5. Bras B, Mistree F: "Designing Design Processes in Decision-Based Concurrent Engineering," SAE Technical Paper 912209, September 1991
6. Anderson D. M.: Profit From DFM: Designing Products to Optimize Cost, Quality, and Time-to-Market, Winter 1991
7. Clark S. N.: The NIST PDES Toolkit: Technical Fundamentals, NISTIR 4335, November 1990
8. Spiby P (editor): EXPRESS Language Reference Manual, ISO 103003: Part 11, March 1991

Data and Process Models for Concurrent Engineering of Mechanical Systems

J.K. Wu and F.N. Choong

Center for Computer Aided Design, Department of Mechanical Engineering, The University of Iowa, Iowa City, IA 52242-1000 USA

Abstract: This paper presents a framework for computer aided engineering (CAE) that has been developed for Concurrent Engineering of mechanical system design and analysis. The framework integrates engineering software systems in a tool-workspace-subenvironment-environment hierarchy. A workspace or a subenvironment represents an engineering application or discipline. The paper discusses (1) an object-oriented tool integration method to integrate engineering applications, (2) global and local data models that are used to integrate product information at different hierarchical levels, and (3) an engineering process modeling method that expresses engineering activities as tool or workspace runstreams. The developments in these three aspects of the framework build a basis to develop activity coordination capabilities. Some considerations that are related to engineering activity coordination are discussed at the end of the paper. The CAE framework presented can be extended for computer aided manufacturing (CAM) applications also.

Keywords: data model / process model / mechanical systems / integration / CAE / CAM / Concurrent Engineering

1 Introduction

Concurrent Engineering, the earliest possible integration of a company's overall knowledge, resources, and experience in design, development, manufacturing, marketing, sales, and disposal, enables a company to create products with high quality and low cost, while meeting customer expectations [1]. Concurrent Engineering shortens the product concept, design, and development process by changing it from a serial to a parallel one [1]. A Concurrent Engineering environment provides a systematic approach to the integrated, concurrent design of products and their related processes, and considers all elements of the product life cycle from design through disposal [2].

Many engineering analysis tools [3-10] and Concurrent Engineering architectures [11-13] have been developed by tool vendors and research institutes. Although some of the architecture developments are not quite completed, the basic functions of the architectures have already been demonstrated. Efforts remaining to develop a Concurrent Engineering environment for design and manufacturing of specific types of products, such as mechanical systems, include the following aspects (1) integrating necessary tools and systems into the architecture, (2) integrating product information for tools and systems, (3) developing a process modeling system to express engineering activities, and (4) developing a coordinating system to manage engineering activities to achieve goals of Concurrent Engineering. The developments in the first two aspects create an integrated environment. With the developments in the last two aspects added to the integrated environment, a concurrent environment is then built.

This paper presents the object-oriented approaches that have been used to develop features of a Concurrent Engineering environment in the first three aspects: tool integration, data models, and process models. The developments in these three aspects construct a basis for the fourth one, which is not included in this paper.

DICE [12] suggests that a Concurrent Engineering environment should consider product, process, and organization of engineering activities. The product perspective defines the products to be shared, analyzed, or derived. The process perspective considers product function evaluating and manufacturing processes. The organization perspective is the management of human and other resources of an enterprise that are involved in product design and manufacturing. The data and process models presented in this paper cover a significant part of the first two perspectives and some of the third.

Before discussing the approaches, some basic terms are introduced here. In the paper, unless noted differently, the term *CAE* means engineering activities in which computers and software systems are used to achieve certain engineering goals. Hence CAE is not limited to the traditional view of CAE--analyzing the engineering functions of products. Using computer programs to analyze the product manufacturing processes and cost is also treated as a CAE activity in this paper.

A *CAE entity* is a software entity used in engineering activities to (1) develop products (in the sense defining and storing the products in computer systems) or analyze functions and costs of the products, or (2) manage the procedures of product development and analysis. A *service* is an engineering functionality or capability that a CAE entity can provide, such as modeling products or analyzing product functions and cost. A *CAE system* is a software system, consisting of CAE entities, that provides high-level, useful engineering services.

In this paper, a *product* of a CAE entity is a mechanical system or a mechanical component. The definition, functions, and manufacturing cost of a product are points of interest in engineering, and are called *product characteristics*.

Integration has two interdependent aspects: tool integration and information integration. *Tool integration* refers to a systematic way to integrate various CAE entities. CAE entities are distinguished into several classes, such as a tool, workspace, and subenvironment (which will be explained later), and are integrated in an engineering environment according to their classes. These CAE entities and relations between the entities form a CAE framework. These relations allow engineer users to navigate in the CAE framework, set up CAE entity invocation sequences, and specify data communication among the entities.

Information integration refers to the common definition and format of product data which every related CAE entity must agree to. The definition of each product includes (1) definitions of its member entities, (2) the hierarchical relationships between the product and its member entities, and (3) relationships between the product and other products, such as is_a_part_of, is_a, and connected_to.

Process modeling refers to a systematic way to express engineering activities that will be carried out in the framework. A process model describes an invocation sequence of CAE entities to accomplish an engineering service according to some engineering theories or principles. The process modeling takes into account the classification and data dependency of CAE entities.

Wasserman [14] introduced five types of integration, namely, data, control, process, presentation, and platform integrations from software integration points of view. The concepts of tool and information integration as well as process modeling that are presented in this paper are similar to those of Wasserman's data, control, and process integration.

Lewis [15] presented seven integration models: three data integration models, two control integration models, and two general data and control integration models. Lewis pointed out that data integration and control integration are tightly coupled and in general cannot be treated separately. The tool and information integration and process modeling that are presented in this paper are closely interdependent and coupled and are more similar to the NSE Link Service [15] than other models that Lewis discussed. However, the integration method and process modeling method presented in this paper take a hierarchical, object-oriented approach and are for mechanical system design and analysis, which are different from the approaches and goals that Lewis and Wasserman had. Using object-oriented approaches, very structured and flexible environments can be developed, and they are easy to extend and modify.

The remainder of the paper is organized as follows. Section 2 discusses the CAE framework and the tool integration method. Section 3 discusses the information integration method. Section 4 discusses the process modeling method. Section 5 presents some elementary developments and techniques that can be used to coordinate engineering activities.

2 CAE Framework and Tool Integration

This section describes (1) an engineering framework for Concurrent Engineering and (2) the object-oriented integration method that is used in developing the framework. A CAE framework is a software system that provides a common, consistent operating and developing environment for CAE entities. In this framework, CAE entities are integrated hierarchically according to a systematic way. The roles of the CAE entities and their interrelations define the framework on which the Concurrent Engineering environment is built.

2.1 Structure of the CAE Framework

CAE entities are structured hierarchically in the CAE framework, as shown in Figure 1. The framework contains six types of CAE entity: tool, workspace, subenvironment, environment, design data server, and blackboard. In the framework, tools that are used by certain type of engineer to accomplish engineering activities, such as modeling a vehicle system and evaluating vehicle performance, are integrated into an *engineering application* in a form of workspace or subenvironment. The workspace or subenvironment is then integrated in a Concurrent Engineering environment.

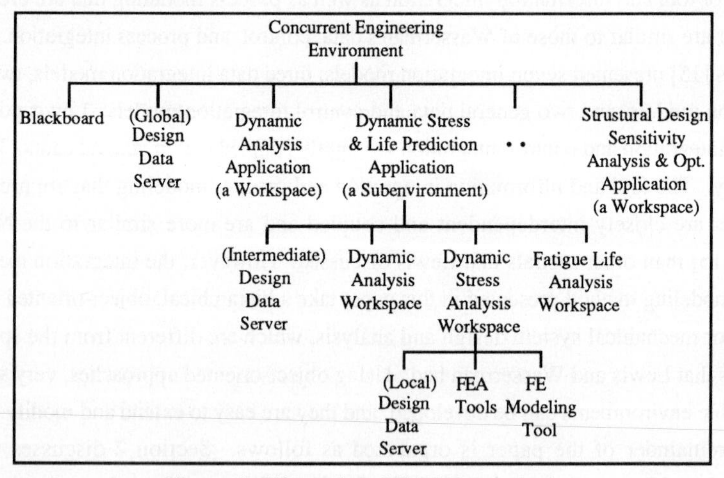

Figure 1. Hierarchy of CAE Framework

An engineering *tool* is a software module or a software package developed for a low-level service or services. Within one invocation of a tool, a service of the tool transforms, displays, determines, or extracts data from one state of the product data (the input to the tool) to another state (the output). Services provided by tools are context insensitive. For instance, a finite element analysis package can be used to analyze structural deformation, structural vibration, and temperature distribution.

Many engineering tools belong to one type, for instance, ANSYS [6], NASTRAN [7], and ABAQUS [8] are instances of a finite element analysis tool. The engineering services that tools of the same type can provide are very similar, however, the input to these tools for the same service varies slightly. The same is true for output. Hence, there exist common features among engineering tools, which allows the object-oriented approach to be used to integrate tools, as described in Section 2.2.

A *functional workspace* or, simply, a *workspace* contains (1) tools, (2) a design data server, and (3) a tool runstream authoring system to provide high-level, context rich (or context specific) engineering services. Each workspace service is carried out by invoking a set of tools sequentially. Each workspace (1) has its abstraction about engineering products, (2) analyzes product characteristics according to certain engineering principles and theories, and (3) has tools and users who have enough knowledge to use the tools to complete the engineering services. Since a functional workspace contains necessary tools for a type of engineering activity, the workspace is corresponding to an agent that is introduced in other Concurrent Engineering research activities [16].

Although the CAE tools that are used by a workspace are context insensitive, it is in a workspace that their use is put into context, because the input data that are generated in the workspace for the tool are for specific engineering services. For instance, a dynamic analysis workspace uses a mechanical system modeling tool to define mechanical system models and uses a dynamic analysis tool, such as the dynamic analysis modules of DADS [9] or ADAMS [10], to analyze the dynamic behavior of the mechanical systems. Note that a CAE tool can be used in several workspaces for the same or different services, and an extreme case of a workspace is that it has one tool. The design data server and runstream authoring system of a workspace will be discussed later.

Both a tool and a workspace are software packages, but the difference is in their intended usage and software contents. When a software package has modeling, analysis, and post processing capabilities (such as ANSYS), and they are used in certain sequence to determine specific engineering information, this software package is a workspace. If only one of the capabilities of the package is used together with other software packages in a workspace, then this software package is a tool in the workspace.

A *subenvironment* is a CAE entity that utilizes workspaces to accomplish higher level engineering services. A subenvironment contains (1) workspaces, (2) a shared design data server, and (3) a workspace runstream authoring system. For instance, fatigue lifes of a mechanical component can be determined based on the dynamic stress histories [17] in the component. These stresses are computed in a dynamic stress workspace, which in turn obtains dynamic loading histories of the component from a dynamic analysis workspace through the shared design data server in the same subenvironment. A subenvironment is an agent in the sense that it is an integral entity and provides services that no other agents are developed for. The workspaces in a subenvironment are subagents, and subagents cannot independently accomplish all of the services that the subenvironment is designed for. The design data server and runstream authoring system of a subenvironment will be discussed later.

Similar to the difference between a tool and a workspace, the difference between a workspace and a subenvironment is their intended engineering usages and their software contents. As an example, the dynamic analysis workspace can be used by a vehicle mobility engineer to model a vehicle system, for instance, as a rigid-body mechanical system model, and analyze the mobility performance of the vehicle model. Hence this workspace will be the engineering application integrated in the Concurrent Engineering environment for mobility engineers. However, a dynamic workspace can also be used by fatigue life prediction engineers to evaluate dynamic load histories of specific components of a vehicle to predict fatigue life. The fatigue life engineers may create slightly different vehicle dynamic models, for instance, flexible-body mechanical system models, but basically consistent with the model used in mobility analysis. In this case, it will be efficient if the fatigue life engineers have an engineering application that is integrated in the Concurrent Engineering environment and allows them to access the dynamic analysis workspace, generate models, and evaluate load histories. Hence such a life prediction application contains dynamic analysis, dynamic stress, and fatigue life prediction workspaces and is called the subenvironment, as shown in Figure 1.

A *Concurrent Engineering environment* is a CAE entity that contains (1) engineering applications (in terms of workspaces and subenvironments), (2) a shared design data server, and (3) a blackboard. The blackboard is developed to coordinate the engineering activities. The major difference between a subenvironment and a Concurrent Engineering environment is that the former is a single discipline system (that means users' activities are organized in advance) and the latter is a multi-disciplinary system (that means disciplines' activities are ordered but determined dynamically). The Concurrent Engineering environment is equipped with a blackboard to coordinate engineering activities. The activity modeling and coordination will be explained in Sections 4 and 5.

The *design data server* of a workspace provides data access services for tools. The design data server needs not be a full-fledged database system; it may use the UNIX file/directory facility for data management. A workspace's design data server is strictly for local use to maintain design/analysis data in transit in the workspace, and its data schema reflects the view of product data that is held by the engineering discipline represented in the workspace. This is why the design data server of a workspace is denoted as the local design data server in Figure 1.

Shared design data servers are provided in a subenvironment and in a Concurrent Engineering environment. The data server in the environment is called the *global design data servers* because it has a shared product definition and a common area for managing design data evolution and access control for any engineering applications (workspaces or subenvironments) in the environment. Similarly, the design data server of a subenvironment, called the *intermediate design data server*, provides shared product definitions and a common area for managing design data evolution and access control for workspaces of the subenvironment. Before a engineering application starts an service, engineers will request necessary product information from the global design data server. When a service of an application is completed, engineers will initiate a data transfer from the design data server of the application (which can be the local design data server of a workspace or a intermediate design data server of a subenvironment) to the global design data server of the environment.

Generally speaking, a global, intermediate, or local design data server controls the access of related CAE entities to the database system of the server. In addition, the design data server maintains design data consistency in a database, regulates design data evolution according to high-level change management policies set up by project leaders, and satisfies each composite object data request by obtaining correct versions of each of its constituent objects. The role of the design data server is similar to the Information Sharing System [18] of the DARPA Initiative in Concurrent Engineering (DICE) architecture [12].

From the engineering point of view, structuring the CAE entities hierarchically in the framework, as shown in Figure 1, has two advantages. First, the product function, product cost, and engineering service that evaluates the product function or cost can frequently be subdivided. The subfunctions, subcosts, and subservices can be further broken down into their elements. It happens very often that several tools are used iteratively to evaluate a subfunction or subcost of a product. These tools are then grouped in a functional workspace for efficiency, and, for the same reason, related workspaces are grouped in a subenvironment.

Second, as pointed out by Paradice [19] that (1) problem-specific model representations can be integrated within levels of the hierarchy and (2) the hierarchy itself can provide a focusing mechanism to help decision makers cope with managerial domain complexity. Hence structuring engineering systems hierarchically provides a systematic way to propagate recommendations from bottom to top and manage engineering activities from top to bottom. In addition, the CAE entity hierarchy reflects the distribution of engineering users' knowledge, sophistication, and responsibility, as discussed in Section 3.3.

2.2 CAE Entity Integration Methods

Tools of the same type, such as finite element analysis codes: ANSYS, NASTRAN, and ABAQUS, have similar features, for instance, they can analyze structural behavior and need finite element models of the structure. However, they describe finite element structural analysis models in their input files differently. The same is true for workspaces and subenvironments. Hence an object-oriented approach is taken to integrate tools, workspaces, and subenvironments. An advantage of an object-oriented approach is that coherent properties of a class of CAE entities can be obtained. The control protocol of entities of the same class is one such coherent property. In addition, replacing CAE entities of the same type requires only the features that vary among the entities be modified, while their common features remain unchanged.

Each tool in a functional workspace belongs to a tool class and is integrated under a tool class hierarchy, as shown in Figure 2. At the highest level, the generic CAE tool class contains general information, such as common attributes, general control, and user interface operations. At the middle level of the hierarchy, each of the derived tool classes contains information specific to a particular subclass of tools, such as how to send requests to a database system to get specific product definitions. At the bottom level of the hierarchy, the specialized derived tool classes would define very specialized information for a tool, for instance, how to convert results of a finite element analysis, such as displacements at finite element nodal points, into a form that is consistent with the schema of the database system which stores finite element analysis results. Such a tool class hierarchy allows tools to inherit information from different levels of abstraction, thus providing a succinct way to present information that is general to all tools as well as to present information that is specific to a particular subclass of tools.

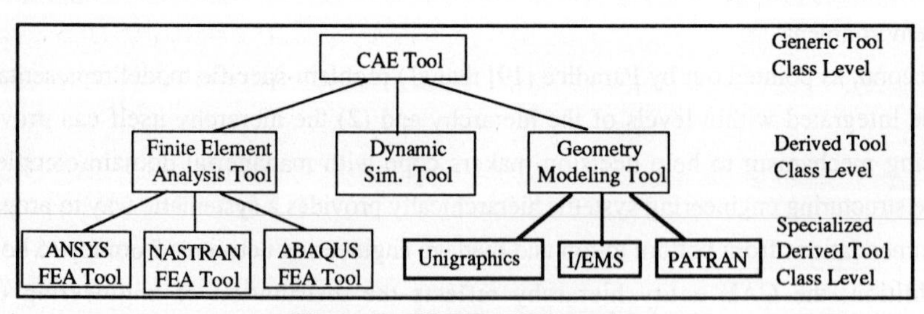

Figure 2. CAE Tool Hierarchy

The significance of this object-oriented approach is that it enables a CAE software developer to (1) follow conceptual maps of objects, (2) utilize information and software templates already built for general classes, and (3) rapidly develop and integrate new engineering tools into the environment.

Class hierarchical concepts are also developed for engineering workspaces as shown in Figure 3. Here different workspaces have different tools, and the engineering objects to be designed/analyzed in the workspaces also differ. For instance, the system analysis workspace analyzes characteristics of a whole product, e.g., a dynamic analysis involves bodies, joints, and force elements of a mechanical system. A component analysis workspace analyzes the characteristics of an element of the product, such as the structural response of a body of a mechanical system. The subsystem analysis workspace analyzes the characteristics of a subset of the product, e.g., an assembly analysis involving two parts and a joint.

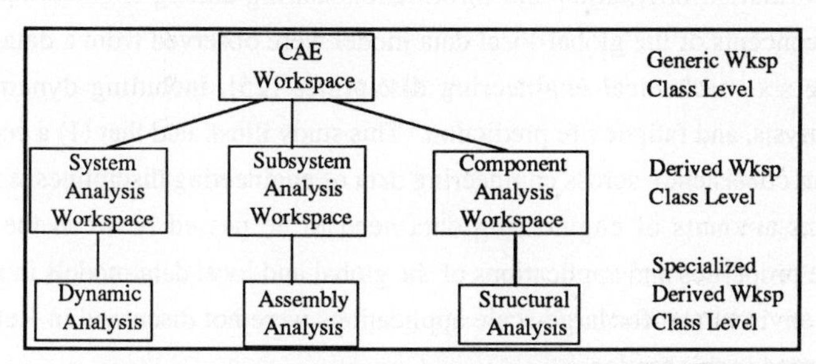

Figure 3. A Partial View of the Workspace Hierarchy

Similarly, there is a subenvironment class hierarchy developed to integrate different subenvironments into the Concurrent Engineering environment. Furthermore, according to the object-oriented approach, an environment class can later be developed to integrate several engineering environments into a super-environment for Concurrent Engineering.

3 Information Integration

This section presents the method used to integrate product informations for engineering applications, such as workspaces and subenvironments, in a Concurrent Engineering environment, and gives a brief review about information integration and data model developments carried out in other research activities.

Tomiyama et. al. [20] defined a *model* as a theory-based set of descriptions about the object world, and defined *modeling* as a process in which observed facts are filtered by a theory to formulate a world which itself is complete in terms of the theory. Shaw et al. [21] defined *product data model* as a class of semantic data models which takes into account the needs of engineering data generated through the product life cycle from specification through design to manufacture. Hardwick and Spooner [22] classified several known data models into four categories: complex objects, objects in fields, hybrid objects, and nontraditional models. They also compared the advantages and disadvantages of these four categories of data model. Bancilhon [23] introduced a semantic data model. Levene and Poulovassilis [24] introduced a formalism for object-oriented data modeling using hypergraphs.

In this paper, the object-oriented information integration is achieved by (1) developing a product information sharing and exchanging model which is called a global-local data model, and (2) providing data access libraries, based on the information model, so that CAE entities can communicate to the product database through a design data server.

In the life cycles of products, many disciplines are involved, including design, analysis, development, and sales/marketing. A global-local data model approach is introduced in this paper for information integration and information sharing among engineering disciplines. Preliminary concepts of the global-local data model were observed from a data dependency study among six mechanical engineering disciplines [25], including dynamic analysis, structural analysis, and fatigue life prediction. This study illustrated that (1) a certain level of coherence and consistency across engineering data of engineering disciplines is required and (2) significant amounts of engineering data need to be passed between the disciplines. However, the properties and applications of the global and local data models in a Concurrent Engineering environment for large-scale applications were not discussed in Reference [25], and are presented in this section.

In this paper, the data model of an engineering application about a class of products is defined as a model of data or an abstraction about the products from the point of view of the application. Generally speaking, the abstraction about a product includes the definition and performance of the product. The former is the input to the application, and the later is the output from the application.

A data model of an engineering product describes data from a static point of view, that is, without considering time or sequence of engineering activities. A data model is part of a conceptual model [26]. Another part of the conceptual model is process models, which describe dynamic behavior of engineering activities. Process models are discussed in Section 4.

The content of the data models of an application in an isolated environment and that of the same application in a Concurrent Engineering environment are basically the same. However, in the isolated environment, entities in the the data model describe the whole engineering world. In a Concurrent Engineering environment, the entities in the data model of an application describe a subset of engineering world that the environment is looking into. When integrating applications into a Concurrent Engineering environment, all data models that have been developed for isolated applications need to be reviewed by engineers from multi-disciplinary points of view to determine the data sharing, dependency, coupling, and communication between engineering applications. Hence, in an integrated environment, some of the product definition attributes are shared, imported from a common database; while in an isolated environment the product definition is always specified by end users.

The relationships and differences among data model, data, database, database schema, design data server, and product model are described below. A *data model* describes the structure and format of sets of data in a database. In other words, the data model defines the database schema. *Data* are stored in the *database* and retrieved from the database according to the structure described in the data model or the schema of the database. A design data server is the data management layer of the database, and applications directly interface with the server to request data manipulation activities. A *product model* is a set of data in the database that describe the product, and the *product data* can be retrieved and updated in the database.

3.1 Global and Local Data Models

A global product data model is developed for an integrated engineering system that contains multiple engineering (sub)applications, such as a Concurrent Engineering environment or a subenvironment. The global data model in an environment (or in a subenvironment) contains product characteristics that are fundamental to and shared among the applications, which are subenvironments or workspaces that are directly integrated in the environment, (or among the workspaces in the subenvironment). The fundamental product characteristics include the product information that is necessary to initiate the activities of some subenvironments or workspaces in the environment (or activities of some workspaces in the subenvironment). The product characteristics that must be consistent among the applications in the environment (or among the workspaces in the subenvironment) are shared product characteristics. Note that the global data model is not intended to include all detailed product characteristics that are used by all applications.

A local data model is developed for a functional workspace in a subenvironment or environment to completely define products from the point of view of the workspace. It is called a local data model, because it is from the workspace point of view, which only covers part of the engineering view of the subenvironment or environment that the workspace is contained in. Hence, the database and design data server of a workspace is called a local database and local design data server, respectively.

The data model of a subenvironment is treated as a global data model relative to the local data models of the workspaces in the subenvironment. However, the data model of the subenvironment is treated as a local data model relative to the global data model of the environment that encloses the subenvironment. Hence, as shown in Figure 1, the design data server of the Concurrent Engineering environment is called the global design data server, and that of a subenvironment is called the intermediate design data server.

3.1.1 Properties of a Global Data Model

A global data model has the following properties:

(1) It contains minimal but sufficient product characteristics to ensure that

  a) some engineering activities can be initiated, once the fundamental and shared definition of the product is available;

  b) the product characteristics that should be consistent across applications are included;

  c) product information that needs to be exchanged between applications are included; and

  d) engineering activities are free within the boundary of information consistency.

(2) It defines product characteristics using a language and primitives that are meaningful to all engineering disciplines.

(3) It needs to be reviewed and possibly modified when

  a) new applications are added, or

  b) changes are made in applications that are related to the global data model.

Examples of fundamental attributes are the geometries of mechanical components and the kinematic and force connectivities between components of a mechanical system. Without these fundamental attributes, it is not possible to initiate the processes to create finite element models for structural analysis and mechanical system models for dynamic analysis.

Examples of shared attributes are the geometry and locations of joint reference frames [27] of each component of a mechanical system. In structural analysis and dynamic stress analysis for fatigue life prediction, engineers need to generate finite element models based on the geometry of the component and the locations where loads apply. These two analyses may generate finite element models that look different but which are somehow consistent, because these models are developed from a common basis--the geometry and reference frame information available in the global database. Without a common definition to refer to, inconsistent analysis models may arise from different applications.

An example of exchanged attributes is that joint reaction forces which are obtained from dynamic analysis are required data for structural analysis. In this case, a copy of the joint reaction force will be exported from the dynamic analysis to the global database and made accessible to the structural analysis.

Product characteristics included in the global database are used to maintain consistency across disciplines and not to restrict the engineer users. These product characteristics, therefore, had better be expressed in a way that is meaningful to engineers, while leaving them substantial engineering freedom. For instance, the global database contains kinematic degrees of freedom between mechanical components (which is a high-level specification), instead of specific joint types (a low-level specification). This is because a rotational degree of freedom between a pair of bodies can be modeled as a revolute joint or a cylindrical joint in dynamic analysis. How the rotational degree of freedom is modeled in dynamic analysis is up to dynamic analysis engineers so long as the degrees of freedom between the components that are specified in the global database are satisfied in the dynamic analysis. If a specific joint type, say a cylindrical joint, were assigned in the global database, dynamic analysis engineers would then be restricted to generate product models using cylindrical joints.

The global data model for a group of applications needs modifications under two conditions. First, when a new application is added to the group, the application may need additional product characteristics that are generated by existing applications or it may pass new product characteristics to existing applications. For instance, adding a thermal analysis application would need the thermal conductivity information to be included in the product definition in the global data model. Second, the global data model needs modifications when changes in the applications require different product characteristics to be shared or obtained from the global database. For instance, when a dynamic analysis application changes from rigid-body based [27] to flexible-body based [28], the product definition in the global database need to be extended to include detail finite element models of bodies of a mechanical system in order to support the flexible-body dynamic analysis.

The level of detail of product characteristics in the global data model determines the levels of consistency, engineering freedom, and data interpretation and communication. Generally speaking, if more content is given in the global data model, there will be tighter consistency across applications, less engineering freedom in applications, and more communication and less interpretation between the global database and applications.

As an example, Figure 4 shows the global data model of mechanical system simulations in the Concurrent Engineering environment that is depicted in Figure 1. The mechanical system definition that is described in Figure 4 is sufficient to initiate the activities of the three applications shown in Figure 1 and maintain consistent product definitions across these applications.

In the global data model, a mechanical system is composed of bodies, kinematic connectors (i.e. kinematic joints), force connectors (i.e., force elements), and subsystems. A subsystem is composed of bodies. The connectivity between bodies are through kinematic or force connectors which are specified in system assembly, as shown in Figure 4.

Each body, joint, and force element is a part assembly. A part assembly contains at least a part. A part is a solid entity that has homogeneous material property, specific geometry and mass property data, and assembly features. A part assembly has composite mass property data, discontinuous material properties, an assembled geometry, finite element models, and internal part assembly information, as described in Figure 4. A part assembly may have several finite element analysis (FEA) models for different applications, such as structural analysis and life prediction.

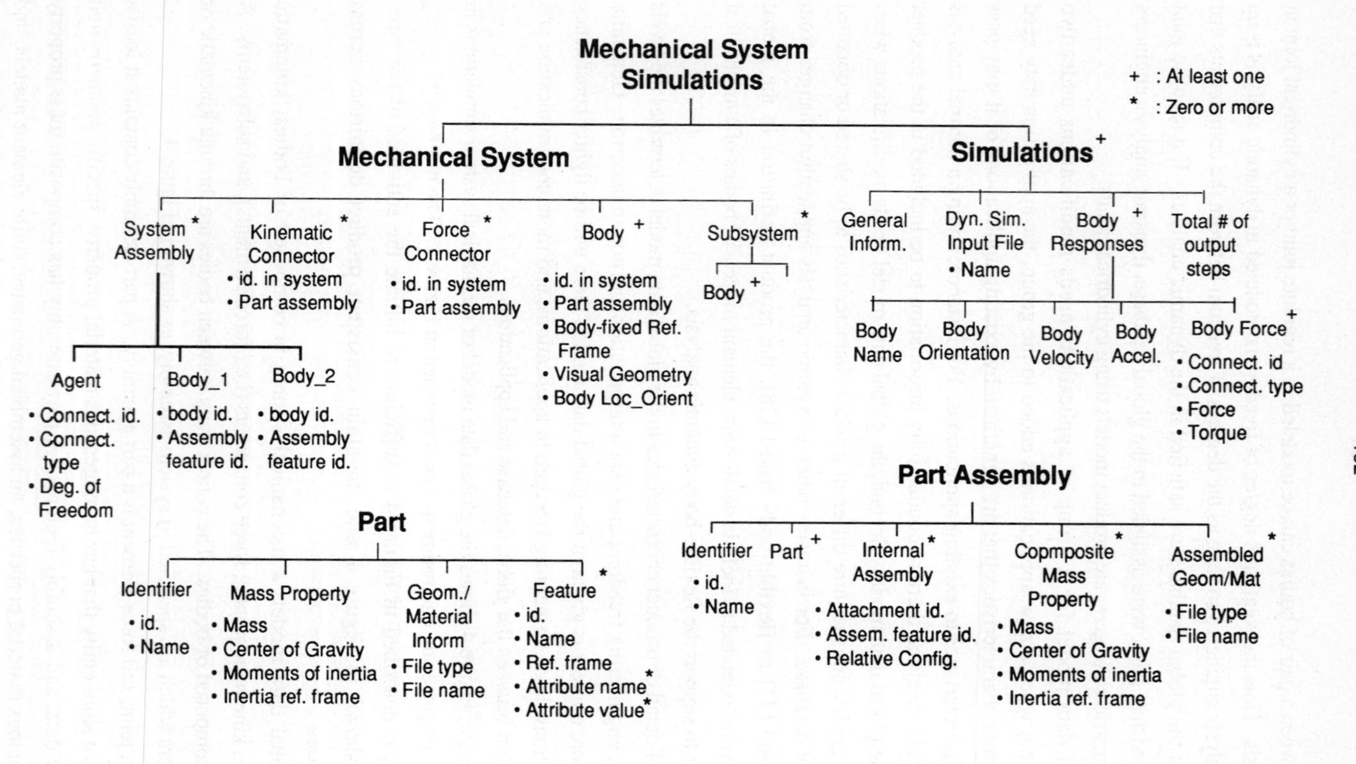

Figure 4. Global Data Model of Mechanical System Simulations

A mechanical system may be used for several dynamic simulations to determine various load histories under different operational scenarios and environments. For each simulation, the global database keeps a brief description about the simulation and body response histories of bodies of the mechanical system that include velocities, accelerations, and body forces (which contain joint reaction forces and force element forces applied on the bodies).

The body response histories are exported to structural analysis and fatigue life prediction applications and used by them as loads on a body. In the global data model each force element or joint is represented as a part assembly, in case their structural behaviors needs to be analyzed.

3.1.2 Properties of Local Data Models

The data model of a workspace in a subenvironment is truly a local data model, while the data model of the subenvironment is a local data model in the subenvironment-environment relationship but a global data model in the workspace-subenvironment relationship. Hence the properties of a local data model discussed here are applicable to a workspace and a subenvironment. However, in this subsection, unless otherwise noted, an application means a workspace. The global data model mentioned in this subsection means the data model of the environment that the workspace belongs to.

Different applications abstract products differently and analyze different kinds of product performance. Such an abstraction is called a local data model in this paper and is called an aspect model by Tomiyama [20].

Local data models have the following properties:

(1) It contains a complete product aspect model from the application point of view, including:
   a) product definition characteristics that are used in the application
   b) product performance characteristics that are estimated by the application

(2) It contains four classes of attributes: imported, interpreted, exported, and resident.

(3) It describes the product characteristics using an application-specific language and primitives.

(4) It needs to be modified when (a) its tools are modified or (b) the related portion of the global data model is changed.

To allow tools to obtain all of data that they need, the local data model defines all product characteristics that are used by the tools and performance characteristics of the product that are evaluated by the tools in an application.

Each application in Concurrent Engineering play certain roles. To define the engineering roles of a workspace in an environment (similarly, those of a subenvironment in a Concurrent

Engineering environment) and to formalize the information sharing and exchange, product data in a local database are classified into four groups.

First, *imported attributes* in a local database are product attributes that are extracted from the global database and used by the application as they are. No changes can be made to the values of this type of attribute in the global database without the agreement from the leader of the engineering environment. Examples of this type of attribute are the masses, moments of inertia, and locations of joints of components of a mechanical system, which are derived from the geometries of the components.

Second, *interpreted attributes* are product attributes in a local database that are interpreted, based on the application's view, from the related product characteristics in the global database. Since product characteristics in the global database are for all applications, these characteristics may not be in a form ready for an application to use directly. Such characteristics need to be accessible to related applications, either through visual displays or by passing them to local memory space. The applications then interpret the attributes into an application-specific language or primitives. The mathematical or engineering meaning indicated by the interpreted attributes in the local database should be consistent with that of attributes in the global database.

For instance, a rotational kinematic freedom between two bodies defined in the global database can be modeled as a revolute joint or a cylindrical joint in a local database for dynamic analysis, so long as the kinematic characteristics between the two bodies that are defined in the global database are maintained in the local database. As another example, the kinematic degree of freedom between a pair of bodies may influence the setting up in boundary conditions for structural analysis in a local database. The boundary conditions of a hole (along the axis of the hole there is a kinematic degree of freedom) may be defined so that the lateral surface of the hole remain circular; hence, the diameter of the hole remains unchanged. In this way, substantial engineering freedom can be kept even when engineering applications are integrated.

Third, *resident attributes* are product characteristics, such as vehicle dynamic analysis models or kinematic joint types, that are defined or generated by an application but which are of no interest to other applications. Generally speaking, all product performance information that is determined by an application belongs to this type of attribute. Examples of resident attributes are mesh densities along certain geometry axes which are used to generate finite element models. Other applications may be interested in using the finite element models but not the mesh density information.

Finally, *exported attributes* are some special product attributes that are generated by an application and also used by other applications. For instance, dynamic analysis computes joint reaction forces, which are used in dynamic stress computation to predict fatigue lifes of mechanical components. So the joint reaction forces are exported attributes in this example, because they are made accessible for other applications in the global database.

The four classes of product attributes define the data relationships between the global database and the applications. These relationships implicitly define the interapplication relationships in a Concurrent Engineering environment. Figure 5 shows the data sharing relationships between an application **k** and the global database. In this figure, for application **k** the attribute set **B** is imported, the attribute set **G** is interpreted to **G'**, the attribute set **H** is resident, and the attribute set **D** is exported. The fact that sets **B** and **D** show up twice in this figure is because these two sets are (1) exported to or generated in the local database of the application and (2) made accessible in the global database. However, when exporting or importing a data set, some data transformation/translation may be needed, because the schema used in the global and local databases may very possibly be different.

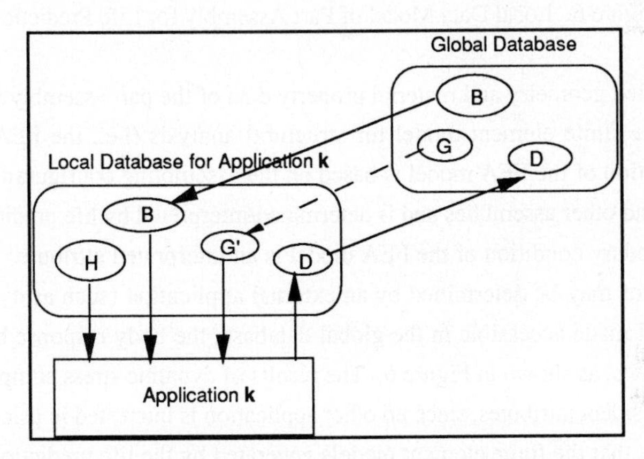

Figure 5. Relationships Between Global Product Characteristics, Local Product Characteristics, and Applications

The language and primitives that are used to define the product in a local database can be application specific, since, in general, the application is the primary user of the product definition stored in its local database. For instance, structural analysis models are defined in terms of finite element primitives, such as nodal points, edges, and elements. These primitives are low level primitives relative to the geometry primitives used in CAD systems.

Attributes in the database of a subenvironment (i.e., the intermediate database) are similarly classified into these four groups relative to the attributes in the global database of the Concurrent Engineering environment.

As an example, Figure 6 shows the local data model of part assemblies for life prediction application. This figure assumes that the dynamic load histories of mechanical components can be determined by the dynamic analysis application (which is an external application relative to the life prediction application in the environment) or by the dynamic analysis workspace in the life prediction subenvironment (see Figure 1). The superscripts # , @ , and ^ highlight imported, interpreted, and exported attributes, respectively. The attributes without any superscript are resident attributes.

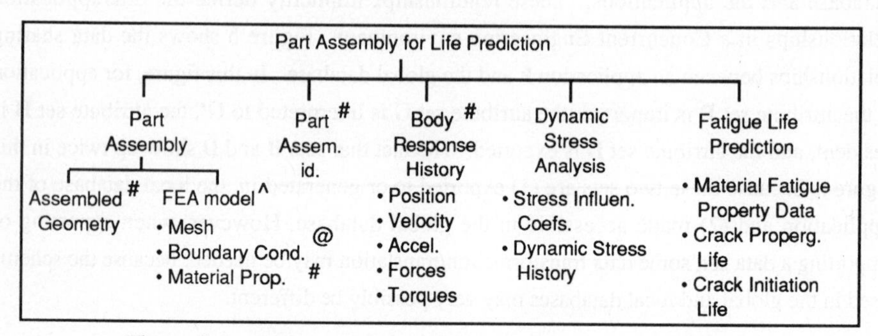

Figure 6. Local Data Model of Part Assembly for Life Prediction

The assembled geometry and material property data of the part assembly is imported and used to generate finite element model for structural analysis (i.e., the FEA model). The boundary condition of the FEA model is based on the assembling configuration between the part assembly and other assemblies and is determined/interpreted by life prediction engineers. Hence, the boundary condition of the FEA model is an interpreted attribute. Since the body response histories may be determined by an external application (such as dynamic analysis application) and made accessible in the global database, the body response histories can be imported attributes, as shown in Figure 6. The results of dynamic stress computation and life prediction are resident attributes, since no other application is interested in using them.

It is possible that the finite element models generated by the life prediction engineers are used in other applications, such as structural sensitivity analysis and optimization. Hence, the finite element model may be exported to the global database for other applications.

3.2 Recursively Applying the Data Model

The global-local data model can be applied recursively to integrate engineering environments and broaden engineering capabilities. Figure 7 depicts such an integration concept. Two environments, computer aided engineering function analysis (CAE) environment and computer aided manufacturing analysis (CAM) environments, share a global database whose schema is defined by a global data model. Note in this subsection and Figure 7, the term CAE means engineering function analysis. In the figure, the CAE environment contains dynamic analysis, fatigue life prediction analysis, and structural analysis applications. The CAM environment has machining and assembly analysis applications.

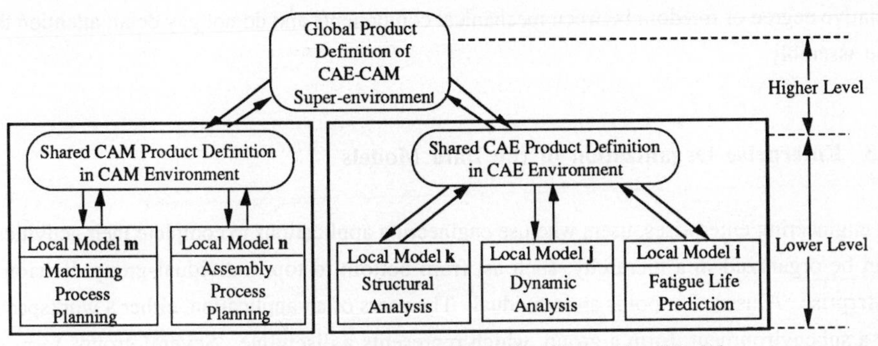

Figure 7. Multi-disciplinary Information Integration Model

Figure 7 shows two hierarchical levels. The higher level contains (1) two environments and (2) a global database for them. The environments share the product definition and put information that is supposed to be passed between them in the global database. Although each of these environments has several subenvironments, workspaces, and tools, the global database at the higher level views each of these environments as a single application. Each environment has a database to integrate applications contained in it. These databases are local databases from the higher level point of view and global databases from the lower level point of view. The lower level of the hierarchy shown in Figure 7 contains two independent environments but no direct communication between them. Each environment has a global database which stores shared CAE (or CAM) product information for CAE (or CAM) applications that are integrated in the environment.

Recall that the language used in a local data model of an application to describe products can be very specific to the application; however, the language used to describe products in a global data model should be general to and understandable by the disciplines that share product information in the global data model. The more diverse the engineering applications that are integrated under a global data model, the more generic the language should be used to described the product in the global data model.

A study [29] shows that Product Data Exchange using STEP (PDES/STEP) [30] can be used to define the product model for selected CAE and CAM applications. This study discusses a way to apply this global-local data model to integrate CAE and CAM environments into a super-environment, as shown in Figure 7, for Concurrent Engineering of mechanical system design, analysis, and manufacturing. Significant object mapping exists between the global data model and the data model of the CAE environment. This is mainly because (1) the granularities of the entities of a mechanical system that are defined in the global data model for CAE and CAM differ from those of the CAE environment, and (2) many CAE environments, for example, the environment discussed in [29], only focus on the relative degree of freedom between mechanical components and do not pay detail attention to the assembly.

3.3 Enterprise Organization of the Data Models

In engineering enterprises, users who use engineering applications to complete their activities can be organized in a hierarchy, such as, from bottom to top, individual-group-division-enterprise. A user of a tool is an individual. The users of an application, either a workspace or a subenvironment, form a group, which represents a discipline. Several groups form a division, i.e., an environment, such as manufacturing division. Finally, several divisions form an enterprise, i.e., a super-environment. By applying the global-local data model one more time, the whole enterprise could be integrated with other enterprises to form a super-super-environment. Hence, this global-local data model can be used to model personnel of enterprises, i.e., it can be used to develop organization models.

4 Process Models

Coordinating engineering activities to achieve common goals of a multi-disciplinary team is one of typical problems faced in a Concurrent Engineering environment. To effectively coordinate engineering activities, the environment must be able to express engineering processes. This section presents such an engineering process modeling capability that can be used when coordinating engineering activities for Concurrent Engineering of mechanical system design and analysis. The process modeling capabilities that are presented in this section are developed within the CAE framework discussed in Section 2, in which engineering activities and engineering software entities are integrated hierarchically using the object-oriented approach.

Research has been carried out on process modeling for software developments. According to Osterweil [31], a "process" is any mechanism used to achieve a goal in an orderly way. Osterweil introduced "process description" as a specification of how the job is to be done. A process model describes who does what, when, and how [32]. Both Humphrey [33] and Osterweil [31] used object-oriented approaches to model software processes. Katayama [34] introduced a formalism for software process description which satisfies three basic requirements: (1) enabling clear description, (2) being able to describe hierarchical process decomposition, and (3) having execution mechanisms.

Humphrey [35] points out that a software process architecture is a framework which establishes the structure, standards, and relationships of various process elements. Process models must (1) represent the way the work is or should actually be performed; (2) provide a flexible, easily understandable, and powerful framework for representing and enhancing processes; and (3) be refinable to the level of detail that is needed for consistent performance.

Results of research on process models for software development are equally applicable to engineering product design, analysis, and development. The process modeling concepts, methodologies, and requirements presented in this section for mechanical system design and analysis are similar to those for software development.

Paris [36] presented process models based on goal-activity relationships to decompose activities. Paris pointed out that when it is impossible to test whether a goal of a complex activity has been achieved, the activity and goal are recursively split up into subactivities and subgoals until a sufficient level of detail is obtained which permits testing of subgoal achievement at the end of a subactivity. This ability to decompose an activity presupposes an activity hierarchy. The CAE framework hierarchy and global and local data models presented in Sections 2 and 3 actually provide an environment to decompose engineering activities into subenvironment, workspace, and tool levels.

Chiueh et. al. [37] presented VSLI design process management models. The process modeling concept presented in this section is similar to Chiueh's approach but is developed independently with mechanical system design and analysis in mind and is based on the CAE framework introduced above.

The purpose of this section is to present a method to model engineering processes, capture engineering knowledge, and manage interprocess relationships.

4.1 Engineering Process Modeling and Knowledge Capturing

In this paper, each engineering analysis process is presented as a sequence of CAE entity invocations, called a runstream. For instance, to accomplish a dynamic analysis for a mechanical system, as shown in Figure 8, a modeling tool is used to generate a dynamic analysis model first. A dynamic analysis tool is then used to analyze the dynamic behavior of the system. Finally, a postprocessor is used to visualize the analysis results and understand the behavior of the system. Each tool runstream is based on an engineering analysis theory or methodology, called background theory. Different engineering analysis methodologies, such as rigid-body or flexible-body dynamic analysis [27, 28], imply different analysis processes.

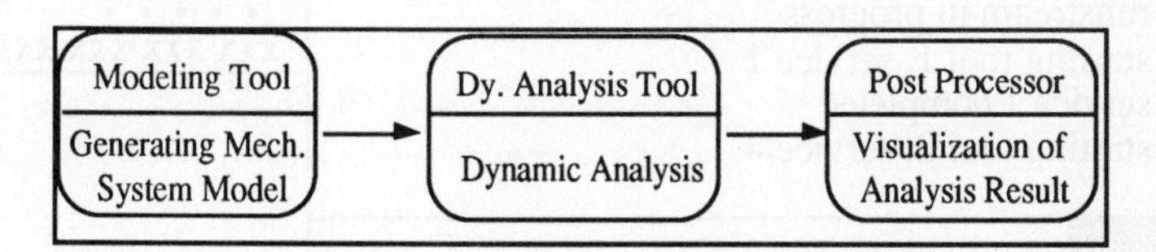

Figure 8. A Dynamic Analysis Runstream

The runstream concept was obtained from observing engineering activities of an engineering group which uses a workspace or a subenvironment to accomplish engineering tasks. A senior engineer in the group knows many tools that are used to design or analyze certain functions of certain products, while a junior engineer knows one or two specific tools. Senior engineers are qualified and may be requested to define engineering processes which describe the CAE entity invocation sequences and user inputs. Junior engineers follow these sequences to accomplish engineering services using the CAE tools.

At the workspace level, a runstream consists of CAE tools and is called a tool runstream; whereas at the subenvironment level, a runstream consists of functional workspaces and is called a workspace runstream. Runstreams give engineers a very flexible means to organize, compose/decompose their design/analysis processes, and save the solutions to the engineering problems. A workspace has a runstream library to store its tool runstreams and the reasons for using them. In this way engineers' knowledge can be kept for reuse and training.

A runstream is represented as oval boxes and arrows between the boxes, as shown in Figure 8 and the upper part of Figure 9. Note that a runstream is not a data flow in that:

1) The arrow between two objects (i.e., tools) indicates the direction of control flow, not data flow;
2) Each oval box in the runstream is a message holder which shows the message to be sent to an object. For example, a vibration analysis message is sent to the CAE tool object ANSYS;
3) Each oval box in the runstream is capable of obtaining mechanical system data from the workspace design data server and will update the workspace database through the design data server.

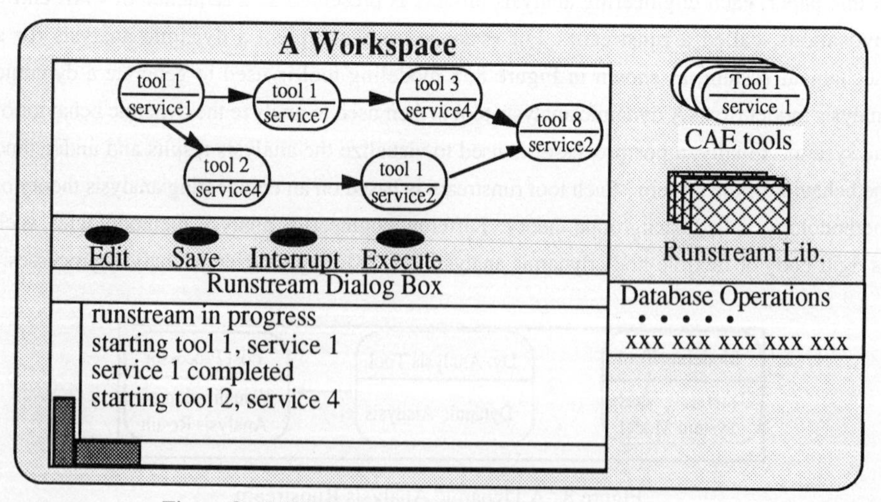

Figure 9. A Workspace and Its Tool Runstream Window

A workspace has a runstream authoring system which has a window to author tool runstreams, as shown in Figure 9. Engineers can use the runstream authoring system to graphically compose tool invocation sequences (i.e., tool service sequences). To carry out a service of a tool, the engineer will create a message holder in the runstream authoring window. Using the message holder, as shown in Figure 10, the engineer specifies the object (i.e., tool) and the service that the object should perform when the control is passed it. The runstream authoring system can send an enquire message to the object in a message holder to find out the preconditions that will have to be satisfied prior to invoking the service of the object in the holder, and the postconditions that will be satisfied upon successful completion of the service.

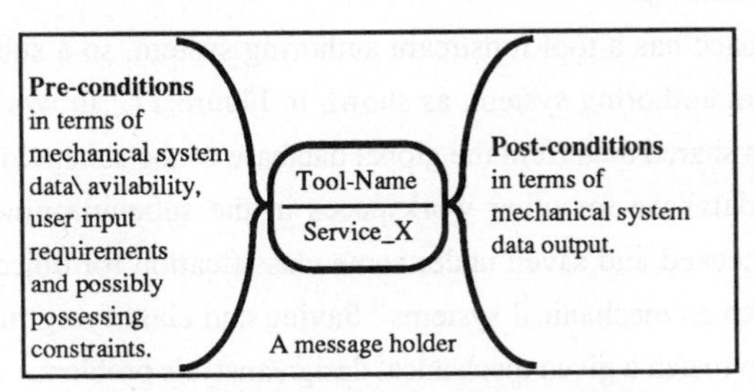

Figure 10. Semantics of a Message Holder

The preconditions and postconditions of a message holder specify data requirements, process constraints, and data generated which provide knowledge to the runstream authoring system. For example, from the preconditions of a service and the postconditions of all the services preceding it, the runstream authoring system can determine whether the data requirements and processing constraint requirements of the service will be met.

The general invocation and control protocol of each tool allow a runstream authoring system to communicate to each tool in a uniform way. For example, each tool will respond to the same request message and reply in a standard way about its input and output requirements, thus making it possible to check data consistency for a specified tool invocation sequence.

Encoding of pre- and postconditions of a CAE tool service opens up the possibility of "knowledge processing" of mechanical system design and analysis. For instance, a runstream's data consistency requirements can be checked against the current database state, or for a given mechanical design/analysis problem its given data can be matched with a runstream.

A service of a tool runstream can be completed when (1) initial conditions (i.e., given product data) of the workspace satisfy the preconditions of the service, (2) the services of tools invoked prior to the service satisfy the preconditions of the service, or (3) both (1) and

(2) hold. When all services of a runstream can be completed, a runstream is completable. Only when the request to execute a service of a tool that can be completed is received from engineer users, the tool will be invoked by the workspace.

The runstream authoring system plays a significant role in defining engineering processes, especially when a type of engineering analysis can be accomplished in several approaches (i.e., different runstreams). For example, dynamic analysis can be carried out by assuming that all mechanical components are rigid bodies or by assuming that some components are rigid, and some are flexible. In the second case, selecting different flexibilities for flexible components in dynamic analysis will produce different analysis results. With the reasons to use a specific runstream stored with the runstream, engineering knowledge can be effectively stored for reuse and training.

Just as a workspace has a tool runstream authoring system, so a subenvironment has a workspace runstream authoring system, as shown in Figure 11. In workspace runstreams, each workspace gets shared data from the global database of the subenvironment and exports data to the global database for other workspaces in the subenvironment. A workspace runstream can be checked and saved under some classification for objects analyzed by the subenvironment, such as mechanical systems. Saving and classifying runstreams allow for intelligent retrieval to match a given mechanical design/analysis problem.

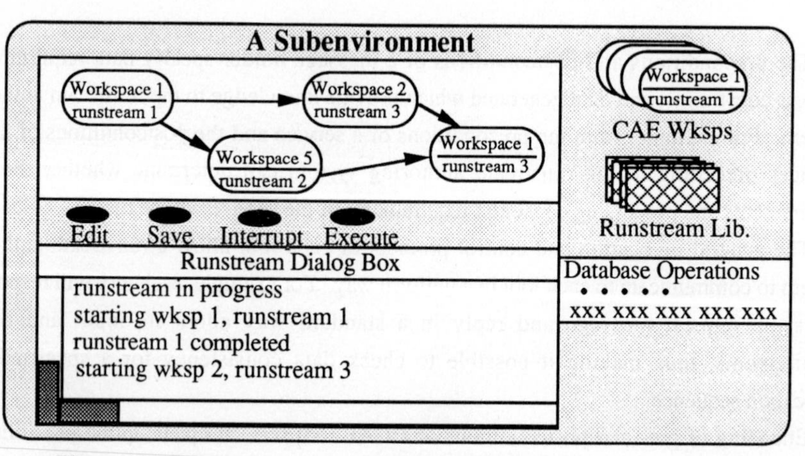

Figure 11. A Subenvironment and Its Workspace Runstream Window

Recall that a workspace runstream is defined using tool runstreams, and the tool runstreams clearly define the activity sequences that engineers of workspaces in a subenvironment should follow. These engineers can be viewed as a single discipline who follows the sequence to play his engineering roles. This is why the subenvironment is treated as a single discipline system.

Subenvironment runstream authoring capabilities are not offered in a Concurrent Engineering environment, instead, a blackboard is used to coordinate engineering activities in the environments. This is because Concurrent Engineering environment is a multi-disciplinary system, and coordinating multi-disciplinary activities is very complex and complicated and will dynamically change depending on the design processes and engineering products to be developed. Before these activities are clearly known and accepted by the whole engineer team, a knowledge-based approach will be more appropriate to coordinate multi-disciplinary activities.

4.2 Interprocess Relationships

Figure 12a shows a workspace runstream consisting of workspaces X, Y, and Z, in which a workspace is represented as a rectangular box. Each workspace in the workspace runstream consists of a tool runstream, as shown in Figure 12b. The integers shown in Figure 12b are tool service identifiers. The arrows that go from services 4 and 6 to service 7 indicate that the preconditions of service 7 would be satisfied by the postconditions of services 4 and 6. The workspace-tool runstream depicted in Figure 12 is hierarchical, and there is no coupling in the runstream.

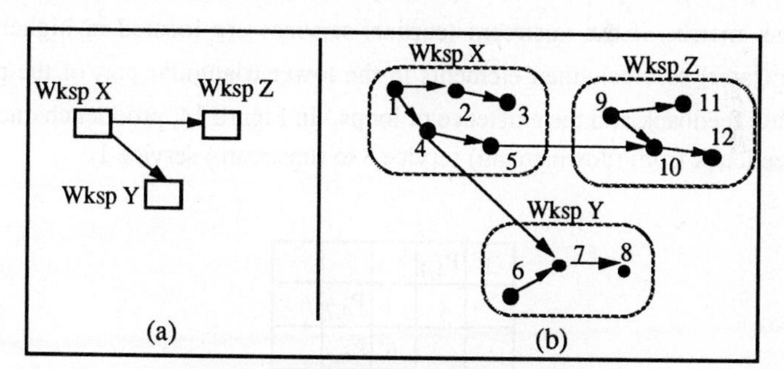

Figure 12. A Hierarchical Runstream

Let us consider a slightly different runstream, as shown in Figure 13; in which there is feedback from tool service 8 to tool service 1. The feedback can be, for instance, (1) a request from a downstream service (service 8) to an upstream service (service 1) for changing a product definition such that certain product performance that is estimated by the downstream service can fulfill product requirements, or (2) product characteristics produced by the downstream service are used by an upstream service, and the downstream service also uses the output of the upstream one.

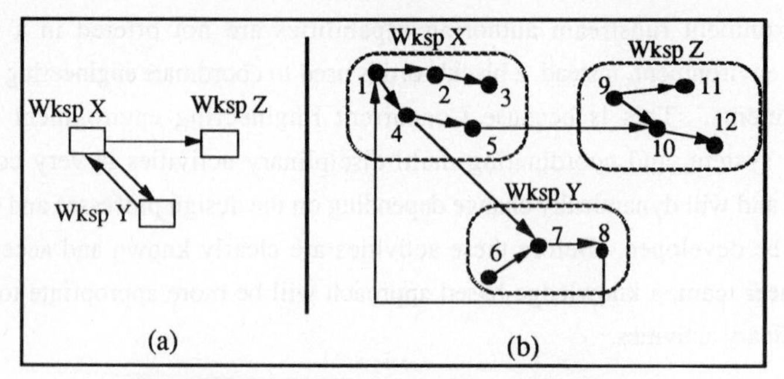

Figure 13. A Coupled Hierarchical Runstream

This feedback forms a loop between workspaces X and Y. As shown in Figure 13, the loop encloses tool services 1, 4, 7, and 8. This workspace-tool runstream is a coupled, hierarchical one. Because of the coupling effect, iterations among services 1, 4, 7, and 8 must take place to obtain reasonable results.

The relationships between runstreams can be captured by using a pseudo task matrix which is similar to the concept of Design Structure Matrix developed by Steward [38]. As an example, the pseudo task matrix of services 1, 4, 6, 7, and 8 that are shown in Figure 13b is shown in Figure 14. In the figure, $P_{i,j}$ of the pseudo task matrix represents a set of product characteristics generated by service i that partially or fully satisfies the preconditions of service j. In the task matrix, if the upstream (earlier) services are located in higher rows than downstream (later) services, then elements in the lower triangular part of the pseudo task matrix indicate feedback and the existence of loops. In Figure 14, product characteristics set $P_{8,1}$ is the feedback from (downstream) service 8 to (upstream) service 1.

1	$P_{1,4}$			
	4		$P_{4,7}$	
		6	$P_{6,7}$	
			7	$P_{7,8}$
$P_{8,1}$				8

Figure 14. Pseudo Task Matrix of Runstream of Figure 13b

Although the tool-runstream of workspace Z shown in Figure 13 is not involved in the loop, service 10 of this runstream cannot be started before reasonable results have been obtained from the iterations of services 1, 4, 7, and 8. This is because the postcondition of service 4 (which is in the loop) is used in service 5 which generates part of the preconditions of service 10.

Such workspace relationships may happen in many computer aided design, analysis, and manufacturing environments or their subenvironments. For instance, in flexible-body dynamic analysis, engineers need to synthesize the flexibility of a body [28] that is modeled as a flexible structure before dynamic analysis can be completed. But, the flexibility of a body cannot be reasonably synthesized until the load characteristics of the body are known. These loadings, however, are obtained from dynamic analysis.

5 Criteria and Methods for Process Coordination

A Concurrent Engineering environment has a blackboard, as shown in Figure 1, to coordinate multi-disciplinary engineering activities to achieve common engineering goals. This section discusses methods and criteria that can be used to coordinate engineering activities that are expressed as runstream within the CAE framework for mechanical system design and analysis.

It is recognized that the process of design involves exploring regions of feasible solutions, and a design problem is a collection of constraint relationships and requirements on attributes of the object to be designed by engineering applications. For example, the feedback $P_{8,1}$, in Figure 14, can be considered as imposing a constraint relationship on the functional workspaces X and Y. Intuitively, such a space of constraint relationships between product attributes and engineering applications can be modeled as a network, as shown in Figure 15, in which rounded rectangles represent applications, such as functional workspaces or subenvironments, and square boxes with a label 'C' represent constraint relationships.

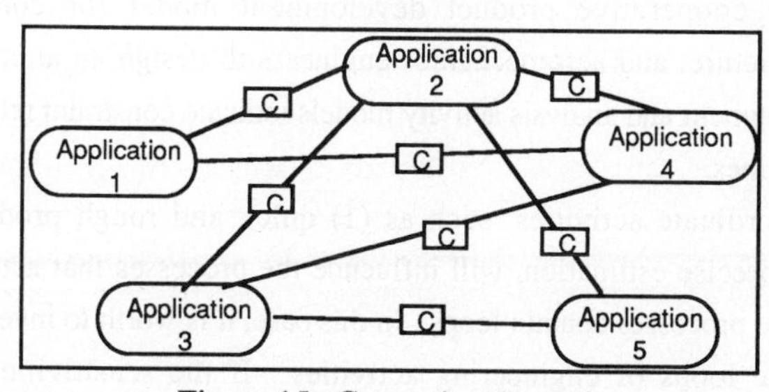

Figure 15. Constraint Network

An important type of constraint is the dependency between product data and processes. The data specified in the pre- and postconditions, which are part of process models, are in global or local databases. Then for a given state of product data, the runstream services that can be completed can be determined, as explained, in the runstream authoring system. However, at the completion of each service, the state of product data changes, and completion of other runstream services becomes feasible. Therefore, product data and process evolve together. Figure 16 briefly shows that: (1) as the product data evolve, the current processes proceed, and new processes may become completable; (2) as processes proceed, product data change, and new data may be produced; and (3) at some point, multiple processes can be carried out parallel.

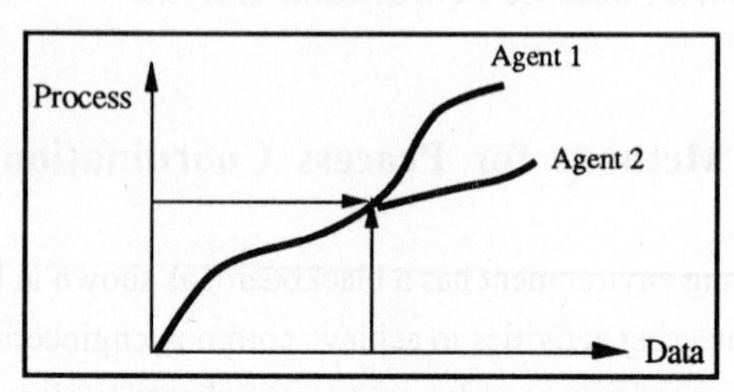

Figure 16. Relationship Between Data and Process Models

Although product data models and process models have developed, to effectively manage engineering processes, authors foresee a critical development in the future is developing a data and process framework to define product data models, process models, and constraint relationships between product data and between data and process models.

When a runstream involves higher level CAE entities, such as workspaces and subenvironments, the constraint relationships between the CAE entities should be modeled according to multi-disciplinary product development methodologies. For example, Sobieski's multi-level system optimization methodology [39, 40] takes into account engineering aspects in several levels, with the most important ones considered first. Similarly, Bond and Ricci [41] presented a cooperative product development model for conceptual design, aerodynamics, structure, and aeromechanics engineers to design an aircraft. Such multi-disciplinary development and analysis activity models indicate constraint relationships among engineering disciplines.

Criteria to coordinate activities, such as (1) quick and rough product performance estimation or (2) precise estimation, will influence the processes that actually carried out, especially when the processes contain loops. In this case, it is worth to investigate the degree of coupling in the loops of engineering activities. If the sensitivities of the product characteristics that are evaluated in the upstream services is sensitive to the change of product characteristics that are fed back from downstream services, the coupling effect (i.e., loops) should not be ignored. On the other hand, if, for instance, in the runstream shown in Figure 13b, the sensitivity of the change of the product parameter set $P_{1,4}$ to the change of the parameter set $P_{8,1}$ is low, and if it is necessary to quickly finish the analysis process, the coupling effect (i.e., the feedback from service 8 to service 1) can be ignored. However, to accurately estimate the product characteristics (such as parameter sets $P_{1,4}$ and $P_{8,1}$), the coupling effect should not be ignored.

For a given set of services and constraints between the services that have taken into account coordination criteria and sensitivities of data dependency in the constraints, Steward's system analysis method [38] and an improved method [42] can be used to determine execution sequences of the engineering services. By using Steward system analysis method and an artificial intelligence system, capabilities to schedule engineering services can be developed [43, 44].

One approach, therefore, to develop process management capabilities for the blackboard to promote cooperation in a Concurrent Engineering environment is to (1) develop the data and process framework; (2) use (i) the object-oriented CAE entity integration hierarchy, (ii) the process modeling methodology, (iii) Steward's system analysis methods, and (iv) artificial intelligence systems as basic elements; and (3) apply multi-level system optimization methodology, sensitivity information, and scheduling criteria to determine and coordinate engineering activities. In this way the blackboard would have the knowledge of the constraint relationships between various disciplines/applications, allow high-level specifications of product attribute changes, and determine and coordinate associated engineering service runstreams.

6 Conclusions

A CAE framework has been developed. The framework uses an object-oriented approach to integrate engineering software systems in a tool-workspace-subenvironment-environment hierarchy, utilizes global and local data models to integrate product informations at different hierarchical levels, and provides a systematic way to model engineering activities as runstreams of tools and workspaces. The developments in engineering tool integration, information integration, and engineering process modeling build a basis to develop an activity coordination system to full support Concurrent Engineering .

While the engineering tools and product data models are for mechanical system design and analysis, the CAE framework can be extended by integrating manufacturing applications, and the product model can be modified for other types of product. The object-oriented approach makes the CAE framework very easy to extend and modify. The framework as well as the global and local data models can also be extended to reflect the personnel organization model of an enterprise. One of the remaining challenges to develop a Concurrent Engineering environment for mechanical systems is to develop a coordination system for the CAE framework.

Acknowledgment: Research supported by NSF-Army-NASA Industry/University Cooperative Research Center for Simulation and Design Optimization of Mechanical Systems.

References

1. Shina, S.G., *Concurrent Engineering and Design for Manufacture of Electronics Products*, Van Nostrand Reinhold, New York, NY., 1991.
2. Carter, D.E. and Baker B.S., *Concurrent Engineering: the Product Development Environment for the 1990s*, Addison Wesley, New York, NY, 1992.
3. PATRAN User Manual, PDA Engineering, Costa Mesa, CA, 1990.
4. Intergraph Engineering Modeling System Reference Manual, Intergraph Co., Huntsville, AL, 1991
5. UG Concept Users' Reference Manual, MacDonnell Douglas Co., Cypress, CA, 1991.
6. DeSalvo, G.J. and Swanson, J.A., ANSYS Engineering Analysis System User's Manual, Vols. I and II, Swanson Analysis System, Inc., Houston, PA, 1990.
7. MSC/NASTRAN, MSC/NASTRAN User's Manual, Vol. I and II, The Macneal-Schwendler Co., 815 Colorado Boulevard, Los Angeles, CA, 90041, 1988.
8. ABAQUS, ABAQUSN User's Manual, Version 4.8, Hibbitt, Karlsson & Sorensen, Inc., 100 Medway Street, Providence, RI, 1989.
9. DADS Reference Manual, Version 6.1, CADSI Co., Oakdale, IA, 1991.
10. ADAMS Users' Manual, Mechanical Dynamics Inc., Ann Arbor, MI, 1981.
11. Wong, A., Sriram, D., and Logcher, R., "SHARED: An Information Model for Cooperative Product Development," Massachusetts Institude of Technology, Intelligent Engineering System Laboratory, Technical Report IESL91-04, Dec. 1991.
12. Cleetus, J., ed., "Red Book of Functional Specification for the DICE Architecture," Concurrent Engineering Research Center, West Virginia University, Morgantown, WV, 1988.
13. Mentor Graphics for Concurrent Design, April, 1990, Mentor Graphics Corp., Beaverton, OR, 97214.
14. Wasserman, A., "Tool Integration in Software Engineering Environments," Software Engineering Environments, Lecture Notes in Computer Science 467, Spinger-Verlag, New York, NY, 1989, pp. 137-149.
15. Lewis, G.R., "CASE Integration Frameworks," Sun Technical Journal, Nov. 1990, pp. 50-51.
16. Sriram, D., Logcher, R., and Fukuda, S., Ed., *Computer-Aided Cooperative Product Development*, Lecture Notes in Computer Sciences, 492, Springer-Verlag, New York, NY, 1991.
17. Baek, W.K. and Stephens, R.I., "Computational Life Prediction Methodology for Mechanical System Using Dynamic Simulation, Finite Element Analysis, and Fatigue Life Prediction Methods," Ph.D. Thesis, Dept. of Mechanical Engineering, The University of Iowa, also in Technical Report R-71, Center for Computer Aided Design, The University of Iowa, Iowa City, IA, May, 1990.
18. Concept of Operations for the Information Sharing System of DICE, Concurrent Engineering Research Center, West Virginia University, Morgantown, WV, 26506, 1992.
19. Paradice, D.B., "An Object-Oriented Approach to Managerial Problem Modeling," ACM, Proceeding of the Twenty-Third Annual Hawaii International Conference on System Sciences, 1990, pp. 353-362.
20. Tomiyama, T., Kirriyama, T., and Takeda, H., "Metamodel: A Key to Intelligent CAD Systems, Research in Engineering," Vol. 1, 1989, pp. 19-34.
21. Shaw, N.K., Bloor, S.M., and de Pennington, A., "Product Data Models, Research in Engineering Design," Vol. 1, 1989, pp. 43-50.
22. Hardwick, M. and Spooner, D.L., "Comparison of Some Data Model for Engineering Objects," IEEE CG&A, March, 1987, pp. 56-66.
23. Bancilhon, F. "Object-Oriented Database Systems," Proceeding of the Seventh ACM Sigact-Sigmod-Sigart Symposium on Principles of Database Systems, March 21-23, 1988, Austin, TX, pp. 152-162.
24. Levene, M., and Poulovassilis, A., "An Object-Oriented Data Model Formalised Through Hypergraphs," Data and Knowledge Engineering, Vol. 6, 1991, pp. 205-244.
25. Wu, J.K., Choong, F.N., Choi, K.K., and Haug, E.J., "A Data Model for Simulation Base Design of Mechanical Systems", International Journal of System Automation Research and Application, Vol. 1, 1991, pp. 67-88.
26. Brodie L.M., "On the Development of Data Model," On Conceptual Modelling, Springer-Verlag, New York, 1982, pp. 19-47.
27. Haug, E.J., *Computer Aided Kinematics and Dynamics of Mechanical Systems*, Allyn and Bacon, Boston, MA, 1989.
28. Kim, S.S. and Haug, E.J., "A Recursive Formulation for Flexible Body Dynamics," Ph.D. Thesis, Dept. of Mechanical Engineering, The University of Iowa, also in Technical Report R-14, Center for Computer Aided Design, The University of Iowa, Iowa City, IA, May, 1988.
29. Wu, J.K., Liu, T.H., and Fischer, G.W., "An Integrated PDES/STEP Based Information Model for CAE and CAM Applications," Proceeding of the Second International Conference on Automation Technology, Taiwan, July 4-6, 1992, pp. 179-188.
30. NIST, "Product Data Exchange Specification First Working Draft," U.S. Department of Commerce, National Institute of Standard and Technology, Gaithersburg, MD, December 1988.

31. Osterweil, L., "Software Process are Software Too," IEEE Proceeding of the Ninth International Conference on Software Engineering, 1987, pp. 2-13.
32 Process Capture and Characterization for DICE, Concurrent Engineering Research Center, West Virginia University, Morgantown, WV, 26506, 1992.
33. Humphrey, W.S. and Keller, M.I., "Software Process Modeling: Principles of Entity Process Models," IEEE Proceeding of Eleventh International Conference on Software Engineering, 1989, pp. 331-342.
34. Katayama, T., "A Hierarchical and Functional Software Process Description and its Enaction," IEEE Proceeding of Eleventh International Conference on Software Engineering, 1989, pp. 343-352.
35. Humphrey, W.S., "Introducing Process Models into Software Organizations," American Programmer, September, 1990, pp. 1-7.
36. Paris, J., "Goal-Oriented Decomposition--Its Application for Process Modeling in the PIMS Project," Software Engineering Environments, Lecture Notes in Computer Science 467, Springer-Verlag, New York, NY, 1989, pp. 69-76.
37. Chiueh, T.F., Katz, R.H., and King, V., "Managing the VLSI Design Process," Computer-Aided Cooperative Product Development, Lecture Notes in Computer Sciences, 492, Springer-Verlag, New York, NY, 1991, pp. 183-199.
38. Steward, D.V., *System Analysis and Management*, Petrocelli Books, New York, NY, 1981.
39. Sobieski, J., "Optimization by Decomposition: a Step From Hierarchic to Non-Hierarchic Systems," in Recent Advances in Multidisciplinary Analysis and Optimization, NASA Conference Publication 3031, Part 1, 1989, pp. 51-78.
40. Adelman, H.M. and Mantay, W.R., "An Initiative in Multidisciplinary Optimization of Rotorcraft," in Recent Advances in Multidisciplinary Analysis and Optimization, NASA Conference Publication 3031, Part 1, 1989, 109-144.
41. Bond, A.H. and Ricci, R.J., "Cooperation in Aircraft Design," Computer-Aided Cooperative Product Development, Lecture Notes in Computer Sciences, 492, Springer-Verlag, New York, NY, 1991, pp. 152-182.
42. Eppinger, S.D., Whitney, D.E., Smith, R.P., and Gebala, D.A., "Organizing the Tasks in Complex Design Projects," Computer-Aided Cooperative Product Development, Lecture Notes in Computer Sciences, 492, Springer-Verlag, New York, NY, 1991, pp. 229-252.
43. Rogers, J., Feyock, S., and Sobieski, J., "Structex-A Prototype Knowledge Based System for Initially Configuring a Structure to Support Point Loads in Two-Dimensions," Artificial Intelligence in Engineering Design, Computational Mechanics Publications, 1988.
44. Rogers, J., "DEMAID: Users' Guide to Design Manager's Aid for Intelligent Decomposition," NASA Langley Research Center, Hampton, VA, 1989.

Storage and Retrieval of Objects for Simulations in Mixed Application Areas

Richard N. Zobel

Department of Computer Science, University of Manchester, Oxford Road, Manchester M13 9PL, United Kingdom

Abstract: Concurrent Engineering of products can benefit from the use of simulation at every stage from the requirements phase through to production, maintenance and update. This paper explores the adaptation of iconic graphical user interface based systems to achieve the necessary rapid prototyping required for parallel product development. To achieve this model definition, interfaces standards and storage in a set of discipline related object oriented databases is proposed.

Keywords: / simulation / graphical user interfaces / concurrent engineering / rapid prototyping / object oriented databases / mixed applications / modelling / object standards / interface standards

1 Introduction

Traditionally, simulation has been used for weapons systems performance studies and more recently for design verification and training simulators. There continues to be expansion of the use of simulation in almost every discipline. No longer is simulation confined to continuous systems, the arrival of discrete event languages and simulations has led to a rapid expansion in this area, and to the development of combined discrete event and continuous languages, models and simulations. Simulation has consequently expanded into almost every activity and discipline.

2 Current Simulation Usage

There is an increasing use of graphical user interface based software systems, and in particular in the simulation area [1]. Further there is an expansion in the use of animation and 2-D and 3-D graphics. Simulation languages are used for part or all of products, but new designs all too frequently require a start again approach. Further, there is relatively little attempt at software re-use except at a low level, and consequently little support for the use of simulation in rapid prototyping.

3 Simulation in Concurrent Engineering

Simulation is not yet generally used to support the parallel concepts of Concurrent Engineering. Indeed, a problem exists in that simulation is not yet integrated into Concurrent Engineering support systems.

In spite of this there is a clear need for the rapid development of the use of simulation in Concurrent Engineering. At the early stages of specification there is a need for concept validation. Simulation should be a powerful tool to aid in this process, as shown by Bradley [2]. Related to this is the need for rapid prototyping. Prototyping by use of simulation models, especially where such models can also be constructed by rapid methods, can save much time and help remove bad design decisions at an early stage.

Moreover, it is clear that simulation should be used to support all stages of product development from specification through design, implementation, testing, production, maintenance, and upgrading.

4 Product Simulation

This section discusses the overall view of product simulation in terms of complex products which involves parts and sub-assemblies from a mix of application areas. The issues involved in simulation of complete products are discussed in terms of the future needs of simulation systems.

4.1 Simulation in Product Design and Manufacture

Simulation is already an important aspect of modern product design and manufacture in some industries. Indeed, in the VLSI area, circuit and system simulation is used to guarantee first time correctness. Extending this desirable concept to complex products requires simulation of parts from many different areas. Few products comprise only mechanical parts. Often electronic and electromechanical assemblies are required along with microcomputers and controllers. In the process industry, items from such diverse disciplines as chemical and control engineering are needed in addition to distributed computers requiring communications. Examples of such systems might be a ducted fan civil aero engine or the distributed control of a chemical plant. A domestic example is the ubiquitous compact disc player, to be found in most homes and illustrated in a graphical based simulation system for mixed application areas [3].

4.2 Complete Product Simulation

An increasing need is perceived for simulation of the entire product or process at an early stage of the specification and design phases as part of the Concurrent Engineering approach. This can be achieved by modular construction of the complete model from a kit of parts, achieved by re-use of software models of the parts obtained from previous projects. The latter may require different parameters and initialization, and sometimes some modification. In principle, complete design should rarely be necessary after an adequate supply of earlier part designs has been assembled.

Work at Manchester is currently concerned with the production of such a design and simulation environment for fast prototyping and full scope simulation. Based on interconnected icons, representing systems parts and signal flows, a current interest concerns the acquisition and storage of a kit of parts or objects [3]. This work has developed from an earlier project to build a system for design and simulation of signal processing systems [4,5]. This was achieved using early versions of C++ [6] and was sufficiently successful to warrant generalization of the techniques to other disciplines and ultimately to a mix of application areas.

5 Storage and Retrieval of Objects

A discussion is presented here of the issues involved principally with the representation of simulation modules using an object oriented approach, and the consequent requirements for the storage and retrieval of such objects in relational data bases and object bases.

5.1 Requirements of an Simulation Object Storage Scheme

Of great significance to the simulation system designer is the use of a system which the designer can employ to select items from a number of application areas such as DSP, control, communications, mechanical engineering, electrical engineering, register transfer logic, instrumentation, etc., to create the particular system specified. Hence, there is a requirement for separate object bases for each discipline, along with browsing systems for user evaluation and selection of suitable parts for the required application.

Figure 1 shows a schema for a system in which the user browses and selects items required for the complete product from a number of such object bases, to be included in a user object base for the particular application at hand. This user application base is then attached to a design and simulation system.

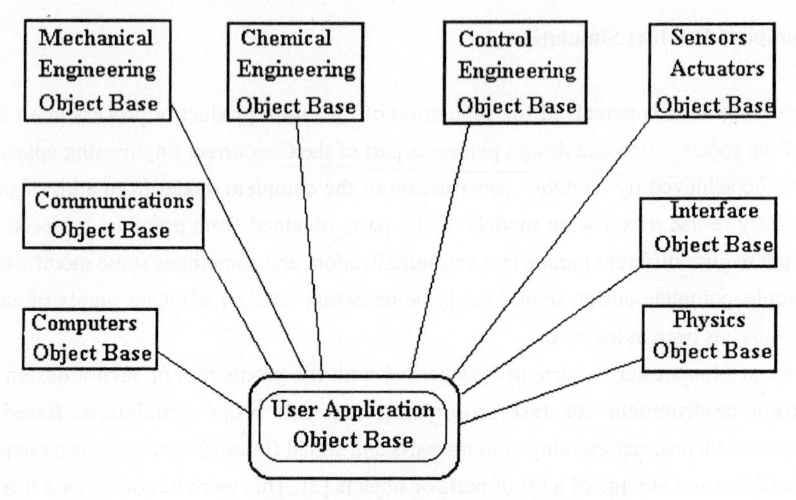

Figure 1. Object Base System

For each object base some structure is required. This is best achieved by adopting the principles of the object oriented approach in which inheritance features strongly and is implemented as a class hierarchy. Difficulties have been experienced in this respect, and much work needs to be done in this area to establish such class hierarchies for some disciplines. Figure 2 illustrates such a hierarchy in a limited way.

In this context one must consider what comprises an object. It is clear that there is a need to at least represent class hierarchy, source and object code, icon, algorithm, and help information. This in turn places significant demands on the storage schema and on the usefulness of existing database technology.

An indication of what is required may be deduced from a provisional list of types of information, data and files that are needed for an object based design and simulation system. In common with modern software thinking, an object requires a specification part and an implementation part. The former relates to specifications of a more general nature related to objects higher in the class hierarchy and to object external interfaces. The latter is concern with implementation of the data structures and methods (functions) associated locally with the object. The code associated with these needs to be stored in both source and object form. The latter may be for more than one hardware processor type in a host-target system, or system comprising a computer with additional hardware accelerators or parallel processors.

Typical initial data and parameter values and ranges are needed to guide the user, plus default values where appropriate.

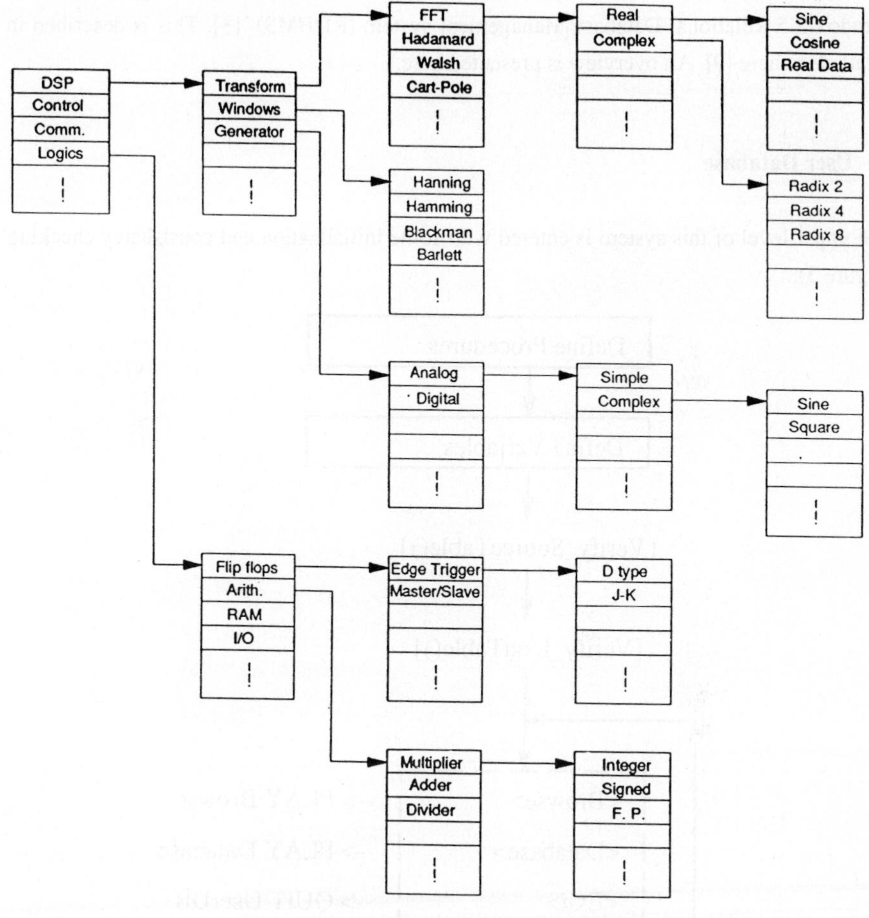

Figure 2. Relation Between Class Objects

In addition, where a suitable object is not available within the system, a mechanism is needed for creating new objects or modifying existing ones, and for integrating them into the object base.

Finally, there can be the problems of linking the user application base to the design and simulation environment. This arises because of inconsistencies between database technologies and the (object oriented) language employed for the design and simulation environment. An early, but useful, discussion of object oriented databases is given by Brown [7].

6 Object Base System

An object base system is currently under development at Manchester, using the Borland Paradox 3.5 Relational Database Management system (RDBMS) [8]. This is described in detail elsewhere [9]. An overview is presented here.

6.1 User Database

The upper level of this system is entered with some initialization and consistency checking (Figure 3).

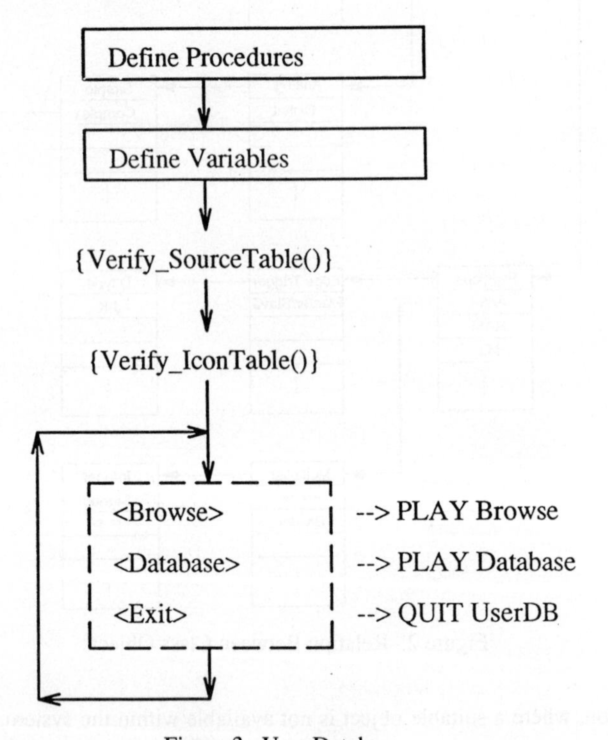

Figure 3. User Database

This is necessary because existing RLDMSs do not readily permit storage of items such as source code, binary object files and icon bit maps. Consequently, the objects are represented by file names instead of actual data. Hence, there is the distinct possibility of inconsistency due to the files being outside of the control of the RDBMS.

After entry, the user may elect to browse the object base system or construct an application object base, by playing the scripts Browse or Database.

6.2 Browsing

This is the subsystem for creation and management of the object bases. The Browse script features a mechanism for navigating through the classes. This gives an overview of the structure of the existing object bases. Subsequently the user may elect to modify the objects and or object bases. The usual insert, extend, and delete, facilities are provided, plus source code and icon viewing. These are subject to some predictable test conditions being applied. In addition the class structure may be modified, and editors are provided for source code and icons.

An important feature of the system is the help facility provided by an additional script Help (Figure 4). This provides a mechanism for navigating through help windows (with graphical help being added at a later stage). In addition, a facility is present to enable users to amend and create help information.

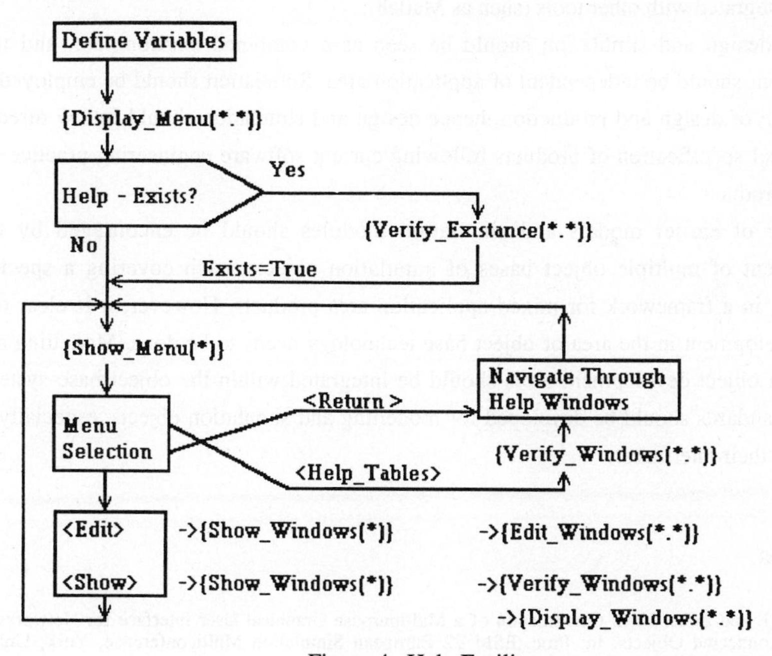

Figure 4. Help Facility

6.3 Application Object Base

This is the subsystem for user creation of the application database for design and simulation of the product under consideration. Again, the user may navigate through the classes for object selection. Objects may be imported from other object bases or externally, and other object bases may also be included for editing. Items may be added or deleted and the resulting object base exported to a design and simulation system.

7 Conclusions

There are many conclusions to be drawn from the considerations of the place that simulation needs to play in Concurrent Engineering and for the consequences for simulation tools and methodologies.

Firstly, modelling and simulation should be integrated into the CE environment in order to encourage support for the parallel product life cycle. This can then used to achieve rapid prototyping by use of simulation to speed up product development.

Secondly, within the CE environment, modelling and simulation should be themselves be fully integrated with other tools (such as Matlab).

Next, design and simulation should be seen as a combined environment, and this environment should be independent of application area. Simulation should be employed at every stage of design and production, hence design and simulation should follow directly from formal specification of products following current software engineering practice for software products.

Re-use of earlier models and simulation modules should be encouraged by the development of multiple object bases of simulation objects, each covering a specific discipline, in a framework for mixed application area products. However, it is clear that much development in the area of object base technology needs to be done. Modelling and simulation object development tools should be integrated within the object base system. Finally, standards should be developed for modelling and simulation objects, especially in relation to their interfaces.

References

1. Lee K.H. and Zobel R.N.: Construction of a Multipurpose Graphical User Interface for Networks of Interconnected Objects. In: Proc. ESM 92 European Simulation Multiconference, York, United Kingdom, pp 455-459 SCSI Ghent. June 1992.
2. Bradley R.: A Method for Specifying Complex Real-Time Systems with Application to an Experimental Variable Stability Helicopter. Ph.D. Thesis, University of Glasgow, United Kingdom. April 1992.
3. Zobel R.N. and Lee K.H.: A Graphical User Interface Based Simulation System for Mixed Application Areas. To be Published in: Proc. SCSC 92 Summer Computer Simulation Conference, Reno, USA. SCSI San Diego. July 1992.
4. Lee K.H. and Zobel R.N.: An Advanced Simulation System for Digital Signal Processing Using C++ and Hardware Accelerators. In: Proc. SCSC91 Summer Computer Simulation Conference, Baltimore, USA, pp 245-250. SCSI San Diego. July 1991.
5. Lee K.H., R.N. Zobel R.N and Tchapda Y.C. Performance of a DSP Design and Simulation System Based on an Enhanced PC/AT. In: Proc. ESM 91 European Simulation Multiconference, Copenhagen, Denmark, pp 958-962. SCSI Ghent. June 1991.

6. Lee K.H. and Zobel R.N.: Experience with the use of C++ for Building Simulation Systems. In: Proc. ESM 91 European Simulation Multiconference, Copenhagen, Denmark, pp 963-967. SCSI Ghent. June 1991.
7. Brown A.B.: Object Oriented Databases. McGraw-Hill, London. 1991.
8. Greenwood D.: Applications Generator Using the Paradox Relational Database Management System. M.Sc. Thesis, Univ. of Manchester, United Kingdom. 1992.
9. Zobel R.N., Greenwood S. and Lee K.H.: Object Bases and Relational Databases for Object Based Simulation Systems. To be published in Proc. EUROSIM 92 European Simulation Congress, Capri, Italy. EUROSIM, Rome. October 1992.

Analysis of Structural Systems Undergoing Gross Motion and Nonlinear Deformations

Jorge A. C. Ambrósio

Institute of Mechanical Engineering, I.S.T., 1096 Lisbon, Portugal

Abstract: Recent developments in flexible multibody dynamics provided new tools for the analysis of structural systems undergoing geometric and material nonlinear deformations. The restrictions that some of the most popular methods present for the analysis of multibody system with flexible components are discussed within the framework of a general methodology. The equations of motion for a flexible body undergoing large overall motion are obtained based on the principles of continuum mechanics and employing nonlinear finite elements. These equations are then simplified based on a lumped mass formulation and referring the nodal accelerations to the inertial reference frame. The resulting equations, which describe completely the coupling between the nonlinear gross motion and the deformation of the flexible components of the system, present a mass matrix that is diagonal and constant. These equations can be further simplified for cases where the level of deformations lies within the elastic range and the geometric nonlinearities are not important. Under these assumptions the modal superposition method can be used to reduce the number of nodal degrees of freedom of the flexible components. If the assumptions of material and geometric linearity are not met the number of nodal degrees of freedom can still be reduced using a static condensation technique. The equations obtained in this manner are implemented in a general purpose program and solved numerically. The study of a light space structure is presented in order to show the effects of the nonlinear geometric deformations in behavior of the system. The application of the methodology described here to crashworthiness and structural impact is illustrated with the study of the rollover of an off-road vehicle.

Keywords: multibody / dynamics / flexibility / plasticity / crashworthiness / vehicle dynamics

1 Introduction

The analysis of of multibody systems experiencing different levels of nonlinear deformations deserved the attention of numerous researchers in recent years. To achieve the goal of concurrent engineering, the dynamic analysis of flexible multibody systems must be convenient to use by an engineer. For this purpose the increasing complexity in the modelling of nonlinear flexible systems must not be transfered directly the design engineer. For each methodology used for the model there is a possible range of applications that must be well known. This paper presents a comprehensive methodology for the analysis of flexible multibody systems that can be used in concurrent engineering.

The initial attempts to describe the motion of flexible bodies experiencing large displacements and rotations was made using the finite element method [3,5,6,16]. However, if the behavior of the system is dominated by such large displacements and rotations and the structural deformations are small or negligible then the finite element method is not adequate. Procedures based on multibody dynamics have shown to be quite efficient to deal with this class of problems [7,8,10]. The combination of multibody dynamics with the finite element method was very successful in describing the motion of systems where the nonlinear gross motion is coupled with the small elastic deformations [18-20,22]. Some researchers have recently extended range of applications of multibody dynamics to problems characterized by the coupling between the gross motion an nonlinear elastic deformations [4,9,21].

In this paper the equations of motion for the multibody system are first obtained in terms of cartesian coordinates. The kinematic constraints between the elements of the system are described by a set of algebraic equations which are introduced in the system by a set of Lagrange multipliers. At a latter stage the cartesian coordinates are replaced by joint coordinates using velocity transformations. This process not only reduces the number of coordinates necessary for the description of the system but also eliminates the constraints from the equations of motion. To describe the deformation of the bodies of the multibody system, an updated Lagrangian formulation based on the work by Bathe et al. [5] is employed. The deformation of the flexible body is always related to a corotated configuration, to which a body fixed referential is attached. The form of the equations of motion obtained is further simplified using a lumped mass formulation for the mass matrix of the flexible body and changing the referential to which the nodal accelerations are referred. Finally this methodology is implemented in a general purpose computer program [11] and it is applied to the study of the rollover and crashworthiness of an off-road vehicle.

2 Multibody System Dynamics Using Joint Coordinates

A multibody system is a collection of rigid and flexible bodies joined together by kinematic joints and force elements as depicted in Figure 1. For the ith body in the system $\mathbf{q}_i$ denotes a vector of coordinates which contains the Cartesian translational coordinates $\mathbf{r}_i$, a set of rotational coordinates $\mathbf{p}_i$, and a set of nodal coordinates $\mathbf{q}'_f$ or δ' (if body i is flexible). A vector of velocities for a rigid body i is defined as $\mathbf{v}_i$, which contains a 3-vector of translational velocities $\dot{\mathbf{r}}_i$ and a 3-vector of angular velocities ω_i (defined in the XYZ coordinate system). If body i is flexible then the vector of velocities $\mathbf{v}_i$ contains $\dot{\mathbf{r}}_i$, ω_i (defined in the $\xi\eta\zeta_i$ coordinate system) and a vector of nodal velocities $\dot{\mathbf{q}}'_f$ or $\dot{\delta}'$. The vector of accelerations for the body is denoted by $\dot{\mathbf{v}}_i$ and it is simply the time derivative of $\mathbf{v}_i$. For a multibody system containing nb bodies, the vectors of coordinates, velocities, and accelerations are $\mathbf{q}$, $\mathbf{v}$ and $\dot{\mathbf{v}}$ which contain the elements of $\mathbf{q}_i$, $\mathbf{v}_i$ and $\dot{\mathbf{v}}_i$, respectively, for i=1, ...,nb.

Let the kinematic joints between rigid bodies be described by mr independent constraints as

$$\mathbf{\Phi}(\mathbf{q}) = \mathbf{0} \tag{1}$$

The first and second derivatives of the constraints yield the kinematic velocity and acceleration equations.

$$\dot{\mathbf{\Phi}} \equiv \mathbf{D}\mathbf{v} = \mathbf{0} \tag{2}$$

$$\ddot{\mathbf{\Phi}} \equiv \dot{\mathbf{D}}\mathbf{v} + \mathbf{D}\dot{\mathbf{v}} = \mathbf{0} \tag{3}$$

where $\mathbf{D}$ is the Jacobian matrix of the constraints. The equation of motion for the system of rigid bodies are written (see reference [12])

$$\mathbf{M}\dot{\mathbf{v}} - \mathbf{D}^{T}\lambda = \mathbf{g} \tag{4}$$

where $\mathbf{M}$ is the inertia matrix, λ is a vector of Lagrange multipliers, and $\mathbf{g} = \mathbf{g}(\mathbf{q}, \mathbf{v})$ contains the gyroscopic terms, and the forces and moments that act on the bodies.

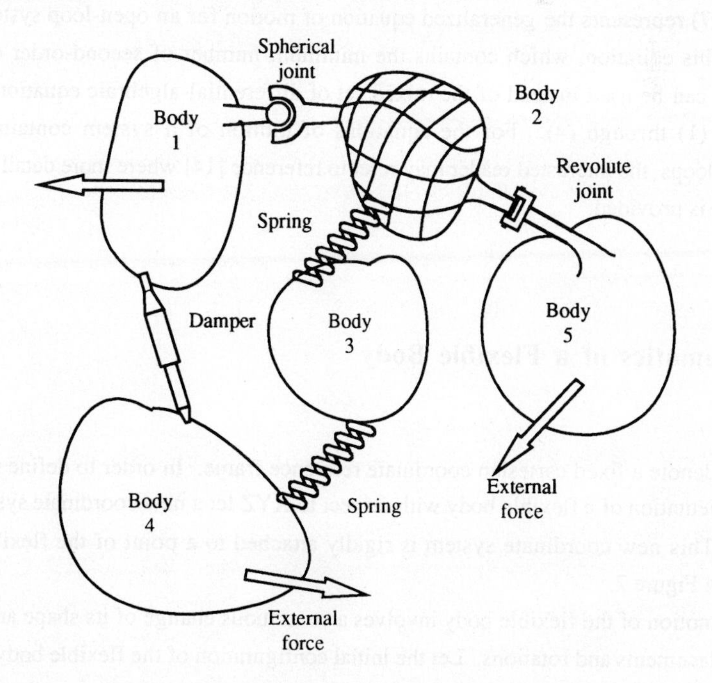

Figure 1 Schematic representation of a multibody system

The constrained equations of motion expressed by equations (1) to (4) can be converted to a smaller set of equations in terms of a set of coordinates known as joint coordinates. Such transformation is briefly discussed here (for more detail refer to Nikravesh and Gim [13]). The relative configurations of two adjacent bodies are described by a set of relative coordinates, known as joint coordinates, equal to the number of relative degrees of freedom between the bodies. The vector of joint coordinates for a system of rigid bodies is denoted by β and it contains all the joint coordinates and the absolute coordinates of the floating base bodies. The vector of joint velocities is defined as $\dot{\beta}$, which is the time derivative of β, and its relation with **v** is given by [13]

$$\mathbf{v} = \mathbf{B}\dot{\beta} \tag{5}$$

Matrix **B** is the velocity transformation matrix and it is orthogonal to the Jacobian matrix **D**. The time derivative of equation (5) provides the formula for the transformation of the accelerations,

$$\dot{\mathbf{v}} = \dot{\mathbf{B}}\dot{\beta} + \mathbf{B}\ddot{\beta} \tag{6}$$

Substituting equation (6) into equation (4), premultiplying by $\mathbf{B}^T$, and using the orthogonality between **B** and **D** yield

$$M\ddot{\beta} = f \tag{7}$$

where

$$M = \mathbf{B}^T\mathbf{M}\mathbf{B} \tag{8}$$

$$f = \mathbf{B}^T\left(\mathbf{g} - \mathbf{M}\dot{\mathbf{B}}\dot{\beta}\right) \tag{9}$$

Equation (7) represents the generalized equation of motion for an open-loop system of rigid bodies. This equation, which contains the minimum number of second-order differential equations, can be used instead of the mixed set of differential-algebraic equations given by equations (1) through (4). For the equations of motion of a system containing closed kinematic loops, the interested reader may refer to reference [14] where more detail and further discussion is provided.

3 Kinematics of a Flexible Body

Let XYZ denote a fixed cartesian coordinate reference frame. In order to define the position and the orientation of a flexible body with respect to XYZ let a new coordinate system $\xi\eta\zeta$ be defined. This new coordinate system is rigidly attached to a point of the flexible body as depicted in Figure 2.

The motion of the flexible body involves a continuous change of its shape and generally large displacements and rotations. Let the initial configuration of the flexible body be denoted as configuration*0*. The current equilibrium configuration, which is generally unknown, is denoted by $t+\Delta t$. t denotes the last known equilibrium configuration. The motion of the flexible body is illustrated in figure 3.

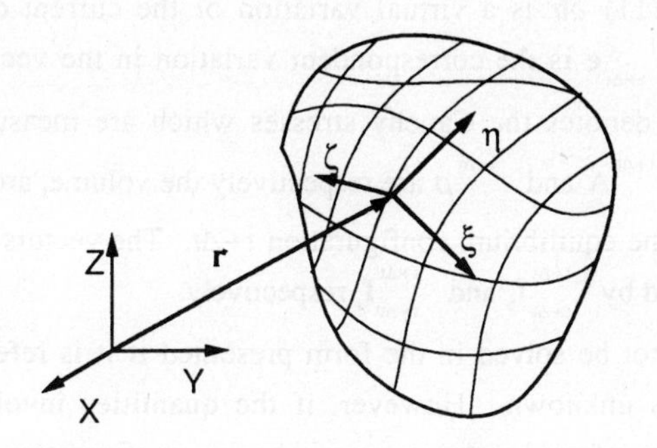

Figure 2 Inertial reference frame XYZ and body fixed coordinate system ξηζ

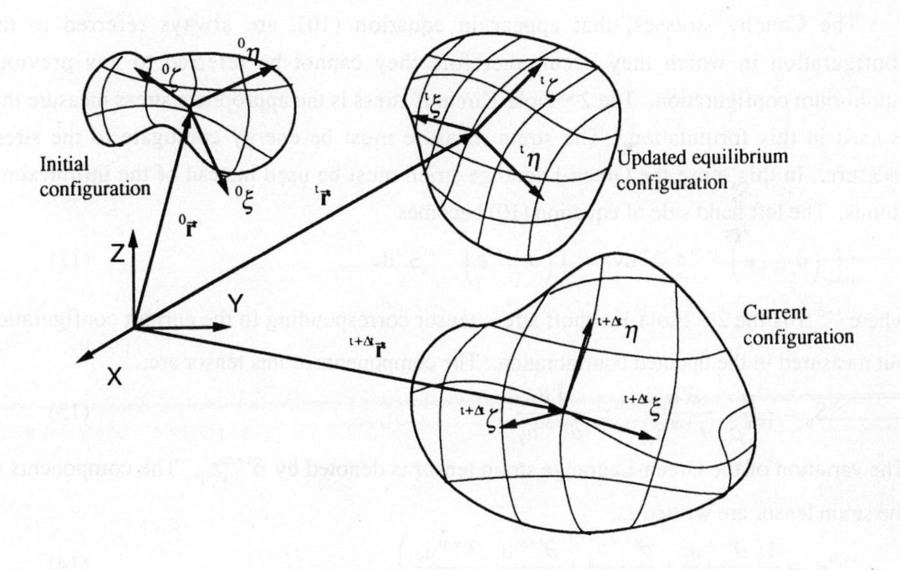

Figure 3 Gross motion of a flexible body.

The principle of the virtual displacements is used to express the equilibrium of the flexible body in the current configuration as

$$\int_{t+\Delta t_V} \left(\partial \,_{t+\Delta t}\mathbf{e} \right)^T \,^{t+\Delta t}\tau \,^{t+\Delta t}dv = \,^{t+\Delta t}\mathbf{R} \tag{10}$$

where the virtual work due to the external forces is given by:

$$ {}^{t+\Delta t}\mathbf{R} = \int_{{}^{t+\Delta t}V} {}^{t+\Delta t}\rho (\delta \mathbf{h})^{T}\, {}^{t+\Delta t}\ddot{\mathbf{h}}\, {}^{t+\Delta t}dv + \int_{{}^{t+\Delta t}V} {}^{t+\Delta t}\rho (\delta \mathbf{h})^{T}\, {}^{t+\Delta t}_{t+\Delta t}\mathbf{f}_{b}\, {}^{t+\Delta t}dv + \int_{{}^{t+\Delta t}A} (\delta \mathbf{h})^{T}\, {}^{t+\Delta t}_{t+\Delta t}\mathbf{f}_{s}\, {}^{t+\Delta t}da \tag{11} $$

In equations (10) and (11) $\delta\mathbf{h}$ is a virtual variation of the current displacements, i.e., $\delta\mathbf{h} = \delta\, {}^{t+\Delta t}\mathbf{h}$, while $\delta\, {}_{t+\Delta t}\mathbf{e}$ is the correspondent variation in the vector of infinitesimal strains. Vector ${}^{t+\Delta t}\tau$ denotes the Cauchy stresses which are measured in the current configuration. ${}^{t+\Delta t}V$, ${}^{t+\Delta t}A$ and ${}^{t+\Delta t}\rho$ are respectively the volume, area and mass density of the flexible body in the equilibrium configuration $t+\Delta t$. The vectors of body forces and surface forces are denoted by ${}^{t+\Delta t}_{t+\Delta t}\mathbf{f}_{b}$ and ${}^{t+\Delta t}_{t+\Delta t}\mathbf{f}_{s}$ respectively.

Equation (10) cannot be solved in the form presented if it is referred to the current configuration, which is unknown. However, if the quantities involved in the virtual displacement equation are referred to the previously known configuration, then a solution can be obtained. In practice, the two candidates for the reference configuration are the initial and the previously known configurations. These options lead to total and updated Lagrangean formulations. In this paper the updated Lagrangean formulation is adopted.

The Cauchy stresses, that appear in equation (10), are always referred to the configuration in which they occur, therefore they cannot be referred to any previous equilibrium configuration. The 2nd Piola-Kirchoff stress is the appropriate stress measure that is used in this formulation. The strain measure must be energy conjugate to the stress measure. In this sense the Green-Lagrange strain must be used instead of the infinitesimal strains. The left hand side of equation (10) becomes

$$ \int_{{}^{t+\Delta t}V} \left(\partial\, {}_{t+\Delta t}\mathbf{e}\right)^{T}\, {}^{t+\Delta t}\tau\, {}^{t+\Delta t}dv = \int_{{}^{t}V} \left(\partial\, {}^{t+\Delta t}_{t}\varepsilon\right)^{T}\, {}^{t+\Delta t}_{t}\mathbf{S}\, {}^{t}dv \tag{12} $$

where ${}^{t+\Delta t}_{t}\mathbf{S}$ is the 2nd Piola-Kirchoff stress tensor corresponding to the current configuration but measured in the updated configuration. The components of this tensor are:

$$ {}^{t+\Delta t}_{t}S_{ij} = \frac{{}^{t}\rho}{{}^{t+\Delta t}\rho} \frac{\partial {}^{t}h_{i}}{\partial {}^{t+\Delta t}h_{s}}\, {}^{t+\Delta t}\tau_{sr} \frac{\partial {}^{t}h_{j}}{\partial {}^{t+\Delta t}h_{r}} \tag{13} $$

The variation of the Green-Lagrange strain tensor is denoted by $\delta\, {}^{t+\Delta t}_{t}\varepsilon_{ij}$. The components of the strain tensor are written as:

$$ {}^{t+\Delta t}_{t}\varepsilon_{ij} = \frac{1}{2}\left(\frac{\partial {}^{t+\Delta t}u_{i}}{\partial {}^{t}u_{j}} + \frac{\partial {}^{t+\Delta t}u_{j}}{\partial {}^{t}u_{i}} + \frac{\partial {}^{t+\Delta t}u_{k}}{\partial {}^{t}h_{j}} \frac{\partial {}^{t+\Delta t}u_{k}}{\partial {}^{t}h_{i}} \right) \tag{14} $$

where the vector of displacements ${}^{t+\Delta t}\mathbf{u}$ is calculated as:

$$ {}^{t+\Delta t}\mathbf{u} = {}^{t+\Delta t}\mathbf{h} - {}^{t}\mathbf{h} \tag{15} $$

It has been shown by several researchers [3,6,16] that referring equation (12) to a corotational coordinate system is more efficient than the traditional updated Lagrangean formulation, in particular if finite elements with rotational degrees of freedom are used to discretize the body. For the purpose of the methodology derived here, the corotational reference frame is the body fixed coordinate frame $\xi\eta\zeta$. Furthermore, let a ghost configuration of the flexible body (equal to the updated configuration but rotated and translated as a rigid body) be associated with the body fixed coordinate frame. The updated ghost configuration, presented in figure 4, is referred to as t'.

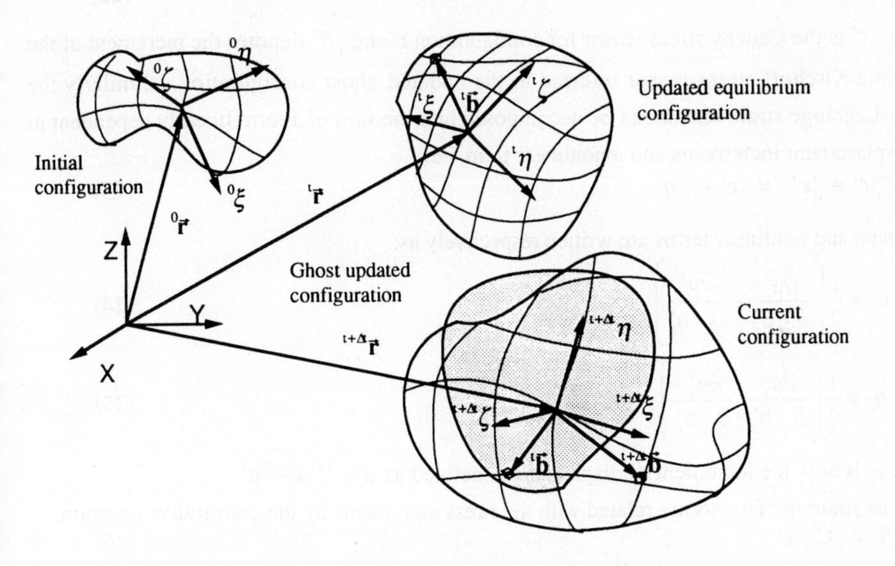

Figure 4 Ghost configuration associated with the body fixed coordinate system

Nygard and Bergan [16] show that the Green-Lagrange strain tensor components ${}^{t+\Delta t}_{t}\varepsilon_{ij}$ and the 2nd Piola-Kirchoff stress tensor components ${}^{t+\Delta t}_{t}S_{ij}$ measured in the current configuration but referred to the updated configuration are respectively equal to the components of the same tensors referred to the corotated ghost configuration and expressed in the body-fixed coordinate system, i.e.,

$$ {}^{t+\Delta t}_{t}\varepsilon_{ij} = {}^{t+\Delta t}_{t'}\varepsilon'_{ij} \tag{16}$$

$$ {}^{t+\Delta t}_{t}S_{ij} = {}^{t+\Delta t}_{t'}S'_{ij} \tag{17}$$

The strain and stress tensor components are written as:

$$ {}^{t+\Delta t}_{t'}\varepsilon'_{ij} = \frac{1}{2}\left(\frac{\partial\, {}^{t+\Delta t}u'_i}{\partial\, {}^{t'}b'_j} + \frac{\partial\, {}^{t+\Delta t}u'_j}{\partial\, {}^{t'}b'_i} + \frac{\partial\, {}^{t+\Delta t}u'_k}{\partial\, {}^{t'}b'_j}\frac{\partial\, {}^{t+\Delta t}u'_k}{\partial\, {}^{t'}b'_i} \right) \tag{18}$$

$$ {}^{t+\Delta t}_{t'}S'_{ij} = \frac{{}^{t}\rho}{{}^{t+\Delta t}\rho}\, \frac{\partial\, {}^{t}b'_i}{\partial\, {}^{t+\Delta t}b'_s}\, {}^{t+\Delta t}\tau'_{sr}\, \frac{\partial\, {}^{t}b'_j}{\partial\, {}^{t+\Delta t}b'_r} \tag{19}$$

where the displacement vector ${}^{t+\Delta t}\mathbf{u}$ is now defined as:

$$ {}^{t+\Delta t}\mathbf{u}' = {}^{t+\Delta t}\mathbf{b}' - {}^{t'}\mathbf{b}' \tag{20}$$

Substituting equations (12) through (19) into equation (10) gives:

$$\int_{{}^{t'}V} \left(\partial\ {}^{t+\Delta t}_{t'}\varepsilon' \right)^T\ {}^{t+\Delta t}_{t'}S'\ {}^{t'}dv\ =\ {}^{t+\Delta t}\mathbf{R} \qquad (21)$$

At this point let the 2nd Piola-Kirchoff stress tensor be decomposed into

$${}^{t+\Delta t}_{t'}\mathbf{S}'\ =\ {}^{t'}\tau'\ +\ {}_{t'}\mathbf{S}' \qquad (22)$$

where ${}^{t'}\tau'$ is the Cauchy stress tensor for configuration t' and ${}_{t'}\mathbf{S}'$ denotes the increment of the 2nd Piola-Kirchoff stress tensor referred to the updated ghost configuration. Similarly the Green-Lagrange strain tensor can be decomposed into the sum of a term linearly dependent in the displacement increments and a nonlinear term, i.e.,

$${}^{t+\Delta t}_{t'}\varepsilon' \equiv {}_{t'}\varepsilon'\ = {}_{t'}\mathbf{e}' + {}_{t'}\eta' \qquad (23)$$

The linear and nonlinear terms are written respectively as:

$${}_{t'}\mathbf{e}'_{ij} = \frac{1}{2}\left(\frac{\partial u'_i}{\partial\ {}^{t'}b'_j} + \frac{\partial u'_j}{\partial\ {}^{t'}b'_i} \right) \qquad (24)$$

$${}_{t'}\eta'_{ij} = \frac{1}{2}\left(\frac{\partial u'_k}{\partial\ {}^{t'}b'_j}\ \frac{\partial u'_k}{\partial\ {}^{t'}b'_i} \right) \qquad (25)$$

where $\mathbf{u}'$ is now the increment of displacement, defined as $\mathbf{u}' = {}^{t+\Delta t}\mathbf{u}' - {}^{t'}\mathbf{u}'$.

The strain increments are related with the stress increments by the constitutive equation:

$${}_{t'}\mathbf{S}' = {}_{t'}\mathbf{C}\ {}_{t'}\varepsilon' \qquad (26)$$

A constitutive equation nonlinear in the displacement increments leads to equations that have no direct solution. An approximate solution is obtained by assuming $\delta\ {}_{t'}\varepsilon' \approx\ \delta\ {}_{t'}\mathbf{e}'$. In addition, an approximate incremental constitutive equation is used:

$${}_{t'}\mathbf{S}' = {}_{t'}\mathbf{C}\ {}_{t'}\mathbf{e}' \qquad (27)$$

The equations of motion for the flexible body are obtained by substituting equations (22) through (27) into equation (21), i.e.,

$$\int_{{}^{t'}V} \left(\delta\ {}_{t'}\varepsilon' \right)^T {}_{t'}\mathbf{C}\ {}_{t'}\varepsilon'\ {}^{t'}dv + \int_{{}^{t'}V} \left(\delta\ {}_{t'}\eta' \right)^T {}_{t'}\tau'\ {}^{t'}dv = {}^{t+\Delta t}\mathbf{R} - \int_{{}^{t'}V} \left(\delta\ {}_{t'}\mathbf{e}' \right)^T {}_{t'}\tau'\ {}^{t'}dv \qquad (28)$$

The vector of the external forces is given by:

$${}^{t+\Delta t}\mathbf{R}\ =\ \int_{{}^{t'}V} {}^{t'}\rho\ {}^{t+\Delta t}\ddot{\mathbf{h}}\ {}^{t'}dv\ +$$

$$\int_{{}^{t'}V} {}^{t'}\rho\ {}^{t+\Delta t}_{t'}\mathbf{f}_b\ {}^{t'}dv\ +\ \int_{{}^{t'}A} {}^{t+\Delta t}_{t'}\mathbf{f}_s\ {}^{t'}da \qquad (29)$$

where it is assumed that the surface and body forces are independent of the configuration in which they are represented, i.e., the loading conditions are independent of the deformation.

A multibody may experience elasto-plastic deformations of one or more of its components. An example of this type of behavior is the structural impact and crash analysis of vehicles. For these problems an elasto-plastic constitutive tensor ${}_{t'}\mathbf{C}$ must be used in the equation (28). For the description of this tensor assume isotropic hardening and isothermal conditions. The material yield condition is written as:

$$f\left({}^{t'}\tau, {}^{t'}\kappa\right) = 0 \tag{30}$$

where ${}^{t'}\tau$ is the Cauchy stress tensor and ${}^{t'}\kappa$ is the hardening parameter (which is a function of the state of strain). Yielding occurs when equation (30) is satisfied. Any further strain increment will be partially elastic and partially plastic. These strain increments are related with the total strain increment by

$$d\ {}_{t'}\mathbf{e} = d\ {}_{t'}\mathbf{e}^{P} + d\ {}_{t'}\mathbf{e}^{E} \tag{31}$$

Furthermore, let associated plasticity be assumed. In these conditions Zienckiewicz [23] shows that the form of the elasto-plastic constitutive tensor is given by

$${}_{t'}\mathbf{C} = {}_{t'}\mathbf{C}^{E} - {}_{t'}\mathbf{C}^{E}\frac{\partial f}{\partial\,{}^{t'}\tau}\left(\frac{\partial f}{\partial\,{}^{t'}\tau}\right)^{T}{}_{t'}\mathbf{C}^{E}\left[H + \left(\frac{\partial f}{\partial\,{}^{t'}\tau}\right)^{T}{}_{t'}\mathbf{C}^{E}\frac{\partial f}{\partial\,{}^{t'}\tau}\right]^{-1} \tag{32}$$

The parameter H is the slope of the plot of the stress versus plastic strain for the uniaxial test if the Huber-Von Mises surface is used in equation (30). The constitutive equation obtained in this fashion is very similar to the one used in small displacement analysis.

4 Equations of Motion of a Flexible Body

In the formulation that follows it is assumed that isoparametric finite elements are used. Furthermore, the matrices describing the flexible body are obtained from the assemblage of the matrices of each finite element in the standard way. In this sense what follows is referred to each finite element.

Referring to figure 5, the position of a material point included inside the boundary of a finite element is written as:

$${}^{t+\Delta t}\mathbf{h} = \mathbf{r} + \mathbf{A}\ {}^{t+\Delta t}\mathbf{b}' \tag{33}$$

where $\mathbf{r}$ is the position of the origin of the body fixed coordinate system ${}^{t+\Delta t}(\xi\eta\zeta)$ associated with the current configuration. In equation (33) $\mathbf{A}$ denotes the transformation matrix from the body fixed coordinate system to the inertial frame XYZ. Since $\mathbf{r}$ and $\mathbf{A}$ are always associated with the current configuration, the left superscript $t+\Delta t$ is ignored for simplicity. The position of the point relative to the body fixed reference frame (denoted as ${}^{t+\Delta t}\mathbf{b}'$) can be decomposed in:

$${}^{t+\Delta t}\mathbf{b}' = {}^{t'}\mathbf{b}' + \mathbf{v}' \tag{34}$$

where $\mathbf{v}'$ is the increment in displacement of the point, from the previously known ghost configuration to the current configuration.

For isoparametric finite elements the coordinates and displacement fields are interpolated by the same shape functions. Denoting by ${}^{t'}\mathbf{x}'$ and $\mathbf{u}'$ the nodal positions and displacements respectively, the position and displacements of a material point in the finite element are given by

$$ {}^{t'}\mathbf{b} = \mathbf{N}\ {}^{t'}\mathbf{x}' \tag{35} $$

$$ \mathbf{v}' = \mathbf{N}\ \mathbf{u}' \tag{36} $$

where $\mathbf{N}$ is a matrix of shape functions. Substituting equations (35) and (36) into equation (34) gives the position of a material point inside the finite element in terms of the nodal displacements and the shape functions as:

$$ {}^{t+\Delta t}\mathbf{h} = \mathbf{r} + \mathbf{A}\mathbf{N}\left({}^{t'}\mathbf{x}' + \mathbf{u}'\right) \tag{37} $$

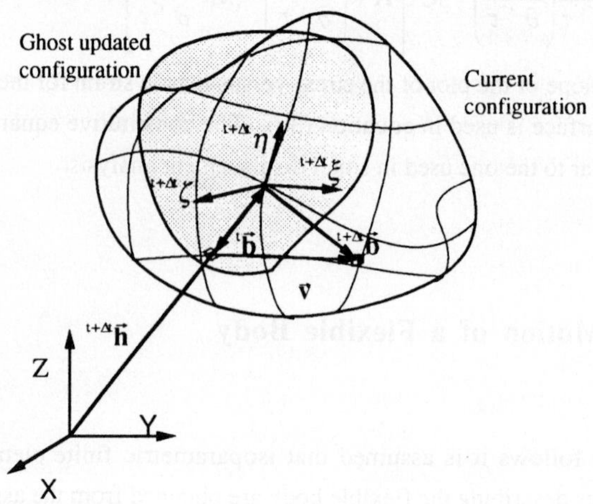

Figure 5 Position of a point in a flexible body

Equation (37) can now be used to evaluate the virtual displacement and acceleration of the point. Substituting these results into equation (28) leads to the equations of motion for a single finite element. For details of this procedure the interested reader is referred to [1]. Assembling the finite element contributions to the flexible body equations of motion results in

$$ \begin{bmatrix} \mathbf{M}_{rr} & \mathbf{M}_{r\phi} & \mathbf{M}_{rf} \\ \mathbf{M}_{\phi r} & \mathbf{M}_{\phi\phi} & \mathbf{M}_{\phi f} \\ \mathbf{M}_{fr} & \mathbf{M}_{f\phi} & \mathbf{M}_{ff} \end{bmatrix} \begin{bmatrix} \ddot{\mathbf{r}} \\ \dot{\omega}' \\ \ddot{\mathbf{u}}' \end{bmatrix} = \begin{bmatrix} \mathbf{g}_r \\ \mathbf{g}'_\phi \\ \mathbf{g}'_f \end{bmatrix} - \begin{bmatrix} \mathbf{s}_r \\ \mathbf{s}'_\phi \\ \mathbf{s}'_f \end{bmatrix} - \begin{bmatrix} \mathbf{0} \\ \mathbf{0} \\ {}^{t}_{t'}\mathbf{F} \end{bmatrix} - \begin{bmatrix} \mathbf{0} & \mathbf{0} & \mathbf{0} \\ \mathbf{0} & \mathbf{0} & \mathbf{0} \\ \mathbf{0} & \mathbf{0} & {}^{t}_{t'}\mathbf{K}_L + {}^{t}_{t'}\mathbf{K}_{NL} \end{bmatrix} \begin{bmatrix} \mathbf{0} \\ \mathbf{0} \\ \mathbf{u}' \end{bmatrix} \tag{38} $$

where the coefficients of the mass matrix are given by

$$ \mathbf{M}_{rr} = \mathbf{I} \int_{{}^0V} {}^0\rho\ {}^0dv \tag{39a} $$

$$ \mathbf{M}_{r\phi} = -\mathbf{A} \int_{{}^0V} {}^0\rho \tilde{\mathbf{b}}'\ {}^0dv \tag{39b} $$

$$\mathbf{M}_{rf} = \mathbf{A}\int_{{}^0V} {}^0\rho \mathbf{N}\ {}^0dv \tag{39c}$$

$$\mathbf{M}_{\phi\phi} = -\int_{{}^0V} {}^0\rho \tilde{\mathbf{b}}'\tilde{\mathbf{b}}'\ {}^0dv \tag{39d}$$

$$\mathbf{M}_{\phi f} = \int_{{}^0V} {}^0\rho \tilde{\mathbf{b}}'\mathbf{N}\ {}^0dv \tag{39e}$$

$$\mathbf{M}_{ff} = \int_{{}^0V} {}^0\rho \mathbf{N}^T\mathbf{N}\ {}^0dv \tag{39f}$$

In these equations the left superscript $t+\Delta t$ of vector **b** is omitted for simplicity. Note that with the exception of $\mathbf{M}_{rr}$ and $\mathbf{M}_{ff}$ the coefficients are not constant because, not only the transformation matrix **A** is not constant, but also the position vector **b** is evaluated at the current configuration. The vector of unknown accelerations is composed by the acceleration of the origin of the body fixed coordinate system $\ddot{\mathbf{r}}$, the angular acceleration of ξηζ denoted by $\dot{\omega}'$, and a vector of nodal accelerations with respect to the body fixed coordinate frame,.$\ddot{\mathbf{u}}'$. The right-hand side of equation (38) is composed by a vector of externally applied forces/moments $\begin{bmatrix}\mathbf{g}_r^T & \mathbf{g}_\phi'^T & \mathbf{g}_f'^T\end{bmatrix}^T$, a vector of gyroscopic forces $\begin{bmatrix}\mathbf{s}_r^T & \mathbf{s}_\phi'^T & \mathbf{s}_f'^T\end{bmatrix}^T$, and internal forces due to the deformation of the flexible body. The vector of the external applied forces/moments is evaluated over the updated configuration and it is written as:

$$_{t'}\mathbf{g} = \begin{bmatrix} \int_{{}^{t'}A} {}^{t+\Delta t}_{t'}\mathbf{f}_s\ {}^{t'}da + \int_{{}^{t'}V} {}^0\rho\ {}^{t+\Delta t}_{t'}\mathbf{f}_b\ {}^{t'}dv \\ \int_{{}^{t'}A} \tilde{\mathbf{b}}'\mathbf{A}^T\ {}^{t+\Delta t}_{t'}\mathbf{f}_s\ {}^{t'}da + \int_{{}^{t'}V} {}^0\rho\tilde{\mathbf{b}}'\mathbf{A}^T\ {}^{t+\Delta t}_{t'}\mathbf{f}_b\ {}^{t'}dv \\ \int_{{}^{t'}A} \mathbf{N}^T\mathbf{A}^T\ {}^{t+\Delta t}_{t'}\mathbf{f}_s\ {}^{t'}da + \int_{{}^{t'}V} {}^0\rho\mathbf{N}^T\mathbf{A}^T\ {}^{t+\Delta t}_{t'}\mathbf{f}_b\ {}^{t'}dv \end{bmatrix} \tag{40}$$

The vector of gyroscopic forces is written as:

$$\mathbf{s} = \begin{bmatrix} \mathbf{A}\tilde{\omega}'\tilde{\omega}'\int_{{}^0V} {}^0\rho\mathbf{b}'\ {}^0dv \\ \int_{{}^0V} {}^0\rho\tilde{\mathbf{b}}'\tilde{\omega}'\tilde{\omega}'\mathbf{b}'\ {}^0dv \\ \int_{{}^0V} {}^0\rho\mathbf{N}^T\tilde{\omega}'\tilde{\omega}'\mathbf{b}'\ {}^0dv \end{bmatrix} + 2\begin{bmatrix} \mathbf{A}\tilde{\omega}'\int_{{}^0V} {}^0\rho\mathbf{N}\ {}^0dv \\ \int_{{}^0V} {}^0\rho\tilde{\mathbf{b}}'\tilde{\omega}'\mathbf{N}\ {}^0dv \\ \int_{{}^0V} {}^0\rho\mathbf{N}^T\tilde{\omega}'\mathbf{N}\ {}^0dv \end{bmatrix}\dot{\mathbf{u}}' \tag{41}$$

In equation (38) vector $\mathbf{u}'$ denotes the increments of displacements from the updated configuration to the current configuration, matrices ${}^t_{t'}\mathbf{K}_L$ and ${}^t_{t'}\mathbf{K}_{NL}$ are respectively the linear and nonlinear stiffness matrices, and ${}^t_{t'}\mathbf{F}$ denotes the vector of equivalent nodal forces due to the actual state of stress. These quantities are given by:

$${}^t_{t'}\mathbf{K}_L = \int_{{}^{t'}V} {}^t_{t'}\mathbf{B}_L^T\ {}_{t'}\mathbf{C}\ {}^t_{t'}\mathbf{B}_L\ {}^{t'}dv \tag{42}$$

$${}^t_{t'}\mathbf{K}_{NL} = \int_{{}^{t'}V} {}^t_{t'}\mathbf{B}_{NL}^T\ {}^t_{t'}\tau'\ {}^t_{t'}\mathbf{B}_{NL}\ {}^{t'}dv \tag{43}$$

$${}^t_{t'}\mathbf{F} = \int_{{}^{t'}V} {}^t_{t'}\mathbf{B}_L^T\ {}^t_{t'}\hat{\tau}'\ {}^{t'}dv \tag{44}$$

In these equations ${}_{t'}^{t}\mathbf{B}_L^T$ and ${}_{t'}^{t}\mathbf{B}_{NL}^T$ denote the linear and nonlinear strain matrices respectively and ${}_{t'}^{t}\tau'$ is the Cauchy stress tensor for the updated configuration. It should be noted that the reference to the linearity of the stiffness matrices ${}_{t'}^{t}\mathbf{K}_L$ and ${}_{t'}^{t}\mathbf{K}_{NL}$ is related to their relation with the displacements. If the constitutive tensor ${}_{t'}\mathbf{C}$ is not linear then both ${}_{t'}^{t}\mathbf{K}_L$ and ${}_{t'}^{t}\mathbf{K}_{NL}$ are not linear.

Equation (38) describes thoroughly the motion of a flexible body. However the form of this equation is not efficient for numerical implementation because not only all the quantities of the right-hand side are not constant but also the mass matrix is variant. A simpler form of the equations of motion for a flexible body is obtained if (a) a lumped mass formulation is used and (b) the accelerations $\ddot{\mathbf{u}}'$ are substituted by a vector of nodal accelerations relative to the nonmoving reference frame $\ddot{\mathbf{q}}_f'$ [1].

The vectors of nodal accelerations can be partitioned into translational and angular accelerations as:

$$\ddot{\mathbf{u}}' = \begin{bmatrix} \ddot{\delta}' \\ \ddot{\theta}' \end{bmatrix} \quad ; \quad \ddot{\mathbf{q}}_f' = \begin{bmatrix} \ddot{\mathbf{d}}' \\ \alpha' \end{bmatrix}$$

The relation between the relative and absolute nodal accelerations for a node k is described by:

$$\begin{bmatrix} \ddot{\delta}' \\ \ddot{\theta}' \end{bmatrix}_k = \begin{bmatrix} \ddot{\mathbf{d}}' \\ \alpha' \end{bmatrix}_k - \begin{bmatrix} \mathbf{A}^T & -\left(\tilde{\mathbf{x}}^k + \tilde{\delta}^k\right)' \\ \mathbf{0} & \mathbf{I} \end{bmatrix} \begin{bmatrix} \ddot{\mathbf{r}} \\ \dot{\omega}' \end{bmatrix} - \begin{bmatrix} \tilde{\omega}'\tilde{\omega}'\left(\mathbf{x}^k + \delta^k\right)' + 2\tilde{\omega}'\left(\dot{\delta}^k\right)' \\ \tilde{\omega}'\left(\dot{\theta}^k\right)' \end{bmatrix} \tag{45}$$

Evaluating equation (45) for all nodes of the flexible body and substituting into equation (38) gives

$$\sum_{k=1}^{n} \left(m\ddot{\mathbf{d}}'\right)_k = \mathbf{g}_r \tag{46a}$$

$$\sum_{k=1}^{n} \left[m\left(\tilde{\mathbf{x}} + \tilde{\delta}\right)' \ddot{\mathbf{d}}' \right]_k = \mathbf{g}_\theta' \tag{46b}$$

$$\mathbf{M}_{ff}\ddot{\mathbf{q}}_f' = \mathbf{g}_f' - {}_{t'}^{t}\mathbf{F} - \left({}_{t'}^{t}\mathbf{K}_L + {}_{t'}^{t}\mathbf{K}_{NL}\right)\mathbf{u}' \tag{46c}$$

Equations (46a) and (46b) are the equations of motion for the center of mass of a system of particles [1]. Equation (46c) is the equation of motion for the nodes of the flexible body, expressed in the body fixed coordinate system. Note that due to the use of the lumped mass formulation the mass matrix $\mathbf{M}_{ff}$ is diagonal and written as:

$$\mathbf{M}_{ff} = \mathrm{Diag}(m_1\mathbf{I}, \mathbf{0}, \cdots, m_k\mathbf{I}, \mathbf{0}, \cdots, m_n\mathbf{I}, \mathbf{0})$$

where m_k is the lumped mass of node k, and $\mathbf{I}$ and $\mathbf{0}$ are 3x3 identity and null matrices associated with the translational and rotational degrees of freedom respectively.

Partially Flexible Body

If the origin of the body fixed coordinate system is coincident with the center of mass of the flexible body, equations (46a) and (46b) are the equations of motion of the origin of referencial

$\xi\eta\zeta$. Very often it is useful to locate the origin in some other point of the flexible body. For this purpose let it be assumed that the flexible body has one rigid part and one flexible part. Let the body fixed coordinate frame be attached to the center of mass of the rigid part as shown in figure 6. The flexible part is attached to the rigid part by the nodes that belong to boundary ψ and the body-fixed coordinate frame is the same for the rigid and flexible parts.

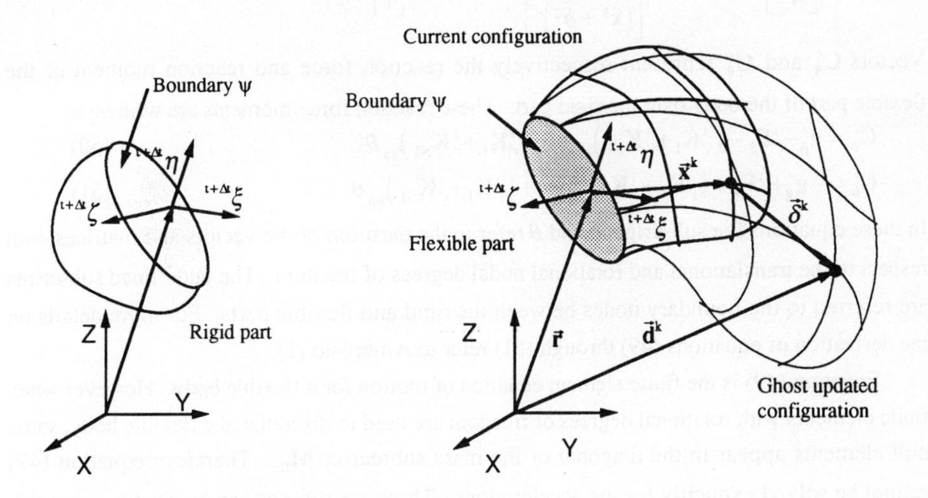

Figure 6 Flexible body with a rigid part

The the Newton-Euler equations of motion for a rigid body are written as

$$m\ddot{\mathbf{r}} = \mathbf{f} \tag{47a}$$

$$\mathbf{J}'\dot{\omega}' = \mathbf{n}' - \tilde{\omega}'\mathbf{J}\omega' \tag{47b}$$

where **f** and **n** are the external forces and moments applied over the center of mass of the rigid part of the body. Equations (47a) and (47b) can be used instead of (46a) and (46b) provided that proper kinematic constraints are introduced between the flexible and rigid parts of the body. These kinematic constraints, that only affect the nodes in the boundary ψ, are described by:

$$\delta' = \dot{\delta}' = \ddot{\delta}' = \theta' = \dot{\theta}' = \ddot{\theta}' = \mathbf{0} \tag{48}$$

These constraint equations can be applied to the equation (45) for each of the boundary nodes and the result introduced into equations (46c) throught (47b) using a Lagrange multiplier technique. At a latter step these multipliers are eliminated from the equations of motion [2] which results in

$$\begin{bmatrix} m + \overline{\mathbf{A}}\underline{\mathbf{M}}^{*}\overline{\mathbf{A}}^{\mathrm{T}} & -\overline{\mathbf{A}}\underline{\mathbf{M}}^{*}\mathbf{S} & \mathbf{0} \\ -(\overline{\mathbf{A}}\underline{\mathbf{M}}^{*}\mathbf{S})^{\mathrm{T}} & \mathbf{J}' + \mathbf{S}^{\mathrm{T}}\underline{\mathbf{M}}^{*}\mathbf{S} & \mathbf{0} \\ \mathbf{0} & \mathbf{0} & \mathbf{M}_{\mathrm{ff}} \end{bmatrix} \begin{bmatrix} \ddot{\mathbf{r}} \\ \dot{\omega}' \\ \ddot{\mathbf{q}}'_{\mathrm{f}} \end{bmatrix} = \begin{bmatrix} \mathbf{f} + \overline{\mathbf{A}}\mathbf{C}'_{\delta} \\ \mathbf{n}' - \tilde{\omega}'\mathbf{J}'\omega' - \mathbf{S}^{\mathrm{T}}\mathbf{C}'_{\delta} - \overline{\mathbf{I}}^{\mathrm{T}}\mathbf{C}'_{\theta} \\ \mathbf{g}'_{\mathrm{f}} - {}^{t}_{t'}\mathbf{F} - \left({}^{t}_{t'}\mathbf{K}_{\mathrm{L}} + {}^{t}_{t'}\mathbf{K}_{\mathrm{NL}}\right)\mathbf{u}' \end{bmatrix} \tag{49}$$

where $\underline{\mathbf{M}}^*$ is a diagonal mass matrix containing the mass of the $\underline{n}$ boundary nodes. Matrices $\overline{\mathbf{A}}^{\mathrm{T}}$, $\mathbf{S}$ and $\bar{\mathbf{I}}$ are made from (3x3) matrices as:

$$\overline{\mathbf{A}}^{\mathrm{T}} = \begin{bmatrix} \mathbf{A}^{\mathrm{T}} \\ \mathbf{A}^{\mathrm{T}} \\ \vdots \\ \mathbf{A}^{\mathrm{T}} \end{bmatrix} \quad ; \quad \mathbf{S} = \begin{bmatrix} \left(\tilde{\mathbf{x}}^{1} + \tilde{\delta}^{1}\right)' \\ \left(\tilde{\mathbf{x}}^{2} + \tilde{\delta}^{2}\right)' \\ \vdots \\ \left(\tilde{\mathbf{x}}^{\underline{n}} + \tilde{\delta}^{\underline{n}}\right)' \end{bmatrix} \quad ; \quad \bar{\mathbf{I}} = \begin{bmatrix} \mathbf{I} \\ \mathbf{I} \\ \vdots \\ \mathbf{I} \end{bmatrix}$$

Vectors $\mathbf{C}'_{\delta}$ and $\mathbf{C}'_{\theta}$ represent respectively the reaction force and reaction moment of the flexible part of the body over the rigid part. These reaction force/moments are written as

$$\mathbf{C}'_{\delta} = \mathbf{g}'_{\underline{\delta}} - {}^{\,t}_{t'}\mathbf{F}_{\underline{\delta}} - \left({}^{\,t}_{t'}\mathbf{K}_{\mathrm{L}} + {}^{\,t}_{t'}\mathbf{K}_{\mathrm{NL}}\right)_{\underline{\delta}\delta} \delta' - \left({}^{\,t}_{t'}\mathbf{K}_{\mathrm{L}} + {}^{\,t}_{t'}\mathbf{K}_{\mathrm{NL}}\right)_{\underline{\delta}\theta} \theta' \tag{50}$$

$$\mathbf{C}'_{\theta} = -\mathbf{g}'_{\underline{\theta}} + {}^{\,t}_{t'}\mathbf{F}_{\underline{\theta}} + \left({}^{\,t}_{t'}\mathbf{K}_{\mathrm{L}} + {}^{\,t}_{t'}\mathbf{K}_{\mathrm{NL}}\right)_{\underline{\theta}\delta} \delta' + \left({}^{\,t}_{t'}\mathbf{K}_{\mathrm{L}} + {}^{\,t}_{t'}\mathbf{K}_{\mathrm{NL}}\right)_{\underline{\theta}\theta} \theta' \tag{51}$$

In these equations the subscripts δ and θ refer to the partition of the vectors and matrices with respect to the translational and rotational nodal degrees of freedom. The underlined subscripts are referred to the boundary nodes between the rigid and flexible parts. For more details on the derivation of equations (49) through (51) refer to Ambrósio [2].

Equation (49) is the finite element equation of motion for a flexible body. However when finite elements with rotational degrees of freedom are used to discretize the flexible body, some null elements appear in the diagonal of the mass submatrix $\mathbf{M}_{\mathrm{ff}}$. Therefore equation (49) cannot be solved explicitly for the accelerations. Three approaches can be used to solve this problem. In the first approach rotational inertias obtained by lumping the off-diagonal terms of the consistent mass matrix $\mathbf{M}_{\mathrm{ff}}$ are used to replace the null coefficients. In the second approach a static condensation of the nodal rotational degrees of freedom is used. In a third approach the modal superposition technique is used to eliminate the explicit use of nodal rotations In what follows, any reference to the use of equation (49) implies the use of the first approach. The second and third approaches will be discussed next.

Static Condensation of Nodal Rotations

In order to use the static condensation of the rotational degrees of freedom let the nodal equations of motion be partitioned into translational and rotational degrees of freedom. The relation between the translational degrees of freedom and the rotational coordinates is described by

$$\theta' = {}^{\,t}_{t'}\mathbf{K}^{-1}_{\theta\theta}\left(\mathbf{g}'_{\theta} - {}^{\,t}_{t'}\mathbf{F}_{\theta} - {}^{\,t}_{t'}\mathbf{K}_{\theta\delta}\delta'\right) \tag{52}$$

Applying equation (52) to equation (49) results in the equations of motion of the reduced system, i.e., without the explicit use of the rotational degrees of freedom. These equations are written as:

$$\begin{bmatrix} m+\overline{\mathbf{A}}\underline{\mathbf{M}}^{*}\overline{\mathbf{A}}^{\mathrm{T}} & -\overline{\mathbf{A}}\underline{\mathbf{M}}^{*}\mathbf{S} & \mathbf{0} \\ -(\overline{\mathbf{A}}\underline{\mathbf{M}}^{*}\mathbf{S})^{\mathrm{T}} & \mathbf{J}'+\mathbf{S}^{\mathrm{T}}\underline{\mathbf{M}}^{*}\mathbf{S} & \mathbf{0} \\ \mathbf{0} & \mathbf{0} & \mathbf{M}_{\delta\delta} \end{bmatrix} \begin{bmatrix} \ddot{\mathbf{r}} \\ \dot{\omega}' \\ \ddot{\delta}' \end{bmatrix} =$$

$$\begin{bmatrix} \mathbf{f}+\overline{\mathbf{A}}\mathbf{C}'_{\delta} \\ \mathbf{n}'-\tilde{\omega}'\mathbf{J}'\omega'-\mathbf{S}^{\mathrm{T}}\mathbf{C}'_{\delta}-\overline{\mathbf{I}}^{\mathrm{T}}\mathbf{C}'_{\theta} \\ \mathbf{g}'_{\delta}-{}^{t}_{t'}\mathbf{F}_{\delta}-{}^{t}_{t'}\mathbf{K}_{\delta\theta}\,{}^{t}_{t'}\mathbf{K}^{-1}_{\theta\theta}\left(\mathbf{g}'_{\theta}-{}^{t}_{t'}\mathbf{F}_{\theta}\right)-\left({}^{t}_{t'}\mathbf{K}_{\delta\delta}-{}^{t}_{t'}\mathbf{K}_{\delta\theta}\,{}^{t}_{t'}\mathbf{K}^{-1}_{\theta\theta}\,{}^{t}_{t'}\mathbf{K}_{\theta\delta}\right)\delta' \end{bmatrix} \quad (53)$$

By a proper choice for the location and orientation of the body fixed coordinate system in the rigid part of the flexible body, the mass matrix in equations (49) and (53) is turned into a diagonal invariant matrix. For this purpose the position of its origin must be coincident with the center of mass of the rigid part plus the boundary nodes of the flexible part of the body. Furthermore the coordinate system must be aligned with the principal directions of inertia of the rigid part plus boundary nodes.

Modal Superposition Technique

In order to achieve computational efficiency in the solution of the flexible body equations of motion, the modal superposition technique has been widely used [17,18]. This method is well suited to reduce the number of degrees of freedom of a flexible body when the mass and stiffness matrix are time invariants and the frequency contents of the external applied forces are of the same order as the lower natural frequencies of the flexible body. This procedure can still be applied for cases where the stiffness matrix shows some level of nonlinearity. Assume that the stiffness matrix is decomposed into an invariant matrix and a displacement dependent matrix. For cases where the material constitutive tensor is constant (linear elastic material) the constant stiffness matrix is ${}^{t}_{t'}\mathbf{K}_{L}$ while the displacement dependent matrix is ${}^{t}_{t'}\mathbf{K}_{NL}$. Moreover, assume that the first two rows of equation (49) or equation (53) have been solved for $\ddot{\mathbf{r}}$ and $\dot{\omega}'$.

Substituting the relation between the global nodal accelerations and the nodal accelerations relative to the body fixed coordinate system, given by equation (45), into the third row of equation (49) gives

$$\mathbf{M}_{ff}\ddot{\mathbf{u}}'+{}^{t}_{t'}\mathbf{K}_{L}\,{}^{t}\mathbf{u}'=\mathbf{g}'_{f}-{}^{t}_{t'}\mathbf{F}-{}^{t}_{t'}\mathbf{K}_{NL}\mathbf{u}'-\mathbf{f}_{\varepsilon} \quad (54)$$

where vector $\mathbf{f}_{\varepsilon}$ represents the inertia forces due to the substitution of global nodal accelerations by local accelerations. This vector is written as:

$$\mathbf{f}_{\varepsilon}=\begin{bmatrix} \mathbf{M}^{*}\left(\overline{\mathbf{A}}^{\mathrm{T}}\ddot{\mathbf{r}}-\mathbf{S}\dot{\omega}'-\mathbf{W}_{2}-2\mathbf{W}_{1}\dot{\delta}'\right) \\ \mathbf{0} \end{bmatrix} \quad (55)$$

In this equation $\mathbf{W}_1$ and $\mathbf{W}_2$ are expressed as

$$\mathbf{W}_{1}=\mathrm{Diag}(\tilde{\omega}',\tilde{\omega}',\cdots,\tilde{\omega}')$$

$$\mathbf{W}_2 = \begin{bmatrix} \tilde{\omega}'\tilde{\omega}'\left(\mathbf{x}^1 + \delta^1\right)' \\ \tilde{\omega}'\tilde{\omega}'\left(\mathbf{x}^2 + \delta^2\right)' \\ \vdots \\ \tilde{\omega}'\tilde{\omega}'\left(\mathbf{x}^n + \delta^n\right)' \end{bmatrix}$$

The solution of the eigenproblem, posed by equating the right-hand side of equation (54) to zero, is a set of natural frequencies and corresponding modes of vibration for the flexible body. The nodal displacements can be expressed as a linear combination of the modes of vibration, i.e.

$$'\mathbf{u}' = \mathbf{X}\mathbf{z} \tag{56}$$

where $\mathbf{X}$ is the modal matrix. The number of modes of vibration envolved in equation (56) is n_m which is normally much smaller than the number of nodal degrees of freedom of the flexible body. Once the modes of vibration are not time dependent, the modal accelerations and velocities are given by

$$\ddot{\mathbf{u}}' = \mathbf{X}\ddot{\mathbf{z}} \tag{57}$$

$$\dot{\mathbf{u}}' = \mathbf{X}\dot{\mathbf{z}} \tag{58}$$

Equations (56) through (58) are now substituted into equation (54) and the result pre-multiplied by $\mathbf{X}^T$. Using the property of orthonormality of the modal matrix with the mass matrix it is found that

$$\ddot{\mathbf{w}} = \mathbf{X}^T\left(\mathbf{g}_f' - {}_{t}'\mathbf{F} - {}_{t}'\mathbf{K}_{NL}\mathbf{u}' - \mathbf{f}_\varepsilon\right) - \Lambda\mathbf{w} \tag{59}$$

where Λ is a diagonal matrix with the squares of the natural frequencies. Equation (59) is the modal equation of motion for the flexible body. The complete set of equations of motion for the flexible body is composed by the two first rows of equation (49) and equation (59).

Application Example

The problem of a canteliver beam attached to a rigid hub, which is spun up from rest to a constant angular speed, is analyzed here. This problem, first proposed by Kane et al. [9] is studied in order to show the preformance of the methodologies presented, namely to show the difference between the application of the different types of coordinates used to describe the deformations of the flexible body.

The canteliver beam, with a lenght of 10 meter and annular cross section is presented in figure 7. The angular speed of the hub is a function of time prescribed as:

$$\omega(t) = \begin{cases} \dfrac{6}{15}\left[t - \dfrac{15}{2\pi}\sin\left(\dfrac{2\pi t}{15}\right)\right] & \text{rad/s} \quad 0 \le t \le 15 \\ 6 & \text{rad/s} \quad t \ge 15 \end{cases}$$

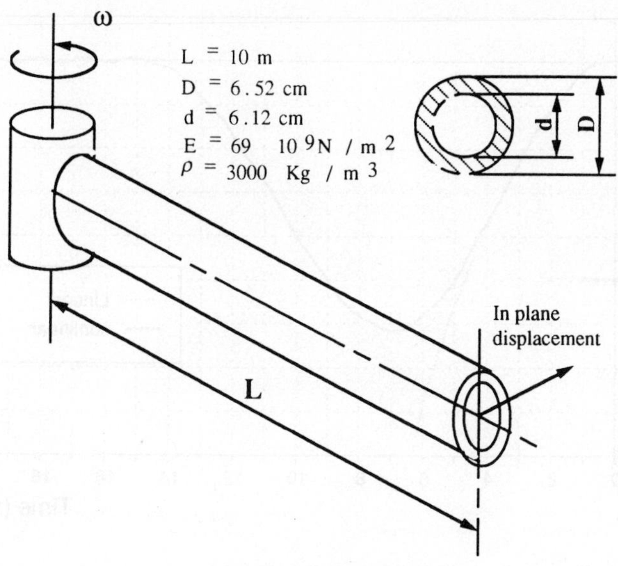

Figure 7 Rotating beam

The results presented in figure 8 show that if a linear behavior is assumed for the beam, i.e., the geometric stiffness is neglected and the deformations are small, the tip displacement of the beam with respect to the body fixed coordinates becomes infinite after 7 seconds of simulation. If equation (53) is used to represent the flexible body, the results are similar to those obtained by Kane et al., i.e., the tip displacement increases while the angular acceleration of the hub is increasing. The tip of the beam ends up oscillating about its undeformed position after the angular speed becomes constant.

Both linear and nonlinear simulations were performed on a DECSTATION 5100 with the three forms of the equations of motion previously derived. The different types of equations of motion are referred here as: equations with no coordinate reduction for equations (49); statically condensed equations for equations (53); and modal equations of motion for equations (59). When a linear behavior for the beam was assumed all forms of the equations of motion provided the same behavior. When the beam was allowed to show a geometric nonlinear behavior the solution of the equations of motion based in the modal superposition had an error of 10% relative to the results obtained with the equations of motion with static condensation or with no coordinate reduction. The computational time spent to simulate the linear behavior of the rotating beam was 13m 15s when no coordinate reduction was used, 13m 56s with static reduction of the nodal rotational degrees of freedom, and 15s for the equations of motion using the modal superposition. For the simulations of the nonlinear behavior of the beam the computational time obtained was 14h 46m, 19h 25m and 15m 12s for each form of the equations of motion respectively.

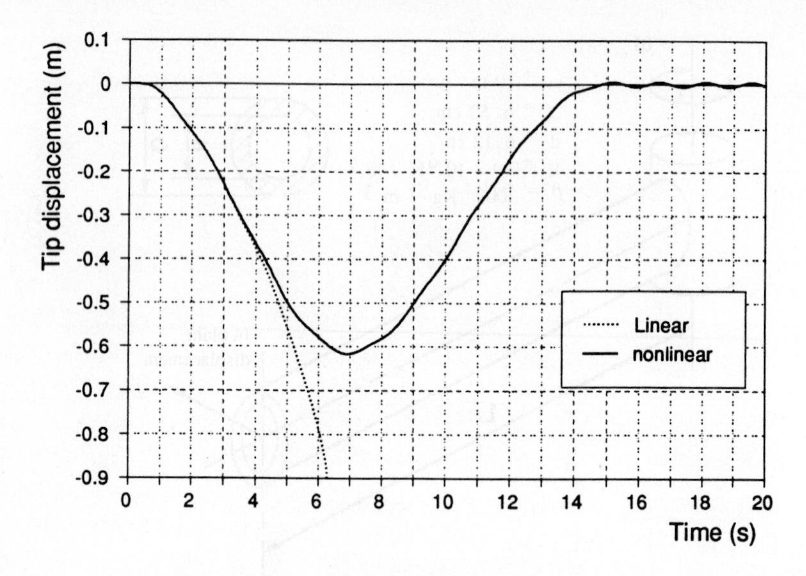

Figure 8 In plane displacement of the tip of the rotating beam with respect to the underformed position

5 Application to a Vehicle Rollover

The vehicle simulated here is an utility truck. Originally this vehicle did not have any protection in case of a rollover. In order to provide that extra protection for passengers, a rollbar cage was attached to the chassis of the truck. The model of the vehicle, excluding the rollbar cage, consists of the main chassis, the complete suspension system, and four wheels. The front wheels are connected to the main chassis by unequal A-arms (double wishbones). The rear wheels are connected to the main chassis by semi-trailing arms. Suspension springs, shock absorbers, and jounce stops are modeled by point-to-point spring-damper elements with nonlinear characteristics. The vehicle model consists of fifteen joint coordinates, equal to the number of degrees of freedom of the system. Six degrees of freedom correspond to the main chassis, four to the four suspension systems, four to the rolling wheels, and one to the steering.

The rollbar cage is a flexible frame mounted over the chassis to protect the passengers in case of a rollover. The actual assembly of the rollbars on the truck is shown in figure 9. The rollbars are made of 1025-1030 steel with a yield strength of 30,000 psi. The cross-sectional area of each bar is annular with an inside radius of 2.14 cm and an outside radius of 2.54 cm.

The finite element model of the cage is composed of 13 beam elements and 12 nodes. In order to simulate the attachment of the cage to the chassis, 6 of the nodes are fixed to body 1. This leads to a finite element model with 36 degrees of freedom.

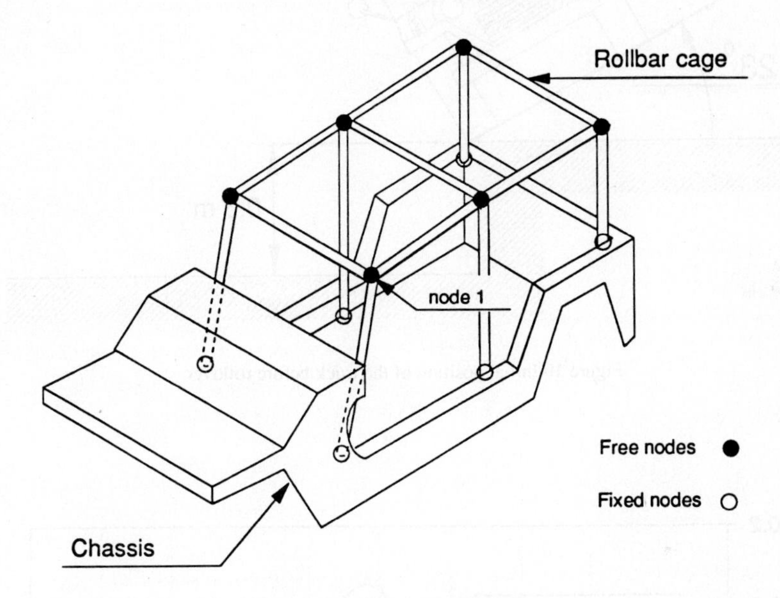

Figure 9 Rollbar cage and chassis

A rollover test for the truck was performed by placing the vehicle over a cart moving at a speed of 30 m.p.h. and impacting a water filled decelerator system, thereby throwing the truck off the cart. The initial roll angle was 23 degrees, and the height of release was 30 cm as shown in figure 10. In order to maintain the total kinetic energy of the vehicle approximately the same as in the experimental test, an initial speed for the truck at the time of departure is assumed to be 25 m.p.h. plus a angular velocity of 1.5 rad/s in the roll direction.

In order to evaluate the performance of the methodology described in this paper, the results of the simulation are compared against the results obtained using a kinetostatic method [15]. Figure 11 shows the vertical displacement of the center of the chassis for both methods. In the kinetostatic method only a linear elastic material behavior is considered. For comparison the first simulation was run with a similar material behavior for the rollbar cage. A second simulation was performed using a material with the yield strength referred above and a tangential plastic modulus given as E^T=E/10.

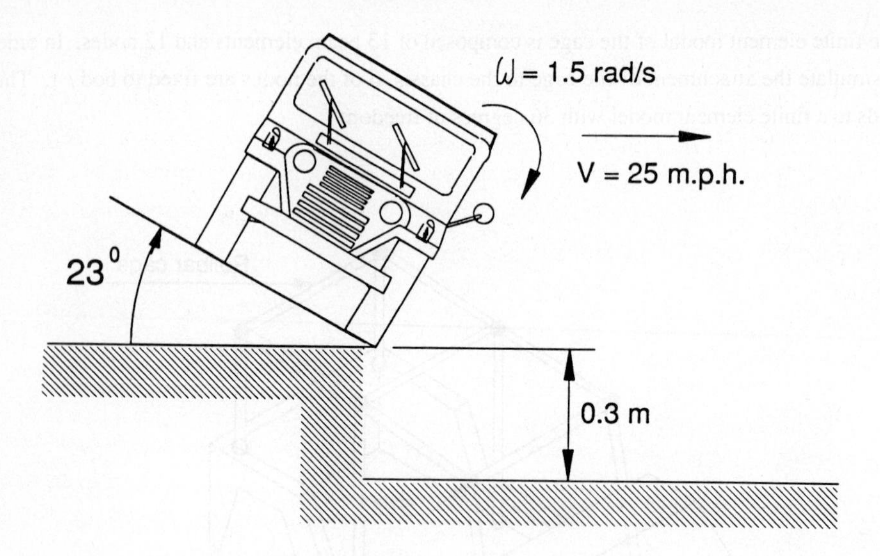

Figure 10 Initial position of the truck before rollover

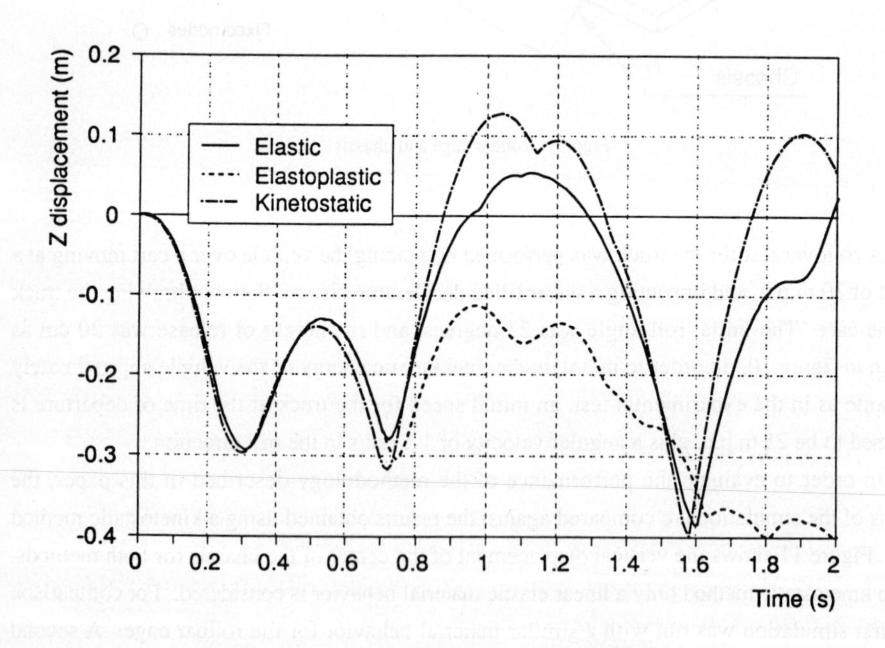

Figure 11 Vertical displacement of the center of mass of the chassis

These simulations show that both methods predict very similar behaviors within the first 1.5 seg. when a linear elastic behavior is assumed for the rollbar cage. The height of the vertical displacement of the chassis for the current method is lower than for the kinetostatic method because: some of the energy of the system is dissipated in the vibration of the cage; there is inertia coupling between the vehicle gross motion and the cage deformations. Due to the extreme nonlinearity of the problem, small initial deviations between the results yield quite different motions after the initial period. When an elasto-plastic behavior is allowed the motion of the vehicle is, as expected, completely different from that of the kinetostatic method.

6 Conclusions

In this paper a methodology for the analysis of structural elements undergoing nonlinear gross motion and nonlinear deformations was presented. It was shown that the use of multibody dynamics together with the finite element method provides a powerful tool to study problems where the nonlinear behavior of the flexible elements is relevant for the system behavior, such as structural impact and crashworthiness.

It was shown that the three forms of the equations of motion for flexible bodies presented in this paper provide similar results if the structural behavior of the flexible components of the system are linear elastic. The use of static condensation to eliminate the nodal rotational degrees of freedom has the same computational cost of not using any reduction at all. The main reason is that the condensation of nodal rotations requires the assembly of the stiffness matrix and the inversion of its sub-matrices. If no coordinate reduction is performed, there is a larger number of coordinates to integrate. However no assembly of the stiffness matrix is required. Clearly the use of the modal superposition is the cheapest. This is because not only the assembly of the stiffness matrix is dispensable but also the number of coordinates to integrate is much smaller than with any of the other forms of the equations of motion. A similar relation for the computational costs is observed when the rotating beam is allowed to exhibit a nonlinear behavior.

For problems that exhibit a strong nonlinear behavior, as in the case of crashworthiness, the number of nodal coordinates cannot be reduced using the modal superposition. However the application of the present methodology to the rollover of the truck show that tthe use of equations (49) and (53) provide good results. Comparing the results obtained with the present method and the kinetostatic method shown the relative importance of the inertia coupling between the gross motion and the structural deformations for the behavior of the system.

References

1. Ambrosio, J.A.C.: Elastic-Plastic Large Deformation of Flexible Multibody Systems in Crash Analysis. Ph.D. Dissertation, University of Arizona, 1991.
2. Ambrosio, J.A.C., Nikravesh, P.E.: Elasto-Plastic Deformations in Multibody Dynamics. Nonlinear Dynamics. 3, 85-104 (1992)
3. Argyris, J.H. and Doltsinis, J.S.: On the Large Strain Inelastic Analysis in Natural Formulation: Dynamic Problems. Comp. Meth. in Appl. Mech. and Engng. 21, 91-126 (1980)
4. Bakr, E.M. and Shabana, A.: Geometrically Nonlinear Analysis of Multibody Systems. Computers and Structures. 23, 739-751 (1986)
5. Bathe, K.-J. and Bolourchi, S.: Large Displacement Analysis of Three-Dimensional Beam Structures. Int. J. Num. Meth. Engng. 14, 961-986 (1979)
6. Belytscko, T. and Hsieh, B.J.: Nonlinear Finite Element Analysis with Convected Coordinates. Int. J. Num. Meth. Engng. 7, 255-271 (1973)
7. Chung, I.S., Nikravesh, P.E. and Arora, J.S.: Automobile Simulation Using a General Purpose Rigid Body Dynamic Analysis Program. ASME Computational Methods in Ground Transportation Vehicles. A. Dm., 50 (1982)
8. Haug, E.J.: Computer Aided Kinematics and Dynamics of Mechanical Systems. New York: Allyn & Bacon 1990
9. Kane, T.R., Ryan, R.R. and Banerjee, A.K.: Comprehensive Theory for the Dynamics of a General Beam Attached to a Moving Rigid Base. Journal of Guidance, Control and Dynamics. 10, 139-151 (1987)
10. Nikravesh, P.E.: Computer Aided Analysis of Mechanical Systems. Englewood Cliffs, New Jersey: Prentice-Hall 1988.
11. Nikravesh, P.E. et al.: MBOSS (MultiBOdy Simulation System). CAEL, Department of Aerospace and Mechanical Engineering, University of Arizona, 1990
12. Nikravesh, P.E.: Systematic Reduction of Multibody Equations of Motion to a Minimal Set Int. J. Non-Linear Mechanics. 25, 143-151 (1990)
13. Nikravesh, P.E. and Gim, G.H.: Systematic Construction of the Equations of Motion for Multibody Systems Containing Closed Kinematic Loops. Proceedings of the ASME Design Automation Conference, Montreal, Canada, 1989
14. Nikravesh, P.E. and Ambrosio, J.A.C.: Systematic Construction of Equations of Motion for Rigid-Flexible Multibody Systems Containing Open and Closed Loops. Int. J. Num. Meth. Engng. 32(8), 1749-1766 (1990)
15. Nikravesh, P.E. and Ambrosio, J.A.C.: Rollover Simulation and Crashworthiness Analysis of Trucks. Forensic Engineering. 2, 387-401 (1990)
16. Nygard, M.K. and Bergan, P.G.: Advances on Treating Large Rotations for Nonlinear Problem. In: State of the Art Surveys on Computational Mechanics (A.K. Noor and J.T. Oden, eds.), pp. 305-333. New York: ASME 1989
17 Pereira, M.S., Proença, P.:Dynamic Analysis of Spacial Flexible Multibody Systems Using Joint Co-ordinates. Int. J. Num. Meth. Engng. 32(8), 1799-1832 (1990)
18. Shabana, A. and Wehage, R.A.: A Coordinate Reduction Technique for Dynamic Analysis of Spatial Substructures with Large Angular Rotations. J. Struct. Mech. 11, 401-431 (1983)
19. Song, J.O. and Haug, E.J.: Dynamic Analysis of Planar Flexible Mechanisms. Computer Methods in Applied Mechanics and Engineering. 24, 359-381 (1980)
20. Sunada, W. and Dubowsky, S.: The Application of the Finite Element Methods to the Dynamic Analysis of Flexible Spatial and Co-Planar Linkage Systems. ASME Journal of Mechanical Design. 103, 643-651 (1981)
21. Walrapp, O., Schwertassek, R.: Representation of geometric Stiffening in Multibody Simulaion. Int. J. Num. Meth. Engng. 32(8), 1833-1850 (1990)
22. Wu, S.C.,Haug, E.J..and Kim, S.S.: A variational Approach to Dynamics of Flexible Multibody Systems. Mech. Struct. Mach. 17, 3-32 (1989)
23. Zienckiewicz, O.C.: The Finite Element Method: McGraw-Hill 1977

Multidisciplinary Simulation

Martin Otter

German Aerospace Research Establishment (DLR), 8031 Oberpfaffenhofen, Germany

Abstract: A new approach is described for multidisciplinary simulation, i.e., the simulation of models, which consist of components from different engineering disciplines. The central part of this approach consists of the definition of a neutral (low level) interface for general event-driven ordinary differential equations and differential algebraic equations, called DSblock (= Dynamic System block). Several preprocessors have been realized that generate DSblocks from models of existing modelling environments from different domains. Furthermore a run time environment is available to simulate DSblocks in an interactive way.

Keywords: multidisciplinary simulation / simulation language / multibody / control

1 Introduction

This paper deals with the multidisciplinary simulation of dynamic systems, i.e., with the simulation of systems consisting of components from different domains, which are connected together in such a way that the overall behavior of the system is strongly influenced by the interaction of all components. The system performance shall be analyzed by dynamic simulation and components shall be developed using simulation as a basic design tool.

The simulation of systems like robots, vehicles or satellites requires for example the modelling of multibody systems, of electronic circuits, of control system devices and of other physical effects such as friction, backlash or heat transfer. If one of these effects is neglected in the model, realistic simulations of the system are often not possible.

A wide variety of commercial programs are available, which allow a comfortable and detailed modelling as well as simulation of systems from usually **one** specific domain as shown by the following examples:

Mechanical systems can be easily modelled via multibody programs like DADS [1], NEWEUL [2], SD-EXACT/SD-FAST [3,4], Simpack [5], which allow a "natural" specification of the system, e.g., by definition of "rigid bodies," "deformable bodies," "joints," and "force-elements." The most difficult and most challenging part of such programs consists of a formalism for building the differential or differential-algebraic equations from the model description.

Electronic circuits can be easily modelled via SPICE [6] or one of its derivates such as I-G SPICE [7], DELIGHT.SPICE [8], since these packages allow a "natural" specification of electronic circuits, e.g., by definition of "resistors," "capacitors," and "MOS-transistors" together with a connection structure. These programs have a lot of know-how in form of large libraries of electronic components consisting of mathematical models of different complexity, the ability to model the connection of a large number of such components and the numerical treatment of the specific differential algebraic equations appearing in the overall model.

Control systems can be easily modelled via a graphical block diagram editor with linear and non-linear function blocks, such as EASY-5 [9], Prosign [10], SimuLink [11] or SYSTEM_BUILD [12], since a block diagram is the basic abstraction mechanism of control theory and therefore reflects directly the user's view.

Physical effects described by mathematical equations which may have discontinuities defined by time or state events, can be easily specified using a general purpose simulation language such as ACSL [13], or one of the more restricted languages such as Desire [14] or Simnon [15], since these languages allow a natural definition of event-driven differential equations or at least of sampled data systems. Model equations can be given in any order, because a sorting algorithm will transform them into a correct (causal) sequence.

All of the mentioned excellent software packages have the disadvantage, that it is difficult or impossible to model components which are not in the single domain, the package was designed for. For example, ACSL or SYSTEM_BUILD are not suited to model (directly) the dynamics of a 6 degree of freedom robot, since the equations of the robot are required as input to these programs and the main difficulty consists in the determination of these equations. However, this task can be easily solved by multibody programs. On the other hand, there is no multibody program available that would allow the simulation of mechanical systems including discrete components, such as a controller realized by a microprocessor. However, the simulation of continuous and discrete elements together is an easy task for ACSL or SYSTEM_BUILD. Furthermore it is noteworthy that most general purpose simulation languages are only able to deal with **explicit** differential equations. In engineering disciplines like mechanics, electronics or hydraulics, systems are often described in a natural way by differential-algebraic equations. In such a case, these **general** packages are only applicable if it is possible to transform the original problem into a problem containing explicit differential equations only.

To summarize, multidisciplinary simulation becomes more and more important for advanced products, but there is no software package available yet for this task. Is it possible to change this situation? A straightforward solution like the development of a new software system from scratch which has the same modelling capabilities as several commercially available domain specific modelling environments together, seems unreasonable. In the next section, a pragmatic solution to this problem is given by **enhancing existing** modelling environments by code generators, which generate the model equations of the environments in the same neutral (low level) format, and by supporting the connection of models described in this format. Another very promising approach is based on the usage of an object oriented modelling language like Dymola [16,17,18] or Omola [19,20]. Such modelling languages **support** the definition of libraries of domain specific components that can be connected together according to the **physical type** of connection (traditional simulation languages like ACSL or Desire do not have this ability). However, it will take some more time to develop domain specific libraries which do have modelling capabilities similar to programs like DADS or SPICE.

2 Multidisciplinary Simulation by Sharing Models of Existing Modelling Environments

In the ANDECS-project of the DLR[1], the following approach was developed to make multidisciplinary simulation available (see Figure 1):

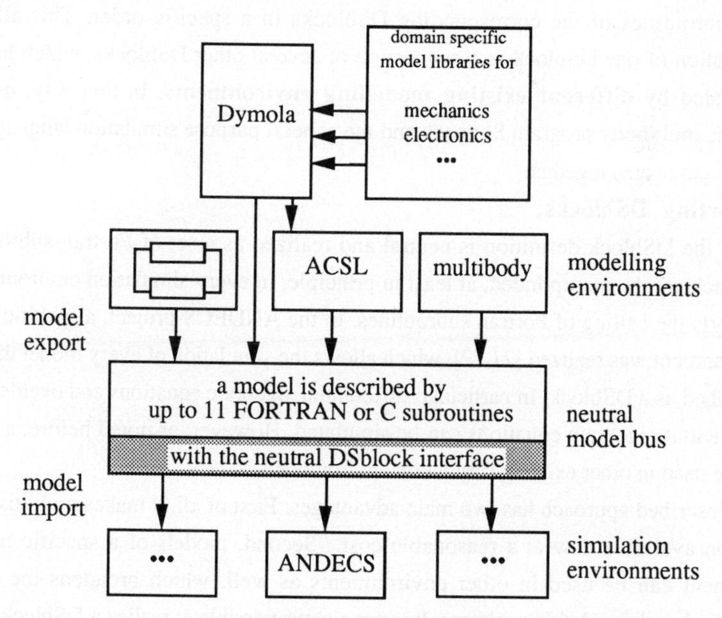

Figure 1. Multidisciplinary Simulation

1 ANDECS stands for Analysis and Design of Controlled Systems. ANDECS is developed by Prof. G. Grübel and his group at the DLR.

1. **Neutral DSblock-interface:**
   Modelling and simulation are separated into two distinct parts of the software. The **interface** between these two parts is defined by a **neutral** description of input/output blocks, which are realized as Fortran-subroutines with **defined** formal arguments.[2] A block of this type is called DSblock (= **D**ynamic **S**ystem **B**lock). A DSblock allows the description of generic, time-delayed, time-, state-, and step-event dependant explicit ordinary differential equations, differential-algebraic equations and overdetermined differential algebraic equations. The latter equations appear if a higher index differential algebraic equation is transformed to an index 1 equation. A DSblock is general enough to describe all models of the software packages mentioned in the introduction.
2. **Exporting DSblocks:**
   Existing modelling environments like ACSL or Prosign are enhanced by a code generator, which generates models of the environment in the standardized DSblock format. Numerical programs like Simpack are enhanced by a subroutine layer in the DSblock format, which allow the calling of the software package at specific time instants to calculate the right hand side of the differential equation.
3. **Connecting DSblocks:**
   The DSblock interface definition supports the connection of different DSblocks by calling the subroutines of the corresponding DSblocks in a specific order. This allows the generation of one DSblock as a connection of several other DSblocks, which have been generated by **different existing modelling environments**. In this way, models of e.g. the multibody program Simpack and the general purpose simulation language ACSL can be connected together.
4. **Importing DSblocks:**
   Since the DSblock definition is neutral and realized as a set of Fortran-subroutines, a DSblock can be incorporated, at least in principle, in every simulation environment that supports the calling of Fortran subroutines. In the ANDECS-project, a specific run time environment was realized [21,22], which allows the simulation of every model that can be described as a DSblock. In particular, differential algebraic equations and overdetermined differential algebraic equations can be simulated. However, as noted before, a DSblock can be used in other existing simulation environments as well.

The described approach has two main advantages: First of all, it makes multidisciplinary simulation available now at a reasonable cost. Second, models of a specific modelling environment can be used in other environments as well, which broadens the scope of previously closed simulation packages. It is not always possible to realize a DSblock interface with reasonable effort. For example in systems like DADS or SPICE, integrators are used that

[2] The DSblock definition was designed in such a way that a C-interface can be directly derived from the Fortran-interface.

make use of the special structure of the specific formulation of the model equations. Solver and right hand side computation are related so closely that it is difficult to extract the pure model equations.

Models described and realized according to the DSblock definition can be used in the following applications:

- A complete model is described by one DSblock. The DSblock can be simulated in different simulation environments, e.g., in a specific real-time system.
- Several models are described by different DSblocks, which are connected together at **compile** time to form a new DSblock. This situation happens if parts of a model are generated by different modelling environments. For example a robot with elastic bodies is modelled by Simpack, and the digital controllers are modelled by ACSL.
- Several models are described by different DSblocks, which are connected together at **run time**. This situation happens if fast experimenting is desired, e.g., to examine different connection structures or to simulate the same system with models of different complexity. If an appropriate graphical user interface is provided, this can be a convenient way to perform simulations.

3 Generating DSblocks From Existing Modelling Environments

Several existing modelling environments have already been enhanced by DSblock code generators. In particular, these are:

ACSL A general purpose simulation language.[3]

Dymola An object oriented modelling language.[4]

Prosign A block-oriented graphical modelling and simulation system.[5]

Simpack A general purpose numerical multibody program.

Code generators for other environments are in discussion. The DSblock interfaces of the mentioned modelling and simulation environments are described in more detail below.

ACSL [13] has become a de facto industrial standard for simulation languages. It allows a descriptive formulation of differential and difference equations and supports a simple type of model libraries via a macro mechanism. Under contract of DLR, a compiler was written by Dr. I. Bausch-Gall (Bausch-Gall GmbH, Munich), and Prof. F. Breitenecker, A. Prinz, G. Schuster (Technical University of Vienna) to generate a DSblock from any ACSL-model of level 10. The compiler is written in C using the parser generator QPARSER+ from QCAD Systems.[6] A few language elements have been added to ACSL, to allow the definition of

---

3 ACSL = Advanced Continuous Simulation Language is a product of Mitchell & Gauthier Associates.

4 Dymola of Hilding Elmquist will be commercially available from DynaSim AB in the near future.

5 Prosign is distributed by R&O, Germering, Germany.

6 QPARSER+ is much more powerful than yacc.

ACSL-variables which are used as input and output signals for the corresponding DSblock. All these language elements start with an ACSL comment character, so that models containing them can also be conveniently simulated in the ACSL run time environment.

Dymola [16,17,18] is an object oriented simulation language for modeling of large continuous systems. The language supports the building of domain specific model libraries. Models are hierarchically decomposed into submodels and can be connected together according to the **physical** coupling of the components. This is possible due to the existence of two features: First of all, Dymola determines by itself whether a variable is an input or an output signal for a model-block and transforms the model equations symbolically if needed. Second, variables can be defined as "across"- and as "through"-variables having different semantics when used in a connection. Dymola can handle explicit differential equations as well as differential algebraic equations (DAE). Higher index DAEs are handled by differentiating specific parts of the model equations symbolically according to the Pantelides algorithm [23]. Dymola is a **modelling** language and does not have a run time environment of its own. At present, Dymola generates code in the ACSL, Desire, Simnon, Simnon-Fortran and DSblock-Fortran format. The last feature, generating a DSblock from a Dymola-model, was realized by Dr. H. Elmqvist very recently. The DSblock-generation, as well as the others, are directly implemented in the Dymola package, using the internal data structure of Dymola as a starting point.

Prosign [10] is a graphical block-oriented simulation system for control and process models. It allows the graphical definition and connection of hierarchically decomposed input/output blocks. Simulations are defined directly on the graphical model definition. By clicking on a connection, the corresponding variable is plotted on-line during the simulation. Since model equations are evaluated in an interpretive fashion, simulations are slow in comparison with a simulation system such as ACSL that uses compiled models. However, the user interface is much better. Under contract of DLR, Dr. Linssen (Linssen & Beese, Fürth, Germany) realized a code generator that produces DSblock code from the internal description of a Prosign model.

Simpack [5] is a numerical general purpose multibody program of the MAN-corporation. Simpack supports the modelling and simulation of rigid and deformable bodies that are connected together by joints into a tree or a closed loop structure. Simpack force elements may depend on time or state events to allow the description of discontinuous components. A multibody system is described in a specific syntax on a file which is read, before a simulation starts. Under contract of DLR, W. Rulka of MAN and M. Otter of DLR realized a subroutine interface for Simpack in the DSblock format. In particular, the right hand side of the differential equations or differential algebraic equations of a Simpack model are computed by a call of the corresponding DSblock subroutine.

4 The Neutral DSblock Model Description

A DSblock describes an input/output "block" of a general nonlinear dynamic system in a neutral way. DSblocks are mathematically described by explicit ordinary differential equations (ODE), differential-algebraic equations (DAE) or overdetermined differential-algebraic equations (ODAE). Furthermore, a general event handling mechanism allows the treatment of discontinuous equations or equations with varying structure. A DSblock is depicted in Figure 2 and consists of:

- Internal signals x(t), i.e. state (ODE) or descriptor variables (DAE/ODAE).
- Input signals u(t), which can be used to connect different DSblocks together or which allow the definition of specific time signals at run time.
- Output signals y(t), which can be used to connect different DSblocks together.
- Indicator signals z(t). The zero crossing of indicator signals define the occurrence of state events, at which the structure of the system may change.
- Auxiliary signals w(t), i.e., "interesting" internal signals not to be used for other purposes. Signals w are computed and stored at communication points only.
- System constants p of type double precision, integer or logical which can be modified at run time before a simulation starts.
- Additional data on a file or on a database to characterize a specific model of a DSblock.
- A mathematical description to compute both internal signals and output signals from input signals, parameters, and initial conditions of the internal signals. The description may depend on time-, state- or step-events, i.e. a dynamic system can be of a variable structure. A time event is defined by a predefined time instant, a state event is defined by the zero crossing of an indicator signal z(t), and a step event is defined by any condition to be checked after each completed internal step of the integrator.[7]

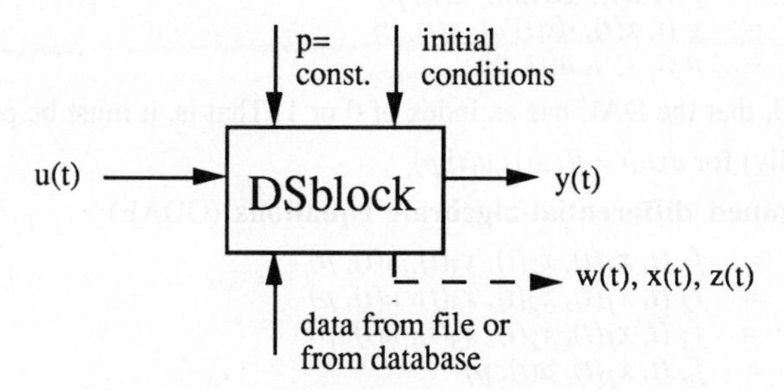

Figure 2. Structure of a DSblock

---

7 Step events can be used to switch efficiently between two sets of state variables, if neither of the two sets is an appropriate description in the whole solution space. E.g., two different 3-parameterizations of a direction cosine matrix in multibody system dynamics.

A DSblock is realized by a subset of at most 11 Fortran subroutines with fixed interfaces but arbitrary (user defined) subroutine names. The structuring of the model is done according to the CSSL-standard. The following subroutines can be provided:

blockB Define a new model of a DSblock and initialize it (e.g., define type of system, define names of signals, read model data from file or database).

blockC Close a model of a DSblock (e.g., close files, free storage).

blockF Compute right hand side f of an ODE, residuum f of a DAE or residua $f_1,...,f_5$ of an ODAE (see below).

blockG Compute output signals y(t).

blockH Compute indicator signals z(t).

blockJ Compute Jacobian matrix.[8]

blockK Compute auxiliary signals w(t).

blockP Check the consistency of system constants p after interactive modification.

blockS To be called before integration starts (initial section).

blockT To be called after integration ends (terminal section).

blockV To be called after a time-, state- or step-events occurred (discrete section).

The details of the interface definition are given in Reference 24. The splitting of the computational part of the model into 5 subroutines blockF, blockG, blockH, blockJ and blockK is motivated by the fact that the tasks of these subroutines have to be performed at different time instants and in different order, depending on the integration method used and whether a single DSblock or several coupled DSblocks are simulated.

The following **mathematical** descriptions of DSblocks are supported:

1. **Explicit ordinary differential equations** (ODE):

$$\begin{aligned} dx/dt &= f(t, x(t), u(t), p) \\ y &= g(t, x(t), u(t), p) \\ z &= h(t, x(t), u(t), p) \end{aligned}$$

2. **Differential-algebraic equations** (DAE):

$$\begin{aligned} 0 &= f(t, x(t), dx(t)/dt, u(t), p) \\ y &= g(t, x(t), dx(t)/dt, u(t), p) \\ z &= h(t, x(t), u(t), p) \end{aligned}$$

It is assumed, that the DAE has an index of 0 or 1. That is, it must be possible to solve f and df/dt (locally) for $dx/dt = f(t,x(t),u(t),p)$.

3. **Overdetermined differential-algebraic equations** (ODAE):

$$\begin{aligned} dx_1/dt &= f_1(t, x_1(t), x_2(t), x_3(t), u(t), p) \\ 0 &= f_2(t, x_1(t), x_2(t), x_3(t), u(t), p) \\ 0 &= f_3(t, x_1(t), x_2(t), x_3(t), u(t), p) \\ 0 &= f_4(t, x_1(t), u(t), p) \\ 0 &= f_5(t, x_1(t), u(t), p) \\ y &= g(t, x_1(t), x_2(t), x_3(t), u(t), p) \\ z &= h(t, x_1(t), x_2(t), x_3(t), u(t), p) \end{aligned}$$

[8] Subroutine blockJ has only to be provided if an analytically derived Jacobian and not a numerically computed Jacobian should be used.

here $\partial f_2/\partial x_2$ and $\partial f_3/\partial x_3$ have to be nonsingular, $f_3 = df_4/dt$, and f_5 has to be an integral invariant of f_1, f_2, f_3, i.e., every exact solution of the first 3 equations must fulfill the last equation. Technically speaking, vector x_2 contains algebraic index-1 variables, which are completely determined by f_2. Vector x_3 contains index-2 variables, which are determined by either f_3 or f_4. Finally f_5 is a constraint that appears, when the original differential algebraic equation is reduced to a lower index form.

There are a substantial number of naturally occurring practical problems, e.g. in multibody systems with closed loops or in transistor models of electronic circuits, which lead to DAEs with an index greater than 1 and can therefore not (directly) be solved by standard DAE-solvers such as DASSL [25,26]. DAE-systems of this type can be reduced to an index-1 DAE by analytically differentiating specific parts of the original DAE according to the algorithm of Pantelides [23]. The resulting equations form an overdetermined set of differential-algebraic equations (= ODAE). Multibody systems are an important application of ODAEs. Here:

f_1 Dynamic equation.
f_2 Friction equation, if friction force depends on constraint force.
f_3 Constraint equations on acceleration level.
f_4 Constraint equations on velocity level.
f_5 Constraint equations on position level.

ODAE-equations can be either solved directly by an ODAE-solver like ODASSL of Führer [27,28,29] or can be transformed to an index-1 DAE by introducing "dummy" derivatives as proposed by Cellier/Elmqvist [17] and Mattsson/Söderlind [30]. In the latter case, the derivatives of some variables are considered to be new variables[9] to get a system with an equal number of equations and unknown quantities.

The DSblock formulation of an ODAE-system was chosen in such a way that several interesting solution strategies of **available** solvers can be carried out with the same model. For example, there exist ODAE-integrators, like MEXX of Lubich [31], which solve directly an index-2 ODAE. Such solvers do not need equation f_3. On the other hand, it is possible to use an index-1 DAE-solver like DASSL to simulate f_1, f_2, f_3, neglecting equations f_4, f_5, if the length of the simulation is sufficiently short and if the initial conditions fulfill f_4, f_5. Finally, DASSL can also be used to solve an index-2 DAE described by equations f_1, f_2, f_4, provided the stepsize control on the index-2 variables x_3 is turned off, i.e. the absolute and relative tolerances of these variables are set to a large value.

To summarize, the ODAE-formulation in the DSblock interface is of a more experimental nature, since the current research on the numerical solution of practical applications of higher index DAE-systems has not yet reached the same level of maturity as for ODE or index-1 DAE systems.

---

9 The fact that a new variable A is the derivative of a variable B is ignored by the solver.

5 The ANDECS Simulation Environment for DSblocks

As already mentioned, a DSblock can be used in any simulation system that allows the calling of Fortran subroutines. In the ANDECS-project of the DLR, a specific simulation system for DSblocks was realized. ANDECS [21,32,33,34] is developed at the DLR on the basis of the engineering database system RSYST [35] of the University of Stuttgart. A preliminary version of ANDECS is already operational and utilized in actual projects at the DLR and some other institutions.

ANDECS is conceived to be a powerful and flexible environment for the analysis and design of controlled dynamic systems. Major components of ANDECS are:

Basic Methods Basic mathematical methods like matrix computation using the Matlab syntax, interpolation of signals or root finding of nonlinear functions.

Linear Methods Analysis and design methods for linear dynamic systems like linear simulation, calculation of poles and zeros, pole placement, LQG or H_∞.

Simulation Simulation environment for (parameterized) DSblocks (see below). Models can be linearized for use with the "Linear Methods." Stationary points of DSblocks can be calculated by use of a nonlinear equation solver.

Optimization Multi-objective parameter optimization and trajectory optimization. Every analysis or design method of ANDECS as well as user defined analysis methods can be used to calculate the optimization criteria. The design history is recorded on an automatically evolving database, allowing new design directions to be started from the actual or from past design steps.

Visualization Standard diagrams like 2-D line, Bode, Nyquist and root locus diagrams. Higher dimensional data, e.g., the n-dimensional criteria space, are visualized by parallel coordinate.

Before simulations of DSblocks can be carried out, the desired DSblocks must be introduced to ANDECS by a configuration module. This module generates a file which has to be compiled together with the corresponding DSblock subroutines. After a new binding run, the defined DSblocks are available in the executable image and can be simulated in the ANDECS simulation environment. Since any number of DSblocks can be kept in the executable image simultaneously, on-line switching between design alternatives or system representations of various complexity is possible.

The result of a simulation experiment is a set of computed signals, which are automatically stored in a database as "signal objects" The "signal objects" can be visualized by different graphic modules. All input data of an experiment, e.g., integration method or length of communication interval, are stored in the database as well. Therefore every simulation run is completely documented and **reproducible** by the data stored in the database.

The ANDECS simulation environment uses well-tested numerical integration routines from various sources. Presently the following solvers are provided:

DEABM — Multistep solver of Shampine/Gordon/Watts [36,37] for non-stiff and moderately stiff ODEs.

LSODE — Multistep solver of Hindmarsh [38] for stiff and non-stiff ODEs.

LSODAR — Multistep solver of Petzold/Hindmarsh [39] which switches automatically between a non-stiff and a stiff integration algorithm along with solution. LSODAR also provides a root finder.

RK45/78 — Runge-Kutta-Fehlberg solvers of Kraft /Führer of orders 5 and 8 with variable stepsize using the Prince-Dormand coefficients according to [40].

DASSL/RT — Multistep solvers of Petzold [25] for DAEs (DASSL) and for DAEs with root finder (DASSLRT).

ODASSL/RT — Multistep solvers of Führer [29] based on DASSL/DASSLRT of Petzold for ODAEs (ODASSL) and for ODAEs with root finger (ODASSLRT).

MEXX — Extrapolation solver of Lubich [31] for a restricted class of index-2 ODAEs (equations f_2 and f_3 of the ODAE are not supported).

There are a wide variety of options available to define a simulation experiment. For example, the communication time grid can be defined as an equidistant grid (defined by initial time, end time, communication stepsize), as an arbitrary user defined grid (defined by a vector of monotonically increasing time instants) or as an automatic grid (the internally used variable stepsize of the integration method is used as stepsize of the communication grid).

In connection with the "optimization" part of ANDECS, the simulation module can be used to calculate criteria in multi-objective parameter optimization and trajectory optimization. In this case, the **parameters** of a DSblock can be used in a parameter optimization, and the **input** signals of a DSblock can be determined by trajectory optimization. The structure of such an optimization together with a hardcopy of a typical ANDECS screen are depicted in Figure 3. Note that the ANDECS modules are independent since they communicate with each other via a fast RSYST database only. The RSYST macro language allows the realization of computational sequences of several modules.

6 Summary

In this paper, a realistic approach to make multidisciplinary simulation available was described. Modelling and simulation are separated into two distinct parts. The interface

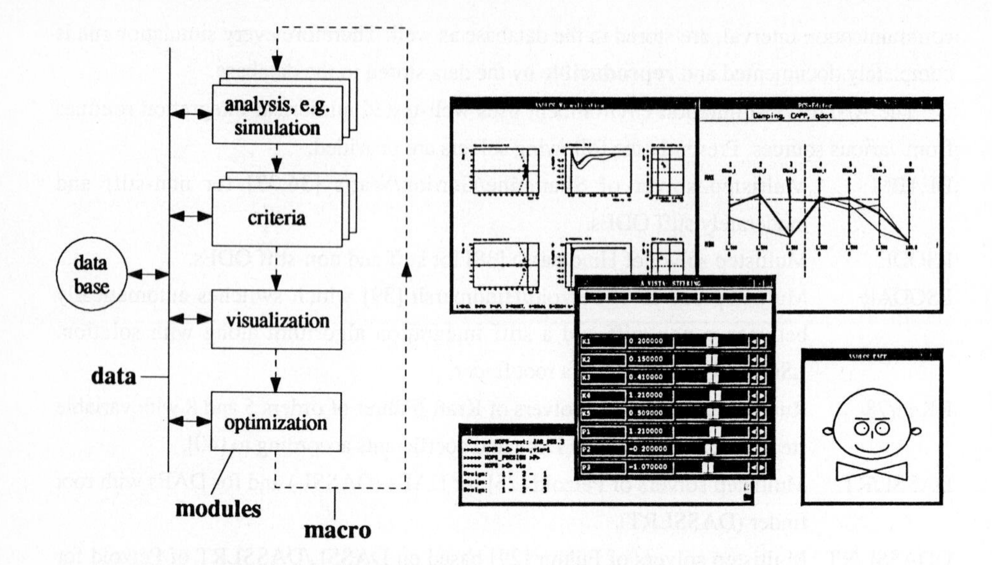

Figure 3. Multi-Objective Optimization within ANDECS

between these two parts is defined by a very general, neutral, low level description of dynamic systems, called DSblock. Some important modelling environments have already been enhanced by code generators, which generate models of the environment in the standardized DSblock format. DSblocks originating from different modelling environments can be connected together to form a new DSblock. DSblocks can be simulated in the ANDECS environment or in any other simulation environment that allows the calling of Fortran subroutines.

References

1. Smith, R.C.; Haug, E.J.: DADS -- Dynamic Analysis and Design System. Multibody Systems Handbook, edited by W. Schiehlen, Springer-Verlag, 1990.
2. Kreuzer, E.; Schiehlen, W.: NEWEUL -- Software for the Generation of Symbolical Equations of Motion. Multibody Systems Handbook, edited by W. Schiehlen, Springer-Verlag, 1990.
3. Rosenthal, R., Sherman, M.A.: High Performance Multibody Simulations via Symbolic Equation Manipulation and Kane's method. The Journal of Astronautical Sciences, Vol. 34, No. 3, pp. 223-239, 1986.
4. Rosenthal, R.: Order N formulation for equations of motion of multibody systems. SDIO/NASA Workshop on Multibody Simulation, JPL, Arcadia, CA, 1987.
5. Rulka, W.: SIMPACK -- A Computer Program for Simulation of Large-motion Multibody Systems. Multibody Systems Handbook, edited by W. Schiehlen, Springer-Verlag, 1990.
6. Nagel, L.W.: SPICE2: A computer program to simulate semiconductor circuits. Berkeley, University of California, Electronic Research Laboratory, ERL -- M 520, 1975.
7. Bowers, J.C.: I-G SPICE -- A circuit designer's dream. Powerconvers. Int., No. 6, pp. 36-40, 1983.
8. Nye, B.; et. al.: DELIGHT.SPICE: An optimization based system for the design of integrated circuits. IEEE Custom Integrated Circuit Conference, pp. 233-238, New York 1983
9. Boeing Computer Services: EASY5/W -- User's Manual. Engineering Technology Applications (ETA) Division, Seattle, Wash, 1988.
10. Linssen; Beese: PROSIGN -- Computer Aided Process Design; User's Guide.

11. Mathworks Inc.: SimuLink -- User's Manual. South Natick, Mass., 1992.
12. Shah, S.C.; Floyd, M.A.; Lehman, L.L: $Matrix_X$: Control Design And Model Building CAE Capability. Computer--Aided Control Systems Engineering, edited by M. Jamshidi and C.J. Herget, Elsevier Science Publishers, pp. 181 - 207, 1985.
13. Mitchell, E.E.L.; Gauthier, J.S.: ACSL: Advanced Continuous Simulation Language -- Reference Manual. Edition 10.0, MGS, Concord., Mass., 1991.
14. Korn, G.A.: Interactive Dynamic System Simulation. McGraw-Hill, New York, 1989.
15. Elmqvist, H.: Simnon -- An Interactive Simulation Program for Nonlinear Systems -- User's Manual. M.S. thesis, Report CODEN:LUTFD2/(TFRT--7502), Department of Automatic Control, Lund Institute of Technology, Lund Sweden, 1975.
16. Cellier, F.E.: Continuous System Modeling. Springer-Verlag, New York, 1991.
17. Cellier, F.E.; Elmqvist, H.: The Need for Automated Formula Manipulation in Object-Oriented Continuous-System Modeling. IEEE Symposium on Computer-Aided Control System Design, CACSD'92, March 17--19, 1992, Napa, California.
18. Elmqvist, H.: A Structured Model Language for Large Continuous Systems. Ph.D. dissertation. Report CODEN:LUTFD2/(TFRT--1015), Department of Automatic Control, Lund Institute of Technology, Lund Sweden, 1978.
19. Anderson, M.: Omola -- An Object-Oriented Language for Model Representation. Licenciate thesis TFRT-3208, Department of Automatic Control, Lund Institute of Technology, Lund, Sweden, 1990.
20. Mattsson, S.E.; Andersson, M.: A kernel for system representation. Preprints of the 11th IFAC World Congress, vol. 10, pp. 91-96, 1990.
21. Gaus, N.; Otter, M.: Dynamic Simulation in Concurrent Control Engineering. IFAC Symposium on Computer Aided Design in Control Systems, Swansea, UK, Preprints pp. 123-126, 15-17 July, 1991.
22. Otter, M.; Gaus, N: Modular Dynamic Simulation with Database Integration. User's Guide, Version 2.1. Technical Report TR R50--91, DLR, Institut für Dynamik der Flugsysteme, D--8031 Oberpfaffenhofen, June 1991.
23. Pantelides, C.C: The consistent initialization of differential-algebraic systems. SIAM Journal of Scientific and Statistical Computing, No. 9, pp. 213-231, 1988.
24. Otter, M.: DSblock: A neutral description of dynamic systems. Version 3.2. Technical Report TR R81--92, DLR, Institut für Dynamik der Flugsysteme, D--8031 Oberpfaffenhofen, May 1992.
25. Petzold, L.R.: A description of DASSL: A differential/algebraic system solver. Proc. 10th IMACS World Congress, Montreal, August 8-13, 1982.
26. Brenan, K.E.; Campbell, S.L.; Petzold, L.R.: Numerical Solution of Initial--Value Problems in Differential--Algebraic Equations. Elsevier Science Publishers, 1989.
27. Führer, C.: Differential-algebraische Gleichungssysteme in mechanischen Mehrkörpersystemen. Ph.D. dissertation, Mathematisches Institut, Technische Universität München, 1988.
28. Eich, E.; Führer, C.; Leimkuhler, B.; Reich, S.: Stabilization and Projection Methods for Multibody Dynamics. Research Report A281, Institut of Mathematics, Helsinki University of Technology, Otakaari 1, SF--02150 Espoo, Finland, August 1990.
29. Führer, C.; Leimkuhler, B.J.: Numerical solution of differential-algebraic equations for constrained mechanical motion. Numerische Mathematik, No. 59, pp. 55-69, 1991.
30. Mattsson, S.E.; Söderlind, G.: A New Technique for Solving High-Index Differential-Algebraic Equations Using Dummy Derivatives. IEEE Symposium on Computer-Aided Control System Design, CACSD'92, March 17--19, 1992, Napa, California.
31. Lubich, C.: Extrapolation integrators for constrained multibody systems. IMPACT Comp. Sci. Eng., No. 3, pp. 213-234, 1991.
32. Grübel, G.; Joos, H.-D.: RASP and RSYST -- Two Complementary Program Libraries for Concurrent Control Engineering. IFAC Symposium on Computer Aided Design in Control Systems, Swansea, UK, Preprints pp. 101-106, 15-17 July, 1991.
33. Joos, H.-D.: Automatic Evolution of a Decision-supporting Design Project Database in Concurrent Control Engineering. IFAC Symposium on Computer Aided Design in Control Systems, Swansea, UK, Preprints pp. 113-117, 15-17 July, 1991.
34. Finsterwalder, R.: A ``Parallel Coordinate" Editor as a Visual Decision Aid in a Multi-objective Concurrent Control Engineering Environment. IFAC Symposium on Computer Aided Design in Control Systems, Swansea, UK, Preprints pp. 118-122, 15-17 July, 1991.
35. Rühle, R.; et. al.: RSYST Unterprogramm- und Modul-Dokumentation, Version 3.5.6, Rechenzentrum der Universität Stuttgart, March 1992.
36. Shampine, L.F.; Gordon, M.K.: Computer Solution of Ordinary Differential Equations. Freeman, San Francisco, 1975.
37. Shampine, L.F.; Watts, H.A.: DEPAC -- Design of a User Oriented Package of ODE Solvers. Sandia National Laboratories, Albuquerque, New Mexico, SAND79--2374, 1980.
38. Hindmarsh, A.C.: LSODE and LSODI, Two New Initial Value Ordinary Differential Equation Solvers. ACM-Signum Newsletter, vol. 15, no. 4, pp. 10-11, 1980.

39. Hindmarsh, A.C.: ODEPACK, a systematized collection of ODE solvers. Scientific Computing, edited by R.S. Stepleman et. al., North-Holland, Amsterdam, 1983.
40. Hairer, E.; Norsett, S.P.; Wanner, G.: Solving Ordinary Differential Equations I. Nonstiff Problems. Springer-Verlag, Berlin, 1987.

Dynamic Analysis of Rigid-Flexible Mechanisms

M. S. Pereira and P. L. Proença

CEMUL – Centro de Mecânica e Materiais da Universidade Técnica de Lisboa,
Instituto Superior Técnico, Av. Rovisco Pais, 1096 Lisboa Codex, Portugal

Abstract: Recent developments in the field of computer methods in rigid-flexible multibody dynamics have been considerable, and have evolved due to the progress of numerical methods and computer technologies.

In this paper, formulations for the dynamic analysis of mechanical systems composed of rigid and flexible bodies are reviewed. Reference and relative kinematics are discussed. The equations of motion are derived using Lagrange multipliers techniques and relative joint velocity methods in order to reduce the number of rigid body and elastic coordinates of the system. An integrated simulation methodology is proposed where all the necessary interface data is generated to carry out a static stress model analysis of the individual bodies using in-house software or commercially available codes. Additional functions of stress history evaluation or stress detailing for fatigue design of complex geometries are presented.

One example of an off-road vehicle is presented where the main body is assumed to be flexible.

Keywords: rigid-flexible mechanisms / multibody dynamics / relative coordinates / finite elements / modal analysis / vehicle simulation and design / structural dynamics

1 Introduction

The dynamic analysis of rigid-flexible mechanical systems has received considerable attention during the last decade. The interest in this area is due to the need to design systems with interconnected rigid and flexible components.

The requirements for efficient mechanisms to work under increased loads and high speeds has been a major incentive to develop mathematical models of flexible mechanisms that accurately predict behavior. Such systems play a major role on the design of machines in general and in the aeronautics, space, vehicle and robot industries, where increasingly lighter components are required to operate under extreme conditions.

This field of study plays a major role in the process of simulation of large-motion global dynamics coupled with the structural behavior of the different mechanical systems components and provides the framework for a simultaneous assessment of the dynamics as well as the strength, fatigue and other related topics in the more conventional structural design areas.

In fact, the dynamic behavior determines reaction forces histories which determine the behavior of the structural components which, in turn, influences the dynamics of the mechanical systems. Unfolding of satellite structures, accurate robot positioning, ride stability and ride comfort of vehicles are typical examples of the importance of these coupling effects.

Much of the work so far has dealt with formulations that allow an automatic generation of the dynamic equations of complex rigid-flexible mechanisms. The dynamic analysis of such systems can be very expensive. However, by careful formulation of the problem considerable efficiency can be gained so that many complex mechanisms can be simulated today using modern computer resources which are now available at reasonable costs.

The major difficulties associated with the analysis of rigid-flexible systems are:

- The coupling of large rigid body motion with the displacements describing the flexibility effects;
- The problem is highly non-linear;
- Considerable increase in the problem size due to the addition of the extra flexible degrees of freedom.

Each one of these aspects have been addressed using different methods: substructuring techniques well developed in structural dynamics have been used to reduce the size of the problem [30]; using modal analysis methods, it is possible [7, 10] to transform the nodal elastic coordinate space in the modal coordinate space by retaining only the more relevant modes; finally, the inertia coupling between the large rigid body motion and the elastic deformation of flexible bodies depend mainly on the choice of body fixed coordinate systems and on the reference conditions for the elastic coordinates. The inertial coupling of large body motion and flexibility effects is strongly dependent on the selection of these reference conditions as will be shown later.

Some authors [24, 27] recognize the potentiality of the finite element method to model the structural behavior of mechanical linkages treating the total mechanisms displacements as the primary unknowns in the dynamic equations of motion. These methods have been extended to generalized finite element formulations [32, 6] where the geometric description of the deformation uses an updated lagrangean formulation with a reference condition corresponding to rigid body motion. Homogeneous transformations describe the kinematics of the system and the coupled rigid elastic equations of motion are derived from the lagrangean of the system.

Approaches on absolute coordinates are more straightforward to develop allowing relatively simple implementations on already existing rigid body cartesian coordinate based formulations [4, 29]. In this approach connectivity between different bodies is described by using a set of non-linear algebraic constraint equations which are adjoined to the system equations of motion using lagrange multipliers techniques. These types of formulation have the advantage of an easy implementation of different types of joints and actuators and user defined driving and forcing functions.

Other formulations use relative joint coordinates to establish a minimum number of generalized coordinates [25, 5]. Recursive formulations for flexible multibody systems using relative joint coordinates have also been developed [28, 15]. These formulations seem to be well suited for implementation in parallel computer platforms since the recursive reduction procedure can be carried out independently along each open-loop chain.

The present paper reviews absolute coordinates and derived methods using a minimum number of equations of motion for spatial flexible multibody systems. Absolute and relative kinematics are developed using absolute and relative joint coordinates.

Computer algorithms are discussed and an example of a flexible vehicle is presented and the results are discussed to illustrate computational efficiency and how the results can be used in the design for fatigue and strength of the vehicle structure.

2 Rigid Body Dynamics

The constrained motion of mechanical systems is often modeled by the Lagrange second order differential equation [20]

$$\mathbf{M}\ddot{\mathbf{q}} - \mathbf{\Phi}_{\mathbf{q}}^{\mathrm{T}}\boldsymbol{\lambda} = \mathbf{Q} \tag{1}$$

If the kinematic constraints for the kinematic joints are expressed in algebraic form

$$\mathbf{\Phi}(\mathbf{q}) = 0 \tag{2}$$

then the first and second time derivatives of the constraints can be written as

$$\mathbf{\Phi}_{\mathbf{q}}\dot{\mathbf{q}} = 0 \tag{3}$$

and

$$\Phi_{\mathbf{q}}\ddot{\mathbf{q}} + \left(\dot{\Phi}_{\mathbf{q}}\right)\dot{\mathbf{q}} = 0 \tag{4}$$

where **M** is the n_c x n_c mass matrix;
q is the n_c generalized coordinate vector;
λ is the n_ϕ Lagrange multiplier vector;
Q is the n_c generalized force vector;
$\Phi_{\mathbf{q}}$ is the n_ϕ x n_c jacobian of the constraints.

The overdots refer to time differentiation and n_c and n_ϕ are the number of system generalized coordinates and the number system constraint equations respectively. Equations 1 and 4 are a mixed set of differential algebraic equations which have to be solved and integrated with respect to time.

These equations can be transformed into a smaller set using a joint velocity transformation [21] which projects the cartesian velocities in the space of a minimum number of joint velocities. Let us define $\dot{\beta}$ as the vector of joint velocities. For any open-loop system a linear relation can be established

$$\dot{\mathbf{q}} = \mathbf{B}\dot{\beta} \tag{5}$$

Differentiating Equation 5 with respect to time

$$\ddot{\mathbf{q}} = \mathbf{B}\ddot{\beta} + \dot{\mathbf{B}}\dot{\beta} \tag{6}$$

and substituting Equation 5 into Equation 3

$$\Phi_{\mathbf{q}}\mathbf{B}\dot{\beta} = \mathbf{0} \tag{7}$$

Equation 7 shows that if the joint velocities are independent, **B** is orthogonal to $\Phi_{\mathbf{q}}$.

Substituting now Equation 6 into Equation 1 and premultiplying both sides by $\mathbf{B}^T$, the corresponding equations of motion can be written in terms of $\ddot{\beta}$ as

$$\overline{\mathbf{M}}\ddot{\beta} = \mathbf{f} \tag{8}$$

where

$$\overline{\mathbf{M}} = \mathbf{B}^{\mathrm{T}}\mathbf{M}\mathbf{B}$$

$$\mathbf{f} = \mathbf{B}^{\mathrm{T}}\left(g - \mathbf{M}\dot{\mathbf{B}}\dot{\beta}\right)$$

Equation 8 is then used with great advantage and efficiency as compared to the previous formulation using Equations 1 and 4.

3 Reference Kinematics

It has been shown [25] that the configuration of a deformable body in a multibody system can be described by a set of global reference coordinates and local elastic coordinates.

First we must consider an appropriate representation of finite rotations in the 3D space. Several techniques have been presented in the literature, such as three independent Euler angles, four dependent Euler parameters or four dependent Rodrigues parameters. The last two representations are more appropriate to computational implementation avoiding the problem of singular positions and transcendental functions evaluation.

Assume an inertial set of axis XYZ and a second set $X^iY^iZ^i$ associated with body i and fixed to it in the undeformed state. The reference generalized coordinates of body i are

$$\mathbf{q}_r^i = \left[\mathbf{r}^{i^T} \mathbf{p}^{i^T}\right]^T \tag{9}$$

where $\mathbf{r}^i$ is the position vector of the set of axis $X^iY^iZ^i$ and $\mathbf{p}^i$ a set of rotational parameters describing the orientation of body i in the reference inertial frame.

Consider now that body i is flexible and is discretized into finite elements. From Figure 1 the reference position of an arbitrary point P^{ij} on the element j of body i can be written as

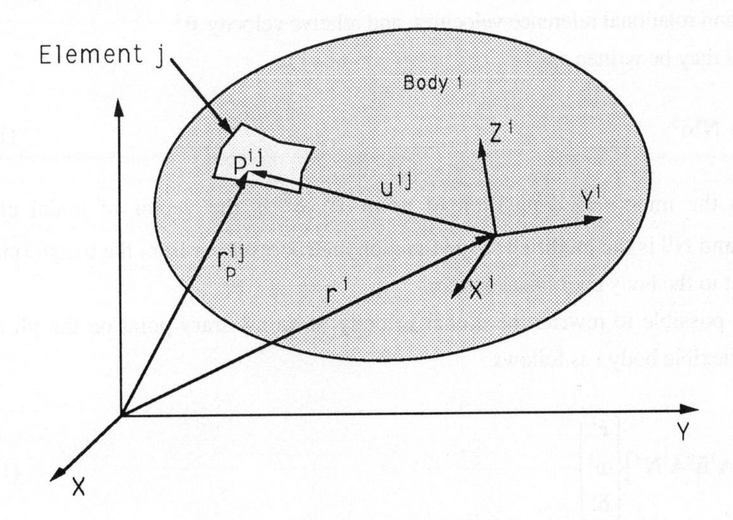

Figure 1. Generalized Reference Coordinates

$$\mathbf{r}_p^{ij} = \mathbf{r}^i + \mathbf{A}^i\mathbf{u}^{ij} \tag{10}$$

where $\mathbf{r}^i$ is the global position of the ith body coordinate system relative to the global inertial frame XYZ, $\mathbf{u}^{ij}$ is the position vector which describes the motion of point P^{ij} in the body coordinate system, and $\mathbf{A}^i$ is the transformation matrix from the ith body coordinate system to the global inertial frame of reference. This transformation matrix is written in terms of orientational parameters $\mathbf{p}^i$.

Differentiating Equation 10 with respect to time, one obtains

$$\dot{\mathbf{r}}_p^{ij} = \dot{\mathbf{r}}^i + \dot{\mathbf{A}}^i\mathbf{u}^{ij} + \mathbf{A}^i\dot{\mathbf{u}}^{ij} \tag{11}$$

$$\dot{\mathbf{r}}_p^{ij} = \dot{\mathbf{r}}^i + \mathbf{A}^i\tilde{\mathbf{u}}^{ij}\omega^i + \mathbf{A}^i\dot{\mathbf{u}}^{ij} \tag{12}$$

where ω^i is the angular velocity of body i expressed in the body attached frame and $\tilde{\mathbf{u}}^{ij}$ is a skew-symmetric matrix given by

$$\tilde{\mathbf{u}}^{ij} = \begin{bmatrix} 0 & -u_z & u_y \\ u_z & 0 & -u_x \\ -u_y & u_x & 0 \end{bmatrix}^i \tag{13}$$

in which u_x, u_y, and u_z are the components of the vector $\mathbf{u}^{ij}$.

Equation 12 represents the velocity of point P^{ij} given as a function of position, translational and rotational reference velocities, and relative velocity $\dot{\mathbf{u}}^{ij}$.

Vector $\mathbf{u}^{ij}$ may be written as

$$\mathbf{u}^{ij} = \mathbf{u}_0^{ij} + \mathbf{N}^{ij}\delta^{ij} \tag{14}$$

where $\mathbf{u}_0^{ij}$ is the undeformed position of point P^{ij}, δ^{ij} is the vector of nodal elastic coordinates, and N^{ij} is the modified shape function matrix resulting from the transformation of the element to the body coordinate system.

It is now possible to rewrite the global velocity of an arbitrary point on the jth finite element of a flexible body i as follows

$$\dot{\mathbf{r}}_p^{ij} = \left[\mathbf{I} - \mathbf{A}^i\tilde{\mathbf{u}}^{ij}\mathbf{A}^i\mathbf{N}^{ij}\right]\begin{Bmatrix} \dot{\mathbf{r}}^i \\ \omega^i \\ \dot{\delta}^i \end{Bmatrix} \tag{15}$$

The spatial kinematic relationships given by Equations 10 and 15 can be used to define the global position and velocity of an arbitrary point P^{ij}

4 Constraint Equations

In this section the nonlinear constraint equations for different kinematic joints between flexible bodies is described. These equations will be considered together with the dynamic equations of motion of flexible bodies using a Lagrange multiplier technique.

4.1 Spherical Joint

Consider two bodies i and j connected through a spherical joint in a common point k, as illustrated in Figure 2.

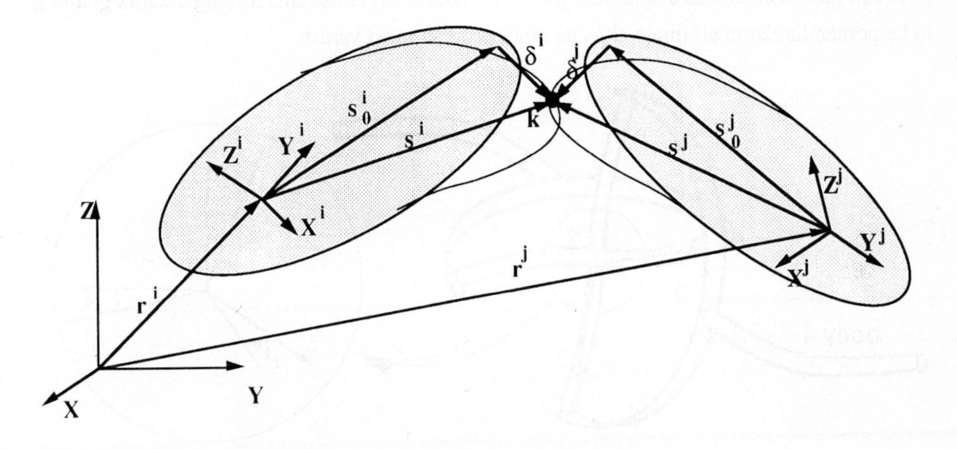

Figure 2. Spherical Joint

The vectorial equation which forces point k to be coincident in both bodies at all times is written in the form

$$\mathbf{r}^i + \mathbf{A}^i\mathbf{s}'^i - \mathbf{r}^j - \mathbf{A}^j\mathbf{s}'^j = \mathbf{0} \qquad (16)$$

where $\mathbf{s}'^i$, $\mathbf{s}'^j$ are position vectors of point k in both bodies i and j respectively;

$\mathbf{A}^i$, $\mathbf{A}^j$ are transformation matrices from the body coordinate systems to the global inertia frame.

If both bodies are flexible s'^i, s'^j depend on the generalized elastic coordinates implying that these vectors have to be calculated at each time for each deformation state.

The previous equation must be then written in the form

$$\mathbf{r}^i + \mathbf{A}^i\left(s_0'^i + \delta_k^i\right) - \mathbf{r}^j - \mathbf{A}^j\left(s_0'^j + \delta_k^j\right) = \mathbf{0} \tag{17}$$

where $s_0'^i$, $s_0'^j$ correspond to the position vectors of point k in the undeformable state; δ_k^i, δ_k^j are the elastic deformations of the connection node at point k of body i and j, respectively.

This joint has three algebraic constraint equations.

4.2 Universal Joint

Consider two bodies i and j connected through a universal joint as shown in Figure 3. This joint can be considered as a spherical joint with additional constrains forcing vectors $\mathbf{g}^i$ and $\mathbf{g}^i$ to be perpendicular at all times, thus its inner product must vanish.

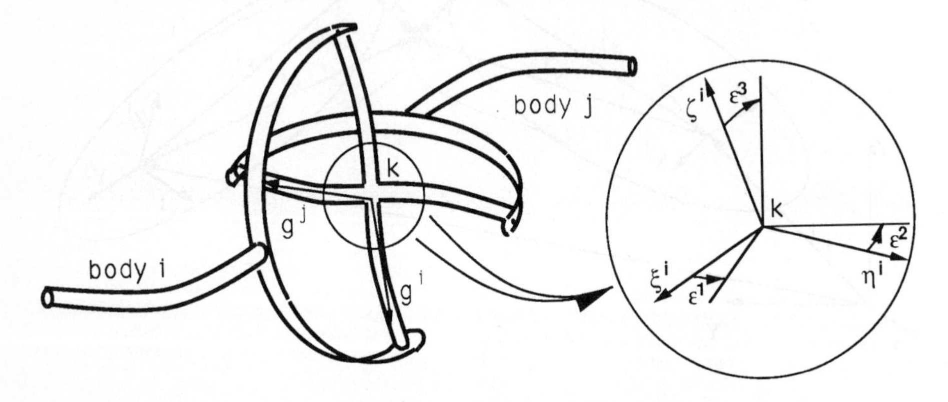

Figure 3. Universal Joint

$$\vec{g}^i \cdot \vec{g}^j = \mathbf{g}^{iT}\mathbf{g}^j = \mathbf{0} \tag{18}$$

When one of the bodies is flexible, say body j, it is necessary to take into account the elastic rotations of the connecting node. Let ε_k^j be the vector of elastic rotations of body j at node k. Assuming that these rotations are sufficiently small to be treated as infinitesimals, it is possible to define a transformation matrix from the $\xi^i\eta^i\zeta^i$ set of axis to the $X^iY^iZ^i$ set of axis as

$$\mathbf{G}^j = \mathbf{I}_3 + \tilde{\varepsilon}^j_k \tag{19}$$

Matrix $\mathbf{G}^i$, as defined, is not orthonormal. This difficulty can be solved assuming for instance that three elastic rotations are applied associated with three Bryant angles.

Equation 19 can now be written as

$$\mathbf{g}^{i^T}\left(\mathbf{A}^j\mathbf{G}^j\mathbf{g}'^j\right) = \mathbf{0} \tag{20}$$

where $\mathbf{g}'^j$ corresponds to $\mathbf{g}^j$ in the undeformed state and defined in the X^iY^iZ set of axis.

When both bodies are flexible Equation 19 will be written as

$$\left(\mathbf{A}^i\mathbf{G}^i\mathbf{g}'^i\right)^T\left(\mathbf{A}^j\mathbf{G}^j\mathbf{g}'^j\right) = \mathbf{0} \tag{21}$$

This joint thus has four algebraic constraint equations.

4.3 Revolute Joint

Consider now two bodies, i and j, connected by a revolute joint as shown in Figure 4. This joint can be considered as a spherical joint with additional constraints forcing vectors $\mathbf{g}^i$ and $\mathbf{g}^j$ to be parallel at all times, thus its vector product must vanish.

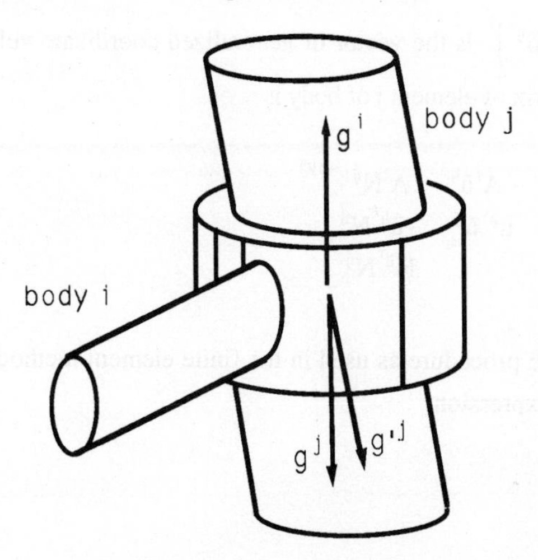

Figure 4. Revolute Joint

$$\tilde{\mathbf{g}}^{i} \times \vec{\mathbf{g}}^{j} = \tilde{\mathbf{g}}^{i}\mathbf{g}^{j} = \mathbf{0} \tag{22}$$

where $\mathbf{g}^{i}$ and $\mathbf{g}^{j}$ are measured in the global system XYZ.

Similarly, if, for example, body j is flexible one has to take into account the elastic rotations of the connecting node and equations (Equation 22) are written in the form

$$\tilde{\mathbf{g}}^{i}\left(\mathbf{A}^{j}\mathbf{G}^{j}\mathbf{g}'^{j}\right) = \mathbf{0} \tag{23}$$

Only two of the above equations are linearly independent, therefore, the total number of equations for this joint is five, which corresponds to one relative degree of freedom.

5 Kinetic Energy

The kinetic energy of element j of body i can be written as

$$\mathrm{T}^{ij} = \frac{1}{2}\int_{V^{ij}} \rho^{ij} \dot{\mathbf{r}}_{P}^{ijT} \dot{\mathbf{r}}_{P}^{ij} dV^{ij} \tag{24}$$

where ρ^{ij} is the material density in the elementary volume dV^{ij}.

Substituting Equation 15 into Equation 24,

$$\mathrm{T}^{ij} = \frac{1}{2}\dot{\mathbf{w}}^{ij^{T}} \mathbf{M}^{ij} \dot{\mathbf{w}}^{ij} \tag{25}$$

where $\dot{\mathbf{w}}^{ij} = \left[\dot{\mathbf{r}}^{i^{T}}, \omega^{i^{T}}, \delta^{ij^{T}}\right]^{T}$ is the vector of generalized coordinate velocities and $\mathbf{M}^{ij}$ is the generalized mass matrix of element j of body i.

$$\mathbf{M}^{ij} = \int_{V^{ij}} \rho^{ij} \begin{bmatrix} \mathbf{I}_3 & -\mathbf{A}^{i}\tilde{\mathbf{u}}^{ij} & \mathbf{A}^{i}\mathbf{N}^{ij} \\ & \tilde{\mathbf{u}}^{ij^{T}}\tilde{\mathbf{u}}^{ij} & -\tilde{\mathbf{u}}^{ij^{T}}\mathbf{N}^{ij} \\ \text{Sim.} & & \mathbf{N}^{ij^{T}}\mathbf{N}^{ij} \end{bmatrix} dV^{ij} \tag{26}$$

By an assemblage procedure as used in the finite element method the kinetic energy of body i results in the expression

$$\mathrm{T} = \frac{1}{2}\dot{\mathbf{w}}^{i^{T}} \mathbf{M}^{i} \dot{\mathbf{w}}^{i} \tag{27}$$

where $\mathbf{M}^i$ is the generalized mass matrix of body i which can be written in a compact notation as

$$\mathbf{M}^i = \begin{bmatrix} \mathbf{M}^i_{rr} & \mathbf{M}^i_{r\omega} & \mathbf{M}^i_{r\delta} \\ & \mathbf{M}^i_{\omega\omega} & \mathbf{M}^i_{\omega\delta} \\ \textbf{Sim.} & & \mathbf{M}^i_{\delta\delta} \end{bmatrix}$$

where: $\mathbf{M}^I_{rr}$ = Diagonal matrix $[m^i, m^i, m^i]$, m^i is the total mass of body i;
$\mathbf{M}^i_{r\omega}$ is zero when the origin of the $X^iY^iZ^i$ is located in the center of mass of body i;
$\mathbf{M}^i_{r\delta}$ and $\mathbf{M}^i_{\omega\delta}$ represent the different inertia coupling effects between the generalized rigid and elastic coordinates;
$\mathbf{M}^i_{\omega\omega}$ for small elastic deformations, is approximately equal to the inertia tensor;
$\mathbf{M}^i_{\delta\delta}$ is constant and corresponds to the usual finite element mass matrix of body i.

6 Elastic Strain Energy

The strain energy of an element j of body i is given by

$$U^{ij} = \int_{V^{ij}} U_0^{ij} dV^{ij} \tag{29}$$

where $U_0^{ij} = \int_0^{\varepsilon^{ij}} \sigma^{ij} d\varepsilon^{ij}$ and σ^{ij} and ε^{ij} are the stress and strain tensors, respectively at the point P^{ij} of the finite element j in body i.

Assuming a linear elastic material with a constitutive law

$$\sigma^{ij} = \mathbf{E}^{ij}\varepsilon^{ij} \tag{30}$$

then the strain energy can be expressed as a quadratic function of the strain tensor

$$U_0^{ij} = \tfrac{1}{2}\varepsilon^{ij^T}\mathbf{E}^{ij}\varepsilon^{ij} \tag{31}$$

For small deformations and small displacements with respect to the set of axis $X^iY^iZ^i$ attached to the body, the strain tensor can be expressed as

$$\varepsilon^{ij} = \mathbf{D}^{ij}\mathbf{u}^{ij} \tag{32}$$

where $\mathbf{D}^{ij}$ is a differential operator normally used in the finite element methodology.

Assuming that body i in its undeformed state has a null strain tensor everywhere

$$\varepsilon^{ij} = \mathbf{D}^{ij}\mathbf{N}^{ij}\delta^{ij} \tag{33}$$

Substituting Equation 33 into Equations 31 and 29,

$$U^{ij} = \tfrac{1}{2}\int_{V^{ij}} \delta^{ij^T}\mathbf{N}^{ij^T}\mathbf{D}^{ij^T}\mathbf{E}^{ij}\mathbf{D}^{ij}\mathbf{N}^{ij}\delta^{ij}dV^{ij} \tag{34}$$

or, using a more compact notation,

$$U^{ij} = \tfrac{1}{2}\delta^{ij^T}\mathbf{K}^{ij}_{\delta\delta}\delta^{ij} \tag{35}$$

where $\mathbf{K}^{ij}_{\delta\delta}$ is the element stiffness matrix.

The elastic strain energy of flexible body i can be written in matrix form after a proper assemblage of the stiffness matrices of all elements in body i:

$$U^{i} = \frac{1}{2}\begin{Bmatrix}\mathbf{r}^{i}\\ \mathbf{p}^{i}\\ \delta^{i}\end{Bmatrix}^{T}\begin{bmatrix}\mathbf{0} & \mathbf{0} & \mathbf{0}\\ \mathbf{0} & \mathbf{0} & \mathbf{0}\\ \mathbf{0} & \mathbf{0} & \mathbf{K}^{i}_{\delta\delta}\end{bmatrix}\begin{Bmatrix}\mathbf{r}^{i}\\ \mathbf{p}^{i}\\ \delta^{i}\end{Bmatrix} \tag{36}$$

Matrix $\mathbf{K}^{i}_{\delta\delta}$ is constant and corresponds to the stiffness matrix usually found in the conventional finite element formulation.

7 Equations of Motion

If, for example, Euler parameters $\mathbf{p}^{i}$ are used, the vector of generalized coordinates for body i can be written in partition form as

$$\mathbf{q}^{i} = \left[\mathbf{r}^{i^T}\mathbf{p}^{i^T}\delta^{i^T}\right]^{T} \tag{37}$$

Using the Lagrange equations, one can write the equations of motion of a system of constrained flexible bodies in terms of generalized coordinates $\mathbf{q}^{i}$ as

$$\frac{d}{dt}\left(\frac{\partial T^{i}}{\partial \dot{\mathbf{q}}^{i}}\right)^{T} - \left(\frac{\partial T^{i}}{\partial \mathbf{q}^{i}}\right)^{T} + \left(\frac{\partial U^{i}}{\partial \mathbf{q}^{i}}\right) - \mathbf{Q}^{i} + \left(\frac{\partial \Phi}{\partial \mathbf{q}^{i}}\right)^{T}\lambda = \mathbf{0} \tag{38}$$

where $\mathbf{Q}^i$ represent the external generalized forces applied in body i associated with the generalized coordinates $\mathbf{q}^i$.

The application of Equation 38 will result in a set of equations of motion involving the time derivatives of the Euler parameters. For the sake of simplicity, assume temporarily that no flexibilities are considered. If the equations of motion corresponding to the reference rotational coordinates are expressed in terms of the angular velocities ω, a modified set of Lagrange equations must be used for each body as [23]:

$$\frac{d}{dt}\left(\frac{\partial T^i}{\partial \omega^i}\right)^T - \frac{1}{2}\mathbf{L}^i\left(\frac{\partial T^{i*}}{\partial \mathbf{p}^i}\right)^T + \tilde{\omega}^i\left(\frac{\partial T^i}{\partial \omega^i}\right)^T - \frac{1}{2}\mathbf{L}\mathbf{Q}^i + \frac{1}{2}\mathbf{L}^i\left(\frac{\partial \Phi}{\partial \mathbf{p}^i}\right)^T \lambda = \mathbf{0} \qquad (39)$$

where $\mathbf{L}^i$ is (3 x 4) matrix which is a function of the Euler parameters. The term $-\frac{1}{2}\mathbf{L}^i\mathbf{Q}^i$ represents the moment components referred to set of axis $X^iY^iZ^i$ of the resultant of the external forces applied on body i.

Substituting the kinetic energy in Equation 27 and the internal energy in Equation 35 into Equations 38 and 39, the equations of motion for body i can be obtained:

$$\begin{bmatrix} \mathbf{M}^i_{rr} & \mathbf{M}^i_{r\omega} & \mathbf{M}^i_{r\delta} \\ & \mathbf{M}^i_{\omega\omega} & \mathbf{M}^i_{\omega\delta} \\ \textbf{Sim.} & & \mathbf{M}^i_{\delta\delta} \end{bmatrix} \begin{Bmatrix} \ddot{\mathbf{r}}^i \\ \dot{\omega}^i \\ \ddot{\delta}^i \end{Bmatrix} + \begin{bmatrix} \mathbf{0} & \mathbf{0} & \mathbf{0} \\ & \mathbf{0} & \mathbf{0} \\ \textbf{Sim.} & & \mathbf{K}^i_{\delta\delta} \end{bmatrix} \begin{Bmatrix} \mathbf{r}^i \\ \mathbf{p}^i \\ \delta^i \end{Bmatrix} = \begin{Bmatrix} \mathbf{Q}^i_r \\ \frac{1}{2}\mathbf{L}^i\mathbf{Q}^i_p \\ \mathbf{Q}^i_\delta \end{Bmatrix} + \begin{Bmatrix} \mathbf{F}^i_r \\ \mathbf{F}^{i*}_\omega \\ \mathbf{F}^i_\delta \end{Bmatrix} - \begin{Bmatrix} \Phi^{iT}_r \\ \frac{1}{2}\mathbf{L}^i\Phi^T_{p^i} \\ \Phi^T_{\delta^i} \end{Bmatrix} \lambda \qquad (40)$$

where the second term in the right-hand side is a vector containing the quadratic velocity terms resulting from the differentiation of the kinetic energy with respect to time and with respect to the system coordinates which correspond to the gyroscopic and Coriolis force components. $\mathbf{Q}^i_r$, $\mathbf{Q}^i_\delta$ are the energy conjugates of the translational and flexible coordinates associated with the externally applied forces.

It is convenient for now to write Equation 40 in the compact notation

$$\mathbf{M}^i\ddot{\mathbf{w}}^i + \Phi^T_{w^i}\lambda = \mathbf{f}^i_e + \mathbf{f}^i_f - \mathbf{f}^i_v \qquad (41)$$

where $\mathbf{f}^i_e$ is the vector of generalized external forces, $\mathbf{f}^i_f$ is the elastic generalized force vector corresponding to the second term in the left hand side of Equation 40 and $\mathbf{f}^i_v$ is a vector containing the quadratic velocity terms resulting from the differentiation of the kinetic energy time. It also contains the gyroscopic and Coriolis force components and other acceleration independent terms resulting from the differentiation of the kinetic energy with respect to the system coordinates.

The system equations of motion can now be obtained by combining equations of motion of each body and kinematical constraint equations describing different kinematical joint between the bodies.

Velocity equations of constraint (3) can be expressed in terms of angular velocities as

$$\mathbf{\Phi}_w \dot{\mathbf{w}} = \mathbf{0} \tag{42}$$

where $\dot{\mathbf{w}} = \left[\dot{\mathbf{r}}^T, \omega^T, \dot{\delta}^T\right]^T$ is the vector of system generalized velocities.

The acceleration equations of constraint are obtained by taking the time derivative of Equation 42

$$\mathbf{\Phi}_w \ddot{\mathbf{w}} - \gamma = \mathbf{0} \tag{43}$$

where

$$\gamma = -\dot{\mathbf{\Phi}}_w \dot{\mathbf{w}} \tag{44}$$

The differential equations of motion for the multibody system can be written as

$$\mathbf{M}\ddot{\mathbf{w}} + \mathbf{\Phi}_w^T \lambda = \mathbf{f}_e + \mathbf{f}_f - \mathbf{f}_v \tag{45}$$

Therefore, the coupled system of differential equations that govern the spatial flexible multibody system can be written in the compact matrix form

$$\begin{bmatrix} \mathbf{M} & \mathbf{\Phi}_w^T \\ \mathbf{\Phi}_w & \mathbf{0} \end{bmatrix} \begin{Bmatrix} \ddot{\mathbf{w}} \\ \lambda \end{Bmatrix} = \begin{Bmatrix} \mathbf{f}_e + \mathbf{f}_f - \mathbf{f}_v \\ \gamma \end{Bmatrix} \tag{46}$$

7.1 Reference Conditions

The problem now is to describe the rigid body motion by use of a coupled set of cartesian and rotational coordinates in order to avoid the difficulties of rigid body modes in describing the large translations and large angular rotations. In fact, in the finite element formulation, the element shape functions $\mathbf{N}^{ij}$, include rigid body modes, thus redundant coordinates have to be eliminated in order to define a unique displacement field. This can be accomplished by imposing in the elastic coordinates a set of reference conditions equal in number to the rigid body modes.

Like a static condensation process, the conditions describing the elimination of rigid body modes for each body i, can be written in the general form

$$\mathbf{R}^i \delta^i = \mathbf{0} \tag{47}$$

where $\mathbf{R}^i$ is a matrix containing *nr* rows corresponding to the number of reference conditions and *nf* columns corresponding to the number of generalized elastic coordinates of flexible body i.

An LU factorization on matrix $\mathbf{R}^i$ allows us the partition

$$\mathbf{R}^i = \left[\mathbf{R}_1^i \mid \mathbf{R}_2^i\right] \tag{48}$$

where $\mathbf{R}_2^i$ is a non-singular (*nr* x *nr*) matrix and $\mathbf{R}_1^i$ is a ((*nf* – *nr*) x *nr*) matrix. The vector δ^i is partitioned accordingly, into the (*nf* – *nr*) dimensional vector ζ^i of independent variables and the *nr* dimensional vector δ_d^i of dependent variables so that

$$\mathbf{R}_1^i \zeta^i + \mathbf{R}_2^i \delta_d^i = \mathbf{0} \tag{49}$$

or

$$\delta^i = \begin{Bmatrix} \zeta^i \\ \delta_d^i \end{Bmatrix} = \begin{bmatrix} \mathbf{I} \\ -\mathbf{R}_2^i \mathbf{R}_1^i \end{bmatrix} \zeta^i \tag{50}$$

Then the relationship between the total elastic coordinates and the independent ones can be written in the compact notation

$$\delta^i = \mathbf{R}_c^i \zeta^i \tag{51}$$

These reference conditions are not a unique set, yet they have to be consistent with the kinematic constraints imposed on the boundary of the deformable body. However, different reference conditions yield different basis for the space of elastic coordinates, so different numerical solutions may be obtained. Probably, the simplest case of reference conditions corresponds to clamping all generalized elastic coordinates in a finite element node located in the origin of the body attached set of reference axes $X^iY^iZ^i$.

It is possible to show that these conditions affect significantly the inertia coupling between the reference motion and elastic deformation of the body. In previous investigations attempts have been made to weaken this coupling by choosing deformable references that satisfy the mean axis condition [3]. However, the deformation modes resulting from the application of these conditions may be suitable only in some specific applications.

7.2 Component Mode Synthesis

In order to reduce the number of elastic degrees of freedom, the component mode synthesis method has been frequently employed [9]. In classical methods only a few of the lowest frequency normal vibration modes are used to represent the elastic deformation. This approach, however, may lead to significant error owing to the inability of normal vibration modes to account for local deformation effects induced by concentrated loads due to kinematic joints.

In the method presented herein, the elastic deformation of a flexible body is represented by a linear combination of the component modes multiplied by time-dependent generalized coordinates, i.e.,

$$\zeta^i = \mathbf{U}^i \chi^i \tag{52}$$

where $\mathbf{U}^i$is a modal matrix whose columns consist of linearly independent deformation modes, and χ^i is the vector of modal coordinates.

Two sets of deformation modes, static and normal modes [11, 32], are used to identify the configuration of flexible bodies.

The static modes are used as interface modes between bodies and are assigned to represent the effects of loads transmitted by interface conditions. Two different types of static modes are possible: constraint modes and attachment modes [17]. A constraint mode is defined as a static deformation due to imposing a unit deformation on one nodal coordinate and zero displacements on the remainder of a specified subset of the nodal coordinates. An attachment mode is obtained as a static deformation due to imposing a unit generalized force in one of the nodal coordinates and zero forces on the remainder of a specified subset of the nodal coordinates. However, the selection of these modes is not easy, and requires the engineer's intuition and judgment, especially in complex mechanisms.

Normal modes are the natural modes of vibration of the flexible body and can be determined by solving a standard eigenvalue problem involving stiffness and mass matrix of the body. This task can be carried out in any commercially available finite element code. It is well known that if the normal modes are normalized with respect to the mass matrix, the modal mass matrix becomes an identity matrix and the model stiffness matrix becomes diagonal. While these vibration modes represent deformation shapes of the body due its mass distribution, static modes represent local deformation shapes due to locally applied forces at the constraints. Therefore, a combination of static modes and vibration normal modes must be employed and most likely improve the model accuracy. Since static modes and normal modes may not be linearly independent, residual static modes are introduced so that they are linearly dependent. A detail discussion of this topic can be found in [33].

7.3 Transformation of Coordinates

In the last two sections, two successive coordinate transformations have been introduced. A first transformation intended to eliminate the redundant rigid body modes and a second transformation which maps physical elastic coordinates into the space of modal coordinates were defined. In both cases there is a dimensional reduction in the number of generalized coordinates which can be drastically significant in the second set of transformations.

Substituting Equation 52 into Equation 51,

$$\delta^i = \mathbf{H}^i\chi^i \tag{53}$$

where $\mathbf{H}^i = \mathbf{R}_c^i\mathbf{U}^i$. Differentiating Equation 53 twice with respect to time,

$$\dot{\delta}^i = \mathbf{H}^i\dot{\chi}^i \tag{54}$$

and

$$\ddot{\delta}^i = \mathbf{H}^i\ddot{\chi}^i \tag{55}$$

Using the preceding transformation rules it is possible to write

$$\dot{\mathbf{w}}^i = \begin{Bmatrix} \dot{\mathbf{r}}^i \\ \omega^i \\ \dot{\delta}^i \end{Bmatrix} = \begin{bmatrix} \mathbf{I}_6 & \mathbf{0} \\ \mathbf{0} & \mathbf{H}^i \end{bmatrix} \begin{Bmatrix} \dot{\mathbf{r}}^i \\ \omega^i \\ \dot{\chi}^i \end{Bmatrix} = \overline{\mathbf{H}}^i\dot{\overline{\mathbf{w}}}^i \tag{56}$$

and in a similar manner

$$\ddot{\mathbf{w}}^i = \overline{\mathbf{H}}^i\ddot{\overline{\mathbf{w}}}^i \tag{57}$$

Applying the transformations in Equations 56 and 57 to Equation 41, the new equations of motion for body i can be written as

$$\overline{\mathbf{M}}^i\ddot{\overline{\mathbf{w}}}^i + \Phi_{\overline{\mathbf{w}}^i}^T\lambda = \overline{\mathbf{F}}^i \tag{58}$$

where

$$\overline{\mathbf{M}}^i = \overline{\mathbf{H}}^{i^T}\overline{\mathbf{M}}^i\overline{\mathbf{H}}^i \tag{59}$$

$$\overline{\mathbf{F}}^{i} = \overline{\mathbf{H}}^{i^{T}}\left(\mathbf{f}_{e}^{i} + \mathbf{f}_{f}^{i} - \mathbf{f}_{v}^{i}\right) \tag{60}$$

$$\mathbf{\Phi}_{\overline{\mathbf{w}}^{i}} = \mathbf{\Phi}_{\mathbf{w}^{i}} \frac{\partial \mathbf{w}^{i}}{\partial \overline{\mathbf{w}}^{i}} = \mathbf{\Phi}_{\mathbf{w}^{i}} \overline{\mathbf{H}}^{i} \tag{61}$$

The acceleration constraint equations have the new form

$$\mathbf{\Phi}_{\overline{\mathbf{w}}^{i}} \ddot{\overline{\mathbf{w}}}^{i} = \mathbf{\Phi}_{\overline{\mathbf{w}}^{i}} \overline{\mathbf{H}}^{i} \ddot{\overline{\mathbf{w}}}^{i} = \gamma^{i} \tag{62}$$

A new system of equations can now be written for the reduced set of coordinates replacing all variables in Equation 45 by those with the overbar notation according to Equations 58–62. However, for the sake of simplicity let us assume, from now on, that Equation 46 without the overbars corresponds to the formulation using the reduced set of elastic coordinates as described by Equations 58 and 61.

8 Numerical Integration Methods for Absolute Dynamics

The main objective in the dynamic analysis consists in the determination of accelerations $\ddot{\mathbf{w}}$ and constraint reactions forces λ for a succession of positions and velocities of a mechanical system subjected to external forces. If Equations 58 and 62 are used, it is well known that the numerical integration of such differential algebraic equations have several drawbacks which may cause the accumulation of numerical errors in the position and velocity constraint equations which are not explicitly used in the numerical integration procedure.

Several procedures have been adopted by different authors to circumvent these problems such as the implementation of a coordinate partitioning method [8] or the introduction of a constraint violation stabilization method [20] and [2].

A different technique is taken herein where, without major efforts, the concept of relative coordinates is used at the numerical integration level giving rise to some computational efficiency, in particular in open-loop systems.

Consider two bodies i and j connected by a kinematic joint. Let θ^{ij} be the relative degrees of freedom of the joint. If θ^{ij} and the flexible coordinates δ^{i} and δ^{j} are known, it is possible to obtain the reference coordinates of body j using a relationship such as

$$\mathbf{q}_{r}^{j} = \mathbf{f}\left(\mathbf{q}_{r}^{i}, \chi^{i}, \chi^{j}, \theta^{ij}\right) \tag{63}$$

Similar expressions can be obtained for the velocities

$$\dot{\mathbf{q}}_r^j = g\left(\dot{\mathbf{q}}_r^i, \dot{\chi}^i, \dot{\chi}^j, \dot{\theta}^{ij}\right) \tag{64}$$

and an inverse expression can be obtained for the accelerations

$$\ddot{\theta}^{ij} = h\left(\ddot{\mathbf{q}}_r^i, \ddot{\mathbf{q}}_r^j, \ddot{\chi}^i, \ddot{\chi}^j, \dot{\chi}^i, \dot{\chi}^j\right) \tag{65}$$

The following algorithm can be implemented:

– At time $t = t^k$, $\mathbf{w}^i = \left[\mathbf{q}_r^{i^T}, \chi^{i^T}\right]^T$, and $\dot{\mathbf{w}}^i$, θ^{ij}, and $\dot{\theta}^{ij}$ are known.

– Using Equations 63 and 64, evaluate $\mathbf{w}^j$ and $\dot{\mathbf{w}}^j$.

– Solve the system of Equations 58 and 62.

– Using Equation 65 calculate $\ddot{\theta}^{ij}$.

– Integrate $\ddot{\chi}^i, \ddot{\chi}^j, \ddot{\theta}^{ij}$ and $\dot{\chi}^i, \dot{\chi}^j, \dot{\theta}^{ij}$ to obtain $\dot{\chi}^i, \dot{\chi}^j, \dot{\theta}^{ij}$ and $\chi^i, \chi^j, \theta^{ij}$ at time $t = t^{k+1}$.

– Using Equations 63 and 64, calculate $\mathbf{q}_r^j$ and $\dot{\mathbf{q}}_r^j$.

– Repeat first step for time $t = t^{k+1}$.

For open-chain mechanical systems all joint coordinates are independent, therefore the derivation of the system equations of motion follows a straightforward recursive procedure yielding a set of differential equations. However, for closed-loop mechanisms the generalized joint coordinates are dependent. In this case, the kinematic relations are based on loop closure conditions which can be stabilized using again the method described in [2].

9 Relative Dynamics

In the preceding sections, the system equations of motion have been derived in terms of absolute reference coordinates and relative modal coordinates. This approach is straightforward because it is easy to formulate the system equations of motion and the kinematical constraint equations for each joint, once the number of generalized coordinates is well defined for a certain finite element discretization. However, this method generates a maximum number of coupled differential and algebraic equations of motion.

A different approach uses relative joint coordinates and relative joint velocities to formulate a minimum number of equations of motion [14]. The numerical integration of these equations is by far more efficient then those expressed in terms of absolute kinematics. For open-chain mechanical systems, all generalized coordinates are independent yielding a set of differential equations for the system. On the other hand, for closed-loop mechanical systems the generalized joint coordinates are no longer independent. thus forcing the establishment of loop closure conditions.

The method herein proposed [22] takes advantage of the simplicity of the absolute coordinate methods and the computational efficiency of the relative joint coordinate formulation based on the velocity transformations as presented in [12]. The system equations of motion are first formulated in terms of the coupled set of absolute reference and elastic modal coordinates. Then these equations are transformed to the joint coordinate space by use of the velocity transformation matrix [18]. This allows an additional advantage of solving for joint relative forces in the case of inverse dynamic problems.

9.1 Relative Coordinates

Consider bodies i and j kinematically connected as shown in Figure 5. Let $X^iY^iZ^i$ and $X^jY^jZ^j$ be, respectively, the body coordinate systems associated with bodies *i* and *j*. To include the effect of flexibility, define two intermediate coordinate systems $\xi^i\eta^i\zeta^i$ and $\xi^j\eta^j\zeta^j$ rigidly attached to nodes k^i and k^j on bodies *i* and *j*, respectively. The relative joint coordinates θ^{ij} are thus defined by the relative angles and distances between the intermediate coordinate systems $\xi^i\eta^i\zeta^i$ and $\xi^j\eta^j\zeta^j$. The number of these coordinates depends on the joint type that connects the two bodies.

From Figure 5, the following relationships between body i and body j can be written:

$$\mathbf{A}^{ij} = \mathbf{B}^{i^T}\mathbf{A}^{i^T}\mathbf{A}^j\mathbf{B}^j \tag{66}$$

$$\mathbf{d}^{ij} = \mathbf{r}^j + \mathbf{A}^j\mathbf{s}^j - \mathbf{r}^i - \mathbf{A}^i\mathbf{s}^i \tag{67}$$

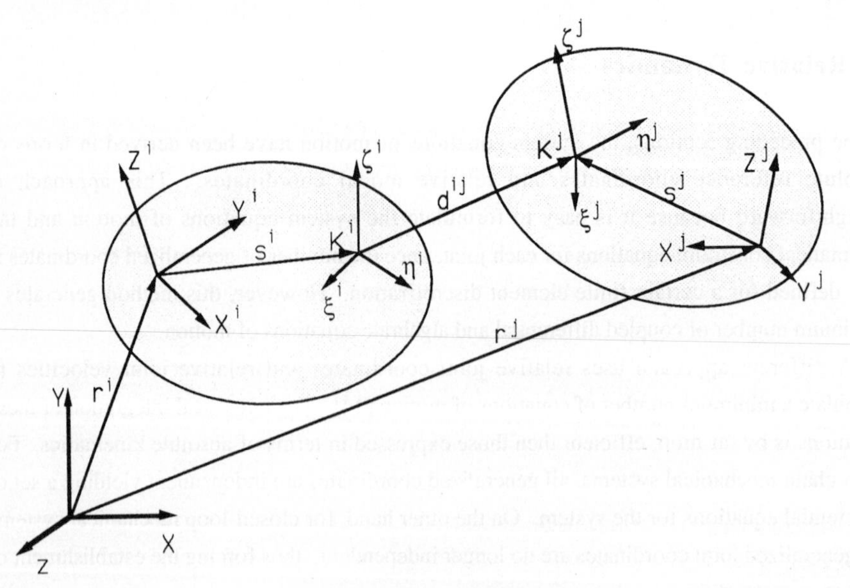

Figure 5. Relative Coordinates

where $\mathbf{A}^{ij}$ is the relative transformation matrix from the intermediate coordinate system $\xi^j\eta^j\zeta^j$ to $\xi^i\eta^i\zeta^i$, $\mathbf{d}^{ij}$ is the relative position vector of body i to body j, $\mathbf{B}^i$ and $\mathbf{B}^j$ are transformation matrices from the intermediate coordinate systems defined in the body coordinate systems. Notice that both $\mathbf{A}^{ij}$ and $\mathbf{d}^{ij}$ depend on the relative joint coordinates $\boldsymbol{\theta}^{ij}$.

Due to local deformation it must be observed that **B** and **s** must be related to the nodal elastic coordinates of the attachment node k, therefore

$$\mathbf{B} = \mathbf{B}(\varepsilon_k) \tag{68}$$

$$s = s_0 + \delta_k \tag{69}$$

where ε_k and δ_k are, respectively, the elastic rotation and displacement vector of the attachment node k. The coefficients of matrix **B** can be obtained by considering three consequential rotations, in any arbitrary order, since it was assumed that small rotations are small enough to be treated as infinitesimal. Thus the elastic rotations of the attachment node can be treated as Bryant angles in the derivation of transformation matrix **B**. This method ensures that **B** remains an orthonormal matrix.

Differentiating Equations 66 and 67 with respect to time one obtains the relationships for the relative velocities:

$$\dot{\phi}^{ij} = \mathbf{A}^j(\omega^j + \dot{\varepsilon}^j_k) - \mathbf{A}^i(\omega^i + \dot{\varepsilon}^i_k) \tag{70}$$

$$\dot{\mathbf{d}}^{ij} = \dot{\mathbf{r}}^j + \mathbf{A}^j(\dot{\delta}^j_k - \tilde{s}^j\omega^j) - \dot{\mathbf{r}}^i - \mathbf{A}^i(\dot{\delta}^i_k - \tilde{s}^i\omega^i) \tag{71}$$

where $\dot{\phi}^{ij}$ and $\dot{\mathbf{d}}^{ij}$ are, respectively, the relative angular and linear velocity vectors, both defined in the global coordinate system. The relative velocities of the intermediate coordinate systems can be written in terms of the modal elastic coordinates, as follows:

$$\dot{\varepsilon}^i_k = \mathbf{N}^i_r\mathbf{H}^i\dot{\chi}^i = \psi^i_r\dot{\chi}^i \tag{72}$$

$$\dot{\delta}^i_k = \mathbf{N}^i_r\mathbf{H}^i\dot{\chi}^i = \psi^i_t\dot{\chi}^i \tag{73}$$

where $\mathbf{N}^i_r$ and $\mathbf{N}^i_t$ are boolean matrices that isolate the terms corresponding to the elastic rotations and displacements of the attachment node k^i.

Differentiating Equations 70 and 71 with respect to time, one obtains the expressions for the relative accelerations:

$$\ddot{\phi}^{ij} = \mathbf{A}^j(\dot{\omega}^j + \ddot{\varepsilon}^j_k + \tilde{\omega}^j\dot{\varepsilon}^j_k) - \mathbf{A}^i(\dot{\omega}^i + \ddot{\varepsilon}^i_k + \tilde{\omega}^i\dot{\varepsilon}^i_k) \tag{74}$$

$$\ddot{\mathbf{d}}^{ij} = \ddot{\mathbf{r}}^{j} + \mathbf{A}^{j}\left(\ddot{\delta}_{k}^{j} - \tilde{\mathbf{s}}^{j}\dot{\omega}^{j} + 2\tilde{\omega}^{j}\dot{\delta}_{k}^{j} + \tilde{\omega}^{j}\tilde{\omega}^{j}\mathbf{s}^{j}\right)$$
$$-\ddot{\mathbf{r}}^{i} + \mathbf{A}^{i}\left(\ddot{\delta}_{k}^{i} - \tilde{\mathbf{s}}^{i}\dot{\omega}^{i} + 2\tilde{\omega}^{i}\dot{\delta}_{k}^{i} + \tilde{\omega}^{i}\tilde{\omega}^{i}\mathbf{s}^{i}\right) \tag{75}$$

9.2 Velocity Transformation

The relative position of vector $\mathbf{d}^{ij}$ is given by

$$\mathbf{d}^{ij} = \mathbf{A}^{i}\mathbf{B}^{i}\mathbf{d}^{ij''}(\theta^{ij}) \tag{76}$$

where $\mathbf{d}^{ij}$ is the relative position vector defined in the local $\xi^i\eta^i\zeta^i$ intermediate coordinate system. Now using Equations 66 and 67, one may express the cartesian elastic coordinates of body j in terms of the cartesian coordinates of body i, elastic coordinates of both bodies and joint relative coordinates

$$\mathbf{A}^{j} = \mathbf{A}^{i}\mathbf{B}^{i}(\chi^{i})\mathbf{A}^{ij}(\theta^{ij})\mathbf{B}^{j}(\chi^{j})^{T} \tag{77}$$

$$\mathbf{r}^{j} = \mathbf{r}^{i} + \mathbf{A}^{i}(\mathbf{s}_{0}^{i} + \psi_{t}^{i}\chi^{i}) - \mathbf{A}^{j}(\mathbf{s}_{0}^{j} + \psi_{t}^{j}\chi^{j}) + \mathbf{A}^{i}\mathbf{B}^{i}\mathbf{d}^{ij''} \tag{78}$$

The relative angular velocity vector can be expressed in terms of joint rotational axis vectors and time derivatives of the relative joint coordinates as

$$\dot{\phi}^{ij} = \mathbf{V}^{ij}\dot{\theta}^{ij} \tag{79}$$

where $\mathbf{V}^{ij}$ is a matrix containing in each column a unit vector coincident with a joint rotational axis, defined in the global coordinate system. If the joint has m relative joint rotational coordinates, the matrix $\mathbf{V}^{ij}$ will have m columns. Rearranging Equations 70 and 71 and using Equation 79, one may write

$$\begin{Bmatrix} \dot{\mathbf{r}}^{j} \\ \omega^{j} \\ \dot{\chi}^{j} \end{Bmatrix} = \begin{bmatrix} \mathbf{D}_{rr}^{ij} & \mathbf{D}_{r\omega}^{ij} & \mathbf{D}_{r\chi}^{ij} & \mathbf{D}_{r\theta}^{ij} & \mathbf{D}_{r\chi}^{ij} \\ \mathbf{0} & \mathbf{D}_{\omega\omega}^{ij} & \mathbf{D}_{\omega\chi}^{ij} & \mathbf{D}_{\omega\theta}^{ij} & \mathbf{D}_{\omega\chi}^{ij} \\ \mathbf{0} & \mathbf{0} & \mathbf{0} & \mathbf{0} & \mathbf{D}_{\chi\chi}^{ij} \end{bmatrix} \begin{Bmatrix} \dot{\mathbf{r}}^{i} \\ \omega^{i} \\ \dot{\chi}^{i} \\ \theta^{ij} \\ \dot{\chi}^{j} \end{Bmatrix} \tag{80}$$

in which

$$\mathbf{D}^{ij}_{rr} = \mathbf{I} \qquad \mathbf{D}^{ij}_{\omega\theta} = \mathbf{A}^{jT}\mathbf{V}^{ij}$$

$$\mathbf{D}^{ij}_{r\theta} = \mathbf{A}^{j}\tilde{\mathbf{s}}^{j}\mathbf{A}^{jT}\mathbf{V}^{ij} + \frac{\partial \mathbf{d}^{ij}}{\partial \theta^{ij}} \qquad \mathbf{D}^{ij}_{r\chi} = \mathbf{A}^{j}\psi^{j}_{t} + \mathbf{A}^{j}\tilde{\mathbf{s}}^{j}\mathbf{A}^{jT}\mathbf{A}^{i}\psi^{i}_{r}$$

$$\mathbf{D}^{ij}_{\omega\chi} = \mathbf{A}^{jT}\mathbf{A}^{i}\psi^{i}_{r} \qquad \mathbf{D}^{ij}_{\omega\omega} = \mathbf{A}^{jT}\mathbf{A}^{i}$$

$$\mathbf{D}^{ij}_{r\omega} = \mathbf{A}^{j}\tilde{\mathbf{s}}^{j}\mathbf{A}^{jT}\mathbf{A}^{i} - \mathbf{A}^{i}\tilde{\mathbf{s}}^{i} \qquad \mathbf{D}^{ij}_{\omega\chi} = -\psi^{j}_{r}$$

$$\mathbf{D}^{ij}_{r\chi} = -\mathbf{A}^{j}\psi^{j}_{t} - \mathbf{A}^{j}\tilde{\mathbf{s}}^{j}\psi^{j}_{r} \qquad \mathbf{D}^{ij}_{\chi\chi} = \mathbf{I} \tag{81}$$

The velocities of body i can be decomposed in the reference velocities $\dot{\mathbf{q}}^{i}_{r}$ and the elastic velocities $\dot{\chi}^{i}$, such that

$$\dot{\mathbf{w}}^{i} = \left[\dot{\mathbf{q}}^{iT}_{r}\dot{\chi}^{iT}\right]^{T} \tag{82}$$

According to this partition, Equation 80 can now be written in the following form

$$\dot{\mathbf{q}}^{j}_{r} = \left[\mathbf{D}^{ij}_{qq}\mathbf{D}^{ij}_{q\theta}\mathbf{D}^{ij}_{q\chi}\mathbf{D}^{ij}_{q\chi}\right]\begin{Bmatrix}\dot{\mathbf{q}}^{i}_{r}\\ \dot{\theta}^{ij}\\ \dot{\chi}^{i}\\ \dot{\chi}^{j}\end{Bmatrix} \tag{83}$$

where

$$\mathbf{D}^{ij}_{qq} = \begin{bmatrix}\mathbf{D}^{ij}_{rr} & \mathbf{D}^{ij}_{r\omega}\\ \mathbf{0} & \mathbf{D}^{ij}_{\omega\omega}\end{bmatrix};\ \mathbf{D}^{ij}_{q\theta} = \begin{bmatrix}\mathbf{D}^{ij}_{q\theta}\\ \mathbf{D}^{ij}_{\omega\theta}\end{bmatrix};\ \mathbf{D}^{ij}_{q\chi} = \begin{bmatrix}\mathbf{D}^{ij}_{r\chi}\\ \mathbf{D}^{ij}_{\omega\chi}\end{bmatrix};\ \mathbf{D}^{ij}_{q\chi} = \begin{bmatrix}\mathbf{D}^{ij}_{r\chi}\\ \mathbf{D}^{ij}_{\omega\chi}\end{bmatrix} \tag{84}$$

Assume for simplicity a first mechanical system composed of three bodies connected to each other in an open-loop topology, as shown in Figure 6.

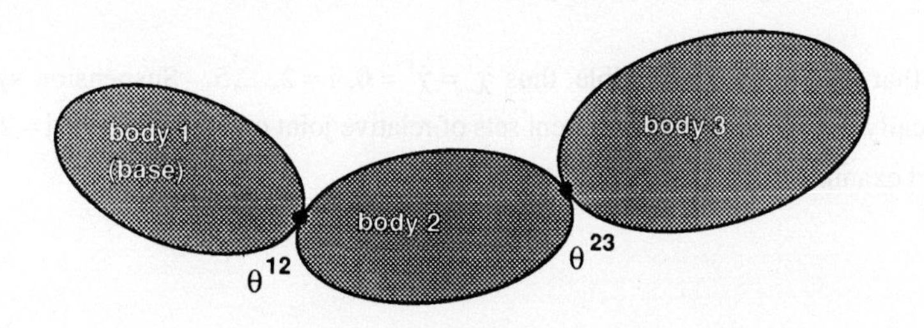

Figure 6. Three Body Open-loop Mechanical System

In this case, Equation 83 takes the form

$$\begin{Bmatrix} \dot{\mathbf{q}}_r^1 \\ \dot{\mathbf{q}}_r^2 \\ \dot{\mathbf{q}}_r^3 \end{Bmatrix} = \begin{bmatrix} \mathbf{I} & \mathbf{0} & \mathbf{0} & \mathbf{0} & \mathbf{0} & \mathbf{0} \\ \mathbf{D}_{qq}^{12} & \mathbf{D}_{q\theta}^{12} & \mathbf{0} & \mathbf{D}_{q\chi}^{12} & \mathbf{D}_{q\chi}^{22} & \mathbf{0} \\ \mathbf{X}_{qq}^{13} & {}^3\mathbf{Y}_{q\theta}^{12} & {}^3\mathbf{Y}_{q\theta}^{23} & \mathbf{Z}_{q\chi}^{13} & \mathbf{Z}_{q\chi}^{23} & \mathbf{Z}_{q\chi}^{33} \end{bmatrix} \begin{Bmatrix} \dot{\mathbf{q}}_r^1 \\ \dot{\theta}^{12} \\ \dot{\theta}^{23} \\ \dot{\chi}^1 \\ \dot{\chi}^2 \\ \dot{\chi}^3 \end{Bmatrix} \tag{85}$$

where

$$\begin{aligned} \mathbf{X}_{qq}^{13} &= \mathbf{D}_{qq}^{23}\mathbf{D}_{qq}^{12} & \mathbf{X}_{q\chi}^{13} &= \mathbf{D}_{qq}^{23}\mathbf{D}_{q\chi}^{12} \\ {}^3\mathbf{Y}_{q\theta}^{12} &= \mathbf{D}_{qq}^{23}\mathbf{D}_{q\theta}^{12} & \mathbf{Z}_{q\chi}^{23} &= \mathbf{D}_{qq}^{23}\mathbf{D}_{q\chi}^{22} + \mathbf{D}_{q\chi}^{23} \\ {}^3\mathbf{Y}_{q\theta}^{23} &= \mathbf{D}_{q\theta}^{23} & \mathbf{Z}_{q\chi}^{33} &= \mathbf{D}_{q\chi}^{33} \end{aligned} \tag{86}$$

A second example is presented related to vehicle technology. A typical vehicle can be assumed to be composed of a main floating base body connected to four wheels as shown in Figure 7.

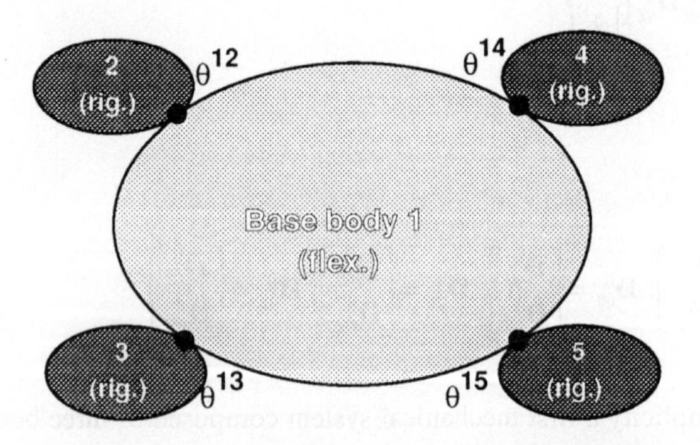

Figure 7. Five Body Four Wheel Vehicle Model

Assume that only body 1 is flexible, thus $\chi^i = \dot{\chi}^i = \mathbf{0}$, $i = 2,\ldots,5$. Suspension systems are kinematically described by four different sets of relative joint coordinates, θ^{1j}, $j = 2,\ldots,5$. In this second example Equation 83 takes the form

$$\begin{Bmatrix} \dot{\mathbf{q}}_r^1 \\ \dot{\mathbf{q}}_r^2 \\ \dot{\mathbf{q}}_r^3 \\ \dot{\mathbf{q}}_r^4 \\ \dot{\mathbf{q}}_r^5 \end{Bmatrix} = \begin{bmatrix} \mathbf{I} & \mathbf{0} & \mathbf{0} & \mathbf{0} & \mathbf{0} & \mathbf{0} \\ \mathbf{D}_{qq}^{12} & \mathbf{D}_{q\theta}^{12} & \mathbf{0} & \mathbf{0} & \mathbf{0} & \mathbf{D}_{q\chi}^{12} \\ \mathbf{D}_{qq}^{13} & \mathbf{0} & \mathbf{D}_{q\theta}^{13} & \mathbf{0} & \mathbf{0} & \mathbf{D}_{q\chi}^{13} \\ \mathbf{D}_{qq}^{14} & \mathbf{0} & \mathbf{0} & \mathbf{D}_{q\theta}^{14} & \mathbf{0} & \mathbf{D}_{q\chi}^{14} \\ \mathbf{D}_{qq}^{15} & \mathbf{0} & \mathbf{0} & \mathbf{0} & \mathbf{D}_{q\theta}^{15} & \mathbf{D}_{q\chi}^{15} \end{bmatrix} \begin{Bmatrix} \dot{\mathbf{q}}_r^1 \\ \dot{\theta}^{12} \\ \dot{\theta}^{13} \\ \dot{\theta}^{14} \\ \dot{\theta}^{15} \\ \dot{\chi}^1 \end{Bmatrix} \tag{87}$$

These examples illustrate the method whereby recursive linear relationships can be obtained to define transformations between the independent reference velocities and the dependent ones.

Let θ be the vector of system relative coordinates containing all the relative joint coordinates, all elastic modal coordinates, and if it is included, the reference coordinates of a base body. Using Equations 80 and 83, it is possible to define a linear transformation between the vector of system generalized velocities and the vector of system relative and independent elastic coordinates

$$\dot{\mathbf{w}} = \mathbf{D}\dot{\theta} \tag{88}$$

It can be shown that matrix $\mathbf{D}$ is orthogonal to the jacobian matrix $\Phi_{\mathbf{w}}$. In fact, substituting Equation 88 into Equation 41 yields

$$\Phi_{\mathbf{w}}\mathbf{D}\dot{\theta} = \mathbf{0} \tag{89}$$

Since $\dot{\theta}$ is a vector of independent coordinates, then

$$\Phi_{\mathbf{w}}\mathbf{D} = \mathbf{0} \tag{90}$$

The system of generalized accelerations can be obtained by taking the time derivative of Equation 88:

$$\ddot{\mathbf{w}} = \mathbf{D}\ddot{\theta} + \dot{\mathbf{D}}\dot{\theta} \tag{91}$$

Substituting the previous equation in the differential equation of motion (Equation 45) and premultiplying by $\mathbf{D}^T$ yields

$$M\ddot{\theta} = f \tag{92}$$

where

$$M = \mathbf{D}^{\mathrm{T}}\mathbf{M}\mathbf{D} \qquad\qquad f = \mathbf{D}^{\mathrm{T}}\left(\mathbf{f}_c + \mathbf{f}_f - \mathbf{f}_v - \mathbf{M}\dot{\mathbf{D}}\dot{\theta}\right)$$

Equation 92 represents the system equations of motion in terms of the system relative and independent accelerations. This system of equations is valid in mechanical systems where the system relative coordinates are all independent. However, in closed-loop mechanisms the system relative coordinates are not independent. Therefore, it is necessary to cut the kinematic closed-loop in one joint and obtain the transformation between the vector of system generalized velocities and the vector of system independent velocities.

9.3 Joint Coordinate Algorithm

The algorithm for the joint coordinate method can be stated as follows:

- For a given time step $t = t^k$, knowing the relative coordinates θ, the absolute coordinates $\mathbf{q}$ for all system can be evaluated using Equations 77 and 78.
- Calculate the velocity transformation matrix $\mathbf{D}$ using Equations 80 and 81.
- Evaluate the absolute velocities $\dot{\mathbf{w}}$ using Equation 88.
- Calculate the reduced system mass matrix M and the right-hand side f.
- Equation 92 yields $\ddot{\theta}$ for the same time step.
- If necessary $\ddot{\mathbf{w}}$ can be calculated from Equation 91.
- The numerical integration of $\dot{\theta}$ and $\ddot{\theta}$ moves the process to the first step with $t = t^k + \Delta t$.

10 Stress Analysis of Multibody Systems

Consider, for example, the automobile industry. The application of an integrated simulator methodology for product design including body stress analysis and vehicle dynamics must include stress analysis histories corresponding to different riding and road conditions. Fatigue and fracture design may eventually result in a byproduct which can be easily dealt with provided the structural stress histories are available.

Various approximate procedures have been used in the past [30, 13], consisting basically in the static analysis of worst case conditions assuming vertical bending loads or torsional conditions, or a combination of both. These loads are normally amplified by dynamic factor experimental programs in testing tracks. The major drawbacks in the use of standard stress analysis codes in such systems stems from the difficulties in the accurate description of realistic boundary and loading conditions, since these structures undergo large rigid body motions which are superimposed on the linear structural deformations. This is particularly true in the case of off-road vehicles.

A procedure for transient stress calculations is derived based on the present rigid-flexible dynamic analysis formulation, allowing thus an automated and accurate description of the large motion and elastic stresses and deformations of flexible mechanisms.

10.1 Reaction Forces Evaluation

In many situations the evaluation of reaction forces is desired for design purposes, once they strongly influence the structural behavior of the different bodies in a mechanical system. These reaction forces need also to be available for the proper use of finite element codes in stress calculations as it will be shown later.

During the dynamic analysis of a mechanical system it has been shown that the elastic coordinates χ^i and the corresponding velocities $\dot{\chi}^i$ and accelerations $\ddot{\chi}^i$ are directly obtained from the integration process. Also, for each body, the reference coordinates $\dot{\mathbf{q}}_r^i$, velocities $\dot{\mathbf{q}}_r^i$, and accelerations $\ddot{\mathbf{q}}_r^i$ can be obtained in a recursive manner from Equations 77, 78, and 91. The nodal elastic coordinates can be extracted using the relations

$$\delta^i = \psi^i \chi^i; \;\; \dot{\delta}^i = \psi^i \dot{\chi}^i; \;\; \ddot{\delta}^i = \psi^i \ddot{\chi}^i \tag{93}$$

where $\psi^i = \mathbf{R}_c^i \mathbf{U}^i$.

Decomposing matrix Mi and vectors ff and fv according to Equation 82 the reaction forces Λ can be extracted from Equation 40 in the following manner:

$$\begin{Bmatrix} \Lambda_q \\ \Lambda_\delta \end{Bmatrix} = \begin{bmatrix} \mathbf{M}_{qq}^i & \mathbf{M}_{q\delta}^i \\ \mathbf{M}_{\delta q}^{i*} & \mathbf{M}_{\delta\delta}^{i*} \end{bmatrix} \begin{Bmatrix} \ddot{\mathbf{q}}_r^i \\ \ddot{\delta}^i \end{Bmatrix} + \begin{bmatrix} \mathbf{0} & \mathbf{0} \\ \mathbf{0} & \mathbf{K}_{\delta\delta}^{i*} \end{bmatrix} \begin{Bmatrix} \mathbf{q}_r^i \\ \delta^i \end{Bmatrix} - \begin{Bmatrix} \mathbf{f}_{f_q} \\ \mathbf{f}_{f*_\delta} \end{Bmatrix} + \begin{Bmatrix} \mathbf{f}_{v_q} \\ \mathbf{f}_{v*_\delta} \end{Bmatrix} \tag{94}$$

where Λ_q and Λ_δ, are the reaction components associated with the reference coordinates and the nodal elastic coordinates, respectively. The overstar denotes the corresponding submatrices which have been stripped of all rows except those associated with the degrees of freedom of the attachment finite element nodes located in the kinematic joints. This simplification considerably speeds up the reaction force evaluation process. Equation 94 assumes no reference conditions for the elastic coordinates, thus care should be taken to provide the off-diagonal mass matrices $\mathbf{M}_{q\delta}^i$ even if mean axis reference conditions have been assumed.

10.2 Stress Calculations

Some approaches have been developed for stress calculations of machine components using quasi-static finite element analysis with rigid body dynamics [19]. Kinetostatic methods can e used by combining rigid body formulations with quasi-static finite element analysis [1].

It is well known that methods for computing stresses from a linear combination of deformation modes using modal generalized coordinates may result in large inaccuracies in he stress calculations. Yoo and Haug [33] have proposed a combination of attachment modes and normal vibration modes to improve the structural model, mainly in situations where locally applied forces will develop due to the kinematic constraints in the joints. Smaller and coarser finite element models may give adequate stiffness and deflection results, which is important to take into account in order to minimize the computing efforts during a ull dynamic analysis of rigid flexible mechanical systems. Complete finite element body models are used primarily for the static behavior, modal analysis and vibration response. Displacements, velocities and acceleration fields must be generated and organized in a data base in order to be used at latter stages for stress calculations. In doing so, the stresses can be recovered only in specific areas of interest which can be remodeled by using refined meshes and providing correct boundary conditions.

Using the partition (Equation 82), a dynamic equation can be derived from Equations 40 and 94:

$$\mathbf{M}^{i}_{q\delta}\ddot{q}^{i}_{r} + \mathbf{M}^{i}_{\delta\delta}\ddot{\delta}^{i} + \mathbf{K}^{i}_{\delta\delta}\delta^{i} = \mathbf{f}_{f_\delta} - \mathbf{f}_{v_\delta} + \Lambda_\delta \tag{95}$$

or, transferring the inertia loads to the right-hand side and using $\dot{\delta}^{i}$ and $\ddot{\delta}^{i}$ from the dynamic analysis, and assuming the generalized elastic coordinates δ^{i*} as unknowns, a structural equilibrium equation call be written:

$$\mathbf{K}^{i}_{\delta\delta}\delta^{i*} = -\mathbf{M}^{i}_{q\delta}\ddot{\delta}^{i} = \mathbf{f}_{f_\delta} - \mathbf{f}_{v_\delta} + \Lambda_\delta \tag{96}$$

The right-hand side of Equation 96 describes a set of external forces including the joint reactions and a set of nodal equivalent point loads consistent with all inertia effects. These two sets of generalized loads are easily available at the end of a dynamic analysis and, according to the D'Alembert's principle are self equilibrated.

Since Equation 96 is written in the body reference frame it can be solved using any commercially available finite element code, regardless of the gross motion of the flexible body. As a result, different loading conditions corresponding to different time steps can be solved simultaneously with one single inversion process of the stiffness matrix $K^i_{\delta\delta}$. A final remark on boundary conditions. Since $K^i_{\delta\delta}$ is singular, a necessary and sufficient set of six boundary conditions must be imposed on the problem in Equation 96 to avoid rigid body motions in the attached body frame.

11 Advanced Integrated Analysis System

A computer oriented methodology is now described based on the formulations developed in the preceding sections. Four main application modules have been developed corresponding to the different phases of the proposed methodology which is schematically shown in Figure 8.

First, a *pre-processor* handles all input data describing the model problem. The data is separated into three sets: i) rigid body data defining the general mass properties of each body, kinematic joint geometric definition data, topology information and initial suggested configuration and kinematically admissible velocities; ii) Flexible body data which corresponds to the finite element model description of each flexible body. Some of the flexible and rigid body data are redundant, allowing several checking procedures for data consistency and giving the user a general format for choosing which bodies are to be considered to be flexible: iii) Special forcing and driving functions can be specified to describe different ride or operational conditions for the mechanical system. In the case of a vehicle a terrain model must be supplied.

General mass and stiffness matrices are assembled at this stage. Modal analysis and the calculation of static correction modes is carried out after imposing reference conditions for the flexible problem in each body. An object oriented data base is then created for the mechanism model to be used at later stages.

A *dynamic analysis* module automatically generates the system dynamic differential equations of motion which are integrated using any appropriate solution method [33]. This module is CPU intensive and can be run in batch mode in a high speed computer platform. It is convenient, at this stage, to generate and save the reaction forces and other forces once $\mathbf{M}^i_{q\delta}$ in Equation 94 is position and velocity dependent.

A *stress analysis* module may use any commercially available finite element code. For this purpose an interface has to be included to use the finite element model definition data and the appropriate generalized force vector describing all external forces, reaction forces and inertia forces which are retrieved from the data base system.

The present analysis tool has been completed with an *animation* program implemented in a CDC910-300 workstation allowing the animation of solid objects made of color filled polygons with hidden line removal. This program is externally linked with the IGL graphics library from Silicon Graphics. This module allows the representation of rigid and flexible bodies including an user defined scaling factor to be applied to the structural deformations.

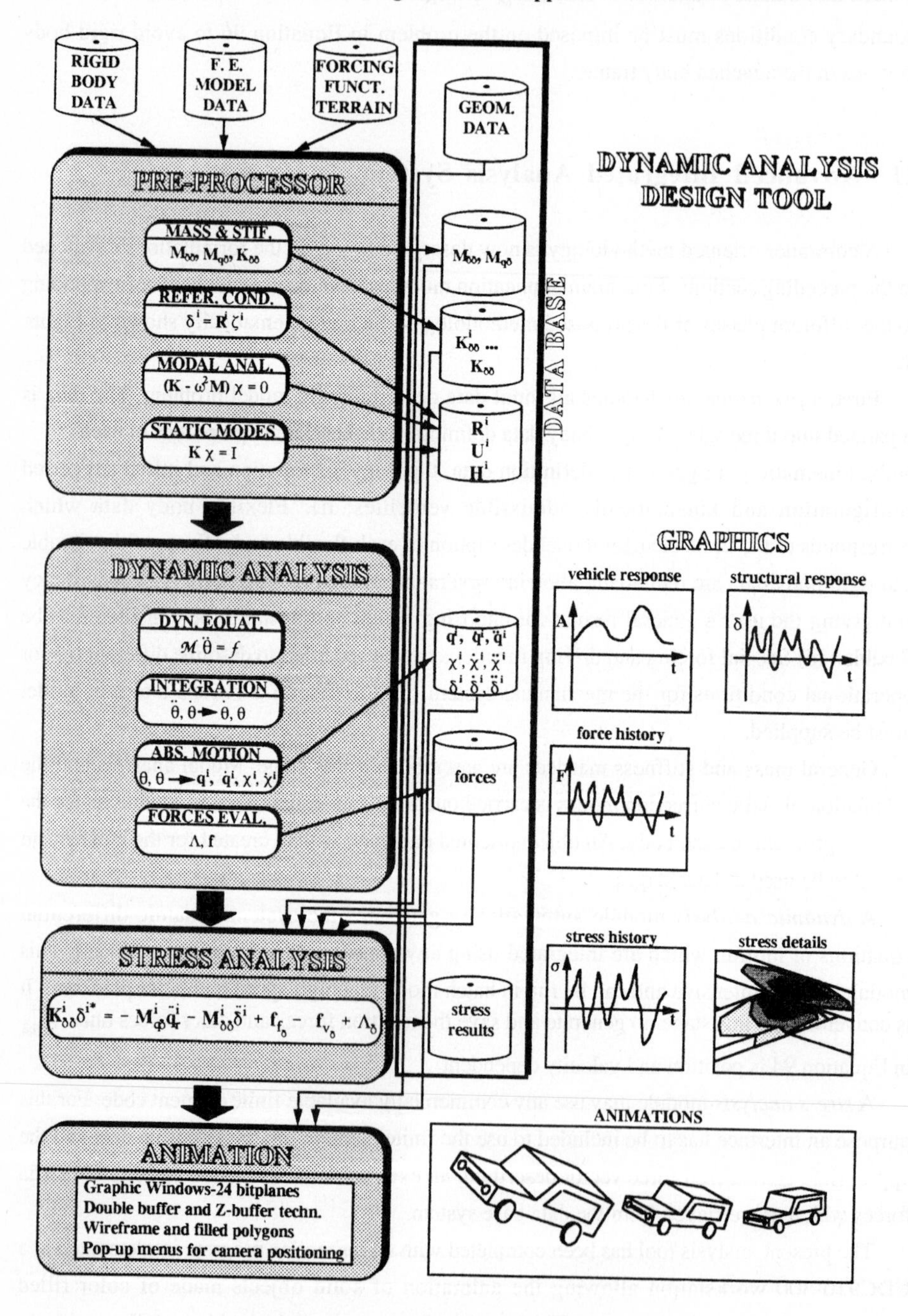

Figure 8. Advanced Integrated Analysis System

Finally, a menu driven XY graphics window can be open to display a wide range of X-t plots of the relevant engineering results, such as position, velocities, forces, stresses and deformations.

12 Example

Figure 9 shows an off-road vehicle UMM atr II. The model of the vehicle consists of the main chassis, the complete suspension system and four wheels. Only the chassis is treated as a flexible body. The two front and back wheels are connected by a rigid axle. Suspension leaf springs and shock absorbers between the wheels and the main chassis are modeled by translational spring dumper actuators with non-linear characteristics.

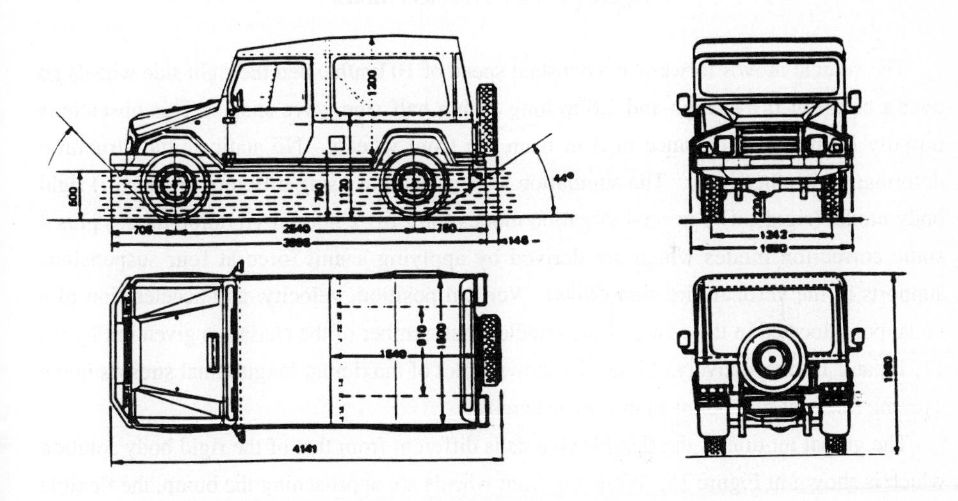

Figure 9. UMM atrII-type Vehicle

As shown schematically in Figure 10, the structure of the chassis has two side rails and three crossmembers. The finite element model consists of 47 nodes and 48 three dimensional beam elements.

Mechanical parts such as engine, transmission, fuel tank, and steering are treated as non-structural lumped masses and are attached to several nodes of the chassis finite element model. Vibration analysis is done with these non-structural masses and with mean axis boundary conditions.

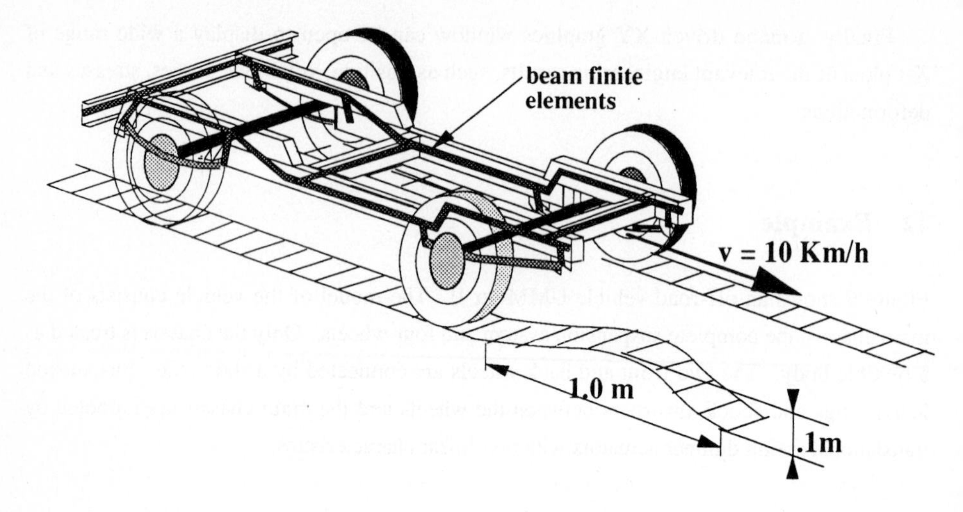

Figure 10. Finite Element Model

The vehicle moves forward at a constant speed of 10 km/h when the right side wheels go over a bump of 0.1 m high and 1.0 m long with a half sine wave shape. This obstacle is initially located at a distance of 3 m from the front wheels. No static initial structural deformation is considered. The simulation was carried out using different models: i) rigid body model (rigid); ii) 6 normal vibration modes (6n); iii) 2 normal vibration modes plus 4 static correction modes which are derived by applying a unit force at four suspensions supports in the vertical direction (2n4s). Vertical position, velocity, and acceleration of a nodal point located in the center of the middle crossmember of the chassis is given in Figures 11, 12, and 13, respectively. Figure 14 shows a plot of maximum longitudinal stresses in the extreme fibers of mid points in the two side rails.

The global motion of the flexible chassis is different from that of the rigid body solution which is shown in Figure 15. While the front wheels are approaching the bump, the flexible chassis tend to oscillate about the static equilibrium position. This behavior is expected since no initial deformation was assumed due to static loads. In this case all velocity and accelerations with higher values as compared with the rigid body solution and some higher frequency vibrations.

The stress results for the right-hand side rail present the same trend of the displacement amplitude whereas the stress history on the symmetric point on the left-hand side rail shows a large peak value of approximately 350 MPa at a latter stage in the simulation. The stress histories of both points exhibit a high frequency content due to the natural modes of vibration of the structure.

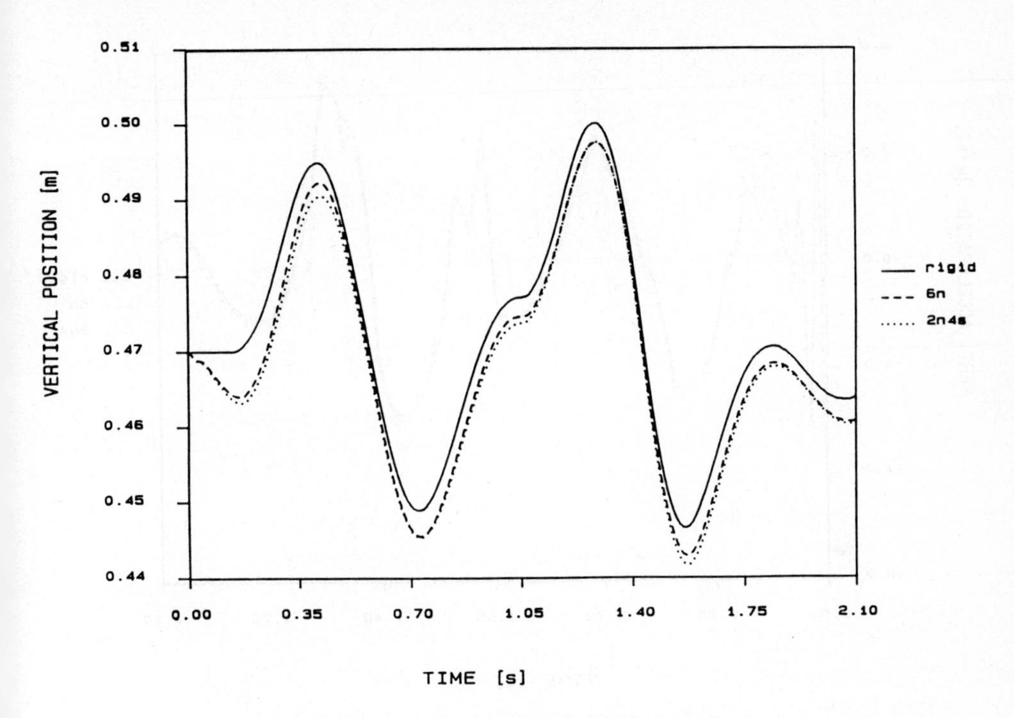

Figure 11. Vertical Position of the Chassis

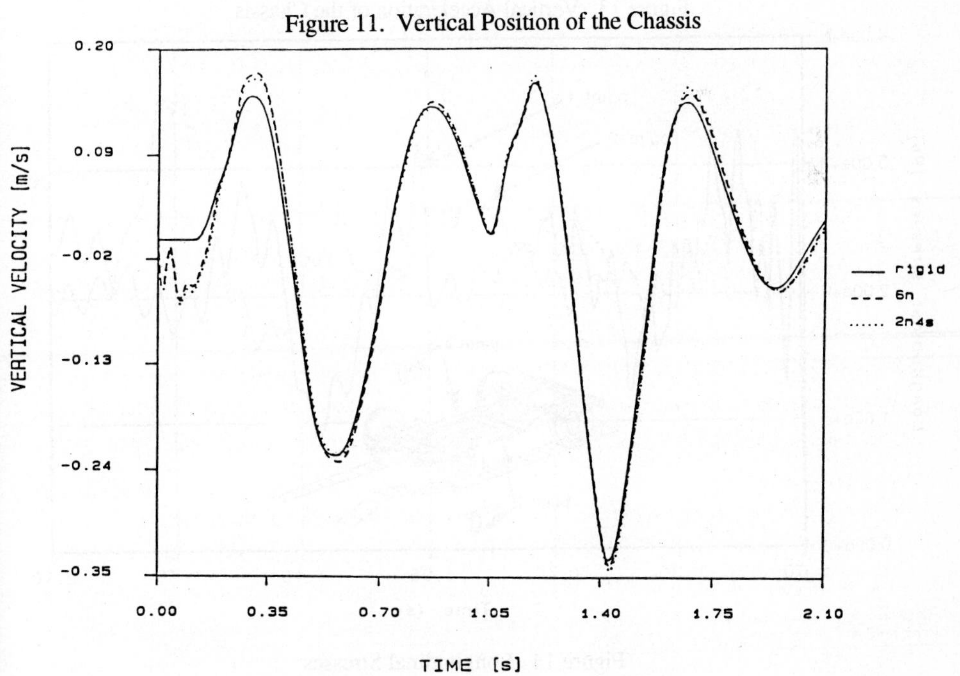

Figure 12. Vertical Velocity of the Chassis

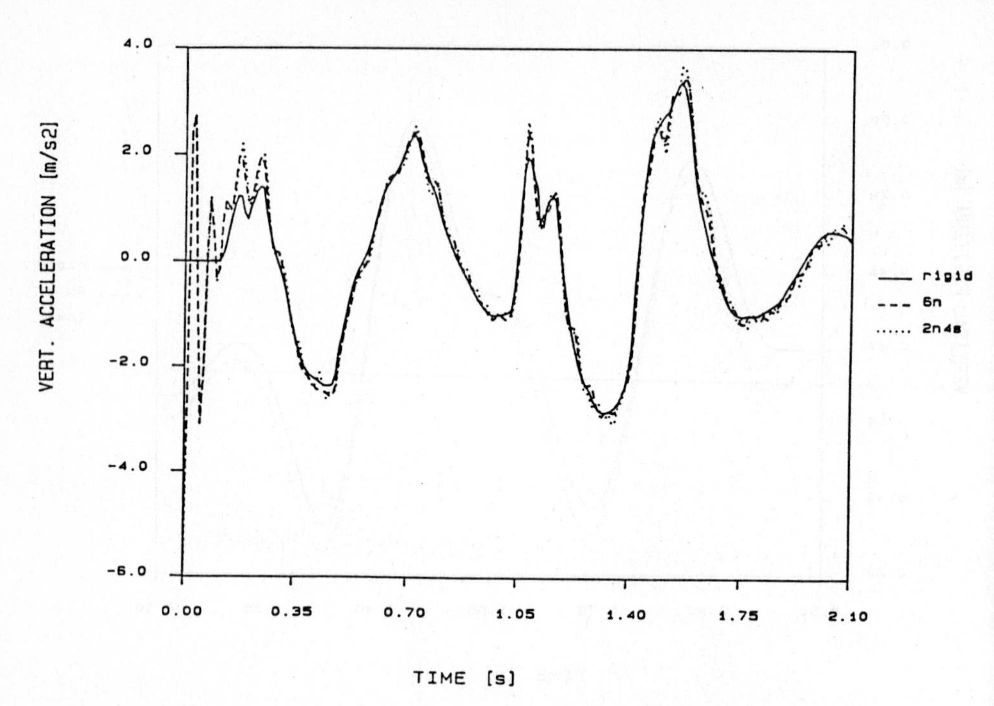

Figure 13. Vertical Acceleration of the Chassis

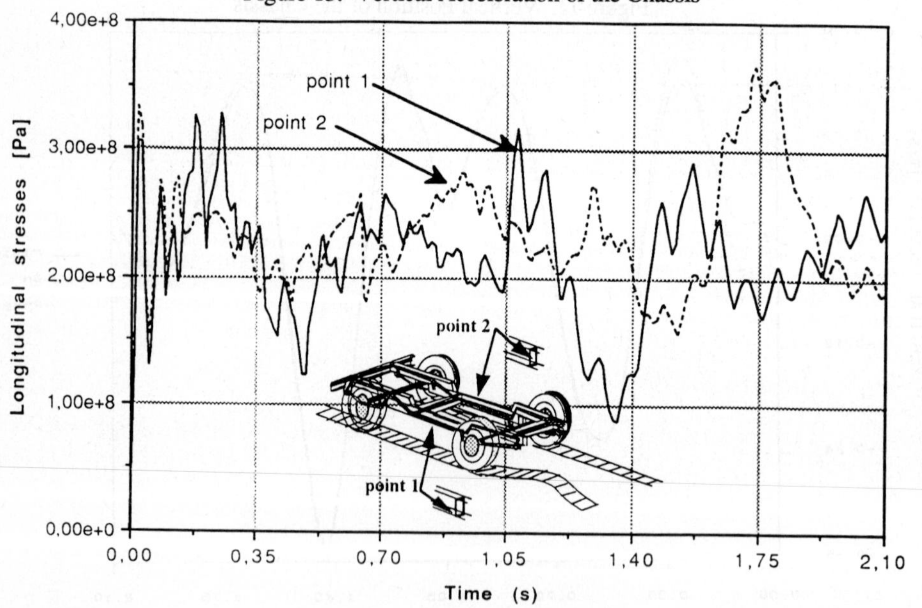

Figure 14. Longitudinal Stresses

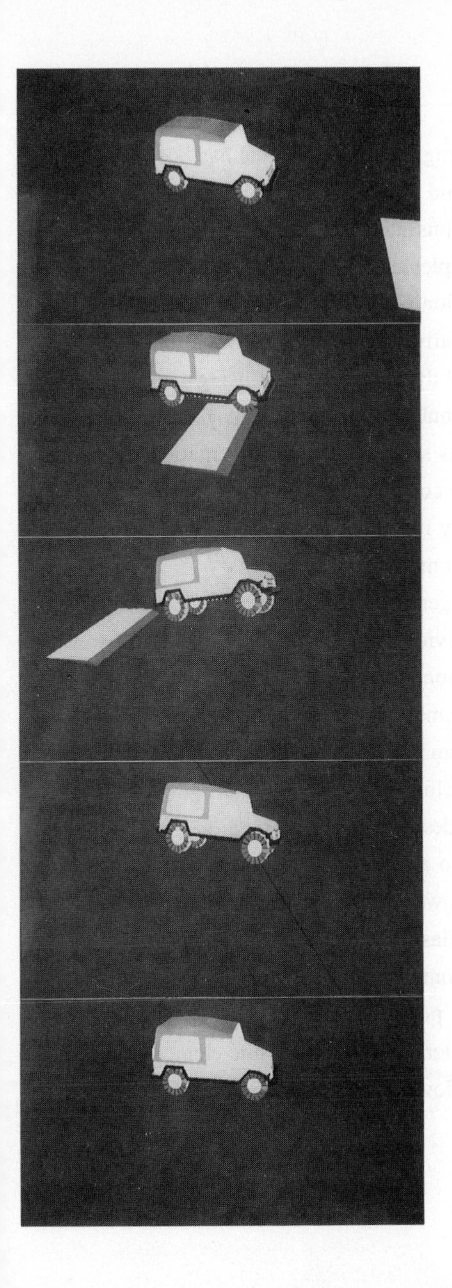

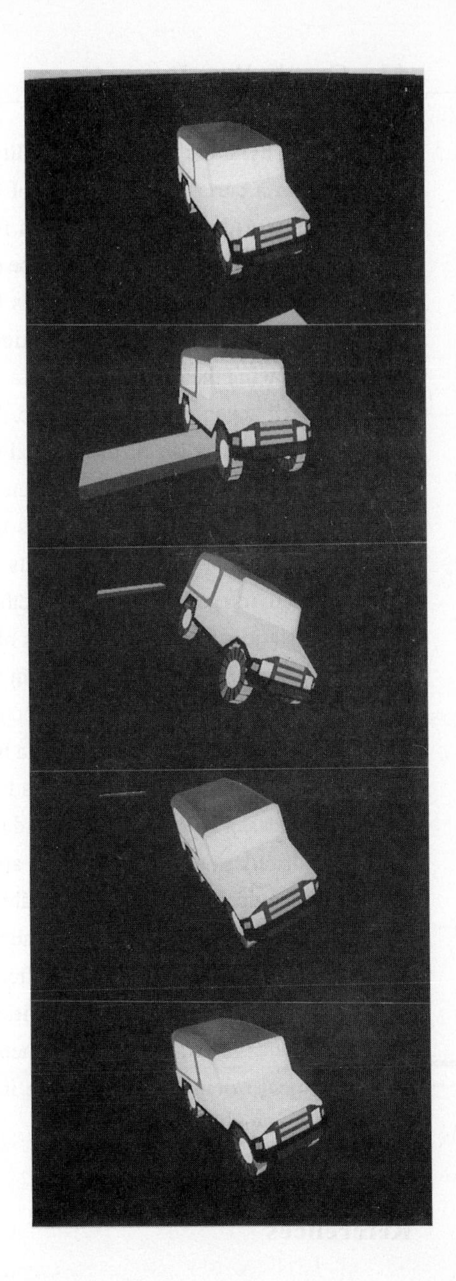

Figure 15. Vehicle Simulation Animations

13 Concluding Remarks

A multibody dynamic analysis capability including flexibility effects has been presented. Starting from a cartesian description of bodies and joints, which is more appealing to the engineering world, absolute dynamics formulations have been reviewed, and integration methods using relative coordinates can be easily implemented with some computational gains.

One of the major drawbacks of this formulation results from the differential algebraic nature of the system dynamics equations. Relative kinematics formulation have been presented allowing the establishment of a system of dynamic equations using a minimal set of rigid and flexible generalized coordinate. This formulation has the advantage of generating in a recursive manner a set of differential equations suitable for implementation in parallel computing platforms, thus drastically increasing the computational efficiency. One of the main features of the present formulations is the ability to model a broad range of mechanical systems with or without flexibility effects with minimum modeling efforts.

The present integrated simulation methodology generates all the necessary interface data to carry out a static stress model analysis of the individual bodies using in-house software or commercially available codes to perform the additional functions of stress history evaluation or stress detailing for fatigue design of complex geometries such as joints and intersections.

A well organized object oriented data base system facilitates the flow between the different modules, gives a more efficient access for the various tasks including all graphical output resulting in a easier and faster response during the design analysis process.

Basic formulations .In the general approach to a design and analysis methodology for flexible mechanical systems is now established. However, some aspects should be addressed in a more systemic manner, such as: number of elastic modes and what modes to be used; robust integration techniques for high frequency contents; different types of finite elements; material and geometric non-linearities with possible applications to impact and crashworthiness; interfacing with geometric modelers and existing finite element codes and modular object oriented programming in C++ allowing a direct mapping of the different engineers concept in software.

References

1. Ambrósio, J.A., Nikravesh, P.E., and Pereira, M.S.: Crashworthiness Analysis of a Truck, Mathl. Comput. Modelling, Vol. 14, pp. 59-964, (1990).
2. Baumgarte, J.: Stabilization of Constraints and Integrals of Motion, Computer Methods in Applied Mechanics in Engineering, 1, (1972).
3. Cavin, R.K. and Dusto, A.R.: Hamilton's Principle: Finite Element Methods in Flexible Body Dynamics, AIAA J. Vol. 15, pp 1684-1690, (1977).
4. Cardona, A., and Geradin, M.: A Beam Finite Element Non-Linear Theory with Finite Rotations, Int. J. Num. Meth. Engng., Vol. 26, pp. 2403-2438, (1988).

5. Djerassi, S. and Kane, T.R.: Equations of Motion Governing the Deployment of Flexible Linkages from a Spacecraft, The J. Astr. Sci., Vol. 33, No. 4, pp. 417-428, (1985).
6. Geradin, M. and Cardona, A.: Kinematics and Dynamics of Rigid and Flexible Mechanisms Using Finite Elements and Quaternion Algebra, Computational Mechanics, Vol. 4, pp. 115-135, (1989).
7. Guyan, R.J.: Reduction of Stiffness and Mass Matrices, AIAA Journal, Vol. 3, p. 380, (1965).
8. Haug, E.J. and Wehage, R.A.: Generalized Coordinate Partitioning for Dimension Reduction in Analysis of Constrained Dynamic Systems, J. Mech. Design, Vol. 104, pp. 247-255, (1982).
9. Hintz, R.M.: Analytical Methods in Component Mode Synthesis, AIAA J., Vol. 15, pp 1007-1016, (1975).
10. Hurty, W.C.: Dynamic Analysis of Structural Systems Using Component Modes, AIAA J. Vol.3, No.4, pp. 687-685, (1965).
11. Imbert, J.F.: Analyse des Structures par Elemnts Finits, Cepadues Ed., Toulouse, 1984.
12. Jerovsky, W.: The Structure of Multibody Dynamic Equations, J. Guid. Control, Vol. 1, pp. 177-182, (1978).
13. Kamal, M.M. and Woll, J.A. Eds.: Modern Automotive Sructural Analysis, Van Nostrand Reinhold.
14. Kane, T.R. and Levinson, D.A.: Dynamics Theory and Applications, McGraw-Hill, New York, 1985.
15. Kim. S.S. and Haug, E.J.: A Recursive Formulation for Flexible Multibody Dynamics, Part 1: Open Loop Systems, Center for Simulation and Design Optimization of Mechanical Systems, The University of Iowa, Techn. Rep. R-21, (1988).
16. Kim, S.S. and Haug, E.J.: A Recursive Formulation for Flexible Multibody Dynamics, Part II: Closed Loop Systems, Center for Simulation and Design Optimization of Mechanical Systems, University of Iowa, Techn. Rep. R-22 (1988).
17. Kim, S.S. and Haug, E.J.: Selection of Deformation Modes for Flexible Multibody Dynamics, T.R. 86-22, Center for Computer Aided Design, University of Iowa, Iowa City, 1986.
18. Kim, S.S. and Vanderploeg M.J.: A General and Efficient Method for Dynamic Analysis of Mechanical Systems Using Velocity Transformations, J. Mech. Trans. Automation Des. ASME, Vol. 108, pp. 176-182, (1986).
19. Liu, T.S. and Haug, E.J.: Computational Methods for Life Prediction of Mechanical Components of Dynamic Systems, Technical Report 86-24, Center for Computer Aided Design, University of Iowa, Iowa Cily, Iowa, 1986.
20. Nikravesh, P.E.: Computer Aided Analysis of Mechanical Systems, Prentice Hall, New Jersey, 1988.
21. Nikravesh, P.E. and Gim, G.H.: Systematic Construction of the Equations of Motion for Multibody Systems Containing Closed Kinematic Loops, Proc. ASME Design Automation Conference, Montreal, Canada, Sept 17-20, (1989).
22. Pereira, M.S. and Proença, P.L.: Dynamic Analysis of Spatial Flexible Multibody Systems Using Joint Coordinates, Int. J. Num. Melh. in Engng., Vol. 32, pp. 1799-1812, (1991).
23. Proença, P.: Análise Tridimensional de Sistemas Mecânicos Rígido-Flexiveis, MSc Thesis, Instituto Superior Técnico, Lisboa, (1990).
24. Shabana, A.A. and Wehage, R.A.: Variable Degree of Freedom Component Mode Analysis of Inertia Variant Flexible Mechanical Systems, ASME J. Mech. Transm. Autom and Des. Vol. 105, pp. 371-378, (1983).
25. Shabana, A.A.: Dynamics of Multibody Systems, John Wiley & Sons, New York, 1989.
26. Shampine, L.F. and Gordon, M.K.: Computer Solution of Ordinary Differential Equations: The Initial Value Problem, Freeman, San Francisco, 1975.
27. Simo, J.C., and Vu-Quoc, L.: On the Dynamics of Flexible Beams under Large Overall Motions – The Plane Case: Parts I and II, ASME J. of Appl. Mech., Vol. 53, pp. 849-863, (1986).
28. Sing, R.P. and Likins, P.W.: Singular Value Decomposition for Constrained Dynamical Systems, J. Appl. Mech., Vol. 52, No. 4, pp. 943-948, (1985).
29. Song, J.0. and Haug, E.J.: Dynamic Analysis of Planar Flexible Mechanisms, Comp. Meth. in Appl. Mech. and Engng.,Vol. 24, pp. 379-381, (1980).
30. Webb, G.G.: Torsional Stiffness in Passenger Cars, Int. Conf. Vehicle Structures, I Mech. E., Cranfield Institute of Technology, 16-18 July 1984.
31. Yae, K.H. and Inman D.H.: Flexible-Body Dynamic Modelling with a Reduced Order Model, Advances in Design Automation, Vol II-Optimal Design and Mechanical Systems, Ed. B. Ravani, ASME, DE-VOL 23, 2, pp. 87-91, 1990.
32. Yang, Z. and Sadler, J.P.: Large Displacement Finite Element Analysis of Flexible Linkages, ASME J. Mech. Design, Vol. 112, pp. 175-182, (1990).
33. Yoo, W.S. and Haug, E.J.: Dynamics of Flexible Mechanical Systems Using Vibration and Static Correction Modes, J. Mech., Transm. and Autom. Design, Vol. 8, pp. 315-322, (1986).

Decoupling Method in Control Design

M. Cotsaftis[1] and C. Vibet[2]

[1]UGRA CEN Fontenay aux Roses, France
[2]Université d'E.V.E., IUT Dépt G.E., Evry, France

Abstract: In the scope of Concurrent Engineering activities, this report introduces a control strategy which allows the systematic construction of controlled multibody systems. The feedback-law derivation is based on the Nonlinear Decoupling Method. This eases the design of hardware and software control loops of mechanical controllers with high-reliability features.

Keywords: controlled mechanisms / high-reliability controllers / nonlinear decoupling method

1. Introduction

Control strategies are playing a crucial role in the design of automated systems. This is especially true in the area of Mechatronics [1] where most mechanical elements, such as the moving parts of manipulators or those of production cells, are governed via feedback-laws by means of digital computers [2]. The increasing need to control mechanisms is due to the fact that stable servo-systems with arbitrary dynamics can be synthesized for most physical systems. As a result, unmodeled dynamics and unwanted disturbances such as the stick-slip effect, for example, are reduced through feedback. At the same time, this means that, on the contrary, uncontrolled mechanisms in which the dynamics result only from a physical arrangement of links and masses of multibody systems, the dynamical features of controlled mechanisms can be easily specified from statements of a real-time computer program. This is such a flexibility which makes that Control enhances at low cost the design versatility.

Though many techniques are available to design controlled mechanical systems, only a few of them allow an automatic generation of feedback-laws in a symbolic form. The Nonlinear Decoupling Method [3-8] presents such an advantage; it allows the construction of codes which automatically formulate the symbolic control laws of dynamical systems. This feature facilitates the use of CAD programs devoted to the analysis of mechanical systems [9] when controlled mechanisms have to be designed [10]. In Section 2 of this report, the concept of servo-mechanisms with graceful-failure capability is introduced in the context of feedback-law design [11]. The third part presents the main goals achieved by the Nonlinear

Decoupling Method when applied to the construction of controlled multibody systems. In particular, this shows that smart servo-systems result from the use of off-line planed trajectory. Section 4 shows how the controllers have to be computed from a state-space description of the dynamics of mechanisms. The simplest case of joint-controlled systems is first analyzed from the mechanism dynamical equations. The method of the feedback-law construction is then given, and the pole allocation problem is solved. A general formulation of the servo-design process is provided in the last part. Next, a methodology to derive the control scheme of a mechanism from its Lagrangian is given in Section 5. This approach reduces the amount of calculations needed to establish the dynamical model as well as its related control laws.

2. Design of Controllers With High-Reliability Features

In the servo-design challenge, many approaches are available to build real-time servo-systems. But few of them provide a feedback-law construction which takes into account the reliability features of servo-loops [11]. In the context of Concurrent Engineering activities, a great advantage of the Decoupling Method is that it allows a servo-design with graceful-failure capabilities. Practically, this ability results from the fact that the hardware part of control circuits which are not computerized must still function and be efficient, even when the servo-computer fails. As shown in Figure 1, the advanced control scheme must be realized in two parts: a hardware part and a computerized part.

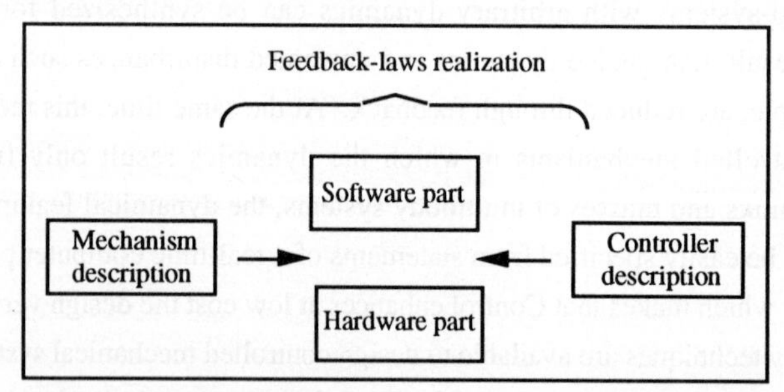

Figure 1. Feedback-Law Implementation in Controllers with High-Reliability Features

There are many advantages to using control systems with such graceful-failure capabilities. In particular, this typical arrangement allows the failure detection of the controller's components via fast automation. Although there is no simple way to decide which part of a feedback-law must be implemented in hardware control circuits, simulation studies can be helpful in solving such a problem. For example, the first step of the work consists of designing via simulation, a simple hardware-servo which sufficiently meets the most essential specifications, such as the closed-loop stability, etc. When designed, the hardware feedback-loop must finally be associated with a computerized control loop in such a way that the resulting effect realizes the decoupling and feedback linearization. Since the Nonlinear Decoupling Method allows the development of such operations through symbolic manipulations or via numerical computations, this control approach appears to be a good candidate to support Concurrent Engineering activities.

3 Nonlinear Decoupling Method

First applied by Freund to robots [3], the Nonlinear Decoupling Method transforms a nonlinear differential system into a linear one via nonlinear feedback . A simple way to show this consists of starting from the dynamical model of a mechanism, such as

$$A(q)\ddot{q} + B(q,\dot{q}) = T \tag{1}$$

where $\mathbf{q}$ represents the $(n \times 1)$ vector of generalized coordinates and $\dot{\mathbf{q}} = d\mathbf{q}/dt$. In Equation 1, $\mathbf{A(q)}$ is the $(n \times n)$ symmetric positive inertia matrix, $B(\mathbf{q},\dot{\mathbf{q}})$ is a $(n \times 1)$ vector which takes into account Coriolis and centrifugal and gravitational forces, while $\mathbf{T}$ represents the $(n \times 1)$ vector of generalized forces (or torques) driving the mechanism. In the case of a joint-controller, the purpose of decoupling is to drive the system from a computed torque such as

$$T(t) = N(q,\dot{q}) + M(q)\,u(t) \tag{2}$$

where the right-hand side represents an algorithm which is computed in real-time, and $\mathbf{u}(t)$ is a n-dimensional control vector. When the feedback algorithms are selected, such as $\mathbf{A(q)} = \mathbf{M(q)}$, and $\mathbf{B}(\mathbf{q},\dot{\mathbf{q}}) = \mathbf{N}(\mathbf{q},\dot{\mathbf{q}})$, the control law implementation reduces Equation 1 to a diagonal linear system, such as

$$\ddot{q} = v(t) \tag{3}$$

when the servo-power input $\mathbf{v}(t)$ is computed from $\mathbf{u}(t) = \mathbf{M(q)}\,\mathbf{v}(t)$. The nonlinear control design is therefore reduced to a linear and decoupled control problem for which the pole assignment is easily realized. In order to do that, the control signal $\mathbf{v}(t)$ must be computed so that Equation 3 becomes

$$\ddot{q} = K_p(q_d - q) + K_v(\dot{q} - \dot{q}) + \ddot{q} \tag{4}$$

where $\mathbf{K}_p$ and $\mathbf{K}_v$ are constant diagonal matrices with positive coefficients. $\mathbf{q}_d$ represents the n-dimensional vector of desired outputs, i.e., the controller input. $\mathbf{q}$ is the actual output vector which comes from instrumented joint measurements. In Equation 4, it is assumed that the kinematic features of the desired trajectory are known in advance for the position, speed, and acceleration profiles. Under these conditions, the controller error equation appears as

$$\ddot{\Phi} + K_v\dot{\Phi} + K_p\Phi = 0 \tag{5}$$

where the static tracking error is defined by $\mathbf{\Phi} = \mathbf{q}_d - \mathbf{q}$. When a stable control system is designed, the tracking error goes to zero with the dynamics specified by Equation 5, e.g., the nth diagonal element k_v^n of the matrix $\mathbf{K}_v$ implies the natural damping coefficient ζ_n of the nth linear controller, since $k_v^n = 2\omega_n\zeta_n$. At the same time, the nth element of $\mathbf{K}_p$ represents the square of the undamped natural frequency ω_n. Note that $\mathbf{K}_p$ and $\mathbf{K}_v$, are solely implied in a real-time computer program so that all the matrix coefficients in Equation 5 can be arbitrarily chosen by the designer. This represents an advantage over the uncontrolled mechanism design. But that leads to massive real-time calculation which is required to construct the feedback-laws from fast hardware systems.

As seen in Equation 4, the design of an ideal mechanical controller needs three *PVA* inputs [12], i.e., the usual position-input $\mathbf{y}_d(t)$, the velocity-input $\dot{\mathbf{y}}_d(t)$ and the acceleration-input $\ddot{\mathbf{y}}_d(t)$. In practice, this requires an off-line construction of the kinematic features of desired motion. When *PV* inputs are used only to drive a second-order servo-system, the step response exhibits a more important overshoot than the *P*-input system does (Figure 2). So, a servo-input with trajectory trapezoidal shapes is then required to reduce the transient overshoot. However, the main advantage of *PV* servo-systems is to reduce the sensitivity of the rise time and peak time with respect to the damping ratio. This effect can be observed in Figures 3 and 4, which provide the normalized response times of *P*-input and *PV*-input linear second-order systems. From these figures, it appears that fast servo-systems result from the use of a *PV*-input scheme. When $\zeta = .8$ for example, the rise time ratio of usual and *PV* systems is about 4, while the peak time ratio drops to 2.5. This means that an additional derivative input transforms a usual controller with positional and speed errors into a servo-system which is twice as fast and has no positional or speed errors. Finally, the most important improvement of servo-features results from the implementation of a PVA-input scheme, which is directly designed by means of the Nonlinear Decoupling Method.

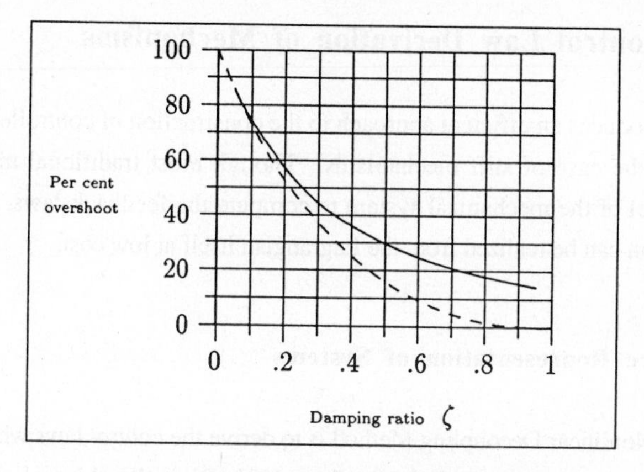

(----) usual controller; (—) with addition derivative input

Figure 2. Maximum Percent Overshoot of Linear Second-Order Systems

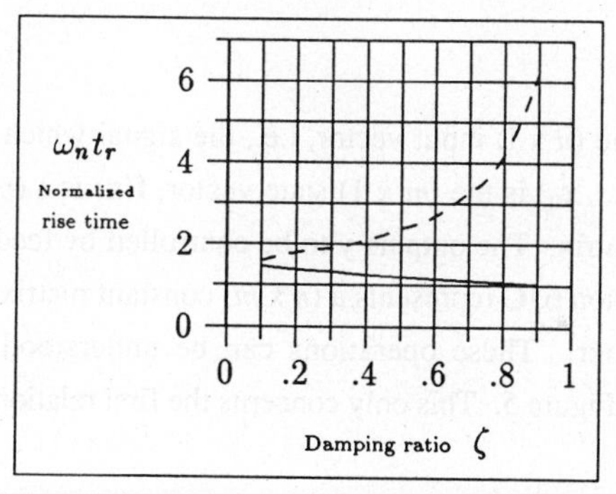

(----) usual controller; (—) with addition derivative input

Figure 3. Normalized Rise-Time

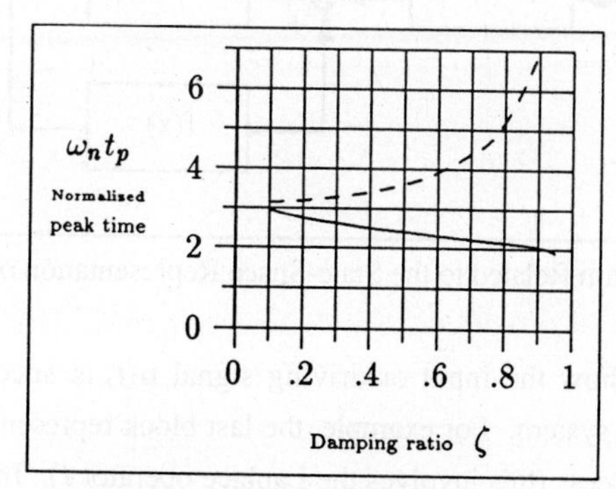

(----) usual controller; (—) with addition derivative input

Figure 4. Normalized Peak-Time

4 Usual Control Law Derivation of Mechanisms

This section introduces an efficient approach to the construction of controlled systems which considers only the case of stiff mechanisms. Though most traditional methods need the dynamical model of the mechanical system to compute the feedback-laws, Section 5 shows that this operation can be realized from the Lagrangian itself at low cost.

4.1 State-space Representation of Systems

The aim of the Nonlinear Decoupling Method is to derive the control laws which decouple and linearize a dynamical system exactly by feedback [13]. Basically, this technique starts from a state-space representation of mechanical systems defined as

$$\dot{x} = f(x) + B(x)\, u(t)$$

$$y = C\, x(t) \tag{6}$$

where $\mathbf{u}(t)$ represents the $(n \times 1)$ input vector, i.e., the signal which governs the dynamical system, $x = col\ (x_1, x_2 ..., x_n)$ is the $(m \times 1)$ state vector, $\mathbf{f(x)}$ is a $(m \times 1)$ nonlinear vector, and $\mathbf{B(x)}$ is a $(n \times m)$ matrix. The outputs $\mathbf{y}$ to be controlled by feedback are specified by a $(n \times 1)$ vector. In Equation 6, $\mathbf{C}$ represents a $(n \times m)$ constant matrix. A more complex case will be investigated later. These operations can be understood from a block scheme representation drawn in Figure 5. This only concerns the first relationship of Equation 6.

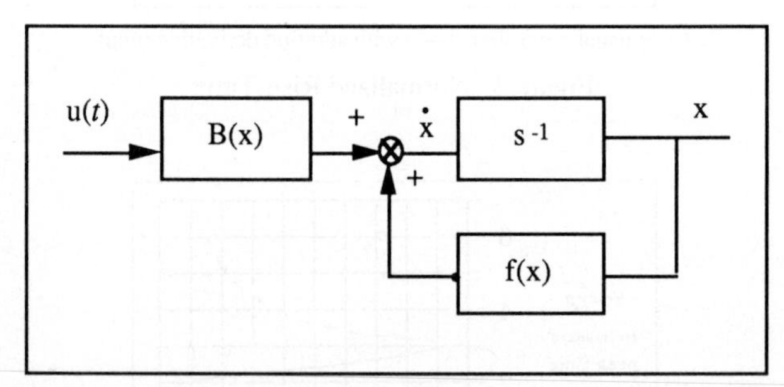

Figure 5. Block-Diagram Related to the State-Space Representation of Dynamical Systems

This figure shows how the input or driving signal $\mathbf{u}(t)$ is successively transformed through the mechanical system. For example, the last block represents the matrix form of a diagonal integration process (this involves the Laplace operator s). In the following section, this scheme will be used to describe a controlled system.

4.2 Background of the Decoupling Method

Let us come back to the goal of the Nonlinear Decoupling Method. It consists of finding which input $\mathbf{u}(t)$ transforms the original nonlinear mechanical system into an exactly linearized and decoupled one. A feedback-law candidate is

$$u(t) = h(x) + G(x)w(t) \tag{7}$$

where $\mathbf{h(x)}$ is a $(n \times 1)$ nonlinear vector which depends on $\mathbf{x}$, $\mathbf{G(x)}$ is a $(n \times n)$ nonlinear matrix which depends on $\mathbf{x}$, and $\mathbf{w}(t)$ represents the new input of the controlled system, i.e., the controller input.

The implementation of the previous feedback-law in the dynamical system described by Equation 6 leads to the typical control scheme drawn in Figure 6. This scheme shows us how the feedback-laws are operating on the system in real-time. Note that the new input of the system is now $\mathbf{w}(t)$ instead of $\mathbf{u}(t)$.

4.3 Usual Feedback-Law Derivation

The feedback-laws defined by Equation 7 are now explicitly derived. Here, the methodology is applied to simple controllers whose output is linear in terms of state variables. The first step is to derive the output $\mathbf{y}$ with respect to the time. The successive time derivatives of the ith output are

$$\begin{aligned}
& y_i = C_i x, \\
& \dot{y}_i = C_i A_o f(x), && \text{when } C_i A_o B(x) = 0 \\
& \dots \quad \dots && \\
& y_i^{d_i]} = C_i A_{d_i-1} f(x), && \text{when } C_i A_{d_i-1} B(x) = 0, \\
& y_i^{1+d_i]} = C_i A_{d_i} f(x) + C_i A_{d_i} B(x) u(t)
\end{aligned} \tag{8}$$

To simplify these expressions, the following operators are used:

$$\begin{aligned}
& A_o(x) = I && A_2(x) = \frac{\partial[A_1 f(x)]}{\partial x} \\
& I = \text{identity matrix} && \dots = \dots \\
& A_1(x) = \frac{\partial[A_o f(x)]}{\partial x} && A_{k+1}(x) = \frac{\partial[A_k f(x)]}{\partial x}
\end{aligned} \tag{9}$$

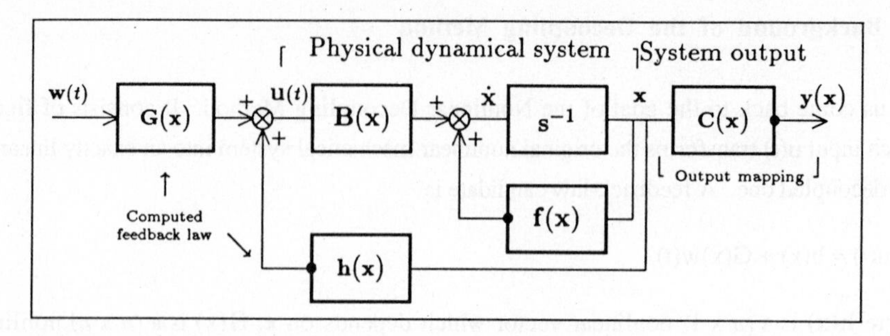

Figure 6. Implementation of Nonlinear Feedback-Laws in Dynamical Systems

Next, integer constants such as $d_1, d_2, ..., d_n$ are computed from the row vector $C_i\mathbf{A}_o\mathbf{B}(\mathbf{x})$, where $C_i(\mathbf{x})$ is the ith row of the matrix $\mathbf{C}$. When such a row equals zero for all allowed $\mathbf{x}$, the next step is to form the row vector $C_i\mathbf{A}_1\mathbf{B}(\mathbf{x})$ and so on, until the first non-zero $C_i\mathbf{A}_j\mathbf{B}(\mathbf{x})$ row vector is obtained. Then, the characteristic number d_i associated with the ith output is the smallest integer j such that the previous row vector is non zero for all the allowed values of $\mathbf{x}$. Finally, a $(n \times n)$ matrix $\mathbf{D}(\mathrm{x})$ and a $(m \times 1)$ vector $\mathbf{f}(\mathrm{x})$ are introduced as

$$D(x) = \begin{bmatrix} \ddots \\ C_i A_{d_i}(x) B(x) \\ \ddots \end{bmatrix}, \qquad f^*(x) = \begin{bmatrix} \cdots \\ C_i A_{d_i}(x) f(x) \\ \cdots \end{bmatrix} \tag{10}$$

Then, the feedback-law defined by Equation 7 is implemented into the last relationship of Equation 8. If $\mathbf{D}$ is a nonsingular matrix, which is the case for mechanisms, a closed-loop linearized and decoupled system results in

$$y_i^{[1+d_i]} = w_i \tag{11}$$

Equation 11 appears as a diagonal and linear relationship. This is schematically depicted in Figure 7, where s is the Laplace operator. This clearly shows that the nonlinear dynamical

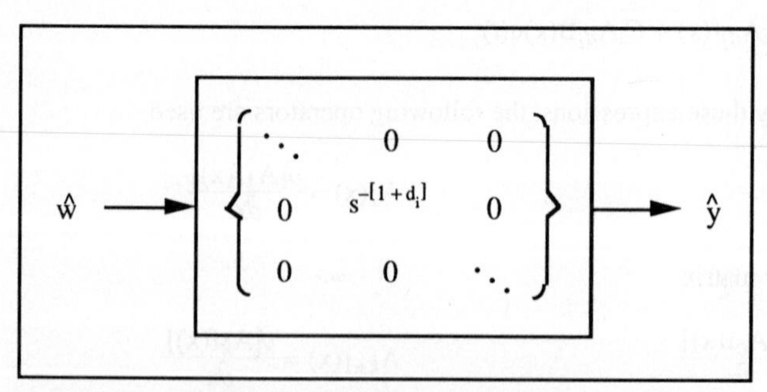

Figure 7. The Goal of the Nonlinear Decoupling Method

system given by Equation 6 becomes, through feedback, a linear and diagonal one. To achieve that, the form of the control law defined by Equation 7 must be taken such as

$$u(t) = D^{-1}(x)[-f^*(x) + w(t)] \tag{12}$$

with

$$h(x) = -D^{-1}(x)f^*(x), \qquad G(x) = D^{-1}(x) \tag{13}$$

To close this part in design terms, the explicit form of the control law has to be derived from Equations 6, 12, and 13. Owing to the disadvantage of writing the state-space representation of a dynamical system, as well as the pole placement formulation, note that these transformations are based on simple algebraic and matrix manipulations.

4.4 Pole Allocation

As shown previously, the Nonlinear Decoupling Method allows the transformation of any set of coupled and nonlinear ordinary differential equations into a linear and decoupled system. It is then observed that the dynamics of the resulting controller obeys linear differential equations. In other words, the input-output relationships are governed by a cascade of $(1 + d)$ integrators. In fact, this does not lead to a stable closed-loop structure. With the aim of finding the controller dynamics described by a stable linear differential equation, a more general form of control laws is to be used, such as

$$u(t) = D^{-1}(x)[\ KQ(x) - f^*(x) + \Gamma z(t)\ ] \tag{14}$$

The implementation of such a control law yields the block scheme drawn in Figure 8. It is observed that the decoupling and linearization effects are due to the inner loop, while the pole assignment is performed by the outer loop. In terms of design steps, the new input of the controller is $\mathbf{z}(t)$. In Equation 14, the vector $\mathbf{Q}(x)$ is defined as

$$Q(x) = \begin{bmatrix} Q_1(x) \\ \vdots \\ Q_n(x) \end{bmatrix} \qquad Q_i(x) = \begin{bmatrix} C_i x \\ C_i\, f(x) \\ \vdots \\ C_i A_{di-1}(x) f(x) \end{bmatrix} \tag{15}$$

where $\mathbf{z}(t)$ represents the $(n \times 1)$ input vector of the resulting new controller, with two return loops (see Figure 8).

In addition, the diagonal matrices are defined as

$$\Gamma = \begin{bmatrix} \lambda_1 & 0 & \cdots & 0 \\ 0 & \lambda_2 & \cdots & 0 \\ 0 & \ddots & \ddots & \vdots \\ 0 & 0 & \cdots & \lambda_n \end{bmatrix} \qquad K = \begin{bmatrix} K_1 & 0 & \cdots & 0 \\ 0 & K_2 & \cdots & 0 \\ 0 & \ddots & \ddots & \vdots \\ 0 & 0 & \cdots & K_n \end{bmatrix} \tag{16}$$

where λ_i, $(i = 1, 2, \cdots, n)$ are constants to be selected by the designer. In Equation 16, the constant row vectors are defined as $K_i = [-k_o^i, -k_1^i, \cdots, -k_{di}^i]$. Combining Equations 14-16, the implementation of the feedback-law into Equation 6 yields a controller in which each output y_i evolves according to the following linear differential equation:

$$y_i^{1+d_i]} + k_{d_i}^i y_i^{d_i]} + \cdots + k_1^i y_i + k_o^i y_i = \lambda_i z_i \tag{17}$$

where the parameters k_j^i and the gains λ_ι are selected to obtain a desired output response from the ITAE criteria, for example.

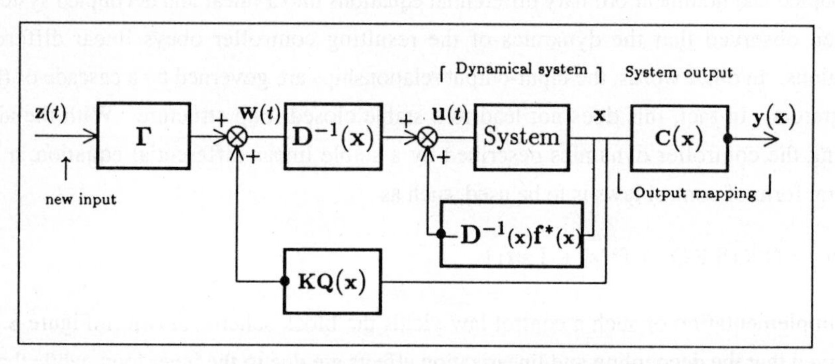

Figure 8. Decoupling and Pole Assignment Through Nonlinear Feedback

4.5 General-Output Design

A generalization of controller design is now given. This concerns systems in which the controlled outputs represent a nonlinear arrangement of state-space variables. Such a case is encountered in designing Cartesian controllers. The dynamical system and the controlled outputs must be specified as follows:

$$\dot{x} = f(x) + B(x)\, u(t)$$

$$y = C(x) \tag{18}$$

where the vector **y** defines the *n*-nonlinear outputs to be controlled by feedback. The problem is to find a feedback-law which produces decoupling and linearization with respect to the outputs. The derivation of the control law consists of computing the successive time derivative of outputs

$$\dot{y}_i = \frac{\partial C_i(x)}{\partial x} f(x) \quad \text{when} \quad \frac{\partial C_i(x)}{\partial x} B(x) = 0$$

$$\ddot{y}_i = \frac{\partial}{\partial x}\left[\frac{\partial C_i(x)}{\partial x}\right] f(x) \quad \text{when} \quad \frac{\partial}{\partial x}\left[\frac{\partial C_i(x)}{\partial x}\right] B(x) = 0 \tag{19}$$

and so on, for higher derivatives. Using Lie's operators, Equation 19 becomes

$$L_i^o(x) = C_i(x), \quad \text{the } i\text{th element of } C(x),$$

$$L_i^k(x) = \frac{\partial L_i^{k-1}}{\partial x} f(x), \quad \text{for } k \geq 1 \tag{20}$$

where $\frac{\partial L_i^k}{\partial x}$ is defined by

$$\frac{\partial L_i^k}{\partial x} = \left(\frac{\partial L_i^k}{\partial x_1}, \frac{\partial L_i^k}{\partial x_2}, \ldots, \frac{\partial L_i^k}{\partial x_n}\right) \tag{21}$$

With the help of Lie's operators, the successive derivatives of output can be written as

$$y_i^{1+di]} = \frac{\partial L_i^{di}}{\partial x} f(x) + \frac{\partial L_i^{di}}{\partial x} B(x)\, u(t) \tag{22}$$

In addition, the characteristic numbers d_i are computed as

$$d_i = min(k), \quad \text{such that} \quad \frac{\partial L_i^k}{\partial x} B(x) \neq 0 \tag{23}$$

It is easily shown that the introduction of the control law into Equation 22 allows the derivation of the generating terms of the nonlinear vector **f*(x)**, and of the matrix **D(x)**:

$$f_i^* = \frac{\partial L_i^{d_i}}{\partial x} f(x), \quad D_i = \frac{\partial L_i^{d_i}}{\partial x} B(x) \tag{24}$$

Finally, the pole assignment is realized by means of Equation 14, with

$$Q(x) = \begin{bmatrix} Q_1(x) \\ \vdots \\ Q_m(x) \end{bmatrix} \qquad Q_i(x) = \begin{bmatrix} L_i^o(x) \\ L_i^1(x) \\ \vdots \\ L_i^{di}(x) \end{bmatrix} = \begin{bmatrix} y_i(x) \\ \dot{y} \\ \vdots \\ y_i^{d_i]}(x) \end{bmatrix} \tag{25}$$

These relationships allow the design of any kind of controller, such as the joint and task spaces controllers. When the kinematics of a desired trajectory of a controlled mechanism are known in advance, a smart servo-system can be designed. The servo-input must then be defined from Equations 3-5 as the usual positional input plus the first and the second time derivatives. This leads to the stable servo-error defined by Equation 5, and to the PSA-control scheme drawn in Figure 9. Finally, when there are more than two higher-order systems, such as the dynamically driven mechanisms, it is suitable to use an additional jerk input.

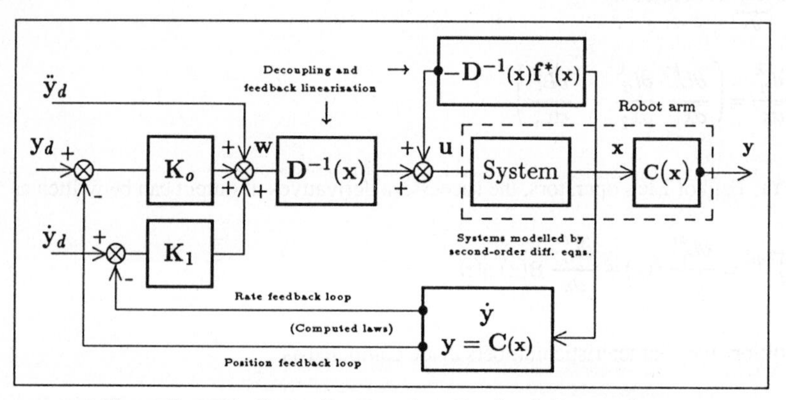

Figure 9. *PSA* Controller Based on Feedback Decoupling

To conclude this part, note that codes which perform the previous manipulations in symbolic form have been developed by researchers with MACSYMA and REDUCE, as well as in C-Language [14].

5 Direct Nonlinear Decoupling Method

In order to avoid a heavy computational scheme for handling the control equations, generating rules [15-18] can be used to compute the feedback-laws of dynamical systems. This simplifies the control design to that of the rule application, which does not require basic knowledge in control theory. The generating rules are derived as follows.

5.1 Resetting of Euler-Lagrange Equations

To make the application of the Nonlinear Decoupling Method on the Lagrange's formalism possible, let us consider a state-space representation of Euler-Lagrange equations [15].

$$\begin{aligned} \dot{q} &= p \\ \dot{P} &= \frac{\partial L}{\partial \dot{q}} - D + T \end{aligned} \tag{26}$$

where L represents the Lagrangian which is a function of $\boldsymbol{n}$ generalized joint coordinates $\mathbf{q}$, and $\mathbf{p}$ is the vector of the generalized velocities $\mathbf{p} = d\mathbf{q}/dt$. In addition, the dissipation function $\mathbf{D}$ has been introduced to take into account friction effects at joints while the vector $\mathbf{T}$ represents the generalized forces driving the mechanical joints. The canonical momentum, or conjugate momentum with respect to the generalized coordinate $\mathbf{q}$ is defined by

$$\mathbf{P} = \frac{\partial L}{\partial \mathbf{q}} \tag{27}$$

To obtain a phase-plane representation of Euler-Lagrange equation from its state-space form, the time derivative of the generalized momentum can be used.

$$\frac{\partial \mathbf{P}}{\partial t} = \frac{\partial \mathbf{P}}{\partial \mathbf{p}} \dot{\mathbf{p}} + \frac{\partial \mathbf{P}}{\partial \mathbf{q}} \dot{\mathbf{p}} \tag{28}$$

The introduction of Equation 28 into Equation 26 yields

$$\begin{aligned} \dot{\mathbf{q}} &= \mathbf{p} \\ \dot{\mathbf{p}} &= \left(\frac{\partial \mathbf{P}}{\partial \mathbf{p}}\right)^{-1} \left(\frac{\partial L}{\partial \mathbf{q}} - \mathbf{D} - \frac{\partial \mathbf{P}}{\partial \mathbf{q}} \mathbf{p} + \mathbf{T}\ (\mathbf{t})\right) \end{aligned} \tag{29}$$

So, Equation 29 presents exactly the form which is needed to apply the Nonlinear Decoupling Method. As shown in the next section, this resumes the application of the decoupling method to a symbolic operation, i.e., a relationship which is valid for all mechanisms.

5.2 Generating Rules Derivation

This section recalls how the generating rules are derived. These equations, which allow a symbolic derivation of control laws from the Lagrangian and from the controller output equation, are computed as follows [17]. The introduction of the new variable $\mathbf{Z} = \boldsymbol{col}(\mathbf{q},\mathbf{p})$ into Equation 29 gives

$$\dot{Z} = F(Z) + H(Z)\,U(t) \tag{30}$$

where

$$F(Z) = \begin{bmatrix} \mathrm{p} \\ \left(\dfrac{\partial \mathbf{P}}{\partial \mathbf{p}}\right)^{-1}\left(\dfrac{\partial L}{\partial \mathbf{q}} - \mathbf{D} - \dfrac{\partial \mathbf{P}}{\partial \mathbf{q}}\,\mathbf{p}\right) \end{bmatrix} \tag{31}$$

with

$$X = \begin{bmatrix} q \\ p \end{bmatrix}, \qquad H(Z) = \begin{bmatrix} 0 & 0 \\ 0 & \left(\dfrac{\partial \mathbf{P}}{\partial \mathbf{p}}\right)^{-1} \end{bmatrix}, \qquad U(t) = \begin{bmatrix} 0 \\ T \end{bmatrix} \tag{32}$$

When considering the design of a position controller, the outputs of this servo-mechanism can be specified by

$$Y = C_p(q) \tag{33}$$

Then, the application of the decoupling method to Equations 28-30 yields generating rules, i.e., a new formalism that can be used to derive the explicit form of the feedback-law $T(t) = h(q) + G(q)\,w(t)$, where

$$\mathbf{h}(\mathbf{q}) = -\frac{\partial \mathbf{P}}{\partial \mathbf{p}}\left(\frac{\partial \mathbf{C}(\mathbf{q})}{\partial \mathbf{q}}\right)^{-1}\left(\frac{\partial}{\partial \mathbf{q}}\left(\frac{\partial \mathbf{C}(\mathbf{q})}{\partial \mathbf{q}}\,\mathbf{p}\right)\mathbf{p}\right) - \frac{\partial L}{\partial \mathbf{q}} + \mathbf{D} + \frac{\partial \mathbf{P}}{\partial \mathbf{q}}\,\mathbf{p}$$

$$\mathbf{G}(\mathbf{q}) = \frac{\partial \mathbf{P}}{\partial \mathbf{p}}\left(\frac{\partial \mathbf{C}(\mathbf{q})}{\partial \mathbf{q}}\right)^{-1} \tag{34}$$

Note that the generating rules require only the Lagrangian, i.e., a scalar expression, the dissipation and friction vector, and the controller output equation. Thus, the dynamical model of mechanisms is not needed in this analysis. As a result, an N-degree-of-freedom system requires the computation of $\sum_n^N \frac{n + n^2}{2}$ traces of matrices while the traditional methods involve up to $\sum_n^N (n+1)\frac{n + n^2}{2}$ traces of matrices. For a six degree-of-freedom mechanism, this leads to 56 and 322 traces respectively. The generalization of this approach to the case of dynamically driven mechanisms can be found in [18].

5.3 Computational Load of Feedback-Laws

In practice, it is suitable to consider the computational load which results from the feedback-law design. This is realized as follows. Since the generating rules also allow the modeling of mechanisms [17], a comparison of this approach can be made with Uicker's modeling method [19-21]. To simplify the analysis, it is assumed that the computational load due to gravity forces is relatively insignificant. Workers [19,20] have proposed an improved Uicker's modeling scheme which takes into account the symmetry of the inertia matrix, and Coriolis and centrifugal vector elements. In terms of trace operations, this reduces the computational load of Coriolis and centrifugal force vectors to $\frac{1}{12}(N-1)N(N+1)(N+2)$.

In contrast, the Lagrangian construction involves the manipulation of $\frac{1}{6}N(N+1)(N+2)$ trace operations. For practical values of N, this is depicted in Table 1. This table shows that even for low values of N, the application of Equation 34 is more computationally efficient than the improved Uicker's method. For higher values of N, for example when a car driver

Traces of Matrices		
No. of Joints N	Lagrangian Design	Uicker's Method
1	1	1
2	4	6
3	10	20
4	20	50
5	35	105
6	56	196
7	84	336
8	120	540
9	165	825
10	220	1210

Table 1. Traces of Matrices Implied in Modeling Mechanisms Through a Direct Lagrangian Construction and Via Uicker's Approach

needs to be modeled from 31 degrees-of-freedom, Uicker's approach involves 87.296 trace operations. At the opposite, a lower computational load of only 5.456 trace operations is required to design the corresponding Lagrangian. Since the operations involved by the generating rules imply the Lagrangian, there is a substantial advantage to modeling both the dynamics of mechanisms and their related control laws from the generating rules.

6 Conclusion

A design methodology based on the Nonlinear Decoupling Method has been introduced to ease the construction of controlled stiff mechanisms. It has been shown that two alternative ways allow the symbolic derivation of control laws from (i) the dynamical model of mechanisms or (ii) directly from the Lagrangian. So, stable schemes with an arbitrary choice of dynamics can be specified by the designer. One advantage of the proposed approach is due to the fact that it can be easily implemented in the codes devoted to the analysis of mechanical system dynamics. This offers a way to overcome the complexity encountered in designing mechanical controllers that in turn would increase the number of potential users of today's CAD programs.

References

1. Ju, M.S. and Hsu, C.J.: Automatic modeling of mechatronic systems via symbolic approach. Mechatronics, 1(2) pp. 157-174,1991
2. Vibet, C.: Research issues associated with the control of mechanism. Int. J. of Robotics and Automation, 5(4) pp. 185-193,1990
3. Freund, E.: A nonlinear control concept for computer-controlled manipulators. Proc. of IFAC Symp. on Multivariable Techn. Systems, Fredericton, Canada, N.Y. Pergamon Press, pp. 395-403,1977
4. Freund, E.: On the design of multirobots systems. Proc. IEEE Conf on Robotics and Automation, pp. 477_490, 1984
5. De Simone, C. and Nicolo, F.: On the control of elastic robots by feedback decoupling. Int. J. of Robotics and Autom., 1(2) pp. 6469,1986
6. Tarn, T.; Bejczy, A. and Yun, X.: Robot arm force control through system linearization by nonlinear feedback. Proc. IEEE Conf. on Robotics and Autom., pp. 1618-1625,1988
7. Tarn, T. and Bejczy A.: Software elements. Int. Encyclopedia of Robotics: Applications and Automation, Edited by R. Dorf, J. Wiley, pp. 1608-1626, 1988
8. Tarn, T.; Bejczy, A.; Ganguly, 5. and Li, Z.: Nonlinear feedback methods of robot arm control: A preliminary experimental study. Proc. IEEE Conf. on Robotics and Autom., pp. 2052-2057, 1990
9. Haug, E.: Computer Aided Analysis and Optimization of Mechanical System Dynamics, NATO ASI, Iowa Univ, Aug 1-12,1983, Springer Verlag, 1984
10. Haug, E. and Dayo, R.: Real-Time Integration Methods for Mechanical System Simulation. NATO ASI, Snowbird, Utha Aug 7-11 1989. Spinger Verlag, 1991
11. Nivens, J.; Whitney, D.; Woodin, A.; Drake, 5.; Lynch, M.; Seltzer, D.; Sturges,R. and Watson, P.: A scientific approach to the design of computer controlled manipulators. Report R837, The Charles Stark Draper Lab., Cambridge, MA, Aug. 1974
12. Vibet, C.: Position-Speed-Acceleration controller design. Electr. Lett., 22(7) pp 382-383,1986
13. Asseo, 5.: Decoupling a class of nonlinear systems and its application to an aircraft control problem. J. of Aircraft, Vol 10 pp 739-747,1973
14. Charbonneau, G.; Vinarnick, 5.; Néel, P.; Evariste, C. and Vibet, C.: Symbolic modeling of controlled mechanisms. Computer Methods in Appl. Mechan. and Engin., Vol 98, pp. 23-40, 1992
15. Cotsaftis, M. and Vibet, C.: A derivation of robot control algorithms from the Lagrange formalism. 2nd Int. Conf. on Robotics and Factories of the Future, San Diego, 1987. Edited by R. Radharamanan, Springer Verlag, pp. 461-465, 1988
16. Cotsaftis, M. and Vibet, C.: Lagrange formalism, decoupling method and control law generation. Int. J. of Robotics and Autom., 3(2) pp. 86-89,1988
17. Cotsaftis, M. and Vibet, C.: Simultaneous derivation of control laws and of dynamical model equations for controlled mechanisms. Real-time Integration Methods for Mechanical System Simulation, Edit. by E. Elaug and R. Dayo. Springer Verlag, pp. 301-327,1991

18. Cotsaftis M. and Vibet, C.: A direct application of the decoupling method on Lagrange formalism. J. of Robotics Research, 11(1), pp. 64-74,1992
19. Murray, J. and Neuman, C.: ARM: an algebraic robot dynamic modeling program. IEEE Conf. on Robotics and Autom., pp. 103-114, 1984
20. Leu, M. and Hemati, N.: Automated symbolic derivation of dynamic equations of motion for robot manipulators. Trans. on the ASME, J. of Dyn. Syst., Measur., and Contr., Vol. 108, pp. 172-179, 1986
21. Muir, P. and Neuman, C.: Dynamic modeling of multibody robotic mechanisms. IEEE Conf. on Robotics and Autom., pp. 1546-1552,1988

Part 2

Application Sectors

Implementation and Applications of Multidisciplinary Concurrent Engineering

Robert G. Vos

Boeing Aerospace and Electronics Company, Seattle, WA 98124, USA

Abstract: Multidisciplinary concurrent implementation is presented from the perspective of practical software development. Model types from several disciplines are described, along with consideration of interfacing models for greater concurrency within an overall CAE integrated system. Supporting tools presented include integrated system simulation; multidisciplinary optimization; model synthesis and reduction; and sharing of tasks and data across a distributed computing network.

Keywords: integrated analysis / concurrent engineering / multidisciplinary analysis / computer software / database / user interface / structures / thermal / controls / optics / multibody

1 Introduction

"Concurrent Engineering is designing and validating a total product, its manufacturing process, and its maintenance process, all at the same time."

Although there are many definitions for Concurrent Engineering (CE), the above is perhaps one of the best. It is concise, and yet states the inclusiveness of CE. It implies that a good CE process ensures all of the "ilities" - analyzability, manufacturability, testability, maintainability, traceability, and reliability.

This study institute emphasizes the mechanical engineering design technologies as a subset of CE. This lecture, in particular, deals with a technology which we will call "multidisciplinary analysis," and which more precisely deals with the integrated modeling, simulation and analysis of a multidisciplinary product. As we deal with the implementation and applications of this technology, we must be guided by the overall characteristics and objectives of CE.

Multidisciplinary analysis is a fairly mature technology, relative to the other supporting CE technologies. Capabilities for this type of analysis have more than a decade of history behind them. It should be noted that multidisciplinary analysis has depended upon, and used, many CE concepts even before they were recognized and promoted as such. In fact, concurrency of the disciplines is required for multidisciplinary analysis, simply to obtain valid results.

It can be expected, therefore, that multidisciplinary analysis will be a major early contributor to CE. Some of these analysis tools have been used long enough to provide considerable experience in feasibility, user issues, and future directions. What remains is to incorporate additional technologies and to achieve a higher level of integration into the overall CE process. At the same time, of course, a broader base of CE acceptance must be achieved within the industry.

The writer's CE experience derives largely from implementation and applications of multidisciplinary analysis software packages. The ISM (Integrated System Modeling) package is currently being developed by Boeing for the U.S. Air Force. Its predecessor, IAC (Integrated Analysis Capability), was originally developed for NASA space structure applications and was later expanded through Boeing in-house development. That work has been described in References 1 and 2. The ISM effort is integrating the structures, controls, thermal, optics, and multibody analysis technologies. Our development in this area now has a 12-year development history and represents a total investment of 10 million dollars.

Figure 1 shows a typical interaction diagram for multidiscipline spacecraft applications generally descriptive of directed energy, kinetic energy, and surveillance systems. Strong potential technology interactions are shown by the heavy arrows, weaker interactions by the lighter arrows. Required concurrency of the disciplines within the analysis process is obvious.

This lecture will discuss several aspects of multidisciplinary analysis against the general backdrop of the ISM related implementation. Alternative approaches to system architecture are mentioned with some details of the approach used for ISM. Individual technologies and associated math models are described to identify differing characteristics and the modeling transformations which must be provided in developing the integrated solution paths. Details of the integrating framework and system simulation concepts are presented, followed by some applications which provide insight into concurrent analysis. Finally, some potential roadblocks to concurrent analysis are discussed, and some thoughts on future directions are given.

2 System Architectures

An ISM type of package can be viewed as providing two different but complementary software products:

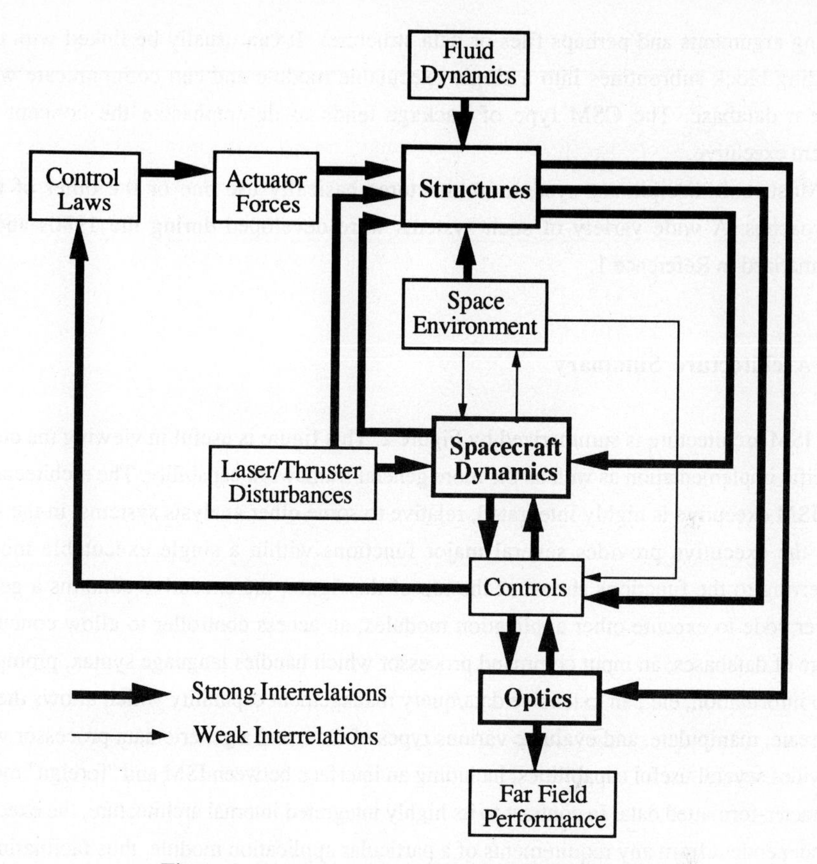

Figure 1. Multidisciplinary Technology Interactions

1. A specific interdisciplinary analysis capability having a set of interfaced technical modules that deal with, for example, the coupling of thermal, structures, and optics disciplines.
2. A general framework product which serves as an "integrating base" whereby user groups can add desired analysis modules. ISM was designed to ease the task of interfacing other analysis modules as they are needed from individual discipline areas.

The key building block for this type of package is the "module." The module is a self contained executable program with defined run-time parameters and using I/O files or data structures. The module can be driven by the system executive, is able to communicate with the system database, and can be incorporated along with other modules into an integrated task runstream.

A somewhat different concept is exemplified by the CSM (Computational Structural Modeling) system under development by NASA/LaRC. The basic building block for CSM is at a lower level, i.e., the "utility." This is generally a computational subroutine with defined calling arguments and perhaps files or data structures. It can usually be linked with other building block subroutines into a single executable module and can communicate with a system database. The CSM type of package tends to de-emphasize the concept of a system executive.

Most multidisciplinary system architectures basically use one or the other of these approaches. A wide variety of such systems were developed during the 1980s and are summarized in Reference 1.

2.1 Architecture Summary

The ISM architecture is summarized by Figure 2. This figure is useful in viewing the current specific implementation as well as the more general framework capability. The architecture of the ISM executive is highly integrated, relative to some other analysis systems, in the sense that the executive provides several major functions within a single executable module. Referring to the functions shown at the top of the figure, the executive contains a general driver code to execute other application modules; an access controller to allow concurrent usage of databases; an input command processor which handles language syntax, prompting, help information, etc.; an extensive data/query management capability which allows the user to create, manipulate, and evaluate various types of data; and a generic data processor which provides several useful capabilities, including an interface between ISM and "foreign" module character-formatted data. In contrast to its highly integrated internal architecture, the executive is independent from any requirements of a particular application module, thus facilitating the substitution or addition of new modules into the system. The ISM concept for overall data organization and storage is summarized by the three data areas shown in the center of the figure. This is discussed more in the section on Data Management.

ISM modules are executed via the executive program with user specified options for run parameters and ISM/module data flow. Groups of technical modules as shown on the left side of Figure 2 have been incorporated to provide a specific current analysis capability. On the right side of the figure are several other groups of modules associated with functions such as graphical data display, model building and viewing, inter-module data flow, mesh interpolation and special data handling, and management support for user owned software. Finally, as indicated at the bottom of the figure, multiple concurrent users may interface with ISM to perform both interactive and batch tasks.

Design of the ISM architecture was driven by several considerations associated with the engineering nature and objectives of ISM. ISM was to be an interactive but in-depth computational package, applicable to the full range of design/analysis activity. There was a need to support a variety of different analysis and user groups, including both technical

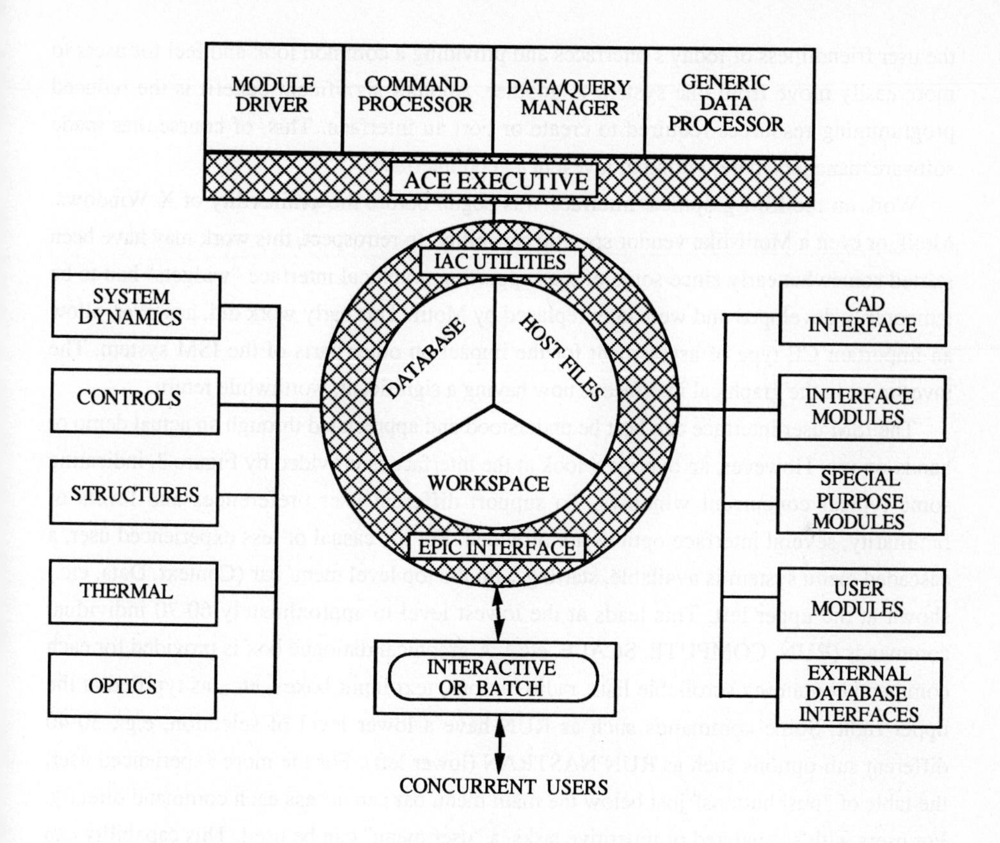

Figure 2. ISM Architecture

specialists as well as multi-disciplinarian types of engineering personnel, and involving diverse technical modules and analysis techniques. In addition, it was felt that the system should be designed to have a long life potential by emphasizing a modular construction and by catering to modem hardware and user interface concepts.

2.2 User Interface

In the past, developers have generally considered three basic user interface approaches - command language, tutorial prompting, and alphanumeric or graphical menu selection. Recently, modern windowing technology has pervaded and complemented all of these techniques, and has also popularized the graphical dialogue box. A powerful force in encouraging improvements has been the emergence of windowing and graphics standards. For workstation oriented engineering analysis, the X-Windows and Motif interface toolkit standards are becoming the most significant. These window based standards are enhancing the user friendliness of today's interfaces and providing a common look and feel for users to more easily move from one system to another. Another significant benefit is the reduced programming resources required to create or port an interface. This, of course, has made software managers more willing to invest in a good interface.

Work on the ISM graphical interface was begun before the availability of X-Windows, Motif, or even a Motif-like vendor specific capability. In retrospect, this work may have been started somewhat early since some of the supporting graphical interface "widgets" had to be temporarily developed and were later replaced by Motif. This early work did, however, allow an important CE type of assessment for the impact on other parts of the ISM system. The investment in the graphical interface is now having a significant, worthwhile return.

The ISM user interface can best be understood and appreciated through an actual demo or hands-on use. However, an overview look at the interface is provided by Figure 3, indicating some typical concurrent windows. To support different user preferences and levels of familiarity, several interface options are provided. For the casual or less experienced user, a cascaded menu system is available, starting with the top level menu bar (Context, Data, etc.) shown at the upper left. This leads at the lowest level to approximately 60-70 individual commands (RUN, COMPUTE, SCALE, etc.). A graphical dialogue box is provided for each command containing scrollable lists, radio buttons, text input boxes, etc., as typified at the upper right. Some commands such as RUN have a lower level of selection, e.g., 30-40 different sub-options such as RUN NASTRAN (lower left). For the more experienced user, the table of "pushbuttons" just below the main menu bar can access each command directly. For users with specialized or repetitive tasks, a "user menu" can be used. This capability can add to, delete from, or modify the list of user commands, and can save or recall a previously defined list. Finally, an experienced user with more complex tasks may choose direct command entry into the command text input window. Command journal files may also be saved and executed to automate the processing of detailed computational sequences.

Other windows may be created, e.g., for informational and error messages or for detail printed output. The database editor may create windows which graphically display ISM data structures (tables, matrices, etc. including data and various labels) as exemplified at the lower right. Data edit options allow scrolling, entry/row/column/block cut and paste, and plotting.

Some might conclude that command language interfaces have been obsoleted by recent windowing and menu/dialogue box techniques. Our experience indicates that this is not the case, for at least two reasons. First, the experienced user will often be able to define a task more quickly in command mode than in graphical mode. Second, and perhaps more importantly, the command language interface allows higher level task sequences to be defined, stored, and accessed in a flexible manner. Automation of repetitive analyses requires that tasks be accessed at a higher level of granularity once the lower levels have been tested and validated. The ISM interface allows a user to combine graphical menu/dialogue box and

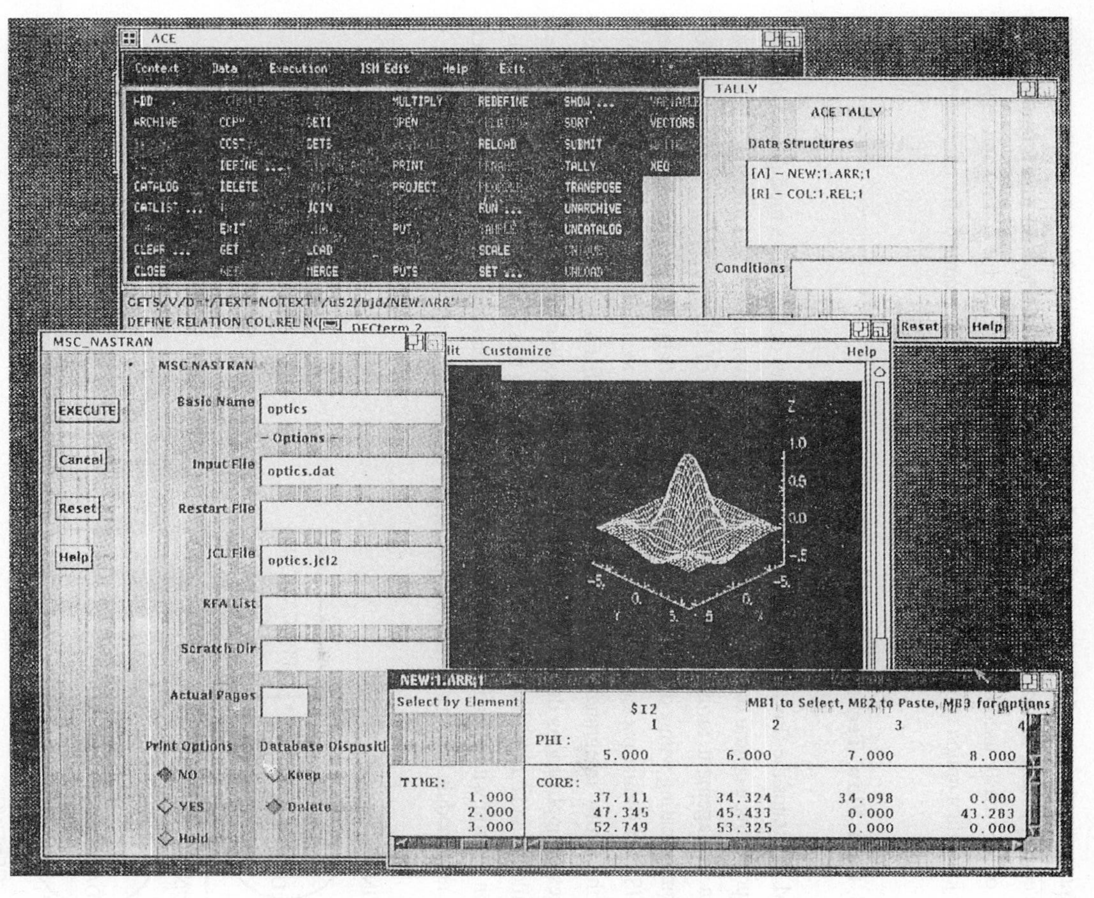

Figure 3. Typical Interface Windows

command techniques. In fact, the graphical definition of a command automatically generates the corresponding textual command and displays it in a small window for user reference. The list of previously used commands is available in another window for editing and reuse, and the journaling option can be activated to permanently record all commands for later reference.

2.3 Data Flow Techniques

ISM facilitates the flow of data between different modules or between a module and the user by providing a central database storage area, standard data structures and formats, and various data management tools. In order to serve the needs of diverse modules and users for communicating with a database, there are three different data flow techniques generally used by ISM or similar packages. Referring to Figure 4, these techniques are denoted respectively by the terms "integrated," "interfaced," and "generic." Each technique is useful for certain types of applications depending upon the particular module characteristics, the resources available, and current and future user needs. In describing these techniques here the emphasis is on dealing with already existing modules, but many of the same considerations apply even if a new module is being developed specifically for use within a multidiscipline system.

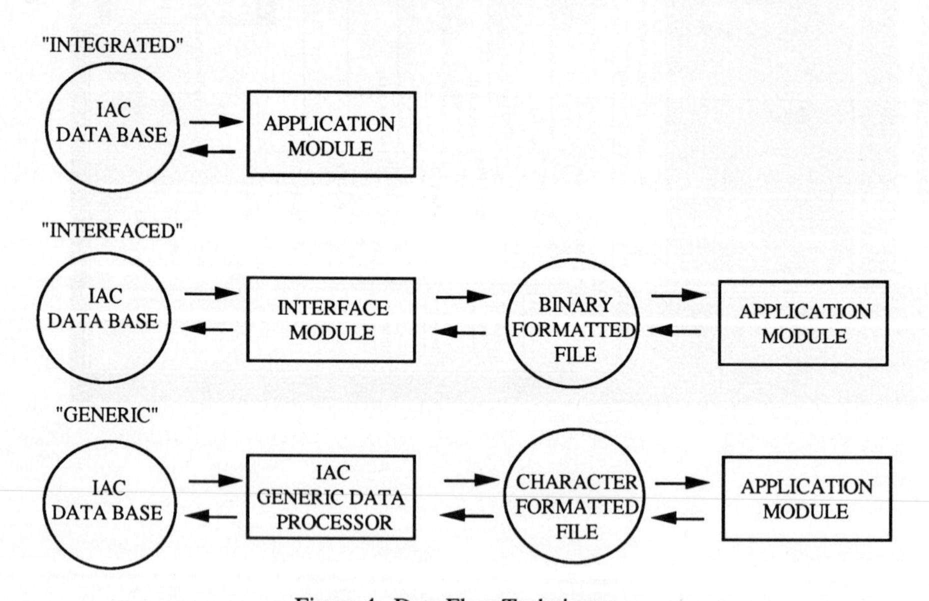

Figure 4. Data Flow Techniques

In the integrated approach, database read/write utilities are linked directly into the module along with any necessary code for conversion of data structures and formats. This approach can be effective if a fairly mature module source code is available, programming personnel are familiar with the module code, and the current and future data flow requirements are well defined. Integrated data flow has been provided for ISM modules such as the DISCOS multibody code and the MIMIC model interpolation code.

The interfaced technique uses an intermediate module, called an interface module, to bridge between the database and a module formatted binary file. (In some cases, especially for database-to-module flow, a character formatted file such as a module card image input file may be utilized.) This technique can be used effectively where module source code is not available or the software is difficult to modify, and the module is capable of reading or writing the required data via a module defined file. It is also helpful where users wish to save significant quantities of module computed data on an output file and then to decide at some later time which selected items of data are to be converted and stored in the database. Interface modules are used to provide data flow for ISM modules such as the SINDA thermal analyzer and the NASTRAN structural analyzer. In the case of NASTRAN, several different interface modules have been developed; some of these modules process NASTRAN generated binary files to extract computed nodal temperatures, static displacements, normal modes data, etc.; other modules use data in an ISM database to generate enhanced bulk input files for NASTRAN containing added data such as thermal load sets or constraint equations.

The generic data flow approach is characterized by a human-readable character formatted file and user-in-the-loop control. The process tends to be less automated than for the integrated or interfaced technique, but provides considerable flexibility and may in some cases prove to be more efficient. As implemented within the ISM executive, the generic data processor provides the user with sophisticated tools for selecting and formatting the data to be read or written. Data may be selected based on criteria such as line numbers or groups, character positions or values, presence or absence of certain character sequences, or similar criteria applied to free-field data items within the line. The ISM generic data flow capability is often used to extract data from printed output files, e.g., NASTRAN stress data, so that the data can be queried and plotted. It is helpful in generating partial input files, e.g., a list of element or node point definitions from data available in the database. The generic technique has many other uses as well, such as in transferring data between two different types of host computers.

Currently, most of the standard ISM modules and solution path operations are based on either the integrated or interfaced technique. The generic technique, however, has been found to be of considerably greater utility than originally envisioned.

2.4 Task Control

The ISM executive was designed to permit both interactive and batch tasks and to serve either as a master controller or as a slave to other application modules. ISM relies upon a host operating system feature to effectively support this design. A spawn feature (VAX/VMS) or fork (Unix) permits an executing module to pause temporarily and request that another module be executed, after which the initial module continues operation as though it had not been interrupted.

ISM task control and relationships between the executive program and other modules are diagrammed in Figure 5. Each circle in the figure represents an execution either of the executive (each execution is associated with a logically independent program, having its own distinct workspace), or of one of the other application modules in the system.

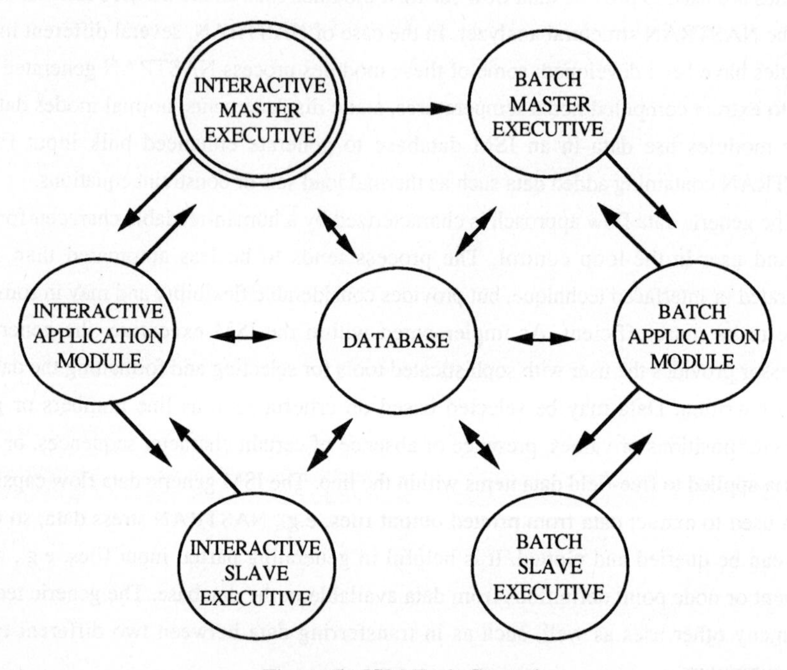

Figure 5. ISM Task Control

The three circles on the left describe an interactive ISM job process. This process could involve a sequence of input data generation steps, technical and interface module executions, query and display of computed results, etc., all managed by the interactive master executive. A module such as the ORACLS controls code might, for example, execute interactively with the user to accomplish several preliminary design cycles for a spacecraft controller, with each cycle involving brief execution of a slave executive in order to query data or to update database parameters and matrices. The master executive might then be used to initiate a batch job, e.g., to perform a more detailed final design analysis requiring more dynamic modes and correspondingly larger matrix computations. A batch job process is described by the circles at the right of the figure. Batch job capabilities are very similar to those of the interactive job, and modules within both types of processes may communicate with one another through concurrent accesses to a shared database.

2.5 Data Management

"Data management" is an old concept inherent in the creation and use of data itself, and data management software has existed in some form since the implementation of the earliest computer programs. Data management has become popularized in more recent years along with terms such as "database" and "query," and there has been a continued evolution in the sophistication of the associated software. The design of an integrated analysis system is influenced by considerations such as typical data group sizes, types, and processing sequences; computational efficiency as well as detailed query requirements; the user interface; and the data management techniques employed by existing modules. The ISM system implementation is characterized by several key data management features:

a. Use of three major areas;
b. A file oriented multi-user database;
c. Standard data structures;
d. A user specific workspace; and
e. Data manipulation and query capabilities.

These features are discussed in the following paragraphs.

Data Areas

As summarized by Figure 2 and the associated overview section on ISM architecture, ISM provides communication between three types of data areas: (l) a special file oriented database; (2) a user specific virtual memory workspace; and (3) the ordinary host computer file system. An ISM database is a permanent storage area containing a cataloged collection of ISM files. The workspace provides for direct user processing of ISM data structures (e.g., for data definition, manipulation and query) as well as for support of the executive processing requirements. An ISM database or a host computer file may be accessible to many different users, while the workspace is a private area accessible to a single user. Utilities provide for various types of data communication between the three major data areas.

Database

A schematic of the overall ISM database strategy is shown in Figure 6. Multi-user concurrent access to the database is supported by an activities file, which contains needed information (user name, access type, etc.) for each database file access currently in progress. This allows for the identification of, and protection against, user access conflicts which might occur. A conflict would occur, for example, if one user were attempting to read data which another user is in the process of writing or deleting. Conflicts between users are reported and, if possible, resolved through an automated repeat/wait cycle algorithm. It should be noted that access conflicts may be the result of erroneous or inappropriate user actions, e.g., an attempt to update a key file without informing other analysts. It is considered important to make conflicts visible by reporting them to the users involved.

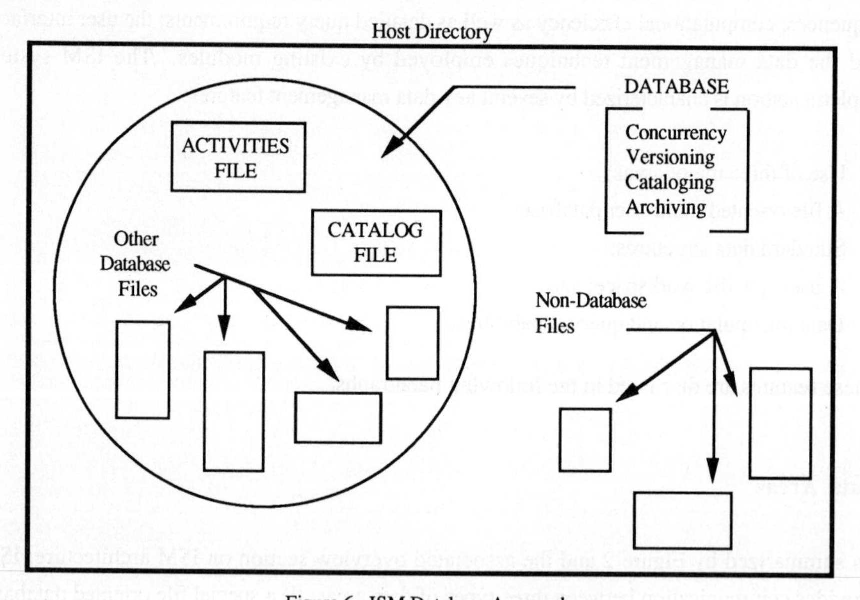

Figure 6. ISM Database Approach

The catalog contains characteristics of all other files in the database, including titles and keyword identifiers, as well as pointers to the data file itself and to any associated user defined textural information. Detailed catalog query capabilities provide an analyst with the ability to regain familiarity with existing data files and to verify the correct identity of a file prior to its use in further computations. As an illustration, a user could make a request to "list the file name, creation date, creating module name, and title; for all database files created by SMITH, after July 1990; which contain the keywords TRUSS, MODES, and/or CONFIGURATION2." Such query capabilities are also valuable in identifying groups of files, prior to performing operations such as copying, deleting, or renaming of the files. The catalog supports an automatic versioning capability for all database files. Versioning encourages backup against data loss, promotes recovery from some types of analysis errors, and helps users to differentiate between obsolete and current data.

The database contains two types of cataloged files:

1. "Unstructured files," i.e., files defined according to individual module formats. Examples are a NASTRAN checkpoint file, a DISCOS plot file, FORTRAN source or object code files, and the input card image or output printer files for any module. In order to maintain computational efficiency, these files are left in their original module defined formats without any ISM modification or appending of additional information.
2. "Structured files," i.e., files defined according to ISM standard formats. These are also referred to as data structures and provide a common basis for communication between different modules or between a module and the executive. There are three classes of ISM data structures--ARRAY, RELATION and USER. These classes are summarized in the Data Structures section below, and associated information is given in the later section of Data Manipulation and Query.

Data Structures

ISM data structures are consistent with the concepts of a file oriented database, a virtual memory workspace, and user hands-on access to data. Data structures are created and stored as individual files, and the system design emphasizes the independent self-contained nature of each file. Transfer of data between a database and a workspace often involves an entire data structure; however, the ISM data handling utilities allow for the option of transferring only selected subsets of data within the structure. Data group types, sizes, and associated identifiers are kept for all data groups but are stored separately from the actual data so as to improve computational efficiency and to simplify any user requirements for dealing directly with the data. The ARRAY, RELATION and USER data structures all share a common record level format definition. This definition is simple enough to allow, if necessary, direct user processing of the data file outside of the ISM system.

The ARRAY is probably the most important ISM data structure, but the RELATION serves many useful purposes. Typical examples for both are shown in Figure 7. The ARRAY is effective for both computational and query purpose and is used in most of the ISM technical module interfaces. This structure is a generalization of vector and matrix type data and can be considered to include the RELATION structure as a subset. The illustrative 2-dimensional ARRAY in the figure contains transient nodal temperature and pressure data. The core of the general structure consists of one or more arrays of arbitrary order (e.g., a matrix may be represented as a second order ARRAY), and each dimension (index) may be of arbitrary size. In order to facilitate user query of the data, each index may have an associated table consisting of one or more labels. Each part of the structure (i.e., label or array) can be referenced by name. An important feature of the ARRAY is the ability to also reference data by index (ISM provides a linear array index as well as an index for each of the dimensions). The ARRAY data structure is used to handle much of the data for ISM modules and solution paths, e.g., nodal temperature data, mode shapes, mass and stiffness matrices, displacement and stress data, and plant/controller definition matrices.

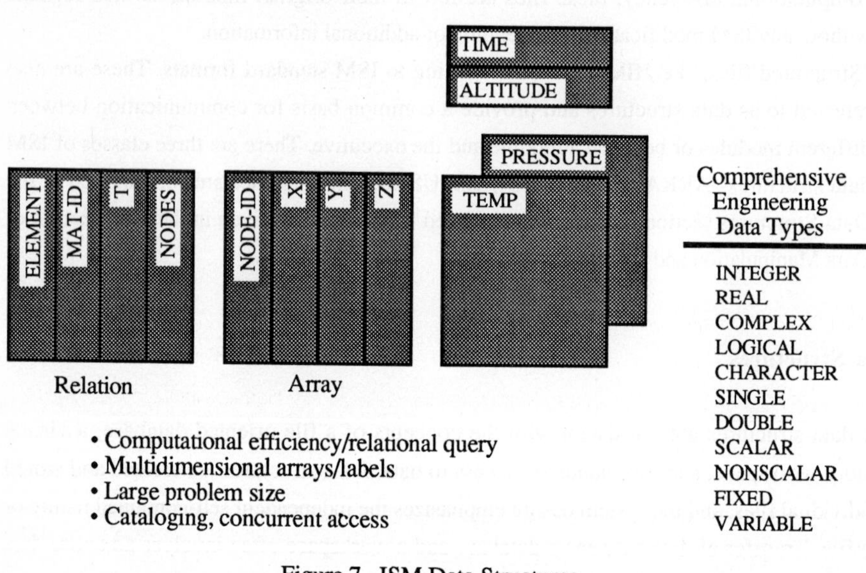

Figure 7. ISM Data Structures

The following are considered to be important operations for an engineering analysis oriented data/query manager, and are provided by ISM.

1. Logical comparison between named variables and constants, e.g., "X greater than 100," or "TIME most equal to 2.5."
2. Logical comparison between different named variables, e.g., "Y most equal to Z" or "NODE less than or equal to 210."
3. Maximum/minimum selection, e.g., "STIFFNESS equals maximum" or "X equals minimum subject to condition that NODE less than 50."

4. Set membership and list comparison, e.g., "KEYWORDS contains all of the substrings MASS, TRUSS, MARCH 1990" or "NODE in the set (25, 30, 71 through 90 incremented by 2, 511, 519, maximum)."

The ISM executive provides on the order of 50 commands associated with data manipulation and query. As discussed in the prior section on Task Control, all of these capabilities can also be accessed from within an application module via the SPAWN type of procedure.

3 Technologies and Models

We include here a brief description of several individual technology modeling characteristics. Not surprisingly, almost every major technology has developed its own preferred methodology. As we move toward a more integrated CE environment, we will need to identify and build on the common characteristics, develop better interfaces between technologies, and consider some new system modeling approaches towards which we can evolve.

3.1 Structures

The predominant characteristics of structural analysis today are derived from several key concepts: finite element modeling; dynamic modal techniques; the requirement for handling very large matrices; and support of CAD geometry, model generation and postprocessing tools. Model refinement for many applications is still limited by capacity and/or cost of available computing hardware. Because of this, less than desired accuracy may have to be accepted and, in fact, the accuracy or correctness of the solution may not be known. The "P-method" is an alternative finite element approach which has recently attracted more interest and recognition. It is based on a sequence of higher order shape functions and can be used to obtain convergence estimates. However, it is not well suited to handling irregular boundaries.

As finite element models become larger and more concurrency is desired between the CAD geometric description and the structural model, ways must be found for defining large models in a more concise manner. One promising approach is to define a 3-dimensional finite element model in terms of only surface and cutting plane discretizations from which the analysis package can then construct its detailed internal math model whenever needed.

3.2 Thermal

Although finite element methods have made some inroads into the thermal analysis technology, finite difference methods are still much more prevalent, at least within the aerospace industry. Actually, thermal analysis methods might better be described as depending on a "nodal network" approach. They do not generally have any kind of spatial metric superimposed on the model. Rather, thermal masses and various types of conductors are defined and lumped by the analyst based on the geometry, but the final model has no explicit knowledge of the spatial distances. This is a serious problem for communication and concurrency with other technologies. The problem will be discussed more in the section on Integrated Solution Paths.

Thermal analysis really includes three subset types of analysis--conduction, convection and radiation. Radiation usually comprises a separate modeling/analysis step, prior to merging its effects into the overall conduction/convection/radiation model. Thermal analysis often deals with very large and highly interconnected models. The biggest cost, which is sometimes a prohibitive one for the desired level of modeling accuracy, is the high analytical "connectivity" of the radiation elements of the model.

3.3 Controls

A basic characteristic of the controls modeling process is that it, like the common thermal approach, does not use a spatial metric. In fact, it is generally considered as not having one. Of course, controller sensors, actuators, gyros, etc. have physical representations, but these representations are usually modeled only in terms of their effects. Physical mass, stiffness, thermal surface, etc. properties are generally left to the other technologies to model. There are several controls modeling techniques, including state space definitions, transfer functions, and a large emphasis on various types of frequency domain plots.

For concurrency considerations, the state space model is most significant. In fact, the state space model probably ranks with finite element methodology in terms of significance to the total multidisciplinary analysts problem. Controls is often the last technology modeled in a sequential analysis process. Controls has often had the role of integrating the other models, implementing required mode selections and truncations, and performing the system simulation. The generality of the state space concept is an important enabling capability for the total system modeling. However, controls analysis tools are generally restricted in effective application by size, i.e., on the order of a one to two hundred degree of freedom. Large sparse matrix methods are absent, and software implementation approaches have obvious limitations evolved from having to consider only this size of problem. A major challenge for

CE is to utilize the general controls modeling techniques in an enhanced form which meets required multidisciplinary and modeling accuracy requirements.

3.4 Optics

Optics modeling deals with two major approaches -- ray-trace or "geometric" analysis, and wavefront or "physical" analysis. As the name implies, ray-trace optics considers light as a series of discrete rays which can be traced from surface to surface in terms of ray direction and path length. Wavefront optics treats light in a more physically represented manner, propagating the light in terms of its intensity and phase. Wavefront optics also can treat additional effects such as diffraction. Transformations between the geometric and physical representations are often performed, at a particular "phase front" or surface. Optics modeling, especially wavefront methodology, generally uses a rectangular grid of evenly spaced mesh points at which appropriate quantities are propagated through a series of reference planes and/or optical elements. Optics interacts with the structure through the motion and deformation of optical elements and with controls through the use of optical sensors to feed information to the controller.

3.5 Multibody

The term multibody refers to a system of two or more bodies, connected in such a way as to allow significant translations and/or rotations to occur between the bodies. Examples are robots, spacecraft with slewing appendages, and aircraft with movable aerodynamic surfaces. The multibody technology is often considered to be a combination of structures and controls. Controls analysts may be more interested in frequency domain simulations for evaluation of controller performance and stability. Structural analysts may be interested in both time and frequency domain simulations for evaluation of dynamic modal and static strength characteristics.

Major concerns in multibody modeling have always been (l) accurate representation of the complex motion effects and structural/control interactions; and (2) high computational cost of the time domain simulations. The accuracy concern has largely been solved through various types of comprehensive and "exact" modeling approaches. The analyst will sometimes drop consideration of certain terms and effects in the formulation in an effort to improve computational speed, although any gains made here generally make only a linear contribution to cost savings. A major reason for increased cost in the time domain solution is that as the system motions take place and the configuration changes, the system modes are constantly changing. This makes it impossible to permanently eliminate any higher frequency modes from the simulation, thus requiring a small and costly time step in the time integration. Major advances have been made in reducing the computational costs, i.e., the recent "order n" or "order n^2" algorithms; and the use of faster computing and/or parallel processing.

From a CE standpoint, a multibody simulation may need to integrate within it the interactions of structures, controls, thermal and optics for spacecraft applications. For aircraft applications, the aerodynamics, structures and controls are interacting technologies. One of the important needs currently being addressed is the use of better graphical methods for visualizing rigid and/or flexible multibody motions.

4 Integrated Solution Paths

There are many solution paths, i.e., multi technology or multi task analysis process, of interest in the CE environment. We describe a few of them in this section. Important considerations are the technical interactions, the availability of appropriate technical modules, communication between the different types of models and meshes, and the practical formatting and flow of data among all of the solution path components. This section is concluded with a discussion of the subject of model synthesis and reduction. This subject is crucial to successful CE analysis and may very well be one of the most important CE research areas yet to be fully developed.

4.1 CAD/Thermal

CAD/thermal might not be considered a true analysis solution path because it explicitly involves only a single analysis technology. However, it does involve the integration of CAD and finite element technology along with the thermal analysis methodology.

The left side of Figure 8 describes the thermal modeling process at issue. There are two main analysis branches here - the radiation analysis, which is sort of a pre-analysis step for the purpose of generating the "radiation conductor" definitions; and the main thermal analysis which includes the combined effects of conduction, convection, and radiation. The basic goal in this solution path is to automate the building of the models and the post processing of the results. Since thermal analysis packages are usually deficient in geometric model building and CAD-type packages are well suited to automatically generating finite element models, the solution path represents a good marriage of the two capabilities.

In the ISM implementation approach, two different but compatible finite element thermal models are constructed. The first model emphasizes radiative surface definitions which can then be mapped onto the representations of the radiation analyzer. This mapping is

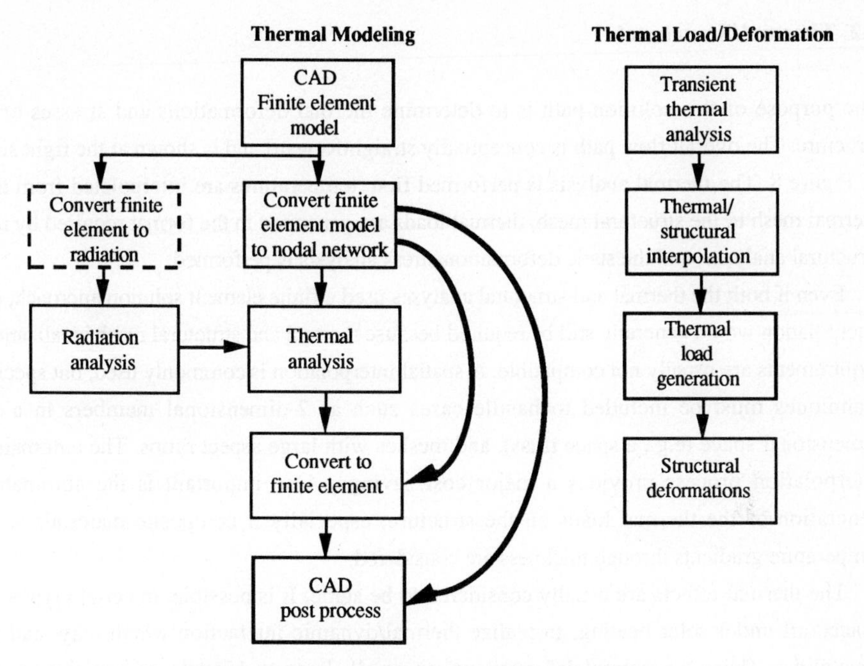

Figure 8. Thermal Modeling and Thermal Loads

conceptually straightforward but requires that the radiation analyzer have a simple quadrilateral type of element in its library. The radiation analyzer then generates lumped node and radiation conductor definitions for the overall analysis. The second model includes both 2-D and 3-D aspects so that conduction and convection characteristics can be represented. A finite-element to nodal-network converter performs the lumping of the finite element material into the required nodes and generates appropriate conductor values. Then, both network models are combined and the overall analysis is performed.

There are two alternative network/finite-element mapping approaches. One approach basically generates nodes at the centroids of the elements; this leads to a narrowly bonded set of equations. The other approach generates nodes at the element vertices and finite element consistent network conductors; this leads to a considerably larger bandwidth in the equations with correspondingly higher solution cost but also additional accuracy.

The final step involves postprocessing to plot thermal contours, etc. Here again, the CAD/ finite-element capabilities are used. Thermal network results are interpolated back to the finite element model using spatial interpolation algorithms and supporting data which was generated in previous mappings.

4.2 Thermal/Structural

The purpose of this solution path is to determine thermal deformations and stresses on a structure. The overall flow path is conceptually straightforward and is shown at the right side of Figure 8. The thermal analysis is performed first, temperatures are interpolated from the thermal mesh to the structural mesh, thermal loads are generated in the format required by the structural analyzer, and the static deformation/stress analysis is performed.

Even if both the thermal and structural analyses used a finite element solution approach, an interpolation would generally still be required because thermal and structural mesh idealization requirements are usually not compatible. A spatial interpolation is commonly used, but special techniques must be included to handle cases such as 2-dimensional members in a 3-dimensional space (e.g., a space truss), and meshes with large aspect ratios. The automated interpolation process provides a major cost savings. Also important is the automated generation of the thermal loads on the structure, especially if composite materials with temperature gradients through thickness are considered.

The thermal effects are usually considered to be static. It is possible, in certain types of spacecraft under solar heating, to realize thermal/dynamic interaction which may lead to instabilities. There, the thermal deformations can usually be treated in a "quasi-static" manner. They are determined a priori, generally converted to modal deformations, and superimposed on the dynamically deforming structure.

4.3 Structural Dynamics/Control

This can be a fairly straightforward path if the system is linear and the controller is simple. The dynamic model is built and mode shapes and frequencies are determined. Desired measurements for the structure, or "plant," are defined and the second order modal representation is converted to the first order state space form:

$$\dot{X} = AX + BU + F(X,U)$$
$$Y = CX + DU + G(X,U)$$

where X is the state vector, $\dot{X}$ are the state derivatives, U are the external disturbances, and Y are the measurements. The matrices A, B, C, U define the linear plant. If nonlinearities are present, terms are added via the functions F and G to represent these effects.

The controller definition is then added. It may be represented as a separate set of matrices, interacting with the plant inputs U (controller outputs) and plant outputs Y (controller inputs). Alternatively, a synthesis of the plant and controller can be accomplished, leading to a single system set of matrices having the same form as those above. In either case, the equations can be used to perform various types of time or frequency domain analysis.

4.4 Structural/Controls/Optics

This path could be applicable to large telescopes, surveillance systems or directed/kinetic energy weapons, and is described referring to Figure 9. The modeling may be considered to begin with the structure, for which modal results are computed and stored in the database. Using the structural geometry, optical sensitivities are computed which provide relationships between structural deformations and optically significant parameters, e.g., sensor inputs. Sensors and actuators, which may have internal dynamics (physical or electrical) of their own, are modeled. Combining the structural, optical sensitivity, sensor and actuator models produces the linear or nonlinear system plant models. The controller is then designed and superimposed, and the required time and/or frequency domain analyses are performed. Detailed performance may be analyzed, usually with the aid of graphical procedures. This is a postprocessing step for directed energy weapons in which the wavefront is accurately propagated to the target.

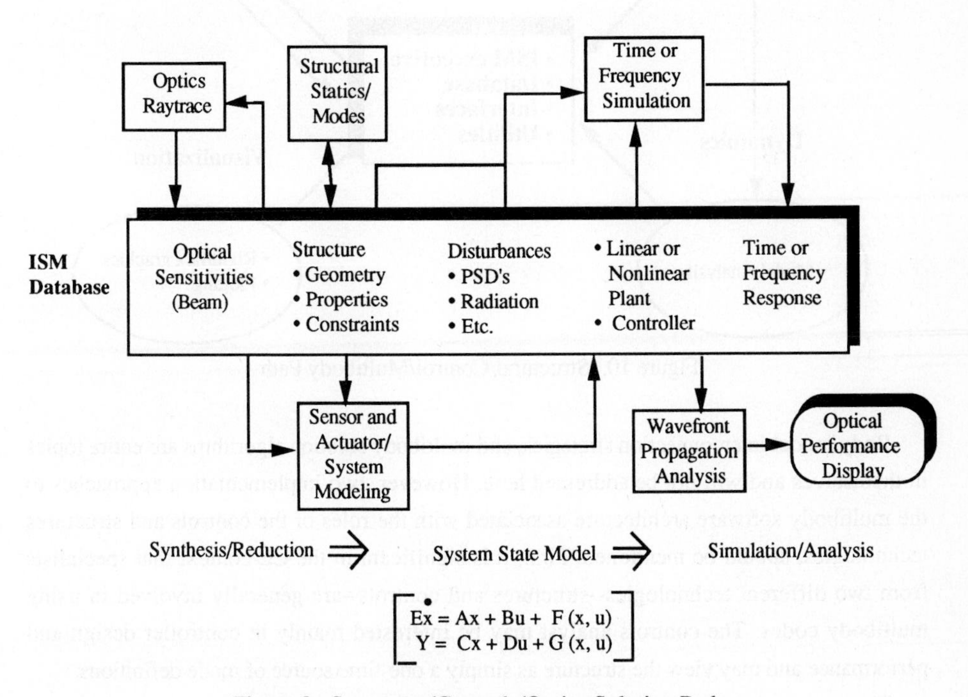

Figure 9. Structures/Controls/Optics Solution Path

The steps in the somewhat complex process must generally be done in an iterative, and hopefully CE, manner to obtain an optimum design. More general accuracy/interaction issues are discussed in Section 4.6, on model synthesis and reduction.

4.5 Structural/Control/Multibody

A brief summary for this path is given here, with the aid of Figure 10. There are four key parts to this path: structural modeling for each individual body in the system; determination of modes and frequencies, again for each separate body; the multibody modeling and simulation, including controller definition and body interconnections; and graphical visualization of the deformed geometry and other results.

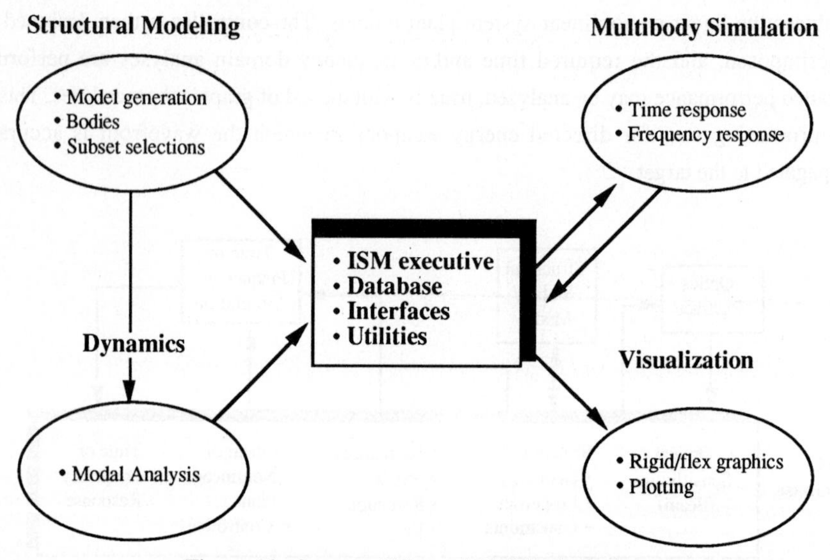

Figure 10. Structural/Control/Multibody Path

Body/modal interconnection strategies, and multibody solution algorithms are entire topics in themselves and will not be addressed here. However, two implementation approaches to the multibody software architecture associated with the roles of the controls and structures technologies, should be mentioned. First, it is significant in the CE context that specialists from two different technologies--structures and controls--are generally involved in using multibody codes. The controls analyst may be interested mainly in controller design and performance and may view the structure as simply a one-time source of mode definitions.

The structural analyst may be interested in strength and dynamic design characteristics and may view the controller simply as a source of external disturbances. Thus, the structural analyst tends to prefer a multibody code emphasizing the dynamic and interconnection effects and finds it useful to have callable access to a few simple controller definition subroutines. The controls specialist, and perhaps the system analyst as well, prefers to have a detailed controls module in charge of the computational and time integration algorithms and would like the dynamic plant definition available at each time step as a callable subroutine. To meet both needs, it is very desirable to have the architecture of a multibody code designed so that it can act either as a self contained overall analysis package or as a cleanly callable subroutine. Without this generality, a multibody code will not reach its full potential user group.

Graphical visualization is especially important in the multibody area because of the complex system/body interactions. Adequate capabilities are only recently becoming available. A key requirement is that both rigid body and flexible deformations be described in a combined fashion. Since flexible deformations tend to be much smaller than the rigid, flexible components must be scaled for visibility. On the other hand, it is desirable, if possible, to avoid having true rigid bodies appear to be deformed. A range of visualization capabilities, from stick-model or wire-frame to full 3-D with hidden surface features, is also useful.

4.6 Model Synthesis/Reduction

A multidisciplinary CE analysis approach, to be successful, must have a common mathematical description in which all technology models can be represented. Since multidiscipline models tend to be large, the approach must also provide an appropriate model reduction process. The first requirement is generally met by a first order state space formulation, which provides a clear, consistent representation for all of the component technology models. An automated way of synthesizing (combining) these models into the system model is also needed. The main task here, aside from organizing the storage and access to all models, is the handling of the model interfaces including connectivities and constraints.

The requirement for model reduction is necessary to maintain practical computational costs. Reduction is typically implemented at the individual technology model level just before these modals are merged into the system model. Guesses and approximations must often be used, trying to avoid truncation of the component models in such a way as cause significant loss of system level effects. This current approach has evolved for two reasons: first, because the technologies do not operate in a CE mode, and therefore the interfaces must be kept simple; and second, because today's methods for handling very large models, even temporarily prior to reduction, are not adequate.

An improved synthesis/reduction approach is badly needed. It must allow the synthesis to be moved upstream and the reduction moved downstream within the total analysis process. Of course, this is a natural strategy for CE to allow more interactions between the disciplines and to achieve a more accurate and optimum result. To accomplish this, there must be several process improvements: more concurrency between the technologies, extending within a practical industry environment to the contractor/subcontractor relationships; more acceptance of common supporting tools, including software, databases, and state space type models; better computational utilities for handling very large models; and a more accurate CE strategy for model synthesis/reduction.

5 Applications

ISM and related Boeing integrated analysis capabilities have been used to support a variety of applications. Although time and space do not allow detailed descriptions, some examples are briefly mentioned here to give further insight into typical solutions paths.

A recent ISM thermal/structural application involved a space-based antenna having severe shape control requirements and subjected to solar radiation. Approximately 20 different thermal radiation load cases were considered, representing different orientations relative to the solar input. The SINDA thermal analysis module was used, and NASTRAN was run to compute thermal stresses and deformations. Automated thermal-to-structural interpolation and NASTRAN thermal load generation resulted in an estimated 60% labor savings over previously available approaches.

An interesting target/decoy discrimination study was performed using a thermal/dynamic solution path. The concept is for a directed-energy weapon to beam energy onto the target or decoy for a very brief time period (less than a millisecond). This causes rapid heating and associated dynamic vibrations. The vibration frequency and/or amplitude content can then be measured using an interferometer type of device. A real-time analysis of the results can theoretically allow discrimination between target and decoy. The ISM solution involved a thermal analysis, thermal load generation on the structure, and dynamic time and frequency response analysis.

An ISM dynamic/optics solution path was used to evaluate the optical performance of a multi-component vehicle under severe external disturbances. This was a typical aerospace contractor/subcontractor application, in which each vehicle component affecting the optical performance had been analyzed separately by the associated contractor. Total performance was estimated by simple addition of terms. When a concurrent analysis of the total vehicle was performed, it was found that there were significant component interaction effects not previously identified due to structural interface flexibilities. The critical location of the flexibilities relative to the optical train resulted in an optical performance 10 times less than previously estimated. These effects were identified early enough in the program to allow them to be included in the vehicle design iterations.

A serious anomaly occurred during an early flight of a key spacecraft. The anomaly was associated with the propulsion system, but the exact cause was unknown. An integrated analysis capability was used to perform approximately 100 different structural configurations, load cases, and "what-if" scenarios within the time span of a very few weeks. The integrated database and graphical plotting capabilities were essential in supporting modeling and evaluation of these scenarios. The results allowed design and manufacturing changes to be made in a timely fashion, minimizing the down time and increasing future quality assurance.

These and other applications indicate the value of integrated, multidisciplinary, and/or CE analysis capabilities. Such applications sometimes follow standard solution paths which can be provided in a well developed, automated form. Other applications tend to be unconventional, such as the previously mentioned thermal/dynamic or anomaly investigations, for which a modular toolkit of capabilities must be provided to the user for developing a specific solution path.

6 Roadblocks to Concurrent Applications

The roadblocks mentioned here have been identified through analysis and training applications, or from experience in developing integrated analysis software. Undoubtedly we will face other challenges as CE implementations evolve further.

A roadblock that immediately comes to mind is the still existing diversity of software tools and "standards." Within the U.S., for example, there are probably at least ten major structural analysis codes, each with a significant following. There are almost that many major thermal analyzers. Costs for the user learning curve, in moving from one analysis package to another, often seem to be a waste of valuable resources. It is well and good to speak of standard formats and generic data flow interfaces but, in the end, interfaces need to be implemented. Existing data flow patterns can be a tremendous help, but the required resources for implementing such interfaces are still significant.

Some of the commercial integrated analysis companies have tended to compound this problem. In their natural desire to create a more consistent and integrated system, to avoid royalty payments for themselves or for their users, or simply to gain more control over future software development, these companies often develop entirely new single-discipline analysis packages. The desire to make product improvements and to avoid dealing with 10 different existing packages is quite understandable. The result, however, is usually the creation of an 11th package.

Some roadblocks remain in dealing with computer operating systems or software development toolkits. We have seen progress in graphics standardization, although much more is desired. Recent developments in the area of operating systems seem to be a step in the right direction. At this time, we must deal with two Unix flavors--Berkeley and AT&T System V. A recent proposed solution is OSF-l Unix. In the near term this will unfortunately result in three Unix flavors, in addition to which each will have sub-varieties depending on the particular vendor's options and system utilities.

Hardware limitations will always be with us. Computers for CE applications will obviously be needing more speed, memory, and external storage. A few years ago, it seemed that speed was the primary roadblock. In the writer's opinion, memory is now the primary limitation, which in turn adversely impacts large-problem computational speed.

Organizational and political roadblocks to CE have always existed. Undoubtedly there are fears that the new philosophy will take away local control over hardware and software selection, modeling decisions, and analysis methods. Even though software tools are costly to develop, organizations often have a pride of ownership in creating and maintaining their own. If integrated tools are to be developed, it is difficult for one technology organization not to be assigned the lead in software development. If participation in the requirements definition and implementation phases is not achieved, ownership roadblocks will discourage use by the other technologies.

There is often a conflict between specific tools and the more general, integrated packages. This is often apparent in coordination between the technical specialist and the more system-oriented multidiscipline analyst. In performing only a single standalone task, the specific tool may very well be justified on the basis of efficiency. In fact, there are always the software "dis-integrators," who have reason to extract pieces from an integrated system and modify them for a specific task. The difficulties occur in a CE environment when the standalone task needs to communicate downstream with other tasks, or when enhancements to the integrated capabilities are developed and the dis-integration must be repeated.

At the company level, proprietary restrictions necessarily inhibit the total implementation of CE. There may be an unwillingness to share data, analysis tools, or process concepts. This roadblock can be somewhat circumvented by careful attention to the contracting relationship and concentration on the "what" of the product rather than the "how." This problem is especially true in government/industry relationships because of the understandable desire by government to establish improved quality at lower cost, detailed product definition, and flexibility in downstream procurement options.

7 Future Directions

In spite of the inherent roadblocks, CE seems destined for continued rapid growth and acceptance. CE is not the answer to all problems, and it remains for each project to implement its processes in an effective manner. The trade offs must be evaluated, and CE and/or sequential approaches must be used where most appropriate.

An obvious future direction is the continuing growth of standards. X-Windows, Motif, Phigs/PEX, and NCS/NFS type capabilities are examples in the graphics and networking areas. OSF-1 may become the standard Unix system, but not in the near future. IGES will continue to be a prominent CAD data exchange standard, and PDES will probably evolve slowly as a more comprehensive tool for product definition. The SQL data query format will probably evolve and become more widely used. A higher level graphics based system for executing modules and establishing task and data flow control is likely to evolve to a standard form. In general, there will be a strong drive for standardization to be applied to any CE related function which is common across hardware platforms, technologies, task executions, data management, or data visualization.

A broader based standard networking approach can be expected, addressing communication protocols, binary data flow, task control, and synchronization across computing hardware.

A higher level of integration and formalization is to be expected as integrated analysis packages are combined with design, manufacturing, testing, and maintenance functions.

Continued commercialization of analysis capabilities can be expected. We will likely see fewer new software packages related to CE analysis, either for individual technologies or for integrated analysis. The high cost of software development and reduced funding levels will encourage the evolution, commercialization, and shakeout of existing packages. There will always be initial room for new software to support new emerging technologies, but such software should be expected to quickly merge with existing broader-scope packages.

A future direction we would certainly hope for is a broader and more effective government/industry partnership in implementing and using CE technology. As the largest customer for CE technology, the government must have a major role in sponsoring joint industry efforts, encouraging standards, and leading the move toward world product economy.

References

1. R.G. Vos, et al, "Development and Use of an Integrated Analysis Capability," AIAA 24th SDM Conference, Lake Tahoe, Nevada, May 1983.
2. H.P. Frisch and R.G. Vos, "The Integrated Analysis Capability (IAC Level 2.0)," ASME/CIE Symposium on Engineering Database Management Critical Issues, New York, August 1987. Also published in Engineering with Computers, Volume 4, 1988, Springer-Verlag.

Simulation Based Design of Automotive Systems

Werner O. Schiehlen

Institute B of Mechanics, University of Stuttgart, W-7000 Stuttgart 80, FRG

Abstract: The design of automotive systems using simulation tools features cost reduction and quality enhancement. This paper presents two basic approaches. The first approach deals with the application of CAD data bases to the evaluation of input data for multibody system formalisms, most adequate for automotive system modeling. An object oriented data model for multibody systems is presented. The second approach covers the development of an integrated simulation tool for automotive vehicles and the corresponding animation facilities. Driving comfort is related to the human perception of mechanical vibration. A companion paper deals with the optimization of automobile parameters using the multibody systems approach.

Keywords: CAD data base / object oriented data model / modeling / multibody systems / equations of motion / vehicle dynamics / simulation / animation / driving comfort

1 Introduction

The strong worldwide competition of automotive industries results in a large variety of automobiles to be developed in shorter and shorter periods. Thus, the classical method of automobile design via intensive experimental testing of prototypes is no longer economically feasible. Therefore, the dynamical behavior of a vehicle has to be simulated during the development process simultaneous with the overall design of the automobile. Most of the data for dynamical modeling and simulation are available in the CAD data base, blueprints and drawings may be omitted. This advanced method is part of Concurrent Engineering (CE) defined as an approach for designing and validating a product, its manufacturing process, and its quality control, all at the same time, see Figure 1. In particular, Concurrent Engineering is superior to the traditional sequential engineering with respect to the time required for the development of a new product. An integrated information processing results in an essential time saving, Figure 2.

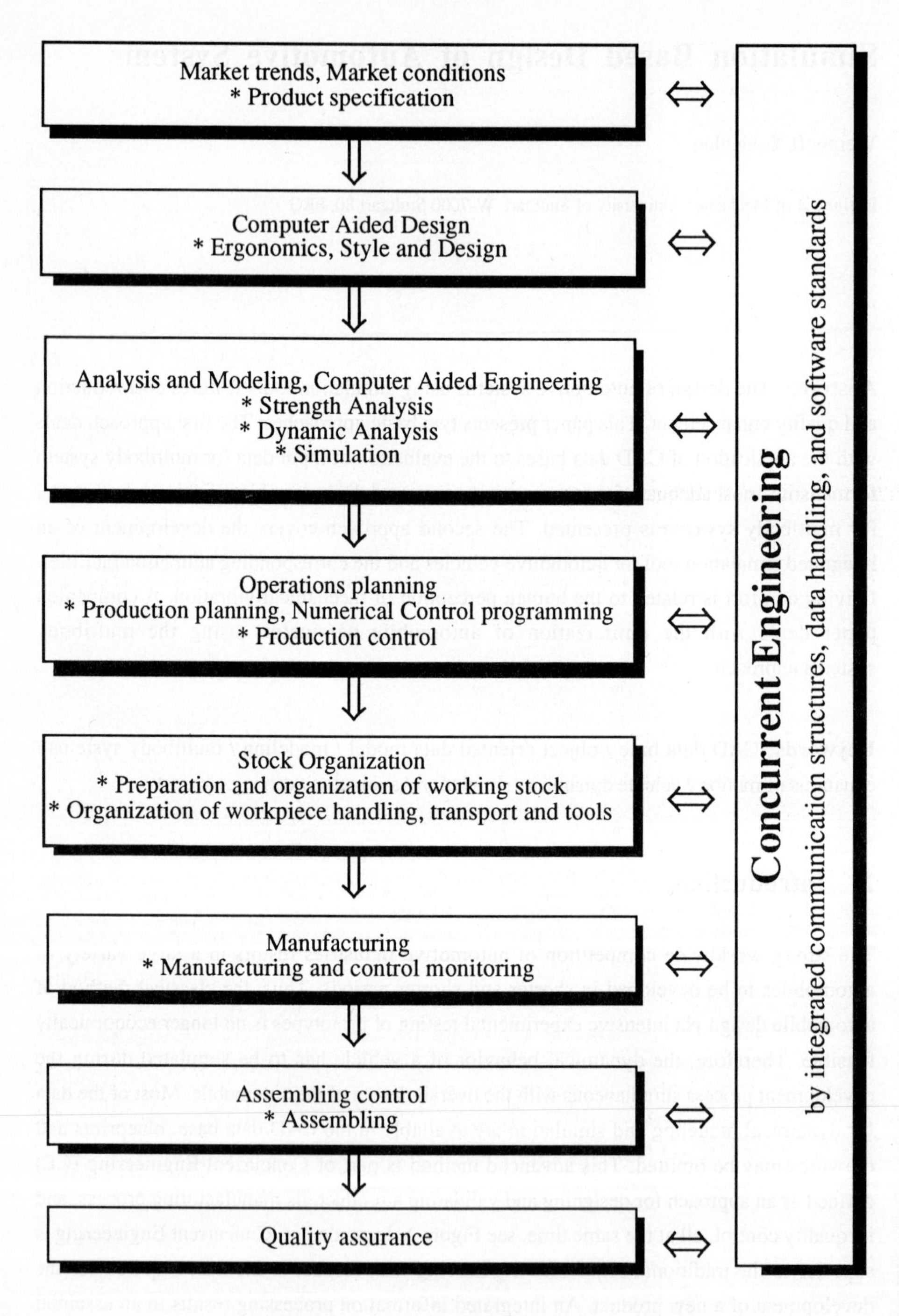

Figure 1. Concurrent Engineering Approach

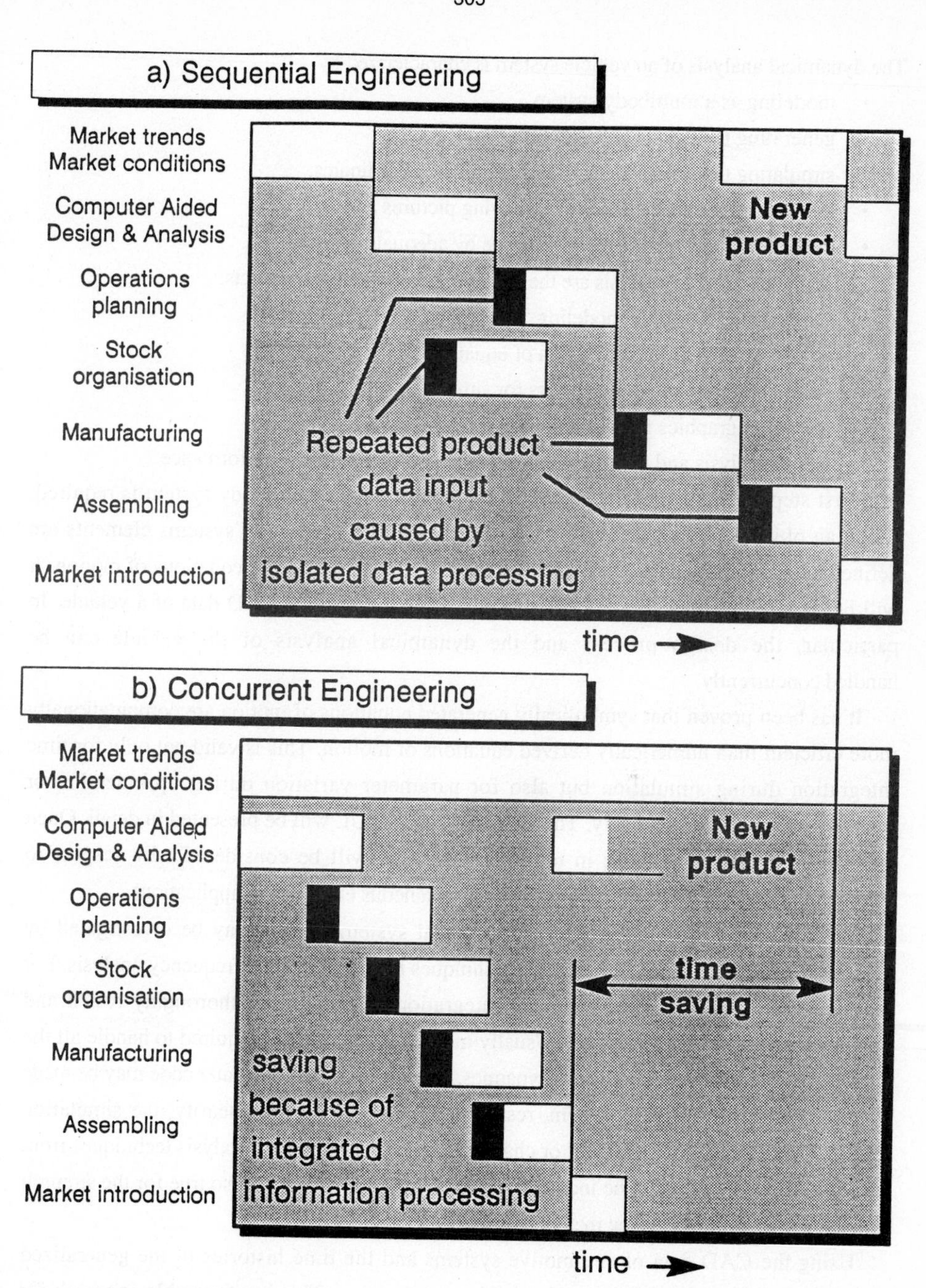

Figure 2. Sequential versus Concurrent Engineering

The dynamical analysis of an vehicle system is characterized by

- modeling as a multibody system,
- generating the equations of motion,
- simulating the trajectories of the generalized coordinates,
- animating the vehicle system by moving pictures and
- evaluating the dynamical performance by adequate criteria.

Related to the dynamical analysis are the following computational aspects:

- CAD software for the modeling,
- CAE formalisms for generation of equations,
- ODE and DAE integration codes for simulations,
- computer graphics for animations and
- signal analysis and optimization codes for evaluation of the performance.

In a first step, a unique description of all the elements of a multibody system is required. Using an object oriented software approach, the data of multibody systems elements are defined independently of the formalism applied for the generation of equations of motion. It will be shown how element data are extracted from the original CAD data of a vehicle. In particular, the design process and the dynamical analysis of the vehicle can be handled concurrently.

It has been proven that symbolically generated equations of motion are computationally more efficient than numerically derived equations of motion. This is valid not only for time integration during simulation but also for parameter variation during optimization or sensitivity analysis, respectively. The formalism NEWEUL will be presented in detail. Open and closed kinematical loops in multibody systems will be considered with respect to computational efficiency most important for simultaneous engineering applications.

Automobiles are highly nonlinear dynamical systems which may be investigated by numerical simulation or by linearization techniques resulting in eigenfrequency analysis. For the numerical simulation the available integration codes have to thoroughly tested and implemented in software packages. Usually more than one code is required to handle all the different problems in vehicle system dynamics. The choice of the computer code may be made by the user or the software system, respectively. Due to the nonlinearity, the simulation results show usually very irregular or chaotic motions. Then, signal analysis techniques from nonlinear dynamics have to be included in the investigation. This is also true for the strength evaluation requiring stochastic methods from material sciences.

Using the CAD data of automotive systems and the time histories of the generalized coordinates, the motion can be made visible by animation. This is a favorable approach for checking the simulation and to obtain a general idea of the motion. However, for an engineering improvement of a highly developed system like an automobile, special criteria for rating are necessary. Two of the most essential criteria are the riding comfort and the riding safety. The comfort is related to the human perception of mechanical vibrations while the safety can be rated via the vertical dynamic tire load. It will be shown how the first criterion may be introduced in the Concurrent Engineering process, too.

The final step, finding optimal parameters for the automotive system, will be treated in the companion paper by Dieter Bestle entitled 'Optimization of Automotive Systems.'

2 Multibody Systems Modeling

Road vehicles can be modeled properly as multibody systems for the design and the analysis of components like suspensions, attitude controllers, shock absorbers, springs, mounts and steering assemblies as well as brakes and antiskid devices. The complexity of the dynamical equations called for the development of computer-aided formalisms a quarter of a century ago. The theoretical background is today available from a number of textbooks authored by Wittenburg [1], Schiehlen [2], Roberson and Schwertassek [3], Nikravesh [4], Haug [5] and Shabana [6]. The state-of-the-art is also presented at a series of IUTAM/IAVSD symposia, documented in the corresponding proceedings, see, e.g., Magnus [7], Slibar and Springer [8], Haug [9], Kortüm and Schiehlen [10], Bianchi and Schiehlen [11], Kortüm and Sharp [12].

In addition, a number of commercially distributed computer codes were developed, a summary of which is given in the Multibody Systems Handbook [13]. The computer codes available show different capabilities: some of them generate only the equations of motion in numerical or symbolical form, respectively, some of them provide numerical integration and simulation codes, too. Moreover, there are also extensive software systems on the market which offer additionally graphical data input, animation of body motions, and automated signal data analysis. There is no doubt that the professional user, particularly in the automotive industry, prefers the most complete software system for dynamical multibody system analysis.

2.1 Elements of Multibody Systems

The method of multibody systems is based on a finite set of elements such as

- rigid bodies and/or particles,
- bearings, joints, and supports,
- springs and dampers,
- active force and/or position actuators.

For more details see Reference [2] or Roberson and Schwertassek [3]. Each vehicle can be modeled for dynamical analysis by these elements as a multibody system, Figure 3.

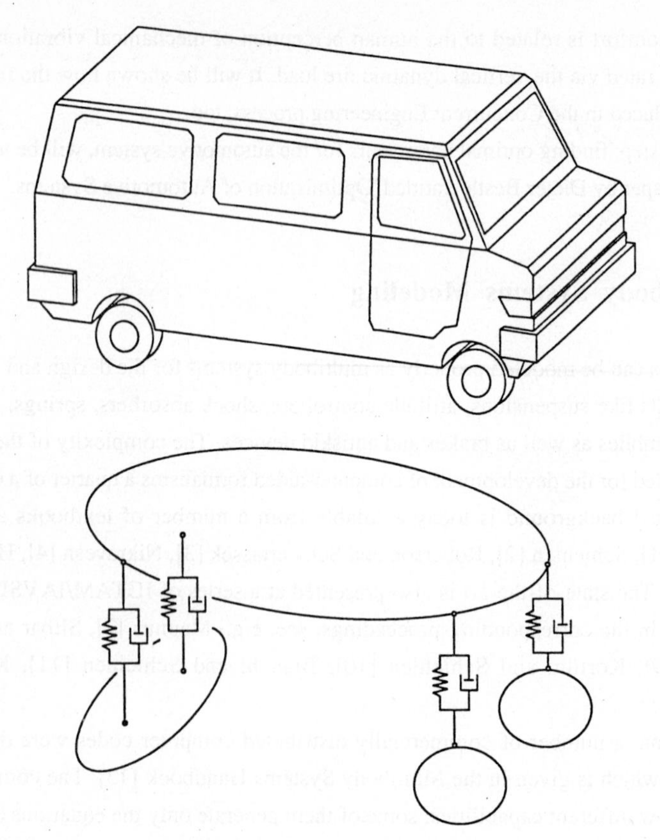

Figure 3. Multibody Model of Vehicle

The elements have to be characterized by body-fixed frames, Figure 4. Then, the absolute and relative motion can be defined by the frame motion using the kinematical quantities of 3x1–translational vectors **r** and 3x3–rotation tensors **S**. The description of joints, Figure 5, requires two frames, one on each of the connected bodies. The joints constrain the relative motion between two rigid bodies and, as a consequence, reaction forces have to be considered. Figure 6 shows the reaction forces $\mathbf{f}^r$ and the reaction $\mathbf{l}^r$ of a revolute joint in a free body diagram. A library of standard joints is shown in Figure 7. For more details see Daberkow [14].

2.2 Multibody System Datamodel

The German Research Council (DFG) sponsored by a nationwide research project the development of a multibody system datamodel, too. In this project, 14 universities and research centers have been engaged and all of them agreed on the datamodel [15].

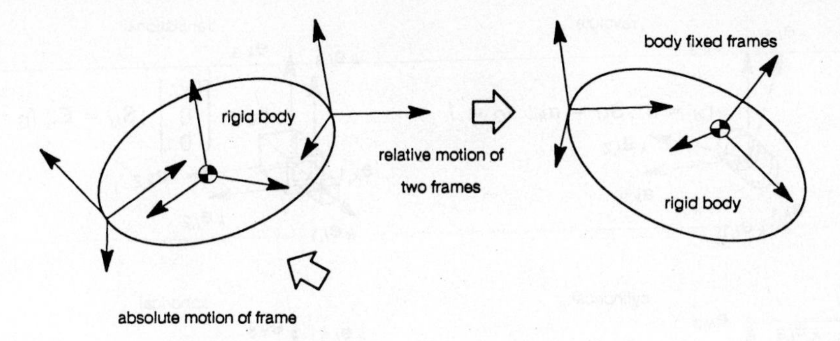

Figure 4. Two Rigid Bodies and Body-Fixed Frames

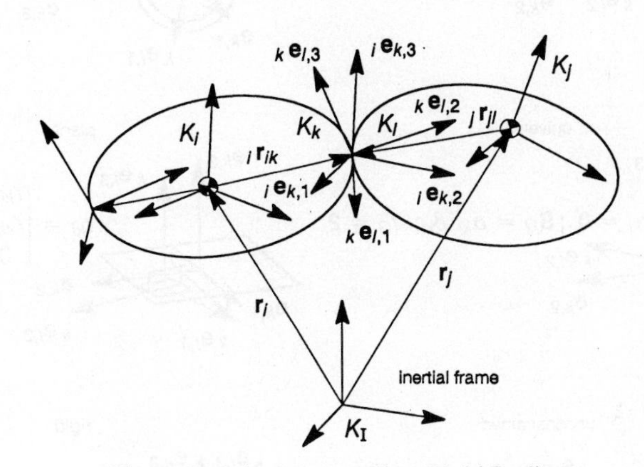

Figure 5. Joint between Two Rigid Bodies

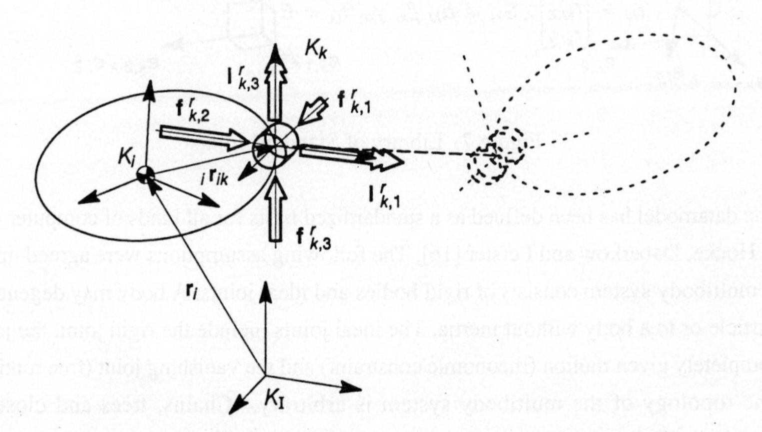

Figure 6. Reaction Forces and Torques of a Revolute Joint

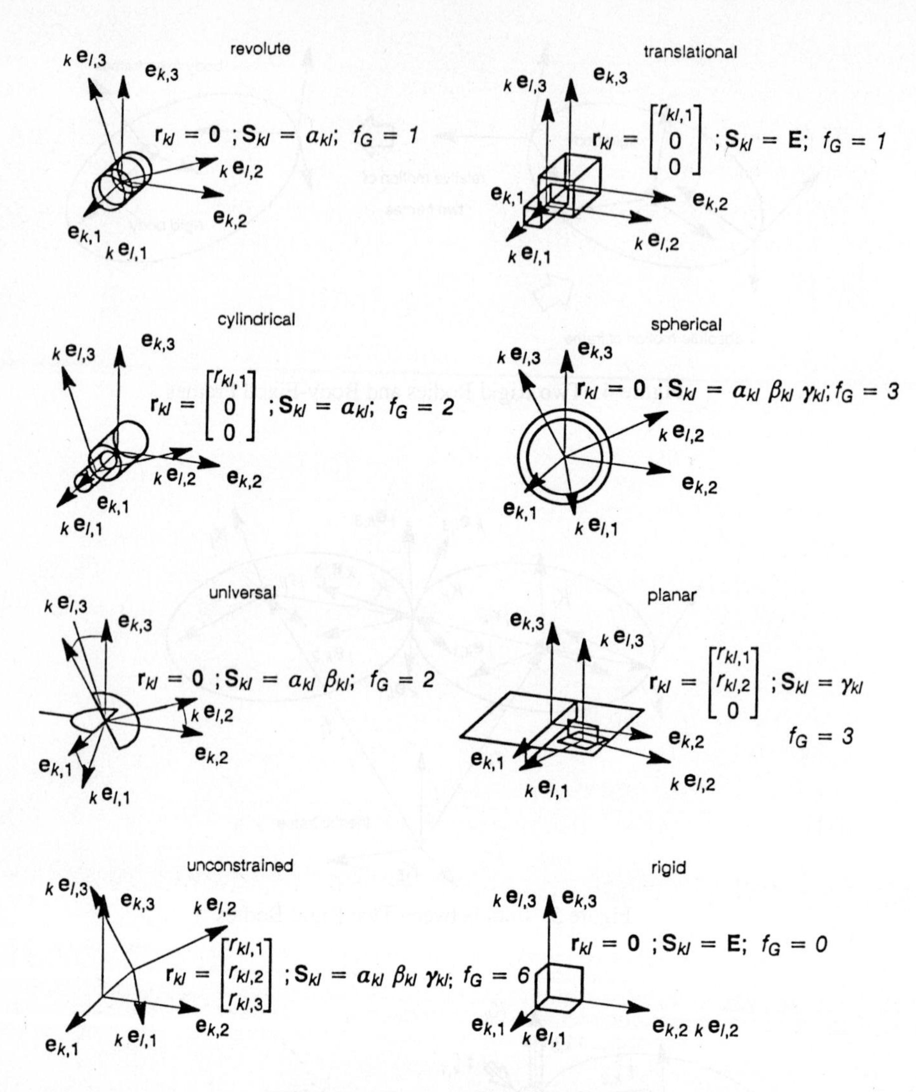

Figure 7. Library of Standard Joints

The datamodel has been defined as a standardized basis for all kinds of computer codes by Otter, Hocke, Daberkow and Leister [16]. The following assumptions were agreed upon:

1. A multibody system consists of rigid bodies and ideal joints. A body may degenerate to a particle or to a body without inertia. The ideal joints include the rigid joint, the joint with completely given motion (rheonomic constraint) and the vanishing joint (free motion).
2. The topology of the multibody system is arbitrary. Chains, trees and closed loops are admitted.

3. Joints and actuators are summarized in open libraries.
4. Subsystems may be added to existing components of the multibody system.

A datamodel for elastic bodies is under development and will be completely compatible with the rigid body datamodel.

A multibody system as defined is characterized by the class *mbs* and consists of an arbitrary number of the objects of the classes *part* and *interact,* see Figure 8.

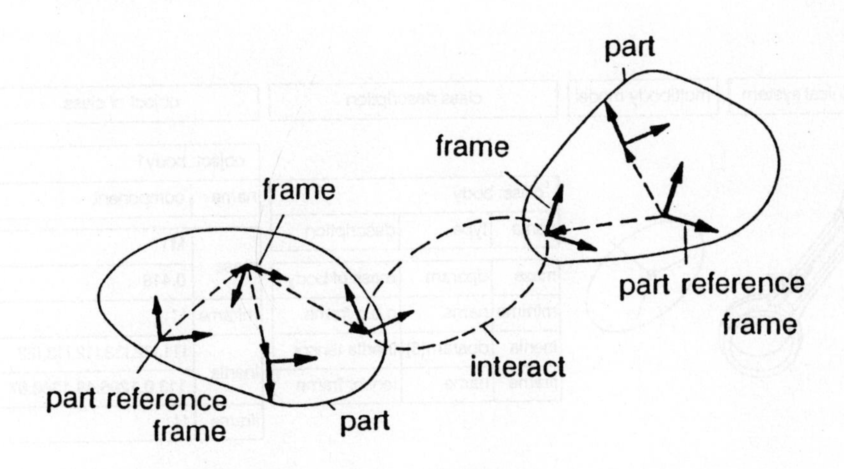

Figure 8. Multibody System to be Represented by the Datamodel

The class *part* describes rigid bodies. Each *part is* characterized by at least one body-fixed *frame,* it may have a mass, a center of mass and a tensor of inertia summarized in the class *body,* Figure 9.

The class *interact* describes the interaction between a frame on *part a* and a frame on *part b*. The interaction may be realized by a joint, by a force actuator or a sensor resulting in the classes *joint, force* or *sensor,* respectively. Thus, the class *interact is* characterized by two types of information: the frames to be connected and the connecting element itself, see Figure 10.

As an example for an element of the class *joint* and the class *force* Figure 11 shows a damper.

The presented classes are the basis of the class *mbs* which means the assembled vehicle. The model assembly using the datamodel is now easily executed. Figure 12 shows the whole procedure. According to the definitions, the datamodel represents holonomic, rheonomic multibody systems.

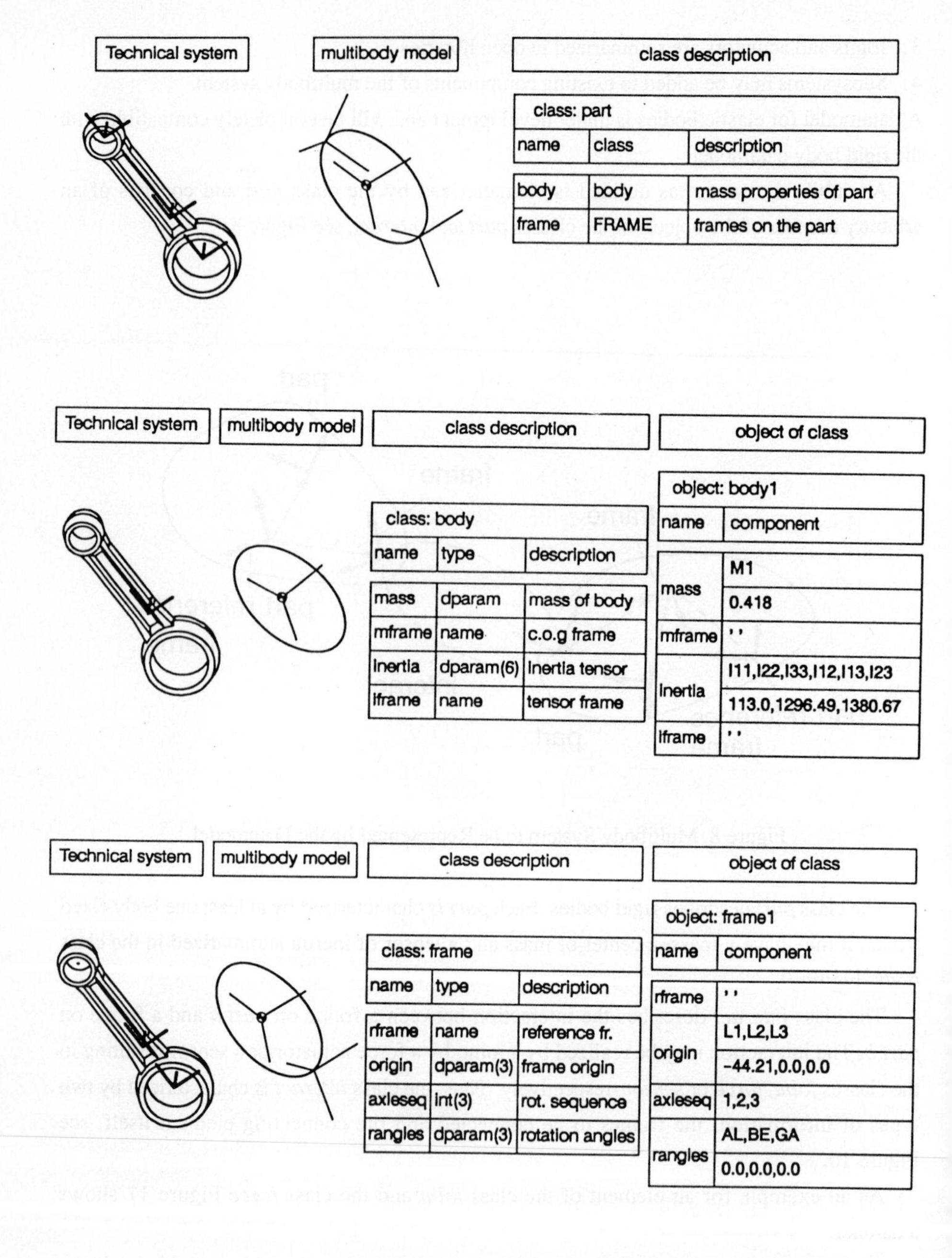

Figure 9. Definition of Class *Part*

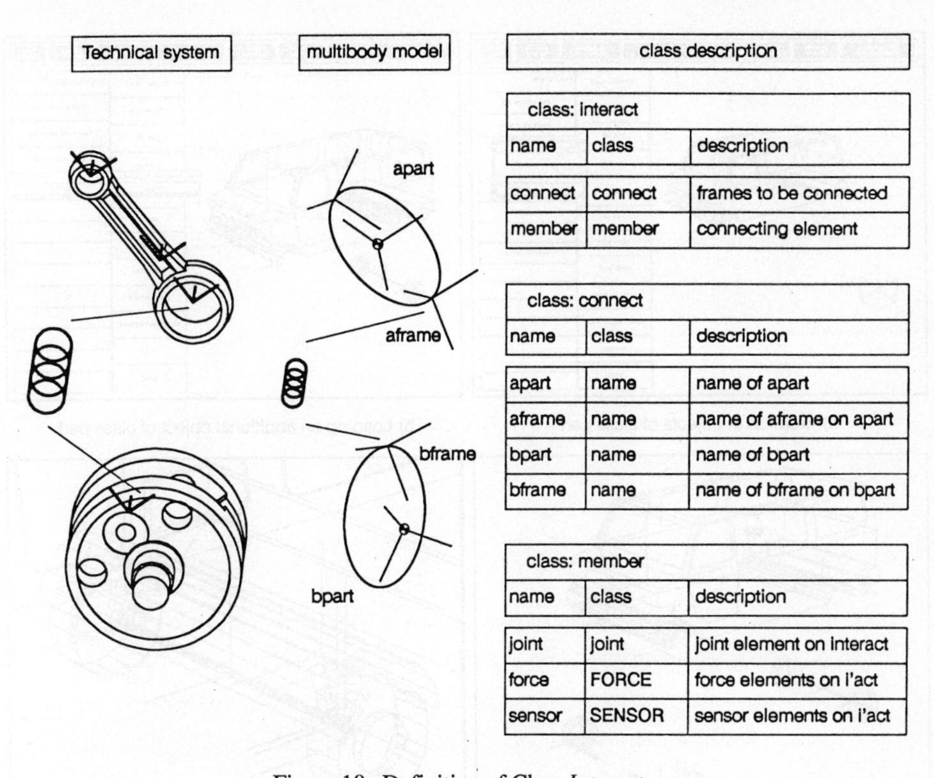

class description

class: interact		
name	class	description
connect	connect	frames to be connected
member	member	connecting element

class: connect		
name	class	description
apart	name	name of apart
aframe	name	name of aframe on apart
bpart	name	name of bpart
bframe	name	name of bframe on bpart

class: member		
name	class	description
joint	joint	joint element on interact
force	FORCE	force elements on i'act
sensor	SENSOR	sensor elements on i'act

Figure 10. Definition of Class *Interact*

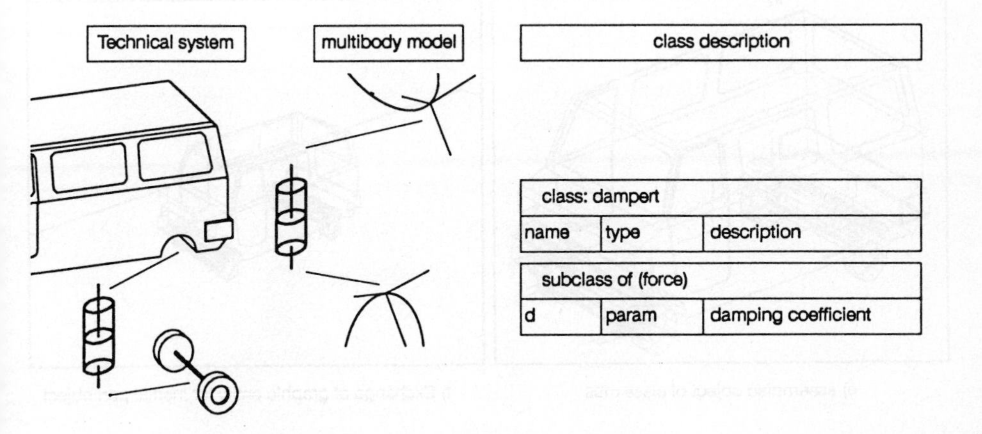

class description

class: dampert		
name	type	description
subclass of (force)		
d	param	damping coefficient

Figure 11. Definition of Class *Dampert*

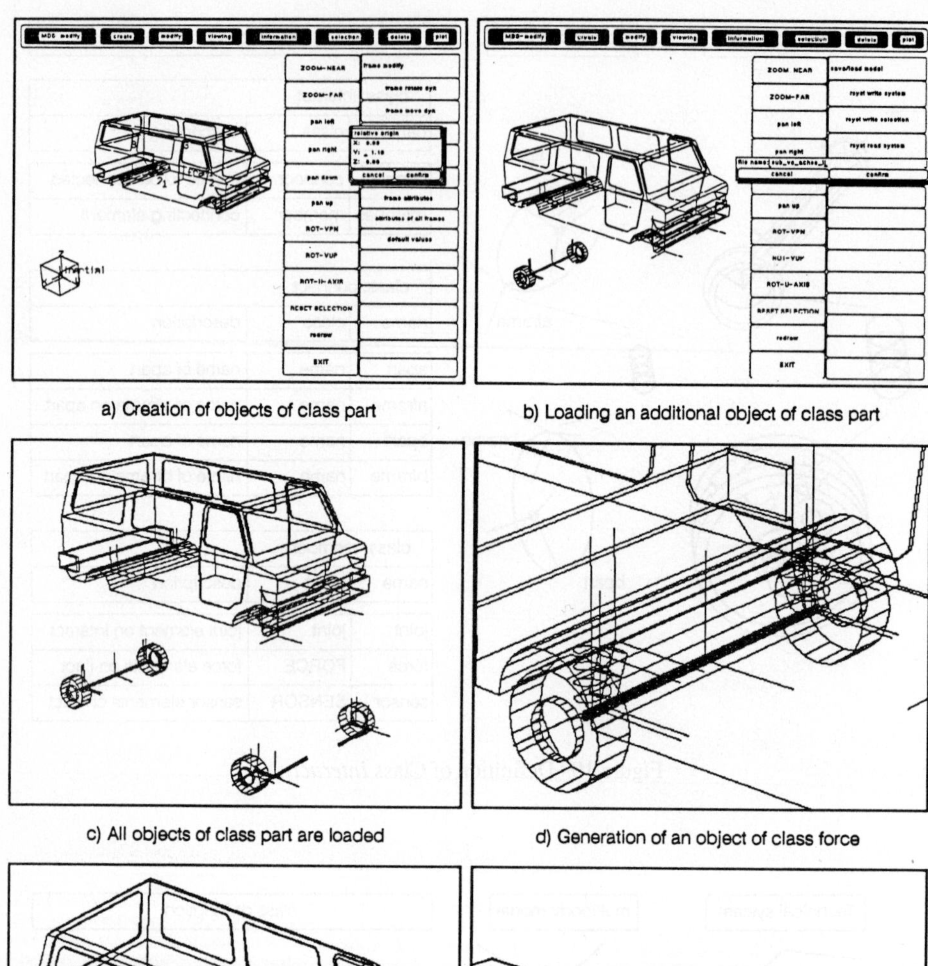

a) Creation of objects of class part

b) Loading an additional object of class part

c) All objects of class part are loaded

d) Generation of an object of class force

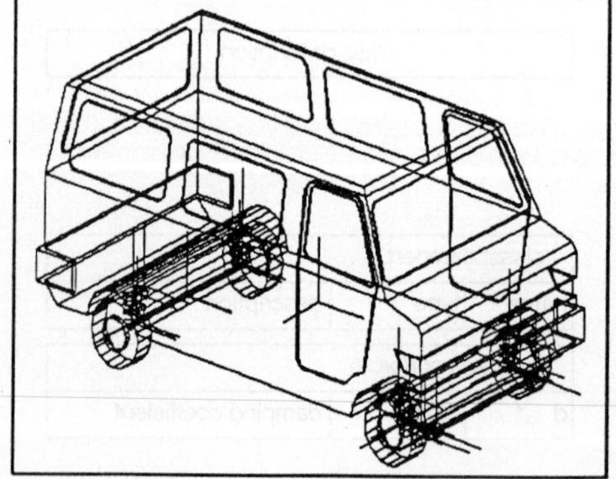

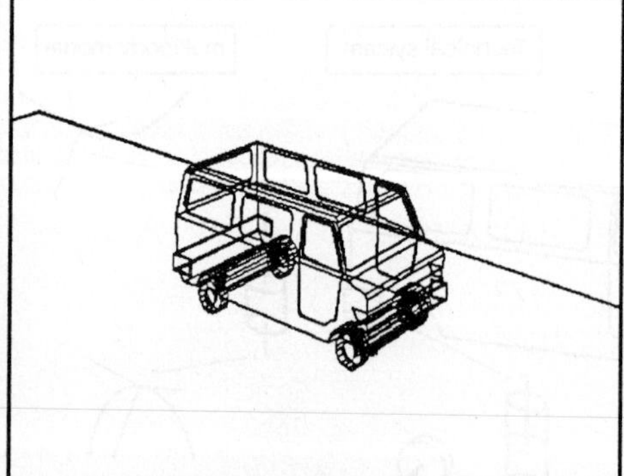

e) assembled object of class mbs

f) Exchange of graphic entity for inertial part object

Figure 12. Model Assembly of a Van

2.3 The Database RSYST

The scientific-engineering database system RSYST is chosen as a software tool. It supports the

- development of user programs,
- development of program packages,
- execution of programs,
- handling of large data sets,
- analysis of data.

One of the main applications of RSYST is the compilation of data and user programs. RSYST is written in FORTRAN 77 and, therefore, it has an excellent portability to all kinds of computers. The fundamental elements of RSYST are the following:

- execution control,
- information system,
- dialogue system,
- output handling,
- dynamic storage handling,
- method and model base,
- database.

The RSYST system has been developed by Rühle and his staff at the Computing Center of the University of Stuttgart. A detailed description is given by Lang [17], Loebich [18] and Rühle [19].

Most important for the multibody system datamodel are the RSYST database and the handling of data objects. All the data in RSYST are considered as objects of a database. Such data objects are, for example, vectors, matrices, sets of parameters, texts, or formally defined objects. The data objects are stored in the RSYST database subject to a very efficient handling, they are identified by special names.

Each data object in the RSYST database is characterized by a data description, identifying the data type. The data description permits a correlation between data objects and possible operations.

RSYST offers the following operations on data objects which are completely internally executed:

- object generating,
- object changing,
- object deleting,
- object listing,
- objects relating to each other,

- objects storing and reading,
- handling of components of objects.

The objects have to be interpreted for the identification of their information. A set of objects with same rules of interpretation are called a class specified by a name. For example, the four components *"rframe, origin, axleseq, rangles"* of the class *"frame,"* are shown in Figure 9.

The database RSYST offers a suitable software engineering concept for multibody system dynamics as shown in Figure 13.

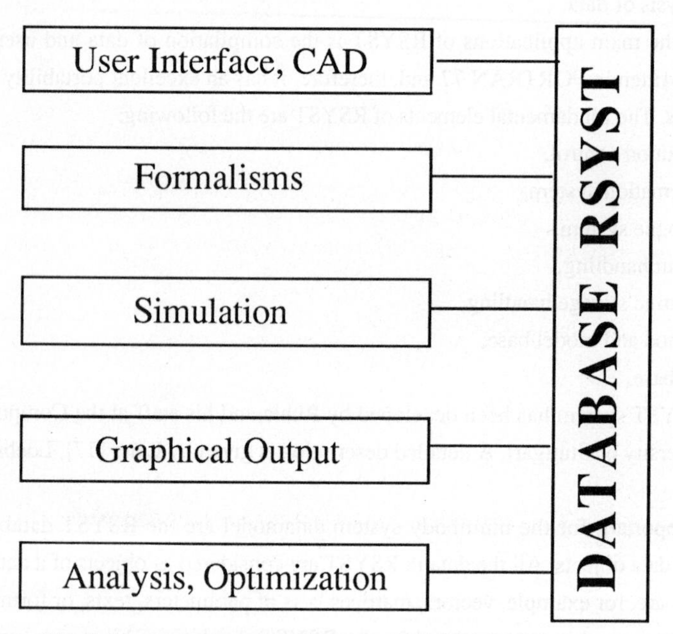

Figure 13. Software Engineering Concept of Multibody System Dynamics

This concept provides the opportunity to use a modular structure of the software, i.e., different multibody formalisms may be combined with different simulation programs via standardized interfaces.

The database structure of the assembled van presented by Figure 12 is shown in Figure 14. It is obvious that the two identical front wheel assemblies result in identical data input. In particular, different designs can be easily implemented and tested.

The object-oriented multibody system modeling can be included in any CAD-3D-software by minor adjustments. The object-oriented multibody modeling kernel, shown in Figure 15, handles the data input and dialog the data output, the storage of *mbs* data and the 3D-graphics. As a result, the CAD-3D and CAD-2D model files are supplemented by a multibody model file.

```
PART
 _inertial
   _FRAME
     _iframe        frame iframe of part inertial
 _vanbody
   _BODY            van-body
   _FRAME
     _jlfwheel      frame of joint left front wheel of part
                    van-body
     _jrfwheel      frame of joint right front wheel of part
                    van-body
     _jraxle        frame of joint rearaxle of part van-body
     _sdlfw         frame of spring-damperforce left front wheel
     _sdrfw         frame of spring-damperforce right front wheel
     _slra          frame of spring rearaxle left
     _dlra          frame of damper rearaxle left
     _srra          frame of spring rearaxle right
     _drra          frame of damper rearaxle right
 _lfrontw
   _BODY            left front wheel  ----------------------+
   _FRAME                                                   I
     _jvanbody      frame of joint vanbody of part left     I
                    front wheel  ---------------------------I-+
     _sdvb          frame of spring-damperforce vanbody  ---I-I-+
     _swheel        frame of springforce left front wheel  -I-I-I-+
 _rfrontw                                                   I I I I
   _BODY            right front wheel ----------------------+ I I I
   _FRAME                                                     I I I
     _jvanbody      frame of joint vanbody of part right      I I I
                    front wheel -------------------------------+ I I
     _sdvb          frame of spring-damperforce vanbody  ---------+ I
     _swheel        frame of springforce right front wheel  -------+
 _rearaxle
   _BODY            rearaxle
   _FRAME
     _bframe        frame bodyframe of part rearaxle
     _jvanbody      frame of joint vanbody of part right front
                    wheel
     _slvb          frame of left springforce vanbody
     _dlvb          frame of left damperforce vanbody
     _srvb          frame of right springforce vanbody
     _drvb          frame of right damperforce vanbody
     _slwheel       frame of springforce right rear wheel
     _srwheel       frame of springforce left rear wheel
```

Figure 14. Database Structure of a Van

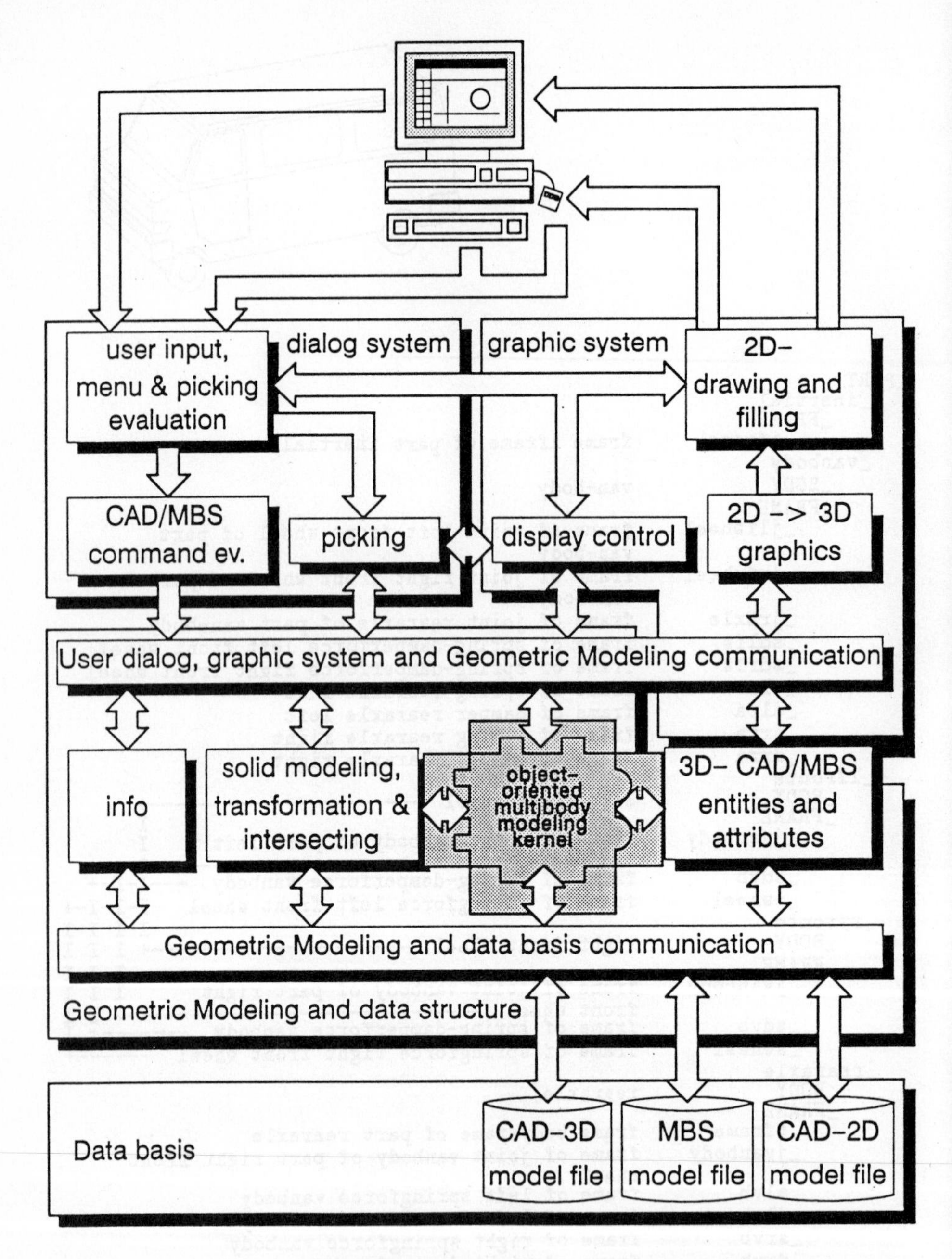

Figure 15. CAD Software and Multibody Modeling Kernel

3 Generation of Equations of Motion

The multibody system model has to be described mathematically by equations of motions for the dynamical analysis. In this chapter, the general theory for holonomic and nonholonomic systems will be presented using a minimal number of generalized coordinates for a unique representation of the motion.

3.1 Kinematics of Multibody Systems

According to the free body diagram of a vehicle system, firstly, all constraints are omitted and the system of p bodies holds 6p degrees of freedom. The position of the system is given relative to the inertial frame by the 3x1–translation vector

$$\mathbf{r}_i = \left[r_{i1}\ r_{i2}\ r_{i3}\right]^T, i = 1(1)p, \tag{1}$$

of the center of mass C_i and the 3x3–rotation tensor

$$\mathbf{S}_i = \mathbf{S}_i(\alpha_i\ \beta_i\ \gamma_i)^T, \tag{2}$$

written down for each body. The rotation tensor $\mathbf{S}_i$ depends on three angles α_i, β_i, γ_i and corresponds with the direction cosine matrix relating the inertial frame I and the body-fixed frame i to each other. The 3p translational coordinates and the 3p rotational coordinates (angles) can be summarized in a 6pxl-position vector

$$\mathbf{x} = \left[r_{11}\ r_{12}\ r_{13}\ r_{21}\ \cdots\ \alpha_p\ \beta_p\ \gamma_p\right]^T. \tag{3}$$

Equations (1) and (2) now read

$$\mathbf{r}_i = \mathbf{r}_i(\mathbf{x}),\ \mathbf{S}_i = \mathbf{S}_i(\mathbf{x})\ . \tag{4}$$

Secondly, the q holonomic, rheonomic constraints are added to the vehicle system given explicitly by

$$\mathbf{x} = \mathbf{x}(\mathbf{y}, t), \tag{5}$$

where the fx1–position vector

$$\mathbf{y} = \left[y_1\ y_2\ y_3\ \cdots\ y_f\right]^T \tag{6}$$

is used summarizing the f generalized coordinates of the system. The number of generalized coordinates corresponds to the number of degrees of freedom, f = 6p–q, with respect to the systems position. Then, translation and rotation of each body follow from (4) and (5) as

$$\mathbf{r}_i = \mathbf{r}_i(\mathbf{y},t), \ \mathbf{S}_i = \mathbf{S}_i(\mathbf{y},t) \tag{7}$$

and the velocities are found by differentiation with respect to the inertial frame:

$$\mathbf{v}_i = \dot{\mathbf{r}}_i = \frac{\partial \mathbf{r}_i}{\partial \mathbf{y}}\dot{\mathbf{y}} + \frac{\partial \mathbf{r}_i}{\partial t} = \mathbf{J}_{Ti}(\mathbf{y},t)\ \dot{\mathbf{y}} + \overline{\mathbf{v}}_i(\mathbf{y},t), \tag{8}$$

$$\boldsymbol{\omega}_i = \dot{\mathbf{s}}_i = \frac{\partial \mathbf{s}_i}{\partial \mathbf{y}}\dot{\mathbf{y}} + \frac{\partial \mathbf{s}_i}{\partial t} = \mathbf{J}_{Ri}(\mathbf{y},t)\ \dot{\mathbf{y}} + \overline{\boldsymbol{\omega}}_i(\mathbf{y},t). \tag{9}$$

The 3xf–Jacobian matrices $\mathbf{J}_{Ti}$ and $\mathbf{J}_{Ri}$ defined by (8) and (9) characterize the virtual translational and rotational displacement of the system, respectively. They are also needed later for the application of d'Alembert's principle. The infinitesimal 3x1–rotation vector $\mathbf{s}_i$ used in (9) follows analytically from the corresponding infinitesimal skew–symmetrical 3x3–rotation tensor. However, the matrix $\mathbf{J}_{Ri}$ can also be found by a geometrical analysis of the angular velocity vector $\boldsymbol{\omega}_i$ with respect to the angles α_i, β_i, γ_i, see e.g., Reference [2].

The accelerations are obtained by a second differentiation with respect to the inertial frame:

$$\mathbf{a}_i = \mathbf{J}_{Ti}(\mathbf{y},t)\ \ddot{\mathbf{y}} + \frac{\partial \mathbf{v}_i}{\partial \mathbf{y}}\dot{\mathbf{y}} + \frac{\partial \mathbf{v}_i}{\partial t}, \tag{10}$$

$$\boldsymbol{\alpha}_i = \mathbf{J}_{Ri}(\mathbf{y},t)\ \ddot{\mathbf{y}} + \frac{\partial \boldsymbol{\omega}_i}{\partial \mathbf{y}}\dot{\mathbf{y}} + \frac{\partial \boldsymbol{\omega}_i}{\partial t}. \tag{11}$$

For scleronomic constraints, the partial time-derivatives in (8), (9) and (10), (11) vanish.

Thirdly, the r nonholonomic, rheonomic constraints, especially due to rigid wheels, are introduced explicitly by

$$\dot{\mathbf{y}} = \dot{\mathbf{y}}(\mathbf{y},\mathbf{z},t) \tag{12}$$

with the gx1–velocity vector

$$\mathbf{z}(t) = \begin{bmatrix} z_1 & z_2 & z_3 & \cdots & z_g \end{bmatrix}^T \tag{13}$$

summarizing the g generalized velocities of the system. The number of generalized velocities characterizes the number of degrees of freedom, g = f – r, with respect to the system's velocity. From (8), (9), and (12), the translational and rotational velocity of each body follow immediately as

$$\mathbf{v}_i = \mathbf{v}_i(\mathbf{y},\mathbf{z},t), \; \boldsymbol{\omega}_i = \boldsymbol{\omega}_i(\mathbf{y},\mathbf{z},t). \tag{14}$$

The accelerations are found again by differentiation with respect to inertial frame I:

$$\mathbf{a}_i = \frac{\partial \mathbf{v}_i}{\partial \mathbf{z}}\dot{\mathbf{z}} + \frac{\partial \mathbf{v}_i}{\partial \mathbf{y}}\dot{\mathbf{y}} + \frac{\partial \mathbf{v}_i}{\partial t} = \mathbf{L}_{Ti}(\mathbf{y},\mathbf{z},t)\;\dot{\mathbf{z}} + \bar{\bar{\mathbf{v}}}_i(\mathbf{y},\mathbf{z},t), \tag{15}$$

$$\boldsymbol{\alpha}_i = \frac{\partial \boldsymbol{\omega}_i}{\partial \mathbf{z}}\dot{\mathbf{z}} + \frac{\partial \boldsymbol{\omega}_i}{\partial \mathbf{y}}\dot{\mathbf{y}} + \frac{\partial \boldsymbol{\omega}_i}{\partial t} = \mathbf{L}_{Ri}(\mathbf{y},\mathbf{z},t)\;\dot{\mathbf{z}} + \dot{\bar{\bar{\boldsymbol{\omega}}}}_i(\mathbf{y},\mathbf{z},t). \tag{16}$$

Here, the 3xg–matrices $\mathbf{L}_{Ti}$ and $\mathbf{L}_{Ri}$ are introduced for the description of the virtual translational and rotational velocity of the system needed also for the application of Jourdain's principle. Further, it has to be mentioned that the partial time–derivatives vanish in (15), (16) for scleronomic systems.

In many applications, a reference frame is given in a natural way. For example, a railway vehicle running on a curved super-elevated track is naturally described in a moving track-related frame. Therefore, the absolute motion may be also presented in a reference frame using the reference motion itself and the bodies' relative motion [20].

3.2 Newton-Euler Equations

For the application of Newton's and Euler's equation to multibody systems, the free body diagram has to be used again. Now the rigid bearings and supports are replaced by adequate constraint forces and torques as discussed later in this section.

Newton's and Euler's equation read for each body in the inertial frame

$$m_i \dot{\mathbf{v}}_i = \mathbf{f}_i^e + \mathbf{f}_i^r, \quad i = 1(1)p, \tag{17}$$

$$\mathbf{I}_i \dot{\boldsymbol{\omega}}_i + \tilde{\boldsymbol{\omega}}_i \mathbf{I}_i \boldsymbol{\omega}_i = \mathbf{l}_i^e + \mathbf{l}_i^r, \quad i = 1(1)p. \tag{18}$$

The inertia is represented by the mass m_i and the 3x3–inertia tensor $\mathbf{I}_i$ with respect to the center of mass C_i of each body. The external forces and torques in (17) and (18) are composed by the 3x1–applied force vector $\mathbf{f}_i^e$ and torque vector $\mathbf{l}_i^e$ due to springs, dampers, actuators, weight, etc., and by the 3x1–constraint force vector $\mathbf{f}_i^r$ and torque vector $\mathbf{l}_i^r$. All torques are related to the center mass C_i . The applied forces and torques, respectively, depend on the motion by different laws, and they may be coupled to the constraint forces and torques in the case of friction.

The constraint forces and torques originate from the reactions in joints, bearings, supports, or wheels. They can be reduced by distribution matrices to the generalized constraint forces. The number of the generalized constraint forces is equal to the total number of constraints (q+r) in the system. Introducing the (q+r)x1–vector of generalized constraint forces

$$\mathbf{g} = \left[g_1 \; g_2 \; g_3 \; \cdots \; g_{q+r} \right]^T \tag{19}$$

and the 3x(q+r)–distribution matrices

$$\mathbf{F}_i = \mathbf{F}_i(\mathbf{y},\mathbf{z},t), \quad \mathbf{L}_i = \mathbf{L}_i(\mathbf{y},\mathbf{z},t) \tag{20}$$

it turns out

$$\mathbf{f}_i^{\,r} = \mathbf{F}_i \mathbf{g}, \quad \mathbf{l}_i^{\,r} = \mathbf{L}_i \mathbf{g}, \quad i = 1(1)p, \tag{21}$$

for each body. The constraint forces or the distribution matrices, respectively, can be found mathematically, or they are derived by geometrical analysis.

The ideal applied forces and torques depend only on the kinematical variables of the system, they are independent of the constraint forces. Ideal applied forces are due to the elements of multibody systems and further actions on the system, e.g., gravity. The forces may be characterized by proportional, differential, and/or integral behavior.

The proportional forces are characterized by the system's position and time-functions

$$\mathbf{f}_i^{\,e} = \mathbf{f}_i^{\,e}(\mathbf{x},t). \tag{22}$$

For example, conservative spring and weight forces, as well as purely time-varying forces, are proportional forces.

The proportional-differential forces depend on the position and the velocity:

$$\mathbf{f}_i^{\,e} = \mathbf{f}_i^{\,e}(\mathbf{x},\dot{\mathbf{x}},t). \tag{23}$$

A parallel spring-dashpot configuration is a typical example for this class of forces. The proportional-integral forces are a function of the position and integrals of the position:

$$\mathbf{f}_i^{\,e} = \mathbf{f}_i^{\,e}(\mathbf{x},\mathbf{w},t), \quad \dot{\mathbf{w}} = \dot{\mathbf{w}}(\mathbf{x},\mathbf{w},t) \, , \tag{24}$$

where the px1–vector $\mathbf{w}$ describes the position integrals. For example, serial spring-damper configurations and the eigendynamics of actuators result in proportional-integral forces. In vehicle systems proportional-integral forces appear, e.g., with modem engine mounts for simultaneous noise and vibration reduction. The same laws hold also for ideal applied torques.

In the case of non ideal constraints with sliding friction or contact forces, respectively, the applied forces are coupled with the constraint forces [20].

The Newton-Euler equations of the complete system are summarized in matrix notation by the following vectors and matrices. The inertia properties are written in the 6px6p–diagonal matrix

$$\overline{\overline{\mathbf{M}}} = \operatorname{diag}\{m_1\mathbf{E} \quad m_2\mathbf{E} \cdots \mathbf{I}_1 \cdots \mathbf{I}_p\}, \tag{25}$$

where the 3x3–identity matrix $\mathbf{E}$ is used. The 6px1–force vectors $\overline{\mathbf{q}}^c$, $\overline{\mathbf{q}}^e$, $\overline{\mathbf{q}}^r$ representing the coriolis forces, the ideal applied forces and the constraint forces, respectively, are given by the following scheme,

$$\overline{\mathbf{q}} = \left[\mathbf{f}^T{}_1\, \mathbf{f}^T{}_2 \cdots \mathbf{1}^T{}_1 \cdots \mathbf{1}^T{}_p\right]^T. \tag{26}$$

Further the 6pxf–matrix $\overline{\mathbf{J}}$ and 6pxg–matrix $\overline{\mathbf{L}}$ as well as the 6px(q+r)–distribution matrix $\overline{\mathbf{Q}}$ are introduced as global matrices, e.g.,

$$\overline{\mathbf{J}} = \left[\mathbf{J}_{T1}{}^T\ \mathbf{J}_{T2}{}^T \cdots \mathbf{J}_{R1}{}^T \cdots \mathbf{J}_{RP}{}^T\right]^T. \tag{27}$$

Now, the Newton-Euler equations can be represented as follows for holonomic systems in the inertial frame

$$\overline{\overline{\mathbf{M}}}\,\overline{\mathbf{J}}\,\ddot{\mathbf{y}} + \overline{\mathbf{q}}^c(\mathbf{y},\dot{\mathbf{y}},t) = \overline{\mathbf{q}}^e(\mathbf{y},\dot{\mathbf{y}},t) + \overline{\mathbf{Q}}\,\mathbf{g} \tag{28}$$

and for nonholonomic systems

$$\overline{\overline{\mathbf{M}}}\,\overline{\mathbf{L}}\,\dot{\mathbf{z}} + \overline{\mathbf{q}}^c(\mathbf{y},\mathbf{z},t) = \overline{\mathbf{q}}^e(\mathbf{y},\mathbf{z},t) + \overline{\mathbf{Q}}\,\mathbf{g}. \tag{29}$$

If the holonomic constraints are omitted, e.g., $\mathbf{z} = \dot{\mathbf{y}}$, Equation (29) reduces to (28), showing a close relation between both representations.

3.3 Equations of Motion

The Newton-Euler equations are combined algebraical and differential equations and the question arises if they can be separated for solution into purely algebraical and differential equations. There is a positive answer given by the dynamical principles. In a first step, the system's motion can be found by integration of the separated differential equations and in a second step the constraint forces are calculated algebraically. For ideal applied forces, both steps can be executed successively while contact forces require simultaneous execution.

Holonomic systems with proportional or proportional-differential forces result in *ordinary* multibody systems. The equations of motion follow from the Newton-Euler equations, applying d'Alembert's principle.

The equations of motion of holonomic systems are found according to d'Alembert's principle by premultiplication of (28) with $\bar{\mathbf{J}}^T$ as

$$\mathbf{M}(\mathbf{y},t)\,\ddot{\mathbf{y}} + \mathbf{k}(\mathbf{y},\dot{\mathbf{y}},t) = \mathbf{q}(\mathbf{y},\dot{\mathbf{y}},t). \tag{30}$$

Here, the number of equations is reduced from 6p to f, the fxf–inertia matrix $\mathbf{M}(\mathbf{y}, t)$ is completely symmetrized $\mathbf{M}(\mathbf{y},t) = \bar{\mathbf{J}}^T\,\bar{\bar{\mathbf{M}}}\,\bar{\mathbf{J}} > 0$, and the constraint forces and torques are eliminated. The remaining fxl–vector $\mathbf{k}$ describes the generalized coriolis forces and the fx1–vector $\mathbf{q}$ includes the generalized applied forces.

The equations of motion following from a moving reference frame agree completely with (30). Thus, the choice of the reference frame doesn't affect the equations of motion at all. However, kinematics and Newton-Euler equations as well as the application of d'Alembert's principle may be strongly simplified by the choice of proper reference frames.

Nonholonomic systems with proportional-integral forces produce *general* multibody systems. The equations of motion are obtained from the Newton-Euler Equations (29) where the proportional-integral forces (24) and Jourdain's principle have to be regarded. However, the equations of motion are not sufficient, they have to be completed by the nonholonomic constraint equation (12). Thus, the complete equations read as

$$\mathbf{M}(\mathbf{y},\mathbf{z},t)\dot{\mathbf{z}} + \mathbf{k}(\mathbf{y},\mathbf{z},t) = \mathbf{q}(\mathbf{y},\mathbf{z},\mathbf{w},t), \tag{31}$$

$$\dot{\mathbf{y}} = \dot{\mathbf{y}}(\mathbf{y},\mathbf{z},t), \quad \dot{\mathbf{w}} = \dot{\mathbf{w}}(\mathbf{y},\mathbf{z},t).$$

Now, the number of equations is reduced from 6p to g and the gxg–symmetric inertia matrix $\mathbf{M}(\mathbf{y},\mathbf{z},t) = \bar{\mathbf{L}}^T\,\bar{\bar{\mathbf{M}}}\,\bar{\mathbf{L}} > 0$ appears. Further, $\mathbf{k}$ and $\mathbf{q}$ are gx1–vectors of generalized coriolis and applied forces. The Equations (31) are in the literature also denoted as Kane's equations.

In addition to the mechanical representation (31) of a multibody system, there also exists the possibility to use the more general representation of dynamical systems, e.g.,

$$\dot{\mathbf{x}} = \mathbf{f}(\mathbf{x},\mathbf{u},t,\mathbf{p}), \;, \quad \mathbf{v} = \mathbf{g}(\mathbf{x},\mathbf{u},t,\mathbf{p}), \tag{32}$$

where $\mathbf{x}$ means in (33) the state vector, $\mathbf{v}$ the output vector, $\mathbf{u}$ the input vector of controls, t the time and $\mathbf{p}$ the vector of mechanical and control parameters or design variables, respectively.

The constraint forces are completely omitted by the dynamical principles. However, they are also of engineering interest for the load in joints, bearings and supports, and they are absolutely necessary for the computation of contact and friction forces. From the 6p coordinates of the constraint force vector $\bar{\mathbf{q}}^r$ there are only (q+r) coordinates linear independent according to (21). Therefore, only the (q+r)x1–vector $\mathbf{g}$ of the generalized constraint forces is needed. The results are given for holonomic systems only, r=0, but they can be transferred to nonholonomic systems without any problem.

The premultiplication of (28) by $\overline{\mathbf{Q}}^T \overline{\overline{\mathbf{M}}}^{-1}$ results according to d'Alembert's principle immediately in the equations of reaction

$$\mathbf{N}(\mathbf{y},t)\,\mathbf{g} + \hat{\mathbf{q}}(\mathbf{y},\dot{\mathbf{y}},t) = \hat{\mathbf{k}}(\mathbf{y},\dot{\mathbf{y}},t) \tag{33}$$

where $\mathbf{N}(\mathbf{y},t) = \overline{\mathbf{Q}}^T \overline{\overline{\mathbf{M}}}^{-1} \overline{\mathbf{Q}} > 0$ is the symmetrical qxq–reaction matrix and $\hat{\mathbf{q}}$ and $\hat{\mathbf{k}}$ are qxl–vectors.

3.4 Formalism NEWEUL

The equations of motion presented are automatically generated by the formalism NEWEUL described in the Multibody Systems Handbook [13], too.

NEWEUL is a software package for the dynamic analysis of mechanical systems with the multibody system method. It comprises the computation of the symbolic equations of motion by the modul NEWEUL and the simulation of the dynamic behavior by the modul NEWSIM. Multibody systems are mechanical models consisting of

- rigid bodies,
- arbitrary constraining elements (joints, position control elements),
- passive coupling elements (springs, dampers), and
- active coupling elements (force control elements).

The topological structure of the models is arbitrary, thus possible configurations are

- systems with chain structure,
- systems with tree structure, and
- systems with closed kinematical loops.

The scleronomic or rheonomic constraints may be

- holonomic or
- nonholonomic.

The software package NEWEUL has been successfully applied in industrial and academic research institutions since 1979. The major fields of application are

- vehicle dynamics,
- dynamics of machinery,
- robot dynamics,
- biomechanics,
- satellite dynamics,
- dynamics of mechanisms.

The software package NEWEUL offers two approaches for multibody system modeling.

These are

- the successive assembly approach using the kinematics of relative motions, and
- the modular assembly approach based on subsystems.

The input data for NEWEUL have to be entered in input files prepared with prompts and comments.

4 Numerical Simulation of Automotive Systems

For the simulation of multibody systems, the computational efficiency is most important. A drawback of the equations of motion (30) is the fact that the inertia matrix has to be inverted

$$\ddot{\mathbf{y}} = \mathbf{M}^{-1}(\mathbf{q} - \mathbf{k}) \tag{34}$$

a procedure of high computational costs. In robot dynamics, however, there has been developed quite a number of methods to circumvent this difficulty see Hollerbach [21], Walker and Orin [22], Brandl, Johanni and Otter [23]. It turns out that for open chains, the second derivative of (5) can be replaced by the recursive kinematic relation

$$\ddot{\mathbf{x}} = \mathbf{C}\ddot{\mathbf{x}} + \mathbf{J}\ddot{\mathbf{y}} + \boldsymbol{\xi} \tag{35}$$

and the Newton-Euler Equations (28) are to be rewritten as

$$\overline{\overline{\mathbf{M}}}\ddot{\mathbf{x}} + \overline{\mathbf{k}} = \overline{\mathbf{q}^{e}} + (\mathbf{E} - \mathbf{C}^{T})\mathbf{Q}\mathbf{g} \tag{36}$$

where $\mathbf{C}$ is a 6px6p–geometry-matrix with submatrices on the lower subdiagonal only and $\mathbf{J}$ and $\mathbf{Q}$ are blockdiagonal matrices of local Jacobian and distribution matrices [24]. Further, the 6px6p–inertia matrix $\overline{\overline{\mathbf{M}}}$ is block-diagonal and time-invariant. Equations (35) and (36) can be solved recursively, avoiding the inversion of the fxf–inertia matrix $\mathbf{M}$ from (30). In Reference [24], it has been shown that the recursive formalisms can be interpreted as a sophisticated backward recursion using the Gaussian algorithm. However, all the recursive approaches are restricted to chain or tree topology of multibody systems, closed kinematical loops cannot be treated.

Equation (35) and (36) represent 12p differential algebraical equations (DAE) for the 12p unknowns summarized in the vector $\mathbf{x}$, $\mathbf{y}$, $\mathbf{g}$ where p is the number of bodies, see Chapter 3. The elimination of the relative accelerations $\ddot{\mathbf{y}}$ in (35) by premultiplication with $\mathbf{Q}^T$ according to the orthogonality condition results in

$$\ddot{\Phi}(\mathbf{x}, \dot{\mathbf{x}}, \ddot{\mathbf{x}}) = \mathbf{Q}^{T}(\mathbf{E} - \mathbf{C}^{T})\,\ddot{\mathbf{x}} - \mathbf{Q}^{T}\boldsymbol{\xi} \tag{37}$$

a set of q equations which have the second integral

$$\Phi(\mathbf{x}) = \mathbf{0} \tag{38}$$

representing the q holonomic, rheonomic constraints (5) of the multibody system implicity. The elimination of the absolute accelerations $\ddot{\mathbf{x}}$ in (35) and (36) results again in the f equations of motion (30).

4.1 Computational Aspects of Simulation

The partially reduced Equations (36) and (38) represent differential algebraical equations (DAE) with the 6p+q unknowns in vectors $\mathbf{x}$, $\mathbf{g}$ and the sparse matrices $\overline{\overline{\mathbf{M}}}$, $\mathbf{C}$ and $\mathbf{Q}$. On the other hand, the fully reduced equations (30) represent ordinary differential equations (ODE) with 6p–q = f unknowns in vector $\mathbf{y}$ and the full matrix $\mathbf{M}$. There are several integration codes for DAE and ODE available which have been tested by Leister [25]. The results presented in Figure 16 show that for closed and open loop structures, the full reduced equations are more efficient. In the case of open loops, the ODE approach can be speeded up recursively if the number of bodies is p>8 as Valasek [26] found.

Due to the implicit formulation of the constraint equations, the closed loop problem does not exist in the DAE approach. However, in the ODE approach it deserves a special treatment. There are two possibilities to overcome this problem:

- automatic choice of an optimal set of generalized coordinates during integration as implemented by Leister [27] and
- preselection of complementary sets of generalized coordinates by kinematical analysis before integration as proposed in Reference [24].

The first method is used in the latest version of the formalism NEWEUL.

4.2 Dynamical Analysis with NEWEUL-NEWSIM Software

NEWEUL generates the equation of motion of multibody systems in symbolic form. The computation is based on the Newton-Euler approach with application of the principles of d'Alembert and Jourdain. The resulting equations of motion may be

- linear,
- partially linearized, or
- nonlinear

symbolic differential equations. Constant parameters can be included in numerical form. Nonlinear coupling elements in kinematically linear models are also permitted.

For the output format of the equations of motion, several options are possible. FORTRAN compatible output allows the equations to be included in commercial software packages for dynamic analysis and simulation such as, ACSL. Another output format allows the processing of the equations with the formula manipulation program MAPLE.

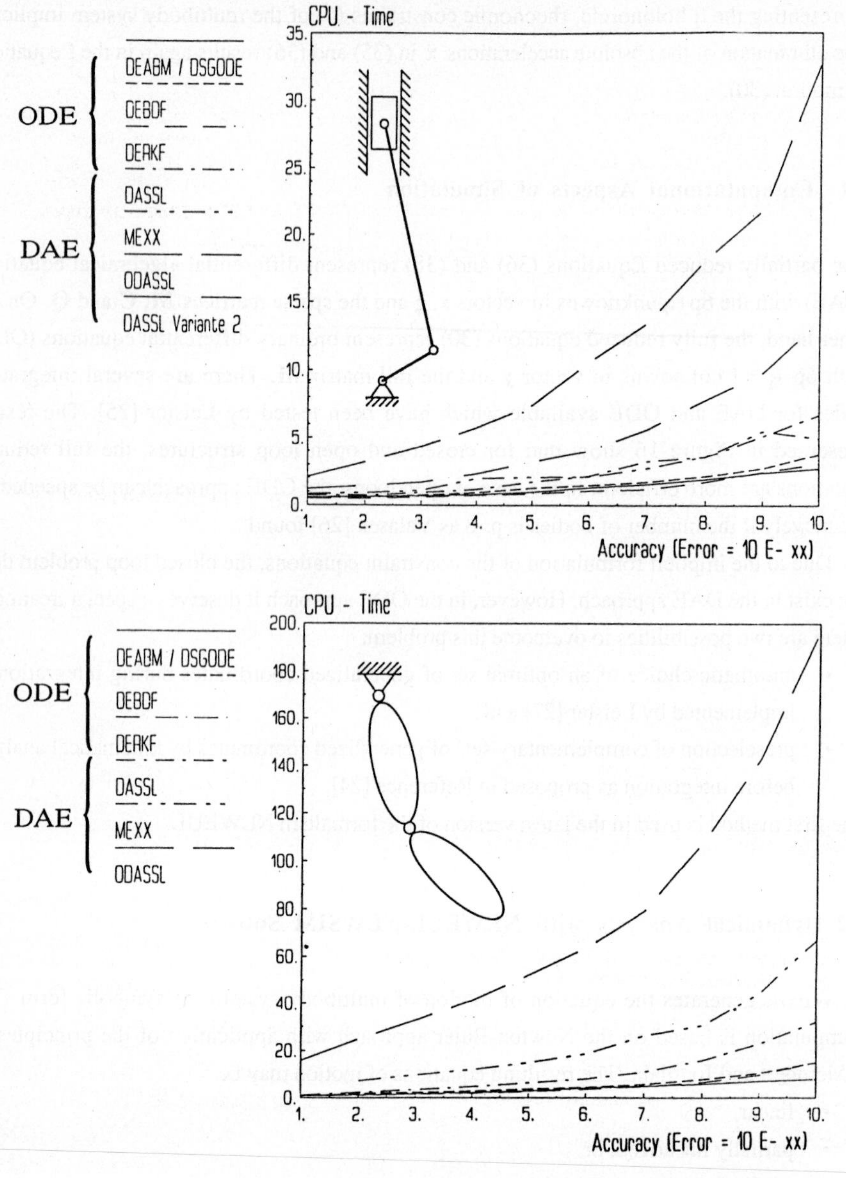

Figure 16. Comparison of Integration Codes for Multibody Simulations

Control parameters for compression and factorization enable the user to change the structure of the output equations, Figure 17. For example, the user may want to obtain fully symbolic equations of motion in order to check the results for modeling and input errors. Later, computationally efficient compressed equations can be generated for the verified model.

Fully symbolic output:

```
C>    Inertia Matrix
       M(1,1)=M1*A**2+I1+
      +        M2*C**2+M3*C**2

       M(2,1)=-M2*B*C*SIN(AL1)*COS(AL2)+
      +         M2*B*C*SIN(AL2)*COS(AL1)
       M(2,2)=M2*B**2+I2

       M(3,1)=M3*B*C*SIN(AL1)*COS(AL3)-
      -        M3*B*C*SIN(AL3)*COS(AL1)
       M(3,2)=0.
       M(3,3)=M3*B**2+I3
```

Factorized output:

```
C>    Inertia Matrix
       M(1,1)=(C*C*(M2+M3)+
      +          (M1*A**2+I1))

       M(2,1)=C*B*M2*SIN(AL2-AL1)
       M(2,2)=(M2*B**2+I2)

       M(3,1)=C*B*M3*SIN(AL1-AL3)
       M(3,2)=0.
       M(3,3)=(M3*B**2+I3)
```

Figure 17. NEWEUL Output of Fully Symbolic and Factorized Inertia Matrix

The software module NEWSIM allows the simulation of the symbolic equations of motion provided by module NEWEUL. It automatically generates a problem specific simulation program. The user simply has to add the specification of

- force laws,
- system parameter values, and
- initial conditions.

The simulation results are stored in ASCII data files that can be visualized with arbitrary graphics packages. The software structure is shown in Figure 18.

The simulation results may contain

- the time history of the state variables,
- the kinematical data of observation points,
- data for animation,
- the time history of the reaction forces, and
- user-defined output data.

Apart from time simulations, additional analyses can be performed with the module NEWSIM. These additional features include:

- the quasi static analysis,
- the computation of the state of equilibrium and
- the treatment of the inverse dynamics problem.

The software package NEWEUL is written in FORTRAN 77 and can be implemented on any workstation or mainframe with a FORTRAN 77 compiler. NEWEUL uses its own formula manipulator.

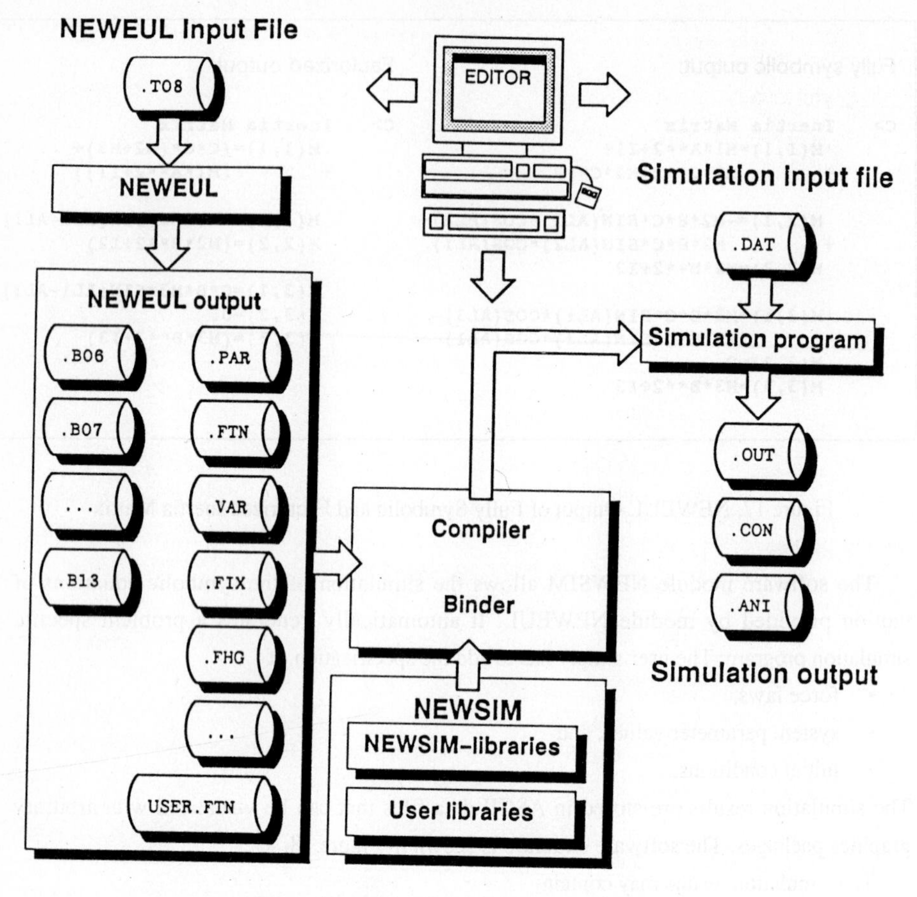

Figure 18. Software Structure of NEWEUL

5 Animation and Human Perception

The simulation produces a large number of data which can be stored in the database. But it is necessary to extract the most valuable information from the simulation data. This can be done by visualization of the motion or by introducing suitable performance criteria, one of them is the human perception of mechanical vibrations.

5.1 Animation of Motion

The visualization of motion can be achieved by the CAD-3D-System or by a separate visualization software tool, respectively, Figure 19. The animation principle is shown in

Figure 20. The information on the 3x1–translation vector $\mathbf{r}_i$ and the 3x3–rotation tensor S_i follows according to (7) from the simulation. This information is used in the 4x4–Denavit-Hardenberg matrix for instancing of the geometric structure at each time step. Daberkow [14] used in the visualization software tool VISANI geometric modeling by polygon sets, as shown in Figure 21, and the PHIGS standard.

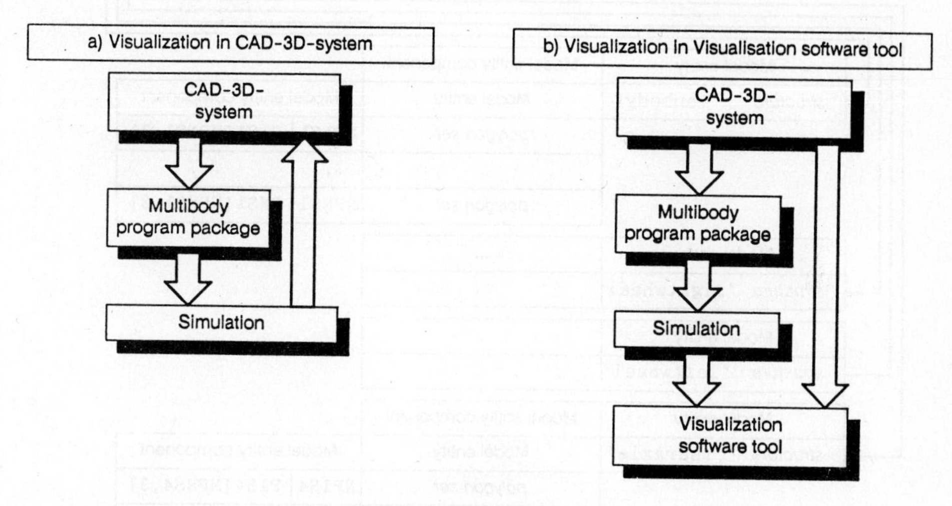

Figure 19. Visualization of Motion

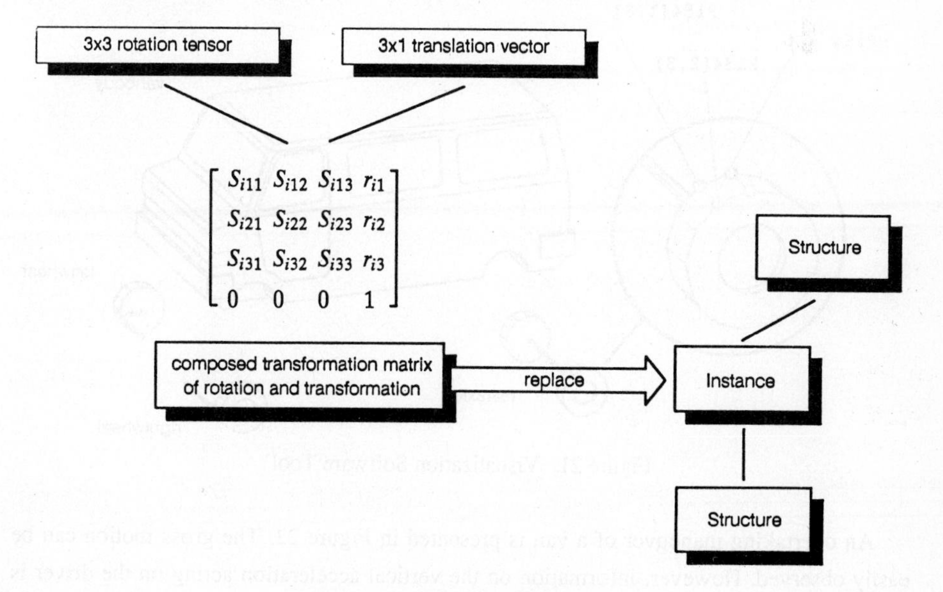

Figure 20. Animation Principle

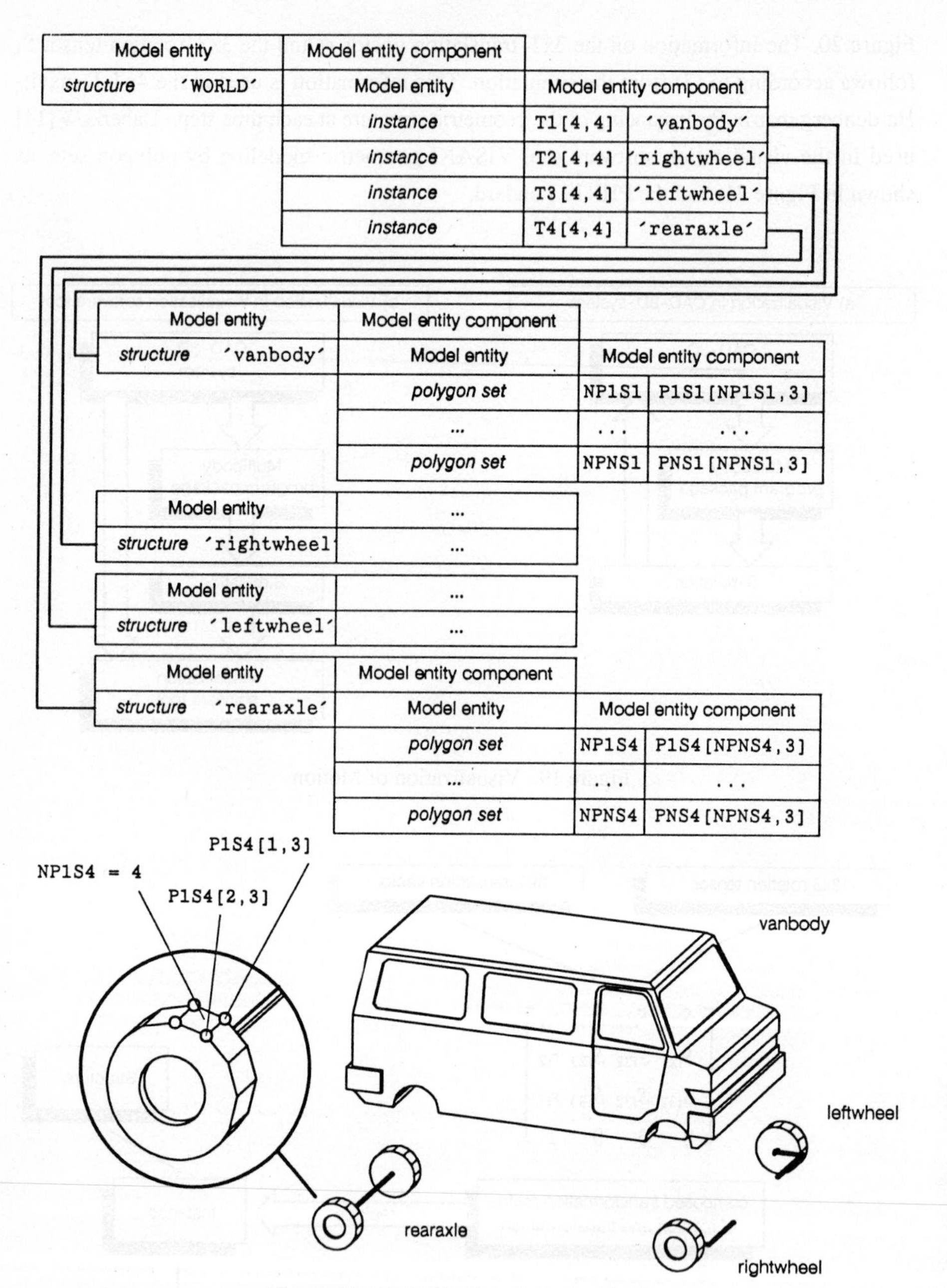

Figure 21. Visualization Software Tool

An overtaking maneuver of a van is presented in Figure 22. The gross motion can be easily observed. However, information on the vertical acceleration acting on the driver is not visible.

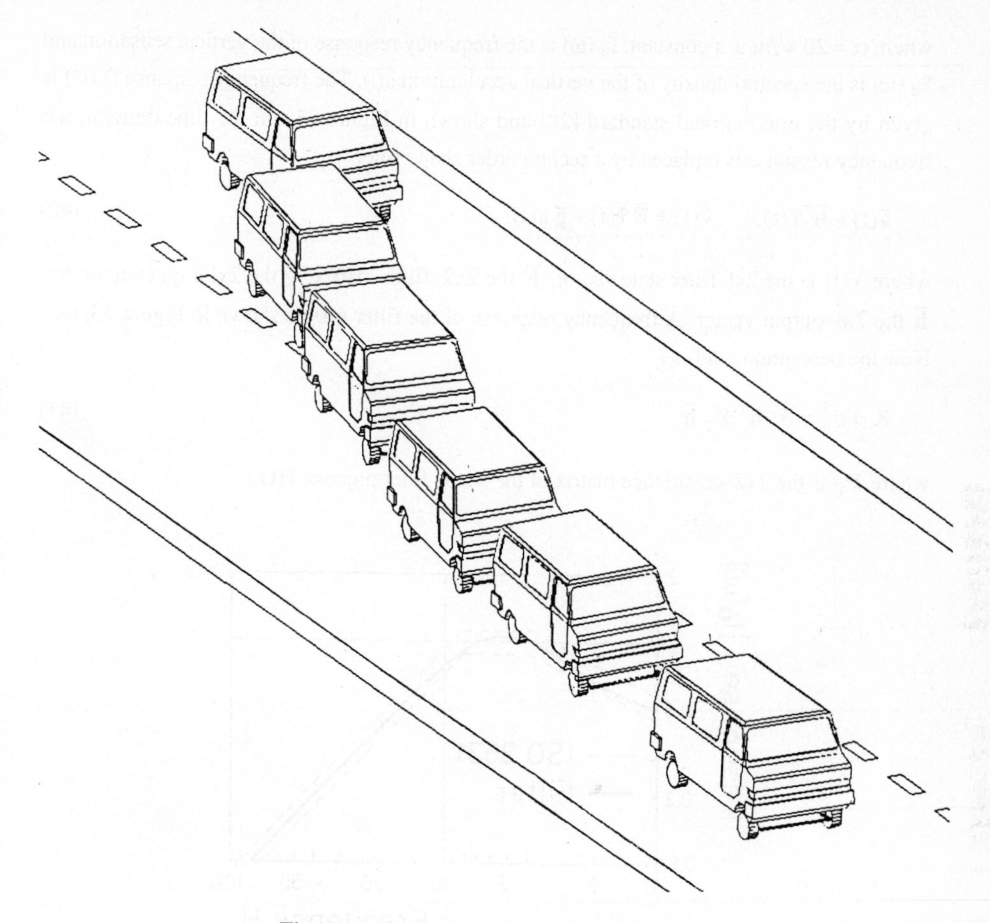

Figure 22. Overtaking Maneuver of a Van

5.2 Human Perception

Even if the road roughness is small and cannot be visualized in an animation, the roughness generates mechanical vibrations of the vehicle acting on the human body. These vibrations depend on the vehicle model, its equations of motion, and the excitation by the road. Since the road roughness represents a stochastic process, the resulting vehicle vibrations are random, too. Therefore, stochastic methods of signal analysis have to be used.

With respect to vehicle vibrations, only the scalar vertical acceleration a(t) will be considered as a criteria for the riding comfort. Then, the perception follows as

$$K = \sigma_{\tilde{a}}^2 = \alpha^2 \int_{-\infty}^{+\infty} \left| f_a(\omega) \right|^2 S_a(\omega)\, d\omega \tag{39}$$

where $\alpha = 20\ s^2/m$ is a constant, $f_a(\omega)$ is the frequency response of the vertical sensation and $S_a(\omega)$ is the spectral density of the vertical acceleration a(t). The frequency response $f_a(\omega)$ is given by the international standard [28] and shown in Figure 23. In the time domain, the frequency response is replaced by a second order shape filter

$$\bar{a}(t) = \bar{\mathbf{h}}^T \bar{\mathbf{v}}(t), \qquad \dot{\bar{\mathbf{v}}}(t) = \bar{\mathbf{F}}\,\bar{\mathbf{v}}(t) + \bar{\mathbf{g}}\,a(t), \tag{40}$$

where $\bar{\mathbf{v}}(t)$ is the 2xl–filter state vector, $\bar{\mathbf{F}}$ the 2x2–filter matrix, $\bar{\mathbf{g}}$ the 2xl–input vector and $\bar{\mathbf{h}}$ the 2xl–output vector. A frequency response of the filter (40) is shown in Figure 23, too. Now the perception reads as

$$K = \sigma_{\bar{a}}^2 = \alpha^2\, \bar{\mathbf{h}}^T\, \mathbf{P}_{\bar{v}}\, \bar{\mathbf{h}} \tag{41}$$

where $\mathbf{P}_{\bar{v}}$ is the 2x2–covariance matrix of the shape filter process $\bar{\mathbf{v}}(t)$.

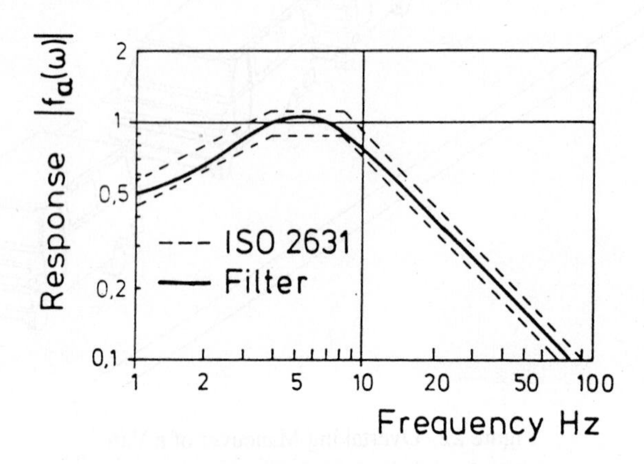

Figure 23. Frequency Response of Human Sensation

The final results (39) and (41) are given in the frequency domain and the time domain, respectively. It turns out that an infinite integral has to be evaluated in the frequency domain while in the time domain, only an algebraic matrix operation is required. Therefore, the covariance analysis using the time domain is preferable [29].

In Figure 24, a complex vehicle is shown consisting of 4 mass points and 7 rigid bodies subject to 35 constraints resulting in f=l9 degrees of freedom. In addition, there are two serial spring–damper–configurations at the engine. Further, four first order excitation shape filters are considered while the sensation shape filter will be neglected. Then, the global system has the order n=44. The human sensation of mechanical vibration will be discussed only with respect to the vertical acceleration of the car body. Numerical results for this complex vehicle have been published by Kreuzer and Rill [30]. Figure 25 shows the RMS value or standard deviation, respectively, of the vertical acceleration normalized by the earth acceleration for each location (C, D) on the car body. It turns out, that the optimal ride characteristics are found in the middle of the car body. Thus, qualitative experience and quantitative measurements are confirmed by theoretical system analysis very well.

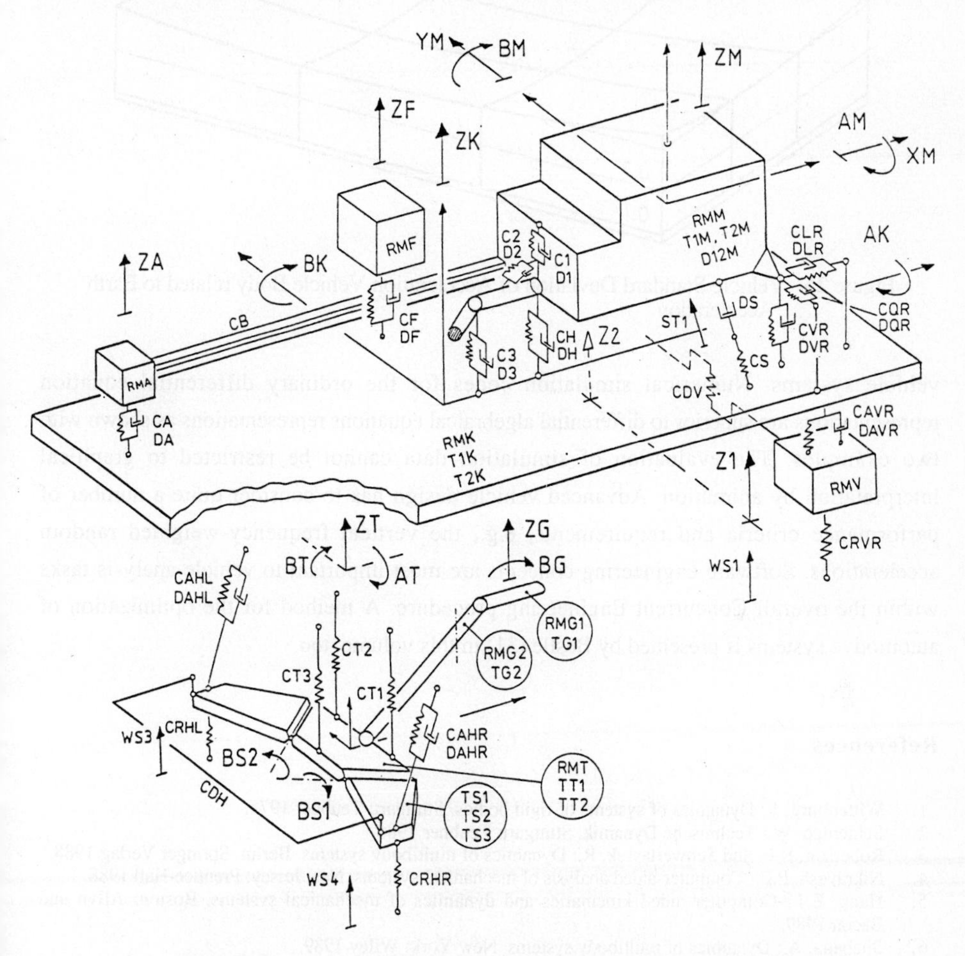

Figure 24. Model of Complex Vehicle under Random Excitation

6 Conclusion

A simulation based design of automotive systems requires modeling using CAD data and well defined multibody system datamodels. The object-oriented approach of modern software engineering is most adequate for multibody system data. Formalisms generating symbolic equations of motion are efficient for dynamical analysis, simulation, and optimization of

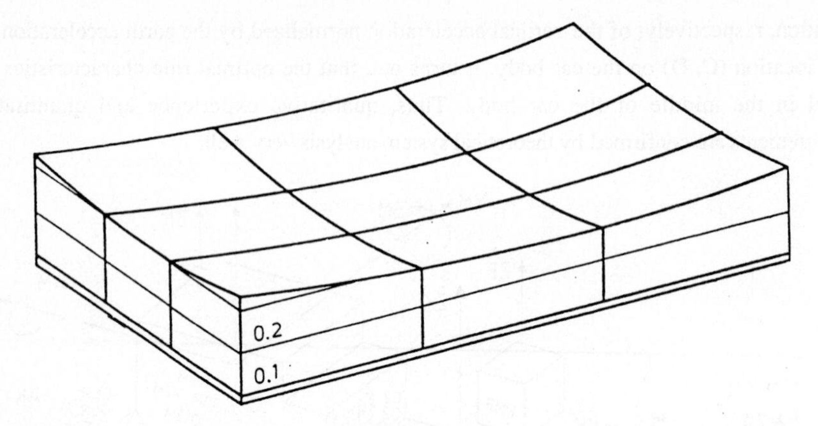

Figure 25. Vehicle Standard Deviation of Acceleration Vehicle Body related to Earth Acceleration

vehicle systems. Numerical simulation codes for the ordinary differential equation representations are superior to differential algebraical equations representations as shown with two examples. The evaluation of simulation data cannot be restricted to graphical interpretation by animation. Advanced vehicle design has to consider quite a number of performance criteria and requirements, e.g., the vertical frequency-weighted random accelerations. Software engineering concepts are most important to vehicle analysis tasks within the overall Concurrent Engineering procedure. A method for the optimization of automotive systems is presented by Bestle [31] in this volume, too.

References

1. Wittenburg, J.: Dynamics of systems of rigid bodies. Stuttgart: Teubner 1977.
2. Schiehlen, W.: Technische Dynamik. Stuttgart: Teubner 1986.
3. Roberson, R.E. and Schwertassek, R.: Dynamics of multibody systems. Berlin, Springer Verlag 1988.
4. Nikravesh, P.E.: Computer-aided analysis of mechanical systems. New Jersey: Prentice-Hall 1988.
5. Haug, E.J.: Computer aided kinematics and dynamics of mechanical systems. Boston: Allyn and Bacon 1989.
6. Shabana, A.: Dynamics of multibody systems. New York: Wiley 1989.
7. Magnus, K (ed.): Dynamics of multibody systems. Berlin: Springer-Verlag 1978
8. Slibar, A; Springer, H. (eds.): Dynamics of vehicles on roads and railway tracks, Swets and Zeitlinger 1978.
9. Haug, E.J. (ed.): Computer aided analysis and optimization of mechanical system dynamics. Berlin: Springer-Verlag 1984.
10. Kortüm, W.; Schiehlen, W.: General purpose vehicle system dynamics software based on multibody formalisms. Vehicle System Dynamics 14(1985), pp. 229-263.
11. Bianchi, G.; Schiehlen, W. (eds.): Dynamics of multibody systems. Berlin: Springer-Verlag 1986.
12. Kortüm, W.; Sharp, R.S.: A report on the state-of-affairs on "Application of multibody computer codes to vehicle system dynamics". Vehicle System Dynamics 20(1991), pp. 177-184.
13. Schiehlen, W. (ed.): Multibody systems handbook. Berlin: Springer-Verlag 1990.
14. Daberkow, A.: Zur CAD-gestützten Modellierung von Mehrkörpersystemen. Ph.D. Thesis. Stuttgart: University of Stuttgart, to appear.

15. Schiehlen, W.: Prospects of the German multibody system research project on vehicle dynamics simulation. In: Dynamics of Vehicles on Road and Tracks. Proc. 12th IAVSD Symposium. Amsterdam: Swet and Zeitlinger 1992, pp. 537-550.
16. Otter, M., Hocke, M., Daberkow, A., Leister, G.: Ein objektorientiertes Datenmodell zur Beschreibung von Mehrkörpersystemen unter Verwendung von RSYST. Institutsbericht IB-16. Stuttgart, Institut B für Mechanik, 1990.
17. Lang, U.: Erstellen von Anwendungsmoduln in RSYST. Stuttgart, Rechenzentrum der Universität Stuttgart, 1988.
18. Loebich, I.: Einführung in RSYST. Stuttgart, Rechenzentrum der Universität Stuttgart, 1988.
19. Rühle, R. et al.: RSYST Unterprogramm– und Moduldokumentation, Version 3.5.0. Stuttgart, Rechenzentrum der Universität Stuttgart, 1988.
20. Schiehlen, W.: Modeling of complex vehicle systems. In: Proc. 8th IAVSD Symposium. Hedrich, J.K. (ed.) Lisse: Swets & Zeitlinger 1984, pp. 548-563.
21. Hollerbach, J.M.: A recursive Lagrangian formulation of manipulator dynamics and a comparative study of dynamics formulation complexity. IEEE Trans. Sys. Man. Cyb. 10(1983), 730-736.
22. Walker, M.W.; Orin, D.E.: Efficient dynamic computer simulation of robot mechanisms. J. Dyn. Sys. Meas., Control 104(1982), 205-211.
23. Brandl, H.; Johanni, R.; Otter, M.: A very efficient algorithm for the simulation of robots and similar systems without inversion of the mass matrix. In: Theory of Robots. Kopacek, P.; Troch, I.; Desoyer, K. (eds.): Oxford: Pergamon Press 1988, pp. 95-100.
24. Schiehlen, W.: Computational aspects in multibody system dynamics. Comp. Meth. Appl. Mech. Eng. 90(1991), pp. 569-582.
25. Leister, G.: Wahl geeigneter Koordinaten zur Dynamikanalyse von Mehrkörpersystemen. Zwischenbericht ZB-49 Stuttgart: Institute B of Mechanik 1990.
26. Valasek, M.: On the efficient implementation of multibody systems formalisms. Institutsbericht IB-17. Stuttgart Institute B of Mechanics 1990.
27. Leister, G.: Beschreibung und Simulation von Mehrkörpersystemen mit geschlossenen kinematischen Schleifen. Fortschr. Ber. VDI Reihe 11 Nr. 169. Düsseldorf: VDI-Verlag 1992.
28. International Standard ISO 2631, Guide for the evaluation of human exposure to whole-body vibrations. Int. Org. Standardization (1974).
29. Schiehlen, W.: Vehicle system dynamics. In: Theoretical and Applied Mechanics. Niordson, F.; Olhoff, N. (eds.): Amsterdam: North-Holland 1985, pp. 387-398.
30. Kreuzer, E.; Rill, G.: Vergleichende Untersuchung von Fahrzeugschwingungen an räumlichen Ersatzmodellen. Ing. Arch. 52(1982), pp. 205-219.
31. Bestle, D.: Optimization of automotive systems. (in this volume)

Simulation-Based Design of Off-Road Vehicles

Ronald R. Beck

U.S. Army Tank-Automotive Research, Development and Engineering Center, Tank-Automotive Technology Directorate, Warren, MI 48397-5000

Abstract: Analytical and physical simulation tools and technologies for the design of off-road wheeled and tracked vehicles are presented. Advanced Computer-Aided Design (CAD) and supercomputer-based analytical tools for off-road vehicle design and evaluation are illustrated via applications. High-capacity hardware-in-the-loop real-time simulators are discussed, with illustrative applications in off-road vehicle development and evaluation. The basic theme of this paper focuses on how the application of simulation technology is becoming an integral part of the Army's combat and tactical vehicle research, development and acquisition process. This paper presents two basic goals to form the basis for simulation-based Concurrent Engineering. The first is use and creation of simulation tools that not only predict vehicle performance, but also subsystem and component reliability. The second is the development of an integrated simulation capability and military vehicle design and performance data base to form the basis for simulation-based Concurrent Engineering.

Keywords: analytical simulation / physical simulation / computer aided engineering (CAE) / computer aided design (CAD) / concurrent engineering / motion base simulators / vehicle systems / real-time dynamics / network computing / data base management

1 Introduction

Simulation of total vehicle and vehicle subsystem performance characteristics has steadily gained acceptance over the past 10 years [1]. Army leadership, especially Program Executive Offices (PEOs), have recognized the value of simulation as a tool for reducing the costs and time associated with traditional approaches to vehicle development. At the U.S. Army Tank-Automotive Research, Development and Engineering Center (TARDEC), in particular, simulation and modeling has been used to the maximum extent possible in support of military vehicle research, design, development, and acquisition. TARDEC has demonstrated that simulation and modeling leads to significant time and cost savings compared to traditional "build-test-break-fix" approaches. Simulation allows analysis of concepts and scenarios which cannot be replicated economically (or not at all) with test-beds.

2 What is Simulation?

Simulation is the coordinated use of analytical and laboratory testing techniques to evaluate off-road mobility, dynamic stability, structural integrity (or other performance aspects of vehicle systems and subsystems) under repeatable, controlled conditions. Simulation is also a tool for screening new technologies or new (or modified) components prior to building expensive prototypes. Most significantly, simulation is a precise and efficient mechanism for evaluating new systems or troubleshooting fielded vehicle problems without having to resort to expensive and time consuming field tests. It is clear that simulation saves time and affords more extensive evaluation than does field testing alone.

Simulation of Army tank and automotive systems is concentrated in two areas: analytical and physical. A state-of-the-art supercomputer-based analytical and physical simulation capability has been created by TARDEC (Figure 1) to reduce the time and high cost of conventional military vehicle prototype-based design and development. These distinct activities encompass a wide-ranging field of tasks in the vehicle development process from analysis of conceptual vehicle systems prior to "bending metal," to evaluation of actual hardware.

3 Analytical Simulation

Analytical simulation involves mathematically modeling vehicle systems and subsystems for the design and engineering analysis of most aspects of combat and tactical vehicle performance. As depicted in Figure 1, engineers and scientists utilize high-performance computing and visualization systems to explore the performance of concepts and vehicle systems under development prior to the fabrication and test of prototypes. The most significant component of this capability is the Army Regional Supercomputer at TARDEC, which is one of a handful of Army sites having this high-performance computing workhorse (a Cray Research, Inc. Cray-2 computer). Augmenting the supercomputer is a growing system of advanced high-performance workstations and networking to perform associated pre- and postprocessing and computer-aided design.

Simulation specialists have developed and implemented the basic methodologies and software tools used to perform analytical simulation. At the same time, increased emphasis is being placed on integration of commercially available analysis software packages. Specifically (Figure 2), simulation codes such as the NATO Reference Mobility Model

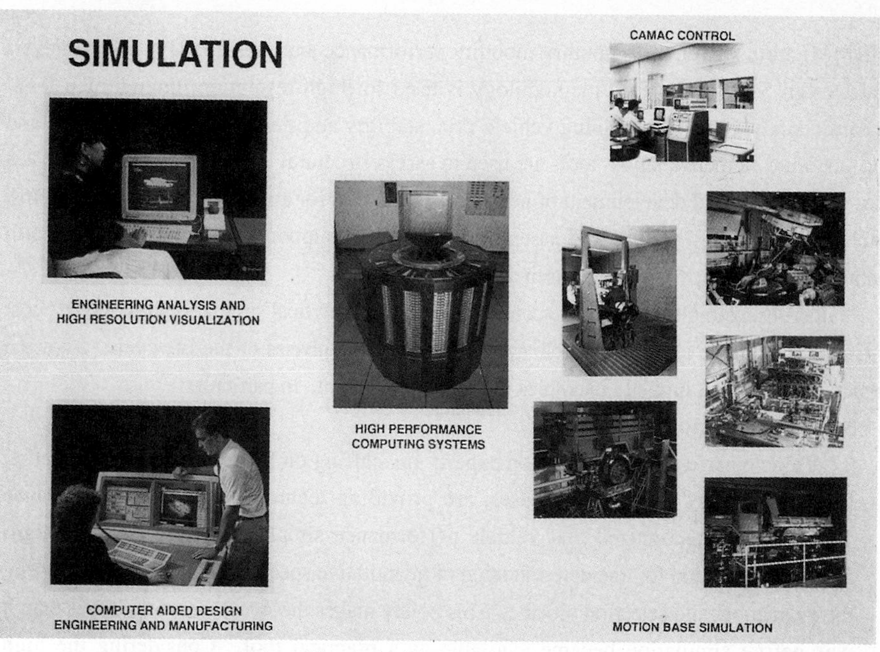

Figure 1. Simulation

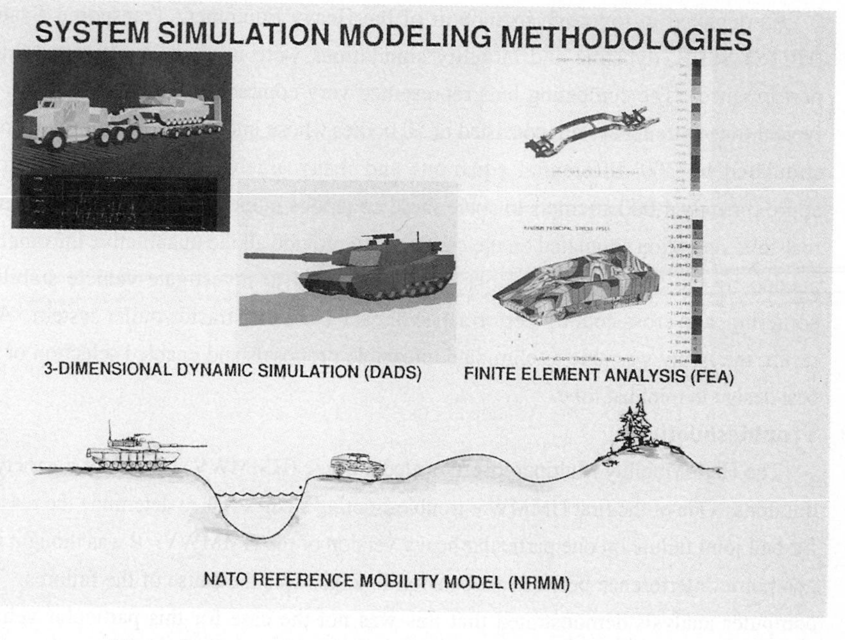

Figure 2. System Simulation Modeling Methodologies

(NRMM) are used for cross-country mobility performance analyses; the Dynamic Analysis and Design System (DADS) methodology is used for high-resolution, three-dimensional dynamic analyses for determining vehicle ride, stability and duty-cycle load histories; and various finite element analysis tools are used to assess structural integrity. Not discussed is a host of existing - and development of new - methodologies for analyzing propulsion systems, track components, signature and assessment models in support of ground vehicle stealth technology research, and vehicle survivability.

Numerous examples have been accumulated where analytical simulation has "saved" cost and time in support of Army vehicle systems through all phases of the life cycle. Figure 3 depicts specific - but limited - examples illustrating this point. In particular:

A. **Source Selection**

Vehicle performance simulation experts are staffing on Source Selection Evaluation Boards (SSEBs), (or in some cases, are providing technical support only) because TARDEC has recognized that vehicle performance simulation is the modern, high-technology method for the determination of quantitative specifications and for supporting the evaluation and selection process. This policy makes the Army a smarter buyer than it was before simulation became available as a practical tool. Considering the high acquisition and subsequent support costs of military vehicle systems, the importance of the ability to avoid the procurement of systems with potential problems cannot be overemphasized.

As depicted in Figure 3, in support of the Heavy Equipment Transporter System (HETS) SSEB, dynamic and mobility simulations were used to investigate vehicle performance. The competing bids represented very complex mechanical systems. A typical tractor-trailer system consisted of 50 bodies whose interactive dynamics had to be simulated by 300 differential equations and many algebraic expressions. It took approximately 1,000 attempts to solve these equations numerically for every second of real-time operation simulated on the computer to provide all the quantitative information needed by the SSEB. Simulations were performed to investigate vehicle stability, cornering, and cross-country performance of each candidate tractor-trailer system. As a result, the Army was able to eliminate infeasible proposals and enabled selection of the best design in minimal time.

B. **Troubleshooting**

The High Mobility Multipurpose Wheeled Vehicle (HMMWV) is used for a variety of missions. One of the first HMMWV troubleshooting studies was to determine the reasons for ball joint failure on one particular heavy version of the HMMWV. It was thought that geometric interference between suspension linkages was the cause of the failures. The computer analysis demonstrated that this was not the case for this particular vehicle configuration and payload. The problem was caused by the fact that the tire (equipped

RECENT APPLICATIONS OF
SUPERCOMPUTER - BASED ANALYTICAL SIMULATION

SOURCE SELECTION
• INTERSECTION NEGOTIATION
• SIDE SLOPE PERFORMANCE
• LATERAL STABILITY
• OFF TRACKING SWING
• CROSS COUNTRY MOBILITY

SAFETY ANALYSIS
• TIRE PRESSURE
• BACK HOE POSITION
• BUCKET LOAD/POSITION
• BUMPS AND HOLES
• RIDE DYNAMICS
• VEHICLE SPEED

TROUBLE SHOOTING
• SUSPENSION SYSTEM DUTY CYCLE
• COMPONENT INTERFERENCE
• SHOCK LOADING

Figure 3. Supercomputer-Based Analytical Simulation

with a run flat device), when hitting a bump, deflects such that the wheel runs on the run flat rim, which then acts as a rigid wheel. This causes the dynamic load on the ball joint to become excessive. The problem was solved by reinforcing certain suspension elements. The PEO was able to make this informed decision once the correct cause was identified. Again, analytical tools were used successfully for the determination of the cause of a serious field problem and for arriving at a solution. This led to substantial savings in time and money over the conventional "build-test-break-fix" approach.

C. **Safety Analysis**

The Small Emplacement Excavator (SEE) is another complex vehicle system which has been simulated to evaluate its dynamic stability both on and off-road. This vehicle is a 4X4 truck with a relatively soft suspension, a back hoe, and a front-end loader. It was not accepted by the user because unstable behavior was observed. The entire fleet of SEE vehicles was grounded temporarily. TARDEC engineers were asked to model the vehicle and establish safe operational limits. Lane change maneuvers, crossing pot holes, step-like obstacles (bumps), and rough terrain were modeled at various tire pressures and back hoe and front-end loader positions. As a result of this analysis, its safe operational limits were established, and the user accepted the SEE vehicle. The point is, however, that the investigation of the same number of different configurations and maneuvers via field tests would have been virtually impossible. Also, in many cases, the driver would have been exposed to dangerously unstable conditions when approaching the boundaries of safe modes of operational limits of the vehicle. It is estimated that the dynamic analyses saved more than $200,000 in build and test costs.

4 Physical Simulation

Physical simulation involves emulating real-time physical motions of actual vehicle systems and subsystems in a computer-controlled laboratory environment to approximate a vehicle's field performance. As hardware components and systems take form, engineers and technicians conduct man- and hardware-in-the-loop motion base simulator studies. Motion base simulators capable of "shaking" complete combat and tactical vehicle systems weighing up to 45 tons are now used to evaluate issues associated with man and machine interaction dynamics, and reliability and other component integration issues associated with the soldier, his displays and controls.

Physical simulation offers accelerated test schedules, repeatable test conditions, and allows for the collection of data otherwise difficult or impossible to obtain. Physical simulation is used to validate analytical simulation models, address man-in-the-loop issues, and determine failure points of a vehicle system or subsystems.

The most significant component of TARDEC's physical simulation capability is its six-degrees-of-motion-freedom Crew Station/Turret Motion Base Simulator (CS/TMBS) (Figure 4), which can accommodate heavy combat vehicle turrets weighing up to 25 tons. The CS/TMBS is used for studying soldier-machine interface problems, gun turret drive stabilization systems, and addressing issues related to the operation of turrets and their components. As shown in Figure 1, other full-scale motion simulators are available at TARDEC. These consist of sets of digitally controlled hydraulic actuators which attach to the wheel spindles of tactical vehicle, or support tracked vehicle road wheels and tracks to simulate the effect of running over specific rough terrain segments. These simulators can be reconfigured and instrumented to isolate and test specific vehicle components.

There are three basic types of motion base simulators: (1) tire/track-coupled simulators, (2) spindle-coupled simulators, and (3) platform simulators. Figures 5, 6, and 7 are pictorial examples of each type, respectively, and briefly summarize their characteristics. Figure 8 is a graphic illustration of the physical simulation process.

There are many documented examples of how laboratory tests can be used in lieu of field testing [2]. Figures 9 and 10 depict the time and costs saved when comparing field vs. laboratory testing of several different trailer systems. Figure 11 is a chart comparing the attributes of field testing to motion base simulator testing. It must be emphasized that each form of testing has its advantages and disadvantages. The best solution is a combination of both field and laboratory testing designed specifically for the situation at hand.

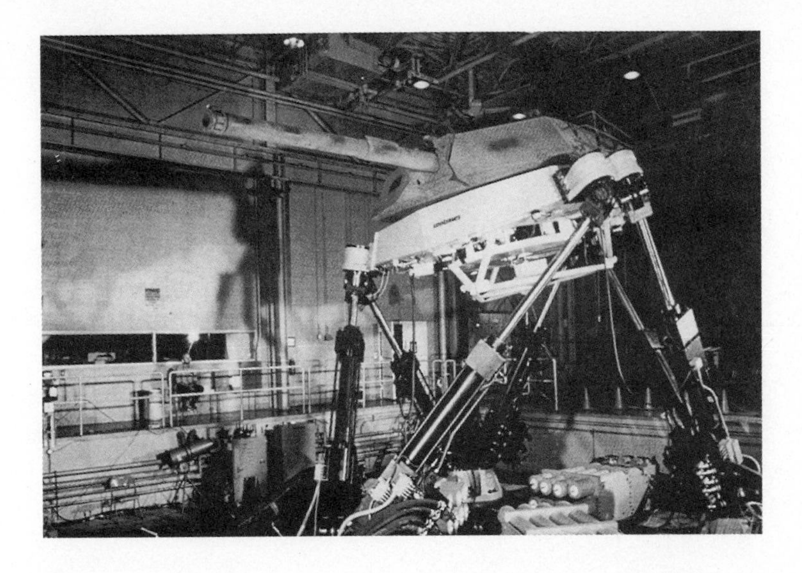

Figure 4. Crew Station/Turret Motion Base Simulator (CS/TMBS)

5 Coordinated Use of Simulation Tools and Field Testing

The ideal integration of analytical and physical simulation coupled with selected field tests, involves each supplying data for, analyzing the results of, and validating the other. For example, detailed analytical evaluations and trade-off analyses of design alternatives are conducted early on to create performance specifications and evaluation criteria to be used later in source selection and subsequent field testing of prototype systems.

As concepts and designs take on definition, analytical simulation can be coupled with laboratory physical simulation for proof-of-principle and man- and hardware-in-the-loop testing. Laboratory tests are conducted under controlled, repeatable, dynamic conditions at the complete system level.

If properly applied, simulation can effectively augment the test planning and validation process. For example, a critical part of vehicle system acquisition is developmental and operational testing. Through modeling and simulation during the development phases, test environments and instrumentation requirements for field testing can be determined in advance with better certainty. Simulation results can also identify potential vehicle problems that may arise during field tests which, therefore, should be addressed in advance of testing. This affords the potential for substantially reducing test costs and time.

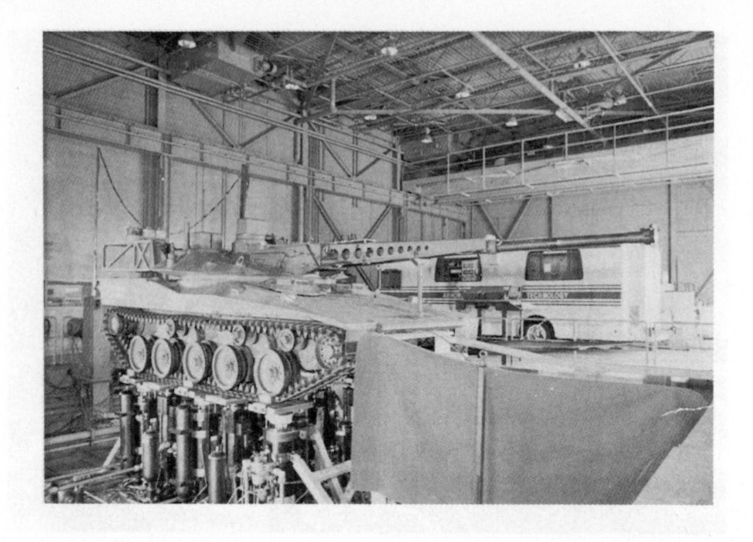

- CHARACTERISTICS:
  - Easy on, easy off vehicle installation
  - Includes tire/track in the test
  - Adaptable to "generic" road/terrain profiles

- APPLICATIONS - DURABILITY EVALUATIONS OF:
  - Body
  - Body-mounted components
  - General noise and vibration analysis
  - Suspension components (limited)

Figure 5. Tire/Track-Coupled Simulators

- CHARACTERISTICS:
  - Lateral, longitudinal and vertical loads included
  - More realistic simulation of multiaxial proving ground environment
  - Provides most accurate simulation for detailed durability assessment
  - Expandable to include:
    - braking torque
    - steering torque

- APPLICATIONS - DURABILITY EVALUATIONS OF:
  - Body/chassis
  - Body/chassis mounted components
  - Suspension, suspension to chassis
  - Engine cradle, engine mounts

Figure 6. Spindle-Coupled Simulators

- CHARACTERISTICS:
  - Can be set up for component/subassemblies
  - Realistic simulation of vibration environment
  - Full vehicle system unnecessary

- APPLICATIONS - DURABILITY AND PERFORMANCE TESTING OF:
  - Hatches
  - Autoloaders
  - Gun/turret drive systems
  - Crew Stations
  - Nearly any subassembly subjected to vibrations

Figure 7. Platform Simulators

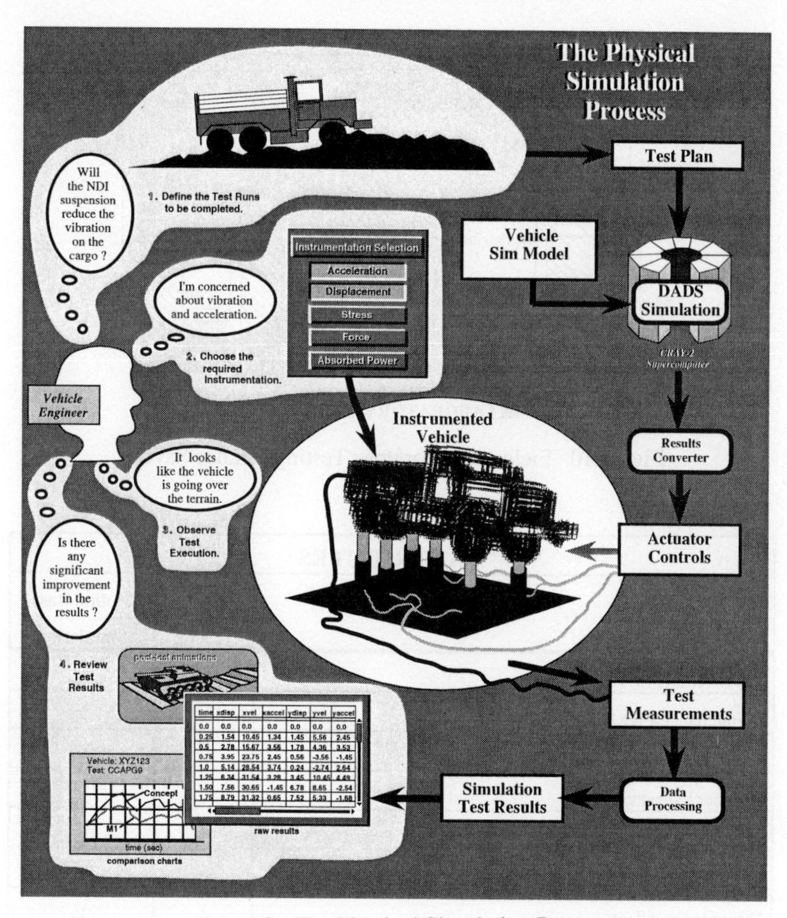

Figure 8. The Physical Simulation Process

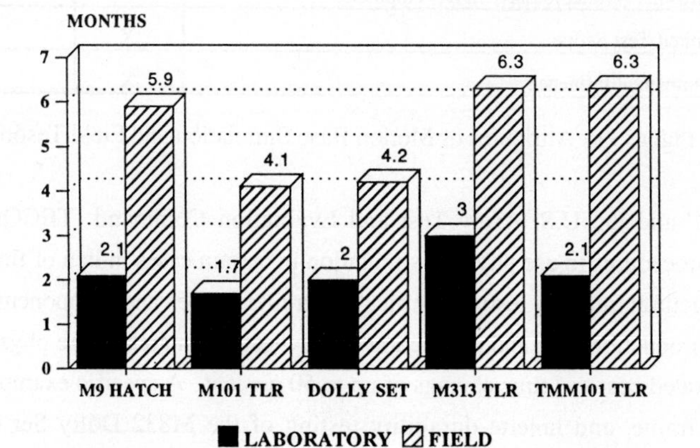

Figure 9. Field vs Laboratory Testing - Time Saved

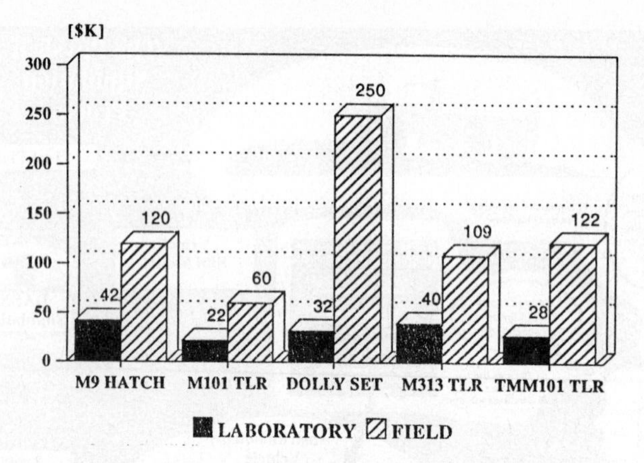

Figure 10. Field vs. Laboratory Testing - Funds Saved

ATTRIBUTES		
	Field Testing	Motion Base Simulation
Wide Variety of Test Courses		X
Maintenance-Free Test Courses		X
Prime Mover Not Required		X
Weather Independent		X
Repeatable		X
Observable		X
Driver Independent		X
Test Results Available Early in Development		X
Unlimited Test Scope	X	
Environmental Influence	X	

Figure 11. Attributes of Motion Base Simulation and Field Testing

TARDEC and the U.S. Army Test and Evaluation Command (TECOM) are jointly developing procedures to use physical simulation in certain cases in lieu of field testing. For example, structural integrity testing of truck and trailer frames and components in support of comparison production tests and production quality tests is an area where physical simulation has demonstrated cost and time savings of up to 50 percent. A specific example of this is the suspension, frame, and lunette durability testing of the M832 Dolly Set (Figure 12) in TARDEC's physical laboratory. In addition, carefully planned test scenarios, vehicle and terrain characteristics, as well as detailed test data, enhances the ability to validate the simulation models. This permits the creation of simulation data base libraries, which can be used for future applications. Relying on simulation during the test-planning process and validating simulation results against carefully controlled tests will greatly enhance the simulation data base libraries and increase confidence in simulation.

Figure 12. Durability Testing of M832 Dolly Set

6 Simulation-Based vs. Conventional Development Approach

Figure 13 is a flowchart comparing the "conventional" and simulation-based development approaches. In the past, the conventional combat and tactical vehicle research and development process relied heavily on building and testing prototype vehicles or test-beds. This allowed engineers to evaluate a limited number of design ideas, based largely on empirical knowledge, under very restrictive field testing conditions. This is what is commonly referred to as "build-test-break-fix" prototype-based development. Also, procurement specifications have often contained vague criteria concerning off-road mobility, ride quality, roll and maneuver stability, as well as many other aspects of military vehicle performance. You might find, for example, such vague specifications as the vehicle has to be able to travel in mud, snow or sand," or, "the vehicle has to be more mobile than the M113 Armored Personnel Carrier." Therefore, decisions were made largely on a judgmental basis using "rules of thumb" and engineering experience. Additionally, since a very limited number of design alternatives could be tried in hardware, advanced technology subsystems with their higher degree of uncertainty and payoff were not given full consideration, because design "rules of thumb" did not apply to them. Advanced component technology was often set aside in favor of better understood conventional methods, which lead to conservative, low-tech designs. In essence, modern engineering analysis tools that permit better design trade-off analyses were not used. Finally, the process was slow and costly.

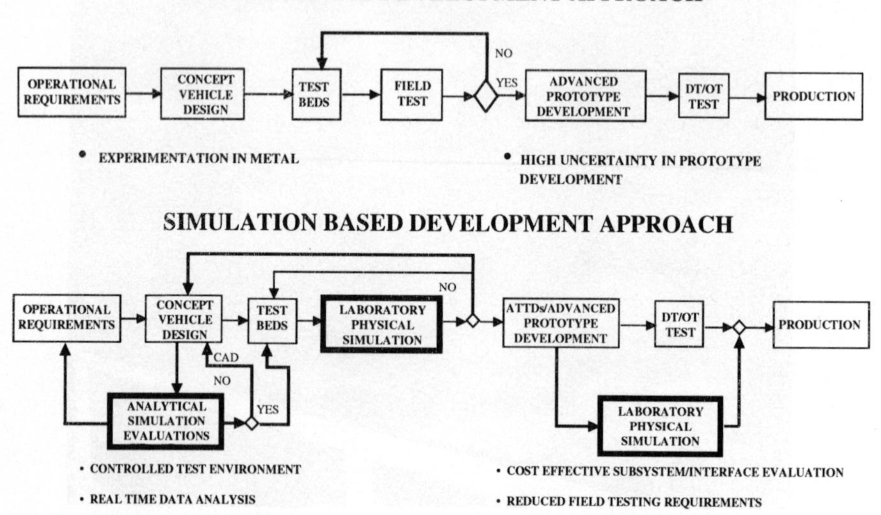

Figure 13. Simulation-Based vs. Conventional Development Approach

Starting in the late 1960s and early '70s, with the availability of computers, theoretically complex mathematical equations that relate the vehicle, the terrain, the human occupants, and the threat were no longer confined to the pages of research reports. Detailed computer simulations became practical tools for analytically predicting vehicle behavior, even in the concept stage. This is especially true with the rapid expansion of the computer and visualization hardware of today. Thus, in contrast to the "prototype-based development approach" where you "build-test-break-fix," the "simulation-based development approach" relies on the early use of high-resolution computerized vehicle engineering design tools and models. It predicts performance of the conceptual system, performs laboratory man- and hardware-in-the-computer-loop technology demonstrations, and finally, fabricates and field test the system to confirm what you know about your system and looks for that 5 percent of unknown capability and possible small performance deficiencies. Additionally, because of the rapid advances in computer hardware and software technology, research is underway to create simulation tools that not only predict vehicle performance, but also subsystem and component reliability.

Simulation can be used to accelerate the component integration process. Clearly, it takes less time to "create" hardware on a computer than in a machine shop. If adequate high-resolution component models are available, their synergistic interaction and performance can be assessed analytically, which leads to a significantly reduced need to build and test hardware.

From the foregoing, it is probably clear why we use simulation. Simulation can also be used to answer questions which cannot be resolved by field testing alone. Engineers can simulate many scenarios, environmental conditions, test sites, and design excursions which would be impractical to attempt via field testing. Also, they can simulate maneuvers which would be dangerous to attempt in reality.

Finally, analytical synthesis is a very cost effective augmentation to field testing. For the reasons discussed in previous sections, simulation saves time and affords much more extensive evaluation than do field tests alone.

7 Simulation-Based Concurrent Engineering

The successful applications and expanded use of simulation over the past decade has clearly demonstrated that analytical and physical simulation are viable tools for drastically improving the military vehicle design, acquisition, and field support process. Additionally, with the rapid expansion and availability of Computer-Aided Design (CAD) and Computer-Aided Engineering tools, every engineering discipline today uses analytical tools in the development of a product. However, each of these tools have been developed and are used in isolation from each other, which results in minimal data sharing. Also, most analytical tools require a high degree of expertise to effectively utilize them.

To neutralize these difficulties, engineers at TARDEC, using advances in network computing and database management systems, are integrating these analytical tools into a "simulation-based Concurrent Engineering environment" which will permit data sharing and close coordination of related tools. Further, the physical man- and hardware-in-the-loop simulators are being linked to this Concurrent Engineering environment to bring engineering-level design considerations into evaluations of prototyped components and new technologies within a simulated full vehicle system. Figure 14 is a pictorial representation of the simulation-based Concurrent Engineering system that is under development.

To improve the ease-of-use for tools within the Concurrent Engineering environment, graphical user interface and knowledge encapsulation techniques are being used. These techniques have been successfully exploited in a Tracked Vehicle Workstation (TVWS) development project that is discussed in more detail in Reference 3. The TVWS, which gives

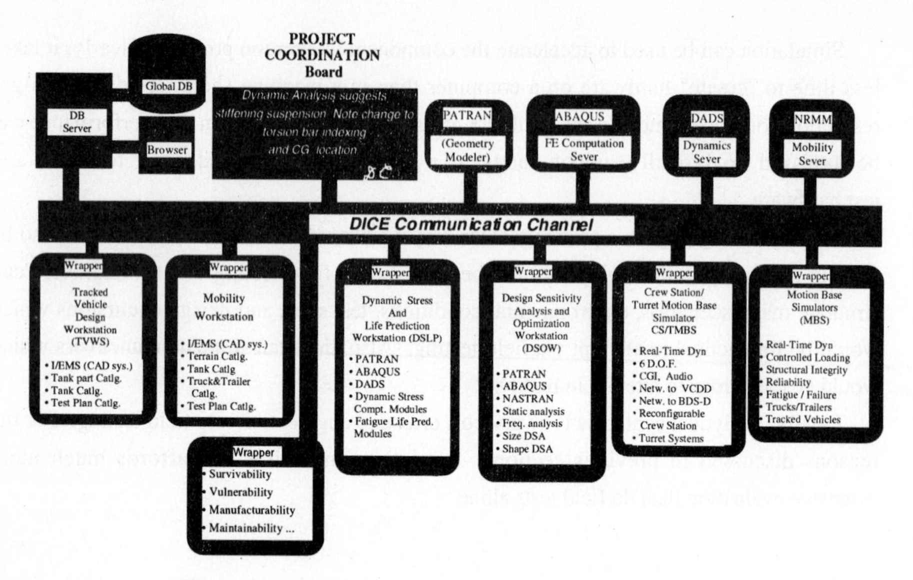

Figure 14. Simulation-Based Concurrent Engineering

combat vehicle design engineers access to advanced vehicle dynamics tools, forms the basis for the TARDEC's Concurrent Engineering environment.

8 Conclusion

The achievements of simulation, in both military and industrial applications, are many. To create a "simulation-based Concurrent Engineering environment" requires dedication of resources and manpower to constantly explore new methodologies and to refine current ones. To properly implement this requires the following:

a. Establish multidisciplinary project teams, foster project team interaction, work in parallel, and share product and simulation data simultaneously.
b. Reorganize the engineering process and create a competitive advantage.
c. Visualize product changes and growth as the design evolves.
d. Increase communication between disciplines, to enable multidisciplinary teams to work in parallel.
e. Create a more productive work environment and foster cross-functional areas of cooperation.

TARDEC has a number of ongoing simulation research thrusts, with one objective in mind: simplifying the simulation process and putting simulation tools in the hands of journeymen engineers. Combined with the ever-expanding proliferation of high-performance computing capabilities, the implementation of complex, theoretical techniques, impractical or impossible before, is now making the engineer's job easier and, most significantly, more productive.

In summary, the following benefits can be realized if simulation becomes routine:

a. Create more designs with fewer people and increase the number of design iterations by well over a factor of 10.
b. Create product designs that are competitive on a "worldwide" basis.
c. Enhance component integration, evaluation of component interaction and verification of component functional and operational performance.
d. Reduce test incidents (surprises) by 90 percent by getting it right the first time.
e. Avoid rework and significantly reduce engineering change orders.

References

1. Beck, R. R. and Schmuhl, J. C. "Role of Simulation at the Army Tank-Automotive Command", Army Bulletin, pp. 33-35, March-April 1992.
2. Beck, R. R. "Real-Time Motion Base Simulation." In *Real-Time Integration Methods for Mechanical System Simulation*, E. J. Haug and R. C. Deyo, (eds.), Computer and Systems Sciences Vol. 69, Springer-Verlag, pp. 45-54, 1991.
3. Ciarelli, K., "Integrated CAE System for Military Vehicle Applications," *Proceedings of the First Annual Symposium on Mechanical System Design in a Concurrent Engineering Environment*, Iowa City, Iowa, pp. 301-318, October 24-25, 1989.

Modeling and Optimization of Aero-Space and Naval Structures

H.A. Eschenauer

Research Center for Multidisciplinary Analyses and Applied Structural Optimization, University of Siegen, D-5900 Siegen, FR Germany

Abstract: After being able to determine structural behavior by means of finite methods, an important goal of engineering activities is to improve and to optimize technical designs, structural assemblies and structural components. The task of structural optimization is to support the engineer in searching for the best possible design alternatives to specific structures. The "best possible" or "optimal" structure is the structure which closely correspond to the designer's desired concept and his objectives while at the same time meeting the functional, manufacturing, and application demands. In comparison to the "trial and error" method generally used in the engineering environment and based on an intuitive empirical approach, the determination of optimal solutions by applying mathematical optimization procedures is more reliable and efficient if correctly applied. These procedures are increasingly entering industrial practice [1].

Keywords: optimization model / multidisciplinary optimization / optimization loop / design model / multicriteria optimization / shape optimization / composite fin / satellite tank / offshore platform

1 Definitions

In order to be able to apply structural optimization methods to an optimization task, it must be possible to express both the design objectives and the constraints by means of mathematical functions. A Scalar Optimization Problem (SO-Problem) is generally defined by the following expression:

Def. 1: Continuous, deterministic SO-problem

$$\min_{x \in \mathbb{R}^n} \{ f(x) \mid h(x) = 0 \,;\, g(x) \leq 0 \} \qquad (1)$$

with $\mathbb{R}^n$ the set of real numbers, f an objective function, $\mathbf{x} \in \mathbb{R}^n$ a vector of n design variables, $\mathbf{g}$ a vector of p inequality constraints, $\mathbf{h}$ a vector of q equality constraints (e.g. system equations for the determination of stresses, deformations, etc.), and X: $\{\mathbf{x} \in \mathbb{R}^n : \mathbf{h}(\mathbf{x}) = \mathbf{0} \, ; \, \mathbf{g}(\mathbf{x}) \leq \mathbf{0} \}$ the "feasible domain", where $\leq$ has to be interpreted for each individual component. The main problem of structural optimization is that the objective function(s) and the constraints are very often nonlinear functions of the design variable vector $\mathbf{x} \in \mathbb{R}^n$; however, the continuity of the functions and their derivations are usually assumed to be given.

For a so-called Vector or Multicriteria Optimization Problem (MCO-Problem), the difficulty lies in finding appropriate solutions taking into account in a given way the multiple objectives. It is significant for such optimization problems with multiple objective functions that an objective conflict exists, i.e. none of the possible solutions permits a simultaneous optimal fulfillment of all objectives, or the individual solutions of the single objective functions are different [2].

Def. 2: Continuous, deterministic MCO-Problem

$$\min_{\mathbf{x} \in \mathbb{R}^n} \{ \mathbf{f}(\mathbf{x}) : \mathbf{h}(\mathbf{x}) = \mathbf{0} \, ; \, \mathbf{g}(\mathbf{x}) \leq \mathbf{0} \} \tag{2}$$

Apart from the objective function vector $\mathbf{f}(\mathbf{x})$, all symbols are the same as in (1). Beside the usual objective "weight" there are other problem-dependent goals of great importance nowadays. These are:

- $f_1(\mathbf{x})$ costs (manufacturing, material, assembly, etc.)
- $f_2(\mathbf{x})$ structural requirements (deformations, stresses, stability, eigenfrequencies, dynamic responses, aeroelastic efficiency, flutter speeds, etc.) [3]
- $f_3(\mathbf{x})$ surface accuracy (optics, astronomy, material manufacturing, etc.) [4]
- $f_4(\mathbf{x})$ quality loss function (environmental conditions, imperfections, etc.) [5]

A vector $\mathbf{x} \in \mathbb{R}^n$ is then and only then called functional-efficient or Pareto-optimal or p-efficient for the problem (2) if no vector $\mathbf{x} \in \mathbb{R}^n$ exists with the characteristics:

$$\begin{aligned} & f_j(\mathbf{x}) \leq f_j(\mathbf{x}^*) \quad \text{for all } i \in \{1,...,m\} \text{ and} \\ & f_j(\mathbf{x}) < f_j(\mathbf{x}^*) \quad \text{for at least one } i \in \{1,...,m\}. \end{aligned} \tag{3}$$

If no vectors are Pareto-optimal, the value of at least one objective function f_j can be reduced without simultaneously increasing the functional values of the remaining components [2]. Consequently, the subject of multicriteria optimization deals with all kinds of "conflicting" problems (therefore the quotation marks in (2)).

A continuous optimization assumes that the objective and constraint values of non-discrete design variables are calculable by interpolation or by means of structural analysis, respectively. If such a possibility is not given, only discrete optimization algorithms, e.g. Integer Gradient Method, must be applied for finding solutions [6,7]. Besides that, for the layout of supporting structures fulfilling several objectives, the degree of reliability of single specified values (objective functions and constraints) will be of increasing interest as a development goal. In the process of fabrication as well as in operation, the constructions are subjected to increasingly stochastic influences (e.g., fabrication tolerances, material characteristics, load characteristics). Hence, in order to receive information about the sensitivity of constructions, it will prove useful to consider the random distribution of objective and constraint functions already in the optimization calculations. Some algorithms are described in [8,9].

2 The "Three Columns" of Structural Optimization Techniques

For all tasks of structural optimization, it is equally essential to note that the application of optimization theories in the design process depends upon the theoretical aspects of the technical problems. When dealing with the structural optimization problem, it is recommendable to proceed according to the "Three Columns Concept," consisting of structural model, optimization algorithms, and optimization model [2,10-12]. Any structural optimization requires the mathematical determination of the physical behavior of the structure. In the case of mechanical structures this refers to the typical structural response subject to static and dynamic loadings such as deformations, stresses, eigenvalues, etc. Furthermore, information on the stability behavior (buckling loads) has to be determined. All state variables required for the objective functions and constraints have to be provided. The computation is carried out using efficient structural analysis procedures such as the finite element method or transfer matrices methods. In order to ensure a wide field of application, it should be possible to adapt several other structural analysis methods, too.

Structural optimization calls for efficient and robust algorithms mathematical programming methods, optimality criteria methods, or hybrid methods. In recent years, mathematical programming algorithms have been introduced more and more for solving nonlinear constrained optimization problems. These algorithms are iterative procedures which, proceeding from an initial design $\mathbf{x}_0$, generally provide an improved design variable vector $\mathbf{x}_k$ as a result of each iteration k. The optimization is terminated if a breaking-off criterion responds during an iteration. Numerous studies have demonstrated that the selection of the optimization algorithm is problem-dependent. This is particularly important for a reliable optimization and a high level of efficiency (computing time, rate of convergence). If all iteration results have to lie within the feasible domain, the Generalized Reduced Gradients (GRG)-Method could be applied [13].

According to the "Three Columns Concept" both the structural model and the optimization algorithms are linked in an optimization loop (Figure 1) via the single models of the **optimization model**. From an engineer's point of view, this column is the most important one of the optimization procedure. First of all, the analysis variables which are to be changed during the optimization process are selected from the structural parameters. The design model including variable linking, variable fixing, shape functions, etc., provides a mathematical link between the analysis and the design variables. In order to increase efficiency and to improve the convergence of the optimization, the optimization problem is adapted to meet the special requirements of the optimization algorithm by transforming the design variables into transformation variables. By using this approach, it is possible to almost linearize the stress constraints of a sizing optimization problem. Additionally, objective functions and constraints have to be determined by procedures that evaluate the structural response or state variables. When formulating the optimization model, the engineer has to regard the demands from the fields of design, material, manufacturing, assembly and operation [12].

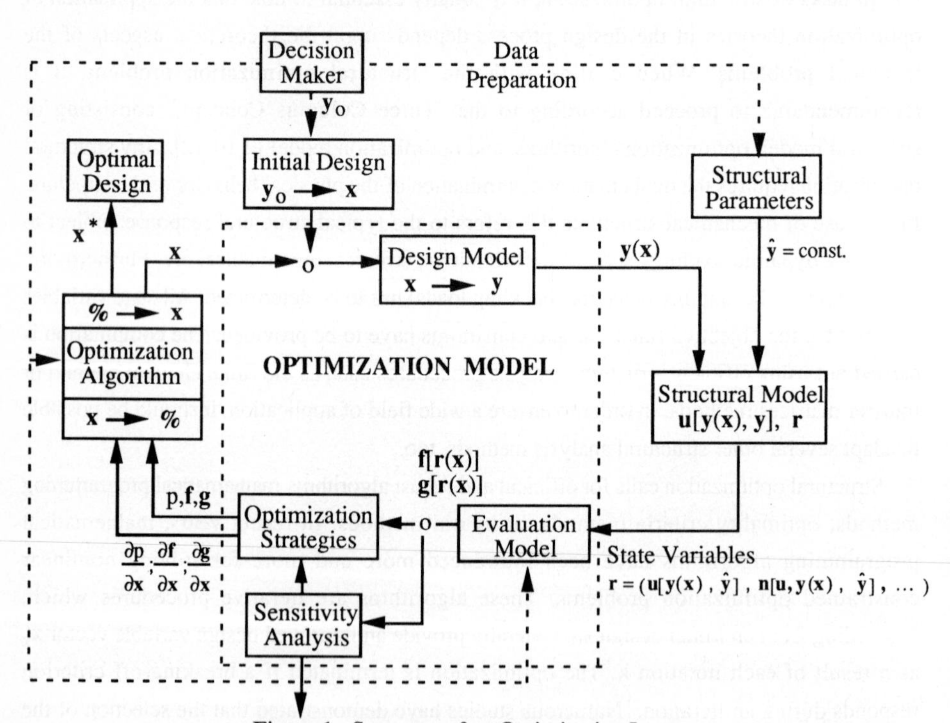

Figure 1. Structure of an Optimization Loop

3 Optimization Modeling

Special attention must be paid to single modules in the optimization modeling which are presented in Figure 2. Therefore, we are going to deal with several aspects of the following models.

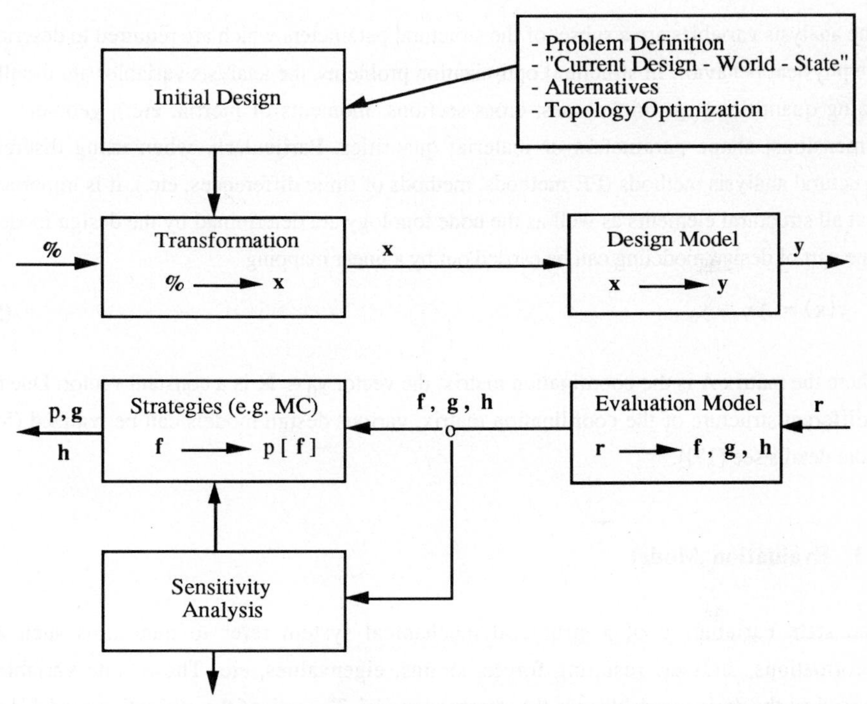

Figure 2. Modules of the Optimization Model

3.1 Initial Design

In order to find the best-possible initial design, a new concept is introduced which incorporates the determination of the general layout or topology of a component into the optimization process. Thus, methods of formal structural optimization should already be applied in the project stage. A methodology, consisting of three phases, which uses the homogenization method [14] as one tool for finding the topology is at several places in progress. This methodology links the homogenization method with a versatile structural optimization procedure via image processing method and geometrical modeling techniques [15]. A further way is the application of expert systems for the synthesis of mechanical and structural components by emulating the design process in design and re-design phases [16].

3.2 Design Model

The task of a design model is to calculate the structural analysis variables **y** from the design variables **x** by using a unique mapping rule:

$$\mathbf{y} = \mathbf{f}(\mathbf{x}) \, , \, \mathrm{x} \in \mathbb{R} \, , \mathbf{y} \in \mathbb{R} \; . \tag{4}$$

The analysis variables are a subset of the structural parameters which are required to describe the physical behavior. In structural optimization problems, the analysis variables are usually sizing quantities (e.g. thicknesses, cross-sections, moments of inertia, etc.), geometrical dimensions, shape parameters or material quantities. Particularly when using discrete structural analysis methods (FE-methods, methods of finite differences, etc.), it is important that all structural elements as well as the node topology are determined by the design model. One part of design modeling can be carried out by a linear mapping

$$\mathbf{y}(\mathbf{x}) = \mathbf{A}\mathbf{x} + \mathbf{y}_0 \, , \tag{5}$$

where the matrix **A** is the coordination matrix, the vector $\mathrm{y}_0 \in \mathbb{R}$ is a constant vector. Due to a different structure of the coordination matrix, various design models can be realized (for more details see [17]).

3.3 Evaluation Model

The state variables **r** of a structural mechanical system refer to quantities such as deformations, stresses, resulting forces, strains, eigenvalues, etc. These state variables depend on the design variables via the structural model. The task of the evaluation model is to formulate the objective function vector **f** and the constraints **g** as functions of these state variables.

As mentioned earlier in (2), the majority of papers published on the examples of structural optimization deal with the minimization of the structural weight. However, for many applications other objectives can be more important than for example the minimization of deformations of highly accurate systems [4], the minimization of stresses in order to reduce stress concentrations, or the maximization of certain eigenvalues of a structure. For this reason, different types of constraints can also be used as an objective. Due to the nonlinearity of these objective functions, they are more difficult to treat than the structural weight.

3.4 Sensitivity Analysis

Gradients of objective function(s) and constraints are determined by the sensitivity analysis with regard to the design variables (see Figure 2). These gradients are necessary for optimization algorithms of first and higher order (determination of search directions, approximation models). Introducing the state variables $\mathbf{r} \in \mathbb{R}^{nj}$ the objective functions and constraints read as follows:

$$\mathbf{f} = \mathbf{f}(\mathbf{r},\mathbf{x}) \text{ resp. } \mathbf{g} = \mathbf{g}(\mathbf{r},\mathbf{x}) \text{ with } \mathbf{r} = \mathbf{r}(\mathbf{x}) = \mathbf{r}[\mathbf{y}(\mathbf{x}), \hat{\mathbf{y}}] \,. \tag{6}$$

The total differentiation yields:

$$\mathbf{df}\,(\mathbf{r},\mathbf{x}) = \frac{\partial \mathbf{f}}{\partial \mathbf{r}}\,\mathbf{dr} + \frac{\partial \mathbf{f}}{\partial \mathbf{x}}\,\mathbf{dx} \text{ with } \frac{\partial \mathbf{f}}{\partial \mathbf{r}} \equiv \left[\frac{\partial f_i}{\partial r_j}\right]_{mxn} . \tag{7}$$

By eliminating $\mathbf{dr}$ it is possible to determine the sensitivity matrix $\mathbf{A_f}$ whose i-th line corresponds to the derivatives $\partial\mathbf{f}/\partial\mathbf{x}$:

$$\mathbf{df} = \mathbf{A_f}\,\mathbf{dx}. \tag{8}$$

Sensitivity information is additionally needed for various purposes like:

- getting more insight into the structural behavior (system identification),
- establishing a design model to be as small as possible by choosing natural design variables,
- sensitivity of the optimal design relating to non-optimized parameters,
- testing of decomposition possibilities.

The derivatives are usually calculated by the following procedures:

- Numerical method by means of finite differences,
- Analytical method,
- Semianalytical method (especially for FE-structural analysis) [2].

In recent years different authors have published numerous works on sensitivity methods (among others [18-20]). Many studies have demonstrated that the selection of the optimization algorithms has to ensue depending on the problem. This is particularly important for a reliable optimization and a high level of efficiency (computing times, rate of convergence etc.).

4 Optimization Strategies

4.1 Multicriteria Optimization

The solution of nonlinear multicriteria optimization problems can be obtained in different ways; one is the transformation by substitutions into scalar optimization problems. Thus, it is possible to determine a compromise solution $\tilde{\mathbf{x}}$ out of the complete solution set $\mathbf{X}^*$, where $\mathbf{X}^*$ is the set of all $\mathbf{x}^*$ [2, 10, 11].

Def. 3: Substitute Problem

$$\min_{x \in X} \; p[\mathbf{f}(\mathbf{x})] \tag{9}$$

is called a substitute problem for a vector optimization problem if there exists an $\tilde{x} \in X^*$ with the characteristic

$$p\,[\mathbf{f}(\tilde{\mathbf{x}})] = \min_{x \in X} \; p[\mathbf{f}(\mathbf{x})] \tag{10}$$

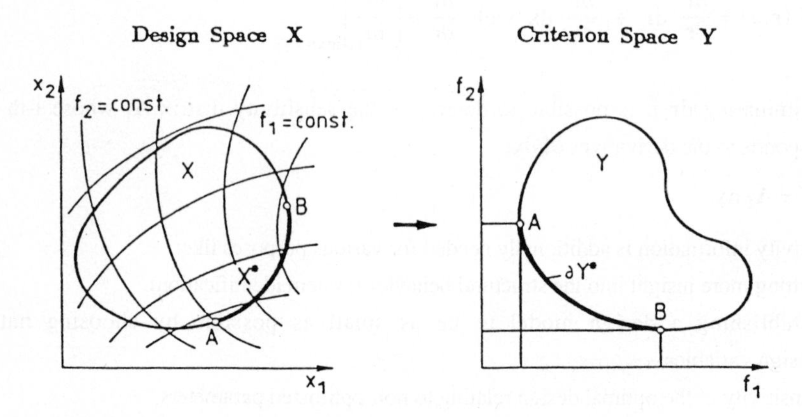

Figure 3. Mapping of a Feasible Design Space into the Criterion Space

The function p is called **preference function** or substitute objective function or quality criterion (the last term being used mainly in control engineering). It is important to prove that the solutions $\tilde{\mathbf{x}}$ of all substitute problems are functional-efficient with regard to X and to the set of objective functions $f_1,\ldots,f_m$; i.e. a point $\mathbf{y} = \mathbf{f}(\mathbf{x})$ really lies on the efficient boundary ∂y^*. At present, there are a number of rules transforming vector optimization problems into substitute problems. Within the scope of the optimization procedure they belong to the "strategies". The problem dependency of different methods can be highly relevant. The following methods can be applied:

- Method of Objective Weighting

$$p_w = \sum_j w_j \frac{f_i(x)}{f_{j0}}, \; f_{j0} = f_j\,(x_0), \; \sum_j w_j = 1, \; w_j \geq 0\,. \tag{11}$$

- Method of Distance Functions

$$p_d^r\,[\mathbf{f}(\mathbf{x}),\bar{\mathbf{f}}] := \left(\sum_{k=1} |\,\Delta\,\mathbf{f}_j\,(\mathbf{x})\,|^r\right)^{1/r}, \; 1 \le r < \infty \; ; \tag{12}$$

$$r = 1 \quad p_d^1 \;=\; \sum_{j=1}^{m} |\,\mathbf{f}_j(\mathbf{x}) - \bar{\mathbf{f}}_j\,| \;, \tag{13}$$

$$r = 2 \quad p_d^2 \;=\; \sqrt{\sum_{j=1}^{m} |\,\mathbf{f}_j(\mathbf{x}) - \bar{\mathbf{f}}_j\,|^2} \qquad \text{Euclidian metric ,} \tag{14}$$

$$r \;\; \infty \quad p_d^\infty \;=\; \max_{k=1,\dots,m} \left\{ |\,\mathbf{f}_j(\mathbf{x}) - \bar{\mathbf{f}}_j\,| \right\} \qquad \begin{array}{l}\text{Chebyshev's metric}\\ \text{Min–max–formulation .}\end{array} \tag{15}$$

- Method of Constraint Oriented Transformations (Trade-Off Method)

$$p = \mathbf{f}_1,\; \mathbf{g}_j \le \bar{\mathbf{f}}_j \quad (j = 2,\dots,m) \;.$$

More details of the single methods are described in [2].

The efficiency of MCD problems can be enhanced by interactive methods. A distinction can be made with respect to the kind of and the stage at which preference information is required from the Decision Maker (DM). As a further distinctive feature the particular organization scheme of an approach can be considered as it prescribes the kind of scalar substitute problems which have been solved during the interactive optimization process.

For the real design process interactive procedures have been mentioned or applied in only a few cases [6, 21, 22]. This, first of all, results from the nonlinearity of the problems, and furthermore, it is due to the fact that the structural analyses such as finite element methods must be carried out numerically which is a time-consuming process. In particular, such problems are considered by Diaz who presents an effective sensitivity analysis to variations in the DM's preferences based on Sequential Quadratic Programming [21].

4.2 Shape Optimization

The definition of a certain design model includes the definition of the required approach functions, their discretization rule, their free function parameters and the assignment of design variables to these function parameters. Thus, one can determine an updated design by means of updated design variables and invariable approach function definitions. Besides this it is possible to define recursive calculations in the geometry model. This option is useful to couple different approach functions or to move control points along arbitrary directions (not only along coordinate directions) [23].

In the following, three typical approach functions of the more complex geometry module are described more precisely. The description of the meridional shape can be carried out by parametric or non-parametric curves. In the case of a parametric description, the independent variables are parameters whereas in the non-parametric description one component of the position vector is independent. The non-parametric formulation is characterized by a major disadvantage in comparison with the parametric formulation: Curves or areas with tangents directed normally to an independent coordinate direction cannot be generated. The parametric formulation, on the other hand, allows this presentation. For this reason, all functions used in the described procedure are formulated parametrically wherever this is possible. The following curves can be used for modeling a meridional shape:

a) Straight line between two points

The simplest shape function implemented is a straight line between two points. The length $\alpha = \xi = s/s_g$ normalized to 1 is taken as a parameter. The shape function describes nothing but the linear interpolation of nodal coordinates between two supporting nodes (Figure 4).

The interpolation can mathematically be formulated as follows:

$$r(\alpha) = \mathbf{r}_1 + \alpha(\mathbf{r}_n - \mathbf{r}_1)\,; \quad 0 \le \alpha \le 1\,. \tag{16}$$

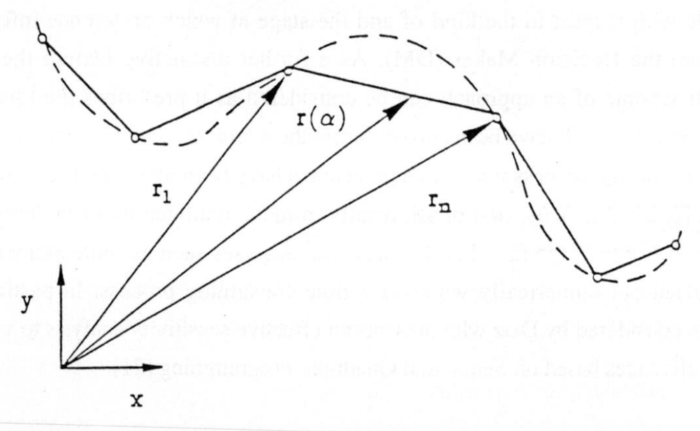

Figure 4. Curve Approximation by Means of Pieces of Straight Lines

Here, the position vector $\mathbf{r}_1$ is the origin and $\mathbf{r}_n$ the end point of the straight line. On the line, points with equidistant or linearly increasing or decreasing, nodal distances can be generated. In the case of this shape function, the coordinates of the origin and the end point can be taken as design variables.

b) B-Spline Curves

The nonlinear, parametric shape function of this type is defined by n + 1 control points which form a control polygon. With the exception of the first and the last point of the polygon, the control points are not placed on the B-spline curve. A B-spline curve $\mathbf{r}(\xi^1)$ with the n + 1 control points $\mathbf{p}_i$ and ξ^1 as independent parameters is defined by

$$\mathbf{r}(\xi^1) = \sum_{i=0}^{n} \mathbf{p}_i \, B_{ik}(\xi^1) \; . \tag{17}$$

The blend functions $B_{ik}(\xi^1)$ of the B-spline curve, the so-called B-spline basic functions, are functions of polynomial parameters of degree k – 1. They can be calculated by means of the following recursive formula:

$$B_{il}(\xi^1) = \begin{cases} 1 \text{ for } t_i \le \xi^1 \le t_{i+1} \, , \\ 0 \text{ otherwise} \; , \end{cases}$$

$$B_{il}(\xi^1) = \frac{(\xi^1 - t_i)}{t_{i+k-1} - t_i} B_{ik-1}(\xi^1) + \frac{(t_{i+k} - \xi^1)}{t_{i+k} - t_{i+1}} B_{i+1\,k-1}(\xi^1) \, , \tag{18}$$

where the definition 0/0 = 0 is assumed for the blend functions. The knots t_j allocate parameters ξ^1 to the control points by means of which the shape of the curve can be influenced. For uniform, non-periodic B-splines with n + 1 control points they are calculated as follows:

$$t_j = \begin{cases} 0 & \text{if} \quad j < k, \\ i-k+1 & \text{if} \quad k \le j \le n, \\ n-k+2 & \text{if} \quad j > n \; . \end{cases} \tag{19}$$

This yields the parametric range for ξ^1 as $0 \le \xi^1 \le n - k + 2$. Figure 5 shows B-spline curves with 9 control points (n = 8) for different degrees (k – 1) of the basic function.

c) Modified Ellipse Curves

A further shape function which is appropriate for special problems is shown in Figure 6. It involves an ellipse function with variable exponents. This can be expressed mathematically as:

$$\left(\frac{x}{a}\right)^{\kappa_1} + \left(\frac{y}{b}\right)^{\kappa_2} = 1 \; , \tag{20}$$

where κ_1, κ_2 are shape parameters and a, b semi-axes.

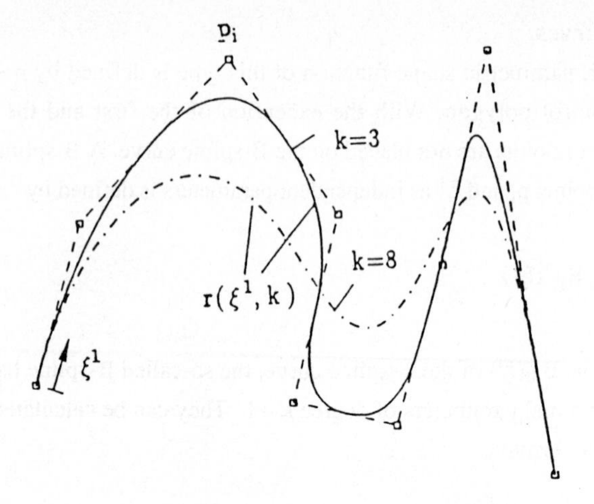

Figure 5. B-Spline Curve with 9 Control Points for Different Degrees (k-l)

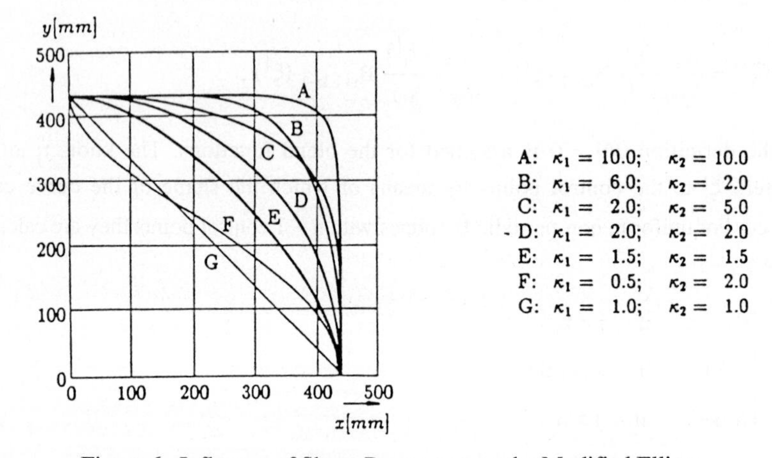

Figure 6. Influence of Shape Parameters on the Modified Ellipse

The use of this modified ellipse has the advantage that meridional shapes always exist for $\kappa_1 > 1$ and $\kappa_2 > 2$ which satisfies the demands made with regard to the curve trend and the tangent position. It is possible to observe the volume constraint for different combinations of the shape parameters κ_1 and κ_2. This equation can be solved for x and y. The modified ellipse equation in a parametric form reads.

$$x = a(\sin\varphi)^{2/\kappa_1}, \quad y = b(\cos\varphi)^{2/\kappa_2}, \tag{21}$$

where φ is the parameter [24].

5 Description of Investigated Problems

Details of the optimization process for the following examples are described in [31].

5.1 Composite Fin

An aircraft fin has to fulfill quite different design requirements with a similar priority. The design of aerodynamic surfaces needs two major design steps, firstly, an aerodynamic design to define the overall geometry like area, span, aspect ratio, taper ratio and profile, and secondly, the structural design to develop the internal structural arrangement of skin, ribs, stringers, spars, rudder support and actuators, to fulfill the following design requirements with minimum weight:

- Static strength due to design loads,
- Aeroelastic efficiencies for performance,
- No flutter inside the flight envelope,
- Manufacturing constraints.

The fin can be subdivided into the main parts stabilizer, tip and rudder (Figure 7). The cover skins of the fin are made of carbon fibre laminate with four different fibre orientations in the stabilizer and three in the rudder. The inner supporting structure is realized by an aluminium honeycomb core. The fin is supported at the connection points to the fuselage and the stiffness of the fuselage is modeled with a general stiffness element. The aerodynamic forces of five different flight conditions are chosen as static load cases. A detailed description of the aerodynamic model and the finite element model is presented in [26, 27].

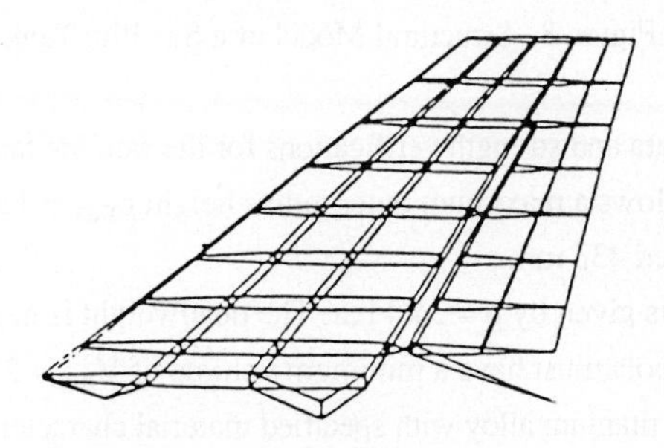

Figure 7. Fin Structural Model

The layer thicknesses and layer angles of a fin have been optimized considering constraints on failure safety, aeroelastic efficiencies and flutter speeds. The objective of this examination was to demonstrate the efficiency of the optimization procedure and also the potential of possible weight savings.

5.2 Satellite Tank [24]

The following example illustrates the requirements for the optimization of a thin-walled, sealed satellite tank subject to constant internal pressure (Figure 8). The main task is to optimize the tank which requires the following demands:

- adherence to the specified design space,
- the weight should be as low as possible,
- the tank must hold a maximum volume.

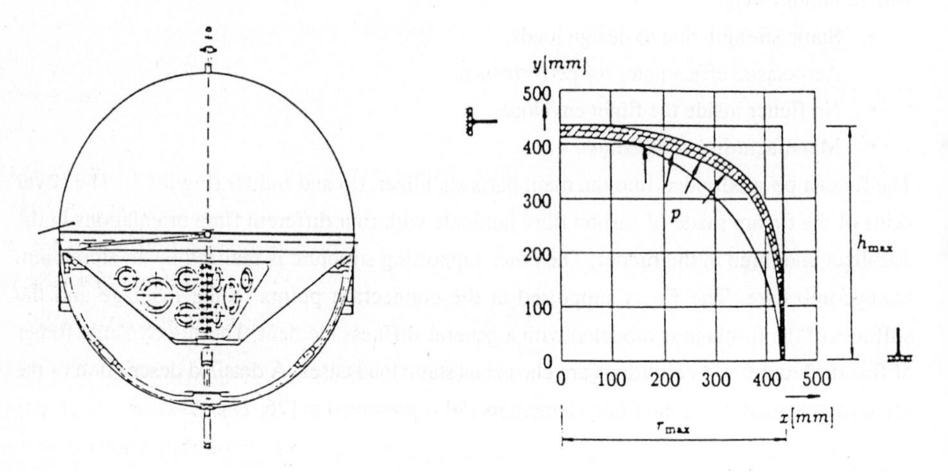

Figure 8. Structural Model of a Satellite Tank

The following design data and strength verifications for the satellite tank are given:

a) The design space allows a maximum outer radius height r_{max} = 436.9 mm; the tank height h_{max} must not exceed 433 mm.

b) The inner pressure is given by p = 34,4 bar. The deadweight is neglected.

c) The half-tank or cupola must have a minimum volume of V_{min}= 215.2 liter.

d) The tank is made of titanium alloy with specified material characteristics:

- density ρ = 4.5g/cm^3,
- Young's modulus E = 1.1 x 10^5 N/mm^2,
- Poisson's ratio ν = 0.3,
- ultimate stress σ_{ult} = 1080N/mm^2.

e) The strength verification is performed depending on the sign of the principal stresses (in meridional and circumferential direction) and according to the following reference-stress hypotheses of the plane stress state.

Case 1. Principal stresses with the same sign imply that the reference stress is calculated on the basis of the maximum stress

$$\max(\sigma_1, \sigma_2) \leq \sigma_{ult} \text{ (for } \sigma_1 \sigma_2 > 0),$$

where σ_1, σ_2 are the principal stresses and σ_{ult} the permissible ultimate strength.

Case 2. Principal stresses with different signs imply that the reference-stress hypothesis according to von Mises should be used:

$$\sqrt{\sigma_1^2 + \sigma_2^2 - \sigma_1 \sigma_2} \quad \leq \sigma_{ult} \quad \text{(for } \sigma_1 \sigma_2 > 0) \, .$$

Functional-efficient solutions of the satellite tank design shall be determined for the objectives "weight minimization" and "volume maximization". As a shape function, we exclusively use a modified ellipsoid with parabola approach in the pole area.

5.3 Offshore Platform [28]

Figure 9 shows the sketch of a semisubmersible which is chosen as the design example for an offshore platform. It consists of a platform, two submerged cylindrical hulls, and several cylindrical columns that connect with them. It has symmetries with respect to its longitudinal and transverse axes. The basic configuration of the semisubmersible is specified by the hull diameter D, the hull length L, the interval between hulls 2b, the submerged depth of a column h, the maximum number of columns per hull (2n + 1), the i-th column diameter d_i (i = 0, 1,..., n), and the distance between the center of the hull and the i-th column l_i (i= 1, 2, ..., n).

In addition to the heave response in irregular waves, it becomes necessary to take account of various design conditions such as building cost, dynamical and statical stabilities, loading weight, platform size, strength, and so forth at the practical design of a semisubmersible. Therefore, relationships among these design conditions become very complex, and it becomes very difficult to formulate the optimization problem and to determine many design variables optimally.

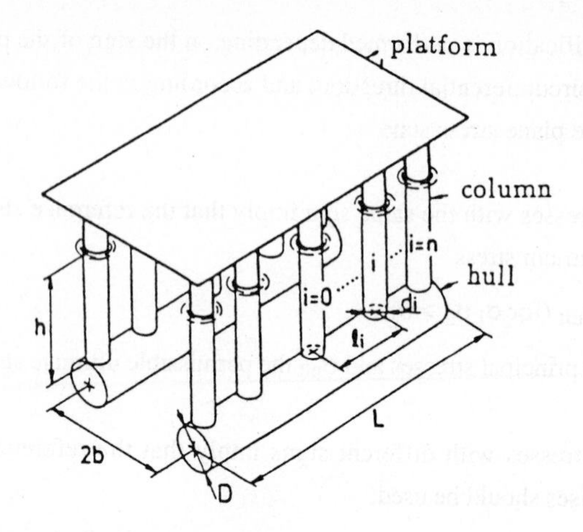

Figure 9. Sketch of a Semisubmersible of Lower-Hull Type

6 Conclusion

In the aerospace and offshore industry, the methods of structural optimization play a more and more important role in the process of engineering design. It is especially for complex design problems that their application leads to better, most possibly optimal layouts which fulfill all requirements in the best possible way. Nowadays, mathematical optimization algorithms and efficient structural analysis packages like finite element methods establish the basis for optimization computations with a high rate of generality and efficiency. The additional inclusion of optimization models does not only lead to a very modular architecture but also to the direct consideration of all relevant practical demands. The problem definitions of some examples of aircraft, spacecraft and naval structures are described.

For such complex structures, the design of high performance structures involves large scale optimization problems with objectives and constraints from different disciplines. These objectives and constraints have to be fulfilled simultaneously, e.g. statics, buckling, aeroelastics and dynamics. In order to treat such problems, a number of optimization procedures have been developed worldwide in the 80s. It has already led to considerable improvements in solving many design problems. Some program architectures like SAPOP [17], LAGRANGE [25], ENGINEOUS [29], and ASTROS [30] are very similarly organized.

The structural weight has often been the most essential and the only objective function for the optimization. But today, more and more problems occur where other objectives have to be optimized with the same priority as well. For the design of aircrafts, the aerodynamic and aeroelastic features are required in addition to the structural weight. Different objectives have to be fulfilled for spacecraft structures during the launching phase and during the operation in orbit. Here, the multicriteria optimization methods are suitable for determining unique and optimal compromise solutions [2].

References

1. Eschenauer, H.A.: Structural Optimization - a Need in the Design Process. In H. Eschenauer; C. Matthek; N. Olhoff: Engineering Optimization. In: Design Processes, pp. 1-13. Berlin-Heidelberg-New York: Springer-Verlag 1991
2. Eschenauer, H.A.; Koski J.; Osyczka, A.: Multicriteria Design Optimization. Berlin-Heidelberg-New York: Springer-Verlag 1990
3. Bergmann, H.W.: Optimization. Berlin-Heidelberg-London: Springer-Verlag 1989
4. Eschenauer, H.A.: Multicriteria Optimization Techniques for Highly Accurate Focusing Systems. In: Multicriteria Optimization in Engineering and in the Sciences (W. Stadler, Ed.), pp. 309-354. Plenum Publishing Corporation 1988
5. Tsai S.-C.; Ragsdell, K.M.: Orthogonal Arrays and Conjugate Directions for Taguchi-Class Optimization. In: Rao, S.S.: Advances in Design Automation - 1988. New York: ASME, 273-278 (1988)
6. Schäfer, E.: Interaktive Strategien zur Bauteiloptimierung bei mehrfacher Zielsetzung und Diskretheitsforderungen. Dissertation, Universitat GH Siegen 1990
7. Gutkowski, W.: Controlled Enumeration with Constraint Variations in Discrete Structural Optimization. Lecture, Minisymposium Structural Optimization, GAMM 1991, Cracow
8. Marti, K.: Stochastic Optimization in Structural Mechanics. Federal Armed Forces University, Munich. Lecture, Miniymposium Structural Optimization, GAMM 1991, Cracow
9. Eschenauer, H.A.; Vietor, T.: Some Aspects on Structural Optimization of Ceramic Structures. In: Eschenauer, H.; Mattheck, C.; Olhoff, N.: Engineering Optimization in Design Process, pp. 145-154. Berlin-Heidelberg-New York: Springer-Verlag 1991.
10. Eschenauer, H.A.: Numerical and Experimental Investigations on Structural Optimization of Engineering Designs. DFG-Research Report, Universitat GH Siegen 1982
11. Sattler, H.-J. (Ed.): Ersatzprobleme fur Vektoroptimierungsaufgaben und ihre Anwendung in der Strukturmechanik. Dissertation, Universitat GH Siegen 1982
12. Eschenauer, H.A.: The "Three Columns" for Treating Problems in Optimum Structural Design. In: H.W. Bergmann: Optimization, pp. 1-21. Berlin-Heidelberg-New York: Springer-Verlag 1989
13. Bremicker, M.: Dekompositionsstrategie in Anwendung auf Probleme der Gestaltsoptimierung. Dissertation, Universitat GH Siegen 1989
14. Bendsøe, M.; Kikuchi, N.: Generating Optimal Topologies in Structural Design Using a Homogenization Method. Computer Methods Applied in Mechanics and Engineering, pp. 197-224 (1988)
15. Bremicker, M.: Ein Konzept zur integrierten Topologie- und Gestaltsoptimierung von Bauteilen. In: Beiträge zur Maschinentechnik, Festschrift Prof. Dr. H. Eschenauer (Müller-Slany, Ed.), pp. 13-39, 1990
16. Hart, P.K.; Rodriguez, J.: A Dual Purpose KBES for Preliminary Structural Design. In: Ravani, B.: Advances in Design Automation - 1989, Vol. 2. Design Automation, pp. 209-216. ASME New York 1989
17. Eschenauer, H.A.; Post, P.U.; Bremicker, M.: Einsatz der Optimierungsprozedur SAPOP zur Auslegung von Bauteilkomponenten. Bauingenieur 63, 515-526 (1988)
18. Haug, E.J.; Arora, J.S.: Design Sensitivity Analysis of Elastic Mechanical Systems. Computer Methods in Applied Mechanics and Engineering 15, 35-62 (1978)
19. Haug, E.J.; Choi, K.K.: Design Sensitivity Analysis of Structural Systems. Orlando-San Diego-New York: Academic Press, Inc. 1986
20. Adelmann, H.M.; Haftka, RT.: Sensitivity Analysis for Discrete Structural Systems A Survey. NASA TM 85333, 1984
21. Diaz, A.: Interactive Solution to Multiobjective Optimization Problems. Inst. f. Num. Methods in Eng. 24, 1865-1877 (1987)
22. Arora, J.S.: Interactive Design Optimization of Structural Systems. In: Eschenauer, H.; Thierauf, G.: Discretization Methods and Structural Optimization Procedures and Application, pp. 10-16. Springer-Verlag 1989

23. Eschenauer, H.; Weinert, M.: Structural Optimization Techniques as a Mathematical Tool for Finding Optimal Shapes of Complex Shell Structures. In: Nonsmooth Optimization - Methods and Applications. by F. Giannessi. Proceedings of an International Conference, Erice, Sicily/Italy, June 19 - July 1, 1991
24. Eschenauer, H.A.: Shape Optimization of Ultra Light Shell Structures in Space Technology. J. Structural Optimization, No. 1, 171-180 (1989)
25. Kneppe, G.: MBB-LAGRANGE: Structural Optimization System for Space and Aircraft Structures. Paper presented at the COMETT-Seminar, Computer Aided Optimal Design. Bayreuth: 1990
26. Eschenauer, H.A.; Schuhmacher, G.; Hartzheim, W.: Optimization of Aircraft FRCP-Structures with Optimization Procedure LAGRANGE. Proceedings of the Computational Structures Technology Conference. Edingburgh: 1991 (to appear 1992).
27. Heinze, P.; Schierenbeck, D.; Hartmann, D.: Structural Optimization of CFRP-Components in Transport Aircraft Design, 4th Conf. of Composite Research in Solid Mechanics. Stuttgart: 1991
28. Akagi, S.; Yokoyama, R.; Ito, K.: Optimal Design of Semisubmersible's Form Based on System Analysis. ASME Paper No. 84-DET-87.
29. Ashley, St.: Engineous Explores the Design Space. Mech. Engineering, Vol. 114/No.2, 49-52 (1992)
30. Venkayya, V.B.: Large Scale Optimization Using New Software Systems like ASTROS. Proceedings of NATO/DFG Advanced Study Institute. Berchtesgaden: l991(to appear 1992)
31. Eschenauer, H.; Mahrenholtz, 0.: Aerospace and Naval Structures. In: Save, M.; Prager, W.: Structural Optimization (Cohn, M.Z.; Frangopol, D.M., Eds.). Plenum Press. New York, London, Vol. 3, Ch. 14 (to appear 1992)

Teleoperation of a Redundant Manipulator

K. Harold Yae

Center for Simulation and Design Optimization of Mechanical Systems, Department of Mechanical Engineering, The University of Iowa, Iowa City, Iowa 52242 USA

Abstract: For an interactive real-time simulation of teleoperation, this paper describes an iterative form of resolved motion rate control in which the constraint Jacobian is constructed on-line in real time and is used in the pseudoinverse method, as the manipulator is teleoperated. The operator's command is interpreted as a series of increments in Cartesian space, and then the constraint Jacobian is developed between two successive increments by viewing the predecessor as the initial configuration and the successor as the target configuration. The Jacobian constructed in this way enables us to treat both free motion and environmental contact in the same way. Although this method requires numerical iterations, its convergence is fast enough to allow real-time control.

Keywords: teleoperation / simulation / recursive dynamics / redundant manipulator / constraint Jacobian

1 Introduction

Teleoperation, one of the challenging domains of robotics [1], involves a human operator, a hand controller, and a manipulator. In typical teleoperation, the manipulator completes, under a human operator's supervision, tasks ranging from simple trajectory following to pick-and-put operation. Consequently, the manipulator intermittently closes the kinematic chain and such closures cause problems in inverse kinematic analysis. The operator's interactions with the manipulator add another problem that is caused by the intrinsic difference between a human operator and a robotic manipulator; that is, a human operator works more efficiently in Cartesian space [2] whereas most manipulators are designed with joint servo control. These problems become even more complicated with a redundant manipulator. In general, teleoperation requires real-time Cartesian space control [3, 4] in the presence of kinematic redundancy and intermittent kinematic loop closure.

In this research the human operator "commands" the end-effector in Cartesian space. The inverse position/orientation problem is solved iteratively using inverse velocity/angular velocity solutions [5]. The difference, however, lies in Jacobian construction. This paper proposes using a *constraint Jacobian* and its pseudoinverse in inverse kinematics analysis. Unlike the Jacobian defined when the end-effector's Cartesian positions are known beforehand, the constraint Jacobian can be constructed on-line as the operator maneuvers the manipulator's end-effector in Cartesian space. In teleoperation, the desired configuration of the end-effector is known only after the operator has commanded through a hand controller, so both Jacobian construction and inverse kinematic analysis must be completed on-line in real time. The constraint Jacobian is derived from the six constraints [6] that are imposed on between the current and the desired end-effector's Cartesian position and orientation. The desired Cartesian position and orientation is viewed as the target configuration that the current position and orientation will eventually have to assume. Constraining the two configurations yields six constraints. This view is applicable to both free motion (unconstrained motion) and environmental contact (constrained motion). For both cases, the constraint Jacobian can be constructed in the same way, which enables a smooth transition between open- and closed-chain kinematics. This Jacobian and its pseudoinverse are used in iterations to yield the joint command angles necessary for the joint controllers. The procedure is illustrated with a positional hand controller, specifically the Kraft 6-dof mini-master, and a 7-dof redundant manipulator [7].

The actual algorithm iterates until the constraint violation reduces to a small quantity. Despite such numerical iterations, the proposed method proves useful to real-time control in teleoperation. Even though the pseudoinverse method has drawbacks as pointed out in [8], it is simple enough to be implemented in real-time control.

2 The Constraint Jacobian

For real-time dynamics, the equations of motion are written in the recursive formulation using the *spatial velocity vector* [9] (also called the *velocity state vector* [10]). This formulation readily produces the constraint Jacobian in spatial vector notation. For use in joint control, however, this Jacobian is then expressed in joint variables. This section introduces some of the key equations in recursive kinematics between two contiguous bodies, the derivation of the constraint Jacobians for position and for orientation separately, and the conversion of the Jacobian in spatial vector notation to that in joint variables.

2.1 Kinematics Between a Pair of Contiguous Bodies

Consistent with the notations in Figure 1, recursive kinematics is developed between a pair of bodies. Primed vectors are expressed relative to the body-fixed coordinate frame (x'-y'-z') whose origin is located at the centroid. Unprimed vectors are relative to the global frame (x-y-z). Subscripts i and j indicate bodies i and j, respectively. The spatial velocity between a pair of bodies [10] is written as

$$\hat{\mathbf{y}}_j = \hat{\mathbf{y}}_i + \mathbf{B}_{ij}\dot{\mathbf{q}}_{ij} \tag{1}$$

where the spatial velocity vector $\hat{\mathbf{y}}_j$ and the velocity transformation matrix $\mathbf{B}_{ij}$ are defined as

$$\hat{\mathbf{y}}_j \equiv \begin{bmatrix} \dot{\mathbf{r}}_j + \tilde{\mathbf{r}}_j \boldsymbol{\omega}_j \\ \boldsymbol{\omega}_j \end{bmatrix} \tag{2}$$

$$\mathbf{B}_{ij} \equiv \begin{bmatrix} \dfrac{\partial \mathbf{d}_{ij}}{\partial \mathbf{q}_{ij}} + (\tilde{\mathbf{r}}_j + \tilde{\mathbf{s}}_j)\mathbf{H}_{ij} \\ \mathbf{H}_{ij} \end{bmatrix} \tag{3}$$

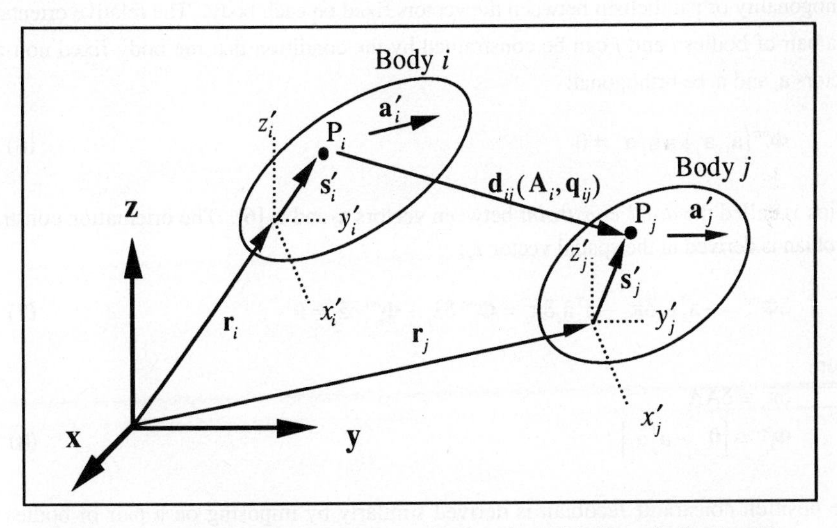

Figure 1. A pair of contiguous bodies: subscripts i and j indicate bodies i and j, respectively; P locates a joint; $\mathbf{r}$ is the vector to the centroid; $\mathbf{s}'$ defines the joint location relative to the body-fixed frame; $\mathbf{a}'$ is an arbitrary vector fixed on the body; $\mathbf{A}$ is the direction cosine matrix; $\mathbf{q}_{ij}$ is the relative coordinate between bodies i and j; and $\mathbf{d}_{ij}$ is the vector between points P_i and P_j, and is a function of $\mathbf{A}_i$ and $\mathbf{q}_{ij}$.

A matrix similar to $\mathbf{B}_{ij}$ is called the *partial velocity matrix for the end-effector* [5]. The transformation matrix $\mathbf{H}_{ij}$ [11, 12, 13] is a function of the orientation matrix $\mathbf{A}_i$ and the joint coordinate $\mathbf{q}_{ij}$. If the connecting joint is revolute, $\mathbf{q}_{ij}$ indicates the joint angle with $\mathbf{d}_{ij} = \mathbf{0}$

and the matrix $\mathbf{H}_{ij}$ becomes a unit vector ($\mathbf{h}_j$) along the axis of rotation. The velocity transformation matrix $\mathbf{B}_{ij}$, therefore, reduces to

$$\mathbf{B}_j = \begin{bmatrix} (\tilde{\mathbf{r}}_j + \tilde{\mathbf{s}}_j)\mathbf{h}_j \\ \mathbf{h}_j \end{bmatrix} \tag{4}$$

Subscripts i and j in $\mathbf{q}_{ij}$ and $\mathbf{B}_{ij}$ indicate relative quantities of the outboard body j with respect to its inboard body i (i.e., j-1). It should be understood that $\mathbf{B}_j$ and $\mathbf{q}_j$ mean $\mathbf{B}_{(j-1)j}$ and $\mathbf{q}_{(j-1)j}$, respectively.

2.2 Constraint Equations and the Constraint Jacobian

The bodies i and j in Figure 1 can also be viewed as being related by the kinematic constraints:

$$\boldsymbol{\Phi}(\mathbf{r}_i, \mathbf{A}_i, \mathbf{r}_j, \mathbf{A}_j) = \mathbf{0} \tag{5}$$

These constraints consist of the position and the orientation constraints that are specified by orthogonality or parallelism between the vectors fixed on each body. The relative orientation of a pair of bodies i and j can be constrained by the condition that the body-fixed non-zero vectors $\mathbf{a}_i$ and $\mathbf{a}_j$ be orthogonal:

$$\boldsymbol{\Phi}^{\text{ori}}(\mathbf{a}_i, \mathbf{a}_j) = \mathbf{a}_i^{\text{T}}\mathbf{a}_j = 0 \tag{6}$$

which is called the *dot-1 constraint* between vectors $\mathbf{a}_i$ and $\mathbf{a}_j$ [6]. The orientation constraint Jacobian is derived in the spatial vector $\hat{\mathbf{z}}_i$:

$$\delta\boldsymbol{\Phi}^{\text{ori}} = -\mathbf{a}_i^{\text{T}}\tilde{\mathbf{a}}_j\delta\boldsymbol{\pi}_j - \mathbf{a}_j^{\text{T}}\tilde{\mathbf{a}}_i\delta\boldsymbol{\pi}_i = \boldsymbol{\Phi}_{\hat{\mathbf{z}}_j}^{\text{ori}}\,\delta\hat{\mathbf{z}}_j + \boldsymbol{\Phi}_{\hat{\mathbf{z}}_i}^{\text{ori}}\,\delta\hat{\mathbf{z}}_i = \mathbf{0} \tag{7}$$

where

$$\begin{aligned} \delta\tilde{\boldsymbol{\pi}}_i &\equiv \delta\mathbf{A}\mathbf{A} \\ \boldsymbol{\Phi}_{\hat{\mathbf{z}}_i}^{\text{ori}} &\equiv \begin{bmatrix} \mathbf{0} & -\mathbf{a}_j^{\text{T}}\tilde{\mathbf{a}}_i \end{bmatrix} \end{aligned} \tag{8}$$

The position constraint Jacobian is derived similarly by imposing on a pair of bodies the condition that points P_i and P_j coincide; that is,

$$\mathbf{d}_{ij} = \boldsymbol{\Phi}^{\text{pos}}(\text{P}_i, \text{P}_j) = \mathbf{r}_j + \mathbf{s}_j - \mathbf{r}_i - \mathbf{s}_i = \mathbf{r}_j + \mathbf{A}_j\mathbf{s}_j' - \mathbf{r}_i - \mathbf{A}_i\mathbf{s}_i' = \mathbf{0} \tag{9}$$

This condition is called the *spherical joint constraint* [6]. The position constraint Jacobian is then found as follows:

$$\delta\boldsymbol{\Phi}^{\text{pos}} = \delta\mathbf{r}_j - \tilde{\mathbf{s}}_j\delta\boldsymbol{\pi}_j - \delta\mathbf{r}_i + \tilde{\mathbf{s}}_i\delta\boldsymbol{\pi}_i = \boldsymbol{\Phi}_{\hat{\mathbf{z}}_j}^{\text{pos}}\,\delta\hat{\mathbf{z}}_j + \boldsymbol{\Phi}_{\hat{\mathbf{z}}_i}^{\text{pos}}\,\delta\hat{\mathbf{z}}_i = \mathbf{0} \tag{10}$$

where

$$\Phi_{\hat{z}_i}^{pos} \equiv \begin{bmatrix} -\mathbf{I} & \tilde{\mathbf{r}}_i + \tilde{\mathbf{s}}_i \end{bmatrix} \tag{11}$$

The constraints of Equations (8) and (11) can be expressed in joint variables

$$\Phi_{\mathbf{q}_j} \equiv \Phi_{\hat{z}_j} \mathbf{B}_j \tag{12}$$

where the followings are used

$$\begin{aligned} \delta\Phi &= \Phi_{\hat{z}_j} \delta\hat{\mathbf{z}}_j + \Phi_{\hat{z}_i} \delta\hat{\mathbf{z}}_i = \Phi_{\hat{z}_j}\left(\delta\hat{\mathbf{z}}_i + \mathbf{B}_j \delta\mathbf{q}_j\right) + \Phi_{\hat{z}_i} \delta\hat{\mathbf{z}}_i \\ &= \underbrace{\left(\Phi_{\hat{z}_j} + \Phi_{\hat{z}_i}\right)}_{=0} \delta\hat{\mathbf{z}}_i + \underbrace{\Phi_{\hat{z}_j} \mathbf{B}_j}_{\Phi_{\mathbf{q}_j}} \delta\mathbf{q}_j = \mathbf{0} \end{aligned}$$

$$\delta\hat{\mathbf{z}}_j = \delta\hat{\mathbf{z}}_i + \mathbf{B}_j \delta\mathbf{q}_j$$

3 Cartesian Space Control

From the human operator's standpoint, teleoperation in Figure 2 is Cartesian space control, involving inverse kinematics, manipulator dynamics and control, and computer graphics. Teleoperation is initiated by the operator's command input through the hand controller. The operator's motion is divided into a series of increments in Cartesian space. These Cartesian increments ($\Delta\mathbf{x}$) are transformed into incremental angles ($\Delta\mathbf{q}$) by the Jacobian $\left(\Delta\mathbf{q} = \Phi_{\mathbf{q}}^{+}\Delta\mathbf{x}\right)$. The incremental joint angles are then input to the joint servo controllers, which in turn produce necessary joint torque. The resulting position and orientation is displayed on a graphics workstation to provide the operator with visual feedback.

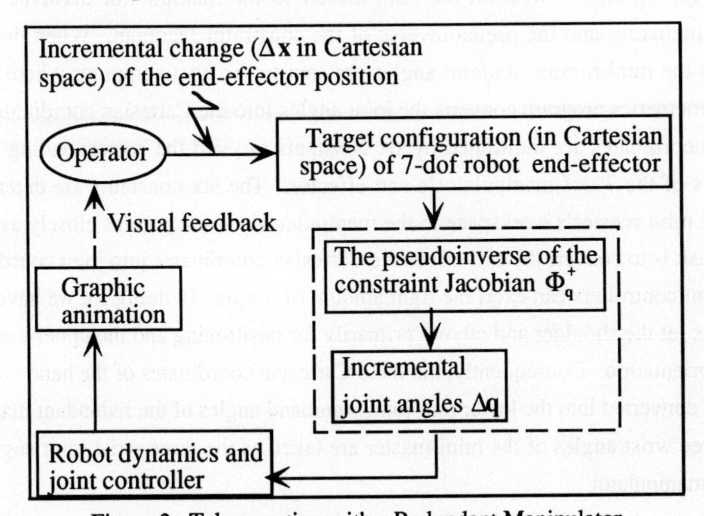

Figure 2. Teleoperation with a Redundant Manipulator

3.1 Kraft Mini-Master and a Position Mapping Algorithm

The manipulator model used here has seven degrees of freedom, shown in Figure 3. The hand controller is the Kraft mini-master (Figure 4), a six degree-of-freedom hand controller interfaced with the manipulator. The mini-master is kinematically similar to the human arm so that the human operator can use it comfortably in Cartesian space with reduced cognitive learning [14]. The manipulator, on the other hand, is joint-controlled. Therefore, the task is to relate six joint angles of the mini-master to seven joint angles of the manipulator in the way that the manipulator's end-effector follows the mini-master's grip.

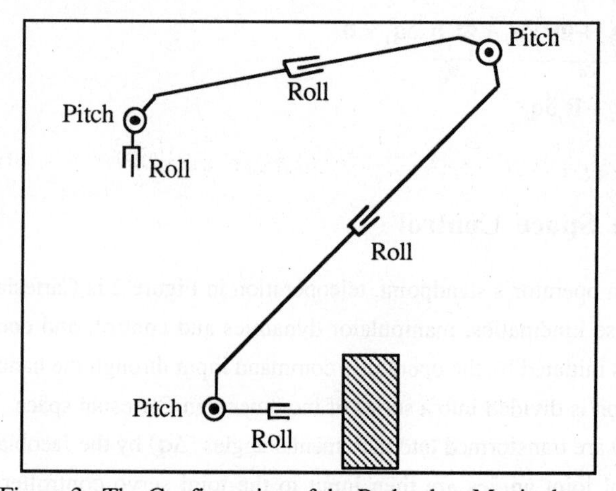

Figure 3. The Configuration of the Redundant Manipulator

The mapping algorithm from the mini-master to the manipulator involves the mini-master's kinematics and the pseudoinverse of the constraint Jacobian. When the operator maneuvers the mini-master, its joint angles are sent to the host computer where the mini-master's kinematics program converts the joint angles into the Cartesian coordinates. These Cartesian coordinates are multiplied by six constants to yield the corresponding Cartesian coordinates of the 7-dof manipulator's end-effector. The six constants are determined to expand the mini-master's workspace to the manipulator's workspace as closely as possible. The next task is to convert the end-effector's Cartesian coordinates into joint coordinates, so that the joint controllers can exert the right amount of torque. In doing so, we have used the lower joints (at the shoulder and elbow) primarily for positioning and the upper joints (at the wrist) for orientation. Consequently, the three Cartesian coordinates of the hand controller's gripper are converted into the lower four joint command angles of the redundant manipulator, and the three wrist angles of the mini-master are taken as the three wrist joint angles of the redundant manipulator.

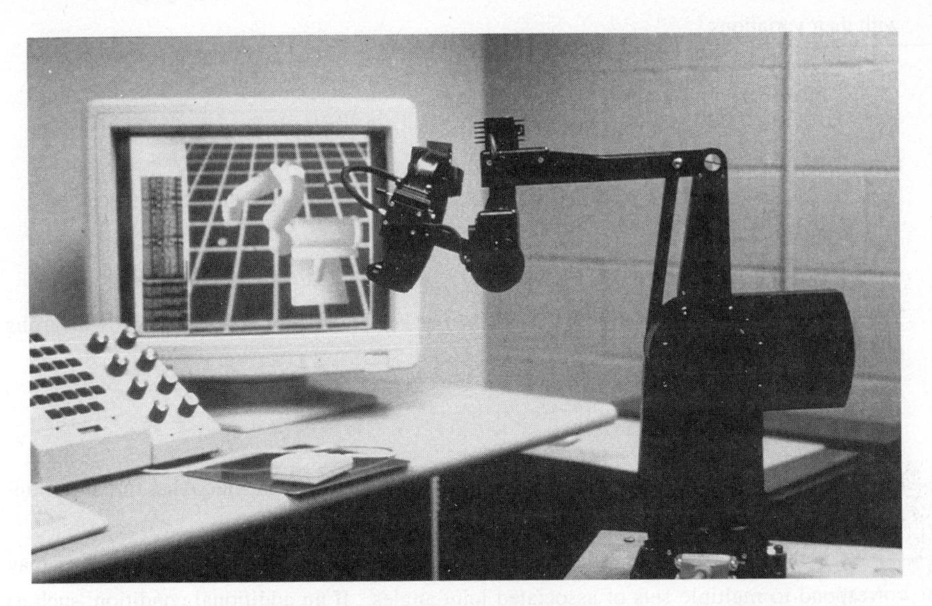

Figure 4. Kraft Mini-Master in the Foreground an the Simulated Manipulator in the Background on the Graphics Screen.

3.2 An Inverse Kinematics Algorithm Based on the Constraint Jacobian

A situation unique to teleoperation is that the target position and orientation is inputted on-line by the operator. Consequently, the control algorithm must first translate incremental Cartesian coordinates ($\Delta\mathbf{x}$) into incremental joint angle change ($\Delta\mathbf{q}$) [5], and then $\Delta\mathbf{q}$ into joint torque. This translation requires inverse kinematics analysis of a manipulator as described in the following.

If the manipulator is non-redundant, the end-effector Cartesian position and orientation has a unique kinematics relation with the joint angles

$$\mathbf{x} = \mathbf{F}(\mathbf{q}) \tag{13}$$

where $\mathbf{x}$ is a 6×1 vector; and $\mathbf{q}$ is also a 6×1 vector, representing the joint angles. For known Cartesian coordinates $\mathbf{x}$, Equation (13) is viewed as constraint equations, instead of an explicit expression for $\mathbf{q}$ being sought, i.e.,

$$\mathbf{\Phi} = \mathbf{x} - \mathbf{F}(\mathbf{q}) = \mathbf{0} \tag{14}$$

with their variations

$$\delta\mathbf{\Phi} = \delta\mathbf{x} - \frac{\partial \mathbf{F}}{\partial \mathbf{q}}\delta\mathbf{q} = \mathbf{0}$$

from which the constraint Jacobian is defined as

$$\mathbf{\Phi}_{\mathbf{q}} \equiv \frac{\partial \mathbf{F}}{\partial \mathbf{q}}$$

The incremental motion in joint space can then be solved, at least conceptually, by inverting the Jacobian,

$$d\mathbf{q} = \mathbf{\Phi}_{\mathbf{q}}^{-1} d\mathbf{x} \tag{15}$$

In fact, Equation (15) can be solved numerically as long as the Jacobian remains non-singular.

When the manipulator is redundant, a single Cartesian position of the end-effector may correspond to multiple sets of associated joint angles. If an additional condition, such as minimization of the Euclidean norm of angular velocity [5, 7, 15], is added, a unique set of joints angles can be identified. Such a method is called the pseudoinverse method, or Moore-Penrose generalized inverse method [16]. When the minimization is applied to the Euclidean norm of angular velocity, the pseudoinverse method defines $\mathbf{\Phi}_{\mathbf{q}}^{+}$ as

$$\mathbf{\Phi}_{\mathbf{q}}^{+} = \mathbf{\Phi}_{\mathbf{q}}^{T}\left(\mathbf{\Phi}_{\mathbf{q}}\mathbf{\Phi}_{\mathbf{q}}^{T}\right)^{-1} \tag{16}$$

If $\mathbf{\Phi}_{\mathbf{q}}$ is of full row rank, then

$$d\mathbf{q} = \mathbf{\Phi}_{\mathbf{q}}^{+} d\mathbf{x} \tag{17}$$

Although using the pseudoinverse for the derivatives ($d\mathbf{q}$ and $d\mathbf{x}$) does not yield an inverse function between the variables themselves [8], the pseudoinverse method turns out to be useful for real-time on-line computation as is required in teleoperation.

The constraint Jacobian is developed by decomposing the commanded path into a series of increments in Cartesian space. For any pair of successive increments, say, *i* and *j* in Figure 5, the predecessor (*i*) is viewed as the current configuration and the successor (*j*) as the target configuration that the end-effector has to assume. Two end-effectors at *i* and *j* can be viewed as a pair of contiguous bodies as defined in Figure 1. The constraint equations result from imposing end-effector *j*'s position and orientation as six constraints on end-effector *i*. In other words, configuration *i* is forced to assume configuration *j*. These constraint equations yield the constraint Jacobian, which in turn gives the joint command angles necessary for configuration i to assume configuration j. This process requires iterations until the constraint violation falls within an acceptable tolerance.

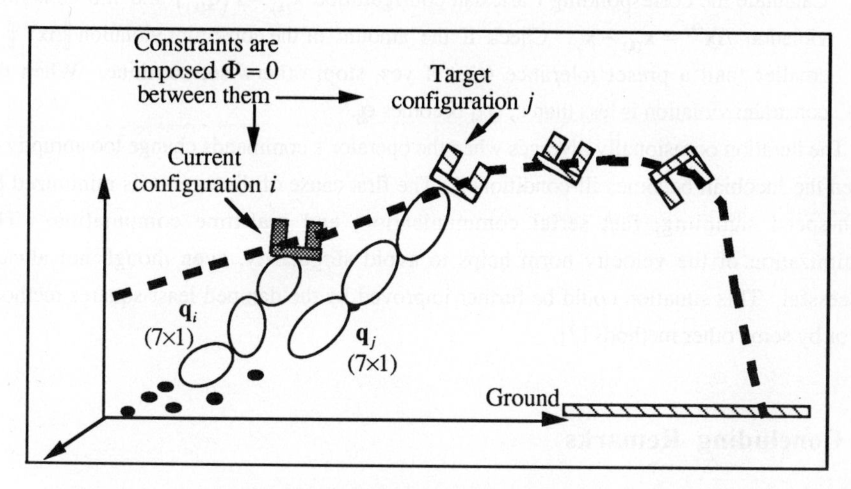

Figure 5. Artificial Constraints Imposed on End-Effector i: End-Effector j is at the Target Position and Orientation.

Numerical iteration is applied to Equation (17) in which the incremental Cartesian motion ($d\mathbf{x}$) is inputted by the operator and the pseudoinverse of the Jacobian is developed. At the start, the current configuration is known both in Cartesian space ($\mathbf{x}_i$) and in joint space ($\mathbf{q}_i$), but the target configuration is known only in Cartesian space ($\mathbf{x}_j$). Also known are the three wrist angles of $\mathbf{q}_i$ because they are taken from the mini-master's wrist angular displacements. The other four components of $\mathbf{q}_i$ are, however, unknown. These four unknowns are determined through the pseudoinverse of the three positional elements of $\mathbf{x}_j$.

(1) Calculate the difference between the initial and the target configuration, $\Delta\mathbf{x}^{(0)} = \mathbf{x}_i - \mathbf{x}_j$.

(2) Determine the position constraint Jacobian in spatial vector space,

$$\Phi^{pos}_{\mathbf{z}_{(k)}} = \left[\mathbf{I} \quad \tilde{\mathbf{r}}_{(k)} + \tilde{\mathbf{s}}_{(k)}\right]_{3\times6}.$$

These vectors $\tilde{\mathbf{r}}_{(k)}$ and $\tilde{\mathbf{s}}_{(k)}$ are already available from dynamics computation. Next, convert this Jacobian into the Jacobian in joint space, $\Phi^{pos}_{\mathbf{q}_{(k)}} = \Phi^{pos}_{\mathbf{z}_{(k)}}\mathbf{B}_{(k)j}$, where $\mathbf{B}_{(k)j}$ is defined in Equation (3) and $k = 0$ indicates the initial configuration i.

(3) Construct the full Jacobian:

$$\Phi_{\mathbf{q}(k)} = \left[\begin{array}{cc} \multicolumn{2}{c}{\Phi^{pos}_{\mathbf{q}_{(k)}}} \\ \hline \mathbf{0}_{3\times4} & \mathbf{I}_{3\times3} \end{array}\right]_{6\times7}$$

The pseudoinverse of the Jacobian is then determined as:

$$\Phi^{+}_{\mathbf{q}_{(k)}} = \Phi^{T}_{\mathbf{q}_{(k)}}\left(\Phi_{\mathbf{q}_{(k)}}\Phi^{T}_{\mathbf{q}_{(k)}}\right)^{-1}$$

(4) Calculate incremental joint angles $\Delta\mathbf{q}^{(k)} = \mathbf{\Phi}^{+}_{\mathbf{q}_{(k)}} \Delta\mathbf{x}^{(k)}$, The new joint angle vector is $\mathbf{q}_{(k)} = \mathbf{q}_i + \Delta\mathbf{q}^{(k)}$.

(5) Calculate the corresponding Cartesian configuration $\mathbf{x}_{(k)} = \mathbf{F}(\mathbf{q}_{(k)})$ and the constraint violation $\Delta\mathbf{x}^{(k)} = \mathbf{x}_{(k)} - \mathbf{x}_j$. Check if the amount of the constraint violation $\|\Delta\mathbf{x}^{(k)}\|$ is smaller than a preset tolerance (ε). If yes, stop; otherwise, continue. When the constraint violation is less than ε, $\mathbf{q}_{(k)}$ becomes $\mathbf{q}_j$.

The iteration occasionally diverges when the operator's commands change too abruptly or when the Jacobian becomes ill-conditioned. The first cause of divergence is minimized by high-speed sampling, fast serial communication, and real-time computation. The minimization of the velocity norm helps to avoid singularity, even though not always successful, This situation could be further improved by the damped least-squares methods [5] or by some other methods [7].

4 Concluding Remarks

To solve the inverse kinematics in teleoperation with a positional hand controller, we iteratively use the resolved motion rate control [4] in which the constraint Jacobian is constructed on-line and used in the pseudoinverse method as the manipulator is teleoperated. The iteration converges within a few cycles, so real-time control is possible. Although the pseudoinverse method has drawbacks, it is simple enough to be implemented for real-time operation. The use of the constraint Jacobian is applicable to both free motion and environmental contact. For the former case, six constraints are "artificially" imposed between two configurations. For the latter, there actually exist six constraints between the end-effector and an object fixed on the environment. Since the constraint Jacobian is constructed in the same way whether in contact or not, the same algorithm applies to both cases, which enables a smooth transition between open-chain (noncontact) and closed-chain (contact) kinematics.

Acknowledgement: Research supported by NSF-Army-NASA Industry/University Cooperative Research Center for Simulation and Design Optimization of Mechanical Systems.

References

1. J. Vertut and P. Coiffet, *Robot Technology, Volume 3A:Teleoperations and Robotics: Evolution and Development*, Englewood Cliffs, NJ: Prentice-Hall, 1986.
2. L. Stark et al., "Telerobotics: Display, Control, and Communication Problem," *IEEE J. of Robotics and Automation*, RA-3(1), pp.67-75, 1987.
3. S. K. Fu, R. C. Gonzalez, and C. S. G. Lee, *Robotics : Control, Sensing, Vision, and Intelligence*, New York: McGraw-Hill, 1987.

4. D. E. Whitney, "Resolved Motion Rate Control of Manipulators and Human Prostheses," *IEEE Trans. on System, Man, and Cybernetics*, MMS-10(2), pp.47-53, 1969.
5. C. W. Wampler II, "Manipulator Inverse Kinematic Solutions Based on Vector Formulation and Damped Lease-Square methods," *IEEE Trans. on Systems, Man, Cybernetics*, SMC-16(1), pp. 93-101, 1986.
6. E. J. Haug, *Computer Aided Kinematic and Dynamics of Mechanical Systems, Vol. 1: Basic Method*, Boston: Allyn and Bacon, 1989.
7. D. N. Nenchev, "Redundancy Resolution Through Local Optimization: A Review," *J. of Robotic Systems*, 6(6), pp.769-798, 1989.
8. C. A. Klein and C. H. Huang, "Review of Pseudoinverse Control for Use with Kinematically Redundant Manipulators," *IEEE Trans. on Systems, Man, and Cybernetics*, SMC-13(3), pp 245-250, 1983.
9. R. Featherstone, *Robot Dynamics Algorithms*, Boston: Kluwer Academic Publishers, 1987.
10. D. S. Bae, R. S. Hwang, and E. J. Haug, "A Recursive Formulation for Real-Time Dynamic Simulation," *ASME Advance in Design Automation*, DE-Vol.14, pp.499-508, 1988.
11. D. S. Bae and E. J. Haug, "A Recursive Formulation for Constrainted Mechanical System Dynamics: Part I, Open Loop System," *Mechanics of Structure and Machines*, 15(3), pp.359-382, 1987.
12. D. S. Bae and E. J. Haug, "A Recursive Formulation for Constrainted Mechanical System Dynamics: Part II, Closed Loop System," *Mechanics of Structure and Machines* , 15(4), pp.481-506, 1987.
13. E. J. Haug, *Intermediate Dynamics*, Boston: Allyn and Bacon, 1990.
14. B. Hannaford and R. Anderson, "Experimental and Simulation Studies of Hard Contact in Force Reflecting Teleoperation," *Proc. IEEE Conf Robotics and Automation*, pp. 584-589, 1988.
15. H. Asada and J. -J. E. Slotine, *Robot Analysis and Control*, Wiley-Interscience Publication, John Wiley and Sons, 1986.
16. C. R. Rao and S. K. Mitra, *Generalized Inverse of Matrices and its Application*, New York: John Wiley & Sons, 1971.

Part 3

Manufacturing

Relationship Between Design for Manufacturing, a Responsive Manufacturing Approach, and Continuous Improvement

J. E. Ashton

FMC Corporation/Naval Systems Division, 4800 E. River Road, Minneapolis, MN 55421-1498 USA

Abstract: Low volume and custom designed products cannot realistically be designed and produced with static or "frozen" design definitions. Rather, design refinement continues throughout the development and production life of the product. Design changes and iterations are a way of life in such an environment, and can be (and should be) a powerful part of "continuous improvement". To take advantage of this powerful potential, design changes must be introduced in a controlled manner, and should be introduced rapidly and efficiently.

In most system job shop environments, the excessive work-in-process inventories and poor in-process quality result in very long times from engineering release through product delivery. These long times and the resulting disruptive effects of engineering changes lead to a variety of "workaround" approaches, including special prototype shops, "skunkworks," and outside fast-turnaround shops. However, dramatic reductions in work-in-process and cycle times can be achieved through aggressive application of just-in-time and total quality management principles. In turn, the much more responsive manufacturing environment greatly facilitates change implementation, which in turn allows direct use of the production shops instead of "special" arrangements or shops. In turn, the iterative learning between design, manufacture, and use can be utilized as a powerful aspect of Continuous Improvement. The approach and results are illustrated with a case example.

Keywords: continuous improvement / concurrent engineering / custom design / job shop / just-in-time / total quality management / design changes

1 The System Job Shop Environment and Design Iterations

Much of the literature and the discussions concerning design for manufacturability or concurrent engineering is oriented to a product design/manufacturing process for the serial production of a high volume of identical items. In this environment an objective to rapidly develop a producible and functional design which is then "frozen" may be appropriate.

However, in many other environments, most notably low volume and custom designed products, such an objective cannot realistically be achieved. Rather, design refinement continues not only through the development cycle, but also during the production life of the product. Design changes are a way of life in such an environment because the evolution of the design over time is required to stay competitive as technologies and environments change. However, manufacturing management and often general management tend to regard such changes as disruptive and harmful leading to attempts to "stop the changes" and slogans such as "do it right the first time". Although some changes are the result of preventable errors, "doing it right" is a naive objective which fails to recognize that design changes are an essential and powerful element of the continuous improvement process. But, to take advantage of the power of continuous improvement, design changes must be introduced in a controlled manner and preferably should be introduced rapidly and efficiently. The attitude and approach of the manufacturing unit are a significant influence in establishing the desirability of the design change process and in developing the ability to handle constructive changes in an iterative manner.

Figure 1 categorizes different kinds of discrete manufacturing businesses ranging along two different axes, one from low volume to high volume and one from low complexity to high complexity. Although it may not be intuitively obvious, high unit volume applications or operations tend to be somewhat easier to manage than low unit volume operations since it is economical to put more intense effort into not only the design but also the management and control aspects of given products. The most complex operations to manage are low volume, high complexity product shops such as capital equipment for the mining and construction environments, aircraft programs, and the like. These operations are characterized by low volumes of individual products, large product mix, and each individual product characterized with multi-level indentured bill of materials involving make parts, buy parts, as well as assemblies.

Figure 2 contrasts the typical design through production cycle of high volume, standard products with that same cycle for low volume and custom designed products. For high volume products, typically a design phase is followed with a prototype phase which is then followed with some early low rate production to check out both the design and also the tooling and the manufacturing processes. After confirmation that the processes as well as the design are satisfactory, the design and tooling and approaches are "frozen" and serial production commences, usually at high rates with little future designs iterations until such time as a major product change is introduced. In contrast, low volume and custom products cannot afford such a prove-out process. For these products, generally design continues throughout the life of the product, involving continuous improvements and iterations, and although there may sometimes be a few development units, the production units themselves continue to evolve and change as technology, customer needs, and manufacturing opportunities present

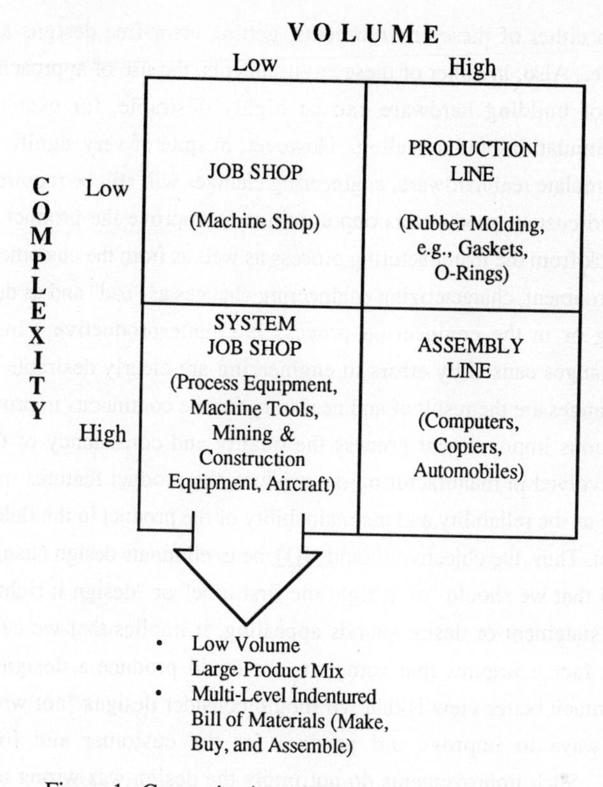

Figure 1. Categorization of Discrete Manufacturing

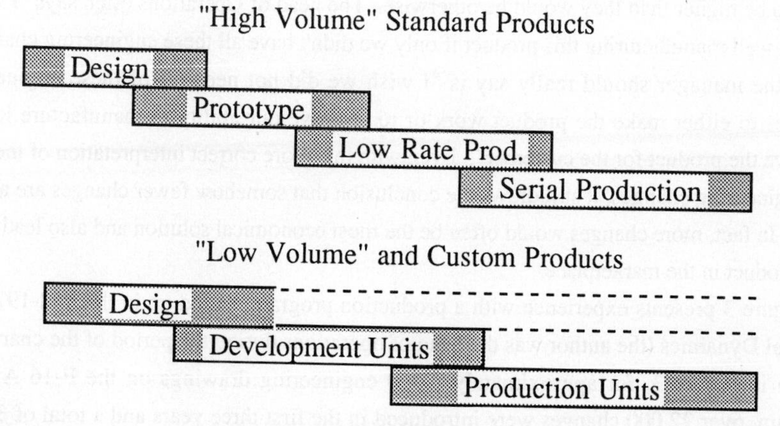

Figure 2. Design Through Production Cycle

themselves. In either of these environments, getting error-free designs accomplished is clearly desirable. Also, in either of these environments, the use of approaches to prove out designs short of building hardware can be highly desirable, for example, the use of mathematical simulations and modeling. However, in spite of very significant advances in our ability to simulate real hardware, engineering changes will still be required in the case of low volume and custom products as opportunities to improve the product are determined through feedback from the manufacturing process as well as from the customer.

In this environment, characterizing engineering changes as "bad" and as due only to errors by engineering or in the engineering process is counterproductive. Instead, although engineering changes caused by errors in engineering are clearly desirable to avoid, many engineering changes are the result of and necessary for the continuous improvement process. With a continuous improvement process the quality and consistency of the product, the productivity involved in manufacturing the product, the product features that the customer desires, as well as the reliability and maintainability of the product in the field, are all subject to improvement. Thus, the objective should NOT be to eliminate design changes. Sometimes we hear it said that we should "do it right the first time" or "design it right the first time". Although this statement or desire sounds appealing, it implies that we can know what is "right" and in fact it implies that somehow we could produce a design that cannot be improved. A much better view is that we should consider designs "not wrong" but should always seek ways to improve our product for the customer and for producibility considerations. Such improvements do not imply the design was wrong to start with, but only that through further experience with the design we can make it better.

Unfortunately, management usually perceives that engineering changes are bad and cause costs to be higher than they would be otherwise. The head of Operations often says "I would do real well manufacturing this product if only we didn't have all these engineering changes". What the manager should really say is "I wish we did not need all of these engineering changes to either make the product work or to improve our ability to manufacture it or to improve the product for the customer". Even with this more correct interpretation of the need for engineering changes, it still draws the conclusion that somehow fewer changes are always better. In fact, more changes would often be the most economical solution and also lead to the best product in the marketplace.

Figure 3 presents experience with a production program started up in the mid-1970s by General Dynamics (the author was the head of operations during the period of the chart). As shown in Figure 3, for approximately 3,000 engineering drawings on the F-16 Aircraft Program, over 22,000 changes were introduced in the first three years and a total of 30,000 changes in the first five years of this development and production program. Conventional wisdom would suggest that with this very high change traffic the program must have been chaotic and probably unsuccessful but, in fact, the opposite was true. Although some of the drawing changes introduced in this period were narrowly the result of engineering errors, much of the change traffic was involved in improving the producibility of the product as well as improving the functionality for the customer.

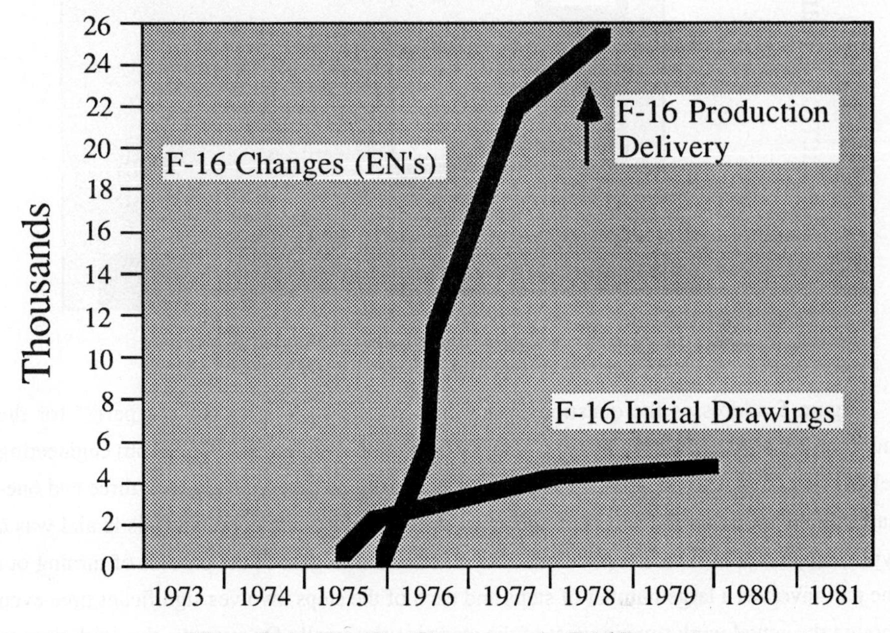

Figure 3. Engineering Changes on the F-16 Program

Figure 4 presents the manhour result for the first development F-16, the first production F-16 (Number 9), the 100th production F-16, and the 200th production F-16. As indicated in the figure, the manhours per aircraft reduced from 183,000 on the first aircraft to 33,000 on the 200th aircraft, a result which represents approximately a 76% learning curve. Although many people would observe that those are very good results in spite of all the engineering changes, a more reasonable interpretation is that those are very good results in part because of the engineering changes.

The example in Figure 4 illustrates that very good results can be achieved with high engineering change traffic and in fact suggests that such engineering change traffic can be desirable. In fact, the objective should not be to reduce the change traffic but rather to manage changes effectively; this means to manage the changes as a controlled process with clear configuration control and to handle the changes rapidly and efficiently. Unfortunately, the way manufacturing operates in most operations, it is difficult to introduce changes either rapidly or efficiently.

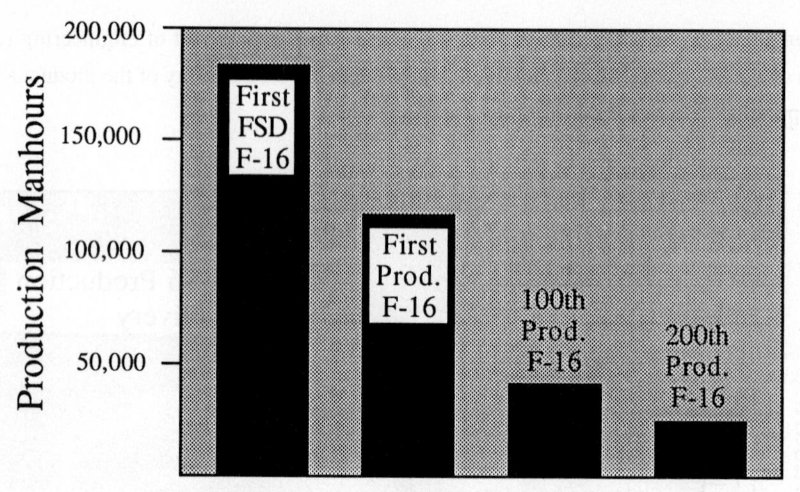

Figure 4. F-16 Production Manhour Improvements

Figure 5 addresses the question "Why does it take so long to make a part?" for the environment at General Dynamics on the F-16 Aircraft Program in 1976. From engineering release of a change to the production of a finished, detailed part typically took three and one-half months based on the way that operation was organized and operated (this is and was a typical approach in such an operation). As illustrated in Figure 5, the process of turning out the part involves a large number of steps and each of the steps involves significant time even though the actual work time in most of the steps is very small. On average, the work time to go through the process from an engineering change to the finished part might represent 20 to 40 hours of work, and yet the elapsed time was typically three and one-half months.

Since in most system job shop environments the time from engineering release through part production, assembly, and test is quite long, either design changes must be introduced slowly or much of the work must be done out of the normal production cycle. Because of these considerations, special engineering prototype shops are often established. Alternatively, "skunkworks" represent another approach to improve and shorten the design/manufacturing cycle and many companies utilize outside, fast turnaround shops for their "changes". Although these approaches may speed up the process and avoid initial disruption of the production shops, they suffer the major disadvantage of "lost learning" by the production shop and often just delay the production disruption of the change implementation.

The primary cause of the long lead times in implementing changes into production are the excessive work-in-process inventories in most such shops, both physically in the shops as well as in the paperwork queues in the office. The large WIP exists for a variety of reasons, especially including quality problems (defects) with the product or process. These long in-process times and large WIP complicate change implementation by requiring either a long time

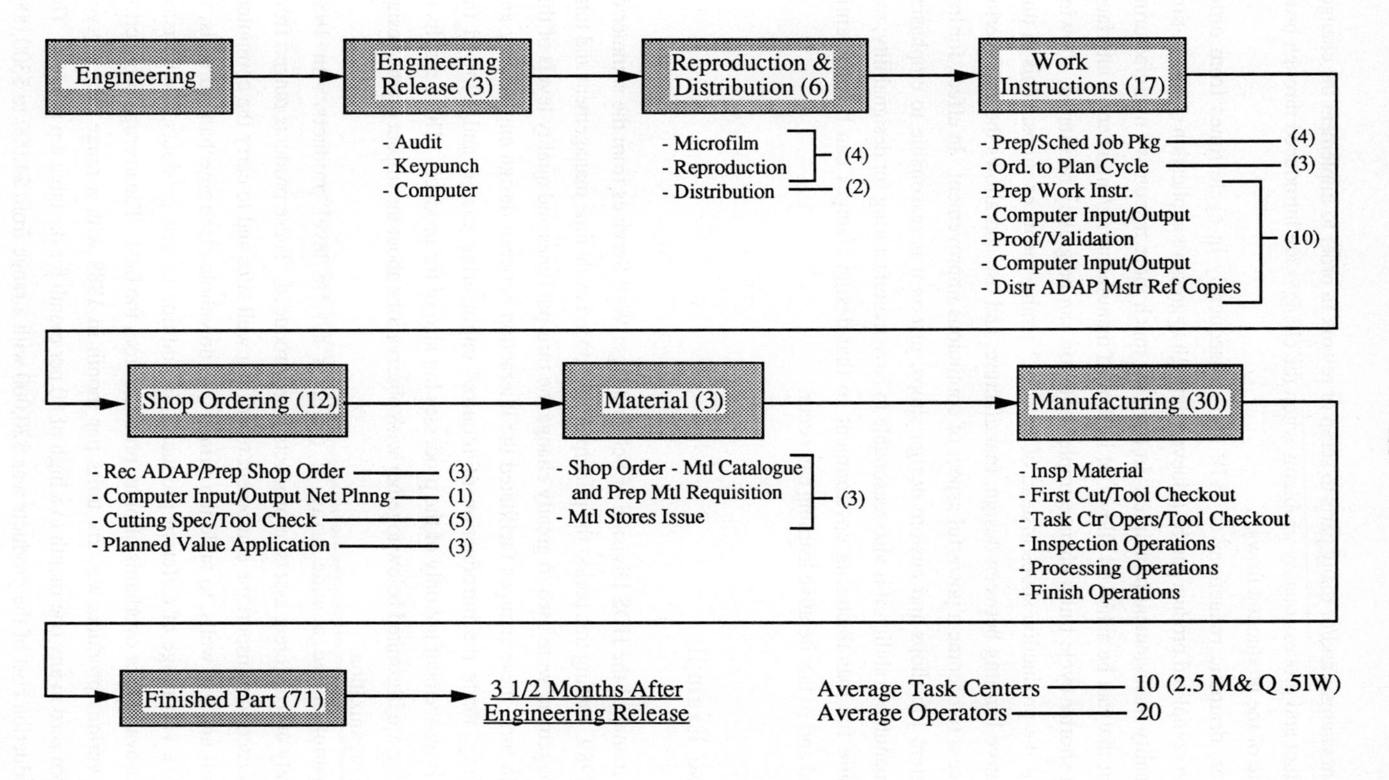
Engineering
Engineering Release (3)
- Audit
- Keypunch
- Computer
Reproduction & Distribution (6)
- Microfilm
- Reproduction
- Distribution
(4)
(2)
Work Instructions (17)
- Prep/Sched Job Pkg
- Ord. to Plan Cycle
- Prep Work Instr.
- Computer Input/Output
- Proof/Validation
- Computer Input/Output
- Distr ADAP Mstr Ref Copies
(4)
(3)
(10)
Shop Ordering (12)
- Rec ADAP/Prep Shop Order
- Computer Input/Output Net Plnng
- Cutting Spec/Tool Check
- Planned Value Application
(3)
(1)
(5)
(3)
Material (3)
- Shop Order - Mtl Catalogue and Prep Mtl Requisition
- Mtl Stores Issue
(3)
Manufacturing (30)
- Insp Material
- First Cut/Tool Checkout
- Task Ctr Opers/Tool Checkout
- Inspection Operations
- Processing Operations
- Finish Operations
Finished Part (71)
3 1/2 Months After Engineering Release
Average Task Centers — 10 (2.5 M& Q .51W)
Average Operators — 20

Figure 5. Why Does it Take so Long to Make a Part?

to implement changes and/or many parts to scrap or rework in order to implement the changes and/or product and process quality problems when the changes are introduced through brute force in spite of the production flows.

However, dramatic reductions in WIP and consequently in cycle times from order placement to completed product can be achieved through aggressive application of just-in-time and total quality management principles. In turn, the much more responsive manufacturing environment that can be achieved allows an improved manufacturing/engineering interface. With much shorter cycle times change implementation can be greatly facilitated, allowing direct use of the production shops instead of "special" arrangements or shops. This in turn allows iterative learning between design, manufacture, and use and allows the engineering change process to become a powerful aspect of continuous improvement. In effect, for low volume system job shops and custom design shops, just as it is reasonable to emphasize design for manufacturability, it is also reasonable to have manufacturing for designability; this is to structure the manufacturing environment so that design changes can be efficiently implemented and so that iterative learning can occur.

2 A Case Example

The author managed the HDS Division of Schlumberger Well Services from the summer of 1985 until 1989. During this period the principles of reduced cycle time management and total quality management were used to greatly change the thru-put times and quality levels of this operation. In turn, these changes facilitated the interaction between design engineering and manufacturing, made the introduction of required engineering easier, and allowed the continuous improvement not only of the processes but also of the products. These results in manufacturing are illustrated below together with observations about the impact on the design engineering organization.

HDS manufactured a wide array of different system level products. In 1988 approximately 400 different assets or products were produced. These products ranged from large trucks used to transport the equipment to drilling well sites and to carry the computers for testing oil and gas wells, to table top printers, to downhole electromechanical tools, to computers - a vast range of different products. In addition to new products, the division produced renovations or overhauls of older products from the field. The average production rate of the various products was six units per month in 1988 with a range (for a given product) from zero in any one month to a high of 80 per month for the table top printers. The average production cost of the products was $20,000 with a range from $4,000 to $300,000. Clearly this operation had a diverse product mix of sophisticated products.

The long lead times to produce already designed products at HDS in 1985 are illustrated in Figure 6. As indicated, the average product took 16 months from the beginning of procurement to the final assembly, test, and delivery. Of this total time, over ten months was involved in value-adding functions within the operation. With these lead times any engineering changes to be introduced had to be introduced many, many months before the delivery of the product. And to be introduced without significant workarounds required typically over a year of elapsed time from design release until such a change could be incorporated into production. In this environment, not surprisingly, changes were resisted and the customer's normal complaint was that HDS was unwilling to improve or even fix product problems that were evident. To avoid disrupting production with its long lead times, HDS had in service in 1985 a dedicated prototype area to build new products. In addition, both engineering and customer support operated their own machine shops and extensive use was made of small outside machine shops to handle engineering prototype efforts. The production machine shop was only used for already designed and developed production machined parts. The division had a high defect rate in ongoing production and had a large number of "use-as-is" dispositions of the defects that were found, suggesting an inability to either make the designs as designed or to change the designs to what was truly needed.

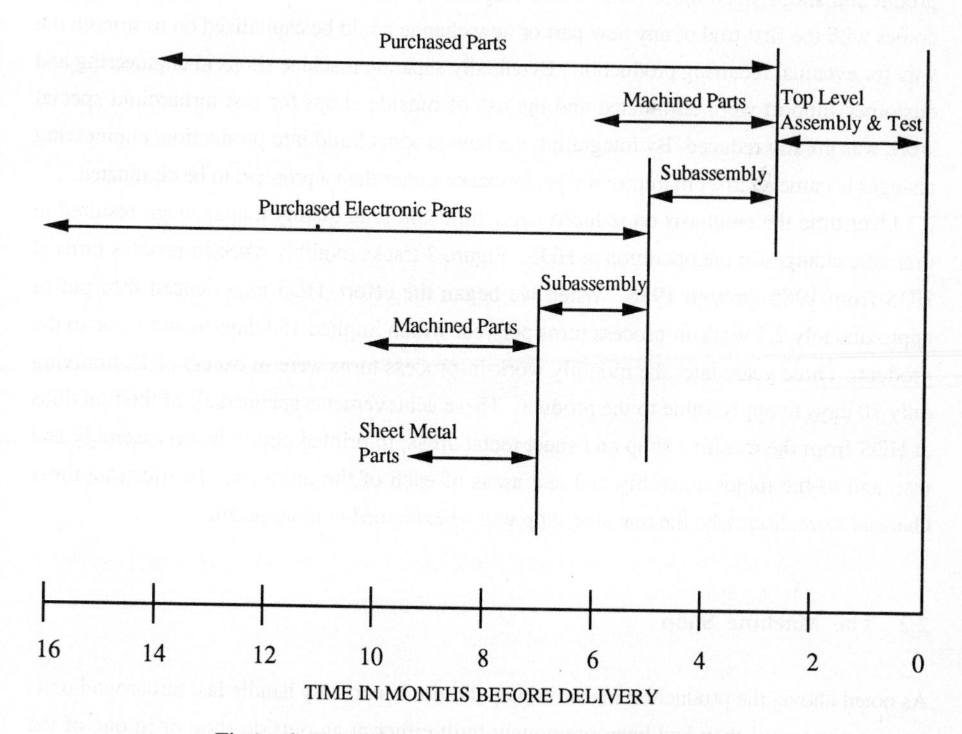

Figure 6. Typical Lead Times in MRP at HDS in 1985

2.1 Actions Taken

In the summer of 1985 a different approach to management and production at HDS was initiated. These efforts are described in more detail in References 1 through 5, wherein the approach and philosophy and many of the techniques undertaken to change the HDS approach are discussed. These changes included a reorganization from a functional organization to an organization by product lines and the introduction of a disciplined project management approach to the management of new products, new product introductions, ongoing manufacture, and improvement of existing products. The key to the approach and the changes was an emphasis on reduced cycle time which was achieved by concentrating on reducing lead times, queues, and lot sizes throughout production. In turn, the emphasis on the reduction in cycle time made an emphasis on improving the quality of the processes mandatory. Defects cannot be allowed to languish or to be oft repeated as less and less inventory exists in the factory or in the office. Thus, an emphasis on reduced cycle time motivates an emphasis on defect reduction and on corrective action to keep defects from recurring. As the cycle times were reduced throughout the operation and the quality improved, it became easier to convince engineering of the advantages of doing prototype work and change introduction through the production shops since these shops could respond in reasonable times and the learning that comes with the first trial of any new part or new change could be capitalized on to smooth the way for eventual recurring production. Eventually separate machine shops in engineering and customer support were eliminated and the use of outside shops for fast turnaround special work was greatly reduced. By integrating the new product build into production, engineering changes became an asset to improving performance rather than a problem to be eliminated.

Over time the emphasis on reduced cycle time and total quality management resulted in dramatic changes in the operation at HDS. Figure 7 tracks monthly work-in-process turns at HDS from 1985 through 1988. When we began the effort, HDS experienced thru-put of approximately 2.3 work-in-process turns per year which implied 150 days to add value to the product. Three years later the monthly work-in-process turns were in excess of 12 implying only 30 days to apply value to the product. These achievements spanned all of the functions at HDS from the machine shop and sheet metal areas, to printed circuit board assembly and test, and to the major assembly and test areas of each of the products. To illustrate these changes more discretely, the machine shop will be examined in more detail.

2.2 The Machine Shop

As noted above, the production machine shop did not historically handle fast turnaround parts or new parts until they had been previously built either at an outside shop or in one of the other machine shops supporting engineering or the customer. Starting in late 1985 we set out

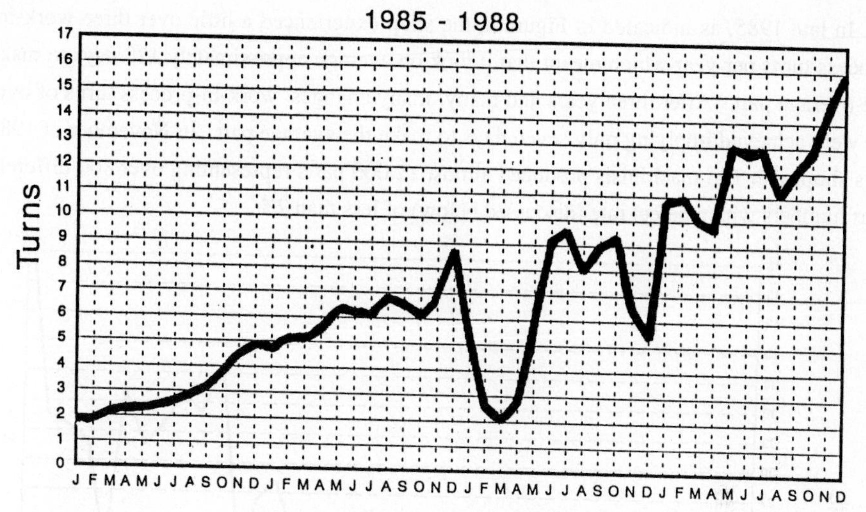

Figure 7. HDS Monthly W-I-P Turns

to produce all of the one-of-a-kind and fast turnaround parts in-house in the production machine shop. The primary motivation for this consolidation of machine shop activity was to be able to capitalize on the learning inherent in manufacturing and in working closely with engineering to refine the designs so they would be manufacturable. To make the production machine shop responsive enough to handle this objective, we put increased emphasis on tooling and on our numerical control programming capability. And, both to achieve this machine shop consolidation and to improve the overall performance and responsiveness of the machine shop, we changed our management approach to achieve greatly reduced cycle times.

This machine shop had 86 operators or machinists when we began the effort and involved both conventional and numerically controlled machines, with parts ranging from simple three-axis parts up to and including sophisticated and very tight tolerance five-axis parts. To handle this variety we established cells. Cellular manufacturing allows a shop to be organized as many little shops instead of one large one; and by grouping similar parts in cells with all of the equipment to produce those parts, reduced setup costs can be achieved within the cell; this in turn allows small lots to be economically produced. With the combination of small lots and cellular manufacturing, extremely small inventories or queues can be maintained and thus very rapid thru-put can be achieved. In order to improve the quality we moved to a "self" inspection process which we believed mandatory so that defects would be found soon after they were created, thus motivating an analysis of what caused the defect and what could be done to the process to avoid them in the future. And, we set out to shorten lead times, reduce lot sizes, reduce queues, and thus greatly reduce cycle time. To accomplish these things we had to conduct extensive training not only of the operators but also of the support people and the management. The results in terms of average cycle times were impressive.

In late 1985, as indicated in Figure 8, the shop experienced a little over three work-in-process turns per year which meant that it took on average approximately 120 days to make the average part. After three years and many, many changes, work-in-process turns of over 40 were achieved implying only seven days to make the average part. In November of 1988 this shop with 106 machinists produced almost 12,000 parts representing over 800 different part numbers with a defect rate (discussed below) of less than 2%.

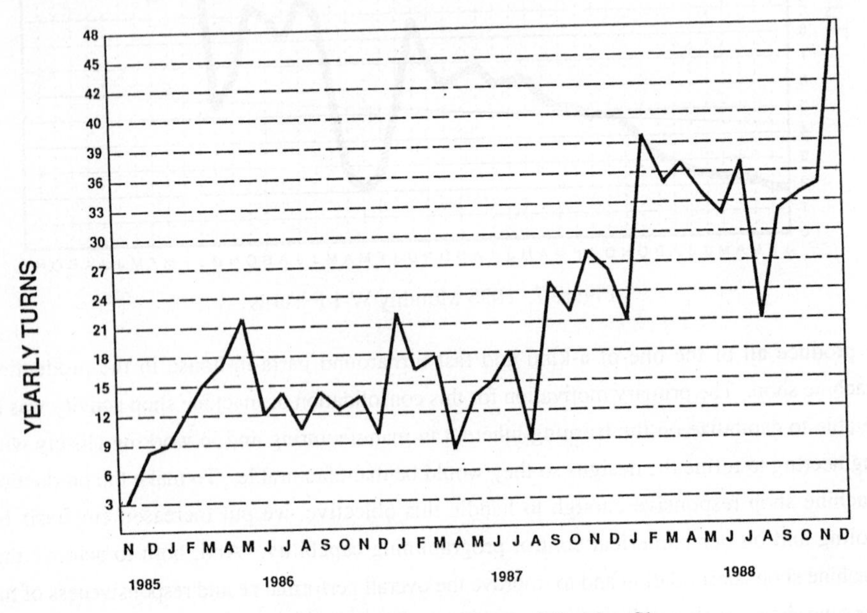

Figure 8. Work-In-Process Turns - Machine Shop

Figures 9 and 10 present the production shipments produced in this machine shop from the beginning of 1987 through May of 1988 and the engineering prototype shipments from this same shop in the same time frame. Because of the oil business collapse in late 1985 and early 1986, production had of necessity been reduced dramatically in 1986. But beginning in 1987, the production volume began to increase. We had reduced employment to 22 machinists at the bottom of the depression in the oil business and we steadily added machinists and increased production throughout 1987 and 1988. As indicated in Figure 9, we increased the number of orders in the machine shop for production products from less than 50 in January 1987 to 500 by November 1988. The past due orders during this time stayed at a low level representing generally no more than 10% to 15% of the orders in the machine shop. At the same time the engineering prototype shipments from this shop stayed in the vicinity of 200 to 400 orders, and in spite of the very large increase in production orders, the prototype orders continued to be completed in a timely way with the past dues held to approximately 10% of the outstanding orders.

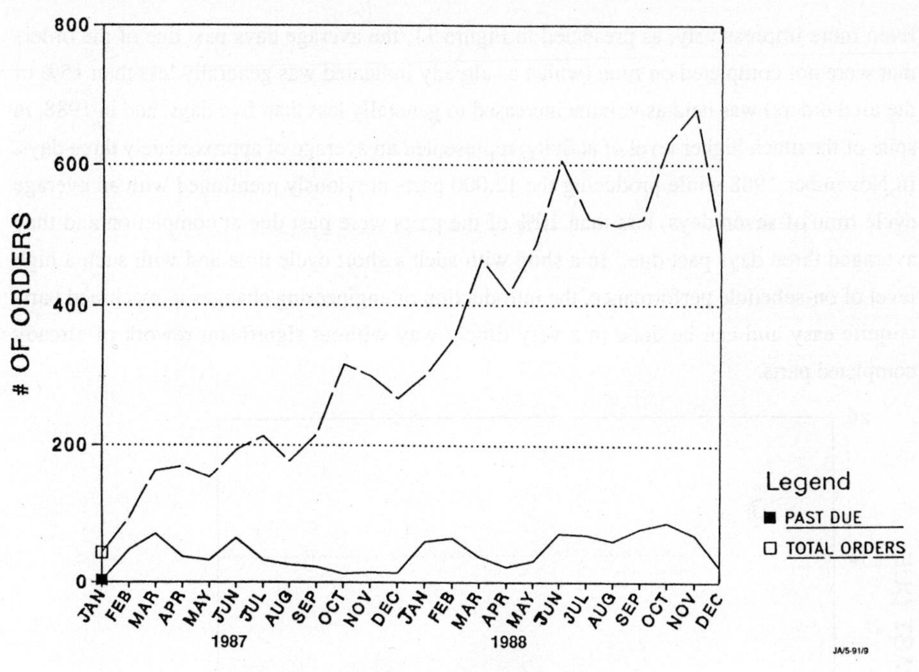

Figure 9. Production Shipments Past Due

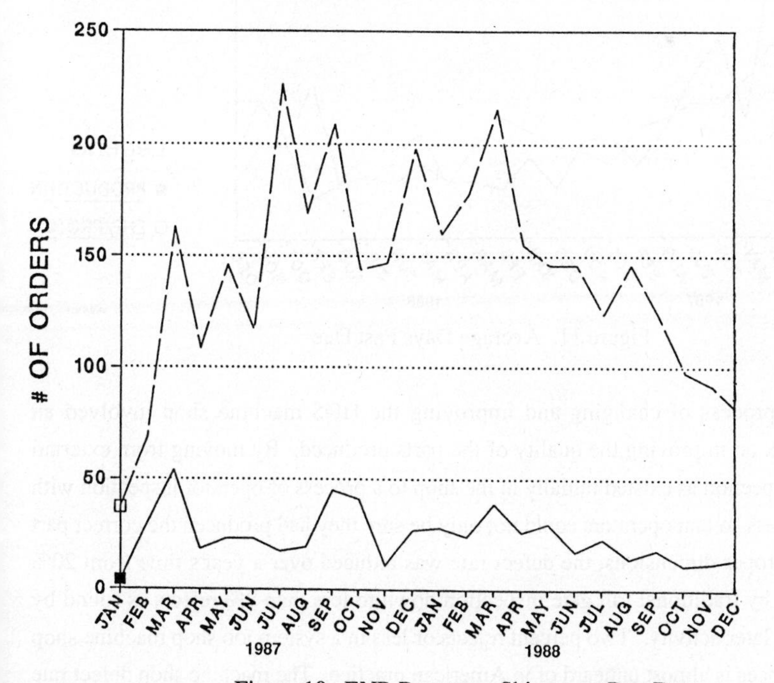

Figure 10. ENP-Prototype Shipments Past Due

Even more impressively, as presented in Figure 11, the average days past due of the orders that were not completed on time (which as already indicated was generally less than 15% of the total orders) was held as volume increased to generally less than five days, and in 1988, in spite of the much higher level of activity, represented an average of approximately three days. In November 1988 while producing the 12,000 parts previously mentioned with an average cycle time of seven days, less than 10% of the parts were past due at completion and they averaged three days past due. In a shop with such a short cycle time and with such a high level of on-schedule performance, the introduction of engineering changes to machined parts is quite easy and can be done in a very timely way without significant rework of already completed parts.

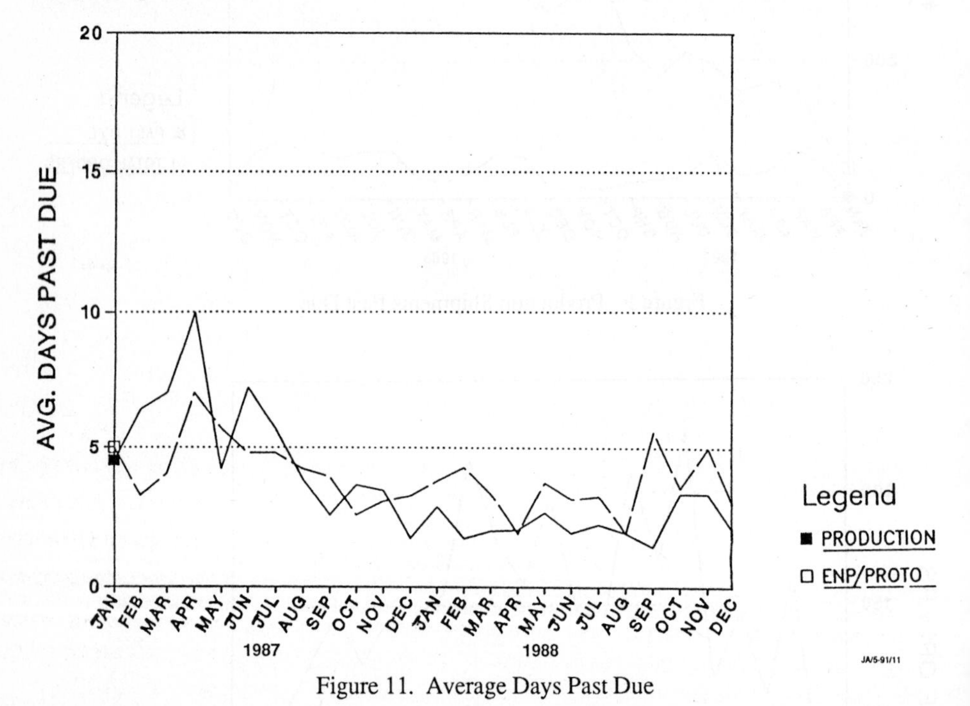

Figure 11. Average Days Past Due

Part of the process of changing and improving the HDS machine shop involved an aggressive attack on improving the quality of the parts produced. By moving from external toll gate type inspection as existed initially in the shop to a process of operator inspection with detailed data sheets so that operators could not only be sure they had produced the correct part but record the proper dimensions, the defect rate was reduced over a year's time from 20% rejects as found by traditional toll gate inspection down to less than 2% rejects as found by operators or any later activity. Two percent rejects or less in a system job shop machine shop with tight tolerances is almost unheard of in American practice. The machine shop defect rate is illustrated in Figure 12. Again, the high quality of this shop made the introduction of new or changed parts easier, since difficulties or problems with newly introduced parts could be rapidly identified, and corrective action rapidly implemented.

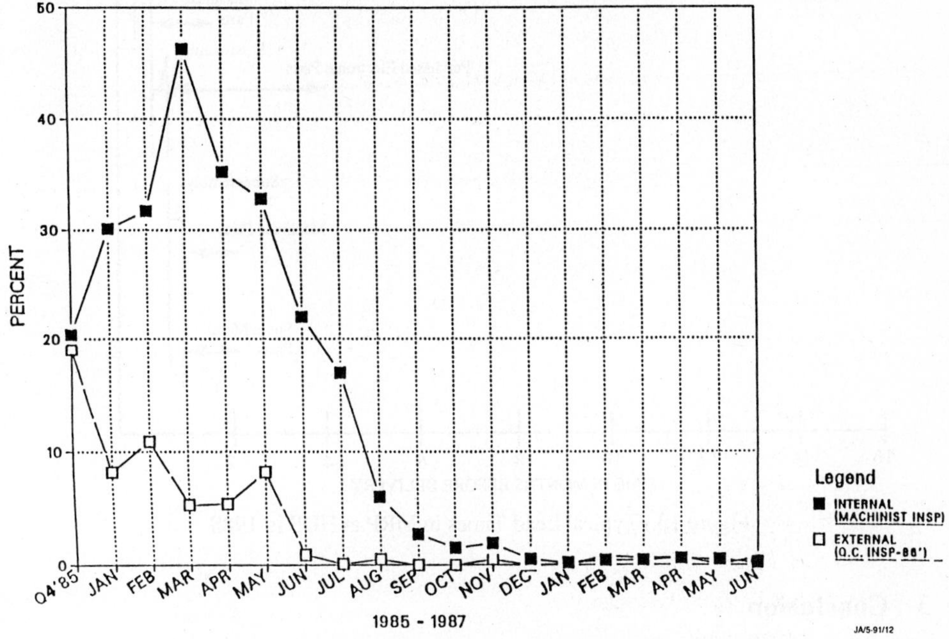

Figure 12. Machine Shop Defect Rate

The changes described in the machined parts area were duplicated in all the feeder shops of HDS and similar reductions in cycle times were achieved in the various assembly areas. In addition, an aggressive effort in purchasing allowed significant reductions in procured part lead times at the same time as we reduced the sizes of the lots we procured. By 1988 the average lead time of the products at HDS had been reduced to approximately five months or less as illustrated in Figure 13 compared to the 16 months for the same or similar products in 1985. With the lead times indicated in Figure 13, engineering changes can be introduced much more rapidly and painlessly because it takes less time to process the change and the changes to the product because there is less product already built to the previous engineering definition and because the quality that is achievable in such an operation allows a smooth assessment of the validity of any design definition. In this operation, manufacturing has become an asset and an ally for designability allowing rapid changes to be tried and proven, improving producibility and also the desirability of the product to the customer.

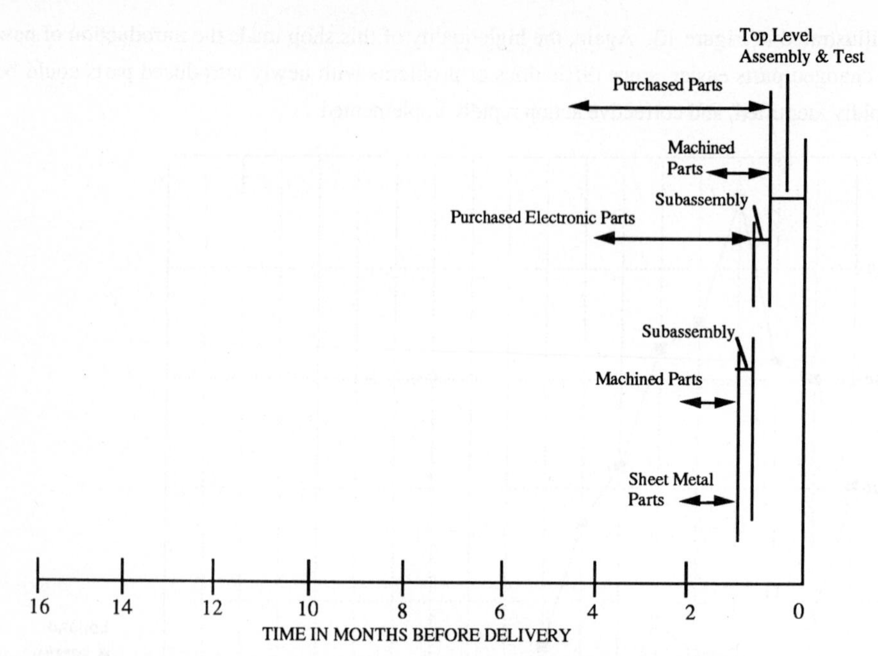

Figure 13. Typical Lead Times in MRP at HDS in 1988

3 Conclusion

The above case example illustrates that reduced cycle time management and TQM in system job shops can dramatically reduce work-in-process, lead times, and defects. In turn, the low inventory and short lead times inherent in the process can greatly facilitate the change process and this in turn allows design and manufacturing to work together to iterate to achieve continuous improvement both of the products and of the processes. Continuous improvement allows better products and better processes at lower costs and is a preferred solution to trying to get the design "right the first time" where right somehow implies the design cannot be improved.

References

1. "It's Time to Reform Job Shop Manufacturing," *Harvard Business Review*, March-April 1989, (with Frank X. Cook).
2. "Relevant Managerial Accounting in the Job Shop Environment," *Manufacturing Review*, December 1988, (with Neal Holmlund).
3. "Shop Floor Control in a System Job Shop: Definitely Not MRP," *Journal of Production and Inventory Management*, Second Quarter 1990, (with M.D. Johnson and F.X. Cook).
4. "Simple and Effective Cellular Approach to a Job Shop Machine Shop," *Manufacturing Review*, June 1989, (with D.L. Stoner and K.J. Trice).
5. "From Status Quo to Continuous Improvement - The Management Process," *Manufacturing Review*, June 1990, (with R.L. Fagan and F.X. Cook).

Defect Preventive Quality Control in Manufacturing

S. M. Wu
S. J. Hu

Department of Mechanical Engineering and Applied Mechanics, University of Michigan, Ann Arbor, MI 48109, U. S. A.

Abstract: Quality control techniques are reviewed, including inspection oriented QC, statistical process control, and design oriented QC. Statistical methods and engineering technologies for next generation quality control methodology – real-time defect prevention during manufacturing – are presented, including (1) real-time sensing to allow the QC system to monitor a process, (2) real-time analysis of process status, and (3) real-time corrective action.

Keywords: statistical process control / defect prevention / time series analysis / forecasting control / variation reduction / precision machining / automobile assembly

1 Quality Assurance Technique Review

Historically, quality assurance has progressed through three development stages, moving from downstream to upstream of the entire manufacturing processes. These three stages are: inspection oriented quality assurance, statistical process control (SPC), and design oriented quality assurance. These methods successfully address the problem of the high cost of data collection and the requirement for human intervention to identify and solve problems. Today, these methods are state-of-the-art techniques and are still being used in industries.

1.1 Inspection Oriented Quality Assurance

Quality assurance started with inspection. Inspection was required for assembly when people started producing interchangeable parts. However, it would be too expensive to inspect every part in mass production. Moreover, if the inspection was destructive, there would be nothing left after inspecting every part. Inspection could also be very tedious for the inspectors during mass production. Therefore, sampling plans were introduced and popularized during World War Two.

A number of problems exist for inspection oriented QA:

(1) Quality cannot be improved through inspection. When defects are found, the only action taken is that of adjusting, reworking, or scrapping the defective parts. In any event, productivity and costs suffer. As pointed out by Ishikawa [1], "the basic notion behind control is the prevention of reoccurrence of errors." Merely finding problems in the manufactured products does not improve product quality.

(2) The concept of Acceptable Quality Level (AQL) in sampling plans is not applicable when the quality level approaches 'zero defects'. For example, if the quality level is 5 parts per million defective, the sampling results will possibly always be zero for a sample size of 100 or 200.

(3) Information from the traditional inspection division to the manufacturing division takes too much time. It is not easy for the manufacturing division to use this information to prevent errors.

Since inspection oriented quality assurance techniques are not enough to improve product quality, other quality assurance techniques are needed.

1.2 Statistical Process Control (SPC)

Statistical Process Control (SPC) based on sampling was introduced by Shewhart from the Bell Lab in the 1920s to reduce the defective rate and balance the control costs. Most of the philosophy behind the use of control charts was laid out in a book published in 1931, Economical Control of Quality of Manufactured Products [2]. The idea is to plot the data in some manner as soon as it becomes available and to observe trends and changes. A typical Shewhart control chart is shown in Figure 1 with both the Upper Control Limit (UCL) and the Lower Control Limit (LCL). Usually a sample of n units is taken periodically, and both the sample mean and sample range are plotted, i.e., $\bar{X}$ and R charts. The idea behind the chart is that when the process is in control, the sample mean should be independently and normally distributed about the target, and the variance should be constant.

In order to obtain sufficiently accurate control limits for the $\bar{X}$ and R charts, a rather large number of subgroups must be taken. According to Grant and Leavenworth [3], a common rule of thumb for $\bar{X}$ and R charts is that at least 25 subgroups should be taken before calculating control limits. A subgroup of 5 is common in industry. This means that a minimum of 125 parts are needed to establish the control limits.

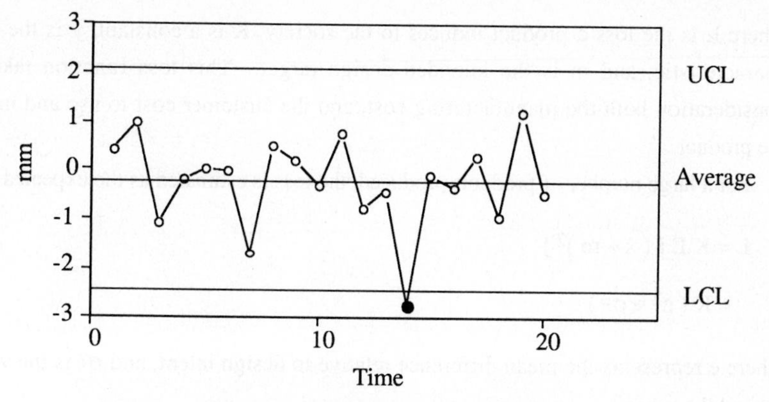

Figure 1. A Shewhart Control Chart

SPC has the following limitations:

(1) Control charts are constructed by inspecting the product. When a process is out of control, defective products have already been produced.

(2) Control charts are based on the assumption of independent data. However, 100% inspection data are usually serially correlated, or autocorrelated. Serial correlation in data results in false alarms, i.e., alarms indicating process out of control when the process is actually in control.

(3) Intentionally making data independent may result in the loss of process information which may be quite useful for studying the process and identifying assignable causes.

1.3 Design Oriented Quality Assurance

Drs. W. Edward Deming [4] and Kaoru Ishikawa [1] have emphasized the importance of continuous, never-ending quality improvement by concentrating on the upstream process, thus making a fundamental break with the past practice of relying on inspection downstream. Dr. Genichi Taguchi developed many useful engineering ideas for quality improvement during the product/process design stage. Several important contributions include:

(1) *Introduction of Loss Function.* When a quality characteristic deviates from its target value, some loss is sustained. According to Taguchi [5], "Quality is the loss a product causes to society after being shipped, other than any losses caused by its intrinsic function." A simple form of the loss function is:

$$L = K\,(y-m)^2$$

where L is the loss a product induces to the society, K is a constant, y is the quality characteristic, and m is the intended design target. This loss function takes into consideration both the manufacturing cost, and the customer cost to use and maintain the product.

For a large number of products produced, the loss is estimated as the expected value:

$$L = K\,E[\,(x-m)^2\,]$$

$$= K\,(\varepsilon^2 + \sigma^2)$$

where e represents the mean difference relative to design intent, and σ^2 is the variance around the mean.

Therefore, two sources contribute to product quality loss:

(a) the quality characteristics not being on target, and

(b) excessive variation around the mean.

Minimizing the loss function implies (a) improving the mean performance relative to design intent,and (b) reducing the variance around the mean. These two aspects of quality improvement are summarized as "being on target with minimum variation." The importance of this minimum variation lies in the very center of continuous quality and productivity improvement.

(2) *Robust Product/Process Design.* Product variation induces loss to society. Statistically designed experiments are used to design the product/process insensitive to environmental noise and component variation [5].

Taguchi's engineering ideas are very useful. However, their application and end results can still be limited because:

(1) Statistical experiment designs are based on the assumption of independence between sampled data, while manufacturing process data are dependent.

(2) A quality design cannot become a quality product without quality assurance measures during the manufacturing process.

Examining the entire product development and manufacturing cycle, one can see that there are QC techniques either before or after a product is manufactured, but not during the manufacturing stage. The reason for missing this important link can be attributed to the fact that manufacturing processes are highly complex and dynamic in nature, and process data is usually serially correlated. Statistically, correlated data is difficult to deal with.

2 Real-Time Defect Prevention

The goal of next generation quality control in manufacturing is to achieve real-time defect prevention. Achieving this goal requires the manufacturing systems to have the following capabilities:

(1) Observation of process variables, i.e., measurement of product quality characteristics in real-time;

(2) Representation of the stochastic properties of the production system, i.e., modeling of the quality characteristics;

(3) Prediction of the future departure of observation from target;

(4) Adjusting the machine to cancel out this departure, i.e., the need for actuators and control.

Either statistics or engineering technologies alone are able to fulfill the above requirements. Therefore, the next generation quality control methodology will focus the fusion of statistics and engineering technologies.

With the development of in-line sensing, product quality characteristics can be measured during manufacturing. For example, on-line machine straightness and flatness measurements provide data for on-line straightness and flatness control [6]. In discrete manufacturing, such as automobile body assembly, the Optical Coordinate Measuring Machine can provide accurate, in-line measurement for every body assembled [7].

In-process measurement data from a continuous manufacturing process, or from a discrete manufacturing process where every product is measured, could be serially correlated. That is, the current behavior of a process variable is influenced by its past behavior plus some new disturbance. Similarly, the present behavior of the variable will affect its future value, thus making prediction possible. Once the future value of a process variable is predicted and the deviation from the target is large, action can be taken by operators or some actuators to move the variable on target, therefore, resulting in minimum variance control.

Therefore, in the envisioned Defect Preventive Quality Control, dependent data analysis based on the Dynamic Data Systems (DDS) methodology [8] will be used to deal with the vast amount of data. The Dynamic Data System (DDS) is a time series based methodology for the estimation of physically meaningful mathematical models of a process from operational or experimental data. It distinguishes between a "best fit" and the "best" physically meaningful model, and seeks the latter. The DDS methodology is based on the fact that many engineering systems can be approximated by a stochastic differential equation of the form shown below, denoted as AM(n,n–1):

$$\frac{d^n x(t)}{dt^n} + b_{n-1} \frac{d^{n-1} x(t)}{dt^{n-1}} + \dots + b_0 x(t) = c_{n-1} \frac{d^{n-1} u(t)}{dt^{n-1}} + \dots + c_0 u(t) \qquad (1)$$

where u(t) and x(t) are the input and output variables of the system, and b_i and c_j are the coefficients of the differential equation.

Corresponding to the AM(n,n–1), there is a stochastic difference equation of the form:

$$x_t = \phi_1 x_{t-1} + \phi_2 x_{t-2} + \ldots + \phi_n x_{t-n}$$
$$+ a_t - \theta_1 a_{t-1} - \ldots - \theta_{n-1} a_{t-n+1} \quad (2)$$

where x_t's are the observed process outputs, a_t's are the Gaussian white noise disturbances to the process, and ϕ_i's and θ_j's are the Autoregressive and Moving Average (ARMA) parameters. This stochastic difference equation is also called ARMA(n,n–1) model.

The problem then becomes: given a set of observations, $x_1, x_2, \ldots$, identify a physically meaningful AM(n,n–1) model that best fit the given data. This problem can be decomposed into: identification, estimation, and transformation. In identification and estimation, ARMA(n,n–1) models are successfully estimated until a statistically adequate representation is obtained. The DDS models are the appropriately constrained ARMA (n,n–1) models that correspond to the AM(n,n–1) processes. A transformation is necessary to convert the ARMA(n,n–1) models to the AM(n,n–1) models based on the criterion of covariance equivalence. Up to this stage, continuous time system parameters, such as natural frequencies and damping ratios, can be determined.

2.1 Features of Dynamic Data System (DDS)

DDS has three features: identification, forecasting, and control. These three features are described below:

(1) *Identification.* Given that an appropriate ARMA (n,n–1) model has been identified, the characteristic equation for the model is given by

$$1 - \phi_1 B - \phi_2 B^2 - \ldots - \phi_n B^n = 0 \quad (3)$$

or

$$(1 - \lambda_1 B)(1 - \lambda_2 B) \ldots (1 - \lambda_n B) = 0 \quad (4)$$

where λ_i's are the eigenvalues of the system, and B is the back-shift operator. Equation (4) can be decomposed as the combination of several first order and second order models. For a pair of complex conjugates, λ and λ^*, the natural frequency and damping ratio can be obtained by the following equations:

$$\omega_n = \frac{1}{2\Delta}\sqrt{[\ln(\lambda\lambda^*)]^2 + 4\left[\cos^{-1}\left(\frac{\lambda+\lambda^*}{2\sqrt{\lambda\lambda^*}}\right)\right]^2} \tag{5a}$$

$$\zeta = \frac{-\ln(\lambda+\lambda^*)}{\sqrt{\left[\ln(\lambda\lambda^*)\right]^2 + 4\left[\cos^{-1}\left(\frac{\lambda+\lambda^*}{2\sqrt{\lambda\lambda^*}}\right)\right]^2}} \tag{5b}$$

(2) *Forecasting*. From estimated time series models as shown in Equation (2), future outputs of the process can be predicted with some confidence. For example, the one-step-ahead forecasting for the ARMA(n,n–1) model is given by:

$$\hat{x}_t(1) = \phi_1 x_t + \phi_2 x_{t-1} + \ldots + \phi_n x_{t-n+1} - \theta_1 a_t - \ldots - \theta_{n-1} a_{t-n+2} \tag{6}$$

(3) *Control.* Once the future outputs of a process can be predicted, some control or compensation action can be taken to make the process on target. Control strategies, such as Forecasting Compensatory Control (FCC) [6] can be applied to result in minimum variance control. FCC has been successfully applied to continuous manufacturing, such as straightness and flatness control in machining. However, it has not been applied to discrete manufacturing processes such as automobile body assembly or door fabrication due to the lack of control mechanisms that activate control instantly. Process adjustment can only be made on a batch-to-batch basis. Therefore, process control is based on the early detection of process faults and the correction of process faults by human interference. Eventually, with the development of adaptive assembly tooling, minimum variance control could be applied to improve process and product quality.

With the availability of on-line sensing techniques, actuators, and control, the Dynamic Data System can be applied for more efficient process control using dependent data. The three features of DDS are related to the three aspects of process control as shown in Figure 2. Not only prediction error, but information contained in the time series model will be used for process monitoring, process parameter identification and process variation reduction.

2.2 Three Aspects of Process Control

(1) *Process Monitoring and Surveillance.* Conventional control charts are based on the assumption of independent data. To satisfy this requirement, Prediction Error Analysis

has been used to remove the correlation, and then apply the control charts the residuals or prediction error [9]. Analysis and simulations showed that the application of Prediction Error Analysis results in the change in the detection speed depending on the parameters of the ARMA model, or the dynamics of the process, and that process damping ratio plays an important role in affecting the detection speed. Therefore, it was suggested that process monitoring be performed together with the identification of process physical characteristics [7].

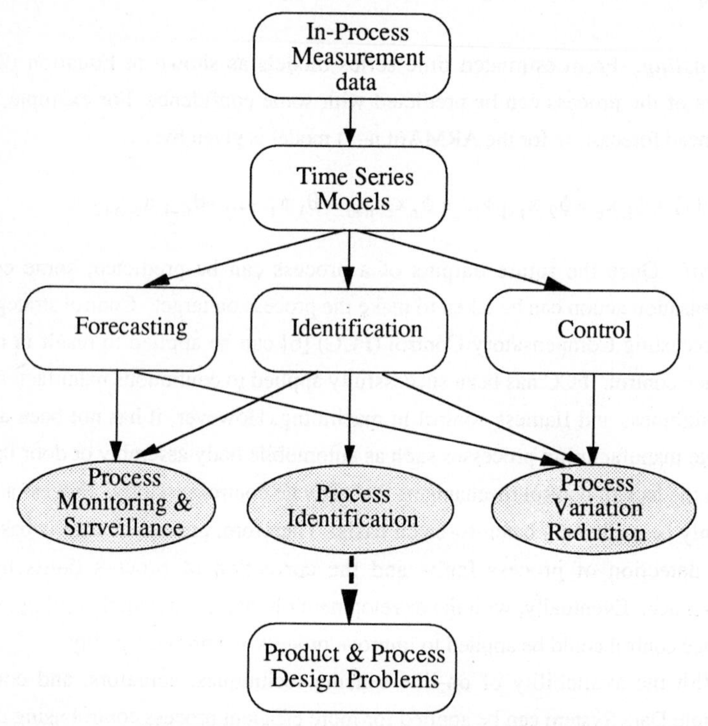

Figure 2. Three Aspects of Process Control

(2) *Process Parameter Identification.* In-process measurement data are usually serially correlated and also cross correlated. Using these correlations, sources of quality variation can be identified. Using serial correlation in the data, process physical characteristics, such as natural frequency and damping ratio, can be identified. Using the cross correlation, inter-sensor relationship can be identified. On many occasions, these sources of dimensional variation can be attributed to the improper product and process design. In the next section, two examples from automobile body assembly will be used to illustrate this point.

(3) *Process Variation Reduction.* Process variation reduction can be achieved in two different ways. The first approach is the identification of variation sources and the removal of these sources through process modification or re-design. The second approach is the application of minimum variance control strategy, such as FCC for on-line process variation reduction.

To fulfill the requirement of real-time defect prevention, engineering technologies, including real-time sensing, signal processing, automated machine monitoring/diagnostics, and control are necessary. These engineering technologies have advanced tremendously. This provides another essential support to the development of NGQC in manufacturing. Some of the engineering technologies in real-time sensing, signal processing, and control will be presented in combination with case studies.

3 Quality Control Case Studies

In this section, work in two areas will be presented. One is precision machining using Forecasting Compensatory Control. The other is variation reduction in automobile body assembly. Case studies from both areas will be used to illustrate the application of statistical techniques and engineering technologies.

3.1 Forecasting Compensatory Control of Machining Accuracy

The objective of this work is to develop methodologies and systems for cost effective precision machining without precise machinery using Forecasting Compensatory Control.

The Forecasting Compensatory Control strategy has been under development over the past 13 years, and has been successfully implemented in the laboratory for straightness control, flatness control, roundness control, and cylindricity control; and in an automobile assembly plant for paint viscosity control. An example of flatness control in disc turning operation is presented here to illustrate the application of real-time sensing, real-time modeling, and control [10].

In manufacturing the Winchester discs, the geometric requirements of the disc require the surface finishing process to be superprecise. Such machining is characterized by the especially stringent demands on the machines, tool, and machining technologies used. Also the properties of the material being processed must be stable and easy to manufacture.

Various methods have been adapted to improve the disc lathe performance. The methods used include ultra precision spindle, increased structural rigidity, improved feeding mechanism, etc. These approaches are effective and straightforward in improving the machining accuracy, but these improvements are related to the extremely high cost for machine tool design and manufacture.

Improving the machining accuracy through error compensation is another technique, which is also the only feasible approach to improve the performance of a built-up machine without changing its structure. Furthermore, this approach may also reduce the statistical errors resulting from ill conditioning of machine components or environmental factors. The compensation techniques can be classified into two approaches, according to the condition of the machine. One is the Direct Compensation Control (DCC), which is usually applied to machines with slow motion. The other method is the Forecasting Compensatory Control (FCC). An error signal, whether deterministic or stochastic, is forecasted with certain confidence based on developed process model. Then, some variables, such as cutting tool tip position, are manipulated to cancel out the predicted error or deviation.

(1) *Flatness and Flatness Measurement.* Flatness is defined as the quality of being flat, or plane. Flatness error or out-of-flatness is considered to be the distance that separates the two parallel planes, within which all points on the surface must fall. Typical methods for flatness measurement include Straightedge, color transfer method, Moir's grid, Autocollimator, and interferometers.

(2) *Flatness Error Sources and In-line Measurement.* Several sources contributed to the surface flatness error in turning operation. As shown in Figure 3, they are:

(a) *Spindle Error Motion* (A_3 and A_4). The spindle is supported by roller bearings during the rotation. The irregular contact motion between the rollers and housing, as well as compliance and damping of the structure in conjunction with the internal and external sources of excitation, will cause the spindle center line to move in radial and axial directions.

(b) *X-slide Error Motion* (A_2). The non-straightness error in the Y-direction of the x-axis linear carriage will cause the surface flatness variation on the turning surface.

(c) *Squareness of the X-slide and Spindle* (A_1). Ideally, the angle between the spindle center line and the x-axis of slide is perpendicular when they are initially assembled. Such a situation is not usually maintained after years of operation due to wear and deterioration. A_1 is the measure of the squareness between the x-slide and the spindle.

(d) *Squareness Between Spindle and Chuck Plate* (A_5). The surface of the chuck plate may not be perpendicular to the center line of spindle rotation. This might be caused by the heavy load of the spindle and the associated deflection of the supporting bearings.

Other factors that will affect the surface flatness include tool wear, machine vibration, etc. To achieve an on-line compensation procedure, it is necessary to

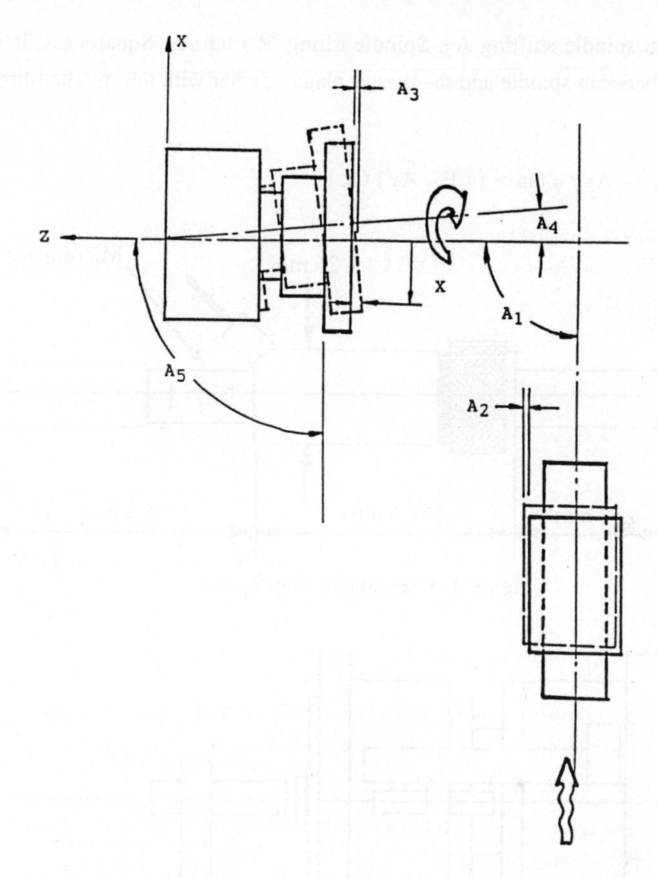

Figure 3. Error Sources Contributing to the Surface Flatness

detect each error source by careful setup of measurement equipment. However, it is very difficult to realize all these because of the limitations of sensor and measurement constraints. These error sources are not considered.

If all the error sources are simultaneously obtained on the XZ plane which includes the motion trajectory of the tool tip, the resultant error in the Z direction will be:

$$E_t = A_2 + A_3 + X \tan A_4 + X \cot A_5 + X \cot A_1 \tag{7}$$

Two probes are used to measure the spindle face error motion. One sensor is fixed in the front of the nose plate whose surface area is twice that of the probe. The nose plate is mounted on the center of the vacuum chuck. The shifting of the spindle will cause the change of the gap, thus, A_3 is obtained.

Another probe is fixed at distance R from spindle center and before the rear surface of the master plate at the same height as the tool tip. The output E from the contains three error sources: spindle shifting A_3, Spindle tilting, $R \bullet \tan A_4$, Squareness, $R \bullet \cot A_5$. The squares between spindle and the master plate together with the spindle tilting can be obtained as:

$$A_4 + (\pi/2 - A_5) = \tan^{-1} [(E - A_3) / R] \tag{8}$$

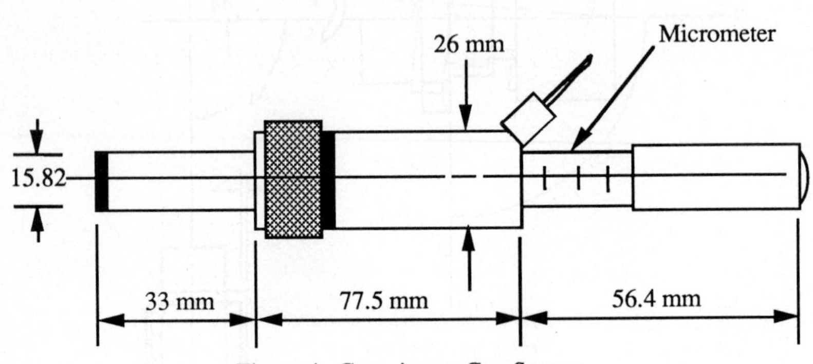

Figure 4. Capacitance Gap Sensor

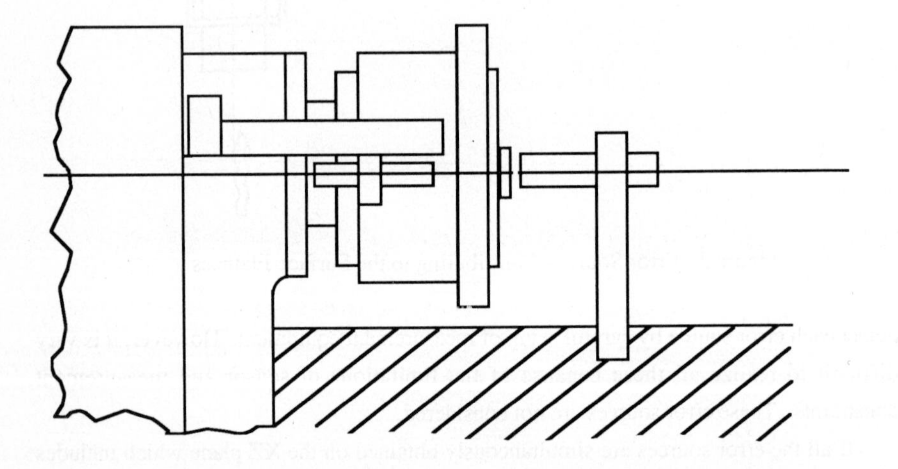

Figure 5. Probe Arrangement for Measuring the Spindle Face-error Motion

A position sensitive lateral type photo-detector and a signal conditioner are also used in this experiment for measuring the straightness of the slide. The setup is shown in Figure 6.

After obtaining each error source on the machine, the resultant flatness error at the tool tip can be simplified as:

$$E_t = A_3 + A_2 + X_t \{ \tan^{-1} [A_4 + (\pi/2 - A_5)] \}$$
$$= A_3 + A_2 + X_t (E - A_3) / R \qquad \cdot (9)$$

where A_3 is the spindle axial motion measured using the first probe; A_2 is the slide straightness measured using photo-detector, x_t is the tool tip position relative to the center of spindle rotation, **E** is the measurement from the second probe, and **R** is the distance of the second probe from the center of spindle.

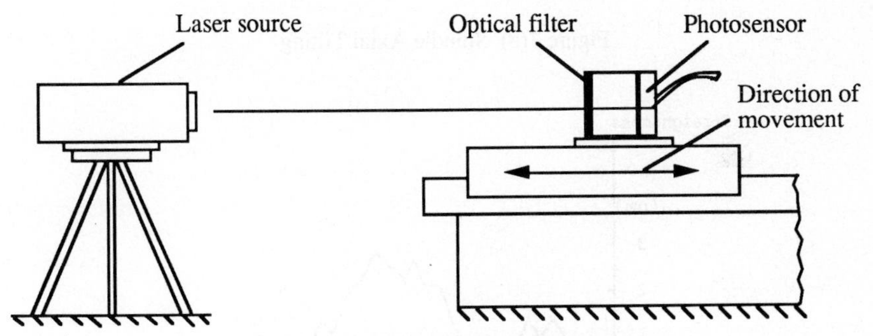

Figure 6. Straightness Measuring Setup

The three error sources measured are shown in Figures 7(a), 7(b), and 7(c), and the resultant errors are shown in Figure 7(d).

(3) *Real-Time Modeling and Forecasting*. After obtaining the flatness error, an initial model was selected off-line. Both the Autoregressive (AR) and the Autoregressive Moving Average (ARMA) models were fitted to the data. It was found that both AR(4) and ARMA (3,2) models fit the data well.

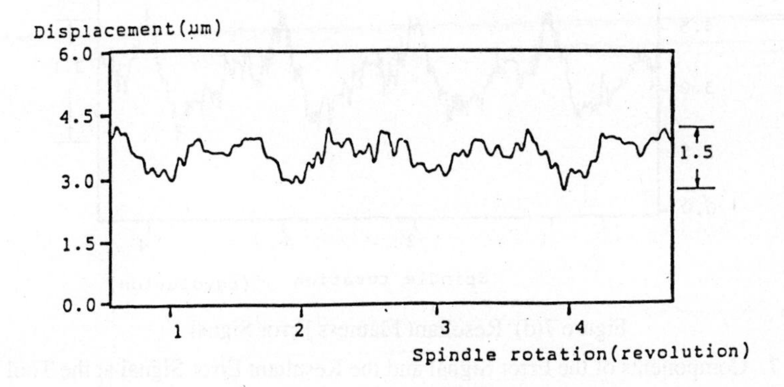

Figure 7(a) Spindle Axial Shifting

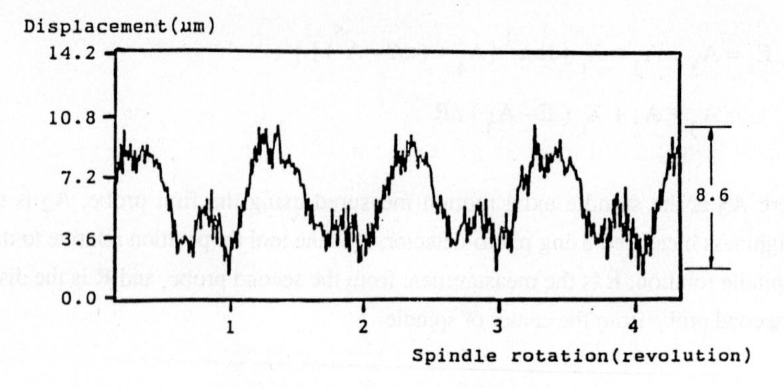

Figure 7(b) Spindle Axial Tilting

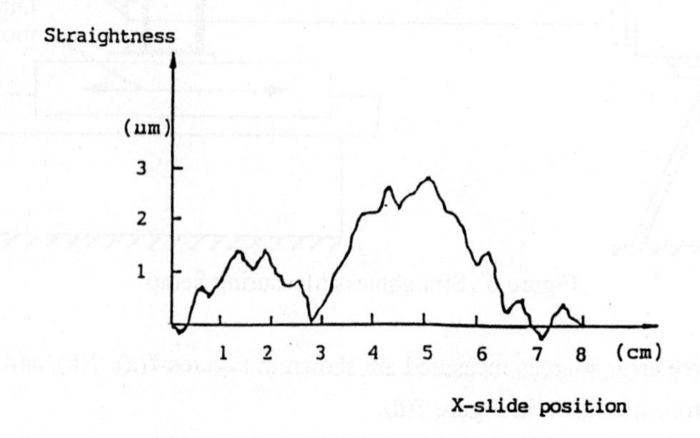

Figure 7(c) Straightness of the X-slide

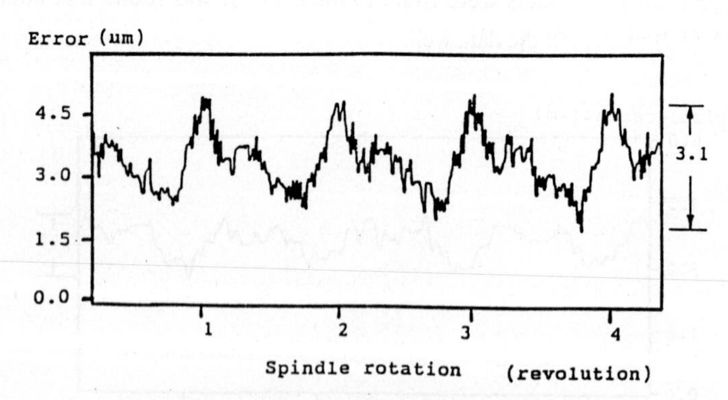

Figure 7(d) Resultant Flatness Error Signal

Figure 7. Components of the Error Signal and the Resultant Error Signal at the Tool Tip

For real-time application, a recursive AR model will be more appropriate. If the modeling and identification routines were implemented on-line using assembly language, the calculation time required for identification and one-step-ahead forecasting for AR(2) was about 0.9 ms, 2.3 ms for AR(3), and 4.7 ms for AR(4). Since the sampling interval for this on-line control system was constrained as 1.2 ms, the AR(2) model was selected for implementation. Though such a decision will reduce the forecasting accuracy, the overall system performance was less affected when all factors are considered.

Once the error source and its model are obtained, the future error signal can be predicted based on the current and previous data by means of forecasting equations, which are another form of the AR model. The one-step-ahead forecasting of the AR(2) model and its forecasting error are shown in Figure 8.

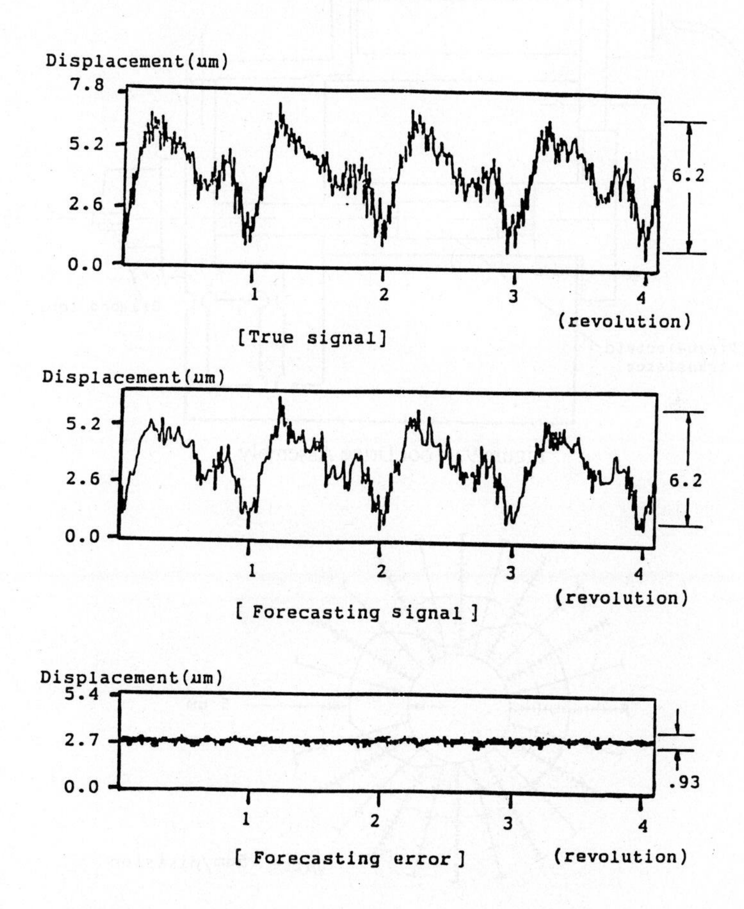

Figure 8. Data, Forecasted Values, and Forecasting Errors

(4) *Real-Time Action.* A piezoelectric actuator is selected for generating the movement to position the diamond cutting tool (Figure 9). The actuator is made of stacked ceramic discs with piezoelectric property. One end of the actuator is connected to the body by one magnetic cylinder, and the other end is fastened on the tool shank. As a high voltage is applied to the actuator, it will expand and push the tool shank and the cutting tool.

(5) *Cutting Experiment.* The effectiveness of the implemented compensatory control system was tested through actual on-line face turning experiments. The uncompensated and compensated disc flatness are shown in Figure 10, and the uncompensated and compensated straightness shown in Figure 11.

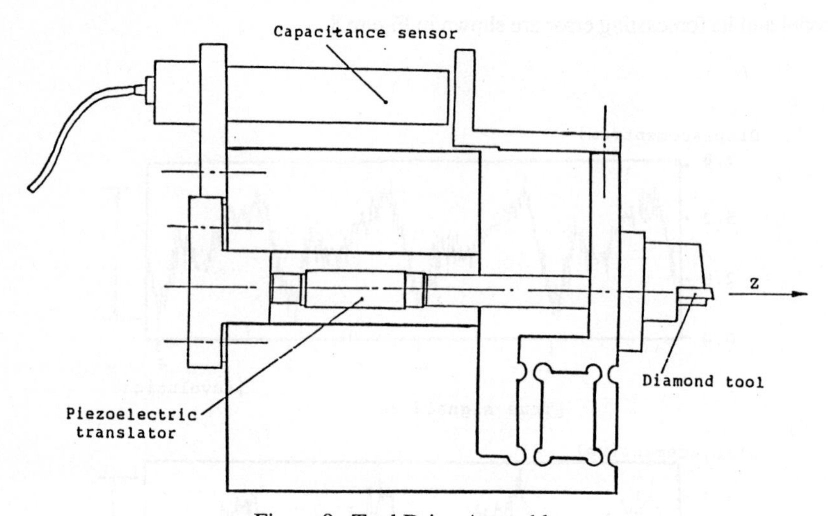

Figure 9. Tool Drive Assembly

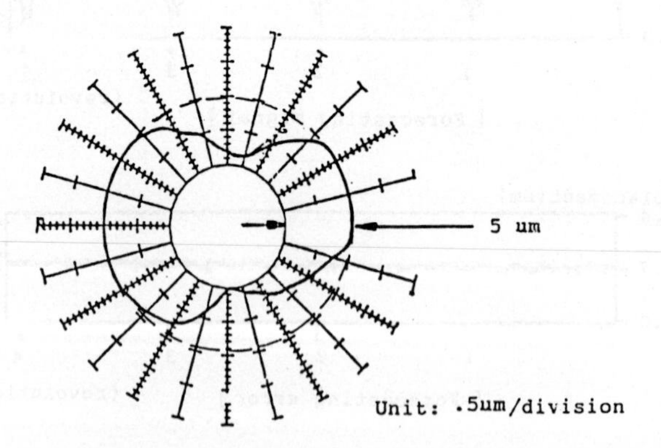

Figure 10(a) Without Compensation: 5μm

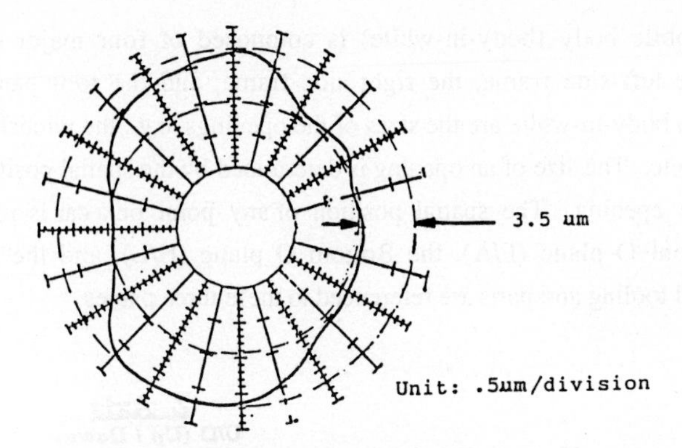

Figure 10(b) After Compensation: 3.5 µm

Figure 10. Circular Flatness of Discs With and Without Compensation

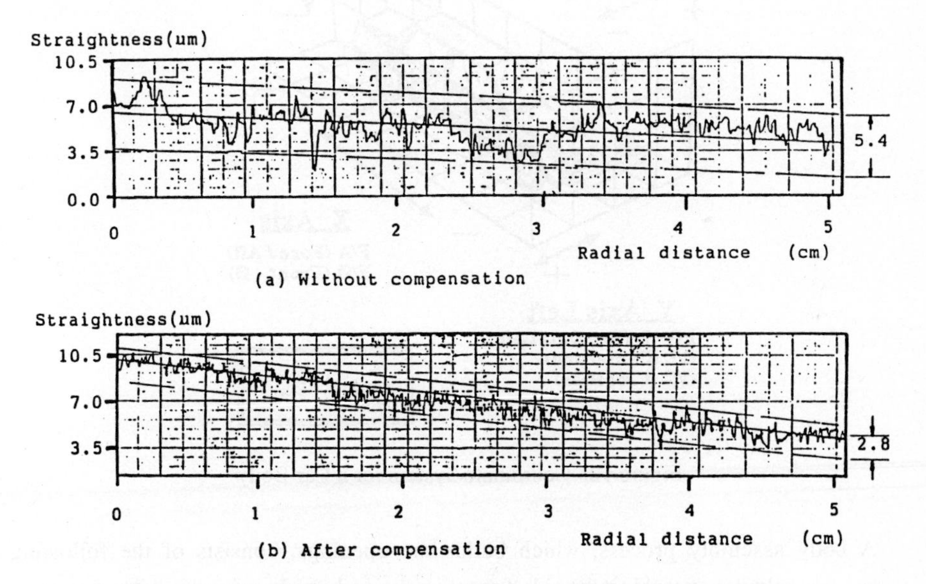

Figure 11. Disc Radial Straightness With and Without Compensation

3.2 Variation Reduction for Automobile Body Assembly

Though automobile body assembly processes vary from corporation to corporation and from plant to plant, the principles behind them are very similar. Presented here is a body assembly process from a major automobile manufacturer.

An automobile body (body-in-white) is composed of four major components: the underbody, the left side frame, the right side frame, and the roof panel. The critical dimensions of a body-in-white are the sizes of the openings, e.g., the windshield opening, the door openings, etc. The size of an opening is determined by the spatial positions of the points surrounding the opening. The spatial position of any point on a car is referenced to three planes, the Front O plane (F/A), the Bottom O plane (H/L), and the Centerline (I/O) (Figure 12). All tooling and parts are referenced to these three planes.

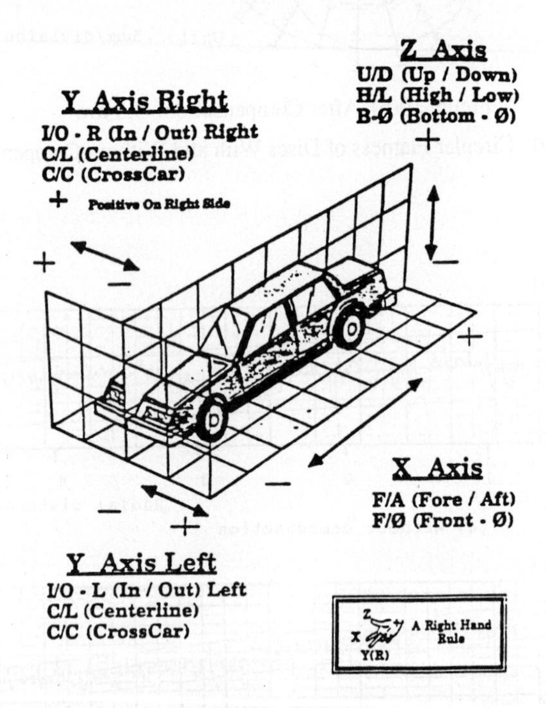

Figure 12. Coordinate System for a Car Body

A body assembly process, which builds the openings, consists of the following stages: toy tabbing--attaching the side frames to the underbody using toy tabs; framing--welding the side frames onto the underbody in a single locating fixture, e.g., robogate; and Re-Spot welding.

These three stages are briefly described as follows:

(1) *Toy Tabbing*. In this stage, the side frames are attached to the underbody using toy tabs. The underbody is positioned and clamped on the pallet. The toy tabs will provide some latitude for the two side frames to move relative to the underbody. Two operators check the positions of the side frames and the toy tabs to ensure that the side frames are attached

to the underbody and that the toy tabs are on. Some other parts, such as the rear end panel, the roof, etc., are also installed onto the unwelded auto body.

(2) *Framing*. The unwelded automobile body, including the underbody, the side frames, the roof, and other small parts, proceeds towards the robogate on a moving pallet. The robogate is a single locating fixture consists of, say, 10 welding robots and two side gates. The pallet can be accurately positioned in the F/A, I/O, and H/L directions inside the robogate. While the positions of the underbody in the F/A, I/O and H/L directions are determined by the pallet, the positions of a side frame relative tothe pallet in the F/A, I/O, and H/L directions are determined by the side gate. Therefore, the repeatability of the side gate will affect the process variance.

Once the side gates position the side frames onto the underbody, clamps are activated to hold these positions. Then weldings are made to fix the relative positions of the side frames relative to the underbody.

(3) *Re-Spot Welding*. After robogate, the partially welded automobile body enters the stage of re-spot welding. In this stage, more welding is done on the body to increase strength and integrity. In general, re-spot welding will not affect the relative position for all the subassemblies. However, it may change the body dimensions due to deformation.

At the end of Re-Spot welding are the dimensional measurement systems: the in-line Optical Coordinate Measuring Machine (OCMM) and the off-line mechanical Coordinate Measuring Machine (CMM). While every body goes through the OCMM being measured, only a selected number of bodies are measured on the CMM. After the OCMM, the bodies go to a hand welding and repair area, and then proceed to the door hanging, fender installation, and hood and deck lid setting operations. A complete body assembly process is shown in Figure 13.

(a) *In-Line Dimensional Measurement*. The in-line Optical Coordinate Measuring Machine (OCMM) is a new dimensional measuring technology introduced in 1988, originated from in-line machine vision gauging. An OCMM checks the dimensions of every car body produced in sync with production rate, resulting in 100% measurement.

Most OCMM sensors are based on the principle of laser triangulation as shown in Figure 14. Feature deviation in space is reflected in the image space of a photo-detector array. A typical OCMM setup for body-in-white consists of about 100 sensors. All these sensors are calibrated to a world coordinate through the use of theodolites, therefore making measurements compatible with CAD data. Sensors work in a serial fashion and are controlled by a logic controller.

(b) *Simulation of Forecasting Compensatory Control*. In Hu [7], the idea of real-time defect prevention using Forecasting Compensatory Control was demonstrated through

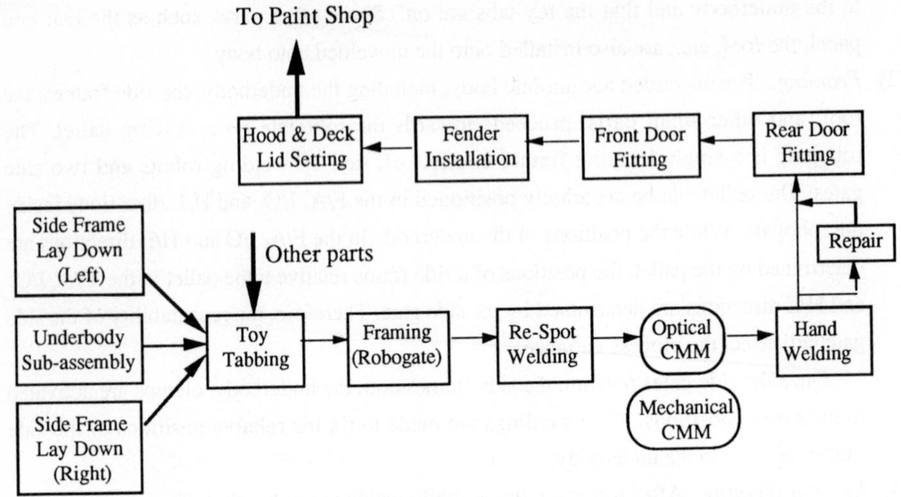

Figure 13. The Body Assembly Process (A Complete Layout)

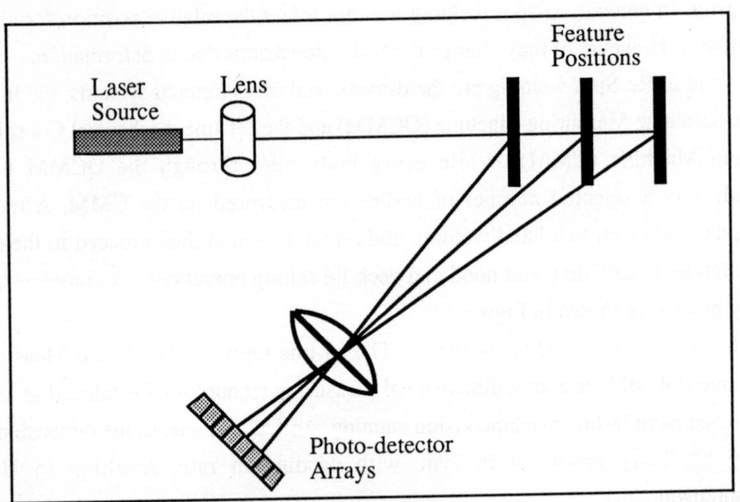

Figure 14. Sensor Principle: Dimensional Measurement by Laser Triangulation

simulation using data from body assembly processes. However, the current design of assembly tooling makes it impossible to adjust the process at every sampling interval. Adjustment can only be made on a batch-to-batch basis. That is, control actions are taken only when the process is stopped. Therefore, the optimal control policy will not be the implementation of Forecasting Control or other minimum variance controls for every body produced because of the lack of control mechanisms that actuate control instantly. However, with the development of flexible and adaptive tooling, minimum variance control could be used to improve process performance and product quality.

If control action cannot be taken at every sampling interval, the process control will be based on the detection of a process fault or a large predicted deviation from the target. Then an action has to be taken to correct the process. In this case, a fast monitoring and diagnostic system becomes a necessity and human operators become an integrated part of the feedback loop in process control. In the sections that follow, two case studies are used to illustrate the procedure of monitoring and control in body assembly.

(c) *Variation Reduction Case Studies.*

Case 1: Side Frame Misalignment in Automobile Body Framing: Figure 15 shows the measurement data in the Fore/Aft (F/A) direction and the positions for the eight points on the left hand side frame. Based on measurement data, a correlation matrix is estimated and is shown in Table 1.

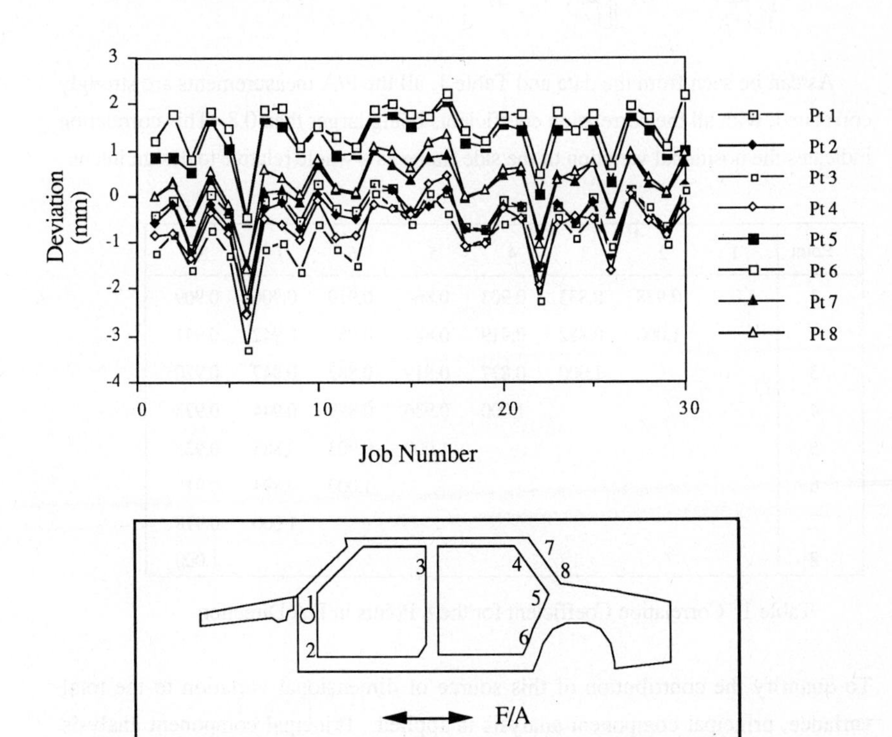

Figure 15. Measurement Data in the F/A Direction and Measurement Point Locations

For a p-dimensional vector time series,

$$\mathbf{x}_t = \begin{bmatrix} x_{1t} \\ \cdots \\ x_{pt} \end{bmatrix}$$

the correlation matrix is defined as:

$$\mathbf{R} = \left[r_{ij} \right]$$

where r_{ij} is the correlation between variable x_i and x_j.

$$r_{ij} = \frac{\sum_{k=1}^{n} (x_i - \bar{x}_i)(x_j - \bar{x}_j)}{\sqrt{\left[\sum_{k=1}^{n} (x_i - \bar{x}_i)^2\right]\left[\sum_{k=1}^{n} (x_j - \bar{x}_j)^2\right]}} \tag{10}$$

As can be seen from the data and Table 1, all the F/A measurements are strongly correlated, with all the correlation coefficients being larger than 0.8. This correlation indicates the positional variation of the side frame as a whole relative to design intent.

Point	1	2	3	4	5	6	7	8
1	1.000	0.978	0.833	0.903	0.864	0.919	0.906	0.909
2		1.000	0.892	0.919	0.883	0.959	0.942	0.941
3			1.000	0.877	0.819	0.882	0.947	0.920
4				1.000	0.936	0.899	0.944	0.973
5					1.000	0.903	0.885	0.925
6						1.000	0.894	0.919
7							1.000	0.978
8								1.000

Table 1. Correlation Coefficient for the 8 Points in F/A Direction

To quantify the contribution of this source of dimensional variation to the total variance, principal component analysis is applied. Principal component analysis involves finding the eigenvalues and eigenvectors of the sample correlation matrix [11] with the variances of the principal components being the eigenvalues of matrix

R, and weight b_{ij} being the normalized eigenvector corresponding to the i-th eigenvalue.

An eigenvalue problem is defined as:

$$\mathbf{R}\,\mathbf{b} = \lambda\,\mathbf{b} \quad (11a)$$

$$[\,\mathbf{R} - \lambda\,\mathbf{I}\,]\,\mathbf{b} = \mathbf{0} \quad (11b)$$

where **b** is the eigenvector, and λ is the eigenvalue. The eigenvalues and eigenvectors are listed in Table 2.

Com ponent	Eigen- Value	%	Eigenvector, coefficients of							
			x_1	x_2	x_3	x_4	x_5	x_6	x_7	x_8
1	7.39	92.37	.350	.407	.429	.378	-.260	-.311	.355	.307
2	.219	2.74	.360	.192	.396	.055	-.042	.156	-.399	-.699
3	.193	2.41	.343	-.695	.177	-.317	-.198	-.443	.141	-.107
4	.111	1.38	.357	.026	-.388	.315	.549	-.487	-.289	.020
5	.041	.51	.345	.327	-.622	-.290	-.544	-.030	-.014	-.074
6	.029	.37	.353	.244	.218	-.664	.420	.172	-.026	.348
7	.011	.14	.359	-.353	-.005	.302	-.218	.465	-.419	.465
8	.007	.09	.362	-.151	-.207	.195	.265	.446	.658	-.250

Table 2. The Eigenvalues and Eigenvectors of the Correlation Matrix for 8 F/A Measurements

The variances of the eight principal components are:

1st component	7.390	92.37%
2nd component	0.219	2.74%
3rd component	0.193	2.41%
4th component	0.111	1.38%
5th component	0.041	0.51%
6th component	0.029	0.37%
7th component	0.011	0.14%
8th component	0.007	0.09%

The first component takes about 92.4% of the total variation. Looking into the coefficients for the first component,

$$z_1 = 0.35\,x_1 + 0.36\,x_2 + 0.343\,x_3 + 0.357\,x_4 + 0.345\,x_5 + 0.353\,x_6 + 0.359\,x_7 + 0.362\,x_8$$

all of the eight coefficients are of almost equal magnitude, which means that the first principal component is basically the average of all the eight measurements. This average is the rigid body movement of the side frame as a whole during the framing stage. Investigation into the manufacturing process identified that this whole body movement was caused by the wheelhouse/underbody interference during framing as showing in Figure 16.

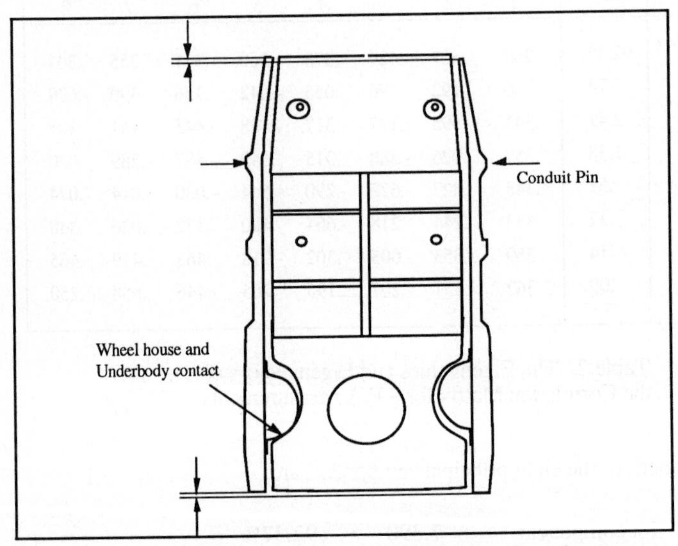

Figure 16. Wheel-house and Underbody Interference Causing the Side Frame Large Variation in the F/A Direction

An alternative process design will be to first weld the wheel house inner to the underbody, instead of to the side frame, and then attaching the side frame to the underbody during framing. In this case, the side frame can move freely relative to the underbody without interference. This reflects the importance of product design in achieving minimum variation.

Case 2: Bi-Modal Distribution in Door Assembly : Manufacturing process design is also of significant importance to product quality. The following example is from a door assembly process.

A typical door assembly consists of the inner panel, the inner belt bar, the outer belt bar, the hinge, the crash bar, and the outer panel. Dimensional data from the in-line Optical Coordinate Measuring Machine revealed that Point A was experiencing large variation compared with other areas on the door (Figure 17).

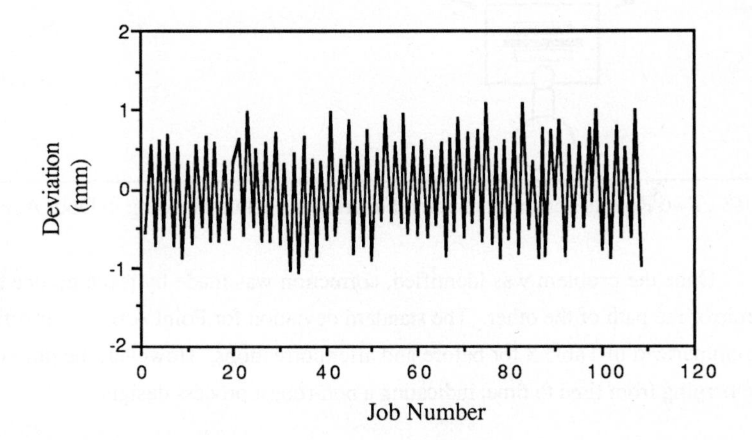

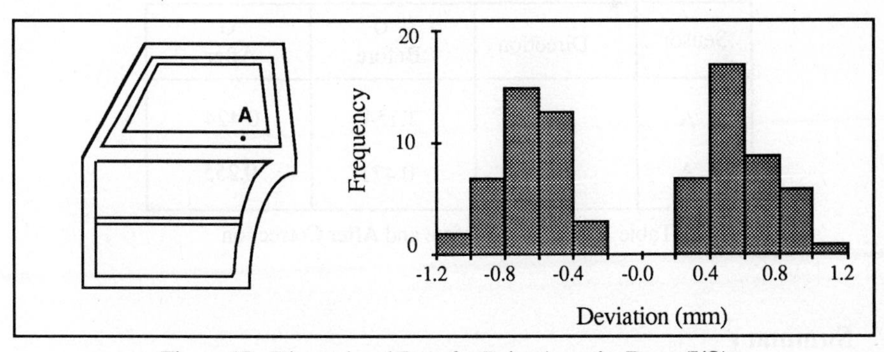

Figure 17. Dimensional Data for Point A on the Door (I/O)

The 6-sigma variation was estimated to be about 4 mm, and the frequency histogram clearly showed bi-modal distribution. Investigation into the production process revealed that the alternating nature in the data was caused in the pick-up welding station for the daylight opening. Two welding robots were used to perform the same function. However, they were not well synchronized, causing bi-modal distribution in the measurement data.

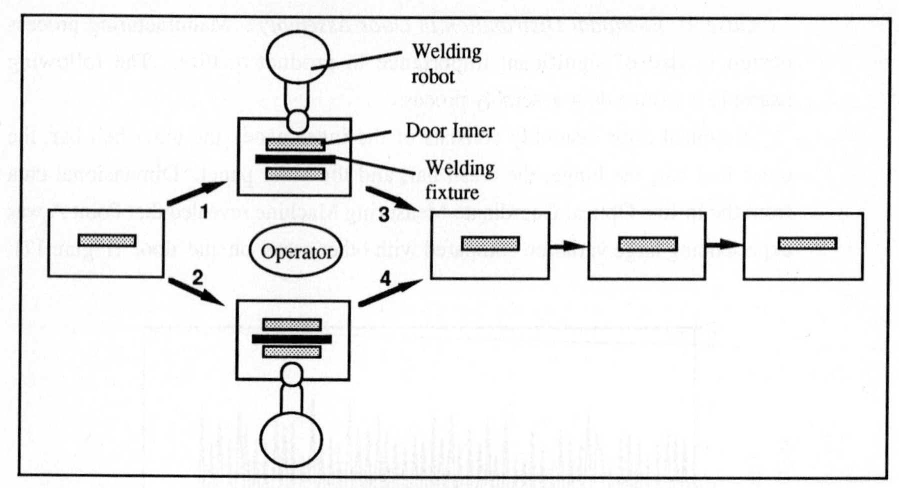

Figure 18. Two Robots Perform the Same Function in Pickup Welding in Door Assembly

Once the problem was identified, correction was made by teaching one robot to mirror the path of the other. The standard deviation for Point A in two directions are summarized in Table 3 for before and after corrections. However, the problem kept emerging from time to time, indicating a non-robust process design.

Sensor	Direction	σ Before	σ After
A	I/O	1.154	0.424
A	H/L	0.471	0.255

Table 3. Variation Before and After Correction

4 Summary

By combining advanced engineering technologies in sensing, computer, and control, and advanced statistical technique, a new generation of quality control methodologies can be developed to achieve real-time defect prevention. Two areas of applications are presented. One is the use of Forecasting Compensatory Control in precision machining. The other is the application of in-process 100% measurement in reducing the dimensional variation of automobile body assembly.

From the viewpoint of Concurrent Engineering, which is a systematic approach to the integrated concurrent design of products and related processes, information obtained during manufacturing and during inspection can contribute to the improved design of products and manufacturing processes.

References

1. Ishikawa, K. 1983. *Guide to Quality Control.* 8th printing, Asian Productivity Organization, Tokyo, Japan.
2. Shewhart, W.A. 1931. *Economic Control of Quality of Manufactured Product*, New York, Van Nostrand.
3. Grant, E.L., and Leavenworth, R.S. 1980. *Statistical Quality Control*, Fifth Edition, New York, McGraw-Hill.
4. Deming, E. 1986. *Out of Crisis.* MIT Center for Advanced Engineering Study, Cambridge, MA.
5. Taguchi, G. 1986. *Introduction to Quality Engineering.* Asian Productivity Organization, Tokyo, Japan.
6. Wu, S.M. and Ni, J. 1989. "Precision Machining without Precise Machinery ", *Annals of CIRP*, Vol. 38-1, pp.533-536.
7. Hu, S. 1990. *Impact of 100% Measurement Data on Statistical Process Control (SPC) in Automobile Body Assembly.* Ph. D. Thesis, University of Michigan
8. Wu, S.M. 1977. "Dynamic Data System: A New Modeling Approach," *Transaction of the ASME, Journal of Engineering for Industry*, August, pp.708-714
9. Alwan, L.C. and Roberts, H.V. 1988. "Time Series Modeling for Statistical Process Control," *American Statistical Association Journal of Business & Economic Statistics*, January, Vol. 6, No.1, pp. 87-95.
10. Your, S.B. 1987, "*Precision Control on the Winchester Disc Flatness in Face-Turning Operations*", Ph.D. Thesis, University of Wisconsin.
11. Manly, B.F.J. 1986. *Multivariate Statistical Methods, A Primer.* Chapman and Hall.

An Approach for PDES/STEP Compatible Concurrent Engineering Applications

Thu-Hua Liu and Gary W. Fischer

Centers for Computer-Aided Design, College of Engineering, The University of Iowa, Iowa City, Iowa, 52242-1000 USA

Abstract: An object-oriented approach to implement PDES/STEP compatible applications is proposed. To demonstrate the approach, a computer processable Assembly Application Protocol (AAP) is developed and documented as an object class structure and implemented in the form of a class library. The approach shows how (1) the PDES/STEP entities are used as building blocks of engineering information, (2) the EXPRESS language is used to model the PDES/STEP compatible data model, and (3) the application oriented data semantics can be embedded in a computer system. The AAP acts as a feature-based information model and persistent object data schema for CE assembly applications. The software tools used to implement the AAP are EXPRESS2C++ translator, C++ programming language, and ROSE database. The use of standard tools are shown to enhance the CE communication and research opportunities, based on the experience with the new generation international product definition and exchange standard, are suggested.

Keywords: PDES/STEP / object-oriented application / feature-based / data model / application protocol / product data communication / concurrent engineering

1 Introduction

1.1 Computer-Aided Concurrent Engineering Using Standard Product Definition

Concurrent Engineering (CE) is a methodology and environment for coordinating multidisciplinary engineering activities, in which design, engineering analysis, and manufacturing analysis of a product can be carried out simultaneously. CE can be achieved by people interacting directly with other people and does not necessarily have to be an automated process; but today's industrial environment often involves complex products and the use of computer-based systems, which tend to be data intensive systems.

The importance of Concurrent Engineering is well known. However, some difficulties must be overcome to implement Concurrent Engineering. Three sources of Concurrent Engineering difficulty are summarized by [1]: (1) the characteristics of the design process, (2) the volume and variety of product data, and (3) the separation of engineering functions. These difficulties can be overcome if "full" communication between multidisciplinary engineers, computer systems, and engineering software is established [2-4].

Full communication to support CE can be achieved when standardized product data are used in computer-based application systems, but two issues need to be addressed [5]:

(1) Being able to obtain or derive product characteristics for discipline specific applications from a generic, shared definition of a product is required in an integrated Concurrent Engineering environment.
(2) The abstractions of product characteristics must be consistent across different applications. Such consistency relies on the existence of a shared standard product definition and the ability to extract product characteristics from the shared definition.

In other words, the key need for a successful Computer-Aided Concurrent Engineering (CACE) environment, is the development of the information modeling techniques and associated standards that define the representation of product data. Today, most product information exchange between CAD, CAE, and CAM applications relies on geometric information standards, such as the Initial Graphics Exchange Standard (IGES) [6]. Unfortunately, IGES either does not convey all required product characteristics, such as form features and tolerances, or loses some of product characteristics in the product information exchange process. Furthermore, when using IGES, product data are exchanged in terms of low level geometry entities, such as curves, lines, and points.

Product Data Exchange using STEP (PDES/STEP) is an international product data standard [7-8]. PDES/STEP provides an avenue for computer-based engineering applications to exchange data between CAD/CAM/CAE/CIM systems [9]. The PDES/STEP standard is a neutral product model data exchange mechanism that is capable of completely representing product definition data throughout the life cycle of a product [7]. PDES/STEP is intended to be informationally complete for all engineering applications and to be directly interpretable by engineering application software [10].

However, PDES/STEP product information models alone cannot ensure the integration of the CACE environment, because inter-relationships and constraints between multidisciplinary applications are not defined in PDES/STEP. Hence, the success of the PDES/STEP depends not only on the process of creating standards, but also on the development of technologies that support the wide variety of automated systems to directly interpret the product definition and to communicate freely with each other.

This research started with the development of a unique conceptual data model that follows the PDES/STEP standard. The data model was constructed by applying object-oriented information modeling as it relates to the manufacturing process. Applying the same data model for both the global database (Concurrent Engineering database) and the local database (multidiscipline engineering function database) has the advantage of avoiding inconsistency, enforcing standards, and ease of maintaining data integrity [5].

1.2 Overview

This paper uses design for assembly to demonstrate a feasible approach to implement the PDES/STEP in a CACE environment. Section 2 describes a data model that contains three major objects: Part, Joint, and Assembly_Model in which the elements of product data are obtained from PDES/STEP. In Section 3 the use of STEP entities to form the feature based information for the assembly application data model is discussed. In Section 4 an approach to implement the PDES/STEP compatible application using the EXPRESS information modeling language and object-oriented ROSE database is proposed. Elements of product definition data from PDES/STEP, the architecture of compatible schemata, and application protocol are discussed. A computer processable Assembly Application Protocol (AAP) is developed and documented as an object class structure and implemented in the form of a class library. The AAP shows (1) how the EXPRESS standard language is used to model the PDES/STEP Compatible data model, (2) how the PDES/STEP entities are used as building blocks of engineering information, and (3) how the application oriented data semantics can be embedded in a computer system. The AAP acts as a feature-based information model and persistent object data schema for CACE assembly applications. Examples are used to demonstrate the capability of the protocol. Section 5 summarizes the methodology and advantages of the approach. In Section 6, the use of standard tools are shown to enhance the CACE communication, and research opportunities are suggested.

2 Product Conceptual Model

2.1 Product and Process Design for Assembly

The manufacture of a product involves the following three steps: (1) product design, (2) manufacturing system design, and (3) manufacturing system operations, e.g., fabrication, assembly, and inspection [11]. Design decisions made at each step determine the constraints, options for manufacturing processes, and costs. One of the main objectives of Concurrent

Engineering is to coordinate the design of products and processes so that effective and efficient manufacturing will be possible. The integration of product design and process design is one of the critical pacing activities for successful Concurrent Engineering.

Following the assumption of morphological analysis [12-14], a mechanical system can be decomposed into subsystems, sub-subsystems and so on until no further breakdown is meaningful. Young et al. define the design process as the separation of the design problem into stages such as conceptual design, detailed design, and analysis and evaluation [1]. Stoll describes a five-step procedure for integrated product/process design that helps encourage parallel development of engineering design [15]. Another procedure for product design that is emphasized for assembly is described by [16].

To explain the concurrent design of products and processes, the production of a mechanical system is divided into six steps (see Figure 1). The concurrent design proceeds from the system to the parts and integrated back to the system. In the first three steps, the conceptual design is separated into functional models and subsystems; then is further decomposed into parts and joints. In the top-down phase, both product function and manufacturing factors influence the breakdown. In steps four and five, the bottom-up phase, mating parts and joints are designed to ensure proper process selection and reliable assembly. Designed groups of parts, joints and subsystems are evaluated for their assemblability and their basis for generating an assembly process plan.

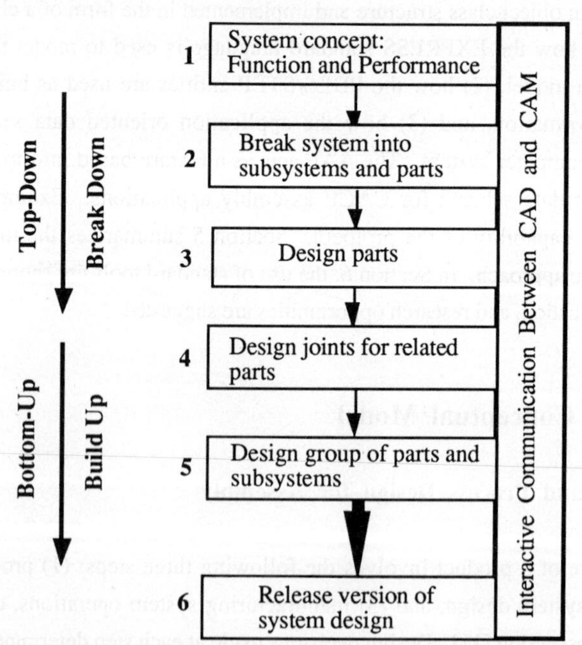

Figure 1. Concurrent Product and Process Design

Figure 1 introduces a common design scenario for emphasizing (1) three important "objects": e.g., Mechanical System (subsystem), Part, and Joint, and (2) the role of assembly engineering activities that are related to the above "objects." The PDES/STEP based Part, Joint, and Mechanical System (called Assembly_Model) definitions are proposed in [17]. Manufacturing related attributes of the Part, Joint, and Assembly_Model objects are also designed in a PDES/STEP compatible fashion and are intended to support the assembly planning and evaluation.

2.2 PDES/STEP Based Information Model

The ordinary practice of information modeling of a computer-aided system is to (1) recognize data elements that are necessary for required functions of the system, and (2) organize data elements into a structure for easier handling and manipulating by engineering applications. For CE practice, additional to the above, the analysis of the data relationships and data flow between multidisciplinary applications also needs to be done.

To establish the successful communication in concurrent engineering, two models must be constructed, namely data model and process model. The data model and process model together are often called the conceptual model [18]. A data model not only contains information to describe a product but also represents the structure of the information for easier retrieving, storing, and revising [19-20]. A process model describes the "dynamic" behavior of application data based on the relationship between data and its pre and post application conditions [21-22].

Unfortunately, the above practices achieve only local CE, because the data model is formed for a specific environment, which makes it difficult to communicate with other environments. Although, in such cases, two data models contain the same information, a translator between each pair of environments or systems is still needed. The need becomes obvious for constructing a data model that follows a standard that will enable CE to be utilized in the dynamic and heterogeneous environments.

This paper does not intend to discuss the details of data and process modeling. The paper concentrates on how to embed a data model in a computer-aided environment that enforces the PDES/STEP standard and provides rapid communication capability (without a translator) for applications that use the data model. A data model for object-oriented CAD/CAE/CAM applications proposed in [17] was refined by [5]. This paper follows the data structure of [5, 17], but reduces the data model to contain only the information that is needed for the assembly application in order to demonstrate a PDES/STEP compatible application.

In [5, 17], the hierarchical structure of mechanical systems is proposed. A mechanical system is composed of parts that are assembled to carry out specific functions. Within a mechanical system, there are two kinds of basic elements: parts and joints. A mechanical system can be defined as a group of related parts and/or subsystems that are assembled by joints. A subsystem is also an assembly of parts and its member subsystems that are connected by joints. Based on the manufacturing and assembly functionalities, the joint is classified into three different types: operational joint, fastener joint, and fusion (welded) joint. The joint in the data model is treated as an information set to describe different assembly relations of mated parts, fastened parts, or welded pieceparts.

2.2.1 Assembly_Model

The process of system decomposition into subsystems is a central aspect of prescriptive and computational models of the design task [23]. The hierarchical model, as suggested by [23], has the advantage of "implicitly incorporating abstraction and refinement." Meunier and Dixon directly assume that the design problem has been decomposed into systems and subsystems for their model of mechanical design [24].

The proposed hierarchical structure of the data model of a mechanical system is shown in Figure 2. From the configuration point of view, a mechanical system can be composed of parts, joints, and subsystems. A subsystem can be further divided into parts, joints, and its member subsystems. This recursive subsystem dividing process is terminated at a subsystem whose subsystem is nil. In this case, the subsystem is composed only of parts and joints. In Figure 2, the composite object (or assembly_model) is used to represent both a mechanical system and its subsystems or subassemblies. The symbol "+" represents one or more elements and the symbol "*" represent zero or more elements.

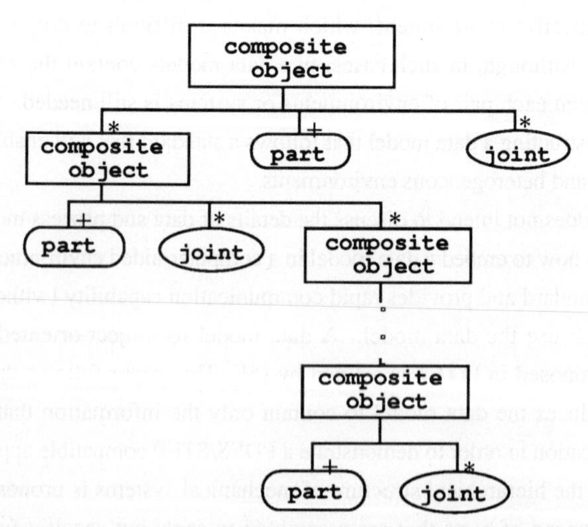

Figure 2. Hierarchical Structure of A Mechanical System

The attributes and functions of the Assembly_Model are listed in Figure 3. The Assembly_Model class carries the is_part_of relationship of a mechanical system and its components. The attributes and methods (functions) of the Assembly_Model are defined to help the designer construct the structure of the mechanical system. The editing functions allow the designer to create specific system configurations. When a designer creates a new system, the system configuration (or decomposition) is based on some special purposes from the designer's viewpoint. Furthermore, based on different application considerations, the designer can edit the configuration or evolve it through experiment to conceive, as a meaningful unit, an analysis that evaluates some aspects of the performance of the system or subsystem.

```
CLASS : Assembly_Model
-----------------------------------------------------
Attributes
        ID
        Name
        Set of Assembly_Models (Nil or Composite IDs)
        Set of Parts (Nil or Part IDs)
        Set of Joints (Nil or Joint IDs)

Methods
        Create Assembly_Model
        Add Part
        Erase Part
        Add Joint
        Erase Joint
        Add Assembly_Model
        Erase Assembly_Model
        Display Component
```

Figure 3. Attributes and Functions of Assembly_Model

A construction backhoe system is used as an example to illustrate concepts presented in the Assembly_Model. In the example, the backhoe system is composed of twelve idealized functional parts, eleven fastener Joints, and three operational Joints (J3, J6, and J10), as shown in Figure 4. A common configuration structure of the backhoe system is shown in Figure 5, which gives the overall data model picture of an example mechanical system. In Figure 5, boxes represent parts; rounded boxes represent systems or subsystems; and ovals represent Joints that specify the assembly relationships of parts. The dashed lines represent the connectivity between parts.

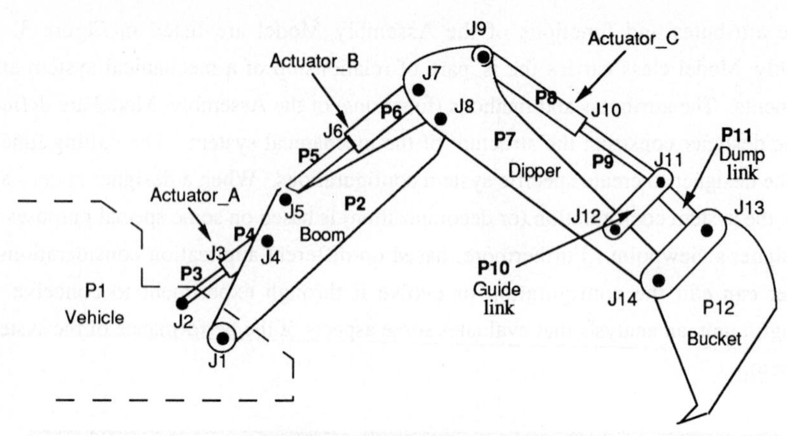

Figure 4. A Backhoe System

2.2.2 Part

A part in a mechanical system is a solid entity that has specific geometry and material properties. Attributes of a part in the data model include a part identification, a name, a geometric representation, and a material property, as shown in Figure 6. The geometric representation and material property of a part are defined primarily according to PDES/STEP. In the paper, the geometric representation of a part contains three components of PDES/STEP [7]: (1) geometric model, (2) form features, and (3) precision features. The geometric model is a formal logical/mathematical representation of the shape and size of a part. Most geometric models are presented as solid models. The geometric model of the Guide link of the backhoe system is a rectangular block, as shown in Figure 7a.

A form feature is a stereotypical portion of a shape; i.e., a portion of the skin of a shape or a dimensionality-2 shape element [7]. A form feature, from the geometry point of view, is a physical portion of a part. From the product definition point of view, a form feature adds detailed geometric characteristics to the geometric model to precisely define the shape of a part. From the manufacturing point of view, a form feature conforms to some manufacturing pattern; e.g., hole, pocket, wall, or chamfer. Whenever form features are appropriate to describe the geometry of a part, they should be applied to the geometric model of the part. In this way, downstream manufacturing applications can more efficiently use the information contained in the geometric representations of parts [25-26]. As an example, the geometry of the Guide link can be defined by applying form features including two holes, an open rectangular section, a slot, and four chamfers to the geometric model of the Guide link as shown in Figure 7b.

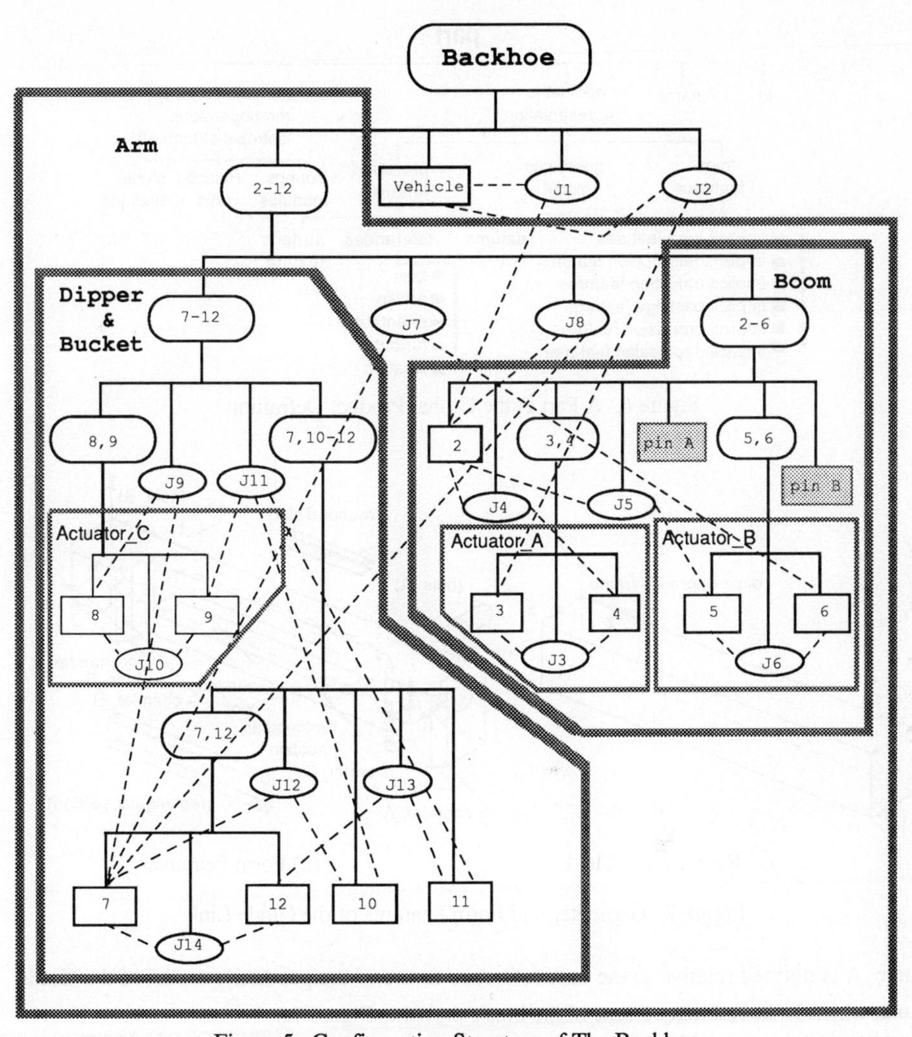

Figure 5. Configuration Structure of The Backhoe

Each part has its own base reference frame (coordinate system). Each volume associated form feature has a feature reference frame. Most reference frames are three dimensional. In PDES/STEP, a reference frame is called an axis-n_placement, where n can be 1 or 2, depending on the type of form feature and/or volume specification procedures. When defining the axis_placement of a form feature of a part, the reference frame of the form feature, which is called the feature reference frame, first needs to be defined relative to the base reference frame of the part. As an example, in Figure 7b the feature reference frame of

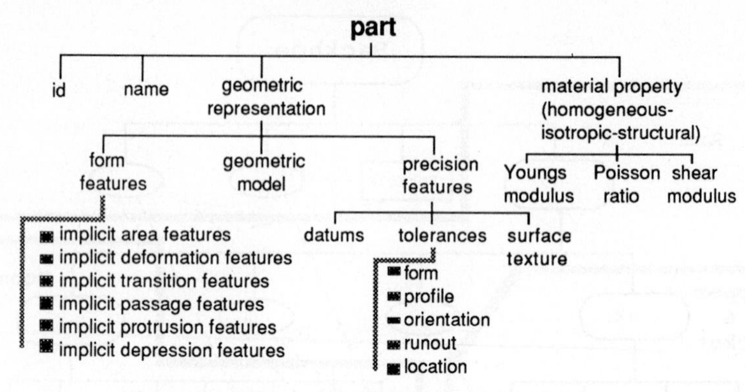

Figure 6. A Part in the Global Product Definition

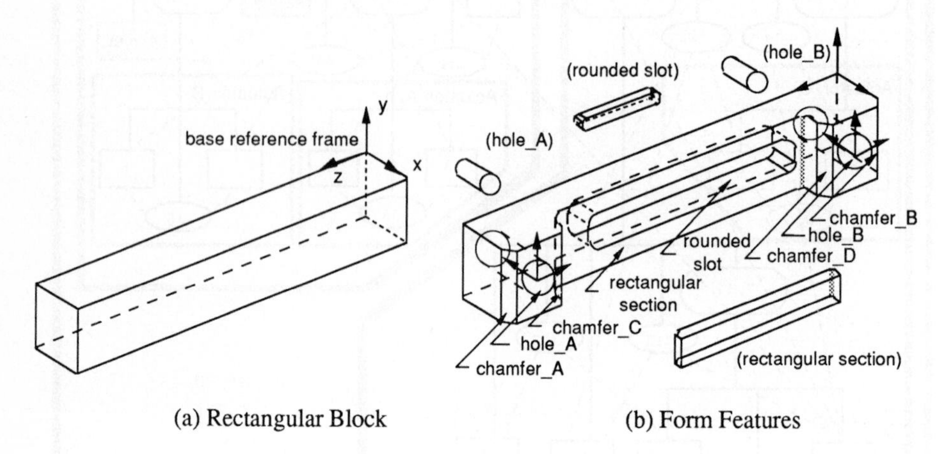

(a) Rectangular Block (b) Form Features

Figure 7. Geometry and Form Features of the Guide Link

hole_A is defined relative to the base reference frame of the geometric model of the Guide link, which is the reference frame of the block shown in Figure 7a.

Precision features such as tolerances and surface texture describe additional geometric characteristics of the existing geometric model and form features of a part. Precision features are used to describe the final product design information for CAM tools to select processes, machine tools, and tooling. In PDES/STEP, part material can be classified into three categories; (1) homogeneous material, (2) composite material, and (3) material table. For detailed classification, readers can refer to [7].

2.2.3 Joint

From the design point of view, a Joint only defines the mating conditions and kinematic constraints between parts. From a CAM point of view, a Joint is an ordered sequence of assembly operations and specifies assembly operations and mating conditions between parts. Hence, a Joint in the assembly application data model defines both connection and assembly process data. Characteristics of a CAM Joint include an identification (id), name, joint type, connectivity, degrees of freedom (d.o.f.), mating operation, and joining operation as shown in Figure 8. Note that a joint agent also is a part.

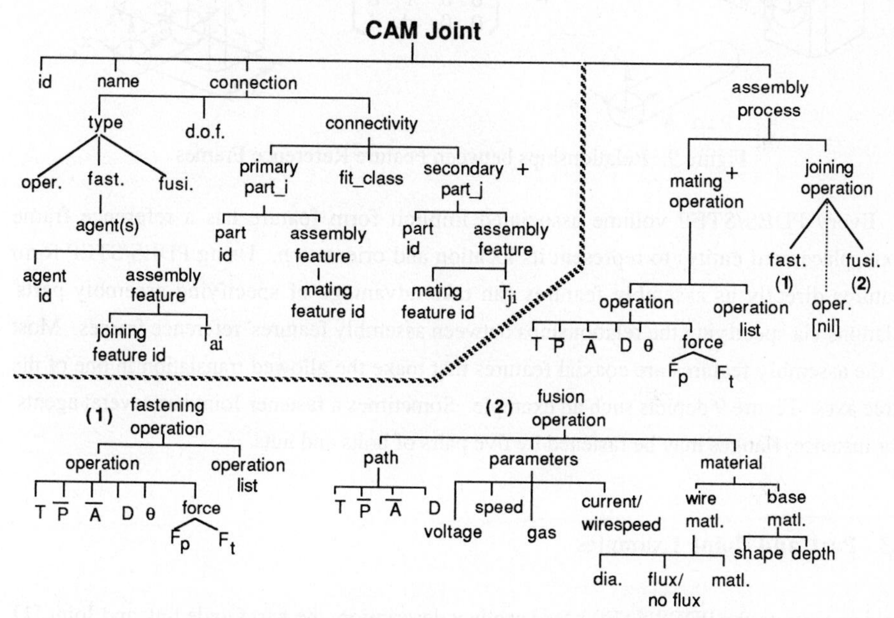

Figure 8. The Definition of Joint in an Assembly Application

Based on assembly processes and the mating conditions, a Joint can be an operational Joint, fastener Joint, or fusion Joint. Beside the joint agent, which is a unique characteristic of a fastener Joint, another difference between these three types of Joints is their assembly processes. The fusion Joint needs a fusion operation to join the contacting parts. The operational Joint requires the mated parts to have assembly related form features to hold each other; e.g., the designed snapping, clipping, gripping, holding, and containing features of mated parts. The fastener Joint needs operations to fix the agents and mated parts.

Since an agent of a fastener Joint is a part to assemble other parts, the agent's geometric features that are involved in assembly should be specified in Joints. The assembly features of an agent include (1) the identification of the assembly form feature and (2) a homogeneous transformation matrix T_{ai}. The homogeneous transformation matrix describes position and orientation of a feature reference frame of the Joint feature of the agent, relative to the reference frame of the assembly feature of the primary part_i. Figure 9 shows an example that illustrates a homogeneous transformation matrix that describes the position and orientation of an agent (pin A) with respect to the primary part (Guide link).

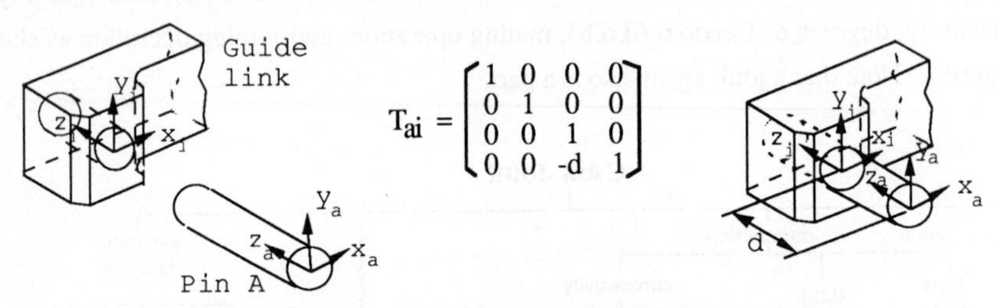

Figure 9. Relationships between Feature Reference Frames

Every PDES/STEP volume associated implicit form feature has a reference frame (axis_placement entity) to represent its location and orientation. Using PDES/STEP form features directly as assembly features can take advantage of specifying assembly parts' relations via specifying the relationships between assembly features' reference frames. Most of the assembly features are coaxial features that make the allowed translation in one of the three axes. Figure 9 depicts such an example. Sometimes a fastener Joint has several agents. For instance, flanges may be fastened by five pairs of bolts and nuts.

2.3 Part and Joint Examples

To demonstrate the PDES/STEP based product description, the part Guide link and Joint J11 are chosen as examples. The geometry of the Guide link (shown in Figure 7b) can be defined according to the data model of a part (shown in Figure 6), which follows the PDES/STEP part definition standard. The geometric representation of the Guide link is defined as shown in Figure 10, by applying form features to the geometric model of the Guide link [5]. In Figure 10, rectangular boxes represent entities introduced in PDES/STEP. Dashed rectangular boxes are the optional entities of the PDES/STEP structure that are not used in the Guide link. Underlined attributes are the fundamental geometric data of a part. Using PDES/STEP to define product data takes advantage of feature-based applications. Because both the abstract information (e.g., Hole_A: Passage feature) and the concrete attributes (e.g., location, length, and radius of Hole_A) are all integrated into PDES/STEP entities, which are used by the proposed data model, the feature recognition procedures can extract the feature information directly from the data model.

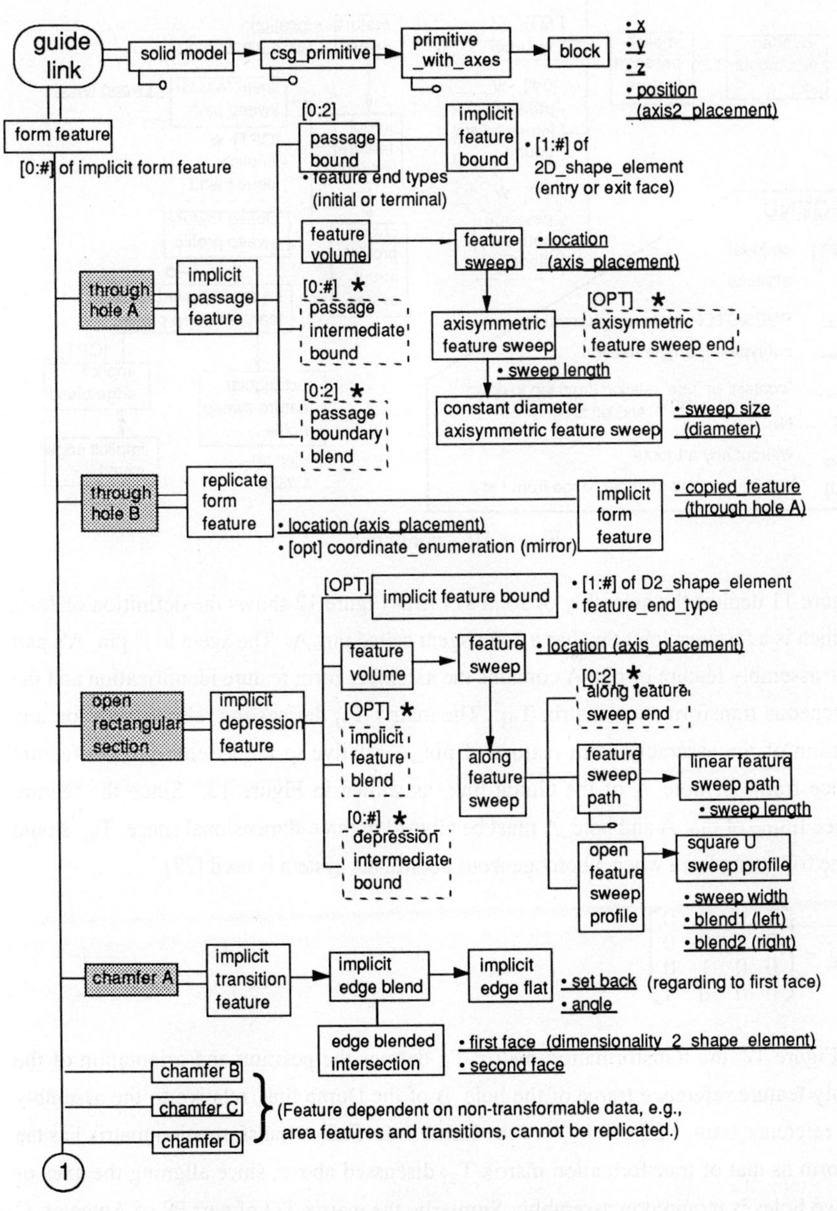

Figure 10. PDES/STEP Definition of the Guide Link

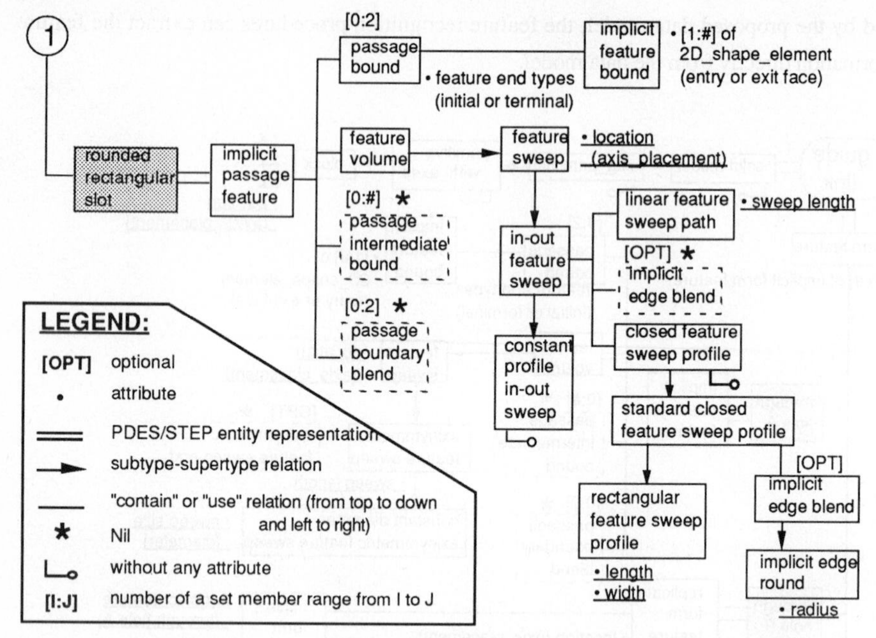

Figure 10. - continued

Figure 11 depicts the assembly of Joint J11 [5]. Figure 12 shows the definition of Joint J11, which is a fastener Joint that has a Joint agent called pin_A. The agent id is pin_A's part id. The assembly feature of pin_A contains the assembly form feature identification and the homogeneous transformation matrix T_{ai}. The matrix T_{ai} defines the relative position and orientation of the assembly form feature of pin_A relative to the assembly form feature reference frame of hole_A of the Guide link, as shown in Figure 12. Since the feature reference frame of pin_A and hole_A must be aligned in three-dimensional space, T_{ai} should be in the following form when a homogeneous coordinate system is used [27].

$$T_{ai} = \begin{bmatrix} 1 & 0 & 0 & 0 \\ 0 & 1 & 0 & 0 \\ 0 & 0 & 1 & 0 \\ 0 & 0 & -d & 1 \end{bmatrix}$$

In Figure 12, the transformation matrix T_{ji} defines the position and orientation of the assembly feature reference frame of the hole_B of the Dump link, relative to the assembly feature reference frame of the hole_A of the Guide link. Such a transformation matrix has the same form as that of transformation matrix T_{ai} discussed above, since aligning the axes of these two holes is required in assembly. Similarly, the matrix T_{ki} of part P9 of Actuator_C defines the position and orientation of hole_C of the part relative to hole_A's feature reference

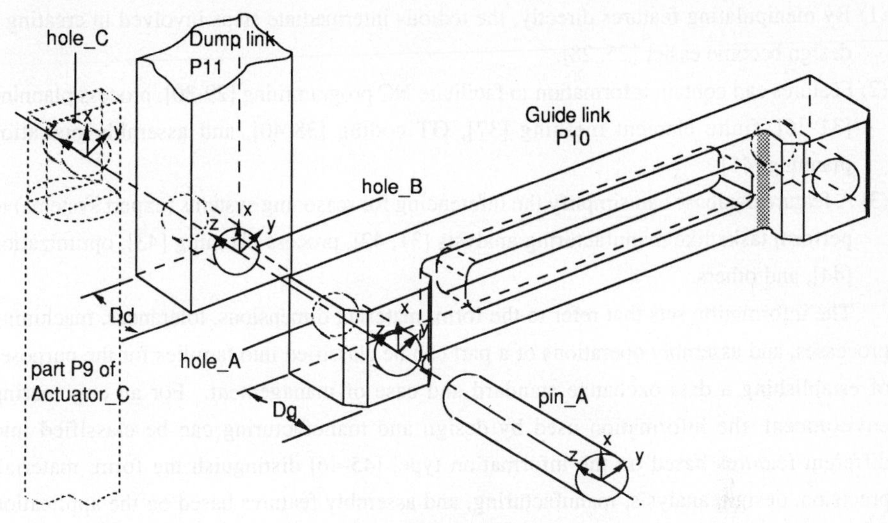

Figure 11. Assembly of Joint J11

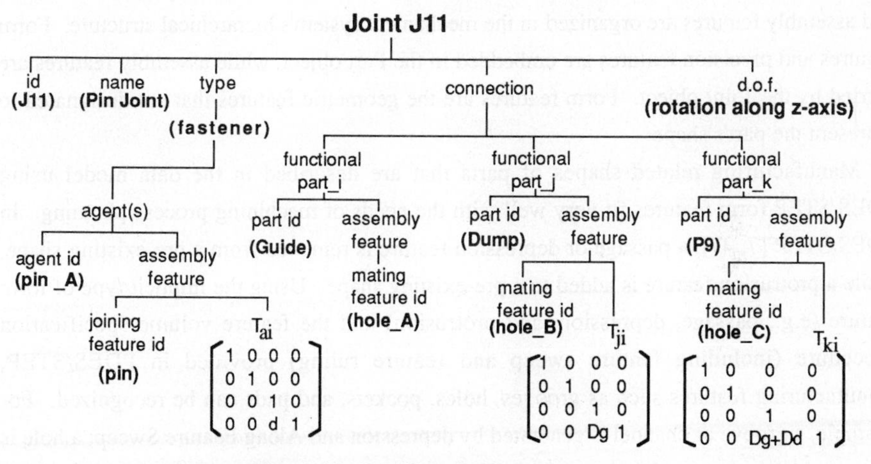

$$T_{ai} = \begin{bmatrix} 1 & 0 & 0 & 0 \\ 0 & 1 & 0 & 0 \\ 0 & 0 & 1 & 0 \\ 0 & 0 & d & 1 \end{bmatrix} \quad T_{ji} = \begin{bmatrix} 1 & 0 & 0 & 0 \\ 0 & 1 & 0 & 0 \\ 0 & 0 & 1 & 0 \\ 0 & 0 & Dg & 1 \end{bmatrix} \quad T_{ki} = \begin{bmatrix} 1 & 0 & 0 & 0 \\ 0 & 1 & 0 & 0 \\ 0 & 0 & 1 & 0 \\ 0 & 0 & Dg+Dd & 1 \end{bmatrix}$$

Figure 12. Detail Definition of Joint J11

frame. The variables Dg and Dd are thicknesses of the Guide link and Dump link, respectively, as shown in Figure 11.

3 Feature-Based Application Using STEP Entities

Features technology has the potential for automating and integrating design, analysis, and manufacturing. Some of the advantages of using features are listed below:

(1) By manipulating features directly, the tedious intermediate steps involved in creating a design become easier [25, 28].

(2) Features can contain information to facilitate NC programming [29-30], process planning [31-36], finite element meshing [37], GT coding [38-40], and assembly operation planning [41].

(3) A feature database can simplify the inferencing for reasoning systems (expert systems) to perform tasks like manufacturing analysis [31, 42], process planning [43], optimization [44], and others.

The information sets that refer to the form, material, dimensions, tolerances, machining processes, and assembly operations of a part can be classified into families for the purposes of establishing a data exchange standard and ease of management. For an engineering environment, the information used by design and manufacturing can be classified into different features based on the information type. [45-46] distinguish the form, material, precision, design, analysis, manufacturing, and assembly features based on the application and abstraction level.

In the data model developed in the present research, the form features, precision features, and assembly features are organized in the mechanical system's hierarchical structure. Form features and precision features are embedded in the Part object, while assembly features are carried by the Joint object. Form features are the geometric features that are designated to represent the part's shape.

Manufacturing related shapes of parts that are described in the data model using PDES/STEP form features fit very well with the needs of machining process planning. In PDES/STEP [7, 47], a passage or depression feature is removed from a pre-existing shape, while a protrusion feature is added to a pre-existing shape. Using the implicit type of form feature (e.g., passage, depression, and protrusion) and the feature volume specification procedure (including feature sweep and feature ruling) provided in PDES/STEP, manufacturing features such as grooves, holes, pockets, and pads can be recognized. For example, a groove or channel is generated by depression and Along Feature Sweep; a hole is generated by passage and Axisymmetric Feature Sweep; a pocket is generated by depression and In-Out Feature Sweep; and a pad or boss is generated by protrusion and In-Out Feature Sweep. The compatibility between manufacturing features and PDES/STEP form features simplifies the standardization and automates decisions for CAPP, GT classification, and NC code generation [48, 49].

Assembly features are particular form features that affect assembly operations. Figure 13 shows an example that depicts the assembly relations described by part assembly features in a Joint. The PDES/STEP form features play a key role in the product information model developed by [50] in which the form features are used as design features in CAD; as machining features in manufacturing process planning; and as assembly features in assembly operation planning. A PDES/STEP form feature can be one of, some of, or all of the above application features, depending on the characteristics and functionality of the form feature. For example, a design feature, say hole_A, can also be a machining feature that requires a drilling operation. Furthermore, hole_A can be an assembly feature, if it is used to receive a cylindrical part that forms a mated or joined connection of an assembly.

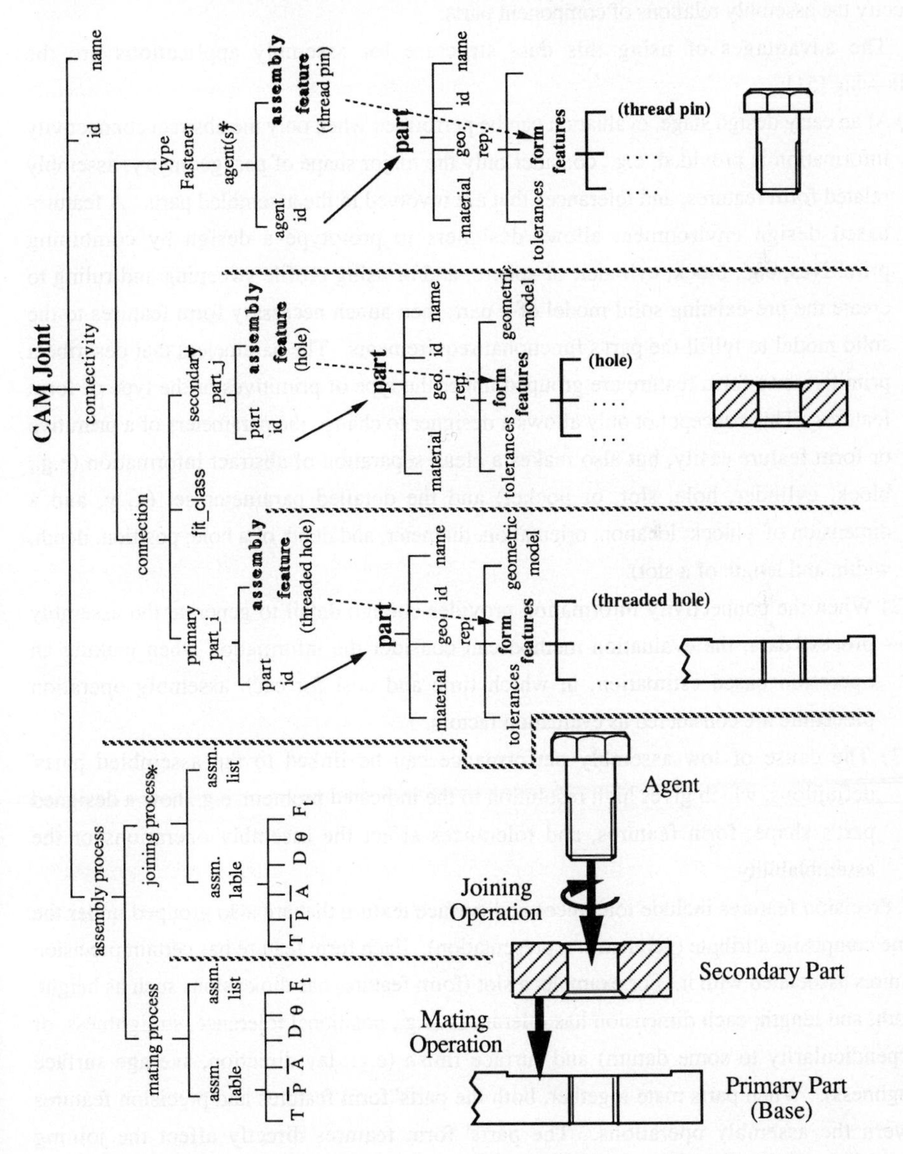

Figure 13. Connectivity Information and Assembly Process Data Carried by Joint

A Joint in CAM applications carries both assembled parts' connectivity information and assembly operation data. The connectivity information, which is decided by the designer, contains mated or joined part identifications and their assembly features. The assembly process data, which are the result of assembly process planning, carry the assembly operation list and assembly operation label [50]. An example of the Joint data structure is shown in Figure 13, which shows the role of Joints in linking the assembly parts' form features to specify the assembly relations of component parts.

The advantages of using this data structure for assembly applications are the following [51]:

(1) At an early design stage, evaluation can be performed when only the abstract connectivity information is provided, e.g., consider only the major shape of part geometry, assembly related form features, and tolerances that are involved in the assembled parts. A feature-based design environment allows designers to prototype a design by combining primitives, e.g., block, cylinder, or sphere, and/or using profile sweeping and ruling to create the pre-existing solid model of a part, then attach necessary form features to the solid model to fulfill the part's functional requirements. The parameters that describe a primitive or a form feature are grouped under the type of primitives or the type of form features. This concept not only allows a designer to change the parameters of a primitive or form feature easily, but also makes a clean separation of abstract information (e.g., block, cylinder, hole, slot, or pocket) and the detailed parameter set (x, y, and z dimension of a block; location, orientation, diameter, and depth of a hole; position, depth, width, and length of a slot).

(2) When the connectivity information provides enough detail to generate the assembly process data, the evaluation method can consider the information when making an operation based estimation, in which time and cost for each assembly operation procedure are considered as evaluation factors.

(3) The cause of low assembly performance can be linked to the assembled parts' definitions, which gives high resolution to the indicated problem, e.g., how a designed part's shape, form features, and tolerances affect the assembly operations or the assemblability.

Precision features include tolerances and surface texture that are also grouped under the same composite attribute (geometric representation). Each form feature has certain precision features associated with it. For example, a slot (form feature) has dimensions such as height, width, and length; each dimension has tolerances (e.g., positional tolerance, straightness, or perpendicularity to some datum) and surface finish (e.g., lay direction, average surface roughness). When parts mate together, both the parts' form features and precision features govern the assembly operations. The parts' form features directly affect the joining conditions, for instance: a hole or pin indicates the fit condition, and a threaded stud or threaded hole suggest a torque operation is needed. Obviously, the precision features of the mating parts also affect the quality and manufacturing processes of the assembly.

Feature based application can simplify the inference for reasoning systems (e.g., process planning) by getting rid of the feature recognition procedures. As mentioned before, the data model is constructed to contain form, assembly, and precision features that are integrated together and can take advantage of feature based applications. However, most of the credit goes to the PDES/STEP. Because PDES/STEP defines the form feature information model (ISO 1030 Part-48) to incorporate the variational tolerance and nominal shape information models, which in turn integrate the product's geometry, form features, dimensions, and tolerances. For example, a form feature size parameters refer to the tolerance entities and the location of a form feature refers to the entities defined in the nominal shape information model (e.g., axis_placement).

4 Implementation of PDES/STEP Compatible Applications

The previous section discusses the information required for assembly applications that is organized as a PDES/STEP based and feature based data model. To embed this data model in a CACE environment and provide communication capabilities for applications and databases, the object-oriented approach is used. At the implementation level, mechanisms are required to support specialization relations concerned with objects and object attributes like "instance-of," "is-a," "reference-to," and "is-part-of" [52].

4.1 Object-Oriented Approach

This research follows the object-oriented concept that uniformly models any real world entity as an object. The advantages of object-oriented modeling are [53]: (1) the introduction of semantics by putting the properties (attributes) and functions (methods) describing the object into the context of that object, (2) the structuring of functions and properties using a unique concept, and (3) the integration of attributes and methods.

The structure of an object is shown in Figure 14. The object attributes can refer to other objects, and inherit from other objects through the super/subtype declaration. The object model represents object relationships, which carry the data semantics of the data model. In Figure 14, the object method is classified into I/O Functions and Algorithms. The I/O Functions define ways the object attributes can be used (e.g., retrieve and update). Algorithms are the special subroutines of the object that perform certain engineering functions of the object. For example, a CAM_Joint may contain a Joint planning method for assembly process planning and a Joint evaluation method to perform an assemblability evaluation.

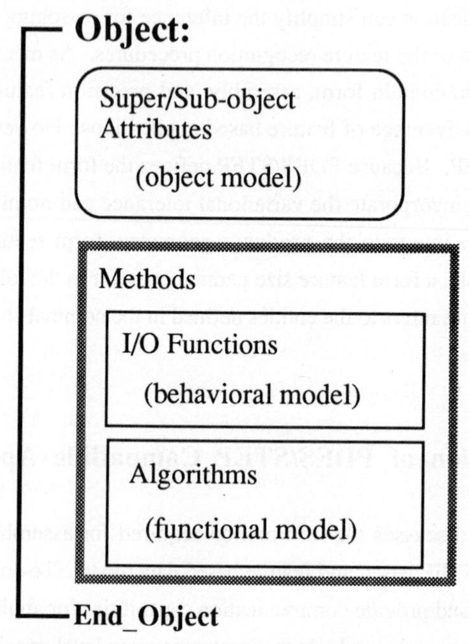

Figure 14. Attribute and Method of Object

The PDES/STEP methodology uses a three layer architecture, including a reference model (concept layer), a formal definition language called EXPRESS (logical layer), and communication file structure called the STEP file (physical layer). The EXPRESS language is used to define the PDES/STEP logic layer, which is designed to satisfy [7, 54]: (a) modeling the object of interest, (b) defining the constraints that are to be imposed on objects, (c) defining the operations that establish how objects are to be used, and (d) modeling in a computer-sensible manner (i.e., processable by computer). EXPRESS is similar to an ordinary computer language like Pascal, C, or C++, but without any input and output statements. The main descriptive elements of EXPRESS are:

Schema
- *Type* (character of data that is used to describe entity)
- *Entity* (describes the object of interest to a UoD (universe of discourse))
- *Algorithm*
  - *Function*
  - *Procedure*
- *Rule* (defines constraints).

The data modeling nature of PDES/STEP information models is exactly object oriented, which is the reason that only an object-oriented approach can take full advantage of both product data and software engineering standardization.

4.2 Application Protocol

The use of PDES/STEP information entities as building blocks for data modeling is the first step toward product data standardization. To standardize the use of STEP to support a particular application is also necessary to communicate between the application and other systems or databases. The documentation of a standard way, for both human and computer systems, to interface with the STEP based data model needs to be accomplished before the full advantage of product data standardization can be achieved.

Application Protocol (AP) aims to document a standard means to interface STEP and a specific application. According to [8], "Application protocols define the content, the use, and the kind of product data for a specific engineering purpose in a product's life cycle." An AP consists of (1) a required product data model to perform particular engineering functions and (2) an application interpreted model to define how the product information is used [55]. This research implements the Assembly AP (AAP) as an object-oriented class library that acts as feature based and STEP based object templates domain and persistent object data schema for computer-aided assembly applications.

4.3 Application and Communication Environment

The implementation approach for the AAP is depicted in Figure 15. The software tools used by this research include EXPRESS language, Express2C++ translator, C++ language, and ROSE database. The assembly application data model is remodeled by the EXPRESS language, which is an object-oriented information modeling language and also is part of STEP (ISO 1030 Part-11). The EXPRESS expression of the Part, Joint, and Assembly_Model (see Figure 15), called application oriented schema, refer to PDES/STEP schemata. All of the EXPRESS schemata are translated by the Express2C++ translator into a class library that keeps the semantics of the data relationships of the data model. The object-oriented class library is easy to use and maintain. A new class can be constructed by (1) inheriting all properties from other existing classes, (2) editing attributes and methods of a class, (3) adding new attributes and/or methods to a class, or (4) combining all the above approaches.

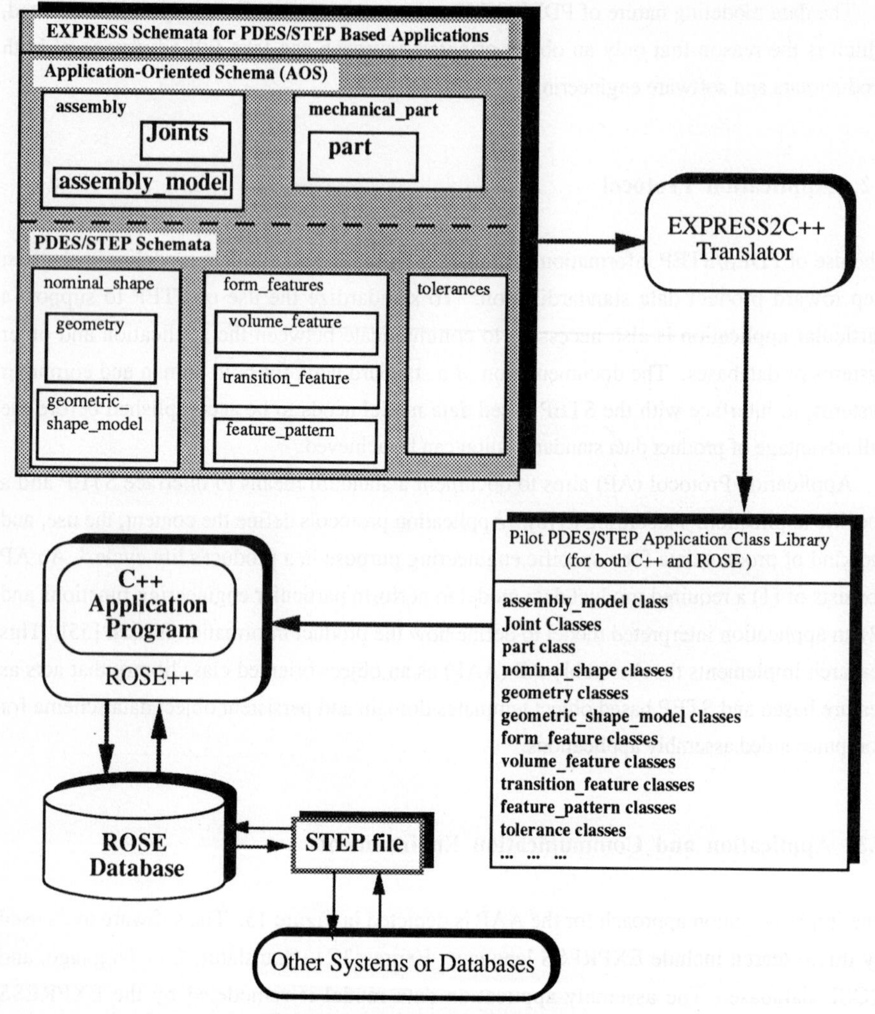

Figure 15. Application Environment

This class library can be used directly by C++ and ROSE database manager. A C++ application program can (1) instance objects from the library to construct user defined objects, (2) used object methods to manipulate object attributes and perform engineering functions, and/or (3) send messages to ROSE++ (ROSE database manager) to save, update, or retrieve user defined objects.

The ROSE database is designed to support CE and PDES/STEP [56], and has the capability to read and write a STEP file. A STEP file represents the physical layer of data communication that is a generic data format for PDES/STEP product information. In the environment shown in Figure 15, system integration and product definition data passing through multi-disciplinary applications to practice concurrent engineering is through the STEP file. But the important issue for a system or an engineering function to understand a STEP file from other systems is to know the original definition and semantic relationships of the information contained in the STEP file. The application protocol is the spirit of the STEP file. Communication between different systems can be done only if those systems have a common data schema to form the foundation to understand each other. PDES/STEP defines a wide range of engineering information entities for applications to construct the application protocol.

4.4 Assembly Application Object Class Library

The class inheritance and reference structure of the Assembly_Model, Part, and Joint objects are proposed in [50] as shown in Figure 16. A set of entities defined in the PDES/STEP ISO 10303-42 (geometry, geometric shape model) [57], ISO 10303-48 (form features) [47], and ISO 10303-47 (shape variational tolerance) [58], as shown in Figure 17, are implemented in the current version of this class library. The overall scope of PDES/STEP should not discourage the implementation of PDES/STEP compatible applications. For example, a C++ class library that implements the STEP model data, called the STEP Class Library (SCL), is under development at the National Institute of Standards and Technology (NIST) as part of the national PDES testbed [59]. The SCL is very likely in the public domain. The users can easily use entities (classes) that are defined in each released PDES/STEP reference model via access to the STEP C++ library.

Figure 16 shows the inheritance and reference structure of the Assembly_Model entity (class), Part entity, and Joint entities. The italic formatted attributes in a class are the attributes inherited from other classes. The plain text formatted attributes are the local attributes of a class. In Figure 16, the solid boxes are classes (entities) that are defined in the assembly schema; the solid lines represent the inheritance relationships from the top down; the shaded arrows represent "refer to;" and the shaded boxes represent the entities defined in other schemata (PDES/STEP reference model). A composite attribute, which itself is an information set, always refers to other classes.

The Nominal_Shape, Form_Feature, and Tolerance entities, shown in Figure 17, are the subsets of entities defined in PDES/STEP (ISO 10303) part 42, part 48, and part 47, respectively. The boxes are PDES/STEP entities; the shaded arrows represent "refer to;" the bold shaded arrows represent "refer to" entities defined in other schemata; the solid lines represent the superclass-subclass relationships.

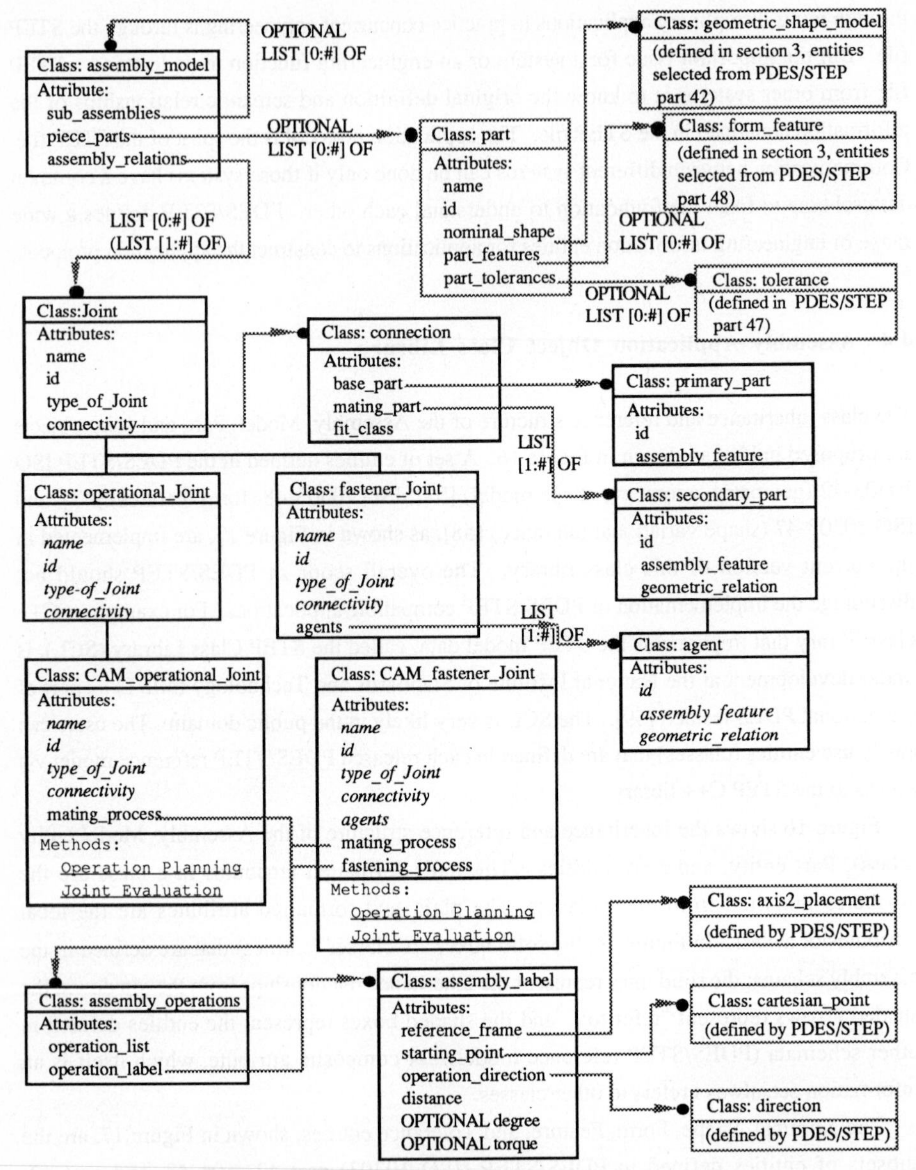

Figure 16. Inheritance and Reference Structure of AAP

In an object-oriented concept, an entity (class) contains not only the data structure to describe the carried information set (static property) of the entity, but also methods (function and procedure) and rules to describe the behavior (dynamic property) of the entity. After the object classes (templates) are established in both the object-oriented programming and database environments, application programs can use the instanced template objects, which

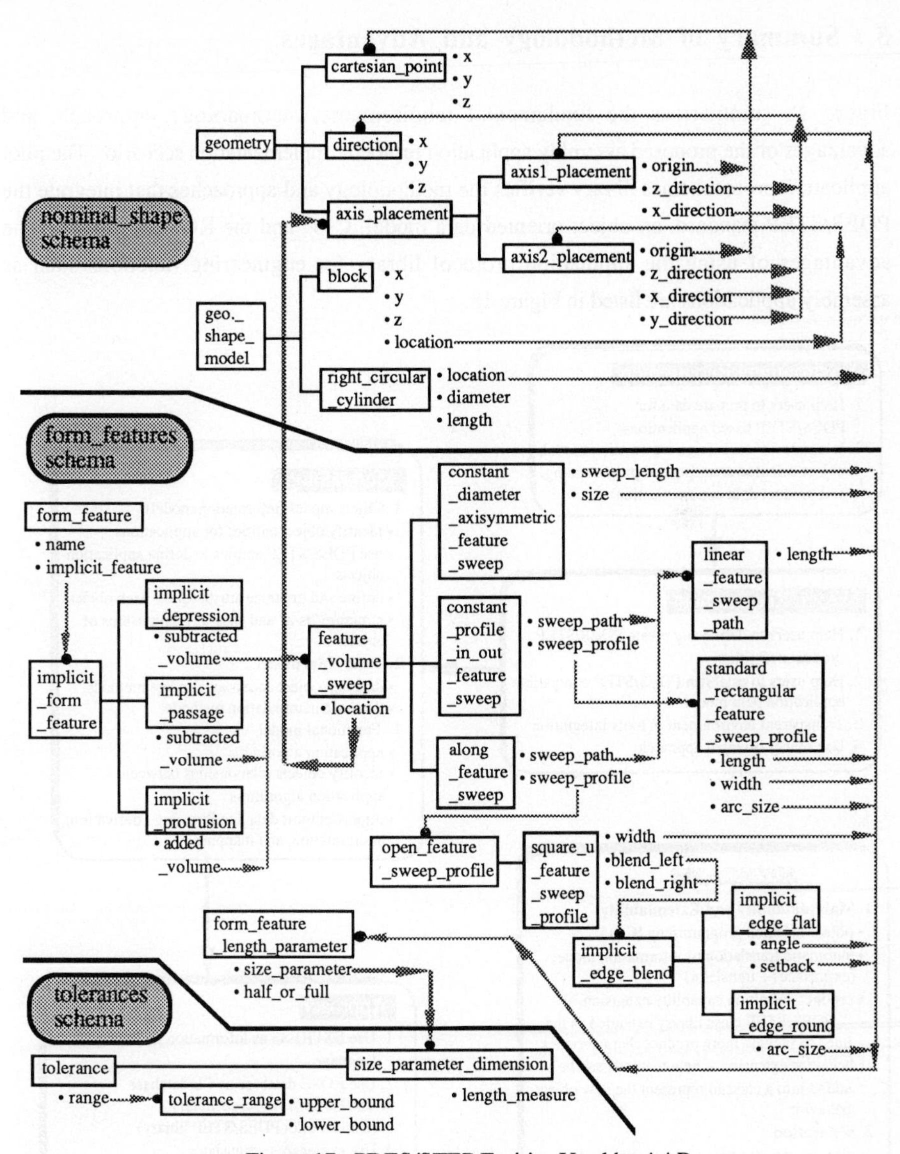

Figure 17. PDES/STEP Entities Used by AAP

follow the PDES/STEP product definition, and can be incorporated into ROSE. The capability of AAP has been demonstrated by coding C++ programs for mechanical system examples. The object lists and STEP files for the examples are available upon request from the authors.

5 Summary of Methodology and Advantages

Figure 18 summarizes the fundamental requirements, methodology, approach, and advantages of the proposed assembly application protocol implementation scenario. The pilot application protocol class library verifies the methodology and approaches that integrate the PDES/STEP standard, an object-oriented data model, C++, and the ROSE database. The advantages of using the application protocol library for engineering functions such as assembly applications are listed in Figure 18.

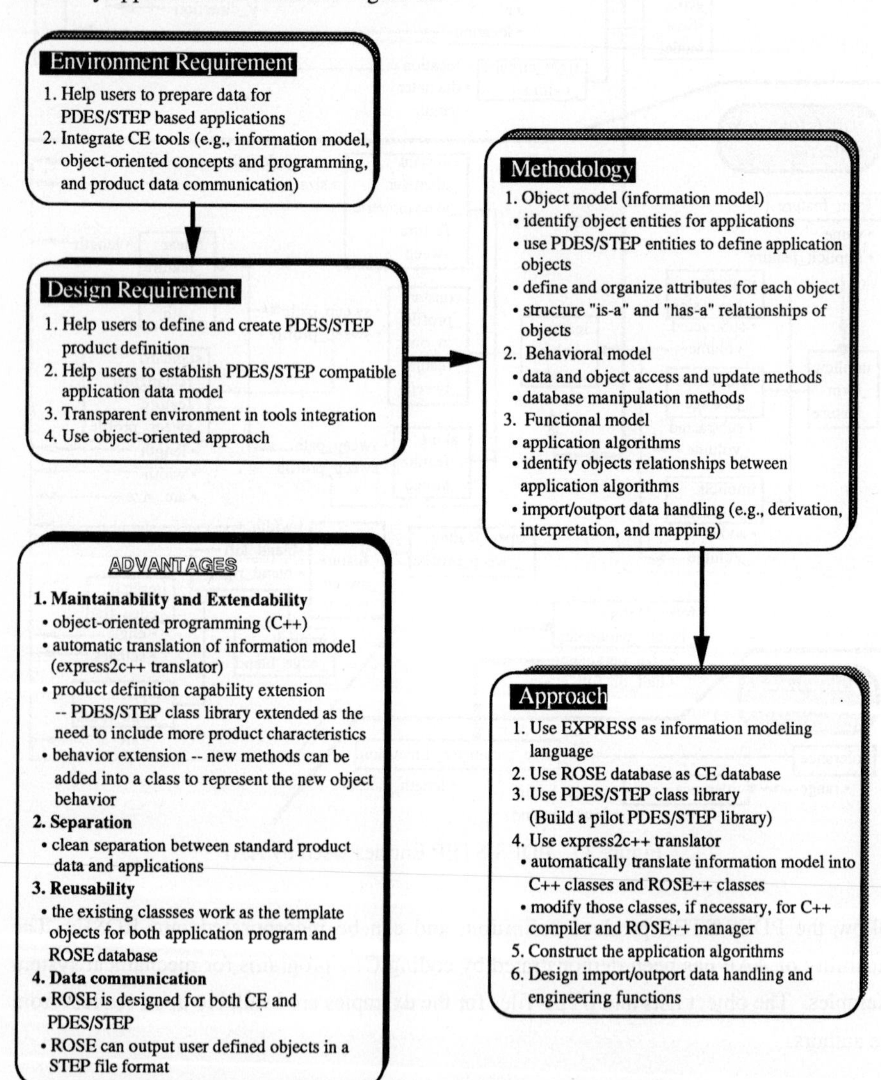

Figure 18. Summary of Requirements, Methodology, Approaches, and Advantages

The benefits of constructing and using such a library in an engineering environment are:

(1) *Maintainability and Extendibility*. The maintainability of using object-oriented programming to implement large and complicated systems is well known. The conceptual PDES/STEP compatible information model can be extended easily by automatically translating it into C++/ROSE++ classes. The PDES/STEP class library for product definition can be efficiently extended as needed to include more product characteristics for a variety of engineering applications. As new engineering functions are integrated into a CACE environment, new methods (procedures or algorithms) can be added into corresponding classes to represent the behavior of the new objects that perform the new engineering functions.

(2) *Separation*. The data relationships and the data model are embedded in the class semantics, thus making a clean separation between standard product definition and applications. Engineers or programmers need not understand the tedious PDES/STEP standard format nor memorize the data semantics of PDES/STEP entities.

(3) *Reusability*. The classes in the pilot library work as template objects for both application programs and the ROSE database. Multi-users can instance objects from the same library and share the database to store their persistent objects (designs).

(4) *Data Communication*. ROSE is designed for both Concurrent Engineering and PDES/STEP. ROSE can output user defined objects in the STEP file format. This capability makes it possible for the applications, which use the pilot library, to communicate with virtually any other systems that follow PDES/STEP.

PDES/STEP is claimed as a generic and neutral product data definition, though PDES/STEP uses the object-oriented concept to document the product definition standard; and most of the tools developed to support the STEP follow or adopt the object-oriented concept. This is not a bias, but instead is a means to guarantee the success of the on-going standard through the use of highly modulated, reusable, maintainable, and extendable object-oriented approaches.

6 Conclusions and Recommendations

Concurrent Engineering (CE) is a process that involves information integration to support data communication between multi-disciplinary applications. Though CE does not necessarily have to be an automated process, most of today's engineering activities are computer-based systems, which tend to be data intensive systems. These automated systems store product information digitally in a database. To enhance CE communication, the standard product definition is necessary for neutral representation of engineering information that can be used directly by different computer-aided automated systems.

This research integrates the following standard tools to enhance the CE communication:

(1) Object-oriented concept, which is a standard approach for software engineering
(2) EXPRESS, which is a standard information modeling language (ISO-10303 Part 11)
(3) PDES/STEP, which is a new generation international standard for product definition
(4) Application protocol, which standardizes the interface with STEP
(5) Object-oriented database, which is the standard database for data intensive applications (e.g., solid modeling) and data centered architecture systems (e.g., CAD system).

The challenges of this research are to (1) embed the data model and data semantics of engineering applications into a computer environment, (2) enforce the engineering data standard, and also (3) provide the communication capabilities between engineering applications. The AAP achieves the above goals from the design for assembly viewpoint. The AAP is PDES/STEP compatible and provides the feature-based product information for both assembly planning and evaluation functions. The AAP is embedded in the computer system in the form of an object class library that works as information templates for the object-oriented programming language (C++) and a database schema for the object-oriented database (ROSE).

The AAP, the virtual spirit of the data model, has the following features: (1) Arbitrary level of structure to represent a mechanical system in an assembly application, which is easy to extend and maintain; (2) Each object is a manipulating unit that enforces the reusability and is highly modulated; (3) Data semantics such as is_part_of, is_a, instance_of, and refer_to are embedded in the class library; and (4) Application features in the data model, such as form feature and precision feature (carried by Part) and assembly feature (carried by Joint), can be used as design, manufacturing, and analysis features. These application features are represented by a well defined parameter set that can describe the geometric shape of a part and have the full capability of dimensioning and tolerancing. Feature-based applications are expected to be one of the keys to real progress in creating a CACE environment [60].

The AAP is also a data map that describes the meaning and relationships of the information contents of a STEP file generated through the integrated environment. The STEP file is the standard communication medium of the PDES/STEP based application activities.

Certainly, there still is much work that can be done in the world of system integration, product data standardization, assembly applications, and concurrent engineering. Figure 19 shows the intent of this research. A methodology for PDES/STEP compatible applications is documented in this paper by using the assembly functions as an example. The same method can be reapplied to augment the AAP into an application protocol for other CAM functions. Hopefully, through the step-by-step extension of a PDES/STEP compatible application protocol, a standardized CACE environment can finally be established in parallel with the development of STEP.

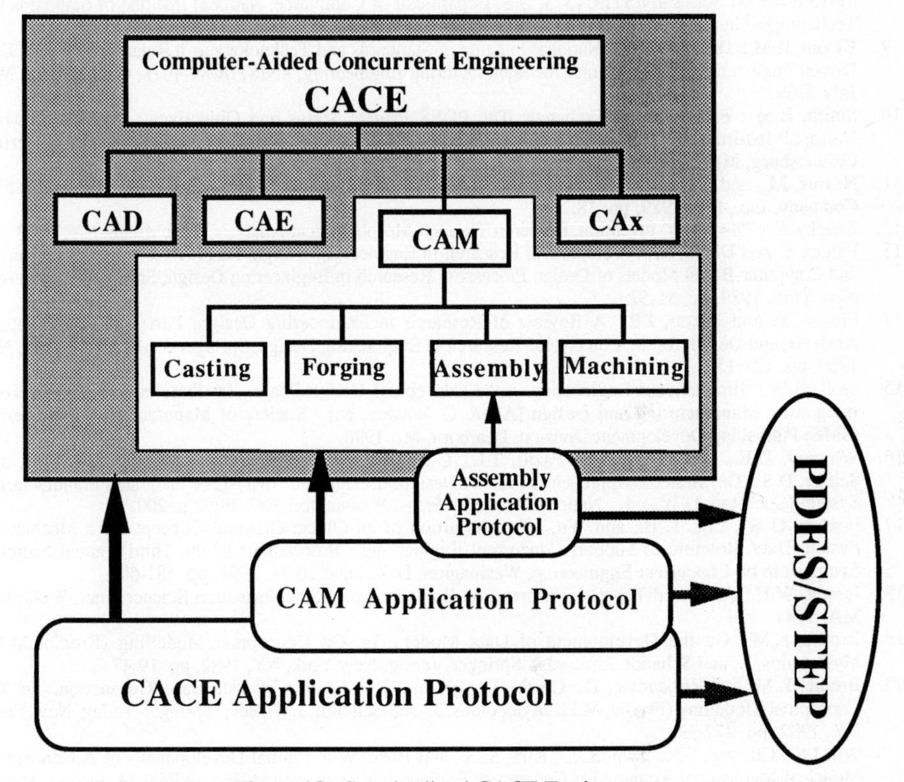

Figure 19. Standardized CACE Environment

References

1. Young, R.E., Greef, A., and O'Grady, P.: An Artificial Intelligence-Based Constraint Network System for Concurrent Engineering. Department of Industrial Engineering Technical Report TR-91-03, North Carolina State University, January 1991.
2. Evans, B.: Simultaneous Engineering. The American Society of Mechanical Engineers (ASME), New York, NY, Vol. 110, No. 2, 1988.
3. Medland, A.J.: The Computer Based Design Process. Springer-Verlag, New York, NY, 1986.
4. Maddux, K.C. and Jane, S.C.: CAE for Manufacturing Engineering: The Role of Process Simulation in Concurrent Engineering. In: Manufacturing Simulation and Processes (Tseng, A. A., Durham, D. R., and Komanduri, R., eds), The American Society of Mechanical Engineers, 1986.
5. Wu, J.-K., Liu, T.-H., and Fischer, G.W.: An Integration of CAE and CAM Applications Using PDES/STEP Information Model. Center of Computer-Aided Design, Technical Report # R-129, The University of Iowa, Iowa City, IA, 1992 (also see Proceedings of The Second International Conference on Automation Technology, Taipei, Taiwan, R.O.C., July 4-6, 1992).
6. Smith, B.M : IGES: A Key to CAD/CAM Systems Integration," IEEE Computer Graphics and Application. Vol. 3, No. 8, 1983, pp. 78-83.

7. NIST: Product Data Exchange Specification First Working Draft. U.S. Department of Commerce, National Institute of Standards and Technology, Gaithersburg, MD, December 1988.
8. Carver, G.P., and Bloom, H.M.: Concurrent Engineering Through Product Data Standards. National PDES Testbed Report Series #NISTIR 4573, U.S. Department of Commerce, National Institute of Standards and Technology, Gaithersburg, MD, May 1991.
9. Bloom, H.M.: The Role of the National Institute of Standards and Technology as it Relates to Product Data Driven Engineering. NIST, Center for Manufacturing Engineering, #NISTIR 89-4078, Gaithersburg, MD, July 1989.
10. Smith, B.M : Product Data Exchange: The PDES Project Status and Objectives. NISTIR 89-4165, National Institute of Standards and Technology (NIST), Center for Manufacturing Engineering, Gaithersburg, MD, September 1989.
11. Nevins, J.L. and Whitney, D.E.: Concurrent Design of Products and Processes. McGraw-Hill Publishing Company, Inc., NY, 1989, pp. 18.
12. Zwicky, F. : Discovery, Invention, Research Through Morphological Analysis. MacMillan, NY, 1969.
13. Finger, S. and Dixon, J.R.: A Review of Research in Engineering Design, Part I: Descriptive, Prescriptive, and Computer-Based Models of Design Processes. Research in Engineering Design, Springer-Verlag, Inc., New York, 1989, pp. 51-57.
14. Finger, S. and Dixon, J.R.: A Review of Research in Engineering Design, Part II: Representation, Analysis, and Design for the Life Cycle. Research in Engineering Design, Springer-Verlag, New York, NY, 1989, pp. 121-137.
15. Stoll, H.W.: Simultaneous Engineering in the Conceptual Design Phase. In: Simultaneous Engineering: Integrating Manufacturing and Design (Allen, C. Wesley, ed). Society of Manufacturing Engineering (SME) Publishing Development Division, Dearborn, MI, 1990.
16. Whitney, D.E., Nevins, J.L., De Fazio, T.L., Gustarson, R.E., Metzinger, R.W., Rourke, J.M., and Seltzer, D.S.: The Strategic Approach to Product Design. Design and Analysis of Integrated Manufacturing System (Compton, D.W., ed). National Academy Press, Washington, DC, 1988, p. 200.
17. Fischer, G.W., Liu, T.-H., and Wu, J.K.: Definition of an Object-Oriented Concept for a Mechanical System Data Structure to Support Concurrent Engineering. Proceedings of the Third Annual National Symposium on Concurrent Engineering, Washington, D.C., June 10-14, 1991, pp. 581-608.
18. Inmon, W.H.: Advanced Topics in Information Engineering. QED Information Science, Inc., Wellesley, MA, 1987.
19. Brodie, L.M.: On the Development of Data Model. In: On Conceptual Modelling (Brodie, M.L., Mylopoulos, J., and Schmidt, J.W., eds), Springer-Verlag, New York, NY, 1982, pp. 19-47.
20. Brodie, L.M. and Ridjanovic, D.: On the Design and Specification of Data Base Transaction. In: On Conceptual Modelling (Brodie, M.L., Mylopoulos, J., and Schmidt, J.W., eds), Springer-Verlag, New York, NY, 1982, pp. 277-331.
21. Wu, J.K., Choong, F.N., Twu, S.-L., Kim, S.-S., and Bake, W.K.: Initial Developments of A Conceptual Model of Mechanical Systems for Concurrent Engineering Environment. Proceedings of Second Annual Symposium on Design of Mechanical Systems in A Concurrent Engineering Environment (Fischer, G.W., ed), Iowa City, IA, October 30-31, 1990.
22. Wu, J.K., Ciarelli, K.J., Kim, S.-S., and Kim, S.-S.: A Generic Data Model for Mechanical System Simulations. Proceedings of 1990 ASME International Computers in Engineering, Boston, MA, August, 1990, pp.113-120.
23. Paz-soldan, J.P. and Rinderle, J.R. : The Alternate Use of Abstraction and Refinement in Conceptual Mechanical Design. Paper #89-WA/DE-8, Proceedings ASME Winter Annual Meeting, San Francisco, CA, December 10-15, 1989.
24. Meunier, K. and Dixon, J.R.: Iterative Respecification: A Computer Model for Hierarchical Mechanical System Design. Proceedings of the ASME Computers in Engineering Conference, The American Society of Mechanical Engineers (ASME), San Francisco, CA, July 31-August 3, 1988.
25. Cunningham, J.J. and Dixon, J.R.: Designing with Features: The Origin of Features. Proceedings of ASME Computers in Engineering Conference, San Francisco, CA, July/August 1988.
26. Cunningham, J.J. and Dixon, J.R.: Design with Feature: The Origin of Features. In Computers in Engineering (Tipnis, V. A. and Patton, E.M., eds), ASME, 1988, pp. 237-243.
27. Harvington, S.: Computer Graphics. McGraw-Hill Publishing Company, New York, NY, 1987.
28. Chung, J.C., Cook, R.C., Patel, D. and Simmons: Feature-Based Geometry Construction for Geometric Reasoning. ASME-Computers in Engineering Conference, San Francisco, CA, August 1988.
29. Wang, H.P. and Liu, C.A.: Automatic Generation of NC Part Programs for Turning Parts Based on 2-D Drawing Files. International Journal of Advanced Manufacturing Technology, Vol. 2, No. 1, 1987.
30. Bobrow, J.E.: NC Machine Tool Path Generation from CGS Part Representations. Journal of CAD, March 1985, pp. 69-76.

31. Fischer, G.W. and Narang R.V.: A Knowledge-Based Approach to Process Selection and Planning: Estimator. Proceedings of Second Annual Symposium on Design of Mechanical Systems in a Concurrent Engineering Environment (Fischer, G.W., ed), Iowa City, IA, October 30-31, 1990.
32. Wei, Y., Fischer G.W. and Santos, José L.T.: A Concurrent Engineering Design Environment for Generative Process Planning Using Knowledge-Based Decisions. The ASME Design Technical Conference 16th Design Automation Conference, DE-Vol. 22, Chicago, IL, September 16-19, 1990, pp. 35-45.
33. Yeh, C.H. and Fischer, G.W.: A Structural Approach to Automotive Planning of Machining Operations for Rotational Parts Based on Computer Integration of Standard Design and Process Data. International Journal of Advanced Manufacturing Technology, No. 6, 1991, pp. 285-298.
34. Chang, T.C. and Wysk, R.A.: Integrating CAD and CAM Through Automated Process Planning. International J. of Production Research, Vol. 22, No. 5, 1984, pp. 877-894.
35. Chang, T.C. and Wysk, R.A.: An Introduction to Automated Process Planning Systems. Prentice-Hall, Englewood Cliffs, NJ, 1985.
36. Chang, T.C., Wysk, R.A., and Wang H.P.: Computer-Aided Manufacturing. Prentice-Hall, Englewood Cliffs, NJ, 1991.
37. Kissil, A. and Kamel, H.A.: An Expert System Finite Element Modeler. Proceedings ASME-Winter Annual Meeting, Anahiem, CA, 1986.
38. Bhadra, A. and Fischer, G.W.; A New GT Classification Approach: A Database with Graphical Dimensions. Manufacturing Review, Vol. 1, No. 1, March 1988, pp. 44-49.
39. Wang, H.P. and Chang, H.: Automated Classification and Coding Based on Extracted Surface Features in a CAD Database. Intl. Journal of Advanced Manufacturing Technology, Vol. 2, No. 1, 1987, pp. 25-35.
40. Bhatnagar, A.: Design and Implementation of Feature Mapping Shell with Application to GT Classification. M.S. Thesis: Mechanical and Aerospace Engineering Department, Arizona State University, Tempe, AZ, June 1988.
41. Padhy, S. and Dwivedi, S.: On the Contact and Geometry of Features in Assembly. Proceedings of Second Annual Symposium on Design of Mechanical Systems in a Concurrent Engineering Environment (Fischer G.W., ed), Iowa City, IA, October 30-31, 1990.
42. Iwata, K. and Sugimura, N.: An Integrated CAD/CAPP System with Know-How on Machining Accuracies of Parts. J. of Engineering For Industry, Vol. 109, May 1987.
43. Wang, H.P. and Wysk, R.A.: Intelligent Reasoning for Process Planning. Computers in Industry, Vol. 8, Jun/July, 1987, pp. 293-309.
44. Opas, J. and Mantyla, M.: Introducing Manufacturing Knowledge into Intelligent CAD Systems. Proc. of 2nd IFIP WG5.2, Workshop on Intelligent CAD, Cambridge UK, September 19-22, 1988.
45. Shah, J.J.: Feature Transformation between Application-Specific Feature Spaces. Computer-Aided Engineering Journal, Vol. 5, No. 6, 1988, pp. 247-255.
46. CAMI: Current Status of Features Technology. CAM-I Revised Report R-88-GM-04.1, Computer-Aided Manufacturing-International, Inc., Arlington, TX, 1988.
47. NIST Part-48: Industrial Automation Systems-Exchange of Product Model Data-Part 48: General Resources: Form Features. National Institute of Standards and Technology, Gaithersburg, MD, June 1991.
48. Ssemakula, M.E. and Satsang, A.: Application of PDES to CAD/CAPP Integration. Computers in Industrial Engineering, Vol. 18, No. 4, 1990, pp. 435-444.
49. Shah, J.J. and Mathew, M.: Experimental Investigation of The STEP Form-Feature Information Model. Computer-Aided Design, Vol. 22, No. 4, May 1991, pp. 282-296.
50. Liu, T.-H. and Fischer, G. W.: Feature-Based Assembly Planning Using PDES/STEP. Working Paper #9201, Department of Industrial Engineering, The University of Iowa, Iowa City, IA, 1992
51. Liu, T.-H. and Fischer, G. W.: Assembly Evaluation Method for PDES/STEP-based Mechanical Systems. Working Paper #9202, Department of Industrial Engineering, The University of Iowa, Iowa City, IA, 1992 (submitted to Journal of Design and Manufacturing, March 1992).
52. Silvestri, M. J.: Product Abstraction Evaluation by Active Process Facilitators. Proceedings of Computer-Aided Cooperative Product Development," (MIT-JSME) Workshop (Sriram, D., Logcher, R., and Fukuda, S., eds), MIT, Cambridge, November 1989, Springer-Verlag, Lecture Notes in Computer Science Series, 1991.
53. Krause, Frank-Lothar: Knowledge Integrated Product Modeling For Design and Manufacture," in Organization of Engineering Knowledge for Product Modelling in Computer Integrated Manufacturing (The 2nd Toyota Conference, Aichi, Japan, 2-5 October 1988) (Sata, Toshio, ed), Elsevier Science Publishers, New York, NY, 1989.
54. NIST Part-11: Industrial Automation Systems-Exchange of Product Model Data-Part 11: The Express Language User Manual. National Institute of Standards and Technology, Gaithersburg, MD, June 1991.
55. Anderson, B.: PDES and STEP. Proceedings of the Army Meeting on Flexible Computer Integrated Manufacturing (Bradham, J.H., ed), South Carolina Research Authority (SCRA), North Charleston, SC, February 20-21, 1992.

56. Hardwick, M., and the RPI DICE Team: ROSE 3.0 User Manual. Rensselaer Design Research Center, Rensselaer Polytechnic Institute, Troy, NY, 1991.
57. NIST Part-42: Industrial Automation Systems-Exchange of Product Model Data-Part 42: Integrated Resources: Geometric and Topological Representation" National Institute of Standards and Technology, Gaithersburg, MD, June 1991.
58. NIST Part-47: Industrial Automation Systems-Exchange of Product Model Data-Part 47: Shape Variation Tolerances. National Institute of Standards and Technology, Gaithersburg, MD, November 1990.
59. McLay, Michael J. and Morris, Katherine C.: The NIST STEP Class Library. National PDES Testbed Report Series #NISTRI 4411, U.S. Department of Commerce, National Institute of Standards and Technology, Gaithersburg, MD, August 1990.
60. Kusiak, A.: Intelligent Manufacturing Systems. Prentice-Hall, Englewood Cliffs, NJ, 1990.

Concurrent Engineering Tools for Forging Die and Process Design

Ramana V. Grandhi and Raghavan Srinivasan

Wright State University, Dayton, Ohio 45435, USA

Abstract: This paper discusses the development of engineering tools for die shape design in the forging process and the utilization of this information for determining optimal operating conditions. The tools discussed here are based on computer graphics, finite element modeling, design optimization and optimal control strategies, and can be run on personal computers, work stations and main frame computers, to meet a variety of customer needs and to provide user friendly tools for conducting "what if" studies. The methods developed here are applicable to several unit processes like extrusion, shape rolling, etc., besides forging. Details about the design variables, design constraints, objectives, analytical sensitivity calculations, condensed states, satisfaction of behavior constraints, and the optimal tracking algorithm are presented here with engineering case studies.

Keywords: concurrent engineering / forging die design / forging process design / optimization / geometric mapping / backward deformation

1 Introduction

Forging is a bulk metalforming process in which a workpiece is deformed between dies by the application of compressive forces. Forged products generally have better mechanical properties than components manufactured through other processes, such as casting. Open-die forging is the term applied to all forging operations in which there is no lateral constraint except for friction, while closed-die forging includes all forging operations involving three-dimensional confinement and control of the deforming material. In closed-die forging, the workpiece is deformed to fill the die cavity representing the final component shape (Figure 1). Metal flow is restricted to fill the closed die cavity, and excess material flows through the gap between the closing dies and forms a flash around the forging at the die parting line (Figure 2). The flash is subsequently trimmed to form the final product.

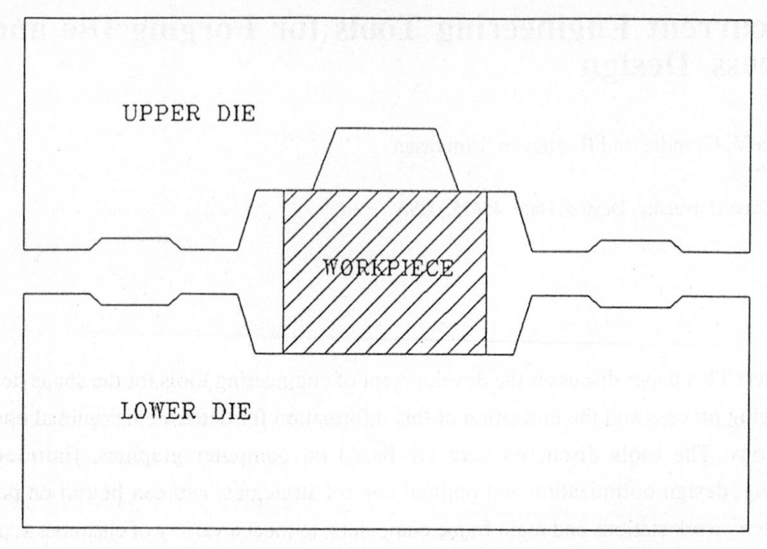

Figure 1. Schematic Diagram of Plane Strain Closed Die Forging

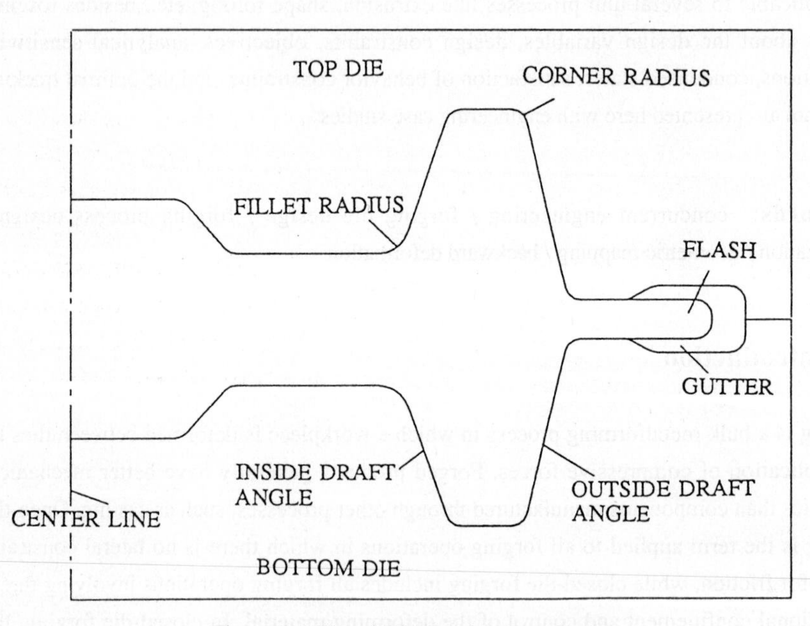

Figure 2. Closed Die Forging Terminology

Engineering components are frequently designed to meet specific functional requirements based on their application. These components are generally of shapes that are not directly forgeable, and their design is generally unalterable by the forging process design engineer.

The design engineer, therefore, has to develop a forging process, taking into account a number of factors, which have been broadly classified into the following two groups:

A. **Determination of the Forging Shape:** The machined or finished part geometry is modified to take into consideration the requirements of forging. This requires the addition of machining allowances, thermal expansion allowances if the part is to be hot forged, draft angles to enable the part to be withdrawn from the forging dies, die and corner radii to reduce stresses on the die, etc. Forging handbooks provide guidelines to the design engineer for making these decisions. In the early 1980s Battelle Columbus Laboratory, with support from the U.S. Air Force Materials Laboratory and several forging companies, demonstrated that the information in handbooks could be coded into a computer based expert system (e.g., AFD) [1,2].

B. **Determination of Processing Conditions:** Since the starting shape of billets for forging operations are generally quite simple (cylindrical, square or rectangular bars) and the forging shape can be quite complex, several decisions are required in order to determine the processing conditions. Metal flow is greatly influenced by the die geometry. If the final component shape is complex and intricate, the billet cannot be deformed to the final shape in a single operation. To avoid problems such as fold-over, localized deformation, excessive die forces and improper die-fill, the workpiece is deformed through several intermediate shapes before a product of the desired shape is formed. These intermediate shapes are generally referred to as the buster, blocker and finisher shapes (Figure 3). The number and shape of intermediate stages, thus, plays an important role in any forging process. Other factors to be considered are the type of forging machine (i.e., press or hammer), the temperature of forging (whether hot, cold or warm), the precision of the forging, location of parting lines, thickness of flash, ram velocity, die material, etc. Design handbooks are available to provide guidelines to the engineer. Again, computer based expert systems like DIE FORGE and BID [3,4] are available to automate some of these tasks.

Conventional design techniques rely heavily on handbooks and on the experience of the design engineer. These techniques provide adequate designs which have been validated through extensive physical modeling - a process entailing considerable expense and long lead times. Established forging processes that have been developed through conventional design methods are generally difficult to alter because of the costs involved in modifying existing equipment, training of personnel, certification of the new process, etc. However, opportunities exist for the application of new design procedures in developing new processes for new materials. Three new methods for the design of processing conditions are discussed in this paper: (l) the use of geometrical mapping to provide a quick view of the change in shape of the workpiece as it deforms from the initial to the final shape, (2) the use of

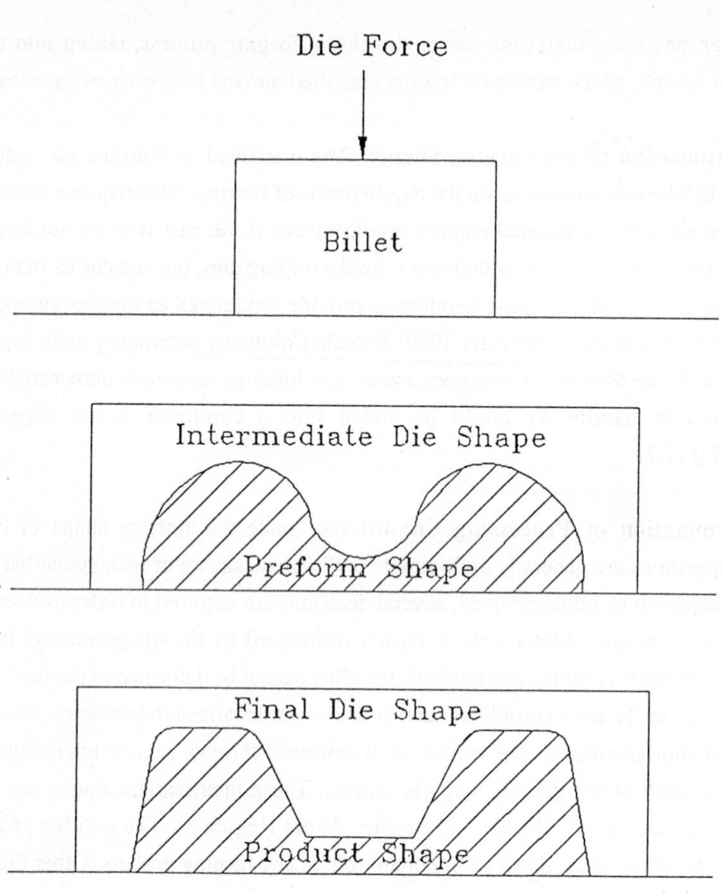

Figure 3. Multi-Stage Forging: Billet, Preform and Final Product Shapes

backward deformation and optimization techniques to determine optimal intermediate shapes, (3) the use of the state space model to design optimal process conditions. This document presents the methodology development for the above three techniques along with example problems to validate the same.

2 Geometric Mapping

The geometric mapping technique has been developed for visualizing the deformation during a forging process. The technique provides the engineer with a sequence of shapes between the billet and the final forging shape. The shapes generated can then be used as initial designs for the intermediate shapes. This would, to some extent, reduce the time consuming and expensive trial-and-error physical modeling.

The transformation of one shape into another in agreement with certain criteria is called conformal mapping. The term conformal mapping, in the present context, is essentially geometrical mapping with adherence to suitable conformance criteria. Since plastic deformation of dense metals occurs without changes in volume, the mapping is done under the condition of volume constancy. In the case of plane strain deformation, this condition translates to the maintenance of constant cross sectional area during deformation.

2.1 Two Dimensional (2-D) Geometric Mapping

This method is explained with reference to a typical forging shown in Figure 4. Due to the symmetry of the 'H' shape, only a quarter section (from the first quadrant) is considered for study. Consider a rectangular element ABCD of the preform (Figure 5). This element is geometrically mapped onto the element $A_1B_1C_1D_1$ of the final shape. The following assumptions are made while mapping the preform to the final shape:

(i) The material of the billet in the element ABCD of the preform is mapped onto the element $A_1B_1C_1D_1$ of the final shape.

(ii) There is no displacement of the centroid of the cross section.

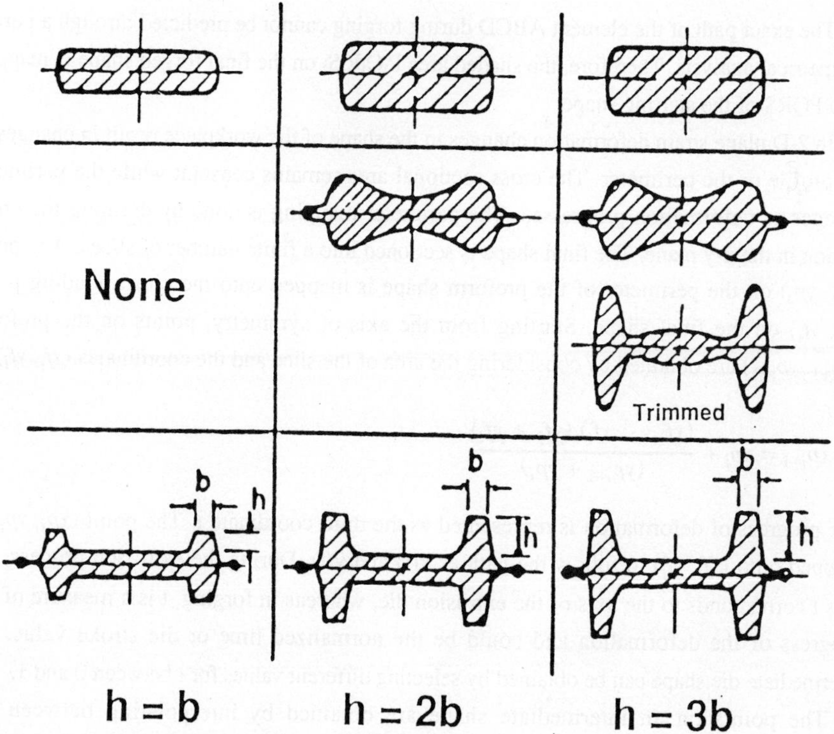

Figure 4. Preforms for Steel Finish Forgings of H Cross-Sections

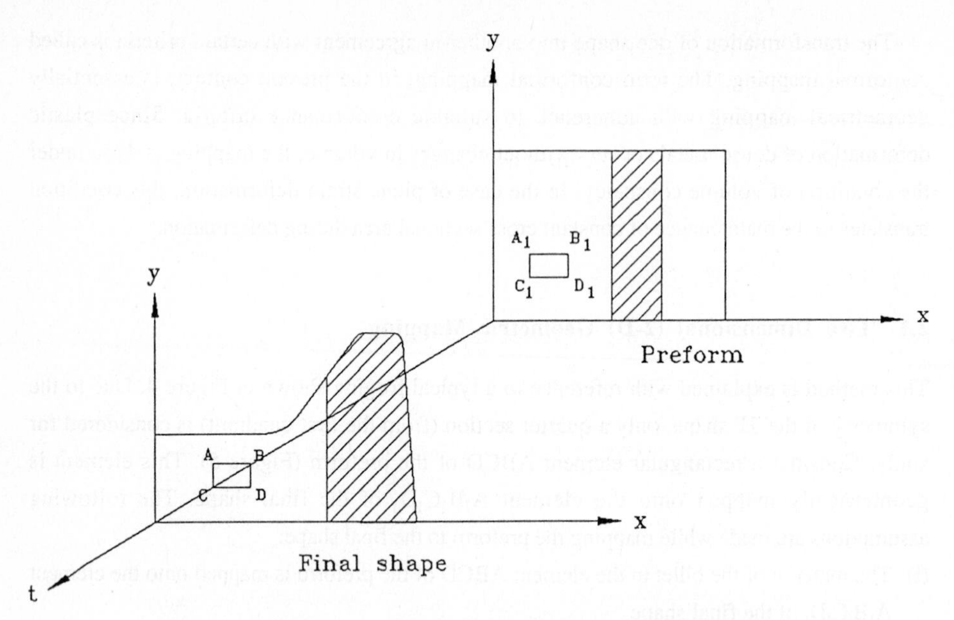

Figure 5. Preform and the Final Forging for Geometric Mapping

The exact path of the element ABCD during forging cannot be predicted through a purely geometrical analysis. Therefore, the shaded area $P_1Q_1R_1S_1$ on the final forged shape is mapped onto PQRS of the starting shape.

In 2-D plane strain deformation changes in the shape of the workpiece result in changes in the profile or the perimeter. The cross sectional area remains constant while the perimeter changes as deformation progresses. The perimeter mapping is done by defining the cross section in the x-y plane. The final shape is sectioned into a finite number of slices. Any point (xp_i, yp_i) on the perimeter of the preform shape is mapped onto the corresponding point (xf_i, yf_i) on the final shape. Starting from the axis of symmetry, points on the preform (xp_{i+1}, yp_{i+1}) are obtained by considering the area of the slice and the coordinates (xf_i, yf_i):

$$xp_{i+1} = xp_i + \frac{(xf_{i+1} - xf_i)\,(yf_i \; + yf_i)}{(yp_{i+1} + yp_i)} \tag{1}$$

The progress of deformation is represented as the third coordinate t. The point (xp_i, yp_i) is mapped onto (xf_i, yf_i) to follow the path as x(t) and y(t). During the extrusion process, the axis t corresponds to the axis of the extrusion die, whereas in forging, t is a measure of the progress of the deformation and could be the normalized time or die stroke value. An intermediate die shape can be obtained by selecting different values for t between 0 and 1.

The points on an intermediate shape are obtained by interpolating between the corresponding points on the perimeter of the preform and the perimeter of the final shape.

During deformation from the initial to the final shape, any point *(x,y)* on the preform initially changes position rapidly and, as it approaches the final shape, it slows down.

The transformation is represented by a logarithmic function and the intermediate shapes are obtained as logarithmic functions along the t direction:

$$x_i = xp_i + (xf_i - xp_i)\ log\ (a + bt) \tag{2}$$

$$y_i = yp_i + (yf_i - yp_i)\ log\ (a + bt) \tag{3}$$

where, $0 \le t \le 1$, ($t = 0$ corresponds to the first intermediate shape and $t = 1$ corresponds to the final shape), a & b are constants, and $a + b = 10$. The fillet radii and the corner radii of the intermediate shapes are critical in order to ensure proper material flow and die fill. General guidelines require that the corner radii and the fillet radii be larger at the preform than at the final shape. The initial radius in the starting billet is assumed to be infinite. The radii at the intermediate shapes are given by:

$$r_i = r_f\left(1 + log\left(\frac{1}{s}\right)\right) \tag{4}$$

where, $0 \le s \le 1$, and $s = 0$ corresponds to the first intermediate shape and $s = 1$ corresponds to the final shape. The areas of the intermediate sections after interpolation are calculated as:

$$A = \sum (x_i y_{i+1} - x_{i+1} y_i) \tag{5}$$

During the mapping process, the cross sectional areas of the intermediate shapes must comply with the constant area requirement. Therefore, correction factors for the coordinates of the intermediate shapes are obtained as:

$$\Delta x = \Delta y = \left(\frac{A_f}{A}\right)^{\frac{1}{2}} \tag{6}$$

The corrected coordinates of the intermediate shapes are calculated as:

$$x_{i_{cor}} = x_i \Delta x \tag{7}$$

$$y_{i_{cor}} = y_i \Delta y \tag{8}$$

The interpolation technique used in this approach does not imply that the material in the shaded area in the initial shape goes to the shaded area in the final shape. An assumption that vertical slices remain vertical made during the mapping process assumes the slices slide along their interfaces without friction. The mapping technique is, therefore, only intended to modify the perimeter and to generate intermediate die shapes having constant cross sectional area.

2.2 Three-Dimensional (3-D) Geometric Mapping

The mapping technique developed for plane strain forgings can be extended to three dimensional cases. The three-dimensional component is assumed to be made of a set of finite number of slabs, and geometric mapping is performed for the two faces of each of the slabs. The volume associated with the component during the transformation is maintained constant.

This method assumes that during the closed die forging process, deformation of the slab takes place in two modes: (i) in the plane of cross section (x-y plane) and (ii) in the direction perpendicular to the cross section (z-axis), as shown in the Figures 6c and 6d. It is also assumed that the planes which are perpendicular to the z-axis remain parallel during deformation. Figures 6a and 6b show the preform and final shapes, respectively. A slab of material (Figure 6c) deforms into the slab of material shown in Figure 6d. During the deformation, changes take place in the perimeter of the cross section and in the thickness of the slab, while adhering to the constant volume criterion.

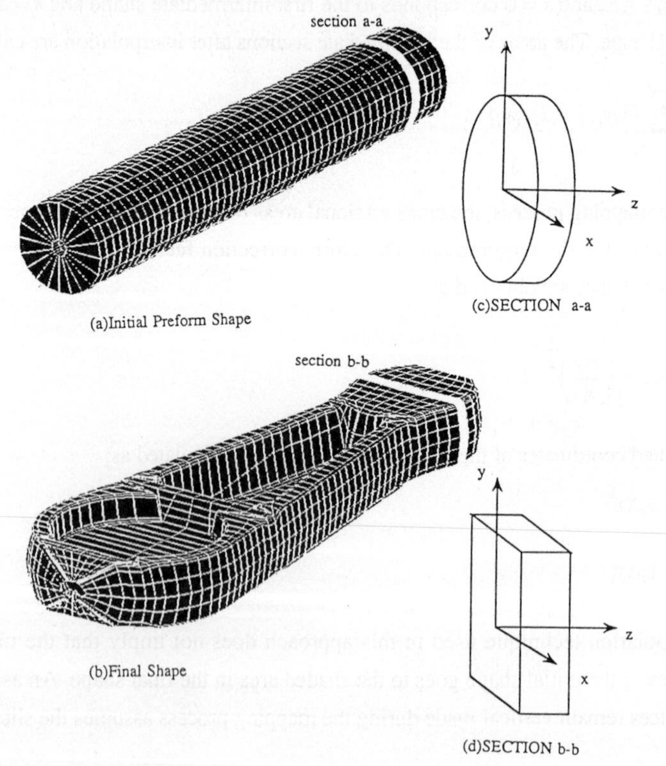

Figure 6. Automobile Steering Arm

This method is explained with reference to an automobile steering arm (Figure 7). The final forging shape of the automotive steering arm is sectioned into a finite number of slabs in a plane perpendicular to the z-axis. The volume of each slab of the final shape is calculated as:

$$V_{slab} = 0.5\ (A_{1i} + A_{2i})\ w_{slab} \tag{9}$$

where, A_{1i} and A_{2i} are the areas of the faces of the slab respectively, and w_{slab} is the thickness of the slab. The total volume of the final shape is obtained by the summation of the volumes of the slabs as:

$$V_{total} = \sum V_{slab} \tag{10}$$

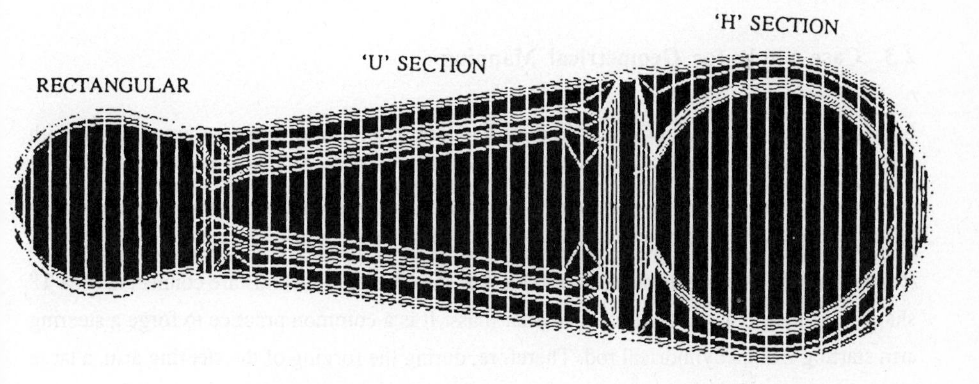

Figure 7. Method of Sectioning into Slabs in 3-D Geometric Mapping

The cross section of the starting billet is selected as a circle, a square, or a rectangle representing the three common bar stocks available. The dimensions of the billet are determined from the total volume of the final component. The billet is sectioned into the same number of slabs as the final component to obtain corresponding slabs of equal volume.

The transformation of the cross section of the profile in the x-y plane is done by geometrically mapping the corresponding cross sections of the initial (billet) and final shapes. During the geometrical mapping, the perimeter of the cross section changes from the initial to the final shape, as discussed for 2-D forgings. The point (xp_i, yp_i) is mapped to (xf_i, yf_i) to follow the path as $x(t)$ and $y(t)$. The variable t has the same properties as in the 2-D case.

During the forging process, the volume of the slab is kept constant while the workpiece transforms from the initial to the final shape. The cross sectional area of each slab in an intermediate shape is calculated as:

$$A_{ji} = \sum (x_i y_{i+1} - x_{i+1} y_i) \tag{11}$$

where j is 1 or 2, depending on the side of the slab. The thickness of the slab at any intermediate stage is determined by the volume constancy criterion and is obtained as:

$$w_{ti} = \frac{V_{ti}}{\frac{(A_{1i}+A_{2i})}{2}} \tag{12}$$

The dimensions of each slab at the intermediate stages are obtained as discussed earlier. The intermediate shapes are then obtained by assembling the slabs obtained for that intermediate stage. The conformal mapping technique thus enables one to model the forging process in three dimensions and to successfully obtain the corresponding intermediate die shapes.

2.3 Case Study for Geometrical Mapping

In this section, the development of intermediate die shapes in forgings is demonstrated with the example of a three-dimensional forging.

The steering arm is a common, elongated forging component. The cross sections perpendicular to the z axis are referred to as the 'H', 'U' and the rectangular sections (Figure 6). The steering arm has a large mass at the ends and the ends are connected by a 'U' shaped channel of a comparatively smaller mass. It is a common practice to forge a steering arm starting from a cylindrical rod. Therefore, during the forging of the steering arm, a large amount of mass needs to be distributed to the ends. The complex variation in the cross section along the axis of this component makes it a good example to demonstrate the concepts developed here.

Figure 7 shows the plan (view) of a steering arm forging. A cylindrical preform of equal volume is used as the starting shape. The final forging shape was sectioned into 50 slabs. The small end of the forging has a rectangular cross section, and the large end has a 'H' cross section. The small end and the big end are connected by a tapering 'U' shaped channel. During the deformation of a cylindrical rod, mass has to be moved from the region of the 'U' channel to the larger 'H' and rectangular regions at the ends. The average cross sectional area of the steering arm is less than that of the cylindrical billet.

The overall length of the workpiece, therefore, increases as deformation progresses (Figure 8). The profiles of cross sections of each of the slabs were digitized, and this data was used as the input for the computer program developed. In this example, the flash associated with the forged shape has not been accounted for. Two intermediate stages of the cross sections were obtained using the geometrical mapping technique. The intermediate shapes of the entire die were obtained by assembling the corresponding slabs at the intermediate stages. Figure 8 shows the intermediate shapes obtained using the geometric mapping technique. Figure 9 gives the intermediate shapes obtained for the 'H', 'U', and rectangular sections, respectively. Other example problems have been successfully analyzed by Lanka et. al [5,6] to validate this design approach.

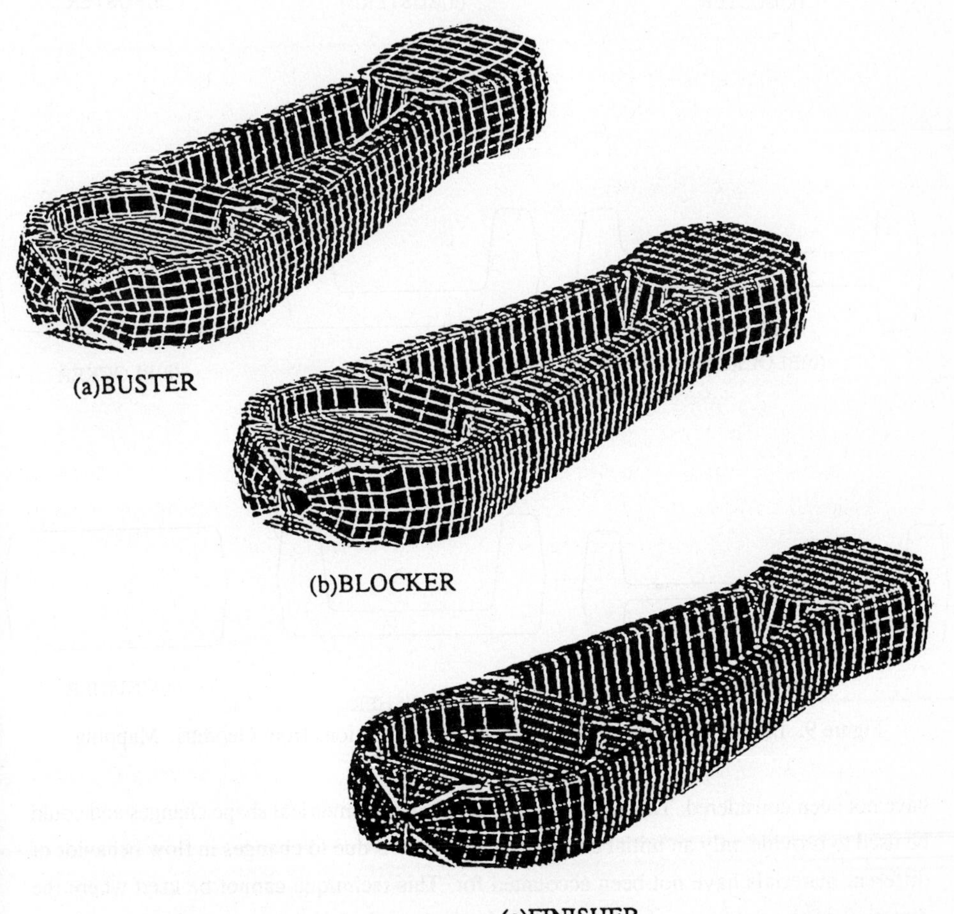

Figure 8. Intermediate Die Shapes for a Steering Arm from Geometric Mapping

2.4 Assumptions and Limitations of the Methodology

This method assumes that the final forging shape, which includes parameters such as machining allowance, draft angles, fillet radii, and parting line, has been determined through other methods, such as by AFD [1,2] or through handbook rules. Some important factors which influence metal flow, like material properties, die-workpiece friction, and flash design

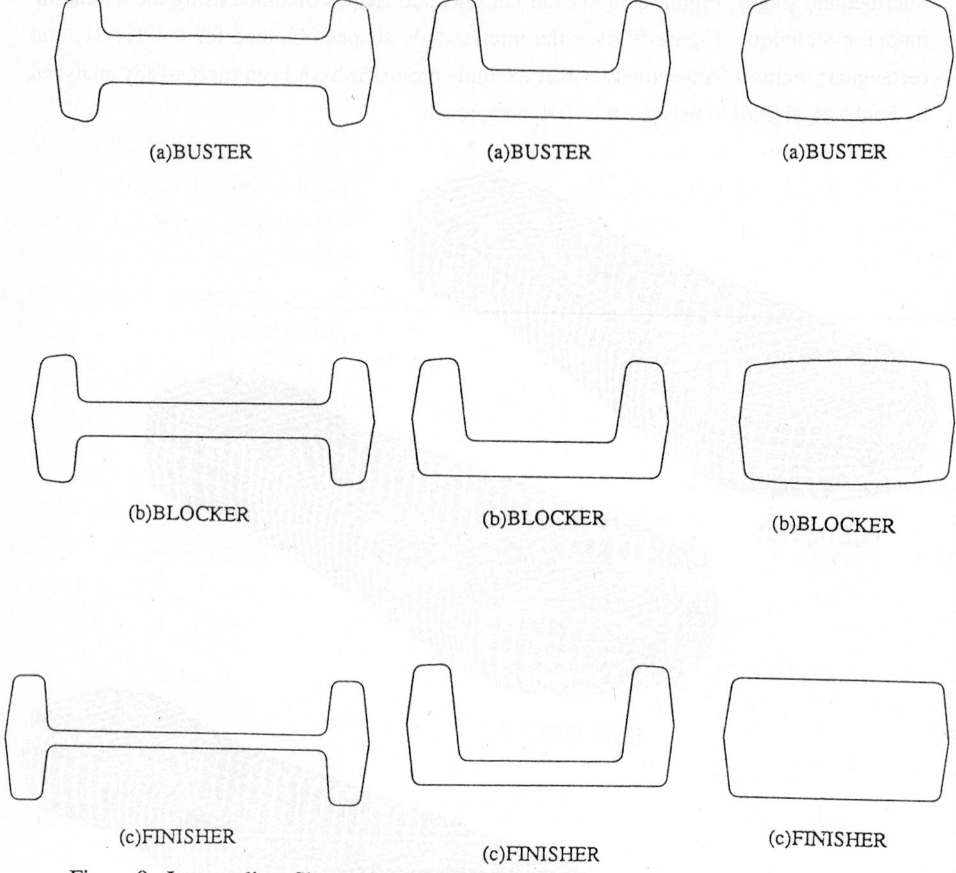

Figure 9. Intermediate Shapes for Different Cross-Sections from Geomtric Mapping

have not been considered. This method is based only on geometrical shape changes and could be used to provide only an initial design, since variations due to changes in flow behavior of different materials have not been accounted for. This technique cannot be used where the direction of the forging or the parting line is changed during the stages of forging. Also, deformation in the planes of the cross section of the slab have been ignored.

3 Backward Deformation Optimization Method

The Backward Deformation Optimization Method (BDOM) presents a methodology based on the rigid viscoplastic deformation control of the workpiece while reversing the deformation process from the final product shape. Since plastic deformation is an irreversible process, no unique path exists between the initial and final shapes. Unlike previous backward tracing methods, the Backward Deformation Optimization Method selects the optimum path based on constraints placed on the deformation of the workpiece. At every backward deformation step in the finite element simulation, the velocity profile of the die contacting workpiece nodal points is determined by minimizing the maximum difference between the effective strain rate and the average effective strain rate. This process gives the information required to judge which workpiece node should be detached from the die to obtain the optimum reverse path in forging. After successfully detaching workpiece nodes from the die, the backward tracing method is used to determine the workpiece geometry at the previous step. The iterative scheme is continued until most of the die contacting nodes become free. The workpiece shape at that point is used as an intermediate die shape. Depending on the complexity of the final product shape, several simulations may be required to determine suitable intermediate die shapes. The resulting optimal intermediate die shapes can provide smooth and complete die fill.

The tracing of a workpiece shape from the final product shape is possible by reversing the die movement as shown in Figure 10. The Backward Tracing Method developed by Hwang and Kobayashi [7,8], iteratively checks whether or not the new workpiece geometry obtained by the backward simulation method would result in the starting shape upon repeating the forward simulation step. The entire procedure is then repeated to trace back the initial shape. A detailed description of the backward tracing procedure is given below.

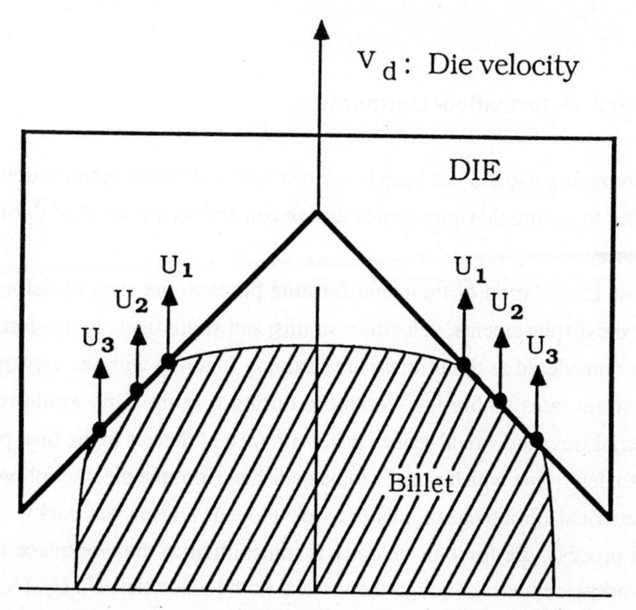

Figure 10. Die Velocity and Design Variables for Backward Deformation Optimization

3.1 Backward Tracing Method [7,8]

Backward tracing refers to the prediction of the part configuration at any stage in the deformation process when the final part geometry and process conditions are given. The concept is illustrated in Figure 11. At time $t = t_0$ the geometrical configuration x_0 of a deforming body is represented by a point Q. The point Q is arrived at from the point P, whose configuration is given as x_{0-1} at $t = t_{0-1}$, through the displacement field during a time step Δt, namely, $x_0 = x_{0-1} + u_{0-1}\Delta t$, where u_{0-1} is the velocity field at $t = t_{0-1}$. Therefore, the problem is to determine u_{0-1}, based on the information (x_0) at point Q. The solution scheme is as follows: taking a loading solution u_0 (forward) at Q, a first estimate of P can be made according to $P^{(1)} = x_0 - u_0\Delta t$. Then, the loading solution $u_{0-1}^{(0-1)}$ can be calculated from the configuration of $P^{(1)}$, with which the configuration x_0 at Q can be compared with $P^{(1)} + u_{0-1}^{(1)}\Delta t = Q^{(1)}$. If Q and $Q^{(1)}$ are not sufficiently close to each other, then $P^{(2)}$ can be estimated by $P^{(2)} = X_0 - u_{(0-1)}^{(1)}\Delta t$. The solution at $P^{(2)}$ is then $u_{0-1}^{(2)}$ and the second estimate of the configuration $Q^{(2)} = P^{(2)} + u_{0-1}^{(2)}\Delta t$ can be made. The iteration is carried out until $Q^{(n)} = P^{(n)} + u_{0-1}^{(n)}\Delta t$ becomes sufficiently close to Q. This algorithm was demonstrated for the shell nosing problem [8] where the die shape was fairly simple and the separation of nodes from the die in going from one step to the previous step was quite straight-forward. But for a complex die shape, a more reliable strategy is needed in releasing the nodes. The following section discusses the Backward Deformation Optimization technique developed for releasing the nodes.

3.2 Backward Deformation Optimization

The backward tracing method has been combined with numerical optimization techniques for releasing nodes to ensure that appropriate design constraints are satisfied during deformation of the workpiece.

In the flow formulation of the metal forming process, the state of deformation is fully described by the displacements, velocities, strains, and strain rates. In this formulation, strain rate has been considered as the primary constraining function with the view to minimize the variation of strain rates within the workpiece during forging. This would result in a more uniform material flow and would reduce the occurrence of defects in the final product.

In order to determine which node is to be detached from the die, the following technique, based on numerical optimization, has been developed. During the backward deformation optimization process, die force or velocity is transmitted to the workpiece through the die contacting workpiece nodes. Though the vertical nodal velocities, U_1, U_2, U_3, etc., as shown in Figure 10 are the same as the die velocity, their effect on the workpiece nodal velocities may be different. Each die-contacting node is considered individually to determine its

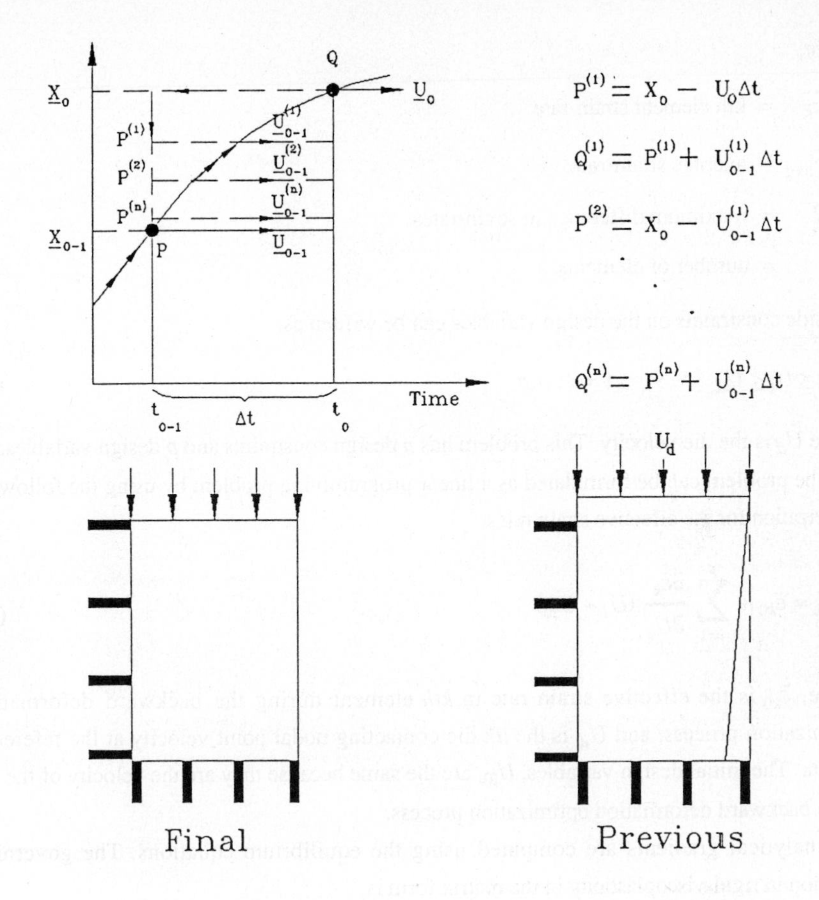

Figure 11. Backward Tracing Method

influence on all the workpiece nodal velocities. This approach is based on the analytical design sensitivity calculation method. The design variables in this problem are U_i, which are initially velocities of the die contacting nodes. Effective strain rates of the elements are used as the design constraints. The design variables, U_i, that minimize the maximum deviation of the effective strain rates from the average strain rates are determined. Mathematically, this mini-max problem is converted into the following minimization problem.

Find U_i to minimize $\boldsymbol{H}$ subject to the following inequality constraints:

$$|\dot{\bar{\varepsilon}}_k - \dot{\bar{\varepsilon}}_{avg}| \le \boldsymbol{H} \qquad k = 1,\cdots,n \tag{13}$$

where,

$\dot{\bar{\varepsilon}}_k$ = kth element strain-rate.

$\dot{\bar{\varepsilon}}_{avg}$ = average strain-rate.

H = maximum difference in strain-rates.

n = number of elements.

The side constraints on the design variables can be written as:

$$0 \leq U_i \leq U_d \qquad i = 1,\cdots,p \tag{14}$$

where U_d is the die velocity. This problem has n design constraints and p design variables.

The problem can be formulated as a linear programming problem by using the following linearization for the effective strain rates:

$$\dot{\bar{\varepsilon}}_k = \dot{\bar{\varepsilon}}_{k0} + \sum_{i=1}^{p} \frac{\partial \dot{\bar{\varepsilon}}_k}{\partial U_i} (U_i - U_{i0}) \tag{15}$$

where, $\dot{\bar{\varepsilon}}_{k0}$ is the effective strain rate in *kth* element during the backward deformation optimization process, and U_{i0} is the *ith* die contacting nodal point velocity at the reference design. The initial design variables, U_{i0}, are the same because they are the velocity of the die in the backward deformation optimization process.

Analytical gradients are computed using the equilibrium equations. The governing equation in rigid-viscoplasticity in the matrix form is,

$$[K]\,\mathbf{V} = \mathbf{F} \tag{16}$$

where,

$$[K] = \frac{\partial \Phi_D}{\partial v_i} + \frac{\partial \Phi_p}{\partial v_i} = \Sigma\left(\int_V \frac{\bar{\sigma}}{\dot{\bar{\varepsilon}}} P_{ij} dV + \int_V rC_iC_j dV\right) \tag{17}$$

$$\mathbf{F} = \frac{\partial \Phi_{S_F}}{\partial v_i} = \Sigma \int_{Sc} mk\frac{2}{\pi}q_j tan^{-1}\left(\frac{q_j u_{sj}}{u_0}\right) dS \tag{18}$$

$[K]$ is the material and process dependent nonlinear stiffness matrix. The solution to Equation 16 is obtained iteratively. The velocity vector, the stiffness matrix, and the traction force vector are implicit functions of the die velocity. These quantities are differentiated with respect to the workpiece die contacting nodal velocities. The resulting equation is:

$$\frac{\partial \mathbf{V}}{\partial U_i} = [K]^{-1}\left(\frac{\partial \mathbf{F}}{\partial U_i} - \frac{\partial [K]}{\partial U_i}\mathbf{V}\right) \tag{19}$$

The stiffness matrix and traction force vector derivatives are given as follows:

$$\frac{\partial [K]}{\partial U_i}\mathbf{V} = \Sigma \int_V \left(\frac{1}{\dot{\bar{\varepsilon}}}\frac{\partial \bar{\sigma}}{\partial \dot{\bar{\varepsilon}}} - \frac{\bar{\sigma}}{\dot{\bar{\varepsilon}}^2}\right)\frac{1}{\dot{\bar{\varepsilon}}} P_{ik}V_kV_mP_{mj}\frac{\partial V_j}{\partial U_i}dV \tag{20}$$

$$\frac{\partial \mathbf{F}}{\partial U_i} = \Sigma \int_S mk\frac{2}{\pi}q_iq_j \frac{u_0}{u_0^2 + (q_ku_{sk})^2}\frac{\partial}{\partial U_i}(V_j - U_j)dS \tag{21}$$

In order to relate the effective strain rates with the design variables, the effective strain rate is defined using strain-rate components as:

$$\dot{\bar{\varepsilon}} = \sqrt{\frac{2}{3}}\,(\dot{\bar{\varepsilon}}_{ij}\dot{\bar{\varepsilon}}_{ij})^{\frac{1}{2}} \tag{22}$$

or, alternatively, in the matrix notation:

$$(\dot{\bar{\varepsilon}})^2 = \dot{\bar{\varepsilon}}^T\,[D]\,\dot{\bar{\varepsilon}} \tag{23}$$

The diagonal matrix $[D]$ has $\frac{2}{3}$ and $\frac{1}{3}$ as its components corresponding to normal strain rate and shear strain-rate, respectively. The strain rate in each element can be derived using the strain rate matrix, $[B]$, and preform nodal velocity vector, $\mathbf{v}$:

$$\dot{\bar{\varepsilon}} = [B]\,\mathbf{v} \tag{24}$$

Therefore,

$$(\dot{\bar{\varepsilon}})^2 = \mathbf{v}^T[B]^T[D][B]\mathbf{v} = \mathbf{v}^T\,[P]\mathbf{v} \tag{25}$$

where,

$$[P] = [B]^T\,[D]\,[B]$$

The derivative of Equation 25 with respect to the design variable U_i, gives:

$$\frac{\partial \dot{\bar{\varepsilon}}_k}{\partial U_i} = \frac{1}{\dot{\bar{\varepsilon}}_{k0}}\mathbf{v_k}^T\,[P]\,\frac{\partial \mathbf{v_k}}{\partial U_i} \tag{26}$$

In Equation 26, $\mathbf{v_k}$ is the nodal velocity vector for the *kth* element, and $\frac{\partial \mathbf{v_k}}{\partial U_i}$, is the nodal velocity sensitivity for the *kth* element, obtained from Equation 19. Equation 26 relates the change of the constraint function to that of the design variable. These gradients are used in approximating the strain-rate $\dot{\bar{\varepsilon}}$.

The linear programming problem is solved by using the simplex method subroutines available in the IMSL package. After solving the optimization problem, the node which shows the minimum velocity is detached from the die because its influence on the billet deformation is minimal. The workpiece geometry of the previous step is determined by using the backward tracing method, after detaching the selected node. In order to ensure that the solution is not far from the linear solution, the time step value is determined such that the strain increment in any element is less than the predetermined maximum time increment. The optimization process and backward tracing process are continued until only one node remains in contact with the die. The flow-chart of the optimization algorithm is shown in Figure 12.

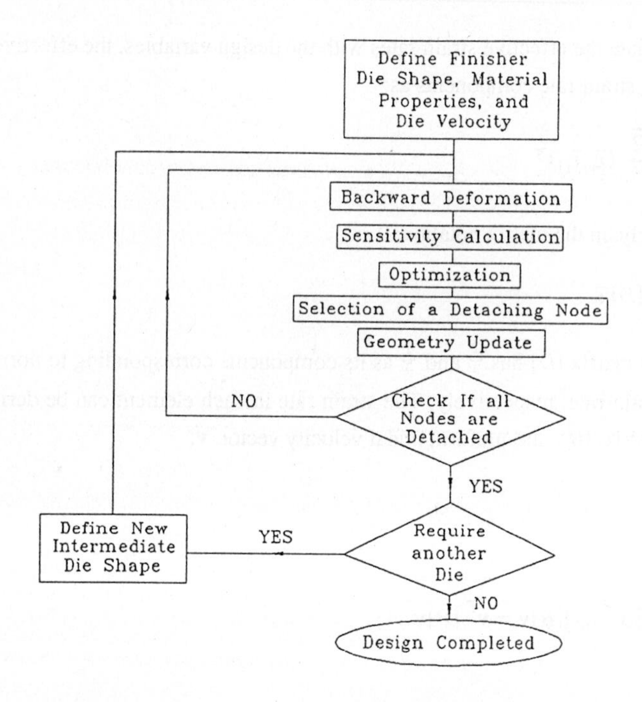

Figure 12. Intermediate Die Shape Optimization

3.3 Case Study for Backward Deformation Optimization

The shape optimization technique used to determine intermediate die shapes is demonstrated using an axisymmetric disk forging problem with 'H' cross-section. A non-strain hardening material having the constitutive relation, $\bar{\sigma} = 10\dot{\bar{\varepsilon}}^{0.3}$ has been used in this work. A constant shear force friction factor of 0.15 is assumed at the workpiece-die interface.

The disk geometry shown in Figure 13a has rotational symmetry about the center line and is also mirror symmetric about the horizontal plane. The deformation of the upper half of Figure 13b is representative of the entire disk forging problem. A cylindrical billet having radius 4.71 inches and height 4.71 inches and of equal volume is taken as the starting shape for the problem. The axisymmetric disk has a web thickness of 0.75 in., a rib height of 1.5 in., and a rib width of 1.5 in. The predicted workpiece shapes during forward simulation using ALPID [9] are shown in Figure 14. These shapes are obtained by considering the final product shape as the die shape. The workpiece material separates from the die in the vicinity of the fillet as seen in Figure 14d. This can result in material fold-over towards the end of deformation as indicated in Figure 14f. The forging of this part should, thus, be conducted in two stages, i.e., the billet should first be deformed to an intermediate shape using a 'blocker' die, and then further deformed to the final shape using the 'finisher' die. The blocker die is designed using the shape optimization technique developed in this work. Figure 15a shows the workpiece completely filling the die. Initially, all surface nodes of the workpiece are in contact with the die. The surface nodes are grouped into three zones marked A, B, and C, respectively. Velocities of the end nodes of these zones are taken as the design variables [10]. The grouping of nodes results in smooth geometric shapes for the die by avoiding local kinks. At the beginning of the backward deformation step, the node on the symmetric plane (marked 1) is assumed to be separated from the die. The sensitivity of the effective strain rates of the workpiece elements is calculated with respect to the design variables. The linearized optimization problem is solved and the new nodal velocities are obtained. The design variable with the least velocity magnitude indicates that the corresponding node separates from the die in that time step. Accordingly, that particular node is detached from the die.

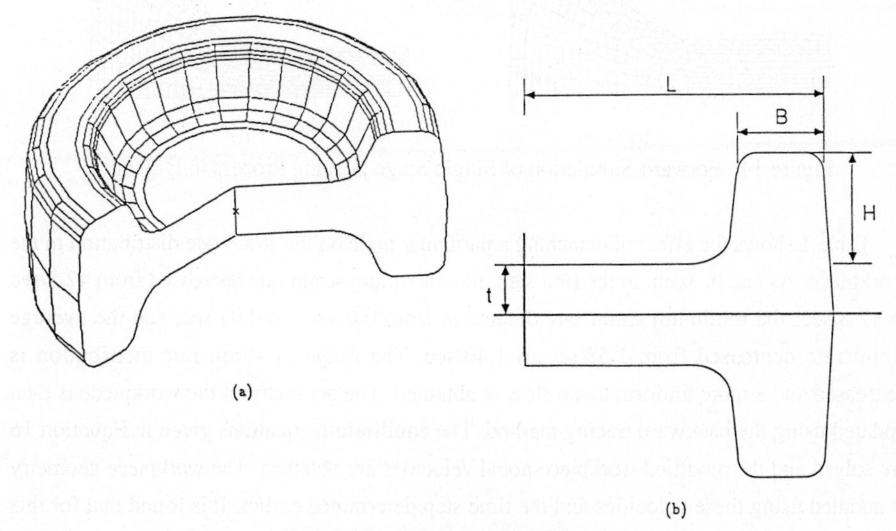

Figure 13. Axisymmetric H-Section

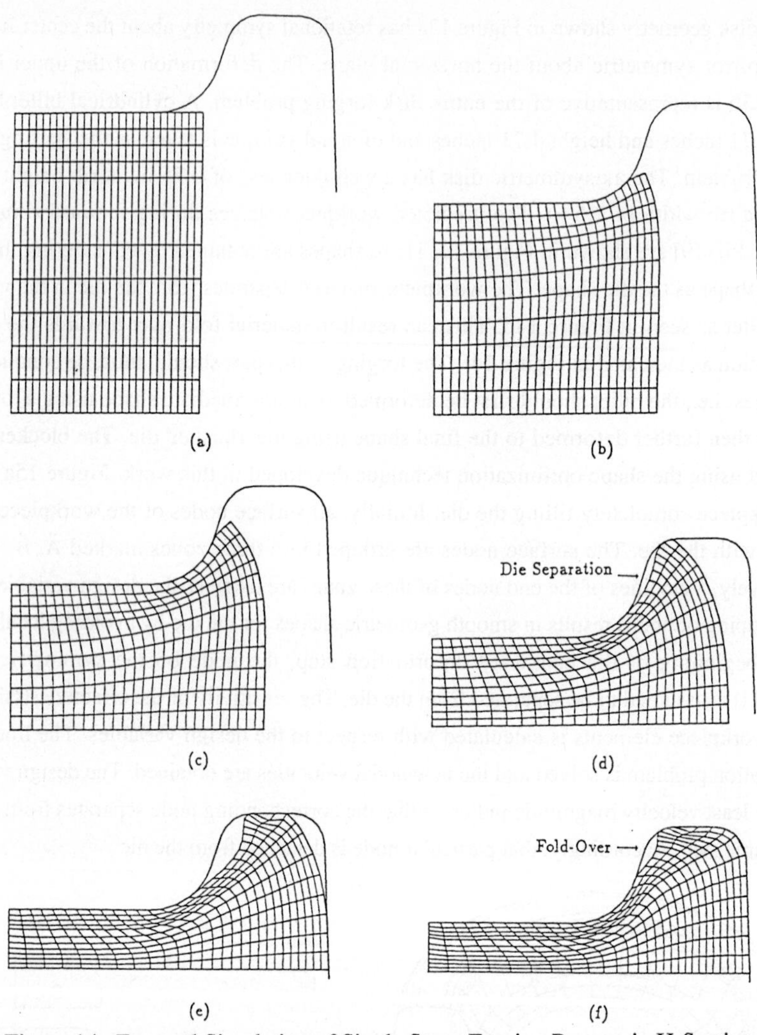

Figure 14. Forward Simulation of Single Stage Forging Process in H-Section

Table 1 shows the effect of detaching a particular node on the strain rate distribution in the workpiece. As can be seen, in the first step, the maximum strain rate decreased from 42.5/sec to 17.8/sec; the minimum strain rate decreased from 0.06/sec to 0.03/sec, and the average strain rate decreased from 1.58/sec to 1.40/sec. The range of strain rate distribution is decreased and a more uniform metal flow is obtained. The geometry of the workpiece is then updated using the backward tracing method. The equilibrium equations given in Equation 16 are solved and the modified workpiece nodal velocities are obtained. The workpiece geometry is updated using these velocities and the time step determined earlier. It is found that for this particular problem the limiting value of maximum strain increment was reached before the

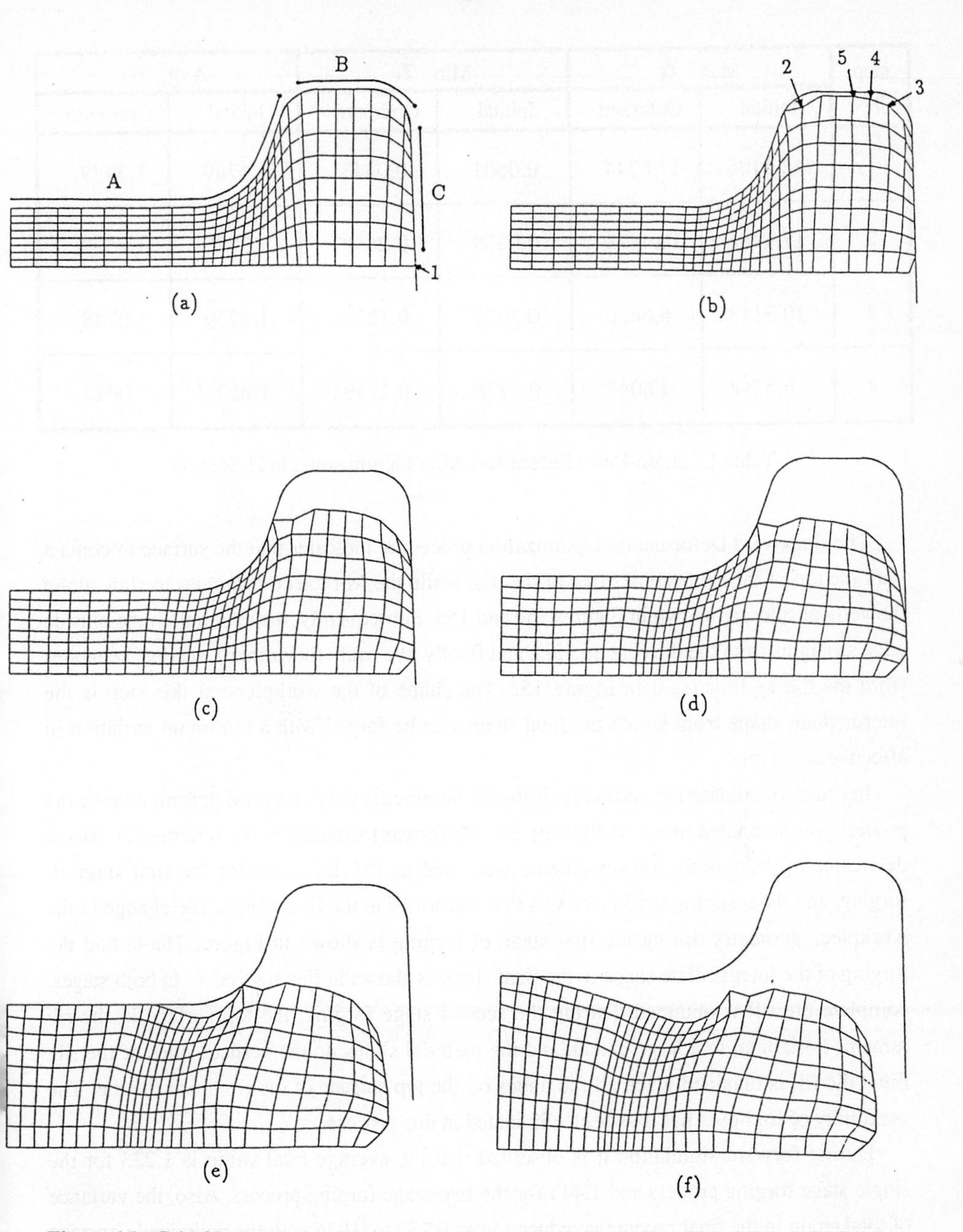

Figure 15. Grid Shapes during Backward Deformation Optimization Simulation in H-Section

maximum allowable time step. Table 1 presents the results obtained for four steps, corresponding to the detachment of nodes 2, 3, 4, and 5 in Figure 15b. In this problem an intermediate die shape was obtained after 30 time steps.

step no.	Max. $\dot{\bar{\varepsilon}}_k$		Min. $\dot{\bar{\varepsilon}}_k$		Avg. $\dot{\bar{\varepsilon}}_k$	
	Initital	Optimum	Initital	Optimum	Initital	Optimum
1	42.5106	17.8744	0.0601	0.0343	1.5789	1.3979
2	17.1486	10.4878	0.0339	0.2652	1.3775	1.2520
3	10.3118	6.6620	0.2025	0.1223	1.2259	1.0748
4	6.5248	4.6047	0.1228	0.1749	1.0591	1.0023

Table 1. Strain-Rates Before and After Optimization in H-Section

The Backward Deformation Optimization procedure indicates that the surface in contact with the top of the rib becomes free of the die, while the workpiece continues to slide along the vertical sides, as shown in Figures 15c and 15d. Subsequently, the material on the outside edge separates from the die (Figure 15e), and finally the web (horizontal) section separates from the die as illustrated in Figure 15f. The shape of the workpiece at this step is the intermediate shape from which the final shape can be forged with a minimum variation in effective strain rate.

In order to validate the optimally designed intermediate die, forward deformation of the process was simulated using ALPID [9]. For the forward simulation, the intermediate shape determined by the optimization scheme was used as the die shape for the first stage of forging, and the resulting workpiece was then deformed to the final shape. The change in the workpiece geometry during the first stage of forging is shown in Figures 16a-c, and the forging of the intermediate shape to the final shape is shown in Figures 16d-f. In both stages, complete die fill is obtained. During the second stage forging, the top surface in the rib remains horizontal to a large extent, and the material slides up the vertical sides of the rib. Final die fill is obtained when all the nodes on the top surface of the rib touch the die. The possibility of fold-over has thus been eliminated in this process.

During forward simulation it is observed that the average total strain is 1.223 for the single stage forging process and 1.445 for the two-stage forging process. Also, the variance of total strain in the final product is reduced from 0.133 to 0.076 with the optimized twostage forging process as compared to the single stage forging process. Thus, two-stage forging would be a better approach for this case, since the optimized die eliminated problems like fold-over, at the same time reducing the variance of total strain in the final product. Other example cases have been analyzed successfully by Han et. al. [11,12] to validate this approach.

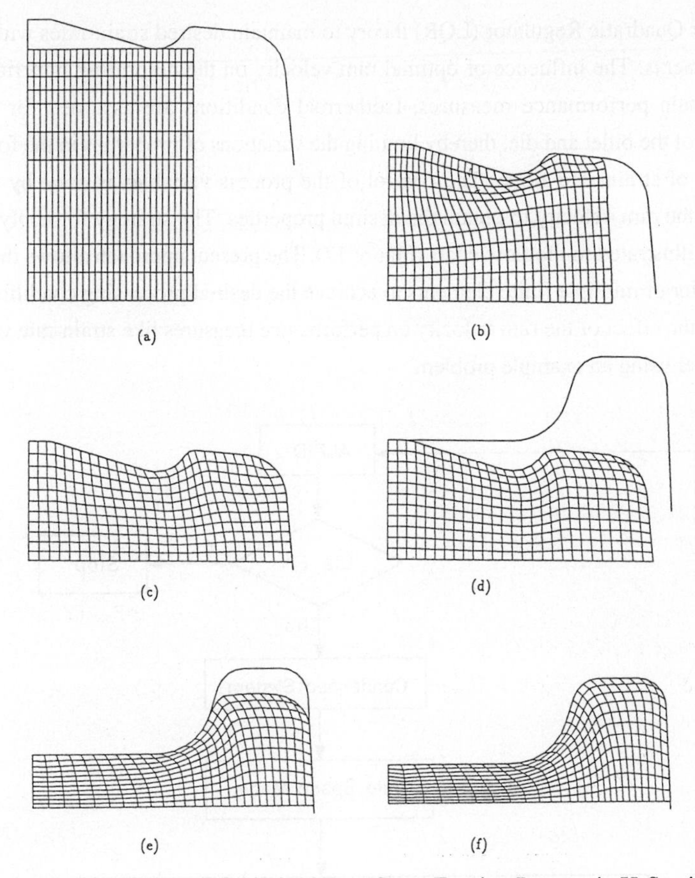

Figure 16. Simulation of Optimized Two-Stage Forging Process in H-Section

4 Process Design

This section presents a state space representation of nonlinear material deformation and an optimal control scheme for obtaining desired process conditions in the deforming material. The state space model is built at discrete time intervals of the process simulation. The state feedback method is used to determine the optimum process inputs. An optimum trajectory of inputs like ram velocity or die load with stroke is determined and supplied to the hydraulic press to obtain a relatively defect free forging.

The main objective in process control is to maintain processing variables such as strain, strain-rate, and temperature within a 'processing window'. The state variables selected in this formulation are the die-billet contact nodal velocities and the nodal velocities of the critical finite elements of the billet. The control input is the ram velocity, which is determined using the Linear Quadratic Regulator (LQR) theory to maintain desired strain-rates within selected finite elements. The influence of optimal ram velocity on the deforming material is studied using certain performance measures. Isothermal conditions are assumed for a specified geometry of the billet and die, thereby limiting the variations of properties in the forged part to the effect of strain-rate alone. The control of the process variables is done by the off-line design of the ram velocity to obtain the desired properties. The strategy for applying process control is illustrated in the flow chart (Figure 17). The present approach allows the control of the behavior of multiple finite elements to achieve the desired processing conditions and also discusses the effect of the ram velocity on performance measures like strain-rate variance and input power using an example problem.

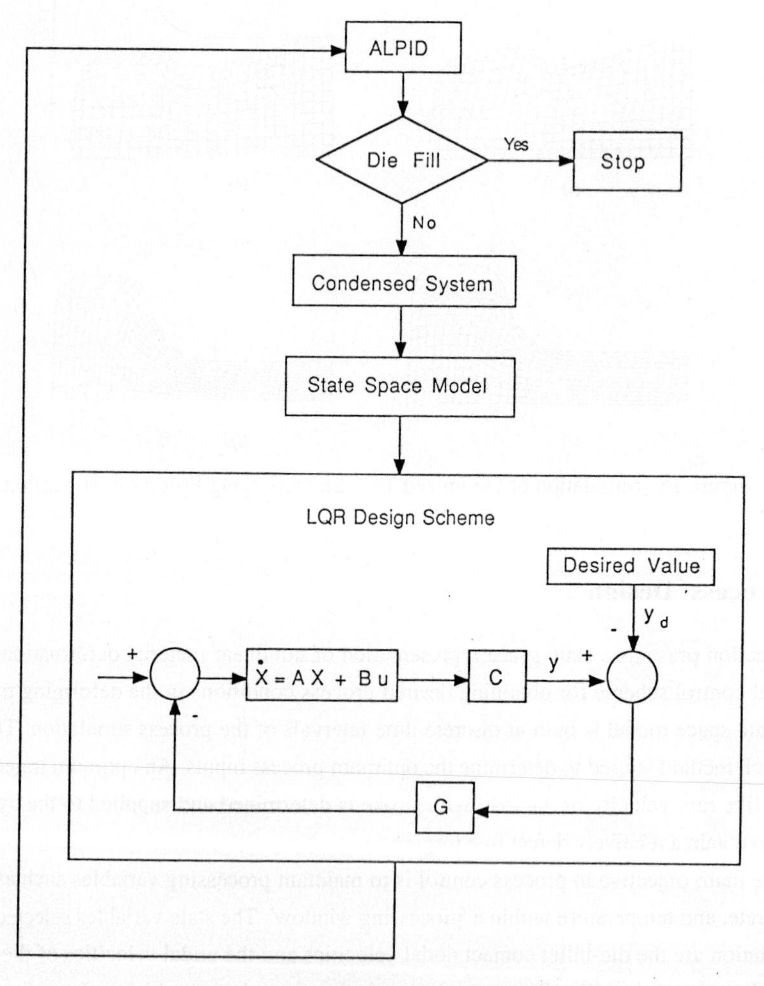

Figure 17. Process Design Flow Chart

4.1 Equilibrium Equations and System Condensation

The finite element equation at the iteration step n, using the direct iteration method is given as,

$$^{n}\mathbf{K}_S(\mathbf{V})\,^{n}\mathbf{V} = {}^{n}\mathbf{F}(\mathbf{V}) \tag{27}$$

where n is the iteration number for convergence at a given time step. The secant stiffness matrix $\mathbf{K}_S(\mathbf{V})$ and the force vector $\mathbf{F}(\mathbf{V})$ are functions of the velocity field $\mathbf{V}$. The billet continuum is discretized into quadrilateral finite elements. Each element has two degrees of freedom per node in the coordinate axes, leading to a system with many degrees of freedom. A reduction in the degrees of freedom is done by defining three zones within the billet continuum (Figure 18):

1. Nodes at the die-billet interface
2. Nodes of elements of interest within the billet continuum
3. Nodes comprising the rest of the billet interior

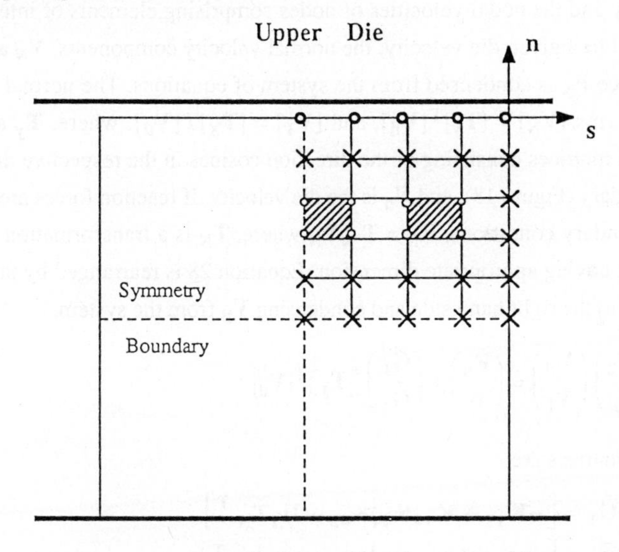

Figure 18. System Condensation

Consider a billet having m nodes, with p nodes at the die contact boundary, q interior nodes for elements of interest and the remaining interior nodes r, as shown in Figure 18. The expanded form of the equilibrium Equation 27 is,

$$\begin{pmatrix} \mathbf{K}_{11} & \mathbf{K}_{12} & \mathbf{K}_{13} & \mathbf{K}_{14} \\ \mathbf{K}_{21} & \mathbf{K}_{22} & \mathbf{K}_{23} & \mathbf{K}_{24} \\ \mathbf{K}_{31} & \mathbf{K}_{32} & \mathbf{K}_{33} & \mathbf{K}_{34} \\ \mathbf{K}_{41} & \mathbf{K}_{42} & \mathbf{K}_{43} & \mathbf{K}_{44} \end{pmatrix} \begin{pmatrix} \mathbf{V}_S \\ \mathbf{V}_N \\ \mathbf{V}_I \\ \mathbf{V}_B \end{pmatrix} = \begin{pmatrix} \mathbf{F}_S \\ \mathbf{F}_N \\ 0 \\ \mathbf{R}_N \end{pmatrix} \tag{28}$$

where $\mathbf{F}_S$ is the frictional force, $\mathbf{F}_N$ is the normal force at the die/billet interface, and $\mathbf{R}_N$ is the reaction force at the symmetry boundary conditions. If no symmetry boundary conditions are specified in the model, $\mathbf{R}_N = \mathbf{0}$. The tangential and normal components of nodal velocities at the die-billet interface are $\mathbf{V}_S$ and $\mathbf{V}_N$, respectively. $\mathbf{V}_I$ comprises the nodal velocities of interior nodes of interest and $\mathbf{V}_B$ the nodal velocities of the rest of the billet domain.

The state variables are identified as the tangential component of nodal velocities at the die-billet boundary and the nodal velocities of nodes comprising elements of interest within the billet domain. For a given die velocity, the normal velocity components, $\mathbf{V}_N$ are known, and the normal force $\mathbf{F}_N$ is condensed from the system of equations. The normal and tangential die velocity terms $[\mathbf{V}_N] = [\mathbf{T}_y]^T[\mathbf{V}_d]$, and $[\mathbf{V}_T] = [\mathbf{T}_X]T[\mathbf{V}_d]$, where, $\mathbf{T}_y$ and $\mathbf{T}_x$ are the transformation matrices consisting of the direction cosines in the respective directions at the die/billet boundary (Figure 18), and $\mathbf{V}_d$ is the die velocity. If reaction forces are present due to symmetry boundary conditions, $\mathbf{R}_N = \mathbf{T}_N^T \mathbf{V}_N$ where, $\mathbf{T}_N$ is a transformation matrix for the reaction forces having appropriate dimension. Equation 28 is rearranged by taking the terms containing $\mathbf{V}_N$ to the right hand side and condensing $\mathbf{V}_B$ from the system.

$$\begin{pmatrix} \bar{\mathbf{K}}_{11} & \bar{\mathbf{K}}_{12} \\ \bar{\mathbf{K}}_{21} & \bar{\mathbf{K}}_{22} \end{pmatrix} \begin{pmatrix} \mathbf{V}_S \\ \mathbf{V}_I \end{pmatrix} = \begin{pmatrix} \mathbf{F}_S \\ 0 \end{pmatrix} + \begin{pmatrix} \bar{\mathbf{G}}_1 \\ \bar{\mathbf{G}}_2 \end{pmatrix} [\mathbf{T}_y]^T[\mathbf{V}_d] \tag{29}$$

where the submatrices are,

$$\bar{\mathbf{G}}_1 = \left[-\mathbf{K}_{12} + \mathbf{K}_{14}\mathbf{K}_{44}^{-1}\mathbf{K}_{42} - \mathbf{H}_1\mathbf{T}_N{}^T\right]$$

$$\bar{\mathbf{G}}_2 = \left[-\mathbf{K}_{32} + \mathbf{K}_{34}\mathbf{K}_{44}^{-1}\mathbf{K}_{42} - \mathbf{H}_2\mathbf{T}_N{}^T\right]$$

$$\bar{\mathbf{K}}_{11} = \left[\mathbf{K}_{11} - \mathbf{K}_{14}\mathbf{K}_{44}^{-1}\mathbf{K}_{41}\right], \ \bar{\mathbf{K}}_{12} = \left[\mathbf{K}_{13} - \mathbf{K}_{14}\mathbf{K}_{44}^{-1}\mathbf{K}_{43}\right]$$

$$\bar{\mathbf{K}}_{21} = \left[\mathbf{K}_{31} - \mathbf{K}_{34}\mathbf{K}_{44}^{-1}\mathbf{K}_{41}\right], \ \bar{\mathbf{K}}_{22} = \left[\mathbf{K}_{33} - \mathbf{K}_{34}\mathbf{K}_{44}^{-1}\mathbf{K}_{43}\right]$$

4.2 State Space Model Development

The linearized form of the reduced system at the (n+l)th iteration for a given time step is,

$$^{n}\bar{\mathbf{K}}_S\, {}^{n}\bar{\mathbf{V}} + {}^{n}\bar{\mathbf{K}}_t\, {}^{n+1}\Delta\bar{\mathbf{V}} = {}^{n+1}\bar{\mathbf{F}} \tag{30}$$

where, $\bar{\mathbf{K}}_t$ is the tangent matrix obtained by linearizing about the velocity field. Here the increment in velocity is, ${}^{n+1}\Delta\bar{\mathbf{V}} = {}^{n+1}\bar{\mathbf{V}} - {}^{n}\bar{\mathbf{V}}$. Substituting without the iteration counters, taking the secant term to the right hand side, multiplying and dividing by Δt on the left side of Equation 30 and then, applying the limit, $\Delta t \rightarrow 0$ we have,

$$[\Delta t\bar{\mathbf{K}}_t]\frac{d\bar{\mathbf{V}}}{dt} = -[\bar{\mathbf{K}}_S]\bar{\mathbf{V}} + \bar{\mathbf{F}} \tag{31}$$

Expanding the right hand side terms according to Equation 30 we get,

$$[\Delta t\bar{\mathbf{K}}_t]\frac{d\bar{\mathbf{V}}}{dt} = -\begin{pmatrix}\bar{\mathbf{K}}_{11} & \bar{\mathbf{K}}_{12} \\ \bar{\mathbf{K}}_{21} & \bar{\mathbf{K}}_{22}\end{pmatrix}\begin{pmatrix}\mathbf{V}_S \\ \mathbf{V}_I\end{pmatrix} + \begin{pmatrix}\mathbf{F}_S \\ 0\end{pmatrix} + \begin{pmatrix}\bar{\mathbf{G}}_1 \\ \bar{\mathbf{G}}_2\end{pmatrix}[\mathbf{T}_y]^T[\mathbf{V}_d] \tag{32}$$

The frictional load vector $\mathbf{F}_S$, is modeled as a function of the sliding velocity [7]:

$$\mathbf{F}_S = -\left[\int_{Sc}\left\{mk\mathbf{N}^T\frac{\mathbf{V}_r}{|\mathbf{V}_r|}\right\}dS\right] \tag{33}$$

where, m is the friction factor, k is the shear strength of the material, and $\frac{\mathbf{V}_r}{|\mathbf{V}_r|}$ is a unit vector along the direction of sliding velocity. The sliding velocity at the die interface $\mathbf{V}_r = \mathbf{V}_S - \mathbf{V}_T$, where $\mathbf{V}_S$ is the tangential nodal velocity at the die interface, and $\mathbf{V}_T = [\mathbf{T}_x]^T[\mathbf{V}_d]$. The modified load vector can be written as,

$$\mathbf{F}_S = -\left[\int_{Sc}\left\{mk\mathbf{N}^T\frac{(\mathbf{V}_S - \mathbf{V}_T)}{|\mathbf{V}_r|}\right\}dS\right] \tag{34}$$

The expanded form of the load vector is,

$$\mathbf{F}_S = -\left[\int_{Sc}\left\{mk\mathbf{N}^T\frac{1}{|\mathbf{V}_r|}\right\}dS\right][\mathbf{V}_S] + \left[\int_{Sc}\left\{mk\mathbf{N}^T\frac{1}{|\mathbf{V}_r|}\right\}dS\right][\mathbf{T}_x]^T[\mathbf{V}_d]$$
$$= -[\mathbf{F}_r][\mathbf{V}_S] + [\mathbf{F}_r][\mathbf{T}_x]^T[\mathbf{V}_d] \tag{35}$$

where $\mathbf{F}_r = \left[\int_{Sc}\left\{mk\mathbf{N}^T\frac{1}{|\mathbf{V}_r|}\right\}dS\right]$. Substituting the load vector in Equation 33 and rearranging the terms according to the velocity vectors we get,

$$[\Delta t\bar{\mathbf{K}}_t]\frac{d\bar{\mathbf{V}}}{dt} = -[\hat{\mathbf{K}}][\bar{\mathbf{V}}] + [\hat{\mathbf{G}}][\mathbf{V}_d] \tag{36}$$

where,

$$\hat{\mathbf{K}} = -\begin{pmatrix}\bar{\mathbf{K}}_{11} & \bar{\mathbf{K}}_{12} \\ \bar{\mathbf{K}}_{21} & \bar{\mathbf{K}}_{22}\end{pmatrix} - [\mathbf{F}_r]$$

and,

$$\hat{\mathbf{G}} = [\mathbf{F_r}][\mathbf{T_x}]^T + \begin{pmatrix} \bar{\mathbf{G}}_1 \\ \bar{\mathbf{G}}_2 \end{pmatrix} [\mathbf{T_y}]^T$$

The state space form is thus given by,

$$\frac{d\bar{\mathbf{V}}}{dt} = -[\Delta t\bar{\mathbf{K}}_t]^{-1}[\hat{\mathbf{K}}][\bar{\mathbf{V}}] + [\Delta t\bar{\mathbf{K}}_t]^{-1}[\hat{\mathbf{G}}][\mathbf{V_d}] \tag{37}$$

This corresponds to the general state space form:

$$\dot{\mathbf{x}} = \mathbf{A}\mathbf{x} + \mathbf{B}\mathbf{u} \tag{38}$$

where the plant matrix $\mathbf{A} = -[\Delta t\bar{\mathbf{K}}_t]^{-1}[\hat{\mathbf{K}}]$,, the input matrix $\mathbf{B} = [\Delta t\bar{\mathbf{K}}_t]^{-1}[\hat{\mathbf{G}}]$, the state vector $\mathbf{x} = \bar{\mathbf{V}}$, and the input vector $\mathbf{u} = \mathbf{V_d}$. Validation of the above state space model has been done earlier by Kumar et al. [13].

4.3 Optimal Control Procedure

The state space model built earlier represents a time invariant system during a particular time step for the nonlinear analysis. The plant matrix and the input matrix do not have a closed form time dependent expression valid for the entire nonlinear FEM simulation. The simulation for plastic deformation is done after reducing the height by 1% at each time step. Hence the state space models built between consecutive time steps are approximately equal. The transient response of the state variables between the time steps is not considered. The constant gain feedback theory is effective in controlling the elemental strain-rates between the time steps of the simulation. The output equation describing the dynamic behavior of the metal forming system is given as,

$$\mathbf{y} = \mathbf{C}\mathbf{x} \tag{39}$$

where y is the output, and $\mathbf{C}$ is the output matrix. Considering the largest strain-rate component $\dot{\varepsilon}_{ij,}$ of the selected finite element, we can write $\dot{\bar{\varepsilon}}_i = \mathbf{D}_v$, where $\mathbf{D}$ is the strain-rate gradient matrix. $\mathbf{D}$ is substituted in the output matrix to transform the state variables into the strain-rate. Since the control objective is posed as a tracking problem, the quadratic performance index $\mathbf{J}$ is minimized as discussed in Anderson and Moore [14].

$$\mathbf{J} = \int_0^\infty \left((\mathbf{y} - \mathbf{y_d})^T \mathbf{Q} (\mathbf{y} - \mathbf{y_d}) + \mathbf{u}^T \mathbf{R} \mathbf{u} \right) dt \tag{40}$$

where **Q** and **R** are the state and control weighting matrices which have to be semidefinite and positive definite, respectively, and $\mathbf{y_d} = \dot{\varepsilon}_\mathbf{d}$, the desired strain-rate. The system matrix is modified by including a state variable $\mathbf{x_{new}} = \dot{\varepsilon}_\mathbf{d}$, and the time derivative $\dot{\mathbf{x}}_\mathbf{new} = \mathbf{0}$. A suitable **Q/R** ratio is selected via simulation to determine the control input. The feedback term is,

$$\mathbf{u} = -\mathbf{G}\mathbf{x} \tag{41}$$

where the gain matrix **G** is,

$$\mathbf{G} = \mathbf{R}^{-1}\mathbf{B}^T\mathbf{P} \tag{42}$$

Here, **P** is the Riccati matrix and it is obtained by the solution of the algebraic equation using the Schur method developed by Laub [15].

$$\mathbf{A}^T\mathbf{P} - \mathbf{P}\mathbf{B}\mathbf{R}^{-1}\mathbf{B}^T\mathbf{P} + \mathbf{P}\mathbf{A} + \mathbf{Q} = \mathbf{0} \tag{43}$$

Additional details on state-space model development and optimal control are presented by Kumar, et. al. [13].

4.4 Case Study for Process Design

This section illustrates the process design procedure by means of an example, wherein, the deformation of a cylindrical billet into a disk is simulated using ALPID. A quarter model of the disk is used for simulation purposes as shown in Figure 19. The effects of friction at the die/billet interface are included to represent a general forging process. The design of ram velocity as a function of time for the non-homogeneous deformation is intended to restrict the process variables within a 'processing window'. The processing window, which defines limits on processing variables such as temperature and strain rate, is generally material specific and is determined either from experimental data or from acceptable industrial practice. Data presented by Prasad et. al. [16] for Ti-6242 has been used in this work. Accordingly, the favorable region for processing the selected material is at 926.667° C temperature and 3.0×10^{-3} /sec strain-rate, obtained from the processing map. The process control objectives are to maintain the desired strain-rate and to ensure die fill. The starting die and billet shapes are shown in Figure 19a. The die has a height to width ratio equal to 0.50. One quarter of the billet, having a radius of 60mm and a height of 40mm, is modeled using 96 elements. The initial velocity of the upper die (ram) is 0.12 mm/sec, and the bottom die is kept stationary. Element number 55 is selected to be the element of interest since it remains within a region which is under a compressive state of stress throughout the simulation. The simulation is carried out until the die-fill condition is reached (Figure 19b).

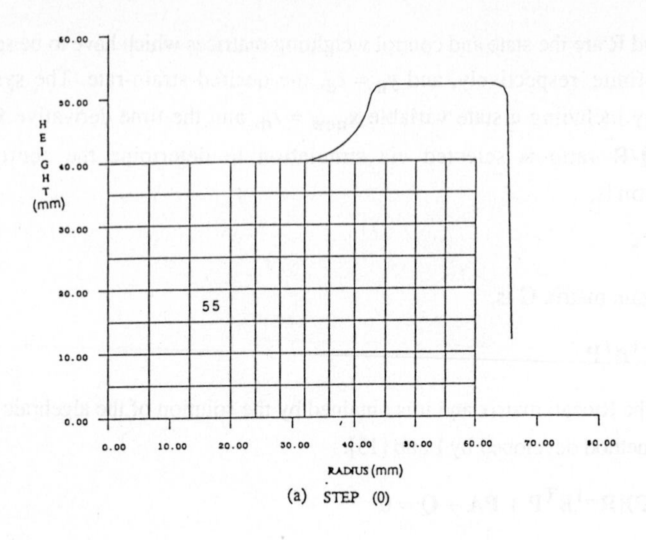

(a) STEP (0)

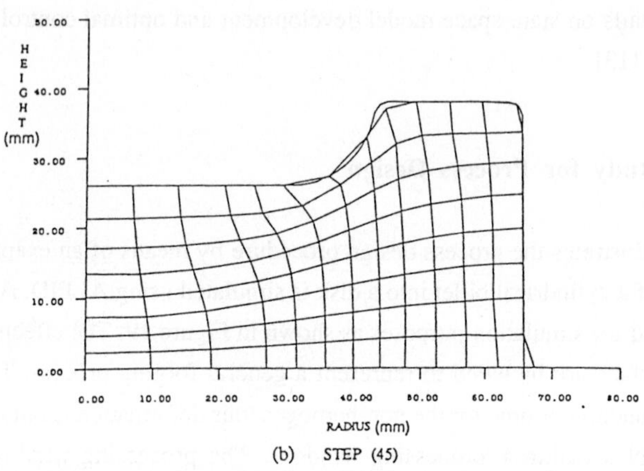

(b) STEP (45)

Figure 19. Disk Forging: Single Element Control

At the start of the simulation, the billet has 7 nodes contacting the die. As a result, the initial plant matrix is of 15 x 15 dimension. Due to the irregular shape of the die, new nodes come into contact with the die with progress in deformation. The state variable vector increases with additional nodes coming in contact with the die, and the final plant matrix size is 27 x 27. The **Q/R** ratio is 10^{10} when the desired strain-rate is greater than the current elemental strain-rate, and is 1 when the elemental strain-rate is lesser than the desired strain-rate. The initial strain-rate at element 55 is lower than the desired value, hence the control algorithm causes the ram velocity to increase until it reaches the desired value (Figure 20). Subsequently, the ram velocity decreases so as to maintain the desired strain-rate value. The controlled ram velocity profile is oscillatory during the die filling stage of the forging process and is reflected in the controlled strain-rate profile (Figure 21). This may be attributed to the nature of the iterative solution scheme for the nonlinear analysis, where the deformed geometry is updated using the product of the converged velocity vector and the time increment. The stiffness matrices are determined for the previous geometry before updating, while the state space model is built using the updated geometry, and the state variable vector includes the additional nodes coming in contact with the die. The error involved is negligible, as the time increment is small, and a smooth curve can be approximated to the die velocity profile for implementation on the hydraulic press.

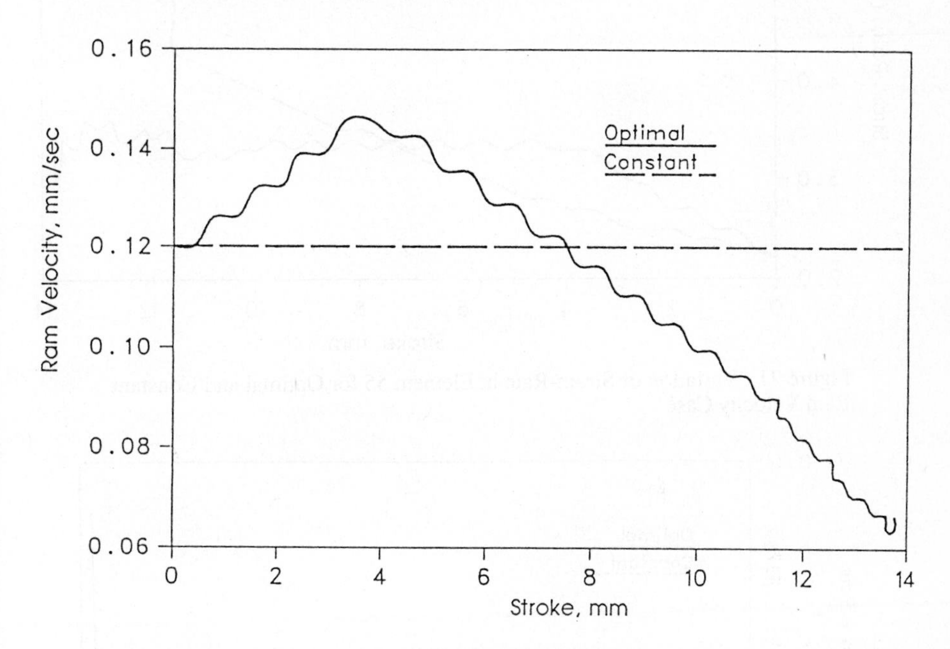

Figure 20. Ram Velocity Profile Using Single Element Control

During the controlled simulation, die-fill occurs in 45 time steps, whereas, the constant velocity simulation takes 43 time steps for die-fill using the same time increment. The variance of strain-rate (Figure 22) indicates a smaller range in the element strain-rates, allowing relatively uniform deformation for 90% of the stroke. The increase in variance during the final stages of die-fill can be altered by changing other design parameters like die geometry, lubrication conditions, temperature, etc. The strain-rate distribution for both the controlled and constant velocity case follow the same pattern. The difference is in the magnitude only. The controlled case gives a better deformation at the center portion of the disk where the desired strain-rate is maintained and reduces the variance in strain-rate at other regions giving a uniform flow of material. The input power requirement (Figure 23) is lower for the optimally designed process, thus improving the efficiency of the process. Also, by controlling the

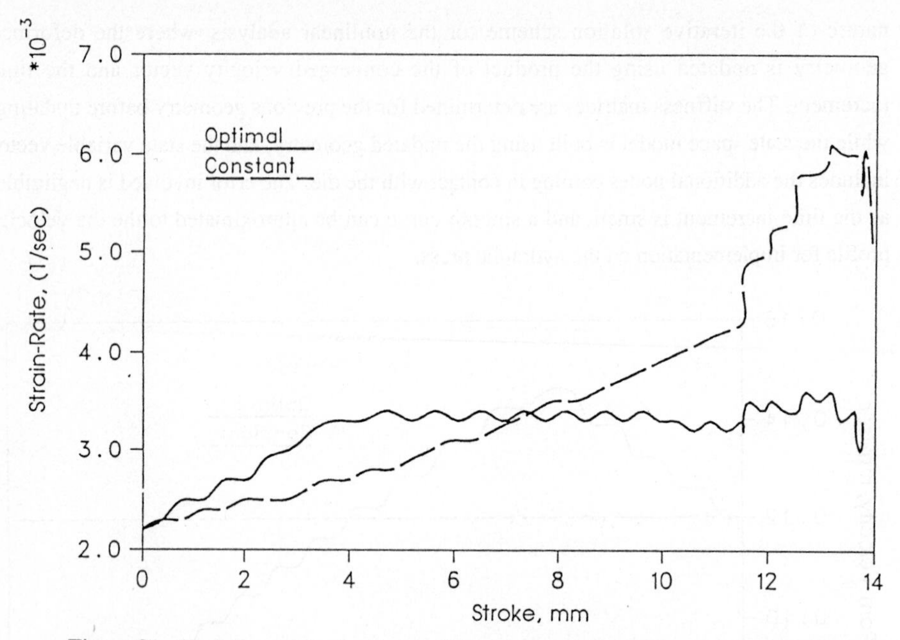

Figure 21. Variation of Strain-Rate in Element 55 for Optimal and Constant Ram Velocity Case

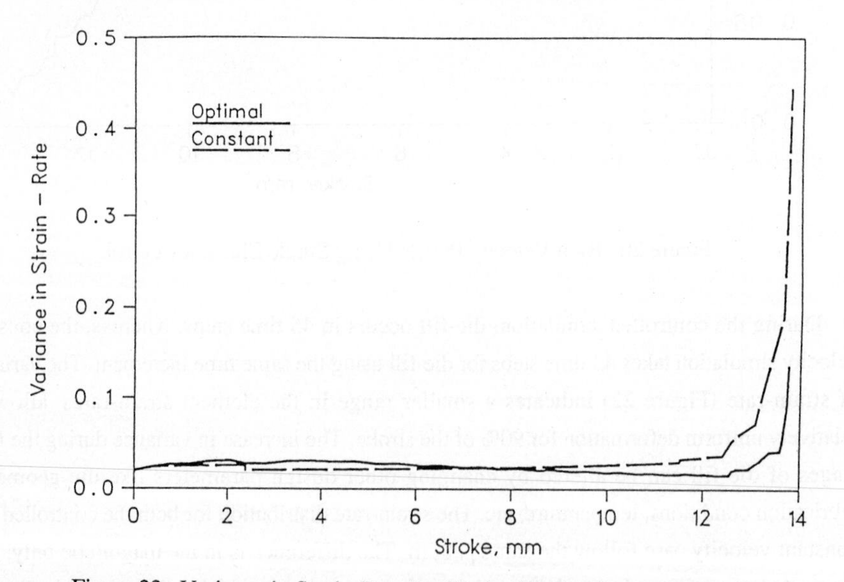

Figure 22. Variance in Strain-Rate Profile for Single Element Control

strain-rate, the required capacity of the forge press is reduced as observed from the load stroke curve (Figure 24). Other example problems have been analyzed successfully by Kumar et. al. [13] to validate this design approach.

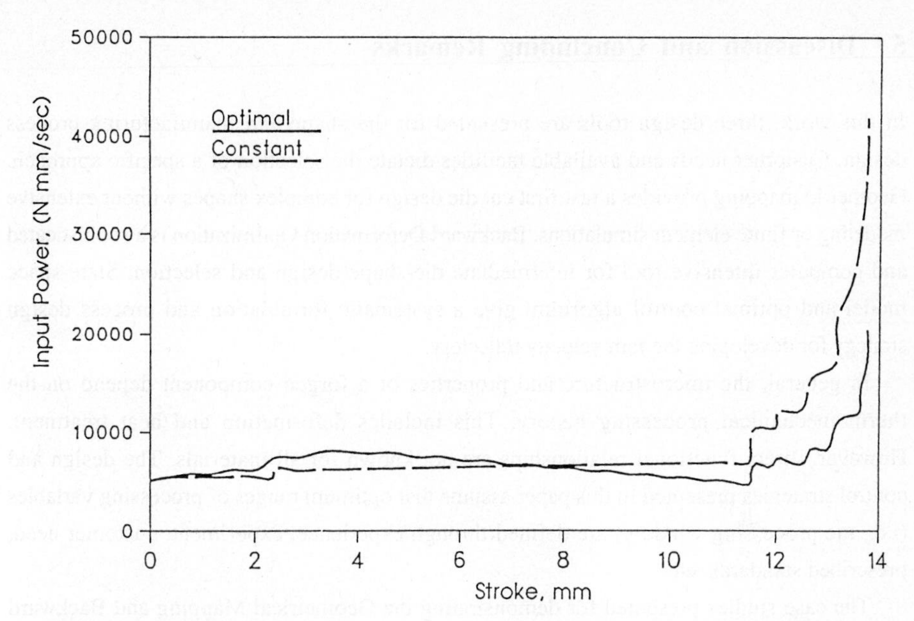

Figure 23. Variation of Input Power for Optimal and Constant Ram Velocity Case Using Single Element Control

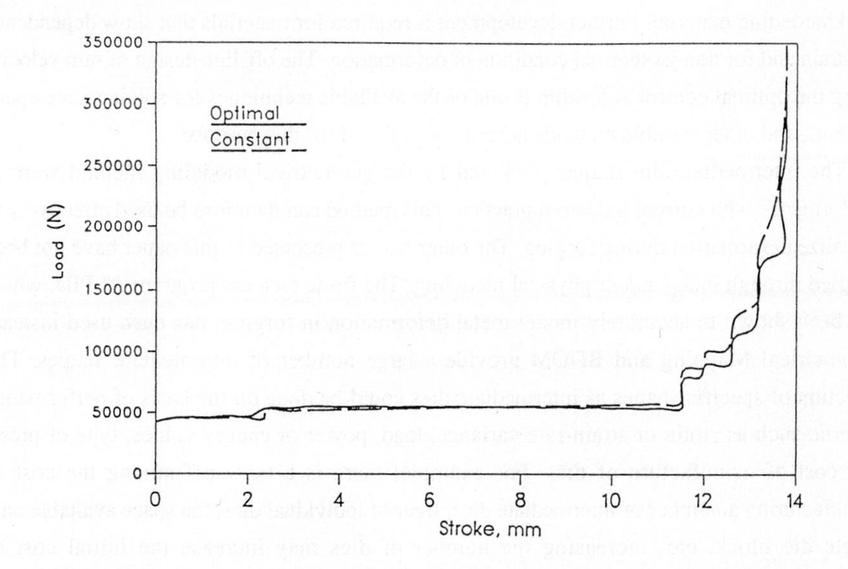

Figure 24. Load-Stroke Curve for Single Element Control Disk Forging

5 Discussion and Concluding Remarks

In this work, three design tools are presented for die shape and manufacturing process design. Customer needs and available facilities dictate the selection of a specific approach. Geometric mapping provides a fast first cut die design for complex shapes without extensive modeling or finite element simulations. Backward Deformation Optimization is a sophisticated and computer intensive tool for intermediate die-shape design and selection. State-space model and optimal control algorithm give a systematic formulation and process design strategy for developing the ram velocity trajectory.

In general, the microstructure and properties of a forged component depend on the thermomechanical processing history. This includes deformation and heat treatment. However, direct functional relationships are not known for all materials. The design and control strategies presented in this paper assume that optimum ranges of processing variables (i.e., the processing window) are defined through experience, experiment, customer need, prescribed standards, etc.

The case studies presented for demonstrating the Geometrical Mapping and Backward Deformation Optimization techniques were directed towards near-net-shape forging. These methods are equally applicable to conventional closed-die forging with in which flash is present. The BDOM and Process Control strategy assume isothermal deformation of a non-workhardening material. Further development is required for materials that show dependence on strain and for non-isothermal condition of deformation. The off-line design of ram velocity using the optimal control algorithm is one of the available techniques for solving state-space systems, and other suitable methods have to be explored for this purpose.

The intermediate die shapes predicted by the geometrical modeling method were in conformance with current industrial practice. This method can therefore be used effectively to visualize deformation during forging. The other results presented in this paper have not been verified through independent physical modeling. The finite element program ALPID, which has been shown to accurately model metal deformation in forging, has been used instead. Geometrical Mapping and BDOM provide a large number of intermediate shapes. The selection of specific shapes as intermediate dies could be done on the basis of performance criteria, such as strain or strain-rate variance, load, power or energy values, type of press, and cost of manufacture of dies. For example, there is a trade-off among the cost of manufacturing a number of intermediate dies, wear of individual dies, the space available on a single die block, etc. Increasing the number of dies may increase the initial cost of manufacture, but these costs may be offset by reduction in the wear of individual dies and the reduction in replacement costs.

Acknowledgements

This research work was sponsored by the Edison Materials Technology Center (EMTEC) through project CT-19 and by the Wright Patterson Air Force Base, Ohio, through the contract F33615-90-C-5950. The authors acknowledge the excellent support of Mr. Shyam Sanjeev Kumar in putting the document together.

References

1. Tang, J.P., Oh, S.I., and Altan, T., "The Application of Expert Systems to Automatic Forging Design", Proceedings of 13th NAMRC, pp. 456-463, 1985.
2. Knight, W.A., "Part Family Methods for Bulk Metal Forming", International Journal of Production Research, Vol. 12, No.2, pp. 209-331,1974.
3. Subramaniam, T.L., Akgerman N. and Altan, T., "Application of Computer-Aided Design and Manufacturing to Precision Isothermal Forging of Titanium alloys", Battelle's Columbus Laboratories, Final Report AML-TR77-108. Vols. I, II, and III, Air Force Materials Lab., July 1977.
4. Subramaniam, T.L., Akgerman, K., and Altan, T., "Computer-Aided Preform Design for Precision Isothermal Forging", Proceedings of 5th NAMRC, pp. 198-213, 1977.
5. Lanka, S.S., Srinivasan, R., and Grandhi, R.V., "A Design Approach for Intermediate Die Shapes in Plane Strain Forgings", Journal of Materials Shaping Technology, Vol. 9, No. 4, pp. 193-206,1991.
6. Grandhi, R.V., Srinivasan, R., and Lanka, S.S., "Design of Intermediate Shapes in Forgings", Vol. I - Development of a Conformal Mapping Technique and an Ideal Material Flow Technique for Forgings, EMTEC Report, EMTEC/CT-19/TR-92-19, June 1992.
7. Kobayashi, S., Soo-Ik Oh, and Altan, T., Metal Forming and the Finite-Element Method, Oxford University Press, 1989.
8. Hwang, S.M. and Kobayashi, S., "Preform Design in Shell Nosing at Elevated Temperatures", International Journal of Machine Tools Manufacture, Vol.27, No.1, pp. 1-14, 1987.
9. Oh, S.I., Lahoti, G.D., and Altan, T., "ALPID - A General Purpose FEM Program for Metal Forming," Proceedings NAMRC-IX, State College, PA, pp. 83-92, 1981.
10. Haftka, R.T., and Grandhi, R.V., "Structural Shape Optimization - A Survey", Computer Methods in Applied Mechanics and Engineering, Vol. 57, No.1, 1986, pp. 91-106.
11. Han, C.S., Grandhi, R.V., and Srinivasan, R., "Optimum Design of Forging Die Shapes for Nonlinear Material Deformation", 33rd AIAA/ASME/ASCE/AHS/ACS Structural Dynamics and Materials Conference, Dallas, Texas, April 1992.
12. Grandhi, R.V., Srinivasan, R., and Han, C.S, "Design of Intermediate Shapes in Forgings", Vol.II - Development of the Backward Deformation Optimization Method, EMTEC Report, EMTEC/CT-19/TR-92-19, June 1992.
13. Kumar, A., Grandhi, R.V., Chaudhary, A., and Irwin, R. "Process Modeling and Control in Metal Forming", Proceedings of American Control Conference, Vol.1, pp. 817-821, June 1992.
14. Anderson, B.D.O. and Moore, J.B., Optimal Control - Linear Quadratic Methods, Prentice Hall, New Jersey, pp. 77-81, 1989.
15. Laub, A.J., "A Schur Method for Solving Algebraic Riccati Equations", IEEE Transactions on Automatic Control, Vol. 24, No. 6, pp. 913-921, 1979.
16. Prasad, Y.V.R.K., Gegel, H.C., Doraivelu, S.M., Malas, J.C., Morgan, J.T., Lark, K.A., and Barker, D.R., "Modelling of Dynamic Material Behaviour in Hot Deformation Forging of Ti-6242", Metallurgical Transactions, Vol. 15A, October, pp. 1883-1892, (1984).
17. Metals Handbook, Forming and Forging, Ninth ed., Vol.14, ASM International, Metals Park, OH, 1988.
18. Biswas, S.K., and Knight, W.A., "Towards an Integrated Design and Production System for Hot Forging Dies", International Journal of Production Research, Vol.14, No.1, pp. 23-49, 1979.
19. Mielnik, E.M., Metalworking Science And Engineering, McGraw-Hill Inc., 1991.
20. Altan, T., Soo-Ik Oh, and Gegel, H.L., Metal Forming, Fundamentals and Applications, 1st ed., American Society for Metals, Metals Park, OH.
21. Park, J.J., Rebelo, N., and Kobayashi, S., "A New Approach to Preform Design in Metal Forming with Finite Element Method", International Journal of Machine Tool Design Research, Vol.23, No.1, pp. 71-79, 1983.
22. Hwang, S.M. and Kobayashi, S., "Preform Design in Disk Forging", International Journal of Machine Tool Design, Vol.26, No.3, 1986, pp. 231-243.
23. Oh, S.I., "Finite Element Analysis of Metal Forming Processes With Arbitrarily Shaped Dies", International Journal of Mechanical Science, Vol. 24, No. 8, pp. 479-493, 1982.

Effects of Structural Dynamics on Chatter in Machine Tools and Its Evaluation at Design Stage

Mehmet S. Tekelioglu

Dokuz Eylul University, Department of Mechanical Engineering, 35101 Bornova-Izmir, Turkey

Abstract: Chatter can be defined as violent vibrations between the workpiece and the cutting tool in machine tools. Its importance in the design of machine tools is discussed in this research. Instead of investigating the chatter probabilities on machine tools of which the prototype has already been developed, these probabilities should be evaluated at the design stage. To do this, a mathematical model has been developed to analyze the structural dynamics of a radial drilling machine. The effects of structural dynamics are pointed out in terms of point and transfer receptances. The receptance properties determine the important factors for design.

Keywords: machine tool vibrations / chatter / structural dynamics / receptance / transfer matrix

1 Introduction

Since machine tools are very important in manufacturing processes, the factors affecting machine tools performance should be evaluated both in design stage and in working conditions.

One of the various factors affecting the machine tools performances is chatter, a violent vibration between the workpiece and the cutting tool. These vibrations have some important effects on dimensional accuracy of workpiece, life of the tool and of the machine, and the number of machined work-pieces per unit time.

Machine tool vibrations can be mainly classified under three groups [16]. First of these is forced vibrations; the second one is free vibrations excited by some shock effects; and the third one is self-excited vibrations. The source of the self-exciting energy is in the cutting process. In metal cutting, these vibrations are referred to as chatter [1,5,13].

Present chatter studies aim at determining stable cutting conditions [10,14]. But these studies give very limited knowledge from the chatter point-of-view, which should be considered in design.

The aim of this research is to study structural dynamics of machine tools in design stage and to investigate its effects on chatter. This study may be useful for determining design factors of machine tools by considering structural dynamics and effects on chatter.

There are three factors which have important effects on chatter. These are the dynamical behavior of the structure, the cutting process, and the effect called as regenerative chatter. The chatter can be shown as a feedback loop considering the above factors [5]. When this loop is studied, it is observed that the part which is under the designer's control is almost only the structure of machine tool. But as a mechanical system, machine tools have a very complex structure, and some simplifications and assumptions must be made when the structure is mathematically modeled. Long and Lemon studied the dynamical behavior of structure considering these facts [4]. Merrit based his stability analysis on this study [5]. It is also possible to determine the dynamical behavior of machine tools by system identification [6,7,11,17].

A mathematical model must be used to determine dynamical behavior of machine tools in design stage. Since it is very difficult to develop only one mathematical model for all kinds of machine tools, this study considers only a certain type of machine tools--namely radial drilling machines. A mathematical model is developed to analyze the structural dynamics of radial drilling machines. The transfer matrix method for continuous system is used and structural damping effect is considered. The dynamical behavior is pointed out in terms of receptances. The results of the mathematical model are compared with experiments on an experimental model of a drilling machine in laboratory.

2 Mathematical Model and Transfer Matrices

2.1 Mathematical Model

Chatter arises in all kinds of machine tools. Different machine tools have different structures. Important static and dynamic properties of machine tools should be investigated from the design point of view.

The mathematical model used for radial drilling machines in this study is shown in Figure 1. It should be noted that Part 2 can slide on the column, and Part 4 can slide on Part 3 in this model.

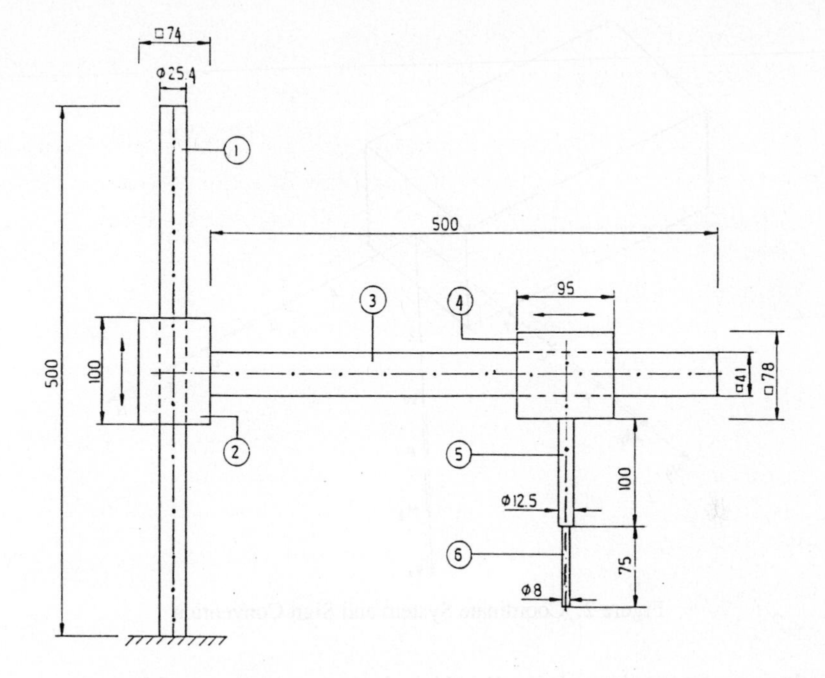

Figure 1. Mathematical Model for Radial Drilling Machines

2.2 Method of Analysis

The mathematical model is analyzed by the transfer matrices obtained by Bernoulli-Euler and St. Venant theories. Structural damping is included in the system by taking Young's modulus in complex form, that is $E(1+js_1)$ instead of E, and $G(1+js_2)$ instead of G, where s_1 and s_2 are loss factors. The coordinate system and the sign convention used throughout the study are shown in Figure 2.

2.3 Transfer Matrices

Since the mathematical model is constructed by considering symmetrical beams, in-plane and out-of-plane vibrations are not coupled. But the bending and longitudinal vibrations in in-plane, and the bending and torsional vibrations in out-of-plane, are coupled. The derivations of transfer matrices can be found in Reference 9.

In the following equation $\{z\}_i$ and $\{z\}_{i-1}$ are complex state vectors at the nodes i and i-1 respectively, and [T] is complex transfer matrix.

$$\{z\}_i = [T]\{z\}_{i-1} \tag{1}$$

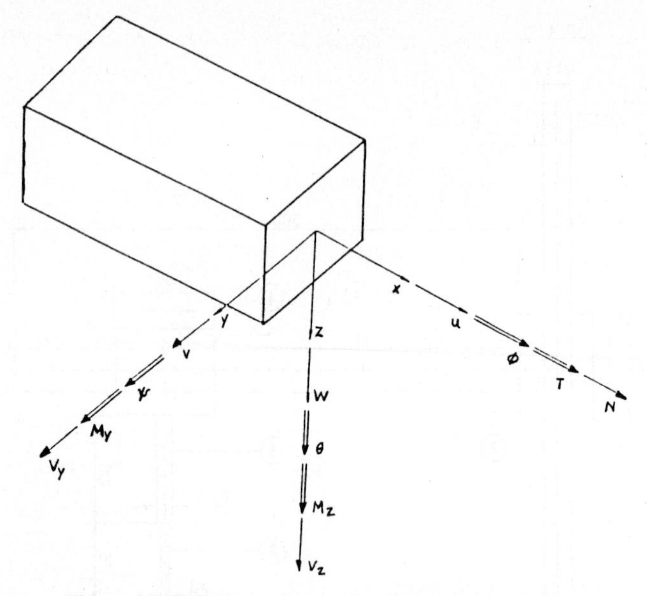

Figure 2. Coordinate System and Sign Convention

When these are separated into their real and imaginary components, the following expression is obtained:

$$\begin{bmatrix} \{z\}^r \\ \{z\}^i \end{bmatrix} = \begin{bmatrix} [T]^r & -[T]^i \\ [T]^i & [T]^r \end{bmatrix} \begin{bmatrix} \{z\}^r \\ \{z\}^i \end{bmatrix} \tag{2}$$

where the superscript r and i denote the real and imaginary components, respectively.

Equation 2 can be used for forced damped vibrations. If the loss factors are taken as zero, the same equation can be used also for free vibrations. The transfer matrices for bending, longitudinal, and torsional vibrations are always in the form of Equation 2.

2.4 Transfer Matrix for the Mathematical Model

Figure 1 is redrawn in Figure 3 schematically for easier interpretation. In Figure 3, a-b-c-d-e-f-g is considered as the main system, and c-h-k as the first branch and e-m-n as the second branch. Figure 3 shows also the coordinate system used for each element in the mathematical model. The expression for the state vectors at the points a and g is given by

$$\{z\}_{3g} = [T]_{a-g}\{z\}_{1a} \tag{3}$$

where subscript 3 and 1 indicate that the state vectors are written in $x_3y_3z_3$ and $x_1y_1z_1$ coordinate system, respectively. The matrix $[T]_{a\text{-}g}$ contains branch effects and required coordinate transformations [9,12].

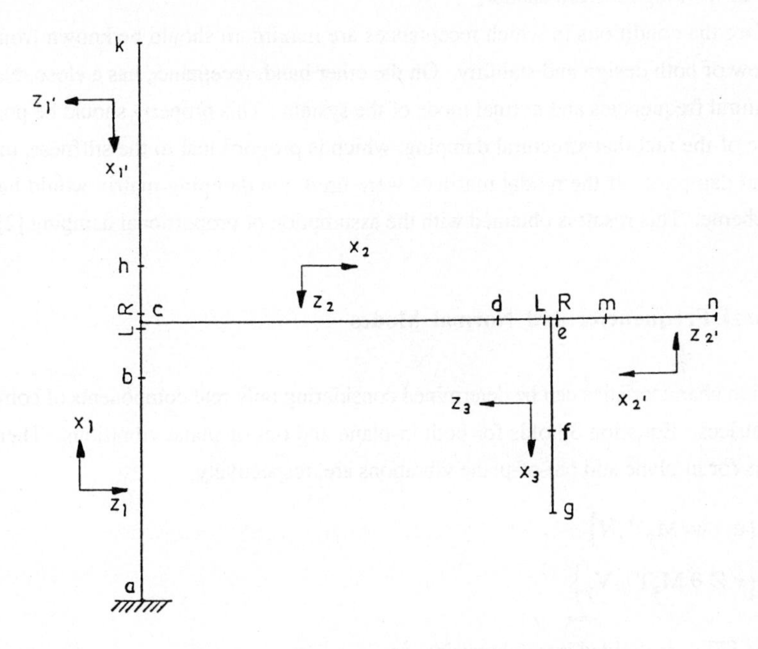

Figure 3. Coordinate Systems for the Elements

3 Dynamical Behavior of Machine Tools

3.1 Chatter Parameters

The number of parameters affecting chatter are many. They are generally classified due to the impossibility of investigating these parameters in only one study. For instance these are classified by Merrit as the parameters in cutting process, the parameters causing regenerative chatter, and the parameters in dynamics of structure [5]. Tlusty also gives a similar classification [13,14]. Tobias, who determines the stability of machine tools by considering whether the damping is positive or negative in the system, studies the subject from the parameters in cutting process point of view but he emphasizes the importance of dynamical characteristics of the structure [16].

The effects of the structure on chatter can be pointed out with reference to its receptance. Receptance is defined as the ratio of a force applied at a point of a mechanical system to displacement of the same or another point. Hence, the smaller the receptance, the larger the probability of working far from chatter.

Therefore the conditions in which receptances are maximum should be known from the point of view of both design and stability. On the other hand, receptance has a close relation with the natural frequencies and normal mode of the system. This property should be pointed out because of the fact that structural damping, which is proportional to the stiffness, means proportional damping. If the modal matrices were used, the damping matrix would have a diagonal scheme. This result is obtained with the assumption of proportional damping [2].

3.2 Natural Frequencies and Normal Modes

Free vibration characteristics can be determined considering only real components of complex transfer matrices. Equation 3 holds for both in-plane and out-of-plane vibrations. Then the state vectors for in-plane and out-of-plane vibrations are, respectively,

$$\begin{aligned} \{z\}_i &= \left\{u \ \text{-}w \ M_y V_z N\right\} \\ \{z\}_i &= \left\{v \ \varnothing \ \theta \ M_z T \ \text{-}V_y\right\} \end{aligned} \tag{4}$$

If the matrix $[T]_{a\text{-}g}$ is divided into submatrices as

$$[T]_{a-g} = \begin{bmatrix} [T_1] & [T_2] \\ [T_3] & [T_4] \end{bmatrix} \tag{5}$$

and it is noted that the displacements at the point a and the forces at the point g vanish, then the frequency equation becomes

$$\det [T_4]=0 \tag{6}$$

The values of ω satisfying the last equation are the natural circular frequencies.

The expression for the natural modes for in-plane and out-of-plane are in the same form. These can be found applying the well-known procedure for the transfer matrix method.

3.3 Dynamic Behavior and Receptance

Machine tools are subjected to a variety of dynamic excitations simultaneously. In addition, the excitation may cause different effects according to the position of the machine tool. Since machine tool positions may change according to the workpiece, it is impossible to define only one dynamic behavior for a machine tool.

A receptance is called point receptance if the displacement considered is at the same point and in the same direction as the force, otherwise it is called transfer receptance. It is possible to obtain the receptance spectrum of a structure by applying a constant force and then determining the displacements in the frequency range considered.

3.3.1 Excitations and Receptance in Machine Tools

Since the mathematical model considered in this work is a linear system, it is possible to analyze the effects of excitations separately. Therefore the displacements caused by in-plane and out-of-plane forces are found separately, and the combined receptance is obtained.

3.3.2 In-Plane Excitations

In the case of in-plane vibrations, bending moment, shear force, and normal force affect the system. If Equation 3 is written in the following form, the expressions may be interpreted more easily. First, it should be noted that,

$$\begin{aligned} \{d\} &= \{u \ \ -w \ \ \psi\} \\ \{p\} &= \{M_y \ V_z \ N\} \end{aligned} \tag{7}$$

Then,

$$\begin{bmatrix} [T_1] & [T_2] \\ [T_3] & [T_4] \end{bmatrix}_{a-g} \begin{bmatrix} \{d\}^r \\ \{d\}^i \\ \{p\}^i \\ \{p\}^r \end{bmatrix}_{1a} = \begin{bmatrix} \{d\}^r \\ \{d\}^i \\ \{p\}^i \\ \{p\}^r \end{bmatrix}_{3g} \tag{8}$$

where $[T_i]$ matrices, (i=1,2,3,4), are 6x6 submatrices of the matrix $[T]_{a\text{-}g}$. An excitation at point g of the system is substituted to the right hand side of Equation 8 as a boundary condition. For example, in the case of a harmonic normal force at point g, the following expression can be written by using Equation 8

$$[T_4]_{a-g}\{p\}_{1a} = \{0\ 0\ 0\ 0\ 0\ N\}_{3g} \tag{9}$$

From this equation,

$$\{p\}_{1a} = [T_4]_{a-g}\{0\ 0\ 0\ 0\ 0\ N\}_{3g} \tag{10}$$

is obtained. Since the displacement at point a is zero, the state vector at this point is completely known. The state vectors at the other points can be obtained easily by using previously explained methods. Receptances are written with respect to the force and displacement at the related points. For example in the case of a harmonic normal force at point g, the point receptance is in the form of

$$R_{u/N} = \left[\left(u_g^r\right)^2 + \left(u_g^i\right)^2\right]^{1/2} / N \tag{11}$$

The necessary change should be made in Equation 10 according to the in-plane excitations.

The formulation gives the real and imaginary components of receptance separately. The analysis of the components gives very useful information about the dynamic characteristic of the system [8].

3.3.3 Out-of-Plane Excitations

Out of plane excitations are in the form of bending moment, shear force and torsional moment. The procedure of analysis is the same as in-plane excitation. But in this case, the following expressions should be used instead of Equation 7:

$$\begin{aligned} \{d\} &= \{v\ \varnothing\ \theta\} \\ \{p\} &= \{M_z\ T\ \text{-}\ V_y\} \end{aligned} \tag{12}$$

4 Quantitative Analysis

Machine tools have different positions according to the workpiece to be machined. Here, four different positions are taken into account in the analysis of the dynamic characteristics of radial drilling machines. These are shown in Figure 4 schematically. The parameters used in the analysis are shown in Table 1. The quantitative analysis is carried out with some programs written in APL programming language [12].

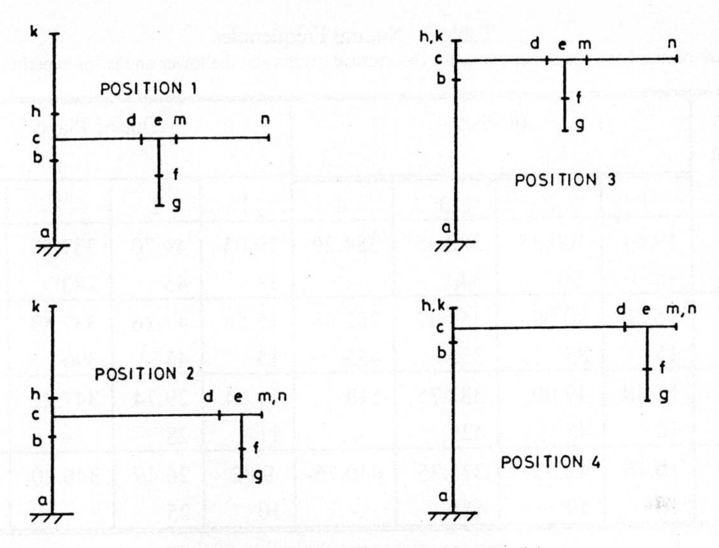

Figure 4. Radial Drilling Machine Positions

Table 1. System Parameters
(Element numbers are given in Figure 1) (E/G=2.6, $I=I_y=I_z$)

Par. \ Element	1	2	3	4	5	6
m (kg/m)	3.925	42.04	13.09	46.16	1.764	0.365
$E \times 10^{-11}$ (Pa)	1.764	2.107	1.764	2.107	1.764	1.764
$A \times 10^4$ (m^2)	5.067	54.76	16.81	60.84	1.277	0.5027
$I \times 10^8$ (m^4)	2.043	250	23.55	308.5	0.1198	0.0201
$J \times 10^8$ (m^4)	4.086	422.8	39.84	522	0.2397	0.0402
$i_x \times 10^3$ (m)	8.98	30.21	16.74	31.84	4.419	2.828

4.1 Natural Frequencies and Mode Shapes

Natural frequencies must be known to determine the dynamic behavior of a mechanical system. In the case of chatter, the maximum values of receptances happen to be important, and these are directly related to the natural frequencies.

Natural frequencies are shown in Table 2 with experimental results. The table shows that in-plane and out-of-plane fundamental frequencies are approximately equal, and this fact yields to the idea that bending modes are predominant at this frequency. The second natural frequencies are different and this may be interpreted that longitudinal and torsional modes are also effective. Experimental results in the table are interpreted below separately.

Table 2. Natural Frequencies
(In every pair of numbers, the upper one is for theorectical results and the lower one is for experimental results)

Position Figure 4	In-Plane				Out-of-Plane			
	1	2	3	4	1	2	3	4
1	19.64 18	100.33 90	380.45 340	384.29 -	19.05 18	49.70 46	337.57 280	383.06 460
2	16.50 15	90.09 85	359.47 330	382.68 435	15.58 15	47.26 45	337.83 390	376.29 -
3	11.40 10	49.00 45	382.75 425	518 -	11.24 11	29.74 28	347.22 -	424.63 -
4	10.16 10	42.53 40	375.35 415	440.76 -	9.82 10	26.49 25	346.20 -	414.25 -

Some of in-plane and out-of-plane mode shapes for Position 1 are presented in Figure 5. The mode shapes for the other positions have similar properties.

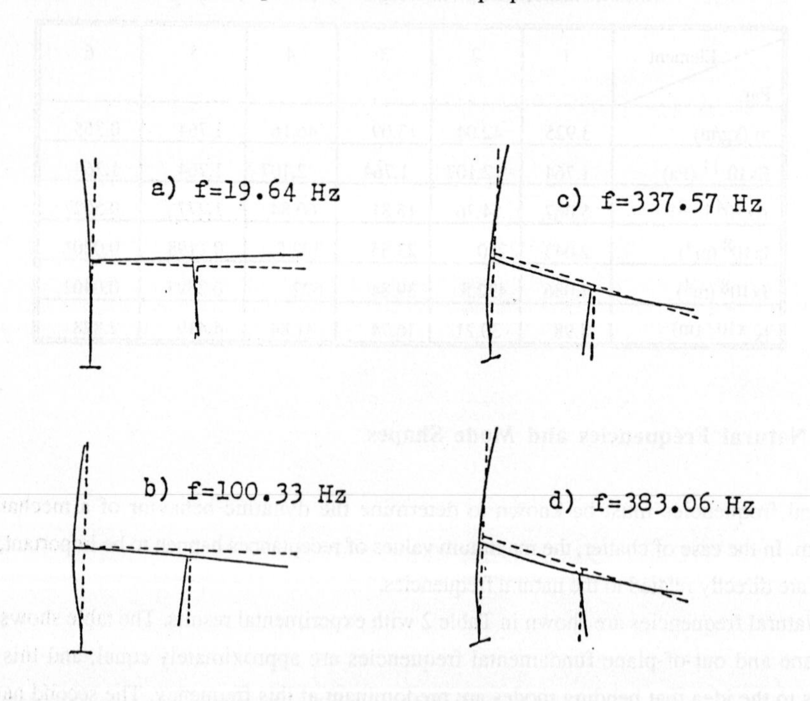

Figure 5. Mode Shapes
a) the first in-plane, b) the second in-plane, c) the third out-of-plane, d) the fourth out-of-plane

4.2 In-Plane and Out-of-Plane Receptances

In the analysis of dynamic behavior of machine tools, the main factor under the designer's control in the chatter loop is receptance of the system [5].

Receptance is important because it carries information about the dynamic stiffness of the system. Naturally, the most important characteristics of the receptance versus frequency graphs are peaks and valleys. These are called resonances and antiresonances, respectively. Resonances indicate the frequencies to which the system is most sensitive, while antiresonances reveal the most insensitive frequencies. Therefore, the maximum values of receptance (the minimum values of the dynamic stiffness) must be known in chatter analysis.

The receptance behavior of the system is presented by considering four different positions of the model shown in Figure 4. One receptance graph is presented for each position. It is possible to observe similar characteristics in the other graphs [12].

In the interpretation of the receptance graphs, some features (such as the position of the machine tool, the excitation being in-plane or out-of-plane, frequency range, and the type of the excitation) are important. So these factors should be considered in analyzing Figures 6 and 7. In these graphs, a value of 0.05 is used for the loss factors. It should be noted that transfer receptances are also important in some cases [12].

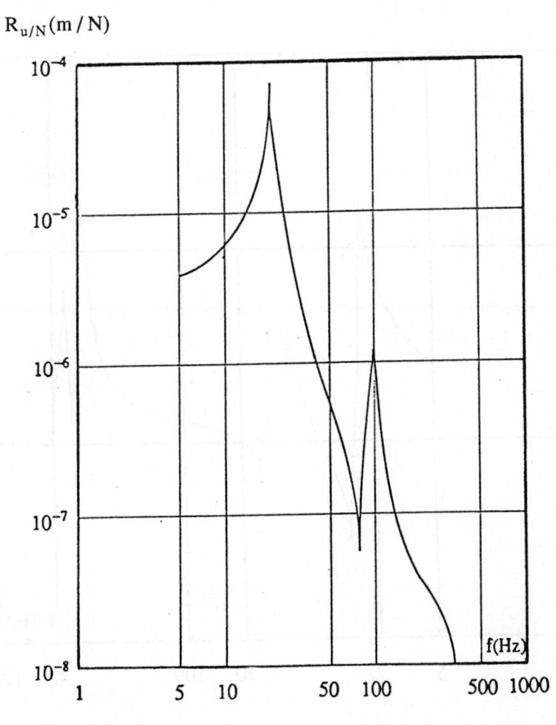

Figure 6a. Point Receptance for Position 1 in the Case of In-Plane Normal Force

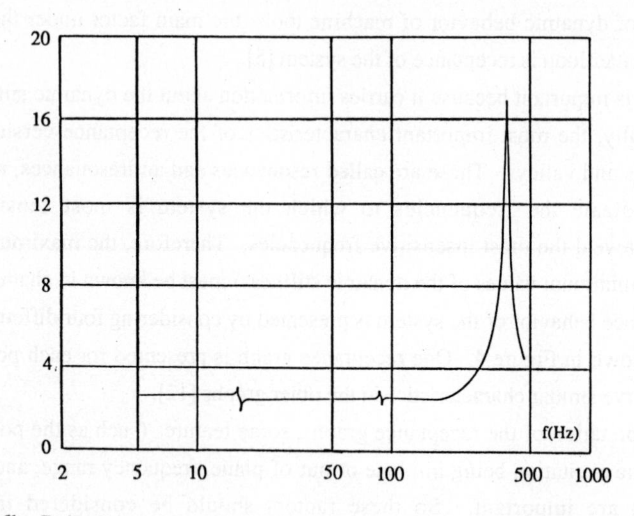

Figure 6b. Point Receptance for Position 2 in the Case of In-Plane Bending Moment

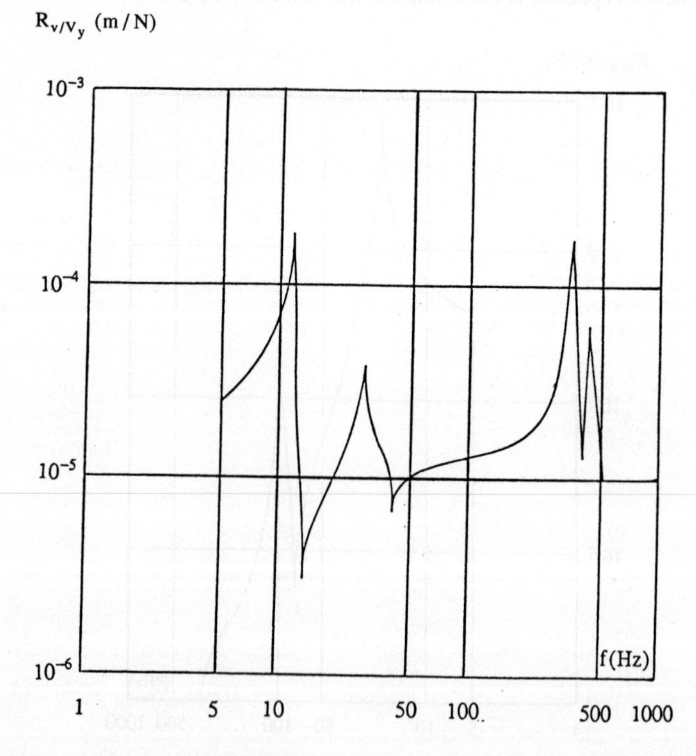

Figure 7a. Point Receptance for Position 3 in the Case of Out-of-Plane Shear Force

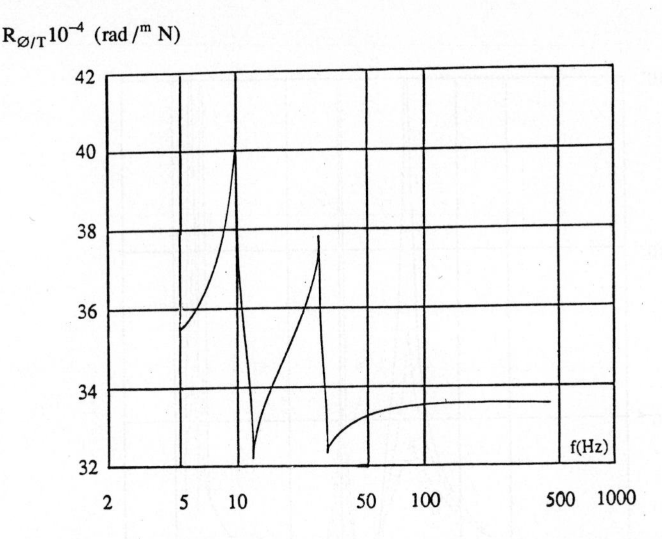

Figure 7b. Point Receptance for Position 4 in the Case of Out-of-Plane Torsional Moment

4.3 The Effects of Damping on Dynamic Behavior

In Figure 8, which is prepared to show the effects of loss factor on receptance, the receptance behavior for Position 1 in the case of out-of-plane harmonic shear force is sketched for loss factors of 0, 0.01, and 0.05. It is observed that the peaks of receptances decrease as the loss factor increases. The effects of damping are less at the points far from the peaks.

5 Experimental Study

A physical model was produced considering the mathematical model for the experimental work. The experimentation set-up and apparatus used is shown in Figure 9. In the experimental work, natural frequencies were measured and receptance graphs were obtained.

The measured natural frequencies are presented in Table 2 together with the theoretical results. The theoretical and experimental results are in good agreement for the first two natural frequencies. Since a regular vibration pattern could not be obtained in general, it was not possible to obtain all of the third and fourth natural frequencies.

The most important point in receptance measurements is to obtain a regular vibration pattern. In these experiments, one side of the force transducer was connected to the shaker and the other side was screwed to the bottom of the shaft. This system was assumed to be a model of radial drilling machine excited by an out-of-plane shear force.

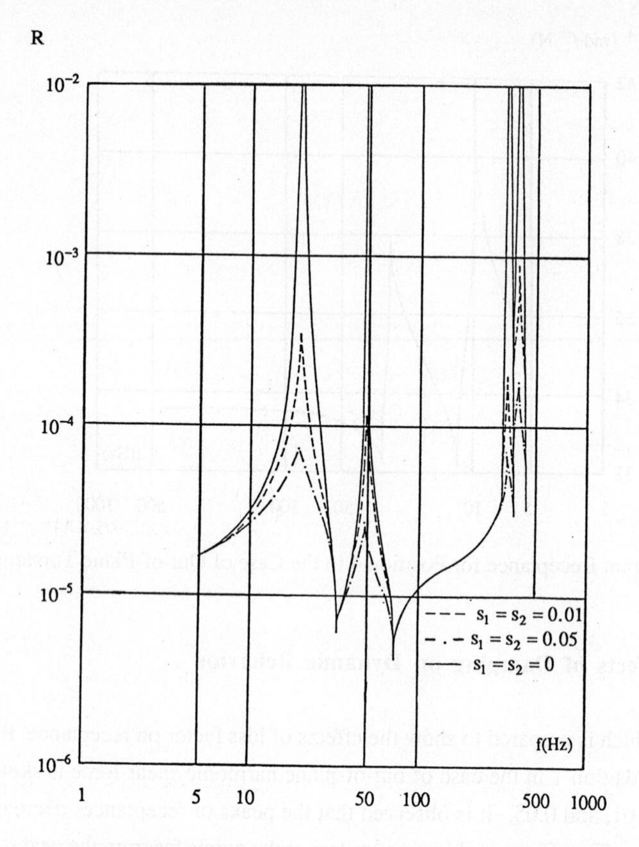

Figure 8. Effects of Damping on Receptance Behavior

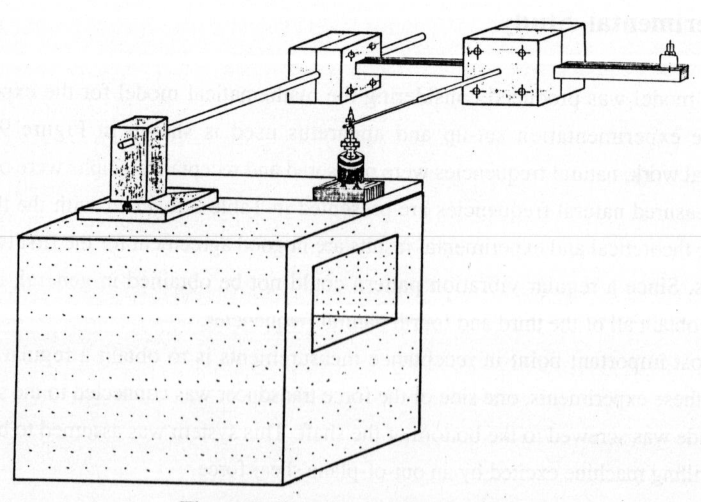

Figure 9. Experimentation Set-Up

Receptance measurements were carried out with respect to two points on which accelerometers are mounted and they are shown in Figure 10. The receptance measurements on the point where accelerometer 1 is mounted gives point receptance and the other one where accelerometer 2 is mounted gives transfer receptance of the related point. The required integration of the acceleration signals was carried out electronically.

The experimental results for receptances are shown in Figure 11. But receptances above 200 Hz could not be obtained accurately due to the various experimental difficulties [12].

6 Comparison of the Results and Discussion

As it is observed in Table 2, the experimental results for the first two natural frequencies are in good agreement with the theoretical results. But the third and the fourth natural frequencies could not be obtained accurately. The difference between the experimental and theoretical results for natural frequencies is around 10%. One reason for this difference may be neglecting the shear deformations and rotary inertia effects.

The receptance values obtained experimentally largely depend on the system loss factor, and the relation of loss factor to frequency should be taken into account. In the theoretical receptance curves shown in Figures 11a and 11b, a value of 0.05 was used as the loss factor. Peak points should be especially noted in comparision of experimental and theoretical results. These results are generally in agreement, and the difference in peak points may be due to the loss factor.

7 Conclusion

In this study, the effects of the dynamics of the structure on chatter, which is one of the most important factors affecting the performance of the machine tool, are analyzed. To show these effects, the dynamic analysis of the machine tool structure is considered from the point of view of chatter.

The chatter possibilities can be shown with respect to receptance. Natural frequencies which have a very important effect on chatter can be obtained without any lumping of mass and elastic property in the mathematical model. Although this work was performed for radial drilling machines, it is possible to apply the present method to other machine tool models. Also, this method can be applied to the model having elements with non-symmetrical sections [3].

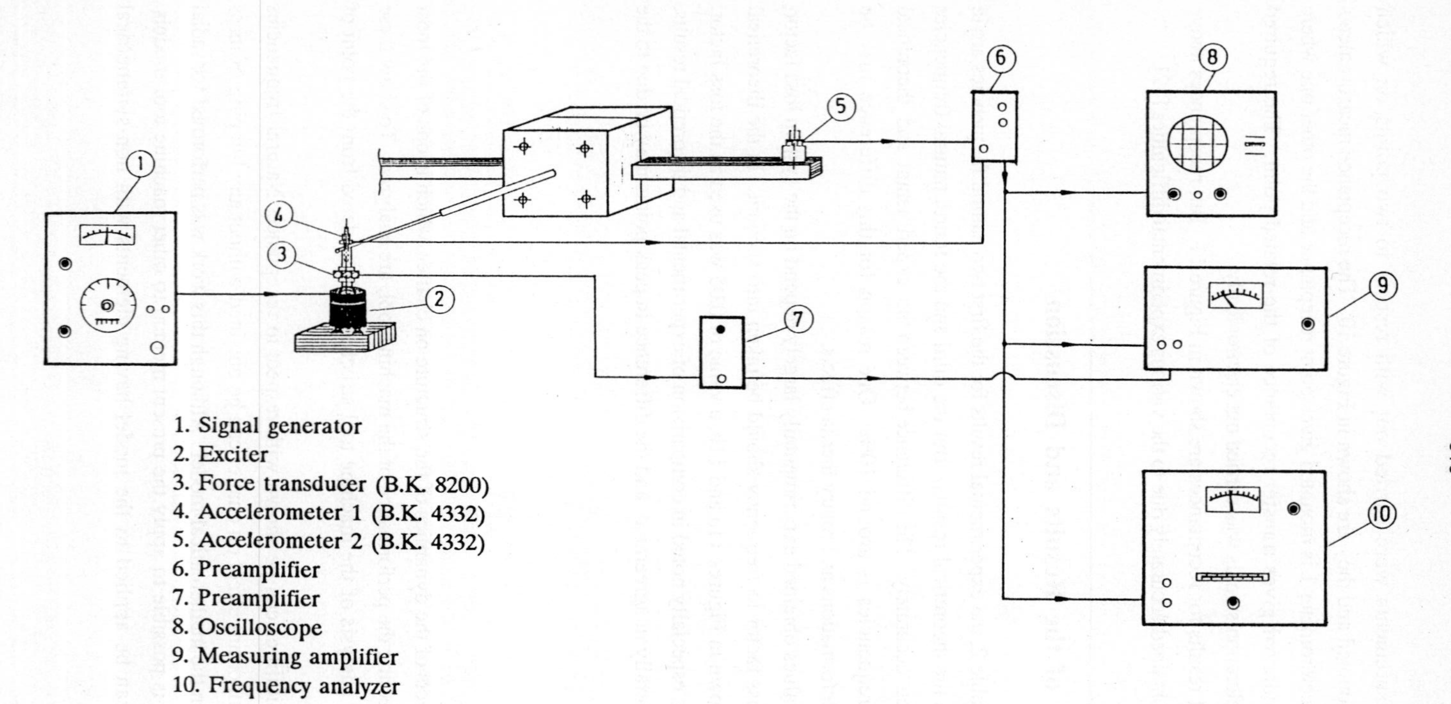

Figure 10. Experimentation Apparatus

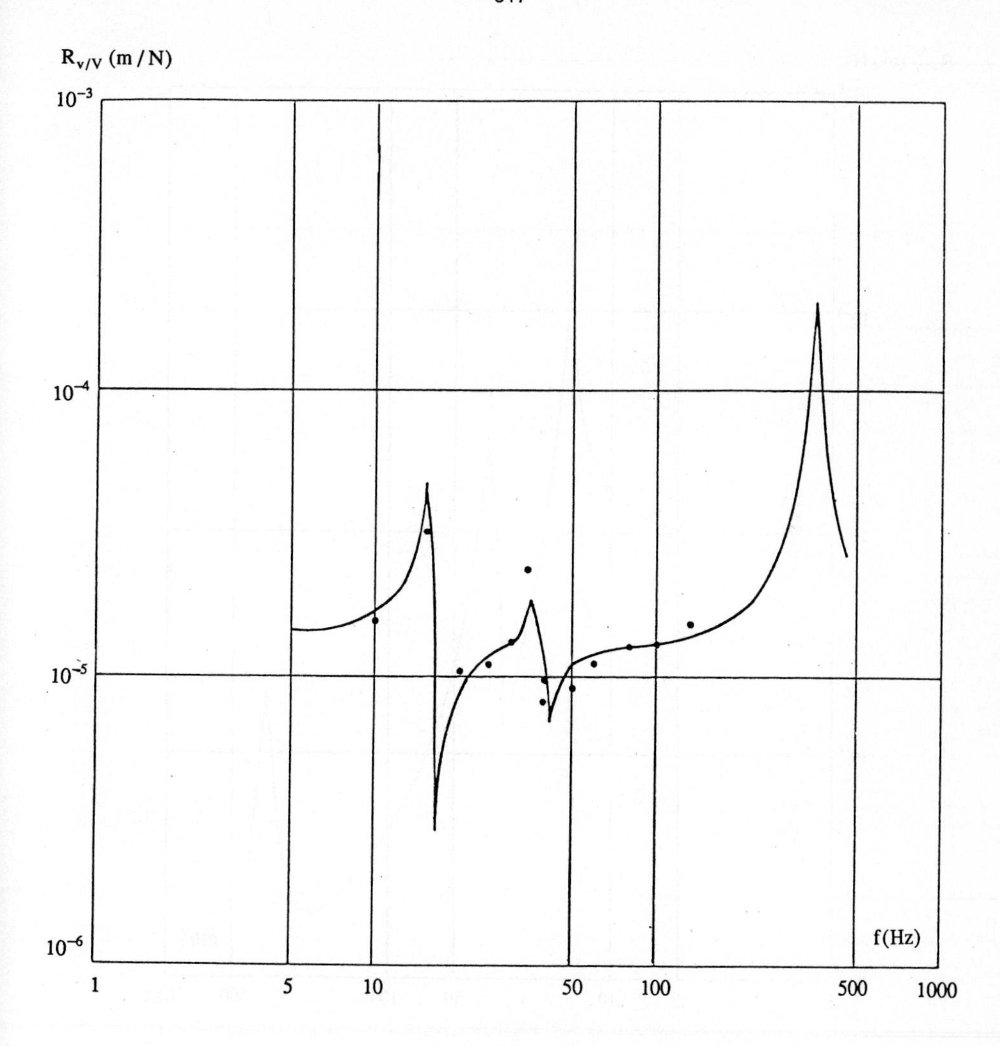

Figure 11a. Theoretical and Experimental Point Receptances in the Case of Out-of-Plane Shear Force

It is observed that structural damping has a very important effect on point and transfer receptances obtained for different radial drilling positions. It is found that transfer receptances are important as much as point receptances. The bending and torsional moments and shear forces are more marked compared to axial excitations for the model considered.

Since almost only one factor under the designer's control in chatter loop is the system receptance, the method of receptance calculations and the importance of being aware of its order of magnitude makes the obtained receptance values more meaningful.

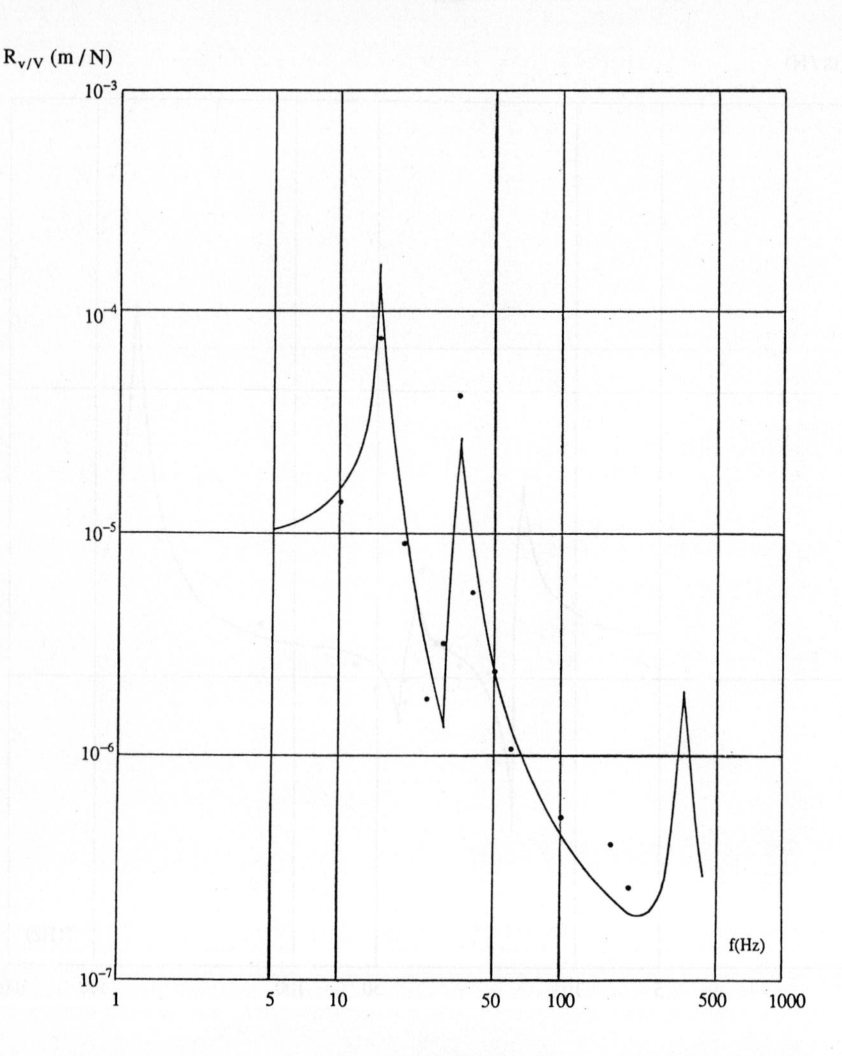

Figure 11b. Theoretical and Experimental Transfer Receptances in the Case of Out-of-Plane Shear Force

References

1. Burney, F.A., Pandit, S.M., Wu, S.M.: A new approach to the analysis of machine-tools system stability under working conditions. J.Eng.Industry, August 1977, pp.585-590.
2. Clough, R.W., Penzien, J.: Dynamics of structures. Mc-Graw Hill, 1975, New York.
3. Dokumaci, E., Tekelioglu, M.S.: Mobility behavior of unsymmetrical beams. Proc. 1st National Machine Theory Symposium. Ankara, 1984, pp.251-59. (In Turkish).
4. Long, G.W., Lemon, J.R.: Structural dynamics in machine-tool chatter. J.Eng.Industry, November 1965, pp.455-463.
5. Merrit, H.E.: Theory of self-excited machine-tool chatter. J.Eng.Industry, November 1965, pp.447-454.
6. Milner, D.A.: System identification of the machining process. J.Eng.Industry, August 1975, pp.1143- 1148.

7. Moriwaki,T., Iwata, K.: In-process analysis of machine tool structure dynamics and prediction of machining chatter. J.Eng.Industry, Feb. 1976, pp.301-305.
8. Ozguven, H.N.: An experimental investigation on the dynamic behavior of damped beams. Dynamics section. Mechanical Engineering, Imperial College, Report No:84001, London.
9. Pestel, E.C., Leckie, F.A.: Matrix Methods in Elastomechanics. Mc-Graw Hill, 1963, New York.
10. Sadek, M.M.: Stability of centre lathes in Orthogonal Cutting. Int.J.Prod.Res., vol.12, No.5, 1974, pp.547-560.
11. Shapton, W.R., Wood, D.M., Morse, I.E.: Shock and vibration identification of machine tool parameters. In:System identification of vibrating structures, ASME publication, 1972, pp.179-200.
12. Tekelioglu, M.S.: The effects of structural dynamics on chatter in machine tools. PhD thesis, DEU, Institute of Science and Engineering, 1985, Izmir-Turkey. (In Turkish).
13. Tlusty, J.: General Features of Chatter. In: Machine Tool Structures, Vol 1. (F.Koenigsberger and J. Tlusty, ed.). Pergamon Press 1970, pp.115-132. London.
14. Tlusty, J.: The theory of chatter and stability analysis. In: Machine Tool Structures, Vol 1. (F.Koenigsberger and J.Tlusty, ed.). Pergamon Press 1970, pp.133-177. London.
15. Tlusty, J., Moriwaki, T.: Experimental and computational identification of dynamic structural models. Annals CIRP, Vol 25, 1976, pp.497-503.
16. Tobias, S.A.:Machine tool vibration. Blackie, 1965, London.
17. Whittaker, A.R., Sadek, M.M.: Mathematical modelling for predicting effects of structural changes in machine tools. ASME paper No.81-DET.39; New York.

Part 4

Design Sensitivity Analysis and Optimization

Concurrent Engineering Design Optimization in a CAD Environment

Niels Olhoff, Erik Lund, and John Rasmussen

Institute of Mechanical Engineering, Aalborg University, DK-9220 Aalborg East, Denmark

Abstract: Concepts, methods and tools for interactive CAD-based concurrent engineering design optimization of mechanical products, systems and components which are critical in terms of cost, development time, functionality and quality, are presented. The emphasis is on formulation, development and implementation of methods and capabilities for finite element analysis, design sensitivity analysis, rational design, synthesis and optimization of mechanical systems and components, and the integration of these methods into a standard CAD modeling environment with a view to develop a concurrent engineering design optimization system. Methods for optimizing the topology of mechanical components are integrated into the system and used as a preprocessor for subsequent shape and sizing optimization. Use of the system for concurrent engineering design of mechanical components is illustrated by examples.

Keywords: concurrent engineering / optimum design system / computer aided design (CAD) / multicriterion optimization / min-max problems / finite element analysis / design sensitivity analysis / shape, sizing, and topology optimization / structures / composites / interactive design / software system / CAD integration / analysis model / design model / design element concept / data structures

1 Introduction

In this paper the Concurrent Engineering Design process is realized as an iterative solution procedure for a multicriterion optimization problem. The problem is defined, possibly redefined during the solution process, and finally solved by the designer in an interactive dialogue with a suitable software system for design optimization. The process can be essentially conceived as a rational search for the optimum spatial distribution of material within a prescribed admissible structural domain subject to multiple criteria and constraints. In the general case, this problem consists in determining both the optimum topology and the optimum design of the structure. Here, the label "optimum design" covers optimum shape or sizing of the design.

Structural optimization is a well-established field of research and, e.g., Haug & Cea (Ref. 1) and Olhoff & Taylor (Ref. 2) have published surveys that include the classical research. The interest in the field has increased considerably during recent decades due to the advent of reliable general analysis methods and the computer power necessary to use them efficiently. This has created the possibility of developing large, practice-oriented optimization systems, see e.g. Refs. 3–18. Ding (Ref. 9) and Hörnlein (Ref. 10) have published extensive overviews of available systems. Furthermore, a number of commercial finite element codes have facilities implemented for design sensitivity analysis and/or structural optimization. Systems like ANSYS, IDEAS, OASIS, MSC–NASTRAN, SAMTECH and NISAOPT are well-known examples.

The work of developing a general CAD-based structural optimization system was initiated at the Institute of Mechanical Engineering of Aalborg University in 1986, and was founded on practical and successful applications of structural optimization by, e.g., Esping, Braibant & Fleury, Bennett & Botkin, Botkin, Young and Bennet, Stanton, and Eschenauer, see Refs. 3–8. Since 1989, our work has been carried out as a project under the "Programme of Research of Computer Aided Design" under the auspices of the Danish Technical Research Council, and we acknowledge inspiring cooperation with profs. M.P. Bendsøe and P. Pedersen of the Technical University of Denmark on this project.

This paper presents concepts and complementary methods for optimization of structural designs, topologies and materials developed and adopted by the authors, and we describe their actual implementation and integration in a user-friendly, interactive, CAD-based optimization system for concurrent engineering design, see also Refs. 15–18.

In Chapter 2 we present the mathematical formulation for multicriterion optimization and the basic developments and capabilities of the optimization system.

Chapter 3 describes the strategies and methods developed for integration of the optimum design facilities into a traditional CAD system. The main strategy behind the integration is the use of two parallel but completely distinct descriptions of the geometry: the design model and the analysis model. Furthermore, full geometric versatility of these models is achieved. Thus, the analysis model is independent of the data structure of the CAD system, and this implies that the optimization system can be easily integrated with several different CAD-systems.

The basic ideas of these methods are discussed along with other important features pertaining to CAD integration. Chapter 3 also presents two systems for optimum structural design, namely the prototype system CAOS and the more general system ODESSY. Finally, the capabilities of the systems are demonstrated by means of examples in Chapter 4.

2 Mathematical Formulation and Basic System Capabilities

The purpose of this chapter is to present part of the mathematical formulation behind the development of the software systems CAOS and ODESSY for concurrent engineering design optimization, and to give an account of the basic capabilities of the systems.

2.1 Basic Concepts

The label *engineering design optimization* identifies the type of design problem where the set of structural parameters is subdivided into so-called *preassigned parameters* and *design variables*, and the problem consists in determining optimum values of the design variables such that they *maximize* or *minimize* a specific function termed the *objective* (or *criterion*, or *cost*) *function*, while satisfying a set of *geometrical* and/or *behavioural requirements* which are specified prior to design, and are called *constraints*.

Design Variables

The design variables will be denoted by

$$a_i, \quad i = 1,..,I, \tag{1}$$

and are assembled in the vector $\boldsymbol{a}$. The design variables can be categorized as follows:

Geometrical design variables:

Sizing design variables: describe cross-sectional properties of structural components like dimensions, cross-sectional areas or moments of inertia of bars, beams, columns and arches; or thicknesses of membranes, plates and shells.

Configurational design variables: describe the coordinates of the joints of discrete structures like trusses and frames; or the form of the center-line or mid-surface of continuous structures like curved beams, arches and shells.

Shape design variables: govern the shape of external boundaries and surfaces, or of interior interfaces of a structure. Examples are the cross-sectional shape of a torsion rod, column or beam; the boundary shape of a disk, plate, or shell; or the shape of interfaces within a structural component made of different materials.

Topological design variables: describe the type of structure, number of interior holes, etc., for a continuous structure. For a discrete structure like a truss or frame, these variables describe the number, spatial sequence, and mutual connectivity of members and joints.

Material design variables: represent constitutive parameters of isotropic materials, or, e.g., stacking sequence of lamina, and concentration and orientation of fibers in composite materials.

Support design variables: describe the support (or boundary) conditions, i.e., the number, positions and types of support for the structure.

Loading design variables: describe the positioning and distribution of external loading which in some cases may be also at the choice of the designer.

Manufacturing design variables: parameters pertaining to the manufacturing process(es), surface treatment, etc., which influence the properties and cost of the structure.

The Notion of Problem Variables

With the aim of developing a versatile mathematical formulation for multicriterion optimization, it has been found convenient to introduce the common notion "problem variables" for functions of the design variables that enter the definition of an optimization problem as part of the multicriterion objective function or as a constrained function. From among these functions we distinguish between *global problem variables* and *local problem variables*.

Global Problem Variables

The global problem variables that enter in the multicriterion objective function are denoted by

$$f_j, \quad j = 1,..,J, \tag{2}$$

and those entering in prescribed constraints are denoted by

$$g_p, \quad p = 1,..,P. \tag{3}$$

The variables may be categorized as *cost variables* and *global behavioural problem variables*, of which the following examples may be given:

Cost problem variables:

Structural volume or weight
Material cost
Manufacturing cost
Life cycle cost
Etc.

Global behavioural problem variables:

Compliance
Buckling load
Plastic collapse load
Eigenfrequency
Dynamic response
Fatigue life
Etc.

Note that the global problem variables are independent of the spatial variables of the structure. A cost problem variable f_j or g_p only depends on the vector $\boldsymbol{a}$ of design variables, i.e.,

$$f_j = f_j(\boldsymbol{a}), \quad g_p = g_p(\boldsymbol{a}) \qquad \text{(cost PVs)} \qquad (4)$$

while a global behavioural problem variable f_j or g_p also depends on the relevant displacement (or state) vector $\boldsymbol{D}_j$ or $\boldsymbol{D}_p$ associated with the behaviour variables,

$$f_j = f_j(\boldsymbol{a}, \boldsymbol{D}_j), \quad g_p = g_p(\boldsymbol{a}, \boldsymbol{D}_p) \qquad \text{(global behavioural PVs)} \qquad (5)$$

Here, $\boldsymbol{D}_j$ or $\boldsymbol{D}_p$ must be first obtained by solution of the relevant finite element equilibrium equations.

The global behavioural problem variables listed above are all of a solid mechanical nature, but extensions into the field of fluid mechanics (e.g., drag in fluid flow) are planned.

Local Problem Variables

Local problem variables entering in the multicriterion objective function will be denoted by

$$f_k^*, \quad k = 1,..,K, \qquad (6)$$

and those entering in prescribed constraints will be denoted by

$$g_s^*, \quad s = 1,..,S. \qquad (7)$$

Examples of this category of variables are the

Local behavioural problem variables:

Stresses
Strains
Displacements
Etc.

These variables depend on the spatial variables of the structure in addition to dependence on the design variables and the relevant displacement field, i.e., we have

$$f_k^* = f_k^* (a, D_k, x) \ , \quad g_s^* = g_s^* (a, D_s, x) \ , \quad x \in \Omega \ , \qquad \text{(local behavioural PVs)} \tag{8}$$

where x designates the coordinates of any point within the structural domain Ω. Note that it is also required to solve the finite element equilibrium equations associated with the given (sets of) external loading in order to determine this type of problem variables.

2.2 Mathematical Formulation for Multicriterion Optimization

The software system for solution of the multicriterion optimization problem is based on the concept of integration of modules of finite element analysis, sensitivity analysis, and mathematical programming. In order to meet practical needs of versatility, the basic mathematical formulation must possess sufficient flexibility such that the system can both handle problems of minimizing cost subject to several constraints, and problems of multicriterion optimization for prescribed resource (and additional constraints). To achieve this goal, the multicriterion problem is cast in scalar form by stating it as minimization of the maximum of a weighted set of the criteria. Such an interpretation of the multicriterion optimization problem can be formulated as a problem of minimizing a variable upper bound on the weighted criteria, see Refs. 19–21, and this *bound formulation* implies the considerable advantage that the multicriterion problem becomes differentiable.

Consider a multicriterion optimization problem in the initial form

Objective:

$$\min_{a_i} \left[\max_{j,k} \left\{ \hat{w}_j \cdot f_j \ ; \ \hat{w}_k^* \cdot \max_{x \in \Omega} f_k^* \right\} \right] \tag{9a}$$

Subject to:

Constraints for problem variables:

$$g_p \le \hat{G}_p \ , \quad p = 1,..,P \qquad \text{(global)} \tag{9b}$$

$$g_s^* \le \hat{G}_s^* \ , \quad s = 1,..,S, \quad \forall \ x \in \Omega \qquad \text{(local)} \tag{9c}$$

Linking constraints for design variables:

$$\sum_{i=1}^{I} \hat{c}_{iq} a_i \leq \hat{c}_{oq} \ , \quad q=1,..,Q \ , \qquad \sum_{i=1}^{I} \hat{b}_{ir} a_i = \hat{b}_{or} \ , \quad r=1,..,R \tag{9d}$$

Side constraints for design variables:

$$\underline{a}_i \leq a_i \leq \overline{a}_i \ , \quad i=1,..,I \tag{9e}$$

Equations for problem variables:

$$f_j = f_j\,(\boldsymbol{a},\boldsymbol{D}_j) \ , \quad g_p = g_p(\boldsymbol{a},\boldsymbol{D}_p) \tag{9f}$$

$$f_k^* = f_k^*\,(\boldsymbol{a},\boldsymbol{D}_k,x) \ , \quad g_s^* = g_s^*(\boldsymbol{a},\boldsymbol{D}_s,x) \ , \quad x \in \Omega \tag{9g}$$

Finite element state equations:

$$(\boldsymbol{KD} = \boldsymbol{F})_t \ , \quad t=1,..,T, \quad etc. \tag{9h}$$

In Eqs. (9a–h), all symbols equipped with carets and upper or lower bars are assumed to be prescribed. Thus, in (9a), $\hat{w}_j$ and $\hat{w}_k^*$ are given weighting factors for the separate design criterion functions f_j, $j=1,..,J$, and f_k, $k=1,..,K$, respectively; in (9b) and (9c) $\hat{G}_p$ and $\hat{G}_s^*$ denote given upper constraint values; $\hat{c}_{iq}$ and $\hat{b}_{ir}$ in (9d) are given linking coefficients; and in (9e) $\underline{a}_i$ and $\overline{a}_i$ denote prescribed lower and upper side constraint values for the design variables a_i, $i=1,..,I$.

The problem defined by Eqs. (9a–h) pertains to determine the set of values of the design variables a_i, $i=1,..,I$, that minimizes the maximum of the set of weighted global and local problem variables f_j, $j=1,..,J$, and f_k^*, $k=1,..,K$, respectively, in the multicriterion objective function (9a), subject to the constraints (9b–h).

It is assumed that each of the problem variables f_j and f_k^* that enter the scalar objective function (9a) have been suitably preconditioned for minimization, and that the finite element state equations in (9h) which may also include eigenproblems etc., contain all the sets of equations that are necessary for computation of the different displacement (or state) vectors required for the variety of behavioural problem variables that needs to be calculated in (9f) and (9g) when the optimization is carried out under consideration of several loading cases and different types of response. Notice that although not stated directly, (9f) also comprise equations for cost variables, and that these generally only depend on the vector of design variables $\boldsymbol{a}$.

Now, the objective (9a) of the multicriterion optimization problem is equivalent to

$$\min_{a_i} \left[\max_{j,(x)_k} \left\{ \hat{w}_j \cdot f_j \; ; \; \hat{w}_k^* \cdot f_k^* \right\} \right] . \tag{10}$$

To circumvent the inherent difficulty that this min–max objective function is generally *not differentiable* with respect to the design variables a_i, $i=1,..,I$, we develop a *bound formulation* of the problem, see Refs. 19–21. By this technique we introduce an additional variable β, which is termed the *bound parameter*, and write (10) in the form of the *differentiable problem*

$$\min_{a_i, \beta} \beta \tag{11a}$$

subject to:

$$\hat{w}_j \cdot f_j \leq \beta \; , \; j=1,..,J \qquad \text{(global constraints)} \tag{11b}$$

$$\hat{w}_k^* \cdot f_k^* \leq \beta \; , \; k=1,..,K, \quad \forall \; x \in \Omega \qquad \text{(local constraints)} \tag{11c}$$

Thus, we simply transform the statement (10) into the constraints (11b) and (11c), and minimize the common upper bound β on these constraints according to (11a).

Notice that in this formulation, the upper bound parameter β is an additional variable which replaces the original non–differentiable functional, and is to be minimized over a constraint set in an enlarged space. The original points of non–differentiability correspond to "corners" in the constraint set of the enlarged space, and arise from intersections of differentiable constraints.

We now in Eqs. (9a–h) replace Eq. (9a) by Eqs. (11a–c) and arrive at the following *bound formulation of the multicriterion optimization problem*:

$$\min_{a_i, \beta} \beta \tag{12a}$$

Subject to:

Variable bound constraints:

$$\hat{w}_j \cdot f_j \leq \beta \; , \; j=1,..,J \qquad \text{(global)} \tag{12b}$$

$$\hat{w}_k^* \cdot f_k^* \leq \beta \; , \; k=1,..,K, \quad \forall \; x \in \Omega \qquad \text{(local)} \tag{12c}$$

Original constraints:

$$g_p \le \hat{G}_p \ , \quad p = 1,..,P \qquad \text{(global)} \tag{12d}$$

$$g_s^* \le \hat{G}_s^* \ , \quad s = 1,..,S, \quad \forall \ x \in \Omega \qquad \text{(local)} \tag{12e}$$

Linking constraints for design variables:

$$\sum_{i=1}^{I} \hat{c}_{iq} a_i \le \hat{c}_{oq} \ , \quad q = 1,..,Q \ , \qquad \sum_{i=1}^{I} \hat{b}_{ir} a_i = \hat{b}_{or} \ , \quad r = 1,..,R \tag{12f}$$

Side constraints for design variables:

$$\underline{a}_i \le a_i \le \overline{a}_i \ , \quad i = 1,..,I \tag{12g}$$

Equations for problem variables:

$$f_j = f_j\,(a, D_j), \quad g_p = g_p(a, D_p) \tag{12h}$$

$$f_k^* = f_k^*\,(a, D_k, x) \ , \quad g_s^* = g_s^*(a, D_s, x) \ , \quad x \in \Omega \tag{12i}$$

Finite element state equations:

$$(KD = F)_t \ , \quad t = 1,..,T, \quad etc. \tag{12j}$$

Notice that the bound approach is a very simple technique for rendering min–max problems differentiable, and that the bound formulation is very versatile in terms of handling different types of problem variables. Thus, it is very simple to switch from, e.g., a prescribed–cost to a cost–minimization formulation of a given type of problem. It is also noteworthy that the sensitivity information concerning problem variables is virtually independent of whether a particular problem variable is an objective or a constraint function.

The problem variables f_j, f_k^*, g_p and g_s^* are determined using necessary finite element software behind (12h–j). Although f_j and g_p are global, and f_k^* and g_s^* are local problem variables, there is no need to make this distinction in practice. Thus, the local problem variables are just evaluated at a number of nodal points of the finite element discretized structure according to some suitable *active set strategy*, and thereby result in a number of inequality constraints that have the same mathematical form as the constraints (12b) and (12d) for the global problem variables.

While Eqs. (12h–j) represent the necessary tools for finite element analysis, Eqs. (12a–g) constitute a standard problem of mathematical programming which is solved sequentially using either a SIMPLEX algorithm, the CONLIN optimizer (Ref. 22) or the Method of Moving Asymptotes (Ref. 23), all of which require first order design sensitivities of the objective and constraint functions to be calculated. For this purpose, a sensitivity analysis is performed for each design variable at each step of redesign. For shape optimization, CAOS and ODESSY use the semi-analytical method which will be briefly discussed in the subsequent section.

2.3 Method of Semi-analytical Sensitivity Analysis

Design sensitivity analysis of a particular behavioural problem variable of a finite element discretized multicriterion structural optimization problem is now considered. Assuming linearly elastic static response, the sensitivity analysis is based on implicit differentiation of the relevant global equilibrium equations

$$\boldsymbol{KD} = \boldsymbol{F} \tag{13}$$

from among the necessary sets of finite element state equations comprised in (12j) for the behavioural problem variable in question. In (13), $\boldsymbol{F}$ is the vector of external nodal loading, $\boldsymbol{K}$ is the reduced global stiffness matrix, and $\boldsymbol{D}$ the vector of nodal displacements associated with the actual behavioural problem variable. The stiffness matrix $\boldsymbol{K}$, and hence $\boldsymbol{D}$, is assumed to depend on the vector $\boldsymbol{a}$ of design variables a_i, $i = 1,..,I$, that may comprise sizing and/or shape design variables. Differentiating Eq. (13) with respect to a_i and rearranging terms, we obtain

$$\boldsymbol{K}(\boldsymbol{a}) \frac{\partial \boldsymbol{D}}{\partial a_i} = - \frac{\partial \boldsymbol{K}(\boldsymbol{a})}{\partial a_i} \boldsymbol{D} + \frac{\partial \boldsymbol{F}}{\partial a_i}, \quad i = 1,..,I, \tag{14}$$

where the right hand side is termed the *pseudo load vector* associated with the design variable a_i.

The first goal of the design sensitivity analysis is to determine the sensitivities $\partial \boldsymbol{D}/\partial a_i$ of the nodal displacements with respect to the design variables. Due to the analogous form of the left hand sides of Eqs. (13) and (14), the sensitivities $\partial \boldsymbol{D}/\partial a_i$ can be solved from Eq. (14) by use of the same factorization of the global stiffness matrix $\boldsymbol{K}$ as is applied when Eq. (13) is solved for the nodal displacement vector $\boldsymbol{D}$ in the initial analysis of a given step of redesign. Thus, with $\boldsymbol{D}$ known at the particular redesign step, and assuming that $\partial \boldsymbol{F}/\partial a_i$ has been computed if $\boldsymbol{F}$ is design dependent, the determination of $\partial \boldsymbol{D}/\partial a_i$ from Eq. (14) only requires that the design sensitivities $\partial \boldsymbol{K}/\partial a_i$ of the global stiffness matrix are known.

If the design sensitivities $\partial \boldsymbol{K}/\partial a_i$ are determined analytically before their numerical evaluation, the approach is called the method of *Analytical* sensitivity analysis, and if they are determined by numerical differentiation, the label *Semi-analytical* sensitivity analysis is used. The latter method was first applied in Refs. 24, 3 and 25.

While the analytical method is expedient and has been applied successfully for problems involving sizing design variables, it is often a quite formidable task to implement the method when shape variables are encountered. Thus, a large amount of analytical work and programming may be required in order to develop analytical expressions for derivatives of various stiffness matrices with respect to a large number of possible shape parameters. For problems involving shape design variables it is much more attractive to apply the semi-analytical method, because this method is easier to implement as it treats different types of finite elements and design variables in a unified way.

Both when the analytical and the semi-analytical methods are applied, usually the differentiation involved in the determination of $\partial K/\partial a_i$ for Eq. (14) is most conveniently carried out at the *element level*. Thus, differentiating with respect to a_i the *element stiffness matrices* $\boldsymbol{k}$ associated with global coordinates, expanding the matrix derivatives $\partial k/\partial a_i$ to the system degrees of freedom, and summing over the number n_e of finite elements of the discretized structure, we obtain

$$\frac{\partial \boldsymbol{K}(\boldsymbol{a})}{\partial a_i} = \sum_{n_e} \frac{\partial \boldsymbol{k}}{\partial a_i} \;, \; i = 1,..,I \tag{15}$$

Notice that this computation of $\partial K/\partial a_i$ is similar to the assembling of the global stiffness matrix $\boldsymbol{K}$ from element matrices $\boldsymbol{k}$, but that only those element stiffness matrices which depend on a particular design variable a_i will contribute to the summation in Eq. (15).

In the following, the symbol $\boldsymbol{k}$ will be used only for the element stiffness matrix operating on the *element degrees of freedom* in global coordinates, and we shall assume that the stiffness matrix $\boldsymbol{k}$ of a particular finite element only depends on a given sub-set from among the total set of design variables a_i, $i = 1,..,I$. Let us re-number this sub-set of the design variables and denote it by a_j, $j = 1,..,J$, where $J < I$.

In the traditional form of the semi-analytical method of sensitivity analysis, the numerical differentiations of the element stiffness matrices $\boldsymbol{k}$ behind the computation of $\partial K(a)/\partial a_i$ in Eq. 15, are carried out by means of the first order forward difference formula

$$\frac{\partial \boldsymbol{k}(a_1,..,a_J)}{\partial a_j} \approx \frac{\Delta \boldsymbol{k}(a_1,..,a_J)}{\Delta a_j} = \frac{\boldsymbol{k}(a_1,..,a_j + \Delta a_j,..,a_J) - \boldsymbol{k}(a_1,..,a_j,..,a_J)}{\Delta a_j} \;, \; j = 1,..,J, \tag{16}$$

where Δa_j is the finite difference increment (perturbation) of the design variable a_j.

<u>Sensitivities of Stresses</u>

For a particular finite element the stress components are assembled in the stress vector function $\boldsymbol{\sigma}(x,y,z) = \{\sigma_x \; \sigma_y \; \sigma_z \; \sigma_{xy} \; \sigma_{yz} \; \sigma_{zx}\}^T$ given by

$$\boldsymbol{\sigma} = \boldsymbol{E}\boldsymbol{\varepsilon} = \boldsymbol{E}\boldsymbol{B}\boldsymbol{d} \tag{17}$$

in terms of the elasticity matrix $\boldsymbol{E}$, strain–displacement matrix $\boldsymbol{B}$ and element nodal displacement vector $\boldsymbol{d}$. Assuming that the material is isotropic, the stress design sensitivity $\partial\sigma/\partial a_i$, with respect to *any* geometric design variable a_i, $i = 1,..,I$, for the structure becomes

$$\frac{\partial \sigma}{\partial a_i} = \boldsymbol{E}\left[\boldsymbol{B}\frac{\partial \boldsymbol{d}}{\partial a_i} + \frac{\partial \boldsymbol{B}}{\partial a_i}\boldsymbol{d}\right], \quad i = 1,..,I. \tag{18}$$

Here, the element nodal displacement sensitivities $\partial \boldsymbol{d}/\partial a_i$, $i = 1,..,I$, are selected from among the displacement design sensitivities $\partial \boldsymbol{D}/\partial a_i$, $i = 1,..,I$, obtained by solution of the system level Eq. (14). The element level derivatives $\partial \boldsymbol{B}/\partial a_i$, $i = 1,..,I$, in (18) are only non-zero if a_i is one of the design variables a_j, $j = 1,..,J$, of the finite element in question, and $\partial \boldsymbol{B}/\partial a_i = \partial \boldsymbol{B}/\partial a_j$ is easily obtained by a standard finite difference approach.

2.4 Method for Elimination of Inaccuracy in Semi-analytical Sensitivity Analysis

The method of finite element based semi-analytical shape sensitivity analysis is fully reliable for usual problems where the structural displacement field entails small rigid body rotations relative to actual deformations of the finite elements, see Refs. 31 and 32.

However, in problems involving linearly elastic bending of long-span, beam-like structures, or of structures like plates and shells that are "long" in two directions, rotations are generally not negligible relative to usual strains, and application of the semi-analytical method may then yield completely erroneous results for lateral displacement design sensitivities with respect to shape design variables associated with the "long" direction(s) of the structure.

The source of the severe error problem is the approximate numerical differentiation of the finite element stiffness matrices, cf. Eqs. (15) and (16), which is the main characteristic of the semi-analytical method, and the inaccuracy problem manifests itself by abnormal dependence of the displacement sensitivity errors on finite element discretization and –refinement, and on the values chosen for perturbations of the shape design variables.

A similar inaccuracy problem is not found if the analytical method or the overall finite difference technique of sensitivity analysis is employed.

It is important to note that the occurrence of severe displacement design sensitivity error is basically *problem dependent rather than element dependent*, i.e., the error problem will manifest itself independently of which type of finite element that is used in the modeling of a long-span structure, see Ref. 32.

The severe error problem readily manifests itself in cases of structures modeled by beam, plate and shell finite elements, see Refs. 26–32. Such elements both possess translational and rotational degrees of freedom whereby the components of their stiffness matrices depend on shape variables in different negative powers. This implies that the use of a forward finite difference formula based on first order polynomial approximation for computation of the element stiffness matrix sensitivities gives rise to different and comparatively large errors of the components of the matrix derivative and hence of the element pseudo loads, see Refs. 27–30.

The non–uniform distribution of pseudo load errors is such that it has a very critical influence on the displacement design sensitivities as the finite element mesh is refined because the values of the latter, which should be mesh–independent, result from subtractions and additions of an increasing number of increasingly large terms as the number of finite elements increases with the mesh–refinement, see Ref. 29.

In order to develop a method for error elimination without a sacrifice of the simple numerical differentiation and other main advantages of the semi–analytical method, the common mathematical structure of a broad range of finite element stiffness matrices has been recently studied in Ref. 32.

This study leads to the result that element stiffness matrices can be generally expressed in terms of a class of special scalar functions and a class of matrix functions of the shape design variables that are defined such that the members of the classes admit "exact" numerical differentiation (exact except for round–off error) by means of a very simple method developed in the paper. The method is based on application of simple correction factors to upgrade usual, computationally inexpensive first order finite differences to "exact" numerical derivatives with respect to the shape design variables. The correction factors can be easily computed once and for all as an initial step of the sensitivity analysis.

Application of the method completely eliminates the problems of severe dependence of semi–analytical design sensitivities on the size of perturbations of design variables and on finite element mesh–size and –refinement, etc. The results are actually equivalent to results that would be obtained by numerical evaluation of corresponding analytical design sensitivities. However, the new method is much more problem independent and easier to implement than the analytical method.

Thus, it is shown in Ref. 32 that the new approach to semi–analytical sensitivity analysis is easily implemented as an integral part of finite element analysis. The method of error elimination by "exact" numerical differentiation can even be implemented in connection with existing computer codes for semi–analytical sensitivity analysis where the different subroutines for computation of element stiffness matrices are only available in the form of black–box routines.

In Ref. 32 the applicability of the method is demonstrated for a broad class of commonly used finite elements. It is also shown that the method is fully compatible with common methods of design boundary parametrization based on master node techniques, and numerical examples are presented for illustration and discussion of the capabilities of the method.

2.5 Boundary Parametrization using Generalized Shape Design Variables

This section presents the mathematical background for important concepts and tools for shape optimization that provide convenient means of describing continuous shapes of structures by a small number of variables, and solves the problem of automatic updating of the finite element model as the shape is changed during the steps of redesign. Also the techniques are of fundamental importance for a suitable CAD–integration of the shape optimization facilities. These aspects will be discussed in Section 3.3 of the subsequent chapter and be illustrated via examples in Chapter 4.

As mentioned in Section 2.3, the present approach to design sensitivity analysis implies the considerable advantage that the factorization of the global stiffness matrix $\boldsymbol{K}$ performed for solution of Eq. (13), can be re–applied when Eq. (14) is solved for the displacement design sensitivities $\partial D / \partial a_i$ associated with a particular behavioural problem variable of the multicriterion problem. For this problem variable and its associated finite element stiffness matrix $\boldsymbol{K}$ and displacement vector $\boldsymbol{D}$, to which our attention will be restricted in the following, solution of Eq. (14) is required for each of the design variables a_i, $i = 1,..,I$, of the structure. However, the total number I of design variables may be large, in particular when a_i are taken to represent shape variables in the form of coordinates of finite element nodes of the discretized structure. In shape optimization, it is therefore customary, in each step of redesign, to only consider the finite element nodes at the design boundaries as design variables, and to only let the coordinates of these nodes be subject to perturbation. This implies that the computation of element stiffness matrices at perturbed values of the design variables is limited as much as possible. This approach is termed *"the design boundary layer technique"*.

In order to both further reduce the number of necessary solutions of Eq. (14), and to introduce smoothness, which improves the convergence properties of the problem, it is advantageous to introduce a *comparatively small number*, M, of *generalized shape design variables* A_m, $m = 1,..,M$, which control the large number of design variables a_i, $i = 1,..,I$ through a suitable set of shape functions for the design boundary. Hereby, Eq. (14) only needs to be solved M times at a given step of redesign.

Mechanically, we may conceive this as an application of the *superposition principle* in terms of pseudo loads and corresponding design displacement sensitivities governed by the linear system, Eq. (14). Thus, instead of solving Eq. (14) separately subject to each pseudo load associated with a design variable a_i, $i = 1,..,I$, we *a priori* superpose properly the *single* pseudo loads into a *resulting* pseudo load that corresponds to the generalized shape design variable A_m, and then just solve Eq. (14) once for the corresponding displacement design sensitivity.

The generalized shape design variables A_m, $m = 1,..,M$, are scalar parameters that will be used to define the positions of *master nodes* which control the shape of the boundary subject to optimization, and thus serve to control the positions of the finite element nodes that are located at the design boundary and follow its changes. The master node technique is well-known, see, e.g. Refs. 4, 6, 15, 16 and 33.

Written in terms of any of the generalized shape design variables A_m, $m = 1,..,M$, the system level Eq. (14) takes the form

$$K \frac{\partial D}{\partial A_m} = - \frac{\partial K}{\partial A_m} D + \frac{\partial F}{\partial A_m}, \qquad m = 1,..,M, \tag{19}$$

and we delimit our interest to the global stiffness matrix derivative $\partial K/\partial A_m$.

Although a particular A_m may not be related with all a_i, for the sake of generality let us assume that $A_m = A_m(a_1,..,a_I)$. Application of the chain rule then yields

$$\frac{\partial K}{\partial A_m} = \sum_{i=1}^{I} \frac{\partial K}{\partial a_i} \frac{\partial a_i}{\partial A_m}, \qquad m = 1,..,M, \tag{20}$$

with summation over all element nodal point coordinates adopted as design variables a_i, $i = 1,..,I$.

Consider now Eq. (15). Here $\boldsymbol{k}$ denotes the element stiffness matrix in global coordinates of any of the finite elements of the discretized structure. The possible nodal point coordinates of this element which play the role as shape design variables, are denoted by a_j, $j = 1,..,J$, and generally constitute a small sub-set of the total set of design variables a_i, $i = 1,..,I$, of the structure. Thus, we may write $\partial K/\partial a_i$ in Eq. (20) as

$$\frac{\partial K}{\partial a_i} = \sum_{n_e}{}' \left. \frac{\partial k}{\partial a_j} \right|_{a_j = a_i} \tag{21}$$

where the prime indicates that the summation is only carried out over those finite elements (from among the total number n_e) that has one or more of the parameters a_i, $i = 1,..,I$, as a nodal point coordinate design variable a_j. Note that the derivatives $\partial k/\partial a_j$ to be substituted into Eq. (21) will be computed by the finite difference formula (16) when the traditional semi-analytical method of sensitivity analysis is used.

2D Boundary Shape Representation

In Eq. (20) we now just need to establish the derivatives $\partial a_i/\partial A_m$. Let us do this for the general case where we are confronted with a 2D design boundary (surface) of a 3D structural domain. Specialization to the more commonly treated cases of boundary *curves* for plane 2D structural domains is straight–forward.

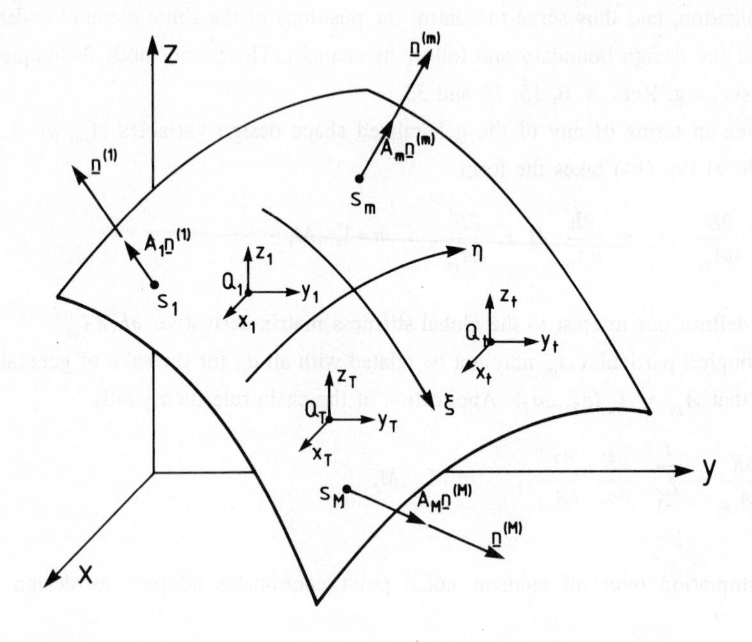

Figure 1. Parametrized 2D Design Boundary Surface of 3D Body

Figure 1 illustrates our 2D design boundary which may be conceived to be a part, or the full surface, of a 3D continuum structure. In the former case, the surface could be the design boundary of a 3D *design element* of the structure (see, e.g., Refs. 4, 6, 15–17 and 33 for similar 2D design elements). The design boundary is described locally by a set of two non–dimensional curvilinear coordinates ξ and η that are embedded in the surface, i.e., a given material point of the surface retains the same values of the coordinates ξ and η independently of shape changes of the surface.

The relationship between the local surface coordinates ξ and η and the global coordinates x, y and z of the surface, see Figure 1, is defined by the following interpolation based on shape functions $R_m(\xi,\eta)$, $m = 1,..,M$:

$$c = \sum_{m=1}^{M} R_m(\xi,\eta)\cdot\left[c_o^{(m)} + A_m n^{(m)}\right]. \tag{22}$$

Here the vector

$$c = \{x\ y\ z\}^T \tag{23}$$

contains the set of global coordinates for any point of the design boundary. The vectors

$$c_o^{(m)} = \{x_o^{(m)} \; y_o^{(m)} \; z_o^{(m)}\}^T \qquad m = 1,..,M \qquad (24)$$

$$n^{(m)} = \{n_x^{(m)} \; n_y^{(m)} \; n_z^{(m)}\}^T$$

refer to M *master nodes* S_m, $m = 1,..,M$, see Figure 1, which serve to control the shape of the design boundary. In Eqs. (24), the vector $c_o^{(m)}$ contains the global coordinates of the initial (or updated) position of the m–th master node S_m, and $n^{(m)}$ is a unit vector in global coordinates that represents an individually assigned move direction to the master node S_m. Finally, the scalar A_m is a *generalized shape design variable* that represents the movement in the direction $n^{(m)}$ of the master node S_m, relative to its initial location given by $c_o^{(m)}$.

In the present context of surface modeling, the shape functions R_m in Eq. (22) may represent Coons patches, 2D Bezier or B–spline surfaces, or special 2D blending functions for the boundary parametrization, depending on the need for accommodation of, e.g., particular geometrical or technological constraints, cf. Refs. 4, 6, 15–17 and 33. Use of smooth shape functions R_m in Eq. (22) implies imposition of smoothness on the variation of the design variables a_i, $i = 1,..,I$, and along these lines it should also be noted that, in general, the parametrization defined by Eq. (22) implies a considerable reduction of the design space for the problem in question.

Let us associate the set of shape design variables a_i, $i = 1,..,I$, with the 2D design boundary under study, and assume, for reasons of generality, that the set a_i, $i = 1,..,I$, comprises *all* the coordinates x_t , y_t and z_t , $t = 1,..,T$, of the finite element nodal points Q_t, $t = 1,..,T$, that belong to the design boundary, see Figure 1. Assign now the following index values to a_i, $i = 1,..,I$, where $I = 3T$:

$$a_t = x_t \; , \quad a_{T+t} = y_t \; , \quad a_{2T+t} = z_t \; , \qquad t = 1,..,T. \qquad (25)$$

From Eqs. (22)–(25) we then easily derive the following expression for the derivatives $\partial a_i / \partial A_m$:

$$\frac{\partial a_i}{\partial A_m} = C_i^{(m)} R_m(\xi_i, \eta_i) \; , \quad i = 1,..,I \; , \quad I = 3T \; , \quad m = 1,..,M, \qquad (26)$$

$$C_i^{(m)} = \begin{cases} n_x^{(m)} & \textit{if } a_i \textit{ is an x-coordinate} \\ n_y^{(m)} & \textit{if } a_i \textit{ is a y-coordinate}. \\ n_z^{(m)} & \textit{if } a_i \textit{ is a z-coordinate} \end{cases} \qquad (27)$$

Here (ξ_i, η_i) are the local coordinates of the reference nodal point for a_i, and indices of the scalar factors $C_i^{(m)}$ correspond to those assigned to a_i in Eq. (25).

Global Stiffness Matrix Derivative

Substitute now Eqs. (21) and (26) into Eq. (20) and obtain the following expression

$$\frac{\partial \boldsymbol{K}}{\partial A_m} = \sum_{i=1}^{I} \left(\left[\sum_{n_e}{}' \left. \frac{\partial \boldsymbol{k}}{\partial a_j} \right|_{a_j = a_i} \right] C_i^{(m)} R_m(\xi_i, \eta_i) \right), \quad m = 1,..,M, \tag{28}$$

for the derivative of the global stiffness matrix $\boldsymbol{K}$ with respect to the generalized shape design variable A_m. Here, $C_i^{(m)}$ is given by Eq. (27), and as a consequence of our initial assumption regarding the design variables, the summation over i, $i = 1,..,I$, $I = 3T$, in Eq. (28) is only over the coordinates x_t , y_t and z_t of the finite element nodes Q_t, $t = 1,..,T$, that belong to the design boundary. If some of these coordinates are not design variables, they are not included in the summation.

Note finally from Eq. (28) that the *same set of element stiffness matrix derivatives* $\partial \boldsymbol{k}/\partial a_j$ computed for the finite element nodes at the design boundary, is used for computation of the global stiffness matrix derivatives with respect to *any of the generalized design variables* A_m, $m = 1,..,M$.

Sensitivities of Stresses with Respect to Generalized Shape Design Variables

Differentiation of Eq. (17) for the stress vector $\boldsymbol{\sigma}$ of a particular finite element with respect to a generalized shape design variable A_m, $m = 1,..,M$, yields

$$\frac{\partial \boldsymbol{\sigma}}{\partial A_m} = \boldsymbol{E} \left[\boldsymbol{B} \frac{\partial \boldsymbol{d}}{\partial A_m} + \frac{\partial \boldsymbol{B}}{\partial A_m} \boldsymbol{d} \right], \tag{29}$$

where the element nodal displacement sensitivities $\partial \boldsymbol{d}/\partial A_m$ are selected from among the displacement design sensitivities $\partial \boldsymbol{D}/\partial A_m$ obtained by solution of Eq. (19) with $\partial \boldsymbol{K}/\partial A_m$ given by Eq. (28). Similarly, $\boldsymbol{d}$ is obtained from the solution $\boldsymbol{D}$ of the global equilibrium equations, Eq. (13).

Consider then the derivative $\partial \boldsymbol{B}/\partial A_m$ of the element strain–displacement matrix $\boldsymbol{B}$ in Eq. (29). Application of the chain rule yields

$$\frac{\partial \boldsymbol{B}}{\partial A_m} = \sum_{i=1}^{I} \frac{\partial \boldsymbol{B}}{\partial a_i} \frac{\partial a_i}{\partial A_m}, \quad m = 1,..,M \tag{30}$$

Now, if we, as assumed earlier, only consider the shape design variables a_i, $i = 1,..,I$, to represent coordinates of the finite element nodes Q_t, $t = 1,..,T$, that belong to the design boundary and are governed by Eqs. (22), (23) and (25) – in other words – if we apply a *design boundary layer technique*, where all other finite element nodal points are held fixed during the sensitivity analysis, then the *derivative* $\partial \boldsymbol{B}/\partial A_m$ in Eqs. (30) and (29) *will only be non-zero for finite elements in the "boundary layer"*, i.e. elements that have nodal points at the design boundary. Moreover, since $\boldsymbol{B}$ is independent of all other design variables a_i, $i=1,..,I$, than those associated with the particular finite element, i.e., a_j, $j=1,..,J$, we can write

$$\frac{\partial \boldsymbol{B}}{\partial A_m} = \begin{cases} \sum\limits_{j=1}^{J}{}' \left[\dfrac{\partial \boldsymbol{B}}{\partial a_j} \dfrac{\partial a_j}{\partial A_m} \right] & \text{for elements in the "boundary layer"} \\ \mathbf{0} & \text{for all other elements} \end{cases} \qquad m = 1,..,M, \tag{31}$$

where the derivatives $\partial a_j/\partial A_m$ are obtained from Eqs. (26) and (27). The prime in Eq. (31) indicates that the summation is only carried out over those shape design variables a_j of the finite element that are associated with a nodal point at the design boundary, and the derivatives $\partial \boldsymbol{B}/\partial a_j$ entering the summation in Eq. (31) may be simply computed by the forward finite difference formula $\Delta \boldsymbol{B}/\Delta a_j = \left(\boldsymbol{B}(a_j + \Delta a_j) - \boldsymbol{B}(a_j)\right)/\Delta a_j$.

2.6 Optimization of Sizing and Configuration of Truss Structures

A capability has been developed for optimization of 2–D and 3–D truss structures under selfweight and multiple loading conditions, using cross–sectional areas of bars and positions of joints as design variables. The system is called SCOTS (Sizing and Configuration Optimization of Truss Structures). The development is inspired by Pedersen (Refs. 34–36). In the current setting, weight minimization is the design objective, and constraints include stresses, displacements, and elastic as well as plastic buckling of bars in compression.

The mathematical programming formulation is:

$$\underset{A_i, x_k}{\textit{Minimize}} \quad \sum_{i=1}^{I} \rho_i A_i \ell_i \tag{32}$$

$$\textit{Subject to} \quad \frac{\sigma_i}{\sigma_i^{yt}} \le 1, \quad \frac{\sigma_i}{\sigma_i^{c}} \le 1, \qquad i = 1,..,I \tag{33a,b}$$

$$\frac{|d_j|}{\overline{d}_j} \le 1, \qquad j = 1,..,J \tag{34}$$

$$\underline{A_i} \le A_i \le \overline{A_i}, \quad i = 1,..,I \;; \quad \underline{x}_k \le x_k \le \overline{x}_k, \quad k = 1,..,K \tag{35a,b}$$

where

$$\sigma_i^{\,c} = -\frac{\pi^2 \alpha_i^{\,2} E_i A_i}{\ell_i^{\,2} S_i} \quad for \quad \ell_i \geq \pi\alpha_i \sqrt{\frac{2E_i A_i}{\sigma_i^{yc} S_i}} \tag{36a}$$

$$\sigma_i^{\,c} = -\sigma_i^{yc} + \frac{(\sigma_i^{yc})^2 S_i \ell_i^{\,2}}{4\pi^2 E_i \alpha_i^{\,2} A_i} \quad for \quad \ell_1 \leq \pi\alpha_i \sqrt{\frac{2E_i A_i}{\sigma_i^{yc} S_i}} \tag{36b}$$

Here, A_i, ℓ_i, ρ_i and E_i denote cross-sectional area, length, specific weight and Young's modulus for the i-th bar of the truss. The design variables of the problem, see Eq. (32), are A_i and x_k, where the latter symbol represents an element of the total set of variable components of position vectors of joints in the structure. Eqs. (35a,b) are side constraints for the design variables. For simplicity in notation, the remaining constraints are written for a single load case. Eqs. (34) are constraints that may be specified for any displacement component for the joints of the truss.

Bar stresses are denoted by σ_i and Eqs. (33a,b) express constraints for tensile and compressive stresses, respectively. In (33a), σ_i^{yt} represents the yield stress or some other specified upper stress limit. When buckling constraints are considered, the lower limit $-\sigma_i^c$ for compressive stresses is design dependent and given by Eq. (36a) or (36b), respectively, where the former represents dimensioning against Euler Buckling, and the latter plastic buckling on the basis of Ostenfeld's formula. In Eq. (36a,b), we have $\alpha_i^2 = r_i^2/A_i$, where r_i is the radius of inertia of the bar cross-section, S_i the factor of safety against buckling, and σ_i^{yc} (>0) the compressive yield stress of the i-th bar. In the literature on optimization of trusses, it is often the case that yielding rather than buckling constraints are considered for compressive bars. This somewhat academic simplification is obtained, if we set $\sigma_i^c = -\sigma_i^{yc}$.

Solution of the mathematical program (Eqs. 32-36) is based on a finite element formulation (the state equations are omitted here for reasons of brevity), and analytical sensitivity analyses. An example is presented in Section 2.7.

2.7 Topology Optimization

The software system possesses the capability of determining the optimum topology of mechanical structures and components by means of a remarkable method recently developed by Bendsøe and Kikuchi, Ref. 37, in which the structure is considered as a domain of space with a high density of material, see also Bendsøe (Ref. 38).

The problem of topology optimization is basically one of discrete optimization, but this difficulty is avoided by introducing relationships between stiffness components and density, based on physical modeling of porous, periodic microstructures whose orientation and density are described by continuous variables over the admissible design domain. The solution of this problem is based on a finite element discretization of the admissible domain, and the optimum values of the design variables (density and orientation of the microstructures) are determined iteratively via an optimality criterion approach.

More precisely, for the topology optimization we minimize compliance for a fixed, given volume of material, and use a density of material as the design variable. The density of material and the effective material properties related to the density is controlled via geometric variables which govern the material with microstructure that is constructed in order to relate correctly material density with effective material property.

The problem is thus formulated as:

$$\begin{aligned} &\textit{Minimize} && L(w) \\ &\textit{Subject to} && a_D(w,v) = L(v) \qquad \forall \; v \in H \\ & && \textit{Volume} \le V \end{aligned} \tag{37}$$

where

$$L(v) = \int_\Omega B_i v_i d\Omega + \int_\Gamma T_i v_i d\Gamma \tag{38}$$

$$a_D(w,v) = \int_\Omega E_{ijkl}(D)\ \varepsilon_{ij}(w)\ \varepsilon_{kl}(v) d\Omega \tag{39}$$

Here, B_i and T_i are the body forces and surface tractions, respectively, and ε_{ij} denote linearized strains. H is the set of kinematically admissible displacements. The problem is defined on a fixed reference domain Ω and the components of the tensor of elasticity E_{ijkl} depend on the design variables used. For a so-called second rank layering constructed as in Figure 2, we have a relation

$$E_{ijkl} \equiv E_{ijkl}(\mu,\gamma,\theta) \tag{40}$$

where μ and γ denote the densities of the layering and θ is the rotation angle of the layering. The relation (40) can be computed analytically (see Ref. 38) and for the volume we have

$$\textit{Volume} = \int_\Omega (\mu + \gamma - \mu\gamma)\ d\Omega \tag{41}$$

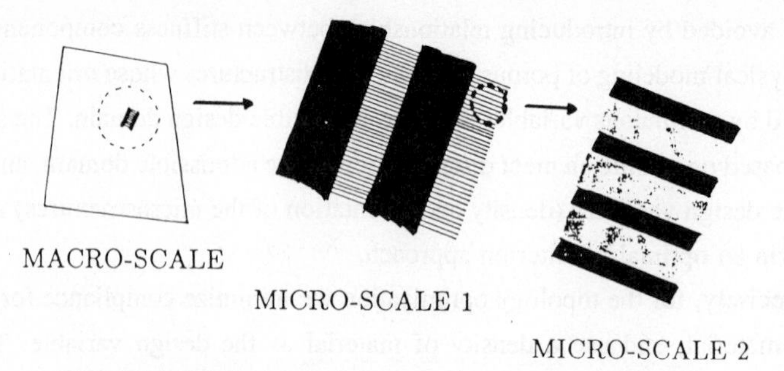

Figure 2. Construction of a Layering of Second Rank

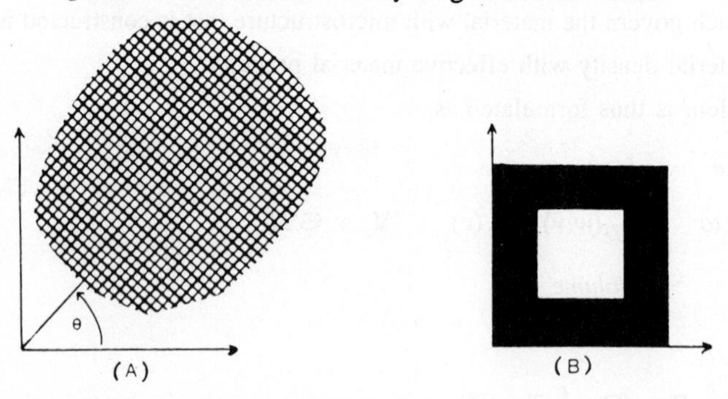

Figure 3. (A): Periodic Microstructure with Square Holes Rotated the Angle θ.
(B): Square Cell with a Square Hole

Layered materials is just one possible choice of microstructure that can be applied. The important feature is to choose a microstructure that allows density of material to cover the complete range of values from zero (void) to one (solid), and that this microstructure is periodic so that effective properties can be computed (numerically) through homogenization (theory of cells). This excludes circular holes in square cells, while square or rectangular holes in square cells, see Figure 3, are suitable choices of simple microstructures. For the case of a rectangular hole in a square cell, the volume is also given by Eq. (41), with μ and γ denoting the amount of material used in the directions of axes of the cell and hole; for this microstructure the angle θ of rotation of the unit cell becomes a design variable.

The optimization problem can now be solved either by optimality criteria methods as in Ref. 38 or by duality methods (Ref. 22), where advantage is taken of the fact that the problem has just one constraint. The angle θ of layer or cell rotation is controlled via the results on optimum rotation of orthotropic materials as presented by Pedersen in Refs. 39, 40, and in these proceedings. The reader is referred to Refs. 41–46 for very recent publications on topology optimization.

As stated above, the optimum topology is determined from the condition of minimum compliance subject to a bound on the total structural volume. Shape and sizing optimization systems, on the other hand, must be able to handle a much larger variety of formulations of which stress and volume minimizations are the most frequent. However, in spite of the incompatible formulations, compliance optimized topologies tend to perform well also from a stress minimization point of view. This is due to the fact that a relatively high amount of energy is stored in possible areas of stress concentration which then become undesirable also in the compliance minimization. The initial topology optimization can therefore in many cases lead to substantial improvements of the final result even though the actual aim is to perform an optimization of a different type.

The topology optimization thus results in a prediction of the structural type and overall lay–out, and gives a rough description of the shape of outer as well as inner boundaries of the structure. This motivates an integration of topology and design optimization, see, e.g., Refs. 15–18 and 43–46. In order to gain the full advantage of these design tools, it is necessary that they be integrated and implemented in a flexible, user–friendly, interactive CAD environment with extensive computer graphics facilities. This type of integration will be discussed in Chapter 3.

Depending on the amount of material available, the generated topology will basically either define the rough shape of a two–dimensional structural domain, possibly with macroscopic interior holes (which shape optimization procedures cannot create), or the skeleton of a truss– or beam–like structure with slender members. The main idea is that the optimum topology can be used as a basis for procedures of refined shape optimization by means of boundary variations techniques as discussed in Section 3.4 and exemplified in Section 4.3, or as a basis for optimization of truss member sizes and positions of joints by means of sizing optimization techniques as presented in Section 2.6.

An example of using topology optimization as a preprocessor for subsequent refined sizing and configuration optimization of a truss structure will be presented here.

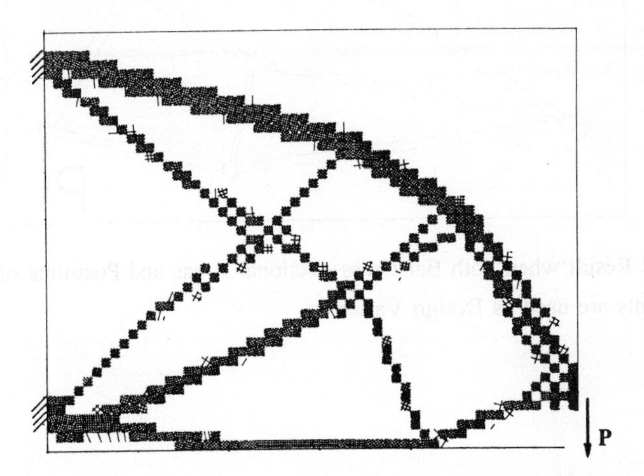

Figure 4. Solution of Topology Optimization Problem

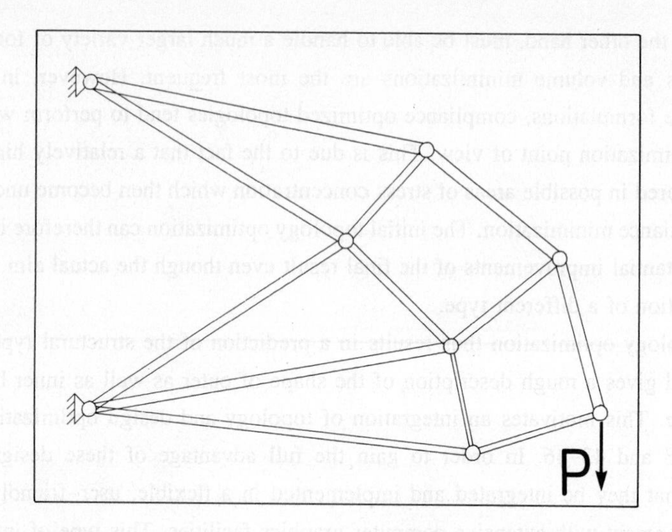

Figure 5. Truss Interpretation of Result in Figure 4 with Bar Areas Determined by Sizing for Fixed Positions of Joints

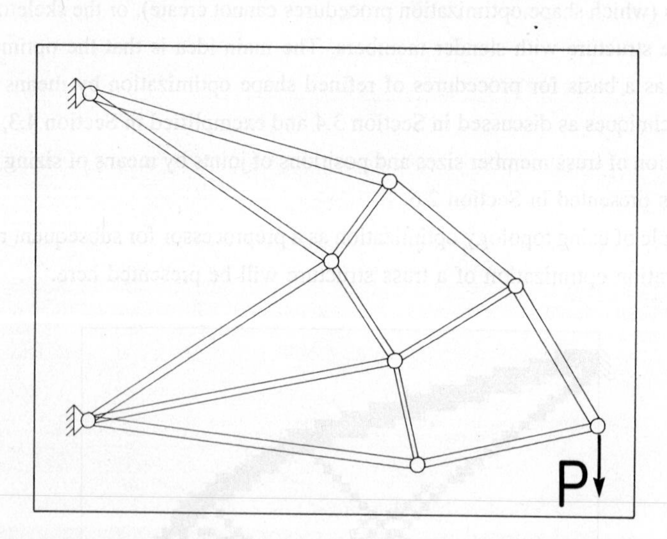

Figure 6. The Result when both Bar Cross-sectional Areas and Positions of Unrestrained Joints are used as Design Variables

Figure 4 presents the result of optimizing the topology of a structure within the rectangular design domain shown in the figure. The structure is required to carry a vertical load P, see Figure 4, and is offered support (displacement constraint) along the two hatched parts of the left hand side of the design domain. The domain is subdivided into 2610 finite elements, and the prescribed solid volume fraction is 16%.

The result is interpreted as a truss with the number and positions of joints shown in Figure 5. This figure also indicates the individual areas of the bars after a sizing optimization (weight minimization) subject to fixed positions of the nodal points and with the value of the compliance constrained to be equal to that obtained by the topology optimization.

Figure 6 shows the result of minimizing the weight at the same compliance value, but using both the bar cross-sectional areas and joint positions as design variables (only the horizontal movement of the load carrying joint and the downward movement of the lower-most joint are restrained). We now obtain a slightly different configuration and distribution of bar areas relative to the result in Figure 5, but the optimum weight/compliance ratio is very close to that obtained by the initial topology optimization (Figure 4).

2.8 Optimization of Fiber Composites

A capability for optimization of fiber composites developed by Thomsen, Ref. 47, is being integrated into the design system. The development pertains to optimization of linearly elastic fiber reinforced composite discs and laminates in plane stress, with variable local orientation and concentration of two mutually orthogonal fiber fields embedded in the matrix material. The thickness and the domain of the discs or laminates are assumed to be given, together with prescribed geometric boundary conditions and in-plane loading.

The problem consists in determining throughout the structural domain the optimum orientations and concentrations of the fiber fields in such a way as to maximize the integral stiffness (minimize the compliance) of the composite disc or laminate under the given loading. The optimization is performed subject to a prescribed bound on the total cost or weight of the composite that for given unit cost factors or specific weights determines the amounts of fiber and matrix materials in the structure.

The mathematical formulation of the problem is identical to that of the topology optimization problem, i.e. Eqs. (37)–(41), except for that the expression for the variable tensor of elasticity E_{ijkl} is different, and that a generalized form of the cost function in (41) is adopted.

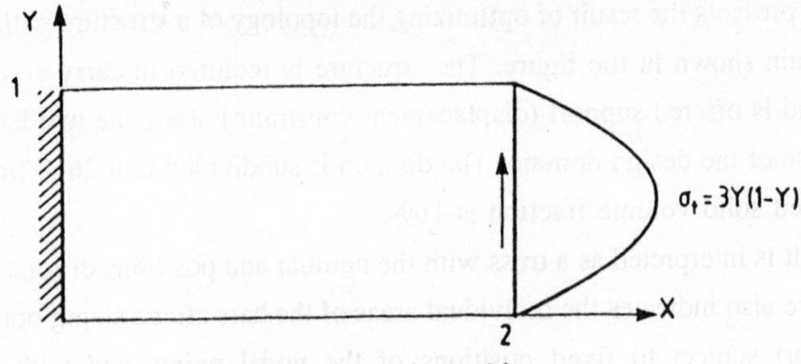

Figure 7. Example Problem

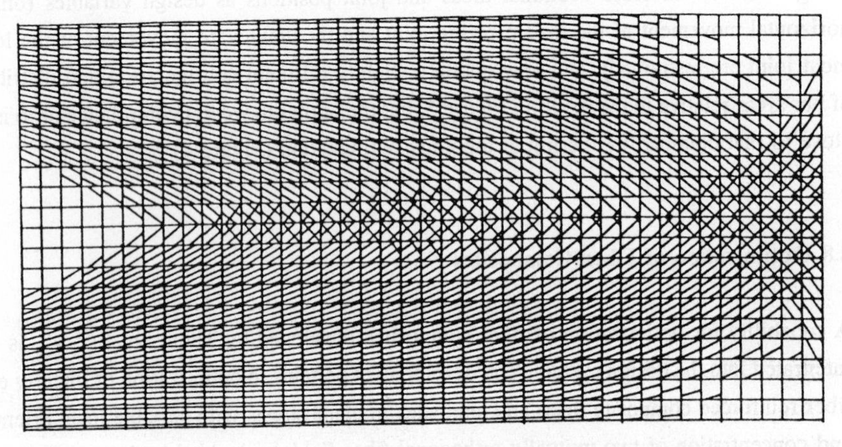

Figure 8. Optimum Distribution and Orientation of Fibers

As an example, consider a rectangular composite disc as indicated in Figure 7. The disc has one of its sides fixed against displacements in both the X and Y directions while the opposite side is subjected to a parabolically distributed shear loading as indicated in the figure.

The result of maximizing the integral stiffness of the disc for given total amounts of the fiber and matrix materials is shown in Figure 8. Here, the direction and density of the hatching (which is shown on top of the applied finite element mesh) illustrate the fiber orientation and concentration, respectively. As is to be expected, the solution contains no fiber reinforcement in lowly stressed sub–domains, only one–way fibers in sub–domains with dominance of a single principal stress component, and a cross–ply fiber arrangement in sub–domains with dominance of shear (i.e., two large principal stresses of opposite signs).

3 Integration of Optimum Design Facilities into a CAD Environment

This chapter presents a set of strategies and methods for integration of optimum design facilities into a traditional Computer Aided Design (CAD) system. The strategy of the work is to solve the optimum design problem using two parallel but completely distinct descriptions of the geometry: the *design model* and the *analysis model*. Furthermore, we seek to achieve full geometric versatility of these models independently of the CAD system's data structure, thus allowing for the integration with several different CAD systems.

The basic ideas of the method are presented and the capabilities of the prototype system CAOS and the more general system ODESSY are demonstrated.

3.1 Basic Ideas and Requirements

The basic idea of integrating optimum design facilities into a CAD environment is that in order to set optimization and rational design facilities at the disposal of the designer, they must be integrated parts of the concurrent engineering design environment.

In the design phase, the technology of Computer Aided Design (CAD) has gained a large influence in mechanical and construction engineering, so the designer is very familiar with this highly developed graphical user interface. From the designer's point of view it would be preferable to be able to perform all stages of design in this environment instead of having to transfer the geometrical information from the CAD environment to another Computer Aided Engineering (CAE) application.

The CAD technology has already been through a rapid development, but the systems of today are still considered as being only the first generation of a long row of increasingly integrated systems handling all stages of design and interfacing with Computer Integrated Manufacturing (CIM) systems. That is a major advantage as optimized designs often involve some rather complex geometries, calling for the use of numerically controlled machines. These CAD systems to come will offer an integrated environment for design, analysis, and manufacture of products of almost any character. Thus, future CAD systems will possibly be regarded as simply databases for geometrical information equipped with a number of tools with the purpose of helping the user in the design process. Among these tools will be facilities for structural analysis, sensitivity analysis and optimization as well as standard CAD features like drawing, modeling and visualization tools.

In short, the designer will be able to perform analysis and redesign of an initial geometry directly in the CAD environment where that geometry was originally defined.

In the case of mere analysis, this problem is close to having found its solution in traditional CAD systems expanding into the field of Computer Aided Engineering and by traditional Computer Aided Engineering systems being equipped with geometrical modeling facilities.

For optimization and redesign, however, the problems are much more involved because the designer will specify ways and limits of possible redesigns which the geometrical model and the analysis model must be able to cope with. The space of possible redesigns is bounded not only by mathematically well-defined criteria like "maximum stress" or "minimum stiffness" but also by considerations pertaining to the functionality and manufacturability of the final design. The system must allow for such constraints to be specified and maintained concurrently during the redesign process, and this requirement further amplifies the need for a highly developed CAD-type user interface.

The state of the art of structural optimization is that a large amount of mathematical techniques is available for the solution of single problems. By implementing collections of the available techniques into general software systems, operational environments for structural optimization have been created. The forthcoming years must bring solutions to the problem of integrating such systems into more general concurrent engineering design environments.

The result of this work should be CAD systems for rational design in which structural optimization is one important design tool among many others, and this chapter presents our strategies and methods for solving this integration problem.

3.2 The Structural Optimization System CAOS

A CAD based structural optimization system called CAOS (Computer Aided Optimization System) has been developed at the Institute of Mechanical Engineering, Aalborg University, Denmark, see Refs. 15-18. The purpose of this development has been to provide a basis for experiments with various solutions to the CAD integration problems outlined in the preceding section as well as to develop a framework where a number of different techniques for structural optimization can be investigated.

The main features of CAOS are:

- The commercial CAD system AutoCAD is used as the basis.
- The CAOS system concept is independent of the AutoCAD data structure.
- It is a fully operational shape design system.
- It has topology optimization facilities due to the integration with the HOMOPT system developed by Bendsøe and Kikuchi.
- It uses finite element modules for the structural analysis.
- It uses semi-analytical sensitivity analysis (see Section 2.3).

- It uses a bound formulation to formulate the multicriterion optimization problem (see Section 2.2).
- It uses mathematical programming techniques to solve the optimization problem.

The above mentioned features of CAOS are also contained in the more general design system ODESSY, which is our new optimization system based on the experiences gained from the development and use of CAOS. The ODESSY system will be described in Section 3.5.

CAOS has been under constant development over a period of four years and it contains the following features of CAD integrated shape optimization:

1. Due to the versatility of the bound formulation and the concept of problem variables, see Sections 2.1 and 2.2, there is a large number of possible formulations of the shape optimization problem. One may choose to minimize weight, stress, compliance, displacement or any other property that can be derived from the geometrical model or the output from an analysis program which in our case is a finite element code. The same set of possibilities is available for specification of constraints. Mathematically, these different formulations lead to very different optimization problems.

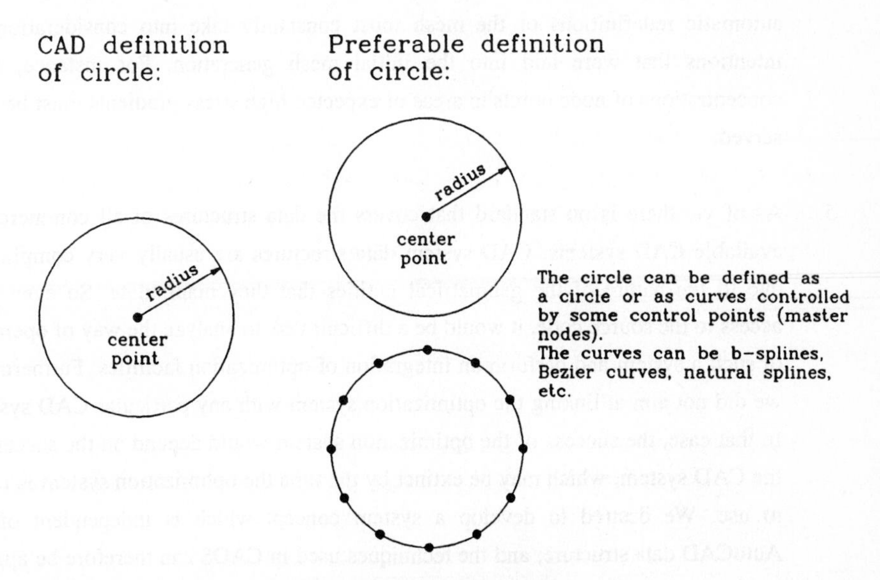
CAD definition of circle:
Preferable definition of circle:
radius
center point
radius
center point
The circle can be defined as a circle or as curves controlled by some control points (master nodes).
The curves can be b-splines, Bezier curves, natural splines, etc.

Figure 9. Definition of Circle

2. In order to use a mathematical programming technique to solve the problem, the continuous shape of the geometry must be described by a finite, preferably small,

number of generalized design variables as discussed in Section 2.5. This problem is closely connected with the data structure of the CAD system which is usually not flexible enough to allow for the shape changes required by the optimization module, as CAD systems seek to store the information at the "highest possible level" in order to minimize the amount of data. For instance, in a CAD system, a circle is defined uniquely by its radius and the coordinates of its center point. In connection with a general optimization system, this data storage scheme has a serious drawback: the circle is bound to remain a circle. It cannot with this data structure be turned into another curve type. Thus, the interface to the optimization system must provide for a conversion of CAD data into a form more convenient for shape optimization. The circle example is shown in Figure 9.

3. Dimensions and shapes that are crucial to the functionality or fabrication can be maintained during the shape optimization process.

4. The initial CAD model of the structure can be converted into an *analysis model*, eg., a finite element model, whose mesh must conform to the changes of the geometry as the optimization process progresses. It is not acceptable for each iteration to call upon the designer to perform a new mesh generation manually. On the other hand, automatic redefinitions of the mesh must constantly take into consideration the intentions that were laid into the initial mesh generation. For instance, local concentrations of node points in areas of expected high stress gradients must be preserved.

5. As of yet there is no standard that covers the data structures of all commercially available CAD systems. CAD system data structures are usually very complicated due to the nature of the geometrical entities that they manipulate. So even with access to the source code, it would be a difficult task to analyze the way of operation of such a system and perform an integration of optimization facilities. Furthermore, we did not aim at linking the optimization system with any particular CAD system. In that case, the success of the optimization system would depend on the success of the CAD system, which may be extinct by the time the optimization system is ready to use. We desired to develop a system concept which is independent of the AutoCAD data structure, and the techniques used in CAOS can therefore be applied in connection with most other CAD systems as well.

6. The geometrical information is automatically updated and interchanged during and after the optimization process.

3.3 Design Model and Analysis Model

CAOS is based on the important distinction between *design model* and *analysis model*. The design model is a variable description of the shape of the structure and it is closely connected with the CAD model and totally distinct from the finite element model that is used for the analysis. The design model consists of so-called *design elements* as presented by Braibant & Fleury in Ref. 4.

In the early days of structural shape optimization system design, attempts were made to use the finite element model directly as the design model, i.e., to use the node coordinates as design variables (see, e.g., Rodrigues, Ref. 40). It turns out that this method has at least four serious drawbacks:

1. The number of design variables can become very large.
2. It is difficult to ensure compatibility and slope continuity between boundary nodes.
3. It is difficult to maintain an adequate finite element mesh during the optimization process.
4. The structural shape sensitivities might not be accurate unless high order finite element types are used.

These results were known when the development of CAOS was initiated, and our investigations have indicated that the concept of design elements provides an acceptable answer to most of the problems mentioned above. The approach used in CAOS is based on a sub-division of the geometry into a number of topologically quadrilateral design elements in case of a 2D structure, and topologically hexahedral design elements in case of a 3D structure. These design elements have a number of attractive features:

1. Design elements lend themselves to a very easy mesh generation with quadrilateral and hexahedral shaped finite elements. A number of randomly placed nodes on the boundaries is the only input needed for a complete mesh generation within the design element. Thus, automatic generation of an analysis model based on the current shape of the design model is achieved. Two examples of this are shown in Figures 10 and 11.

   The method used for creating the finite element mesh is an isoparametric mapping technique as illustrated in Figure 12.

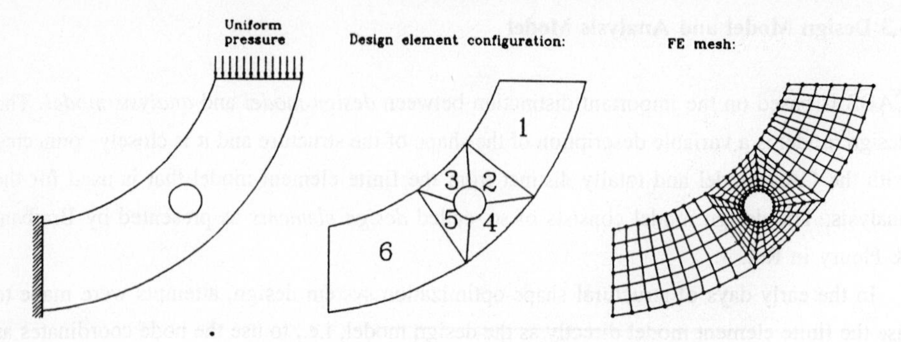

Figure 10. 2D Model of a Mechanical Component

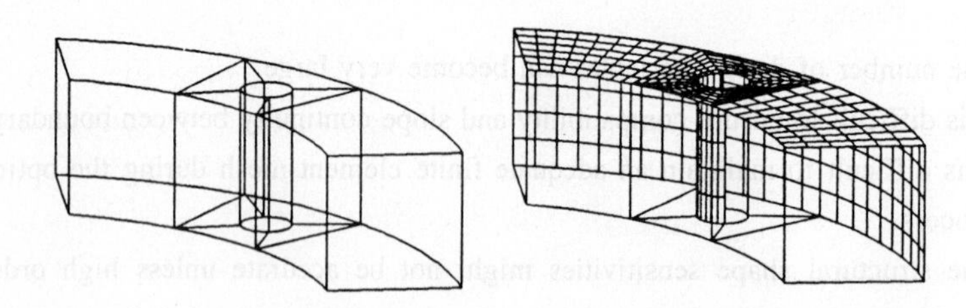

Figure 11. 3D Model of the Mechanical Component shown in Figure 10

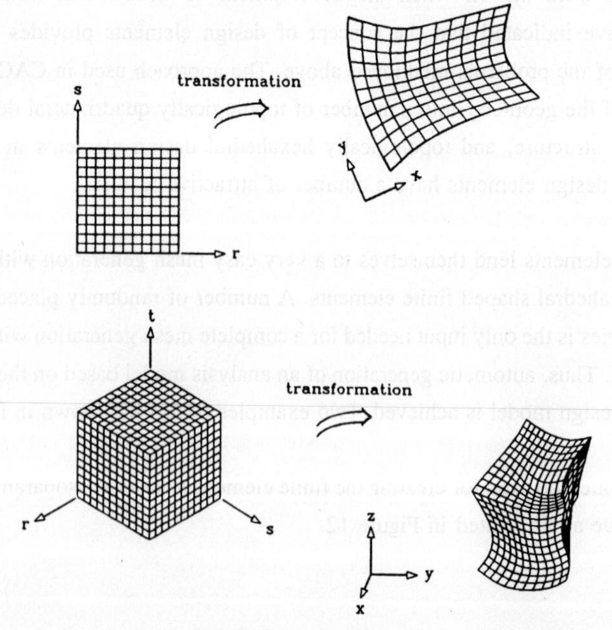

Figure 12. Isoparametric Mapping Technique in 2D and 3D

2. The boundaries of the design elements can be curves of almost any character, i.e., piecewise straight lines, arcs, b–splines with specified degree of continuity, bezier curves, etc. It is therefore very simple to generate relatively complicated geometries with a small number of design elements. Furthermore, geometrical requirements to the final shape are easily specified by assigning specific curve types to the boundaries in question.

3. The shapes of the boundaries are controlled by a number of master nodes, cf. Section 2.5. In case of defining a shape optimization problem, this creates an evident connection between the generalized design variables (the movements of the master nodes) and the shape of the geometry. The direction and size of the movement of each master node is constrained to follow some predefined move directions specified by the designer (see Section 2.5). This gives the designer further possibility to control the outcome of the optimization process in order to meet functional or fabricational requirements. The finite element analysis model is automatically updated as the modified positions of the master nodes produce improved geometries. A variable model of the component shown in Figure 10 can be seen in Figure 13, where the arrows represent the move directions.

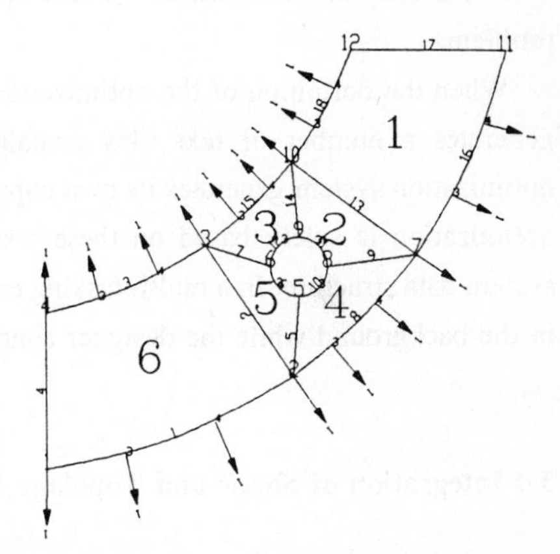

Figure 13. Variable Design Model

4. With the drawing aids of the CAD system it is very easy to draw the design elements in a separate drawing layer on top of the original drawing.

5. Design elements provide a way of liberating the geometrical description from the data structure of the CAD system as illustrated by the circle example in Figure 9.

6. Full geometric versatility of the design model is achieved. Through changes of the boundary shapes, design elements have the capability to represent several different shapes with the same element configuration.

The connection between the CAD system and the optimization system is as follows: Design elements are drawn on top of the original geometry. This is performed with a number of predefined geometrical entities (named "blocks" in most systems). These entities have attributes assigned to them. Attributes are variable numbers or text strings that describe the entity. This is a very common facility which is normally used to store prices, types, dimensions, etc., of standard components on a drawing. The CAD system automatically scans the drawing for attributes of a given type and generates a text file containing the data. In this particular case, the attributes describe curve types, number of finite element nodes, state variables etc.; in other words, all the specifications necessary to define an optimization problem.

When the definition of the optimization problem is completed, the system automatically generates a number of text files containing the information. Based on these files, the optimization system generates its own copy of the geometry and starts optimizing. The actual optimization is solely based on these text files and is therefore independent of the CAD system data structure. In a multi-tasking environment the optimization process can be started in the background while the designer continues working with the CAD-system.

3.4 Integration of Shape and Topology Optimization

The design models of CAOS can be subjected to either shape or topology optimization.

The optimization process is initiated by inserting the necessary specifications, ie., master nodes, curve types, design element definitions, loads, boundary conditions etc., into the CAD model. This is all done interactively using the drawing and visualization facilities already available in the CAD system. This also means that the specifications can be erased, moved, redefined or otherwise modified as any other entity in the CAD model.

In the case of a shape optimization problem the generalized design variables identify the movements of the master nodes as mentioned earlier, and the designer defines a variable design model by assigning move directions with prescribed upper and lower limits on the allowable movement, to the variable master nodes. The outcome of the shape optimization process can be controlled by the designer in three ways:

- Design boundaries with a predefined function can be forced to take on a certain shape as, e.g., an arch or a piecewise straight line.
- If the designer desires a boundary to locally or globally maintain its shape, one or several of the master nodes can be fixed. This is accomplished simply by not assigning the particular master node any design variables.
- The move directions can be joined together, i.e., linked, in order to fulfil specific symmetry or shape requirements.

When the variable design model has been defined in the CAD environment, a number of simple text files specifying the optimization problem are generated, and from these files the optimization system generates a finite element analysis model. Then the shape optimization is performed automatically by using the semi-analytical sensitivity analysis as outlined in Section 2.5, the bound formulation for the definition of the mathematical programming problem (see Section 2.2) and the SIMPLEX algorithm as an optimizer.

The topology optimization is performed by the HOMOPT system developed by Bendsøe and Kikuchi, Ref. 37, which represents an important break–through in structural optimization. As discussed in Section 2.7, the topology optimization minimizes the compliance of the structure subject to an upper bound on the available volume, and HOMOPT works directly on the finite element analysis model generated by CAOS. Furthermore, CAOS has been equipped with facilities for visualization of the optimized topologies that come out of HOMOPT. The examples in Section 2.7 and in Section 4.3 show how the user, dependent on whether the resulting topology is a skeleton–like or continuum–like structure, can continue with refined sizing or shape optimization. This way, the topology optimization can be used as a preprocessor for subsequent sizing or shape optimization. For further information on the homogenization method, the reader is referred to Bendsøe, Ref. 38, and to Refs. 41–46.

3.5 Fully Parametrical Design Model (the ODESSY System)

The design model of CAOS was much inspired by the works of Braibant & Fleury and others (Refs. 3–8) and reflected the desire to perform shape optimization using a mathematical programming algorithm. Thus, the method of controlling shape by the use of master nodes as discussed in Section 2.5, solves the problem of parametrizing the geometry for general shape optimization.

In practical design, shapes of the final products are very often limited by the interaction of the component in question with its surroundings and/or by fabricational considerations. Modern solid modeling systems reflect this situation by allowing the user to build models from predefined simple geometrical shapes like boxes, cylinders, cones, etc., as well as more freely shaped volumes. Thus, in order to integrate optimization facilities with contemporary modeling systems, we need to provide a design model that enables description of the geometry by the same simple parameters as used by the CAD system, i.e., radii, heights, angles of rotation, etc. Furthermore, the design model must be so well–constructed and self–contained that changes of the parameters of the model will automatically lead to a full update of all entities depending on the parameter in question.

The **Optimum Design System** (ODESSY) is a new development founded on the experiences of CAOS but with a more general design model allowing for precisely the type of geometrical description explained above. The following is a short description of the data structure underlying the geometrical description within ODESSY.

The design model of ODESSY is built in a strictly hierarchical way inspired by solid modeling techniques. Data is separated into three levels depending on complexity:

Level 0 is the "ground" level. This level contains geometrical entities which are independent of other entities in the model. This would typically be points, vectors and scalar numbers. In CAOS, master nodes would be level 0 entities. Level 0 is characterized by the property that it is the only level containing real numbers.

Level 1 contains entities which depend only on level 0 information. The boundary curves of CAOS are a typical example. Their shapes are defined solely by the position of the master nodes. This level could contain

- circles
- curves
- boxes
- tetrahedrons
- cylinders
- cones
- etc.

The actual shapes of all entities in level 1 are defined by pointers to level 0 information. Level 1 thus consists only of integer values.

The expansion of geometrical facilities from the curves of CAOS to the set of geometrical entities of level 1 represents a drastic improvement of the modeling capabilities of the system. However, for practical purposes, it is very often necessary to be able to define geometries by boolean operations between predefined, so-called primitives like the entities mentioned above. For instance, the union of two cylinders as shown in Figure 14 contains an intersection curve that depends on the dimensions and relative positions of the two cylinders. Thus, if the dimensions of the cylinders are used directly as design variables, we cannot model the intersection curve independently.

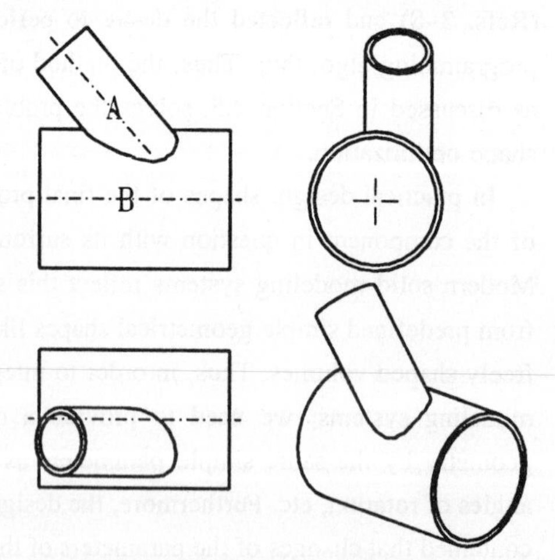

Figure 14. Fusion of Two Cylinders

For purposes like this we need a **Level 2** which is reserved for more complex entities resulting from interaction between other entities.

The parametrization of the geometry is obtained through the subdivision of the information into these three levels. We simply control the geometry by modifying the entities of level 0. Because of the hierarchical construction of the model, modifications of level 0 entities automatically lead to corresponding changes in levels 1 and 2.

In order to improve the possibility of limiting the design space, data in level 0 are controlled by the use of so-called **modifiers**. CAOS, for instance, contains only one type of modifier, namely the move directions which may also be thought of as translation transformations. Other possible modifiers may be point scaling, line scaling, and rotation. This way, level 0 entities may be subjected to one or several transformations thus controlling the overall design. The **design variables** of the problem will then simply be the amount of transformation, i.e. the rotation angle or the scale factor.

In this framework, the move directions of the CAOS system are realized as translation modifiers coupled with generalized design variables which specify the magnitude of the translation. Each master node is level 0 information, and the entire geometry therefore changes with relocation of the master nodes.

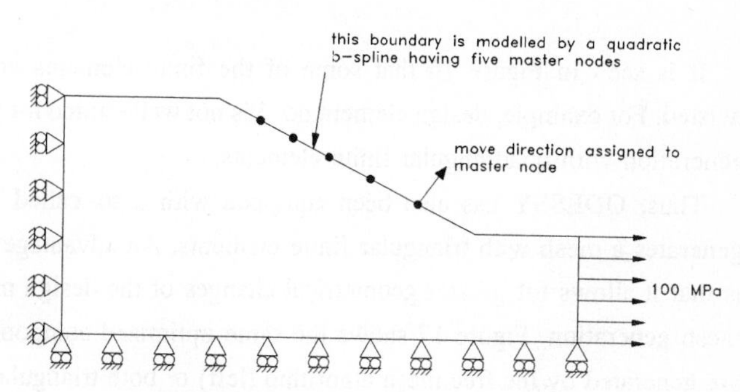

Figure 15. ODESSY Definition of the Well-known Fillet Problem

3.6 Mesh Generation

The mapping mesh generation method of CAOS forced the user into subdividing the geometry into quadrangular design elements. For most geometries this is not a disadvantage, but in some cases, mainly in connection with overall triangular shapes, it may cause problems. This is illustrated by the example shown in Figure 16, where the mechanical component from Figure 10 has been shape optimized.

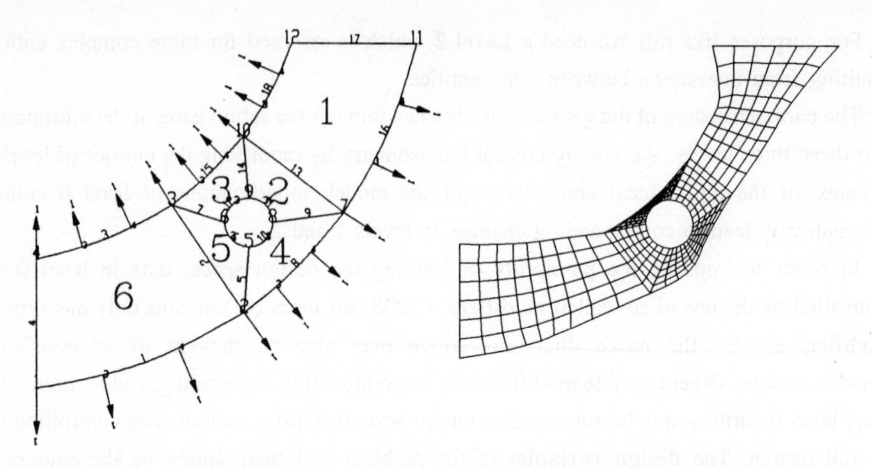

Figure 16. Possible Result of Shape Optimization

It is seen in Figure 16 that some of the finite elements are very distorted and nearly twisted. For example, design element no. 3 is not well-suited for isoparametric mapping mesh generation with quadrangular finite elements.

Thus, ODESSY has also been equipped with a so-called free mesh algorithm, which generates a mesh with triangular finite elements. An advantage of the free mesh generation is that it allows for greater geometrical changes of the design model than does the mapping mesh generation. Figure 17 shows the same optimized component when either all elements are generated by the free mesh algorithm (left) or both triangular and quadrangular elements have been used (right). The quality of the finite element mesh is maintained during the redesign process when this algorithm is used.

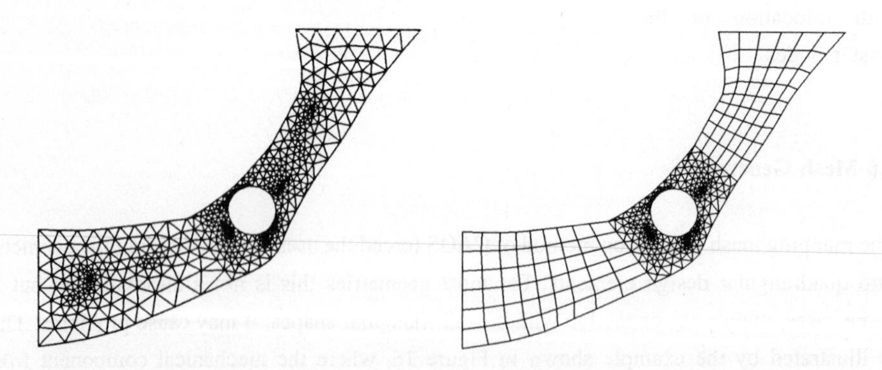

Figure 17. Optimized Design when the Free Mesh Algorithm has been used

The free mesh algorithm allows for automatic mesh generation of any simply connected domain described by a set of boundaries. Therefore, instead of using six design elements for the design model description of the mechanical component, only one is necessary if we use the free mesh algorithm as shown in Figure 18.

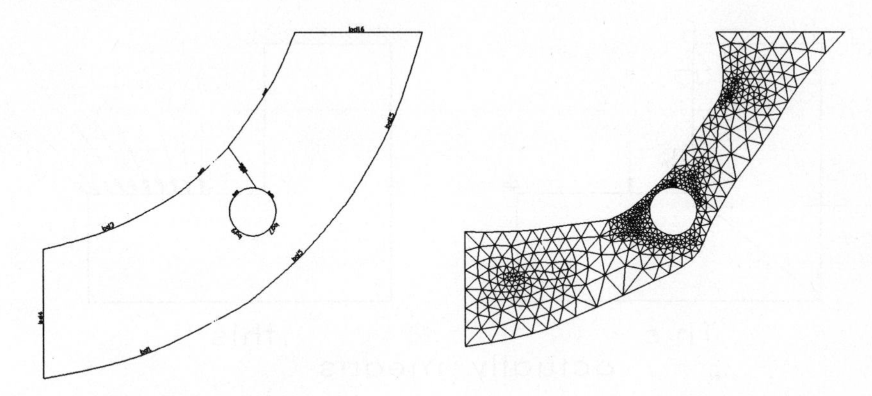

Figure 18. Possible Design Element Configuration and Final Design when the Free Mesh Algorithm is used

The free mesh facilities are currently only available in 2D, but a 3D extension is being developed.

4 Examples

The following examples are extracts from Ref. 17 which contains descriptions of a number of problems solved with CAOS. The examples are all practical real-life problems.

The optimization of the subproblem at each iteration is performed by sequential linear programming (SLP) based on a SIMPLEX subroutine. Other more sophisticated optimizers have previously been used, but extensive tests have shown that SLP provides stable results for all problems if a proper move-limit strategy is employed. The examples presented here have all been subjected to an initial move-limit of 10% of the range of each design variable in the first three iterations. In subsequent iterations, the move-limit is adjusted according to the iteration history for each variable such that oscillating behavior leads to a tightening of the move-limit and stable convergence gives more slack to the variable in question. This strategy has worked for all problems and there has been no attempt to evaluate the optimum initial move-limit or alternative adjustment strategies for particular problems.

The problems are presented by means of figures which are in most cases direct hardcopies of the screen when working with the system. These figures should give a feeling of working with the system. In particular, the system's way of depicting boundary conditions and loads require a little explanation. When fixing or loading a boundary segment, CAOS draws corresponding icons at the two end points of the segment even though the condition actually applies to all points between the two end points as shown in Figure 19.

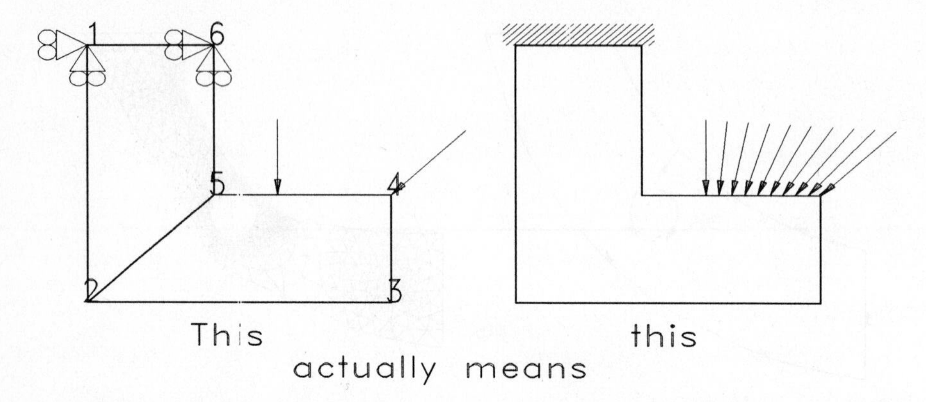

Figure 19. Icons Representing Boundary Conditions and Loads and Their Interpretation

The result of optimization is in most cases presented by drawings of the finite element models and/or stress fields for the initial and final designs, together with iteration histories for design criteria and constraints. The actual values of the design variables and sensitivities at each iteration constitute too much information to be presented here. All stress fields, finite element geometries and other results shown in this article are produced by the CAOS postprocessor, CAOSPOST. Unfortunately, the reproducible black and white pictures presented here are not doing justice to the colourful plots which CAOSPOST can produce on the screen or a plotter.

4.1 Turbine Wheel

In this example we seek to optimize the shape of the turbine wheel in Figure 20. The wheel is a part of a turbocharger.

Two different formulations of the problem are attempted:

Minimize the mass moment of inertia
Subject to an upper bound on the von Mises stress of 524 MPa.

Minimize the volume
Subject to an upper bound on the von Mises stress of 524 MPa.

Figure 21 shows a section in the turbine wheel with the initial dimensions.

The fast rotation of the wheel calls for inclusion of the volume centrifugal forces in the problem. CAOS performs this calculation automatically and takes the change of forces with changing geometry into account in the solution of the problem.

We consider only the shape of the turbine wheel itself and therefore replace the blades by a uniformly distributed load on the outer boundary of the wheel. This load is an approximation of the centrifugal forces from the blades.

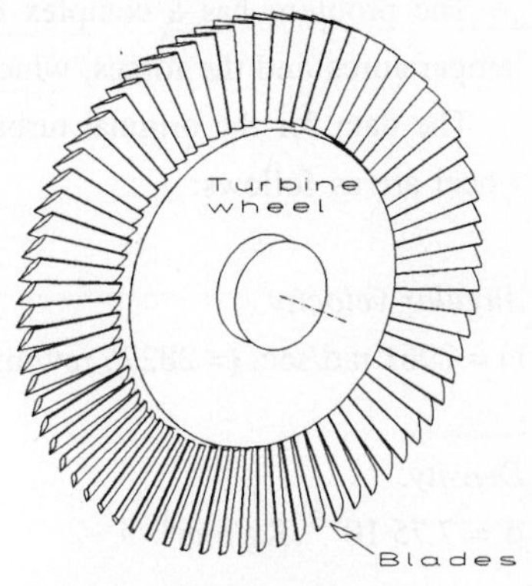

Figure 20. Turbine Wheel

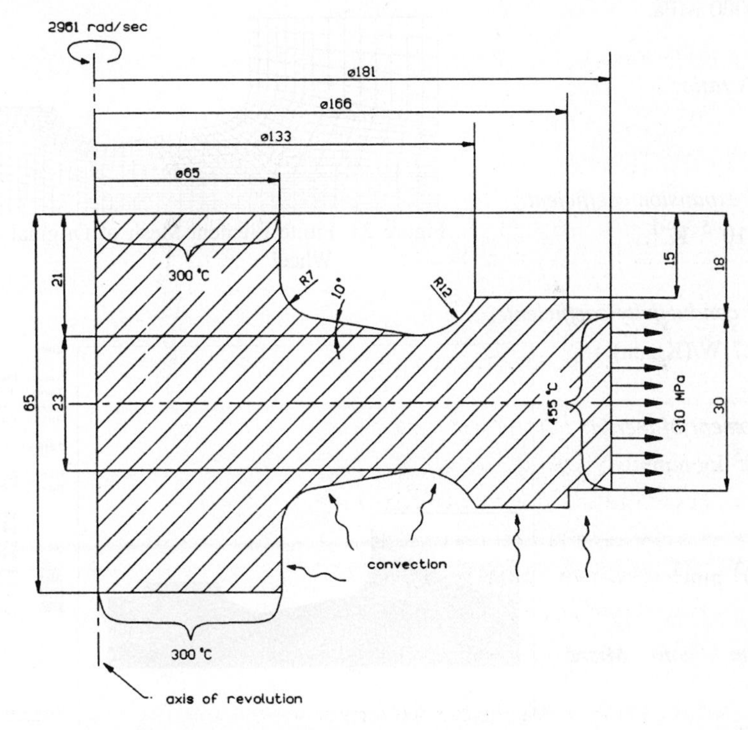

Figure 21. Dimensions of Original Turbine Wheel

Furthermore the turbine wheel is influenced by temperatures (also shown in Figure 21). These temperatures derive from the hot exhaust gas which drives the wheel. Part of the boundary is subjected to a convection condition, which also is due to the exhaust gas. At these boundaries the environmental temperature is specified to 455 °C and the convection coefficient to 0.0005 W/(mm^2·K).

The problem has a complex character, because the stresses depend on the design, the temperatures and the forces, which again depend on the design.

The data for the original turbine wheel are as follows:

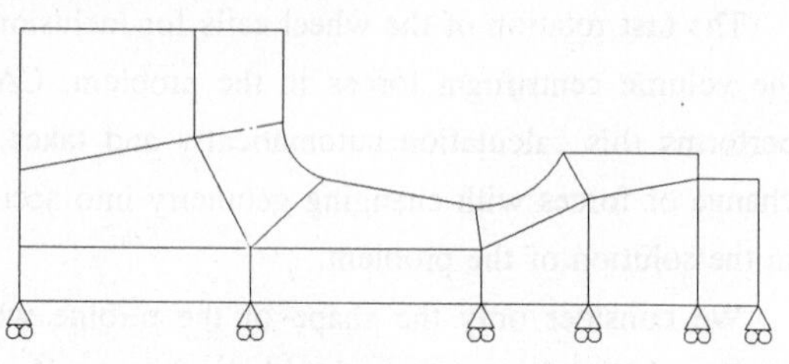

Figure 22. AutoCAD Definition of the Problem

Angular velocity:
$\omega = 2961$ rad/sec. (= 28275 rev/min)

Density:
$\rho = 7.75 \cdot 10^{-6}$ Kg/mm^3.

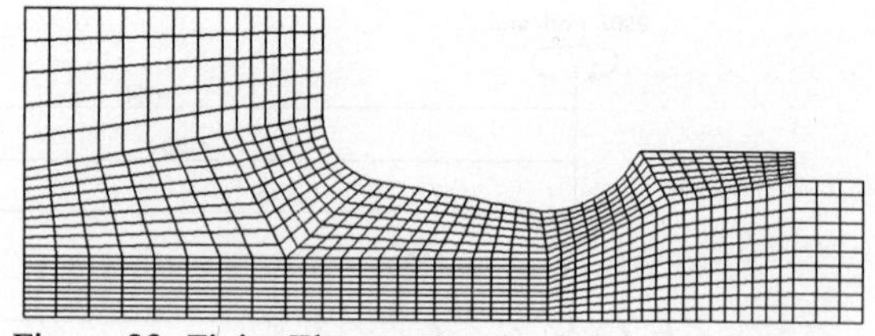

Figure 23. Finite Element Mesh of Original Turbine Wheel

Young's modulus:
E = 180000 MPa.

Poisson's ratio:
$\nu = 0.3$.

Thermal expansion coefficient:
$\alpha = 1.2 \cdot 10^{-5}$ K^{-1}.

Thermal conductivity coefficient:
k = 0.027 W/(K·mm).

Mass moment of inertia:
$13.41 \cdot 10^3$ kg·mm²

Volume:
$451.9 \cdot 10^3$ mm^3

Maximum von Mises stress:
806 MPa

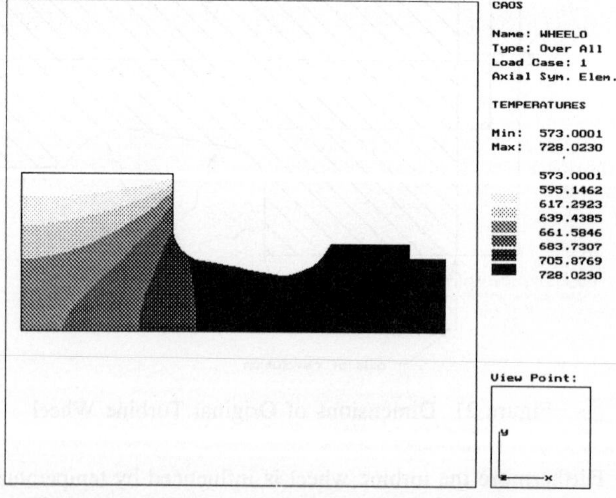

Figure 24. Temperature Field in Original Turbine Wheel

The analysis of the initial geometry reveals that the stress constraint is violated by 54%. The initial temperature and stress fields are shown in Figures 24 and 25.

The initial geometry is unnecessarily complicated as a basis for the shape optimization process, and we therefore define a new model for this purpose as shown in Figure 26. This model consists of 432 4-node isoparametric elements for axial symmetry.

The iteration histories for mass moment of inertia and volume are shown in Figures 27 and 28, respectively. The final finite element mesh, temperature and stress fields in the case of mass moment of inertia minimization are illustrated in Figures 29, 31, and 32.

In the case of minimization of mass moment of inertia, the final data are the following:

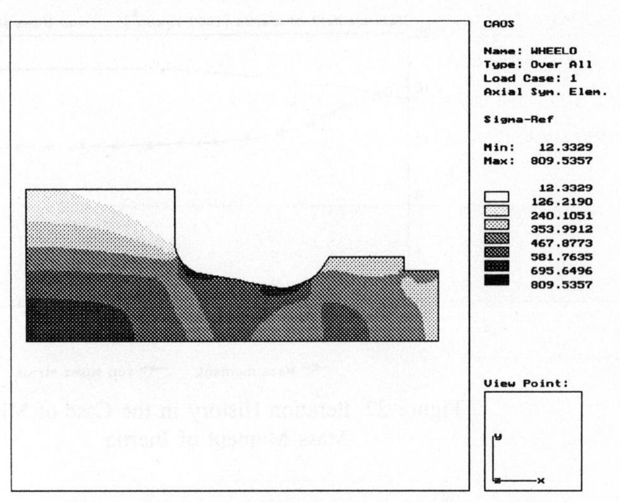

Figure 25. Stress Field in Original Turbine Wheel

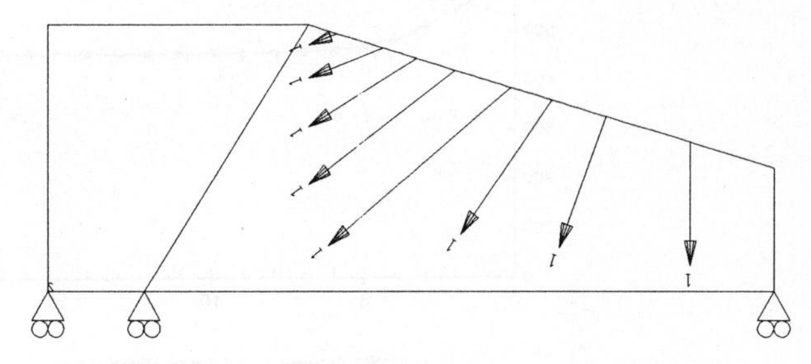

Figure 26. AutoCAD Definition of the Problem

Mass moment of inertia:
$11.74 \cdot 10^3$ kg·mm²
Maximum von Mises stress:
524 MPa

In the case of volume minimization the following result is achieved:

Volume:
$454 \cdot 10^3$ mm³
Maximum von Mises stress:
524 MPa

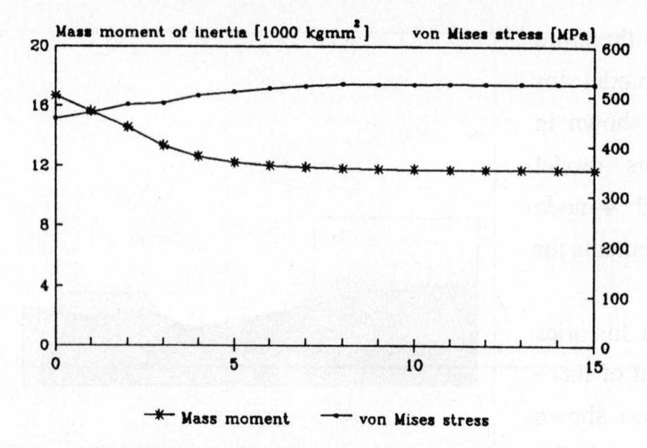

Figure 27. Iteration History in the Case of Minimization of Mass Moment of Inertia

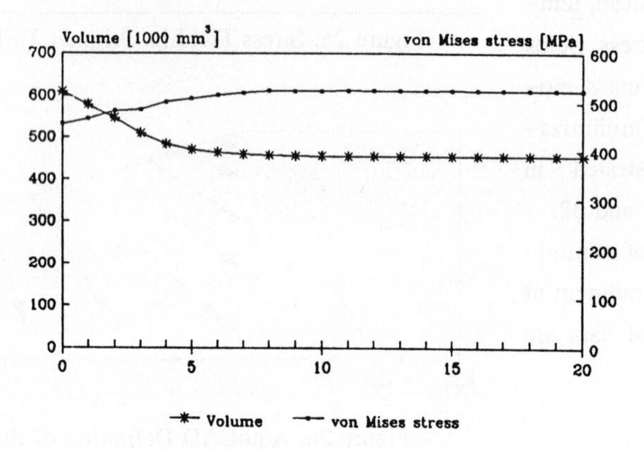

Figure 28. Iteration History in the Case of Minimization of Volume

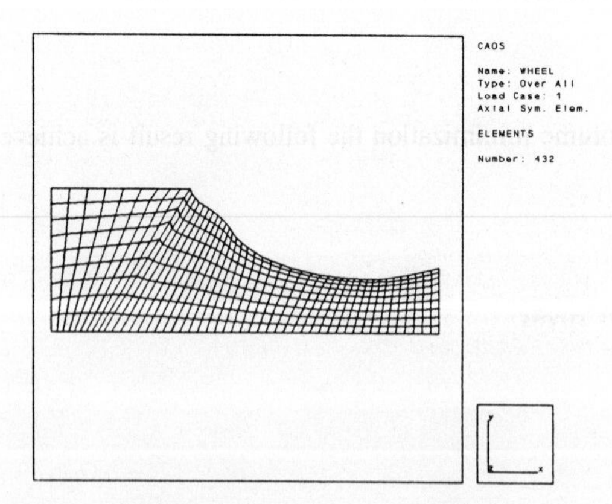

Figure 29. Final Finite Element Geometry in the Case of Minimization of Mass Moment of Inertia

The geometric results in the two cases are very similar. Figure 30 shows a comparison with the volume optimized geometry shown by dotted lines.

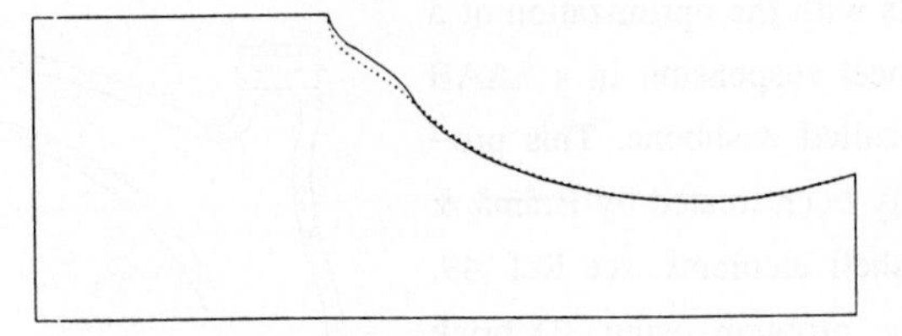

Figure 30. Comparison of Final Geometries.

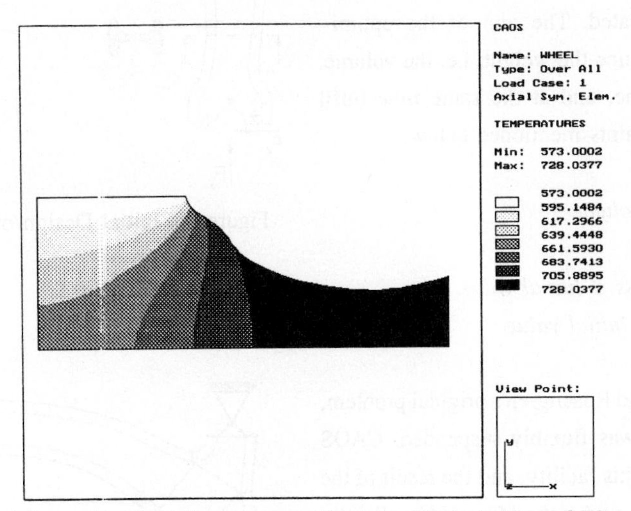

Figure 31. Final Temperature Field for Minimization of Mass Moment of Inertia

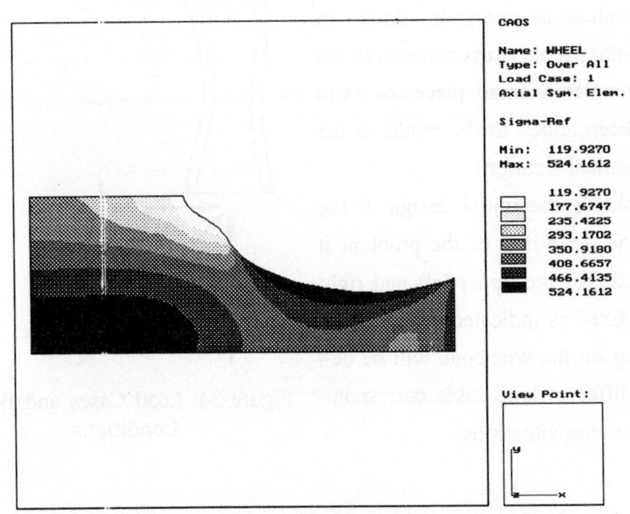

Figure 32. Final von Mises Stress Field for Minimization of Mass Moment of Inertia

4.2 Wishbone

This example deals with the optimization of a part of a front wheel suspension in a SAAB 9000 car, the so–called wishbone. This problem has previously been treated by Bråmå & Rosengren using shell elements, see Ref. 49. We shall solve the problem using 3D brick elements, and using a design model which is rather complicated. The aim of the optimization is to reduce the weight, i.e. the volume, of the wishbone, and at the same time fulfil the two constraints mentioned below.

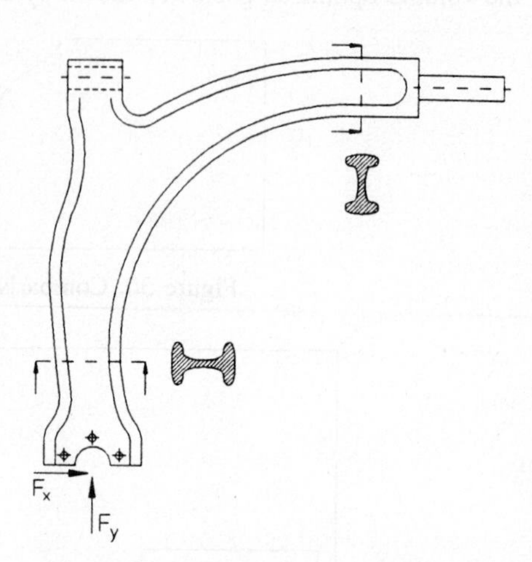

Figure 33. Initial Design of Wishbone

Minimize the volume
Subject to
von Mises stress ≤ Initial value.
Compliance ≤ Initial value.

In Bråmå and Rosengren's original problem, the wishbone was flexibly suspended. CAOS does not have this facility, and the result of the two optimizations can therefore not be directly compared. This is also the reason why we choose initial values as constraint limits. In order to avoid undesirable stress concentrations in the actual geometry, small pieces of extra material have been added to the model at the points of mechanical fixation.

Figure 33 shows the initial design of the wishbone. In the definition of the problem it will be assumed that the upper left and right boundaries are fixed as indicated in Figure 34. The loads acting on the wishbone will be defined in three different load cases corresponding to the following situations:

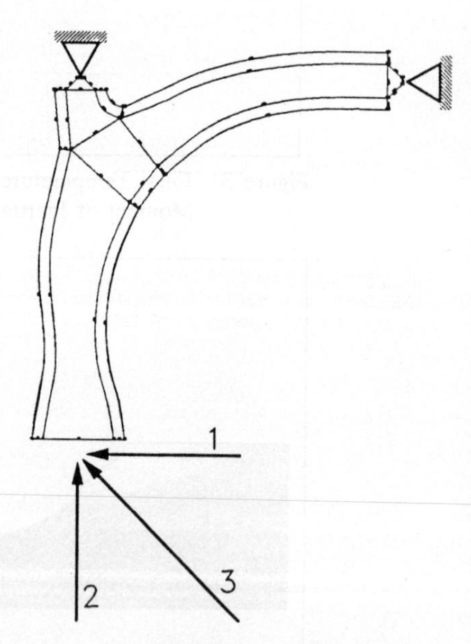

Figure 34. Load Cases and Boundary Conditions

1. Maximum straight line braking, i.e., F_x is non zero.
2. Maximum lateral acceleration, i.e., F_y is non zero.
3. Maximum combined braking/lateral acceleration, i.e., both F_x and F_y are non zero.

Figure 35 illustrates the rather complex CAD model. The geometry is modeled by 202 boundaries, 30 design element and 308 move directions. The number of independent design variables is reduced to 112 by linking some of the move directions. It is evident that the user has to rely heavily on good interactive graphical facilities to maintain the overview of a model as complicated as this. The lengths of the move directions in Figure 35 also indicate the geometrical constraint that, due to physical limitations, the wishbone has to remain within a predetermined space.

Figure 35. Analysis Model including Design Elements and Move Directions

The wishbone is made of aluminum and has the following material parameters:

Young's modulus: 71000 MPa.
Poisson's ratio: 0.3.

The initial finite element mesh is illustrated in Figure 36 and is modeled by 776 8-node isoparametric solid elements. The corresponding von Mises stress fields for the three load cases are shown in Figures 37 – 39.

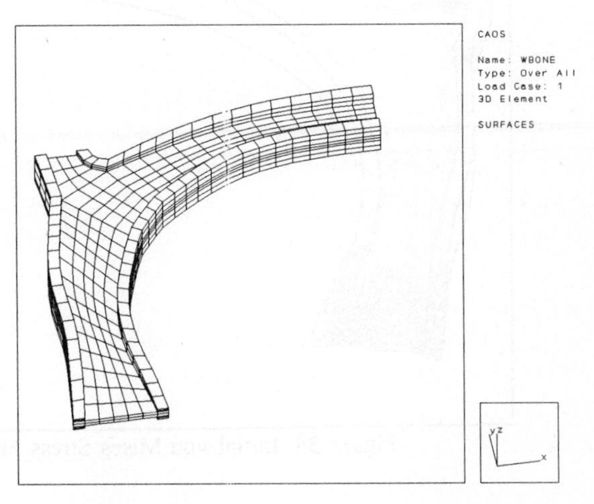

Figure 36. Initial Finite Element Model

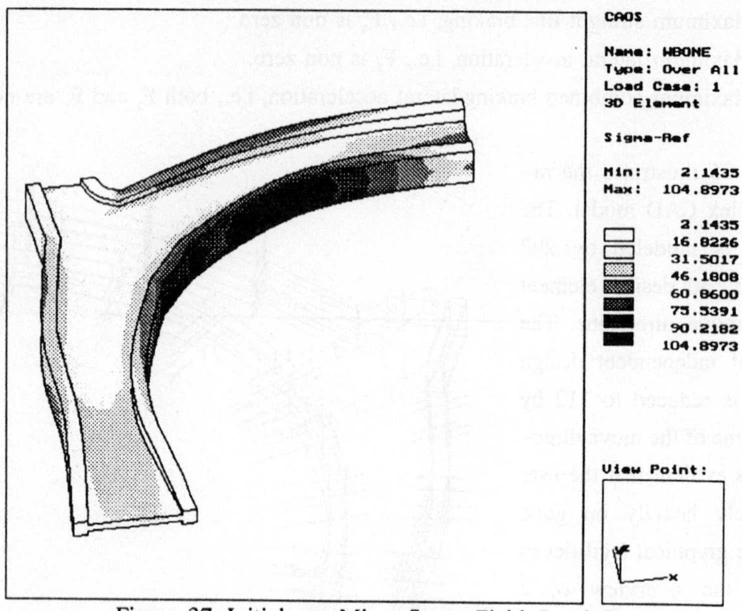

Figure 37. Initial von Mises Stress Field, Load Case 1

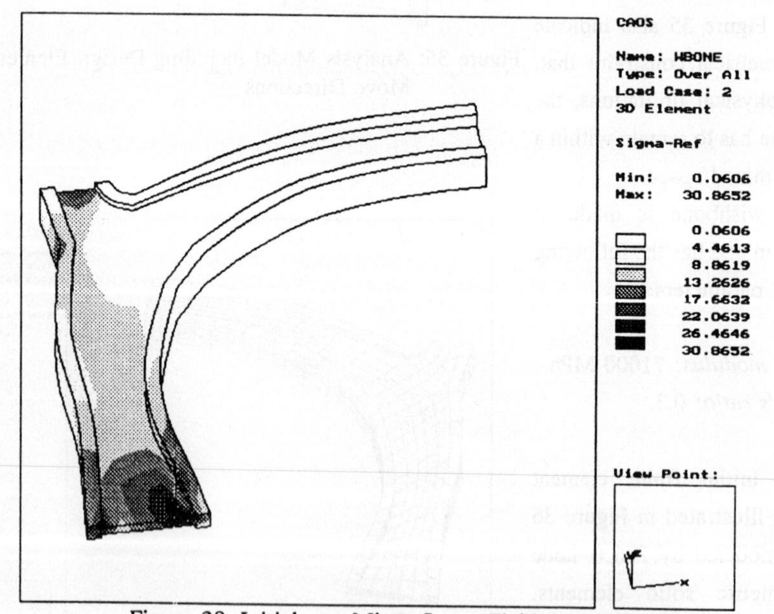

Figure 38. Initial von Mises Stress Field, Load Case 2

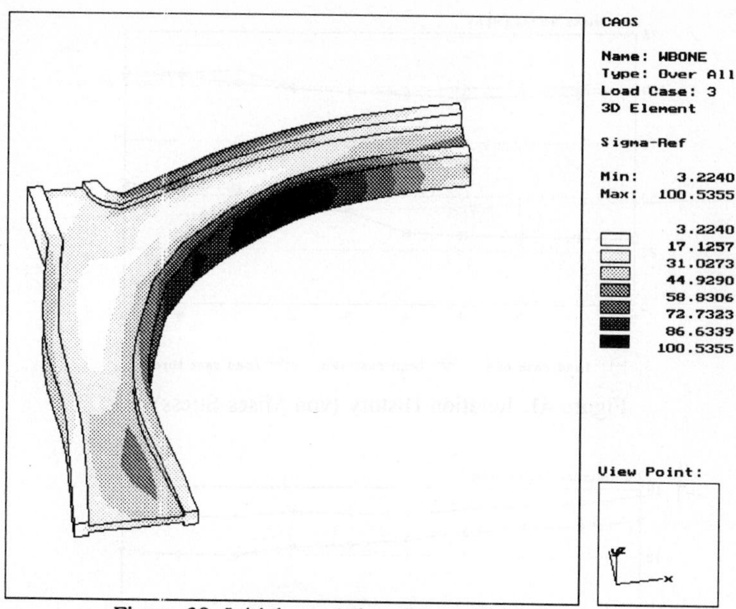

Figure 39. Initial von Mises Stress Field, Load Case 3

The iteration history proceeds steadily (Figures 40 – 42) and leads to a considerable volume reduction of 40%. The final finite element model, which is feasible for all load cases, is shown in Figure 43. In spite of the considerable volume reduction, the visual geometrical alterations are rather small. The final stress fields are illustrated in Figures 44 – 46. It is evident that the optimization has caused a much smoother stress distribution in the two critical load cases 1 and 3.

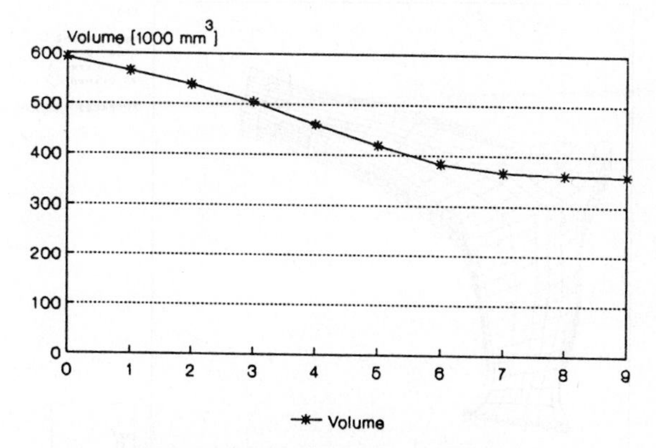

Figure 40. Iteration History (Volume)

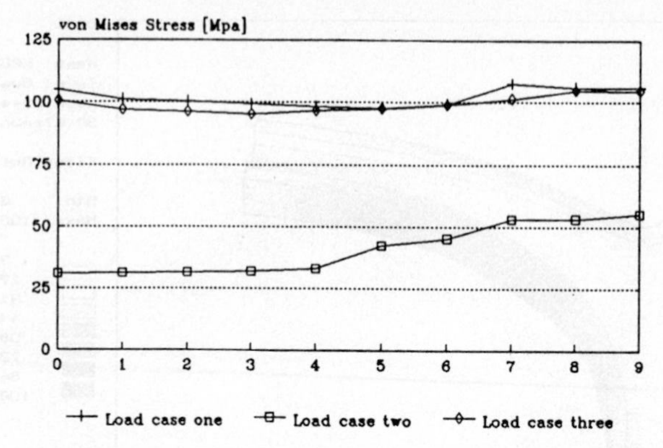

Figure 41. Iteration History (von Mises Stress)

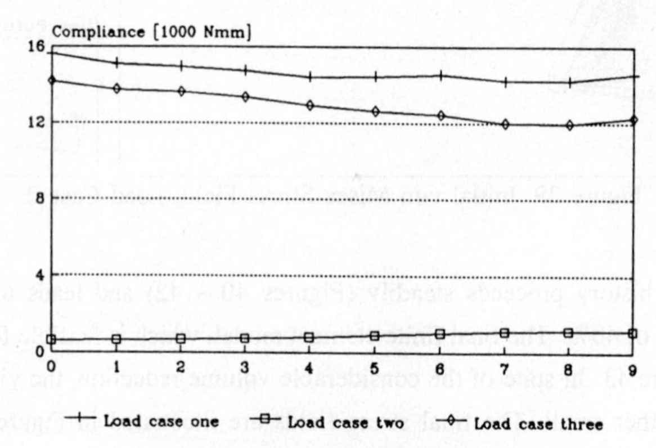

Figure 42. Iteration History (Compliance)

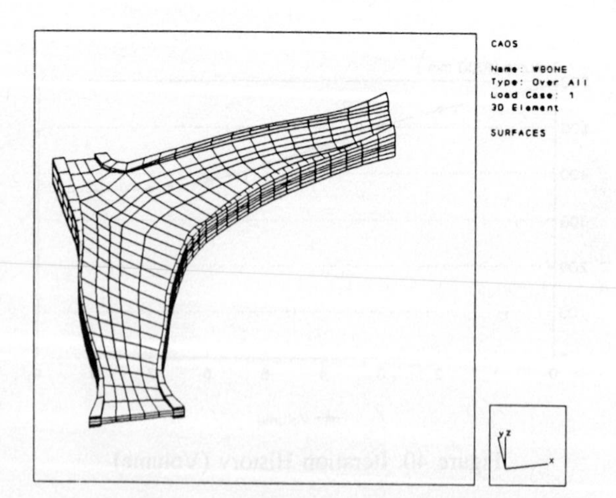

Figure 43. Final Finite Element Mesh

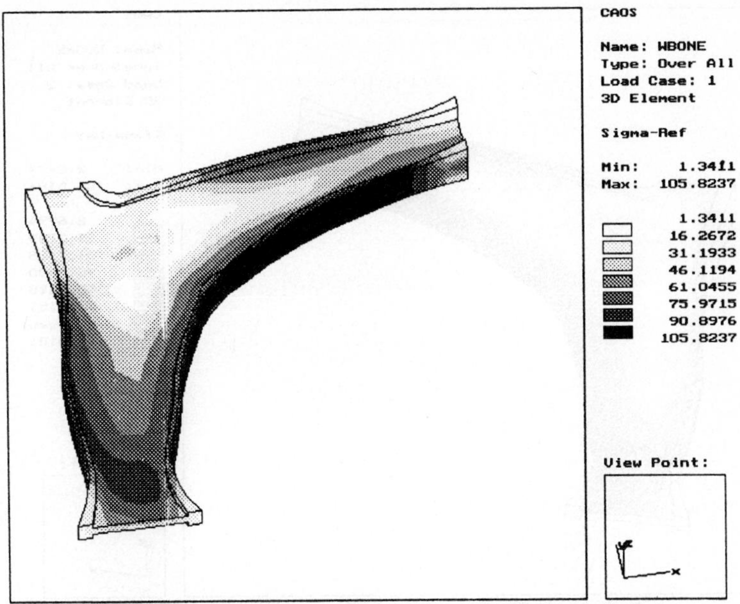

Figure 44. Final von Mises Stress Field, Load Case 1

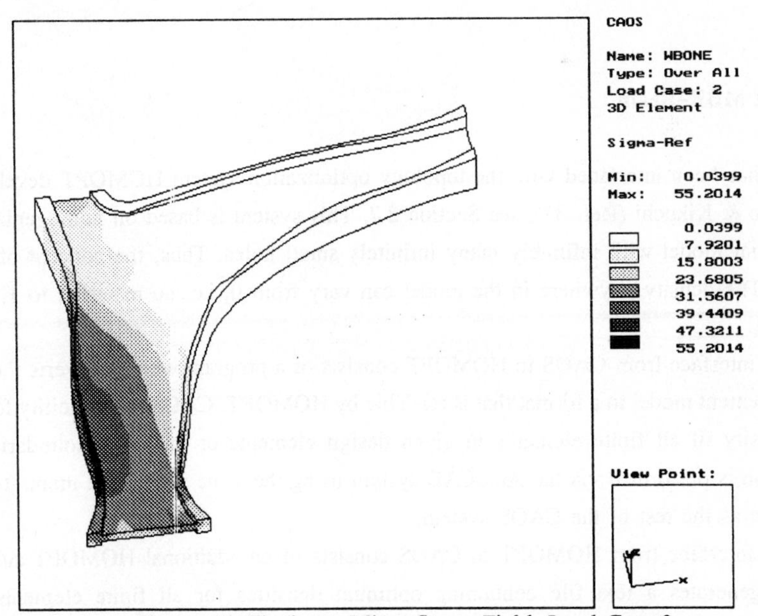

Figure 45. Final von Mises Stress Field, Load Case 2

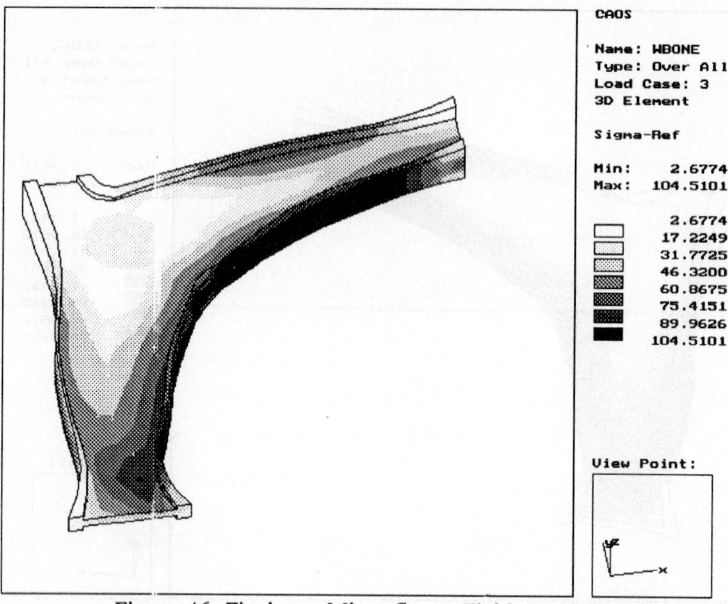

Figure 46. Final von Mises Stress Field, Load Case 3

4.3 The MBB-Beam

CAOS has been interfaced with the topology optimization system HOMOPT developed by Bendsøe & Kikuchi (Ref. 37), see Section 2.7. This system is based on homogenization of a material model with infinitely many infinitely small holes. Thus, the concept of density arises. The density anywhere in the model can vary from 0, i.e. no material, to 1, i.e. full material.

The interface from CAOS to HOMOPT consists of a program which converts the CAOS finite element model to a format that is readable by HOMOPT. CAOS has a facility for fixing the density of all finite elements in given design elements or on given boundaries. This operation is performed via the AutoCAD system using the same interactive menu-type user interface as the rest of the CAOS system.

The interface from HOMOPT to CAOS consists of an additional HOMOPT subroutine which generates a text file containing optimum densities for all finite elements in the structure. This file is read by a CAOS function which generates a picture of the optimized model. Finite elements are drawn on the screen and filled according to their density. Based on this picture, the user can easily generate a new design element configuration and perform a traditional shape optimization based on the optimized topology.

HOMOPT works on different variations of the formulation:

Minimize *Compliance*

Subject to *a bound on the total volume*

This section describes the optimization of a support beam from a civil aircraft produced by Messerschmitt–Bölkow–Blohm GmbH, München, BRD. The structure in question (Figure 47) has the function of carrying the floor in the fuselage of an Airbus passenger carrier and must meet the following requirements:

1. The upper and lower surfaces must be planar and the distance between them cannot be changed.
2. The maximum deflection of the beam must not exceed 9.4 mm under the given load.
3. The maximum von Mises stress should not exceed 385 MPa.
4. There must be a number of holes in the structure to allow for wires, pipes etc. to pass through.

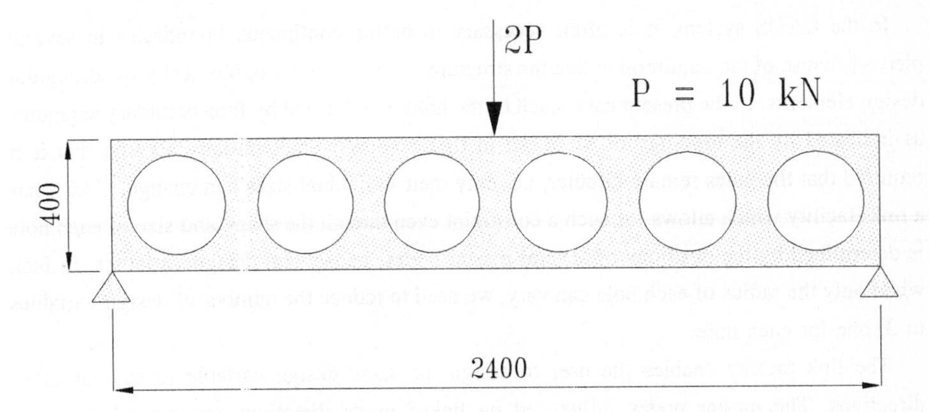

Figure 47. Initial Geometry with Loads and Boundary Conditions

The purpose of the optimization is to find the shapes of the holes that minimize the weight of the beam while not violating any of the requirements mentioned above. Because of the symmetry of the structure, we shall analyze only one half of the beam with boundary conditions as indicated in Figure 48. With the purpose of demonstrating the various possibilities of the CAOS system, we shall try several different formulations of the problem which lead to very different solutions.

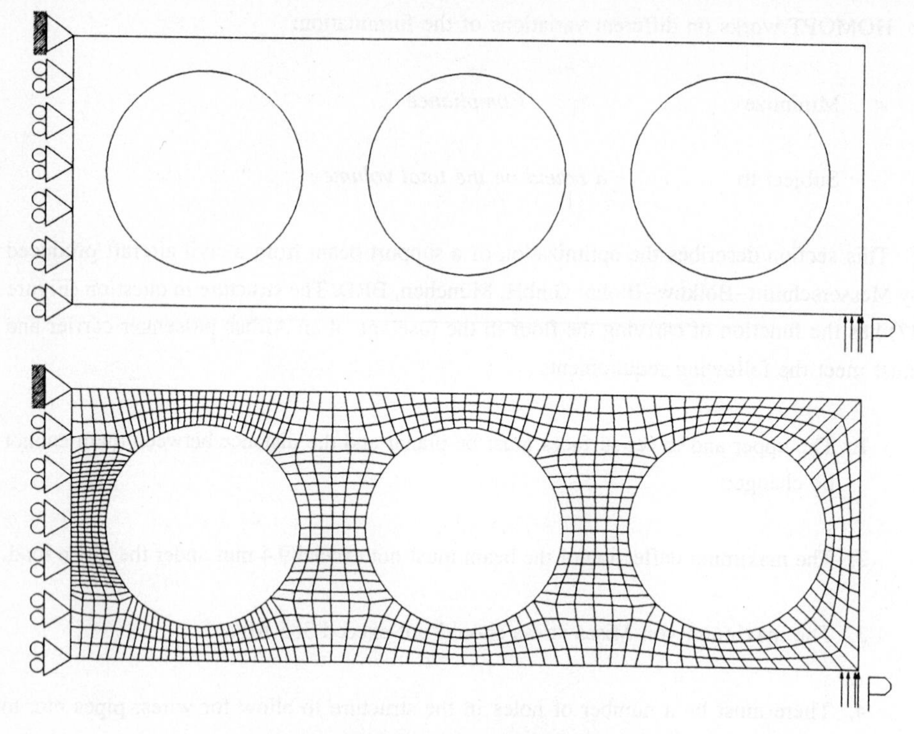

Figure 48. Initial Model with Circular Holes

In the CAOS system, it is often necessary to define continuous boundaries in several pieces because of the requirement that the structure be divided into topologically quadrangular design elements. In the present case, each of the holes are defined by four boundary segments as indicated by the topology of the mesh in figure 48. We shall initially assume that it is required that the holes remain circular, i.e. only their individual sizes can change. CAOS has a link–facility which allows for such a constraint even though the shape and size of each hole is determined by the positions of several master nodes, i.e. several design variables. In fact, when only the radius of each hole can vary, we need to reduce the number of design variables to 3, one for each hole.

The link facility enables the user to assign the same design variable to several move directions. The master nodes influenced by linked move directions are moved the same distance in their individual directions. In the present case, the boundary of each of the three holes is composed from four circular arcs, totally 3 times 8 master nodes. The 8 master nodes of each hole are assigned radial move directions. For each hole, a master is chosen, a master–master node so to speak, and the remaining master nodes are slaves of the first one, i.e. they will all move the same distance in their individual directions when the system starts to change the design. This reduces the total number of design variables to 3 and ensures that the holes remain circular. This system enables the use of the center position of the holes as a design variable also. However, in the present case, there is little room left in the structure to move the holes around.

Figure 48 shows the finite element model of the initial geometry which has the following data:

$$Volume = 1.07 \cdot 10^6\ mm^3$$
$$Deflection = 10.1\ mm$$
$$Max.\ Stress = 292\ MPa$$

The constraint on deflections in the vertical direction is an upper limit of 9.4 mm, and the initial design is therefore infeasible by at least 7.4% because the displacement based finite element method overestimates the stiffness of the structure.

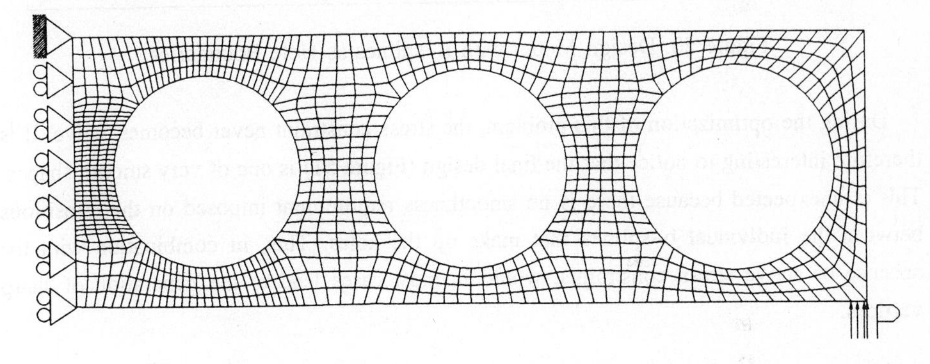

Figure 49. Final Finite Element Model of Geometry with Circular Holes

Figure 49 is the finite element model of the final geometry. This geometry has the following data:

$$Volume = 1.10 \cdot 10^6\ mm^3$$
$$Deflection = 9.4\ mm$$
$$Max.\ Stress = 248\ MPa$$

We see that the final geometry satisfies the initially violated deflection constraint at the cost of a volume increase by 3%. The three holes initially had a radius of 150 mm. In the final design, the radii are (counting from the left on Figure 49):

1: 141 mm

2: 148 mm

3: 151 mm

i.e, the first two holes decrease in size and the last one gets a little bigger. However, as evident from Figure 49, the geometrical alterations are marginal.

As a second attempt, the circular arcs making up the hole boundaries were replaced by b–

splines, and additional master nodes were introduced in order to give the system more design parameters for the optimization. We shall require symmetry about the horizontal mid-axis of the geometry and thus link the movements of master nodes above this line to the correspon-ding master nodes below. This new design model is illustrated in Figure 50.

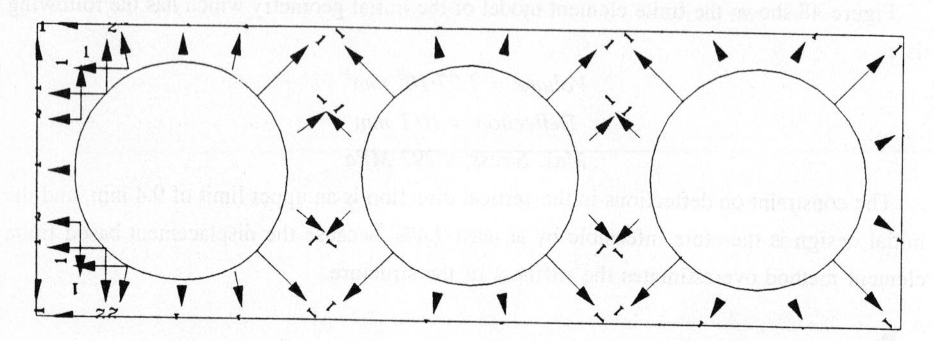

Figure 50. Design Model with B-splines as Hole Boundaries

During the optimization of this problem, the stress constraint never becomes active. It is therefore interesting to notice that the final design (Figure 51) is one of very smooth shapes. This is unexpected because there is no smoothness requirement imposed on the transitions between the individual b-splines that make up the holes. This, in combination with the absence of active stress constraints, would in most cases lead to the generation of sharp vertices.

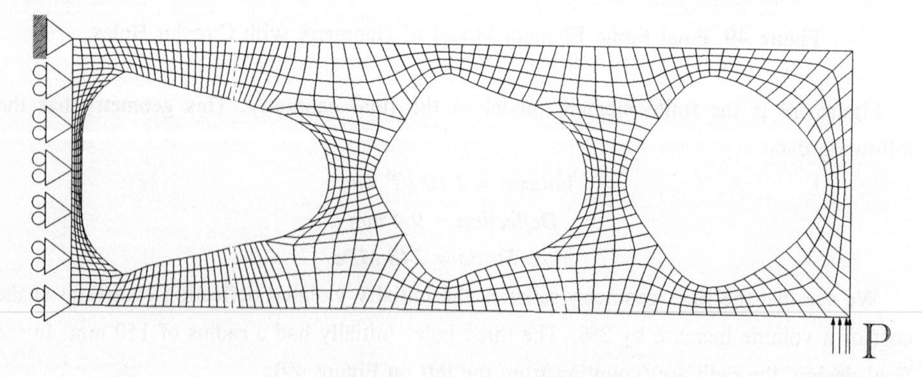

Figure 51. Final Finite Element Model

The final design has the following data:

$$Volume = 1.02 \cdot 10^6 \ mm^3$$

$$Deflection = 9.4 \ mm$$

$$Max. \ Stress = 372 \ MPa$$

i.e., with this model, we managed to create a feasible design and save 5.2% of the volume.

The fact that the possible volume reduction even with a b-spline model is rather modest leads to the suspicion that the three-hole topology is not well suited for a structure of this type. It is therefore tempting to start the redesign procedure by a topology optimization.

As previously explained, the topology optimization requires a volume constraint to be defined. The topology optimization system will then distribute the given material volume in the available domain such that the stiffness is maximized. The system enables the user to specify regions or boundaries which are required to be solid, that is, of density 1. We shall use this facility in the present example because the function of the structure requires that the outer contour, except for the left vertical symmetry boundary, remains unchanged.

The original geometry with three circular holes of radius 150 mm has a volume of $1.07 \cdot 10^6$ mm^3. A full beam has a volume of $1.92 \cdot 10^6$ mm^3, i.e., the volume of the initial geometry is 56% of that of a full beam.

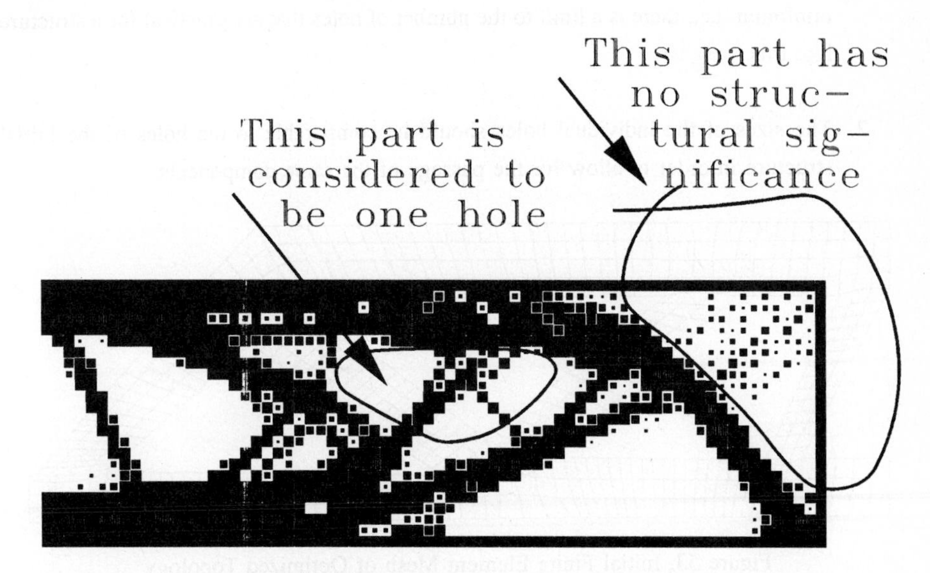

Figure 52. Result of Topology Optimization

A topology optimization with a bound on the volume corresponding to the volume of the initial structure with three circular holes and no additional requirements was performed. In addition to the volume constraint, we require the rim of the structure to remain solid. The resulting topology is shown in Figure 52. Graphically, each element is filled with a lump of material corresponding to its final density. This creates the impression that, in some regions, isolated lumps of material remain outside the solid part of the structure. This is not necessarily the case. The lumps are merely a convenient way of illustrating the porous material. When using CAOS for the actual shape optimization based on an optimized topology, the user has to decide upon the actual position of interfaces between material and void based on the filling of elements by lumps of material.

It is evident from the figure that a number of holes allowing for the necessary passage of wires, pipes etc. have emerged.

We shall now attempt a shape optimization based on this topology. We therefore return to the original definition of the problem, i.e. minimize volume with a bound on displacement and stress. The problem is a difficult one because the overall impression of the type of the structure is that it is on the interface between a disk and a frame or truss structure. Thus, the division of the geometry into design elements is relatively complicated. Furthermore, while creating the shape optimization model, we shall have to take a number of practical considerations into account:

1. Due to the cost of manufacture, the complexity of the geometry should be kept at a minimum, i.e., there is a limit to the number of holes that are practical for a structure like this.

2. The sizes of the individual holes should be comparable to the holes of the initial structure in order to allow for the passage of the same components.

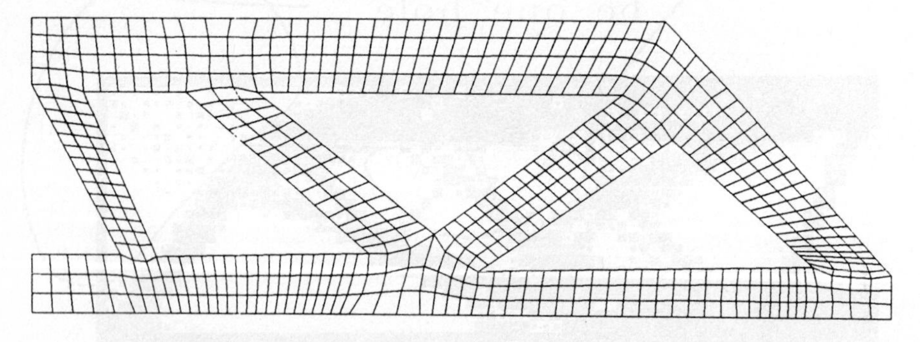

Figure 53. Initial Finite Element Mesh of Optimized Topology

The upper right corner of the frame has been removed. This part of the geometry has a function, but it is structurally insignificant and can therefore be excluded from the shape optimization and added to the modified structure afterwards. This simplification greatly facilitates the generation of the design model. Figure 52 illustrates the modifications that have been imposed on the optimized topology and the resulting initial finite element model is shown in Figure 53. This structure has the data:

$$\textit{Volume} = 1.10 \cdot 10^6 \ mm^3$$

$$\textit{Deflection} = 6.0 \ mm$$

$$\textit{Max. Stress} = 227 \ MPa$$

The volume of this geometry is slightly larger than the volume of the initial geometry with three circular holes. However, due to the topology optimization, this geometry has significantly larger stiffness. The maximum stress has also been reduced, but this value is unreliable because of the vertices of this geometry. Mathematically, the stress is infinite at sharp concave vertices, but the stress functions of the finite element model in question are unable to model such a state correctly. From a physical point of view, neither vertices of infinite curvature nor infinite stresses exist.

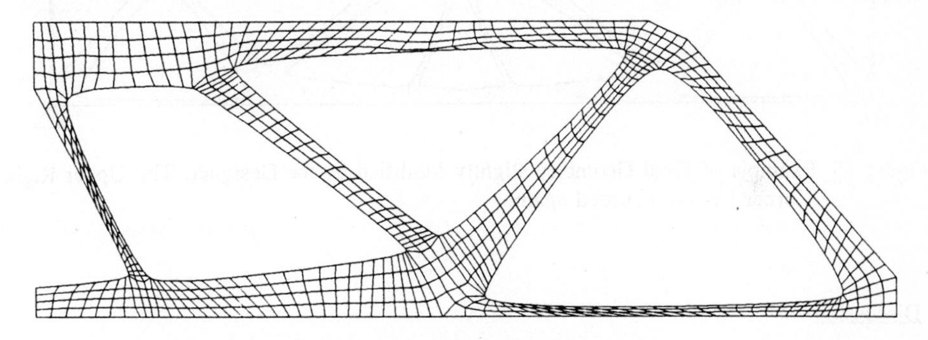

Figure 54. Final Finite Element Model

CAOS proceeds by reducing the volume significantly. Because of the structure's resemblance with a truss structure, small relocations of the master nodes may lead to large distortions of the finite element mesh. It has therefore been necessary to perform a redefinition of the finite element mesh topology on the half way between the initial and the final designs. The final design is illustrated in Figure 54. It has the data:

$$Volume = 0.624 \cdot 10^6 \ mm^3$$
$$Deflection = 9.4 \ mm$$
$$Max. \ Stress = 305 \ MPa$$

The volume is reduced by 42% in comparison with the initial design of circular holes. The final geometry is a frame-like structure. The stress constraint is not active because, for practical reasons, a minimum thickness is specified for the members of the resulting geometry.

Unfortunately, the geometry of the solution introduces the problem of stability which is not covered by CAOS. It is also a problem that the generation of a suitable finite element mesh for a very thin-webbed structure like this is difficult and often requires the topology of the mesh to be redefined. However, the use of membrane elements for a problem like this creates a result that is rich in the sense that it has details that could never have been found by the use of a dedicated system for truss or frame optimization.

Based on the final design in Figure 54 the designer can update his geometrical model and perform the final adjustments, e.g. add the structurally insignificant upper right corner that was removed in order to facilitate the generation of an analysis model, and thereby yield the final design in Figure 55.

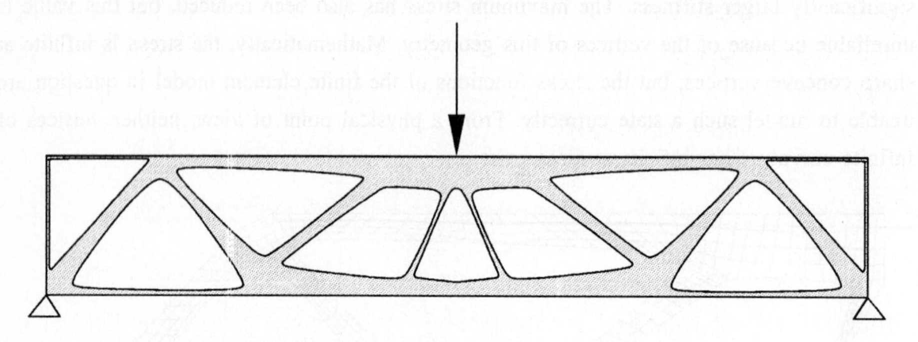

Figure 55. Example of Final Geometry slightly Modified by the Designer. The Upper Right Corner has been Added again

Discussion

It is evident that the initial topology optimization allows the shape optimization to arrive at a much better final result. For problems like this, where there are large possibilities for geometrical variations, the topology optimization is a very valuable tool in the design process. It is the experience from the present example that topology optimization should be used in the early stages of the development in order to inspire the designer and lead him/her in a beneficial direction. The result of the topology optimization is merely a crude guess and can therefore safely be modified by the designer to meet practical requirements, before the more detailed shape optimization is performed. In fact, one of the properties of the homogenization method is that very small changes in the problem formulation may lead to substantial geometrical changes of the final solution. Fortunately, these different solutions, in terms of stiffness, tend to be very similar in quality. The general impression from working with topology/shape interfacing is that the solutions tend to be rather insensitive to minor geometrical alterations, thus providing the designer with some degree of freedom to take considerations other than stiffness into account in the design.

The present example also shows the importance of setting a toolbox of various facilities at the disposal of the designer. Structural optimization enforces rather than removes the creative aspect of designing, and the final result is therefore very difficult to predict. The collection of structural optimization facilities must be versatile enough to allow the designer to continue work no matter what type of structure emerges. The final design must be a product of creativity rather than availability or lack of analysis facilities.

5 Closure

This article reflects a coordinated research effort towards the creation of a concurrent engineering design environment for rational design and optimization of traditional as well as high–technology mechanical products, systems, and components. The work embraces development, implementation and integration of methods and techniques for computer aided design, geometric modeling and parametrization, finite element analysis, sensitivity analysis, and multicriterion optimization. Design with the use of advanced materials is given considerable attention.

As a result of the work, two modular, interactive, CAD–integrated software systems for shape optimization have been created, the prototype system CAOS and the more general system ODESSY. These systems include important preprocessing capabilities for generation of optimum structural topologies, and the concurrent engineering design environment comprise facilities for optimum sizing and configuration of truss structures and optimization of composite structures as well.

The software systems are based on a specially developed mathematical programming formulation for multicriterion optimization which, in notable contrast to the traditional "trial and error" approach, ensures that multiple design criteria and constraints of various types are treated simultaneously, and that the design process converges towards the optimum product through a sequence of successively improved solutions. At the same time, the design phase becomes well–documented, and the concurrency implies a fast design process that considerably reduces the design to product lead time, which, along with the optimum characteristics of the product itself, is an important competition parameter.

Our current efforts are devoted to further extensions of the engineering design facilities in terms of methods as well as design criteria and objectives, with a view of allowing for solution of as broad a spectrum of practical engineering design problems as possible. This to a large extent implies expansion of the system analysis capabilities into other fields of engineering, and will hopefully result in a truly multidisciplinary concurrent engineering design system.

Acknowledgement

The work presented in this paper received support from the Danish Technical Research Council (Programme of Research on Computer Aided Design).

References

1. Haug, E.J.: A review of distributed parameter structural optimization literature. In: Optimization of distributed parameter structures (E.J. Haug & J. Cea, Eds.). Vol. 1, pp. 3–74. Alphen aan den Rijn: Sijthoff & Noordhoff 1981.
2. Olhoff, N. & Taylor, J.E.: On structural optimization. J. of Appl. Mech. 50, 1139 – 1151 (1983).
3. Esping, B.J.D.: The OASIS structural optimization system. Computers & Structures, 23, 365–377 (1984).
4. Braibant, V. & Fleury, C.: Shape optimal design using B–splines. Comp. Meths. Appl. Mech. Engrg. 44, 247 – 267 (1984).
5. Bennett, J.A. & Botkin, M.E.: Structural shape optimization with geometric description and adaptive mesh refinement. AIAA J. 23, 458 – 464 (1985).
6. Botkin, M.E., Yang, R.J. & Bennet, J.A.: Shape optimization of three–dimensional stamped and solid automotive components. In: The optimum shape. Automated structural design (J.A. Bennet & M.E. Botkin, Eds.). New York: Plenum Press 1986.
7. Stanton, E.L.: Geometric modeling for structural and material shape optimization. In: Loc. cit. 6.
8. Eschenauer, H.: Numerical and experimental investigations of structural optimization of engineering designs. Siegen: Bonn+Fries, Druckerei und Verlag 1986.
9. Ding, Y.: Shape optimization of structures: A literature survey. Computers & Structures 24, 985–1004 (1986).
10. Hörnlein, H.R.E.M.: Take–off in optimum structural design. In: Computer aided optimal design. Structural and mechanical systems. NATO ASI Series F: Computer and System Sciences (C.A. Mota Soares, Ed.). Vol. 27, Springer–Verlag 1987.
11. Sobieszanski–Sobieski, J. & Rogers, J.L.: A programming system for research and applications in structural optimization. In: New directions in optimum structural design (E. Atrek et.al., Eds.), pp. 563–585. Chichester: Wiley 1984.
12. Kneppe, G., Hartzheim, W. & Zimmermann, G.: Development and application of an optimization procedure for space and aircraft structures. In: Discretization methods and structural optimization – procedures and applications (H.A. Eschenauer & G. Thierauf, Eds.), pp. 194–201, Berlin: Springer–Verlag.
13. Arora, J.S.: Interactive design optimization of structural systems. In: Loc. cit. 12, pp. 10–16.
14. Choi, K.K. & Chang, K.H.: Shape design sensitivity analysis and what–if workstation for elastic solids. The University of Iowa, TH Report R–105, April 1991.
15. Rasmussen, J.: The structural optimization system CAOS. Structural Optimization 2, 109–115 (1990).
16. Rasmussen, J.: Shape optimization and CAD. Int. J. Systems Automation 1, 35–47 (1991).
17. Rasmussen, J., Lund, E. & Birker, T.: Collection of examples. CAOS optimization system. 3rd edition. Special Report No. 13, Institute of Mechanical Engineering, Aalborg University, April 1992.
18. Olhoff, N., Bendsøe, M.P. & Rasmussen, J.: On CAD–integrated structural topology and design optimization. Comp. Meths. Appl. Mech. Engrg. 89, 259–279 (1991).
19. Bendsøe, M.P., Olhoff, N. & Taylor, J.E.: A variational formulation for multicriteria structural optimization. J. Struct. Mech. 11, 523–544 (1983).
20. Taylor, J.E. & Bendsøe, M.P.: An interpretation of min–max structural design problems including a method for relaxing constraints. Int. J. Solids Struct. 20, 301–314 (1984).

21. Olhoff, N.: Multicriterion structural optimization via bound formulation and mathematical programming. Structural Optimization 1, 11–17 (1989).
22. Fleury, C. & Braibant, V.: Structural optimization: A new dual method using mixed variables. Int. J. Num. Meths. Engrg. 23, 409–428 (1986).
23. Svanberg, K.: The method of moving asymptotes – a new method for structural optimization. Int. J. Num. Meths. Engrg. 24, 359–373 (1987).
24. Zienkiewicz, O.C. & Campbell, J.S.: Shape optimization and sequential linear programming. In: Optimum structural design – theory and applications (R.H. Gallagher & O.C. Zienkiewicz, Eds.), pp. 109–126. London: Wiley and Sons 1973.
25. Cheng, G. & Liu, Y.: A new computation scheme for sensitivity analysis. Engrg. Opt. 12, 219–235 (1987).
26. Barthelemy, B. & Haftka, R.T.: Accuracy analysis of the semi-analytical method for shape sensitivity analysis. Mech. Struct. & Mach. 18, 407–432 (1990).
27. Pedersen, P., Cheng, G. & Rasmussen, J.: On accuracy problems for semi-analytical sensitivity analysis. Mech. Struct. & Mach. 17, 373–384 (1989).
28. Cheng, G., Gu, Y. & & Zhou, Y.: Accuracy of semi-analytical sensitivity analysis. Finite Elements in Analysis and Design 6, 113–128 (1989).
29. Olhoff, N. & Rasmussen, J.: Study of inaccuracy in semi-analytical sensitivity analysis – a model problem. Structural Optimization 3, 203–213 (1991).
30. Fenyes, P.A. & Lust, R.V.: Error analysis of semi-analytical displacement derivatives for shape and sizing variables. AIAA J. 29, 217–279 (1991).
31. Cheng, G. & Olhoff, N.: New method of error analysis and detection in semi-analytical sensitivity analysis. Report no. 36, Institute of Mechanical Engineering, Aalborg University, May 1991 (to appear in Compters and Structures).
32. Olhoff, N., Rasmussen, J. & Lund, E.: Method of "exact" numerical differentiation for error elimination in finite element based semi-analytical shape sensitivity analyses. Special Report no. 10, Institute of Mechanical Engineering, Aalborg University, February 1992 (to appear in Mech. Struct. & Mach.).
33. Braibant, V.: Shape sensitivity by finite elements. J. Struct. Mech. 14, 209–228 (1986).
34. Pedersen, P.: On the minimum mass layout of trusses. Proc. AGARD Symposium 1970, Istanbul, AGARD–CP–36–70, 1970.
35. Pedersen, P.: On the optimal layout of multi-purpose trusses. Computers & Structures 2, 695–712 (1972).
36. Pedersen, P.: Optimal joint positions for space trusses. J. Struct. Div., ASCE 99, 2459–2476 (1973).
37. Bendsøe, M.P. & Kikuchi, N.: Generating optimal topologies in structural design using a homogenization method. Comp. Meths. Appl. Mech. Engrg. 171, 197–224 (1988).
38. Bendsøe, M.P.: Optimal shape design as a material distribution problem. Structural Optimization 1, 193–202 (1989).
39. Pedersen, P.: On optimal orientation of orthotropic materials. Structural Optimization 1, 101–106 (1989).
40. Pedersen, P.: Bounds on elastic energy in solids of orthotropic materials. Structural Optimization 2, 55–63 (1990).
41. Diaz, A.R. & Bendsøe, M.P.: Shape optimization of structures for multiple loading conditions using a homogenization method. Structural Optimization 4, 17–22 (1992).

42. Thomsen, J.: Topology optimization of structures composed of one or two materials. (To appear in Structural Optimization).

43. Bendsøe, M.P. & Rodrigues, H.C.: Integrated topology and boundary shape optimization of 2-D solids. Comp. Meths. Appl. Mech. Engrg. 87, 15-34 (1991).

44. Suzuki, K. & Kikuchi, N.: A homogenization method for shape and topology optimization. Comp. Meths. Appl. Mech. Engrg. 93, 291-318 (1991).

45. Papalambros, P.Y. & Chirehdast, M.: An integrated environment for structural configuration design. J. Engrg. Design 1, 73-96 (1990).

46. Bremicker, M., Chirehdast, M., Kikuchi, N. & Papalambros, P.: Integrated topology and shape optimization in structural design. Mech. Struct. & Mach. 19, 551-587 (1991).

47. Thomsen, J.: Optimization of composite discs. Structural Optimization 3, 89-98 (1991).

48. Rodrigues, H.C.: Shape optimal design of elastic bodies using a mixed variational formulation. Comp. Meths. Appl. Mech. Engrg. 69, 29-44 (1988).

49. Bråmå, T. & Rosengren, R.: Application of the structural optimization program OPTSYS. Proc. ICAS Conf. 1990. ICAS-90-2.1.3, 1990.

Design Sensitivity Analysis and Optimization Tool for Concurrent Engineering

Kyung K. Choi and Kuang-Hua Chang

Center for Simulation and Design Optimization, College of Engineering, The University of Iowa, Iowa City, Iowa 52242, U.S.A.

Abstract: This paper together with the following two papers present emerging technologies for development of design sensitivity analysis (DSA) and optimization that can be used for concurrent engineering. A summary of recently developed unified continuum DSA methods for linear and nonlinear structural systems is presented. Design sensitivities of static and dynamic responses of elastic solids and built-up structures with respect to material property, sizing, shape, and configuration design variables are considered. For DSA of acoustic response, acousto-elastic systems are treated. For DSA of nonlinear structural systems, both geometric and material nonlinearities are considered. The adjoint variable and direct differentiation methods are used to derive explicit design sensitivity expressions that can be evaluated numerically using analysis results from established finite element analysis (FEA) codes. It is demonstrated that the continuum based DSA method allows design sensitivity computations to be carried out using established FEA codes with respect to geometric design parameters that are employed in computer-aided design (CAD) tools, so that industry standard tools can be exploited in concurrent engineering. A DSA and optimization (DSO) tool with a visually driven user interface is developed to allow design engineers to easily create geometric, design, and analysis models; define performance measures; perform DSA; and carry out a four-step interactive design process that includes visual display of design sensitivity, what-if study, trade-offs, and interactive design optimization.

Keywords: sensitivity analysis / optimization / concurrent engineering / finite element analysis / linear structures / nonlinear structures / acoustic / continuum sensitivity theory / sizing design / CAD / user interface / numerical implementation

1 Introduction

In design of structural systems that are made of truss, beam, membrane, shell, and elastic solid members, there are five different kinds of design variables: material property design variables such as Young's modulus; sizing design variables such as thickness and cross-sectional area; shape design variables such as length and geometric shape; configuration design variables such as orientation and location of structural components; and topological design variables. Substantial literature has emerged in the field of sizing and shape DSA and optimization of structural systems [1-7]. DSA of structural systems and machine components has emerged as a much needed design tool, not only because of the role of design sensitivity information in optimization algorithms, but also because design sensitivity information can be used in a computer-aided engineering environment for early product trade-offs in a concurrent design process.

Developments in sizing and shape DSA methods have been made using two fundamentally different approaches. In the first approach, which is called the discrete method, design derivatives of a discretized structural finite element equation are taken to obtain design sensitivity information. In the second approach, which is called the continuum method, design derivatives of the variational governing equation of the structural system are taken to obtain explicit design sensitivity expressions in integral form with integrands written in terms of variations of material property, sizing, shape, and configuration design variables, and natural physical quantities such as displacements, stresses, and strains [7]. The explicit design sensitivity expressions are then numerically evaluated using analysis results of finite element analysis codes [7,8]. The numerical method provides accurate design sensitivity information without the lack of numerical accuracy associated with the selection of finite difference perturbations. Also, the method does not require derivatives of stiffness, mass, and damping matrices as in the discrete method. Another advantage of the continuum method is that it provides a unified structural DSA capability, so it is possible to develop one DSA software system that works with a number of well-established FEA codes. The unified computational DSA method has been implemented in the DSO so it can provide design sensitivity for ANSYS, MSC/NASTRAN, and ABAQUS FEA codes [9-11].

This paper presents continuum sizing DSA methods for the dynamic frequency of linear acousto-elastic structures and the static response and critical load of nonlinear structural systems. The shape DSA method for elastic solids and the configuration DSA method for built-up structures are summarized in the following two papers in this volume. Continuum DSA methods for the static and transient dynamic responses of linear structural systems are also presented in Refs. 7, 8, and 12-14.

The continuum DSA method makes it possible to compute design sensitivity of analysis results of the established FEA codes with respect to the geometric design parameters employed in CAD tools. That is, connections to CAD tools can be made by providing design sensitivity of performance measures with respect to the design parameters defined on the CAD tool. Using the same CAD design parameters in manufacturing tools as well lays the foundation for concurrent engineering. Once models are based on the same CAD tool, an integrated CAD-FEA-DSA can be used to develop a design tool for concurrent engineering such that the design and manufacturing engineers can carry out early trade-offs. A connection can also be made to dynamic simulation and other computer-aided engineering (CAE) tools, if these tools use the same geometric CAD modeler [15].

The Design Sensitivity Analysis and Optimization tool (DSO), developed at the Center for Simulation and Design Optimization, link CAD-DSA-FEA and makes the technology accessible to design engineers, with a visually driven user interface. From a functional point of view, three design stages, pre-processing, design sensitivity computation, and post-processing, have been developed in the DSO. In the pre-processing design stage, the design engineer can easily create geometric and finite element analysis models, and a design model by defining design parameters and performance measures. In the design sensitivity computation stage, at the command of the design engineer, the DSO will execute complicated runstreams to compute design sensitivity information. The post-processing stage provides a four-step interactive design process that includes visual display of design sensitivity information, what-if study, trade-offs, and interactive design optimization, has been implemented in the post-processing design stage, so that the design engineer can better control the design process in concurrent engineering.

In Section 2, a continuum based design sensitivity analysis method for dynamic frequency response of acousto-elastic built-up structures is presented. For numerical example, a simple vehicle system is used. In Section 3, continuum design sensitivity analyses methods for the static response and critical load of nonlinear structural systems are presented. In Section 4, the DSO tool is described and the conclusions are given in Section 5.

2 DSA of Dynamic Frequency Response of Acousto-Elastic Built-Up Structures

Dynamic frequency response is used in various industries for noise, vibration, and harshness (NVH) analysis of mechanical and structural systems that are subject to harmonically varying external loads caused by reciprocating powertrains or other rotating machineries such as motors, fans, compressors, and forging hammers [16]. For example, airplane body and wing structures are subject to a harmonic load transmitted from the propulsion system. Also, ship vibrations resulting from propeller and engine excitations can cause noise problems, cracks, fatigue failure of the tailshaft, and discomfort to crew. When a machine or any structure oscillates in some form of periodic or random motion, the motion generates alternating pressure waves that propagate from the moving surface at the velocity of sound. For instance, interior sound pressure in an automobile compartment can occur when the input forces transmitted from road and power train excite the vehicle compartment boundary panels.

In Refs. 17 and 18, the continuum DSA method for the static response of Refs. 7 and 8 is extended to develop a continuum DSA method of dynamic frequency response of acousto-elastic built-up structures using the adjoint and direct differentiation methods. A typical acousto-elastic built-up structure is shown in Fig. 1. All members of the built-up structure are assumed to be plate (two-dimensional components) enclosures and/or beam (one-dimensional components) stiffeners in three-dimensional space. The built-up structure encloses a three-dimensional cavity that contains a medium (fluid) that transmits linear acoustic waves.

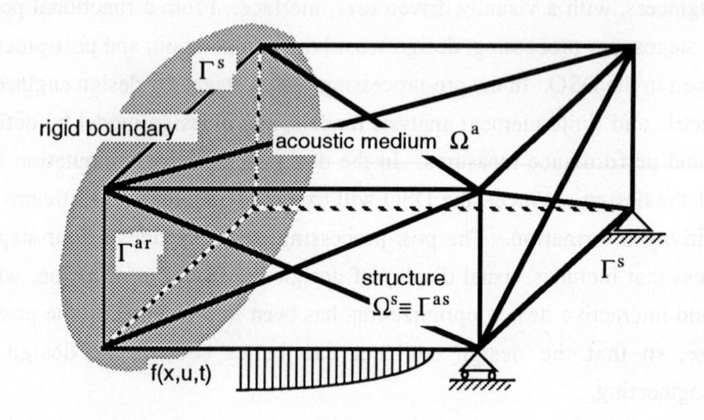

Figure 1 Acousto-Elastic Built-up Structure

The coupled dynamic motion of the built-up structure and acoustic medium can be described using the following system of partial differential equations:

Structure

$$m(x,u)z_{tt}(x,u,t) + C_u z_t(x,u,t) + A_u z(x,u,t) = f(x,u,t) + p^s(x,u,t)n, \quad x \in \Omega^s,\ t \geq 0 \tag{1}$$

with boundary and initial conditions

$$Gz = 0, \quad x \in \Gamma^s \tag{2}$$

$$z(x,u,0) = z_t(x,u,0) = 0, \quad x \in \Omega^s \tag{3}$$

Fluid

$$\frac{1}{\beta}p_{tt}(x,u,t) + Tp_t(x,u,t) - \frac{1}{\rho_0}\nabla^2 p(x,u,t) = 0, \quad x \in \Omega^a,\ t \geq 0 \tag{4}$$

with boundary and initial conditions

$$\nabla p^T n = 0, \quad x \in \Gamma^{ar} \tag{5}$$

$$p(x,u,0) = p_t(x,u,0) = 0, \quad x \in \Omega^a \tag{6}$$

Interface Conditions

$$p^s = p, \quad x \in \Gamma^{as} \equiv \Omega^s \tag{7}$$

$$\nabla p^T n = -\rho_0 z_{tt} n, \quad x \in \Gamma^{as} \equiv \Omega^s \tag{8}$$

where Ω^s is the domain of the structure; m(x,u) is the mass effect of the structure; C_u is the linear partial differential operator that corresponds to the damping effect of the structure; A_u is the fourth-order partial differential operator that corresponds to the stiffness effect of the structure; f(x,t,u) is the time-dependent applied load; p^s is the acoustic pressure at the interface between the structure and the acoustic medium; and n is the outward unit normal vector on the boundary of the acoustic medium. The design variable u(x) that can be defined using a CAD tool is independent of time, and the dynamic response $z(x,u,t)=[z_1, z_2, z_3]^T$ is displacement field of the structure. Boundary conditions, Eq. 2, are imposed on the structural boundary Γ^s using the trace operator G [19].

In Eq. 4, Ω^a is the domain of the acoustic medium; β is the adiabatic bulk modulus; and ρ_0 is the equilibrium density. The dynamic response p(x,u,t) is the acoustic or excessive pressure, and the linear operator T corresponds to dissipation of the acoustic energy. The normal gradient of the pressure vanishes on the rigid wall Γ^{ar} as shown in Eq. 5.

Structure-fluid interaction between two systems can be seen in Eq. 7 in the form of structural load p^s and, in Eq. 8, in the form of the acoustic boundary condition that relates the structural acceleration z_{tt} and the gradient ∇p of the acoustic pressure at the interface. Note that, as can be seen in Fig. 1, the acoustic-structure interface Γ^{as} is the domain Ω^s of the structure.

When the harmonic force f(x,u,t) with a single frequency ω is applied to the built-up structure of the coupled linear system, the corresponding dynamic responses z(x,u,t) and p(x,u,t) are also harmonic functions with the same frequency ω. These can be represented using complex harmonic functions as

$$\left.\begin{aligned} f(x,u,t) &= \mathrm{Re}\,\{\,\mathbf{f}(x,u)\,e^{i\omega t}\,\} \\ z(x,u,t) &= \mathrm{Re}\,\{\,\mathbf{z}(x,u)\,e^{i\omega t}\,\} \\ p(x,u,t) &= \mathrm{Re}\,\{\,\mathbf{p}(x,u)\,e^{i\omega t}\,\} \end{aligned}\right\} \tag{9}$$

where **f**, **z**, and **p** are complex phasors that are independent of time. Equations 1-8 can then be reduced to the following time-independent system of equations:

Structure

$$D_u\mathbf{z} \equiv -\,\omega^2 m(x,u)\mathbf{z} + i\omega C_u\mathbf{z} + A_u\mathbf{z} = \mathbf{f}(x,u) + \mathbf{p}^s n, \qquad x \in \Omega^s \tag{10}$$

with boundary conditions

$$G\mathbf{z} = 0, \qquad x \in \Gamma^s \tag{11}$$

Fluid

$$B\mathbf{p} \equiv -\frac{\omega^2}{\beta}\mathbf{p} + i\omega T\mathbf{p} - \frac{1}{\rho_0}\nabla^2\mathbf{p} = 0, \qquad x \in \Omega^a \tag{12}$$

with boundary conditions

$$\nabla\mathbf{p}^T n = 0, \qquad x \in \Gamma^{ar} \tag{13}$$

Interface Conditions

$$\mathbf{p}^s = \mathbf{p}, \qquad x \in \Gamma^{as} \equiv \Omega^s \tag{14}$$

$$\nabla\mathbf{p}^T n = \omega^2\rho_0\mathbf{z}^T n, \qquad x \in \Gamma^{as} \equiv \Omega^s \tag{15}$$

The differential operator D_u in Eq. 10 depends on design u, while the differential operator B in Eq. 12 does not, because an acoustic medium such as air is assumed to be fixed during the design process.

A variational form of Eqs. 10 and 12 can be obtained by multiplying both sides of Eqs. 10 and 12 by the transpose of complex conjugates $\bar{\mathbf{z}}^*$ and $\bar{\mathbf{p}}^*$ of kinematically admissible virtual states $\bar{\mathbf{z}} \in Z$ and $\bar{\mathbf{p}} \in P$, respectively, integrating by parts over each physical domain, adding them, and using boundary and interface conditions; that is,

$$b_u(\mathbf{z}, \bar{\mathbf{z}}) - \iint_{\Gamma^{as}} \mathbf{p}\bar{\mathbf{z}}^{*T} n \, d\Gamma + d(\mathbf{p}, \bar{\mathbf{p}}) - \omega^2 \iint_{\Gamma^{as}} \bar{\mathbf{p}}^* \mathbf{z}^T n \, d\Gamma = \ell_u(\bar{\mathbf{z}}) \tag{16}$$

which must hold for all kinematically admissible virtual states $\{\bar{\mathbf{z}}, \bar{\mathbf{p}}\} \in Q$ where

$$Q = \{ \mathbf{z} \in Z, \ \mathbf{p} \in P \mid \mathbf{p}^s = \mathbf{p} \ \text{and} \ \nabla \mathbf{p}^T n = \omega^2 \rho_o \mathbf{z}^T n, \ \ x \in \Gamma^{as} \} \tag{17}$$

and

$$\left. \begin{aligned} Z &= \{ \mathbf{z} \in [H^2(\Omega^s)]^3 \mid G\mathbf{z} = 0, \quad x \in \Gamma^s \} \\ P &= \{ \mathbf{p} \in H^1(\Omega^a) \mid \nabla \mathbf{p}^T n = 0, \quad x \in \Gamma^a \} \end{aligned} \right\} \tag{18}$$

and $\mathbf{H}^1$ and $\mathbf{H}^2$ are Sobolev spaces of orders one and two, respectively [19]. In Eq. 16, the sesquilinear forms b_u and d, and the semilinear form ℓ_u, are defined [20], using the L_2-inner product on complex function spaces, as

$$b_u(\mathbf{z}, \bar{\mathbf{z}}) \equiv (D_u \mathbf{z}, \bar{\mathbf{z}}) = -\iint_{\Omega^s} \omega^2 m \bar{\mathbf{z}}^{*T} \mathbf{z} \, d\Omega + i\omega c_u(\mathbf{z}, \bar{\mathbf{z}}) + a_u(\mathbf{z}, \bar{\mathbf{z}}) \tag{19}$$

where

$$c_u(\mathbf{z}, \bar{\mathbf{z}}) \equiv \iint_{\Omega^s} \bar{\mathbf{z}}^{*T} C_u \mathbf{z} \, d\Omega \quad \text{and} \quad a_u(\mathbf{z}, \bar{\mathbf{z}}) \equiv \iint_{\Omega^s} \bar{\mathbf{z}}^{*T} A_u \mathbf{z} \, d\Omega \tag{20}$$

$$d(\mathbf{p}, \bar{\mathbf{p}}) \equiv (B\mathbf{p}, \bar{\mathbf{p}}) = \iiint_{\Omega^a} \left[-\frac{\omega^2}{\beta} \mathbf{p} \, \bar{\mathbf{p}}^* + i\omega T \mathbf{p} \, \bar{\mathbf{p}}^* + \frac{1}{\rho_0} \nabla \mathbf{p}^T \nabla \bar{\mathbf{p}}^* \right] d\Omega \tag{21}$$

and

$$\ell_u(\bar{\mathbf{z}}) \equiv \iint_{\Omega^s} \mathbf{f}^T \bar{\mathbf{z}}^* \, d\Omega \tag{22}$$

If the built-up structure does not have an acoustic medium, then the variational form, Eq. 16, can be simplified by dropping all terms corresponding to the acoustic medium, including interface conditions. In this way, the results of Ref. 17 will be obtained.

2.1 Direct Differentiation Method of DSA

To develop the direct differentiation method of DSA, take the first variation of Eq. 16 with respect to the design variable u and rearrange to obtain

$$b_u(\mathbf{z}', \bar{\mathbf{z}}) - \iint_{\Gamma^{as}} \mathbf{p}' \bar{\mathbf{z}}^{*T} n \, d\Gamma + d(\mathbf{p}', \bar{\mathbf{p}}) - \omega^2 \iint_{\Gamma^{as}} \bar{\mathbf{p}}^* \mathbf{z}'^T n \, d\Gamma = \ell'_{\delta u}(\bar{\mathbf{z}}) - b'_{\delta u}(\mathbf{z}, \bar{\mathbf{z}}) \tag{23}$$

which must hold for all kinematically admissible virtual states $\{\bar{z}, \bar{p}\} \in Q$. In Eq. 23,

$$z' \equiv \frac{d}{d\tau} z(x, u + \tau\delta u)\Big|_{\tau=0} = \lim_{\tau \to 0} \frac{z(x, u + \tau\delta u) - z(x, u)}{\tau} \tag{24}$$

and

$$p' \equiv \frac{d}{d\tau} p(x, u + \tau\delta u)\Big|_{\tau=0} = \lim_{\tau \to 0} \frac{p(x, u + \tau\delta u) - p(x, u)}{\tau} \tag{25}$$

are the first variations of **z** and **p** with respect to the design variable u in the direction δu of the design change. Also, the first variations of the sesquilinear form b_u and semilinear form ℓ_u with respect to explicit dependence on the design variable u are

$$b'_{\delta u}(z, \bar{z}) \equiv \frac{d}{d\tau} b_{u+\tau\delta u}(\tilde{z}, \bar{z})\Big|_{\tau=0} \tag{26}$$

and

$$\ell'_{\delta u}(\bar{z}) \equiv \frac{d}{d\tau} \ell_{u+\tau\delta u}(\bar{z})\Big|_{\tau=0} \tag{27}$$

where $\tilde{z}$ denotes the state **z** with dependence on t (design variable u) suppressed, and $\bar{z}$ is independent of t. Equation 23 is a variational equation in which the design sensitivities z' and p' are unknowns. If the solution **z** of Eq. 16 is obtained, the fictitious load, i.e., right side of Eq. 23, can be computed using the shape functions of the finite element to evaluate integrands at Gauss points for numerical integration.

2.2 Adjoint Variable Method of DSA

Harmonic performance measures of the acousto-elastic built-up structure can be expressed in terms of complex phasors of the structural displacement and the acoustic pressure. For the adjoint variable method, first consider the pressure at a point $\hat{x}$ in the acoustic medium enclosed by the built-up structure under harmonic excitation

$$\psi_p = \iiint_{\Omega^a} \delta(x - \hat{x}) p \, d\Omega \tag{28}$$

The pressure can be correlated to noise audible to the human ear in passenger vehicles, aircraft, or ships. The first variation of the performance measure is

$$\psi_p' = \iiint_{\Omega^a} \delta(x - \hat{x}) p' \, d\Omega \tag{29}$$

To use the adjoint variable method, define non-selfadjoint operators D_u^a and B^a corresponding to the operators D_u and B of Eqs. 10 and 12, respectively, as

$$(D_u z, \lambda) \equiv (z, D_u^a \lambda), \qquad \text{for all } z, \lambda \in Z \tag{30}$$

and

$$(Bp, \eta) \equiv (p, B^a \eta), \qquad \text{for all } p, \eta \in P \tag{31}$$

Then, by the definition of $b_u(\bullet,\bullet)$ of Eq. 19 and $d(\bullet,\bullet)$ of Eq. 21,

$$b_u(\bar{\lambda}, \lambda) = (\bar{\lambda}, D_u^a \lambda) = -\iint_{\Omega^s} \omega^2 m \bar{\lambda}^T \lambda^* \, d\Omega + i\omega c_u(\bar{\lambda}, \lambda) + a_u(\bar{\lambda}, \lambda) \tag{32}$$

and

$$d(\bar{\eta}, \eta) = (\bar{\eta}, B^a \eta) = \iiint_{\Omega^a} \left[-\frac{\omega^2}{\beta} \bar{\eta}\, \eta^* + i\omega T \bar{\eta}\, \eta^* + \frac{1}{\rho_0} \nabla \bar{\eta}^T \nabla \eta^* \right] d\Omega \tag{33}$$

Define an adjoint equation for the performance measure of Eq. 28 by replacing $\mathbf{p}'$ in Eq. 29 by a virtual pressure $\bar{\eta}$ and equating the term to the sesquilinear forms as

$$b_u(\bar{\lambda}, \lambda) - \iint_{\Gamma^{as}} \bar{\eta} \lambda^{*T} n \, d\Gamma + d(\bar{\eta}, \eta) - \omega^2 \iint_{\Gamma^{as}} \eta^* \bar{\lambda}^T n \, d\Gamma = \iiint_{\Omega^a} \hat{\delta}(x - \hat{x}) \bar{\eta} \, d\Omega \tag{34}$$

which must hold for all kinematically admissible virtual states $\{\bar{\lambda}, \bar{\eta}\} \in Q$. Note that the solution of Eq. 34 is the pair of complex conjugates $\{\lambda^*, \eta^*\}$. To take advantage of the adjoint equation, we may evaluate Eq. 34 at $\bar{\lambda}=\mathbf{z}'$ and $\bar{\eta}=\mathbf{p}'$ to obtain

$$b_u(\mathbf{z}', \lambda) - \iint_{\Gamma^{as}} \mathbf{p}' \lambda^{*T} n \, d\Gamma + d(\mathbf{p}', \eta) - \omega^2 \iint_{\Gamma^{as}} \eta^* \mathbf{z}'^T n \, d\Gamma = \iiint_{\Omega^a} \hat{\delta}(x - \hat{x}) \mathbf{p}' \, d\Omega \tag{35}$$

which is the term on the right side of Eq. 29 that we would like to write explicitly in terms of du. Similarly, evaluate Eq. 23 at $\bar{\mathbf{z}}^*=\lambda^*$ and $\bar{\mathbf{p}}^*=\eta^*$ to obtain

$$b_u(\mathbf{z}', \lambda) - \iint_{\Gamma^{as}} \mathbf{p}' \lambda^{*T} n \, d\Gamma + d(\mathbf{p}', \eta) - \omega^2 \iint_{\Gamma^{as}} \eta^* \mathbf{z}'^T n \, d\Gamma = \ell'_{\delta u}(\lambda) - b'_{\delta u}(\mathbf{z}, \lambda) \tag{36}$$

Since the left sides of Eqs. 35 and 36 are equal, the desired explicit design sensitivity expression can be obtained from Eqs. 29, 35, and 36,

$$\psi_p' = \ell'_{\delta u}(\lambda) - b'_{\delta u}(\mathbf{z}, \lambda)$$

$$= \iint_{\Omega^s} f_u^T \lambda^* \delta u \, d\Omega + \iint_{\Omega^s} \omega^2 m_u \lambda^{*T} \mathbf{z} \delta u \, d\Omega - i\omega c'_{\delta u}(\mathbf{z}, \lambda) - a'_{\delta u}(\mathbf{z}, \lambda) \tag{37}$$

Note that, to evaluate the design sensitivity of Eq. 37, the solution λ^* of Eq. 34, which is the complex conjugate of λ, must be obtained.

Another performance measure of the acousto-elastic built-up structure is the structural displacement at a point $\hat{x}$. For instance, the performance measure could be the vibration amplitude of the seat of a passenger vehicle, aircraft, or ship. The performance measure can be written as

$$\psi_{z_i} = \iint_{\Omega^s} \hat{\delta}(x - \hat{x}) \mathbf{z}_i \, d\Omega, \qquad i = 1,2,3 \tag{38}$$

The adjoint equation for this performance measure is defined as

$$b_u(\bar{\lambda}, \lambda) - \iint_{\Gamma^{as}} \bar{\eta} \lambda^{*T} n \, d\Gamma + d(\bar{\eta}, \eta) - \omega^2 \iint_{\Gamma^{as}} \eta^* \bar{\lambda}^T n \, d\Gamma = \iint_{\Omega^s} \hat{\delta}(x - \hat{x}) \bar{\lambda}_i \, d\Omega \tag{39}$$

which must hold for all kinematically admissible virtual states $\{\bar{\lambda}, \bar{\eta}\} \in Q$. Once the complex conjugate λ^* of the adjoint response is obtained from Eq. 39, the design sensitivity expression of Eq. 37 can be used to obtain design sensitivity information. Another point to note is that, since the sizing design variable u is defined only on the structural part, Eq. 37 requires only the structural response λ^* of adjoint Eqs. 34 or 39.

In the adjoint variable method, the adjoint load for each performance measure needs to be computed. For the displacement performance measure at a node, a unit harmonic load is applied at the node in the direction of the displacement for which the design sensitivity information is to be computed. For the pressure performance measure, the adjoint load is the second time derivative of a unit volumetric strain at the point in the acoustic medium at which the pressure performance measure is defined. The complex conjugates of the adjoint structural responses; i.e., solutions of Eqs. 34 and 39, can be obtained efficiently by the restart option of established FEA codes. Using the original response and complex conjugate of adjoint structural response, the design sensitivity information of Eq. 37 can be obtained by evaluating the integrands of Eq. 37 at Gauss points using the shape functions of the finite element for numerical integration. Computational procedures for continuum design sensitivity analysis can be found in Refs. 7 and 8. If the direct differentiation method is used, the fictitious load on the right side of Eq. 23 is computed using the shape functions of the finite element for numerical integration. An efficient solution of Eq. 23 can also be obtained using the restart option of FEA codes.

2.3 Example - Simple Vehicle System

The low frequency noise in the passenger compartment of a vehicle occurs over a wide range of vehicle speeds and is known to be dominant at frequencies between 20 and 200 Hz. Considerable interaction between the motion of the body structure and the vibration of the acoustic medium at critical frequencies has been observed, and it is suggested that the noise may be amplified by the resonance of the acoustic medium [21].

The simple vehicle model, which can be used to identify the system characteristics prior to a practical engineering model analysis of a vehicle system, is shown in Fig. 2. The body structure is made of thin plates of uniform thickness that enclose the acoustic medium (air) and is mounted on a simplified suspension system consisting of springs and dampers. The mass density and Young's modulus of the plates are ρ=10 kg/m^3 and E=0.5$\times$10^9 N/m^2, respectively. The adiabatic bulk modulus and equilibrium density of the acoustic medium are β=141700 N/m^2 and ρ_0=1.20236 kg/m^3, respectively. The finite element model consists of 232 triangular shell elements for the body structure, 84 hexagonal and 8 tetrahedral elements for the acoustic medium, and 12 spring elements and 12 dampers for the suspension system. The body structure has 118 grid points and 708 degrees of freedom, and the acoustic medium has 160 grid points with 160 degrees of freedom.

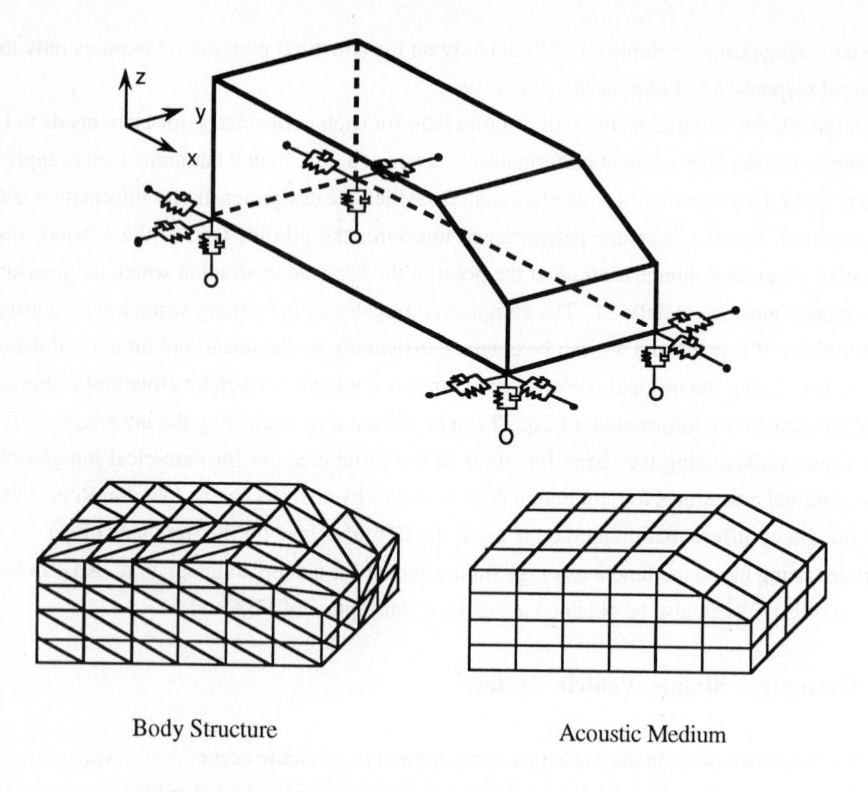

Figure 2 Simple Vehicle System and Finite Element Models

This model is used to investigate effects of body structure design changes on vibration amplitudes of acoustic pressure at the driver's and passenger's ear levels and vibration amplitudes at the driver and passenger seats. The design variable is the thickness t of the body panels, which is 0.009m at the current nominal design. Two rear suspension supports are harmonically excited in the x_3-direction with an amplitude of 0.0001m and frequency of 60 Hz.

The direct frequency response method of ABAQUS [11] is used to analyze the original and adjoint systems. The complex conjugates of the adjoint responses are obtained by solving the coupled system of equations with harmonic adjoint loads. For the acoustic pressure performance measure, the adjoint load is the second time derivative of a unit volumetric strain at the point of the performance measure. Also a unit force is applied for the displacement performance measure at the point where the performance measure is defined.

The accuracy of design sensitivity results has been checked using the finite difference results. For the test, $\pm 0.01\%$ perturbations of the uniform thickness of the body panels are taken. Table 1 shows accuracy of design sensitivity results for the acoustic pressure amplitudes. Table 2 shows the accuracy of design sensitivity results for the structural displacement amplitude in the x_3-direction. In Tables 1 and 2, R, I, and D are the real and imaginary components, and magnitude of complex responses. Good agreement is obtained between the design sensitivity predictions φ' and finite differences Δφ. In addition to the structural displacement amplitude, Table 2 presents design sensitivities of the velocity and acceleration amplitudes, V and A, respectively.

Table 1 Accuracy of Design Sensitivity of Acoustic Frequency Responses at Driver's and Passenger's Ear Levels, in pascals

Location		ψ(u-δu)	ψ(u+δu)	Δψ	ψ'	ψ'/Δψ(%)
Driver	R	-.2316D-01	-.2299D-01	0.8600D-04	0.8944D-04	104.0
	I	-.1294D-01	-.1311D-01	-.8420D-04	-.8776D-04	104.2
	D	0.2653D-01	0.2647D-01	-.3349D-04	-.3547D-04	105.9
Passenger	R	0.1133D-01	0.1125D-01	-.4395D-04	-.4613D-04	105.0
	I	0.7174D-02	0.7265D-02	0.4549D-04	0.4831D-04	106.2
	D	0.1341D-01	0.1339D-01	-.1253D-04	-.1284D-04	102.5

Table 2 Accuracy of Design Sensitivity of Frequency Responses at Driver and Passenger Seats in x_3-Direction, in meters

Loaction		ψ(u-δu)	ψ(u+δu)	Δψ	ψ'	ψ'/Δψ(%)
Driver	R	0.3681D-05	0.3650D-05	-.1579D-07	-.1667D-07	105.6
	I	0.3459D-05	0.3488D-05	0.1436D-07	0.1536D-07	107.0
	D	0.5052D-05	0.5048D-05	-.1583D-08	-.1530D-08	96.7
	V	0.1904D-02	0.1903D-02	-.5969D-06	-.5769D-06	96.7
	A	0.7179D+00	0.7175D+00	-.2250D-03	-.2174D-03	96.7
Passenger	R	-.3979D-05	-.3941D-05	0.1885D-07	0.1991D-07	105.6
	I	-.4586D-05	-.4620D-05	-.1708D-07	-.1829D-07	107.1
	D	0.6071D-05	0.6072D-05	0.6539D-09	0.8835D-09	135.1
	V	0.2289D-02	0.2289D-02	0.2465D-06	0.3330D-06	135.1
	A	0.8628D+00	0.8630D+00	0.9294D-04	0.1256D-03	135.1

3 DSA of Nonlinear Structural Systems

Design requirements for structures that are expected to perform under severe loading conditions and the need to use materials with nonlinear properties necessitate development of a unified DSA method for nonlinear structural systems. A unified continuum DSA method for nonlinear structural systems with sizing design variables has been developed in Refs. 22 and 23 using the total and updated Lagrangian formulations of nonlinear analysis. This continuum DSA method developed is valid for large displacements, large rotations, small strains, and nonlinear elastic material properties when appropriate kinematic and constitutive descriptions are used for analysis. Using the adjoint variable method, as in the case of linear structural systems, a linear adjoint equation is obtained for each performance measure.

The DSA method of Ref. 22 is extended in Ref. 24 to develop a method for DSA of the critical load with respect to sizing and shape design variables of a nonlinear structural system subject to a conservative static loading. A stability equation for the critical load has been developed using the total Lagrangian formulation. The design derivatives of the stability equation are used to obtain the design sensitivity of the critical load. This section summarizes the DSA methods for the static response and critical load presented in Refs. 22 and 24.

3.1 Equilibrium and Stability Equations of Nonlinear Structural Systems

In this section, the nonlinear equilibrium equation, linearizations of the incremental nonlinear equation, linearized eigenvalue equation, and nonlinear stability equation are introduced. These equations are used to obtain first variations of the static governing and stability equations. Using the principle of virtual work, the equilibrium equation for a body with domain ${}^{t+\Delta t}\Omega$ and boundary ${}^{t+\Delta t}\Gamma$ in the equilibrium configuration at time t+Δt can be expressed as [25-27]

$$\iint_{{}^{t+\Delta t}\Omega} {}^{t+\Delta t}\sigma_{ij}\ {}_{t+\Delta t}\bar{e}_{ij}\ {}^{t+\Delta t}d\Omega = {}^{t+\Delta t}\bar{R}\,, \quad \text{for all } \bar{z} \in Z \tag{40}$$

where ${}^{t+\Delta t}\sigma_{ij}$ are Cartesian components of the Cauchy stress tensor at time t+Δt and ${}_{t+\Delta t}e_{ij}$ are components of the infinitesimal strain tensor, both referred to the configuration at time t+Δt. Also, $\bar{z}=[\bar{z}_1,\bar{z}_2,\bar{z}_3]^T$ is a kinematically admissible virtual displacement, Z is the space of kinematically admissible virtual displacement, and an overbar " $\bar{\ }$ " indicates the variation of the quantity under the bar. In Eq. 40, ${}_{t+\Delta t}\bar{e}_{ij}$ are components of the infinitesimal strain tensor that corresponds to the virtual displacement and referred to the configuration at time t+Δt; i.e.,

$${}_{t+\Delta t}\bar{e}_{ij} = \frac{1}{2}\left({}_{t+\Delta t}\bar{z}_{i,j} + {}_{t+\Delta t}\bar{z}_{j,i}\right) \tag{41}$$

On the right side of Eq. 40, ${}^{t+\Delta t}\bar{R}$ is the virtual work done on the body by externally applied force through a kinematically admissible virtual displacement $\bar{z}$,

$${}^{t+\Delta t}\bar{R} = \iint_{{}^{t+\Delta t}\Omega} {}^{t+\Delta t}f_i\, \bar{z}_i\ {}^{t+\Delta t}d\Omega + \int_{{}^{t+\Delta t}\Gamma} {}^{t+\Delta t}T_i\, \bar{z}_i\ {}^{t+\Delta t}d\Gamma \tag{42}$$

where ${}^{t+\Delta t}f_i$ and ${}^{t+\Delta t}T_i$ are the components of externally applied body and surface force vectors, respectively. In the following development, forces ${}^{t+\Delta t}f_i$ and ${}^{t+\Delta t}T_i$ are assumed to be conservative static loadings, i.e., deformation independent.

Incremental Equilibrium Equation

Equation 40 cannot be solved directly, since the final equilibrium configuration at time t+Δt is unknown. A solution can be obtained by referring all variables to a previously known equilibrium configuration and linearizing the resulting incremental equation. Two formulations, the updated and the total Lagrangian formulations, can be used according to the reference configuration. In this paper, the total Lagrangian formulation is summarized. For design sensitivity analysis of the updated Lagrangian formulation, see Refs. 22 and 24.

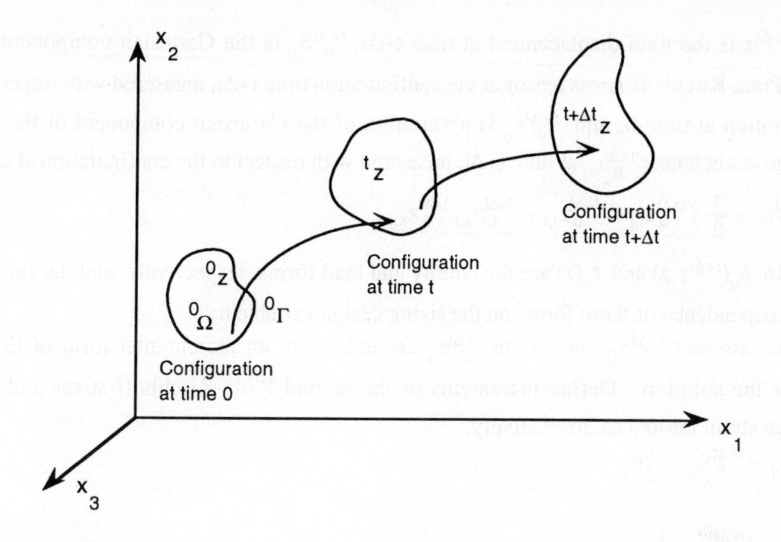

Figure 3 Motion of the Body in a Stationary Cartesian Coordinate System

Consider the motion of a body in a fixed Cartesian coordinate system, as shown in Fig. 3. The body can experience large displacements, large rotations, large strains, and nonlinear material behavior. The coordinates of a generic point P in the body at time 0 are $[{}^0x_1, {}^0x_2, {}^0x_3]^T$, and at time t they are $[{}^tx_1, {}^tx_2, {}^tx_3]^T$, where the left superscript refers to the configuration of the body, and the right subscript refers to the coordinate axis. For derivatives of displacements with respect to the coordinates, the left subscript indicates the configuration in which the coordinate is measured; thus, we have

$$ {}^{t+\Delta t}_{\ \ 0}z_{i,j} = \frac{\partial\, {}^{t+\Delta t}z_i}{\partial\, {}^0x_j} \tag{43} $$

As an exceptional case, if the quantity under consideration occurs in the same configuration in which it is also measured, then the left subscript is omitted [27]; e.g.,

$$ {}^{t+\Delta t}\sigma_{ij} \equiv {}^{t+\Delta t}_{t+\Delta t}\sigma_{ij} \tag{44} $$

Increments in the displacements from time t to time t+Δt are denoted as

$$ z_i = {}^{t+\Delta t}z_i - {}^tz_i\,, \qquad i = 1,2,3 \tag{45} $$

In the total Lagrangian formulation, all static and kinematic variables are referred to the initial configuration at time 0, and Eq. 40 is transformed to [22, 27]

$$ \begin{aligned} a_u({}^{t+\Delta t}z, \bar{z}) &\equiv \iint_{{}^0\Omega} {}^{t+\Delta t}_{\ \ 0}S_{ij}\; {}^{t+\Delta t}_{\ \ 0}\bar{\varepsilon}_{ij}\; {}^0d\Omega \\ &= \iint_{{}^0\Omega} {}^{t+\Delta t}_{\ \ 0}f_i\, \bar{z}_i\; {}^0d\Omega + \int_{{}^0\Gamma} {}^{t+\Delta t}_{\ \ 0}T_i\, \bar{z}_i\; {}^0d\Gamma \equiv \ell_u(\bar{z})\,, \quad \text{for all } \bar{z} \in Z \end{aligned} \tag{46} $$

where ${}^{t+\Delta t}z$ is the total displacement at time t+Δt; ${}^{t+\Delta t}_{0}S_{ij}$ is the Cartesian component of the second Piola-Kirchhoff stress tensor at the configuration time t+Δt, measured with respect to the configuration at time 0; and ${}^{t+\Delta t}_{0}\bar{\varepsilon}_{ij}$ is a variation of the Cartesian component of the Green-Lagrange strain tensor ${}^{t+\Delta t}_{0}\varepsilon_{ij}$ at time t+Δt, measured with respect to the configuration at time 0,

$$ {}^{t+\Delta t}_{0}\varepsilon_{ij} = \frac{1}{2}\left({}^{t+\Delta t}_{0}z_{i,j} + {}^{t+\Delta t}_{0}z_{j,i} + {}^{t+\Delta t}_{0}z_{k,i}\, {}^{t+\Delta t}_{0}z_{k,j} \right) \tag{47} $$

In Eq. 46, $a_u({}^{t+\Delta t}z, \bar{z})$ and $\ell_u(\bar{z})$ are the energy and load forms, respectively, and the subscript u denotes dependence of these forms on the sizing design variable u.

Since stresses ${}^{t+\Delta t}_{0}S_{ij}$ and strains ${}^{t+\Delta t}_{0}\varepsilon_{ij}$ are unknown, an incremental form of Eq. 46 is used for the solution. Define increments of the second Piola-Kirchhoff stress and Green-Lagrange strain tensors as, respectively,

$$ {}_{0}S_{ij} = {}^{t+\Delta t}_{0}S_{ij} - {}^{t}_{0}S_{ij} \tag{48} $$

and

$$ \begin{aligned} {}_{0}\varepsilon_{ij} &= {}^{t+\Delta t}_{0}\varepsilon_{ij} - {}_{0}\varepsilon_{ij} \\ &= \frac{1}{2}\left({}_{0}z_{i,j} + {}_{0}z_{j,i} + {}^{t}_{0}z_{k,i}\, {}_{0}z_{k,j} + {}_{0}z_{k,i}\, {}^{t}_{0}z_{k,j} + {}_{0}z_{k,i}\, {}_{0}z_{k,j} \right) \end{aligned} \tag{49} $$

A variation of the Green-Lagrange strain tensor is

$$ \begin{aligned} {}^{t+\Delta t}_{0}\bar{\varepsilon}_{ij} &= {}_{0}\bar{\varepsilon}_{ij} \\ &= \frac{1}{2}\left({}_{0}\bar{z}_{i,j} + {}_{0}\bar{z}_{j,i} + {}^{t}_{0}z_{k,i}\, {}_{0}\bar{z}_{k,j} + {}_{0}\bar{z}_{k,i}\, {}^{t}_{0}z_{k,j} + {}_{0}\bar{z}_{k,i}\, {}_{0}z_{k,j} + {}_{0}z_{k,i}\, {}_{0}\bar{z}_{k,j} \right) \end{aligned} \tag{50} $$

since ${}^{t}\bar{z}_i=0$ gives ${}^{t}_{0}\bar{\varepsilon}_{ij}=0$.

Using these definitions, the nonlinear equation of Eq. 46 can be written in incremental form as

$$ \iint_{{}^0\Omega} {}_{0}C_{ijrs}\, {}_{0}\varepsilon_{rs}\, {}_{0}\bar{\varepsilon}_{ij}\; {}^{0}d\Omega + \iint_{{}^0\Omega} {}^{t}_{0}S_{ij}\, {}_{0}\bar{\eta}_{ij}\; {}^{0}d\Omega = \ell_u(\bar{z}) - \iint_{{}^0\Omega} {}^{t}_{0}S_{ij}\, {}_{0}\bar{e}_{ij}\; {}^{0}d\Omega, \quad \text{for all } \bar{z} \in Z \tag{51} $$

where ${}_{0}C_{ijrs}$ is the incremental material property tensor, and ${}_{0}e_{ij}$ and ${}_{0}\eta_{ij}$ are the linear and nonlinear parts of ${}_{0}\varepsilon_{ij}$, respectively [22, 27]. If Eq. 51 is linearized by assuming ${}_{0}\varepsilon_{ij}={}_{0}e_{ij}$, an approximate equation can be obtained,

$$ \iint_{{}^0\Omega} {}_{0}C_{ijrs}\, {}_{0}e_{rs}\, {}_{0}\bar{e}_{ij}\; {}^{0}d\Omega + \iint_{{}^0\Omega} {}^{t}_{0}S_{ij}\, {}_{0}\bar{\eta}_{ij}\; {}^{0}d\Omega = \ell_u(\bar{z}) - \iint_{{}^0\Omega} {}^{t}_{0}S_{ij}\, {}_{0}\bar{e}_{ij}\; {}^{0}d\Omega, \quad \text{for all } \bar{z} \in Z \tag{52} $$

with the incremental constitutive law

$$ {}_{0}S_{ij} = {}_{0}C_{ijrs}\, {}_{0}e_{rs} \tag{53} $$

The right side in Eq. 52 represents the out-of-balance virtual work results from the linearization performed to derive Eq. 52. The solution step with the incremental Eq. 52 is repeated until the out-of-balance virtual work is zero, to within a certain convergence measure [27].

Define the first and second terms on the left side of Eq. 52 as the linear strain energy form A_u and the negative of the nonlinear strain energy form D_u, respectively. Also, define the linearized total strain energy form a_u^* as [22]

$$a_u^*({}^tz; z, \bar{z}) \equiv A_u({}^tz; z, \bar{z}) - D_u({}^tz; z, \bar{z}) \tag{54}$$

Then Eq. 52 can be rewritten as

$$a_u^*({}^tz; z, \bar{z}) \equiv A_u({}^tz; z, \bar{z}) - D_u({}^tz; z, \bar{z})$$

$$= \ell_u(\bar{z}) - \iint_{{}^0\Omega} {}^t_0S_{ij}\, {}_0\bar{e}_{ij}\, {}^0d\Omega \quad \text{for all } \bar{z} \in Z \tag{55}$$

and, from Eqs. 49, 50, and 52,

$$A_u({}^tz; z, \bar{z}) \equiv \iint_{{}^0\Omega} {}_0C_{ijrs}\big({}_0z_{r,s}\, {}_0\bar{z}_{i,j} + {}^t_0z_{k,i}\, {}_0z_{r,s}\, {}_0\bar{z}_{k,j} + {}^t_0z_{k,r}\, {}_0z_{k,s}\, {}_0\bar{z}_{i,j}$$

$$+ {}^t_0z_{k,r}\, {}^t_0z_{l,i}\, {}_0z_{k,s}\, {}_0\bar{z}_{l,j} \big)\, {}^0d\Omega \tag{56}$$

and

$$D_u({}^tz; z, \bar{z}) \equiv -\frac{1}{2} \iint_{{}^0\Omega} {}^t_0C_{ijrs}\big({}^t_0z_{r,s} + {}^t_0z_{s,r} + {}^t_0z_{m,r}\, {}^t_0z_{m,s} \big)\, {}_0z_{k,i}\, {}_0\bar{z}_{k,j}\, {}^0d\Omega \tag{57}$$

The fact that the incremental material property tensor ${}_0C_{ijrs}$ and the material property tensor ${}^t_0C_{ijrs}$ are symmetric with respect to their indices has been used in Eqs. 56 and 57. Thus, the linearized total strain energy bilinear form $a_u^*({}^tz; \bullet, \bullet)$ of Eq. 55 is symmetric in its arguments [27]. The constitutive law

$${}^t_0S_{ij} = {}^t_0C_{ijrs}\, {}^t_0\varepsilon_{rs} \tag{58}$$

and the Green-Lagrange strain-displacement relationship in Eq. 47 are used in Eq. 57.

Stability Equation

Consider a structural system with the equilibrium path as shown in Fig. 4. The critical limit point in Fig. 4 is a relative maximum point in the nonlinear load-displacement curve. This point defines the boundary between the prebuckling and the postbuckling equilibrium paths.

Eigenvalue Problem

Various eigenvalue problems have been suggested to evaluate the stability of nonlinear structural systems [28-30]. The estimated critical load factor of nonlinear structural systems can be obtained by solving the eigenvalue problems at a prebuckling equilibrium configuration. The mathematical derivation of the eigenvalue problem follows from the incremental equilibrium equations; i.e., the left side of Eq. 55 vanishes at the critical limit point since the out-of-balance virtual work is zero. With the total displacement ${}^{cr}z$ at the critical limit point t=cr, the stability equation becomes

$$A_u({}^{cr}z; y, \bar{y}) - D_u({}^{cr}z; y, \bar{y}) = 0, \quad \text{for all } \bar{y} \in Z \tag{59}$$

where the incremental displacement z in Eq. 47 is replaced with y to distinguish this from a real incremental displacement z. The existence of a nontrivial incremental solution y of this nonlinear equation serves to identify a point of instability. That is, if the final equilibrium configuration is not the critical limit point, the solution y of Eq. 59 must be zero.

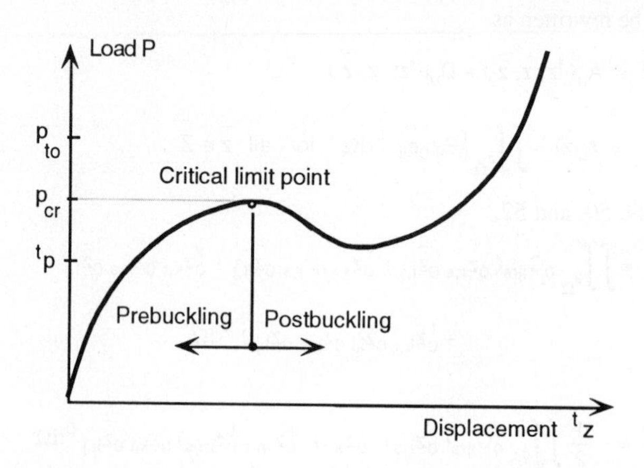

Figure 4 Equilibrium Path of a Nonlinear Structural System

To estimate the critical load, it is necessary to evaluate the left side of Eq. 59 at the projected critical limit point, using available information at a prebuckling equilibrium configuration at time t≤cr. Utilizing information at the prebuckling equilibrium configuration at time t, Eq. 59 can be rewritten as an eigenvalue problem. Two common approaches - one- and two-point linear eigenvalue problems - are formulated in variational forms [28-30]. In this paper, the one-point eigenvalue problem is used for design sensitivity analysis [28, 29]. For DSA that uses the two-point eigenvalue problem, see Ref. 30. By linearizing the nonlinear relationship between the negative of the nonlinear strain energy form D_u and the additional load increment, D_u at the critical limit point is approximated by

$$D_u({}^{cr}z; y, \bar{y}) = {}^t\zeta \, D_u({}^tz; y, \bar{y}) \tag{60}$$

with ${}^t\zeta$ representing the estimated critical load factor at the prebuckling equilibrium configuration time t. By neglecting variations in the linear strain energy form A_u due to a loading change,

$$A_u({}^{cr}z; y, \bar{y}) = A_u({}^tz; y, \bar{y}) \tag{61}$$

Equation 59 then becomes the one-point linear eigenvalue problem,

$$A_u({}^t z; y, \bar{y}) - {}^t\zeta\, D_u({}^t z; y, \bar{y}) = 0\,, \qquad \text{for all } \bar{y} \in Z \tag{62}$$

where ${}^t\zeta$ is the lowest eigenvalue of the eigenvalue problem, and y is the corresponding eigenvector. Note that ${}^t\zeta$ is a function of design variable u and that ${}^t\zeta \geq 1$. If the total applied load vector ${}^t p$ is equal to the critical load vector $p_{cr} = {}^t p$, then ${}^t\zeta = 1$ and Eq. 62 is the same as Eq. 59.

Stability Equation

To obtain the actual critical load factor, assume that the magnitude of the intended total applied load vector p_{to} is larger than the magnitude of the critical load vector p_{cr} and that the critical load vector p_{cr} is equal to the applied load vector ${}^t p$ at the critical limit point at time t. Note that the load vectors p_{cr}, p_{to}, and ${}^t p$ have the same directions, since they are assumed to be proportional loadings. The magnitude of the critical load is unknown before the system is analyzed. With the total strain energy a_u at the prebuckling equilibrium configuration at time t=cr and the load form ℓ_u, which is the virtual work done by the intended total applied load vector p_{to}, the equilibrium Eq. 46 can be rewritten as (note that it is assumed that the critical limit point occurs at t)

$$a_u({}^t z, \bar{z}) = \beta_u\, \ell_u(\bar{z}), \qquad \text{for all } \bar{z} \in Z \tag{63}$$

where the actual critical load factor β_u is defined as the ratio of the magnitude of the critical load vector $p_{cr} = {}^t p$ to the magnitude of the intended total applied load vector p_{to}, i.e.,

$$p_{cr} = \beta_u\, p_{to} \tag{64}$$

The actual critical load factor β_u can be evaluated only after the critical load is known. Note that β_u is a function of design variable u and that $\beta_u \leq 1$. If the total applied load vector p_{to} is equal to the critical load vector $p_{cr} = {}^t p$, then $\beta_u = 1$, and equilibrium Eq. 24 is the same as equilibrium Eq. 7.

3.2 Sizing DSA of Static Response

In this section, the linearized incremental Eq. 55 is used to obtain the first variation of the nonlinear equilibrium equation with respect to sizing design variables such as cross-sectional geometry of truss/beam and thickness of membrane/plate. Both the adjoint variable and direct differentiation methods are presented for the static response. Using the adjoint variable method that parallels the method presented in Ref. 7 for linear structural systems, a linear adjoint equation is obtained for each performance measure of which the design sensitivity is to be computed. If the finite element method is used for numerical evaluation, the stiffness matrix of the adjoint system is the tangent stiffness matrix of the nonlinear system at the final equilibrium configuration. For the direct differentiation method, the first variation of the state is obtained by solving the linear equation resulting from differentiation of the nonlinear equilibrium equation with respect to the design variable. If the finite element method is used, the stiffness matrix of the linear equation is the same tangent stiffness matrix as in the adjoint variable method. Equations of DSA include the effects of large displacements, large strains, and material nonlinearities, with appropriate kinematic and constitutive relations.

Assume the structural system is in the equilibrium configuration at time t for a given design u. When the design is perturbed to $u+\tau\delta u$, the structural system may reach an equilibrium configuration at time $t+\Delta t$. Using Eq. 46, equilibrium equations can be written as

$$a_u({}^t z, \bar{z}) = \ell_u(\bar{z}), \qquad \text{for all } \bar{z} \in Z \tag{65}$$

and

$$a_{u+\tau\delta u}({}^{t+\Delta t}z, \bar{z}) = \ell_{u+\tau\delta u}(\bar{z}), \qquad \text{for all } \bar{z} \in Z \tag{66}$$

As the magnitude of the design perturbation $\tau\delta u$ becomes smaller, the difference between the equilibrium states of the original and perturbed designs will become smaller. Hence, as the design perturbation vanishes, so does the difference in state; i.e., $\Delta t \to 0$ as $\tau \to 0$.

Define the first variations of the nonlinear energy and load forms in Eq. 66 with respect to their explicit dependence on the design variable u as [22]

$$a'_{\delta u}({}^t z, \bar{z}) \equiv \frac{d}{d\tau} a_{u+\tau\delta u}({}^t\tilde{z}, \bar{z}) \Big|_{\tau=0} \tag{67}$$

and

$$\ell'_{\delta u}(\bar{z}) \equiv \frac{d}{d\tau} \ell_{u+\tau\delta u}(\bar{z}) \Big|_{\tau=0} \tag{68}$$

where ${}^t\tilde{z}$ denotes the state ${}^t z$ with dependence on τ (design variable u) suppressed, and $\bar{z}$ is independent of τ. Define the first variations of the solutions of the equilibrium equations at $\tau=0$ with respect to the design variable u as

$$z' \equiv \frac{d}{d\tau} {}^{t+\Delta t}z(u+\tau\delta u) \Big|_{\tau=0} = \lim_{\tau\to 0} \frac{{}^{t+\Delta t}z(u+\tau\delta u) - {}^t z(u)}{\tau} \tag{69}$$

Using the above definition, the order of taking the first variation and the partial derivative of the response can be interchanged; i.e., $({}_0 z_{i,j})' = ({}_0 z'_i)_{,j}$, i,j=1,2,3 [22].

Since the energy form $a_{u+\tau\delta u}$ in Eq. 66 is nonlinear in ${}^{t+\Delta t}z$, the chain rule of differentiation can be used to obtain [22]

$$\frac{d}{d\tau} \{ a_{u+\tau\delta u} [{}^{t+\Delta t}z(u+\tau\delta u), \bar{z}] \} \Big|_{\tau=0} = a'_{\delta u}({}^t z, \bar{z}) + a^*_u({}^t z; z', \bar{z}) \tag{70}$$

For the second term on the right of Eq. 70, the linearization process from Eq. 46 to Eq. 55 has been used. By taking the first variations of both sides of Eq. 66 and using Eqs. 68 and 70,

$$a^*_u({}^t z; z', \bar{z}) = \ell'_{\delta u}(\bar{z}) - a'_{\delta u}({}^t z, \bar{z}), \qquad \text{for all } \bar{z} \in Z \tag{71}$$

where explicit expression for $a_u^*({}^tz;z',\bar{z})$ can be obtained by replacing z with z' in Eq. 54. Presuming that the state ${}^tz$ is known as the solution of Eq. 65, Eq. 71 is the variational equation for the first variation z'. As mentioned before, since the total Lagrangian formulation includes large displacements, large rotations, large strains, and material nonlinearities when appropriate kinematic and constitutive descriptions are used, Eq. 71 is valid for these cases. However, in this paper, only infinitesimal strains are considered.

For the adjoint variable method, consider a structural performance measure that can be written in integral form, at the final equilibrium configuration time t+Δt with reference configuration time 0,

$$\psi_\tau = \iint_{{}^0\Omega} g({}^{t+\Delta t}z,\ {}_0\nabla^{t+\Delta t}z,\ u+\tau\delta u)\,{}^0d\Omega \tag{72}$$

where

$${}_0\nabla^{t+\Delta t}z = [\,{}_0\nabla^{t+\Delta t}z_1,\ {}_0\nabla^{t+\Delta t}z_2,\ {}_0\nabla^{t+\Delta t}z_3]$$

Taking the first variation of the functional of Eq. 72 with respect to the sizing design variable,

$$\psi' = \iint_{{}^0\Omega} (g_{{}^tz}\, z' + g_{{}_0\nabla^tz}\ {}_0\nabla z' + g_u\,\delta u\,)\,{}^0d\Omega \tag{73}$$

where $g_{{}_0\nabla^tz} = \left[\dfrac{\partial g}{\partial\, {}^t_0 z_{i,j}}\right]$. Since explicit expression for ψ' is desired in terms of δu, the first two terms of the integrand of Eq. 73 must be rewritten explicitly in terms of δu.

Introducing an adjoint equation in the same manner as for the linear structural systems in Ref. 7, i.e., by replacing z' in Eq. 73 by a virtual displacement $\bar{\lambda}$ and equating terms involving $\bar{\lambda}$ in Eq. 73 to the energy bilinear form $a_u^*({}^tz;\lambda,\bar{\lambda})$ defined in Eq. 54, yields the adjoint equation for the adjoint variable λ,

$$a_u^*({}^tz;\,\lambda,\,\bar{\lambda}) = \iint_{{}^0\Omega} (g_{{}^tz}\,\bar{\lambda} + g_{{}_0\nabla^tz}\ {}_0\nabla\bar{\lambda}\,)\,{}^0d\Omega, \qquad \text{for all } \bar{\lambda}\in Z \tag{74}$$

where a solution λ is sought. Since z'∈Z, evaluate Eq. 74 at $\bar{\lambda}$=z' to obtain

$$a_u^*({}^tz;\,\lambda,\,z') = \iint_{{}^0\Omega} (g_{{}^tz}\, z' + g_{{}_0\nabla^tz}\ {}_0\nabla z'\,)\,{}^0d\Omega \tag{75}$$

Since $\bar{z}$ and λ belong to the same space of kinematically admissible virtual displacement Z, evaluate Eq. 71 at z=λ to obtain

$$a_u^*({}^tz;\,z',\,\lambda) = \ell'_{\delta u}(\lambda) - a'_{\delta u}({}^tz,\,\lambda) \tag{76}$$

Using symmetry of the energy bilinear form $a_u^*({}^tz;\,\bullet,\,\bullet)$ in its arguments,

$$\iint_{{}^0\Omega} (g_{{}^tz}\, z' + g_{{}_0\nabla^tz}\ {}_0\nabla z'\,)\,{}^0d\Omega = \ell'_{\delta u}(\lambda) - a'_{\delta u}({}^tz,\,\lambda) \tag{77}$$

where the right side is linear in δu and can be evaluated once the state ${}^tz$ and the adjoint variable λ are determined as solutions of Eqs. 65 and 74, respectively. Substituting this result in Eq. 73,

$$\psi' = \iint_{{}^0\Omega} g_u\,\delta u\ {}^0d\Omega + \ell'_{\delta u}(\lambda) - a'_{\delta u}({}^tz,\,\lambda) \tag{78}$$

which expresses the dependence of design sensitivity on the design variation, and the form of the last two terms on the right depends on each design component. Explicit expressions for these two terms are given in Refs. 22-24 for some prototype structural components. An interesting fact is that, even though the original governing equation is nonlinear, the sensitivity Eq. 74 and adjoint Eq. 35 are linear and have the same tangent stiffness matrix at the final equilibrium configuration. This means that effort to compute design sensitivity will be the same for both linear and nonlinear structural systems [23].

3.3 Sizing DSA of Critical Load

In this section, DSA is carried out for the actual critical load factor by taking the first variation of the stability equation with respect to sizing design variables. Design sensitivity of the estimated critical load factor is given in Ref. 30. Suppose that the design is perturbed by $\tau\delta u$. The new stability equation of the structural system at the critical configuration at time $t+\Delta t$ for the perturbed design $u+\tau\delta u$ becomes

$$a_{u+\tau\delta u}({}^{t+\Delta t}z, \bar{z}) = \beta_{u+\tau\delta u}\, \ell_{u+\tau\delta u}(\bar{z}), \qquad \text{for all } \bar{z} \in Z \tag{79}$$

The first variation of both sides of Eq. 79 can be taken to obtain

$$a'_{\delta u}({}^{t}z, \bar{z}) + a^{*}_{u}({}^{t}z; z', \bar{z}) = \beta'\, \ell_u(\bar{z}) + \beta_u\, \ell'_{\delta u}(\bar{z}), \qquad \text{for all } \bar{z} \in Z \tag{80}$$

where a^*_u is the first variation of a_u with respect to design variables, implicitly through the total displacement ${}^tz$, and is defined in Eq. 54. Since Eq. 80 holds for all $\bar{z}\in Z$, this equation may be evaluated at $\bar{z}=y$, which is the eigenfunction corresponding to the lowest eigenvalue of the linear eigenvalue problem of Eq. 62. Then Eq. 80 becomes

$$a'_{\delta u}({}^{t}z, y) + a^{*}_{u}({}^{t}z; z', y) = \beta'\, \ell_u(y) + \beta_u\, \ell'_{\delta u}(y) \tag{81}$$

Using Eq. 54 and the symmetry of energy forms $A_u({}^tz; \bullet, \bullet)$ and $D_u({}^tz; \bullet, \bullet)$ in their arguments, a^*_u in Eq. 81 can be rewritten as

$$a^{*}_{u}({}^{t}z; z', y) \equiv A_u({}^{t}z; y, z') - D_u({}^{t}z; y, z') \tag{82}$$

At the critical limit point with ${}^t\zeta=1$ and $z'\in Z$, the right side of Eq. 82 vanishes using Eq. 62. Thus, in Eq. 80 with $\bar{z}=y$, the second term of the left hand side vanishes and, using Eq. 64, the sizing design sensitivity expression of the critical load becomes

$$p'_{cr} = p_{to}\, \beta' = \frac{p_{to}}{\ell_u(y)} [\, a'_{\delta u}({}^{t}z, y) - \beta_u\, \ell'_{\delta u}(y)] \tag{83}$$

In Eq. 83, ${}^tz$ is the displacement at the final prebuckling equilibrium configuration. Note that, for evaluation of the design sensitivity expression in Eq. 83, no adjoint system is required, unlike the design sensitivity analysis of the estimated critical load in Ref. 30.

3.4 Example - Vehicle Passenger Compartment Frame

Consider a vehicle passenger compartment that is modeled as a planar, closed frame structure whose outline matches the projections of the two side-frame assemblies on the vertical fore-and-aft body central plane, as shown in Fig. 5 [31]. This finite element model is composed of 45 equivalent beam members. The moment of inertia (about the axis perpendicular to the x_1-x_2 plane) and cross-sectional area are equal to the sum of the corresponding members of both side frames. The six flexible joints at nodes 5, 11, 17, 26, 33, and 38 in Fig. 5 correspond to the six major joint assemblies in each side frame of a passenger compartment. An important characteristic concerning the flexibility of a body joint is its nonlinear moment-rotation (M-θ) relationship. The M-θ relationship of joints is given as a piecewise linear curve, as shown in Fig. 6.

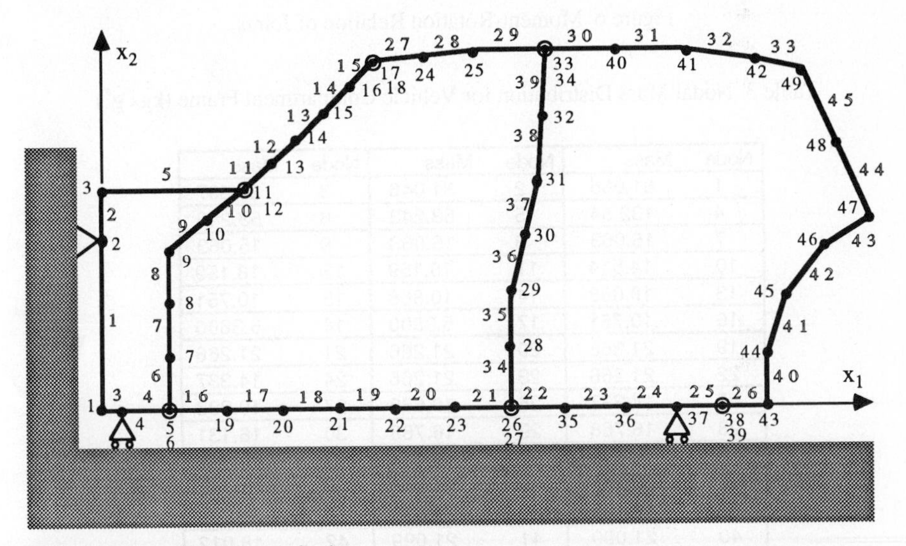

◉ ; Nonlinear Rotational Spring

Figure 5 Vehicle Passenger Compartment Frame

For the loading condition, consider the case of vehicle frontal impact by collision with a barrier at node 2. During deceleration from a certain speed to zero, a part of the kinetic energy of the vehicle is dissipated by the work resulting from a plastic deformation of the front portion of the automobile passenger compartment. In this example, it is assumed that the nonstructural weight of 1045 kg (75% of the vehicle sprung mass) and the structural frame weight are imposed on the frame as an inertia load in the horizontal direction during the deceleration time. Out of nonstructural mass, 35% of the sprung mass is distributed over the bottom rail as nodal masses, while 65% is distributed over the other nodal points. The engine/transmission mass is distributed at nodes 1, 4, and 5. The mass distribution at each nodal point is shown in Table 3 for the initial design.

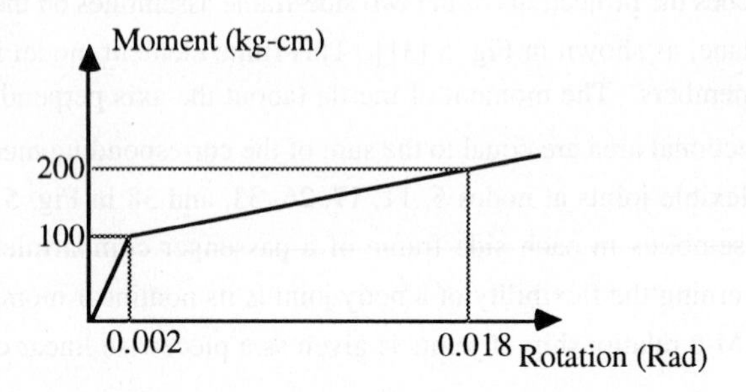

Figure 6 Moment-Rotation Relation of Joints

Table 3 Nodal Mass Distribution for Vehicle Compartment Frame (kg× g*)

Node	Mass	Node	Mass	Node	Mass
1	61.005	2	31.046	3	28.567
4	132.54	5	53.243	6	53.243
7	15.063	8	15.063	9	15.063
10	14.514	11	18.159	12	18.159
13	10.868	14	10.868	15	10.751
16	10.751	17	5.3800	18	5.3800
19	21.266	20	21.266	21	21.266
22	21.266	23	21.266	24	14.337
25	14.337	26	10.633	27	10.633
28	16.768	29	16.768	30	16.131
31	16.313	32	18.581	33	20.306
34	20.306	35	21.266	36	21.266
37	18.424	38	7.8010	39	7.8010
40	21.099	41	21.099	42	18.012
43	7.8010	44	14.965	45	16.885
46	18.512	47	15.964	48	22.687
49	20.355				

* The gravitational acceleration (9.8 m/sec^2)

Selected design variables are 5 cross-sectional areas of beam members. Element groups for each design variable are shown in Table 4. For a fixed cross-sectional type, the moment of inertia can be uniquely defined as $I = cA^2$, where c is a constant that depends on the cross-sectional type, and A is the cross-sectional area. In this example, a rectangular hollow cross section is used with two geometric ratios fixed between width b and height h, and width b and thickness t. In Table 5, properties of 3 rectangular cross-sectional types are given with corresponding design variable numbers. The value of the design vector is [100, 100, 100, 100, 100 $]^T$ cm^2.

Table 4 Design Variable Linking for Vehicle Compartment Frame

Design	Elements Linked	Number of Element
A_1	6-15	10
A_2	16-26	11
A_3	27-33	7
A_4	34-39	6
A_5	40-45	6

Table 5 Properties of Rectangular Hollow Cross-Sections for Vehicle Compartment Frame

Cross-section #	Design id.	h/b	t/b	$c=I/A^2$
1	A_1, A_4, A_5	1.5	0.075	0.8609
2	A_2	0.5	0.05	0.2866
3	A_3	1.0	0.05	0.7939

For structural analysis, STIF3 (2-dimensional beam element), STIF39 (nonlinear force-deflection element), and STIF21 (general mass) of ANSYS [9] are used for beam members, flexible joints, and nodal masses, respectively. For linear elastic material of the beam members, an aluminum alloy with Young's modulus of $E=7.4 \times 10^5$ kg/cm^2 is used. The specific weight of 2.69 $gram/cm^3$ is used for the self-weight of beam members. Using the incremental analysis method of ANSYS for the initial design, the critical deceleration d_{cr} is found to be between 3.41278g and 3.41279g, where g is the gravitational acceleration. In other words, the vehicle frame buckles when deceleration d_{cr}=3.41279g.

The sizing design sensitivity coefficients of the critical deceleration are evaluated using design sensitivity expressions in Eq. 83. In structural analysis of the frame, inertia loads of structural masses are imposed with the applied deceleration. In particular, the inertia loads due to self-weight of the frame are proportional to the cross-sectional areas. In other words, the load form ℓ_u in Eq. 63 depends on cross-sectional areas.

The last column of Table 6 presents design sensitivity coefficients of the critical deceleration with a uniform design for several deceleration levels between d=3.0g and 3.41278g. The second and third columns represent the lowest eigenvalue and the design sensitivity coefficient of the actual critical deceleration factor. As in Eq. 83, the design sensitivity of the critical deceleration in the last column is the result of multiplying the applied deceleration d/g in the first column with the design sensitivity of the actual critical deceleration factor $\partial\beta/\partial A_{uniform}$ in the third column.

The design sensitivity results are verified to be accurate using the finite difference method. Design sensitivity coefficients with uniform design at the applied deceleration level d=3.41278g are compared with the results of the finite difference method for 3 design perturbations: 5%, 1%, and 0.1%. In Table 7, $\Delta d_{cr}/g$ is the finite difference of the critical deceleration divided by g, and d'_{cr}/g is the change predicted by the proposed design sensitivity method for the corresponding design perturbation. The agreements between Δd_{cr} and d'_{cr} in the last column of Table 7 approach 101.4% with 0.1% design perturbation. As Table 7 shows, the results of the finite difference method converge to the results of this work.

Table 6 Design Sensitivity of Critical Deceleration for Vehicle Compartment Frame

d/g	$^t\zeta$	$\partial\beta/\partial A_{uniform}$	$\partial d_{cr}/\partial A_{uniform}$
3.00000	2.2622	0.018158	0.054474
3.40000	1.1532	0.018110	0.061574
3.41000	1.0694	0.018103	0.061731
3.41200	1.0362	0.018100	0.061757
3.41270	1.0117	0.018097	0.061759
3.41278	1.0024	0.018096	0.061758

Table 7 Verification of Design Sensitivity of Critical Deceleration Using Finite Difference Method for Vehicle Compartment Frame (Applied Deceleration = 3.41278g)

Area	Perturbation	d_{cr}/g	$\Delta d_{cr}/g$	d'_{cr}/g	$d'_{cr}/\Delta d_{cr}$
100.0	-	3.41279	-	-	-
110.0	10.0%	4.00490	0.59211	0.61758D-0	104.3%
101.0	1.0%	3.47290	0.06012	0.61758D-1	102.7%
100.1	0.1%	3.41888	0.00609	0.61758D-2	101.4%

The design sensitivity vector for the critical deceleration factor b with respect to the areas A_i, i = 1, 5 at the final deceleration d = 3.41278g is [0.22141D-5, 0.13299D-1, 0.10683D-2, 0.83548D-4, 0.36434D-2 $]^T$. At the initial design, the cross-sectional area A_2 of the bottom rail members (elements 16-26) is most effective for controlling the buckling behavior. The cross-sectional area A_2 of the wind shield pillar (A-pillar) has no significant effect on the critical deceleration.

4 Design Sensitivity Analysis and Optimization (DSO) Tool

In the past few years, the advent of powerful graphics-based engineering workstations with increasing computational power has created an ideal environment for making interactive design optimization a viable alternative to more monolithic batch-based design optimization. The Center for Simulation and Design Optimization at The University of Iowa exploits the multi-window environment provided by engineering workstations to develop a highly interactive menu-driven system for structural sizing design sensitivity analysis and optimization -- the Design Sensitivity Analysis and Optimization (DSO) tool [32-34]. The DSO integrates design procedures by letting the designer create a geometrical model, build a finite element model, parameterize the geometric model, perform finite element analysis, visualize finite element analysis results, characterize performance measures, and carry out DSA. To improve designs, design engineers will use the DSO to carry out a four-step interactive design process: (1) visually display design sensitivity information, (2) carry out what-if studies, (3) make trade-off determinations, and (4) execute interactive design optimization [35].

Design parameterization, which allows the designer to define geometry properties for each design component of the structural system being designed, is treated as a vital design step in the DSO [34]. Design parameterization forces the design, analysis, and manufacturing engineering groups to interact at an early stage, and supports a unified design parameter set to be used as the common ground for these engineering groups to carry out analysis, design, and manufacturing processes. As a result of design parameterization, CAD, CAE, and CAM procedures are tied to form a concurrent engineering design environment. Currently, the DSO integrates CAD and CAE disciplines.

4.1 Three Design Stages

Three design stages, pre-processing, design sensitivity computation, and post-processing, have been implemented in the DSO to support structural sizing design. The major goals in the pre-processing design stage are to formulate a design problem by creating a geometric model using PATRAN [36], converting the geometric model to a design model through the DSO's design parameterization process, generating a finite element model for analysis, and defining structural performance measures. In the DSA stage, a design sensitivity coefficient matrix is computed for the design problem defined in the pre-processing stage. In the post-processing design stage, a four-step interactive design process helps the designer understand structural behaviors at the current design and improve the design by utilizing the design sensitivity information computed in the previous stage. The three design processes have been integrated using a MOTIF-based menu-driven user interface [37].

Pre-processing Design Stage

The pre-processing design stage consists of geometry and finite element model generation, design parameterization, finite element analysis, analysis result evaluation, and performance measure definition. Figure 7 shows the methodology used in the DSO to support the pre-processing design stage.

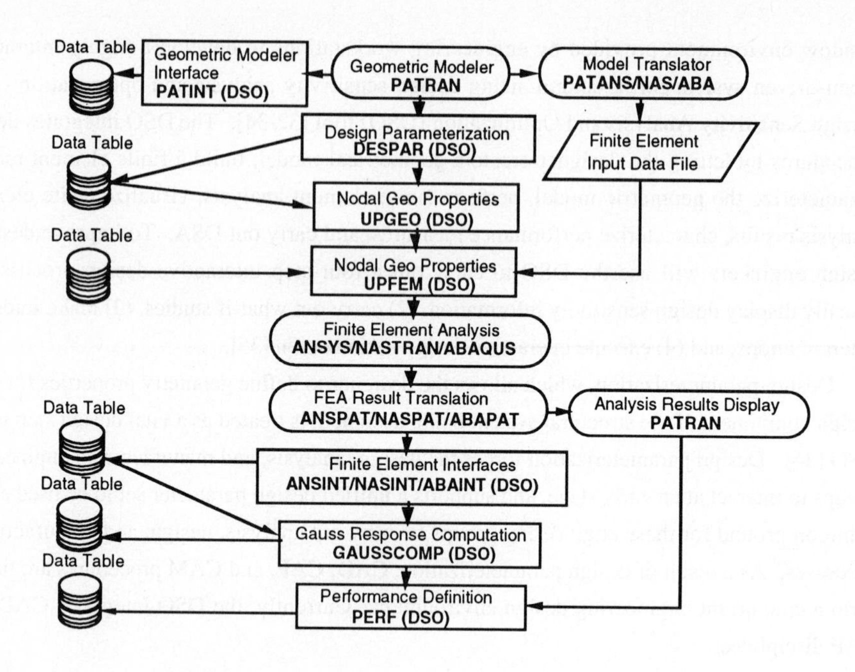

Figure 7 Pre-Processing Stage of the Design Process

Geometric Model Creation

PATRAN [37], from PDA Engineering, is used for geometry modeling and finite element meshing in the DSO. Line and patch geometry entities are used to model line (truss, beam) and surface (membrane, plate) design components, respectively, both of which can be meshed for finite element analysis. Material properties as well as boundary conditions are also defined in PATRAN. Geometric properties can be defined in PATRAN for the design components with fixed geometric parameters. However, design components with thickness or cross-sectional size that can be changed during the design process need to be parameterized using the design parameterization method provided in the DSO.

PATRAN allows the designer to access various finite element analysis codes for design evaluation by generating analysis input files in specific formats through its model translation programs, such as PATANS [38] for ANSYS [9], PATNAS [39] for MSC/NASTRAN [10], and PATABA [40] for ABAQUS [11].

Design Model Creation

Design parameterization is one of the most important steps in the structural design process. The principal functionality of the design parameterization procedure is to define the geometry parameters which characterize the geometry model and to collect a subset of the geometry parameters as design parameters. These geometry parameters are defined at end grid points of a line design component and at corner grid points of a surface design component. Design parameters can be defined among the geometry parameters that characterize dimensions of cross-sectional geometry and thickness distribution for line and surface design components, respectively. In the DSO, constant and linear design parameterizations, as shown in Fig. 8, can be defined. Note that, each dimension that defines the cross-sectional geometry, such as width or height in Fig. 8(a), could be treated as a design parameter and is allowed to vary together with the same value at both ends (constant parameterization) or different values at the ends (linearly interpolated in between). A bilinear thickness distribution can be used to characterize a surface design component, as shown in Fig. 8(b).

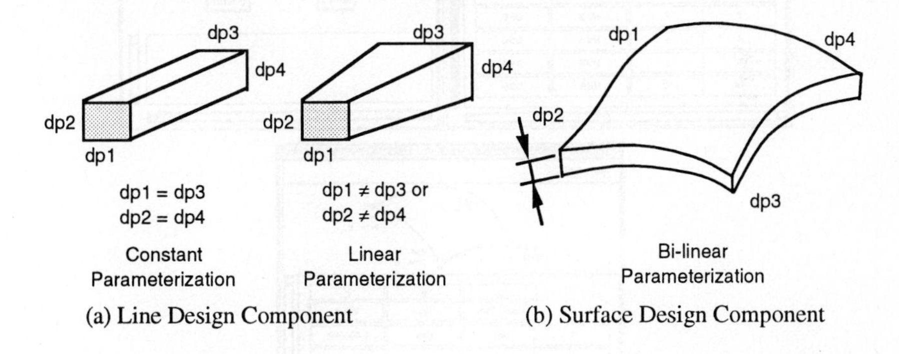

Figure 8 Constant and Linear Parameterization for Line and Surface Design Components

By interacting with the user interface menu of the DSO shown in Fig. 9, and the geometric model displayed in the PATRAN window, the user can carry out design parameterization for all design components of the structural system. The DSO will read the design parameterization information from the user's input and generate design specifications for the structure. The design specification is then written to the DSO data table for analysis model creation, design sensitivity computation, and design improvement in the post-processing stage.

<u>Analysis Model Creation</u>

In the DSO, a design model is transferred to an analysis model using the UPGEO and UPFEM modules. The UPGEO module maps the finite element type in the commercial analysis codes to the DSO element type, computes element nodal point geometric properties, and computes geometric properties at element Gauss points for design sensitivity computation. The geometric properties are computed in accordance with design parameterization. The UPFEM module then updates the finite element input data file using the nodal geometric properties computed by UPGEO to perform finite element analysis. The UPFEM module supports analysis model updates for ANSYS, MSC/NASTRAN, and ABAQUS. UPGEO, UPFEM, and finite element interface (FEINT, which is described later) make it easy to extend the DSO to accommodate other existing finite element codes.

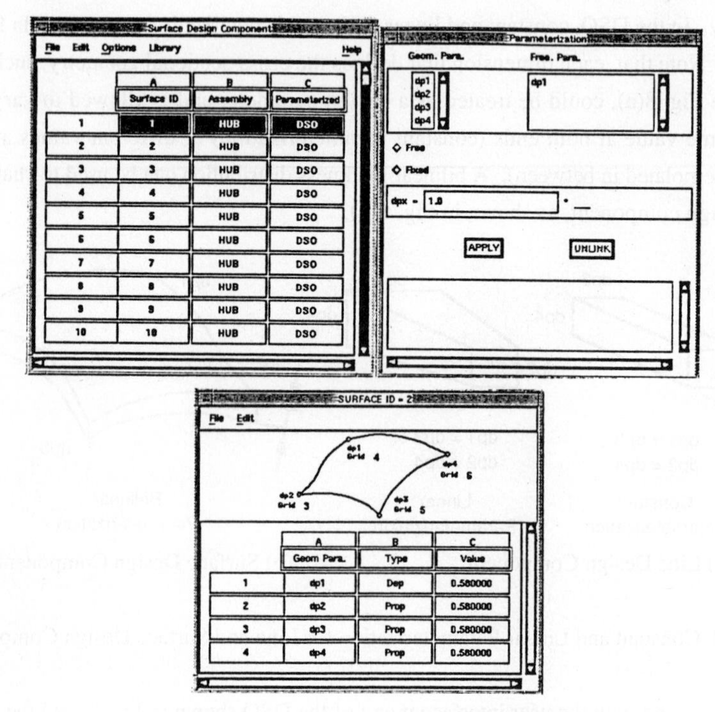

Figure 9 Design Parameterization Menu

The UPGEO and UPFEM modules also facilitate the design optimization process, in which design parameters are changed in each design iteration. With these modules, the analysis input data file can be updated automatically and efficiently.

To visualize the finite element analysis results, such as stress contour or deformed shape, the result translators provided by PATRAN, such as ANSPAT [38] for ANSYS, NASPAT [39] for MSC/NASTRAN, ABAPAT [40] for ABAQUS, are utilized in the DSO. Visualization of the analysis results in PATRAN plots helps the designer define performance measures.

Once the finite element analysis is done, the Finite Element Interface (FEINT) module is executed to retrieve finite element nodal point responses from the data file of the analysis code. Data retrieved by the interface are: (1) nodal displacements, (2) nodal stresses (if possible), and (3) eigenvalues and eigenvectors for modal analysis. The interfaces consist of ANSINT for ANSYS, NASINT for MSC/NASTRAN, and ABAINT for ABAQUS. Element Gauss point responses are then computed using GAUSSCOMP.

Performance Measure Definitions

Performance measures can be defined using finite element responses. Based on the evaluation of analysis results, engineering concerns, such as high stress spots, clearance, natural frequency, or mass, can be identified as performance measures for design improvement. Seven different types of performance measures are supported by the DSO: mass, volume, displacement, stress, compliance, buckling, and frequency [7]. Performance measure definition permits the design engineer to specify the structural performance for which design sensitivity information must be computed.

A subset of structural performance measures can be selected for visual display of design sensitivity and what-if studies after the design sensitivity computation is completed. Also, the cost and constraints can be defined by combining certain performance measures, with appropriate constraint bounds, for trade-off determination and interactive design optimization. The DSO user interface menu allows performance measures and cost and constraint functions to be defined easily and conveniently.

Design Sensitivity Computation

Using the continuum theory of DSA presented in this paper and Refs. 7 and 8, sizing design sensitivity expressions can be evaluated numerically. To evaluate design sensitivity information, the original and adjoint structural responses obtained from the finite element analysis are needed. The procedure of computing design sensitivity information is shown in Fig. 10.

The ADJLOAD module reads the performance measure definition from the DSO data tables and computes the adjoint loads for displacement and stress performance measures. The FECOM module is used to create an adjoint restart input file using the adjoint loads as additional loading cases. Since stiffness matrices of the original and adjoint structures are the same, once a decomposed stiffness matrix is obtained as the result of solving the original structural responses in the pre-processing stage, only backward and forward substitutions are required to obtain adjoint structural responses. Compared to the original structural analysis, the adjoint analyses are very efficient [7, 8]. However, as the number of performance measures increases, the amount of computational resources necessary to carry out the adjoint response computations increases significantly. To meet the turnaround requirements associated with an interactive design environment, design sensitivity computations can be distributed to a network of computers. To distribute the computation, a computation algorithm has been developed [43] that relies on a client-server model supported by Apollo's Network Computing System [41], Current Programming Support System [42], and standard UNIX features.

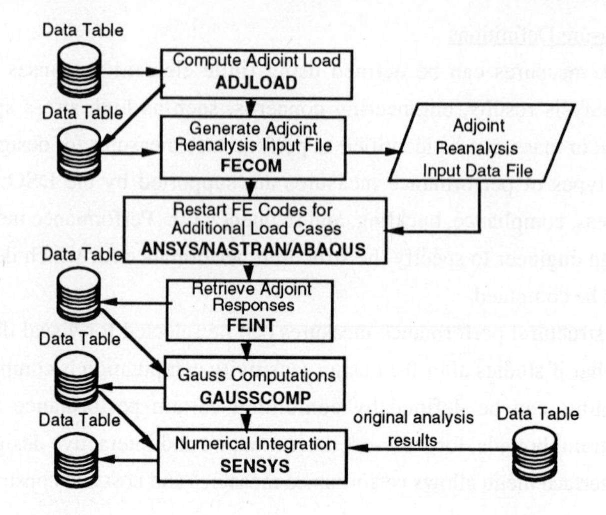

Figure 10 Procedure of Design Sensitivity Computation

The finite element interface and Gauss computation modules are executed to compute the adjoint responses at element Gauss points. Finally, the SENSYS module reads the performance measure definition and performs numerical integration to compute design sensitivity information for the structural system.

Since design sensitivity computation is carried out outside the finite element analysis code using finite element post-processing data only, various finite element codes, such as ANSYS, MSC/NASTRAN, and ABAQUS, can be accommodated for design sensitivity computation in the DSO without modifying the finite element analysis codes.

Post-processing Design Stage

The principal objective of the post-processing design stage is to utilize the design sensitivity information to improve design through the four-step design process: visual display of design sensitivity, what-if study, trade-off determination, and interactive design optimization. Figure 11 shows the four-step interactive design process used in the DSO to support the post-processing stage.

The first three design steps help the design engineer obtain a better design by understanding structural behaviors at the current design. The last design step launches commercial optimization codes to perform design optimization. Depending on the design problem, the design engineer will use some or all of the four design steps to improved the design at each design iteration. That is, new designs could be obtained from what-if, trade-off, or interactive optimization design steps. Once a new design is obtained, the UPGEO module will be executed to compute the geometry properties at nodal points for the new design, and UPFEM will update the analysis model for analysis. After finite element analysis results are obtained, design sensitivity computation can be performed only for the active constraints of the new design.

When an optimum design is obtained using the optimization algorithm, the design engineer might again display the design sensitivity or repeat the what-if study to identify sensitive design parameters that are helpful in defining manufacturing tolerance.

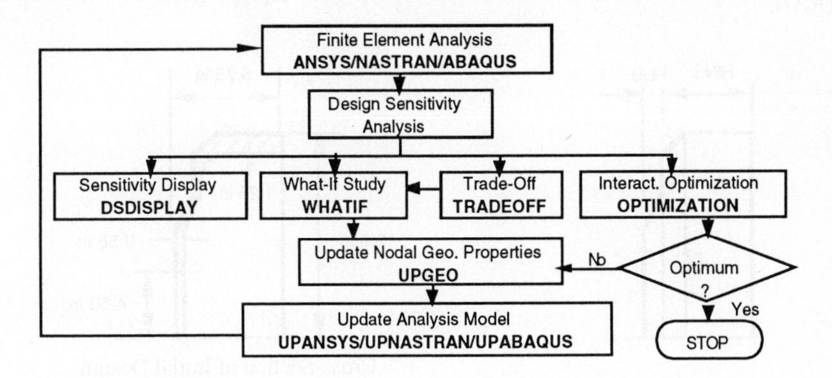

Figure 11 Post-processing Design Stage In the DSO

4.2 Example - Road Wheel

A tracked vehicle road wheel shown in Fig. 12 is utilized as an example to demonstrate the capabilities and design methodologies developed in the DSO. The objective of this application is to vary the thickness of the road wheel in order to minimize the structural volume while allowing a certain deformation at the contact point.

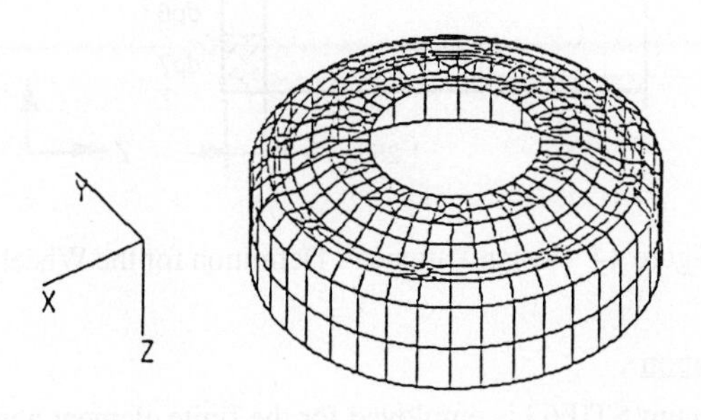

Figure 12 A Tracked Vehicle Road Wheel

Geometric Model Generation

Because the wheel is symmetric, only half of it is taken for design and analysis. The outer diameter of the wheel is 25 in, with two cross-sectional thicknesses; 1.25 in at the ring section and 0.58 in at the hub section, as shown in Fig. 13. To model the wheel, 216 patches and 432 triangular finite elements are defined in the geometric and finite element models using PATRAN.

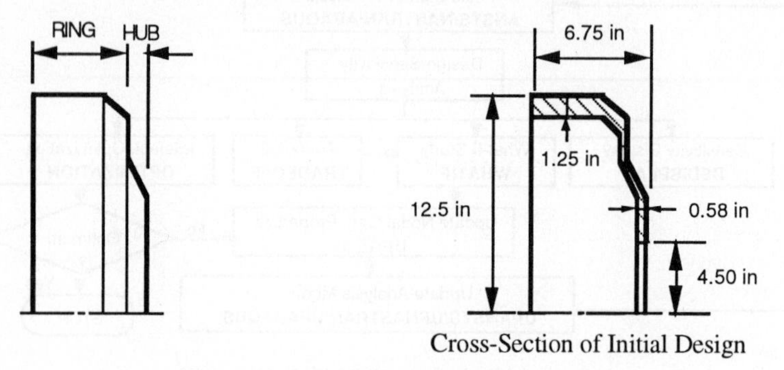

Figure 13 Road Wheel Geometric Model

Design Parameterization

Thicknesses of the wheel are defined as design parameters, which are linked along the wheel circumferential direction to maintain a symmetric design. Seven design parameters are defined for the wheel along the radial direction as shown in Fig. 14.

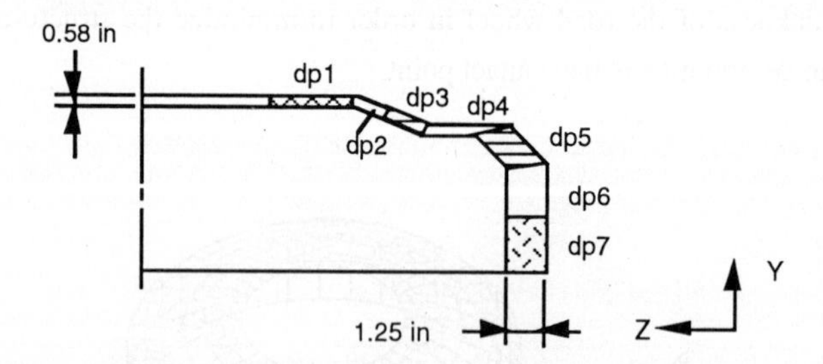

Figure 14 Design Parameter Definition for the Wheel

Analysis Model Generation

ANSYS plate element STIF63 is employed for the finite element analysis. There are 432 triangular plate elements and 1,650 degrees of freedom defined in the model, as shown in Fig. 15 (a). This wheel is made of aluminum with Young's modulus, $E = 10.5 \times 10^6$ psi, shear modulus, $G = 3.947 \times 10^6$ psi, and Poisson's ratio, $\nu = 0.33$.

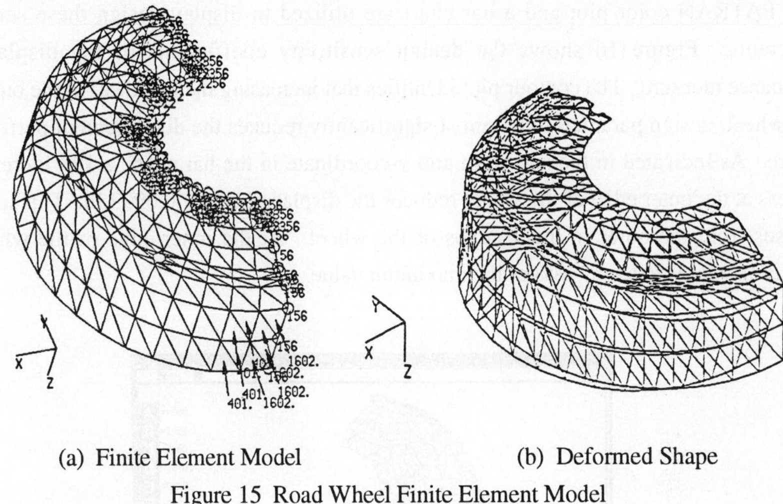

(a) Finite Element Model (b) Deformed Shape

Figure 15 Road Wheel Finite Element Model

A deformed wheel shape, obtained from the ANSYS analysis result, is displayed for evaluation using PATRAN, as shown in Fig. 15 (b). The analysis results show that the maximum displacement occurs at the contact point, node 266, in the y-direction with magnitude 0.108173 in. The volume of the structure computed by the DSO is 361.94 in^3.

Performance Measure Definition

The maximum displacement at node 266 in the y-direction and structural volume are defined as performance measures for the wheel.

Design Sensitivity Results and Display

Table 8 lists design sensitivity coefficients which were computed by the DSO using the adjoint variable method and analysis results obtained from ANSYS.

Table 8 Design Sensitivity Coefficients

Performance	dp1	dp2	dp3	dp4	dp5	dp6	dp7
Displacement	-.045865	-.01553	-.011015	-.019515	-.019420	-.056306	-.099343
Volume	35.1301	26.2068	29.8408	34.8080	47.7519	93.8567	92.4915

As shown in Table 8, sensitivity coefficients of the displacement performance measure are all negative. This result means that increasing wheel thicknesses will reduce the displacement measure. However, all sensitivity coefficients of the volume are positive, i.e., increasing wheel thicknesses will also increase wheel volume.

A PATRAN color plot and a bar chart are utilized to display design these sensitivity coefficients. Figure 16 shows the design sensitivity coefficients of the displacement performance measure. The contour plot identifies that increasing the thickness at the outer edge of the wheel, design parameter dp7, most significantly reduces the displacement performance measure. As indicated in the spectrum and y-coordinate in the bar chart, a 1 in increment of thickness at the outer edge of the wheel reduces the displacement by 0.0993 in. This influence decreases from the outer to inner edges of the wheel. At the inner edge of the wheel, the influence increases to about 40% of the maximum value.

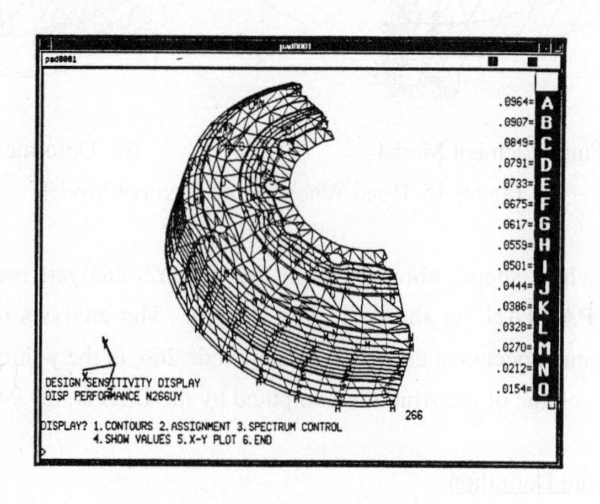

(a) PATRAN Contour Plot

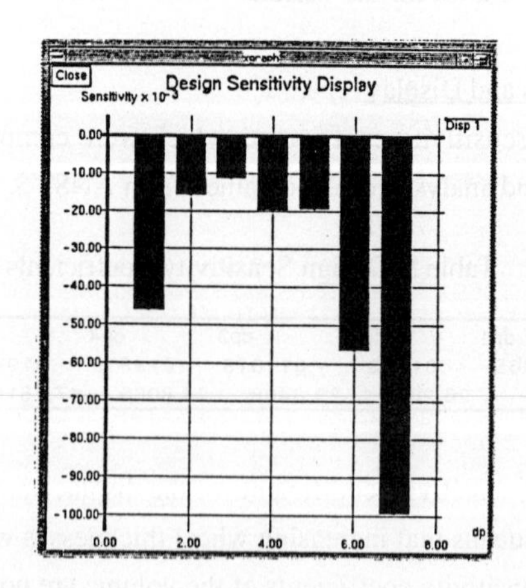

(b) Bar Chart

Figure 16 Design Sensitivity of Displacement Performance Measure

Trade-Off Determination

As the design sensitivity results show, a conflict exists between reducing the structural volume and maximum deformation. A trade-off study is therefore performed to find the best design direction. For this study, the volume performance measure is selected as the cost function, and the displacement performance measure is used to define the constraint function, with an upper bound of 0.1 in. Notice that at the current design, the displacement performance measure is 0.108173 in, which is greater than the upper bound. Therefore, the current design is infeasible. The side constraints are defined for all the design parameters with bounds 0.1 and 10.0 in, as listed in Table 9.

Table 9 Side Constraint Definitions for the Road Wheel (Unit: in.)

dpid	Lower Bound	Current Value	Upper Bound
1	0.1	0.58	10.0
2	0.1	0.58	10.0
3	0.1	0.58	10.0
4	0.1	0.58	10.0
5	0.1	1.25	10.0
6	0.1	1.25	10.0
7	0.1	1.25	10.0

Since the design is infeasible, the second algorithm that neglects cost (constraint correction) is selected for the trade-off study [44]. The DSO determines a design direction by forming forms QP (Quadratic Programming) subproblem, which is solved using the Stanford QP Solver [45]. Table 10 shows the design direction obtained from the QP Solver.

Table 10 Design Direction Obtained from Trade-Off Determination

dpid	Current Value	Direction	Perturbation	%
1	0.58	0.3596	0.03596	6.20
2	0.58	0.1218	0.01218	2.10
3	0.58	0.08638	0.008638	1.49
4	0.58	0.1530	0.01530	2.64
5	1.25	0.1523	0.01523	1.22
6	1.25	0.4415	0.04415	3.53
7	1.25	0.7790	0.07790	6.23

What-if Study

Following the design direction suggested by the trade-off determination (Table 10), a what-if study is carried out again using a step size of 0.1 in. Table 11 shows the approximate cost and constraint values using the design sensitivity coefficients and design perturbation. Also, as shown in Table 11, finite element analysis is carried out at the perturbed design to verify accuracy of the what-if predictions.

Table 11 What-if Results and Verification

Cost/Constraint	Current Value	Predicted Value	FEA Results	Accuracy
Cost	361.941	376.345	376.387	100.3
Constraint	0.108173	0.095420	0.096571	91.0

Design Optimization

The same cost, constraint, and side constraint definitions used in trade-off determination are utilized in the design optimization, which combines DOT [46], ANSYS, and local sensitivity computation and model update programs. After seven iterations, a local minimum is obtained. The optimization histories for cost, constraint, and design parameters are shown in Figs. 17(a), (b), and (c), respectively.

Figure 17(a) shows that the cost function starts around 362 in^3 and immediately jumps to 382 in^3 to correct the constraint violation. Then the cost is reduced until it achieves a minimum at 353 in^3. Also, the constraint function history plot, Fig. 17(b), shows that an 80% violation is found at the initial design, but that the violation is reduced to 65% below the upper bound at the first iteration. The constraint function then stabilizes and stays feasible for the remaining iterations. At optimum, the constraint is 4% below the upper bound, the maximum displacement becomes 0.09950 in, and the design is feasible. Interestingly, the design parameter history, Fig. 17(c), shows that all design parameters are descending in the design process, except dp1 and dp7. Design parameter dp1 increases from 0.58 to 0.65 in at the optimum, although the most significant design change is dp7, which increases from 1.25 to 1.44 in to resist deformation. Decrement of the remaining design parameters contributes to the wheel volume reduction. The design parameter values at the optimum are listed in Table 12.

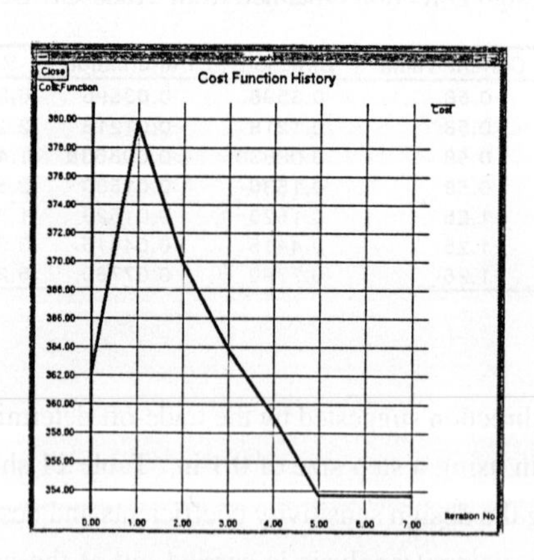

(a) Cost Function History

Figure 17 Design Optimization History

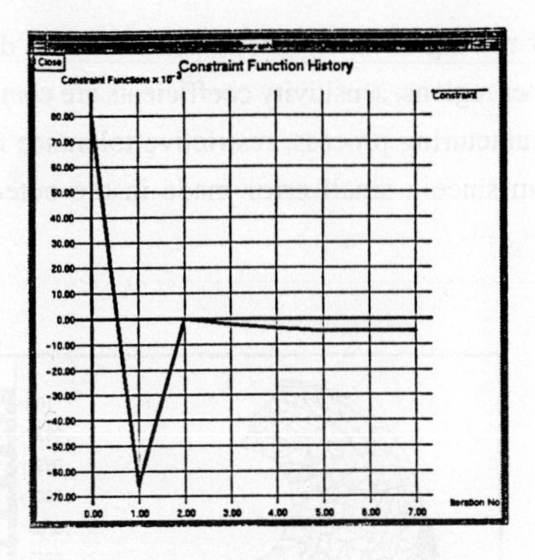

(b) Constraint Function History

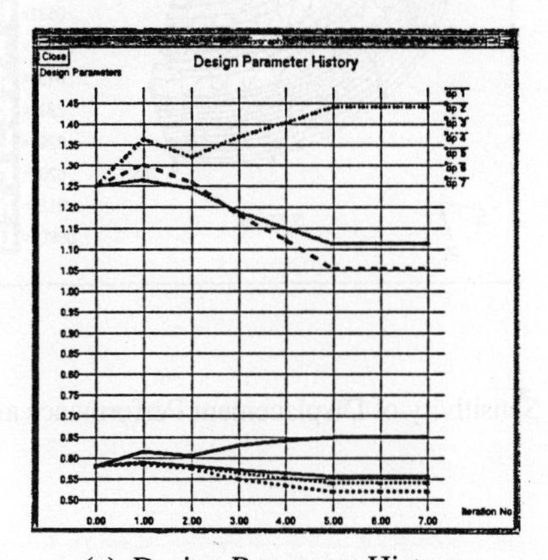

(c) Design Parameter History

Figure 17 Design Optimization History (Cont)

Table 12 Design Parameter Values at Optimum

dpid	Initial Design	Optimum Design
1	0.58	0.650
2	0.58	0.556
3	0.58	0.512
4	0.58	0.540
5	1.25	1.113
6	1.25	1.053
7	1.25	1.442

<u>Post-Optimum Study</u>

At the optimum point, the designer can still use the four-step design process to get significant information about the structural behavior. The design sensitivity plot for the displacement performance measure at the optimum design is shown in Fig. 18. The plot shows that thickness at the outer rim significantly affects the maximum displacement at the contact point. However, at the other regions, sensitivity coefficients are comparatively small. This plot suggests that, in the manufacturing process, restrictive tolerance needs to be applied to the thickness at the outer rim since a small error made in the outer rim will exaggerate the displacement measure.

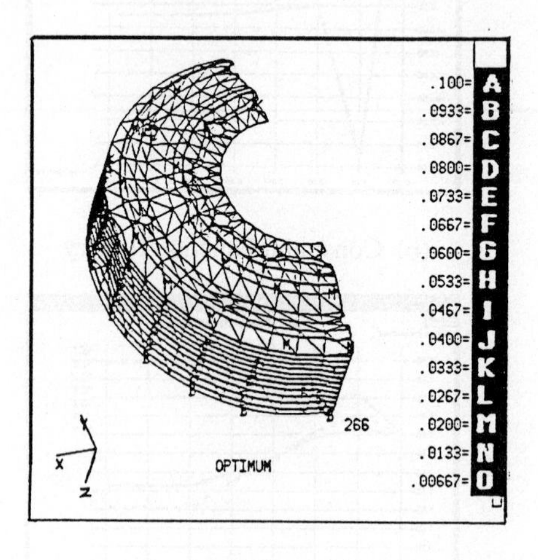

Figure 18 Design Sensitivity of Displacement Performance at Optimum Design

5 Conclusions

In this paper, a summary of a recently developed unified continuum DSA method for linear and nonlinear structural systems has been presented. Design sensitivities of static and dynamic responses of structural systems with respect to sizing design variables are summarized. It is demonstrated that the continuum DSA method allows development of a unified structural DSA capability that works with a number of well-established FEA codes. Also, it is demonstrated that the continuum based DSA method makes it possible to compute design sensitivity of analysis results of established FEA codes with respect to geometric design parameters that are employed in CAD tools, so that industry standard tools can be exploited in concurrent engineering. A DSA and optimization (DSO) tool with a visually driven user interface is developed to allow designers to easily perform structural optimum design. A road wheel example has been presented to demonstrate the DSO capabilities.

Acknowledgement: Research supported by NSF-Army-NASA Industry/University Cooperative Research Center for Simulation and Design Optimization of Mechanical Systems.

References

1. Adelman, H.M. and Haftka, R.T., "Sensitivity Analysis for Discrete Structural System," *AIAA*, Vol. 24, No. 5, 1986, pp. 831-900.
2. Haftka, R.T. and Grandhi, R.V., "Structural Shape Optimization-A Survey," *Computer Methods in Applied Mechanics and Engineering*, Vol. 57, No. 1, 1986, pp. 91-106.
3. Bennet, J.A. and Botkin, M.E., Eds., *The Optimum Shape:Automated Structural Design*, Plenum Press, New York, 1986.
4. Mota Soares, C.A., Ed., *Computer Aided Optimal Design*, Springer-Verlag, Heidelberg, 1987.
5. Eschenauer, H.A., Mattheck, C., and Olhoff, N., Eds., *Engineering Optimization in Design Processes*, Lecture Notes in Engineering 63, Springer-Verlag, 1990.
6. Rozvany, G., Ed., *Optimization of Large Structural Systems*, Kluwer Academic Publishers, to appear 1992.
7. Haug, E.J., Choi, K.K., and Komkov, V., *Design Sensitivity Analysis of Structural Systems*, Academic Press, New York, N.Y., 1986; also translated to the Russian by N.V. Banichuk and published in the USSR, 1988.
8. Choi, K.K., Santos, J.L.T., and Frederick , M.C., "Implementation of Design Sensitivity Analysis with Existing Finite Element Codes," *ASME J. of Mechanisms, Transmissions, and Automation in Design*, Vol. 109, No. 3, 1987, pp. 381-395.
9. DeSalvo, G.J., and Swanson, J.A., *ANSYS Engineering Analysis System, User's Manual Vols. I and II*, Swanson Analysis Systems, Inc., P.O. Box 65, Houston, PA, 1989.
10. MSC/NASTRAN, *MSC/NASTRAN User's Manual*, Vol. I and II, The Macneal-Schwendler Co., 815 Colorado Boulevard, Los Angeles, CA 90041, 1988.
11. ABAQUS, *ABAQUS User's Manual*, Version 4.8, Hibbitt, Karlsson & Sorensen, Inc., 100 Medway Street, Providence, Rhode Island, 1989.
12. Wang, S. and Choi K.K., "Continuum Design Sensitivity Analysis of Transient Responses Using Ritz and Mode Acceleration Methods," *AIAA Journal*, Vol. 30, No. 4, 1992, pp. 1099-1109.
13. Choi, K.K. and Wang, S., "Continuum Design Sensitivity of Structural Dynamic Response Using Ritz Sequence," *AIAA/ASME/ASCE/AHS/ASC 31st Structures, Structural Dynamics and Materials Conference*, Paper No. 90-1137, 1990, pp. 385-393.
14. Wang, S. and Choi, K.K., "Continuum Design Sensitivity Analysis of Eigenvectors Using Ritz Vectors," *AIAA/ASME/ASCE/AHS/ASC 32nd Structures, Structural Dynamics and Materials Conference*, Paper No. 91-1092, 1991, pp. 406-412.
15. Wu, J.K., Choong, F.N., Choi, K.K., and Haug, E.J., "A Data Model for Simulation Based Concurrent Engineering of Mechanical Systems," *International Journal of Systems Automation, Research & Applications*, Vol. 1, 1991, pp. 67-87.
16 Crede, C., *Shock and Vibration Concepts in Engineering Design*, Prentice-Hall, Englewood Cliffs, N.J., 1965.
17. Choi, K.K. and Lee, J.H., "Sizing Design Sensitivity Analysis of Dynamic Frequency Response of Vibrating Structures," *ASME Journal of Mechanical Design*, Vol. 114, No. 1, 1992, pp. 166-173; also presented at the 15th ASME Design Automation Conference, September 1989.
18. Choi, K.K., Shim, I., Lee, J.H., and Kulkarni, H.T., "Design Sensitivity Analysis of Dynamic Frequency Responses of Acousto-Elastic Built-up Structures," *Optimization of Large Structural Systems*, (Ed. G. Rozvany), to appear, 1992.
19. Adams, R.A., *Sobolev Spaces*, Academic Press, N.Y., 1975.
20. Horvath, J., *Topological Vector Spaces and Distributions*, Addison Wesley, Reading, MA, 1966.
21. Kamal, M.M. and Wolfe, J.A., Jr. ed., *Modern Automotive Structural Analysis*, Van Nostrand Reinhold Company, New York, 1982.
22. Choi, K.K. and Santos, J.L.T., "Design Sensitivity Analysis of Nonlinear Structural Systems. Part I: Theory," *International Journal for Numerical Methods in Engineering*, Vol. 24, No. 11, 1987, pp. 2039-2055.
23. Santos, J.L.T. and Choi, K.K., "Sizing Design Sensitivity Analysis of Nonlinear Structural Systems. Part II: Numerical Method," *International Journal for Numerical Methods in Engineering*, Vol. 26, No. 9, 1988, pp. 2097-2114.
24. Park, J.S. and Choi, K.K., "Design Sensitivity Analysis and Optimization of Nonlinear Structural Systems with Critical Loads," *ASME Journal of Mechanical Design*, to appear, 1991; also presented at the 16th ASME Design Automation Conference, September 1990..
25. Hibbit, H.D., Marcal, P.V., and Rice, J.R., "A Finite Element Formulation for Problems for Large Strain and Large Displacement," *International Journal of Solids and Structures*, Vol. 6, 1970, pp. 1069-1086.
26. Bathe, K.J., Ramm, E., and Wilson, E.L., "Finite Element Formulations for Large Deformation Dynamic Analysis," *International Journal for Numerical Methods in Engineering*, Vol. 9, 1975, pp. 353-386.

27. Bathe, K.J., *Finite Element Procedures in Engineering Analysis*, Prentice Hall, Inc., Englewood Cliffs, NJ, 1982.
28. Brendel, B. and Ramm, E., "Linear and Nonlinear Stability Analysis of Cylindrical Shells," *Computers and Structures*, Vol. 12, 1980, pp. 549-558.
29. Rammerstofer, F.G., "Jump Phenomena Associated with the Stability of Geometrically Nonlinear Structures," *Recent Advances in Nonlinear Computational Mechanics*, Eds. E. Hinton, D.R.J. Owen, and C. Taylor, University College of Swansea, U.K., 1982, Ch. 5, pp. 119-153.
30. Park, J.S. and Choi, K.K., "Design Sensitivity Analysis of Critical Load Factor for Nonlinear Structural Systems," *Computers and Structures*, Vol. 36, No. 5, 1990, pp. 823-838.
31. Chang, D.C. and Ni, C-M., "Plastic Deformation Analysis," *Modern Automotive Structural Analysis*, Eds. M.M. Kamal and J.A. Wolf, Van Nostrand Reinhold Co., 1981, Chapter 10.
32. Chang, K.H. and Choi, K.K. "Structural Design Sensitivity Analysis and Optimization Workstation," in preparation, 1992.
33. Santos, J.L.T., Godse, M.M. and Chang, K.H., "An Interactive Post-Processor for Structural Design Sensitivity Analysis and Optimization : Sensitivity Display and What-If Study," *Computers and Structures*, Vol. 35, No. 1, pp. 1-13, 1990.
34. Stone, T.A., Santos, J.L.T. and Haug, E.J., "An Interactive Pre-Processor for Structural Design Sensitivity Analysis and Optimization," *Computers and Structures*, Vol. 34, No. 3, pp. 375-386, 1990.
35. Santos, J.L.T. and Choi, K.K., "Integrated Computational Considerations for Large Scale Structural Design Sensitivity Analysis and Optimization," *GAMM Conference on Discretization Methods and Structural Optimization*, University of Siegen, FRG, October, 1988.
36. PDA Engineering, *PATRAN Plus User's Manuals Vols. I and II*, Software Products Division, 1560 Brookhollow Drive, Santa Ana, CA, 1988.
37. Open Software Foundation, "OSF/Motif Programmer's Guide," Prentice Hall, Englewood Cliffs, New Jersey 07632, 1988.
38. PDA Engineering, *PATRAN - ANSYS Interface Guide*, Software Products Division, 1560 Brookhollow Drive, Santa Ana, CA, 1985.
39. PDA Engineering, *PATRAN - MSC/NASTRAN Interface Guide*, Software Products Division, 1560 Brookhollow Drive, Santa Ana, CA, 1988.
40. PDA Engineering, *PATRAN - ABAQUS Interface Guide*, Software Products Division, 1560 Brookhollow Drive, Santa Ana, CA, 1988.
41. Apollo Computer Inc., "Network Computing System (NCS) Reference," 330 Billerica Road, Chelmsford, MA, 1987.
42. Apollo Computer Inc., "Current Programming Support (CPS) Reference," 330 Billerica Road, Chelmsford, MA, 1987.
43. Chang, K.H. and Santos, J.L.T., "Distributed Design Sensitivity Computations on a Network of Computers," *Computers and Structures*, Vol. 37, No. 3, pp. 265-275, 1990.
44. Arora, J.S., *Introduction to Optimum Design*, McGraw-Hill, 1989.
45. Gill, P.E., Murray, W., Saunders, M.A., and Wright, M.H., User's Guide for SOL/QPSOL, *Technical Report SOL 83-7*, System Optimization Laboratory, Stanford University, July 1983.
46. Vanderplaats, G.N. and Hansen, S.R., *DOT User's Manual*, VMA Engineering, 5960 Mandarin Avenue, Suite F, Goleta, CA 93117, 1990.

Concurrent Engineering Design with and of Advanced Materials

Pauli Pedersen

Department of Solid Mechanics, The Technical University of Denmark, Lyngby, Denmark

Abstract: Advanced materials are now used frequently in engineering design and that has opened for the possibility of material design. A general characteristic of these materials is that they are anisotropic, and this puts new demands on the analysis capabilities and optimization methods. In recent years a number of questions have been clarified, and the intention of the present notes is to distribute the knowledge gained. Active research areas are also commented on, and the concurrent design with a number of different design parameters is put forward.

Keywords: optimal design / anisotropic materials / laminates / sensitivity analysis / shape design / material orientation

1 Introduction

Often design parameters are classified under the headings of size, shape and topology; mentioned here in order of increasing importance. Design parameters related to materials, to some extent, fall outside these classes and are now extensively dealt with due to the rapid evolution of a number of new materials such as fibre-reinforced materials and ceramics. Two aspects of optimization are then of practical interest -- the influence on size, shape and topology design *with these materials*, and the more or less detailed design *of a material* for a specific purpose.

The two aspects are not uniquely defined; for example, we may argue that design of ply angles in a laminate is design *with plies*, but if the laminate is defined as a material, then it is design *of laminate material*. Anyhow, it is found valuable to recognize the different aspects. A general characteristic of the new materials is that they are anisotropic, and we shall start these notes *with a classification of constitutive matrices*. Although limited to two-dimensional problems, this classification includes information that is not commonly available. We shall restrict the possible materials only by the constraints of a positive definite, constitutive matrix.

Research related to material design has also focused on the very basic sensitivity analysis related to elastic strain energy. Many important results are not well known and as it is possible to prove these results at the energy level we shall do that. This analysis holds for one- ,two- and three-dimensional problems, holds for different models, and holds for analytical as well as different numerical solutions. The most important result is the localized determination of sensitivities that we often encounter.

With this basic knowledge we then, in Section 4, treat optimal material orientation for plates and discs. Here too, a number of rather general results are available, and we shall focus on these rather than on the specific problems. We may say that two-dimensional problems are to a large extent solved, but the three-dimensional problems are still a subject of intensive research. An interesting result is the co-alignment of principal strains and stresses.

Size optimization, such as distribution of plate thickness, is naturally influenced by the use of anisotropic material but in general the methods of isotropic design are applicable. Therefore, in Section 5, we focus on the concurrent design of orientation and thickness. Comments on non-linear elasticity and on the limitations of optimality criteria methods are given here.

A short account of the detailed material design, in terms of design of constitutive matrices, is given in Section 6. This very active research area already has a number of important results available, and we go through a specific example. The idea here is to see this as the third step after having first solved the orientational problem and then the total density (thickness) problem.

Finally, in Section 7, the shape optimization based on anisotropic materials are discussed and recent solutions shown. The influence from numerical modeling is focused on, i.e., the number of elements and the element type in a finite element model. This naturally leads to the concurrent mesh design, a research area we only touch upon.

The paper is not written as a review paper. It concentrates mainly on the results of personal research. However, many groups are presently active in this area and their reports should be studied in addition to the present notes.

2 Classification of Constitutive Matrices for 2D Linear Elasticity

The reaction to the heading of this section might be that this information must be contained in the classical textbooks. However, material anisotropy is seldom treated in great depth and even in the well-known reference of Lekhnitskii [1], we cannot find all the results needed for optimal design based on anisotropic behavior. Hence, it is our intention to make this section self-contained.

First of all, what is a constitutive law? Is it described by material parameters alone?, and what is a material? The notions are often used as synonyms, and in fact we should add the notion of model. A two-dimensional constitutive law is not based on material behavior alone, but also depends on the mathematical modeling, i.e., the reduction from three to two dimensions, as exemplified by the difference between plane strain and plane stress models. Furthermore, the definition of the concepts of strain and stress influences the constitutive law.

To be specific, let us define the constitutive law by the constitutive matrix [C] (or by its inverse - the compliance matrix)

$$\{\sigma\}_x = [C]_x \{\varepsilon\}_x \;;\; \{\varepsilon\}_x = [C]_x^{-1} \{\sigma\}_x$$
$$\begin{Bmatrix} \sigma_{11} \\ \sigma_{22} \\ \sigma_{12} \end{Bmatrix}_x = \begin{bmatrix} C_{11} & C_{12} & C_{13} \\ C_{12} & C_{22} & C_{23} \\ C_{13} & C_{23} & C_{33} \end{bmatrix}_x \begin{Bmatrix} \varepsilon_{11} \\ \varepsilon_{22} \\ 2\varepsilon_{12} \end{Bmatrix}_x \tag{2.1}$$

As stated, [C] is symmetric and it is also non-singular. Furthermore, we have indicated in Eq. (2.1) that reference to a specific Cartesian x-coordinate system is assumed.

The concepts of strain energy density u and stress energy density u^C (complementary strain energy) are defined by their first variations as

$$\delta u := \{\sigma\}^T \{\delta\varepsilon\} \;;\; \delta u^C := \{\delta\sigma\}^T \{\varepsilon\} \tag{2.2}$$

If linear elasticity is assumed, the definitions (2.2) result in

$$u = u^C = \frac{1}{2}\{\sigma\}^T\{\varepsilon\} = \frac{1}{2}\{\varepsilon\}^T[C]\{\varepsilon\} = \frac{1}{2}\{\sigma\}^T[C]^{-1}\{\sigma\} \tag{2.3}$$

From the definition of energy density (2.2), it follows directly that [C] must be symmetric and positive definite, i.e., all its eigenvalues must be positive. So for all 2D constitutive models, we have

[C] symmetric, i.e., max. 6 parameters and

$$\begin{aligned} & C_{11} > 0 \;;\; C_{22} > 0 \;;\; C_{33} > 0 \\ & C_{11}C_{22} - C_{12}^2 > 0 \;;\; C_{22}C_{33} - C_{23}^2 > 0 \;; \\ & C_{11}C_{33} - C_{13}^2 > 0 \;;\; |\,[C]\,| > 0 \end{aligned} \tag{2.4}$$

The conditions for being positive definite can be written in alternative forms.

To familiarize ourselves with these conditions, let us write the well-known constitutive matrix for isotropic models with only 2 parameters

$$[C]=\begin{bmatrix} C_{11}\ ; & C_{12}\ ; & 0 \\ C_{12}\ ; & C_{11}\ ; & 0 \\ 0 & ;\ 0 & ;\ \frac{1}{2}(C_{11}-C_{12}) \end{bmatrix} \tag{2.5}$$

for which the conditions (2.4) give $C_{11}>0\ ;\ C_{11}^2-C_{12}^2>0\ ;\ C_{11}-C_{12}>0$ or, alternatively,

$$0 \le |C_{12}| < C_{11} \text{ (subset } 0<C_{12}<C_{11}) \tag{2.6}$$

from which we see that $C_{12}<0$ is possible, although often not the case. In plane stress $C_{11}=E/(1-\upsilon^2)\ ;\ C_{12}=\upsilon C_{11}$ and in plane strain $C_{11}=E(1-\upsilon)/((1+\upsilon)(1-2\upsilon))\ ;$ $C_{12}=\upsilon C_{11}/(1-\upsilon)$, so with $E>0$ we get $-1<\upsilon<0.5$.

Let us then characterize the most important case of orthotropic models with 4 parameters and a specified direction. We note that only along the orthotropic directions are there no couplings between shear and normal stresses/strains

$$[C]_x=\begin{bmatrix} C_{11} & C_{12} & 0 \\ C_{12} & C_{22} & 0 \\ 0 & 0 & C_{33} \end{bmatrix}_x \tag{2.7}$$

To be more precise, we choose the larger modulus direction of the two orthogonal directions, i.e., $C_{11}>C_{22}$. Then the conditions (2.4) give additional relations $C_{22}>0\ ;$ $C_{33}>0\ ;\ C_{11}C_{22}-C_{12}^2>0$, and from the last condition follows $\frac{1}{2}(C_{11}+C_{22})-C_{12}>0$. It is often assumed that $C_{12}>0$, but this is not generally a real condition. Summing up, we have

$$\begin{aligned} &|C_{12}|<\sqrt{C_{11}C_{22}}\ ;\ 0<C_{22}<C_{11}\ ;\ 0<C_{33} \\ &\left(\text{subset } 0\le C_{12}<\sqrt{C_{11}C_{22}}<\tfrac{1}{2}(C_{11}+C_{22})\right) \end{aligned} \tag{2.8}$$

for the four parameters of the orthotropic models with a uniquely chosen reference axis.

Based on the quantities of the orthotropic constitutive matrix (2.7), we -- in agreement with the literature on composite materials (e.g., Jones [2]) -- define **practical parameters** (often termed invariants, which is misleading)

$$\begin{aligned}
C_1 &:= \tfrac{1}{8}(3(C_{11}+C_{22})+2(C_{12}+2C_{33})) \\
C_2 &:= \tfrac{1}{2}(C_{11}-C_{22}) \\
C_3 &:= \tfrac{1}{8}((C_{11}+C_{22})-2(C_{12}+2C_{33})) \\
C_4 &:= \tfrac{1}{8}((C_{11}+C_{22})+2(3C_{12}-2C_{33})) \\
C_5 &:= \tfrac{1}{8}((C_{11}+C_{22})-2(C_{12}-2C_{33}))
\end{aligned} \tag{2.9}$$

Two of these are of major importance and have a distinct physical interpretation, i.e., C_2 and C_3. The difference in modulus between the two orthotropic directions is measured by C_2. The parameter C_3 is of the utmost importance for the optimal design. It has a physical interpretation as relative shear modulus and, according to its sign, we classify the constitutive law as low or high shear modulus, respectively. From (2.8) we know that for "small" C_{33} we have $C_3 > 0$ and from (2.9) we see that for "large" C_{33} we get $C_3 < 0$. Thus

$$\begin{aligned}
&\text{Low shear modulus:} \\
&C_3 > 0 \quad \text{i.e.,} \quad 2C_{33} < \tfrac{1}{2}(C_{11}+C_{22})-C_{12} \\
&\text{High shear modulus:} \\
&C_3 < 0 \quad \text{i.e.,} \quad 2C_{33} > \tfrac{1}{2}(C_{11}+C_{22})-C_{12}
\end{aligned} \tag{2.10}$$

(Isotropic $C_3 = 0$)

This classification is not well known, and it is sometimes argued that $C_3 < 0$ is not possible. Most ordinary "materials" give $C_3 > 0$, but constructed materials like laminates can easily be designed to have $C_3 < 0$.

In a broader perspective, the notion of constitutive matrices is also used for plate

stiffnesses-in the plane:
$$\begin{Bmatrix} N_{11} \\ N_{22} \\ N_{12} \end{Bmatrix} = [A] \begin{Bmatrix} \varepsilon_{11}^0 \\ \varepsilon_{22}^0 \\ 2\varepsilon_{12}^0 \end{Bmatrix} \tag{2.11}$$

as well as out of the plane:

$$\begin{Bmatrix} M_{11} \\ M_{22} \\ M_{12} \end{Bmatrix} = [D] \begin{Bmatrix} \kappa_{11} \\ \kappa_{22} \\ 2\kappa_{12} \end{Bmatrix} \tag{2.12}$$

where [A] is the matrix of extensional stiffnesses, {N} is a vector of plate forces per unit length, [D] is the matrix of bending stiffnesses, {M} is a vector of plate moments per unit length, and $\left\{\varepsilon^0\right\}$, $\{\kappa\}$ are midplane strains and curvatures, respectively.

In order to deal with non-orthotropic constitutive matrices, we define, in addition to the parameters $C_1 - C_5$ in (2.9),

$$\begin{aligned} C_6 &:= \tfrac{1}{2}(C_{13} + C_{23}) \\ C_7 &:= \tfrac{1}{2}(C_{13} - C_{23}) \end{aligned} \tag{2.13}$$

two parameters, that vanish for orthotropic laws. However, the orthotropic directions may be unknown and then we need a criterion to test for existence of orthotropic directions. This is derived in Pedersen [3] and is expressed using the parameters C_2, C_3, C_6 and C_7 (based on any reference axis) as

$$C_7C_2^2 - 4C_7C_6^2 - 4C_6C_3C_2 = 0 \tag{2.14}$$

If (2.14) is satisfied, we choose the direction of orthotropy for which $C_2 > 0$. If (2.14) is not satisfied, we have a non-orthotropic constitutive matrix but still need a specific reference axis for comparison of equal constitutive matrices which may, in reality just be mutually rotated. No clear tradition for selection of this direction seems to be available in the literature.

In our studies, we have chosen the direction which maximizes C_{11}, i.e.,

$$\underset{\gamma}{\text{Max}}\ (C_{11} = C_{11}(\gamma)) \text{ for } 0 \le \gamma \le \pi \tag{2.15}$$

and in most (but not all) practical cases, this is uniquely determined by

$$(C_{13})_x = 0 \text{ , i.e., } (C_6)_x = -\,(C_7)_x \tag{2.16}$$

from which follows a general constitutive matrix

$$[C] = \begin{bmatrix} C_{11} ; & C_{12} ; & 0 \\ C_{12} ; & C_{22} ; & C_{23} \\ 0 \ ; & C_{23} ; & C_{33} \end{bmatrix} \text{ with } C_2 = \tfrac{1}{2}(C_{11} - C_{22}) > 0 \tag{2.17}$$

This constitutive matrix is positive definite with the conditions

$$0 < C_{22} \; ; \; 0 < C_{33} \; ; \; |C_{12}| < \sqrt{C_{11}C_{22}}$$
$$|C_{23}| < \sqrt{C_{22}C_{33} - C_{12}^2 C_{33} / C_{11}} \tag{2.18}$$

to simplify (2.1) and (2.4).

The results of this section as a whole are shown in Figure 2.1.

Before ending the section, we shall list the transformation formulas for constitutive laws from x-coordinates to y-coordinates as defined in Figure 2.2. Generally we have

$$
\begin{aligned}
(C_{11})_y &= \tfrac{1}{2}(C_{11}+C_{22})_x + (C_2)_x \cos(2\gamma) \\
&\quad - (C_3)_x(1-\cos(4\gamma)) + (C_6)_x 2\sin(2\gamma) + (C_7)_x \sin(4\gamma) \\
(C_{22})_y &= \tfrac{1}{2}(C_{11}+C_{22})_x - (C_2)_x \cos(2\gamma) \\
&\quad - (C_3)_x(1-\cos(4\gamma)) - (C_6)_x 2\sin(2\gamma) + (C_7)_x \sin(4\gamma) \\
(C_{12})_y &= (C_{12})_x + (C_3)_x(1-\cos(4\gamma)) - (C_7)_x \sin(4\gamma) \\
(C_{33})_y &= (C_{33})_x + (C_3)_x(1-\cos(4\gamma)) - (C_7)_x \sin(4\gamma) \\
(C_{13})_y &= -\tfrac{1}{2}(C_2)_x \sin(2\gamma) - (C_3)_x \sin(4\gamma) \\
&\quad + (C_6)_x \cos(2\gamma) + (C_7)_x \cos(4\gamma) \\
(C_{23})_y &= -\tfrac{1}{2}(C_2)_x \sin(2\gamma) + (C_3)_x \sin(4\gamma) \\
&\quad + (C_6)_x \cos(2\gamma) - (C_7)_x \cos(4\gamma)
\end{aligned}
\tag{2.19}
$$

If the constitutive matrix is orthotropic in the x-coordinates it simplifies to

$$
\begin{aligned}
(C_{11})_y &= (C_1)_x + (C_2)_x \cos(2\gamma) + (C_3)_x \cos(4\gamma) \\
(C_{22})_y &= (C_1)_x - (C_2)_x \cos(2\gamma) + (C_3)_x \cos(4\gamma) \\
(C_{12})_y &= (C_4)_x - (C_3)_x \cos(4\gamma) \\
(C_{33})_y &= (C_5)_x - (C_3)_x \cos(4\gamma) \\
(C_{13})_y &= -\tfrac{1}{2}(C_2)_x \sin(2\gamma) - (C_3)_x \sin(4\gamma) \\
(C_{23})_y &= -\tfrac{1}{2}(C_2)_x \sin(2\gamma) + (C_3)_x \sin(4\gamma)
\end{aligned}
\tag{2.20}
$$

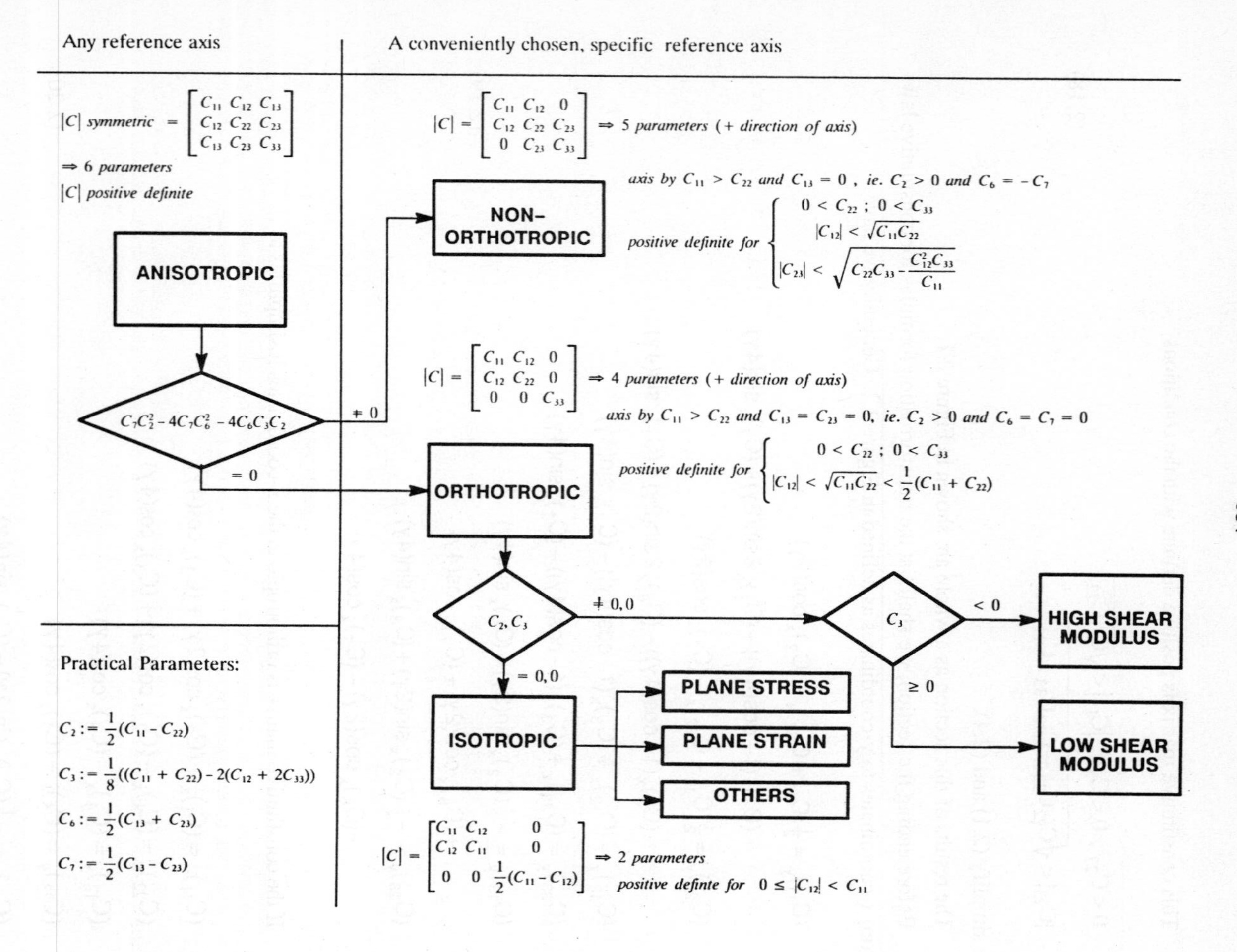

Figure 2.1. Overview of the Results

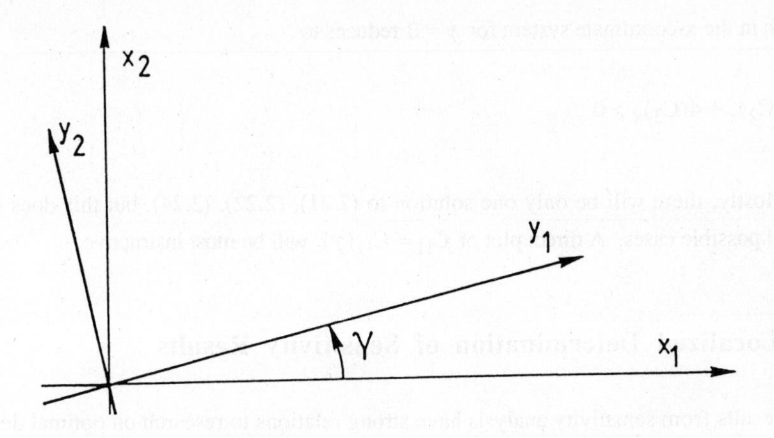

Figure 2.2. The Specific Reference Axis x_1 and a Relatively Rotated Coordinate System $y_1;y_2$. Angle γ defined from x_1 to y_1, positive anti-clockwise.

Note that the definition is from the orthotropic direction and that the rotation angle γ is measured positive anti-clockwise. Alternative definitions are found in the literature.

Based on the transformations (2.19) we can prove the postulate (2.16) and see its limitations. The condition of $C_{11} > C_{22}$ gives

$$2C_2 \cos 2\gamma + 4C_6 \sin 2\gamma > 0 \text{ ; i.e., } \tan 2\gamma < -\frac{C_2}{2C_6} \tag{2.21}$$

with C_2, C_6 from the arbitrary coordinates. (From directions x we have $\gamma = 0$, i.e., $(C_2)_x > 0$). The condition of stationarity for C_{11} gives $\partial C_{11} / \partial\gamma = 0$, i.e.,

$$-2C_2 \sin 2\gamma - 4C_3 \sin 4\gamma + 4C_6 \cos 2\gamma + 4C_7 \cos 4\gamma = 0 \tag{2.22}$$

and this also holds from the system x, for which $\gamma = 0$ results in

$$(C_6)_x = -(C_7)_x \text{ ; i.e., } (C_{13})_x = 0 \tag{2.23}$$

The condition of maximum C_{11} with $\partial^2 C_{11} / \partial\gamma^2 < 0$ gives the relation

$$-4C_2 \cos 2\gamma - 16C_3 \cos 2\gamma - 8C_6 \sin 2\gamma - 16C_7 \sin 2\gamma < 0 \tag{2.24}$$

which in the x-coordinate system for $\gamma = 0$ reduces to

$$(C_2)_x + 4(C_3)_x > 0 \tag{2.25}$$

Mostly, there will be only one solution to (2.21), (2.22), (2.24), but this does not hold for all possible cases. A direct plot of $C_{11} = C_{11}(\gamma)$ will be most instructive.

3 Localized Determination of Sensitivity Results

The results from sensitivity analysis have strong relations to research on optimal design but are in reality of much wider importance and applicability. Often the results are derived for specific models, and the generality is lost or at least not visible.

When the quantity for which we seek the sensitivity is related to a global energy quantity, we have very simplified results of localized nature. These results will be derived here without reference to a specific model and are thus valid for one-, two- and three-dimensional models, for analytical calculation, and for numerical model and are valid independent of the numerical method chosen, say finite difference, finite element, or more global Galerkin approaches.

We shall concentrate on static problems for which the sensitivity results are less intuitive. In general this is due to the different nature of the work of the external forces compared with the kinetic energy. Let us start with the work equation

$$W + W^C = U + U^C \tag{3.1}$$

where W, W^C are physical and complementary work of the external forces, and U, U^C are physical and complementary elastic energy, also named strain and stress energy, respectively. The work equation (3.1) holds for any design $\mathbf{h}$ and therefore for the total differential quotient with respect to $\mathbf{h}$

$$\frac{dW}{d\mathbf{h}} + \frac{dW^C}{d\mathbf{h}} = \frac{dU}{d\mathbf{h}} + \frac{dU^C}{d\mathbf{h}} \tag{3.2}$$

Now in the same way as $\mathbf{h}$ represents the design field generally, ε represents the strain field and σ represents the stress field. Remembering that as a function of $\mathbf{h}, \varepsilon$ we have W,U, while the complementary quantities W^C, U^C are functions of $\mathbf{h}, \sigma$, we then get (3.2) in greater detail by means of

$$\frac{\partial W}{\partial \mathbf{h}}+\frac{\partial W}{\partial \varepsilon}\frac{\partial \varepsilon}{\partial \mathbf{h}}+\frac{\partial W^C}{\partial \mathbf{h}}+\frac{\partial W^C}{\partial \sigma}\frac{\partial \sigma}{\partial \mathbf{h}}=\frac{\partial U}{\partial \mathbf{h}}+\frac{\partial U}{\partial \varepsilon}\frac{\partial \varepsilon}{\partial \mathbf{h}}+\frac{\partial U^C}{\partial \mathbf{h}}+\frac{\partial U^C}{\partial \sigma}\frac{\partial \sigma}{\partial \mathbf{h}} \tag{3.3}$$

The principles of virtual work, which hold for solids/structures in equilibrium, are

$$\frac{\partial W}{\partial \varepsilon}=\frac{\partial U}{\partial \varepsilon} \tag{3.4}$$

for the physical quantities with strain variation and for the complementary quantities with stress variation, we have

$$\frac{\partial W^C}{\partial \sigma}=\frac{\partial U^C}{\partial \sigma} \tag{3.5}$$

Inserting (3.4) and (3.5) in (3.3) we get

$$\frac{\partial U^C}{\partial \mathbf{h}}-\frac{\partial W^C}{\partial \mathbf{h}}=-\left[\frac{\partial U}{\partial \mathbf{h}}-\frac{\partial W}{\partial \mathbf{h}}\right] \tag{3.6}$$

and for design-independent loads

$$\boxed{\left[\frac{\partial U^C}{\partial \mathbf{h}}\right]_{\substack{\text{fixed}\\ \text{stresses}}}=-\left[\frac{\partial U}{\partial \mathbf{h}}\right]_{\substack{\text{fixed}\\ \text{strains}}}} \tag{3.7}$$

as stated by MASUR [4]. Note that the only assumption behind this is the design-independent loads $\partial W/\partial \mathbf{h}=0$, $\partial W^C/\partial \mathbf{h}=0$.

To get further into a physical interpretation of $(\partial U/\partial \mathbf{h})_{\text{fixed strains}}$ (and by (3.7) of $(\partial U^C/\partial \mathbf{h})_{\text{fixed stresses}}$) we need the relation between external work W and strain energy U. Let us assume that this relation is given by the constant c

$$W = cU \tag{3.8}$$

For linear elasticity and dead loads we have $c = 2$, and in general we expect $c > 1$, i.e., we lose energy.

Parallel to the analysis from (3.1) to (3.3) and on the basis of (3.8), we obtain

$$\frac{\partial W}{\partial \mathbf{h}}+\frac{\partial W}{\partial \varepsilon}\frac{\partial \varepsilon}{\partial \mathbf{h}}=c\frac{\partial U}{\partial \mathbf{h}}+c\frac{\partial U}{\partial \varepsilon}\frac{\partial \varepsilon}{\partial \mathbf{h}} \tag{3.9}$$

which for design-independent loads $\partial W/\partial \mathbf{h}=0$ with virtual work (3.4), gives

$$\frac{\partial W}{\partial \varepsilon}\frac{\partial \varepsilon}{\partial \mathbf{h}}=\frac{\partial U}{\partial \varepsilon}\frac{\partial \varepsilon}{\partial \mathbf{h}}=\frac{c}{1-c}\frac{\partial U}{\partial \mathbf{h}} \tag{3.10}$$

and thereby

$$\frac{dU}{d\mathbf{h}}=\frac{\partial U}{\partial \mathbf{h}}+\frac{\partial U}{\partial \varepsilon}\frac{\partial \varepsilon}{\partial \mathbf{h}}=\frac{1}{1-c}\left[\frac{\partial U}{\partial \mathbf{h}}\right]_{\substack{\text{fixed}\\ \text{strains}}} \tag{3.11}$$

Note in this important result that with $c > 1$ we have different signs for $dU/d\mathbf{h}$ and $(\partial U/\partial \mathbf{h})_{\text{fixed strains}}$.

For the case of linear elasticity and dead loads we have, with $c = 2$ and adding (3.7),

$$\frac{dU}{d\mathbf{h}}=-\left[\frac{\partial U}{\partial \mathbf{h}}\right]_{\substack{\text{fixed}\\ \text{strains}}}=\left[\frac{\partial U}{\partial \mathbf{h}}\right]_{\substack{\text{fixed}\\ \text{stresses}}} \tag{3.12}$$

For the case of nonlinear elasticity modeled one dimensionally by

$$\sigma = E\varepsilon^{p} \tag{3.13}$$

and still "dead loads" ($W^C = 0$) we get $c = 1+p$ and thereby

$$\boxed{\frac{dU}{d\mathbf{h}}=-\frac{1}{p}\left[\frac{\partial U}{\partial \mathbf{h}}\right]_{\substack{\text{fixed}\\ \text{strains}}}=\frac{1}{p}\left[\frac{\partial U}{\partial \mathbf{h}}\right]_{\substack{\text{fixed}\\ \text{stresses}}}} \tag{3.14}$$

This result is applied in Pedersen and Taylor [29], with examples for two-dimensional cases.

Now let us discuss the localized determination of sensitivities as given generally by (3.11). The strain energy is summed over all domains, and the design parameter $\mathbf{h}_e$ is local. Then we get

$$\frac{dU}{d\mathbf{h}_e} = \frac{dU_e}{d\mathbf{h}_e} + \sum_{i \neq e} \frac{dU_i}{d\mathbf{h}_e} = \frac{\partial U_e}{\partial \mathbf{h}_e} + \frac{\partial U_e}{\partial \varepsilon_e}\frac{\partial \varepsilon_e}{\partial \mathbf{h}_e} + \sum_{i \neq e} \frac{\partial U_i}{\partial \varepsilon_i}\frac{\partial \varepsilon_i}{\partial \mathbf{h}_e} \tag{3.15}$$

where ε_e describes the strain field of domain e. The results (3.11) tells that we need not calculate $\partial U_i / \partial \varepsilon_e$, because

$$\frac{dU}{d\mathbf{h}_e} = \frac{1}{1-c}\left[\frac{\partial U_e}{\partial \mathbf{h}_e}\right] \tag{3.16}$$

and from (3.15-16) we can again determine the "indirect" effect

$$\frac{\partial U_e}{\partial \varepsilon_e}\frac{\partial \varepsilon_e}{\partial \mathbf{h}_e} + \sum_{i \neq e} \frac{\partial U_i}{\partial \varepsilon_i}\frac{\partial \varepsilon_i}{\partial \mathbf{h}_e} = \sum_{\text{all } i} \frac{\partial U_i}{\partial \varepsilon_i}\frac{\partial \varepsilon_i}{\partial \mathbf{h}_e} - \frac{c}{1-c}\frac{\partial U_e}{\partial \mathbf{h}_e} \tag{3.17}$$

Eq. (3.17) is the same as Eq. (3.11) for the localized parameter $\mathbf{h}_e$. For linear elasticity and dead loads we have c = 2, and the indirect effect is then twice the direct effect (with opposite signs).

We will often determine the strain energy by the strain energy density u_e and the domain volume V_e, i.e.,

$$U_e = u_e V_e \tag{3.18}$$

We then naturally treat two groups of design parameters, i.e., the ones without influence on V_e and the ones without explicit influence on u_e. For the first group, say $\mathbf{h}_e$ is a parameter of the constitutive matrix, the sensitivity determination (3.16) simplifies to

$$\frac{dU}{d\mathbf{h}_e} = \frac{V_e}{1-c}\left[\frac{\partial u_e}{\partial \mathbf{h}_e}\right]_{\substack{\text{fixed}\\ \text{strains}}} \tag{3.19}$$

and for the second group, say $\mathbf{h}_e$ is a thickness or area parameter, the sensitivity determination (3.16) simplifies to

$$\frac{dU}{d\mathbf{h}_e} = \frac{u_e}{1-c}\frac{\partial V_e}{\partial \mathbf{h}_e} \tag{3.20}$$

In the following we shall apply these general sensitivity results to a number of specific problems.

4 Optimal Material Orientation for Plates and Discs

In this section our general design parameter **h** is taken to be the material orientation θ_e in the domain e. We assume linear elasticity and dead loads (W = 2U) and thus have directly from (3.19)

$$\frac{dU}{d\theta_e} = -V_e\left[\frac{\partial u_e}{\partial \theta_e}\right]_{\substack{\text{fixed}\\ \text{strains}}} \tag{4.1}$$

For coupled plate/disc problems using traditional symbols from laminate analysis, the energy density per plate area $u_e t_e$ is given by

$$\begin{aligned} u_e t_e &= \tfrac{1}{2}\{\varepsilon^0\}^T\{N\} + \tfrac{1}{2}\{\kappa\}^T\{M\}; \\ \{N\} &= [A]\{\varepsilon^0\} + [B]\{\kappa\}; \\ \{M\} &= [B]\{\varepsilon^0\} + [D]\{\kappa\}, \end{aligned} \tag{4.2}$$

and the combined result is

$$u_e t_e = \tfrac{1}{2}\{\varepsilon^0\}^T[A]\{\varepsilon^0\} + \{\varepsilon^0\}^T[B]\{\kappa\} + \tfrac{1}{2}\{\kappa\}^T[D]\{\kappa\} \tag{4.3}$$

with t_e for plate thickness, $\{\varepsilon^0\}$, [A] for midsurface strains and extensional stiffnesses; and $\{\kappa\}$, [D], [B] for curvatures, bending stiffnesses and coupling stiffnesses. Applying the result (4.1) we get

$$\frac{dU}{d\theta_e} = -a_e\left[\tfrac{1}{2}\{\varepsilon^0\}^T\left[\frac{\partial A}{\partial \theta_e}\right]\{\varepsilon^0\} + \{\varepsilon^0\}^T\left[\frac{\partial B}{\partial \theta_e}\right]\{\kappa\} + \tfrac{1}{2}\{\kappa\}^T\left[\frac{\partial D}{\partial \theta_e}\right]\{\kappa\}\right] \tag{4.4}$$

with a_e for domain area. Even for the fully coupled problems, this result can be written

$$\frac{dU}{d\theta_e} = U_1 \sin 2\theta + U_2 \cos 2\theta + U_3 \sin 4\theta + U_4 \cos 4\theta \tag{4.5}$$

This follows from the fact that all the matrices [A], [B] and [D] originate from the constitutive matrix [C], which, as seen from (2.19), contains only the trigonometric functions of Eq. (4.5).

Before treating the specific and simplified problems, it should be appreciated that according to (4.5) we, in the general case, can find at most four different solutions to $dU / d\theta_e = 0$. This follows from rewriting $dU / d\theta_e$ as a fourth order polynomial. However, analytical solutions are too complicated to be shown here.

For orthotropic materials and models where only the cosine terms are involved, like $C_{13} = C_{23} \equiv 0$ in Eq. (2.20), analytical solutions to $dU / d\theta_e$ are obtainable. Keeping in Eq. (4.5) only the sine terms, we have for these (specially orthotropic/balanced) models

$$\frac{dU}{d\theta_e} = 2U_3 \sin 2\theta \left[\frac{U_1}{2U_3} + \cos 2\theta\right] \tag{4.6}$$

Stationarity is then obtained for

$$\theta = 0 \; ; \; \theta = \pi / 2 \; ; \; \theta = \pm\tfrac{1}{2} \arccos\left[-\frac{U_1}{2U_3}\right] \tag{4.7}$$

and, furthermore, supplementary angles will return the same energy density

$$u(\pi - \theta) = u(\theta) \tag{4.8}$$

when only $\cos 2\theta$ and $\cos 4\theta$ appear in the stiffness expressions. Thus, the orientational dependence is described completely by the interval $0 \le \theta \le \pi / 2$.

Now let us discuss some specific results. First, simply supported plates in pure bending, for which the Navier analytical response solutions are available. With a load pressure by

$$p = p_{mn} \sin \frac{m\pi x_1}{a} \sin \frac{n\pi x_2}{b} \tag{4.9}$$

the transverse displacement is

$$w = p\frac{b^4}{\pi^4 n^4}\frac{1}{\Phi_{mn}}$$

$$\Phi_{mn} := \eta_{mn}^4 D_{11} + D_{22} + \eta_{mn}^2\, 2(D_{12} + 2D_{33}) \tag{4.10}$$

$$\eta_{mn} := (mb)/(na)$$

with the mode parameter η_{mn} defined as the ratio between the two actual half-wavelengths. The same functional Φ give us eigenfrequencies by

$$\omega_{mn} = \frac{\pi^2 n^2}{b^2}\sqrt{\frac{\Phi}{\rho t}} \tag{4.11}$$

where ρ is the mass density of the plate. Also buckling load, say $(N_{x1})_{cr}$, is described by Φ as

$$\left(N_{x1}\right)_{cr} = \frac{\pi^2}{b^2}\frac{\Phi}{\eta_{mn}^2} \tag{4.12}$$

Thus, maximizing Φ, we at the same time maximize buckling load, maximize frequency, and minimize displacements for the same mode, i.e., the same η_{mn}.

The solution which maximizing $\Phi = \Phi(\theta)$ gives as a specific case of (4.7)

$$\theta_{opt} = 0 \text{ or } \theta_{opt} = \pm\pi/2 \text{ or } \theta_{opt} = \pm\tfrac{1}{2}\arccos(-\delta)$$

$$\delta := \frac{C_2}{4C_3}\frac{\left(\eta^4 - 1\right)}{\left(1 - 6\eta_2 + \eta^4\right)} = \frac{(C_{11} - C_{22})}{(C_{11} + C_{22} - 2(C_{12} + 2C_{33}))}\frac{\left(\eta^4 - 1\right)}{\left(1 - 6\eta^2 + \eta^4\right)} \tag{4.13}$$

where the global best extremum can be located by second order derivatives. The result (4.13) is derived in Pedersen [6] and can also be found in the paper by Bert [7] or the more recent paper by Muc [8]. We note from the definition of the optimization parameter δ that two inverse cases of η_1 and $\eta_2 = \eta_1^{-1}$ have opposite signs of δ and thereby give rise to complementary angles. The results are thus fully explained in Figure 4.1.

The result in Figure 4.1 can be directly explained in physical terms. When the deformation wavelength is large in a certain direction, then the fibredirection should be perpendicular to that direction, i.e., $\theta = 0°$ or 90°, respectively. At a certain wavelength ratio, depending upon material but about 1:1.8, use of skew fibredirections starts, and with a square deformation pattern, the fibredirection should be in the diagonal direction of the square.

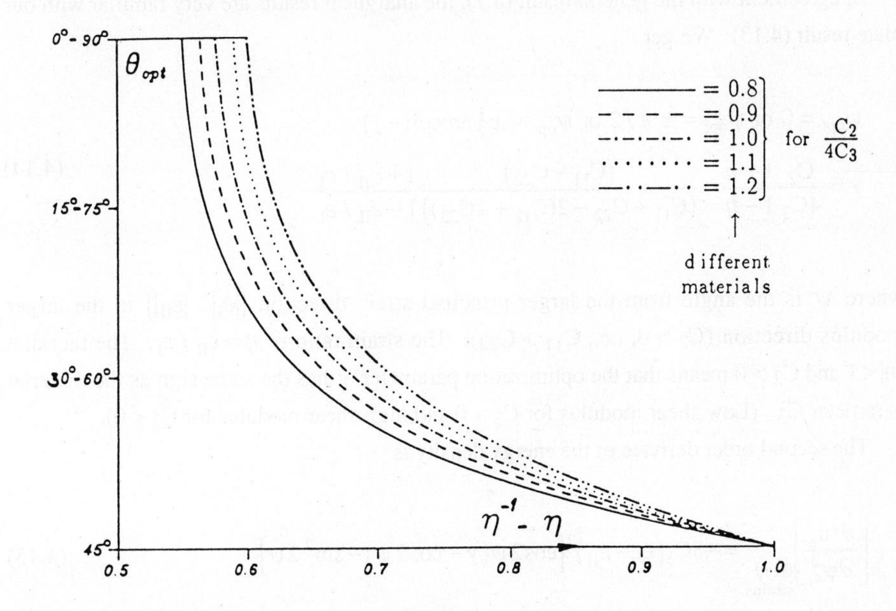

Figure 4.1. Optimal Orientation as a Function of the Mode Parameter η for Plate Bending

From an engineering point of view, the main conclusions to be drawn from Figure 4.1 are:

- The optimal orientation depends mainly on the mode parameter η. Thus if the deformation pattern is known, the optimal fibredirection can be estimated directly.
- Cases of inverse mode parameters $\eta_1 = \eta_2^{-1}$ have complementary solutions $\theta_2 = \pi/2 - \theta_1$.
- For "extreme" values of η the optimal orientation is perpendicular to the long wavelength.
- The change of optimal fibredirection to a skew direction is very sensitive to the mode parameter.
- The optimal orientation is rather insensitive to the material parameters.
- The optimal orientation is independent of the position of the ply in the laminate, and thus the same for all plies.
- Local optima exist.

By simple numerical analysis, problems with combined modes - as exemplified by adding terms (4.9) - can also be optimized; Pedersen [9] shows such solutions.

We shall now move on to the disc problem and specifically treat optimal orientation for maximum extensional stiffness. For details see Pedersen [10] and [11]. The description is limited to orthotropic material, i.e., $C_{13} = C_{23} = 0$ in the principal material directions.

In agreement with the general result (4.7), the analytical results are very familiar with our plate result (4.13). We get

$$\psi_{\text{opt}} = 0 \text{ or } \psi_{\text{opt}} = \pm\pi/2 \text{ or } \psi_{\text{opt}} = \pm\tfrac{1}{2}\arccos(-\gamma)$$
$$\gamma := \frac{C_2}{4C_3}\frac{1+\eta}{1-\eta} = \frac{(C_{11}-C_{22})}{(C_{11}+C_{22}-2(C_{12}+2C_{33}))}\frac{1+\varepsilon_{II}/\varepsilon_I}{1-\varepsilon_{II}/\varepsilon_I} \tag{4.14}$$

where ψ is the angle from the larger principal strain direction $(|\varepsilon_I| > |\varepsilon_{II}|)$ to the larger modulus direction ($C_2 > 0$, i.e., $C_{11} > C_{22}$). The strain ratio is $\eta = \varepsilon_{II}/\varepsilon_I$. The fact that $|\eta| < 1$ and $C_2 > 0$ means that the optimization parameter γ has the same sign as the material parameter C_3. (Low shear modulus for $C_3 > 0$ and high shear modulus for $C_3 < 0$).

The second order derivate of the energy density is

$$\left[\frac{\partial^2 u_e}{\partial \psi_e^2}\right]_{\substack{\text{fixed}\\ \text{strains}}} = -8C_3(\varepsilon_I - \varepsilon_{II})^2\left[\cos 2\psi(\gamma + \cos 2\psi) - \sin^2 2\psi\right] \tag{4.15}$$

and with this we obtain the results given in Table 4.1 for solutions of global maximum or global minimum. For non-orthotropic materials, analytical solutions are difficult and extremum are found numerically, say with Newton-Raphson iterations. A more extended analysis than can be shown here, Poulsen [12] offers information about appropriate starting points for such iterations.

Angle ψ of stationarity	Low shear modulus material $C_3 > 0$		High shear modulus material $C_3 < 0$	
	$0 < \gamma < 1$	$\gamma < 1$	$\gamma < -1$	$-1 < \gamma < 0$
0°	Global min.	Global min.	Global min.	Local max.
±90°	Local min.	Global max.	Global max.	Global max.
$\cos 2\psi = -\gamma$	Global max.			Global min

Table 4.1. Table for Selection of Optimality Criterion with Respect to Global Minimum or Global Maximum of Total Strain Energy

The numerical procedure for solving a specific problem is, briefly stated, as follows:

1) For a given design a finite element analysis gives the actual strain field, i.e., the principal strains with the ε_I direction in each element.
2) For each element, the optimization parameter γ_e is evaluated by (4.14).
3) Based on Table 4,.1 the new material angle (relative to the ε_I direction) is determined.
4) If actual changes are not within a given convergence criterion, return to 1) for a new analysis.

The total number of necessary iterations is normally about 5-10, provided that an extreme convergence criterion is not specified. Convergence in terms of total energy is much faster than in terms of design variables.

It is shown in Pedersen [11] that for the optimal design, the principal strain directions are aligned with the principal stress directions. Numerically it is found more efficient to redefine the material angle relative to the larger principal stress direction as an alternative to the larger principal strain direction.

As our first example we analyze the short cantilever in Figure 4.2, where the results for uniform material orientation throughout the model are shown. Two different materials are used as illustrative examples. If we optimize with the possibility of local orientation in each of the 72 finite elements, we can for the low shear modulus material further minimize the energy with a factor 0.51 or maximize with a factor 1.37. For the high shear modulus material, the corresponding numbers are 0.86 and 1.53.

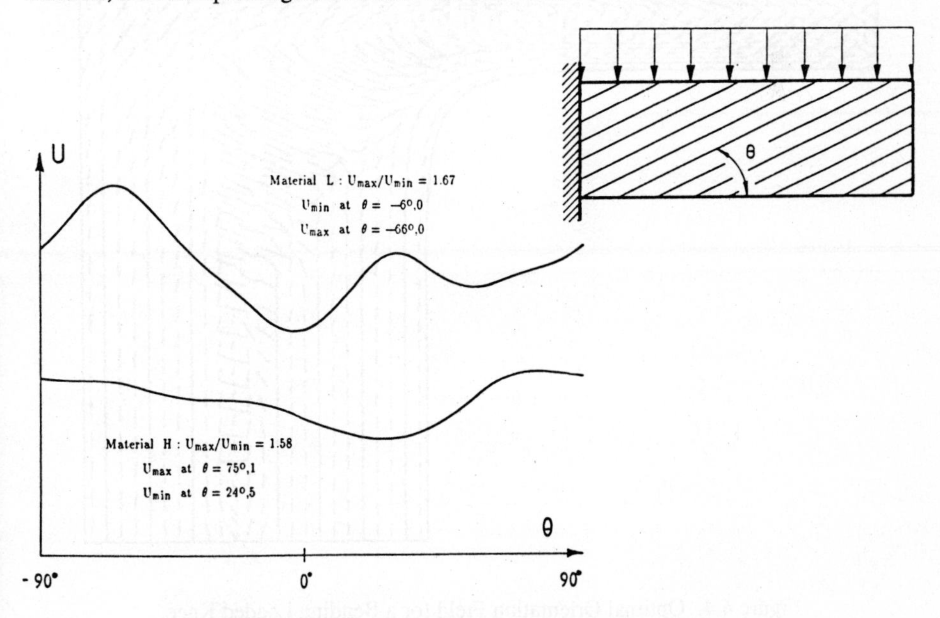

Figure 4.2. Elastic Energy for Different Material Orientations

Upper curve for a low shear modulus material ($C_{11} = 8$; $C_{22} = 4$; $C_{12} = 1$; $C_{33} = 0.5$) and lower curve for a high shear modulus material ($C_{11} = 8$; $C_{22} = 4$; $C_{12} = 3$; $C_{33} = 3.5$).

As always, we optimize the model, and detailed models should also be analyzed. Two such solutions are now shown. In Figure 4.3, we show the results of a cantilever example and in Figure 4.4 the results for a bending loaded knee example. For the 720 design parameters (cantilever), the stiffness is improved by a factor 1.8, and for the 1408 design parameters (knee), by a factor 1.6.

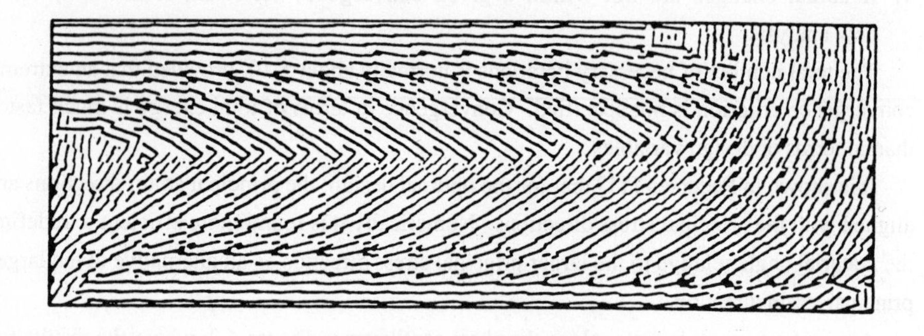

Figure 4.3. Optimal Orientation Field for a Uniformly Loaded Cantilever, Based on a 720 Element Model
Relative material parameters: (C_{11} = 80; C_{22} = 40; C_{12} = 10; C_{33} = 5).

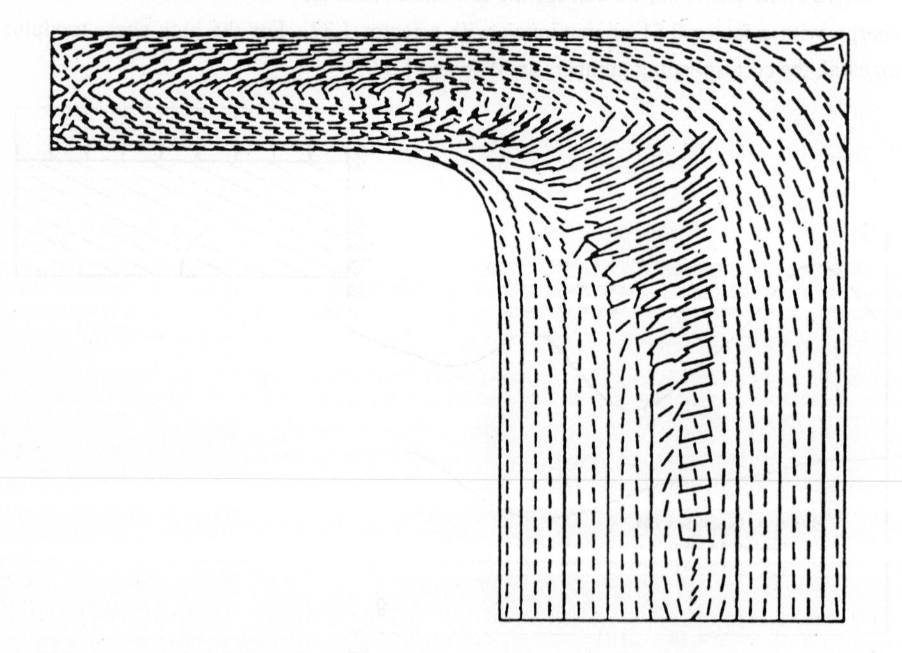

Figure 4.4. Optimal Orientation Field for a Bending Loaded Knee, Based on a 1408 Element Model
Relative material parameters: (C_{11} = 400; C_{22} = 100; C_{12} = 30; C_{33} = 75).

5 Thickness Distribution and Optimality Criterion

After the unconstrained optimization problem of orientation, we now treat the also simple class of optimization problems, characterized by only a single constraint. The actual constraint is a given volume $\overline{V}$, i.e.,

$$V - \overline{V} = \sum_e V_e - \overline{V} = \sum_e a_e t_e - \overline{V} = 0$$

$$\frac{\partial(V - \overline{V})}{\partial t_e} = \frac{\partial V}{\partial t_e} = \frac{\partial V_e}{\partial t_e} = a_e = \frac{V_e}{t_e} \tag{5.1}$$

The domain volume, area and thickness are V_e, a_e and t_e, respectively.

The necessary optimality criterion with only a single constraint is proportional gradients for all design variables. When the objective is to minimize strain energy U by thickness design this criterion gives

$$\frac{\partial U}{\partial t_e} = A \frac{\partial V}{\partial t_e} \quad \text{for all } e \tag{5.2}$$

where A is the common factor of proportionality. From the general sensitivity results, say (3.20), we have

$$\frac{\partial U}{\partial t_e} = -\left[\frac{\partial U_e}{\partial t_e}\right]_{\substack{\text{fixed}\\ \text{strain}}} = -\frac{u_e V_e}{t_e} \tag{5.3}$$

and inserting (5.1) and (5.3) in (5.2) we get directly the well-known criterion of constant energy density, equal to the mean energy density $\overline{u}$

$$u_e = \overline{u} \text{ for all } e \tag{5.4}$$

This optimality criterion is derived by Wasiutynski [5] as early as 1960.

Solutions which satisfy (5.4) may correspond to maximum or minimum or just stationarity. Furthermore, the extremum may be local or global. Lastly, the existence of a design satisfying (5.4) is not proven, and a procedure for obtaining such a possible solution still has to be described.

How is a thickness distribution that fulfills (5.4) obtained? We shall discuss a practical procedure, Rozvany [13], which is based on a number of approximations. Firstly, we neglect the mutual influences from element to element, i.e., each element is redesigned independently (but simultaneously)

$$(t_e)_{next} = t_e + (\Delta t_e) \tag{5.5}$$

Secondly, the optimal mean energy density $\bar{u}$ is taken as the present mean energy density $\tilde{u}$. Thirdly, the element energy U_e is assumed to be constant through the change Δt_e and then, from (5.4), we get

$$\begin{gathered} \frac{U_e}{V_e(1+\Delta t_e / t_e)} = \tilde{u}\text{ , i.e.,} \\ \Delta t_e = t_e(u_e - \tilde{u}) / \tilde{u} \text{ or } (t_e)_{next} = t_e\, u_e / \tilde{u} \end{gathered} \tag{5.6}$$

A relaxation power, say 0.8, is normally introduced in the formulation. It is natural to ask why the gradient of element energy is not taken into account

$$(U_e)_{next} = U_e + \frac{\partial U_e}{\partial t_e} \Delta t_e \tag{5.7}$$

but this is explained by the fact that although $\partial U / \partial t_e$ is known from (5.3), the gradient of the local energy (the element strain energy)

$$\frac{\partial U_e}{\partial t_e} = \left[\frac{\partial U_e}{\partial t_e}\right]_{\text{fixed strain}} + \left[\frac{\partial U_e}{\partial \varepsilon}\right]\frac{\partial \varepsilon}{\partial t_e} \tag{5.8}$$

is more difficult to determine. The two terms in (5.8) have different signs, and the other neglected terms $\partial U_e / \partial t_i$ for $e \neq i$ may also be of the same order.

Although the procedure (5.6) mostly works rather satisfactory, we shall now extend our analysis to the coupled problem. The redesign procedure provided by (5.6) neglects the mutual sensitivities, i.e., the change in element energy due to change in the thickness of the other elements. These sensitivities can be calculated by classical sensitivity analysis. Assume that the analysis is related to a finite element model

$$[S]\{D\} = \{A\} \tag{5.9}$$

where $\{A\}$ are the given nodal actions, $\{D\}$ the resulting nodal displacements and $[S] = \sum_e [S_e]$ the system stiffness matrix accumulated over the element stiffness matrices $[S_e]$ for $e = 1,2,...,N$.

Let t_e be without influence on $\{A\}$, we then get

$$[S]\frac{\partial\{D\}}{\partial t_e} = -\frac{\partial[S]}{\partial t_e}\{D\} = \{P_e\} \tag{5.10}$$

where the right-hand side $\{P_e\}$ is a pseudo load, equivalent to thickness design change. Knowing $\partial\{D\}/\partial t_e$ it is straightforward to calculate $\partial U_i/\partial t_e$. Generally, the computational effort involved corresponds to one additional load for each design parameter.

Then, with all the gradients $\partial U_i/\partial t_e$ available, we can formulate a procedure for simultaneous redesign of all element thicknesses

$$\{t\}_{next} = \{t\} + \{\Delta t\} \tag{5.11}$$

that takes the mutual sensitivities into account. In agreement with the optimality criterion (5.4) we change towards equal energy density $\tilde{u}$ in all elements. Formulated in terms of strain energy per area, we want

$$u_e t_e + \sum_{i=1}^{N} \frac{\partial(u_e t_e)}{\partial t_i}\Delta t_i = \tilde{u}(t_e + \Delta t_e) \quad \text{for } e = 1,2,... \tag{5.12}$$

or in matrix notation

$$\{ut\} + [\nabla(ut)]\{\Delta t\} = \tilde{u}[\{t\} + \{\Delta t\}] \tag{5.13}$$

with solution

$$\{\Delta t\} = [[\nabla(ut)] - \tilde{u}[I]]^{-1}\{(\tilde{u} - u)t\} \tag{5.14}$$

The gradient matrix $[\nabla(ut)]$ consists of the quantities $\partial(u_e t_e)/\partial t_i$. Note that with the assumption of fixed element energy, the strain energy per area is unchanged, i.e., $[\nabla(ut)] = [0]$ and we get the simple redesign formula (5.6).

An alternative formulation would be Newton-Raphson iterations directly on energy densities

$$(u_e - \tilde{u}) + \sum_{i=1}^{N} \frac{\partial(u_e - \tilde{u})}{\partial t_i} \Delta t_i = 0 \text{ for } e = 1,2,... \tag{5.15}$$

or, in matrix notation,

$$[\nabla u]\{\Delta t\} = \tilde{u}\{1\} - \{u\} \tag{5.16}$$

Here, the gradient matrix $[\nabla u]$ constitutes $\partial u_e/\partial t_i$. An interesting formulation is obtained when we multiply every row e with area a_e and get

$$[\nabla(ua)]\{\Delta t\} = \{(\tilde{u} - u)a\} \tag{5.17}$$

The present matrix is now symmetric, which, to the author's knowledge, is not well known. Remembering that $u_e a_e = U_e / t_e$, we prove this directly from (5.3)

$$\frac{\partial^2 U}{\partial t_e \partial t_i} = -\frac{\partial(U_e / t_e)}{\partial t_i}$$
$$\frac{\partial^2 U}{\partial t_i \partial t_e} = -\frac{\partial(U_i / t_i)}{\partial t_e} \tag{5.18}$$

Therefore, as $\partial^2 U / (\partial t_e \partial t_i) = \partial^2 U / (\partial t_i \partial t_e)$, we have

$$[\nabla(ua)]^T = [\nabla(ua)] \tag{5.19}$$

The two specific examples optimized with respect to material orientation shall now be optimized, with respect to only thickness distribution, and then simultaneously with respect to thickness and orientation. These examples are taken from Pedersen [14]. In all the figures we have used the same way of visualizing the results. The design is characterized by thickness and orientation, which are shown by hatching the triangular finite elements in the direction of the larger modulus direction and with the hatch density proportional to the thickness. Dark areas are therefore areas with large thicknesses. The response is characterized by the distribution of the strain energy density and by the direction of the larger principal stress. The technique of hatching is again used with hatch direction equal to principal stress direction and with hatch density proportional to strain energy density. Dark areas are therefore areas with energy concentration.

The cantilever example shown in Figure 5.1 is based on a 720 element model, with constant thickness and orientation in each element. For the uniform cantilever the mean and the maximum values of energy density are 787, 32919 (relative measures of stiffness and stress concentration). Only thickness optimization gives 414, 450, which means stiffness improved by a factor 1.9 and almost no energy concentration. With simultaneous thickness, orientational optimization we obtain 181, 199, i.e., a stiffness improved by a factor of 4.3.

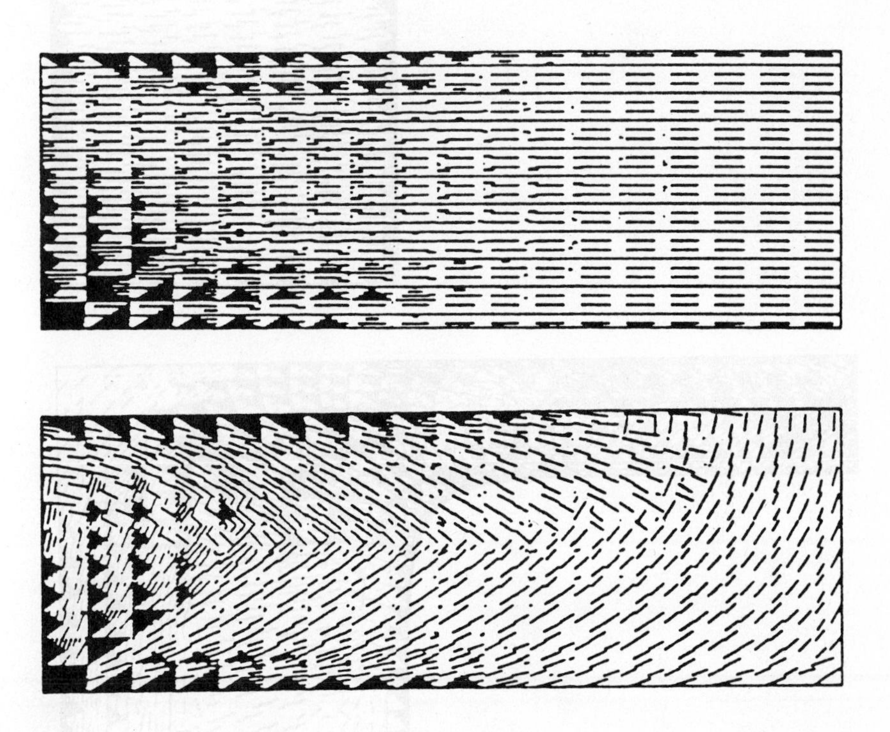

Figure 5.1. Upper: The Optimal Thickness Distribution for the Uniformly Loaded Cantilever, Which in Figure 4.3 is Shown with Orientational Optimization Alone. Lower: The Combined Thickness and Orientational Optimization.

The results for the bending loaded knee example are illustrated in Figure 5.2 and again we conclude that both thickness and orientational optimization need to be performed. The specific values for the uniform model are u_{mean}, u_{max} = 726, 5002; optimized only for thickness distribution we get 287, 328 and simultaneous thickness; orientational optimization gives 151, 161. Totally, we can thus gain a factor of 4.8 for stiffness and almost eliminate energy concentration.

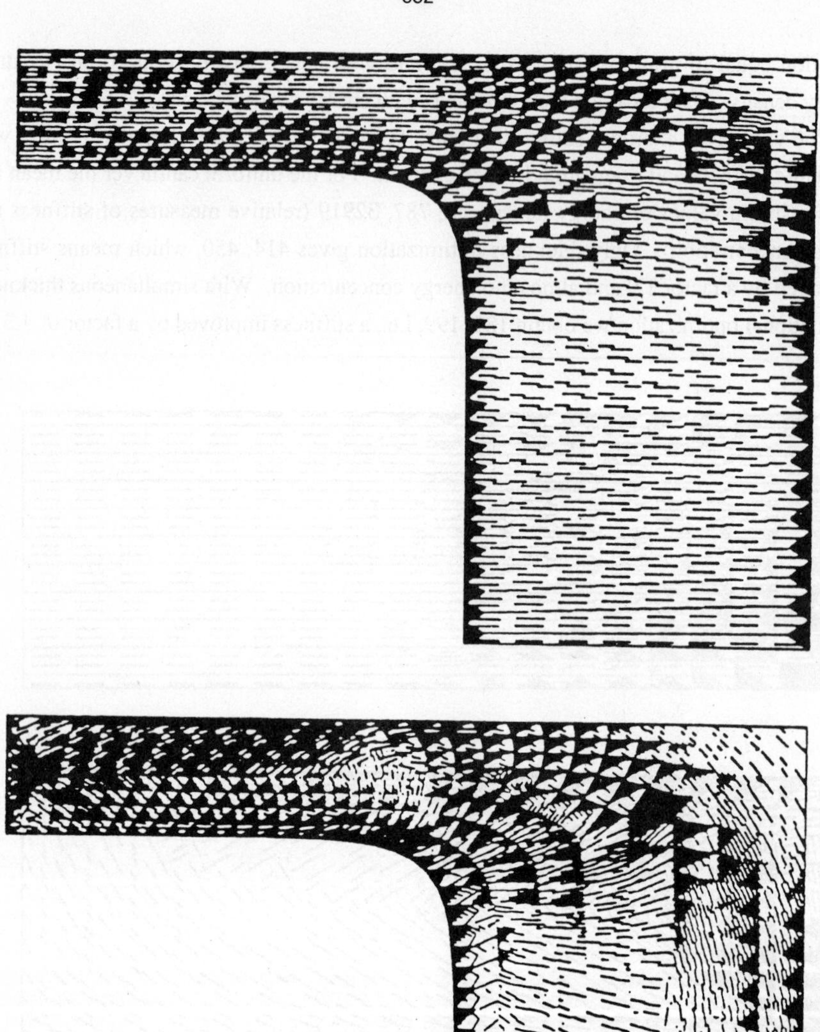

Figure 5.2. Upper: The Optimal Thickness Distribution for the Bending Loaded Knee, which in Figure 4.4 is Shown with Orientational Optimization Alone. Lower: The Combined Thickness and Orientational Optimization.

6 Detailed Material Design

The orientational design in Section 4 and the thickness design in Section 5, in fact show how to use advanced materials. The detailed design of the constitutive matrix is also a very active research area, as documented by the works of Lurie, Fedorov & Chekaev [15], Kohn [16], Bendsøe [17], Diaz and Bendsøe [18], among others.

We shall here focus only on the objective of maximize stiffness (minimize strain energy) based on orthotropic materials classified as low shear modulus materials. In reality we shall go directly to the material model from the above-mentioned references, as illustrated in Figure 6.1.

The total, relative volume densities ρ_e give the total volume V by

$$V = \sum_e V_e = \sum_e \tilde{V}_e \rho_e$$
$$\frac{\partial V}{\partial \rho_e} = \frac{\partial V_e}{\partial \rho_e} = \tilde{V}_e = \frac{V_e}{\rho_e} \tag{6.1}$$

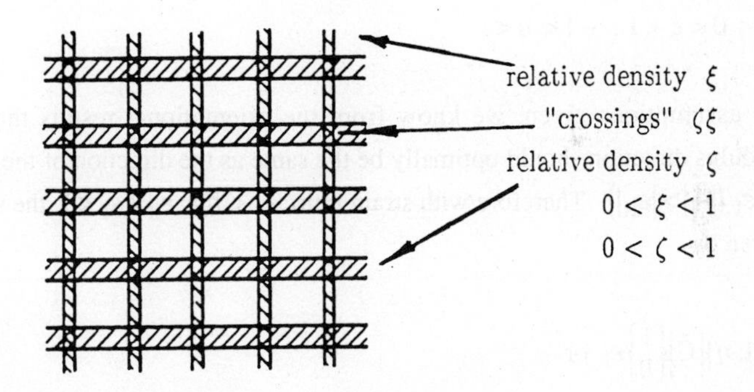

Figure 6.1. A Material Model with Two Directional Densities

where $\tilde{V}_e$ is the maximum volume of domain e. In principle, ρ_e acts like the thickness in Section 5. Often we shall omit the domain index e, and the relations of the directional relative densities ζ, ξ are

$$\rho = \zeta + \xi - \zeta\xi$$
$$\frac{\partial \rho}{\partial \zeta} = 1 - \xi \; ; \; \frac{\partial \rho}{\partial \xi} = 1 - \zeta \tag{6.2}$$

From Bendsøe [17] we take the orthotropic constitutive matrix, expressed in known modulus E and Poissons ratio υ and in the directional parameters ζ, ξ, as

$$[\mathrm{C}]=\begin{bmatrix} [\tilde{\mathrm{C}}] & ; & \begin{matrix}0\\0\end{matrix} \\ 0\,;\,0 & ; & \mathrm{C}_{33}\end{bmatrix}$$

$$\left[\tilde{\mathrm{C}}\right]=\frac{\mathrm{E}\xi}{\alpha-\beta^2}\begin{bmatrix}1 & ; & \beta\\ \beta & ; & \alpha\end{bmatrix};\left[\tilde{\mathrm{C}}\right]^{-1}=\frac{1}{\mathrm{E}\xi}\begin{bmatrix}\alpha & ; & -\beta\\ -\beta & ; & 1\end{bmatrix} \tag{6.3}$$

$$\alpha=\xi^2+\xi(1-\xi)/\zeta\,;\,\beta=\xi\upsilon$$

and we assume the major principal axis chosen by

$$1>\alpha \Rightarrow \zeta>\xi/(1+\xi) \tag{6.4}$$

The conditions of positive definite and limited $\left[\tilde{\mathrm{C}}\right]$ are satisfied for

$$0<\zeta<1\,;\,0<\xi<1\,;\,-1<\upsilon<1 \tag{6.5}$$

With the assumptions given, we know from the orientational results that the larger principal modulus direction should optimally be the same as the direction of the numerically larger strain ε_{I} $\left(|\varepsilon_{\mathrm{I}}|>|\varepsilon_{\mathrm{II}}|\right)$. Therefore with strain ratio $-1<\eta:=\varepsilon_{\mathrm{II}}/\varepsilon_{\mathrm{I}}<1$, the strain energy density is given by

$$\begin{aligned}&\mathrm{u}=\tfrac{1}{2}\varepsilon_{\mathrm{I}}\{1;\eta\}\left[\tilde{\mathrm{C}}\right]\begin{Bmatrix}1\\ \eta\end{Bmatrix}\varepsilon_{\mathrm{I}} \quad \text{or}\\ &\mathrm{u}=\tfrac{1}{2}\varepsilon_{\mathrm{I}}^2\frac{\mathrm{E}\xi}{(\alpha-\beta^2)}\left(1+2\beta\eta+\alpha\eta^2\right)\end{aligned} \tag{6.6}$$

We know that principal stresses are generally also in the same direction, and for the present case it is proved that the numerically larger principal stress $\sigma_{\mathrm{I}}\left(|\sigma_{\mathrm{I}}|>|\sigma_{\mathrm{II}}|\right)$ match with ε_{I}. Thus expressed in stresses and stress ratio $-1<\mu:=\sigma_{\mathrm{II}}/\sigma_{\mathrm{I}}<1$ we have from (6.3)

$$\mathrm{u}=\mathrm{u}^{\mathrm{C}}=\tfrac{1}{2}\sigma_{\mathrm{I}}^2\frac{1}{\mathrm{E}\xi}\left(\alpha-2\beta\mu+\mu^2\right) \tag{6.7}$$

These are the results from the orientational optimization. Next, in the same way as in the thickness optimization, we have

$$U = \sum_{e} U_e = \sum u_e \tilde{V}_e \rho_e$$

$$\frac{\partial U}{\partial \rho_e} = -\left[\frac{\partial U_e}{\partial \rho_e}\right]_{\substack{\text{fixed} \\ \text{strains}}} = -\frac{u_e V_e}{\rho_e} \tag{6.8}$$

and thus, with (6.1), again the result of uniform energy density equal to mean energy density $\bar{u}$

$$u_e = \bar{u} \tag{6.9}$$

Solution by optimality criterion iterations can thus give us ρ_e for $e = 1,2,...$

Lastly, we come to the detailed design. Knowing ρ_e, then how to optimize $\zeta_e \xi_e$?

We again have a single constraint problem (omitting index e)

$$\begin{aligned} &\text{Minimize } U = u\tilde{V}\rho \\ &\text{over } \zeta, \xi \\ &\text{with constraint } \rho = \zeta + \xi - \zeta\xi \end{aligned} \tag{6.10}$$

Derivatives with respect to the constraint are stated in (6.2) and derivatives with respect to u can be found from (6.6) to be:

$$\begin{aligned} &\partial u / \partial \zeta = (1-\xi)(1+\xi\upsilon\eta)^2 / N \\ &N = \left(1-\xi+\zeta\xi(1-\upsilon^2)\right)^2 \\ &\partial u / \partial \xi = \left[\zeta\left(1-\zeta(1-\upsilon^2)\right)+2\zeta\upsilon\eta \right. \\ &\left. +\eta^2\left[(1-\xi)^2+\zeta\xi(2(1-\xi)+\xi\upsilon^2)+\zeta^2\xi^2(1-\upsilon^2)\right]\right] / N \end{aligned} \tag{6.11}$$

Then, by means of (6.11) and (6.2), the optimality criterion of constant ratios between objective and constraint gradients

$$\frac{\partial u / \partial \zeta}{\partial \rho / \partial \zeta} = \frac{\partial u / \partial \xi}{\partial \rho / \partial \xi} \tag{6.12}$$

gives the equation of optimality

$$(1-\zeta)(1+\xi\upsilon\eta)^2 = \zeta\left(1-\zeta\left(1-\upsilon^2\right)\right) + 2\zeta\upsilon\eta$$
$$+\eta^2\left[(1-\xi)^2 + \zeta\xi\left(2(1-\xi)+\xi\upsilon^2\right) + \zeta^2\xi^2(1-\upsilon^2)\right] \tag{6.13}$$

With the help of the MATHEMATICA programme, it is possible to find that (6.13) correspond to the product of two linear equations

$$\left[(1-\zeta)+\upsilon\zeta+\eta(1+\xi(\zeta-1)(1+\upsilon))\right] \times$$
$$\left[(1-\zeta)-\upsilon\zeta-\eta(1+\xi(\zeta-1)(1-\upsilon))\right] = 0 \tag{6.14}$$

and thus we have with (6.2) and (6.4)-(6.5) an analytical solution, see Jog, Haber and Bendsøe [30].

In the stress formulation (6.7) the results are more simple to derive, and we get the final results after little algebra:

$$\zeta = \frac{\rho}{1+|\mu|} \;;\; \xi = \frac{|\mu|\rho}{1+|\mu|-\rho} \tag{6.15}$$

in terms of the total relative density ρ and stress ratio $\mu := \sigma_{II} / \sigma_I$.

In the paper by Thomsen [19] concurrent density and orientational design is solved with constraints on a cost defined functional.

7 Shapes for Minimum Energy Concentration

The problem of shape design for minimum energy concentration is highly nonlinear and must be solved iteratively. Thus, we can immediately convert the problem into a sequence of problems of optimal redesign, i.e., how do we change a given shape into a better "neighboring" shape? The solution to this involves three steps: finite element strain analysis for the given shape - sensitivity analysis with respect to the parameters c_i describing the design - and optimal decision of redesign.

In mathematical terms, the objective of our shape design is to

$$\underset{\text{(over feasible shapes)}}{\text{Minimize}} \qquad \underset{\text{(over the structural space (x) and load cases)}}{\text{Maximum } u} \tag{7.1}$$

where u is the strain energy density. Choosing u as our objective can naturally be questioned, but alternative objectives can be treated in a similar way.

Converting (7.1) to a Min-problem, concentrating on redesign by design parameters Δc_i, and including -- for computational reasons -- an area (volume) constraint $A = \overline{A}$, we have the actual redesign formulation

$$\underset{\text{(within move-limits on } \Delta c_i)}{\text{Minimize}} \quad u_{max} \text{ subject to}$$

$$u(x) + \sum_i \frac{\partial u(x)}{\partial c_i} \Delta c_i - u_{max} \le 0 \ , \ A + \sum_i \frac{\partial A}{\partial c_i} \Delta c_i - \overline{A} = 0 \tag{7.2}$$

where u_{max} is a further unknown to be determined.

Specifically, we shall concentrate in this section on the two dimensional problem of a biaxially loaded hole. The extensions relative to earlier works [20-22] are to study the influence from orthotropic material behavior, although still assuming linear elasticity. The studies by Lee, Kikuchi & Scott [23] and by BAcklund and Isby [24] are also related to shape optimization with anisotropic materials, but in all very few papers are published on the subject. We also shall study the influence of finite element modeling and compare the solutions based on elements with constant and linear strain assumptions, respectively. Extensive details can be found in the M.Sc. thesis [25], written in Danish, and the present section is closely related to the paper Pedersen, Tobiesen and Jensen [26].

The techniques used for the design description, for the finite element modeling, and for the sensitivity analysis, have proved accurate, robust and effective. Therefore, a number of detailed comments on the methods are included.

Design Parameters and FE Model. The success of shape optimization depends to a large extent on the chosen design parameters, i.e., the description. Many possibilities exist and we use a global description that enforces smoothness and desired connections to neighboring shapes, as exemplified by symmetric requirements. Other possible descriptions are by local methods, for which such requirements are enforced as constraints.

In the early works [20-22], cylindrical coordinates for 2D problems and polar coordinates for 3D problems were applied. These descriptions, although successfully applied, certainly have some limitations, so a generalized description was introduced in [27]. In two dimensions, the shape to be designed is a curve C, mathematically described by

$$C = C_0 + \sum_{i=1}^{I} c_i f_i(s) \tag{7.3}$$

The basic curve C_0 may be given or perhaps depend on some (preliminary) design parameters, as in the case of a superelliptic curve C_0 in a x_1,x_2 Cartesian system

$$C_0 : (x_1 / a)^n + (x_2 / b)^n = 1 \tag{7.4}$$

where the half-axes a, b and the power n are design parameters. The further design parameters for C are the factors c_i that give the linear combination of the preselected modification functions $f_i(s)$, where s is the natural parameter for the curve C_0.

We choose the functions $f_i(s)$ to be mutually orthogonal; on the basis of their physical background (beam vibrational modes), we can control smoothness, slopes, connections and know clearly the details that we are unable to describe. Mostly four-five functions, i.e., I = 5 in (7.3), are enough.

For the quarter hole models to be shown later, the natural functions to choose are

$$\begin{aligned} &f_i(s) = \sin(\gamma_i s) - \sinh(\gamma_i s) + k_i(\cos(\gamma_i s) - \cosh(\gamma_i s)) \\ &k_i := (\cos(\gamma_i S) - \cosh(\gamma_i S)) / (\sin(\gamma_i S) + \sinh(\gamma_i S)) \\ &\gamma_i \text{ given by } \cos(\gamma_i s)\cosh(\gamma_i s) = 1 \end{aligned} \tag{7.5}$$

where S is the length of C_0. These functions satisfy the boundary conditions

$$f_i(0) = f_i(S) = \frac{df_i}{ds}(0) = \frac{df_i}{ds}(S) = 0 \tag{7.6}$$

In Figure 7.1, we illustrate the description with only two functions f_1 and f_2.

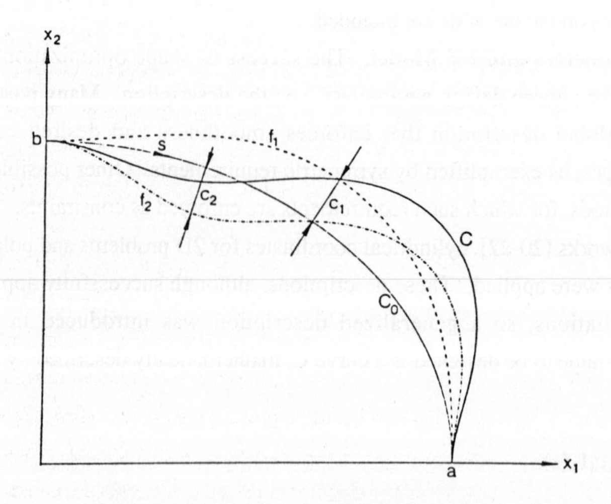

Figure 7.1. Description of a Shape of Design by Basic Curve C_0 and Two Modification Functions, f_1 and f_2

The finite element modeling is by no means straightforward, either. A good finite element model is strongly related to the actual problem, which constitutes the design as well as the load case. We need to choose elements and a mesh and have to be prepared to perform mesh adaption for the individual designs as well as for the changed designs. Automatic mesh generation is a necessity.

In general, our analysis is based on a "deformed" rectangular mesh with each quadrangle divided into four triangular basic elements, illustrated in Figure 7.2. The node concentrations can be controlled by a few parameters (exponential powers), and a simple scheme for automatic mesh adaption (uniform element total energy) has also been worked out. Convergence tests confirm the accuracy, and we shall present a comparative study between the three-node constant strain triangles and the six-node linear strain triangles.

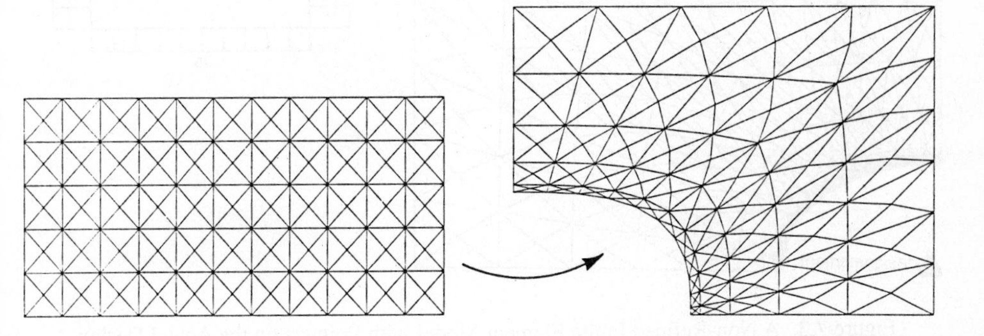

Figure 7.2. Illustration of the Mesh Generations

In Figure 7.3 we show the total model, the quarter part for analysis, and a representative finite element model.

Sensitivity Analysis and Optimization Procedure. The key information needed for the optimal design process is how the strain energy-density u changes with the parameters of the shape design, here symbolically denoted by c. With the constitutive matrix [E] from $\{\sigma\} = [E]\{\varepsilon\}$ (in this section symbol [E] is used because c were chosen for the design parameters) and assuming linear elasticity, the strain energy density at a given point is

$$u = \tfrac{1}{2}\{\varepsilon\}^T[E]\{\varepsilon\} \tag{7.7}$$

We emphasize here that orthotropy enters only through the matrix [E] and that the following derivations are completely analogous to the analyses needed for isotropic materials. The derivations are repeated for the sake of completeness of presentation. From (7.7) follows

$$\frac{\partial u}{\partial c} = \{\varepsilon\}^{T}[E]\frac{\partial\{\varepsilon\}}{\partial c} \tag{7.8}$$

with [E] symmetric and independent of c .

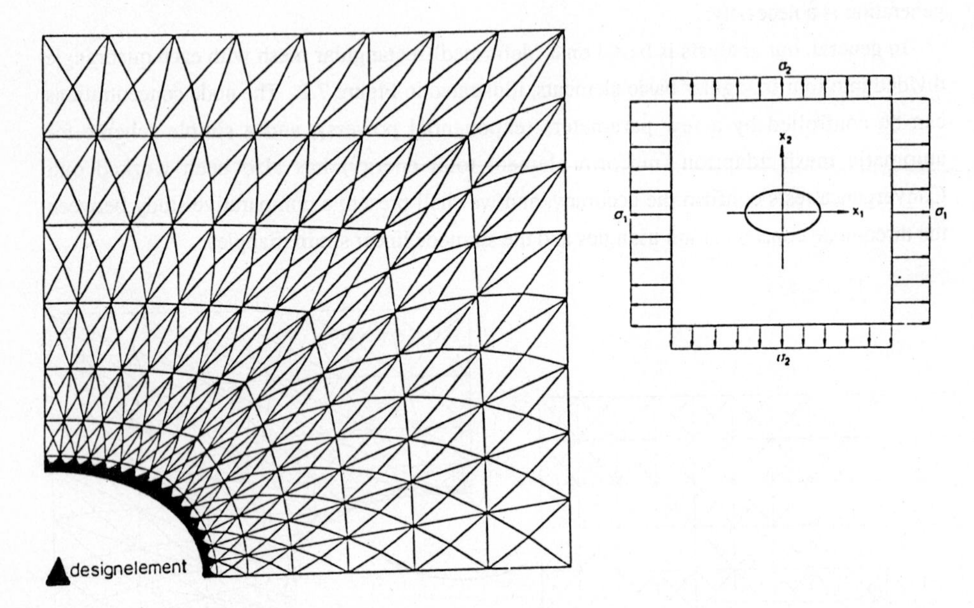

Figure 7.3. A Non-Refined Finite Element Model with Pointers on the Actual Design Elements, i.e., Elements Influenced Directly by Boundary Design. The total model with loads shown on the right.

From the strain/displacement relation $\{\varepsilon\} = [B]\{D_e\}$ follows

$$\frac{\partial\{\varepsilon\}}{\partial c} = \frac{\partial[B]}{\partial c}\{D_e\} + [B]\frac{\partial\{D_e\}}{\partial c} \tag{7.9}$$

The partial derivative $\partial[B]/\partial c$ is local, meaning that this term will only be non-zero if c changes the nodal positions of the actual element. The partial derivative $\partial\{D_e\}/\partial c$ is global and we traditionally determine it by the pseudo load approach, i.e., from the system equilibrium $[S]\{D\} = \{L\}$, which, with design-independent loads $\partial\{L\}/\partial c = \{0\}$, gives

$$[S]\frac{\partial\{D\}}{\partial c} = -\frac{\partial[S]}{\partial c}\{D\} \tag{7.10}$$

Then we get the gradient of the design element nodal displacements $\partial\{D_e\}/\partial c$, directly contained in $\partial\{D\}/\partial c$.

Very simply, we apply the semi-analytic approach, see Haftka, Gürdal and Kamat [28],

$$\frac{\partial[S]}{\partial c} \approx \frac{[S(c+\Delta c)]-[S(c)]}{\Delta c} \tag{7.11}$$

This technique is a powerful tool that does not require the details of the analytical evaluation of $\partial[S]/\partial c$ while it does not require the extreme computational efforts of an overall finite difference approach by $\partial\{D\}/\partial c \approx (\{D(c+\Delta c)\}-\{D(c)\})/\Delta c$.

We also use the difference technique to obtain the gradient of the strain/displacement matrix

$$\frac{\partial[B]}{\partial c} \approx \frac{[B(c+\Delta c)]-[B(c)]}{\Delta c} \tag{7.12}$$

In many cases, central differences are preferred to (7.11) and/or (7.12). The relative changes of c are of the order 10^{-3}.

Only at the critical points, i.e., points of large u, do we perform the analysis to determine $\partial u/\partial c$. It should also be remembered that only elements connected to the shape to be designed are included in the semi-analytical analyses (7.11) and (7.12), see Figure 7.3. So what may at first seem overwhelming is in fact a fast and simple computer analysis.

Lastly, in the sensitivity analysis, we need the gradient of model area $\partial A/\partial c$. Again the difference approach is applied

$$\frac{\partial A}{\partial c} \approx \frac{A(c+\Delta c)-A(c)}{\Delta c} = \frac{1}{\Delta c}\sum_e (A_e(c+\Delta c)-A_e(c)) \tag{7.13}$$

with element summation only for connections to the shape of design.

The optimization procedure used for each design improvement is linear programming with move limits. The general technique of converting the Min-max problem into a pure min problem is to introduce the further unknown u_{max} and then constraint the strain energy density everywhere to be less than u_{max}. Details are given in [20] and in many other papers, so we will omit them here.

The Model Problem. A number of interesting parameter studies can be performed with the programs developed. We shall concentrate on the biaxial single-load case illustrated in Figure 7.2. External stress ratios σ_1/σ_2 of 3/2 and 3/1 are taken as examples. Material with the x_1,x_2 axes as orthotropic axes are assumed, and we shall then vary the degree of orthotropy. From the law of mixture (see JONES [2]) the material parameters in Table 7.1 are taken as examples.

Material #	$E_1/10^{11}$Pa	$E_2/10^{11}$Pa	υ_{12}	$G_{12}/10^{11}$Pa	E_1/E_2
I, Isotropic	1.0	1.0	0.3	0.3846	1.0
II, 5% fiber	3.450	1.052	0.3	0.4044	3.281
III, 10% fiber	5.900	1.109	0.3	0.4264	5.322
IV, 20% fiber	10.80	1.244	0.3	0.4784	8.683
V, 30% fiber	15.70	1.416	0.3	0.5448	11.08
VI, Extreme	60.914	1.4503	0.3	0.13778	42.00

Table 7.1. Applied Material Data

The Isotropic Case. To test the method and programs, the well known result for the isotropic materials was firstly recalculated. The both analytically, as well as numerically, proven result from [20] is an elliptic shape with half axes proportional to the external stress ratio, i.e., for the design parameters (7.1)–(7.2)

$$n = 2 \; ; \; a / b = \sigma_1 / \sigma_2 \; ; \; \text{all } c_i = 0 \tag{7.14}$$

This result was also obtained with our energy objective, because we have a uniaxial stress state at the shape of design. Therefore the squared von Mises stress and strain energy densities are proportional here.

In this case, as for all the following cases, we find almost uniform energy density along the designed shape. This is, however, not presumed in our formulation and it cannot be expected to be the result with, say, multiple load cases.

Dependence on Element Model. It is well known and should always be borne in mind that we are optimizing the finite element model. Thus, not only the analysis result but also the resulting optimal shape could be expected to change with the model. We therefore performed a study of the influence of model refinement, in terms of "better" mesh and/or "better" elements. A better mesh is obtained with more elements, i.e., more degrees of freedom and more computer time. Better elements relative to the simple three-node, constant strain triangles are the six-node, linear strain triangles.

No unexpected results were obtained and the shape information from simple models was not changed in principle. Thus, refined models are not needed, at least not before the final iterations. More detailed optimal shapes were obtained with six-node triangles but, compared at equal computer time, the three-node triangles seems to be "the winner." In Figure 7.4, we show the results of comparative studies based on the strongly orthotropic material VI in Table 7.1. Note that even with the same number of nodes, the computer time for the model with six-node elements will be about four times greater.

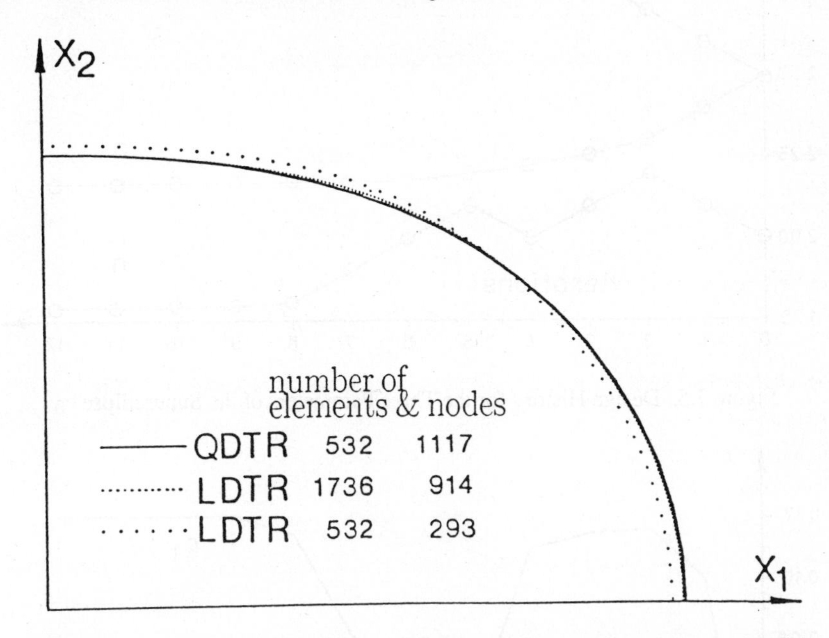

Figure 7.4. Resulting Optimal Shapes with Three Different Finite Element Models

The Design History. For all the examples the history of iteration is very much the same. The necessary number of iterations is between 5 and 10, naturally depending on the convergence test. Without comment, Figures 7.5-7.8 illustrate the stability of the method. The example relates to material VI, i.e., the strongly orthotropic material and the applied stress ratio is 3/1. The only "oscillations" are related to the fact that the design parameters n in (7.1) and c_1 in (7.2) can describe almost the same changes. Being aware of this fact, it creates no complications but merely focuses attention on the importance of choosing mutually orthogonal design functions as is the case for the f_i functions (7.4).

Influence of Degree of Orthotropy. In Figure 7.9, we show the optimal shape designs corresponding to the materials I-V in Table 7.1. The external stress ratio σ_1 / σ_2 is 3/2. The corresponding values of the design parameters are given in Table 7.2. These values relate to the modeling with linear strain elements, 532 elements and 1117 nodes. Refined models confirm the results. (For these results the power parameter n could in fact be fixed to 2, i.e., a normal elliptic basic curve). In Table 7.3 we compare the optimal shape design with the initial designs, which for all cases were the isotropic optimal shape design.

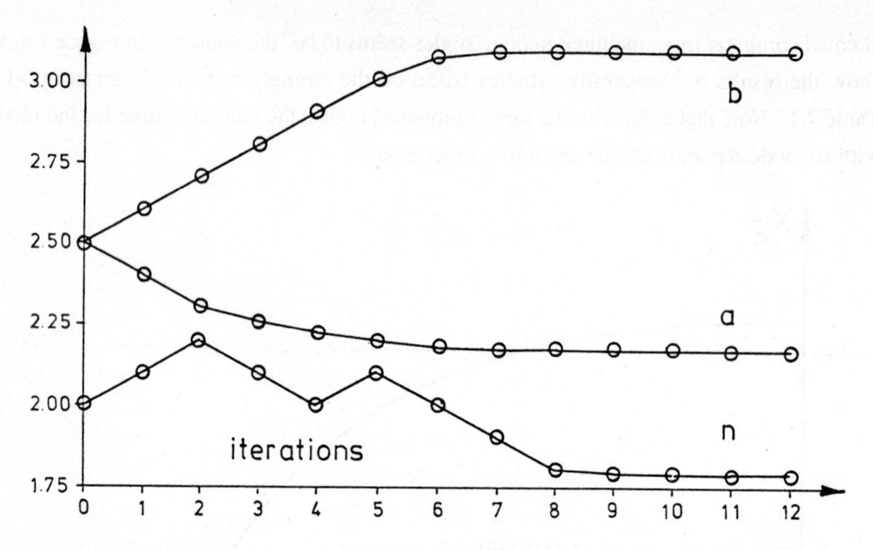

Figure 7.5. Design History for the Three Parameters of the Superellipse c_0

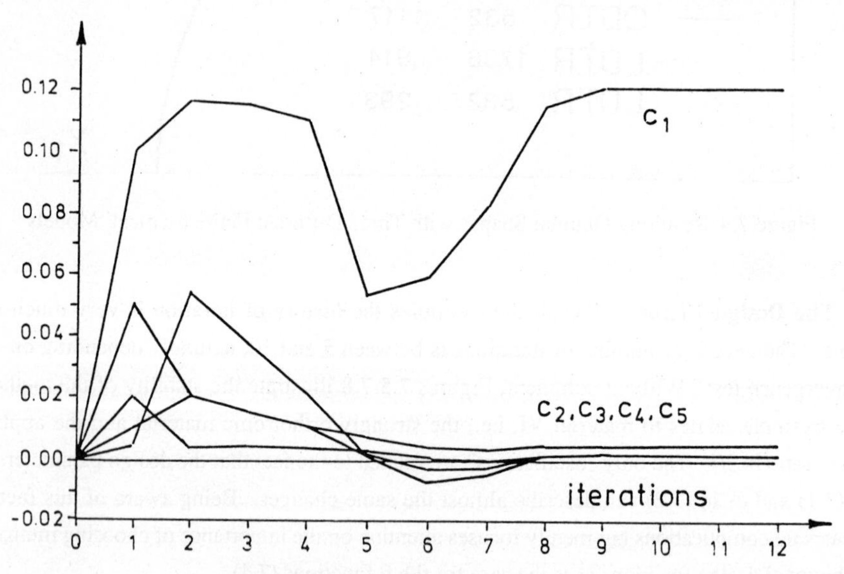

Figure 7.6. Design History for the Five Parameters of the Modification Functions

Through the present study, we have seen that the earlier methods for optimal shape design with isotropic materials also work effectively when the design is based on the use of orthotropic materials. The recommended methods are: global design description finite element analysis, semianalytical sensitivity analysis, and linear programming.

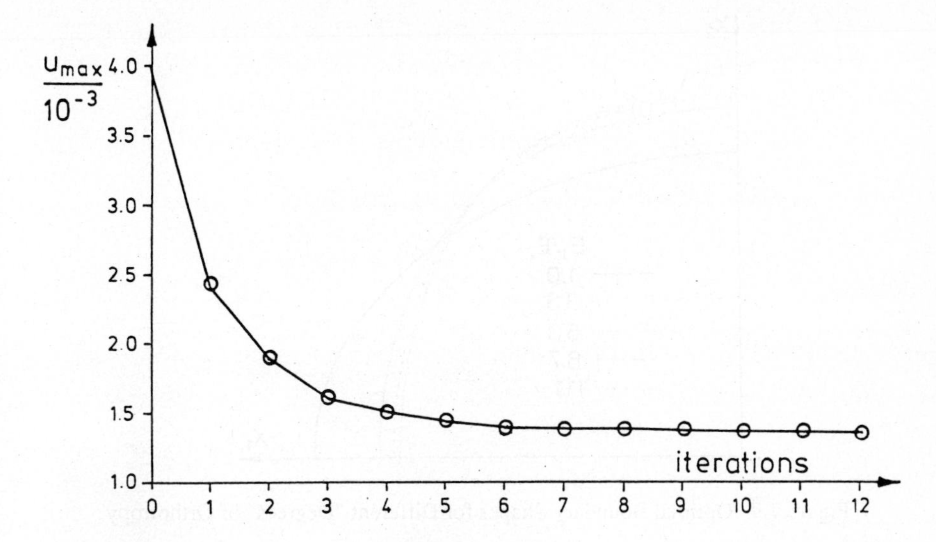

Figure 7.7. Design History for the Objective, i.e., Maximum Strain Energy Density

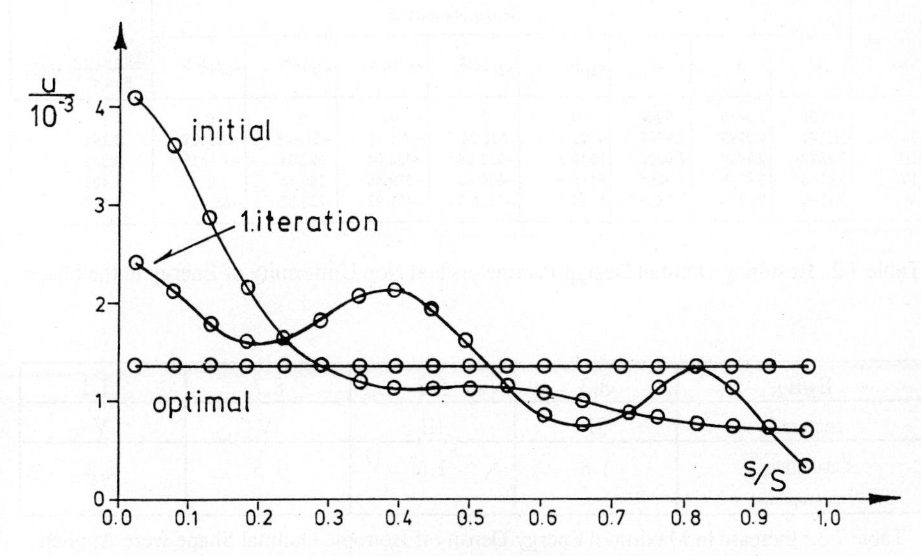

Figure 7.8. Distribution of Strain Energy Density Along the Shape of Design for Three Different Designs

Comparisons related to the refinement of the analysis show that even a course finite element analysis gives a rather accurate optimal design. The design description with a few global design parameters is the basis for this possibility, and is generally of major importance.

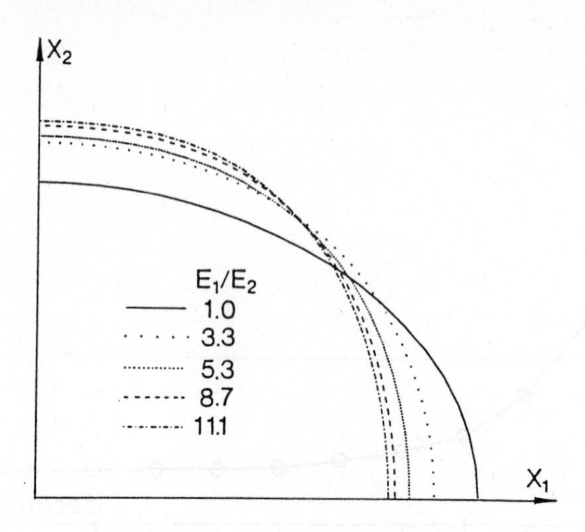

Figure 7.9. Optimal Boundary Shapes for Different "Degrees" of Orthotropy

Material	Design parameters								$\frac{\Delta u_{SHAPE}}{u_{max}} \cdot 100\%$
	a	b	n	$c_1/10^{-5}$	$c_2/10^{-5}$	$c_3/10^{-5}$	$c_i/10^{-5}$	$c_5/10^{-5}$	
I	3.0092	2.0415	1.9908	0	0	0	0	0	
II	2.7141	2.2999	1.9987	4721.2	331.24	−231.46	−31.108	−2.8422	0.194
III	2.5529	2.3420	2.0306	5349.2	−212.04	−433.04	32.934	−7.7518	0.335
IV	2.4563	2.4095	1.9688	8115.8	−870.03	−379.46	146.15	0	1.403
V	2.4109	2.4387	2.0895	5324.7	−1326.5	−451.93	223.33	−68.25	0.376

Table 7.2. Resulting Optimal Design Parameters and Non Uniformity of Energy at the Shape

E_1/E_2	3.3	5.3	8.7	11.1
material	II	III	IV	V
$\frac{u_{max\ initial}}{u_{max\ optimal}}$	1.8	1.6	1.5	1.2

Table 7.3. Increase in Maximum Energy Density if Isotropic Optimal Shape were Applied

The objective of minimum-maximum energy density could easily be substituted by an alternative objective, say an experimentally verified strength criterion. Problem formulations with multiple load cases and with constraints on displacements as well as on geometry could also be incorporated, but the results would then naturally be of only specific interest.

It should be noted that the resulting uniform energy density along the optimal shape is not enforced in the formulation and cannot be expected with additional load cases and/or additional constraints.

8 Conclusion

In this tutorial introduction to research results from optimal design with and of anisotropic material, we have tried to focus on the more general aspects.

Our goal is to design concurrently structural shape, pointwise density or thickness, material orientation, and the detailed constitutive behavior. However, certain aspects, such as material orientation and material density, are given by optimality criteria that hold generally. Thus, advantageous decouplings are possible and should be utilized.

The description is restricted to two-dimensional problems, which to a large extent are now clarified. Interesting extensions to three-dimensional problems are the subject of active research, and connections to topology design are being pointed out.

It would be fair to say that the use of advanced materials has put optimal design into a higher level of application. The present paper is a rather subjective point of view and appropriate reference to other research groups is not given.

References

1. Lekhnitskii, S.G.: Theory of Elasticity of an Anisotropic Body, Mir Publishers, Moscow, 1981, p. 430.
2. Jones, R.M.: Mechanics of Composite Materials, 2, McGraw-Hill, New York, N.Y., 1975, p. 355.
3. Pedersen, P.: Combining Material and Element Rotation in One Formula, Comm. in Appl. Num. Meth., Vol. 6, 549-555, 1990.
4. Masur, E.F.: Optimum Stiffness and Strength of Elastic Structures, J. of the Engineering Mechanics Div., ASCE, EM5, 621-649, 1970.
5. Wasiutynski, Z.: On the Congruency of the Forming According to the Minimum Potential Energy with that According to Equal Strength, Bull. de L'Academie Polonaise des Sciences, Serie des sciences technique, Vol. VIII, No. 6, 259-268, 1960.
6. Pedersen, P.: On Sensitivity Analysis and Optimal Design of Specially Orthotropic Laminates, Eng. Opt., Vol. 11, 305-316, 1987.
7. Bert, C.W.: Optimal Design of a Composite-Material Plate to Maximize its Fundamental Frequency, J. Sound and Vibration, Vol. 50, 229-239, 1977.
8. Muc, A.: Optimal Fibre Orientation for Simply-Supported, Angle-Ply Plates under Biaxial Compression, Composite Structures, Vol. 9, 161-172, 1988.
9. Pedersen, P.: Minimum Flexibility of Non-Harmonic Loaded Laminated Plates, In "Mechanical Characterization of Fibre Composite Materials" (ed. R. Pyrz), AUC, Denmark, 182-196, 1986.
10. Pedersen, P.: On Optimal Orientation of Orthotropic Materials, Structural Optimization, Vol. 1, 101-106, 1989.
11. Pedersen, P.: Bounds on Elastic Energy in Solids of Orthotropic Materials, Structural Optimization, Vol. 2, 55-63, 1990.
12. Poulsen, H.H.: Optimal Material-Orientation, Thesis for M.Sc., Solid Mechanics, DTH, 1990 (in Danish).
13. Rozvany, G.I.N.: Structural Design via Optimality Criteria, Kluwer, 1989, p. 463.

14. Pedersen, P.: On Thickness and Orientational Design with Orthotropic Materials, Structural Optimization, Vol. 3, 69-78, 1991.
15. Lurie, K.A., Fedorov, A.V. and Chekaev, A.V.: Regularization of Optimal Design Problems for Bars and Plates, J. Optim. Theory Appl., Vol. 37(4), 499-513, 1982.
16. Kohn, R.: Composite Materials and Structural Optimization, to appear in Proc. Workshop on Smart/Intelligent Materials, Honolulu, Technomic Press, 1990.
17. Bendsøe, M.P.: Optimal Topology Design and Homogenization, to appear in "Mechanics, Numerical Modelling and Dynamics of Materials" (ed. Blanc, Raous, Suquet), CNRS, 1991.
18. Diaz, A.R. & Bendsøe, M.P.: Shape Optimization of Structures for Multiple Loading Conditions using a Homogenization Method, Structural Optimization, Vol. 4, 17-22, 1992.
19. Thomsen, Jan: Optimization of Composite Discs, Structural Optimization, Vol. 3, 89-98, 1991.
20. Kristensen, E.A. and Madsen, N.F.: On the optimum shape of fillets in plates subjected to multiple in-plane loading cases. Int. J. Numer. Meth. Engng. 10, 1007-1019, 1976.
21. Pedersen, P. and Laursen, C.L.: Design for minimum stress concentration by finite elements and linear programming. J. Struct. Mech. 10, 243-271, 1982-83.
22. Dybbro, J.D. and Holm, N.C.: On minimization of stress concentration for three-dimensional models. Comp. & Struct. 4, 637-643, 1986.
23. Lee, M.S., Kikuchi, N. and Scott, R.A.: Shape optimization in laminated composite plates. Comp. Meth. in Appl. Mech. & Eng. 72, 29-55, 1989.
24. Bäcklund, J. and Isby, R.: Shape optimization of holes in composite shear panels. Proc. IUTAM Symposium on Structural Optimization, Melbourne, Australia, 9-16, 1988.
25. Tobiesen, L. and Jensen, S.H.: Optimal hulfacon i ortotrop skive (Optimal hole shape in orthotropic plate), Solid Mechanics, DTH, 1990 (in Danish).
26. Pedersen, P., Tobiesen, L. and Jensen, S.H.: Shapes of Orthotropic Plates for Minimum Energy Concentration, Mechanics of Structures and Machines, Vol. 20, No. 4, 1992.
27. Pedersen, P.: Design for minimum stress concentration--some practical aspects, in Rozvany & Karihaloo eds. "Structural Optimization," Kluwer, 225-232, 1988.
28. Haftka, R.T., Gürdal, Z. and Kamat, M.P.: Elements of structural optimization. 2nd ed., Kluwer, p. 396, 1990.
29. Pedersen, P. and Taylor, J.E.: Optimal design based on power-law nonlinear elasticity, IUTAM Symposium: Optimal Design with Advanced Materials, Lyngby, Denmark, 1992.
30. Jog, C.S., Haber, R.B. and Bendsøe, M.P.: Topology design using a material with self-optimizing microstructure, IUTAM Symposium: Optimal Design with Advanced Materials, Lyngby, Denmark, 1992.

Latin symbols:

A	common factor of proportionality
A, A_e	total area and area of domain e
a	length of plate (in x_1-direction)
a_e	area of domain e
b	width of plate (in x_2-direction)
C_{ij}	modulus from the constitutive matrix
C_n	practical combination of C_{ij} (n=1,2,...,7)
c	constant factor between external work and elastic energy
c_i	design parameter
C, C_0	design shapes
D_{ij}	stiffnesses from the bending stiffness matrix
E	modulus of elasticity
f_i	design modification function

h, h_e	design parameter and local design parameter of domain e
m	number of halfwavelength in x_1-direction
n	number of halfwavelength in x_2-direction
n	power of superelliptic curve
N_{cr}	buckling load
p	power of non-linear elasticity
p, p_{mn}	pressure and pressure coefficients
s,S	length parameter and total length of design boundary
t_e	plate thickness of domain e
U, U_e	total strain energy and same of domain e
u	strain energy density
U^C, U^C_e	total stress energy (complementary energy) and same of domain e
u^C	stress energy density
V, V_e	total volume and volume of domain e
v	displacement field
W	work of the external forces
W^C	complementary work of the external forces
w	transverse displacement
x	structural space

Greek Symbols:

γ	angle of material orientation
γ	combined optimization parameter for extensions
Δ	incremental prefix
δ	variational prefix
δ	combined optimization parameter for bendings
ε	strain field
$\varepsilon_I, \varepsilon_{II}$	principal strains $(\lvert\varepsilon_I\rvert > \lvert\varepsilon_{II}\rvert)$
ζ	relative directional volume density
η, η_{mn}	mode parameter of bending
η	ratio of principal strains
θ	angle of material or plate orientation
μ	ratio of principal stresses
υ	Poissons ratio
ξ	relative directional volume density
ρ, ρ_e	relative volume density in domain e
σ	stress field

σ_I, σ_{II}	principal stresses
Φ_{mn}	response functional of bending
ψ	angle of material from the direction of the larger principal strain

Vectors and Matrices:

[A]	extensional plate stiffnesses
{A}	finite element load vector
[B]	coupling plate stiffnesses
[B]	strain/displacement matrix
[C]	constitutive matrix (3 by 3 for 2D)
$[\tilde{C}]$	constitutive submatrix (2 by 2 for 2D)
[D]	bending plate stiffnesses
{D}	nodal displacement vector
[E]	constitutive matrix (3 by 3 for 2D)
$\{P_e\}$	pseudo load vector
$[S],[S_e]$	total stiffness matrix and element stiffness matrix
$\{\varepsilon\}$	strain vector $\left(\{\varepsilon_{11} ; \varepsilon_{22} ; 2\varepsilon_{12}\}\text{ for 2D}\right)$
$\{\varepsilon^0\}$	midsurface strain vector
$\{\kappa\}$	curvature/bending vector
$\{\sigma\}$	stress vector $\left(\{\sigma_{11} ; \sigma_{22} ; \sigma_{12}\}\text{ for 2D}\right)$

Optimization of Automotive Systems

Dieter Bestle

University of Stuttgart, Institute B of Mechanics, Pfaffenwaldring 9, W-7000 Stuttgart 80, Germany

Abstract: As an example for the optimal design of automotive systems, the problem of suspension systems with passive and actively controlled elements is formulated as an integrated modeling and design problem. For the modeling part, multibody formalisms are applied; the resulting nonlinear programming problem is solved by standard optimization algorithms. The missing link is an efficient and reliable procedure for computing gradients. Advantages and drawbacks of three different approaches are discussed in detail. Optimization of a vehicle with active and passive suspension is performed with respect to riding comfort, riding safety, and relative displacements between the wheels and the car body.

Keywords: multibody system / optimization / sensitivity analysis / simulation / vehicle dynamics / active suspension

1 Introduction

Passive vehicle suspensions using springs and dampers have been steadily improved over the last century. In the past, this was achieved by experimental studies of prototypes and test vehicles. With help of experience and intuition, rather good trade-offs between the two conflicting tasks of suspension systems, providing comfort and safety, have been found. But, due to the time-consuming and costly construction of prototypes, development cycles were rather long.

Only recently, automobile companies have started to switch to a computer-aided design process for suspension systems. Analyzing the dynamic behavior and performance of a vehicle in a an early design stage can help to shorten the development time significantly. Furthermore, introducing active elements in suspension systems is a new challenge which cannot be solved by intuition of mechanical engineers only. Tools for a systematic computer-aided design of vehicle suspensions have to be developed.

In the classical optimal theory approach, active elements of vehicle suspensions are designed independently of the passive components using very simple quarter-car models, Figure la, e.g., [1]. Applied to a full-vehicle model, such controllers may behave less than optimal due to the missing modeling of roll- and pitch-motion [2]. Only an integrated modeling and control design process may help to achieve an optimal behavior of the overall vehicle, Figure lb. More realistic models have to be used for adjusting parameters of both active and passive elements of suspension systems.

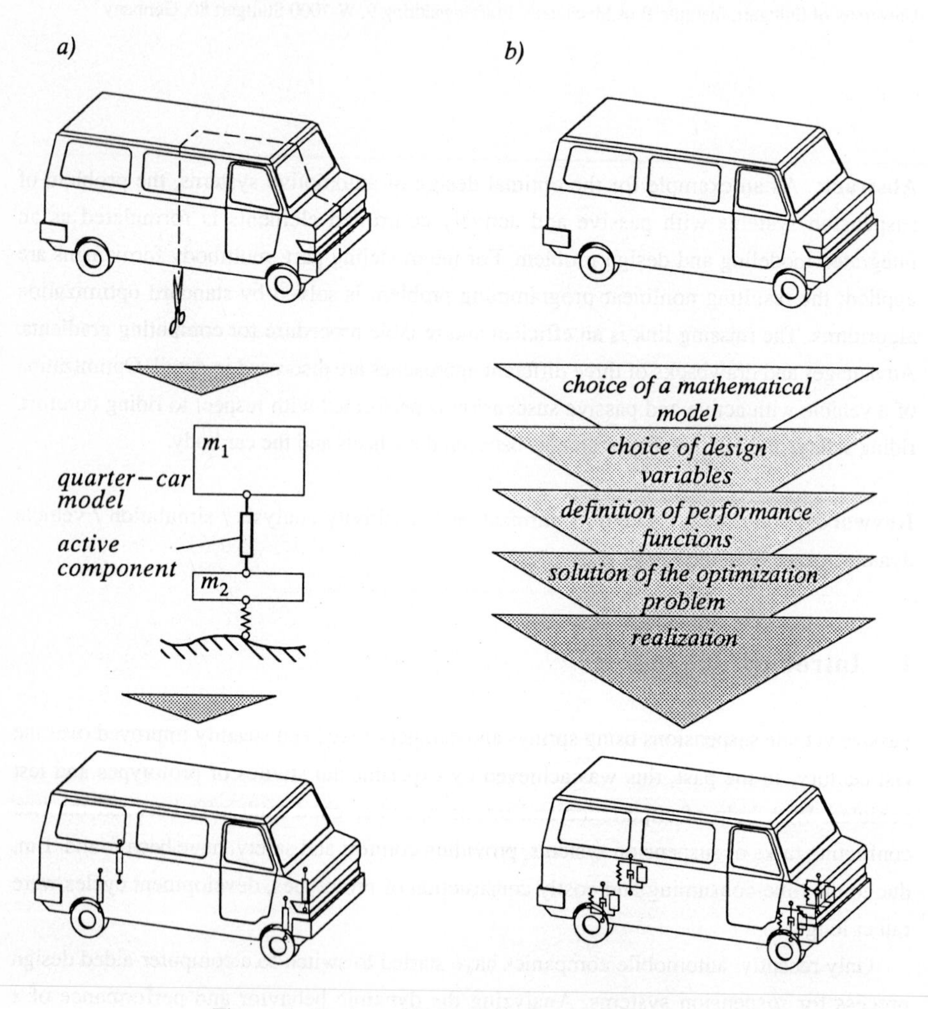

Figure 1. Design of Vehicle Suspension Systems.
Optimal control theory approach (a) and integrated modeling and control design approach (b)

Vehicle systems are only one example for a broad class of dynamic systems which can be modeled by multibody systems. In this paper, tools for a computer-aided design of such systems are described. The tools are integrated in a software-package which may be used for performance evaluation and optimization in an early design stage to support Concurrent Engineering.

2 Actively Controlled Vehicle Systems

For low frequency motions, the car body, the driver, and the axles can be considered as rigid bodies resulting in a multibody system model. The bodies are connected by ideal links and coupled by force elements like springs, dampers, and active elements, Figure 2.

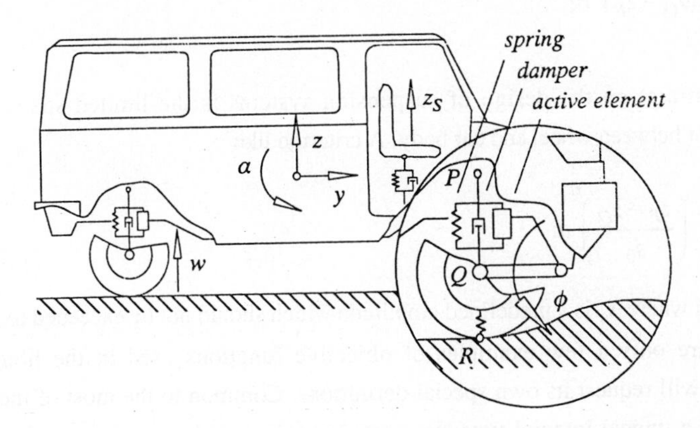

Figure 2. Planar Vehicle Model

The coefficients of the laws for the force elements may serve as design variables which can be varied within given ranges for achieving optimal dynamic behavior. For example, the stiffness coefficient c of the spring acting on the wheelset and car body with the force $F_c = c(z_P - z_Q)$, the damping coefficient d of a linear or nonlinear damper resulting in a force $F_d = d(\dot{z}_P - \dot{z}_Q)$ or $F_d = d(\dot{z}_P - \dot{z}_Q)^3$, respectively, or the control parameter d_a of an active element generating the force $F_a = d_a \dot{z}_P$. But also geometrical data, masses and moments of inertia of some bodies may be chosen as design variables.

For optimization purposes the performance of the suspension system has to be expressed by objective functions. This is perhaps one of the most critical parts of formulating the design problem, and only experience and experiments can help to achieve coincidence of the objectives with the human perception.

A frequently used measure for comfort is the vertical acceleration $\ddot{z}_s$ of the driver. If the driving over a bump is considered as a test, accelerations may be penalized by time to avoid long term vibrations:

$$\psi_C := \int_{t^0}^{t^1} (t\,\ddot{z}_s)^2\,dt \tag{1}$$

where $[t^0,\ t^1]$ is the time interval of interest. Optimal comfort is then expressed by a minimal value of ψ_C.

Riding safety is related to the dynamic variation of the load between the wheels and the road. If the tire is considered as a linear spring, Figure 2, the load is proportional to the relative displacement between the wheel and road surface. Thus, a potential performance function evaluating safety is

$$\psi_S := \int_{t^0}^{t^1} (z_Q - z_R)^2 \, dt. \tag{2}$$

A constraint on the design of suspension systems is the limited space for relative displacement between wheel and car body. A criterion like

$$\psi_D := \int_{t^0}^{t^1} \left(\frac{z_P - z_Q}{s_0} \right)^6 dt \tag{3}$$

may be used where s_0 is a predefined amplitude which should not be exceeded to much.

These are only a few examples of objective functions used in the literature; each application will request its own special definitions. Common to the most of the objectives, however, is a typical integral type structure depending on kinematic quantities. This will enable us to define a rather general type of objective function.

3 Optimization of Multibody Systems

The problem as stated above can be reduced to a nonlinear programming problem [3]. The optimum has to be found by an iterative process where each evaluation step involves a time-consuming numerical simulation of the dynamic behavior of the vehicle. In order to keep the number of iterations small, highly developed optimization algorithms using gradient information should be applied. Therefore, sensitivity analysis is an essential part of the design process, bridging the gap between multibody system formalisms for simulation and optimization procedures.

3.1 Design Problem

The design variables can be summarized in a vector $\boldsymbol{p} \in I\!R^h$ The dynamic behavior of a multibody system with f degrees of freedom can be described by generalized coordinates $\boldsymbol{y} \in I\!R^f$ and generalized velocities $\boldsymbol{z} \in I\!R^f$ [4]. If values are assigned to the design variables, the dynamic behavior of the system is uniquely determined by the equations of motion

$$\begin{aligned} \dot{\boldsymbol{y}} &= \boldsymbol{z}, \\ \boldsymbol{M}(t, \boldsymbol{y}, \boldsymbol{p})\, \dot{\boldsymbol{z}} + \boldsymbol{k}(t, \boldsymbol{y}, \boldsymbol{z}, \boldsymbol{p}) &= \boldsymbol{q}(t, \boldsymbol{y}, \boldsymbol{z}, \boldsymbol{p}), \end{aligned} \tag{4}$$

and the initial conditions

$$y^0 := y(t^0): \quad \Phi^0(t^0, y^0, p) = 0, \quad \det \frac{\partial \Phi^0}{\partial y^0} \neq 0,$$

$$z^0 := z(t^0): \quad \dot{\Phi}^0(t^0, y^0, z^0, p) = 0, \quad \det \frac{\partial \dot{\Phi}^0}{\partial z^0} \neq 0. \tag{5}$$

This also determines the values of the objective functions (1)-(3). A more general form of this type of criterion is

$$\psi(p) = G^1(t^1, y^1, z^1, p) + \int_{t^0}^{t^1} F(t, y, z, \dot{z}, p)\, dt \tag{6}$$

where the final time can be given explicitly or depend on the final state $y^1 = y(t^1)$, $z^1 = z(t^1)$:

$$t^1: H^1(t^1, y^1, z^1, p) \overset{!}{=} 0. \tag{7}$$

Thus, the objective functions can also be considered as functions depending on the design variables only, i.e., $\psi_C = \psi_C(p)$, $\psi_S = \psi_S(p)$, $\psi_D = \psi_D(p)$.

Minimizing all three criteria simultaneously results in a multicriteria optimization problem, e.g., [5,6]. Due to the high computational costs for evaluating the objective functions, it is unrealistic to assume that the whole Pareto-optimal set can be found. Thus, the problem has to be scalarized resulting in a nonlinear programming problem for finding some of the Pareto-optimal solutions. There are several possibilities to scalarize the problem. The most popular are the weighting objectives method, i.e.,

$$\begin{aligned} &\textit{minimize} \quad \psi \\ &\textit{where} \quad \psi = w_C \frac{\psi_C}{\psi_C^*} + w_S \frac{\psi_S}{\psi_S^*} + w_D \frac{\psi_D}{\psi_D^*}, \quad w_C + w_S + w_D = 1, \end{aligned} \tag{8}$$

and the trade-off method, e.g.,

$$\begin{aligned} &\textit{minimize} \quad \psi_C \\ &\textit{where} \quad \psi_S \le \hat{\psi}_S, \ \psi_D \le \hat{\psi}_D. \end{aligned} \tag{9}$$

With the first method, a weighted sum of the normalized criteria is minimized where the factors ψ_C^*, ψ_S^* and ψ_D^* can be the values of the objectives for the initial design or the individual minima. It is up to the designer to choose the weighting coefficients w_C, w_S and w_D; different choices will result in different Pareto-optimal solutions.

With the second method, one of the criteria is minimized, e.g., the criterion for comfort, where the others are limited by user-defined upper bounds. Variation of the bounds or interchange of the objective and the constraints yields different Pareto-optimal solutions.

The decision on a particular Pareto-optimal solution to be realized has to be made by the designer. The search for Pareto-optimal solutions can be performed numerically by general purpose optimization codes which are highly developed for solving the general nonlinear programming problem

$$\underset{p \in P}{\textit{minimize}}\ f(p) \quad \textit{where}\ \ P = \left\{ p \in I\!R^h \,\middle|\, g(p) = 0,\ h(p) \leq 0 \right\}. \tag{10}$$

Sequential quadratic programming (SQP) algorithms have shown to be superior to most other methods, but they require reliable information on the gradients [7].

3.2 Sensitivity Analysis

In principal, the gradient can be computed by three different methods: numerical differentiation, the direct differentiation method, and the adjoint variable method. These approaches will be discussed with respect to their application to multibody systems.

The numerical differentiation is a purely numerical method using additional evaluations of the objective function for slightly perturbed design variables. The components of the gradient are approximated by finite differences, e.g.,

$$\nabla \psi_k := \frac{d\psi}{dp_k} \approx \frac{\psi(p + \Delta p_k\, e_k) - \psi(p)}{\Delta p_k}, \quad k = 1(1)h, \tag{11}$$

where $\Delta p_k\, e_k$ is a small perturbation of the k-th design variable. The variations have to be chosen carefully with respect to approximation errors and errors in the function evaluation. If the function under consideration would be algebraic, the function value for specific values of the design variables could be computed within the machine precision. In this case, Δp could be chosen very small, e.g., $\Delta p_k / p_k = \sqrt{\varepsilon}$, $k = 1(1)h$, where ε is the machine epsilon. But the objective function (6) can be evaluated by numerical integration with moderate accuracy only. Increasing the parameter perturbation adequately will then lead to large errors due to the 0-order approximation of the gradient which can cause slow convergence of the iterative optimization process or a break down. Even for an optimal perturbation the results for the gradient are much less accurate than the function evaluation itself.

Furthermore, applying finite differences to dynamic systems has a second disadvantage. Approximating the whole gradient by finite differences requires h additional evaluations of the function ψ for perturbed parameters, and each evaluation is a time consuming dynamic analysis, i.e., a numerical integration of the equations of motion. Therefore, it is desirable to have more analytical information on the gradient. This is performed by the direct differentiation method and the adjoint variable method.

For deriving additional information on the gradient it will be advantageous to use the variational theory and index notation with the major rule that summation over all possible values of an index is automatically assumed if an index occurs twice in an expression.

Regarding the function ψ as a function of $\boldsymbol{p}$ only, i.e., $\psi = \psi(\boldsymbol{p})$, the first variation is

$$\delta\psi = \frac{d\psi}{dp_k}\delta p_k = \nabla\psi_k \delta p_k. \tag{12}$$

On the other hand, the structure of the objective function can be considered in more detail. A variation of the design variables will yield a variation of the dynamic behavior and the final time. The value of the objective function (6) will then vary by

$$\delta\psi = \delta G^1 + F^1\delta t^1 + \int_{t^0}^{t^1} \delta F dt = \left(\frac{\partial G^1}{\partial t^1} + F^1\right)\delta t^1 + \frac{\partial G^1}{\partial y_i^1}\delta y_i^1 + \frac{\partial G^1}{\partial z_j^1}\delta z_j^1$$
$$+ \int_{t^0}^{t^1}\left(\frac{\partial F}{\partial y_i}\delta y_i + \frac{\partial F}{\partial z_j}\delta z_j + \frac{\partial F}{\partial \dot{z}_j}\delta \dot{z}_j\right)dt + \left[\frac{\partial G^1}{\partial p_k} + \int_{t^0}^{t^1}\frac{\partial F}{\partial p_k}dt\right]\delta p_k \tag{13}$$

where $\boldsymbol{F}^1 := \boldsymbol{F}(t^1, \boldsymbol{y}^1, \boldsymbol{z}^1, \dot{\boldsymbol{z}}^1, \boldsymbol{p})$. Without the additional variations of the generalized coordinates, velocities and final time, a comparison of the two formulations (12) and (13) would immediately yield coincidence of $\nabla\psi_k$ with the coefficients of δp_k due to the independence of the variations δp_k. This is not valid in our case because of the dependence of δy_i, δz_j , and δt^1 on the design variations which is expressed by the first variations of the equations of motion, the initial conditions, and the final condition:

$$\delta\dot{\boldsymbol{y}} - \delta\boldsymbol{z} = \boldsymbol{0}, \quad \boldsymbol{M}\,\delta\dot{\boldsymbol{z}} + \delta\boldsymbol{M}\,\dot{\boldsymbol{z}} + \delta\boldsymbol{k} - \delta\boldsymbol{q} = \boldsymbol{0},$$
$$\delta\boldsymbol{\Phi}^0 = \boldsymbol{0}, \quad \delta\dot{\boldsymbol{\Phi}}^0 = \boldsymbol{0}, \quad \delta\mathrm{H}^1 = 0. \tag{14}$$

Therefore, terms with dependent variations have to be eliminated. This can be achieved in two different ways: elimination of the dependent variations themselves or elimination of the corresponding coefficients.

Elimination of the dependent variations is achieved by substituting the definitions

$$\delta y_i = \frac{dy_i}{dp_k}\delta p_k, \quad \delta z_j = \frac{dz_j}{dp_k}\delta p_k \tag{15}$$

for them in Eq. (13). The (f x h)-sensitivity matrices $\boldsymbol{Y}_p := d\boldsymbol{y}/d\boldsymbol{p}$ and $\boldsymbol{Z}_p := d\boldsymbol{z}/d\boldsymbol{p}$ are computed from an initial value problem resulting from analogous substitutions in Eqs. (14). Figure 3 shows the final result and procedure for computing the gradient.

Gradient

$$\nabla\psi = Y_p^{1T}\left(\frac{\partial G^1}{\partial y^1} - \frac{\dot{G}^1 + F^1}{\dot{H}^1}\frac{\partial H^1}{\partial y^1}\right) + \int_{t^0}^{t^1} Y_p^T \frac{\partial F}{\partial y}\, dt$$
$$+ Z_p^{1T}\left(\frac{\partial G^1}{\partial z^1} - \frac{\dot{G}^1 + F^1}{\dot{H}^1}\frac{\partial H^1}{\partial z^1}\right) + \int_{t^0}^{t^1} Z_p^T \frac{\partial F}{\partial z}\, dt + \int_{t^0}^{t^1} \dot{Z}_p^T \frac{\partial F}{\partial \dot{z}}\, dt$$
$$+ \left(\frac{\partial G^1}{\partial p} - \frac{\dot{G}^1 + F^1}{\dot{H}^1}\frac{\partial H^1}{\partial p}\right) + \int_{t^0}^{t^1} \frac{\partial F}{\partial p}\, dt$$

Sensitivity differential equations

$$\dot{Y}_p = Z_p$$
$$M\,\dot{Z}_p = -\frac{\partial(M\dot{z} + k - q)}{\partial y}\,Y_p - \frac{\partial(M\dot{z} + k - q)}{\partial z}\,Z_p - \frac{\partial(M\dot{z} + k - q)}{\partial p}$$

Initial conditions

$$Y_p^0:\ \frac{\partial\phi^0}{\partial y^0}\,Y_p^0 = -\frac{\partial\phi^0}{\partial p} \qquad Z_p^0:\ \frac{\partial\dot{\phi}^0}{\partial z^0}\,Z_p^0 = -\frac{\partial\dot{\phi}^0}{\partial y^0}\,Y_p^0 - \frac{\partial\dot{\phi}^0}{\partial p}$$

Figure 3. Direct Differentiation Method

This approach is called the direct differentiation method, since the results can also be obtained by direct differentiation of the equations of motion and initial conditions with respect to the design variables [8]. The equations representing the gradient are exact and can be solved simultaneously with the equations of motion with the same accuracy as the objective function evaluation. The major drawback of the direct differentiation method is the computational effort which is proportional to the number of design variables similar to the numerical differentiation.

The second possibility for eliminating the terms with dependent variations is the elimination of their coefficients leading to the adjoint variable method. This is done by including the implicit constraints (14) in Equation (13) by means of Lagrange multipliers or adjoint variables [3]. For a special choice of these adjoint variables the coefficients of dependent variations vanish and a comparison with Equation (12) yields the desired result for the gradient. The conditions for vanishing of the coefficients result in algebraic and differential equations for the adjoint variables, Figure 4.

The equations for computing the gradient are also exact and can be solved by numerical integration with the same accuracy as the evaluation of the corresponding performance function. The major advantage of the adjoint variable method is that the dimension of the adjoint differential equations is independent of the number of design variables and identical to the number of degrees of freedom. The only drawback is the integration of the adjoint differential equations backward in time which requires storage and interpolation of the state

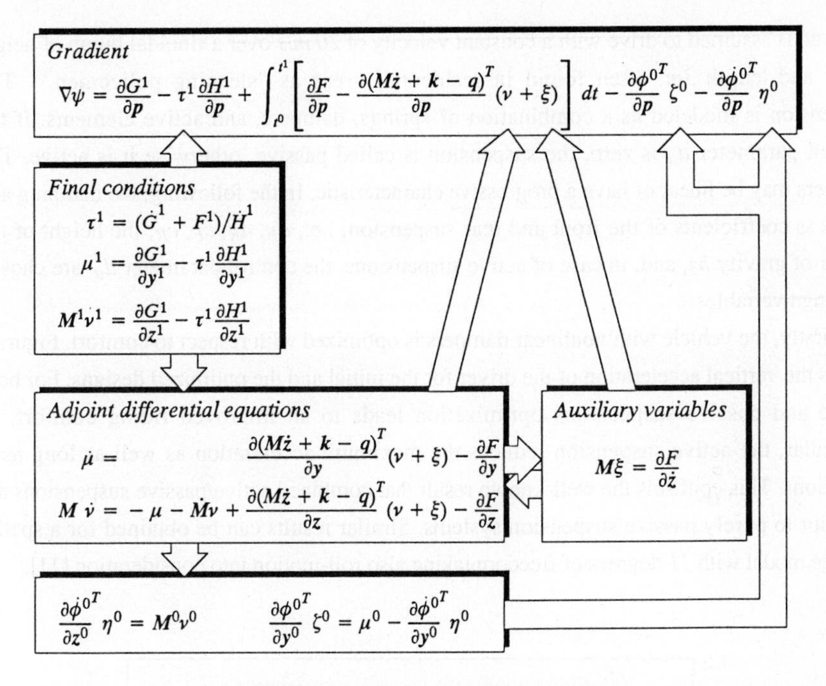

Figure 4. Adjoint Variable Method

variables due to the coupling. By using multistep integration algorithms this can be performed without loss of accuracy [3].

With respect to computational effort the adjoint variable method has shown to be superior to the other methods if more than three to four design variables are considered. The computed gradients are accurate and reliable even in the neighborhood of minima. Therefore, it has become the heart of a design tool AIMS (Analyzing and Improving Multibody Systems) being developed at the Institute B of Mechanics, University of Stuttgart. It is based on NEWEUL [9] for generating symbolic equations of motion and MAPLE [10] for generating the partial derivatives used in the adjoint equations. These symbol manipulation packages help to avoid human errors and allow to apply the design optimization procedures to more complex systems.

4 Some Numerical Results

The planar vehicle model in Figure 2 with 6 degrees of freedom can be described by the generalized coordinates

$$y = (y,\ z,\ a,\ \phi,\ w,\ z_S)^T. \tag{16}$$

The van is assumed to drive with a constant velocity of *20 m/s* over a sinoidal bump of height *0.1m* and length *3m,* often found in residential areas as "sleeping policemen." The suspension is modeled as a combination of springs, dampers, and active elements. If the control parameter d_a is zero, the suspension is called passive, otherwise it is active. The dampers may be linear or have a progressive characteristic. In the following, the damping and stiffness coefficients of the front and rear suspension, i.e., d_F, d_R, c_F, c_R, the height of the center of gravity h_3, and, in case of active suspensions, the control parameter d_a are chosen as design variables.

Firstly, the vehicle with nonlinear dampers is optimized with respect to comfort. Figure 5 shows the vertical acceleration of the driver for the initial and the optimized designs. For both active and passive suspensions optimization leads to an improved riding comfort. In particular, the active suspension reduces the maximum acceleration as well as long term vibrations. This confirms the well-known result that combined active/passive suspensions are superior to purely passive suspension systems. Similar results can be obtained for a spatial vehicle model with *11* degrees of freedom taking also roll-motion into consideration [11].

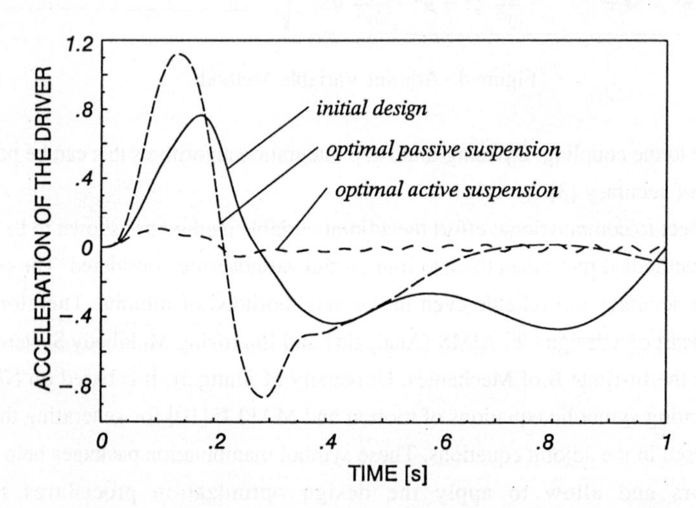

Figure 5. Optimized Vehicle with Respect to Riding Comfort

It turns out that the reduction of the performance function related to an improvement of the design is very high during the first few steps of optimization, and then it remains nearly constant, Figure 6. Although the value of the performance function does not change very much, the optimization algorithm does not converge due to significant changes of some design variables which have low influence on the performance function. A realized system at such an optimal design point will be robust with respect to variations of these parameters which may result from manufacturing problems.

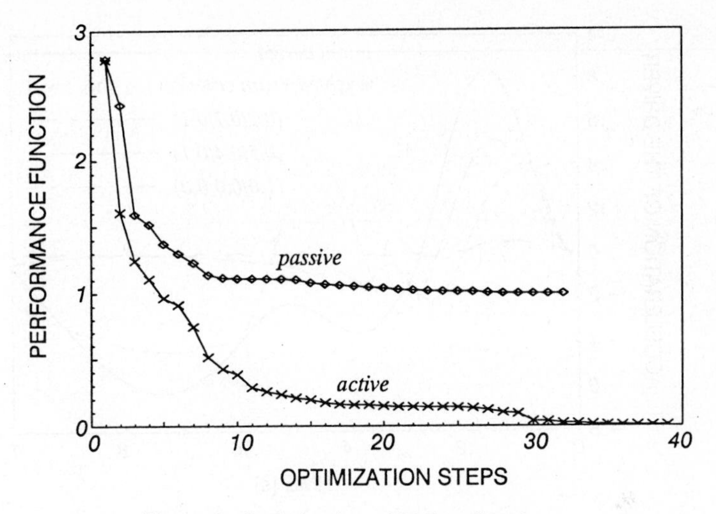

Figure 6. Optimization of Riding Comfort

Figure 7 shows the riding comfort of an optimized vehicle with active suspension and linear dampers using single criteria (1)-(3). Considering the relative displacement leads to high maximal accelerations while optimization of the riding safety yields low but poorly damped accelerations. Such an analysis of conflicting optimization criteria shows that the improvement of one criterion worsens the other criteria. Thus, only a multicriteria approach will give an acceptable trade-off, Figure 8. If, for instance, a weighted-sum criterion is used instead of the comfort criterion, the improvement in riding comfort is not as high as in Figure 7. Depending on the weighting coefficients more or less improvement of the riding comfort can be gained without much pay-off in riding safety or relative displacement. It will be a matter of engineering intuition to make a good choice on the weighting coefficients w_C, w_S and w_D.

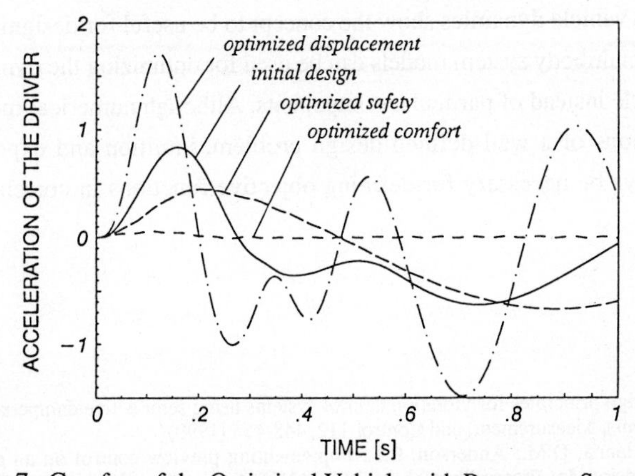

Figure 7. Comfort of the Optimized Vehicle with Respect to Several Criteria

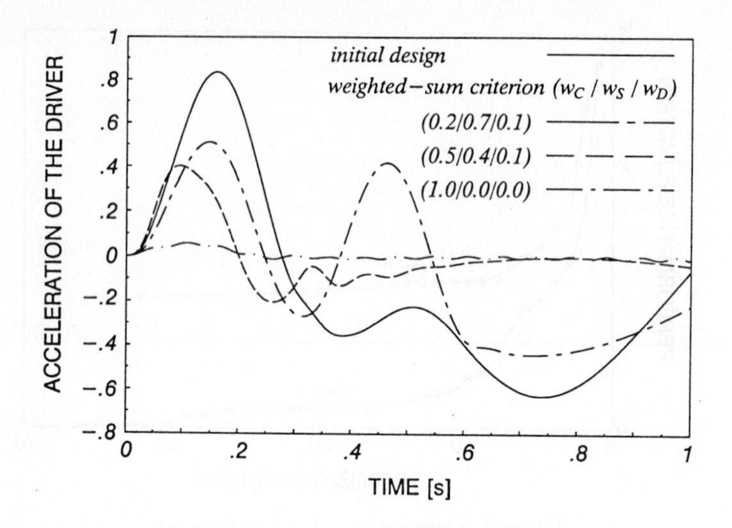

Figure 8. Multicriteria Optimization of the Vehicle

5 Conclusions

The optimization of multibody systems is formulated as a nonlinear programming problem for rather general assumptions. The missing link between multibody system codes for analyzing the dynamic behavior and optimization codes is the sensitivity analysis of the objective and constraint functions with respect to parameter perturbations. The adjoint variable method is more reliable and efficient than the often applied numerical differentiation for gradient evaluation. Additional analytical computations can be largely automated by symbolic manipulation packages.

Applications to vehicle dynamics show the concept to be useful for designing suspension systems. Realistic multibody system models can be used for optimizing the dynamic behavior of the overall vehicle instead of particular components. Although numerical methods help to find optimal solutions of a well-defined design problem, intuition and experience of the designer will always be necessary for defining objective functions in correlation with the human perception.

References

1. Karnopp, D.: Design principles for vibration control systems using semi-active dampers. ASME Journal of Dynamic Systems, Measurement, and Control 112, 448-455 (1990)
2. Langlois, R.G.; Hanna, D.M.; Anderson, R.J.: Implementing preview control on an off-road vehicle with active suspension. In: Proceedings of the 12th IAVSD-Symposium, Lyon 1991 (G. Sauvage, Ed.) Supplement to Vehicle System Dynamics 20, 340-353 (1991)

3. Bestle, D.; Eberhard, P.: Analyzing and optimizing multibody systems. Mechanics of Structures and Machines 20, 67-92 (1992)
4. Schiehlen, W.: Simulation based design of automotive systems (in this volume)
5. Osyczka, A.: Multicriteria optimization in engineering. New York: Ellis Horwood 1984
6. Hwang, Ch.-L.; Masud, A.S.: Multiple objective decision making – Methods and applications. Lect. Notes in Economics and Math. Sys., Vol. 164. Berlin: Springer 1979
7. Fletcher, R.: Practical methods of optimization. Chichester: Wiley 1987
8. Haug, E.J.: Design sensitivity analysis of dynamic systems. In: Computer aided design: Structural and mechanical systems (C.A Mota-Soares, Ed.) Berlin: Springer 1987
9. Kreuzer, E.; Leister, G.: Programmsystem NEWEUL'90, User's Guide AN-24. Stuttgart: University, Institute B of Mechanics 1991
10. Char, B.W. et. al.: MAPLE – Reference manual. Waterloo: Waterloo Maple Publ. 1990
11. Bestle, D.; Eberhard, P.; Schiehlen, W.: Optimization of an actively controlled vehicle system (to appear)

Multicriterion Optimization of Large Scale Mechanical Systems

Stefan Jendo[1] and Witold M. Paczkowski[2]

[1] Institute of Fundamental Technological Research, Polish Academy of Sciences, Swietokrzyska 21, 00-049 Warsaw, Poland
[2] Civil Engineering Dept., Technical University of Szczecin, Al. Piastow 50, 70-311 Szczecin, Poland

Abstract: The paper deals with discussion of optimization problems in engineering structural design. The following questions are discussed: continuous or discrete optimization, single-or multicriteria optimization, one-or multilevel optimization. The paper is illustrated with example of multicriterion discrete optimization of large scale truss systems.

Keywords: multicriterion optimization / discrete optimization / large scale systems / multilevel optimization

1 Introduction

The first work on multiobjective (multicriterion, vector) optimization was presented by Pareto [74]. After at least fifty years, problems of multiobjective optimization were again considered by von Neumann and Morgenstern [70] and by Debreu [14]. A relatively modern formulation of the multiobjective optimization problem was presented by Zadeh [98]. But wider interest in optimization theory, operation research, and control theory did not occur until the late 1960s, and since then numerous papers on the subject have been published. Most are concerned with the theory and applications of multiobjective decision making from a general viewpoint, and few applications to engineering design can be found. Comprehensive bibliographies on multicriterion decision-making and related areas are in Lin [68], Keeney and Raiffa [57], Hwang and Masud [35], Stadler [89,90,91], Chankong and Heimes [11], and Dauer and Stadler [13]. Cohon [12], Homenjuk [34], Yu [97], Gerasimov, Potchman and Skalozub [27], and Dubov, Travkin and Jakimec [15] and Stadler [92] present the foundations of modern multiobjective optimization.

Continuous optimization is based on the assumption that the design variables can take any values from the continuous sets. In discrete optimization due to technological and/or other requirements the design (decision) variables can take only discrete values from finite sets. The following design variables can be treated as continuous: height of truss, dimensions of cross-sectional area of reinforcement beam, sag of cable, rise of arch or camber of beam, value of prestressing force, plate thickness and so on. However, in practice only certain discrete values are permissible, for example, type of cross-sectional area of bar, series of types of cross-sectional areas, partition of structure on stiffness zones, support properties, stiffness of joints, number of bridge spans, number of storeys in building, number of girders in hall, number of prestressing cables and so on. Discrete variables occur very often in engineering design. Rounding the continuous solution to the nearest discrete solution can lead to the solution which is far from optimal discrete solution (see Jendo and Paczkowski [45], Paczkowski [73] and Bauer, Gutkowski and Iwanow [4]). In such cases discrete optimization is recommended.

Multicriterion (multiobjective, vector) optimization takes into account the vector objective function contrary to single-criteria (scalar) optimization, where only one criterion is considered. The choice of criterion of optimization is essential to the result of solution. The most popular criteria used in structural optimization are as follows: minimum weight or cost, maximum stiffness, minimum displacement at certain point of structures, maximum frequency of free vibration an so on (see Brandt [6,7]). These criteria are very often in conflict. In such case it is necessary to formulate a multicriteria optimization problem and look for the set of compromise solutions in objective space. Next, the so called preferable solution should be chosen taking into account an additional criterion or using fuzzy set theory (see Kickert [58]) or so called global criterion like utility function, distance function or hierarchical method (see Eschenauer [17], Jendo [40] and Peschel and Riedel [75]) .

Multilevel optimization can be applied to solve large scale optimization systems using decomposition and coordination method. Large scale systems can be decomposed into a few subsystems which can be optimized separately. Coordination level allows to get the solution of decomposed system which converges to the solution of original system (see Jendo and Stachowicz [54,55]).

Many engineering structures can be decomposed into subsystems. For example the hall structure can be divided into: roofing panels, trusses, columns, walls and foundations. In optimization process it is necessary to choose the local design variables concerning with particular subsystems and global design variables common for at least two or more subsystems. The particular subsystems are optimized independently of each other with respect to local variables. Next, coordination is performed according to global objective function with respect to coordination (global) variables.

2 Applications of Multiobjective Optimization in Engineering Design

Stadler [86,87,88] was the first who applied multicriterion optimization in structural mechanics. He considered the bicriterion optimization problem, with weight and stored energy as criteria, and obtained analytical solutions for some structural elements, calling the results "natural structural shapes". Leitmann [66] and Stadler [89] applied control theory to solve some multicriterion optimization problems in viscoelasticity. Baier [2,3] studied multicriterion optimization of structures from a general point of few, taking weight and energy stored under separate loading conditions as optimization criteria. Truss optimization with weight and displacement criteria were considered by Koski [61,62,63] and Koski and Silvennoinen [64]. Osyczka [71,72] and Rao and Hati [76] considered some conflicting objectives in solving mechanical optimization problems. Eschenauer [16,17,18], and Sattler [78] applied the multicriterion optimization approach to spatial structures that support radiotelescopes assuming the following criteria: minimum weight and minimum displacements of the radiotelescope surface from its initial configuration under selected loading conditions. Also, Eschenauer [19] considered the multiobjective optimization of fiber-reinforced plates and disks, choosing weight and deformation as optimization criteria. Carmichael [8] solved a multicriterion optimization problems using the method of constrained objective functions. Jendo and Marks [41,42,43] applied multicriterion optimization in solving some civil engineering structures like flexural elements and truss structures, using weight and displacements or potential energy as optimization criteria. Jendo, Marks and Thierauf [44] considered multicriterion optimization of an isostatc truss using dynamic programming. Jendo [36,37,38,40] considered multiobjective optimization of single-layer cable systems, choosing minimum weight and maximum lowest frequency of free vibration as optimization criteria. Some aspects of these problems, including dynamic plasticity of cable systems, were considered also by Eschenauer and Jendo [20]. Eschenauer, Kneppe and Stenvers [21] compared deterministic and stochastic multiobjective optimization of beams and shell structures. Watkins and Morris [941 applied a multicriterion optimization for laminated composites. Stadler [92] and Eschenauer, Koski and Osyczka [22] presented many applications of multicriterion structural design in diffrent mechanical and civil engineering structures. Jendo [39], Jendo and Putresza [49,50 51,52,53] presented a multicriterion optimization of bar structures with random parameters using weight of structure and reliability of structure as optimization criteria.

3 Multilevel Optimization and Decomposition Techniques

It is well known that optimization of large-scale structures by mathematical programming methods can present some difficulties because enormous computing time and computer memory are needed. We can avoid such difficulties by multilevel optimization and decomposition methods. In general, an integrated optimization problem involving many subsystems and variables cannot be decomposed into independent subproblems that can be independently optimized. The methods described in this section do permit decomposition of optimization problems into subproblems, each of which when solved independently yields the overall system optimum. The interconnection of the subsystems may take many forms, but one of the most common is the hierarchical form in which the second-level unit controls or coordinates the units on the level below, called the *first-level* or *optimization level.* The subsystems are solved independently at the first level. Our goal at the second level is to coordinate the action of the first-level units so that we obtain the solution of the original problem.

The first step in the hierarchical approach is to separate the set of design variables Z and the set of constraints H into two classes. The first class contains the coordination variables ZA and the constraints HA, containing only the design variables ZA that are used to create the second, or coordination, level. The second class contains the rest of the design variables ZB=Z\ZA and the constraints HB=H\HA, and is used to create the first, or optimization, level. After separating design variables and constraints, we can write the optimization problem as follows:

$$\min_{x \in Z} C(\mathbf{x}) = \min_{x \in ZA} [\min_{x \in ZB} C(\mathbf{x})]. \quad (1)$$

Now, the basic problem is to decompose the optimization level. Two classes of decomposition methods can be distinguished: intuitive and formal. *Intuitive methods* construct iterative schemes that are reasonable but are not guaranteed to lead to a true optimum. *Formal methods,* on the other hand, converge to a true optimum under well-defined assumptions.

Intuitive optimization methods are based on the use of substructures. Each substructure is optimized assuming that the loads on the substructure do not change as a result of the resizing. The entire structure is analyzed periodically to obtain the new loads acting on each substructure. Giles [28] applied this approach to aircraft wing design under static loading, where each substructure (cover panel-shear webs) was optimized using the fully stressed design approach. Sobieszczanski-Sobieski [82] and Sobieszczanski-Sobieski and Leondorf [84] treated fuselage and wing structures, using mathematical programming for the substructure optimization.

Sobieszczanski-Sobieski, James, and Dovi [83] described a method for decomposing an optimization problem into a set of subproblems and a coordination problem that preserves coupling between the subproblems. The decomposition is achieved by separating the structural element optimization subproblems from the assembled structural optimization problem. Each element optimization and optimum sensitivity analysis yields the cross-sectional dimensions that minimize the cumulative measure of the element constraint violation as a function of the elemental forces and stiffness. The assembled structural optimization produces the overall mass and stiffness distributions optimized for minimum total mass subject to constraints that include the cumulative measures of the element constraint violations extrapolated linearly with respect to the element forces and stiffnesses. The method is introduced as a special case of a multilevel, multidisciplinary system optimization and its algorithm is fully described for two-level optimization for structures assembled of finite elements of arbitrary type. Numerical results are given for an example of a framework to show that the decomposition method converges and yields results comparable to those obtained without decomposition. They pointed out that optimization by decomposition should reduce the design time by allowing groups of engineers using different computers to work concurrently on the same large problem.

Kirsch, Reiss, and Shamir [60] and Kirsch [59] proposed analyzing the total structure after each substructure optimization to improve the convergence of the method. This approach is ideally suited for local constraints such as stress constraints. For global constraints we need to devise ad hoc procedures. Sobieszczanski-Sobieski and Leondorf [84] described such approach for aerospace applications. Schmit and Ramanatham [80] and Schmit and Mehrinfar [79] used a two-level approach to design trusses and aircraft wing structures. Ho [32, 33] applied limit design techniques to multistage planar trusses. He took advantage of the staircase structure of the constraint matrix to apply a nested decomposition technique that is guaranteed to converge to the optimum design. Chang [10] used the Rosen [77] decomposition algorithm to solve the problem of limit design of a planar truss structure subject to vulnerability constraints. Woo and Schmit [95] and Kaneko and Cu Dong Ha [56] applied a decomposition method to large-scale optimum plastic design problems. Jendo and Stachowicz [54, 55] and Stachowicz and Ziobron [85] applied multilevel optimization to double-layer cable systems and to underground water tank systems.

4 Optimization with Discrete Design Variables

Optimization problems in structural design must often be considered to be of discrete type in which the variables can assume only certain predetermined values. Among the problems that commonly arise in structural design is the selection of standard steel sections for continuous or frame structures. Thus, from a practical standpoint, it is more convenient to formulate the design problem with discrete rather then continuous variables.

Many researchers have attempted to develop methods for discrete optimization of structures. Toakley [93] formulated the optimal plastic design of frame structures as a discrete problem, in which the solution space consisted of the available standard sections, and solved it by using a mixed integer-continuous programming algorithm. Cella and Soosar [9] presented a state-of-the-art paper on discrete variables in structural optimization. Lai and Achenbach [65] presented a direct search algorithm to solve nonlinear optimization problems in which the variables were discrete. Glankwahmdee, Liebman, and Hogg [29] presented a method specifically for nonlinear discrete unconstrained optimization problems. Next, Liebman, Khachaturian, and Chanaratna [67] using the last paper as a starting point, developed a method of discrete optimization that uses concepts well-known for problems with continuous variables. Bauer, Gutkowski, and Iwanow [4] presented a semianalytical approach to the discrete optimization of large space-truss structures. Sepulveda and Cassis [81] developed a method for efficiently solving the problem of minimum weight design of plane or space trusses with discrete or mixed variables.

Following Grierson and Lee [30], we present a formulation of a minimum weight design problem of planar frameworks under service and ultimate performance conditions, using discrete section members. Service load conditions ensure acceptable elastic stresses and displacements, and ultimate load conditions ensure adequate safety against plastic collapse. We assume that all loads are static and the specified service loads are proportional to the specified ultimate loads. The framework is discretized into an assemblage of n prismatic members. For each member, the type of cross-section is specified a *priori,* and a corresponding set of discrete sections from which its design cross-section is to be selected is identified. The design variables for member i is its cross-sectional area a_l. The minimum-weight design problem can be stated as follows:

$$\min W = \sum_{i=1}^{n} w_i a_i \tag{2}$$

subject to

$$\underset{\sim}{\delta}_j \le \delta_j \le \tilde{\delta}_j, \qquad j=1,2,\ldots,d \tag{3}$$

$$\underset{\sim}{\sigma}_k \le \sigma_k \le \tilde{\sigma}_k, \qquad k=1,2,\ldots,s \tag{4}$$

$$\underset{\sim}{\alpha}_m \le \alpha_m \le \tilde{\alpha}_m, \qquad m=1,2,\ldots,p \tag{5}$$

$$a_i \in A_i, \quad i=1,2,\ldots,n \tag{6}$$

Equation (2) defines the weight of the structure (w_i is the weight coefficient for member i); inequality (3) defines the d service load constraints on displacements (undertilde and overtilde denote specified lower and upper bounds, respectively); inequality (4) defines the s service load constraints on stresses σ_k; inequality (5) defines the p ultimate load constraints on plastic collapse load factors α_m; and condition (6) requires each member cross-sectional area a_i to belong to a specified set of discrete section areas $A_i = \{a_1, a_2, \ldots\}_i$.

This discrete-weight optimization problem was solved by an iterative method, using an efficient generalized optimality criteria technique due to Fleury [23]. The iterative process terminates with the final design when there is no change in the structural weight from one design stage to the next (for details see Grierson and Lee [30]).

5 Discrete Optimization of Large Scale Truss Systems

5.1 Object of Optimization

This section is concerned with optimization of steel hall structure covered by space truss. This kind of structure can be considered as large-scale system (see Makowski [69]). The hall structure can be divided into roofing space truss, cantilever structure, roof covering, walls and foundations (Figure 1). The local and global design variables should be distinguished. The local design variables describes the particular elements (subsystems) contrary to global variables which describes at least two elements (Figure 2). This approach allows to consider a few optimization subsystems instead one large scale system (Figure 3). Each subsystem is optimized separately (independently). The total objective function and global design variables play a role of coordination between subsystems (see Sobieszczński-Sobieski et.al. [83], Jendo and Stachowicz [54,55], and Jendo and Paczkowski [45,46,47,48]).

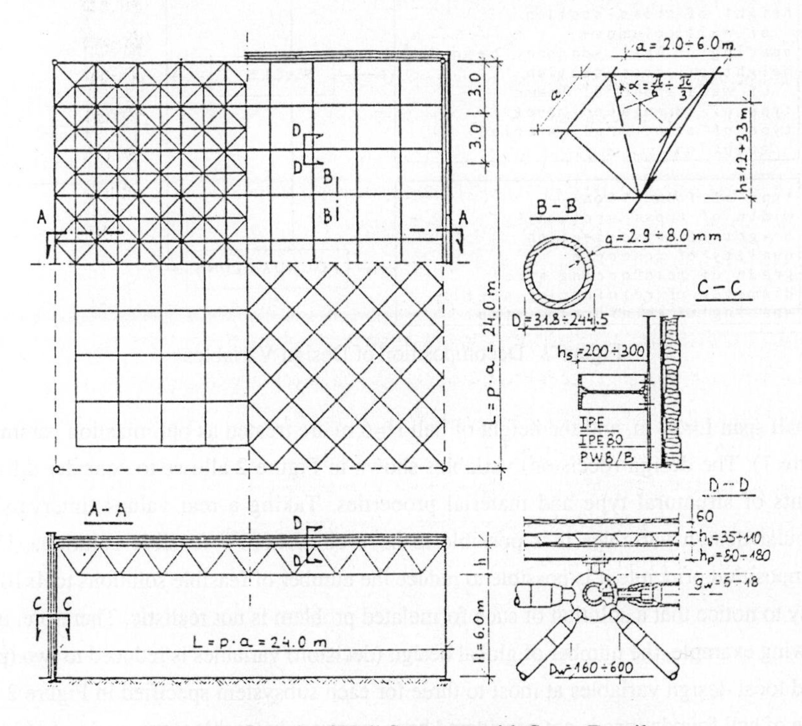

Figure 1. Sketch of Hall Structure with Some Details

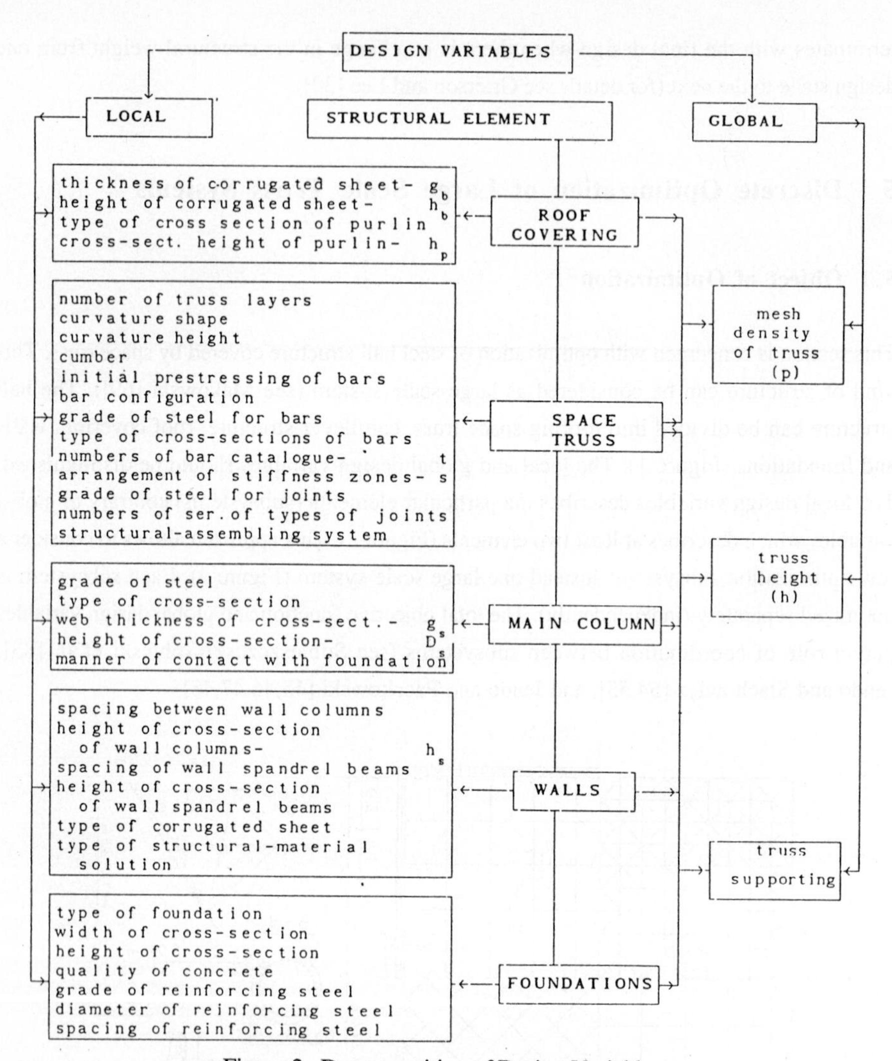

Figure 2. Decomposition of Design Variables

The hall span L=24 m, and the height of hall H=6 m are treated as optimization parameters (Figure 1). The design (decision) variables shown in Figure 2 allows to consider different variants of structural type and material properties. Taking a real values (intervals) for particular design variables it is possible to have circa 4×10^{20} feasible solutions. Using decomposition principle it is possible to reduce the number of feasible solutions to 4×10^{10}. It is easy to notice that a solution of such formulated problem is not realistic. Therefore, in the following example, the number of global design (decision) variables is reduced to two (p and h) and local design variables at most to three for each subsystem specified in Figure 2. The design of hall foundations is not considered here, as strongly problem oriented task (type of

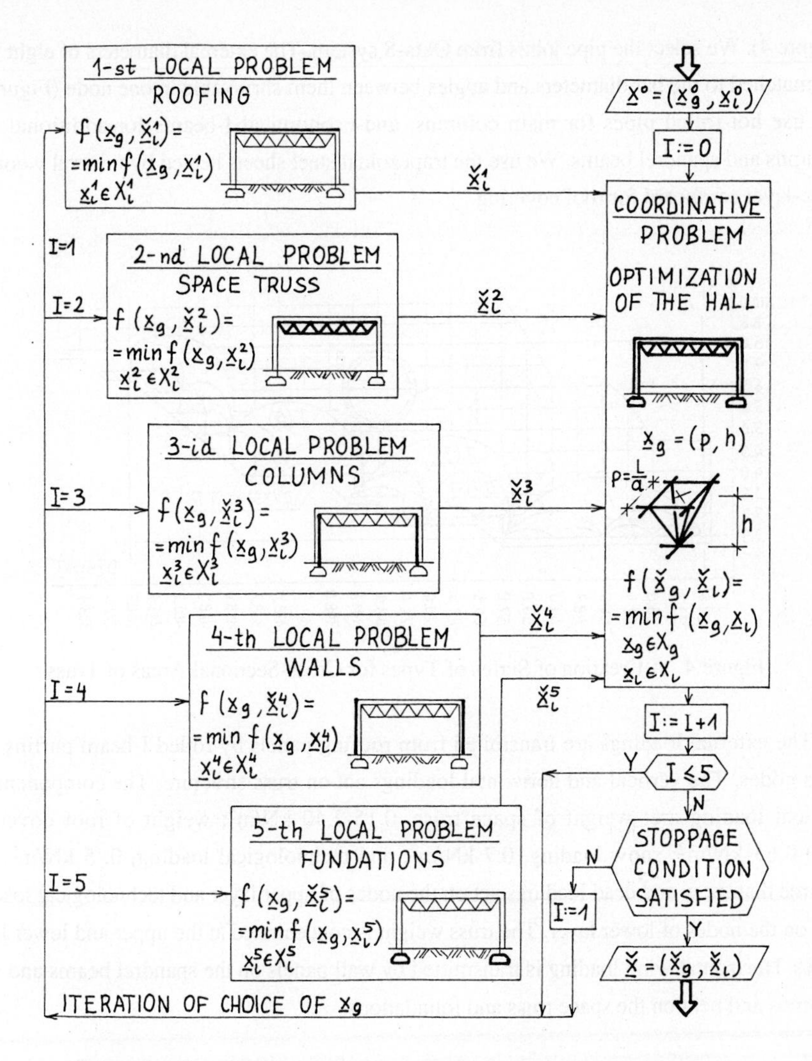

Figure 3. Flow Diagram and Decomposition of Hall into Subsystems

soil). The number of feasible solutions of this problem is reduced to 3200. Using decomposition principle and looking for the efficient solutions this number can be reduced to tens.

We use an the orthogonal double-layer space truss simply supported at the corners of the upper layer for hall covering. We use cylindrical pipes made from R35 hot-rolled steel for truss bars. We order the series of types of bars, containing 33 elements, according to increasing cross-sectional areas and simultaneously increasing values of limit loads

(Figure 4). We select the pipe joints from Okta-S system. The external diameters of eight balls are matched to the bar diameters and angles between them shrinkable in one node (Figure 1). We use hot-rolled pipes for main columns. and economical I-beams for additional wall columns and spandrel beams. We use the trapezoidal steel sheets heated by mineral wool and three-layer cardboard for roof covering.

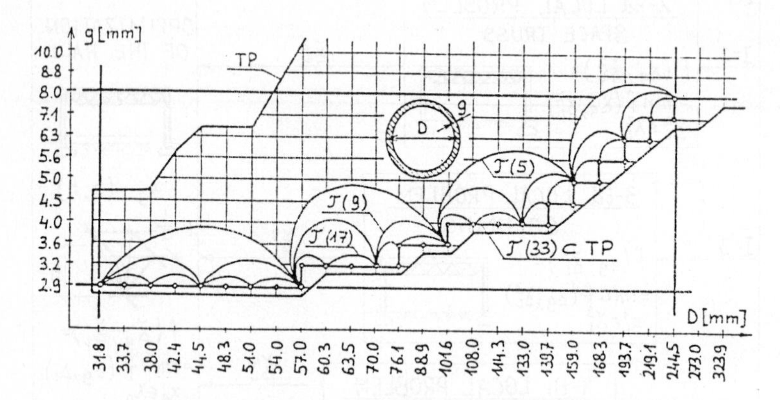

Figure 4. A Creation of Series of Types for Cross-Sectional Areas of Truss

The external loadings are transmited from roofing panels by rolled I-beam purlins into truss nodes. The vertical and horizontal loadings act on truss structure. The components of vertical loading are: weight of space truss, 0.15-2.40 kN/m^2; weight of roof covering, 0.50-0.65 kN/m^2; snow loading, 0.7 kN/m^2; and technological loading, 0. 5 kN/m^2. We assume that snow and dead loadings act on the nodes of upper layer and technological loading acts on the nodes of lower layer. The truss weight is concentrated at the upper and lower layer nodes. Horizontal wind loading is transmitted by wall panels on the spandrel beams and wall columns and next on the space truss and foundations.

5.2 Problem Formulation

We formulate the optimization problem using decomposition principle as follows: find the vectors of global and local variables $\hat{\mathbf{x}}_g$ and $\hat{\mathbf{x}}_l$ which minimize the objective function

$$f(\hat{\mathbf{x}})=f(\hat{\mathbf{x}}_g,\hat{\mathbf{x}}_l) = \min_{\mathbf{x}_g \in X_g,\ \mathbf{x}_l \in X_l} f(\mathbf{x}_g,\mathbf{x}_l) = \min_{\mathbf{x}_g \in X_g} \sum_{i=1}^{4} \min_{\mathbf{x}_l^{(i)} \in X_l^{(i)}} f(\mathbf{x}_g,\mathbf{x}_l^{(i)}), \tag{7}$$

where X_g and X_l are global and local domains of feasible solutions:

$$X_g=\{\mathbf{x}_g: c_g(\mathbf{x}_g) \le 0, \mathbf{x}_g \in C\} \tag{8}$$

$$X_1^{(i)}=\{\mathbf{x}_1^{(i)}: c_1^{(i)}(\mathbf{x}_g,\mathbf{x}_1^{(i)}) \le 0 , \mathbf{x}_1^{(i)} \in C\}, \tag{9}$$

$$\bigcup_{i=1}^{4} \mathbf{x}_1^{(i)} = \mathbf{x}_1, \bigcup_{i=1}^{4} X_1^{(i)} = X_1, (X_g \cup X_1) \subset X. \tag{10}$$

We perform the optimization process in two stages. In the first stage, we consider eight single-criteria optimization problems. In the second stage, we choose three objective functions and we perform multiobjective optimization. We find the sets of compromise and preferable solutions.

In the first stage, we minimize the following objectives:

$$f_1(\mathbf{x}) = \frac{\gamma}{L^2} \sum_{kp=1}^{Lp} l_{kp} . A_{kp}, \tag{11}$$

$$f_2(\mathbf{x}) = \frac{\gamma}{L^2} \sum_{kw=1}^{Lw} V_{kw}, \tag{12}$$

$$f_3(\mathbf{x}) = \frac{1}{L^2} \sum_{k=1}^{6} M_k, \tag{13}$$

$$f_4(\mathbf{x}) = f_1(\mathbf{x}) + f_2(\mathbf{x}), \tag{14}$$

$$f_5(\mathbf{x}) = f_1(\mathbf{x}) + f_2(\mathbf{x}) + f_3(\mathbf{x}), \tag{15}$$

$$f_6(x) = \Delta_{max}, \tag{16}$$

$$f_7(x) = \frac{1}{2} \sum_{kw=1}^{Lw} P_{kw} . \Delta_{kw}, \tag{17}$$

$$f_8(\mathbf{x}) = \frac{p^2 + (p+1)^2}{40} a^2 \left[1 + \left|\sin\left(\frac{\Pi}{4} - \alpha\right)\right|\right] [1 - 0.4 \exp(-0.35t)].$$

$$\cdot (0.005\, b + 0.9)(0.005\, c + 0.94)(0.005\, d + 0.96), \tag{18}$$

where: $f_1(\mathbf{x})$ is mass of truss bars, $f_2(\mathbf{x})$ is mass of truss joints, $f_3(\mathbf{x})$ is mass of the steel elements in supporting structure and hall walling, $f_4(\mathbf{x})$ is mass of truss bars and joints, $f_5(\mathbf{x})$ is total mass of truss structure and mass of steel elements of hall walling, $f_6(\mathbf{x})$ is maximal vertical displacement of truss node, $f_7(\mathbf{x})$ is elastic energy of truss, $f_8(\mathbf{x})$ is conventional labor cost of structure, $\mathbf{x}$ is vector of design variables $\mathbf{x}=(\mathbf{x}_g, \mathbf{x}_1)$, γ =7850 kg/m^3 is the bulk density of steel, A_{kp} is cross-sectional area of k-th truss bar, l_{kp} is the length of k-th truss bar, L=24 m is the truss span, $L_p \in$ <128, 1152> denotes the number of truss bars, V_{kw} denotes material volume of k-th truss joint, $L_w \in$ <41, 313> denotes the number of truss nodes, M_k

is mass of purlins for k=l, or mass of trapezoidal sheet of covering for k=2, or mass of hall corner columns for k=3, or mass of wall columns for k=4, or mass of spandrel beams for k=5, or mass of steel in heated wall panels PW8/B for k=6, respectively, Δ_{max} is maximal vertical displacement of truss midnode, Δ_{kw} is vertical displacement of k-th truss node, p=L/a is a mesh density of truss, a is a distance between truss nodes, α is the angle of inclination of web members to upper or lower layers, t is a numbers of catalogue of cross-sectional areas, b is a number of different cross-sectional areas of bars in i-th truss layer, c is a maximal number of different cross-sectional areas of bars in i-th truss layer, d is a maximal number of different cross-sectional areas of bars shrinkable in one truss node.

The domain of feasible solutions is determinated by the set of optimization constraints specified in Table 1. We use the following notation: $\sigma_k, \sigma_{ks}, \sigma_{kr}$, denote the stresses, compressive and tensile stresses in k-th truss bar, respectively, N_k is an internal force in k-th truss bar, m_{wk} is a buckling coefficient for k-th compressed truss bar, A_k, A_{kn}, A_{kg} are the cross-sectional area of k-th truss bar, net and gross, respectively, λ_k is a slenderness ratio for k-th compressed truss bar, i_k is a radius of inertia of cross-section area of k-th truss bar, R is a calculated strength of material, n is a number of iterations in truss design process, T(t) is the series of types of cross-sectional areas with numbers t, C is the set of integer numbers. Additional notations are shown in Figures 1, 2, 4, and 5.

Table 1. Optimization Constraints

Design constraints	Technological constraints
for truss	for truss
	$L/12 \le a \le L/4$ (31)
$\sigma_{ks} = \frac{N_k m_{wk}}{A_{kn}} \le R = 210$ MPa (19)	$p = L/a \in \rho = \{4, 5, 6, \ldots, 12\}$ (32)
	$L/20 \le h \le L/7$ (33)
	$h \in \mathscr{h} = \{1.2, 1.5, 1.8, \ldots, 3.0, 3.3\}$ (34)
$\sigma_{kr} = \frac{N_k}{A_{kg}} \le R = 210$ MPa (20)	$5 \le t \le 33$ (35)
	$t \in \mathscr{t} = \{5, 9, 17, 33\}$ (36)
	$s \in \mathscr{s} = \{1, 2, 3, \ldots, 12\}$ (37)
$\lambda_k = \frac{l_k}{i_k} \le 250$ (21)	$A_{kp} \in \mathcal{T}(t) = \{T(5), T(9), T(17), T(33)$ (38)
	$\mathcal{T}(5) \subset \mathcal{T}(9) \subset \mathcal{T}(17) \subset \mathcal{T}(33) \subset TP$ (39)
$\Delta_{max} \le \frac{\sqrt{2}\ L}{300} = 113.1$mm (22)	$3.2 \le A_{kp} \le 80.1$ cm^2 (40)
	$2.9 \le g \le 8.0$ mm (41)
for other elements	$g \in \mathscr{g} = \{g_1, g_2, \ldots, g_{10}\}$, (Fig. 4) (42)
	$31.8 \le D \le 244.5$ mm (43)
$\sigma = \frac{Nm_w}{A} + \frac{M}{W} \le 1.05\ R$ (23)	$D \in \mathcal{D} = \{D_1, D_2, \ldots, D_{33}\}$, (Fig. 4) (44)
$\sigma = \frac{M}{W} \le R = 210$ MPa (24)	$g_w \in \mathscr{g}_w = \{6, 7, 9, 12, 14, 18$ mm$\}$ (45)
$\Delta \le \frac{l}{150}$ (25)	$D_w \in \mathcal{D}_w = \{160, 210, 270, 340, 42$ $510, 600$ mm$\}$ (46)
Computational constraints	for other elements
$K_u \Delta_u = P$ (26)	$h_b \in \mathscr{h}_b = \{35, 43.5, 55, 80, 110$ mm$\}$ (47)
$1 \le u \le 15$ (27)	$g_b \in \mathscr{g}_b = \{0.75, 0.88, 1.0, 1.25mm\}$ (48)
$120 \le L_p \le 1200$ (28)	$h_p \in \mathscr{h}_p = \{80, 100, 120, 140, 160$ 180 mm$\}$ (49)
$40 \le L_w \le 320$ (29)	$g_s \in \mathscr{g}_s = \{10, 12, 14$ mm$\}$ (50)
$0.5 \cdot p \in C \subset R^n$ (30)	$D_s \in \mathcal{D}_s = \{580, 600, 620, 640$ mm$\}$ (51)
	$h_s \in \mathscr{h}_s = \{200, 220, 240, 270,$ 300 mm$\}$ (52)

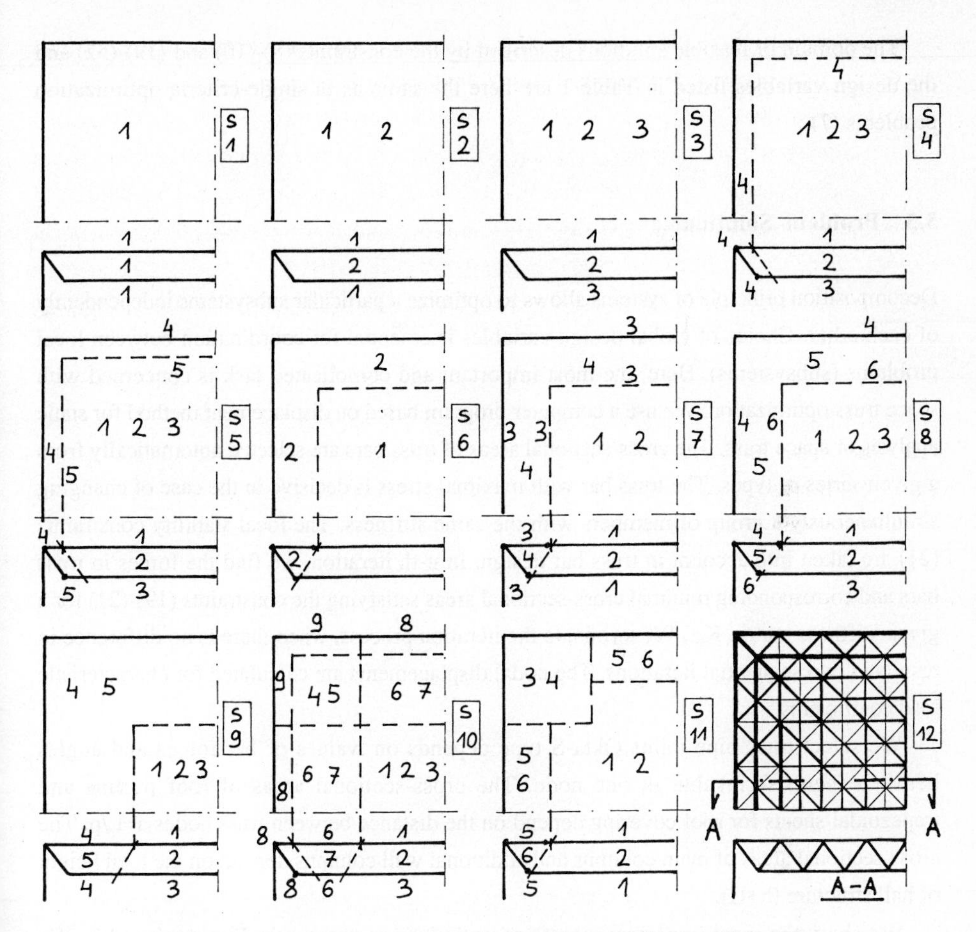

Figure 5. Numbering of Zones with the Same Stiffness of Truss Bars

We create the series of types of cross-sectional areas for truss bars as shown in Figure 4. We choose, for example, from basic series of types with a numbers t=33, every second (t=17), every third (t=9), and every eighth (t=5) cross sectional area.

We show numbering of zones with the same stiffness of truss bars in Figure 5. The bars in solution s=l have the constant cross-sectional areas. We choose individually the cross-sectional areas of bars in solution s=13 without partition on stiffness zones.

We choose in multiobjective optimization approach three objectives: mass of steel hall elements, the vertical displacement of midnode of truss and conventional labor cost of structure, i.e.,

$$\min \mathbf{f}(\mathbf{x}) = \min \left[f_5(\mathbf{x}), f_6(\mathbf{x}), f_8(\mathbf{x}) \right] \tag{53}$$

The domain of feasible solutions described by the constraints (8)-(10) and (19)-(52) and the design variables listed in Table 1 are here the same as in single-criteria optimization problems (7).

5.3 Problem Solution

Decomposition principle of systems allows to optimize a particular subsystems independently of each other. Choice of global design variables is essential for coordination between local problems (subsystems). Here, the most important and complicated task is concerned with space truss optimization. We use a computer program based on displacement method for static analysis of space truss. The cross-sectional areas of truss bars are selected automatically from a given series of types. The truss bar with maximal stress is decisive in the case of changing simultaneously a group of members with the same stiffness. The local stability constraints (21) are taken into account in truss bar design. In u-th iteration, we find the forces in truss bars and corresponding minimal cross-sectional areas satisfying the constraints (19)-(21) for a given stiffness matrix K_u . We terminate the iteration process, when there is no difference in results in two sequential iterations. The nodal displacements are calculated for characteristic load coefficients.

We select truss pipe joints Okta-S type depends on .values of bar forces and angles between them, shrinkable in one node. The cross-sectional areas of roof purlins and trapezoidal sheets for roof covering depend on the distance between truss nodes a=L/p. The cross-sectional areas of main columns and additional wall columns depend on the total height of hall structure (h+H).

We show the most important results of optimization process in Figures 6 and 7. We present the contours of the objective functions on global design variables plane (p,h) in Figure 6, for a given

$$\mathbf{x}_1 = [g_b, h_b, h_p, t, s, g_s, D_s, h_s\} =$$

$$= [0.75, 43.5, 80, 33, s12, 12, 620, 270] \tag{54}$$

We present the normalized values (interval [0,1]) of the objective functions $f_5(\mathbf{x})$, $f_6(\mathbf{x})$ and $f_8(\mathbf{x})$ in Figure 7, for a given vector of global variables.

$$\mathbf{x}_g = [p, h] = [8, 2.1] \tag{55}$$

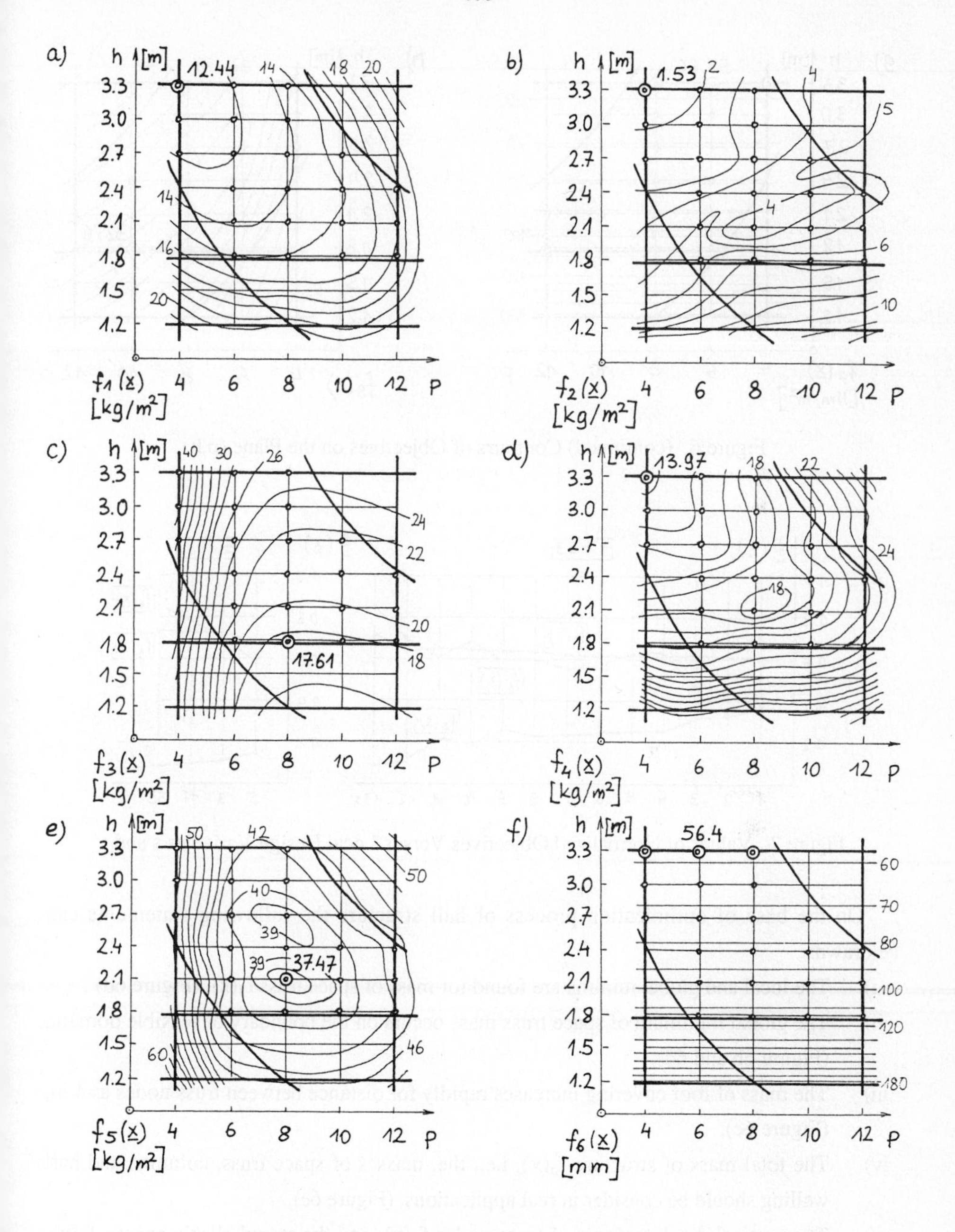

Figure 6. Contours of Objectives on the Plane (p,h)

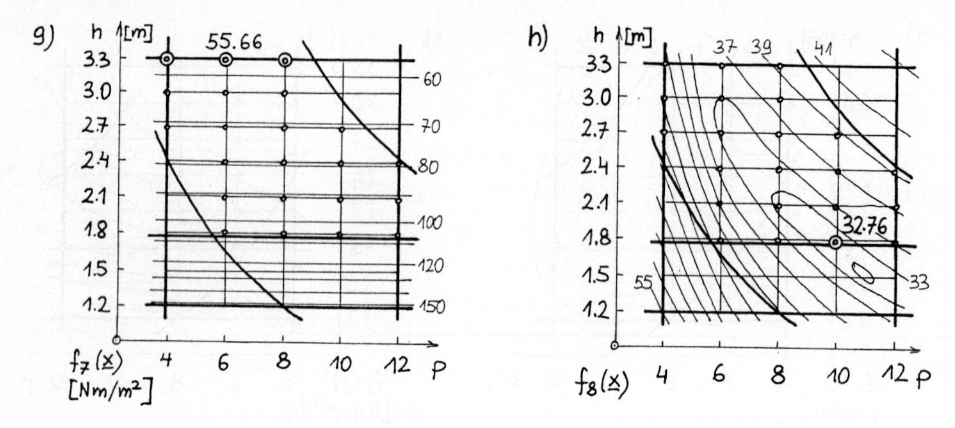

Figure 6. (continued) Contours of Objectives on the Plane (p,h)

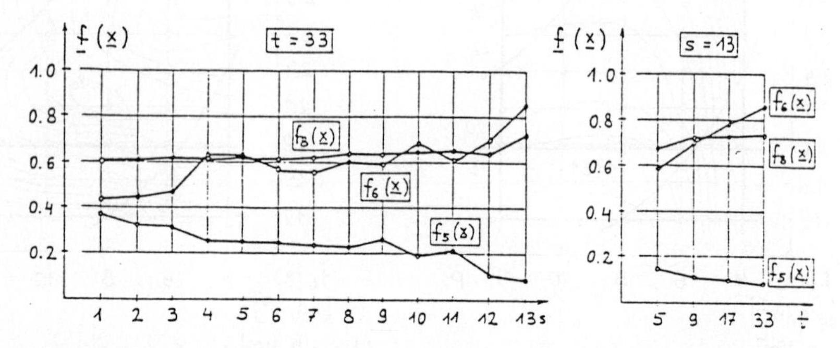

Figure 7. Values of Normalized Objectives Versus Local Design Variables s and t

On the base of optimization process of hall structure the following statements can be drawn:

i) The local and global minima are found for mass of space truss $f_3(\mathbf{x})$, (Figure 6d) ,

ii) The global minimum of space truss mass occurs on the boundary of feasible domain, (Figure 6b and 6d),

iii) The mass of roof covering increases rapidly for distance between truss nodes a>4 m, (Figure 6c),

iv) The total mass of structure $f_5(\mathbf{x})$, i.e., the. masses of space truss, columns and hall walling should be consider in real applications, (Figure 6e),

v) The vertical displacements of truss nodes $f_6(\mathbf{x})$, and the stored elastic energy $f_7(\mathbf{x})$, for a given local variable t=33 do not depend on mesh density of truss p. This is due to the fact that the bending stiffness in fully stress truss is inversely proportional to the distance between truss nodes a=L/p.

5.4 Multiobjective Optimization of Hall Structure

We assume in problem formulation (53), that all objective functions are minimized. Vector $\hat{\mathbf{x}}_{ND}$ is a Pareto optimal (compromise) solution of problem (53). This means, that $\hat{\mathbf{x}}_{ND}$ belongs to the set of feasible (compromise) solutions $\hat{\mathbf{x}}_{ND}$ if the corresponding vector objective function $\hat{\mathbf{y}}_{ND}=f(\hat{\mathbf{x}}_{ND})$ is a minimal (efficient) vector, in a partially ordered sense (see Peschel and Riedel [75] and Amelianczyk [1]). In other words, $\hat{\mathbf{x}}_{ND}$ is a Pareto optimal solution if there is no feasible solution x that would decrease some criterion without causing a simultaneous increase in at least one criterion. We denote by $\prec_\Lambda$ a cone relation order, generated by convex cone $\Lambda \subset R^n$, i.e., such set of the pairs (**y**, **z**) that ($\mathbf{y}$-$\mathbf{z}$) $\in \Lambda$. In other words, $\mathbf{z} \in (y + \Lambda) \Rightarrow (\mathbf{y}, \mathbf{z}) \in \prec_\Lambda$ [1] determinated by positive orthant R_0^+ of objective space R^J, i. e.,

$$\Lambda = \{(\lambda_1,...\lambda_j,...\lambda_J) \in R^J : \lambda_j \geq 0, j \in \overline{1,J}\} = R_0^+ \subset R^J, \tag{56}$$

then, the set of efficient solutions can be described as follows:

$$Y_{ND} = \{\hat{\mathbf{y}}_{ND} \in Y : \bigwedge_{z \in Y} (\hat{\mathbf{y}}_{ND}, \mathbf{z}) \in \prec_\Lambda \Rightarrow \hat{\mathbf{y}}_{ND} = \mathbf{z}\}. \tag{57}$$

The set Y_{ND} corresponds to the set of Pareto optimal solutions X_{ND} in design space, i. e.,

$$X_{ND} = \mathbf{f}^{-1}(Y_{ND}) = \{\hat{\mathbf{x}}_{ND} \in X : \hat{\mathbf{y}}_{ND} = f(\hat{\mathbf{x}}_{ND}) \in Y_{ND}\}. \tag{58}$$

The sets X_{ND} and Y_{ND} are finite (because of discretization of design variables) with dimesionality zero. A convex hull of the set of efficient solutions of problem (53), denoted by conv(Y_{ND}), is a surface with dim conv(Y_{ND})=2, in objective space. Dimensionality of convex hull of the set of compromise solutions, i.e., dim conv(X_{ND})=10, and after decomposition, dim conv(X^g_{ND})=2, and dim conv($X^{l(i)}_{ND}$) = {3(i=1), 2(i=2,3), l(i=4)}.

We use a property of monotonicity of objectives and design variables to calculate the sets X_{ND} and Y_{ND} (see [5,24, 25,26,75,96]). The minimal values of objectives, the bicriteria and three-criteria efficient sets belong to the set of efficient solutions Y_{ND}. Adding the (j+l) objective, where $j \in \overline{1,(J-1)}$, can only increase a numbers of existing set Y_{ND}(j) and dimensionality of the set conv{Y_{ND}(j)}. The points $\hat{\mathbf{y}}_{ND}$ belonging to Y_{ND}(j), belong also to the set $Y_{ND}(j+l) \subset Y_{ND}$. The monotonicity of design variables gives similar result, i.e., adding the (n+l) variable, where $n \in \overline{1,(N-1)}$, can only improve the actual set Y_{ND}(n). This means, that $Y_{ND}(n) \subset Y_{ND}(n+l)$, and there exist such value of active design variable x(n), that the product of sets $X_{ND}(n) \cap X_{ND}(n+l)$ is nonempty set.

We show the minimal set Y_{ND} for multiobjective optimization of hall structure in Figure 8. We mark the influence of taking into account the sequential design variables on minimal set.

Y^I_{ND} is the minimal set coming from variation of global variables $\mathbf{x}_g$, for a given vector of local variables $\mathbf{x}_l$, according to (54). Among local design variables, only t and s have a crucial influence on objectives. We investigate the influence of t and s on objective function, for a given global variables $\mathbf{x}_g$ =[8, 2.1]. These values of $\mathbf{x}_g$ corresponds to the minimal value of global criterion (60). Y^s_{ND} and Y^t_{ND} denote the sets of minimal (efficient) points for different stiffness zones of truss bars and different numbers of catalogue elements. Minima of $f_6(\mathbf{x})$ and $f_8(\mathbf{x})$ occur in the feasible domain on the opposite constraints, (Figure 6). The set X^g_{ND} runs across the feasible domain X_g, (Figure 9a). The functions $f_6(\mathbf{x})$ and $f_8(\mathbf{x})$ are cooperative (no conflict) for local design variables t and s. The solution x_l(t,s)=[5,1] dominates all other solutions. Finally, we find the efficient set Y^g_{ND} for the functions $f_6(\mathbf{x})$ and $f_8(\mathbf{x})$ with respect to global variables p and h for a given $\mathbf{x}_l$(t,s)=[5,1], (Figure 8).

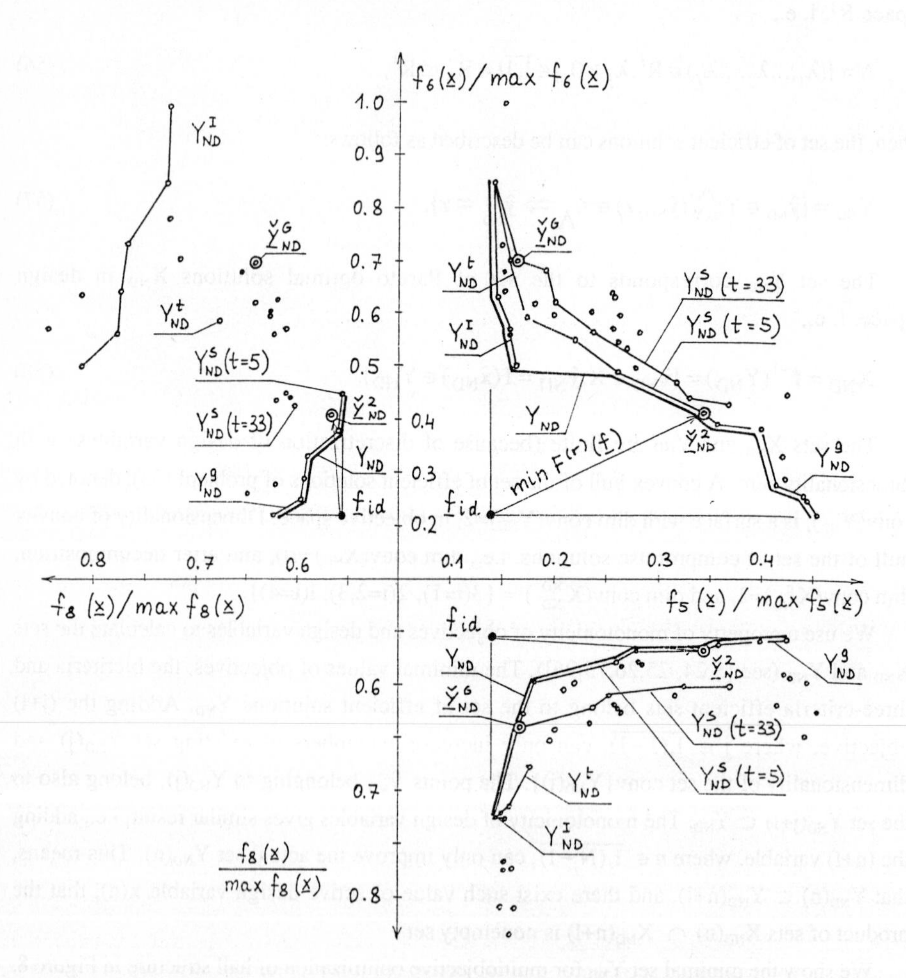

Figure 8. The Sets of Minimal (Efficient) Solutions for Hall Structure

The set of Pareto optimal (compromise) solutions is shown in Figure 9. The corner points (denoted by circles) correspond to minima of particular objectives. X^{56}_{ND}, X^{58}_{ND}, X^{68}_{ND} denote the sets of compromise solutions for particular bicriteria optimization problems, respectively.

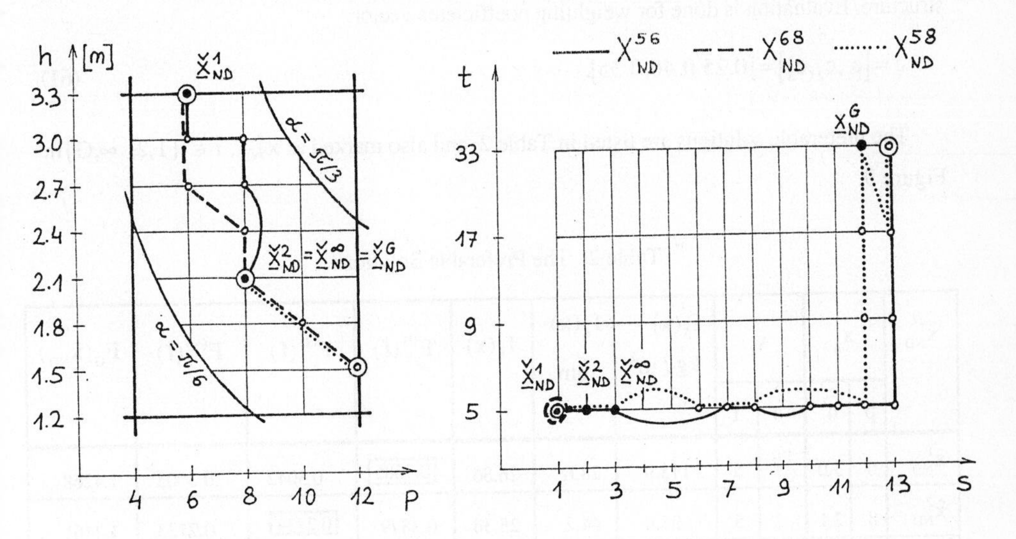

Figure 9. The Sets of Pareto Optimal (Compromise) Solutions

5.5 Choice of Preferable Solution

We choose the preferable solution using two methods: metric and utility function methods. Metric function method is based on determination of distance between ideal point, f_{id} and points belonging to the set of efficient solutions Y_{ND}. We find coordinates of ideal point minimizing particular objectives (Figure 8). The metric function has a form [1]:

$$F^{(r)}(\mathbf{f})=\left\{\sum_j\left[\left|\frac{f_j(\hat{\mathbf{x}}_{ND})}{\max f_j(\hat{\mathbf{x}}_{ND})}-\min\frac{f_j(\hat{\mathbf{x}}_{ND})}{\max f_j(\hat{\mathbf{x}}_{ND})}\right|\right]^r\right\}^{\frac{1}{r}} \tag{59}$$

$$j\in\{5,6,8\},\quad r\in\{1,2,\infty\}.$$

$F^{(2)}(\mathbf{f})$ for r=2 denotes geometrical distance between the ideal point and the set of efficient solutions. The minimal distance $F^{(2)}(\mathbf{f})$ is marked in Figure 8.

Cost function including material cost, $K_m(\mathbf{x})$; labor cost, $K_w(\mathbf{x})$; and maintain cost, $K_u(\mathbf{x})$; is taken as utility function with weighting coefficients vector $\mathbf{c}$, i.e.,

$$F_G(\hat{\mathbf{x}}_{ND}) = c_1\frac{K_m(\mathbf{x})}{K_m(\mathbf{x}^o)} + c_2\frac{K_w(\mathbf{x})}{K_w(\mathbf{x}^o)} + c_3\frac{K_u(\mathbf{x})}{K_u(\mathbf{x}^o)}, \qquad \sum_{i=1}^{3} c_i = 1.0. \tag{60}$$

We evaluate the compromise solutions $\hat{\mathbf{x}}_{ND} \in X_{ND}(n) \subset X$. A hall structure with local design variables given by (54) and global variables, x_g^0 = [10, 1.8]; is taken as a reference structure. Evaluation is done for weighting coefficients vector:

$$c = [c_1, c_2, c_3] = [0.25, 0.40, 0.35]. \tag{61}$$

The preferable solutions are listed in Table 2. and also marked as $\mathbf{x}_{ND}^r$, $r \in \{1,2,\infty,G\}$ in Figure 9.

Table 2. The Preferable Solutions

$\hat{\mathbf{x}}_{ND}^r$	$\mathbf{x}_g$		$\mathbf{x}_l$		$f_5(\mathbf{x})$ kg / m^2	$f_6(\mathbf{x})$ mm	$f_8(\mathbf{x})$	$F^{(1)}(\mathbf{f})$	$F^{(2)}(\mathbf{f})$	$F^{(\infty)}(\mathbf{f})$	$F_G(\hat{\mathbf{x}}_{ND})$
	p	h	s	t							
$\hat{\mathbf{x}}_{ND}^1$	6	3.0	1	5	115.9	28.6	26.86	0.3685	0.3042	0.3003	1.4268
$\hat{\mathbf{x}}_{ND}^2$	8	2.1	2	5	93.0	44.2	25.30	0.3879	0.2734	0.2133	1.2461
$\hat{\mathbf{x}}_{ND}^\infty$	8	2.1	3	5	90.0	47.0	25.57	0.4072	0.2811	0.2013	1.2296
$\hat{\mathbf{x}}_{ND}^G$	8	2.1	12	33	43.0	78.9	28.83	0.5812	0.4854	0.4780	0.9694

6 Conclusions

i) The real structure should satisfy all technological requirements. Also, the final solution should be a compromise between the minimization of weight of structure and minimization of labor cost. Decomposition principle allows to choose the best solution. Decomposition may concern with design variables as well as with vector objective function. The appropriate optimization procedure allows to find a preferable solution, after initial structural design, and evaluation (comparison) of tens of structures.

ii) The intervals of changing values for particular objectives in feasible domain are usually large. In this example, we have maximal rates of changing values for objectives: 270% for $f_5(\mathbf{x})$, 78% for $f_6(\mathbf{x})$ and 35% for $f_8(\mathbf{x})$. It does not mean that similar effects can be obtained in optimization process. Usually, the improvement of objectives is within the range of (2-8)%, compare with a solution proposed by an experienced designer. In this example, we obtained 3.06% improvement for competitive cost function.

iii) Structural optimization is particulary important in two cases, namely: for systems applied many times and for new large-scale structures designed without previous engineering experience.
iv) Structural optimization problems should be formulated as discrete, nonlinear, multicriteria and multilevel problems.
v) Sometimes it is enough to evaluate different variants of system and choose one which is close to optimum then to look for strictly optimum solution.

References

1. Ameliańczyk, A.: Multicriteria optimization in the control and management problems (in Polish), Ossolineum, Wroclaw 1984
2. Baier, H.: Uber Algorithmen zur Emittlung und Charakterisierung Pareto-Optimaler Losungen bei Entwurfsaufgaben elastischer Tragwerke, ZAMM, 57, 318-320 (1977)
3. Baier, H.: Mathematische Programmierung zur Optimierung von Tragwerken inbesondere bei mehrfachen Zielen, Dissertation D17, Darmstadt 1978
4. Bauer, J., Gutkowski, W., Iwanow. Z.: Optimum design of regular space structures, in: Nooshin H., (ed.), Proc. Third Int. Conf. on Space Structures, Elsevier, London and New York, 672-676 (1984)
5. Benson, H.P.: An improved definition of efficiency for vector maximization with respect to cones. Journal of Mathematical Analysis and Applications, Vol. 71, 232-241 (1979)
6. Brandt, A.M., (ed.): Criteria and Methods of Structural Optimization, PWN-Polish Scientific Publishers (Warszawa) and Martinus Nijhoff Publishers (The Hague-Boston-Lancaster) 1984
7. Brandt, A.M., (ed.): Foundations of Optimization of Structural Elements, PWN-Polish Scientific Publishers (Warszawa) and Martinus Nijhoff Publishers, Dordrecht/Boston/London 1989
8. Carmichael, D.G.: Computation of Pareto optima in structural design, Int. J.Numer.Meth.Engng., 15, 925-934 (1980)
9. Cella, A. and Soosar, K.: Discrete variables in structural optimization, In: Optimum Structural Design; Theory and Applications, (Eds. Gallagher, R.H. and Zienkiewicz, O.C.) 201-222, John Wiley and Sons, London 1973
10. Chang, C.: Minimum weight design of trusses subject to damage tolerance constraints, M.S.Thesis, Illinois Institute of Technology, 1981
11. Chankong, V. and Haimes, Y.Y.: Multiobjective Decision-making Theory and Methodology, Elsevier, North-Holland, New York 1982
12. Cohon, J.L.: Multiobjective Programming and Planning, Academic Press, New York 1978
13. Dauer, J.P. and Stadler, W.: A survey of vector optimization in infinite-dimensional spaces, Part 2, JOTA, 51, 2, 205-241 (1986)
14. Debreu, G.: Theory of Value, John Wiley, New York 1959
15. Dubov, Ju.A., Travkin, S.I. and Jakimec, V.N.: Multiobjective Models for Creating and Choosing a System of Variants (in Russian), Nauka, Moscow 1986
16. Eschenauer, H.: Eine Vektoroptimierungsaufgabe der Strukturmechanik, Karl-Marguerre-Gedachtniskolloquium, THD Schriftenreihe Wissenschaft und Technik, Bd. 16 (ISSN 0172-2085) 89-101 (1980)
17. Eschenauer, H.: Anwendung der Vektoroptimierung bei raumlichen Tragstrukturen, Der Stahlbau, 4, 110-115 (1981)
18. Eschenauer, H.: Vector optimization in structural design and its application on antenna, Proc. of Euromech-Colloquium 164 on Optimization Methods in Structural Design, (Eds.) Eschenhauer, H. & Olhoff, N., Bibl.Institut, Mannheim, 146-155 (1983)
19. Eschenauer, H.: Optimierung ebener Flachentragwerke aus Verbundwerkstoff, Z. Flugwiss. Weltraumforsch., Vol. 8, 6, 367-378 (1984)
20. Eschenauer, H. and Jendo, S.: Multiobjective optimization of cable structures, ZAMM, 66, 5, T342-T343 (1986)
21. Eschenauer, H., Kneppe, G. and Stenvers, K.H.: Deterministic and stochastic multiobjective optimization of beam and shell structures, J. of Mech. Transm.& Autom. in Design, Vol. 108, 31-37 (1986)

22. Eschenauer, H, Koski, J. and Osyczka, A. (eds.): Multicriterion Design Optimization: Procedures and Applications, Springer Verlag, Berlin 1990
23. Fleury, C.: Structural weight optimization by dual methods of convex programming, Int. J. Numer. Methods Engng., 14, 1761-1783 (1979)
24. Galas, Z. and Nykowski, J. (eds.): Mathematical Programming Exercises, Vol. 1: Linear Programming (in Polish), PWN-Polish Scientific Publishers, Warszawa 1986
25. Galas, Z. and Nykowski, J. (Eds.): Mathematical Programming Exercises, Vol. 2: Nonlinear Programming (in Polish), PWN-Polish Scientific Publishers, Warszawa 1988
26. Galas, Z., Nykowski,J. and Zólkiewski, Z.: Multicriterion Programming (in Polish), PWE-Polish Economical Publishers, Warszawa 1987
27. Gerasimov, E.N., Potchman, Ju.M. and Skalozub, V.V.: Multicriterion Structural Optimization (in Russian). Vistcha Schola, Kiev 1985
28. Giles, G.L.: Procedure for automated aircraft wing structural design, J. Struct. Div. ASCE, 97, 99-113 (1971)
29. Glankwahmdee, A., Liebman, J.S. and Hogg, G.L.: Unconstrained discrete nonlinear programming, Engng. Optim., 4, 95-107 (1979)
30. Grierson, D.E. and Lee, W.H.: Optimal synthesis of frameworks under elastic and plastic performance constraints using discrete sections, J. Struct. Mech., 14, 4, 401-420 (1986)
31. Hauk, V.: The Mannesmann Okta-S joint for tubular space structures. Second Int. Conf. on Space Structures, Reports, University of Surrey, Sept. 1975
32. Ho, J.K.: Optimal design of multi-stage structures - a nested decomposition approach, Technical Report 72-4, System Optimization Lab., Dept. of Operations Research, Stanfort University 1974
33. Ho, J.K.: Optimal design of multi-stage structures: a nested decomposition approach, Comp. and Struct., 5, 4, 249-255 (1975)
34. Homenjuk, V.V.: Elements of Theory of Multiobjective Optimization (in Russian), Nauka, Moscow 1983
35. Hwang, C.L. and Masud, A.S.M.: Multiple Objective Decision Making; Methods and Applications, Lecture Notes in Economics and Mathematical Systems, Vol. 164, Springer-Verlag, New York 1979
36. Jendo, S.: Some problems of multicriterion optimization of cable structures, IFTR Reports, Warsaw, 43, 1985
37. Jendo, S.: Multicriteria optimization of cable systems, Proc. of the IASS Symp. on Membrane Structures and Space Frames, Sept. 1986, Osaka, Japan, Elsevier, Amsterdam, Vol. 2, 71-78 (1986)
38. Jendo, S.: Multicriteria optimization of single-layer cable systems, Arch. of Mech., 38, 3, 219-234 (1986)
39. Jendo, S.: Multiobjective reliability-based optimization of plastic frames, in: Lecture Notes in Engineering, 42, 171-177, Edited by C.A. Brebbia and Orszag, Proc. of GAMM Seminar on Discretization Methods and Structural Optimization - Procedures and Applications, October 5-7, 1988, Siegen, FRG, H.A. Eschenauer and G. Thierauf (eds.), Springer-Verlag, Berlin 1989
40. Jendo, S.: Multiobjective Optimization. In: Save M. and Prager W. (eds.): Structural Optimization, Vol. 2: Mathematical Programming, Plenum Press, New York/London, 311-342 (1990)
41. Jendo, S. and Marks, W.: Multiobjective structural optimization, In: Lecture Notes in Control and Information Science, Vol. 84, pp. 365-374, Edited by Thoma, M. and Wyner, A., System Modelling and Optimization, Proc. of the 12th IFIP Conference, Budapest, Hungary, Sept. 1985, Edited by Prekopa, A., Szelezsan, J. and Strazicky, B., Springer-Verlag, Berlin (1986)
42. Jendo, S. and Marks, W.: Formulation of structural optimization problems (in Polish), Rozp. Inzyn. (Engng. Trans.) 34, 4, 457-481 (1986)
43. Jendo, S. and Marks, W.: Multicriteria optimization in civil engineering design, Proc. of the Int. Conf. on Optimization Techniques and Applications, Singapore, 998-1006 (1987)
44. Jendo, S., Marks, W. and Thierauf, G.: Multicriteria optimization in optimum structural design, In: Sage,A.P.(ed.), Large Scale Systems, Elsevier Science Publishers B.V.(North-Holland), 9, 141-150 (1985)
45. Jendo, S. and Paczkowski, W.M.: Some problems of optimization in civil engineering structural design. Proc. X Polish Conf. "Computer Methods in Mechanics", Swinoujscie, 301-308 (1991)
46. Jendo, S. and Paczkowski, W.M.: Multicriteria discrete optimization of large scale truss systems, Proc. of the NATO/ASI, Berchtesgaden/Germany 1991
47. Jendo, S. and Paczkowski, W.M.: Multicriteria discrete optimization of large scale truss systems, Structural Optimization, 1993 (in print).
48. Jendo, S. and Paczkowski, W.M.: Multicriterion discrete optimization of large scale frame structures, Proc. of the ICOTA-92, Singapore, 343-351 (1992)
49. Jendo. S. and Putresza. J.: Vector optimization of bar structures according to minimum cost and maximum reliability, Transactions of WSI-Koszalin, 13, 61-72 (1988)

50. Jendo, S. and Putresza, J.: Multicriteria optimization of bar structures with random parameters, pp. 1085-1097, in: Proc. of Inter. Conf. on Multiple Criteria Decision Making: Application in Industry and Service, M.T. Tabucanon and V. Chankong (eds.), Bangkok, Thailand 1989
51. Jendo, S. and Putresza, J.: Reliability-based multicriteria optimization of bar structures (Monte-Carlo Simulation), Proc. of Conf. on Computer Methods in Mechanics, Swinoujscie/Poland, 309-315 (1991)
52. Jendo, S. and Putresza, J.: Multicriteria optimization of elasto-plastic bar structures, Proc. of IFIP Conf. on Reliability and Optimization of Structural Systems, Munich/Germany, 239-249 (1992)
53. Jendo, S. and Putresza, J.: Reliability-based multicriteria optimization of bar structures (failure-tree method), ZAMM, 72, 6, T 561 - T 563 (1992)
54. Jendo, S. and Stachowicz, A.: Optimum structural design by multilevel optimization, ZAMM, 66,5, T347-T349 (1986)
55. Jendo, S. and Stachowicz, A.: Multilevel optimization in civil engineering structural design, ZAMM, 68, 5 (1988)
56. Kaneko, I. and Cu Duong Ha: A decomposition procedure for large-scale optimum plastic design problems, Int. J. Numer. Methods Eng., 19, 873-889 (1983)
57. Keeney, R.L. and Raiffa, H.: Decision with Multiple Objectives: Preferences and Value Trade-offs, John Wiley & Sons, New York 1976
58. Kickert, W.J.M.: Fuzzy theories on decision-making. A critical review, In: Frontiers in System Research, Vol. 3, Martinus Nijhoff Social Sciences Division, VIII, Leiden, Boston, London 1978
59. Kirsch, U.: Multilevel approach to optimum structural design, Proc. ASCE, 101, ST 4, J. Struct. Div., 957-974 (1975)
60. Kirsch, U., Reiss, M. and Shamir, U.: Optimum design by partitioning into substructures, J. Struct. Div. ASCE, 98, 249-261 (1972)
61. Koski, J.: Truss optimization with vector criterion, Tampere University of Technology, Tampere/Finland, Publications, No. 6, 1979
62. Koski, J.: Multicriteria optimization in optimum structural design, In: New Directions in Optimum Structural Design, Ed. by ATREK, E., et.al., John Wiley & Sons, Chichester, 483-503 (1984)
63. Koski, J.: Bicriterion optimum design method for elastic trusses, Acta Polytechnica Scandinavica, Mech. Engng. Series, No. 86, Helsinki 1984
64. Koski, J. and Silvennoinen, R.: Pareto optima of isostatic trusses, Comp. Meth. Appl. Mech. Engrg., 31, 3, 265-279 (1982)
65. Lai, Y.S. and Achenbach, J.D.: Direct search optimization method, J. of Struct. Div., ASCE, 98, STl, 119-131 (1973)
66. Leitmann, G.: Some problems of scalar and vector-valued optimization in linear viscoelasticity, JOTA., 23, 1, 93-99 (1977)
67. Liebman, J. S., Khachaturian, N. and Chanaratna, V.: Discrete structural optimization, J. of Struct. Div., ASCE, 107, STll, 2177-2197 (1981)
68. Lin, J.G.: Maximal vector and multi-objective optimization, JOTA, 18, 1, 41-63 (1976)
69. Makowski Z.S.: Räumliche Tragwerke aus Stahl, Verlag Stahleisen, MBH, Dusseldorf 1963
70. Von Neumann, J. and Morgenstern, O.: Theory of Game and Economic Behaviour, 2nd Edition, Princeton University Press, Princeton, New Jersey 1947
71. Osyczka, A.: An approach to multicriterion optimization problems for engineering design, Comput. Meth. Appl. Mech. Engng.,15, 309-333 (1978)
72. Osyczka, A.: Multicriterion Optimization in Engineering, Ellis Horwood Publishers, Chichester, England 1984
73. Paczkowski, W.M.: OPTIM-nonlinear discrete optimization program, Proc. of X Polish Conf. on Computer Methods in Mechanics, Swinoujscie/Poland, 607-614 (1991)
74. Pareto, V.: Cours D'Economie Politique, Vol. I and II, F. Rouge, Lausanne 1896
75. Peschel, M., Riedel, C.: Polioptimierung eine Entscheidungshilfe fur ingenieurtechnische Kompromisslösungen, VEB Verlag Technik, Berlin 1976
76. Rao, S.S. and Hati, S.K.: Game theory approach in multi-criteria optimization of function generating mechanisms, Trans. ASME, J. of Mech. Engrg., 101, 398-405 (1979)
77. Rosen, J.B.: Primal partition programming for block diagonal matrices, Num. Mathem., 6, 250-260 (1964)
78. Sattler, H.J.: Ersatzprobleme fur Vektoroptimierungsaufgaben und ihre Anwendung in der Strukturmechanik, VDI-Verlag, GmbH, Dusseldorf, Series 1, No. 88, 1982
79. Schmit, L.A., Jr. and Mehrinfar, M.: Multilevel optimum design of structures with fiber-composite stiffened panel components, AIAA Paper 80-0723, 1980 and AIAA J., 20, 138-147 (1982)
80. Schmit, L.A., Jr. and Ramanatham, R.K.: Multilevel approach to minimum weight design including buckling constraints, AIAA J., 16, 97-104 (1978)
81. Sepulveda, A. and Cassis, J.H.: An efficient algorithm for the optimum design of trusses with discrete variables, Int. J. Num. Meth. in Engng., 23, 1111-1130 (1986)

82. Sobieszczański-Sobieski, J.: An integrated computer procedure for sizing composite airframe structures, NASA TP-1300, 1979
83. Sobieszczanski-Sobieski, J., James, B.B., and Dovi, A.R.: Structural optimization by multilevel decomposition, AIAA J., 23, 1775-1782 (1985)
84. Sobieszczański-Sobieski, J. and Leondorf, D.D.: A mixed optimization method for automated design of fuselage structures, J. Aircraft, 9, 805-811 (1972)
85. Stachowicz, A. and Ziobroń, W.: Underground Water Tank Systems (in Polish) Arkady, Warsaw 1986
86. Stadler, W.: Preference optimality and applications of Pareto optimality, In: Multi-Criteria Decision Making, Ed. by Leitmann, G. and Marzollo, A., CISM Courses and Lectures, No. 211, Springer-Verlag, Wien and New York 1975
87. Stadler, W.: Natural structural shapes of shallow arches, J. Appl. Mech., Trans. ASME., 44, 2, 291-298 (1977)
88. Stadler, W.: Natural structural shapes (the static case), Quart. J. Mech. Appl. Math., 31, 2, 196-217 (1978)
89. Stadler, W.: Preference optimality in multicriterion control and programming problems, Nonlinear Anal. Theory, Methods and Applications, 4, No. 1, 51-65 (1980)
90. Stadler, W.: A comprehensive bibliography on MCDM, In: MCDM-Past Decade and Future Trends; A Source Book of Multiple Criteria Decision Making, Ed. by Zeleny, M., Greenwich, Conn., JAI-Press, 1984
91. Stadler, W.: Multicriteria optimization in mechanics (a survey), AMR, 37, 3, 277-286 (1984)
92. Stadler, W. (ed.): Multicriteria Optimization in Engineering and in the Sciences, Plenum Press, New York, 1988
93. Toakley, A.R.: Optimum design using available sections, J. Struct. Div., ASCE, 94, ST5, 1219-1241 (1968)
94. Watkins, R.I. and Morris, A.J.: A multicriterion objective function optimization scheme for laminated composites for use in multilevel structural optimization schemes, Comp. Meths. in Appl. Mech. Eng., 60, 233-251 (1987)
95. Woo, T.H. and Schmit, L.A.: Decomposition in optimal plastic design of structures, Int. J. Solids Struct., 17, 39-56 (1981)
96. Yu, P.L.: Cone convexity, cone extreme points and nondominated solutions in decision problems with multiobjectives. IOTA, Vol. 14, 319-377 (1974)
97. Yu, P.L.: Multiple-Criteria Decision Making: Concepts, Techniques, and Extensions, Plenum Press, New York and London 1985
98. Zadeh, L.: Optimality and non-scalar-valued performance criteria, IEEE Trans., Automatic Control, AC-8, 1963

Design Sensitivity Analysis for Coupled Systems and Their Application to Concurrent Engineering

Daniel A. Tortorelli

University of Illinois at Urbana-Champaign, Urbana, Illinois 61801 USA

Abstract: Design sensitivities are derived via the direct and adjoint methods for fully coupled nonlinear systems. These sensitivity derivations may then be used to create a concurrent engineering design environment where the individual systems are the product analyses and manufacturing process analyses. The sensitivity information, in turn, may be used in a trade-off study to quantify how product or manufacturing design changes affect any performance functional, e.g. cost, mass, stress, or cooling rate. Finally, it is noted that the necessary operator required for the sensitivity analyses is, in fact, the tangent operator of the coupled system analysis; and that this operator may be partitioned to take advantage of existing software capabilities.

Keywords: concurrent engineering / multidisciplinary analysis / design sensitivity analysis

1 Introduction

In a truly concurrent engineering environment, the design of the product and its manufacturing process are a unified effort. In practice, however, this is seldom the case as the product is first designed and then given to the manufacturing engineer who, in turn, designs its manufacturing process. Thus, it is often the case that the original product design is not conducive to the manufacturing process and the manufacturer's redesign recommendations adversely affect the product's performance. Fortunately, with today's computer simulation capabilities, design flaws are identified early and the necessary redesigns are implemented before costly production begins. Indeed, both production and manufacturing engineers have access to powerful computer analysis tools, in particular the finite element method, which enable them to create simulation models (albeit a time consuming task), perform the analysis, post-process the simulation results, and evaluate the effects of the proposed design changes. Still, the process is arduous due to the many redesign iterations which pass amongst and between the product and manufacturing design teams. The excessive number of design iterations is due to the engineer's inability to access the effects which design changes have on both the product performance and product manufacturability.

We propose a scheme in which design sensitivity analysis is incorporated into the design loop. In this way, the effects which the proposed design changes have on the the product performance and product manufacturability may be quantified. These sensitivities may then be used to drive a formal optimization [21] or to perform a trade-off study. The proposed sensitivity analysis approach assumes that both the production and manufacturing design teams have access to numerical simulation capabilities such as the the finite element method, that the input data of the simulation package can be directly related to the set of continuous design parameters, and that the simulation results can be used to measure the quality of the design. This approach is extends the work of Sobieszczanski-Sobieski et. at. [13, 14] for multi-disciplinary optimization.

In the following, design sensitivities are derived for coupled systems in which the output of one system analysis is the input to the other, and vice versa. Both adjoint [10] and direct [9] approaches of sensitivity analysis are presented. Further, the sensitivity analyses are valid for nonlinear and transient systems. The sensitivity derivations are preceded by the system analysis and the problem statement.

2 System Analysis

Herein, we consider two systems which are implicitly defined through,

$$\begin{aligned} A(u(\phi);(v(\phi),\phi)) &= 0 \\ B(v(\phi);(u(\phi),\phi)) &= 0 \end{aligned} \tag{1}$$

where $u(\phi)$ and $v(\phi)$ represent the state of systems A and B (and may be vector valued), respectively; $(v(\phi),\phi)$ and $(u(\phi),\phi)$ are the control of systems A and B, respectively; and $\phi \in \Re^L$ represents the L-dimensional set of design parameters. The operators A and B, are typically differential operators and may be nonlinear and transient. Note that coupling occurs in these analyses in that the state $u(\phi)$ of system A is defined on the state $v(\phi)$ of analysis B, and vice versa. Finally note that both u and v are implicit functions of the design parameters ϕ, and we write $u(\phi)$ and $v(\phi)$. Henceforth, the above systems are referred to as the local systems.

As an example of the above system, one may consider dynamic thermoelasticity. The state field u, of the elasticity analysis A, is the displacement, stress and strain and the state field v, of the thermal analysis B, is the temperature, heat flux, and temperature gradient. The geometry of the structure, the material properties, and the prescribed boundary conditions for both systems are expressed as functions of the design parameters ϕ. Coupling occurs due to the presence of a strain rate dependent heat source in the energy equation and the inclusion of thermal strains which lead to body type loads in the momentum equation. Obviously, both states are ultimately determined by the design. Sensitivities for this class of problems are derived in [17].

The solution to the above system (for a fixed design) generally requires an iterative procedure. We suggest the use of the Newton-Raphson iteration in which the global residual is defined

$$R(u(\phi), v(\phi); \phi) \equiv \left\{ \begin{array}{c} A(u(\phi); (v(\phi), \phi)) \\ B(v(\phi); (u(\phi), \phi)) \end{array} \right\} = 0 \tag{2}$$

Note that the A and B may viewed as the local residuals of the respective systems. The system R as defined above, is deemed the global system.

Suppose we have an existing guess for the solution of the above, say $u^I(\phi)$ and $v^I(\phi)$, which does not satisfy equation 2. The objective then, is to determine the appropriate state changes, i.e. the Δu and Δv, which when added to the current solution guess, will satisfy the above equation. To this end, we perform a first-order Taylor series expansion about the current solution (iterate) which gives

$$\begin{aligned} 0 &= R(u^I(\phi) + \Delta u, v^I(\phi) + \Delta v; \phi) \\ &= R(u^I(\phi), v^I(\phi), \phi) + \left[\begin{array}{cc} \frac{\partial A(u^I(\phi);(v^I(\phi),\phi))}{\partial u} & \frac{\partial A(u^I(\phi);(v^I(\phi),\phi))}{\partial v} \\ \frac{\partial B(v^I(\phi);(u^I(\phi),\phi))}{\partial u} & \frac{\partial B(v^I(\phi);(u^I(\phi),\phi))}{\partial v} \end{array} \right] \left\{ \begin{array}{c} \Delta u \\ \Delta v \end{array} \right\} \end{aligned} \tag{3}$$

The above may be solved for the incremental response Δu and Δv by inverting the global tangent stiffness matrix, i.e.,

$$\left\{ \begin{array}{c} \Delta u \\ \Delta v \end{array} \right\} = - \left[\begin{array}{cc} \frac{\partial A(u^I(\phi);(v^I(\phi),\phi))}{\partial u} & \frac{\partial A(u^I(\phi);(v^I(\phi),\phi))}{\partial v} \\ \frac{\partial B(v^I(\phi);(u^I(\phi),\phi))}{\partial u} & \frac{\partial B(v^I(\phi);(u^I(\phi),\phi))}{\partial v} \end{array} \right]^{-1} R(u^I(\phi), v^I(\phi), \phi) \tag{4}$$

after which the response is updated according to

$$\begin{aligned} u^{I+1}(\phi) &= u^I(\phi) + \Delta u \\ v^{I+1}(\phi) &= v^I(\phi) + \Delta v \end{aligned} \tag{5}$$

The global residual is again evaluated, and if it is not sufficiently small, the process is repeated. It should be noted that this process experiences quadratic convergence; however, uniqueness of the solution is not guaranteed as the solution is highly dependent on the initial solution guess, $u^0(\phi)$ and $v^0(\phi)$. Further note, that existence of the solution is not necessarily ensured.

In numerical simulations, such as the finite element method, $u(\phi)$ and $v(\phi)$ form the local solution vectors, $A(u(\phi); (v(\phi), \phi))$ and $B(v(\phi); (u(\phi), \phi))$ form the local residual vectors, and $\frac{\partial A(u(\phi);(v(\phi),\phi))}{\partial u}$ and $\frac{\partial B(v(\phi);(u(\phi),\phi))}{\partial v}$ form the local tangent stiffness matrices of the respective system analyses.

In general, the local tangent stiffness matrices are banded, whereas, the global tangent stiffness matrix is not. Further, the tangent stiffness matrices may be computed in different parts of the analysis code or from different analysis codes, entirely. To this end, we propose the following partitioning scheme where we assume that the local systems A and B are of dimension M and N, respectively, and $M > N$. First we expand equation 3 to give

$$\begin{aligned} \Delta u &= \frac{\partial A}{\partial u}^{-1}[-A - \frac{\partial A}{\partial v}\Delta v] \\ \frac{\partial B}{\partial v}\Delta v &= [-B - \frac{\partial B}{\partial u}\Delta u] \\ &= [-B - \frac{\partial B}{\partial u}\frac{\partial A}{\partial u}^{-1}[-A - \frac{\partial A}{\partial v}\Delta v]] \end{aligned} \tag{6}$$

where A and B and their derivatives in the above are understood to be the values of the respective functions evaluated at $u^I(\phi)$, $v^I(\phi)$ and ϕ. This abuse of notation sporadically appears in the sequel. The above may be solved via

$$\begin{aligned} [\frac{\partial B}{\partial v} - \frac{\partial B}{\partial u}\frac{\partial A}{\partial u}^{-1}\frac{\partial A}{\partial v}]\Delta v &= [-B + \frac{\partial B}{\partial u}\frac{\partial A}{\partial u}^{-1} A] \\ \Delta u &= \frac{\partial A}{\partial u}^{-1}[-A - \frac{\partial A}{\partial v}\Delta v] \end{aligned} \tag{7}$$

where we first solve for Δv and then subsequently solve for Δu. Note that the above scheme makes use of the decomposed tangent stiffness matrix from the larger local analysis, thus a majority of the banded structure is utilized. The $[\frac{\partial B}{\partial v} - \frac{\partial B}{\partial u}\frac{\partial A}{\partial u}^{-1}\frac{\partial A}{\partial v}]$ must be assembled and decomposed. This term is generally a full matrix and requires the decomposed matrix from the larger analysis. Thus, for each iteration of the Newton-Raphson process, we 1) assemble the local residuals A and B and 2) perform a convergence check. If the solution has not converged we proceed to 3) form the matrices $\frac{\partial A}{\partial u}$, $\frac{\partial A}{\partial v}$, $\frac{\partial B}{\partial u}$, and $\frac{\partial B}{\partial v}$, 4) decompose $\frac{\partial A}{\partial u}^{-1}$, 5) assemble and decompose $[\frac{\partial B}{\partial v} - \frac{\partial B}{\partial u}\frac{\partial A}{\partial u}^{-1}\frac{\partial A}{\partial v}]$, 6) compute Δv and then Δu via equation 7, and 7) update the solution via equation 5. Most of the terms in steps 1 - 4 are computed in standard finite element codes, [1] with the possible exception of the $\frac{\partial A}{\partial v}$ and $\frac{\partial B}{\partial u}$ quantities; which are only included if the analysis is performed within a single code which simulates fully coupled problems. Thus, finite element code enhancements may be required to perform the Newton-Raphson iteration for fully coupled problems.

[1] We assume that the finite element code incorporates the Newton-Raphson procedure, that implicit (or partially implicit) time integration schemes are employed, and that direct solvers are utilized.

3 Sensitivity Problem Definition

Once the necessary analyses are performed, the design may be judged via a cost and a series of constraint functions. Presumably, these cost/constraint functions are expressed in terms of the system state, i.e.

$$G(\phi) = F(u(\phi), v(\phi), \phi) \tag{8}$$

In the above, the generalized function G is defined on only the design space as ultimately, the design determines the response, i.e. $\phi \mapsto G(\phi)$. The function F is defined on the design through the state, $u(\phi)$ and $v(\phi)$. Typically, F is an integral function over the spatial and time domains. However, local quantities, may be described by utilizing the appropriate weighting functions, c.f. [17] for details. For example, we again refer to the dynamic thermoelastic problem. The function G may be used to measure the total strain energy of the body integrated over time, the average temperature at a point in the body, or the stress in the body at a distinct point and time.

The objective of the sensitivity analysis is to quantify how changes in the design parameters affect the value of the cost/constraint function $G(\phi)$. These changes may be approximated via a first-order Taylor series expansion, $G(\phi + \Delta\phi) \approx G(\phi) + \frac{\partial G(\phi)}{d\phi}\Delta\phi$. Hence, the goal the of the sensitivity analysis is to evaluate the derivative $\frac{dG(\phi)}{d\phi}$,

$$\begin{aligned} \frac{dG(\phi)}{d\phi} &= \frac{\partial F(u(\phi), v(\phi), \phi)}{\partial u}\frac{du(\phi)}{d\phi} + \frac{\partial F(u(\phi), v(\phi), \phi)}{\partial v}\frac{dv(\phi)}{d\phi} \\ &\quad + \frac{\partial F(u(\phi), v(\phi), \phi)}{\partial \phi} \end{aligned} \tag{9}$$

where we invoked the chain-rule. The difficulty in evaluating the above expression is due to the presence of the state derivatives, $\frac{du(\phi)}{d\phi}$ and $\frac{dv(\phi)}{d\phi}$. These terms are not explicitly known quantities, because u and v are implicitly defined on the design through equation 1.

The most obvious way to evaluate the sensitivity $\frac{dG(\phi)}{d\phi}$ is to perform a finite difference approximation where the system analysis is performed for a given design ϕ and G is evaluated. Then one of the design parameters ϕ_α is perturbed, the system is re-analyzed, G is re-evaluated, and the sensitivity $\frac{\partial G}{\partial \phi_\alpha}$ is approximated via the difference of the perturbed and original G values divided by the perturbation. This procedure must then be repeated for each of the L design parameters. This method is costly due to the additional system analysis and may be unreliable. Indeed, if the perturbation is too small or too large, then round-off and truncation will erode the results, respectively [16].

In the following, we describe two methods to compute the sensitivity analytically, the direct and adjoint methods. These methods are both accurate and efficient.

4 Direct Differentiation Method

In the direct differentiation method we evaluate the derivatives $\frac{du(\phi)}{d\phi}$ and $\frac{dv(\phi)}{d\phi}$ by defining and solving a series of *pseudo* problems. Once these quantities are known, equation 9 may be readily evaluated.

To evaluate the state derivatives, we merely differentiate equation 1 with respect to each of the design parameters, $\phi_\alpha, \alpha = 1, L$, to evaluate the partial derivatives $\frac{\partial u(\phi)}{\partial \phi_\alpha}$ and $\frac{\partial v(\phi)}{\partial \phi_\alpha}$. The total derivatives $\frac{du(\phi)}{d\phi}$ and $\frac{dv(\phi)}{d\phi}$ are then formed by assembling the partial derivatives.

Differentiation of equation 2 (the equivalent of equation 1) with respect to ϕ_α gives

$$0 = \begin{bmatrix} \frac{\partial A(u(\phi);(v(\phi),\phi))}{\partial u} & \frac{\partial A(u(\phi);(v(\phi),\phi))}{\partial v} \\ \frac{\partial B(v(\phi);(u(\phi),\phi))}{\partial u} & \frac{\partial B(v(\phi);(u(\phi),\phi))}{\partial v} \end{bmatrix} \begin{Bmatrix} \frac{\partial u(\phi)}{\partial \phi_\alpha} \\ \frac{\partial v(\phi)}{\partial \phi_\alpha} \end{Bmatrix} + \begin{Bmatrix} \frac{\partial A(u(\phi);(v(\phi),\phi))}{\partial \phi_\alpha} \\ \frac{\partial B(v(\phi);(u(\phi),\phi))}{\partial \phi_\alpha} \end{Bmatrix} \tag{10}$$

The above is similar to equation 3, where we interchange $R(u^I(\phi), v^I(\phi), \phi) = \{A(u(\phi); (v(\phi), \phi)), B(v(\phi); (u(\phi), \phi))\}^T$ with $\{\frac{\partial A(u(\phi);(v(\phi),\phi))}{\partial \phi_\alpha}, \frac{\partial B(v(\phi);(u(\phi),\phi))}{\partial \phi_\alpha}\}^T$ and interchange $\{\Delta u, \Delta v\}^T$ with $\{\frac{\partial u(\phi)}{\partial \phi_\alpha}, \frac{\partial v(\phi)}{\partial \phi_\alpha}\}^T$. Using this analogy, we may compute the pseudo state $\{\frac{\partial u(\phi)}{\partial \phi_\alpha}, \frac{\partial v(\phi)}{\partial \phi_\alpha}\}^T$ by solving the pseudo problem

$$\begin{aligned} [\frac{\partial B}{\partial v} - \frac{\partial B}{\partial u}\frac{\partial A}{\partial u}^{-1}\frac{\partial A}{\partial v}]\frac{\partial v}{\partial \phi_\alpha} &= [-\frac{\partial B}{\partial \phi_\alpha} + \frac{\partial B}{\partial u}\frac{\partial A}{\partial u}^{-1}\frac{\partial A}{\partial \phi_\alpha}] \\ \frac{\partial u}{\partial \phi_\alpha} &= \frac{\partial A}{\partial u}^{-1}[-\frac{\partial A}{\partial \phi_\alpha} - \frac{\partial A}{\partial v}\frac{\partial v}{\partial \phi_\alpha}] \end{aligned} \tag{11}$$

Note that the above problems use the same operators as the incremental problems c.f. equation 7. Thus, the partial derivatives $\frac{\partial u(\phi)}{\partial \phi_\alpha}$ and $\frac{\partial v(\phi)}{\partial \phi_\alpha}$ may be efficiently computed as the matrices associated with the terms $[\frac{\partial B}{\partial v} - \frac{\partial B}{\partial u}\frac{\partial A}{\partial u}^{-1}\frac{\partial A}{\partial v}]$ and $\frac{\partial A}{\partial u}$ have already been formed and decomposed. To compute the sensitivities, we merely 1) assemble $[-\frac{\partial B}{\partial \phi_\alpha} + \frac{\partial B}{\partial u}\frac{\partial A}{\partial u}^{-1}\frac{\partial A}{\partial \phi_\alpha}]$ and solve for $\frac{\partial v(\phi)}{\partial \phi_\alpha}$ via equation 11.1 and 2) assemble $[-\frac{\partial A}{\partial \phi_\alpha} - \frac{\partial A}{\partial v}\frac{\partial v}{\partial \phi_\alpha}]$ and solve for $\frac{\partial u(\phi)}{\partial \phi_\alpha}$ via equation 11.2. These steps are repeated for each of the $\alpha = 1, L$ design parameters after which equation 9 may be computed for any number of functions. This method is cost efficient in that only pseudo load vectors need be assembled and back substituted into the existing decomposed tangent stiffness matrices; whereas the *primal* analysis used to compute $u(\phi)$ and $v(\phi)$ may require numerous iterations to converge in which the tangent stiffness matrices must be formed and decomposed. Note that the above results are consistent with those obtained by Sobieszczanski-Sobieski et. at. [13, 14].

5 Adjoint Method

In the adjoint method we eliminate the derivatives $\frac{du(\phi)}{d\phi}$ and $\frac{dv(\phi)}{d\phi}$ by defining and solving the appropriate *adjoint* problem. This method relies on the Lagrange multiplier method [2].

Following the Lagrange multiplier method, we define the augmented functional G^*

$$G^*(\phi) \;=\; F(u(\phi), v(\phi), \phi) + \lambda(\phi) R(u(\phi), v(\phi); \phi) \tag{12}$$

Note that $G^*(\phi) = G(\phi)$ since the augmented term is identically zero. The augmented functional is next differentiated with respect to the design, which after some manipulation gives

$$\begin{aligned}\frac{dG^*(\phi)}{d\phi} \;=\; & \{\frac{\partial F(u(\phi), v(\phi), \phi)}{\partial u}, \frac{\partial F(u(\phi), v(\phi), \phi)}{\partial v}\}\left\{\begin{array}{c}\frac{du(\phi)}{d\phi}\\ \frac{dv(\phi)}{d\phi}\end{array}\right\}\\ & + \frac{\partial F(u(\phi), v(\phi), \phi)}{\partial \phi}\\ & + \lambda(\phi)\left\{\left[\begin{array}{cc}\frac{\partial A(u(\phi);(v(\phi),\phi))}{\partial u} & \frac{\partial A(u(\phi);(v(\phi),\phi))}{\partial v}\\ \frac{\partial B(v(\phi);(u(\phi),\phi))}{\partial u} & \frac{\partial B(v(\phi);(u(\phi),\phi))}{\partial v}\end{array}\right]\left\{\begin{array}{c}\frac{du(\phi)}{d\phi}\\ \frac{dv(\phi)}{d\phi}\end{array}\right\}\right.\\ & \left. + \left\{\begin{array}{c}\frac{\partial A(u(\phi);(v(\phi),\phi))}{\partial \phi_\alpha}\\ \frac{\partial B(v(\phi);(u(\phi),\phi))}{\partial \phi_\alpha}\end{array}\right\}\right\}\end{aligned} \tag{13}$$

where $\frac{dG^*(\phi)}{d\phi} = \frac{dG(\phi)}{d\phi}$ since the derivative of the augmented term is identically zero, c.f. equation 10. Further manipulation of the above yields

$$\begin{aligned}\frac{dG^*(\phi)}{d\phi} \;=\; & \{\frac{du(\phi)}{d\phi}, \frac{dv(\phi)}{d\phi}\}\left\{\left\{\begin{array}{c}\frac{\partial F(u(\phi),v(\phi),\phi)}{\partial u}\\ \frac{\partial F(u(\phi),v(\phi),\phi)}{\partial v}\end{array}\right\}\right.\\ & \left. + \left[\begin{array}{cc}\frac{\partial A(u(\phi);(v(\phi),\phi))}{\partial u} & \frac{\partial A(u(\phi);(v(\phi),\phi))}{\partial v}\\ \frac{\partial B(v(\phi);(u(\phi),\phi))}{\partial u} & \frac{\partial B(v(\phi);(u(\phi),\phi))}{\partial v}\end{array}\right]^T\left\{\begin{array}{c}\lambda_u(\phi)\\ \lambda_v(\phi)\end{array}\right\}\right\}\\ & + \frac{\partial F(u(\phi), v(\phi), \phi)}{\partial \phi}\\ & + \{\frac{\partial A(u(\phi);(v(\phi),\phi))}{\partial \phi_\alpha}, \frac{\partial B(v(\phi);(u(\phi),\phi))}{\partial \phi_\alpha}\}\left\{\begin{array}{c}\lambda_u(\phi)\\ \lambda_v(\phi)\end{array}\right\}\end{aligned} \tag{14}$$

where we have partitioned the adjoint state λ according to $\lambda = \{\lambda_u, \lambda_v\}^T$.

To eliminate the state derivatives $\frac{du(\phi)}{d\phi}$ and $\frac{dv(\phi)}{d\phi}$ from equation 14, we equate the coefficient of this term, given by the top two rows of the right-hand side, to zero. Again, we note the similarities of this system and that defined by the incremental problem, c.f. equation 3, where we interchange
$R(u^I(\phi), v^I(\phi), \phi) = \{A(u(\phi); (v(\phi), \phi)), B(v(\phi); (u(\phi), \phi))\}^T$ with $\{\frac{\partial F(u(\phi),v(\phi),\phi)}{\partial u}, \frac{\partial F(u(\phi),v(\phi),\phi)}{\partial v}\}^T$ and interchange $\{\Delta u, \Delta v\}^T$ with $\{\lambda_u(\phi), \lambda_v(\phi)\}$. Using

this analogy, we may compute the adjoint state $\{\lambda_u, \lambda_v\}$ by solving the adjoint problem

$$
\begin{aligned}
[\frac{\partial B}{\partial v} - \frac{\partial B}{\partial u}\frac{\partial A}{\partial u}^{-1}\frac{\partial A}{\partial v}]^T \lambda_v &= [-\frac{\partial F}{\partial v} + \frac{\partial A^T}{\partial v}\frac{\partial A}{\partial u}^{-T}\frac{\partial F}{\partial u}] \\
\lambda_u &= \frac{\partial A}{\partial u}^{-T}[-\frac{\partial F}{\partial u} - \frac{\partial B^T}{\partial u}\lambda_v]
\end{aligned} \tag{15}
$$

Note that the above adjoint problems use the adjoint operators of the incremental problems c.f. equation 7. Thus, like the pseudo problem, the adjoint state $\{\lambda_u, \lambda_v\}$ may be efficiently computed. To compute the sensitivities, we 1) assemble $[-\frac{\partial F}{\partial v} + \frac{\partial A^T}{\partial v}\frac{\partial A}{\partial u}^{-T}\frac{\partial F}{\partial u}]$ and solve for λ_v via equation 15.1 and 2) assemble $[-\frac{\partial F}{\partial u} - \frac{\partial B^T}{\partial u}\lambda_v]$ and solve for λ_u via equation 15.2.

Once the adjoint problem is solved, the implicit response derivatives are annihilated and the explicit sensitivity may be computed from

$$
\begin{aligned}
\frac{dG^*(\phi)}{d\phi} &= \frac{\partial F(u(\phi), v(\phi), \phi)}{\partial \phi} \\
&+ \{\frac{\partial A(u(\phi); (v(\phi), \phi))}{\partial \phi_\alpha}, \frac{\partial B(v(\phi); (u(\phi), \phi))}{\partial \phi_\alpha}\} \left\{ \begin{array}{c} \lambda_u(\phi) \\ \lambda_v(\phi) \end{array} \right\}
\end{aligned} \tag{16}
$$

As opposed to the direct formulation, to compute the sensitivities for each cost/ constraint functional this method requires the assembly and back substitution of one adjoint load into the transpose of the existing decomposed tangent stiffness matrices. Whereas in the direct method we solve one pseudo problem for each of the design variables. Hence, if the design variables outnumber the number of cost/constraint functionals, the adjoint method if preferred and vice versa.

6 General Remarks

We conclude this document with several miscellaneous remarks. If the operators are linear, then only one iteration is required to perform the analysis. In addition, the analysis is valid for any number of systems, however, we only consider two systems as any additional systems may be combined by following the above procedures to form the above two system format. The derivative terms, e.g. $\frac{\partial A}{\partial \phi}$, may be computed analytically or by using the semi-analytical method. When using the semi-analytical method these terms are computed via finite difference approximations; and caution must be exercised to obtain reliable results, c.f. [6]. Finally, note that the tangent operators must be exact (i.e. consistent with the numerical formulation) or the sensitivities will be erroneous.

As previously mentioned, A and B are typically differential operators. If the design parameters are used to define the geometry of the domain complications arise in the sensitivity analysis, as the spatial domain of the problem is now considered to be variable.

Several approaches, notably the material derivative method [3, 4, 5, 10] and the domain parameterization method [1, 3, 4, 8, 12, 16, 17] have been utilized to obtain these so called shape sensitivities.

For weakly coupled systems equation 1.2 becomes

$$\begin{aligned} A(u(\phi);(v(\phi),\phi)) &= 0 \\ B(v(\phi);\phi) &= 0 \end{aligned} \tag{17}$$

so that $\frac{\partial B}{\partial u} = 0$. In this situation, equation 7 becomes

$$\begin{aligned} \frac{\partial B}{\partial v}^{-1}\Delta v &= -B \\ \Delta u &= \frac{\partial A}{\partial u}^{-1}[-A - \frac{\partial A}{\partial v}\Delta v] \end{aligned} \tag{18}$$

so the B system is completely independent of the A system. Thus, the system analysis is greatly simplified and follows the current order, in that we first solve for Δv and then determine Δu. Similarly, for the uncoupled system the pseudo problem reads

$$\begin{aligned} \frac{\partial B}{\partial v}\frac{\partial v}{\partial \phi_\alpha} &= -\frac{\partial B}{\partial \phi_\alpha} \\ \frac{\partial u}{\partial \phi_\alpha} &= \frac{\partial A}{\partial u}^{-1}[-\frac{\partial A}{\partial \phi_\alpha} - \frac{\partial A}{\partial v}\frac{\partial v}{\partial \phi_\alpha}] \end{aligned} \tag{19}$$

Again the analysis is greatly simplified and follows the current order. The uncoupled adjoint problem becomes

$$\begin{aligned} \frac{\partial B}{\partial v}\lambda_v &= [-\frac{\partial F}{\partial v} + \frac{\partial A^T}{\partial v}\frac{\partial A^{-T}}{\partial u}\frac{\partial F}{\partial u}] \\ &= [-\frac{\partial F}{\partial v} - \frac{\partial A^T}{\partial v}\lambda_u] \\ \lambda_u &= -\frac{\partial A^{-T}}{\partial u}\frac{\partial F}{\partial u} \end{aligned} \tag{20}$$

where we used the result of equation 20.3 to obtain equation 20.2. Note here that the simplified analysis requires the evaluation of λ_u prior to the evaluation of λ_v, thus the order of the adjoint problems has reversed. These results are consistent with references [7, 11, 18] for weakly coupled thermoelasticity.

7 Conclusion

Sensitivities for coupled problems have been derived via the direct and adjoint approaches. It is believed that these formulations may formalize a concurrent engineering environment in which the effects of design variations by the production design team are readily assessed by the manufacturing design team and vice versa. The approach here is not without its limitations, it assumes that analysis capabilities exist, that the analyses utilize implicit (or partially implicit) time integration schemes and direct solvers, and that the state is differentable, i.e. discrete design parameters are not allowed.

In the future, we will utilize these approaches to demonstrate the possibility of performing design optimization within the concurrent engineering environment. Several pieces are currently in place, namely the ability to optimize products [19] and their manufacturing process [15, 20].

Finally, the above approach may be used to compute sensitivities for other classical coupled problems, such as advection-diffusion and thermoelasticity; and as noted in [13, 14], they may be used to compute the sensitivities for large finite element models with sub-structuring.

References

1. J.S. Arora, T.H. Lee, J.B. Cardoso, Structure Shape Design Sensitivity Analysis: A Unified Viewpoint, AIAA/ASME/ASCE/ASC 32nd Structure, Structural Dynamics, and Materials Conference, Baltimore, Maryland, pp. 675-683, 1991.
2. A.D. Belegundu, Lagrangian Approach to Design Sensitivity Analysis, Journ. Engrg. Mech. Div. ASCE, 111(5), pp 680-695, 1985.
3. J. Cea, Problems of Shape Optimal Design, Optimization of Distributed Parameter Structures, (E.J. Haug, J. Cea, eds), Vol. II, pp. 1005-1048, Sijthoff & Noordhoff, Alphen aan den Rijn, The Netherlands, 1981.
4. J. Cea, Numerical Methods in Shape Optimal Design, Optimization of Distributed Parameter Structures, (E.J. Haug, J. Cea, eds), Vol. II, pp. 1049-1087, Sijthoff & Noordhoff, Alphen aan den Rijn, The Netherlands, 1981.
5. K.K. Choi, E.J Haug, Shape Design Sensitivity Analysis of Elastic Structures, J. Struct. Mech., Vol. 11(2), pp. 231-269, 1983.
6. G. Cheng, N. Olhoff, New Method of Error Analysis and Detection in the Semi-Analytical Sensitivity Analysis, NATO/DFG ASI Optimization of Large Structural Systems, Berchtesgaden, Ed. G. Rozvany, Sept. 23 - Oct. 4, pp. 234-267.
7. K. Dems, Z. Mroz, Variational Approach to Sensitivity Analysis in Thermoelasticity, J. Therm. Stresses, Vol. 10, pp. 283-306, 1987.
8. R.B. Haber, A New Variational Approach to Structural Shape Design Sensitivity Analysis, NATO ASI Series, Vol. F27, Computer Aided Optimal Design: Structural and Mechanical Systems. Edited by C.A. Mota Soares, Springer-Verlag Berlin Heildelberg 1987.
9. R.T. Haftka, Z. Gurdal, Elements of Structural Optimization, Third Edition, Kluwer Academic Publishers, Boston, 1991.
10. E.J. Haug, K.K. Choi, V. Komkov, Design Sensitivity Analysis of Structural Systems, Academic Press, New York, 1986.
11. R.A. Meric, Material and Load Optimization of Themoelastic Solids. Part I: Sensitivity Analysis, J. Therm. Stresses, Vol, 9, pp. 359-372, 1986.
12. D.G. Phelan, C. Vidal, R.B. Haber, Explicit Sensitivity Analysis of Nonlinear Elastic Systems, Computer Aided Optimum Design: Recent Advances, Brebbia & Hernandez, Eds., Comp. Mechanics Publications/Springer-Verlag, pp. 357-366, 1989.
13. J.Sobieszczanski-Sobieski, Sensitivity of Complex, Internally Coupled Systems, AIAA J., Vol. 28(1), pp. 153-160, 1990.

14. J.Sobieszczanski-Sobieski, C.L. Bloebaum, P. Hajela, Sensitivity of Control-Augmented Structure Obtained by a System Decomposition Method, AIAA J., Vol. 29(2), pp. 264-270, 1991.
15. D.A. Tortorelli, Design Sensitivity Analysis for Nonlinear Dynamic Thermoelastic Systems, Ph.D Thesis, University of Illinois, Urbana-Champaign, 1988.
16. D.A. Tortorelli, R.B. Haber, First Order Design Sensitivities for Transient Conduction Problems by an Adjoint Method, Int. J. Numer. Methods Eng., Vol. 28(4), pp. 733-752, 1989.
17. D.A. Tortorelli, G.S. Subraamani, C.Y. Lu, R.B. Haber, Sensitivity Analysis for Coupled Thermoelastic Systems. Int. J. Solids Structures , Vol. 27(12), pp. 1477-1497, 1991.
18. D.A. Tortorelli, R.B. Haber, S. C-Y Lu, Adjoint Sensitivity Analysis for Nonlinear Dynamic Thermoelastic Systems, AIAA J., Vol. 29(2), pp. 253-263, 1991.
19. D.A. Tortorelli, A Geometric Representation Scheme Suitable for Shape Optimization, Mech. of Struct. Mach., (to appear).
20. D.A. Tortorelli, J.A. Dantzig, Optimal Design of Advanced Materials, IUTAM Symposium on Optimal Design With Advanced Materials, Lyngby, Denmark, Aug. 18-20, 1992.
21. G.N. Vanderplaats, Numerical Optimization Techniques for Engineering Design - With Applications, (McGraw-Hill, New York, 1984).

Configuration Design Sensitivity Analysis for Design Optimization

Sung-Ling Twu[1] and Kyung K. Choi[2]

[1] Chiou Technical Service, Columbus, Indiana 47201, U.S.A.
[2] Center for Simulation and Design Optimization, College of Engineering, The University of Iowa, Iowa City, Iowa 52242, U.S.A.

Abstract: A unified configuration design sensitivity analysis (DSA) is developed for built-up structures that include truss, beam, plane elastic solid, and plate design components. Taking the total variation of the energy equation and using an adjoint variable or direct differentiation method, configuration design sensitivity results for static and eigenvalue response are formulated in terms of the design velocity fields. Displacement, stress, and eigenvalue performance measures are considered. A computational procedure for configuration design optimization is presented, using an established finite element analysis (FEA) code, the continuum DSA method, and an optimization code. A domain displacement method is presented to compute both the domain velocity and the angular velocity. A configuration design optimization of a crane structure is demonstrated using the FEA code ANSYS, continuum DSA, and Pshenichny's linearization method.

Keywords: configuration / sensitivity analysis / built-up structures / finite element analysis / optimization / continuum sensitivity theory / material derivative / shape design variable / orientation design variable / numerical implementation

1 Introduction

Because engineering design is intricate, the design of a mechanical system is specified using parameters of the geometry, material, manufacturing process, simulation, and environment. The layout of a structure is one of the most challenging areas in which to improve system performance because it involve several engineering disciplines and the design of other structural components. Previous research has shown that performance improves more when the configuration of structural components is altered than when the geometry is assumed to be fixed [1-3].

A comprehensive review of the literature in configuration design optimization of skeletal structures has been published by Topping [4]. Also, a general review of the field of structural shape design sensitivity analysis and optimization can be found by Haug [5], Ding [6], and

Haftka [7]. A list of the references connected with configuration design optimization of structural systems is given [8-23].

This paper discusses a unified configuration design sensitivity analysis for design optimization of built-up structures using distributed parameter (continuum) structural theory. Although only the static response is discussed in this paper, the development discussed here has been applied to eigenvalue problems [24] and can be extended to other engineering problems. A numerical implementation of the configuration design sensitivity analysis is shown by using an established finite element code ANSYS [25]. Design sensitivity analysis of a swept wing structure is demonstrated. Finally, configuration design optimization of a crane structure is presented.

2 Derivatives for Configuration Design Sensitivity Analysis

One of the key differences between shape and configuration design sensitivity analyses is the orientation change of the design component. In shape design problems, the domain shape is treated as the design variable, and the orientation of the design component remains fixed. In configuration design, on the other hand, both the domain shape and the orientation of design components are changed. The configuration design change of a design component can be viewed as a dynamic process of moving the design component in three steps: translation, rotation, and shape variation. Since translations, rotations, and shape variation are three independent design changes, configuration design sensitivity can be obtained by adding the design sensitivity results that are obtained from each design perturbation. In fact, since translation of the design component does not contribute to the design sensitivity result of a performance measure [26], configuration design sensitivity can be obtained by adding contributions from the shape variation and rotation of each individual design component in a built-up structure.

For shape variation, a unified shape design sensitivity analysis method has been developed in Ref. 27 using the material derivative idea of continuum mechanics. For the completeness of discussion, the basic definition and results are written here. The process of shape variation, as shown in Fig. 1, is viewed as the process of deforming a continuum medium form Ω to $\Omega_\tau = T_\Omega(\Omega,\tau)$, with τ playing the role of time. A shape design velocity field $\mathbf{V}_\Omega$ is considered as the perturbation of the shape design. Suppose the displacement $\mathbf{z}_\tau(\mathbf{x}_\tau)$ is a smooth solution of the boundary value problem on the perturbed domain Ω_τ. The existence of pointwise material derivative $\dot{\mathbf{z}}_{V_\Omega}$ at $\mathbf{x} \in \Omega$ is shown in Ref. 27 and is defined as

$$\dot{\mathbf{z}}_{V_\Omega}(\mathbf{x}) \equiv \frac{d}{d\tau}\mathbf{z}_\tau(\mathbf{x} + \tau\mathbf{V}_\Omega(\mathbf{x}))\Big|_{\tau=0} = \lim_{\tau\to 0}\frac{\mathbf{z}_\tau(\mathbf{x}+\tau\mathbf{V}_\Omega(\mathbf{x})) - \mathbf{z}(\mathbf{x})}{\tau} \quad (1)$$

If $\mathbf{z}_\tau$ has a regular extension to a neighborhood U_τ of the closed domain $\bar{\Omega}_\tau$, denoted again as $\mathbf{z}_\tau$, then

$$\mathbf{z}_{V_\Omega}(\mathbf{x}) = \mathbf{z}'_{V_\Omega}(\mathbf{x}) + \nabla\mathbf{z}^T\mathbf{V}_\Omega(\mathbf{x}) \tag{2}$$

where

$$\mathbf{z}'_{V_\Omega}(\mathbf{x}) \equiv \lim_{\tau\to 0}\frac{\mathbf{z}_\tau(\mathbf{x}) - \mathbf{z}(\mathbf{x})}{\tau} \tag{3}$$

is the partial derivative of $\mathbf{z}$ due to the shape variation $\mathbf{V}_\Omega$, and $\nabla\mathbf{z} = [\mathbf{z}_{,1}\ \mathbf{z}_{,2}\ \mathbf{z}_{,3}]^T$, where subscript i, i = 1, 2, 3, denotes the derivative with respect to x_i.

Consider a domain functional, defined as an integral over Ω_τ,

$$\psi = \int_{\Omega_\tau} f_\tau(\mathbf{x}_\tau)\, d\Omega_\tau \tag{4}$$

where f_τ is a regular function defined on Ω_τ. If Ω has C^k regularity, the material derivative of ψ at Ω is

$$\begin{aligned}\psi'_{V_\Omega} &= \int_\Omega [f'_{V_\Omega}(\mathbf{x}) + (\nabla f^T\mathbf{V}_\Omega) + f(\nabla^T\mathbf{V}_\Omega)]\, d\Omega \\ &= \int_\Omega [f'_{V_\Omega}(\mathbf{x}) + \mathrm{div}(f\mathbf{V}_\Omega)]\, d\Omega\end{aligned} \tag{5}$$

Similar to shape variation, the orientation change of a design component (as shown in Figs. 2 and 3) is viewed as a process of rotating a continuum medium form Ω to $\Omega_\tau = T_\theta(\Omega,\tau)$, with τ playing the role of time. An orientation design velocity field $\mathbf{V}_\theta$ is considered as the perturbation of the orientation and is normal to the domain of the design component.

Suppose the displacement $\mathbf{z}_\tau(\mathbf{x}_\tau)$ is a smooth solution on the perturbed domain Ω_τ. The pointwise derivative $\dot{\mathbf{z}}_{V_\theta}$ at $\mathbf{x} \in \Omega$ due to the orientation change, if it exists, is defined as

$$\mathbf{z}_{V_\theta}(\mathbf{x}) \equiv \frac{d}{d\tau}\mathbf{z}_\tau(\mathbf{x} + \tau\mathbf{V}_\theta(\mathbf{x}))\bigg|_{\tau=0} = \lim_{\tau\to 0}\frac{\mathbf{z}_\tau(\mathbf{x} + \tau\mathbf{V}_\theta(\mathbf{x})) - \mathbf{z}(\mathbf{x})}{\tau} \tag{6}$$

Next, define a regular extension of $\mathbf{z}_\tau$ in the original local coordinate system x_1-x_2-x_3 as

$$\mathbf{z}_\tau(\mathbf{x}) \equiv \mathbf{z}_\tau(\mathbf{x}_\tau) \tag{7}$$

if $\mathbf{x}_\tau = \mathbf{x} + \tau\mathbf{V}_\theta$. Then, we have the relationship

$$\mathbf{z}_\tau(\mathbf{x}_\tau(\mathbf{x})) = \mathbf{A}_1(\delta\alpha, \delta\beta, \delta\gamma)\mathbf{z}_\tau(\mathbf{x}) \tag{8}$$

for a line design component, and

$$\mathbf{z}_\tau(\mathbf{x}_\tau(\mathbf{x})) = \mathbf{A}_2(\delta\alpha, \delta\beta)\mathbf{z}_\tau(\mathbf{x}) \tag{9}$$

for a surface design component. In Eqs. 8-9, $\mathbf{z}_\tau(\mathbf{x}_\tau(\mathbf{x}))$ denotes evaluating the perturbed solution $\mathbf{z}_\tau$ at location $\mathbf{x}_\tau$ in the original local coordinate system. Using Eqs. 8-9, Eq. 6 becomes

$$\begin{aligned} \mathbf{z}_{V_\theta}(\mathbf{x}) &= \lim_{\tau\to 0}\frac{\mathbf{z}_\tau(\mathbf{x}) - \mathbf{z}(\mathbf{x})}{\tau} + \lim_{\tau\to 0}\frac{\mathbf{z}_\tau(\mathbf{x}_\tau) - \mathbf{z}_\tau(\mathbf{x})}{\tau} \\ &= \mathbf{z}'_{V_\theta} + \lim_{\tau\to 0}\frac{\mathbf{A}_i\mathbf{z}_\tau(\mathbf{x}) - \mathbf{z}_\tau(\mathbf{x})}{\tau} \\ &= \mathbf{z}'_{V_\theta} + \left.\frac{d\mathbf{A}_i}{d\tau}\right|_{\tau=0}\mathbf{z}(\mathbf{x}) \\ &= \mathbf{z}'_{V_\theta} + \tilde{\mathbf{V}}_{i_\theta}\mathbf{z}(\mathbf{x}), \qquad \text{for } i = 1, 2 \end{aligned} \tag{10}$$

where $\mathbf{A}_i$, i = 1, 2 is the rotational transformation matrix and $\tilde{\mathbf{V}}_{i\theta}$, i = 1, 2 contains derivatives of the orientation design velocity field,

$$\mathbf{A}_1(\delta\alpha,\delta\beta,\delta\gamma) = \begin{bmatrix} \cos\delta\alpha & -\sin\delta\alpha & 0 & 0 & 0 & 0 \\ \sin\delta\alpha & \cos\delta\alpha & 0 & 0 & 0 & 0 \\ 0 & 0 & 1 & \sin\delta\alpha & 0 & 0 \\ 0 & 0 & 0 & \cos\delta\alpha & -\sin\delta\alpha & 0 \\ 0 & 0 & 0 & \sin\delta\alpha & \cos\delta\alpha & 0 \\ 0 & 0 & 0 & 0 & 0 & 1 \end{bmatrix} \begin{bmatrix} \cos\delta\beta & 0 & \sin\delta\beta & 0 & 0 & 0 \\ 0 & 1 & 0 & -\sin\delta\beta & 0 & 0 \\ -\sin\delta\beta & 0 & \cos\delta\beta & 0 & 0 & 0 \\ 0 & 0 & 0 & \cos\delta\beta & 0 & \sin\delta\beta \\ 0 & 0 & 0 & 0 & 1 & 0 \\ 0 & 0 & 0 & -\sin\delta\beta & 0 & \cos\delta\beta \end{bmatrix} \begin{bmatrix} 1 & 0 & 0 & 0 & 0 & 0 \\ 0 & \cos\delta\gamma & -\sin\delta\gamma & 0 & 0 & 0 \\ 0 & \sin\delta\gamma & \cos\delta\gamma & 0 & 0 & 0 \\ 0 & 0 & 0 & 1 & 0 & 0 \\ 0 & 0 & 0 & 0 & \cos\delta\gamma & -\sin\delta\gamma \\ 0 & 0 & 0 & 0 & \sin\delta\gamma & \cos\delta\gamma \end{bmatrix} \tag{11}$$

where

$$\left.\begin{aligned} \delta\alpha &= \tau V_{2,1} \\ \delta\beta &= -\tau V_{3,1} \\ \delta\gamma &= -\tau V_{2,3} \text{ or } \tau V_{3,2} \end{aligned}\right\} \tag{12}$$

and

$$\tilde{\mathbf{V}}_{1_\theta} = \begin{bmatrix} 0 & -V_{2,1} & -V_{3,1} & 0 & 0 & 0 \\ V_{2,1} & 0 & V_{2,3} & V_{3,1} & 0 & 0 \\ V_{3,1} & -V_{2,3} & 0 & V_{2,1} & 0 & 0 \\ 0 & 0 & 0 & 0 & -V_{2,1} & -V_{3,1} \\ 0 & 0 & 0 & V_{2,1} & 0 & V_{2,3} \\ 0 & 0 & 0 & V_{3,1} & -V_{2,3} & 0 \end{bmatrix} \tag{13}$$

for a line design component, and

$$\mathbf{A}_2(\delta\alpha,\delta\beta) = \begin{bmatrix} 1 & 0 & 0 & 0 & 0 & 0 \\ 0 & \cos\delta\alpha & -\sin\delta\alpha & 0 & 0 & 0 \\ 0 & \sin\delta\alpha & \cos\delta\alpha & 0 & 0 & -\sin\delta\alpha \\ 0 & 0 & 0 & 1 & 0 & 0 \\ 0 & 0 & 0 & 0 & \cos\delta\alpha & -\sin\delta\alpha \\ 0 & 0 & 0 & 0 & \sin\delta\alpha & \cos\delta\alpha \end{bmatrix} \begin{bmatrix} \cos\delta\beta & 0 & \sin\delta\beta & 0 & 0 & 0 \\ 0 & 1 & 0 & 0 & 0 & 0 \\ -\sin\delta\beta & 0 & \cos\delta\beta & 0 & 0 & \sin\delta\beta \\ 0 & 0 & 0 & \cos\delta\beta & 0 & \sin\delta\beta \\ 0 & 0 & 0 & 0 & 1 & 0 \\ 0 & 0 & 0 & -\sin\delta\beta & 0 & \cos\delta\beta \end{bmatrix} \tag{14}$$

where

$$\left.\begin{aligned} \delta\alpha &= \tau V_{3,2} \\ \delta\beta &= -\tau V_{3,1} \end{aligned}\right\} \tag{15}$$

and

$$\tilde{\mathbf{V}}_{2_\theta} = \begin{bmatrix} 0 & 0 & -V_{3,1} & 0 & 0 & 0 \\ 0 & 0 & -V_{3,2} & 0 & 0 & 0 \\ V_{3,1} & V_{3,2} & 0 & 0 & 0 & -V_{3,1}-V_{3,2} \\ 0 & 0 & 0 & 0 & 0 & -V_{3,1} \\ 0 & 0 & 0 & 0 & 0 & -V_{3,2} \\ 0 & 0 & 0 & V_{3,1} & V_{3,2} & 0 \end{bmatrix} \tag{16}$$

for a surface design component. Note that, like $\mathbf{z}'_{V_\Omega}$, $\mathbf{z}'_{V_\theta}$ commutes with the derivative with respect to x_i. Also, the assumption of a small design perturbation has been used to obtain Eqs. 12 and 15. Using the regular extension of a displacement function and the fact that the determinant of the Jacobian is independent of the orientation change, the first variation of the general functional in Eq. 4 due to the orientation change is

$$\begin{aligned} \psi'_{V_\theta} &= \frac{d}{d\tau}\int_\Omega f_\tau(\mathbf{x}_\tau)\ |\mathbf{J}|\ d\Omega \Big|_{\tau=0} \\ &= \int_\Omega \frac{d}{d\tau} f_\tau(\mathbf{x})\ d\Omega \Big|_{\tau=0} \\ &= \int_\Omega f'_{V_\theta}(\mathbf{x})\ d\Omega \end{aligned} \tag{17}$$

3 Configuration Design Sensitivity Analysis

The variational equation of a boundary value problem for a built-up structure can be written as

$$a_\Omega(\mathbf{z},\bar{\mathbf{z}}) = \ell_\Omega(\bar{\mathbf{z}}), \qquad \text{for all } \bar{\mathbf{z}} \in Z \tag{18}$$

where $a_\Omega(\mathbf{z},\bar{\mathbf{z}})$ is the energy bilinear form, $\ell_\Omega(\bar{\mathbf{z}})$ is the load linear form, and Z is the space of kinematically admissible displacements. Assuming that the energy bilinear and load linear forms are differentiable with respect to the configuration and noting $\dot{\mathbf{z}} = \dot{\mathbf{z}}_{V_\Omega} + \dot{\mathbf{z}}_{V_\theta}$, the first variation of both sides of Eq. 18 is

$$\begin{aligned}[a_\Omega(\mathbf{z},\bar{\mathbf{z}})]' &\equiv a_\Omega(\dot{\mathbf{z}},\bar{\mathbf{z}}) + a_\Omega(\mathbf{z},\dot{\bar{\mathbf{z}}}) + a'_{V_\Omega}(\mathbf{z},\bar{\mathbf{z}}) + a'_{V_\theta}(\mathbf{z},\bar{\mathbf{z}}) \\ &= \ell_\Omega(\dot{\bar{\mathbf{z}}}) + \ell'_{V_\Omega}(\bar{\mathbf{z}}) + \ell'_{V_\theta}(\bar{\mathbf{z}}) \equiv [\ell_\Omega(\bar{\mathbf{z}})]'\end{aligned} \tag{19}$$

Using the fact that $\dot{\bar{\mathbf{z}}} \in Z$ and $a_\Omega(\mathbf{z},\dot{\bar{\mathbf{z}}}) = \ell_\Omega(\dot{\bar{\mathbf{z}}})$, Eq. 19 becomes

$$a_\Omega(\dot{\mathbf{z}},\bar{\mathbf{z}}) = \ell'_{V_\Omega}(\bar{\mathbf{z}}) + \ell'_{V_\theta}(\bar{\mathbf{z}}) - a'_{V_\Omega}(\mathbf{z},\bar{\mathbf{z}}) - a'_{V_\theta}(\mathbf{z},\bar{\mathbf{z}}), \qquad \text{for all } \bar{\mathbf{z}} \in Z \tag{20}$$

Consider a performance measure in integral form as

$$\psi = \int_{\Omega_\tau} g(z_{i\tau}, \nabla z_{i\tau}, z_{i\tau,jk})\, d\Omega_\tau \tag{21}$$

where function g is continuously differentiable with respect to its arguments; $z_{i\tau}$ is the ith component of the displacement vector $\mathbf{z}_\tau$; and $z_{i\tau,jk} = \partial^2 z_{i\tau}/\partial x_{j\tau}\partial x_{k\tau}$, j, k = 1, 2, 3, denotes the second derivative of $z_{i\tau}$. If only the first derivative of $z_{i\tau}$ appears in the performance measure, the second derivative of $z_{i\tau}$ in Eq. 21 can be ignored.

Using Eqs. 5 and 17 and the fact that $\dot{z}_i = \dot{z}_{iV_\Omega} + \dot{z}_{iV_\theta}$, the first variation of Eq. 21 becomes

$$\begin{aligned}\psi' = \int_\Omega [&g_{z_i}\dot{z}_i + g_{\nabla z_i}\nabla\dot{z}_i + g_{z_{i,jk}}\dot{z}_{i,jk} - g_{z_i}(\tilde{\mathbf{V}}_\theta\mathbf{z})_i - g_{\nabla z_i}\nabla(\tilde{\mathbf{V}}_\theta\mathbf{z})_i - g_{z_{i,jk}}(\tilde{\mathbf{V}}_\theta\mathbf{z})_{i,jk} \\ &- g_{z_i}(\nabla z_i^T\mathbf{V}_\Omega) - g_{\nabla z_i}\nabla(\nabla z_i^T\mathbf{V}_\Omega) - g_{z_{i,jk}}(\nabla z_i^T\mathbf{V}_\Omega)_{,jk} + \nabla g^T\mathbf{V}_\Omega + g(\nabla^T\mathbf{V}_\Omega)]\, d\Omega\end{aligned} \tag{22}$$

where $(\tilde{\mathbf{V}}_\theta\mathbf{z})_i$ denotes the ith component of the vector $\tilde{\mathbf{V}}_\theta\mathbf{z}$. In the direct differentiation method, Eq. 20 is solved for $\dot{\mathbf{z}}$ with the given design velocity fields $\mathbf{V}_\Omega$ and $\mathbf{V}_\theta$. Once the original response $\mathbf{z}$ and the first variation $\dot{\mathbf{z}}$ are obtained, the design sensitivity expression in Eq. 22 can be evaluated.

In the adjoint variable method, an adjoint equation is defined as

$$a_\Omega(\lambda,\bar{\lambda}) = \int_\Omega [g_{z_i}\bar{\lambda}_i + g_{\nabla z_i}\nabla\bar{\lambda}_i + g_{z_{i,jk}}\bar{\lambda}_{i,jk}]\, d\Omega, \qquad \text{for all } \bar{\lambda} \in Z \tag{23}$$

The adjoint equation is solved for the adjoint response λ. Since $\dot{\mathbf{z}}$ is in the space of kinematically admissible displacements, Eq. 23 can be evaluated at $\bar{\lambda} = \dot{\mathbf{z}}$ and Eq. 20 at $\bar{\mathbf{z}} = \lambda$, to obtain

$$\begin{aligned} \psi' = {} & \ell'_{V_\Omega}(\lambda) + \ell'_{V_\theta}(\lambda) - a'_{V_\Omega}(\mathbf{z},\lambda) - a'_{V_\theta}(\mathbf{z},\lambda) - \int_\Omega [g_{z_i}(\tilde{\mathbf{V}}_\theta \mathbf{z})_i + g_{\nabla z_i}\nabla(\tilde{\mathbf{V}}_\theta \mathbf{z})_i \\ & + g_{z_{i,jk}}(\tilde{\mathbf{V}}_\theta \mathbf{z})_{i,jk} + g_{z_i}(\nabla z_i^T \mathbf{V}_\Omega) + g_{\nabla z_i}\nabla(\nabla z_i^T \mathbf{V}_\Omega) + g_{z_{i,jk}}(\nabla z_i^T \mathbf{V}_\Omega)_{,jk}]\, d\Omega \\ & + \int_\Omega [\nabla g^T \mathbf{V}_\Omega + g(\nabla^T \mathbf{V}_\Omega)]\, d\Omega \end{aligned} \tag{24}$$

Once the design velocity fields $\mathbf{V}_\Omega$ and $\mathbf{V}_\theta$ are defined, with the original response $\mathbf{z}$ and the adjoint response λ, the configuration design sensitivity expression in Eq. 24 can be calculated.

4 Numerical Implementation of Configuration Design Sensitivity Analysis

One virtue of the continuum design sensitivity analysis method is that its numerical implementation is independent of the computer program for finite element analysis [28]. The analysis results obtained from an established finite element code can be used to carry out the design sensitivity analysis. The computational procedure of the configuration design sensitivity analysis is shown in Fig. 4. The overall procedure is similar to the computational procedure for sizing and shape design sensitivity analyses. Design parameterizations of both the domain shape and the orientation of the design component are required for configuration design sensitivity analysis. As mentioned in the previous section, the regularity requirements have to be considered in the selection of the design velocity fields. The domain shape design velocity $\mathbf{V}_\Omega$ may be generated using the boundary displacement method [29], and the orientation design velocity $\mathbf{V}_\theta$ can be calculated using the orientation design parameterization, such as the grid point locations of the design component.

To demonstrate the accuracy of configuration design sensitivity, the swept wing model shown in Fig. 5 is treated in this section. The wing is made of aluminum with Young's modulus E = 10.6 Mpsi, and Poisson's ratio n = 0.3 is subjected to a uniform pressure of

0.556 psi acting on top of the skin panels. The cross sectional areas are 0.02 in^2 for longitudinal spar caps and 0.2 in^2 for vertical spar caps. The thickness of the skin panels on the first half of the wing and all of the shear panels (vertical panels) is 0.2 in. The thickness of the skin panels on the second half (wing tip) of the wing is 0.1 in. Because of the symmetry of the structure and loading, only half of the wing box is analyzed. The finite element model is created using the ANSYS finite element code. The model consists of 60 3-D truss elements, STIF8, and 130 membrane elements, STIF41. This model has 88 nodal points and 160 degrees of freedom.

For a configuration design change, the tip of the swept wing is moved forward as shown in Fig. 5. The design velocity fields are defined so that all ribs (shear panels) that are parallel to the y-axis remain parallel while moving. The orientation of the spars, shear panels in the x-direction, and the skin panels will then be rotated accordingly so that the shear panels which are plane will remain as planes. Based on the perturbation of nodal points, linear and bilinear shape functions are used to obtain the shape design velocity field for the line and surface design components, respectively.

The displacements at the tip of the wing, the averaged axial stress on the spar caps, and the averaged von Mises stress on the skin panels and the shear panels are selected as the performance measures. For averaged stress performance measures, the general performance measure given in Eq. 21 can be written explicitly as

$$\psi = \iint_{\Omega} g(\sigma^{ij}(\mathbf{z}))m_p \, d\Omega \tag{25}$$

where σ^{ij} denotes the stress tensor, and m_p is the characteristic function. For the averaged axial stress performance measure,

$$g(\sigma^{ij}(\mathbf{z})) = \sigma^{11} = Ez_{1,1} \tag{26}$$

and for the averaged von Mises stress performance measure,

$$g(\sigma^{ij}(\mathbf{z})) = \sqrt{(\sigma^{11})^2 - \sigma^{11}\sigma^{22} + (\sigma^{22})^2 + 3(\sigma^{12})^2} \tag{27}$$

The adjoint equation of Eq. 25 is

$$a_{\Omega}(\lambda,\bar{\lambda}) = \iint_{\Omega_p} g_{\sigma^{ij}}[\sigma^{ij}(\bar{\lambda})]m_p \, d\Omega, \qquad \text{for all } \bar{\lambda} \in Z \tag{28}$$

and the final design sensitivity expression of Eq. 25 is

$$\psi' = \ell'_{V_\Omega}(\lambda) + \ell'_{V_\theta}(\lambda) - a'_{V_\Omega}(\mathbf{z},\lambda) - a'_{V_\theta}(\mathbf{z},\lambda)$$
$$- \iint_{\Omega_p} g_\sigma{}^{ij}[\sigma^{ij}(\nabla \mathbf{z}^T \mathbf{V}_\Omega)] m_p \, d\Omega + \iint_{\Omega_p} (\nabla g^T \mathbf{V}_\Omega + g \, \mathrm{div} \mathbf{V}_\Omega) m_p \, d\Omega$$
$$- \iint_{\Omega_p} g \, m_p \, d\Omega \iint_{\Omega_p} \mathrm{div} \mathbf{V}_\Omega m_p \, d\Omega - \iint_{\Omega_p} g_\sigma{}^{ij}[\sigma^{ij}(\tilde{\mathbf{V}}_\theta \mathbf{z})] m_p \, d\Omega \qquad (29)$$

Note that the differentials $a'_{V_\Omega}(\mathbf{z},\lambda)$, $a'_{V_\theta}(\mathbf{z},\lambda)$, $\ell'_{V_\Omega}(\lambda)$, and $\ell'_{V_\theta}(\lambda)$ in Eq. 29 are obtained by adding the contributions of each truss and membrane design component. An explicit expression of these differentials can be found in Ref. 24. The configuration design sensitivity results of displacement, and averaged axial stress, averaged von Mises stress are presented in Table 1. The results of Table 1 show an accurate prediction of $\psi'(\mathbf{b})$ compared with the prediction of the central finite difference method $\Delta\psi(\mathbf{b})$.

5 Configuration Design Optimization

The crane structure shown in Fig 6 is considered for configuration design. The element cross-sectional areas are linked as follows: $A_1 = A_4 = A_8 = A_{12} = A_{16} = 10.71$ in^2; $A_2 = A_6 = A_{10} = A_{14} = A_{18} = 15.19$ in^2; $A_3 = A_7 = A_{11} = A_{15} = 1.94$ in^2; $A_5 = A_9 = A_{13} = A_{17} = 5.19$ in^2. Young's modulus is E = 10 Mpsi. The material density is given as $\rho = 0.1$ lb/in^3. Finite element analysis is performed using the ANSYS 2-D truss element, STIF1, which has 18 truss elements, 11 nodal points, and 18 degrees of freedom.

The design problem of the crane structure is stated as follows: "Design the layout of the structural members, subject to the geometric limitations, such that the weight of the crane will be minimized and none of the structural members will yield under a loading condition." Based on this statement, the optimal design problem can be formulated as

$$\min_{\mathbf{b}} \psi_0(\mathbf{b}) \qquad (30)$$

subject to

$$\sigma_i \le \sigma_0, \quad i = 1 \text{ to } 18 \qquad (31)$$

and

$$\mathbf{b}_\ell \le \mathbf{b} \le \mathbf{b}_u \qquad (32)$$

where ψ_0 is the total weight of the crane; $\mathbf{b} = [X_1(3), X_2(3), X_1(5), X_2(5), X_1(7), X_2(7), X_1(9), X_2(9)]^T$ is the design variable, which consists of the X_1 and X_2 coordinates of nodal points 3, 5, 7, and 9; σ_i is the axial stress for the ith element; $\sigma_0 = 20$ ksi is the given allowable stress; and $\mathbf{b}_\ell = -10$ in and $\mathbf{b}_u = 1500$ in are geometric limitations.

The design problem of Eqs. 30-32 is optimized using the LINRM linearization method [30]. For numerical design sensitivity analysis, a linear shape design velocity field is used for each truss design component, and derivatives of orientation design velocity are computed based on the orientation change of the design component. The results of the optimal crane structure are summarized in Table 2. The weight of the crane is reduced from 4322.77 lbs to 3904.90 lbs in the final design, which is 9.7% lower than the initial design. Also, large constraint violations which exist initially have been removed in the final design. This indicates that a much better design has been obtained with a significant saving of material.

The convergence histories of the objective function and the ℓ^2 norm of the design perturbation are shown in Figs. 7 and 8. A slow convergence near the optimal point is shown in Figs. 7 and 8 using LINRM. About 8.2% of the cost reduction is achieved in the first 7 iterations. Between the 8th and 75th iteration, only 1.4% of the cost reduction is obtained. After the 75th iteration, the value of the objective function stays almost the same. This kind of slow convergence has been revealed in other studies of configuration design optimization [2, 17]. Hansen and Vanderplaats [22] presented an optimal configuration of the same crane structure using an approximate structural analysis based on first-order Taylor series expansions of the member forces. They used the super-DOT optimization packages and, as shown in Table 2, obtained faster convergence than was obtained in this section using LINRM. The values of the objective function at the optimal point are nearly identical.

The optimal configurations are similar in Figs. 9(b) and 9(e). The configurations obtained from LINRM at the 7th, 20th, and 95th iterations are shown in Figs. 9(c), 9(d), and 9(e), respectively. As shown in Figs. 9(d) and 9(e), the shape at the 20th iteration is close to the final configuration, and yet the program takes another 75 iterations to get the final design. The results obtained from LINRM tend to have a sharp tip with evenly distributed cross members.

Acknowledgement: Research supported by NSF-Army-NASA Industry/University Cooperative Research Center for Simulation and Design Optimization of Mechanical Systems.

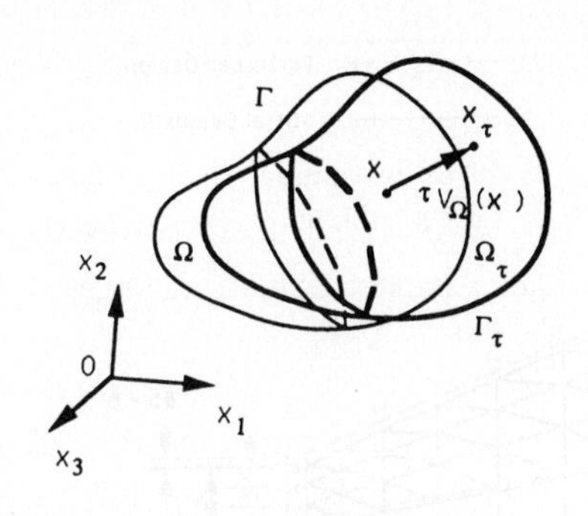

Figure 1 Domain Shape Variation

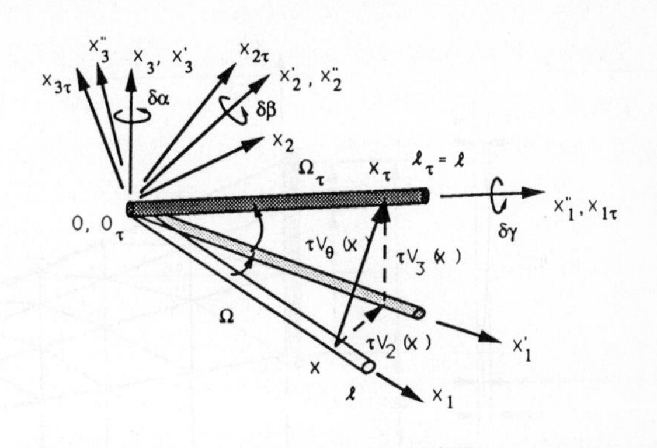

Figure 2 Orientation Change of a Line Design Component

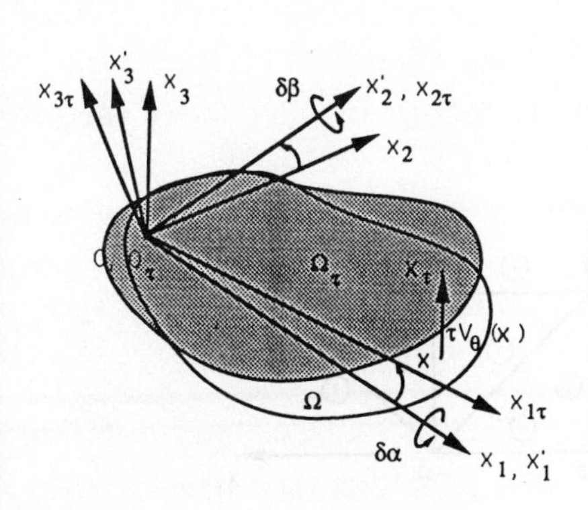

Figure 3 Orientation Change of a Surface Design Component

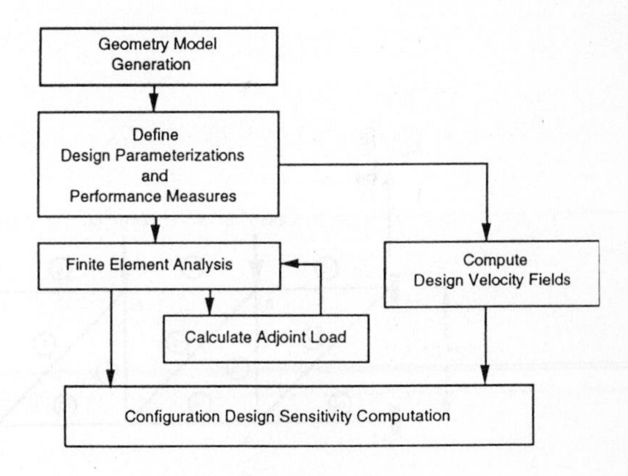

Figure 4 Flow Chart of Configuration Design Sensitivity Analysis

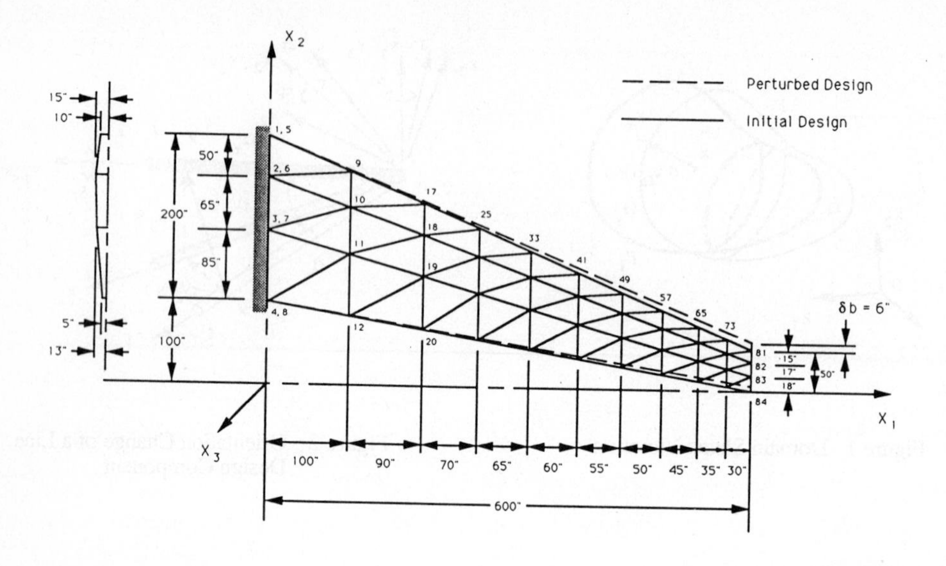

Figure 5 Swept Wing Model

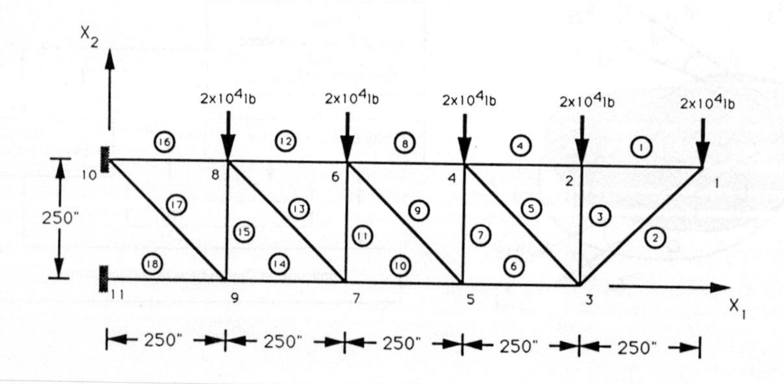

Figure 6 18-Bar Crane Structure

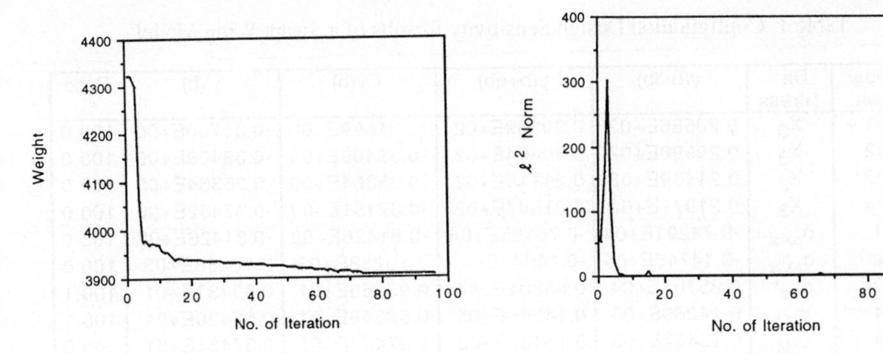

Figure 7 Convergence History of the Objective Function

Figure 8 Convergence History of ℓ^2 Norm of the Design Perturbation

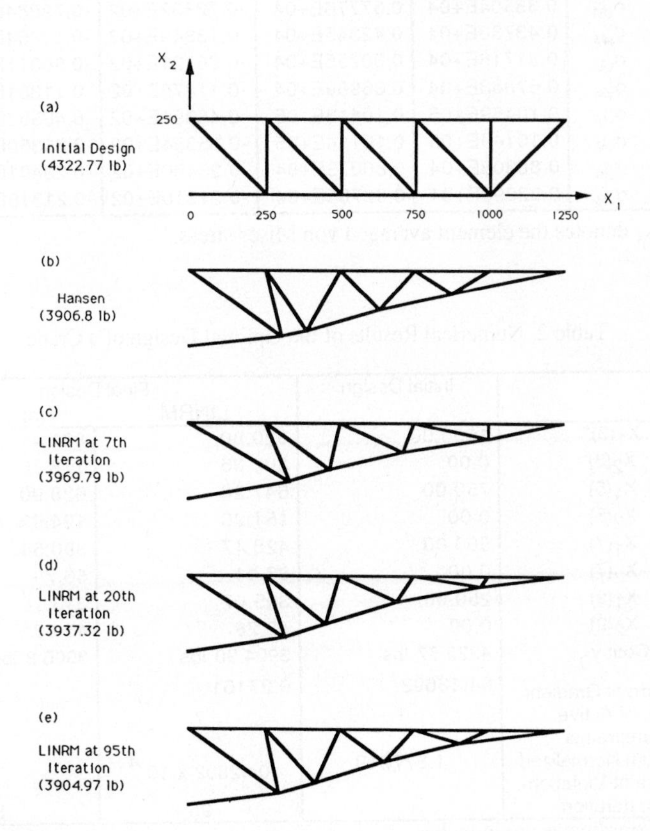

Figure 9 Comparison between Initial and Final Configuration for Optimal Design of a Crane Structure

Table 1 Configuration Design Sensitivity Results of a Swept Wing Model

Node/ Elem.	Dir./ Stress	y(b-db)	y(b+db)	Dy(b)	y'(b)	Ratio %
81	X_3	0.20586E+02	0.20269E+02	-0.31699E+00	-0.31700E+00	100.0
82	X_3	0.20998E+02	0.20664E+02	-0.33409E+00	-0.33409E+00	100.0
83	X_3	0.21469E+02	0.21115E+02	-0.35364E+00	-0.35364E+00	100.0
84	X_3	0.21971E+02	0.21597E+02	-0.37451E+00	-0.37452E+00	100.0
1	σ_{axial}	-0.74291E+04	-0.75105E+04	-0.81426E+02	-0.81426E+02	100.0
2	σ_{axial}	-0.14745E+05	-0.14643E+05	0.10288E+03	0.10288E+03	100.0
3	σ_{vM}*	0.85701E+04	0.85801E+04	0.99366E+01	0.99431E+01	100.1
4	σ_{vM}	0.14260E+05	0.14266E+05	0.52348E+01	0.52420E+01	100.1
5	σ_{vM}	0.13482E+05	0.13486E+05	0.37437E+01	0.37431E+01	100.0
6	σ_{vM}	0.14530E+05	0.14424E+05	-0.10514E+03	-0.10514E+03	100.0
7	σ_{vM}	0.14479E+05	0.14395E+05	-0.84104E+02	-0.84108E+02	100.0
8	σ_{vM}	0.75624E+04	0.74249E+04	-0.13746E+03	-0.13747E+03	100.0
9	σ_{vM}	0.38504E+04	0.37776E+04	-0.72837E+02	-0.72864E+02	100.0
10	σ_{vM}	0.43730E+04	0.42345E+04	-0.13849E+03	-0.13854E+03	100.0
11	σ_{vM}	0.31716E+04	0.30755E+04	-0.96121E+02	-0.96311E+02	100.2
63	σ_{vM}	0.67083E+04	0.66969E+04	-0.11379E+02	-0.11381E+02	100.0
64	σ_{vM}	0.10459E+05	0.10413E+05	-0.46390E+02	-0.46393E+02	100.0
65	σ_{vM}	0.10149E+05	0.10114E+05	-0.35354E+02	-0.35360E+02	100.0
66	σ_{vM}	0.88308E+04	0.88073E+04	-0.23486E+02	-0.23491E+02	100.0
67	σ_{vM}	0.92998E+04	0.92785E+04	-0.21310E+02	-0.21315E+02	100.0

*Note: σ_{vM} denotes the element averaged von Mises stress.

Table 2 Numerical Results of the Optimal Design of a Crane

	Initial Design	Final Design	
		LINRM	Hansen [22]
$X_1(3)$*	1000.00	920.80	881.42
$X_2(3)$	0.00	202.36	178.76
$X_1(5)$	750.00	647.28	628.90
$X_2(5)$	0.00	151.20	124.92
$X_1(7)$	500.00	428.17	390.54
$X_2(7)$	0.00	83.11	66.79
$X_1(9)$	250.00	305.07	313.16
$X_2(9)$	0.00	42.28	45.03
Cost y_0	4322.77 lbs	3904.90 lbs	3906.8 lbs
ℓ^2 Norm of Gradient	54.13692	0.27161	-
No. of Active Constraints	1	7	-
Maximum Normalized Constraint Violation	1.577319	0.42802×10^{-4}	-
No. Iteration		95	8

*Note: Coordinates are in inches.

References

1. Saka, M.P. and Attili, B., "Shape Optimization of Space Trusses," *The Proceedings of the International Conference on the Design and Construction of Non-Conventional Structures* (Ed. B.H.V. Topping), Civil-Comp Press, London, 1987, pp. 115-121.
2. Felix, J. and Vanderplaats, G.N., "Configuration Optimization of Trusses Subject to Strength, Displacement, and Frequency Constraints," *ASME Journal of Mechanisms, Transmissions, and Automation in Design*, Vol. 109, No. 2, June 1987, pp. 233-241.
3. Sandren, E., Lee, H., and El-Sayed, M., "Optimization Design of Mechanical Components with Sizing, Configurational and Topological Consideration," The 1989 ASME Design Technical Conferences - 15th Design Automation Conference, *Advances in Design Automation*, Vol. 2, 1989, pp. 101-110.
4. Topping, B.H.V., "Shape Optimization of Skeletal Structures: A Review," *ASCE Journal of the Structural Division*, Vol. 109, 1983, pp. 1933-1952.
5. Haug, E.J., "A Review of Distributed Parameter Structural Optimization Literature," *Optimization of Distributed Parameter Structures*, (Ed. E.J. Haug and J. Cea), Sijthoff & Noordhoff, Alphen Aan Den Rijn, Netherlands, 1980.
6. Ding, Y.L., "Shape Optimization of Structures: A Literature Survey," *Computers and Structures*, Vol. 24, 1986, pp. 985-1004.
7. Haftka, R.T. and Adelman, H.M., "Recent Developments in Structural Sensitivity Analysis," *Structural Optimization*, Vol. 1, 1989, pp. 137-151.
8. Mitchell, A.G.M., "The Limits of Economy of Materials in Frame Structures," *Philosophical Magazine*, Series 6, Vol. 8, No. 47, 1904, pp. 589-597.
9. Maxwell, J.C., "On Reciprocal Figures, Frames and Diagrams of Forces," *Scientific Papers*, Cambridge University Press, United Kingdom, Vol. 2, 1980, pp. 175-177.
10. Cox, H.L., "The Design of Structures of Least Weight," Pergamon Press, London, United Kingdom, 1965.
11. Owen, J.B.B., "The Analysis and Design of Light Structures," Edward Arnold, London, United Kingdom, 1974.
12. Chan, H.S.Y., "Optimum Structural Design and Linear Programming," College of Aeronautics, Cranfield, United Kingdom, Report No. 175, September, 1964.
13. Hemp, W.S., "Optimum Structures," Clarendon Press, Oxford, United Kingdom, 1973.
14. Dorn, W.S., Gomory, R.E., and Greenberg, H.J., "Automatic Design of Optimal Structures," *Journal de Mecanique*, Vol. 3, No. Mars, France, 1964.
15. Dobbs, M.W. and Felton, L.P., "Optimization of Truss Geometry," *ASCE Journal of the Structural Division*, Vol. 95, No. ST10, Oct., 1969, pp. 2105-2118.
16. Pedersen, P., "Optimal Joint Positions for Space Trusses," *ASCE Journal of the Structural Division*, Vol. 99, No. ST12, Dec., 1973, pp. 2459-2476.
17. Imai, K., *Configuration Optimization of Trusses by the Multiplier Method*, Report No. UCLA-ENG-7842, Mechanics and Structures Department, School of Engineering and Applied Science, University of California, Los Angeles, 1978.
18. Imai, K. and Schmit, L.A., "Configuration Optimization of Trusses," *ASCE Journal of the Structural Division*, Vol. 107, 1981, pp. 745-756.
19. Lin, J.H., Che, W.Y., and Yu, Y.S., "Structural Optimization on Geometrical Configuration and Element Sizing with Static and Dynamical Constraints," *Computers and Structures*, Vol. 15, No. 5, 1982, pp. 507-515.
20. Kapoor, M.P. and Kumarasamy, K., "Optimum Configuration of Transmission Towers in Dynamic Response Regime," *Proceedings, International Symposium on Optimum Structural Design*, Tucson, AZ, Oct., 1981.
21. Kuritz, S.P., *Configuration Optimization of Trusses Using Convex Linearization Techniques*, Master Thesis, University of California, Los Angeles, 1986.
22. Hansen, S.R. and Vanderplaats, G.N., "An Approximation Method for Configuration Optimization of Trusses," *AIAA/ASME/ASCE/AHS 29th Structures, Structural Dynamics, and Materials Conference, Part 3*, 1988, No. 88-2432.
23. Zhou, M., "Geometrical Optimization of Trusses by a Two-Level Approximation Concept," *Structural Optimization*, Vol. 1, 1989, pp. 235-240.
24. Twu, S.L. and Choi, K.K., *Configuration Design Sensitivity Analysis and Optimization of Built-up Structures*, Technical Report R-91, Center for Simulation and Design Optimization, The University of Iowa, January 1991.
25. DeSalvo, G.J. and Swanson, J.A., *ANSYS Engineering Analysis System User's Manual, Vol. I and II*, Swanson Analysis System, Inc., P.O. Box 65, Houston, PA, 1985.

26. Lee, H.G., Choi, K.K., and Haug, E.J., *Design Sensitivity Analysis and Optimization of Built-Up Structures*, Technical Report 84-12, Center for Computer Aided Design, The University of Iowa, December 1984.
27. Haug, E.J., Choi, K.K., and Komkov, V., *Design Sensitivity Analysis of Structural Systems*, Academic Press, New York, 1986.
28. Choi, K.K., Santos, J.L.T., and Frederick M.C., "Implementation of Design Sensitivity Analysis with Existing Finite Element Codes," *ASME Journal of Mechanisms, Transmissions, and Automation in Design*, Vol. 109, No. 3, 1987, pp. 385-391; also presented at the 11th ASME Design Automation Conference, September 1985, Paper No. 85-DET-70.
29. Yao, T.M. and Choi, K.K., "3-D Shape Optimal Design and Automatic Finite Element Regridding," *Int. J. of Numerical Methods in Engineering*, Vol. 28, No. 2, 1989, pp. 369-384.
30. Pshenichny, B.N. and Danilin, Y.M., *Numerical Methods in External Problems*, MIR Publishers, Moscow, 1978.

Shape Design Sensitivity Analysis and What-if Tool for 3-D Design Applications

Kuang-Hua Chang and Kyung K. Choi

Center for Simulation and Design Optimization of Mechanical Systems and Department of Mechanical Engineering, The University of Iowa, Iowa City, Iowa 52242, USA

Abstract: Shape design parameters that govern the geometric shape of a structural component are the most effective way to improve design of 3-D elastic solid components. Four major characteristics, however, that are unique to the shape design problem make it more complicated than the traditional sizing problem; (1) it is difficult to retain the accuracy of finite element analysis results for a design model whose geometry changes during the design process, (2) it is difficult to handle the sophisticated shape design parameterization and update geometric shape, (3) efficient computation of shape design sensitivity information for a large scale problem is difficult to achieve, and (4) visualization of important design sensitivity information and automation of shape design processes to provide an effective design environment is not available.

To support Concurrent Engineering activities, design parameters defined in the CAD model are the most important common data shared by various engineering disciplines. Design sensitivity analysis that computes structural responses with respect to design parameters defined in the CAD model is a critical step to support Concurrent Engineering activities.

This paper presents a methodology to support structural shape design for 3-D elastic solid components, using the geometric modeler PATRAN [1]. The proposed methodology overcomes four major difficulties of shape design; accuracy, integration, efficiency, and effectiveness. A clevis and a turbine blade examples are given to demonstrate capabilities of the design tool.

Keywords: shape design / sensitivity analysis / concurrent engineering / finite element analysis / material derivative / shape design variable / numerical implementation / user interface / CAD

1 Introduction

Shape design problems in which the structural geometry is to be determined have attracted the attention of both academic and industrial researchers during the past fifteen years. The reason for this interest is that traditional sizing design variables have proven less effective than shape design variables for many applications [2]. Four major characteristics that are unique to shape design make it more complicated than traditional sizing design. The first difficulty is to maintain the accuracy of finite element analysis results during the design process. In analysis of a 3-D structural component with a sophisticated geometric shape, a refined finite element mesh is necessary to accurately capture stress concentration. Thus, the dimension of finite element models tends to be very large for 3-D structural components. Furthermore, the finite element mesh is continuously changing since the geometry of the design model is changing during the design process [2,3]. Maintaining accuracy of finite element analysis is difficult since finite element meshes may become distorted due to a change in structural shape. Currently, most of the CAD modelers in the market have automatic mesh generation capability that reduces the finite element modeling effort. However, lack of error analysis and mesh adaptation capability and supporting small class of finite element types present major disadvantages of moving shape DSA to CAD modelers.

Second, shape design parameterization is more complicated and difficult to handle than sizing design parameterization. The optimum shape is highly dependent on the geometric parameterization method. An inappropriate design parameterization may result in an impractical design. On the other hand, changing the geometric shape of the design model to reflect successive changes in design parameters is a tedious, complicated, and inefficient process. Manual updates of geometric shape and finite element meshes are quite impractical. An efficient automatic shape updating procedure is needed for supporting the shape design process.

The intensive computations associated with shape design sensitivity analysis constitute the third difficulty. Design sensitivity evaluation and function updates, which involve generation of design velocity fields for all shape design parameters and finite element analysis, respectively, are computationally intensive tasks. An efficient and reliable computational algorithm that fully utilizes the computational power of a computer network is necessary to speed up computations.

The fourth and most difficult problem is the development of an interactive design process that provides effective visualization of design sensitivity information and supports automation of the shape design process. An integrated and automated design environment must be developed to minimize difficulty in handling tedious routine operations, such as control of execution of programs and data manipulation. This shape design environment must also provide tools for easy interpretation of design results.

The proposed methodology overcomes four major difficulties of shape design; accuracy, integration, efficiency, and effectiveness, using the following strategies: (1) implementation of error analysis and interactive mesh adaptation methods to ensure accuracy of finite element analyses in the shape design process, (2) development of a systematic design method, including shape design parameterization method to handle a large class of structural shape design problems, (3) implementation of an integrated shape design computation procedure to compute design sensitivity information for a large class of problems, (4) development of an efficient shape design sensitivity analysis method, utilizing the shape design component approach, to solve large scale problems, (5) implementation of an efficient distributed computational algorithm for shape design sensitivity analysis, (6) demonstration of a design process that utilizes "sensitivity display" and "what-if study" design steps to obtain improved designs, and (7) development of a menu-driven user interface to integrate and ease the shape design process.

Development of a shape design tool involves many different disciplines. Capabilities that are necessary to develop an integrated shape design tool include: (1) geometric modeling, (2) finite element mesh generation, (3) finite element error analysis and adaptive mesh refinement, (4) shape design parameterization, (5) velocity field and design sensitivity computation, (6) geometric shape and finite element mesh updates, (7) computational speedup, (8) system integration, and (9) data management. The major challenge is to bring up-to-date technologies in each discipline and appropriate methodologies that are developed or adapted in the design process to overcome the four difficulties noted above.

A major concern for shape design is that the concept, methodology, and data structure in CAD communities are not yet unified. A single methodology for shape design that will support several CAD modelers is not known at this moment. Moreover, lack of CAD connection to dedicate finite element analysis codes creates another difficulty in developing design sensitivity analysis capabilities directly to the CAD modelers. In spite of these difficulties, bringing the various disciplines to bear on development of a structural shape design methodology using PATRAN, which has strong connection to major finite element analysis codes, is the major goal of this study.

2 Shape Design Sensitivity Analysis and What-if Tool

From a functionality point of view, three design stages: pre-processing, design sensitivity analysis, and post-processing, have been developed in the Shape Design Sensitivity Analysis and What-if Tool (SDSW) to support shape design process. The pre-processing design stage includes geometric and finite element modeling, design parameterization, and definition of performance measures. In this design stage, finite element error analysis and mesh adaptation methods are developed and implemented to ensure accuracy of finite element analysis results. A design parameterization method is developed to support the design process.

In the design sensitivity analysis stage, the SDSW will compute design derivatives of pre-defined structural performance measures with respect to shape design parameters that are defined in the PATRAN geometric model. For this stage, a systematic and general method for boundary and domain velocity field computations is developed. The SDSW uses velocity field information, post-processing finite element data, and adjoint or sensitivity reanalysis results to compute design sensitivity information. Generation of design velocity fields and computation of design sensitivity information are distributed over a network of computers to speed up computation [4]. All design sensitivity computation processes are integrated and automated, using UNIX shell scripts [5]. Consequently, the sophisticated and complicated shape design sensitivity computation can be launched by simply clicking on a menu icon from the user interface.

By providing "sensitivity display" and "what-if study" capabilities, the post-processing stage will allow the design engineer to manipulate and display design sensitivity information and assist in finding improved designs. The entire shape design process is integrated by employing a menu hierarchy system using engineering spreadsheet [6] and OSF-Motif [7].

3 Pre-Processing Design Stage

The major goals to be achieved in the pre-processing design stage are: (1) constructing a design model based on the geometric concept, (2) bringing the geometric concept into the SDSW through design parameterization, (3) translating the design model into a finite element analysis model for design evaluation, and (4) defining performance measures for which design sensitivity information is to be computed. To achieve these goals, the following steps must be performed: geometric modeling, design parameterization, mesh generation, finite element modeling, finite element analysis, finite element error analysis, visualization of finite element error, mesh adaptation, visualization of the finite element mesh, performance measure definition, and visualization of structural performance. The basic design process and computation procedures of the pre-processing design stage are shown in Figure 1.

3.1 Shape Design Parameterization

The shape design parameterization method developed in this study deals with geometric features. A geometric feature is a subset of the geometric boundaries of a structural component. For example, a fillet or a circular hole is a geometric feature that has certain

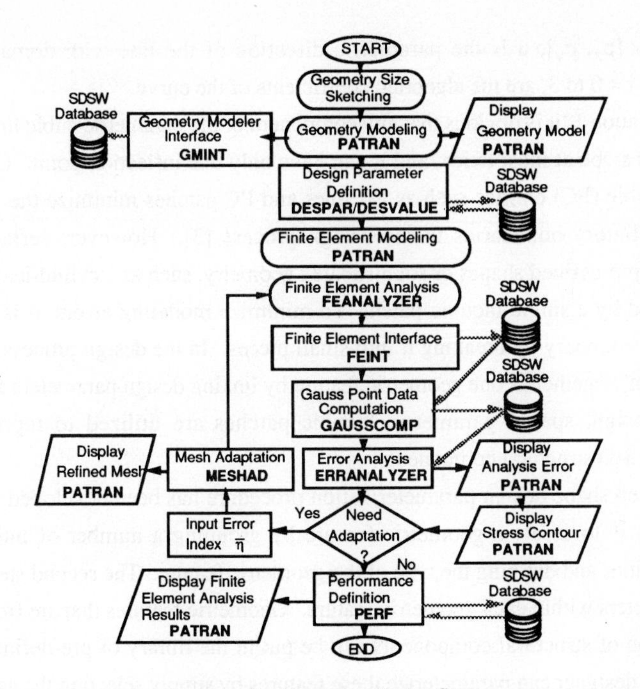

Figure 1. Pre-Processing Design Stage

characteristics associated with it and is likely to be chosen as the design. A geometric feature with design parameters defined is a parameterized geometric feature. A parameterized geometric feature is treated as a single entity in the shape design process. For example, a circular hole, with the radius and location of its center defined as design parameters, is a parameterized geometric feature. Such a parameterized circular hole can be moved around in the structure, with its size varied due to design changes. However, the shape of the circular hole is retained.

The design parameterization method developed in this study uses the basic PATRAN capability. In PATRAN, all geometric entities are represented using parametric cubic (PC) lines, patches (surfaces), and hyperpatches (solids). A planar parametric cubic line is represented as [8]

$$\begin{aligned} \mathbf{p}(u) &= \mathbf{a}_3 u^3 + \mathbf{a}_2 u^2 + \mathbf{a}_1 u + \mathbf{a}_0 \\ &= [u^3 \; u^2 \; u \; 1] \begin{bmatrix} \mathbf{a}_3 \\ \mathbf{a}_2 \\ \mathbf{a}_1 \\ \mathbf{a}_0 \end{bmatrix}_{4\times 2} \\ &= \mathbf{U}\,\mathbf{A}\,, \qquad u \in [0,1] \end{aligned} \tag{1}$$

where $\mathbf{p}(u) = [p_x, p_y]$, u is the parametric direction of the line with domain [0,1], and $\mathbf{a}_i = [a_{ix}, a_{iy}]$, i = 0 to 3, are the algebraic coefficients of the curve.

From Equation 1, it is obvious that any component of the parametric cubic line can change the sign of its slope at most twice, and it can have only one inflection point. Consequently, parametric cubic (PC) entities such as PC lines and PC patches minimize the possibility of yielding oscillatory boundaries in the design process [3]. However, certain geometric features with pre-defined shapes or sophisticated geometry, such as a cylindrical hole, cannot be represented by a single bicubic patch. To minimize modeling errors, it is necessary to model such a boundary by breaking it into small pieces. In the design process, these pieces must be "glued" together as one geometric feature by linking design parameters appropriately. For shape design, spatial parametric bicubic patches are utilized to represent design boundaries of 3-D structural components.

A three-step shape design parameterization procedure has been developed in this study. The first step is to create a geometric feature by grouping a number of inter-connected geometric entities and defining the type of the geometric feature. The second step is to define design parameters within each geometric feature. Geometric features that are frequently used in construction of structural components can be put in the library of pre-defined geometric features. The designer can parameterize these features by simply selecting the associated pre-defined shape design parameters from the user interface menu. On the other hand, geometric features that are not included in the library must be defined as user-defined features. To generate a parameterized user-defined geometric feature, the designer can use the design parameter definition within the geometric entities and link design parameters across the entities. The third step is to link design parameters across parameterized geometric features, if necessary.

As described above, the fundamental shape design parameterization is defined within geometric entities, and parameterized geometric features are created using geometric entities. Hierarchy of the design parameterization method is shown in Figure 2.

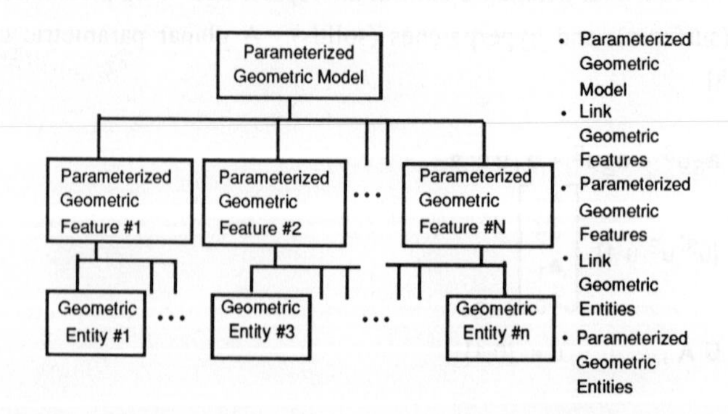

Figure 2 . Hierarchy of Shape Design Parameterization

In general, there are forty eight degrees of freedom for a parametric bicubic surface. The mathematical expression for a bicubic parametric surface is

$$\mathbf{p}(u,w) = \sum_{i,j=0}^{3} \mathbf{a}_{ij}\, u^i\, w^j$$

$$= [u^3\ u^2\ u\ 1] \begin{bmatrix} \mathbf{a}_{33} & \mathbf{a}_{32} & \mathbf{a}_{31} & \mathbf{a}_{30} \\ \mathbf{a}_{23} & \mathbf{a}_{22} & \mathbf{a}_{21} & \mathbf{a}_{20} \\ \mathbf{a}_{13} & \mathbf{a}_{12} & \mathbf{a}_{11} & \mathbf{a}_{10} \\ \mathbf{a}_{03} & \mathbf{a}_{02} & \mathbf{a}_{01} & \mathbf{a}_{00} \end{bmatrix}_{4\times4\times3} \begin{bmatrix} w^3 \\ w^2 \\ w \\ 1 \end{bmatrix}_{4\times1}$$

$$= \mathbf{U}\,\mathbf{A}\,\mathbf{W}^T, \quad (u,w) \in [0,1]\times[0,1] \tag{2}$$

where $\mathbf{p}(u,w) = [p_x, p_y, p_z]$, $\mathbf{a}_{ij} = [a_{ijx}, a_{ijy}, a_{ijz}]$ are the algebraic coefficients of the surface, and u and w are the parametric directions of the geometric entity. The above bicubic patch in algebraic format can be translated into other formats, such as Bezier, geometric, etc., using linear transformation to support various design applications [4].

A bicubic surface (patch) in geometric format is represented by the positions, tangent vectors, and twist vectors at the four corner points of the surface, as shown in Figure 3. To parameterize the geometric surface, all 48 geometric coefficients in the **B** matrix can be defined as shape design parameters; i.e.,

$$\mathbf{B} = \begin{bmatrix} \mathbf{p}_{00} & \mathbf{p}_{01} & \mathbf{p}_{00}^{w} & \mathbf{p}_{01}^{w} \\ \mathbf{p}_{10} & \mathbf{p}_{11} & \mathbf{p}_{10}^{w} & \mathbf{p}_{11}^{w} \\ \mathbf{p}_{00}^{u} & \mathbf{p}_{01}^{u} & \mathbf{p}_{00}^{uw} & \mathbf{p}_{01}^{uw} \\ \mathbf{p}_{10}^{u} & \mathbf{p}_{11}^{u} & \mathbf{p}_{10}^{uw} & \mathbf{p}_{11}^{uw} \end{bmatrix}_{4\times4\times3} \tag{3}$$

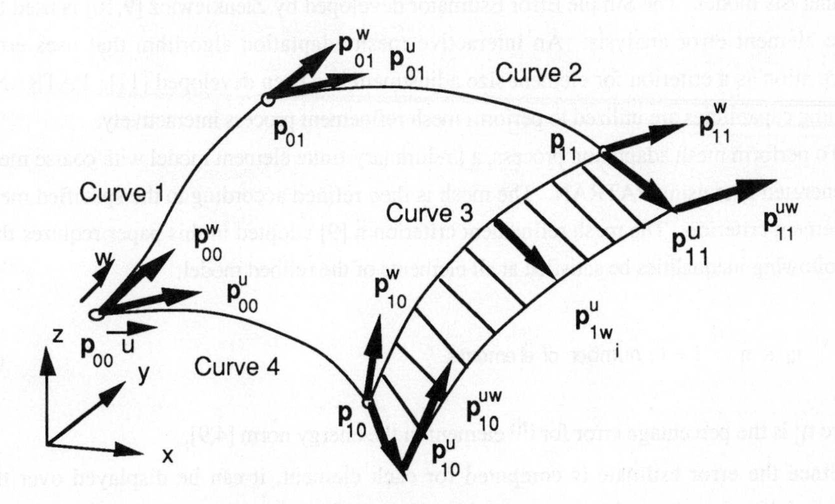

Figure 3. Geometric Surface

In this study, algebraic, geometric, 16-point, Bezier, plane surface, cylindrical surface, ruled surface, and surface of revolution are developed for handling 3-D shape design problems. To parameterize the geometric curve, a design parameterization menu shown in Figure 4 is implemented to allow the designer interact with the SDSW conveniently.

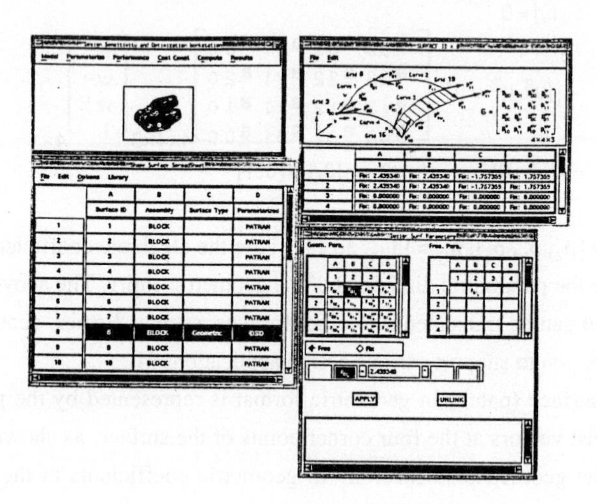

Figure 4. Design Parameterization Menu

3.2 Error Analysis and Mesh Adaption

Finite element error analysis and mesh adaptation procedures are used to ensure accuracy of the analysis model. The Simple Error Estimator developed by Zienkiewicz [9,10] is used for finite element error analysis. An interactive mesh adaptation algorithm that uses error information as a criterion for element size adjustment has been developed [11]. PATRAN's meshing capabilities are utilized to perform mesh refinement process interactively.

To perform mesh adaptation process, a preliminary finite element model with coarse mesh is generated first using PATRAN. The mesh is then refined according to the specified mesh refinement criterion. The mesh refinement criterion $\bar{\eta}$ [9] adopted in this paper requires that the following inequalities be satisfied at all elements of the refined model,

$$\eta_i \leq \bar{\eta}, \quad i = 1, \text{number of elements} \tag{4}$$

where η_i is the percentage error for i^{th} element in the energy norm [4,9].

Since the error estimate is computed for each element, it can be displayed over the structural domain using color contours to identify regions with high error. The ratio between element percentage error η_i and refinement criterion $\bar{\eta}$ determines elements that need further refinement, and the amount of refinement [4,9].

High stress concentration areas are of critical concern for design engineers. To ensure adequate accuracy of FEA stress results, a refined mesh must be used in these critical regions. To what extent the mesh should be refined is difficult to determine beforehand. Large errors may occur in areas with low stress. Only regions of high stress concentration and large error accumulation need to be refined. In this study, both stress and error contour plots are displayed in a PATRAN window, to help the designer in refining the mesh.

4 Design Sensitivity Analysis

In continuum shape design sensitivity analysis, shape of the structural domain is treated as the design variable. The relationship between a shape variation of a continuous domain and the resulting variation in structural performance measures can be described by the material derivative of continuum mechanics [16].

The continuum form of the governing equation on a deformed domain Ω_t can be written as

$$a_{\Omega_\tau}(z_\tau, \bar{z}_\tau) = \ell_{\Omega_\tau}(\bar{z}_\tau) \quad \text{for all } \bar{z}_\tau \in Z_\tau \tag{5}$$

where $\bar{z}_\tau$ is a virtual displacement, Z_τ is the space of kinematically admissible virtual displacements, and $a_{\Omega_\tau}(z_\tau, \bar{z}_\tau)$ and $\ell_{\Omega_\tau}(\bar{z}_\tau)$ are the energy bilinear and load linear forms, respectively. The subscript Ω_τ in Equation 5 is used to indicate the dependency of the equilibrium equation on the domain of the structure.

Performance measure such as displacements and stresses can be written in integral form as

$$\Psi = \iint_{\Omega_\tau} g\,(z_\tau, \nabla z_\tau)\, d\Omega_\tau \tag{6}$$

Using the adjoint variable method of shape design sensitivity analysis, the design derivative of the performance measure Ψ of Equation 6 can be expressed as

$$\Psi' = \ell_V'(\lambda) - a_V'(z,\lambda)$$
$$- \iint_{\Omega} [\, g_z(\nabla z^T V) + g_{\nabla z} \nabla(\nabla z^T V)\,]\, d\Omega + \int_{\Gamma} g(V^T n)\, d\Gamma \tag{7}$$

where λ is the solution of the adjoint equation

$$a_\Omega(\lambda,\bar{\lambda}) = \iint_\Omega [\, g_z\bar{\lambda} + g_{\nabla z}\nabla\bar{\lambda} \,]\, d\Omega \qquad \text{for all } \bar{\lambda} \in Z \tag{8}$$

For the direct differentiation method, the design derivative of the performance measure Ψ can be written as

$$\Psi' = \iint_\Omega [\, g_z\dot{z} + g_{\nabla z}\nabla\dot{z} - g_z(\nabla z^T V) - g_{\nabla z}\nabla(\nabla z^T V) \,]\, d\Omega + \int_\Gamma g(V^T n)\, d\Gamma \tag{9}$$

where $\dot{z}$ is the solution of the equation

$$a_\Omega(\dot{z},\bar{z}) = \ell_V'(\bar{z}) - a_V'(z,\bar{z}) \qquad \text{for all } \bar{z} \in Z \tag{10}$$

The subscript V on the right sides of Equations 7 and 10 is used to indicate the dependency of the terms on design velocity fields.

Numerical evaluation for Equations 7 and 10 requires knowledge of the original structural responses z, adjoint responses λ, material derivative $\dot{z}$, and the design velocity field V. The solution z of Equation 5 is obtained from finite element analysis carried out in the pre-processing design stage. The solutions λ and $\dot{z}$ of Equations 8 and 10, respectively, are obtained by restarting the finite element analysis code with different loading vectors, i.e., right sides of Equations 8 and 10. Proper generation of design velocity fields is an important step in obtaining accurate shape design sensitivity information. Velocity field and sensitivity computation methods are discussed in the following sections.

4.1 Velocity Field Computation

For design sensitivity analysis, boundary velocity field that occurs due to shape design perturbations is first computed. Domain velocities are then computed, using either the Boundary Displacement [12-14] or isoparametric mapping method [15]. In order to support the Boundary Displacement method, a general scheme has been developed to identify boundary curves and surfaces of the geometric model [4]. Once the velocity fields are computed, sensitivity coefficients can be evaluated using either direct differentiation or adjoint variable methods. Note that the velocity field computation is decoupled from the sensitivity computation.

4.2 Boundary Velocity Field Computation Using the Isoparametric Mapping Method

Computation of the boundary velocity field for the discretized finite element model is directly related to the design parameterization that is defined for the geometric model. As discussed previously, the most primitive design parameters are defined on the geometric entities. In order to parameterize the geometric features, design parameter linking must be carried out across geometric entities. Consequently, the boundary velocity field must be computed for all the geometric entities on which the associated shape design parameters, either independent or dependent, are defined. Using the isoparametric mapping method, the boundary velocity field can be expressed as

$$
\begin{aligned}
V_{1\times 3} &= [u^3\ u^2\ u\ 1] \begin{bmatrix} \delta a_{33} & \delta a_{32} & \delta a_{31} & \delta a_{30} \\ \delta a_{23} & \delta a_{22} & \delta a_{21} & \delta a_{20} \\ \delta a_{13} & \delta a_{12} & \delta a_{11} & \delta a_{10} \\ \delta a_{03} & \delta a_{02} & \delta a_{01} & \delta a_{00} \end{bmatrix}_{4\times 4\times 3} \begin{bmatrix} w^3 \\ w^2 \\ w \\ 1 \end{bmatrix}_{4\times 1} \\
&= \mathbf{U}\,\delta\mathbf{A}\,\mathbf{W}^T, \quad (u,w) \in [0,1]\times[0,1]
\end{aligned}
\tag{11}
$$

where matrix $\delta\mathbf{A}$ is the algebraic form of shape design perturbations of the geometric entity that represents the design boundary. Vectors $\mathbf{U}$ and $\mathbf{W}$ are locations of nodes in the parametric directions of the geometric entity of the design boundary. The matrix $\delta\mathbf{A}$ can be obtained from the perturbed design parameter matrix $\delta\mathbf{B}$ following a linear transformation. Matrix $\delta\mathbf{B}$ is defined as the variation of the design parameter matrix $\mathbf{B}$ of the geometric entity.

4.3 Design Sensitivity Computation

To compute design sensitivity information, both the adjoint variable and direct differentiation methods of continuum design sensitivity analysis have been implemented to achieve a more efficient computation, together with the domain method of shape design sensitivity analysis [4,16,17]. To provide a transparent design environment, a UNIX shell script has been written to automate complicated shape design sensitivity computation procedures, as shown in Figure 5.

4.4 Shape Design Sensitivity Analysis Using Submodeling Techniques

In engineering structural analysis, where the geometric shape of the structure is complicated and accurate stress analysis is required, the structural model may be very large. For such

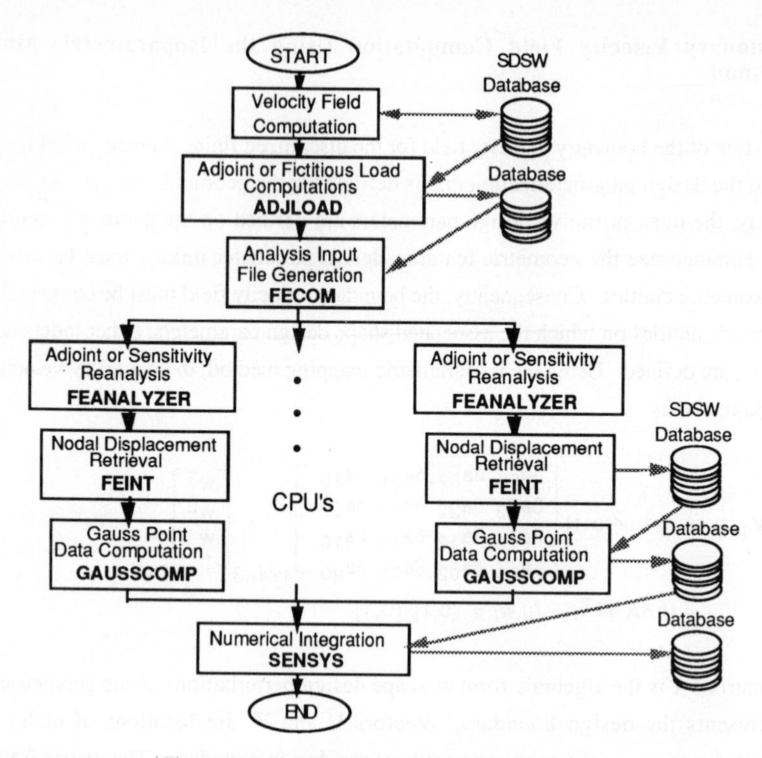

Figure 5. Design Sensitivity Computation Procedure

models, shape design sensitivity computations require that velocity field computations, adjoint or sensitivity reanalyses, and domain integration be carried out on the entire domain of the model. Consequently, intensive computations, large memory space, and large disk space are necessary. In SDSW, shape design sensitivity computations using a design component method incorporating PATRAN's "assembly" capability for easy submodeling are implemented. Using the submodeling approach, significant saving in computation and accurate design sensitivity information are obtained [4].

4.5 Distributed Computational Algorithm

To speed up design sensitivity computations, boundary and domain velocity computations, adjoint and sensitivity reanalyses, and numerical integration are distributed over a network of computers [4]. To support the distributed design sensitivity computation, a client-server model built on top of the Apollo Network Computing System (NCS) [18] is used. With the distributed design sensitivity computation, an almost linear performance speedup has been obtained [4,19].

5 Post-Processing Design Stage

The post-processing design stage supports two capabilities: design sensitivity display and what-if study [20,21], as shown in Figure 6. A method that utilizes PATRAN visualization capabilities for display of shape design sensitivity information has been implemented. For design sensitivity display, the steepest descent direction of a performance measure is displayed using color plots. First order predictions of performance measures with respect to a specific perturbation of design parameters are provided to assist the designer in the design decision making process. Predicted values of performance measures for perturbed designs can be computed in the what-if study step, using design sensitivity information and performance values of the current design. A general algorithm that uses design velocity fields and design parameter perturbations for computing new node positions at perturbed designs has been developed. An automatic geometric shape updating method, using PATRAN's geometric database, has also been developed [4]. Both the finite element mesh and the geometric shape are updated directly in PATRAN's database, using PATRAN commands or database I/O routines [22]. Predicted performance values of the what-if study can be displayed on the perturbed geometric or finite element model.

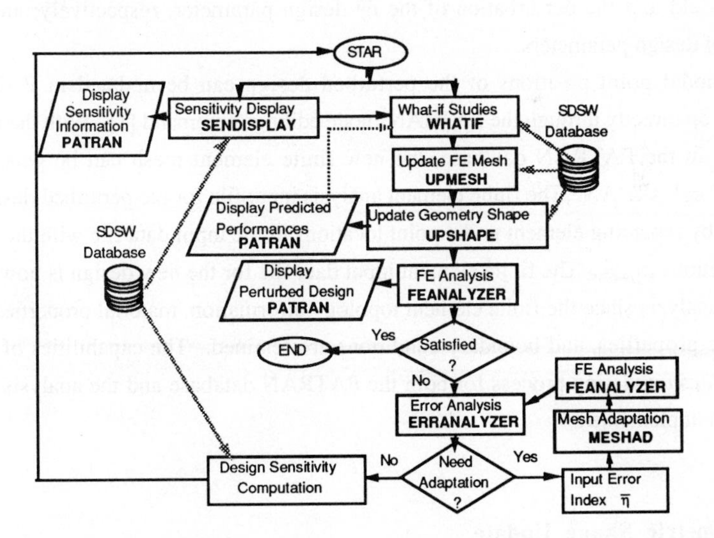

Figure 6. Post-Processing Design Stage

The advantage of using the what-if study, compared with reanalysis, is that the former does not require the designer to go through geometric modeling, model translation, and finite element analysis to evaluate new designs. Indeed, the post-processing capabilities of the SDSW permit the designer to perform many design tryouts in a short time, with the minimum requirement of both human and machine efforts.

To allow the design engineer to perform finite element analysis for the next design iteration, nodal point positions are updated directly in the analysis input command file. Since the finite element mesh can be updated using the velocity field, mesh generation is not required for the perturbed design unless the finite elements become distorted during the design process.

5.1 Finite Element Mesh Update

The finite element mesh of the perturbed design can be generated using velocity field information. The locations of nodes in the perturbed design can be computed by

$$\mathbf{p}_{b+\delta b} = \mathbf{p}_b + \delta\mathbf{p} = \mathbf{p}_b + \sum_{i=1}^{n} \mathbf{V}_i \, \delta b_i \tag{12}$$

where $\mathbf{p}_{b+\delta b}$ and $\mathbf{p}_b$ are the locations of the nodes of the perturbed and the current designs, respectively; $\delta\mathbf{p}$ is the nodal point movement due to design perturbations; $\mathbf{V}_i$ and δb_i are the velocity field and the perturbation of the ith design parameter, respectively; and n is the number of design parameters.

The nodal point positions of the perturbed design can be updated in PATRAN by inputting $\delta\mathbf{p}$ directly through the PATRAN "node editing" command [1]. With the new nodal positions in the PATRAN database, the new finite element mesh can be generated and displayed in PATRAN. The finite element analysis input file for the perturbed design can be obtained by replacing element nodal point locations in the input data file with the perturbed node locations $\mathbf{p}_{b+\delta b}$. The finite element input data file for the new design is now ready for the next analysis since the finite element topology information, material properties, element geometric properties, and boundary conditions are retained. The capabilities of the finite element mesh updating process for both the PATRAN database and the analysis input file have been implemented.

5.2 Geometric Shape Update

The geometric shape of the perturbed design can also be determined by manipulating the design parameter perturbation and geometric data of the structure. Using the above information, the geometric coefficient matrix of the perturbed geometric entity can be obtained [4]. The geometric entity can then be updated by changing the PATRAN database, using PATRAN commands or database I/O routines [22].

With the capabilities developed in the post-processing design stage, the structural performance measures of the current and perturbed designs can be displayed on their respective finite element models in PATRAN. Also, finite element analysis of the perturbed design can now be carried out. Furthermore, the geometric shape of both the current and the perturbed designs can be displayed in PATRAN. Visualization of this design information will help the designer relate predicted performance measures with the perturbed geometric shape.

6 Examples

Two design applications, a 3-D clevis and a turbine blade, are presented to demonstrate the design methodologies developed in the SDSW. In the turbine blade example, difficulty encountered in model translation from CAD modeler to PATRAN is discussed.

6.1 3-D Clevis

A redesigned clevis model [23] shown in Figure 7(b) is utilized to demonstrate the SDSW capabilities. This clevis connects a tow bar and the top eye hookup of the M-1 tank. When the M-1 tank is subjected to certain maneuvers, interference exists between the clevis and the hookup. In order to eliminate such an interference, a decision was made to place a chamfer on the lower surface of the clevis to increase vertical motion of the tow bar, as shown in Figure 7.

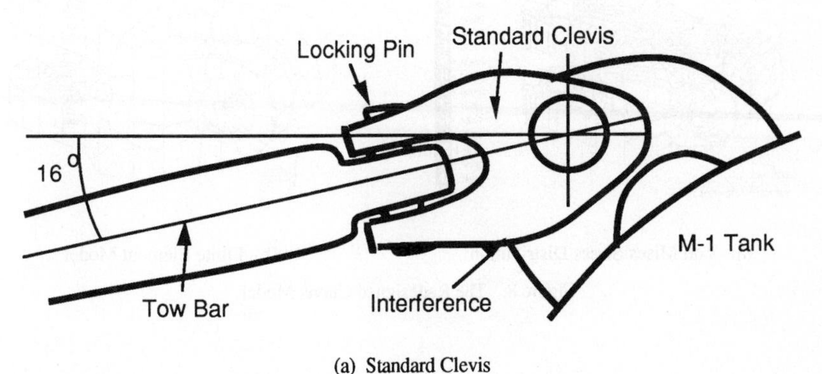

(a) Standard Clevis

Figure 7. 3-D Clevis

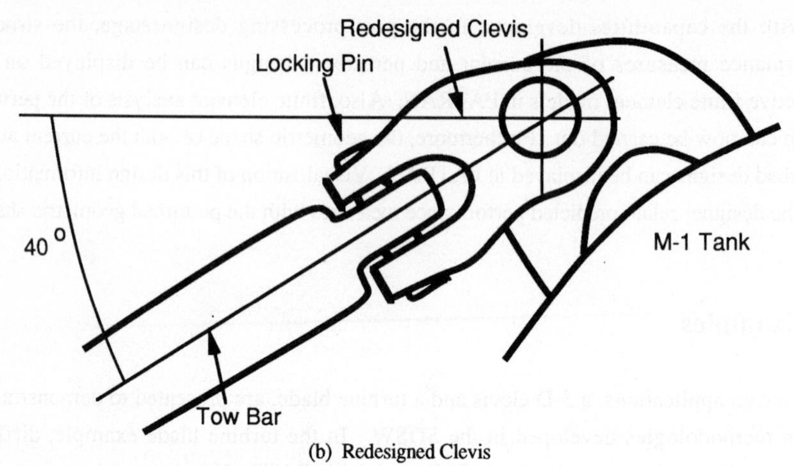

(b) Redesigned Clevis

Figure 7. 3-D Clevis (con't)

However, the redesigned clevis failed in the field test, due to stress concentration on its lower surface. Finite element analysis results confirmed the stress concentration in the lower surface, as shown in Figure 8(a) (at the upper edge of the semi-cylindrical surface with stress label G), under boundary conditions applied at the four half pins, as shown in Figure 8(b). The material properties are Young's modulus $E = 10.5\times10^6$ psi and Poisson's ratio $\nu = 0.3$. The finite element model has 247 20-node isoparametric elements ANSYS STIF 95 [24], with 5,000 degrees of freedom.

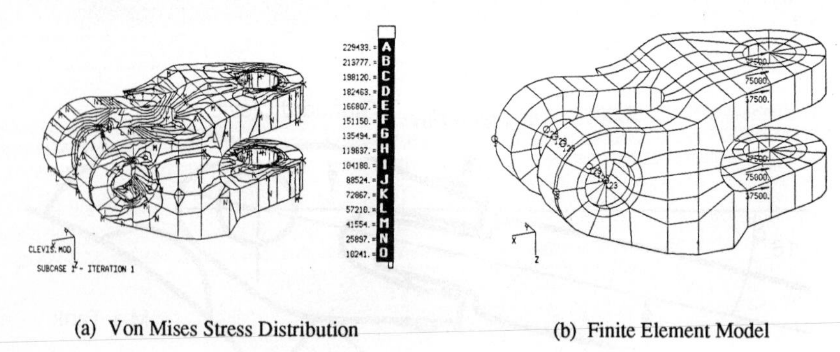

(a) Von Mises Stress Distribution (b) Finite Element Model

Figure 8. The Redesigned Clevis Model

Design Parameterization

The objective of modifying the redesigned clevis is to reduce the stress concentration by altering certain boundary surfaces without significantly increasing its weight. To parameterize the redesigned clevis model, five geometric features, front surface, rear surface, semi-cylindrical surface 1, semi-cylindrical surface 2, and top surface are identified as design boundaries, as shown in Figures 9(a) to (e), respectively. Notice that a surface (patch) in PATRAN is represented by a 3×3 lattice, not a 3×3 finite element mesh, as shown in Figure 9.

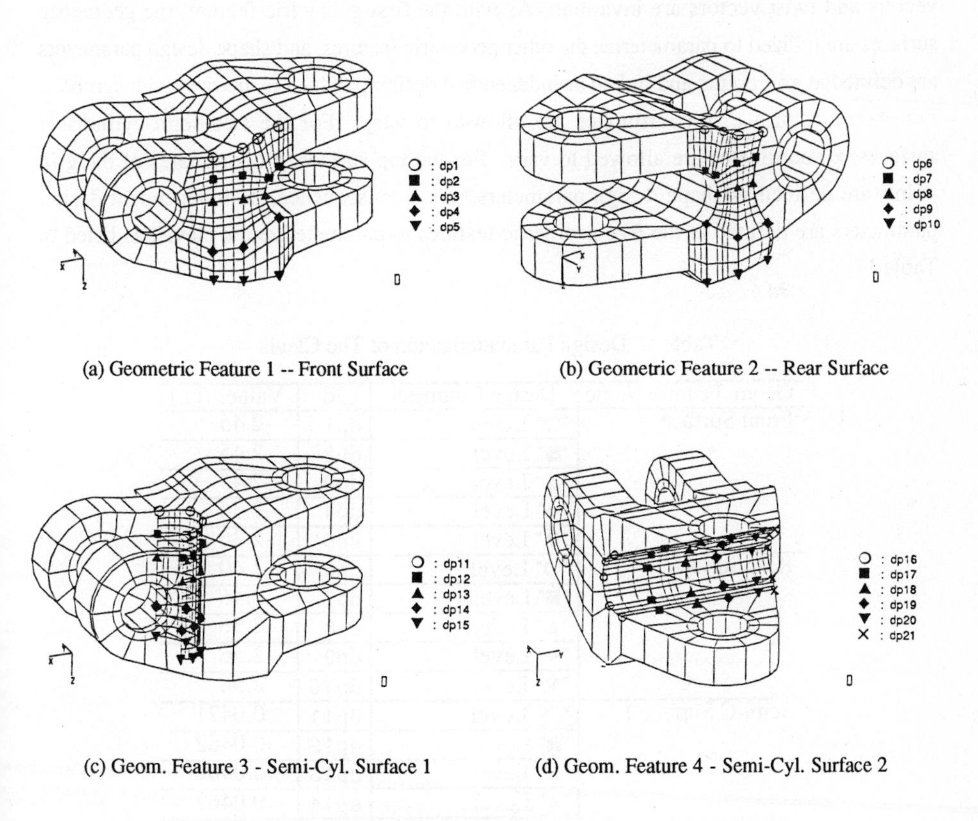

(a) Geometric Feature 1 -- Front Surface

(b) Geometric Feature 2 -- Rear Surface

(c) Geom. Feature 3 - Semi-Cyl. Surface 1

(d) Geom. Feature 4 - Semi-Cyl. Surface 2

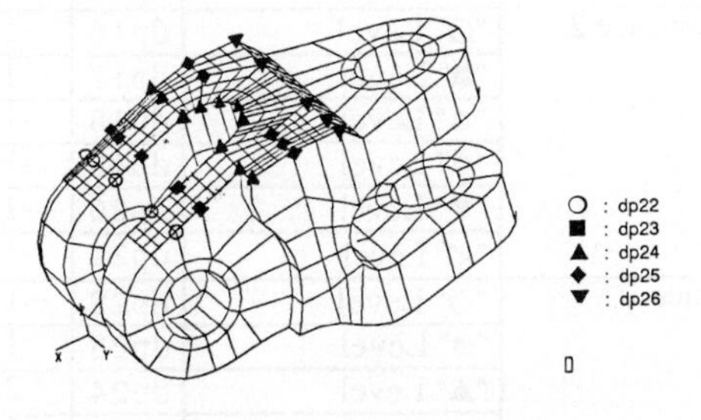

(e) Geometric Feature 5 -- Top Surface

Figure 9. Design Parameterization of The Redesigned Clevis

To parameterize the first geometric feature, the fifteen patches are parameterized as geometric patches, and the y-coordinates of the seventeen grid points are linked as five independent shape design parameters, marked with five symbols, as illustrated in Figure 9(a). For this geometric feature, C^0-continuity is retained by the design parameter linking at the grids of inter-surface boundaries. Moreover, C^1-continuity is retained since the tangent vectors and twist vectors are invariant. As with the first geometric feature, the geometric surfaces are utilized to parameterize the other geometric features, and shape design parameters are defined at grid points and linked as independent design parameters, marked with symbols. For the rear surface, y-coordinates are allowed to vary. For the two semi-cylindrical surfaces, x-coordinates are allowed to vary. For the top surface, z-coordinates of the grid points are defined as shape design parameters. As a consequence, 26 independent design parameters are defined at the five geometric features to parameterize the clevis, as listed in Table 1.

Table 1. Design Parameterization of The Clevis

Geom. Feature Name	Design Parameter	Id	Values (in.)
Front Surface	"○" Level	dp1	-2.45
	"■" Level	dp2	-2.45
	"▲" Level	dp3	-2.45
	"◆" Level	dp4	-2.36
	"▼" Level	dp5	-2.36
Rear Surface	"○" Level	dp6	2.40
	"■" Level	dp7	2.46
	"▲" Level	dp8	2.56
	"◆" Level	dp9	2.56
	"▼" Level	dp10	2.56
Semi-C Surface 1	"○" Level	dp11	0.0471
	"■" Level	dp12	0.0462
	"▲" Level	dp13	0.0462
	"◆" Level	dp14	0.0462
	"▼" Level	dp15	0.0462
Semi-C Surface 2	"○" Level	dp16	-1.25
	"■" Level	dp17	-1.25
	"▲" Level	dp18	-1.25
	"◆" Level	dp19	-1.25
	"▼" Level	dp20	-1.25
	"×" Level	dp21	-1.25
Top Surface	"○" Level	dp22	1.73
	"■" Level	dp23	1.98
	"▲" Level	dp24	2.40
	"◆" Level	dp25	2.56
	"▼" Level	dp26	2.56

Design Sensitivity Analysis

Boundary Displacement and direct differentiation methods are used to evaluate velocity fields and sensitivity coefficients, respectively. Accuracy of the sensitivity coefficients has been verified using central finite difference of finite element analysis results [15] and a what-if study result discussed next.

Shape Design Sensitivity Display

The influence of design parameters on the stress performance measure is displayed in Figure 10. The performance measure is defined as the von Mises stress at a Gauss point in the element 219, where stress is concentrated, as shown in Figure 8. The contours of nodal point movements of the steepest descent design sensitivity direction are given in Figures 10(a) - (e), and the sensitivity coefficients are listed in Table 2.

Table 2. Design Sensitivity Coefficients and Design Perturbation

dpid	Sensitivity (219)	Current Values	Perturbation
1	5469.8	-2.45	-0.0804
2	3258.0	-2.45	-0.0479
3	734.7	-2.45	-0.0108
4	1434.6	-2.36	-0.0210
5	640.3	-2.36	-0.0094
6	-8396.9	2.40	0.1235
7	-5599.7	2.46	0.0824
8	-3417.3	2.56	0.0493
9	-1265.3	2.56	0.0156
10	293.2	2.56	-0.0043
11	193.3	0.0471	-0.0028
12	-421.4	0.0462	0.0062
13	-5270.9	0.0462	0.0775
14	-13625.1	0.0462	0.2000
15	-4438.0	0.0462	0.0654
16	1798.0	-1.25	-0.0272
17	5304.0	-1.25	-0.0797
18	8595.1	-1.25	-0.1265
19	12102.7	-1.25	-0.1765
20	10131.9	-1.25	-0.1485
21	3225.75	-1.25	-0.0471
22	217.7	1.73	-0.0032
23	2149.8	1.98	-0.0316
24	2469.6	2.40	-0.0363
25	680.3	2.56	-0.0100
26	299.3	2.56	-0.0044

Figure 10(a) shows that, on the front surface, moving those grid points marked with "O"; i.e., design parameter 1, a unit magnitude in the negative y-direction to make the front surface thicker decreases the stress in element 219 by an amount of 5,470 psi. Design

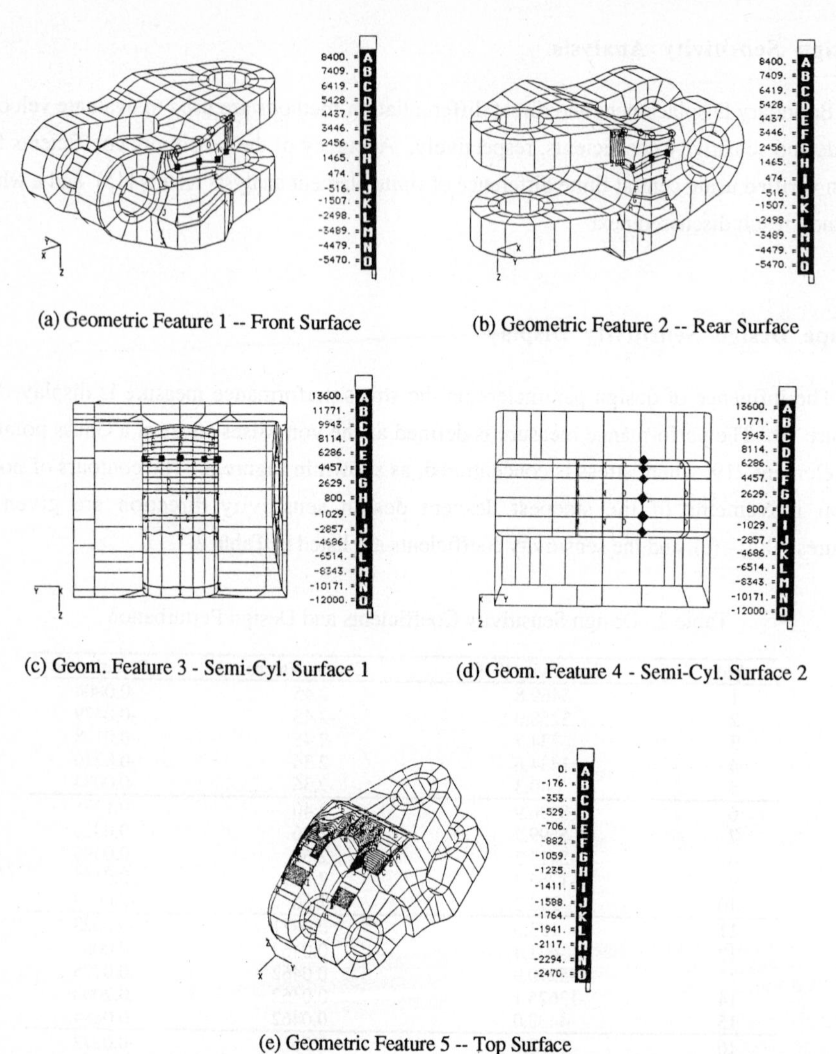

(a) Geometric Feature 1 -- Front Surface

(b) Geometric Feature 2 -- Rear Surface

(c) Geom. Feature 3 - Semi-Cyl. Surface 1

(d) Geom. Feature 4 - Semi-Cyl. Surface 2

(e) Geometric Feature 5 -- Top Surface

Figure 10. Contours of Sensitivity Display for 3-D Clevis

parameter 2, which is defined as the y-movement of the grid point marked with "■" similarly decreases the stress, but by a lesser amount. The other design parameters are not influential. Figure 10(b) shows that, on the rear surface, perturbing design parameter 6, which are defined at grid points marked with "O", a unit magnitude in the positive y-direction causes the rear surface to be thicker decreases the stress concentration an amount of 8,400 psi. The other design parameters defined on the rear surface similarly decrease the stress, but by a lesser amount.

Figure 10(c) shows that design parameter 14, which is defined as the x-movement of grids marked as "■", has the most influence on the stress performance. Perturbing design parameter 14 a unit magnitude in the x-direction to make the upper cylindrical surface thicker decreases the stress concentration 13,600 psi. The other design parameters defined on the cylindrical surface similarly decrease the stress, but by a lesser amount. For the other cylindrical surface, moving the grids marked as "◆"; i.e., design parameter 19, a unit magnitude in the negative x-direction decreases the stress concentration 12,000 psi, as shown in Figure 10(d). The other design parameters defined on the cylindrical surface similarly decrease the stress, but by a lesser amount. Among the five geometric features, the design perturbations on the top surface have the least influence on the stress concentration. However, the interesting observation is that, instead of adding material to the top surface, moving the grids inward reduces the stress concentration. This is because by cutting out the material on the surface, the overall stress field is redistributed so that the stress concentration is reduced.

In summary, the most influential design parameters are located on the two cylindrical surfaces. The closer the grids are to the stress concentration area, the more influential they are. Moreover, the variations of the top surface have little influence.

What-if Study

Based on the design sensitivity display, a design perturbation that perturbs the design in the steepest descent direction, but allows a maximum 0.2 in. boundary movement is selected to carry out the what-if study. Design perturbations are summarized in Table 2. Figure 11 shows that, following such a design perturbation, the von Mises stress of the Gauss point defined at element 219 is reduced from labels E to F (127,618.5 to 116,369.1 psi), without introducing other stress concentrations. For such design changes, the volume of the clevis increases from 89.674 to 91.566 in^3; i.e., a 2.1 % increment. Finite element analysis results for the perturbed design show that the maximum von Mises stress of the Gauss point is reduced from 127,618.5 to 117,522.1 psi, a 7.9% stress reduction, as listed in Table 3.

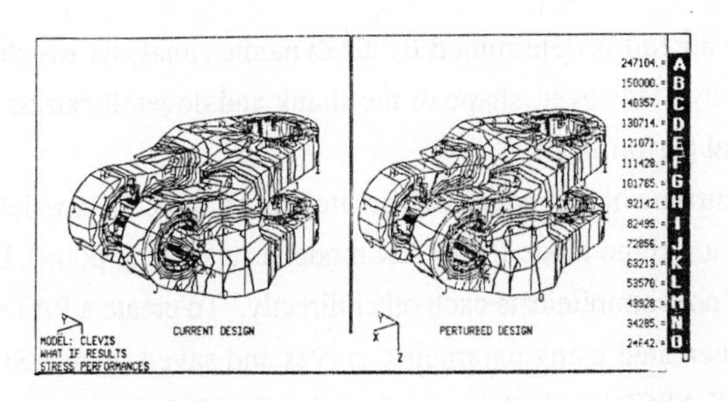

Figure 11.. What-if Study for 3-D Clevis

Table 3. What-if Results and Verification

Performance	Current Value	Predicted Value	% change	FEA Results	Accuracy%
Stress 219	127618.5	116369.1	-8.81	117522.1	89.8

6.2 Turbine Blade

Physically, a blade is inserted to a slot in a disc mounted on a rotating shaft, as shown in Figure 12. The blade can be divided into four major parts, air foil, platform, shank, and dovetail. Due to shaft rotation, fluid pressure is applied on the surface of the air foil, and a centrifugal force is applied to the whole blade structure. By analyzing the blade model, the centrifugal force is found dominant in contributing to blade structural deformation, therefore the fluid pressure is ignored in the finite element and sensitivity analyses. Moreover, as found in the finite element analysis, since platform does not contribute significantly to the blade structural behaviors, it is removed in the modeling process. Note that shank is the part that sustains a major stress flow from the dovetail to the air foil, due to rotation.

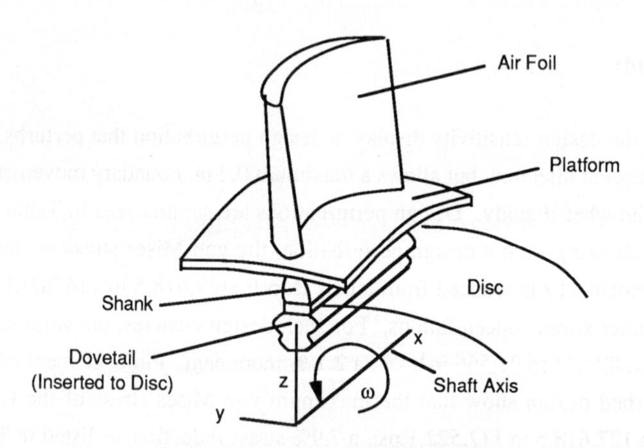

Figure 12. Turbine Blade Physical Model

Profile of the air foil is determined by air dynamics analysis which is not considered to change in this study. However, shape of the shank and dovetail can be modified to improve the blade structural performance.

Initially, the turbine blade model was created using the CAD modeler ICAD. However, the SDSW needs a turbine blade PATRAN model as a starting point. Unfortunately, ICAD and PATRAN do not communicate each other directly. To create a PATRAN blade model, an ICAD model is recreated using parametric curves and saved in an ASCII file in IGES [25] format. The IGES ASCII file is then transferred to PATRAN, using PATIGE [25], to create a wireframe blade model shown in Figure 13(a).

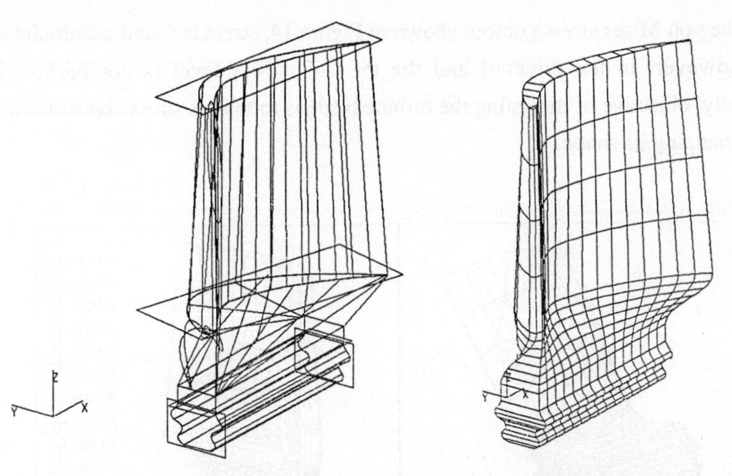

(a) Wireframe Model Translated From ICAD (b) PATRAN Finite Element Model

Figure 13. Turbine Blade PATRAN Model

Notice that the wireframe model is accurate in the air foil and the dovetail, however, straight lines are connected between the air foil and the dovetail, which is not appropriate to form the shank. Instead of straight lines, smooth surfaces are required to be put between the air foil and the dovetail, with C^0- and C^1-continuity across the upper and lower interfaces, i.e., interface between the air foil and the shank, and interface between the shank and the dovetail.

To create the air foil and the dovetail, the wireframe model is cleaned up by removing unnecessary curves and points by engineers, and a number of patches are created using parametric curves in the wireframe model. After patches are formed, a number of hyperpatches are constructed using the patches.

To create the shank, Geometric patches [1] are found appropriate to connect the air foil and the dovetail with specified tangent vectors across interfaces to maintain C^0- and C^1-continuity. Once the Geometric patches are created for the shank, hyperpatches are formed using the surrounding patches.

After geometric model is completed, finite elements are meshed in each hyperpatches to generate finite element model shown in Figure 13(b). The model translation and creation processes took two engineer-weeks to accomplish. The PATRAN blade model has 315 3-D 20-node elements (ANSYS STIF95) and 2,388 nodes. The material properties are Young's Modulus $E = 2.99938 \times 10^7$, Possion's ratio $\nu = 0.29$, and mass density $\rho = 7.317313 \times 10^{-4}$. For boundary condition, displacement of the nodes at two sides of the dovetail are fixed in all directions, and a constant angular velocity w = 1,570.8 rad/sec (15,000 rpm) is applied in x-direction to create centrifugal force in the whole blade model, which is a body force type loading.

From the von Mises stress contour shown in Figure 14, stress is found accumulated in the shank. However, in the dovetail and the air foil, stress level is comparatively low. Consequently, objective of designing the turbine blade is to relieve stress concentration in the shank by changing its shape.

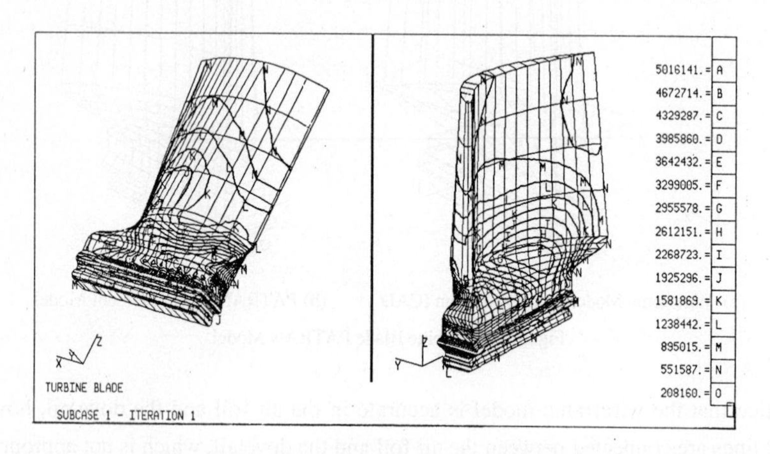

Figure 14. von Mises Stress Contour of the Turbine Blade Model

Design Parameterization

Five design parameters are defined in the turbine blade model to explore the blade design trend, i.e., offset of the dovetail in x-, y-, and z-directions, and rotation of the dovetail along z-direction, as shown in Figure 15.

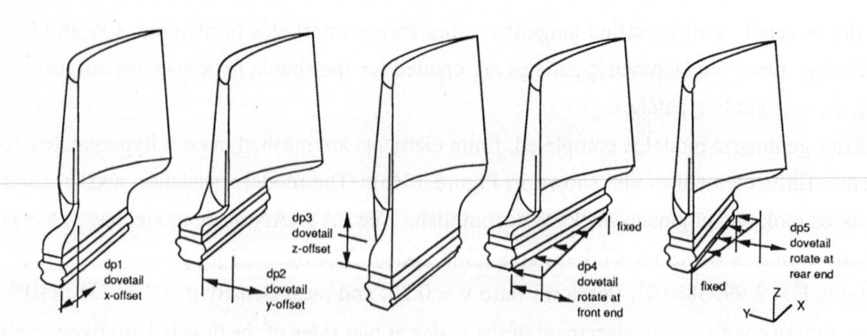

Figure 15. Design Parameterization For The Blade

The first three design parameters characterize the design changes by repositioning the dovetail. In such designs, the dovetail is translating with constant shape, however, shape of the shank is changing. Both the shape changes of the shank and translation of the dovetail affect the blade structural performance since mass is redistributed. The last two design parameters characterize the dovetail rotation design. In such a design, the dovetail is tilted sideways at ends, however, its cross section shape and normal direction are maintained. Again, both changes in the shank and the dovetail contribute to the blade structural performance variations. In fact, effect of design change due to dovetail y-offset is equivalent to the sum of dovetail rotation with same amount of changes at ends.

Design Consideration

As discussed earlier, design of the turbine blade is focusing on reducing stress concentration in the shank area by changing shapes of the shank and repositioning the dovetail, without significantly increasing its mass. Since blade mass and volume differ only by a factor of mass density ρ, volume is considered. Consequently, structural volume and high stress measures (over 4.0 MPa) are considered as performance measures for design improvements. At current design, structural volume is 15268.3041 mm^3, and high stresses occur in six elements, as listed in Table 4. Locations of these elements in the blade are shown in Figure 16.

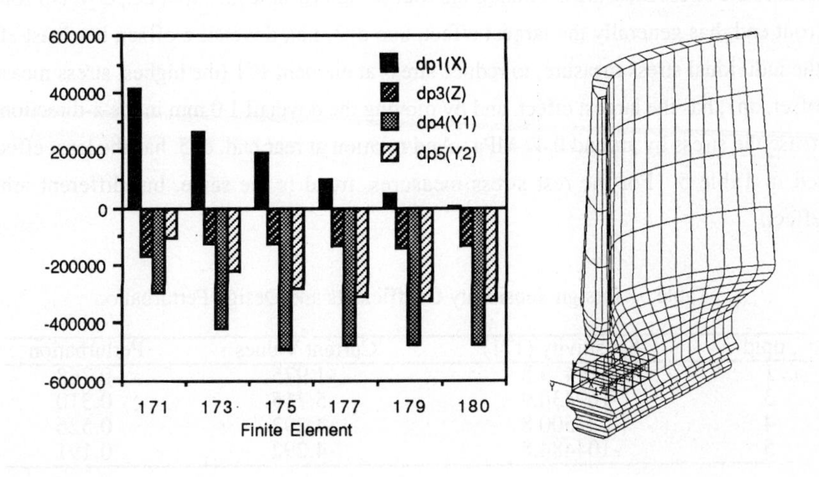

Figure 16. Stress Sensitivity Displayed in Bar Chart

Table 4. Gauss Stress Measures Over 4.0 MPa

Element	Gauss Point	Stress
171	2	0.481950993E+07
173	13	0.422697428E+07
175	6	0.419180741E+07
177	6	0.415963585E+07
179	6	0.410238461E+07
180	6	0.401685113E+07

Design Sensitivity Analysis

Isoparametric mapping and direct differentiation methods are used to evaluate velocity fields and sensitivity coefficients, respectively. Accuracy of the sensitivity coefficients can be verified using the what-if study and finite element analysis results described below.

Shape Design Sensitivity Display

To display design sensitivity coefficients, a bar chart is employed. For stress performance measures, design sensitivity coefficients for the six high stress measures are displayed using a bar chart in Figure 16. It shows that increasing design parameter dp1, i.e., moving the dovetail in the x-direction, increases the stress measures, and the effect is decreasing along the x-direction. However, increasing the other three design parameters decreases the stress measures. Among the four design parameters, dp4, i.e., dovetail rotation at front end, has generally the largest effect, and dp3, i.e., dovetail z-offset, has least effect. In the individual stress measure, to reduce stress at element 171 (the highest stress measure), x-offset, dp1, has the largest effect, and by moving the dovetail 1.0 mm in the x-direction will increase the stress by around 0.42 MPa. And rotation at rear end, dp5, has the least effect, as listed in Table 5. For the rest stress measures, trend is the same, but different amount of effect.

Table 5. Design Sensitivity Coefficients and Design Perturbation

dpid	Sensitivity (171)	Current Values	Perturbation
1	419459.8	-1.975	-0.768
3	-169430.9	-5.715	0.310
4	-287500.8	-4.292	0.526
5	-104484.8	-4.292	0.191

What-if Study

As discussed earlier, stress concentration is the engineer's highest concern. A design change that focuses on reducing the highest stress measure, identified in element 171, is performed using what-if design step. A steepest descent design direction for stress in element 171 and a step size 1.0 mm, that yield design perturbations shown in Table 5, have been selected to perform the what-if study. Prediction of volume and stress measures using design sensitivity coefficients is listed in Table 6. Also, accuracy of the predicted performance measures are verified to be excellent using the finite element analysis results obtained at the perturbed design. Reduction of stress measures due to such design change is displayed in Figure 17.

Table 6. What-if Results and Verification

Performance	Current Value	Predicted Value	% change	FEA Results	Accuracy%
Volume	15268.3041	15182.7667	-0.56	15182.7573	100.0
Stress 171	4819509.93	4273413.35	-11.33	4289771.25	97.0
Stress 173	4226974.28	3722203.13	-11.94	3734589.77	97.5
Stress 175	4191807.41	3695253.89	-11.85	3704051.12	98.2
Stress 177	4159635.85	3732537.12	-10.27	3740478.96	98.1
Stress 179	4102384.61	3714012.11	-9.49	3723099.27	97.7
Stress 180	4016851.13	3657743.19	-8.94	3666848.63	97.5

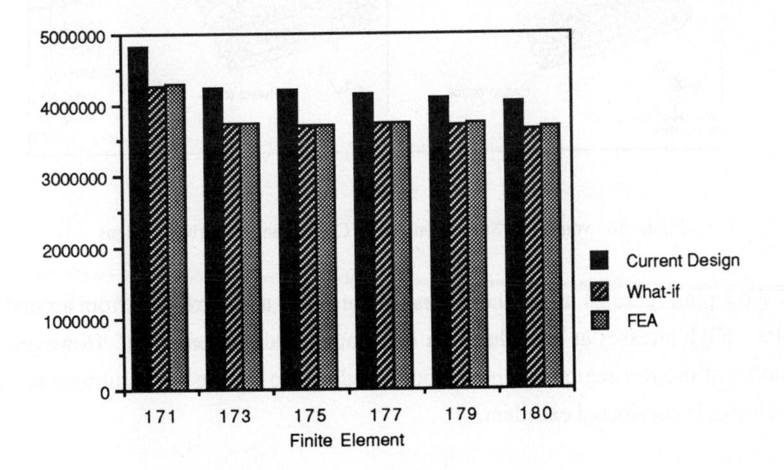

Figure 17. Stress What-if Study Results Displayed in Bar Chart

From Figure 17, stresses at the six elements are reduced by approximately 10% due to the design change. However, blade volume is reduced by 0.56%. Such a design change is not only relieving stress concentration at the shank but reducing structural volume. Based on such a design change, blade finite element mesh is updated, using the velocity field and design perturbation.

The predicted performance measures and finite element mesh update can be obtained in few cpu minutes which is extremely efficient compared with finite element analysis at the perturbed design. Moreover, using the SDSW, design engineers can try out several design changes conveniently and efficiently, without going through modeling and finite element analysis processes. The element von Mises stress contours at current and perturbed designs are also displayed in Figure 18 for stress in all elements. Sensitivity coefficients are computed for all stress measures using direct differentiation method. Note that contour shown in Figure 18 is obtained from what-if prediction which has been verified using finite element analysis results.

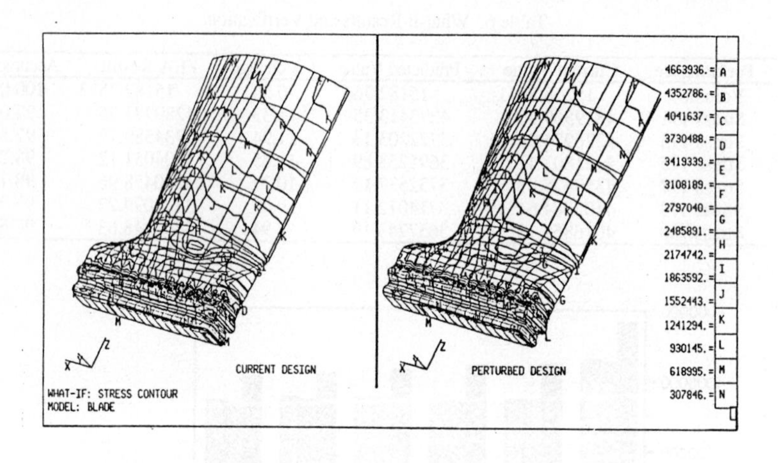

Figure 18. von Mises Stress Contour At Current and Perturbed Designs

As shown in Figure 18, the highest stress in the shank are decreasing from around 4.8 to 4.2 MPa. High stresses at the other five elements are also decreasing. However, stress distribution at the rest region is almost unchanged due to the design perturbation. Such a design change is considered excellent.

7 Conclusions

A methodology that supports structural shape design for 3-D elastic solid components using the geometric modeler PATRAN is presented in this paper. This methodology overcomes four major difficulties: accuracy, integration, efficiency, and effectiveness, of structural shape design problems.

To ensure accuracy of finite element analyses in the shape design process, an error analysis and interactive mesh adaptation method has been implemented. The method utilizes both Zienkiewicz's Simple Error Estimator and PATRAN's meshing capabilities. To support the mesh adaptation process, PATRAN's result visualization capabilities have been incorporated to display important analysis error and mesh refinement information. However, to automate the mesh adaptation process for 3-D components is difficult. Moreover, large size finite element models are very often resulted from the mesh adaptation process.

To integrate the shape design process, a shape design parameterization method, a velocity field computation method, a menu-driven user interface, and UNIX scripts are developed. This shape design parameterization method, which utilizes PATRAN's geometric data and modeling capabilities is capable of handling a large class of structural shape design problems. Through the menu-driven user interface, the design parameterization and other design processes can be carried out easily and conveniently. The boundary and domain velocity field computation method that links the velocity field with the shape design parameterization has also been implemented. Furthermore, a hybrid shape design sensitivity computation algorithm that utilizes both the adjoint variable and direct differentiation methods has been implemented to reduce computational effort. Integration of the shape design process has been demonstrated using the menu-driven user interface and UNIX scripts that automate sophisticated shape design sensitivity computation runstreams.

For handling large size problems, an efficient shape design sensitivity analysis method that uses the shape design component method has been implemented. Computational efficiency is also achieved by a new computational algorithm, which distributes computationally intensive tasks over a network of computers.

Visualization of design sensitivity information has been implemented using PATRAN's visualization tools. Furthermore, methodologies for automatic mesh regridding, geometric shape update, and performance measure prediction are developed and implemented to support the what-if study design step. These two design steps, together with visualization of design information, are demonstrated as a useful tool for the designer to carry out the shape design process effectively.

Finally, a three-stage shape design process that consists of pre-processing, shape design sensitivity analysis, and post-processing has been proposed. The three design stages have been integrated by the menu-driven user interface. Capabilities of the SDSW have been demonstrated using a clevis and a turbine blade examples.

Acknowledgement: Research supported by NSF-Army-NASA Industry/University Cooperative Research Center for Simulation and Design Optimization of Mechanical Systems.

References

1. PDA Engineering, *Patran Plus User's Manuals Vols I and II*, Software Products Division, 1560 Brookhollow Drive, Santa Ana, CA, 1987.
2. Haftka, R. T. and Grandhi, R. V., "Structural Shape Optimization - A Survey," *Computer Methods in Applied Mechanics and Engineering*, Vol. 57, pp. 91-106, 1986.
3. Ding, Y., "Shape Optimization of Structures - A Literature Survey," *Computers & Structures*, Vol. 24, No. 6, pp. 985-1004, 1986.
4. Chang, K. H. and Choi, K. K., "Shape Design Sensitivity Analysis and A What-if Design Workstation For Elastic Structural Components," *Technical Report R-93*, Center for Simulation and Design Optimization, The University of Iowa, 1991.
5. Anderson, G. and Anderson, P., *The UNIX C SHELL Field Guide*, Prentice-Hall, Inc., Englewood Cliffs, NJ 07632, 1986.
6. Godse, M. M., Koch, P., and Santos, J. L. T., "Development of A User Interface For A Design Sensitivity Analysis and Optimization Workstation," *Advances in Design Automation*, Vol. II Optimal Design and Mechanical Systems Analysis, Ed. by Ravani, B., pp. 117-126, ASME Conference, 1990.
7. Open Software Foundation, *OSF/Motif Programmer's Guide*, Prentice Hall, Englewood Cliffs, New Jersey 07632, 1988.
8. Mortenson, M. E., *Geometric Modeling*, John Wiley & Sons, Inc., New York, 1985.
9. Zienkiewicz, O. C. and Zhu, J. Z., "A Simple Error Estimator and Adaptive Procedure for Practical Engineering Analysis," *Int. J. for Numerical Methods in Engineering*, Vol. 24, pp. 337-357, 1987.
10. Rank, E. and Zienkiewicz, O. C., "A Simple Error Estimator in the Finite Element Method," *Comm. in Applied Numerical Methods*, Vol. 3, pp. 243-249, 1987.
11. Chang, K. H. and Choi, K. K., "An Error Analysis and Mesh Adaptation Method for Shape Design of Structural Components," *Computers and Structures*, to appear, 1992.
12. Yao, T. M. and Choi, K. K., "3-D Shape Optimal Design and Automatic Finite Element Regridding," *Int. J. for Numerical Methods in Engineering*, Vol. 28, pp. 369-384, 1989.
13. Yao, T. M. and Choi, K. K., "Shape Optimal Design of An Arch Dam," *ASCE Journal of Structural Engineering*, Vol. 115, No. 9, pp.2401 - 2405, 1989.
14. Choi, K. K. and Yao, T. M., "3-D Shape Modeling and Automatic Regridding in Shape Design Sensitivity Analysis," *Sensitivity Analysis in Engineering*, NASA Conference Publication 2457, pp. 329-345, 1987.
15. Chang, K. H. and Choi, K. K., "A Geometry Based Shape Design Parameterization Method For Elastic Solids," *Mechanics of Structures and Machines*, Vol. 20, No. 2, pp. 215-252, 1992
16. Haug, E. J., Choi, K. K. and Komkov, V., *Design Sensitivity Analysis of Structural Systems*, Academic Press, New York, 1986.
17. Choi, K. K. and Seong, H. G., "A Domain Method for Shape Design Sensitivity Analysis," *Computer Methods in Applied Mechanics and Engineering*, Vol. 57, pp. 1-15, 1986.
18. Apollo Computer Inc., *Network Computing System (NCS) Reference*, 330 Billerica Road, Chelmsford, MA, 1987.
19. Chang, K. H. and Santos, J. L. T., "Distributed Design Sensitivity Computations on a Network of Computers," *Computers & Structures*, Vol. 37, No. 3, pp. 265-275, 1990.
20. Santos, J. L. T. and Choi, K. K., "Integrated Computational Considerations for Large Scale Structural Design Sensitivity Analysis and Optimization," *GAMM Conference on Discretization Methods and Structural Optimization*, University of Siegen, FRG, October 5-7,1988.
21. Santos, J. L. T., Godse, M. M., and Chang, K. H., "An Interactive Post-Processor for Structural Design Sensitivity Analysis and Optimization: Sensitivity Display and What-If Study," *Computers and Structures*, Vol. 35, No. 1, pp. 1-13, 1990.
22. PDA Engineering, *G/DB - ACCESS GATEWAY UTILITY - Release 2.3*, Software Products Division, 1560 Redhill Avenue, Costa Mesa, CA, July 1988.
23. Peterson, S. and Stone, T. A., "Finite Element Analysis of M88 Tow Bar Clevis," *TACOM Report*, May 15, 1987.
24. DeSalvo, G. J. and Swanson, J. A., ANSYS Engineering Analysis System, *User's Manual Vols I and II*, Swanson Analysis Systems, Inc., P.O. Box 65, Houston, PA, 1989.
25. PDA Engineering, *PAT/IGES Application Interface*, Software Products Division, 1560 Brookhollow Drive, Santa Ana, CA, 1989.

Optimal Design of Vibrating Structures

Tomasz Lekszycki

Institute of Fundamental Technological Research, Polish Academy of Sciences, 00 049 Warsaw, Poland

Abstract: The problem of optimal design of vibrating elastic systems with damping is discussed. Linear constitutive relations in differential form for viscoelastic materials are cited with remarks concerning some experimental results of optimal identification of material damping. These relations are used in the derivation of selected variational principles in complex formulation, together with analysis of forced vibrations and eigenproblems. It is demonstrated that material damping, described in commonly used linear models of viscoelastic materials, significantly changes mechanical properties of a system with even an arbitrarily small amount of dissipated energy. It is shown that a continuous viscoelastic system, in contrast to a perfectly elastic one, may have a finite number of eigenfrequencies and resonant frequencies. Moreover, some of the eigenfrequencies associated with high order eigenmodes may be shifted down, due to dissipation of energy. Therefore the eigenvalue optimization problem is directly related to material characteristics and frequently cannot be formulated in such a manner as for perfectly elastic systems. The formulation of sensitivity analysis and optimization of structures for two important classes of problems, namely forced steady state harmonic vibration and eigenproblems, are presented and discussed.

Keywords: design variable / eigenproblem / experimental verification / material damping / models of viscoelastic materials / objective function / overdamping phenomenon / sensitivity analysis / steady state vibrations

1 Introduction

The problems of optimization of forced vibrations or eigenfrequencies of elastic structures have been discussed in various contexts in numerous papers within the last three decades, and many interesting results were published [11,12]. But in the real world, perfectly elastic materials do not exist and motion is always associated with dissipation of energy.

The aim of the present work is to report and discuss selected results of recent investigations concerning vibrating systems with damping, in the context of optimization problems. The question of proper description of damping properties in modern structural materials should be answered before doing serious dynamic analysis of contemporary structures. Therefore, experimental investigations are sometimes necessary, especially because results of optimization are strongly dependent on the material model and values of the material parameters. In the present work, an assumption is made that in many cases the linear model of viscoelasticity is good enough for analysis and optimization of vibrating mechanical systems. The reason for taking such an assumption is that in complex computer calculations, especially these associated with optimization, the mathematical model should be simple enough to make possible practical computations. On the other hand, it follows from the author's experimental experience that for many materials with high dissipation of energy, linear models of viscoelasticity are satisfactory [8,10].

Relevant to isotropic viscoelastic materials, the following general differential form of a relation between the deviatoric components of stress $\underline{s}_{ij}(t)$ and strain $\underline{e}_{ij}(t)$ is used [1].

$$p_{10}\underline{s}_{ij}(t)+p_{11}\frac{d\underline{s}_{ij}(t)}{dt}+p_{12}\frac{d^2\underline{s}_{ij}(t)}{dt^2}+\ldots=p_{20}\underline{e}_{ij}(t)+p_{21}\frac{d\underline{e}_{ij}(t)}{dt}+p_{22}\frac{d^2\underline{e}_{ij}(t)}{dt^2}+\ldots \tag{1}$$

This equation can be rewritten in a compact form as follows

$$P_1(D)\underline{s}_{ij}(t) = P_2(D)\underline{e}_{ij}(t) \tag{2}$$

where P_1 and P_2 are differential operators

$$P_1(D) = \sum_{k=0}^{N_1} p_{1k}D^k \tag{3}$$

$$P_2(D) = \sum_{k=0}^{N_2} p_{2k}D^k \tag{4}$$

and D denotes differentiation with respect to time.

Sometimes, similar relationships but with fractional derivatives are used, but here consideration is restricted to the case of non-negative integer values of k.

In an entirely similar manner, the dilatational part of the isotropic stress strain relation can be introduced,

$$P_3(D)\underline{\sigma}_{ij}(t) = P_4(D)\underline{\varepsilon}_{ij}(t) \tag{5}$$

with differential operators

$$P_3(D) = \sum_{k=0}^{N_3} p_{3k} D^k \tag{6}$$

$$P_4(D) = \sum_{k=0}^{N_4} p_{4k} D^k \tag{7}$$

In the above relations the following notation has been used,

$$\underline{s}_{ij} = \underline{\sigma}_{ij} - \frac{1}{3}\underline{s}\delta_{ij} \tag{8}$$

$$\underline{e}_{ij} = \underline{\varepsilon}_{ij} - \frac{1}{3}\underline{\Theta}\varepsilon_{ij} \tag{9}$$

$$\underline{s} = \underline{\sigma}_{11} + \underline{\sigma}_{22} + \underline{\sigma}_{33} \tag{10}$$

$$\underline{\Theta} = \underline{\varepsilon}_{11} + \underline{\varepsilon}_{22} + \underline{\varepsilon}_{22} \tag{11}$$

$$\delta_{ij} = \begin{cases} 1 & \text{for} \quad i = j \\ 1 & \text{for} \quad i \neq j \end{cases} \tag{12}$$

Here, consideration is restricted to the class of motions described by the relations,

$$\begin{aligned} \underline{s}_{ij}(t) &= Re\left[s_{ij} e^{\Omega t} \right] \\ \underline{e}_{ij}(t) &= Re\left[e_{ij} e^{\Omega t} \right] \\ \underline{\sigma}_{kk}(t) &= Re\left[\sigma_{kk} e^{\Omega t} \right] \\ \underline{\varepsilon}_{kk}(t) &= Re\left[\varepsilon_{kk} e^{\Omega t} \right] \end{aligned} \tag{13}$$

where Ω is the so called complex frequency,

$$\Omega = \alpha + \iota\omega, \tag{14}$$

the imaginary unit $\iota = \sqrt{-1}$ should be distinguished from the index i, and stress and strain fields s_{ij}, e_{ij}, σ_{kk}, ε_{kk} are defined in complex space,

$$s_{ij} = s_{ij}^{R} + \imath s_{ij}^{I} \quad , \quad e_{ij} = e_{ij}^{R} + \imath e_{ij}^{I}$$
$$\sigma_{kk} = \sigma_{kk}^{R} + \imath \sigma_{kk}^{I} \quad , \quad \varepsilon_{kk} = \varepsilon_{kk}^{R} + \imath \varepsilon_{kk}^{I} \tag{15}$$

and are time independent.

Two independent classes of problems are covered by the above description. For Ω imaginary (i.e., $\alpha = 0$), $\Omega = \imath\omega$, the first of the Equations 13 can be rewritten in the form

$$\underline{s}_{ij}(t) = Re\left[s_{ij} e^{\imath\omega t} \right] = Re[(s_{ij}^{R} + \imath s_{ij}^{I})\,(\cos\omega t + \imath \sin\omega t)]$$
$$= s_{ij}^{R} \cos\omega t - s_{ij}^{I} \sin\omega t \tag{16}$$

and the three others in a similar manner. Equation 16 represents the dependence on time during forced steady state harmonical vibrations with circular frequency ω.

In the other case Ω, in general, has a full complex value and an eigenproblem can be considered. Thus Equation 13, describing damped free vibrations, can be expanded as follows,

$$\underline{s}_{ij}(t) = Re\left[s_{ij} e^{(\alpha + \imath\omega)t} \right] = Re[e^{\alpha t}\,(s_{ij}^{R} + \imath s_{ij}^{I})\,(\cos\omega t + \imath \sin\omega t)]$$
$$= e^{\alpha t}\,(s_{ij}^{R}\cos\omega t - s_{ij}^{I}\sin\omega t) \tag{17}$$

An expanded form of stress-strain relation can be obtained by combining Equations 2 and 5,

$$P_1(D)P_3(D)\,\underline{\sigma}_{ij}(t) = P_2(D)P_4(D)\,\underline{\varepsilon}_{ij}(t) + \frac{1}{3}\delta_{ij}[P_1(D)P_4(D) - P_2(D)P_3(D)]\underline{\varepsilon}_{kk}(t) \tag{18}$$

Equation 13 can now be used to obtain, instead of Equations 2, 5, and 18, equations analogous to these that are valid for perfect elastic materials. After elimination of time dependence,

$$s_{ij} = 2G^{*}\, e_{ij} \tag{19}$$

$$\sigma_{kk} = 3K^{*}\, \varepsilon_{kk} \tag{20}$$

$$\sigma_{ij} = 2G^{*}\, \varepsilon_{ij} + \lambda^{*}\, \Theta\, \delta_{ij} \tag{21}$$

The material parameters in the above relations are complex and dependent on the complex frequency Ω,

$$G^{*} = \frac{1}{2}\,\frac{\Sigma_{k=0}^{N_2}\, p_{2k}\, \Omega^{k}}{\Sigma_{k=0}^{N_1}\, p_{1k}\, \Omega^{k}} \tag{22}$$

$$K^* = \frac{1}{3} \frac{\Sigma_{k=0}^{N_4} p_{4k} \, \Omega^k}{\Sigma_{k=0}^{N_3} p_{3k} \, \Omega^k} \tag{23}$$

and

$$\lambda^* = K^* - \frac{2}{3} G^* \tag{24}$$

2 Analysis of Damped Vibrations

In the present section, the formulation and analysis of harmonic forced vibrations and an eigenproblem in complex variational form are discussed, important properties of considered systems are presented, and selected experimental results of optimal identification of viscoelastic models of materials are mentioned.

2.1 Forced Steady State Harmonic Vibrations

Consider steady state vibrations of linearly damped mechanical structure excited by harmonically varying external loads with a circular frequency ω. Equations 19 - 21 can be used in a similar manner for perfect elastic case in order to derive complex element stiffness matrices. The global stiffness matrix can then be assembled. Let us assume the displacement vector $\underline{X}(t)$ varies in time according to the relation similar to the Equation 13,

$$\begin{aligned} \underline{X}(t) &= Re\left[X e^{\iota \omega t} \right] = Re[(X^R + \iota X^I)\,(\cos\omega t + \iota \sin\omega t)] \\ &= X^R \cos\omega t - X^I \sin\omega t \end{aligned} \tag{25}$$

Application of Equation 25 leads to the following equation of motion which, after elimination of time dependence, corresponds to the statical equation of equilibrium but is rewritten in complex form,

$$KX - \omega^2 MX = F \tag{26}$$

where the term $\omega^2 MX$ represents the complex body forces and F denotes complex vector of external forces,

$$\begin{aligned} \underline{F}(t) &= Re\left[F e^{\iota \omega t} \right] = Re[(F^R + \iota F^I)\,(\cos\omega t + \iota \sin\omega t)] \\ &= F^R \cos\omega t - F^I \sin\omega t \end{aligned} \tag{27}$$

The stiffness matrix K is comprised of element matrices derived with use of complex constitutive relations Equations 19 - 21. Therefore, the elements of matrix K are also complex and frequency (ω) dependent.

Equation 26 implies a virtual work equality occurring for any kinematically admissible variation of displacement vector δX, namely

$$\langle KX,\ \delta\bar{X}\rangle - \omega^2\langle MX,\ \delta\bar{X}\rangle = \langle F,\ \delta\bar{X}\rangle \tag{28}$$

where $\langle x, y\rangle$ denotes the scalar product of complex vectors x, y

$$\langle x,\ y\rangle = x^T\bar{y} = \sum_{j=1}^{N} x_j\bar{y}_j \tag{29}$$

and $x_j = x_j^R + \iota x_j^I$, $\bar{y}_j = y_j^R - \iota y_j^I$.

Consider the mutual potential energy dependent on displacement fields X and Y

$$\Pi(X, Y) = \frac{1}{2}\langle LX - F, Y\rangle - \frac{1}{2}\Psi(X) \tag{30}$$

where $\Psi(X)$ is an arbitrary function of X, and L is a linear operator. Its variation due to changes of vectors X and Y yields

$$\delta\Pi(X, Y) = \frac{1}{2}\langle LX - F, \delta Y\rangle + \frac{1}{2}\langle \delta X, L^a Y\rangle - \frac{1}{2}\langle \delta X, \overline{\frac{d\Psi}{dX}}\rangle \tag{31}$$

where L^a denotes an adjoint operator to L.

Condition $\delta\Pi(X, Y) = 0$ for arbitrary variations δX and δY implies

$$LX - F = 0 \tag{32}$$

and

$$L^a Y - \overline{\frac{d\Psi}{dX}} = 0 \tag{33}$$

The last two equations establish stationarity conditions of mutual potential energy. The first constitutes the equilibrium equation, Equation 26, for the considered mechanical structure for L defined as

$$L = K - \omega^2 M \tag{34}$$

The second provides the state equation for an adjoint system. Since $\Psi(X)$ is an arbitrary function of X, it can be assumed that $\Psi(X) = X^T F$ and then

$$\overline{\frac{d\Psi}{dX}} = \bar{F} \tag{35}$$

Since matrices M and K are symmetric, as only self-adjoint problems are considered here, the adjoint matrix $L^a = \bar{L}^T = \bar{L}$ and the equilibrium equation for the adjoint system is

$$\bar{L}Y - \bar{F} = 0 \tag{36}$$

It follows from the above relations that

$$Y = \bar{X} \tag{37}$$

and the mutual potential energy is expressed as follows:

$$\Pi(X) = \frac{1}{2}\langle KX - \omega^2 MX, \bar{X}\rangle - \langle F, \bar{X}\rangle \tag{38}$$

Therefore, for self-adjoint systems, the complex potential energy is built only on the displacement field X of the considered (primary) system, and the stationarity condition of $\Pi(X)$ implies the equilibrium equation, Equation 26. In a similar manner, the appropriate relations for non- self-adjoint systems can be derived [5].

2.2 Eigenproblem

Consider now the free motion of linearly damped structures. The material model described by Equations 2 and 5 is used here, but similar study can be done for finite memory model considered in [2]. Assuming the motion of the system to take the form

$$\begin{aligned} \underline{X}(t) &= Re\left[Xe^{\Omega t}\right] = Re[e^{\alpha t}(X^R + \iota X^I)\,(\cos\omega t + \iota\sin\omega t)] \\ &= e^{\alpha t}\,(X^R\cos\omega t - X^I\sin\omega t) \end{aligned} \tag{39}$$

which leads to the state equation in the form

$$KX + \Omega^2 MX = 0 \tag{40}$$

The stiffness matrix K is now dependent on the complex eigenvalue Ω. The associated complex potential energy has the form

$$\Pi(X) = \frac{1}{2}\langle KX + \Omega^2 MX, \bar{X}\rangle \tag{41}$$

The stationary condition of Equation 41 implies Equation 40. In fact, for any kinematically admissible variation of displacement vector,

$$\delta\Pi(X) = \frac{1}{2}\langle K\delta X + \Omega^2 M\delta X, \bar{X}\rangle + \frac{1}{2}\langle KX + \Omega^2 MX, \delta\bar{X}\rangle = \langle KX + \Omega^2 MX, \delta\bar{X}\rangle \quad (42)$$

Thus the following theorem can be formulated: Among all kinematically admissible displacement vectors X, the eigenvectors satisfying Equation 40 correspond to a stationary value of the complex potential energy Π specified by Equation 41.

Assume now that both X and Ω in the functional of Equation 41 can vary, so $\Pi = \Pi(X, \Omega)$. The variation of Π yields

$$\delta\Pi(X, \Omega) = \langle KX + \Omega^2 MX, \delta\bar{X}\rangle + \frac{1}{2}\langle 2\Omega MX + K_{,\Omega}X, \bar{X}\rangle\delta\Omega \quad (43)$$

The stationarity requirement for $\Pi(X, \Omega)$ implies

$$\langle 2\Omega MX + K_{,\Omega}X, \bar{X}\rangle\delta\bar{\Omega} = 0 \quad (44)$$

so either

$$\Omega = -\frac{\langle K_{,\Omega}X, \bar{X}\rangle}{2\langle MX, \bar{X}\rangle} \quad (45)$$

or

$$\delta\Omega = 0 \quad (46)$$

Equation 45 provides the condition for critical damping. On the other hand, Equation 46 describes the stationarity condition of functional $\Omega(X)$, if one exists, for an eigenvector X satisfying the state equation. In fact for simple models of viscoelasticity, such as Voigt or Zener models, these functionals can be expressed in explicit form. For the simple illustration the relations for Voigt model will be quoted here. In such a case, the complex stiffness matrix can be expressed as follows:

$$K = S + \Omega C \quad (47)$$

where S denotes elastic real stiffness matrix and C is a damping matrix. The complex kinetic T, dissipative D, and elastic U energies can be defined, respectively, as follows:

$$T = \langle MX, \bar{X}\rangle \quad (48)$$

$$D = \langle CX, \bar{X}\rangle \quad (49)$$

$$U = \langle SX, \bar{X}\rangle \quad (50)$$

and the functional $\Omega(X)$ has a form,

$$\Omega(X) = \frac{-D + \sqrt{D^2 - 4TU}}{2T} \tag{51}$$

or

$$\Omega'(X) = \frac{-D - \sqrt{D^2 - 4TU}}{2T} \tag{52}$$

The stationarity condition $\delta\Omega = 0$ implies the state equation, provided that damping differs from the critical damping of the system.

Selected important properties of systems with damping described with use of the simplest models of viscoelasticity covered by above relations are presented in the following part of this section. In order to accomplish that, consideration is restricted to the very simple case of so called "proportional damping." The complex stiffness matrix K can be rewritten in the form

$$K(\Omega) = S\mu(\Omega) \tag{53}$$

where S is a real elastic stiffness matrix and $\mu(\Omega)$ follows from the constitutive equation, Equation 21,

$$\mu(\Omega) = \frac{\Sigma_{k=0}^{k=k_N} \Omega^k b_k}{\Sigma_{k=0}^{k=k_M} \Omega^k a_k} \tag{54}$$

and a_k and b_k are material parameters.

With such assumptions, the state equation takes the following form:

$$SX + \frac{\Omega^2}{\mu(\Omega)} MX = 0 \tag{55}$$

On the other hand, for elastic systems, $a_k = 0$ $(k = 1, ..., k_N)$, $b_k = 0$ $(k = 1, ..., k_M)$, and Equation 55 can be reduced to the simple form

$$SX - \omega_e^2 MX = 0 \tag{56}$$

where ω_e is a circular eigenfrequency of a perfect elastic system with no damping.

Comparison of the two last equations leads to the conclusion that eigenfunctions of perfect elastic and associated viscoelastic systems are equal, and the direct relation between eigenvalues of both systems can be found in the following form:

$$-\omega_e^2 = \frac{\Omega^2}{\mu(\Omega)} \tag{57}$$

This relation is used in the next part of this section to investigate selected dynamical properties of structural elements built of viscoelastic materials that are described by simple models, using the constitutive relation assumed here and satisfying assumptions of "proportional damping."

Consider Equation 57 for the simplest viscoelastic solid, the so called Voigt - Kelvin model. This model is often used in numerical calculations, for the sake of its simplicity, in spite of the fact that it does not offer good approximation of real structural materials. Although it has defects [8], the fundamental relations and their effects for Voigt-Kelvin material are discussed here in detail because the basic conclusions have important consequences in optimization of structures. They also allow better understanding of complex behaviors of systems that behave in accord with more advanced models. Additionally, certain designed structural members, such as dampers or vibration absorbers, can sometimes be described by the Voigt-Kelvin relation.

The stiffness matrix has the form

$$K(\Omega) = S(1 + \Omega b_1) \tag{58}$$

One of the weak points of the Voigt - Kelvin model can be observed when the forced harmonic vibrations with a given circular frequency ω_f are under consideration. In such a case, the complex frequency $\Omega = i\omega_f$ and the real part of the complex stiffness is constant, while its imaginary part is linearly depends linearly on ω_f,

$$Re(K) = S \quad , \qquad Im(K) = \omega_f b_1 S \tag{59}$$

This conclusion is in disagreement with experimental results, where diminishing values of $Im(K)$ are observed for values of ω_f growing to infinity.

The relation of Equation 57 is now examined, in order to present some interesting and important properties of viscoelastic vibrating systems. It follows from this equation that

$$\Omega^2 + \gamma\omega_e^2\Omega + \omega_e^2 = 0 \tag{60}$$

where γ, instead of b^1, has been used for the damping coefficient.

The solution of Equation 60 provides the real and imaginary parts of Ω as functions of ω_e. These relations are illustrated in Figure 1 for different values of the damping coefficient ($\gamma = 0$ and $\gamma = 0.4$). It is clear there exists some critical value of ω_e such that, for all greater eigenfrequencies of elastic systems, the associated complex eigenvalues of the viscoelastic one have imaginary parts equal to zero. This means the viscoelastic continuous system has only a finite number of eigenfrequencies, all of the other eigenvalues being associated with non-periodic motion because of overdamping phenomenon.

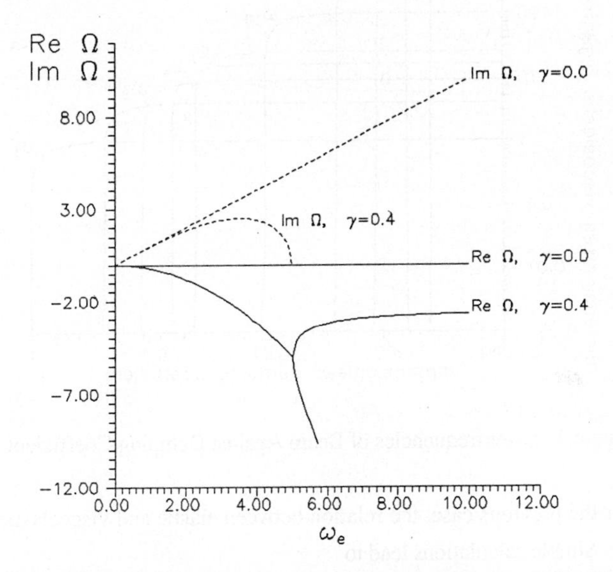

Figure 1. Complex Eigenvalues of Viscoelastic System (Voigt Model)

The eigenfrequencies ω_k of a simply supported viscoelastic beam have been examined, in order to discuss the consequences of the above conclusion. The plots for selected first eigenfrequencies (k = 1, 2, ..., 10, ..., 50) and different values of dimensionless damping coefficient (γ/γ_{cr}), where γ_{cr} is a critical damping over which the free vibrations of the beam are not possible, are presented in Figure 2. Assuming that damping coefficient falls between the values 0.01 and 0.1, it follows from this picture that the considered beam has at most ten eigenfrequencies, and all the others are overdamped. In addition, there exist domains where the order of eigenfrequencies is changed; i.e., the low frequencies are associated with higher-order eigenmodes. This situation sometimes appears for perfect elastic systems, but this results from a particular shape of structure and mass distribution giving priority for a specified eigenmode of vibration. In the present example, this is not the case. The beam is uniform, and it exhibits a damping phenomenon that results in changed order of eigenfrequencies.

The next model of viscoelastic material investigated here is a Zener model. This is a much more realistic assumption to choose for materials description. The stiffness matrix has the following form:

$$K(\Omega) = S\,\frac{1+\Omega b_1}{1+\Omega a_1} \tag{61}$$

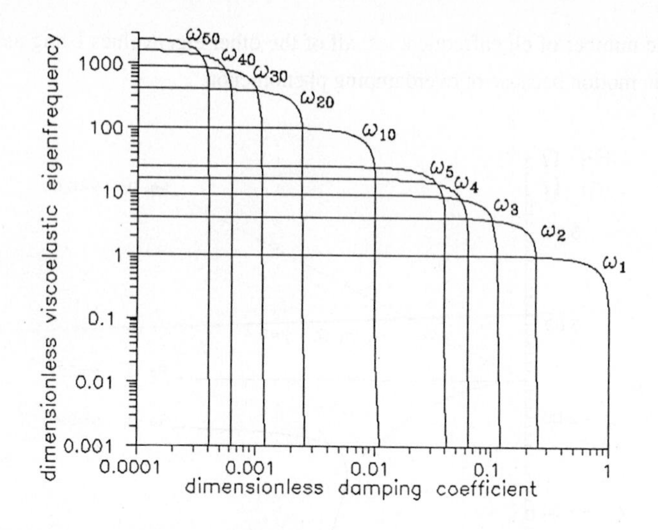

Figure 2. Eigenfrequencies of Beam Against Damping Coefficient

Again, as for the previous case, the relation between elastic and viscoelastic eigenvalues can be examined. Simple calculations lead to

$$a_1\Omega^3 + \Omega^2 + b_1\omega_e^2\Omega + \omega_e^2 = 0 \tag{62}$$

It follows from Equation 62 there are three eigenvalues associated with one eigenmode, three multiple or distinct real, or one real and two conjugate complex. Similarly to Figure 1 for the Voigt model, the function $\Omega(\omega_e)$ is plotted in Figures 3, 4, and 5, for different values of material parameters. It can be observed, depending on the material, that some local minima are possible, as well as the domains where the system is overdamped. This conclusion is important and should be taken into consideration in formulation of synthesis problems [6,9].

3 Optimization and Sensitivity Analysis of Vibrating Viscoelastic Structures

The focus of this section is on a problem of sensitivity analysis and optimization of vibrating viscoelastic structures. Two important classes of problems are discussed; optimization of structures under forced harmonic vibrations and optimization of complex eigenvalues.

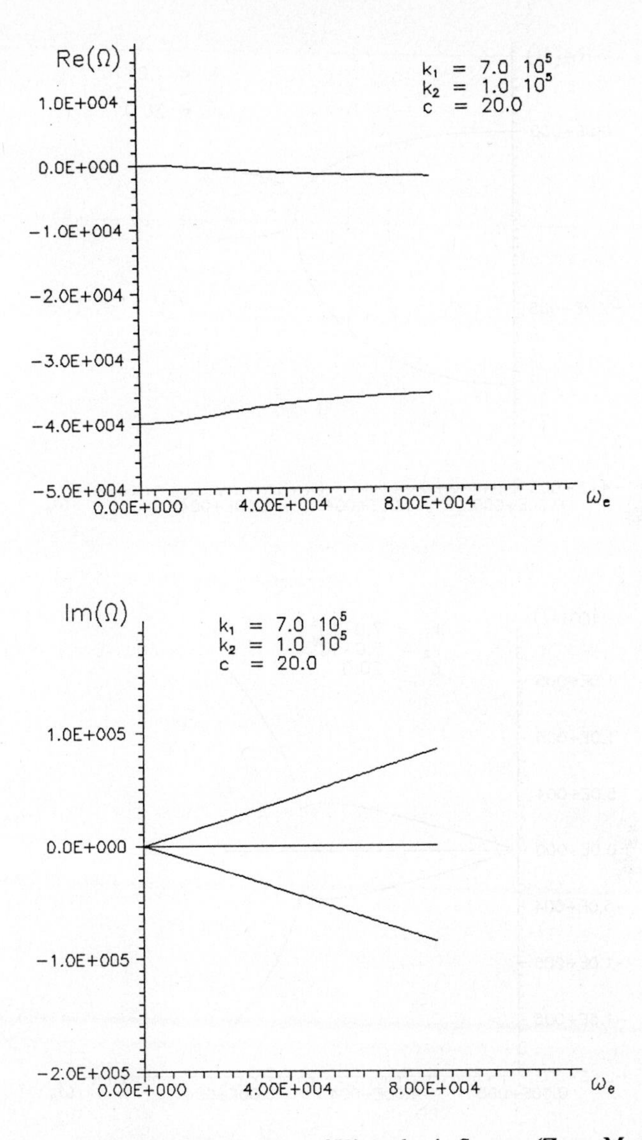

Figure 3. Complex Eigenvalues of Viscoelastic System (Zener Model)

3.1 Optimal Design and Sensitivity Analysis of Structures Under Forced Vibrations

The stationarity condition of an arbitrary objective function is derived in this section. In a particular case, the relations obtained can be simplified in order to perform a sensitivity analysis.

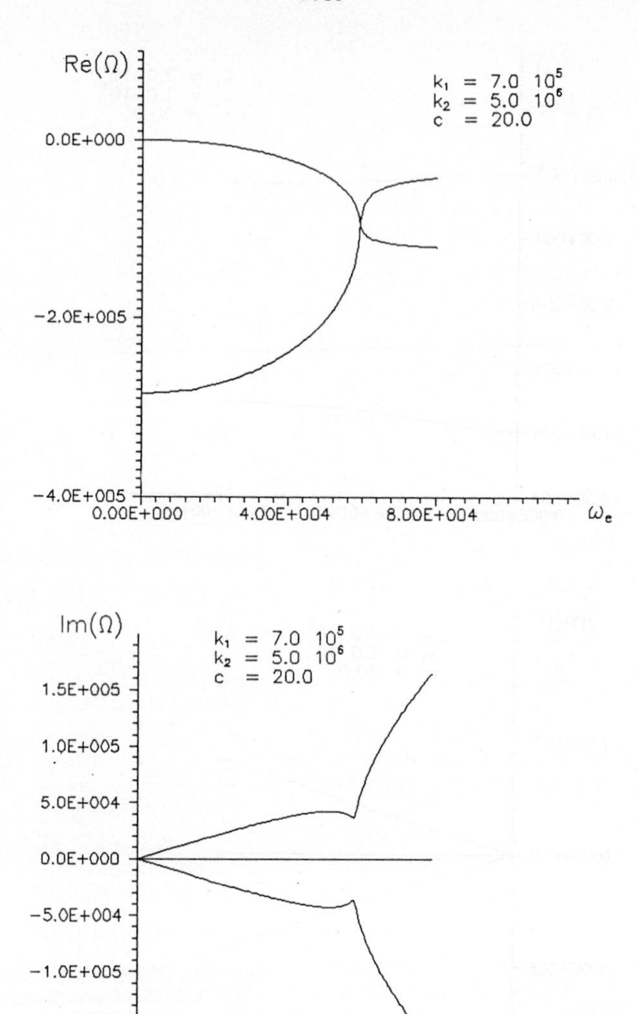

Figure 4. Complex Eigenvalues of Viscoelastic System (Zener Model)

Consider the optimal design of a mechanical structure, with r design parameters constituting the design vector d. Assume the objective function G is dependent on the design vector d and the displacement vector X,

$$G = G(X, d) \tag{63}$$

As there are constraints imposed on vectors d and X, they can be introduced into the objective function G by means of Lagrange multipliers. Therefore, define the new design

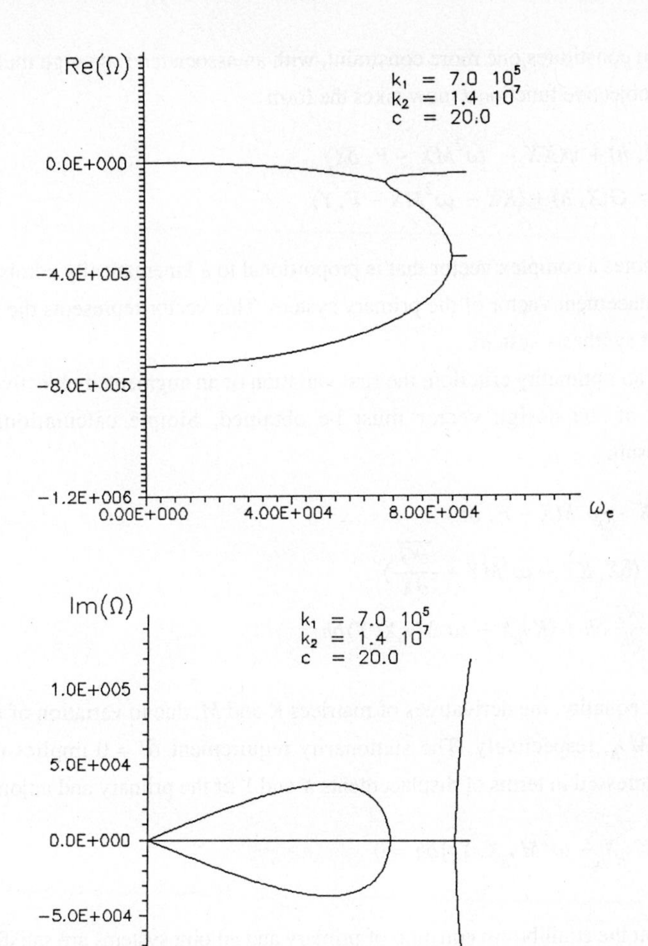

Figure 5. Complex Eigenvalues of Viscoelastic System (Zener Model)

vector h by extending d by Lagrange multipliers associated with these constraints. The extended objective function then becomes

$$G^* = G^*(X, h) \tag{64}$$

The complex equilibrium equation follows from the condition $\delta\Pi = 0$, which is equivalent to the virtual work equality,

$$\langle KX - \omega^2 MX - F, \delta X \rangle = 0 \tag{65}$$

This relation constitutes one more constraint, with an associated Lagrange multiplier ψ. The augmented objective function J now takes the form

$$J = G(X, h) + \psi\langle KX - \omega^2 MX - F, \delta X\rangle$$
$$= G(X, h) + \langle KX - \omega^2 MX - F, Y\rangle \tag{66}$$

where Y denotes a complex vector that is proportional to a kinematically admissible variation δX of a displacement vector of the primary system. This vector represents the displacements of an adjoint synthesis system.

To state an optimality criterion, the first variation of an augmented objective function due to changes of the design vector must be obtained. Simple calculations lead to the following result:

$$\delta J = \langle KX - \omega^2 MX - F, \delta Y\rangle$$
$$+ \langle \delta X, \bar{K}Y - \omega^2 MY + \overline{\frac{\partial G}{\partial X}}\rangle$$
$$+ \frac{\partial G}{\partial h}\delta h + \langle K_{,h}X - \omega 2M_{,h}X, Y\rangle\delta h \tag{67}$$

In the above equality, the derivatives of matrices K and M, due to variation of h, are denoted by $K_{,h}$ and $M_{,h}$, respectively. The stationarity requirement $\delta J = 0$ implies the optimality condition expressed in terms of displacements X and Y of the primary and adjoint structures,

$$\left(\frac{\partial G}{\partial h} + \langle K_{,h}X - \omega^2 M_{,h}X, Y\rangle\right)\delta h = 0 \tag{68}$$

providing that the equilibrium equation of primary and adjoint systems are satisfied,

$$KX - \omega^2 MX - F = 0$$
$$\bar{K}Y - \omega^2 MY + \overline{\frac{\partial G}{\partial X}} = 0 \tag{69}$$

To perform the sensitivity analysis, the relation of Equation 67 can be used with the vector h replaced by the design parameters d.

The above general relations can be used in the solution of specific problems. In particular, the problem of shape optimization of layers of materials attached to viscoelastic beams with high dissipation of energy damping due to forced vibrations, the optimization of support conditions in frames, and sensitivity analysis of vibrating plates on viscoelastic foundation have been solved successfully [3,4,5,7].

3.2 Eigenproblem Optimization

The optimization of complex eigenvalues is a subject of the present section. There exists another important class of problems associated with eigenproblem optimization, namely optimization of eigenmodes but it is not considered here.

Consider free vibrations of a viscoelastic structure, with the constitutive law for the structural material given in the complex form Equation 21. According to the previous considerations, the equation of equilibrium describing an eigenproblem takes the form

$$KX + \Omega^2 MX = 0 \tag{70}$$

where the complex stiffness matrix K is dependent on the complex frequency Ω. The general relation for the variation of a complex eigenvalue is derived here. The application of its general form can be difficult or time consuming in numerical computations. For specific particular problems, it can be simplified to very simple formulas that can be used in sensitivity analysis or optimization of structures.

Consider the variation of the Equation 70,

$$\begin{aligned}(K_{,d}X + \Omega^2 M_{,d}X)\delta d + (K_{,\Omega}X + 2\Omega MX)\delta\Omega \\ + K\delta X + \Omega^2 M\delta X = 0\end{aligned} \tag{71}$$

Multiplication of Equation 71 by $\bar{X}$ implies

$$\begin{aligned}\langle K_{,d}X + \Omega^2 M_{,d}X,\bar{X}\rangle\, \delta d + \langle K_{,\Omega}X + 2\Omega MX, \bar{X}\rangle\, \delta\Omega \\ + \langle KX + \Omega^2 MX, \overline{\delta X}\rangle = 0\end{aligned} \tag{72}$$

It follows from Equation 72 that the variation of an eigenvalue, due to variation of design parameters, is

$$\delta\Omega = -\frac{\langle K_{,d}X + \Omega^2 M_{,d}X, \bar{X}\rangle}{\langle K_{,\Omega}X + 2\Omega MX, \bar{X}\rangle}\delta d \tag{73}$$

For the particular case of the Voigt material discussed earlier, this relation takes the form

$$\delta\Omega = -\frac{\langle K_{,d}X + \Omega^2 M_{,d}X, \bar{X}\rangle}{\sqrt{D^2 - 4TU}}\delta d \tag{74}$$

As a simple example of an eigenvalue optimization, the maximization of the lowest eigenfrequency ($Im[\Omega]$) of a simply supported viscoelastic beam made of Voigt material is presented here. Assume the cross-sectional area of a beam is given and the area moment of inertia is a design parameter. The relation between the design parameter and eigenfrequencies associated with ten lowest eigenmodes are displayed in Figure 6. It can be observed that the objective function is not continuous, and the problem cannot be solved by a typical procedure. Moreover, additional constraints imposed on the design variable are necessary.

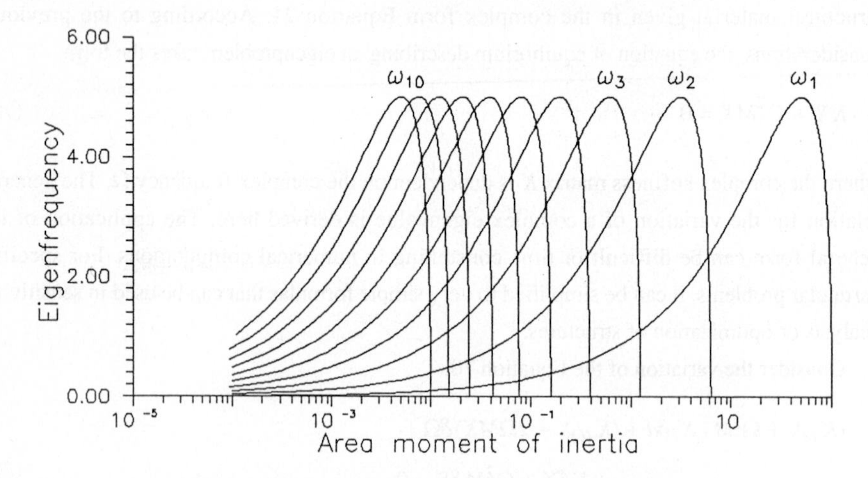

Figure 6. Eigenfrequencies Against Stiffness of a Beam

This simple example shows some of possible difficulties arising during optimization of viscoelastic structures.

4 Conclusions

Very simple examples have been considered to show that for optimization of real structures, the dissipation of energy should be taken into consideration. The results of optimization are strongly dependent on the model of viscoelastic material and values of material parameters used for calculations. Therefore the analysis and optimization in various cases should be preceeded by identification of structural material. On the other hand, it is now possible to design a material with required optimal behavior. The three analysis functions; identification, analysis, and synthesis of structures, should thus go together in the design process to achieve the advantages created by modern structural materials and computer facilities.

References

1. Christensen R.M., "*THEORY OF VISCOELASTICITY: An Introduction*", Academic Press, Ed.1, 1971
2. Frischmuth K., Kosiński W. and Lekszycki T. "Free Vibrations of Finite Memory Material Beams", *Int. J. Eng. Sc.*, 1992
3. Lekszycki T. and Olhoff N., "Optimal Design of Viscoelastic Structures Under Forced Steady-State Vibration", *J. Struct. Mech.*, Vol.9, No 4, pp.363-387, 1981
4. Lekszycki T. and Mróz Z., "On Optimal Support Reaction in Viscoelastic Vibrating Structures ", *J. Struct. Mech.*, Vol.11, No 1, pp. 67-79, 1983
5. Lekszycki T. and Mróz Z., "Variational principles in analysis and synthesis of elastic systems with damping", *Solid Mech. Arch.*, Vol.14, No 3-4, pp. 181-202, 1989, (The special issue dedicated to the memory of Prof. H. Leipholz)
6. Lekszycki T., "Eigenvalues optimization - new view about the old problem", *Control and Cybernetics*, Vol. 19, 3-4, pp. 189-201, 1990
7. Lekszycki T., "Application of Variational Methods in Analysis and Synthesis of Viscoelastic Continuous Systems", *J. Mech. of Struct. and Machines*, Vol.19, No 2, pp.163-192, 1991
8. Lekszycki T., Olhoff N. and Pedersen J.J., "Modelling and identification of viscoelastic properties of vibrating sandwich beams", *Composite Structures*, Vol. 22, 1992
9. Lekszycki T., "Optimization of Vibrating Structures with Damping", in preparation for publication in *Structural Optimization*
10. Lekszycki T., Pedersen J.J. and Chaika V., "Attempt to combine analytical solution of free vibration of elastic cantilever beam with experimental results from viscoelastic beam", in preparation for publication
11. Olhoff N., "A survey of the optimal design of vibrating structural elernents", *Shock Vib. Dig.*, Vol.8, No.8, pp. 3-10, (Part 1: Theory), Vol.8, No.9, pp. 3-10, (Part 11: Applications), 1976
12. Olhoff N. "Variational methods in structural eigenvalue optimization" in Cinquini C. and Rovati M. editors, *Computer Aided Optimal Design of Structures*, Vol.2, pp. 1-56, 1989

The Problem of Additional Load of Minimum Work in a Thin Plate

Salvatore Sergio Ligaró

Istituto di Scienza delle Costruzioni, Università di Pisa, 56126 Pisa, Italy

Abstract: A rectangular elastic plate is simply supported along its boundary. A flexural couple is exerted on one of its sides, while the others remain unloaded. The question is posed, assuming the ability to produce couples along the side opposite the already loaded one, of determining the values of these latter couples which would minimize the elastic strain energy of the system.

The solution is obtained in explicit form by representing the transverse displacement in the form of a Fourier series and directly calculating the minimum of the strain energy.

Keywords: incomplete information / optimization of loads

1 Introduction

When dealing with structural optimization problems, the functionals usually considered for optimization are either the structure's total weight or stiffness under designated load conditions, or some parameters indicating the stress level. However, the variety of problems posed in optimization is a great deal wider. At times loads are not unconditionally defined; all that is known is that they must produce a certain resultant and act on a specific part of the structure. The problem therefore arises, given known resultant and region of action, of determining the point distribution necessary in order to optimize a certain effect. This type of problem is commonly called of extremum with "incomplete information" [1]. The most typical case, described by Banichuk, is that of a thin plate of variable thickness **h** on which the load per unit surface **q** is of the same sign at all points, for instance $\mathbf{q} \geq 0$, the resultant $\mathbf{R} = \iint_{\Omega} q\,dx\,dy$ (Ω is the area) is assigned and the function h(x,y), which minimizes the overall weight of the structure, is sought.

Unfortunately these problems can not be dealt with in a systematic fashion even in their simplest formulation, that is, a one-dimensional structure loaded perpendicularly on its longitudinal axis. Moreover, all conceivable numerical calculation procedures turn out to be highly unstable and therefore compromise all efforts at precision in the early stages of calculation.

The present note addresses a further, more specific problem, still regarding the equilibrium of thin plates. A thin rectangular plate, simply supported on its sides, is loaded along one of its edges with a known distribution of flexural couples. These couples, which might simulate, for instance, the effect of elastic connection of the side to another structure, generate on the plate a state of stress with an associated elastic strain energy. It is assumed that flexural couples can now be applied along the side opposite that of the loaded one. These additional couples generate new stresses which are superimposed on the initial ones. The problem at hand, therefore, is calibrating these second couples such that the final strain energy of the system is not only reduced, but minimized. According to the terminology already adopted in elasticity theory (see [2]), the added couples will serve to improve the "apparent stiffness" of the structure.

An alternative version of the problem is one in which the plate is supported on two opposite sides only, with the other sides free. On one of these latter, known shear forces are applied, and the transverse load which needs to be applied on the other in order to minimize strain energy is sought.

Both problems can be dealt with explicitly through Fourier series representation of data and unknowns. Since the functions of the series are mutually orthogonal, the strain energy is then represented by the sum of squares and its minimum can be calculated directly without having to solve a system of infinite linear equations.

The results, quite unexpected, are illustrated in the following two numerical examples.

2 The Simply Supported Plate

A rectangular plate with sides **a** and **b** (Fig. 1) occupies the region $0 \le x \le a$, $0 \le y \le b$ of the plane x,y. It is simply supported on its four sides and initially loaded on side $y = 0$ with flexural couples distributed according to the analytic expression

$$\overline{m_y}(x) = \sum_{n=1}^{\infty} Y_n \sin \alpha_n x, \qquad \alpha_n = \frac{n\pi}{a}, \tag{2.1}$$

where Y_n are constants and n are integer numbers. It is implicit that the function $\overline{m_y}$ in (2.1) can be developed in a Fourier sine series within the interval $0 \leq x \leq a$.

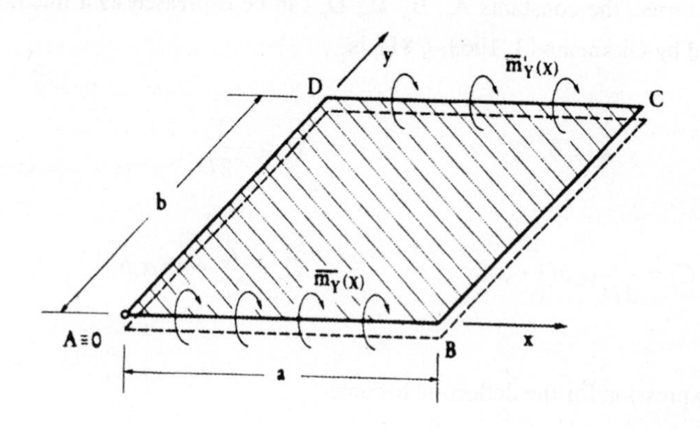

Fig. 1. A simply supported rectangular plate.

Calculation of the deflection produced by the boundary couples $\overline{m_y}$ can be done simply by separation of variables, which leads to the expression

$$w(x,y) = \sum_{n=1}^{\infty} \frac{1}{\alpha_n^2}(A_n \cosh \alpha_n y + \alpha_n y B_n \sinh \alpha_n y + C_n \sinh \alpha_n y + \alpha_n y D_n \cosh \alpha_n y) \sin \alpha_n x, \quad (2.2)$$

in which A_n, B_n, C_n, D_n are unknown constants which are determined on the basis of the boundary conditions

$$\left.\begin{array}{llll} w = 0, & m_x = 0, & \text{for} & x = 0, \quad x = a, \\ w = 0, & m_y = \overline{m_y}, & \text{for} & y = 0, \\ w = 0, & m_y = 0, & \text{for} & y = b. \end{array}\right\} \quad (2.3)$$

In (2.3) m_x and m_y must be written as a function of **w** by means of the constitutive equations

$$m_x = -D\left(\frac{\partial^2 w}{\partial x^2} + \sigma \frac{\partial^2 w}{\partial y^2}\right), \qquad m_y = -D\left(\frac{\partial^2 w}{\partial y^2} + \sigma \frac{\partial^2 w}{\partial x^2}\right), \quad (2.4)$$

where D is the plate's flexural stiffness and σ is the Poisson factor.

Substituting (2.2) into the four boundary conditions (2.3) and imposing term to term equality in the series, the constants A_n, B_n, C_n, D_n can be expressed as a function of Y_n. The result, furnished by Girkmann [3, 1963, § 81], is

$$A_n = 0, \qquad B_n = -\frac{Y_n}{2D}, \tag{2.5}$$

$$C_n = \frac{Y_n}{2D}\alpha_n b(1-\coth^2\alpha_n b), \qquad D_n = \frac{Y_n}{2D}\coth\alpha_n b,$$

thus, the full expression for the deflection becomes

$$w(x,y) = \frac{1}{2D}\sum_{n=1}^{\infty}\frac{Y_n}{\alpha_n}\left[y\cosh\alpha_n(b-y) - b\frac{\sinh\alpha_n y}{\sinh\alpha_n b}\right]\frac{\sin\alpha_n x}{\sinh\alpha_n b}. \tag{2.6}$$

From this, all other characteristics of deformation and stress can be determined, in particular the rotation of the loaded boundary y = 0

$$\frac{\partial w}{\partial y}(x,0) = \sum_{n=1}^{\infty} Y_n f_{yn}\sin\alpha_n x, \tag{2.7}$$

and that of the opposite side y = b

$$\frac{\partial w}{\partial y}(x,b) = \sum_{n=1}^{\infty} Y_n f'_{yn}\sin\alpha_n x, \tag{2.8}$$

having put

$$f_{yn} = \frac{1}{2D\alpha_n}\left[\coth\alpha_n b - \frac{\alpha_n b}{\sinh^2\alpha_n b}\right], \qquad f'_{yn} = \frac{\alpha_n b\coth\alpha_n b - 1}{2D\alpha_n\sinh\alpha_n b}. \tag{2.9}$$

Formula (2.6) represents the plate's "initial" load state. Let

$$\overline{m'_y}(x) = \sum_{n=1}^{\infty} Y'_n \sin \alpha_n x, \tag{2.10}$$

be a new couple distributed on the side at y = b ; it is to be regarded as an unknown, but we assume that $\overline{m_y}$ and $\overline{m'_y}$ are equally oriented for every x within the interval $0 \leq x \leq a$. The solution can again be reached by separation of variables and the application of the new boundary conditions. Repetition of this calculation, although trivial, is however unnecessary since the rotations of the supports y = 0, and y = b can be determined directly using formulas (2.7) and (2.8) in which Y_n has been replaced by Y'_n and f_{yn} e f'_{yn} are inverted, as are the signs. Having designated **w'** as the new deflection, the rotations at the two boundaries y = 0, y = b, take the form

$$\frac{\partial w'}{\partial y}(x,0) = -\sum_{n=1}^{\infty} Y'_n f'_{yn} \sin \alpha_n x, \tag{2.11}$$

$$\frac{\partial w'}{\partial y}(x,b) = \sum_{n=1}^{\infty} Y'_n f_{yn} \sin \alpha_n x. \tag{2.12}$$

When the two couples act simultaneously, the above information is sufficient to calculate the strain energy since, by virtue of Clapeyron's theorem, it is equal to half the work done by the external loads, that is

$$W = \frac{1}{2}L = \frac{1}{2}\int_0^a \overline{m_y}\left[\frac{\partial w}{\partial y}(x,0) + \frac{\partial w'}{\partial y}(x,0)\right]\mathrm{d}x + \frac{1}{2}\int_0^a \overline{m'_y}\left[\frac{\partial w}{\partial y}(x,b) + \frac{\partial w'}{\partial y}(x,b)\right]\mathrm{d}x. \tag{2.13}$$

From this, by substituting the expressions which give the moments and rotations, and integrating with respect to x, we obtain

$$W = \frac{a}{4}\sum_{n=1}^{\infty} Y_n(f_{yn}Y_n - f'_{yn}Y'_n) + \frac{a}{4}\sum_{n=1}^{\infty} Y'_n(f_{yn}Y'_n - f'_{yn}Y_n). \tag{2.14}$$

This is a quadratic form of the coefficients Y_n e Y'_n, the first of which is known and the second to be determined. The nth term of the series contains only the coefficients Y_n e Y'_n, and is positive, as the quadratic expression

$$f_{yn}Y_n^2 - 2f'_{yn}Y_nY'_n + f_{yn}Y'^2_n$$

is positive definite, its discriminant being strictly positive. It is therefore a minimum for those values of Y'_n for which

$$Y'_n = \frac{f'_{yn}}{f_{yn}}Y_n, \tag{2.15}$$

or alternatively

$$Y'_n = \frac{\alpha_n b\cosh\alpha_n b - \sinh\alpha_n b}{\sinh\alpha_n b\cosh\alpha_n b - \alpha_n b}Y_n. \tag{2.16}$$

Placing $z = \alpha_n b$, the function

$$f(z) = \frac{z\cosh z - \sinh z}{\sinh z\cosh z - z},$$

defined for $z \geq 0$, is positive, decreasing and with an upper limit of the value $f(0) = 1/2$, by which, if the series ΣY_n^2 is converging, then so is $\Sigma Y'^2_n$. This means that, if the moment $\overline{m_y}$ is square summable then so will $\overline{m'_y}$.

Furthermore, the optimality condition (2.15) requires satisfaction of the condition

$$\frac{\partial w}{\partial y}(x,b) + \frac{\partial w'}{\partial y}(x,b) = 0, \tag{2.17}$$

and this means that the distribution of the opposing couple $\overline{m'_y}$ which minimizes the elastic strain energy of the plate will coincide with that of the clamped edge. This result can be considered an extension to a two-dimensional system of Castigliano's theorem [4] in the inverse formulation and of Menabrea's theorem of minimum work [5].

Having determined the coefficient Y'_n, a comparison can be made of the strain energy due to the presence of the couple $\overline{m_y}$ with that produced by the combined action of $\overline{m_y}$ and $\overline{m'_y}$. In the former case, as the values of Y'_n are all zero, from (2.14) we can obtain

$$W = \frac{a}{4}\sum_{n=1}^{\infty} f_{yn} Y_n^2; \tag{2.18}$$

in the second, placing $Y'_n = f'_{yn}/f_{yn}\, Y_n$, we instead find

$$W = \frac{a}{4}\sum_{n=1}^{\infty} f_{yn} Y_n^2 \left[1 - \frac{f'^2_{yn}}{f^2_{yn}}\right]. \tag{2.19}$$

which is clearly less than the former.

3 The Case of a Single Concentrated Couple

In order to have a numerical check of the gain in strain energy obtained by applying opposite couples on the y = b side, it is convenient to consider the case in which the distribution $\overline{m_y}$ is made up of a single concentrated couple, M_y, applied at the mid point of the side AB at coordinates $x = a/2$, $y = 0$. Although such conditions do not satisfy the assumptions made above, as the function $\overline{m_y}$ is not square summable, this case can be dealt with by utilizing formal Fourier series, since the strain energy associated to it is still finite.

The couple concentrated at the mid point of the $y = 0$ side can be developed in a Fourier series when, beginning with its representation in the form of a series of flexural couples of intensity $\overline{m_y}(x) = M_y/2$ distributed uniformly in the interval $a/2-\varepsilon \leq x \leq a/2+\varepsilon$, the limit is reached by making ε tend towards zero, so that

$$\lim_{\varepsilon\to 0} \int_{\frac{a}{2}-\varepsilon}^{\frac{a}{2}+\varepsilon} \overline{m_y}\, \mathrm{d}x = \lim_{\varepsilon\to 0} 2\overline{m_y}\varepsilon = M_y. \tag{3.1}$$

Since each coefficient Y_n is by definition given by the integral

$$Y_n = \frac{\int_0^a \overline{m_y} \sin \alpha_n x \, \mathrm{d}x}{\int_0^a \sin^2 \alpha_n x \, \mathrm{d}x},$$

it is easy to show that

$$Y_n = \frac{2}{a} \lim_{\varepsilon \to 0} \int_{\frac{a}{2}-\varepsilon}^{\frac{a}{2}+\varepsilon} \overline{m_y} \sin \alpha_n x \, \mathrm{d}x = 2\frac{M_y}{a} \sin\frac{n\pi}{2}, \qquad (n = 1,3,5,...). \qquad (3.2)$$

In this particular case $\Sigma Y_n{}^2$ is not converging, but instead the series

$$\frac{\partial w}{\partial y}\left(\frac{a}{2},0\right) = \sum_{n=1}^{\infty} Y_n f_{yn} \sin\frac{n\pi}{2}, \qquad (n = 1,2,3,...), \qquad (3.3)$$

which represents the rotation of point x = a/2, y = 0, where the couple acts, is converging.

As a result, the work of deformation produced by this couple is finite, and we can therefore proceed with determination of the coefficients of the opposite couple $\overline{m'_y}$ to apply along the y = b side. By directly applying (2.15), we obtain

$$\overline{m'_y}(x) = 2\frac{M_y}{a} \sum_{n=1}^{\infty} \frac{f'_{yn}}{f_{yn}} \sin\frac{n\pi}{2} \sin \alpha_n x, \qquad (n = 1,2,3,...). \qquad (3.4)$$

The formal result is that the distribution of the opposing couples $\overline{m'_y}$ need not necessarily be concentrated, but is quite regular, which can be illustrated with some numerical examples.

If, for the sake of simplicity, we consider the case of a square plate and place a = b, then α_n b = nπ, and therefore from (2.16)

$$Y'_n = \frac{n\pi \cosh n\pi - \sinh n\pi}{\sinh n\pi \cosh n\pi - n\pi} Y_n,$$

where now $Y_n = 2\, M_y/a \sin n\pi/2$.

The plot of the function $\overline{m'_y} = \Sigma Y'_n \sin \alpha_n x$, presented in Figure 2 for several values of the ratio b/a, confirms the aforementioned law of distribution.

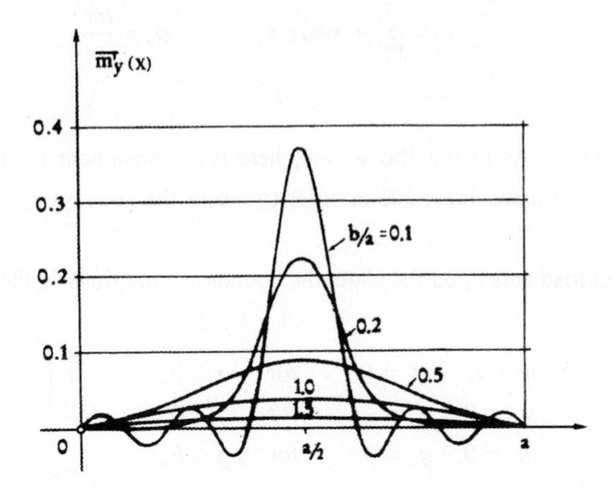

Fig. 2. Optimal distribution of couples.

4 A Plate Supported on Two Opposite Sides

An instance which is in some respects complementary to the one considered above is that in which a rectangular plate is supported on two opposite edges, say for instance x = 0 and x = a, while the other two remain free (see Fig. 3).

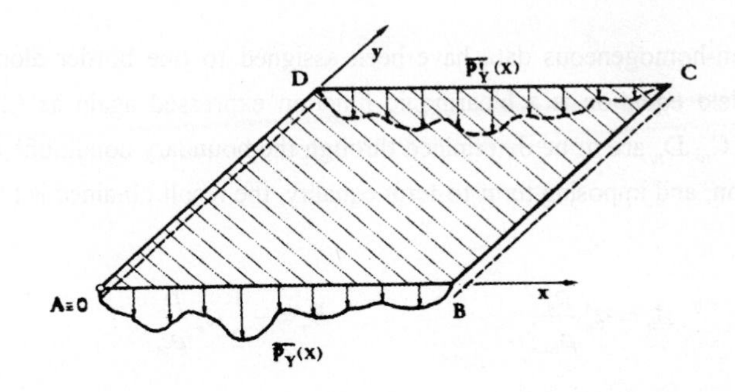

Fig. 3. A rectangular plate supported on two opposite sides.

The $y = 0$ edge is loaded by forces perpendicular to the middle plane given by the expression

$$\overline{p_y}(x) = \sum_{n=1}^{\infty} p_n \sin \alpha_n x, \qquad \alpha_n = \frac{n\pi}{a}, \tag{4.1}$$

where p_n are constants. As in the above case, here it is convenient to impose the *a priori* condition that $\overline{p_y}(x)$ is square integrable in order to insure the convergence of the series $\Sigma\, p_n^2$ (see [6]).

If only the $\overline{p_y}(x)$ load acts upon the plate, the boundary conditions of the problem are

$$\left.\begin{array}{llll} w = 0, & m_x = 0, & \text{for} & x = 0, \quad x = a, \\ m_y = 0, & q_y = -\overline{p_y}, & \text{for} & y = 0, \\ m_y = 0, & q_y = 0, & \text{for} & y = b, \end{array}\right\} \tag{4.2}$$

where m_y, the bending moment along the y axis, and q_y, the generalized shear force on the elements perpendicular to the y axis, can both in turn be expressed as functions of **w**. Assuming for simplicity's sake that $\sigma = 0$, the expressions which link these characteristics to **w** take the form

$$m_y = -D\frac{\partial^2 w}{\partial y^2}, \qquad q_y = -D\left(\frac{\partial^3 w}{\partial y^3} + 2\frac{\partial^3 w}{\partial x^2 \partial y}\right). \tag{4.3}$$

Since the non-homogeneous data have been assigned to one border alone, the general solution to the field equation is a biharmonic function expressed again as (2.2), where the constants A_n, B_n, C_n, D_n are to be determined through the boundary conditions (4.2). Carrying out the substitution, and imposing term-to-term equality, the result obtained is (see [3, § 85])

$$A_n = \alpha_n^2 \frac{p_n}{\rho_{1n}}, \qquad B_n = -\frac{1}{2}\alpha_n^2 \frac{p_n}{\rho_{1n}},$$

$$C_n = -\frac{3\sinh^2 \alpha_n b - 1/2\alpha_n^2 b^2}{3/2\sinh 2\alpha_n b + \alpha_n b}\frac{p_n}{\rho_{1n}}, \qquad D_n = \frac{3/2\sinh^2 \alpha_n b}{3/2\sinh 2\alpha_n b + \alpha_n b}\frac{p_n}{\rho_{1n}}, \tag{4.4}$$

in which

$$\rho_{1n} = \frac{1}{2} D\alpha_n \frac{9\sinh^2 \alpha_n b - \alpha_n^2 b^2}{3/2\sinh 2\alpha_n b + \alpha_n b}.$$

As the constants are known, so is the deflection w(x,y), and furthermore, the displacement w(x,0) can be calculated

$$w(x,0) = \sum_{n=1}^{\infty} \frac{p_n}{d_{n0}} \sin \alpha_n x, \tag{4.5}$$

as well as that of the side y = b

$$w(x,b) = \sum_{n=1}^{\infty} \frac{p_n}{d_{nb}} \sin \alpha_n x, \tag{4.6}$$

in which we have placed

$$d_{n0} = \rho_{1n}\alpha_n^2 = \frac{1}{2} D\alpha_n^3 \frac{9\sinh^2 \alpha_n b - \alpha_n^2 b^2}{3/2\sinh 2\alpha_n b + \alpha_n b}, \tag{4.7}$$

$$d_{nb} = \rho_{1n}\alpha_n^2 \left[\cosh \alpha_n b - \frac{1}{2}\alpha_n b \sinh \alpha_n \right.$$

$$\left. - \frac{3\sinh^2 \alpha_n b - 1/2\alpha_n^2 b^2}{3/2\sinh 2\alpha_n b + \alpha_n b} \sinh \alpha_n b + \frac{3/2\sinh^2 \alpha_n b}{3/2\sinh 2\alpha_n b + \alpha_n b} \alpha_n b \cosh \alpha_n b \right]^{-1}. \tag{4.8}$$

If we now distribute onto side y = b the unknown loads,

$$\overline{p_y}(x) = \sum_{n=1}^{\infty} p_n^{'} \sin \alpha_n x, \tag{4.9}$$

the displacements on sides y = 0 and y = b are immediately calculable by means of formulas (4.5) and (4.6), where p_n is now replaced by p'_n, and d_{n0} and d_{nb} must be inverted. Indicating the new deflection with **w'**, the displacements of the two edges y = 0 and y = b are expressed as

$$w'(x,0) = \sum_{n=1}^{\infty} \frac{p'_n}{d_{nb}} \sin \alpha_n x, \tag{4.10}$$

$$w'(x,b) = \sum_{n=1}^{\infty} \frac{p'_n}{d_{n0}} \sin \alpha_n x, \tag{4.11}$$

so that, if $\overline{p_y}$ and $\overline{p'_y}$ act simultaneously, the resulting strain energy is

$$W = \frac{1}{2} L = \frac{1}{2}\int_0^a \overline{p_y}\left[w(x,0) + w'(x,0)\right] \mathrm{d}\,x + \frac{1}{2}\int_0^a \overline{p'_y}\left[w(x,b) + w'(x,b)\right] \mathrm{d}\,x. \tag{4.12}$$

Furthermore, carrying out substitution of the expressions that yield loads and displacements and integrating with respect to x, we obtain

$$W = \frac{a}{4} \sum_{n=1}^{\infty} \left[\frac{p_n^2}{d_{n0}} + 2\frac{p_n p'_n}{d_{nb}} + \frac{p'^2_n}{d_{n0}} \right], \tag{4.13}$$

a quadratic form for the coefficients p_n and p'_n, the values p_n being known and the p'_n to be determined. The nth term of the series contains the coefficients p_n and p'_n alone, and results positive, as the quadratic form

$$\frac{p_n^2}{d_{n0}} + 2\frac{p_n p'_n}{d_{nb}} + \frac{p'^2_n}{d_{n0}}$$

is positive defined since its discriminant is strictly positive. It is a minimum for those values of p'_n for which

$$p_n' = -\frac{d_{n0}}{d_{nb}} p_n, \tag{4.14}$$

or, that is for

$$p_n' = \left[\frac{1}{2}\alpha_n b \sinh \alpha_n b - \cosh \alpha_n b\right.$$

$$\left. + \frac{3\sinh^2 \alpha_n b - 1/2\alpha_n^2 b^2}{3/2\sinh 2\alpha_n b + \alpha_n b}\sinh \alpha_n b - \frac{3/2\sinh^2 \alpha_n b}{3/2\sinh 2\alpha_n b + \alpha_n b}\alpha_n b \cosh \alpha_n b\right] p_n. \tag{4.15}$$

Placing again $z = \alpha_n b$, the function

$$g(z) = \cosh z - \frac{1}{2} z \sinh z - \frac{3\sinh^2 z - 1/2 z^2}{3/2 \sinh 2z + z} + \frac{3/2\sinh^2 z}{3/2\sinh 2z + z} z \cosh z,$$

defined for $z \geq 0$, results positive and decreasing with a maximum $g(0) = 1$, by which, if the series Σp_n^2 is converging, then so is $\Sigma p'^2_n$. This means that, if the load $\overline{p_y}$ is square summable, so does $\overline{p'_y}$.

The optimality condition (4.14) furthermore provides

$$w(x,b) + w'(x,b) = 0, \tag{4.16}$$

which means that the distribution of the opposite forces, $\overline{p'_y}$, which minimizes the plate elastic strain energy coincides with that of the simply supported edge.

As the coefficients p'_n are known, it is possible to carry out a comparison of the strain energy resulting from the presence of the forces $\overline{p_y}$ with that consequent to the combined action of $\overline{p_y}$ and $\overline{p'_y}$. In the former case, as the values of p'_n are all zero, from (4.12) we can obtain

$$W = \frac{a}{4}\sum_{n=1}^{\infty}\frac{p_n^2}{d_{n0}}, \tag{4.17}$$

while in the second, by putting $p'_n = -d_{n0}/d_{nb}\,p_n$, we instead obtain

$$W = \frac{a}{4}\sum_{n=1}^{\infty}\frac{p_n^2}{d_{n0}}\left[1-\frac{d_{n0}^2}{d_{nb}^2}\right]. \tag{4.18}$$

5 The Case of a Single Concentrated Load

With the aim of numerically assessing the gain in strain energy obtained by applying opposite loads on the y = b side, it is convenient to consider the case in which the distribution $\overline{p_y}$ is made up of a single concentrated load, P_y, applied at the mid point of the side AB at coordinates x = a/2, y = 0. Although the function $\overline{p_y}$ is not square summable, a formal Fourier series can still be used since the strain energy associated to it is finite.

The load concentrated at the mid point of the y = 0 side can be developed in a Fourier series when, beginning with its representation in the series form of a uniform load of intensity $\overline{p_y}(x) = P_y/2\varepsilon$ acting in the interval $a/2-\varepsilon \le x \le a/2+\varepsilon$, the limit is reached by making ε tend towards zero, so that

$$\lim_{\varepsilon\to 0}\int_{\frac{a}{2}-\varepsilon}^{\frac{a}{2}+\varepsilon}\overline{p_y}\,\mathrm{d}\,x = \lim_{\varepsilon\to 0} 2\overline{p_y}\varepsilon = P_y. \tag{5.1}$$

Since each coefficient p_n is, by definition given by the integral expression

$$p_n = \frac{\int_0^a \overline{p_y}\sin\alpha_n x\,\mathrm{d}\,x}{\int_0^a \sin^2\alpha_n x\,\mathrm{d}\,x},$$

it can be easily shown that

$$p_n = \frac{2}{a}\lim_{\varepsilon\to 0}\int_{\frac{a}{2}-\varepsilon}^{\frac{a}{2}+\varepsilon} p_y \sin\alpha_n x \,\mathrm{d}x = 2\frac{P_y}{a}\sin\frac{n\pi}{2}, \qquad (n = 1,3,5,\ldots). \tag{5.2}$$

The series Σp_n^2 is not converging, while instead the series

$$w(a/2,0) = \sum_{n=1}^{\infty}\frac{p_n}{d_{n0}}\sin\frac{n\pi}{2}, \qquad (n = 1,2,3,\ldots), \tag{5.3}$$

which represents the displacement of the point x = a/2, y = 0 where the load acts, is converging.

Consequently, the work of deformation produced by P_y is bounded, and we can therefore proceed again to calculation of the coefficients of the opposite load $\overline{p'_y}$ which is to be applied along the y = b side. Directly applying (4.15), we obtain

$$p_n' = 2\frac{P_y}{a}\Bigg[\frac{1}{2}\alpha_n b \sinh\alpha_n b - \cosh\alpha_n b$$

$$+\frac{3\sinh^2\alpha_n b - 1/2\alpha_n^2 b^2}{3/2\sinh 2\alpha_n b + \alpha_n b}\sinh\alpha_n b - \frac{3/2\sinh^2\alpha_n b}{3/2\sinh 2\alpha_n b + \alpha_n b}\alpha_n b\cosh\alpha_n b\Bigg]\sin\frac{n\pi}{2}. \tag{5.4}$$

The formal result is that $\overline{p'_y}$ is not necessarily a concentrated force, but is instead quite regular. In the case of a square plate, by placing a = b, and α_n b = n π, then (5.4) provides

$$p_n' = 2\frac{P_y}{a}\Bigg[\frac{1}{2}n\pi\sinh n\pi - \cosh n\pi$$

$$+\frac{3\sinh^2 n\pi - 1/2n^2\pi^2}{3/2\sinh 2n\pi + n\pi}\sinh n\pi - \frac{3/2\sinh^2 n\pi}{3/2\sinh 2n\pi + n\pi}n\pi\cosh n\pi\Bigg]\sin\frac{n\pi}{2}. \tag{5.5}$$

The plot of the function $\overline{p'_y} = \Sigma p'_n \sin\alpha_n x$, presented in Figure 4 for several values of the ratio b/a, confirms the above property.

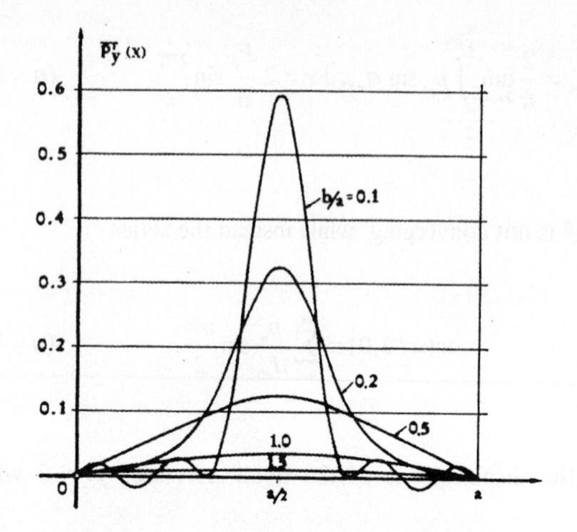

Fig. 4. Optimal distribution of transversal loads.

References

1. N. V. Banichuck, *Problems and Methods of Optimal Structural Design*, Plenum Press, New York, (1983).
2. H. Neuber, *Theory of Nocth Stresses*, Edwards, Ann Arbor, Michigan, (1946).
3. K. Girkmann, *Flächentragwerke*, Springer-Verlag, Wien, (1963).
4. A. Castigliano, Nuova teoria intorno all'equilibrio dei sistemi elastici, R. Acc. d. Scienze, Torino, (1875).
5. L. F. Menabrea, Principe d'èlasticité ou principe du moindre travail, R. Acc. d. Scienze, Torino, (1871).
6. H. F. Weinberger, *A First Course in Partial Differential Equations*, Blaisdell, New York, (1965).

Layout Optimization of Large FE Systems by New Optimality Criteria Methods: Applications to Beam Systems

O. Sigmund, M. Zhou and G.I.N. Rozvany

FB 10, Essen University, W-4300 Essen 1, Germany

Abstract: The aim of this paper is to demonstrate the optimization capability of recently developed optimality criteria methods (COC, DCOC) by solving layout problems for beam systems (grillages) involving many thousand potential members. Although the method presented can include all practical design constraints, its validity and accuracy is also verified by comparing the numerical output with closed form analytical solutions for some layout problems.

Keywords: optimality criteria / layout optimization / topology optimization / grillages / finite-elements / elastic design / plastic design / stress constraints / deflection constraints

1 Introduction–COC/DCOC

One of the difficulties in the optimization of large structural systems is the discrepancy between analysis capability (up to ~10^5 DF's) and optimization capability (~10^3 variables and ~10^3 active behavioral constraints if, respectively, primal and dual programming or traditional OC methods are used). It was demonstrated recently by the second and third authors [1,2] that the above discrepancy can be eliminated, and even reversed, by using new, continuum type optimality criteria (COC) methods. A modified version of the above technique (termed DCOC) was formulated by Zhou [3] directly for discretized systems using matrix notation, which makes this method readily applicable to finite elements.

The almost unlimited capability of COC/DCOC in terms of the number of design variables *and* active behavioral constraints is due to the fact that in these methods, the computer time requirement depends essentially only on the number of active global (e.g., deflection) constraints and the number of such constraints for typical large structural systems is very small in comparison to the active local (e.g., stress) constraints.

In DCOC the Lagrange multipliers for active stress constraints are determined explicitly at the element level and their effect on the optimal design is ensured by using suitable prestrains in a modified virtual load system termed *adjoint system*. These two operations require a rather negligible amount of computer time in comparison to the repeated analyses of the considered structure. The analysis of the adjoint system also requires little computational effort because it only involves the forward and backward substitution of the decomposed stiffness matrix of the real system.

In Section 2, general aspects of layout optimization by the proposed methods are discussed, and in Section 3 the treatment of beams by DCOC is reviewed. Section 4 is concerned with an extension of DCOC to non-prismatic beam elements. In Section 5 some results are compared with known analytical solutions, while in Section 6 it is shown how DCOC results can lead to new analytical solutions. Finally, Section 7 discusses practical applications including stress and deflection constraints .

2 Layout Optimization by the COC/DCOC Methods

COC/DCOC are particularly useful in layout optimization problems because the latter always involve a very large number of potential members (elements).

Layout optimization by OC methods is based on the so-called layout theory, developed in the seventies by Prager and Rozvany (e.g., [4]), which uses optimality criteria and the concept of a so-called "structural universe" (or "ground structure"), the union of all potential members. The idea of a "ground structure" was used already in the sixties by Dorn, Gomory and Greenberg [5] who, however, combined it with a linear programming method. Since continuum based optimality criteria give adjoint strains also for vanishing members (of zero cross-sectional area), in convex problems the fulfillment of these criteria for all potential members represents a necessary and sufficient condition for layout optimality and also provides the optimal cross-sections for non-vanishing members. The same criteria have been used successfully for nonconvex problems, for which they constitute only a necessary condition, provided that certain additional requirements are satisfied. While in *analytical layout solutions* the structural universe consists of an *infinite number of potential members* and the *cross-sectional area can take on a zero value,* numerical COC/DCOC solutions are based on a *finite* but large number (up to many thousand) *of potential members* and a *finite* but small (down to 10^{-12} times the average) *prescribed minimum cross-sectional area.* Owing to the large number of potential members, COC/DCOC effectively achieves a simultaneous optimization of the topology, geometry and cross-sectional sizes [2, Part II]. These methods thereby complement favorably the two-stage procedure for layout optimization (e.g., Kirsch [6]).

3 Formulation for Prismatic Beam Elements

Although the DCOC method [7,3] has been developed for most design conditions used in practice, a simplified version of it is used here within the following restrictions:

- one single load condition;
- stress constraints and one deflection constraint;
- only nodal loads are applied;
- the objective function is the structural weight;
- there is only one design variable per element, which is not linked with other design variables.

The optimality criteria for the above problem take the following form:

$$\frac{\partial W^e}{\partial x^e} + \nu\left(\{\bar{\mathbf{F}}\}^T\left[\frac{\partial \mathbf{f}}{\partial x^e}\right]\{\mathbf{F}\}\right) + \lambda_j^e\left\{\frac{\partial \mathbf{S}_j^e}{\partial x^e}\right\}^T\{\mathbf{F}\} - \beta^e = 0\,, \tag{1}$$

where W^e is the weight of the element e, x^e the design variable for the element e, $\{\bar{\mathbf{F}}\}$ the vector of virtual nodal forces (in flexibility formulation), $[\mathbf{f}]$ the flexibility matrix, $\{\mathbf{F}\}$ the vector of real nodal forces, $\{\mathbf{S}_j^e\}$ a vector converting the nodal forces into a stress at location j of element e, and ν, λ_j^e and β^e are Lagrange multipliers for the displacement constraint, stress constraints and lower side constraints ($x^e \geq \underline{x}^e$), respectively. These multipliers are nonzero only if the corresponding constraints are active.

For the so-called adjoint structure

- the flexibility matrix is the same as for the real structure;
- the load is the virtual load (e.g., unit "dummy" load at the constrained displacements), subject to
- the initial nodal displacement (prestrains) in flexibility formulation

$$\{\bar{\mathbf{u}}_i^e\} = \frac{1}{\nu}\sum_j \lambda_j^e\{\mathbf{S}_j^e\}\ . \tag{2}$$

The actual analysis of the real and adjoint structure is carried out by the stiffness method and hence the initial displacements in (2) are converted into fixed end forces in the actual computational procedure. It is assumed in this paper that the torsional stiffness is very low and hence torsion can be neglected in (1). This neglect causes no significant error in the final result because optimal beam directions are known to coincide with the principal directions with zero torsion.

In this paper we consider beam elements of constant depth and variable width (x^e) clamped at the left end. If F_1 and F_2 are the vertical force and bending moment at the free end of the beam, we have the following matrices (in flexibility formulation).

Flexibility matrix:

$$[\mathbf{f}] = \frac{1}{x^e}\begin{bmatrix} \dfrac{4L_e^3}{Eh^3} & \dfrac{6L_e^2}{Eh^3} \\ \dfrac{6L_e^2}{Eh^3} & \dfrac{12L_e}{Eh^3} \end{bmatrix} = \frac{[\mathbf{f}_0]}{x^e} \quad . \tag{3}$$

Bending stress constraint:

$$\left\{\mathbf{S}_1^e\right\}\left\{\mathbf{F}^e\right\} \le \sigma_a \text{ with} \left\{\mathbf{S}_1^e\right\} = \begin{Bmatrix} \dfrac{6\xi^e}{h^2 x^e} \\ \dfrac{6}{h^2 x^e} \end{Bmatrix} \text{sgn } M_{\max} \, . \tag{4}$$

Shear stress constraint:

$$\left\{\mathbf{S}_2^e\right\}\left\{\mathbf{F}^e\right\} \le \tau_a \text{ with} \left\{\mathbf{S}_2^e\right\} = \begin{Bmatrix} \dfrac{1.5}{hx^e} \\ 0 \end{Bmatrix} \text{sgn } V_{\max} \, , \tag{5}$$

where h is the depth of all beams, $M_{\max}$ and $V_{\max}$ are the moment and shear force with the greatest absolute value, $\xi^e = 0$ and $\xi^e = L_e$, respectively, if the maximum moment is at the same end as or opposite to the nodal forces $\{\mathbf{F}^e\}$ considered and L_e is the length of element e.

On the basis of (1) and (3), the redesign formula for beam elements with inactive stress constraints ($\lambda_j = 0$ for $j = 1, 2$) becomes

$$x^e = \sqrt{\nu \{\bar{\mathbf{F}}^e\} [\mathbf{f}_0^e] \{\mathbf{F}^e\}/w^e} \tag{6}$$

where the superscript "e" indicates that the corresponding symbol is restricted to element e and w^e is the factor giving the weight W^e of the beam element e:

$$W^e = w^e x^e \text{ with } w^e = \gamma h L_e \, , \tag{7}$$

where γ is the specific weight. Elements governed by (6) will be termed *active elements* (A) and those governed by stress or side constraints termed *passive elements* (P). The value of the Lagrange multiplier can be calculated from (6) and the deflection constraint (as an equality):

$$\Delta = \sum_e \{\bar{\mathbf{F}}^e\} [\mathbf{f}^e] \{\mathbf{F}^e\} = \sum_A \frac{\{\bar{\mathbf{F}}^e\} [\mathbf{f}_0^e] \{\mathbf{F}^e\}}{\sqrt{\nu \{\bar{\mathbf{F}}^e\} [\mathbf{f}_0^e] \{\mathbf{F}^e\}/w^e}} + \sum_P (\{\bar{\mathbf{F}}^e\} [\mathbf{f}^e] \{\mathbf{F}^e\}) \, , \tag{8}$$

implying

$$\sqrt{\nu} = \frac{\sum_{A} \sqrt{w^e \{\bar{\mathbf{F}}^e\} [\mathbf{f}_0^e] \{\bar{\mathbf{F}}^e\}}}{\Delta - \sum_{P} (\{\bar{\mathbf{F}}^e\} [\mathbf{f}^e] \{\mathbf{F}^e\})} . \tag{9}$$

Each iteration in the DCOC algorithm consists of the following steps:

- analysis of the real and adjoint systems;
- calculation of the Lagrange multiplier from (9);
- resizing of beam elements on the basis of (4), (5), (6), and the side constraints.

4 Extension to Nonprismatic Elements

Denoting the widths at the two ends of a linearly varying element by x_1 and x_2 and its length by L, the four entries in the flexibility matrix in (3) are replaced by:

$$f_{11} = \frac{L_e^3}{EI_1} \frac{2x_2^2 \ln\left(\frac{x_2 L_e}{x_1}\right) - 2x_2^2 \ln L_e - (x_1 - 3x_2)(x_1 - x_2)}{2(x_1 - x_2)^3} ,$$

$$f_{12} = f_{21} = \frac{L_e^2}{EI_1} \frac{x_2 \ln\left(\frac{x_2 L_e}{x_1}\right) - x_2 \ln L_e - x_2 + x_1}{2(x_1 - x_2)^2} ,$$

$$f_{22} = \frac{L_e}{EI_1} \frac{\ln\left(\frac{x_2 L_e}{x_1}\right) - \ln L_e}{x_1 - x_2} , \tag{10}$$

where $I_1 = h^3/12$. To ensure computational stability, the flexibility matrix for prismatic members (3) was used if $|x_1 - x_2| < 2.5 \times 10^{-5}$.

Non-prismatic elements have only been used so far for either a single stress constraint for each member or a compliance (total external work) constraint. For either of these, the COC method [l] gives the redesign formula

$$x_i^e = \nu \, |M_i| \quad (i = 1, 2) . \tag{11}$$

The flexibility coefficients in (10) are not valid if the moment changes sign within a beam element because in that case the width varies bi-linearly over that element. However, this error is insignificant if the number of beam elements is large. Moreover, such elements do not usually occur in the final optimal solution.

5 Comparisons of FE Solutions with Known Analytical Solutions

Figure 1 shows some simple analytical layout solutions for grillages with various boundary conditions (e.g., [8]). Double thin lines denote simple supports, thick lines clamped edges, and single thin lines free edges. Arrows indicate the optimal beam directions and the signs refer to the sign of the optimal beam moments. Small circles with a sign indicate that all beam directions are equally optimal.

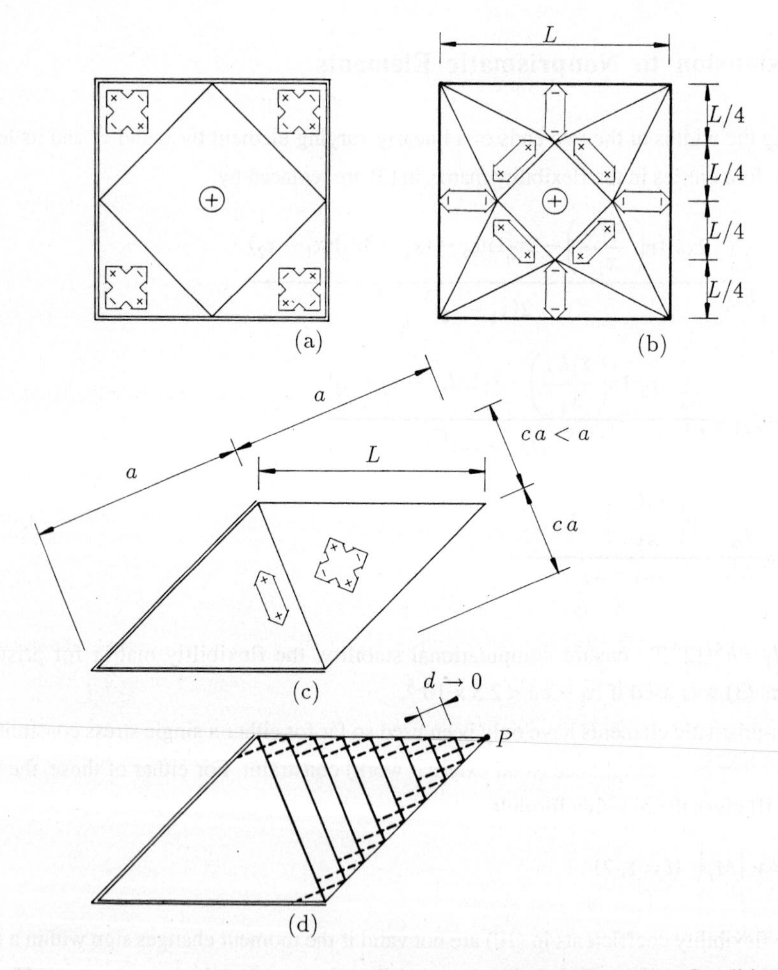

Figure 1. Analytical Grillage Layout Solutions for Various Boundary Conditions

In comparing the optimal weight of analytical and numerical solutions, two types of design problems are considered:

(a) stress (strength) constraint (elastic or plastic design);

(b) compliance constraint (elastic design).

The total weight for these two classes of problems will be denoted by W_S and W_C, respectively. It can be shown ([2], Part II, Equation 15) that for *normalized* problems with the same boundary and loading conditions, we have

$$W_C = W_S^2 \ . \tag{12}$$

5.1 Simply Supported Square Grillage

The layout based on the analytical solution for this problem is given in Figure la. Some discretized solutions were generated with the very simple ground structure (structural universe) with 80 elements shown in Figure 2a. In this and subsequent figures, a small circle in the middle shows the position of a unit point load. It can be seen from Figure la that for the central region in this solution, all beam directions are optimal so long as the beam moments are positive. The COC program with non-prismatic beams gave first the solution in Figure 2b. After removing from the ground structure (Figure 2a), in turn, beams with a slope of 2:1 and 1:1 to the sides, the solutions in Figures 2c and 2d were obtained. For a side length of $L = 2$, all three optimal layouts (Figures 2b to 2d) gave the same normalized weight ($W_S = 0.5$, $W_C = 0.25$) as the known analytical solution. This also shows that in the considered region, the optimal beam directions are non-unique.

5.2 Square Grillage with Built-In Edges

Figure lb shows the analytically derived optimal layout for this problem [8]. Using a structural universe with 624 members (Figure 3a), the discretized solutions shown in Figures 3b and c were obtained for the nine point loads (circles in Figure 3b, all point loads are unit forces, except that the central load has a value of 4.0). The latter are in complete agreement with the above analytical solution. In the central region of Figure lb, where the optimal beam direction is indeterminate, the discretized solution gave a rather dense grid of beams (which is one of the possible optimal solutions). For a side length of $L = 2$, the analytical optimal weights are $W_S = 31/64 = 0.484375$, $W_C = W_S^2 = 0.234619$. The above numerical solution (Figures 3b and c) gave $W_C = 0.234620$.

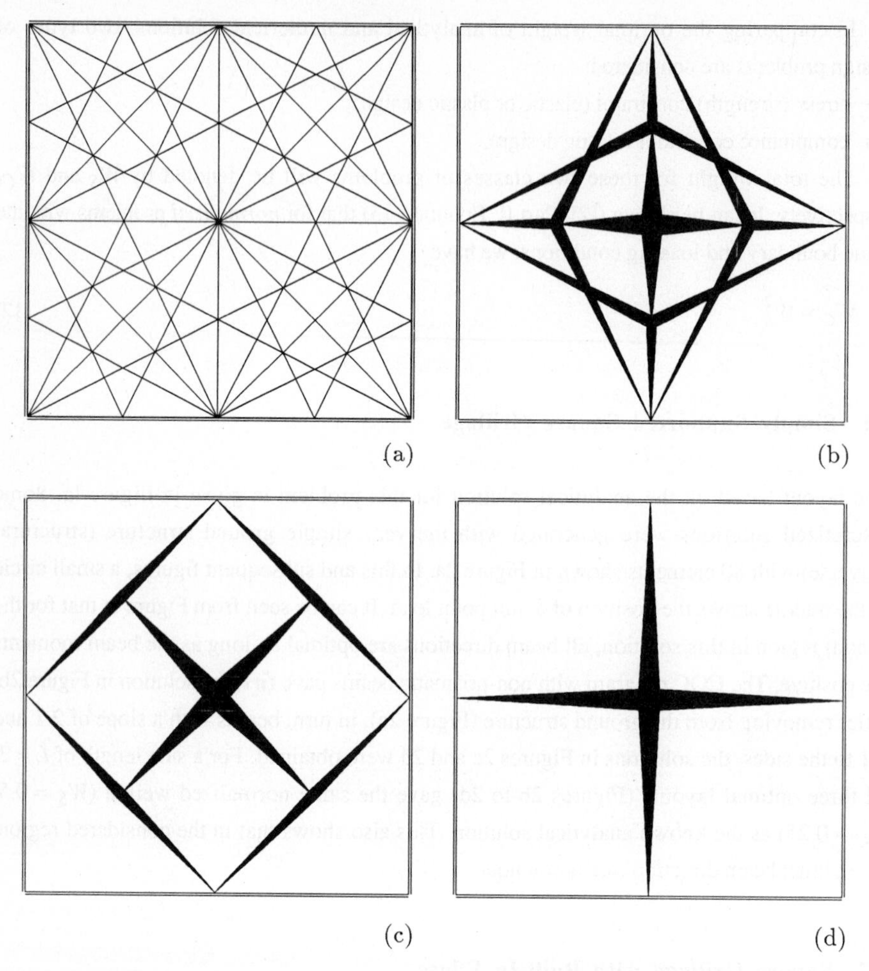

Figure 2. Discretized Solutions for Square, Simply Supported Grillages

5.3 Gillage with Two Simple Supports and Two Free Edges

It was shown about fifteen years ago by Rozvany, Hill and Prager [9,10] that along free edges, the optimal grillage layout contains a so-called "beam-weave" consisting of long beams with positive moments and short beams with negative moments (Figure ld). The length of the short beam is theoretically infinitesimal. The above solution has now been confirmed by discretized numerical computations (Figure 4b) using 620 elements in the ground structure (Figure 4a). For a finite number of beams, the optimal grillage weight is given by [10]

$$W_S = Pa^2[1 + 3c^2 + (1 - c^2)/n]/2 , \tag{13}$$

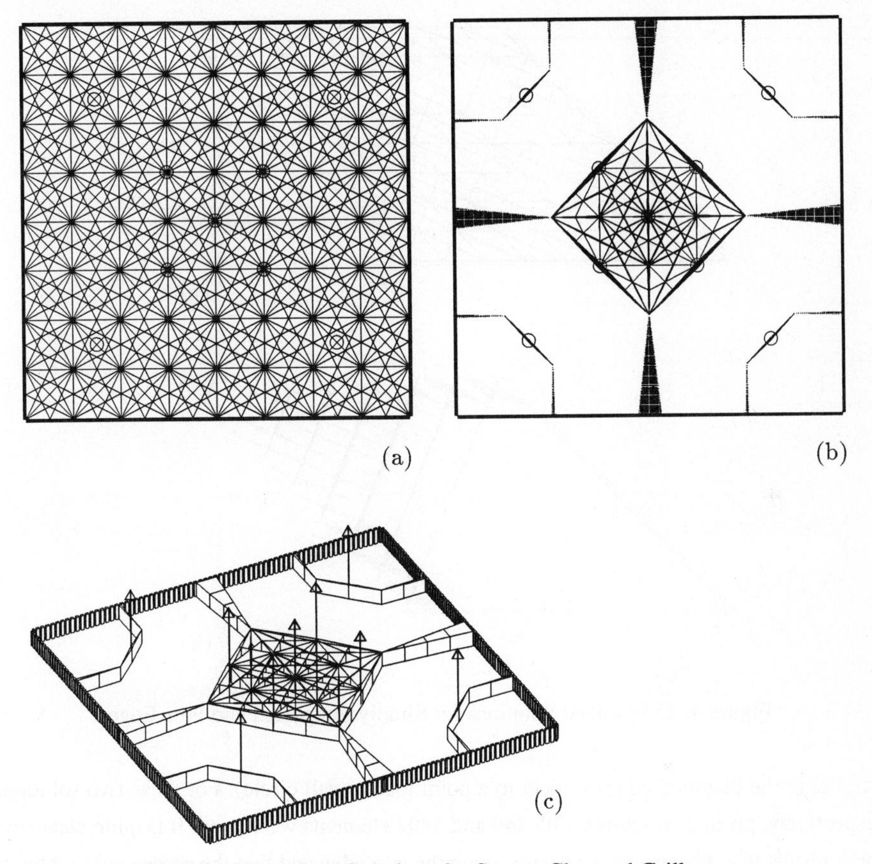

Figure 3. Discretized Solutions for Square Clamped Grillages

where n is the number of long beams and the meaning of other symbols is explained in Figure lc. On the basis of a side length of L = 1.0 in Figure lc, we have the values $a = \sqrt{1/(4-2\sqrt{2})} = 0.92387953$, $c = \sqrt{2} - 1 = 0.41421356$, n = 10, P = 1. Then (13) gives the weight $W_S = 0.68180195$, $W_C = W_S^2 = 0.46485390$. The latter was fully confirmed (to eight digits accuracy) by the discretized numerical results.

6 New Analytical Solutions Prompted by Numerical Results

6.1 Grillages with Simply Supported, Clamped, and Free Edges

For combinations of all the above three boundary conditions, analytical solutions were not available until recently. Figures 5a and b show solutions for a rectangular grillage with a clamped edge (thick line), a simply supported edge (double line) and a free edge (top edge

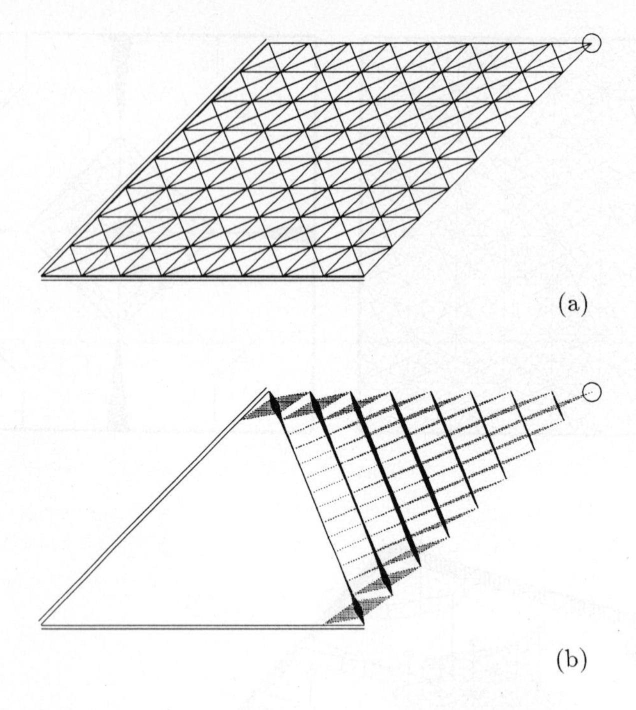

Figure 4. Discretized Solutions for Simply Supported and Free Edges

parallel to the clamped edge) subject to a point load (small circle). For these two solutions, respectively, ground structures with 466 and 1892 elements were used. It is quite clear from these results that a beam-weave occurs along the free edge and that the angle enclosed by the beams and the free edge gradually becomes smaller with the distance from the free edge.

On the basis of the above topology and the general theory of least-weight grillages [8-10], the differential equations representing the exact solution for the above problem were derived. The unknown variables for the assumed topology are (Figure 6) the angle α enclosed by the long beams and the free edge, the distance t along the free edge, the distance (y) between the clamped edge and the boundary separating beams in positive and negative bending, the angle (ε) enclosed by the tangent of this boundary and the free edge, and the adjoint deflection ($\bar{u}$) along the free edge.

The optimal values of the above parameters have been shown to be governed by the following equations:

$$dt = -\frac{L-y}{\sin^2\alpha} \cdot \frac{1+\sin^2\varepsilon}{2\sin^2(\alpha+\varepsilon)-1-\sin^2\varepsilon}\, d\alpha \;, \quad -d^2\bar{u}/dt^2 = \cos(2\alpha),$$

$$y = L - \sin\alpha\sqrt{(L^2-2\bar{u})/(1+\sin^2\alpha)}\;, \qquad \tan\varepsilon = dy/dx\;. \tag{14}$$

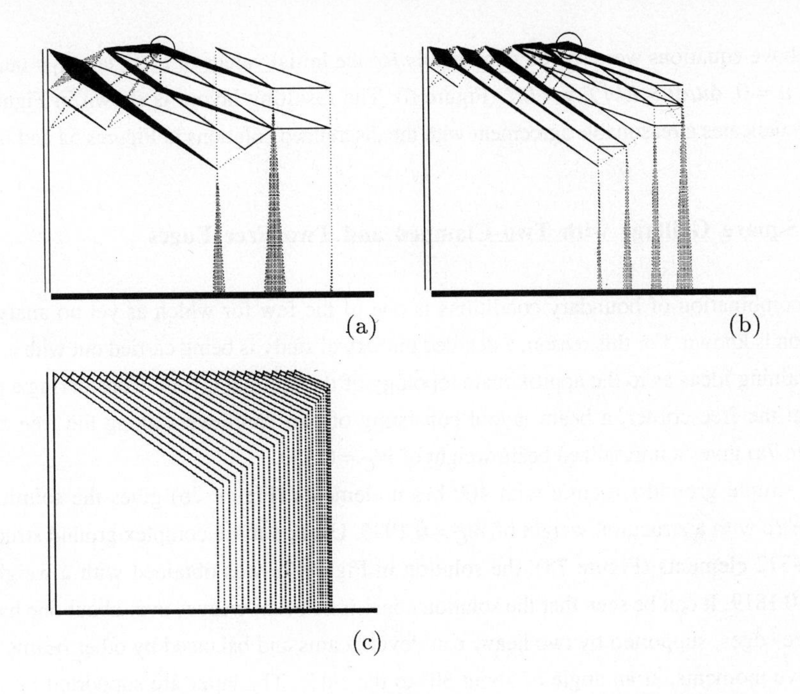

Figure 5. Discretized Solutions for a Combination of Clamped, Simply Supported and Free Edges

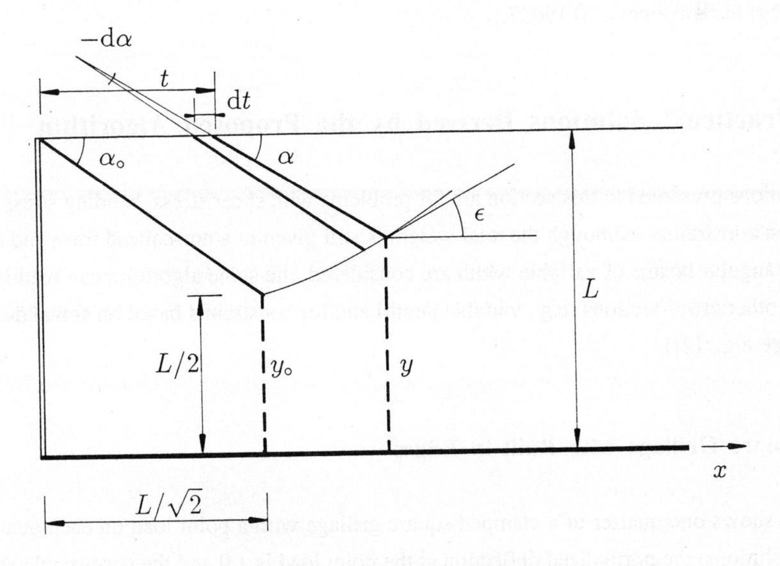

Figure 6. Analytical Solution for the Problem in Figure 5

The above equations were solved numerically for the initial values of $t = 0$, $\tan \alpha_0 = \tan \varepsilon_0 = 1/\sqrt{2}$, $u = 0$, $du/dt = L/\sqrt{2}$, $y = L/2$ (Figure 6). The resulting layout is shown in Figure 5c which indicates a reasonable agreement with the discretized solutions in Figures 5a and b.

6.2 Square Grillage with Two Clamped and Two Free Edges

This combination of boundary conditions is one of the few for which as yet no analytical solution is known. For this reason, a detailed numerical study is being carried out with a view to obtaining ideas as to the approximate topology of the optimal layout. Considering a point load at the free corner, a beam layout consisting of two cantilevers along the free edges (Figure 7a) gives a normalized beam weight of $W_C = 0.250$.

A simple ground structure with 400 beam elements (Figure 7b) gives the solution in Figure 7c with a structural weight of $W_C = 0.1995$. Using a more complex ground structure with 9312 elements (Figure 7d), the solution in Figure 7e was obtained with a weight of $W_C = 0.1819$. It can be seen that the solution consists of beam weaves over about one half of the free edges, supported by two heavy cantilever beams and balanced by other beams with negative moments, at an angle of about 30° to the sides. The latter are supported by short beams with positive moments which are probably normal to the clamped edges.

Whereas in Figures 6b and c the beam directions are restricted to slopes of 0 and 1:1 to the sides and Figures 7d and e include also slopes of 1:2, the ground structure in Figure 7f has additional slopes of 1:3 and 2:3 but only 3752 beam elements. The corresponding solution (Figure 7g) has a weight of 0.19057.

7 "Practical" Solutions Derived by the Proposed Algorithm

The solutions presented in this section are for problems with shear stress, bending stress and deflection constraints. Although the total weight is still given in a normalized form and here only rectangular beams of variable width are considered, the same algorithm can readily be used for other cross-sections (e.g., variable depth) and for constraints based on actual design codes (see e.g., [2]).

7.1 Square Grillage with Built-In Edges

Figure 8 shows one quarter of a clamped square grillage with a point load on each quarter. For all solutions, the normalized deflection at the point load is 1.0 and the permissible shear stresses and corresponding structural weights are: Figure 8a: $\tau_a = 10.6$, $W_C = 0.445804$;

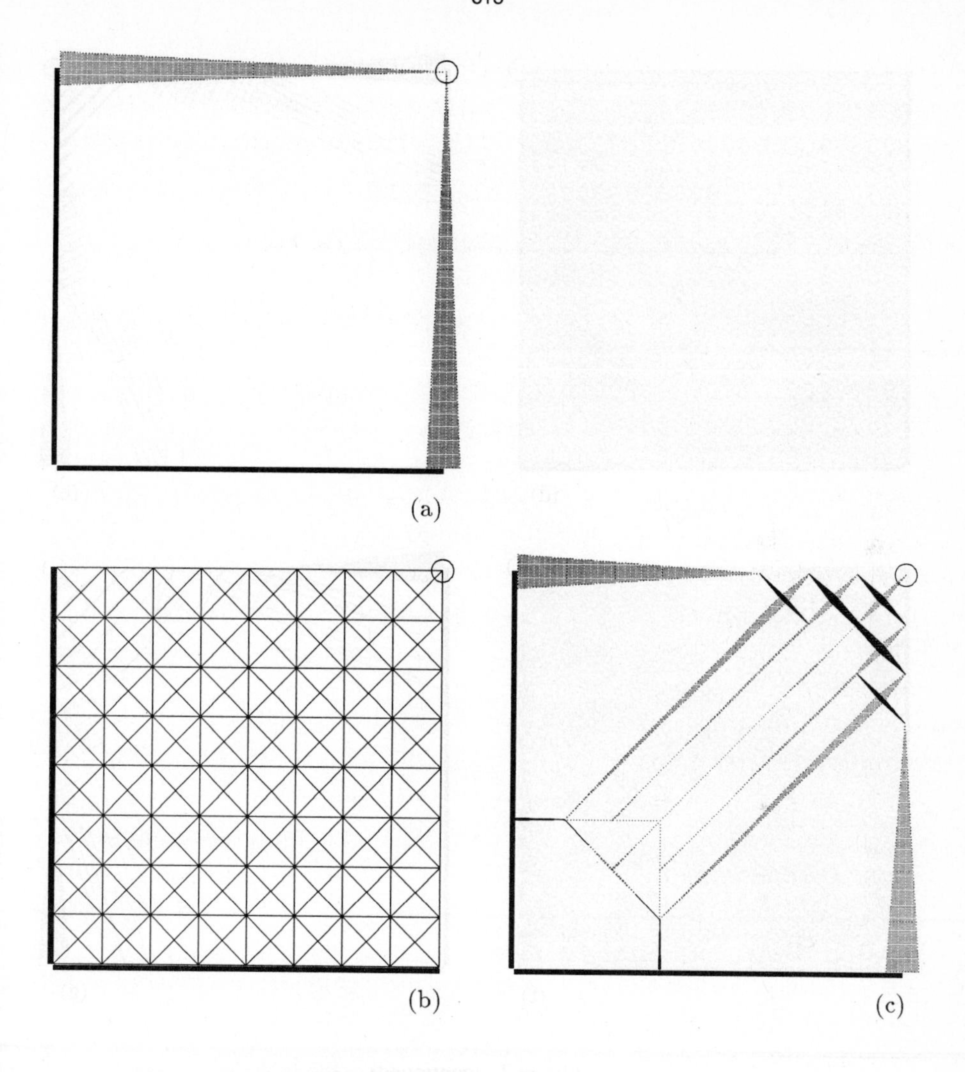

Figure 7. Discretized Solutions for a Problem for which the Analytical Solution is not yet Known

Figure 8b: $\tau_a = 5.0$, $W_C = 0.456674$; Figure 8c: $\tau_a = 3.0$, $W_C = 0.503007$; Figure 8d: $\tau_a = 2.0$, $W_C = 0.640169$. In Figure 8a, the shear stress constraint is inactive. Figure 8d is governed almost entirely by the shear stress constraint. The bending stress constraint is not active for this problem because the deflection constraint reduces to a compliance constraint which also provides optimal material distribution for a permissible bending stress [4, 8].

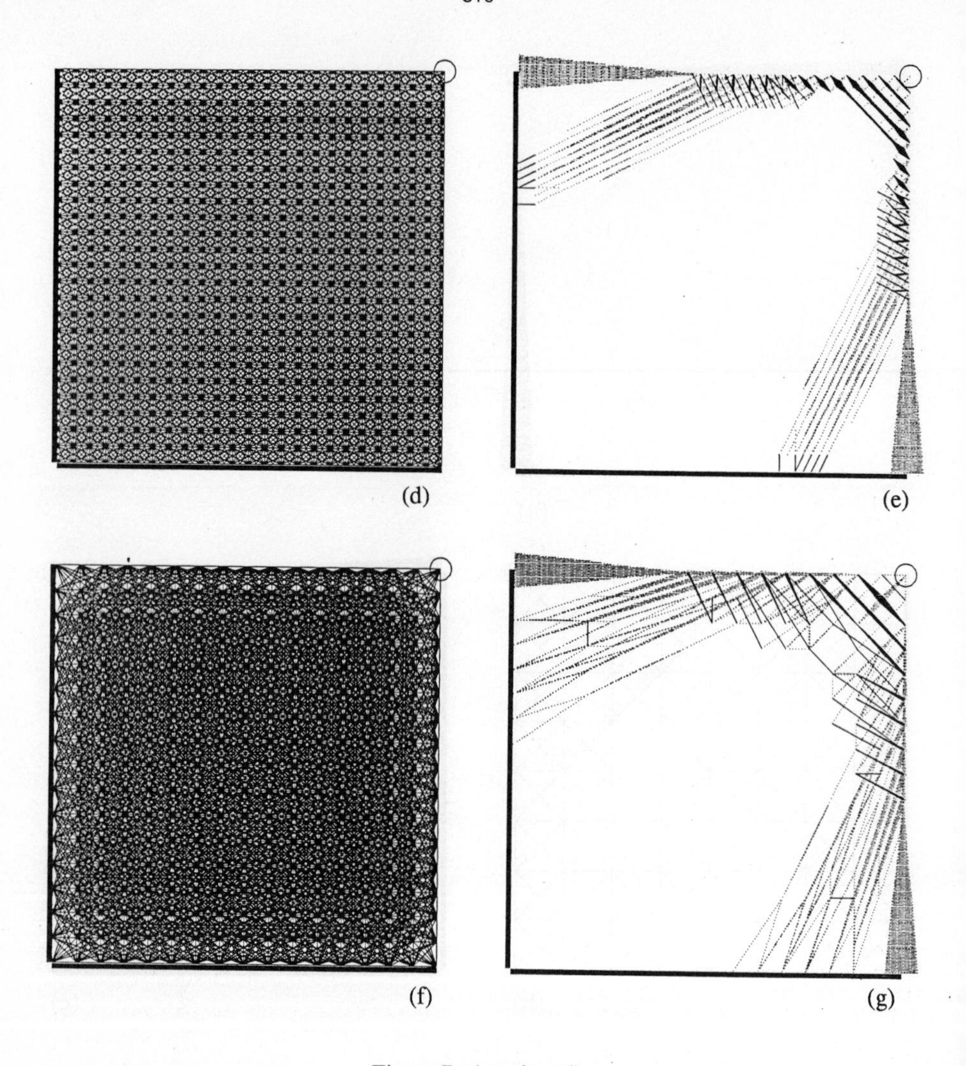

Figure 7. (continued)

7.2 Triangular Grillage with One Free Edge and Two Simply Supported Edges

The length of the free edge in Figure 9a is 2.0 and its distance from the simply supported corner is 0.5. The ground structure (299 elements) for this problem is shown in Figure 9a and solutions with various permissible shear stresses in Figures 9b-d with the following optimal weights: Figure 9b: $\tau_a = 17.79$, $W_C = 0.026653$; Figure 9c: $\tau_a = 10.00$, $W_C = 0.027801$; Figure 9d: $\tau_a = 6.00$, $W_C = 0.035282$. If we restrict all beams parallel to the free edge, then we obtain the solution in Figure 9e, with a weight of $W_C = 0.161626$ ($\tau_{max} = 4.82$ is

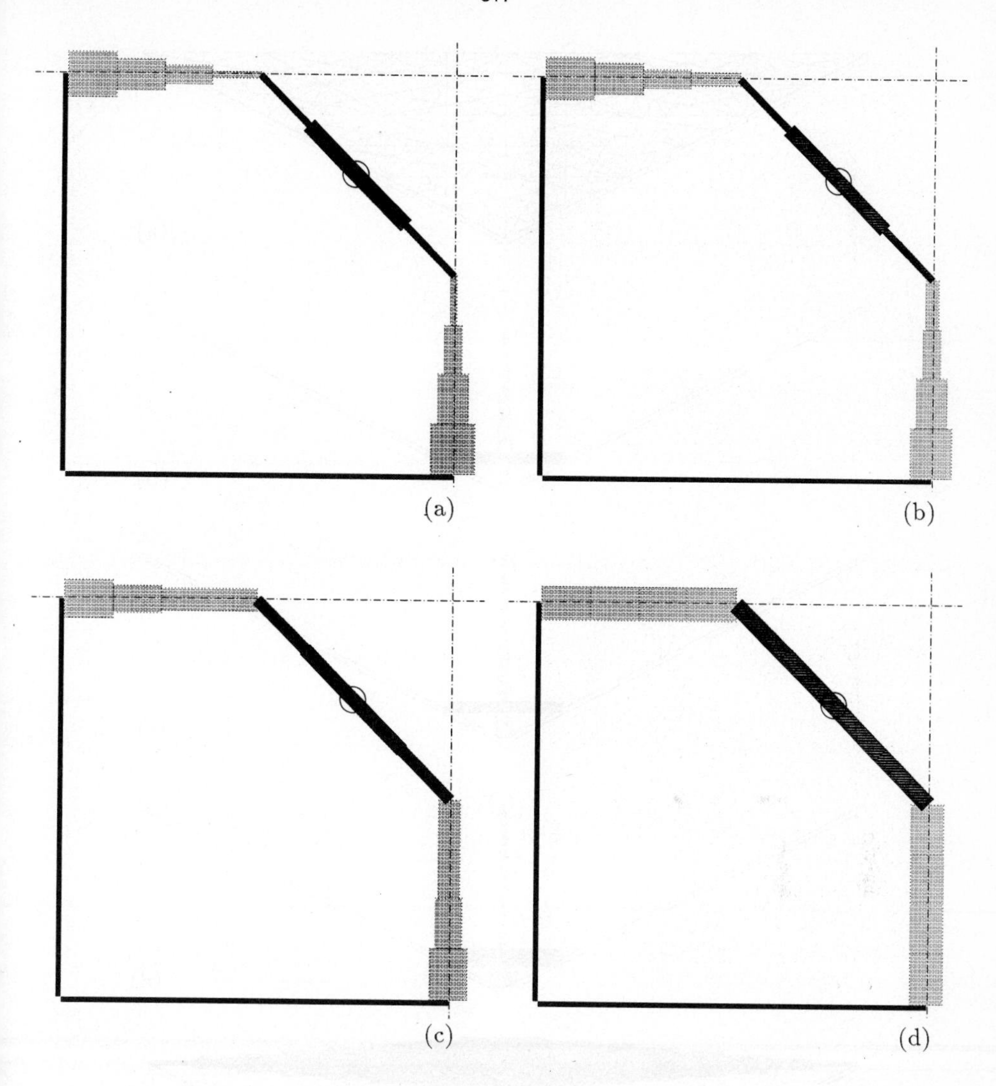

Figure 8. Discretized Solutions for Square Clamped Grillage with Active Deflection and Shear Stress Constraints

inactive). The latter represents over 400% of the optimal weight, although some authors (e.g., [11]) still claim that this topology is relatively economical.

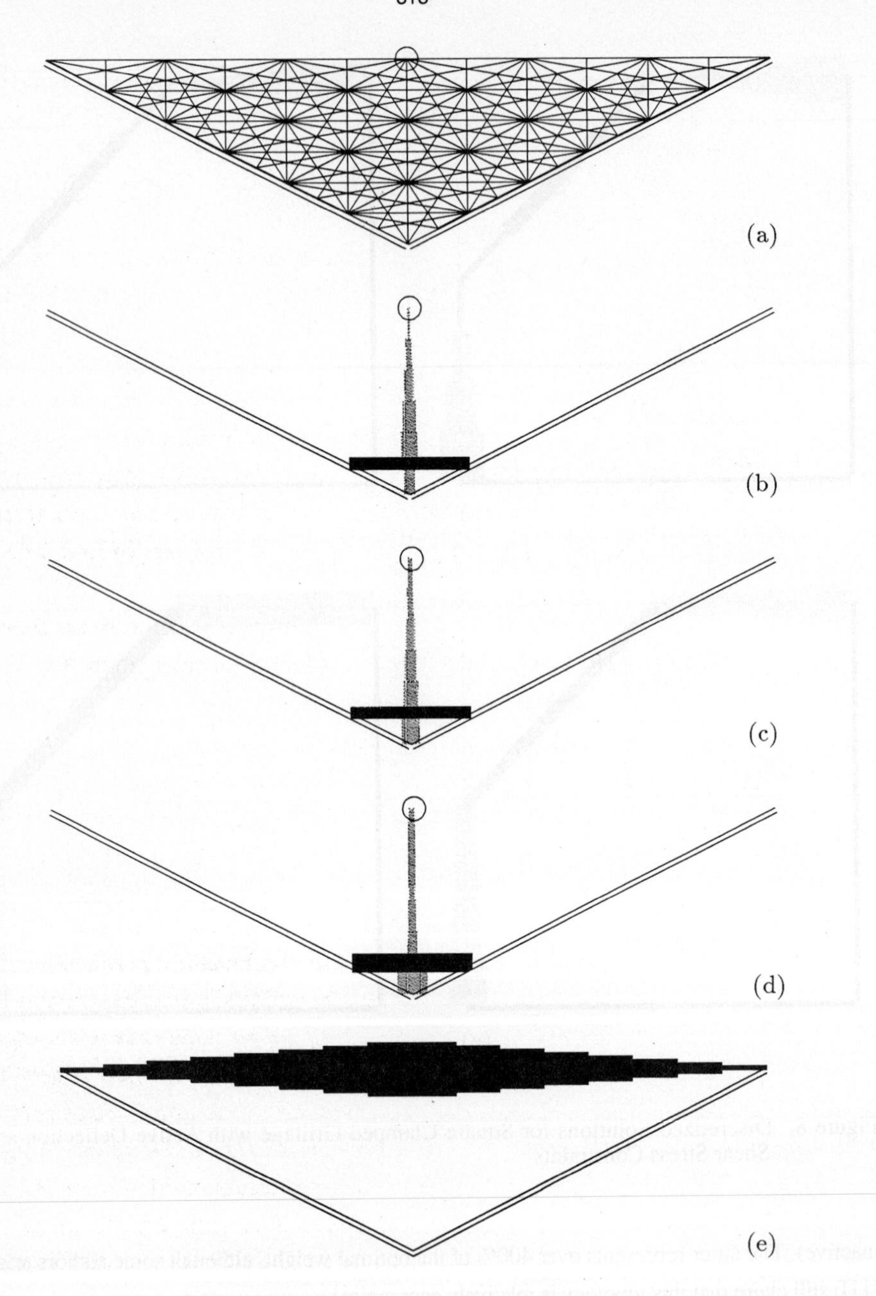

Figure 9. Discretized Solutions for a Simply Supported Corner with Active Deflection and Shear Stress Constraints

References

1. Rozvany, G.I.N.; Zhou, M. *et al.*: Continuum-type optimality criteria methods for large finite element systems with displacement constraints, Parts I and II, *Struct. Optim.* **1**, 47-72, 1989; **2**, 77-104, 1990.
2. Rozvany, G.I.N.; Zhou, M.: The COC algorithm, Part I: Cross-section optimization or sizing. Zhou, M.; Rozvany, G.I.N.: The COC algorithm, Part II: Topological, geometrical and generalized shape optimization, *Comp. Meth. Appl. Mech. Eng.* **89**, 281-308, 309-336, 1991.
3. Zhou, M.; Rozvany, G.I.N.: A new discretized optimality criteria method in structural optimization, *Proc. AIAA/ASME/ASCE/AHS/ASC Structures, Dynamics and Materials Conf.*, Dallas, 1992, pp. 3106-3120.
4. Rozvany, G.I.N.: *Structural Design via Optimality Criteria*, Kluwer, Dordrecht, 1989. Chapt. 8.
5. Dorn, W.S.; Gomory, R.E.; Greenberg, H.J.: Automatic design of optimal structures, *J. Mec.* **3**, 25-52, 1964.
6. Kirsch, U.: Optimal topologies of structures, *Appl. Mech. Rev.* **42**, 223-239, 1989.
7. Zhou, M.: *A new discretized optimality criteria method in structural optimization.* Doctoral Dissertation, FB 10, Essen University, 1992.
8. Rozvany, G.I.N.: *Optimal design of flexural systems,* Pergamon, Oxford, 1976.
9. Hill, R.H.; Rozvany, G.I.N.: Optimal beam layouts: the free edge paradox, *J. Appl. Mech.* **44**, 696-700, 1977.
10. Prager, W.; Rozvany, G.I.N.: Optimal layout of grillages, *J. Struct. Mech.* **5**, 1-18, 1977.
11. Lowe, P.G.: Optimization of systems in bending - conjectures, bounds and estimates relating to moment volume and shape. In: Rozvany, G.I.N. (Ed.): *Proc. IUTAM Symp. on Structural Optimization* (held Melbourne, Australia, 1988), Kluwer, Dordrecht, 1988, pp. 169-176.

Analysis and Design of Structural Sandwich Panels Against Denting

Ole Thybo Thomsen

Institute of Mechanical Engineering, Aalborg University, Pontoppidanstræde101, DK-9220 Aalborg East, Denmark

Abstract: The problem of local bending often encountered in laterally loaded sandwich panels is analyzed by considering the deflection of the loaded facing against the not loaded facing as being governed by an elastic foundation model. This is achieved by the development of a two-parameter elastic foundation formulation which takes into account the existence of shear interactions between the loaded facing and the supporting medium (core material). The results obtained by application of the developed approximate solution procedure, i.e., the local stress field in the near vicinity of the area of application of the external load, are compared with finite element solution results and very good comparative results are obtained.

Keywords: structural sandwich panels / indentation / local bending effects / elastic foundation / classical sandwich théory / comparative study / finite element analysis / parametric effects / design tool / concurrent engineering.

1 Introduction

The structural member, known as a sandwich panel, is a special form of a laminated composite which consists of three principal parts; thin, strong, and relatively high density facings separated by a thick, light, and weaker core. The faces and the core of the sandwich panel are adhesively joined in order to transfer the load between the components. This provides a structure which is very efficient for resisting bending and buckling loads since the sandwich panel is much stronger and stiffer, in most respects, than the sum of the individual strengths and stiffnesses.

Although sandwich structures have been accepted as an excellent way to design strong, durable and lightweight structures, a number of important problems have been left more or less unattended, even by classical textbooks as the ones by Allen [1], Plantema [2] and Stamm & Witte [3]. An area of significant practical importance, and belonging to this class of more or less unattended problems, is the so called denting problem associated with local bending effects in structural sandwich panels. These local bending effects appear more frequently than is usually expected, and two typical examples of areas/positions where local bending can be observed are: 1. various joints between sandwich panels (T-joints, corner-joints); 2. inserts and fasteners for introduction or "take-out" of external forces as well as for mechanical fastening of various types of equipment.

It is a well known fact that sandwich panels are notoriously sensitive to failure by the application of strongly localized external lateral loads, i.e., point loads, line loads and distributed loads of high intensity. This pronounced sensitivity toward the application of concentrated external loads is due to the associated inducement of significant local deflections of the facing into the core material. The result of this may be a premature failure of the sandwich panel.

The present paper is exclusively devoted to local bending analysis of sandwich panels (primarily foam-cored) which can fail in several distinctly different ways [1-5]. Possible failure modes in the present context could be crushing (denting failure) of the core material [4-7]; delamination at the interface between the loaded face and the core material due to the concentration of interlaminar stresses in the region adjacent to the area of external load application; shear failure of the core material; tensile or compressive failure of the facings by yielding or fracturing dependent on the properties of the face-materials; and finally delamination due to interlaminar shear and transverse normal stresses for the case of laminated FRP-faces.

Relatively few references have been treating the problems associated with local bending of the faces of sandwich panels subjected to concentrated loads, and no one gives an explicit description of the onset and development of actual failure modes. They restrict themselves to the development of structural analyses based on the somewhat unrealistic assumption of linear elastic behavior of the core material as well as the faces of the considered sandwich panel. Thus, the results obtained do not reflect the actual sequence of events experienced by the constituent materials during the onset and development of failure, but they do have the potential of giving valuable information about the fundamental mechanics of the local bending problem as well as the parameters controlling the onset of irreversible failure.

In the modelling of the problem, it seems reasonable to consider the relative deflection of the loaded face of the sandwich panel against the not loaded face as being governed by an appropriate elastic foundation formulation.

This approach was used by Reference 6, who considered the foam-core as continuously distributed linear tension/compression springs supporting the loaded face of the sandwich panel. Thus, the elastic response of the foam-core material was assumed to be governed by the "classical" Winkler foundation model, which is extensively treated by, among others, Hetényi [8].

However, the Winkler foundation model suffers a serious drawback, since it does not account for the possible existence of shear interactions between the loaded face and the supporting medium (core material). This feature of the Winkler foundation model suggests that it becomes inadequate for deformations of short wave-length, in which the shear deformations of the foam-core of the sandwich beam become important.

2 Elastic Foundation Analogy

2.1 Formulation of Two-Parameter Foundation Model

The present formulation is based on the assumption of an elastic foundation model, which does take into account the existence of shear interactions between the loaded face and the supporting medium (core material). This additional consideration, compared to the Winkler foundation model, is necessary in order to account for the build-up of interfacial shear stresses occurring adjacent to the local area of external load application.

Figure 1 illustrates the considered problem with its constituent parts: the loaded facing of a sandwich beam and the supporting medium (core material). The considered facing can be subjected to arbitrary distributed external transverse normal and shear loads $p_Z(x)$, $p_X(x)$ along its upper surface.

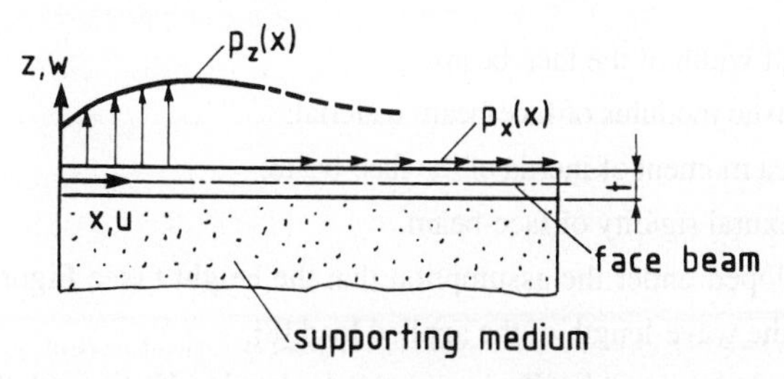

Figure 1. Loaded Facing Supported by Core Material.

It is assumed that the face as well as the core material can be satisfactory described by use of linear elastic constitutive relations, even though this assumption in general has to be considered as being unrealistic as mentioned earlier.

The elastic response of the supporting medium (core material) is suggested expressed by the following equations, which relates the deflections of the loaded facing to the interfacial stress components measured per unit length of the face-beam (Figure 1):

$$q_x(x) = K_x\, u\left(x, -\frac{t}{2}\right) \;; \tag{1}$$

$$q_z(x) = K_z\, w(x) \;;$$

where K_x, K_z : foundation moduli;

$q_x(x)$, $q_z(x)$: interfacial shear and normal stress resultants;

$u(x,-t/2)$: longitudinal displacement of lower fibre of the facing;

$w(x)$: lateral displacement of the loaded facing;

x : independent variable (longitudinal coordinate).

It should be emphasized that $q_x(x)$, $q_z(x)$ represents the foundation normal and shear stress resultants per unit length of the beam (unit: N/mm), i.e., they can be obtained by multiplying the stress components $\tau_{zx}(x)$, $\sigma_z(x)$ at the interface with the width of the beam (2b). In the forthcoming derivations $q_x(x)$, $q_z(x)$ will be referred to as the interlaminar stress distribution functions.

The foam-core and the faces are assumed to be homogeneous, isotropic and linear elastic, and the foundation modulus K_z is suggested related to the elastic coefficients of the core (E_c,ν_c) and the loaded facing (E_f) by the following expression [8][1]:

$$K_z = 0.71\, E_c \sqrt[3]{\frac{E_c\, b^4}{E_f\, I}} \;; \tag{2}$$

where b : half width of the face-beam;

E_f : elastic modulus of face-beam material;

I : area moment of inertia of the face-beam;

$E_f I$: flexural rigidity of face-beam.

Equation 2 is developed under the assumption that the height t (see Figure 1) of the face is small compared to the wave-length of the applied load [8].

The second foundation modulus K_x is suggested related to K_z through the relation:

$$K_x = \frac{K_z}{2(1+\nu_c)} \;. \tag{3}$$

The loaded facing is modelled by application of the classical theory of bending of beams. Thus, it is assumed that normals to the undeformed neutral axis of the face-beam remains straight, normal and inextensional during deformation so that transverse normal and shearing strains may be neglected in deriving the beam kinematic relations. The displacement

[1] Obtained by comparison with 2-D elasticity solution to the problem of an infinitely long beam attached to an elastic medium which extends to infinity on one side.

components u, w of an arbitrary point in the face-beam may be expressed by the neutral axis quantities u_0, w_0 by the equations:

$$u = u_0 - z\,\frac{dw}{dx}\,; \qquad w = w_0 \tag{4}$$

The first of Equations 1 together with Equation 4 yields:

$$q_x(x) = K_x\left(u_0 + \frac{t}{2}\,\frac{dw}{dx}\right). \tag{5}$$

From Equation 5 it is seen that the interfacial shear stress distribution function $q_x(x)$ is proportional to the angular rotation dw/dx of the neutral axis of the face-beam. The interfacial stress distribution functions $q_z(x)$, $q_x(x)$ are unknown, and the objective of the proceeding derivations is to formulate and solve the equations determining these functions.

Referring to Figure 2 for notation and sign convention, the three equilibrium equations for the elastically supported face-beam can be written in the form:

$$\frac{dN}{dx} = q_x - p_x\,; \qquad \frac{dQ}{dx} = q_z - p_z\,; \qquad \frac{dM}{dx} = Q - \frac{t}{2}\,(q_x + p_x)\,. \tag{6}$$

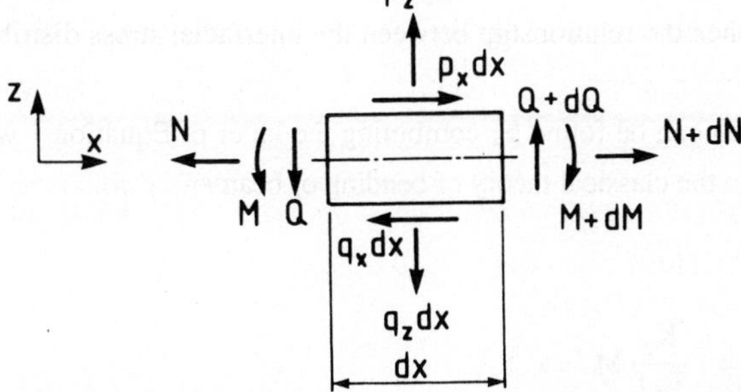

Figure 2. Notation and Sign Convention for Equilibrium of Loaded Face-Beam.

The longitudinal normal strain ε_x at the interface between the facing and the core material is expressed as follows, where a unit width (2b=1) sandwich beam is considered:

$$\varepsilon_x\left(x,-\frac{t}{2}\right)=\frac{d}{dx}\left(u\left(x,-\frac{t}{2}\right)\right)=\frac{1}{E_f}\,\sigma_x\left(x,-\frac{t}{2}\right)\Rightarrow$$

$$\frac{d}{dx}\left(u\left(x,-\frac{t}{2}\right)\right)=\frac{1}{E_f\,t}\left(N-\frac{6\,M}{t}\right). \tag{7}$$

From the first of Equation 1 the following is derived:

$$\frac{dq_x(x)}{dx}=K_x\,\frac{d}{dx}\left(u\left(x,-\frac{t}{2}\right)\right). \tag{8}$$

By combination of Equations 7 and 8 followed by successive differentiation as well as introduction of the equilibrium Equation 6, the following expressions appear:

$$\frac{dq_x}{dx}=\frac{K_x}{E_f\,t}\left(N-\frac{6\,M}{t}\right)\Rightarrow$$

$$\frac{d^2q_x}{dx^2}=\frac{K_x}{E_f\,t}\left(4\,q_x-\frac{6}{t}\,Q+2\,p_x\right)\Rightarrow \tag{9}$$

$$\frac{d^3q_x}{dx^3}=\frac{K_x}{E_f\,t}\left(4\,\frac{dq_x}{dx}+2\,\frac{dp_x}{dx}-\frac{6}{t}\,(q_z-p_z)\right).$$

By rearrangement, the latter of Equation 9 can be written in the form:

$$\frac{d^3q_x}{dx^3}-\frac{4\,K_x}{E_f\,t}\,\frac{dq_x}{dx}=-\frac{6\,K_x}{E_f\,t^2}\,(q_z-p_z)+\frac{2\,K_x}{E_f\,t}\,\frac{dp_x}{dx}. \tag{10}$$

Equation 10 establishes the relationship between the interfacial stress distribution functions $q_x(x)$, $q_z(x)$.

A similar equation can be found by combining the latter of Equation 1 with the "flexure formula" known from the classical theory of bending of beams:

$$q_z=K_z\,W\Rightarrow$$

$$\frac{d^2q_z}{dx^2}=K_z\,\frac{d^2w}{dx^2}=-\frac{K_z}{E_f\,I}\,M\Rightarrow$$

$$\frac{d^3q_z}{dx^3}=-\frac{K_z}{E_f\,I}\left(Q-\frac{t}{2}\,(q_x+p_x)\right)\Rightarrow \tag{11}$$

$$\frac{d^4q_z}{dx^4}=-\frac{K_z}{E_f\,I}\left(q_z-p_z-\frac{t}{2}\left(\frac{dq_x}{dx}+\frac{dp_x}{dx}\right)\right).$$

Rearrangement of the latter of Equation 11 gives:

$$\frac{d^4 q_z}{dx^4} + \frac{K_z}{E_f I} q_z = \frac{t}{2} \frac{K_z}{E_f I} \left(\frac{dq_x}{dx} + \frac{dp_x}{dx} \right) + \frac{K_z}{E_f I} p_z \,. \tag{12}$$

Equations 10 and 12 represent a set of two coupled ordinary constant-coefficient differential equations expressed in the two unknown interfacial stress distribution functions.

In order to obtain two differential equations expressed solely in the two unknown functions $q_x(x)$ and $q_z(x)$, Equation 10 can be employed to eliminate $q_z(x)$ and its derivatives from Equation 12. Similarly Equation 12 can be employed to eliminate the derivatives of $q_x(x)$ from Equation 10.

From Equation 10 the following relations are obtained:

$$q_z = -\frac{E_f t^2}{6 K_x} \left(\frac{d^3 q_x}{dx^3} - \frac{4 K_x}{E_f t} \frac{dq_x}{dx} \right) + \frac{t}{3} \frac{dp_x}{dx} p_z \Rightarrow$$

$$\frac{d^4 q_z}{dx^4} = -\frac{E_f t^2}{6 K_x} \left(\frac{d^7 q_x}{dx^7} - \frac{4 K_x}{E_f t} \frac{d^5 q_x}{dx^5} \right) + \frac{t}{3} \frac{d^5 p_x}{dx^5} + \frac{d^4 p_z}{dx^4} \,. \tag{13}$$

Similarly Equation 12 gives:

$$\frac{dq_x}{dx} = \frac{2 E_f I}{K_z t} \left(\frac{d^4 q_z}{dx^4} + \frac{K_z}{E_f I} q_z \right) - \frac{2}{t} p_z - \frac{dp_x}{dx} \Rightarrow$$

$$\frac{d^3 q_x}{dx^3} = \frac{2 E_f I}{K_z t} \left(\frac{d^6 q_z}{dx^6} + \frac{K_z}{E_f I} \frac{d^2 q_z}{dx^2} \right) - \frac{2}{t} \frac{d^2 p_z}{dx^2} - \frac{d^3 p_z}{dx^3} \,. \tag{14}$$

Inserting Equations 13 and 14 into Equations 10 and 12, yields a set of two ordinary non-homogeneous differential equations of seventh and sixth order respectively, where each of the two differential equations is expressed solely in one of the unknown interfacial stress distribution functions $q_x(x)$, $q_z(x)$:

$$\frac{d^7 q_x}{dx^7} + \alpha_2 \frac{d^5 q_x}{dx^5} + \alpha_1 \frac{d^3 q_x}{dx^3} + \alpha_0 \frac{dq_x}{dx} = -\frac{3 \alpha_2}{2 t} \frac{d^4 p_z}{dx^4} - \frac{\alpha_2}{2} \left(\frac{d^5 p_x}{dx^5} - \frac{\alpha_1}{2} \frac{dp_x}{dx} \right); \tag{15}$$

$$\frac{d^6 q_z}{dx^6} + \alpha_2 \frac{d^4 q_z}{dx^4} + \alpha_1 \frac{d^2 q_z}{dx^2} + \alpha_0 q_z = \alpha_1 \left(\frac{d^2 p_z}{dx^2} + \frac{\alpha_2}{4} p_z \right) + \frac{\alpha_1 t}{2} \left(\frac{d^3 p_x}{dx^3} + \frac{\alpha_2}{2} \frac{dp_x}{dx} \right); \tag{16}$$

$$\alpha_2 = -\frac{4\,K_x}{E_f\,t}\;;\;\alpha_1 = \frac{K_z}{E_f\,I}\;;\;\alpha_0 = \frac{\alpha_1\,\alpha_2}{4}\;. \tag{17}$$

The complete solution to the governing set of two ordinary non-homogeneous constant-coefficient differential, Equations 15 and 16 can be written in the general form:

$$q_x(x) = (q_x(x))_h + (q_x(x))_p\;;$$
$$q_z(x) = (q_z(x))_h + (q_z(x))_p\;; \tag{18}$$

where subscript "h" denotes the solutions to the homogeneous parts of Equations 15 and 16 and subscript "p" denotes the particular solutions to Equations 15 and 16.

The solution of the homogeneous part of Equations 15 and Equation 16 is straightforward, and it can be shown [9] that the homogeneous solution to Equation 15 can be written in the form:

$$\begin{aligned}(q_x(x))_h = {} & A_0 + A_1\cosh(\phi_1 x) + A_2\sinh(\phi_1 x)\\ & + A_3\cosh(\xi x)\cos(\eta x) + A_4\sinh(\xi x)\cos(\eta x)\\ & + A_5\sinh(\xi x)\sin(\eta x) + A_6\cosh(\xi x)\sin(\eta x)\;;\end{aligned} \tag{19}$$

where ϕ_1, ξ, η are coefficients defined in terms of α_2, α_1, α_0 given by Equation 17. ϕ_1, ξ, η are given in explicit form in Reference 9 to which the reader is referred for further details. A_j (j=0,1,...,6) are seven integrational constants which have to be determined from the statement of the boundary conditions of the problem.

From the first of Equation 13 (omitting the non-homogeneous terms) it follows, that the homogeneous part of the transverse normal stress response $(q_z(x))_h$ can be expressed by the first and third derivatives of $(q_x(x))_h$:

$$(q_z(x))_h = \frac{2\,t}{3\,\alpha_2}\left(\frac{d^3(q_x(x))_h}{dx^3} + \alpha_2\,\frac{d(q_x(x))_h}{dx}\right)\;. \tag{20}$$

Thus, the general solution to the homogeneous parts of the governing differential Equations 15 and 16 has been established.

The particular solution parts of Equations 15 and 16 still need to be determined, and this is accomplished by assuming suitable particular solution functions by "qualified quessing". The derivation of the particular solutions $(q_x(x))_p$, $(q_z(x))_p$ is of course strongly influenced by the form of the external loads, since these load distribution functions $p_x(x)$, $p_z(x)$ determine the non-homogeneous terms appearing in the governing differential equations.

For further information about the explicit expressions for the particular solution parts, for various types of external loading, the reader is referred to Thomsen [10,11].

2.2 Specification of Boundary Conditions

In order to determine the seven integrational constants A_j (j=0,1,...,6), it is necessary to formulate seven boundary conditions. To exemplify the formulation of the necessary boundary conditions the simplest possible example is considered: the loaded facing of a simply supported sandwich beam in 3-point bending (shown in Figure 3). The necessary boundary conditions for the loaded facing can be expressed as follows:

$$\left[q_x\right]_{x=0} = 0\,; \qquad \left[\frac{dw}{dx}\right]_{x=0} = 0\,;$$
$$\left[\frac{d^2w}{dx^2}\right]_{x=L} = 0\,; \qquad \left[\frac{d^3w}{dx^3}\right]_{x=L} = 0\,; \tag{21}$$
$$\int_0^L q_x(x)\,dx = 0\,; \qquad \int_0^L q_z(x)\,dx = \frac{P}{2}\,.$$

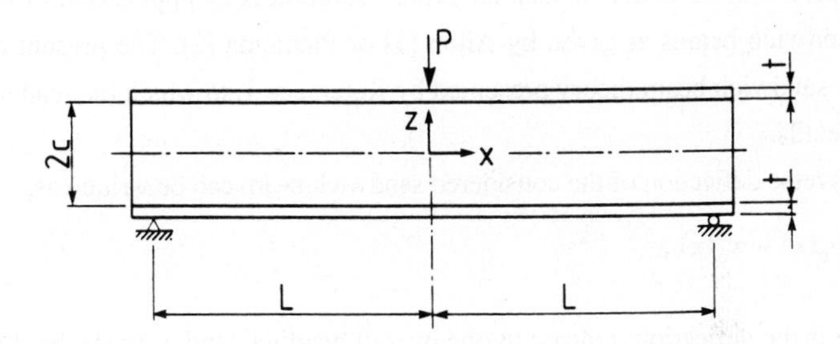

Figure 3. Unit Width Sandwich Beam in 3-Point Bending.

The first and the second of Equation 21 represent the conditions of symmetry about the center of the beam span of the facing (x=0); the third and the fourth of Equation 21 represent the conditions at the free end of the facing (x=L); and the last two of Equation 21 represent the conditions of horizontal and vertical equilibrium, respectively, for the loaded facing of the considered sandwich beam, where $p_x(x)=p_z(x)=0$ (see Figure 3).

The last of the seven boundary conditions is derived by combination of the second of Equations 9 and the third of Equation 11. The resulting equation can be written in the form (Q is eliminated):

$$\frac{d^3q_z}{dx^3} + \frac{2\,t\,\alpha_1}{3\,\alpha_2}\frac{d^2q_x}{dx^2} + \frac{t\,\alpha_1}{6}\,q_x = \frac{t\,\alpha_1}{6}\,P_x\,. \tag{22}$$

The last boundary condition, supplementing the already established boundary conditions for the considered problem, is then specified by requiring fulfillment of Equation 22 at some position, say x=0 (center of beam span).

Thus, the complete set of boundary conditions necessary in order to determine the seven integrational constants has been completed, and the solution of the problem of the face-beam elastically supported by the core material has been achieved for the case of a sandwich beam in 3-point bending (Figure 3).

3 Superposition with Classical Sandwich Theory Solution

In order to obtain a complete solution to the specific problem of a sandwich beam in 3-point bending, it is necessary to investigate the overall bending and shearing effects; i.e., to establish an overall solution supplementing the local bending solution derived in the preceding sections of the present paper.

The simplest way to establish such an overall solution is by application of the classical theory of sandwich beams as given by Allen [1] or Plantema [2]. The present approach is based on the sandwich beam theory presented by Reference 1, to which the reader is referred for further details.

The transverse deflection of the considered sandwich beam can be written as:

$$w(x) = w_b(x) + w_s(x) \; ; \tag{23}$$

where $w_b(x)$ is the deflection induced by the overall bending, and $w_s(x)$ is the displacement induced by the overall shearing of the sandwich beam.

The displacement $w_b(x)$ is calculated by means of the ordinary theory of bending of beams, and for the considered problem (see Figure 3) the following is obtained:

$$w_b(x) = -\frac{P}{12\,D}\,(x^3 - 3\,L\,x^2 + 2\,L^3),\; 0 \le x \le L \; ; \tag{24}$$

where D is the flexural rigidity of the sandwich beam. D is given by (for the present case: 2b=1.0 mm):

$$D = E_f\,(2b)\left(\frac{t^3}{6} + \frac{t\,(2c+t)^2}{2}\right) + E_c\,(2b)\,\frac{2\,c^3}{3} \; . \tag{25}$$

The displacement $w_s(x)$ is expressed as:

$$w_s(x) = -\frac{P}{2\,A\,G_c}\,(L - x)\,,\; 0 \le x \le L : \tag{26}$$

where G_c is the shear modulus of the foam-core, and A is expressed by (2b=1.0 mm):

$$A = \frac{2b\,(2c+t)^2}{2c}\ . \tag{27}$$

The term AG_c, appearing in Equation 26, is referred to as the shearing stiffness of the sandwich beam.

The stresses (overall solution) in the faces and the core of the considered unit width sandwich beam (Figure 3) are expressed in the form; $0 \le x \le L$:

$$(\sigma_x(x,z))_f = -\frac{P\,(L-x)\,z}{2\,D}\,E_f, \qquad \begin{pmatrix} (c+t) \ge z \ge c \\ -c \ge z \ge -(c+t) \end{pmatrix};$$

$$(\sigma_x(x,z))_c = \frac{P\,(L-x)\,z}{2\,D}\,E_c\,, \qquad c \ge z \ge -c\,; \tag{28}$$

$$(\tau_{zx}(x,z))_c = \frac{P}{2\,D}\left(E_f\,\frac{t\,(2c+t)}{2} + \frac{E_c}{2}\,(c^2 - z^2)\right), \; c \ge z \ge -c\,.$$

The overall solution (Equations 23-28) can be superposed to the approximate local bending solution derived earlier. Thus, the total transverse displacement of the loaded facing due to overall bending, shearing, as well as local bending can be written:

$$w^{total}(x) = w_b(x) + w_s(x) + w_{local}\,(x)\,. \tag{29}$$

The local transverse displacement w_{local} is defined by the latter of Equation 1. The local displacement component w_{local} only defines the displacement of the "elastic line" of the loaded facing; i.e. the superposition of the overall and local deflection components suggested by Equation 29 will only describe the local bending effects at the interface between the loaded facing and core. It is not possible to decide, in explicit terms, the decay through the thickness of the core material of the deflections induced by local bending.

The stress field induced by the overall bending and shearing as well as local bending of the sandwich beam is another subject of major interest. The normal stress distribution in the loaded face-beam can be written in the form:

$$[(\sigma_x)_f]^{total} = [(\sigma_x)_f]_{overall} + [(\sigma_x)_f]_{local}\,; \tag{30}$$

where $[(\sigma_x)_f]_{local}$ for a unit width beam can be expressed in terms of $q_z(x)$; $0 \le x \le L$

$$[(\sigma_x(x,z_{local}))_f]_{local} = \frac{M_{local} z_{local}}{I} \Rightarrow$$

$$[(\sigma_x(x,z_{local}))_f]_{local} = -\frac{E_f \, z_{local}}{K_z} \frac{d^2 q_z(x)}{dx^2} \; ; \; \frac{t}{2} \geq z_{local} \geq -\frac{t}{2} \, . \tag{31}$$

The coordinate z_{local} referred to in Equation 32 corresponds to the local face-beam coordinate shown in Figure 1.

The transverse normal stress component σ_z at the interface between the loaded face and the core is given by dividing the elastic response function $q_z(x)$ with the width of the beam (unit width), i.e., the interface transverse normal stress component can be written as:

$$[\sigma_z(x,z{=}c)]^{total} = \frac{q_z(x)}{2b} \; ; \; (2b = 1 \text{ mm}) \, . \tag{32}$$

Similarly the shear stress component at the interface between the loaded face and the core can be expressed as:

$$[\tau_{zx}(x,z{=}c)]^{total} = [\tau_{zx}(x,z{=}c)]_{overall} + \frac{q_x(x)}{2b} \; (2b = 1 \text{ mm}) \, . \tag{33}$$

As mentioned earlier for the total transverse displacements, the superposition of the overall and local stress field components only describes the state of stress in the loaded face and in the interface between the loaded face and the core of the considered sandwich beam. This area however, is by far the most interesting part of the structure, as it is here the peak values of the stress components will appear.

Thus, a complete (approximate) solution to the problem of a sandwich beam loaded in 3-point bending (Figure 3) has been derived by superposition of the local and overall displacement and stress fields respectively.

4 Results and Discussion

The applicability of the developed solution procedure will be demonstrated by an example, and the results obtained will be compared to a finite element solution in order to ascertain the quality of the suggested approach.

The example chosen in the present paper is the earlier described unit width sandwich beam in 3-point bending shown in Figure 3. The geometry, the material properties and the external point load P are as follows:

GEOMETRY:	L=40.0 mm; t=1.0 mm; c=2.5 mm; 2b=1.0 mm;
FACE-BEAMS:	E_f=15.0 GPa ("E-glass/epoxy");
FOAM-CORE:	E_c=0.1 GPa; ν_c=0.35 ("PVC-foam": ρ_c=100.0 kg/m^3);
POINT LOAD:	P=10.0 N/mm (force per unit width).

For the results presented in the following parts of the paper, the stress quantities will be normalized with respect to the "average" shear stress of the considered sandwich beam, which can be calculated by the expression [1]:

$$\tau_{avg} = \frac{P}{2\,(2c + t)} \,. \tag{34}$$

4.1 Comparison with Finite Element Solution

The software used for the model generation was the general finite element (FEM) program ANSYS, version 4.4A, developed and distributed by Swanson Analysis Systems, Inc. The sandwich beam shown in Figure 3, with geometrical and material properties as quoted above, was modeled using 2-D isoparametric 4-node plane stress/strain elements (element type referred to as STIF42 in the ANSYS 4.4A code) for the upper and lower facings as well as the core material. The upper and lower facings were modeled with two layers of elements, and the core was modeled using ten layers of elements. Due to the symmetry of the problem, only one half of the sandwich beam was generated. The total number of elements in the half-beam model was 1120 [9].

Figure 4 shows the distribution of normalized normal stresses σ_x/τ_{avg} along the upper and lower boundaries of the loaded facing of the sandwich beam. The results observed from Figure 4 are that the local bending effects are of significant importance near the point of external load application (x=0). At the upper surface of the facing a compressive state of stress is present (curves (1) and (2) in Figure 4) and the peak compressive value of σ_z/τ_{avg} is about -105. It is observed that the results obtained by the analytical and the FEM solutions, respectively, are very close to each other.

At the lower boundary of the facing a tensile state of stress is present at x=0, and this phenomenon is explained by the significant local bending contribution. Again a very close resemblance between the two solutions is observed. As the x-coordinate is increased, the local bending contribution is seen to fade out, and from x/L=0.3-0.4 the stress state corresponds to the linear variation of the overall bending moment which is a characteristic feature of the classical beam theory solution for the sandwich beam in 3-point bending.

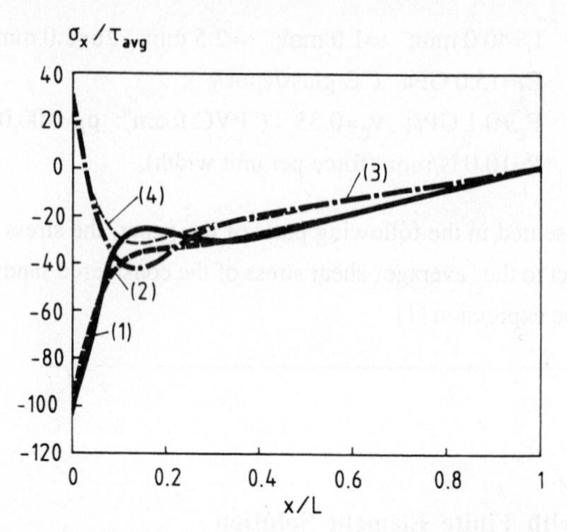

Upper Fibre: 1) ANALYT; 2) FEM ; Lower Fibre: 3) ANALYT; 4) FEM.

Figure 4. Longitudinal distribution of σ_x/τ_{avg} in upper and lower fibre of the loaded facing obtained by the analytical and FEM solutions.

Figure 5 shows the distribution through the thickness at x=0 of the sandwich beam of the normalized stress component σ_x/τ_{avg}. Curve (1) shows the results obtained by the analytical solution, and curve (2) shows the corresponding FEM results. The overall tendency is again a very close match of the results: the upper and lower facings of the sandwich beam are transferring the overall bending load and significant stresses are present. The stress state in the upper facing is determined by the overall bending together with the local bending contribution: at x=0 a compressive state of stress is present at the upper boundary, and at the lower boundary a tensile stress state is present. The stress state of the lower facing at x=0 is exclusively tensile since its is determined by the overall bending alone.

At this point, it should be mentioned that the very close match between the two solutions for the lower facing, shown in Figures 4 and 5, is a special feature of the example considered in the present paper, i.e., in general it cannot be expected that the classical sandwich theory solution will give accurate results for the lower facing. This circumstance can be explained by the fact (experimental results; [12]) that the decay of the transverse normal stress σ_z down through the core-material may not be complete, which again implies that transverse normal stresses have to be transferred between the core-material and the lower facing. As mentioned earlier, however, the question mark posed on the accuracy of the results obtained for the lower facing is no serious drawback for the suggested approximate solution procedure, as the loaded facing (where reliable results are obtained) will always be the most interesting part of the sandwich structure (most severely loaded).

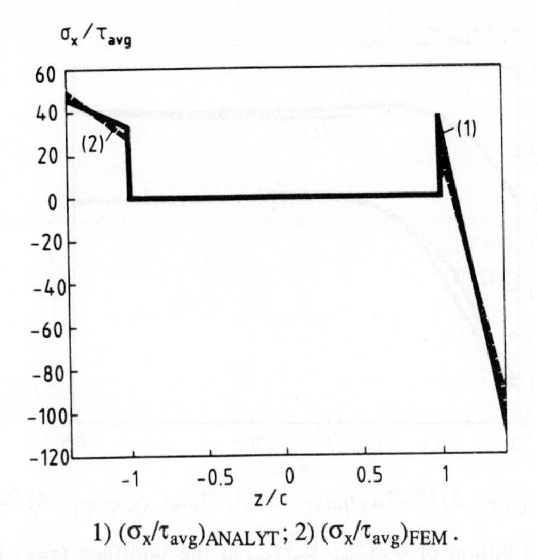

1) $(\sigma_x/\tau_{avg})_{ANALYT}$; 2) $(\sigma_x/\tau_{avg})_{FEM}$.

Figure 5. Transverse distribution at x=0 of σ_x/τ_{avg} obtained by the analytical and FEM solutions, respectively,

Figure 6 shows the distribution of the normalized transverse normal and shear stress components, σ_z/τ_{avg}, τ_{zx}/τ_{avg} at the interface between the loaded face-beam and the core material. Focusing at first on the curves (1) and (2) of Figure 6, showing the distribution of σ_z/τ_{avg}, it is observed that the results obtained by the analytical and FEM solutions, respectively, show a very close match. The predicted peak values (x=0) are very close to each other, and the characteristic decay with increasing values of x is also observed for both solutions. Furthermore, the wavy harmonic character of the "decay function" predicted by the elastic foundation formulation is confirmed by the finite element solution.

Curves (3) and (4) of Figure 6 show the interfacial distribution of τ_{zx}/τ_{avg} obtained by the two solution methods, and again a very close resemblance of the results is observed.

Other results regarding the comparison between the developed approximate solution procedure and the FEM model of the sandwich beam could be shown, but they all show the same close match as observed from Figures 4, 5 and 6, and no further comparative results will be given in the present paper (for further information see [9]).

4.2 Parametric Study

In order to illustrate the influence of certain characteristic parameters on the local interfacial stress distribution in the near vicinity of the point of external load application, the results obtained from a brief parametric study will be presented. The characteristic material and

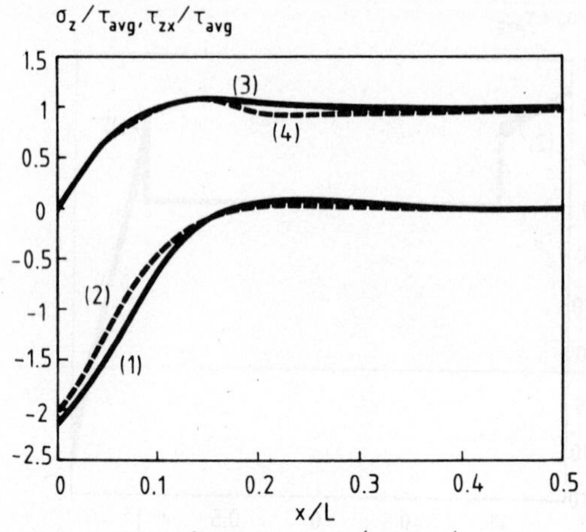

1) $(\sigma_z/\tau_{avg})_{ANALYT}$; 2) $(\sigma_z/\tau_{avg})_{FEM}$; 3) $(\tau_{zx}/\tau_{avg})_{ANALYT}$; 4) $(\tau_{zx}/\tau_{avg})_{FEM}$.

Figure 6. Distribution of σ_z/τ_{avg}, τ_{zx}/τ_{avg} at the interface (z=c) between the loaded facing and the core material.

geometrical parameters included in the analysis are: the ratio between the elastic moduli E_f/E_c, and the thickness t of the loaded facing.

The parameters L and c (see Figure 3) also influence the stress distribution within the sandwich beam, but they only exert influence on the overall bending and shearing of the sandwich beam, and therefore they will not be included in the presented parametric study.

Figures 7 and 8 show the effects of altering the stiffness ratio E_f/E_c on the interfacial distribution of the normalized transverse normal stress, σ_z/τ_{avg}, and the normalized local shear stress, τ_{zx}/τ_{avg}. The thickness of the loaded face is chosen to t=1.0 mm, and E_f/E_c is given the values 50, 150 and 450.

From Figure 7 it is seen that the lower the value of the stiffness ratio, E_f/E_c, the higher the peak value of σ_z/τ_{avg}. Furthermore, it is observed that the wave-length of the elastic response is increased significantly as E_f/E_c is increased.

Similar results are observed from Figure 8, except that the peak value of τ_{zx}/τ_{avg} (representing only the shear contribution induced by local bending) is located some distance away from x=0. Again it is observed that the wave-length of the elastic shearing response increases with increasing values of E_f/E_c.

The results shown in Figure 8 also indicates when it is advisable to use the two-parameter elastic foundation model, suggested in the present approach, instead of the simpler Winkler foundation model (one-parameter elastic foundation model) used by Weissman-Berman et al. [6]. It is observed that the peak value of τ_{zx}/τ_{avg} decreases very rapidly as E_f/E_c is increased, and for large values of the stiffness ratio the build up of interfacial shear stresses due to local bending is negligible in comparison with the other stress components. In this case the results

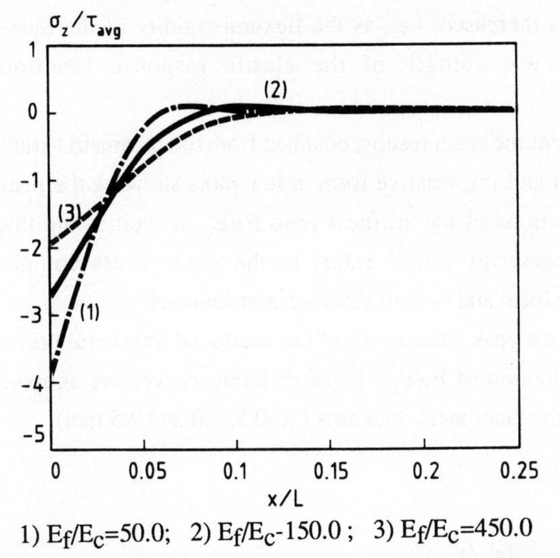

1) E_f/E_C=50.0; 2) E_f/E_C-150.0 ; 3) E_f/E_C=450.0

Figure 7. σ_z/τ_{avg}. vs. x/L at the interface between the loaded facing and the core material (E_C fixed to E_C=0.1 GPa).

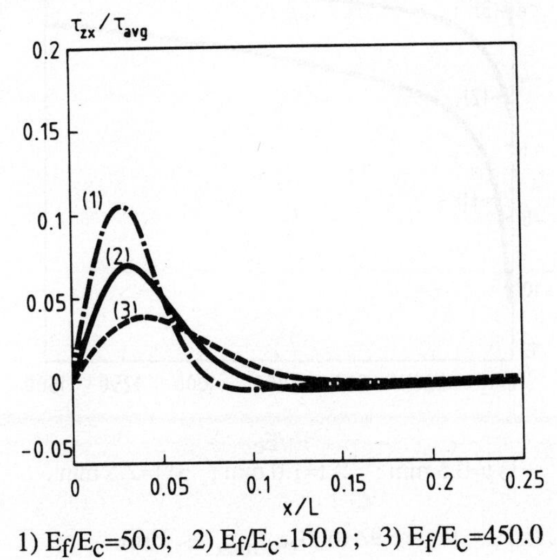

1) E_f/E_C=50.0; 2) E_f/E_C-150.0 ; 3) E_f/E_C=450.0

Figure 8. τ_{zx}/τ_{avg} vs. x/L at the interface between the loaded facing and the core material (E_C fixed to E_C=0.1 GPa).

obtained by the two-parameter elastic foundation model will be nearly identical to the results predicted by use of the Winkler foundation model (further treatment in Section 4.3).

Curves similar to the ones shown in Figures 7 and 8 can be obtained by altering the thickness t of the loaded facing for a fixed value of E_f/E_C, but they will not be shown in the present paper, since the conclusions drawn are similar to the conclusions drawn from Figures 7 and 8. The overall tendencies are that the peak values of σ_z/τ_{avg}, τ_{zx}/τ_{avg} decreases significantly as t is increased, i.e., as the flexural rigidity of the face-beam is increased. Furthermore, the wave-length of the elastic response functions increases with increasing t-values.

In order to present the main results, obtained from the parametric studies quoted above, in a more compressed and informative form, a few plots showing the peak values of σ_z/τ_{avg}, $\tau_{zx}^{total}/\tau_{avg}$ as functions of the stiffness ratio E_f/E_c as well as the thickness t have been prepared. The superscript "total" refers to the shear stress components obtained by superposition of the local and overall shearing contributions.

Figure 9 shows the peak value (x=0) of the interfacial transverse normal stress component $(\sigma_z/\tau_{avg})_{max}$ as function of E_f/E_c. Three different curves are shown, each representing different values of the face-beam thickness t (t=0.5, 1.0 and 2.5 mm).

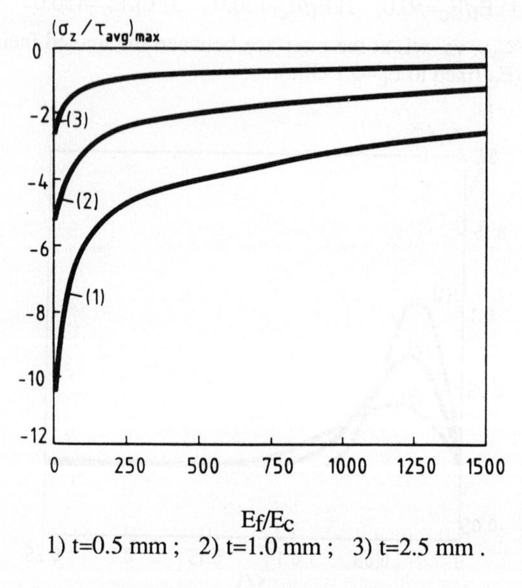

1) t=0.5 mm ; 2) t=1.0 mm ; 3) t=2.5 mm .

Figure 9. $(\sigma_z/\tau_{avg})_{max}$ vs. EfEc.

It is seen that $(\sigma_z/\tau_{avg})_{max}$ shows a strong dependency of E_f/E_c as expected from the results shown in Figure 7. A significant feature of the plots shown is that the peak value of σ_z/τ_{avg} attains very large values for very small values of E_f/E_c, even though it should be recalled that the elastic foundation approach, on which the results are based, becomes inadequate for very small values of E_f/E_c (corresponding to deflections of short wave-length).

It is also observed that the peak value of σ_z/τ_{avg} is influenced by the face-beam thickness t in the way that the larger the value of t (i.e. the larger the flexural rigidity of the face-beam) the lower the peak values of σ_z/τ_{avg}.

Figure 10 shows the peak values of the interfacial shear stress τ_{zx}/τ_{avg} as function of the stiffness ratio E_f/E_C for three different values of the thickness t of the loaded facing (t=0.5, 1.0 and 2.5 mm). It is recognized that the peak values of τ_{zx}/τ_{avg} are strongly dependent on the E_f/E_C-values in the way, that very large peak shear stresses are obtained for very small values of the stiffness ratio (it is emphasized that the calculated peak stress values for low values of E_f/E_C are questionable as mentioned earlier), and that the peak values of $\tau_{zx}^{total}/\tau_{avg}$ decreases strongly as the stiffness ratio is increased. The peak value of $\tau_{zx}^{total}/\tau_{avg}$ approaches unity asymptotically as E_f/E_C goes to infinity, i.e., the interfacial shear stress contribution induced by local bending diminishes as the stiffness ratio becomes large.

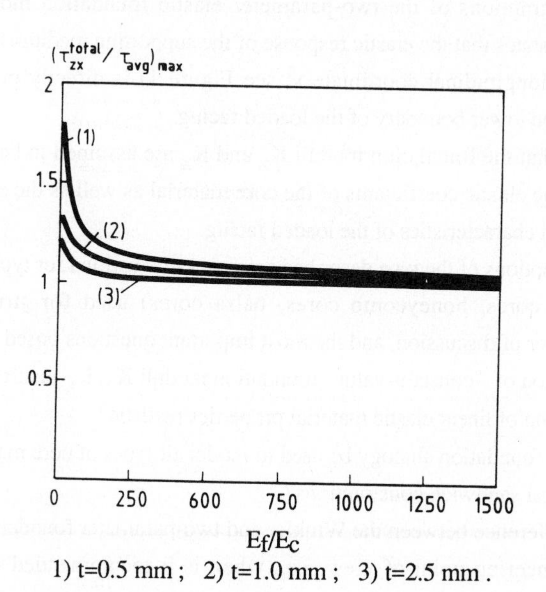

1) t=0.5 mm ; 2) t=1.0 mm ; 3) t=2.5 mm .

Figure 10. $(\tau_{zx}/\tau_{avg})_{max}$ vs. E_f/E_C.

Again it is clear, that the peak value of $\tau_{zx}^{total}/\tau_{avg}$ depends on the face thickness t in the way, that the larger the value of t, the lower the peak value of $\tau_{zx}^{total}/\tau_{avg}$.

4.3 Discussion of Range of Applicability

The introduction of the elastic foundation analogy, in the context of local bending of sandwich panels subjected to localized loads, has been carried out in order to present a simple method of obtaining detailed information about the displacement and stress fields induced locally (loaded facing and interface between loaded facing and core material) as a result of localized lateral loads.

The basic idea behind the application of an elastic foundation analogy is, as described earlier, to approximate the supporting medium (core material of sandwich beam) by continuously distributed linear tension/compression and shear springs (two-parameter foundation model). The quality of the approximation, however, is strongly dependent on the quality of the basic assumptions of the elastic foundation model employed.

The basic assumptions of the two-parameter elastic foundation model are given by Equation 1, which states that the elastic response of the supporting medium at a given position (specified by the longitudinal coordinate x; see Figure 1) is directly proportional to the displacements of the lower boundary of the loaded facing.

This implies, that the foundation moduli K_Z and K_X are assumed to be constants, which can be related to the elastic coefficients of the core material as well as the elastic coefficients and the geometrical characteristics of the loaded facing.

Whether assumptions of the type described are generally justified for typical core materials (polymeric foam cores, honeycomb cores, balsa cores) used for structural sandwich structures is a matter of discussion, and the most important questions posed are:

1. Is the assumption of "constant-value" foundation moduli K_Z, K_X justified?
2. Is the assumption of linear elastic material properties realistic?
3. Can the elastic foundation analogy be used to model all types of core materials typically in use for structural sandwich constructions?
4. What is the difference between the Winkler and two-parameter foundation models from a practical engineering point of view, i.e., when is it recommended to use one model instead of the other?

Concerning the first of the questions posed, the answer demands some additional considerations. The assumption of constant value elastic foundation moduli K_X, K_Z, is obviously not generally justified, but can, at the most, be legalized if some, as yet unspecified, further restrictions are imposed on the class of problems for which the proposed elastic foundation analogy can be used successfully.

The considerations leading to these further restrictions, on the range of applicability of the elastic foundation analogy, will not be given in detail in the present paper, but the reader is instead referred to the analysis presented in Reference 1 (pp. 169-171) concerning the bounds of application on the Winkler hypothesis in the context of analysis of wrinkling (local instability) of structural sandwich panels. Furthermore, the reader is referred to Reference 9 (pp. 21-24) in which the bounds on the two-parameter elastic foundation model are discussed.

It is shown in References 1 and 9, that the key to the specification of the necessary restrictions on the applicability of elastic foundation models (as the ones discussed) is the wave-length (denoted by λ) of the deflections of the elastically supported loaded facing. The conclusion drawn from the reflections presented in the quoted references is that the elastic foundation models (Winkler as well as two-parameter foundation model) become inadequate for deformations of short wave-length. This is caused by the fact that the shearing deformations of the supporting medium becomes increasingly important as the wave-length of the deflections decreases. Thus, it is recognized that the simplifications introduced by the application of an elastic foundation model, instead of a continuum model for the core material, are not justified for deformations of short wave-length.

A natural question in this context is, whether it is possible to give an explicit quantification of the concept of "deformations of short wave-length". Unfortunately the answer to this question turns out to be negative if the term "quantification" means definition of very precise bounds on the applicability of the foundation models. This is caused by the circumstance that the wave-length of the deflections is related to the characteristic material and geometrical parameters of the problem in a very complex manner.

For practical sandwich panels, however, the bounds imposed by the vaguely formulated concept of "deformations of short wave-length" are not likely to be active, since the typical face thicknesses ($0.5 \text{ mm} < t < 10.0 \text{ mm}$), and the typical stiffness ratios ($25 < E_f/E_c < 1500$), will usually ascertain sufficiently large deflectional wave-lengths to ensure the justification of the simple elastic foundation approach instead of a very complex two or three dimensional continuum formulation.

The second question posed was whether it is reasonable to assume linear elastic properties of the constituent materials of typical structural sandwich panels. The answer to this question is not a very direct and simple one, as some face materials as well as some core materials can be said to behave reasonably linear elastically, and some do not.

However, the service conditions, under which structural sandwich panels are employed, are usually sought to be well within the safe domain specified by the proportional limit of the constituent materials. Thus, it is expected that the suggested approach, based on linear elastic assumptions, is capable of giving a fairly good estimate of the magnitude of the stress concentrations induced by local bending effects.

The third question posed was whether the elastic foundation approach can be used to model all types of core materials in use for structural sandwich panels, and in order to give a proper answer to this question it is necessary to include some additional comments.

There are three major types of core materials used for structural sandwich panels: 1. polymeric foam core materials, 2. honeycomb core materials, and 3. balsa cores. The first and third of the listed core material types can be considered as being homogeneous materials (at least from a macroscopic point of view which is usually preferred in the context of engineering applications), since the cellular structure of these materials is usually at least one order of magnitude smaller than the characteristic dimensions of the considered sandwich panel (face thickness, core thickness, width of sandwich panel, sandwich panel span width etc.). Thus, it seems reasonable to model the foam and balsa core materials respectively by using a continuum or an elastic foundation formulation as suggested.

With respect to the honeycomb type cores, it is a bit more complicated to present an answer to the posed question. Honeycomb cores are discrete by nature, i.e., they do not support the facings of a sandwich panel continuously, but rather in a discrete manner along the edges of the honeycomb cells. This means, that continuum formulations or continuous spring support formulations (Winkler and two-parameter models) may be incapable of modelling the core material realistically. Whether this is the case, is determined from the cell size of the honeycomb cell structure: if the characteristic dimensions (width and edge length of cells) of the cells are small compared to the thickness of the facings, continuum or continuous spring support formulations will be capable of giving a fair estimate of the mechanical behavior from an engineering point of view. If, on the other hand, the cell size of the honeycomb structure is comparable to, or even larger than, the face thicknesses, it is not likely that good results can be obtained, as the facings will tend to act like plates within the boundaries of each cell in the honeycomb structure.

The above mentioned considerations deals with the justification of elastic foundation models in general as opposed to elastic continuum models, with a discussion about the assumption of linear elastic material properties, and with a discussion about the capability of continuum or elastic foundation formulations to describe the mechanical behavior of various types of core materials. No distinction between the very simple Winkler foundation model and the more complicated two-parameter model has been presented.

It is clear that none of the foundation models will give satisfactory answers for "deformations of short wave-length", as it can be shown [1,9], that it is not possible to select constant values of the foundation moduli K_z and K_x which are appropriate for displacements of any wave-length. The reason for this is, as described earlier in this section, the fact that the shearing deformations of the core material becomes very influential for deformations of short wave-length.

This last statement, however, makes it possible to distinguish between the "quality" of the Winkler and two-parameter foundation models, because the latter actually does take into account the possible shearing interaction (although in very simple form) between the loaded facing and the core material of a sandwich beam subjected to strongly localized lateral loads.

Thus, it is recognized that the two-parameter foundation model is superior to the Winkler foundation model, as it predicts the existence of the interfacial shear stress distribution function $q_x(x)$, although the formulation of the shearing interaction effects is not sophisticated enough to handle problems characterized by displacements of very short wave-length.

The application of the two-parameter foundation model instead of the simpler, from a mathematical point of view, Winkler foundation model can be recommended for structural sandwich panels characterized by the following approximate relationship:

$$\lambda \approx 5.03\, t \sqrt[3]{\frac{E_f}{E_c}} < 50\text{–}60 \text{ [mm]} ; \qquad (35)$$

where λ is the wave-length of the elastic deformations (elastic line). The guidelines defined by the inequality (35) should only be considered as a rough estimate of the recommended bounds of applicability of the two foundation models under consideration. Thus, it is possible to imagine structural sandwich panels characterized by extreme values of the thickness t or the stiffness ratio E_f/E_c giving rise to short wave-length elastic responses, even though λ lies well within the "safe domain" specified by Equation 35, i.e., in the domain where the Winkler hypothesis should be able to supply sufficiently accurate results.

The values of λ suggested by the inequality (35) as an appropriate applicational separation between the Winkler and two-parameter foundation models, however, will usually ensure that the peak value of the interface shear stress component $(\tau_{zx})_{local}$ is negligible compared to the interface peel stress component $(\sigma_z)_{local}$.

5 Concluding Remarks

A method of analyzing the local stress and displacement fields in the near vicinity of strongly localized external loads applied to sandwich beams (or plates in cylindrical bending) has been presented. The developed formulation is based on the assumption that the deflection of the loaded facing against the not loaded facing of a laterally loaded sandwich beam can be properly modeled by use of an elastic two-parameter foundation model. Furthermore, it is assumed that the facings as well as the core material behave linear elastic.

The local stress and displacement fields derived by application of the two-parameter elastic foundation model are superimposed to the stress and displacement fields derived by use of the classical theory of sandwich beams (or a rough FEM model), and an overall solution is obtained including overall bending and shearing as well as local bending effects.

The applicability of the derived solution procedure has been investigated by analyzing a test-example: a sandwich beam in 3-point bending. The results obtained by use of the developed method have been compared with a finite element solution, and an excellent match of the computed solutions has been observed.

Thus, it is has been shown that it is possible to estimate (with a high degree of accuracy) the severity of the stress concentrations in the loaded facing and in the interface between the loaded facing and the core of sandwich beams by application of a suitable elastic foundation formulation.

A brief parametric study, based on the developed solution procedure, has shown that the local bending effects are very strongly influenced by the stiffness ratio E_f/E_c and the thickness t of the loaded facing. The peak values of the interfacial transverse normal and shear stresses attain very large values for small values of both E_f/E_c and t, corresponding to elastic responses of short wave-lengths. Furthermore, it is observed that these peak values of the interface stresses decreases rapidly as E_f/E_c and t are increased.

From the results obtained in the parametric study, it is recognized that the only way to reduce the locally induced stress concentrations is to increase either of E_f/E_c or t. From a practical point of view, however, it is not possible to change E_f/E_c locally. Thus, it is seen that t is the only parameter which (realistically) can be adjusted locally in order to accomplish a less severe distribution of interface stresses.

On the basis of the elastic foundation analogy presented in the present paper, graphical "design-charts" have been prepared for different load cases (point load, uniformly distributed load, concentrated bending moment [11]).

For a specific load case it is possible, from these graphical "design-charts" (similar to Figures 9 and 10) to find peak values of the interface stress components, as well as the peak value of the bending stress in the loaded facing, for specified values of the stiffness ratio E_f/E_c and the face-thickness t. These peak stresses, which are induced by local bending effects, can be superposed to the stresses obtained by an overall solution, i.e., a classical sandwich theory solution or a rough finite element solution obtained by use of sandwich beam or plate elements.

Thus, graphical "design-charts" [11] can be used in a design process, as a very cost effective Concurrent Engineering tool, to give a fairly accurate estimate of the stress concentrations induced by local bending, and among the type of questions which can be answered are:

1. Considering a sandwich with given face and core materials, given face- and core-thicknesses; what magnitude of external load could be applied?
2. Considering a sandwich with given face material and face-thickness, given core material with given compressive strength; what face-thickness is required to prevent core-failure by crushing?

3. Considering a sandwich with given face and core materials, given face- and core-thicknesses, given total external load; over how large an area should the external load be distributed to prevent the maximum allowable stresses (core and facings) to be exceeded?

The approximate solution procedure presented is only strictly valid for sandwich beams, but it is easily extendable to the case of sandwich plates in cylindrical bending. Plate analysis in general, however, cannot be accomplished directly with the solutions presented so far, but general sandwich plate solutions, based on an extended version of the two-parameter foundation model, will be presented in the near future.

Acknowledgement: The author wishes to thank the Danish Technical Research Council, under the "Programme of Research on Computer Aided Engineering Design", for the financial support received during the period of the present work.

Symbols Used

Latin Symbols

A	:	$b(2c+t)^2/c$ (mm^3);
A_j	:	$(j=0,...,6)$ integrational constants;
b	:	half width of sandwich beam (mm);
c	:	half thickness of core material (mm)
D	:	flexural rigidity of sandwich beam (Nmm^2)
E_c,E_f	:	elastic moduli of core and face materials respectively (GPa);
G_c	:	shear modulus of core material (GPa);
I	:	area-moment of inertia of facing (mm^4);
K_x,K_z	:	elastic foundation moduli (GPa);
L	:	half length of sandwich beam (mm);
M	:	bending moment (Nmm);
N	:	normal force (N);
p_x,p_z	:	external surface loads (N/mm);
P	:	external point load measured per unit width (N/mm);
q_x,q_z	:	interfacial stress distribution functions (N/mm);
Q	:	shear force (N);
t	:	thickness of faces (mm);
u	:	longitudinal displacement (mm);
w	:	lateral displacement (mm);
x	:	longitudinal coordinate (mm);
z	:	lateral coordinate (mm).

Greek Symbols

α_j	:	(j=0,1,2) coefficients;
ε_x	:	longitudinal normal strain;
η	:	coefficient (root to characteristic equation);
λ	:	wave-length of elastic response (mm);
ν_c	:	Poisson's ratio of core material;
ξ	:	coefficient (root to characteristic equation);
ρ_c	:	density of core material (kg/m^3);
σ_x,σ_z	:	normal stress components (MPa);
t_{zx}	:	shear stress component (MPa);
ϕ_1	:	coefficient (root to characteristic equation).

Superscripts

total	:	superposition of local and overall bending contributions.

Subscripts

avg	:	average;
b	:	bending;
c	:	core;
f	:	facing;
local	:	contribution induced by local bending;
0	:	neutral axis (loaded face-beam) quantity;
overall	:	contribution induced by overall bending;
s	:	shear.

References

1. Allen, H.G.: Analysis and design of structural sandwich panels. Pergamon Press, Oxford, 1969
2. Plantema, F.J.: Sandwich construction. John Wiley & Sons, New York, 1966
3. Stamm, K. and Witte, H.: Sandwichkonstruktionen (in German). Springer-Verlag, Wien, Austria, 1974.
4. Triantafillou, T.C. and Gibson, L.J.: Failure mode maps for foam core sandwich beams. Materials Science and Engineering, Vol.95, 1987, pp.37-53
5. Allison, I.M.: Localised loading of sandwich beams. Proceedings of the 9th International Conference on Experimental Mechanics, 20-24 August, 1990, Copenhagen, Denmark, pp.1604-1609
6. Weismann-Berman, D., Petrie, G.L. and Wang, M.-H.: Flexural response of foam-cored frp sandwich panels. The Society of Naval Architects and Marine Engineers (SNAME), November 1988

7. Meyer-Piening, H.-R.: Remarks on higher order stress and deflection analyses. Sandwich Construction 1 (K.-A. Olsson and R.P. Reichard; Eds.), Proceedings of the First International Conference on Sandwich Construction, 19-21 June, Stockholm, Sweden, 1989, pp.107-127
8. Hetényi, M.: Beams on elastic foundations. The University of Michigan Press, Ann Arbor, Michigan, 1946
9. Thomsen, O.T.: Flexural response of sandwich panels subjected to concentrated loads. Special Report No.7, Institute of Mechanical Engineering, Aalborg University, May 1991
10. Thomsen, O.T.: Further remarks on local bending analysis of sandwich panels using a two-parameter elastic foundation model. Institute of Mechanical Engineering, Aalborg University, March 1992
11. Thomsen, O.T.: Localised loads. Contributing chapter in Handbook of Sandwich Constructions (Dr. Dan Zenkert, Royal Institute of Technology, Stockholm, Sweden; Ed.), Nordic Fund for Industrial Development, to be published in 1992
12. Thomsen, O.T.: Photoelastic investigation of local bending effects in sandwich beams, Report No. 41, Institute of Mechanical Engineering, May 1992

Part 5

Virtual Prototyping and Human Factors

Virtual Prototyping for Mechanical System Concurrent Engineering

Edward J. Haug, Jon G. Kuhl, and Fuh Feng Tsai

Center for Computer Aided Design, The University of Iowa, Iowa City, Iowa 52242-1000 USA

Abstract: The emergence of high-speed computers, new mechanical system dynamic simulation formulations, and a broad range of operator-in-the-loop simulators is shown to provide a revolutionary new virtual prototyping tool to support Concurrent Engineering of mechanical systems. The state-of-the-art of operator-in-the-loop simulation and projections regarding its refinement for use in a broad range of engineering applications is outlined, with emphasis on providing a virtual prototyping capability that accounts for the operator-machine interaction, prior to fabrication and test of prototypes. Examples of advanced ground vehicle simulators, telerobotic simulators, and construction equipment simulators are used to illustrate virtual prototyping applications that hold the potential to revolutionize the process of mechanical system design for the human operator. The potential now exists to routinely investigate trade-offs involving mechanical system design and operator effectiveness that will permit the engineering community to optimize the design of mechanical systems for the human operator, beginning early in the design and development process and continuing through commercialization and product improvement. Bringing the human factor into design consideration using virtual prototyping before design decisions are finalized is projected to be one of the greatest advances in Concurrent Engineering of Mechanical Systems to occur in the decade.

Keywords: virtual prototyping / operator-in-the-loop simulation / real-time dynamic simulation

1 Introduction

Dynamic simulation of mechanical systems has seen a renaissance during the 1980s, due to a number of synergistic developments. The rapid increase in digital computer power has permitted an international community concerned with mechanical system dynamics to create new analytical and computational formulations that take advantage of emerging computer power and automate the process of forming and solving the differential-algebraic equations of mechanical system dynamics, using only engineering model data that can be naturally and effectively provided by the engineer. With this computational burden transferred from the engineer and analyst to the computer, the creative process of model development, design concept formulation and analysis, and testing of designs prior to fabrication of a prototype has revolutionized the process of mechanical system design for dynamic performance [1-6]. As an illustration of the explosive growth in the field of mechanical system dynamic simulation, six textbooks and advanced research monographs on the topic have been published since 1988 [7-12], whereas only two such books had been published prior to 1988 [13,14].

As impressive as has been the advancement in computer–based dynamic simulation of mechanical systems, computer times required for realistic simulation of dynamic performance of mechanical systems have been extremely high. Even on the most powerful computers available in the late 1980s, the computer time required has typically been a factor of 10 to 100 greater than the clock-time that transpires during actual motion of the mechanical system. As a result, only off-line (non-real-time) dynamic simulation could be carried out in support of design applications, precluding applications in which the operator must interact with the mechanical system to control performance. Projections of increased computer performance and the emergence of revolutionary new dynamics formulations in the late 1980s suggested the potential for real-time simulation; i.e., computing the motion of a mechanical system in one unit of the computer time that corresponds to the same unit of time required for actual performance of the system. This led to a vision for operator-in-the-loop simulation for a broad range of applications in the late 1980s that is only now coming to fruition. The purpose of this paper is to summarize the development of enabling technologies for operator-in-the-loop simulation, providing references for more detailed development. The role of operator-in-the-loop simulation is defined, to permit concurrent consideration of the human operator in design of mechanical systems, beginning in the conceptual design phase and continuing through production and product improvement. This new capability might be thought of as prototyping and testing designs on the simulator with the operator-in-the-loop, suggesting the term "virtual prototyping."

A precise definition of the term "virtual prototyping," especially as it applies to mechanical system design, is needed to avoid confusion with other concepts in design. As a foundation for such a definition, consider the following dictionary definitions:

Prototype: A first full-scale functional form of a new type or design of a construction (such as an airplane).

Virtual: Being such in essence or effect, although not in actual fact.

Reality: The quality or fact of being real.

While not yet in the dictionary, the term "virtual reality" has taken on the meaning of the "computer generated perception of reality on the part of an involved human." It is believed that the term "virtual reality" motivated the emerging use of the term "virtual prototype," suggesting both computer and human involvement in virtual prototyping.

A key concept that is implicit in each of the above definitions, but not explicitly stated, is that the functionality of the system or environment being addressed is clearly understood. The functionality of a prototype is central to the purpose for which it is fabricated and tested; e.g., assessment of dynamic performance, maintainability, manufacturability, and supportability. The expression "being such" in the definition of the word "virtual" implies some well understood form of functionality. The essence of the concept of "reality" is that some form of functionality should be, or appear to be, real. Thus, the central issue in defining and using the term "virtual prototyping" is making explicit the intended functionality of the prototype that is to be realized virtually.

With this background, the following definitions are proposed:

Virtual Prototype: A computer based simulation of a prototype system or subsystem with a degree of functional realism that is comparable to that of a physical prototype.

Virtual Prototyping: The process of using a virtual prototype, in lieu of a physical prototype, for test and evaluation of specific characteristics of a candidate design.

These definitions are intended to **include** the following:

1. The intended functionality of the prototype that is to be created virtually is clearly defined and realistically simulated; e.g., vehicle dynamic performance, vehicle maintainability functions, engine reliability, and vehicle component manufacturability.
2. If human action is involved in the intended functionality of the prototype, then the human functions involved must be realistically simulated, or the human must be included in the simulation; i.e., real-time operator-in-the-loop simulation.
3. If no human action is involved in the intended functionality of the prototype, then either off-line (non-real-time) computer simulation of the functions can be carried out; e.g., dynamic performance of an engine, stresses in its connecting rods, and fabriction of the connecting rods, or a combination of computer and hardware-in-the-loop simulation can be carried out; e.g., vehicle dynamic performance prediction, laboratory durability testing for difficult to model failure modes, and manufacturing process analysis or critical components.

These definitions are intended to **exclude** the following:

1. Partial simulation that does not include the full functionality intended for the prototype; e.g., geometric modeling with a CAD system that does not simulate dynamic performance, finite element stress analysis of a component that does not include system or subsystem performance simulation that defines loads on the component, and manufacturing process planning that does not consider component performance or design constraints.

2. Show-and-tell exercises that lack a prototype level of functional reality; e.g., goggles and gloves simulation with no underlying physical or mathematical simulation at an engineering or manufacturing level of reality.

To provide background on developments that have occurred during the past decade in dynamic simulation, a brief summary of its evolution for off-line (non-real-time) applications is provided. A vision is suggested for real-time operator-in-the-loop simulation that creates the opportunity for virtual prototyping in a broad range of mechanical system design and development applications, with emphasis on ground vehicles and construction equipment applications. Real-time recursive dynamics formulations that are well-suited to exploit the emerging capabilities of shared memory parallel processors are summarized, with references to further developments. Emerging technologies, including recursive dynamics, for advanced driving simulation and virtual prototyping applications are summarized, culminating in the current implementation of a number of advanced ground vehicle driving simulators that will support a broad range of human factors research, including highway safety and vehicle and highway system design. Finally, other operator-in-the-loop applications, primarily telerobotics and construction equipment are outlined, to provide an indication of the potential that exists for virtual prototyping in a broad range of Concurrent Engineering applications.

2 Off-Line Dynamic Simulation

To illustrate the capabilities that have evolved in off-line, or non-real-time dynamic simulation, consider the tractor-trailer roll stability analysis suggested by the scenario shown in Figure 1. A tractor-trailer vehicle drives along a road surface and encounters a depression that is at an angle with the roadway, causing roll motion of the vehicle as it transits the irregular road surface. The objective is to create a dynamic simulation of the tractor-trailer and its contact with the road surface via its tires, to predict roll motion of the vehicle as it moves through the depression.

To model the tractor-trailer using a commercial dynamic simulation program such as DADS [15], the vehicle kinematics and internal force characteristics are modeled using a library of joints and force elements that are offered within the dynamic simulation computer program. Shown in Figure 2 is a sampling of standard kinematic connections between pairs of bodies [9] that can be used to make up the model of a mechanical system, such as the tractor-trailer vehicle in question. Also shown is an illustration of a force element [9] that can be used to account for forces acting internal to the mechanical system due to springs, dampers, hydraulic actuators, electrical motors, and a variety of related force generating devices.

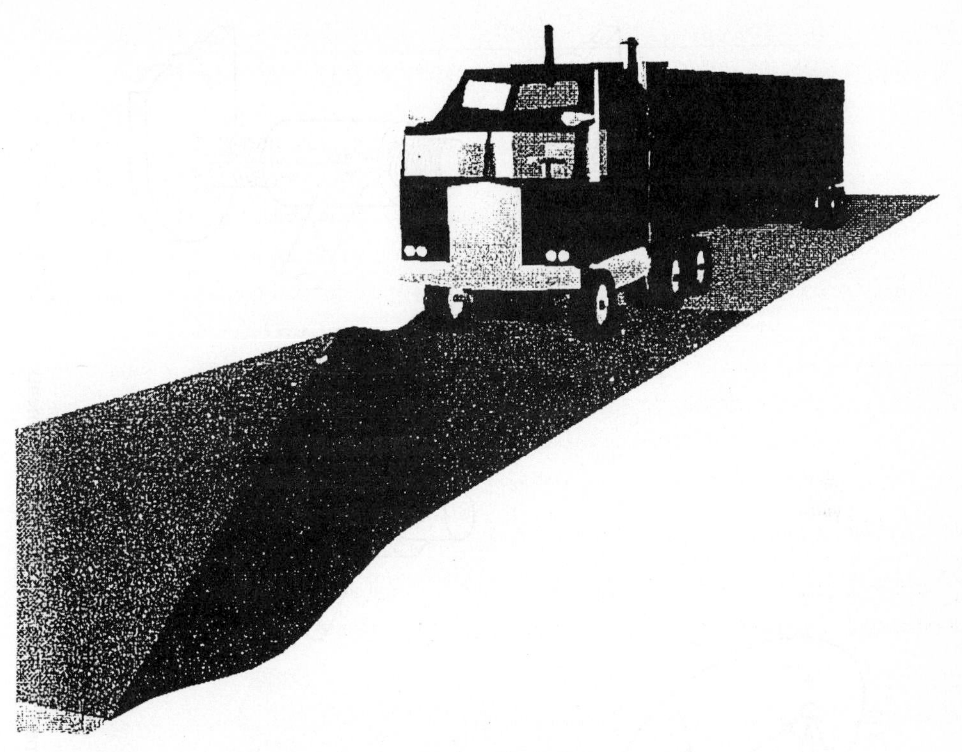

Figure 1. Tractor-Trailer with Depression in Road

To illustrate the use of kinematic and force building blocks illustrated in Figure 2, a tractor-trailer model suitable for roll stability analysis in the situation shown in Figure 1 is described in Figure 3. Rotational and translational joints are used to permit roll and heave (vertical relative to chassis) motion of each of the five axles and associated wheel sets that make up the model of a tractor-trailer. Suspension springs and shock absorbers are accounted for by force elements between the axles and the chassis of the tractor and the trailer, as shown. The load leveling effect of leaf springs in the tractor and trailer suspension subsystems is accounted for by modeling the leaf spring as bodies that are pivoted relative to the chasses of the tractor and trailer, as in the real application, with spring effects concentrated at the ends of the bodies. This permits the intended nearly equal distribution of loads across pairs of axles in the tractor and trailer. Vertical and lateral forces due to tire contact with the road surface are calculated using empirical models of the tires and are transfered to the appropriate axle. The tractor and trailer are coupled to represent the effect of the "fifth wheel" connection between the tractor and trailer.

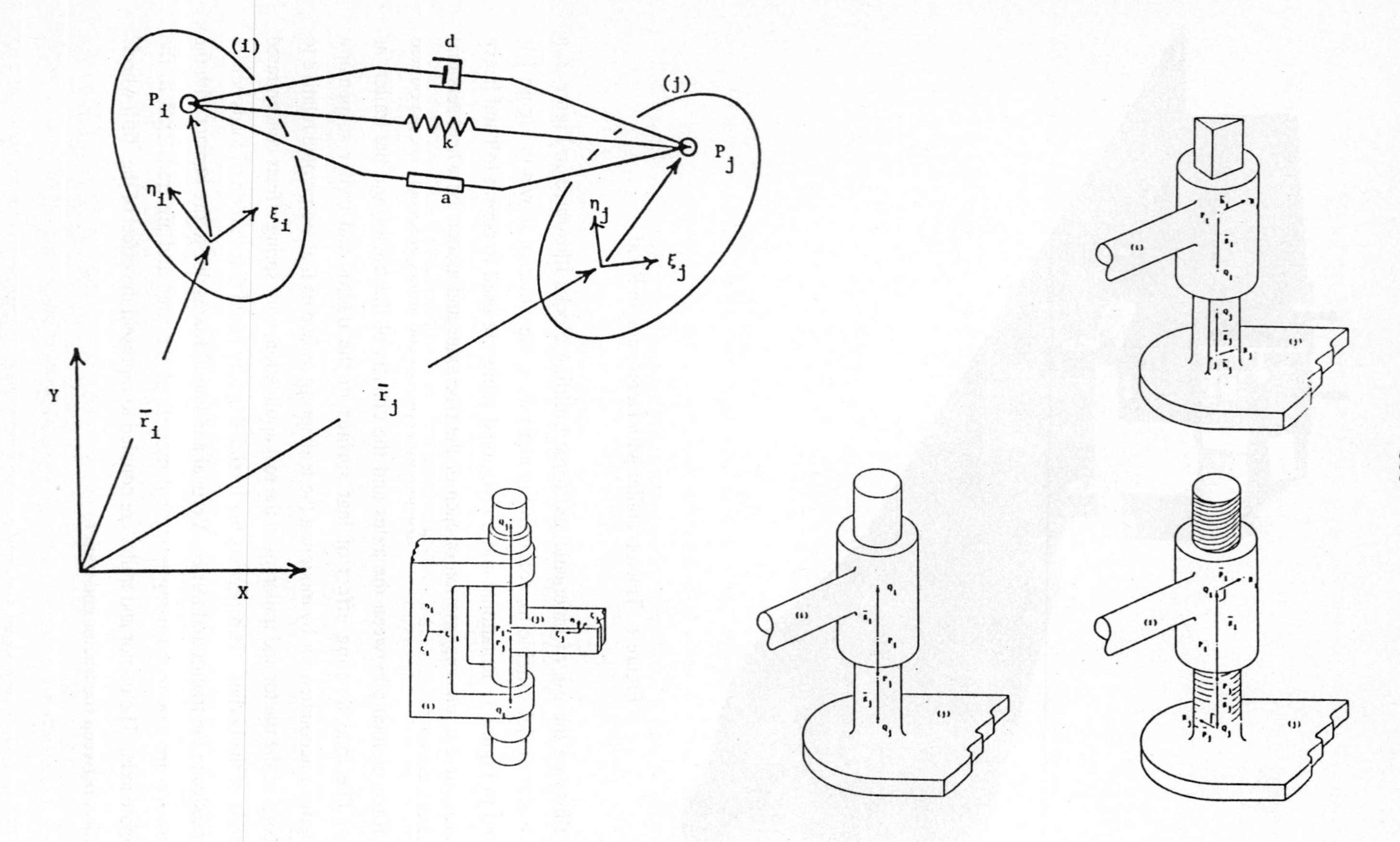

Figure 2. Library of Kinematic Connections and Force Elements

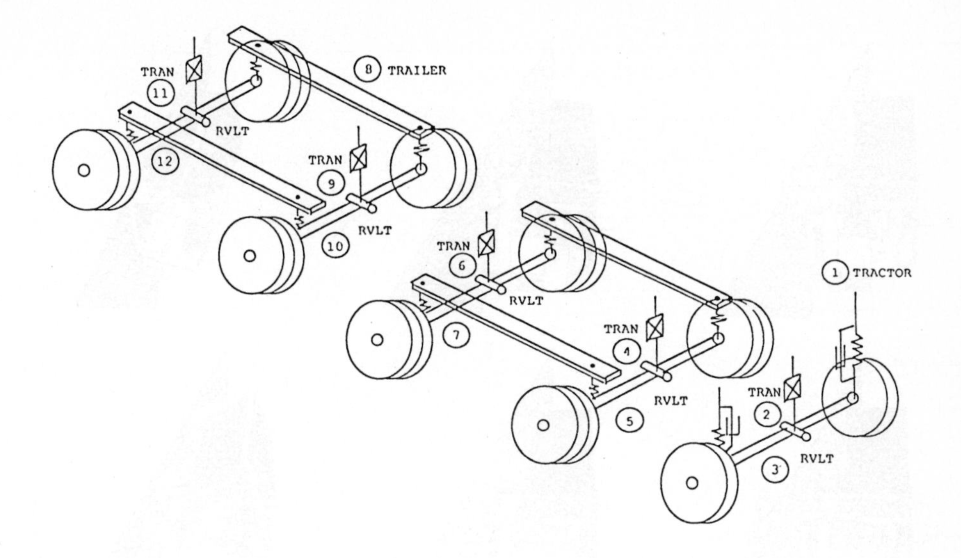

Figure 3. Tractor-Trailer Kinematic Model

Simulations may now be carried out at varying speeds as the vehicle encounters and traverses the irregular road surface shown in Figure 1. A multi-frame report of results presented in Figure 4 illustrates the results of simulation, showing that at a critical case the trailer may actually roll over during operation.

It is interesting to note that simulations of this complexity in 1982, on a then-modern super minicomputer, required seven CPU hours to carry out the simulation sequence shown in Figure 4, which represented approximately one minute of actual operational time. Furthermore, the rendering of the approximately 1,200 individual images required for animation at 20 frames per second for the single minute required another seven CPU hours on the same computer. Thus, 14 CPU hours were required in 1982 to precompute quantities needed to create an animation of one minute of actual vehicle motion illustrated in Figure 4. While this is a great deal of computing time, the value of the result was significant and has led to adoption of this form of analysis in a broad cross section of vehicle and other mechanical system development and manufacturing communities [1,5].

3 Recursive Dynamics

In order to motivate concepts that underlie recursive dynamics formulations, the Army's High Mobility Multipurpose Wheeled Vehicle (HMMWV) shown in Figure 5 is used as an illustration. This heavy duty multipurpose off-road vehicle represents challenges in ground

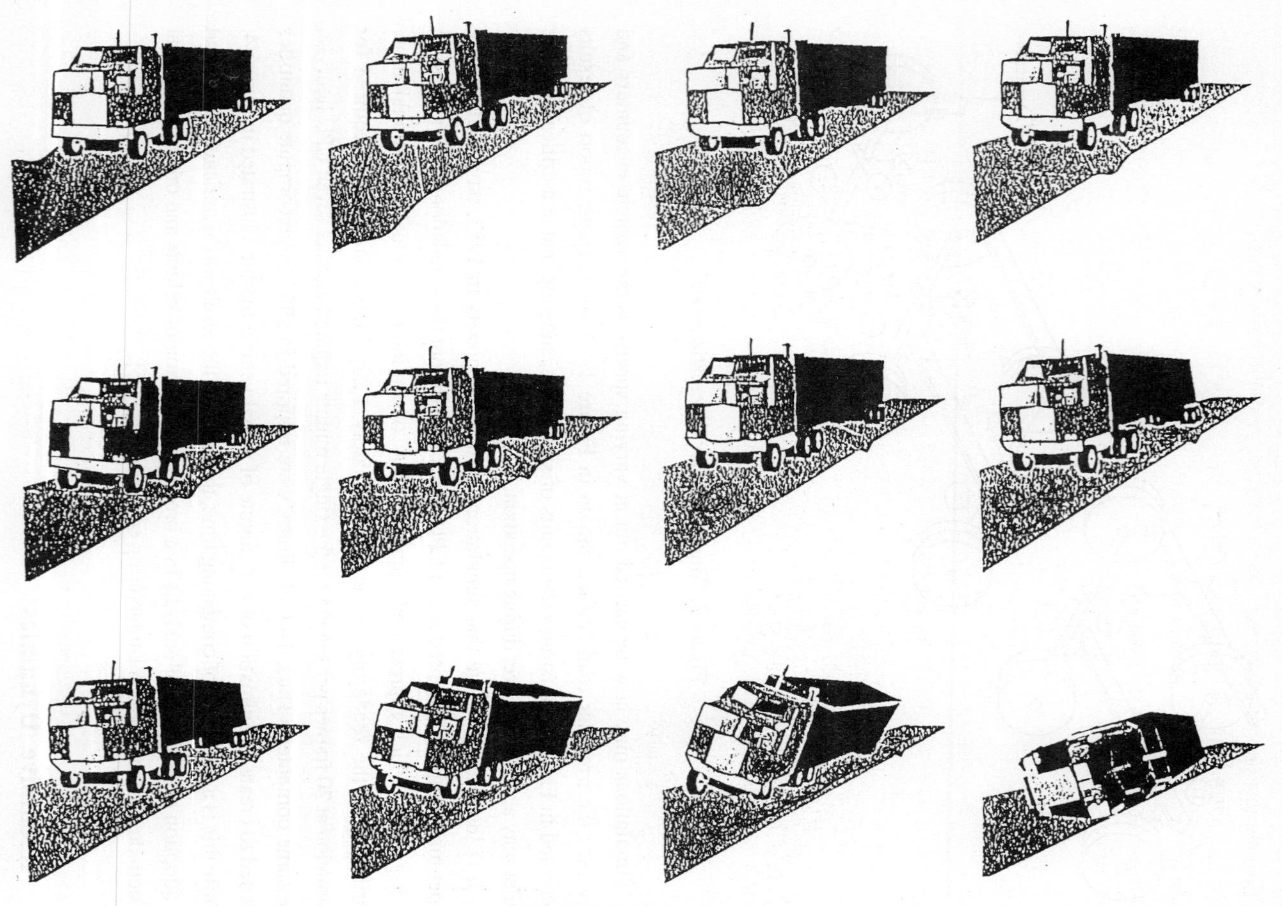

Figure 4. Animation of Trailer Roll-Over

vehicle simulation that are typical of those found in automobiles and heavy trucks. It has been used extensively in comparison of alternative algorithms and benchmarking on alternate computer platforms. It is used in this paper to provide a concrete example of the class of algorithms being considered and to serve as the basis for computational efficiency comparisons between algorithms and alternate computer implementations.

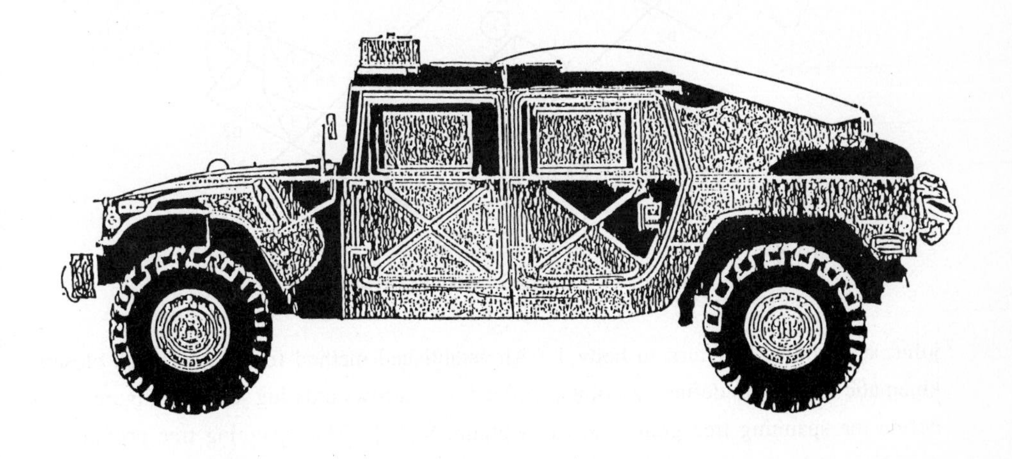

Figure 5. High Mobility Multipurpose Wheeled Vehicle (HMMWV)

The schematic representation of a fourteen body model of the HMMWV is shown in Figure 6. Rigid bodies that may move in space and relative to each other are shown schematically as circled numbers representing bodies 1 through 14. Body 1 is the chassis of the vehicle and body 2 is the steering rack. The right front suspension subsystem is comprised of the lower control arm (body 3), the wheel assembly (body 4), and the upper control arm (body 5). Each of the other three suspension subsystems is similarly constructed.

Translational and rotational joints allow only one relative degree of freedom, translation and rotation, respectively, between bodies they connect. Spherical joints permit three relative rotation degrees of freedom between bodies they connect. Finally, tie rod distance constraints serve to constrain the distance between points on bodies they connect.

A graph theoretic representation of the HMMWV model is shown in Figure 7. Numbered nodes (circles) of the graph represent bodies identified in Figure 6. Edges of the graph that connect bodies represent joints and tie rod distance constraints between the bodies connected. It may be observed that there are eight independent closed kinematic loops in this vehicle mechanism; i.e., paths that may be traversed beginning from body 1 and crossing successive

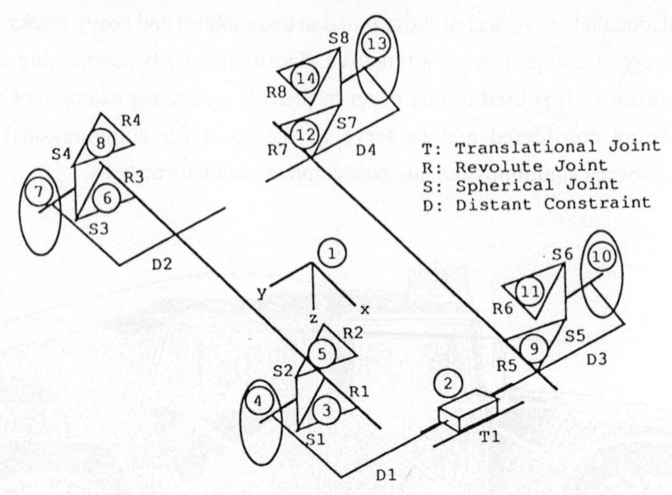

Figure 6. Kinematic Model of HMMWV

joints and bodies to return to body 1. An established method for treating such closed kinematic loops is to define cut-joints, denoted with arrows crossing joints in Figure 7, to define the spanning tree graph shown in Figure 8 [13]. This spanning tree provides a definition of kinematic and dynamic computational sequences that are used to create the equations of motion of the system. As will be noted in the development that follows, the cut-joints identified in Figure 7 must be treated as constraints in the formulation of the equations of motion, using the Lagrange multiplier method of dynamics [9].

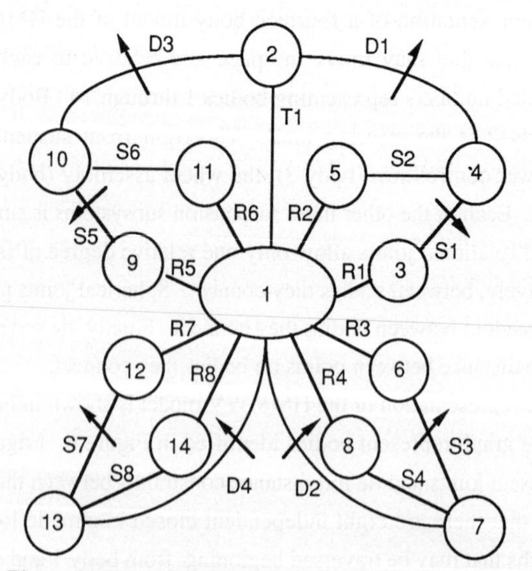

Figure 7. Graphical Representation of the HMMWV

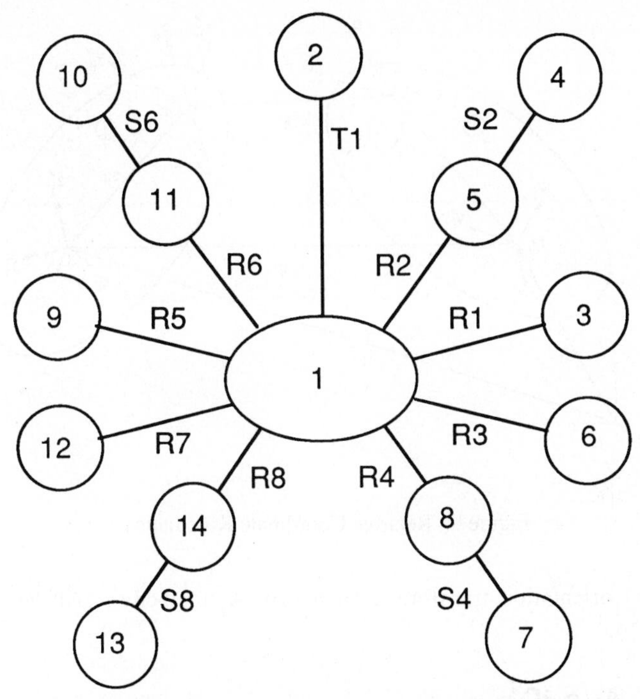

Figure 8. Spanning Tree Corresponding to Figure 4

The basic concept of relative coordinate kinematics between bodies that are connected by a joint is illustrated in Figure 9. The pair of bodies shown, designated by indices i and j, each have associated x'-y'-z' body reference frames, with origins at O_i and O_j. In addition, joint x"-y"-z" reference frames are located on the bodies at joint definition points O_{ij} and O_{ji}. The vectors $\mathbf{s}_{ij}$ and $\mathbf{s}_{ji}$ locate the origins of the joint reference frames in the respective bodies. The orientations of the joint reference frames relative to the body reference frames are defined by constant orthogonal rotation transformation matrices [9] $\mathbf{C}_{ij}$ and $\mathbf{C}_{ji}$ on bodies i and j, respectively.

Denoting the vector (column matrix) of joint relative coordinates between bodies i and j as $\mathbf{q}_{ij}$, which depend on the type of joint selected, the following vector relationship can be written to define the vector $\mathbf{r}_j$ that locates body j in space, as a function of $\mathbf{r}_i$ and the relative coordinates $\mathbf{q}_{ij}$

$$\mathbf{r}_j = \mathbf{r}_i + \mathbf{s}_{ij} + \mathbf{d}_{ij}(\mathbf{q}_{ij}) - \mathbf{s}_j \tag{1}$$

where the vector $\mathbf{d}_{ij}$ depends on joint relative coordinates. Similarly, the orthogonal orientation transformation matrix $\mathbf{A}_j$ for the body j reference frame can be written in terms of

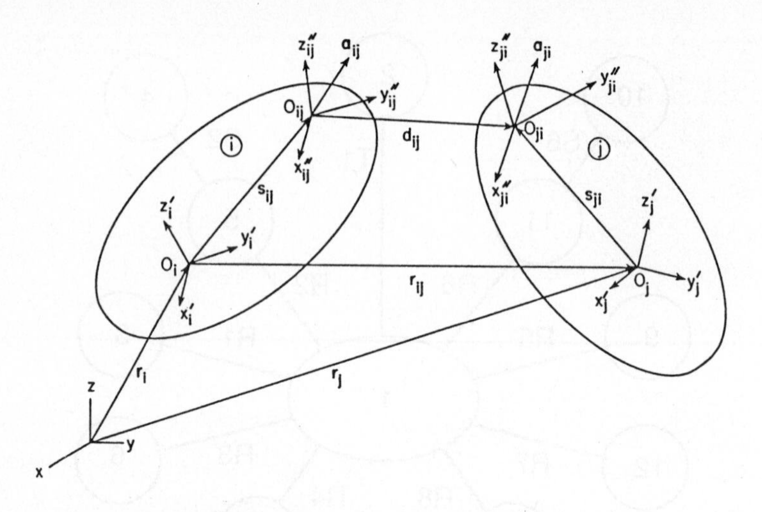

Figure 9. Relative Coordinate Kinematics

the orthogonal orientation transformation matrix $\mathbf{A}_i$ for body i and the joint relative coordinates as [9]

$$\mathbf{A}_j = \mathbf{A}_i \mathbf{C}_{ij} \mathbf{A}_{ij}(\mathbf{q}_{ij}) \mathbf{C}_{ji}^T \tag{2}$$

where $\mathbf{A}_{ij}$ is the orthogonal orientation transformation matrix from the body j joint reference frame to the body i joint reference frame, which depends on the joint relative coordinates $\mathbf{q}_{ij}$.

As a concrete illustration of relative coordinate kinematic relationships, consider the chassis and upper control arm of the HMMWV model shown in Figure 6. As shown in Figure 10, for the rotational joint between bodies 5 and 1, $\mathbf{d}_{ij} = \mathbf{0}$ and Eq. 1 specializes to

$$\mathbf{r}_5 = \mathbf{r}_1 + \mathbf{s}_{15} - \mathbf{s}_{51} \tag{3}$$

Noting that, in the case of this rotational joint, the relative coordinate q_{15} is a rotation about the unit vector $\mathbf{u}_{15}$ along the axis of relative rotation, Eq. 3 may be differentiated to obtain a relative velocity relationship. From geometric considerations, an angular velocity relationship between bodies 5 and 1 can similarly be written [16]. The combined result is the matrix relationship

$$\begin{bmatrix} \dot{\mathbf{r}}_5 \\ \omega_5 \end{bmatrix} = \begin{bmatrix} \mathbf{I} & (\tilde{\mathbf{s}}_{51} - \tilde{\mathbf{s}}_{15}) \\ \mathbf{0} & \mathbf{I} \end{bmatrix} \begin{bmatrix} \dot{\mathbf{r}}_1 \\ \omega_1 \end{bmatrix} + \begin{bmatrix} \tilde{\mathbf{s}}_{51} \\ \mathbf{u}_{15} \end{bmatrix} \dot{\mathbf{q}}_{15} \tag{4}$$

where the operator ~ denotes vector product [9]. Defining coefficient matrices in this relationship as $\mathbf{B}_{15}$ and $\mathbf{D}_{15}$, Eq. 4 may be written in the form

$$\mathbf{Y}_5 = \mathbf{B}_{15}\mathbf{Y}_1 + \mathbf{D}_{15}\dot{\mathbf{q}}_{15} \tag{5}$$

where $\mathbf{Y} \equiv \left[\dot{\mathbf{r}}^T, \boldsymbol{\omega}^T\right]^T$ is the composite vector of Cartesian velocity and angular velocity, relative to the inertial reference frame.

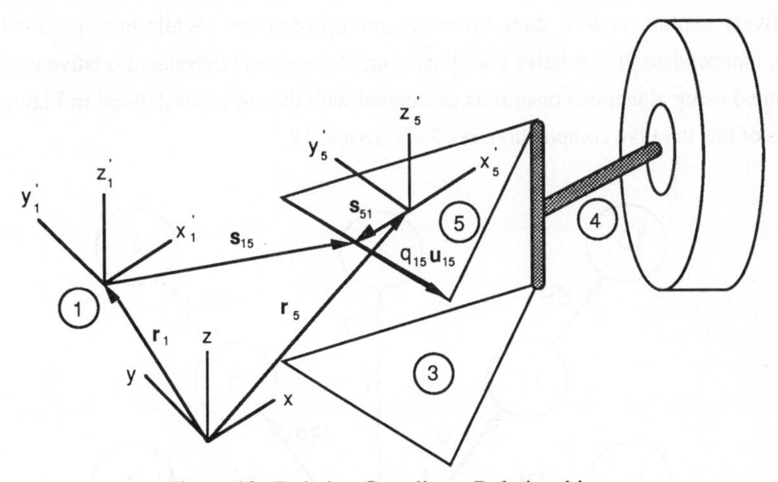

Figure 10. Relative Coordinate Relationships

Denoting $\delta\mathbf{Z} \equiv \left[\delta\mathbf{r}^T, \delta\boldsymbol{\pi}^T\right]^T$ as a composite vector of virtual displacement and virtual rotation, an analogous relationship to Eq. 5 [16] is obtained as

$$\delta\mathbf{Z}_5 = \mathbf{B}_{15}\delta\mathbf{Z}_1 + \mathbf{D}_{15}\delta\mathbf{q}_{15} \tag{6}$$

where δq_{15} is a variation in the joint relative coordinate q_{15}.

Differentiating Eq. 5 with respect to time yields the acceleration relationship

$$\dot{\mathbf{Y}}_5 = \mathbf{B}_{15}\dot{\mathbf{Y}}_1 + \mathbf{D}_{15}\ddot{\mathbf{q}}_{15} + \mathbf{E}_{15} \tag{7}$$

where $\mathbf{E}_{15}$ is a term that is quadratic in velocities. For details of the derivation of these equations and construction of the associated matrices, see Refs. 16 to 18.

One of the key computational steps in dynamic simulation is the calculation of the position and velocity of each body in the system, relative to the inertial reference frame, once all relative coordinates and their time derivatives are known. This computation proceeds systematically, using Eqs. 1, 2, and 4, along each branch of the spanning tree shown in Figure 8. As shown schematically in Figure 11, for each branch in the spanning tree, computations begin with the chassis and proceed outward toward the extreme bodies in each branch, called tree end bodies. In each branch, computations cross a joint from body 1 to the next body and, if there is a subsequent body in the chain, carrying out the computation across that joint. The graph shown in Figure 11 serves as a guide for efficient use of a parallel computer, illustrating that computations may proceed in parallel along each of the nine branches in the spanning tree. This serves as a guide to coarse-grain parallelism that can effectively exploit modern shared-memory multiprocessors. While not discussed in this paper, independent joint relative coordinates are defined, and dependent relative coordinates computed using algebraic constraints associated with the cut joints defined in Figure 7. For details of this iterative computation, see Refs. 16 and 18.

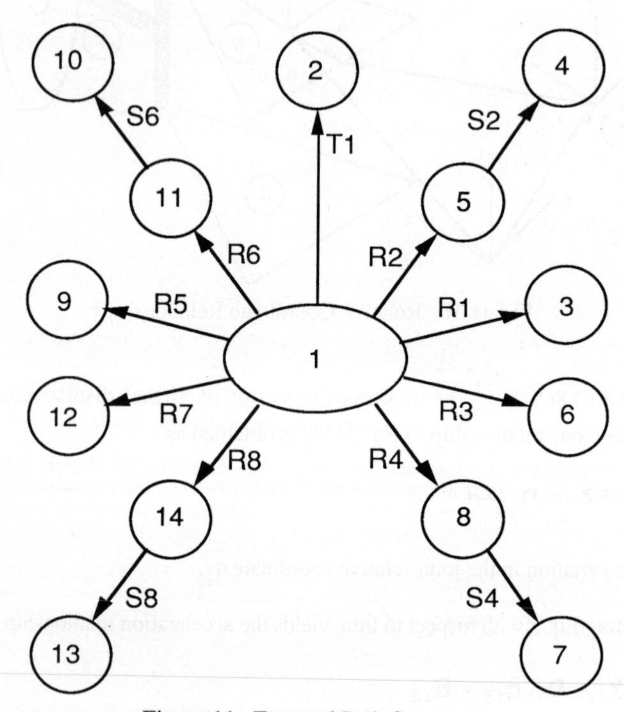

Figure 11. Forward Path Sequence

Denoting the cut joint algebraic constraints as

$$\boldsymbol{\Phi}(\mathbf{q}_{ij}) = \mathbf{0} \tag{8}$$

the variational form of the equations of motion of the entire system, and of the right front suspension subsystem of the HMMWV, can be written as

$$\delta \mathbf{Z}_3^T(\mathbf{M}_3\dot{\mathbf{Y}}_3 - \mathbf{Q}_3 + \mathbf{\Phi}_{\mathbf{Z}_3}^T\lambda) + \delta \mathbf{Z}_1^T(\mathbf{M}_1\dot{\mathbf{Y}}_1 - \mathbf{Q}_1)$$

$$+ \delta \mathbf{Z}_5^T(\mathbf{M}_5\dot{\mathbf{Y}}_5 - \mathbf{Q}_5) + \delta \mathbf{Z}_4^T(\mathbf{M}_4\dot{\mathbf{Y}}_4 - \mathbf{Q}_4 + \mathbf{\Phi}_{\mathbf{Z}_4}^T\lambda) = 0 \tag{9}$$

which must hold for all kinematically admissible virtual displacements and rotations $\delta \mathbf{Z}_i$. Using equations analogous to Eqs. 6 and 7 that relate the virtual displacement and acceleration of body 4 to those of body 5 into Eq. 9 yields

$$\delta \mathbf{Z}_3^T(\mathbf{M}_3\dot{\mathbf{Y}}_3 - \mathbf{Q}_3 + \mathbf{\Phi}_{\mathbf{Z}_3}^T\lambda) + \delta \mathbf{Z}_1^T(\mathbf{M}_1\dot{\mathbf{Y}}_1 - \mathbf{Q}_1)$$

$$+ \delta \mathbf{Z}_5^T\{(\mathbf{M}_5 + \mathbf{K}_5)\dot{\mathbf{Y}}_5 + \mathbf{R}_5\ddot{\mathbf{q}}_{54} - (\mathbf{Q}_5 + \mathbf{L}_5) + \mathbf{P}_5^T\lambda\}$$

$$+ \delta \mathbf{q}_{54}^T(\mathbf{G}_5\dot{\mathbf{Y}}_5 + \mathbf{H}_5\ddot{\mathbf{q}}_{54} + \mathbf{V}_5 + \mathbf{W}_5^T\lambda) = 0 \tag{10}$$

where coefficient matrices are products of those appearing in Eqs. 6, 7, and 9. Note that Eq. 10 holds for all kinematically admissible virtual displacements of bodies 3 and 5 and arbitrary values of $\delta \mathbf{q}_{54}$. Thus, the coefficient of $\delta \mathbf{q}_{54}$ must be **0**, yielding

$$\ddot{\mathbf{q}}_{54} = -\mathbf{H}_5^{-1}(\mathbf{G}_5\dot{\mathbf{Y}}_5 + \mathbf{V}_5 + \mathbf{W}_5^T\lambda) \tag{11}$$

This observation [16,17] permits reduction of the equations of motion and solution for relative coordinate accelerations between bodies 5 and 4, as functions of inboard body accelerations and Lagrange multipliers.

The above process is continued by substituting from Eqs. 6 and 7 to eliminate $\delta \mathbf{Z}_5$ and $\dot{\mathbf{Y}}_5$, yielding expressions that involve only chassis accelerations and Lagrange multipliers. Carrying out similar reductions along other branches of the spanning tree, beginning with the outermost bodies and moving in toward the chassis, yields the matrix equation

$$\begin{bmatrix} \mathbf{K}_1 & \mathbf{\Phi M}_1 \\ \mathbf{\Phi M}_1 & \mathbf{\Phi L}_1 \end{bmatrix} \begin{bmatrix} \dot{\mathbf{Y}}_1 \\ \lambda \end{bmatrix} = \begin{bmatrix} \mathbf{L}_1 \\ \mathbf{RHS} \end{bmatrix} \tag{12}$$

which involves only the chassis acceleration and Lagrange multipliers associated with cut-joint constraints. The second line of Eq. 12 is obtained by differentiating the constraint equations of Eq. 8 twice. For details of this reduction, see Refs. 16 and 17.

A key characteristic of this recursive formulation of the equations of motion, based on the spanning tree graph and elimination of joint relative accelerations, is that it eliminates all relative coordinate accelerations from the reduced equations of motion of Eq. 12. This algorithm is thus called the recursive algorithm "with elimination". The number of computations required for its implementation is proportional to the number of relative coordinates in the longest chain in the mechanism. For a single-chain mechanism with n joints, the number of calculations is proportional to n. The algorithm is thus called "order-n".

Rather than eliminating the relative coordinate acceleration using Eq. 11, which involves the inversion of a matrix, the last term in Eq. 10 may be retained in the equations of motion and the recursive process of eliminating $\delta \mathbf{Z}_5$ may be applied to obtain

$$\begin{aligned}
&\delta \mathbf{Z}_3^T(\mathbf{M}_3\dot{\mathbf{Y}}_3 - \mathbf{Q}_3 + \mathbf{\Phi}_{\mathbf{Z}_3}^T\lambda) \\
&+ \delta \mathbf{Z}_1^T\{(\mathbf{M}_1 + \mathbf{K}_1)\dot{\mathbf{Y}}_1 + \mathbf{R}_1\ddot{\mathbf{q}}_{15} + \mathbf{B}_{15}^T\mathbf{R}_5\ddot{\mathbf{q}}_{54} - [\mathbf{Q}_1 + \mathbf{L}_1] + \mathbf{B}_{15}^T\mathbf{P}_5^T\lambda\} \\
&+ \delta \mathbf{q}_{15}^T(\mathbf{G}_1\dot{\mathbf{Y}}_1 + \mathbf{H}_1\ddot{\mathbf{q}}_{15} + \mathbf{V}_1 + \mathbf{D}_{15}^T\mathbf{P}_5^T\lambda) \\
&+ \delta \mathbf{q}_{54}^T\{\mathbf{G}_5\mathbf{B}_{15}\dot{\mathbf{Y}}_1 + \mathbf{G}_5\mathbf{D}_{15}\ddot{\mathbf{q}}_{15} + \mathbf{H}_5\ddot{\mathbf{q}}_{54} + (\mathbf{G}_5\mathbf{E}_{15} + \mathbf{V}_5) + \mathbf{W}_5^T\lambda\} = 0
\end{aligned} \tag{13}$$

After this process is complete, equations analogous to Eq. 7 are used to write all Cartesian accelerations in terms of relative coordinate accelerations. Coefficients of relative coordinate variations must then be **0** [18], yielding

$$\begin{bmatrix} \bar{\mathbf{M}} & \mathbf{\Phi}_\mathbf{q}^T \\ \mathbf{\Phi}_\mathbf{q} & \mathbf{0} \end{bmatrix} \begin{bmatrix} \ddot{\mathbf{q}} \\ \lambda \end{bmatrix} = \begin{bmatrix} \bar{\mathbf{Q}} \\ \mathbf{rhs} \end{bmatrix} \tag{14}$$

where the last row is the second time derivative of Eq. 8, written in terms of joint relative coordinates. This formulation is fundamentally different from the recursive algorithm with elimination that resulted in Eq. 12. First, it typically involves more variables, hence larger matrices, so that the number of calculations in solving Eq. 14 is proportional to the cube of the number of relative generalized coordinates. This algorithm has come to be called the recursive algorithm "without elimination" and is designated as "order-n^3". For details of this algorithm, see Refs. 18 and 19.

For more complex mechanical systems that consist of multiple, closed kinematic chains, computational complexity issues are somewhat more involved than indicated by the order-n versus order-n^3 designation for single chain systems. For both the formalations "without elimination" and "with elimination," all independent kinematic chains can be traversed simultaneously, once loop cuts have been applied. Thus in both cases, provided that sufficient parallel processing is used, the complexity of forming the linear equations of motion (Eq. 12 and Eq. 14, respectively) can be kept proportional to the length of the longest such chain, regardless of the number of chains in the overall model. In the formulation "with elimination," solving Eq. 12 will have complexity proportional to the cube of the overall number of Lagrange multipliers (cut constraints) in the model. In the latter case, the complexity of solving Eq. 14 will be proportional to the cube of the overall number of generalized coordinates in the system.

The question of which formulation will be more efficient for a given mechanical system model is dependent on the characteristics of the model; e.g., the total number of generalized coordinates, number of chains, maximum length of chains, and number of cut constraints. A third variation of the recursive dynamics formulations allows elimination of Lagrange multipliers from each kinematically decoupled loop. This may provide computational advantages for systems that contain a significant number of decoupled loops. Full discussion of this method is beyond the scope of this paper, but details can be found in Refs. 18 and 19.

Much as the forward path sequence of Figure 11 identified parallelism in kinematic computations, the backward path sequence in Figure 12 illustrates that for either the order-n or order-n^3 algorithms, formation of the equations of motion proceeds along each branch of the spanning tree, beginning with the outermost body and moving back to the chassis, as illustrated in Figure 12. Since each of these computations is independent, this diagram provides a guide to coarse-grain parallelism for parallel computer implementation.

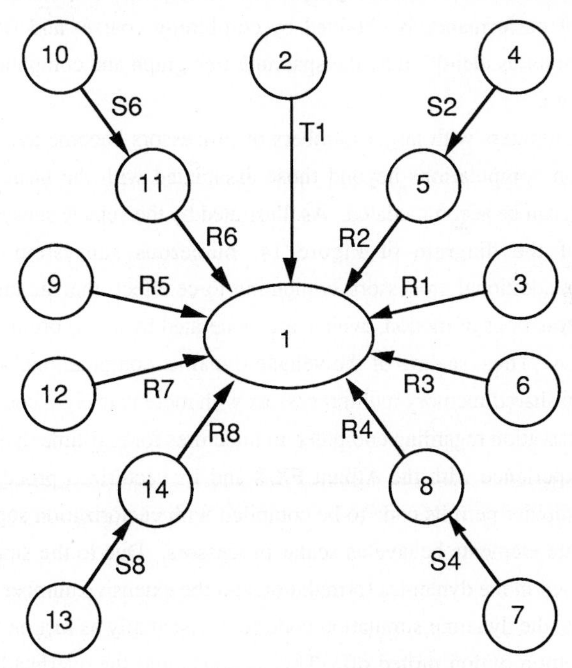

Figure 12. Backward Path Sequence

4 Parallel Processing Real-Time Dynamic Simulation

Parallel processing algorithms that exploit the coarse-grain parallelism outlined in the preceding section, for both kinematic and kinetic computations, have been developed in

Refs. 20 and 21. A number of refinements in parallel computational implementation have been developed and demonstrated in Ref. 22, to identify fine-grain parallel computation opportunities that exploit emerging shared-memory multiprocessor computer architectures. Benchmark parallel computer implementations of the recursive algorithms, both with elimination and without elimination, have been made on an eight-processor Alliant FX/8 parallel computer. In order to achieve real-time simulation of the HMMWV vehicle illustrated in the preceeding section, a total computation time per integration time step of 6.7 msec is required for explicit integration with constant time step. This figure is based on an objective of capturing 15 Hz behavior of the vehicle suspension and a rule-of-thumb estimate of ten integration time steps per Hz.

The parallel task graph for the recursive algorithm without elimination shown in Figure 13, which is explained in detail in Refs. 21 and 22, yielded a 6.4 msec per integration time step performance. This represents real-time simulation of a realistic ground vehicle and achieves 75 percent utilization of the eight-processor Alliant FX/8 parallel computer. This enhanced level of performance is obtained by combining coarse- and fine-grain parallel processing opportunities identified by the spanning tree graph and computational sequences within the algorithm.

As parallel computers with larger numbers of processors become available, additional vehicle simulation computations beyond those associated with the basic suspension and chassis subsystem can be accommodated. As illustrated by the vehicle subsystem modules on the periphery of the diagram of Figure 14, numerous subsystem models can be accommodated on additional processors, computing force effects that are incorporated in the right side of the equations of motion, which are generated by the algorithms outlined in the preceeding section. Thus, scaling of the vehicle dynamic computational load is relatively straightforward on shared-memory multiprocessors with more than eight compute elements.

As a final observation regarding computer architectures for real-time dynamic simulation, computational experience with the Alliant FX/8 and its vectorized processors is of some interest. This computer permits code to be compiled with vectorization suppressed. In this mode, the compute elements behave as scalar processors. Due to the small dimension of vectors that are used in the dynamics formulation and the extensive number of computations with 3x3 matrices, the dynamic simulation code runs essentially as fast on the Alliant FX/8 with the vectorization option turned off. This suggests that the overhead associated with starting up pipeline operations with the small vectors and matrices that are encountered in dynamics exceeds the benefits gained. The conclusion that can be drawn from this computational experience is that parallel computers and workstations with high-speed scalar RISC processors, functioning with a shared-memory, are ideally suited for high-speed dynamic computation. In contrast, there appears to be little gain to be achieved with these algorithms in the use of pipelined supercomputers. The emergence of modest-cost parallel

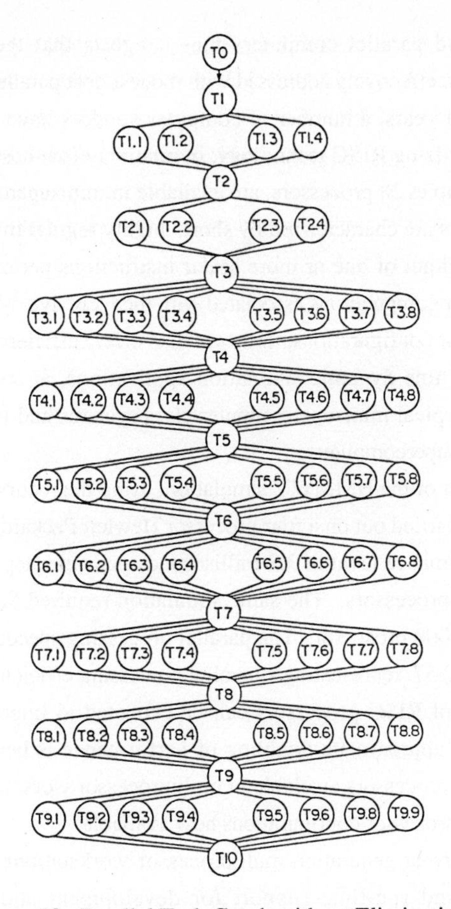

Figure 13. Parallel Task Graph without Elimination

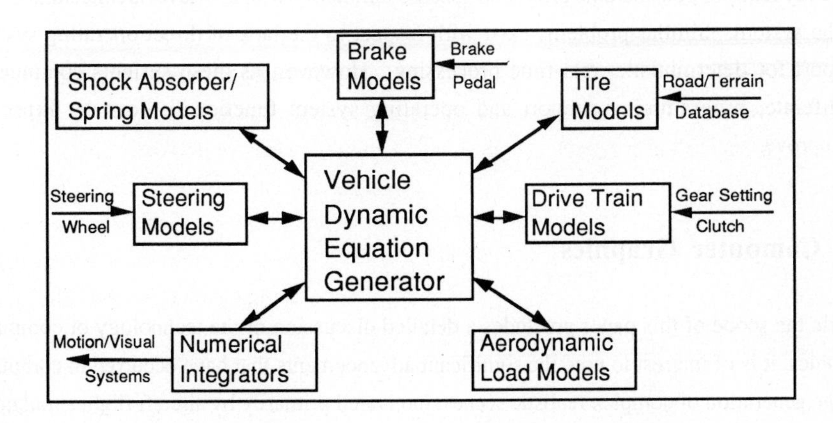

Figure 14. Structure of Real-Time Vehicle Simulation

superworkstations and parallel computers thus suggests that there is a broad class of applications that can be effectively addressed with modest-cost parallel computers.

In the past several years, a number of computer vendors have begun to offer multiple processor systems, utilizing RISC technology, in relatively low-cost workstation platforms. Larger systems, with up to 28 processors, are available in minisupercomputer configurations. These RISC processors are characterized by short, highly regular instruction pipelines and a sustained CPU throughput of one or more scalar instructions per cycle. As such, they are ideally suited for scalar computations associated with the recursive dynamics formulations. In four- to eight-processor configurations, these systems offer sufficient computational capacity to support some real-time dynamic simulation applications, at costs that are an order of magnitude less than typical minisupercomputer class systems and two orders of magnitude less than full-fledged supercomputers.

An implementation of the HMMWV simulation, using the recursive formulation without elimination, has been carried out on a four processor Hewlett Packard/Apollo DN10000 RISC workstation. A performance level of 3.3 milliseconds per time step was achieved, using all four of the DN10000 processors. The same simulation required 8.5 milliseconds per time step on a single DN10000 processor. The parallel processing speedup factor for the parallel version was therefore 2.57, representing a parallel processing efficiency of over 64 percent.

The performance of RISC processors can be expected to improve dramatically in the future. Currently, an approximate doubling of performance is being observed every two years. The number of processors available in multiprocessor workstations is also increasing, with eight- to sixteen-processor configurations now available.

A limitation of current generation multiprocessor workstations is the lack of adequate programming tools and run-time support for development and execution of parallel applications. In the absence of such support, constructing parallel dynamics implementations currently requires considerable effort and specific familiarity with low-level architectural detail of the system. Similar problems exist with respect to the lack of direct operating system support for deterministic, real-time processing. However, as these systems continue to proliferate, programming support and operating system functionality can be expected to improve.

5 Computer Graphics

While the scope of this paper precludes a detailed discussion of the technology of computer graphics, it is of interest to note the significant advancements that have occurred in computer image generation of complex realistic scenes, motivated primarily by aircraft flight simulators. More pertinent to the ground vehicle applications discussed thus far in this paper, the scene shown in Figure 15 indicates the level of textural detail that can be accommodated in scenes through which a driver can function [23]. The revolutionary developments that have occurred in high-performance computer image generation provide extraordinarily realistic visual feedback to the driver of the vehicle, with realistic motion predicted using the dynamics methods outlined in the previous two sections.

Figure 15. Ground Vehicle Visual Imagery

The type of high-quality, textured graphics capability currently provided only by specialized, multi-million dollar image generation systems is rapidly evolving into lower-cost graphics workstation platforms. High-end graphics workstations, such as the Silicon Graphics IRIS 4D, offer features such as texture mapping and can provide a significant frame rate capability. Such systems are not currently capable of supporting the demands of real-time image generation for highly realistic operator-in-the-loop simulation. However, as current rates of performance increase, the highest-end workstation systems can soon be expected to achieve this level of capability. By the mid-1990s, it can be expected that multiprocessor graphics workstation platforms will be available that will be sufficiently powerful to support both real-time dynamic simulation and reasonably high-quality real-time image generation. This should result in a dramatic reduction in the cost of achieving low and mid-range vehicle simulation capabilities.

6 Motion Generation

To complete the realism of the operator's experience in driving a vehicle, it is important that the platform on which the driver sits while driving the vehicle moves so that the motion cues experienced during driving are replicated. In the area of aircraft flight simulation, one of the most advanced simulators operated by NASA at Moffett Field, California is shown schematically in Figure 16. This major flight simulator has a motion base that moves sixty feet vertically, forty feet laterally, and eight feet longitudinally, with substantial acceleration capability. The pilot thus feels motion cues associated with flying the aircraft that is being simulated, in addition to seeing a visual display of the motion that would be experienced in flying the actual aircraft. While this motion envelope is well suited to advanced aircraft simulation, the basic motion envelope is not suitable for ground vehicle applications in which the vehicle experiences sustained longitudinal and lateral accelerations. Under conditions of high acceleration, only modest vertical displacement is required for the ground vehicle. Nevertheless, this motion generation technology has been developed for aircraft applications.

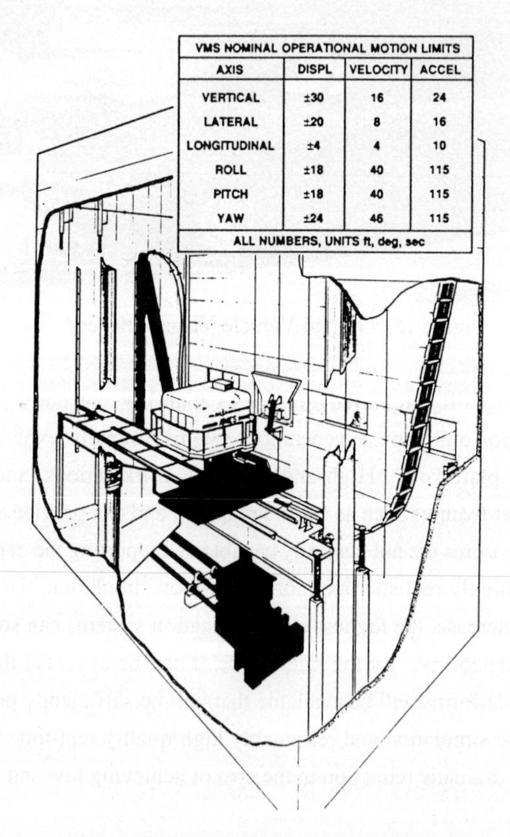

VMS NOMINAL OPERATIONAL MOTION LIMITS

AXIS	DISPL	VELOCITY	ACCEL
VERTICAL	±30	16	24
LATERAL	±20	8	16
LONGITUDINAL	±4	4	10
ROLL	±18	40	115
PITCH	±18	40	115
YAW	±24	46	115

ALL NUMBERS, UNITS ft, deg, sec

Figure 16. NASA Vertical Motion Simulator

At the other extreme of motion generation, a massive hexapod motion base discussed in Ref. 1 has recently been installed at the US Army Tank-Automotive Command in Warren, Michigan. This high-capacity motion base can move a 25 ton turret, with up to 5 g acceleration, in precision motion. This and the aircraft simulator motion base shown in Figure 16 clearly illustrate that the technology for motion generation in vehicle simulation is in hand.

7 Ground Vehicle Virtual Prototyping

The most advanced ground vehicle driving simulator in existence to date is operated by Daimler-Benz in Berlin [24]. This system, shown schematically in Figure 17, consists of a thirty-foot-diameter dome on a platform that supports the vehicle cab in which the driver functions. Graphic imagery is displayed on the interior of the dome, wrapped 180 degrees around the driver's vehicle. The dome and platform are moved by a six-degree-of-freedom hexapod system that provides approximately two Hz motion response, with substantial roll and pitch. This simulator utilizes 1985 vintage graphics that are not textured, but provide a sharp scene at high frame rate to the driver of the vehicle. Experience with this simulator has attracted a great deal of attention to the potential that exists for this new class of advanced ground vehicle driving simulators.

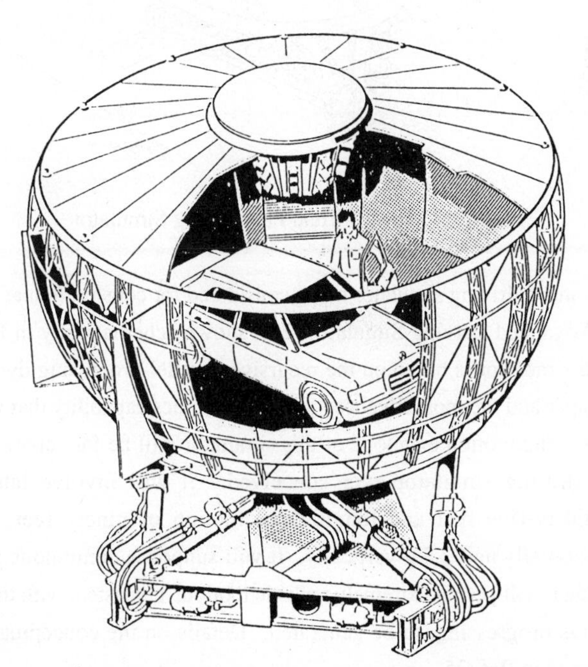

Figure 17. Daimler-Benz Driving Simulator

A new virtual prototyping simulator that is under construction at The University of Iowa, using advanced textured graphics and the recursive dynamics algorithms outlined in Sections 2 and 3, is shown schematically in Figure 18. This simulator employs a small hexapod motion base with frequency response up to approximately ten Hz and represents the most advanced vehicle virtual prototyping simulator in the US.

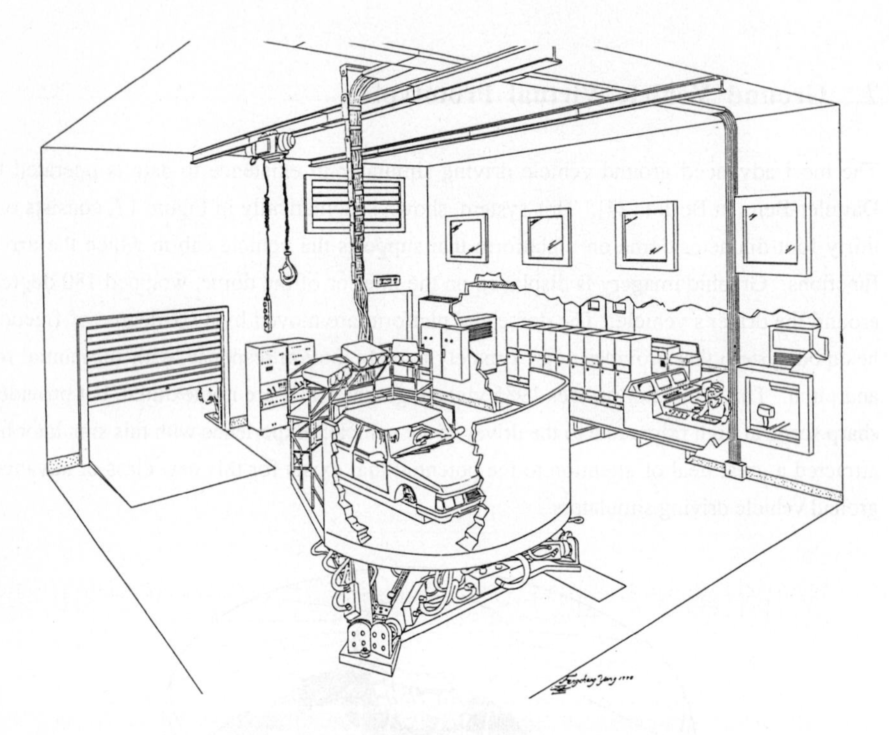

Figure 18. Iowa Virtual Prototyping Simulator

The most advanced driving simulator being considered for construction at the present time is the National Advanced Driving Simulator [25], shown schematically in Figure 19. This advanced driving simulator is based on the recursive parallel processing dynamics methods outlined in this paper and the most advanced textured graphics capability that will be available in the mid-1990s. The motion envelope of this simulator will be far superior to that of any ground vehicle driving simulator ever conceived. It will involve lateral motion of approximately thirty-five feet and longitudinal motion of ninety feet, with one g of acceleration horizontally and 2.5 g vertically. It will support a continuous yaw ring on the motion platform that will permit extremely realistic motion, consistent with the scene through which the driver is progressing, to be generated. Details on the conceptual design of this device may be found in Ref. 25.

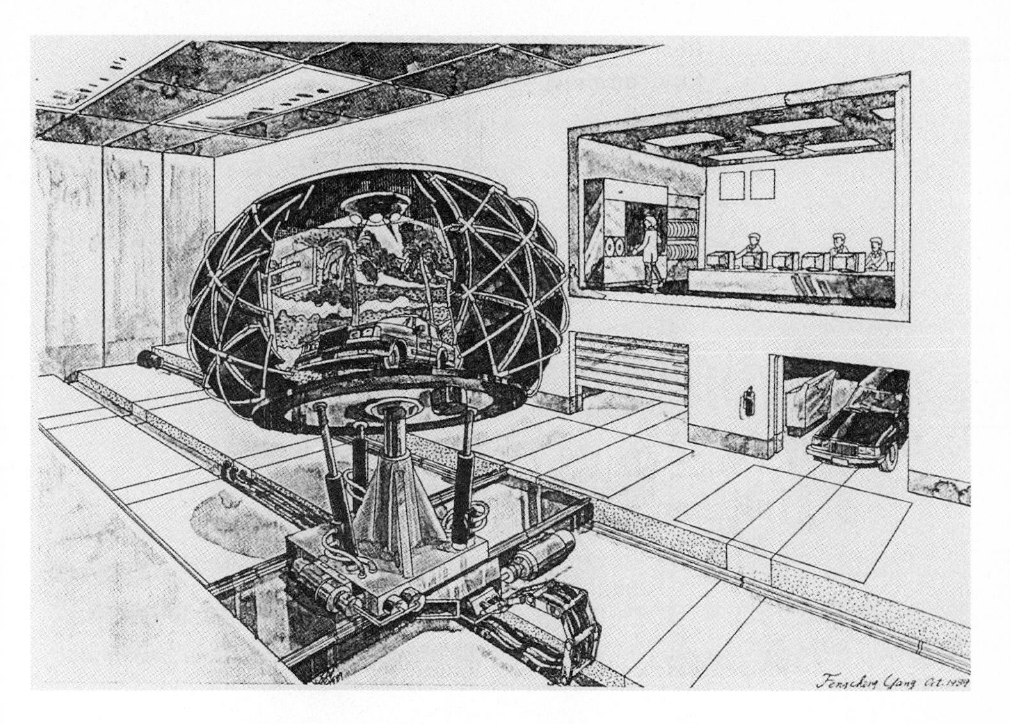

Figure 19. National Advanced Driving Simulator

8 Telerobotic and Construction Equipment Virtual Prototyping

The concept of a virtual prototyping simulator for a remotely operated robot shown in Figure 20 illustrates the concept of creating capability to simulate both the performance and visual environment of a manipulator or robot that is controlled by a human operator using video feedback. This concept has been studied extensively with NASA for remote teleoperation of robots in space. Implementation of this concept on an advanced graphics work station, using a six degree-of-freedom force-feedback manipulator controller shown in Figure 21 (Kraft mini-master with robot on screen) used by the operator to input desired motion and receive force feedback indicating level of effort by actuator on the robot. The level of simulation detail incorporated in this application includes dynamic performance of high-gear ratio special purpose drives in the actual robot in [26].

Motivated by the robot application, developments have taken place in construction equipment operator-in-the-loop simulation; e.g., a construction backhoe. Actual construction backhoe operator consoles have been implemented with a large screen visual display shown in Figure 22 for simulation of backhoe operation. The real-time computer simulation in this application includes the kinematics and dynamics of the backhoe construction equipment as

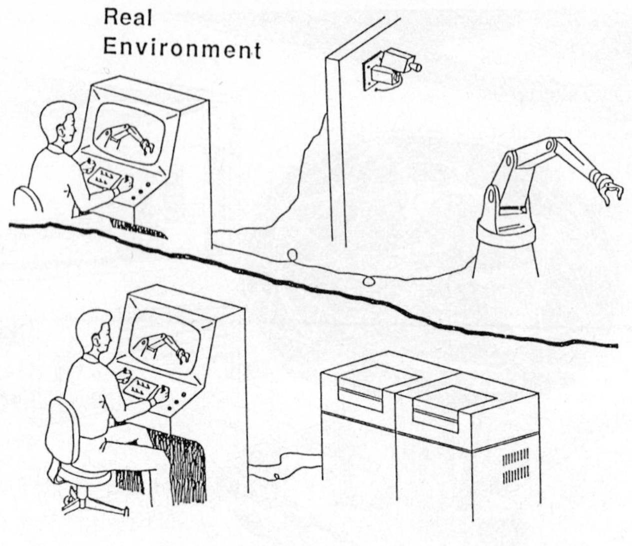

Figure 20. Schematic of Robot Virtual Prototyping Simulator

Figure 21. Kraft Mini-Master for Robot Control

well as dynamics of the hydraulics system that drives the backhoe. This simulation, which functions on a two processor workstation that also drives the graphics projector can now be used to investigate alternative operator interfaces and control algorithms [26]. The breadth of such applications for both training and design for optimum performance of equipment in the hands of the operator is now feasible and finding its way into engineering and training applications.

Figure 22. Backhoe Simulator

9 Conclusions

The technology for operator-in-the-loop virtual prototyping, as regards graphics and motion subsystems, has been developed over the past two decades for aircraft flight simulation. Dynamic simulation of aircraft motion for pilot-in-the-loop aircraft flight simulation is, however, much less complex and demanding than simulation of the extremely nonlinear dynamic effects of vehicle suspensions and tire-road surface interaction, and construction equipment hydraulics. Major new applications in ground vehicle driving and robot/construction equipment virtual prototyping are only now feasible, as a result of the advancements in recursive dynamics algorithms and parallel computer implementations outlined in this paper. These developments, combined with available computer graphics and motion base technologies, create a unique opportunity for tailoring the design of vehicles, robots, and construction equipment to the capabilities of the operator, investigating the influence of human conditions and capabilities on the operator's ability to carry out complex task of equipment operation, and for numerous other important applications that influence the lives of virtually every citizen of the world on a daily basis. These advances have been made possible by mathematical and computational developments in the theory of dynamics and its parallel implementation on emerging high-speed RISC-based parallel computers. It is interesting that this mathematical development has been felt very quickly in the field of ground vehicle virtual prototyping.

Acknowledgment: Research supported by NSF-Army-NASA Industry/University Cooperative Research Center for Simulation and Design Optimization of Mechanical Systems.

References

1. Beck, R.R., "Simulation Based Design of Off-Road Vehicles," *Concurrent Engineering Tools and Technologies for Mechanical System Design* (E. J. Haug ed.), Springer-Verlag, Heidelberg, 1993.
2. Ciarelli, K., "Integrated CAE System for Military Vehicle Applications," *Proceedings of the First Annual Symposium on Mechanical System Design in a Concurrent Engineering Environment*, Iowa City, Iowa, pp. 301-318, October 24-25, 1989.
3. Frisch, H.P., "Man/Machine Interaction Dynamics and Performance Analysis," *Concurrent Engineering Tools and Technologies for Mechanical System Design* (E. J. Haug ed.), Springer-Verlag, Heidelberg, 1993.
4. Kuhl, J.G., Papelis, Y.E., Romano, R.A., "An Open Software Architecture for Operator-in-the-Loop Simulator Design and Integration," *Concurrent Engineering Tools and Technologies for Mechanical System Design* (E. J. Haug ed.), Springer-Verlag, Heidelberg, 1993.
5. Schiehlen, W.O., "Simulated Based Design of Automotive Systems," *Concurrent Engineering Tools and Technologies for Mechanical System Design* (E. J. Haug ed.), Springer-Verlag, Heidelberg, 1993.
6. Bestle, D., "Optimization of Automotive Systems," *Concurrent Engineering Tools and Technologies for Mechanical System Design* (E. J. Haug ed.), Springer-Verlag, Heidelberg, 1993.
7. Roberson, R.E., and Schwertassek, R., *Dynamics of Multibody Systems*, Springer-Verlag, Berlin, 1988.
8. Nikravesh, P.E., *Computer-Aided Analysis of Mechanical Systems*, Prentice-Hall, Englewood Cliffs, New Jersey, 1988.
9. Haug, E.J., *Computer Aided Kinematics and Dynamics of Mechanical Systems, Vol. I: Basic Methods*, Allyn and Bacon, Boston, 1989.
10. Shabana, A.A., *Dynamics of Multibody Systems*, John Wiley & Sons, New York, 1989.
11. Huston, R.L., *Multibody Dynamics*, Butterworth-Heinemann, Boston, 1990.
12. Amirouche, F.M.L., *Computational Methods in Multibody Dynamics*, Prentice-Hall, Englewood Cliffs, New Jersey, 1992.
13. Wittenburg, J., *Dynamics of Systems of Rigid Bodies*, Teubner, Stuttgart, 1977.
14. Haug, E.J.(ed.), *Computer Aided Analysis and Optimization of Mechanical System Dynamics*, Springer-Verlag, Berlin, 1984.
15. *DADS User's Manual Rev. 5.0*. Computer Aided Design Software, Inc., Oakdale, IA, 1988.
16. Bae, D.S., and Haug, E.J., "A Recursive Formulation for Constrained Mechanical Systems, Part I - Open Loop," *Mechanics of Structures and Machines*, 15:3, pp. 359-382, 1987.
17. Bae, D.S., and Haug, E.J., "A Recursive Formulation for Constrained Mechanical Systems, Part II - Closed Loop," *Mechanics of Structures and Machines*, 15:4, 1987.
18. Tsai, F.F., and Haug, E.J., "Real-time multibody system dynamic simulation, Part I - A modified recursive formulation and topological analysis," *Mechanics of Structures and Machines*, 19:1, 1991.
19. Bae, D.S., Hwang, R.S., and Haug, E.J., "A Recursive Formulation for Real-Time Dynamic Simulation," *Proceedings of the 1988 ASME Design Automation Conference*, pp. 499-508, 1988.

20. Bae, D.S., Haug, E.J., and Kuhl, J.G., A Recursive Formulation for Constrained Mechanical Systems, Part III - Parallel Processor Implementation, *Mechanics of Structures and Machines*, 16:2, 1988.
21. Hwang, R.S., Bae, D.S., Kuhl, J.G., and Haug, E.J., "Parallel Processing for Real-Time Dynamic Simulation," submitted to *Journal of Mechanisms, Transmissions, and Automation in Design*.
22. Tsai, F.F., and Haug, E.J., "Real-Time Multibody System Dynamic Simulation, Part II - A Parallel Algorithm and Numerical Results," *Mechanics of Structures and Machines*, 19:2, 1991.
23. *ESIG-4000 Image Generator Specification*, Simulation Division, Evans & Sutherland, 1990.
24. Drosdol, J., and Panik, F., *The Daimler-Benz Driving Simulator, A Tool for Vehicle Development*, Society of Automotive Engineers, 1986.
25. Haug, E.J., et. al., *Feasibility and Conceptual Design of National Advanced Driving Simulator*, DOT HS 807 596, US Department of Transportation, National Highway Traffic Safety Administration, March 1990.
26. Yae, K.H., "Teleoperation of Redundant Manipulator," *Concurrent Engineering Tools and Technologies for Mechanical System Design* (E. J. Haug ed.), Springer-Verlag, Heidelberg, 1993.

An Open Software Architecture for Operator-in-the-Loop Simulator Design and Integration

Jon G. Kuhl, Yiannis E. Papelis, Richard A. Romano

Center for Computer-Aided Design, The University of Iowa, Iowa City, Iowa 52242-1000 USA

Abstract: This paper describes the design and implementation of an open software architecture for operator-in-the-loop simulators to support virtual prototyping applications. This framework provides properties that are critical to the cost-effective design of complex simulators, including uniform software specification techniques, modular design methodologies, enhanced reliability and maintainability, ease of reconfigurability, and reusability of simulator subsystems. The open architecture has already served as a basis for development of the Iowa Driving Simulator (IDS) and will, with further enhancement, be used to support development of tracked vehicle virtual prototyping simulation under support from the Defense Advanced Projects Research Agency (DARPA). The architecture is suitable for general application to the design and integration of a wide variety of simulation systems.

Keywords: operator-in-the-loop simulation / software engineering / real-time / software integration / virtual prototyping

1 Introduction

The overall complexity of an operator-in-the-loop simulator grows rapidly as a function of increasing system fidelity and realism. This growth in complexity is only partially due to the greater sophistication of the individual cueing systems (visual, motion, etc.). The major source of complexity lies in the integration of interacting subsystems that comprise the simulator. Most of this integration process is performed by software components. Thus, the design of a high fidelity simulator is, in large measure, a software engineering challenge. Since virtual prototyping simulators must embody a high degree of functional validity, they are subject to this high level of software complexity. At the same time, however, a virtual prototyping simulator must be rapidly deployable, easily reconfigurable and adaptable to changes in the prototype definition, and reasonably cost effective to design and implement.

The software architecture of a simulator can be viewed as a set of communicating subsystems, as pictured in Figure 1. Typically, each subsystem must operate at some fixed iteration rate, which may be different for each subsystem. Potentially large amounts of data must be communicated among subsystems at a fixed rate that is compatible with the requirements of various subsystems. Subsystem execution must be scheduled in a constrained temporal ordering that is consistent with system data-flow requirements. This scheduling must explicitly account for subsystem execution times, inter-subsystem data communication latencies, and other system overheads. As fidelity increases, the accuracy of underlying computational simulation models and subsystem control algorithms must become proportionally greater, and they typically must operate at higher iteration rates. This results in greater computational demands and implies the need for parallel computers and/or the distribution of computational demands among multiple computing platforms. In addition, the frequency and extent of inter-subsystem data communication may increase dramatically.

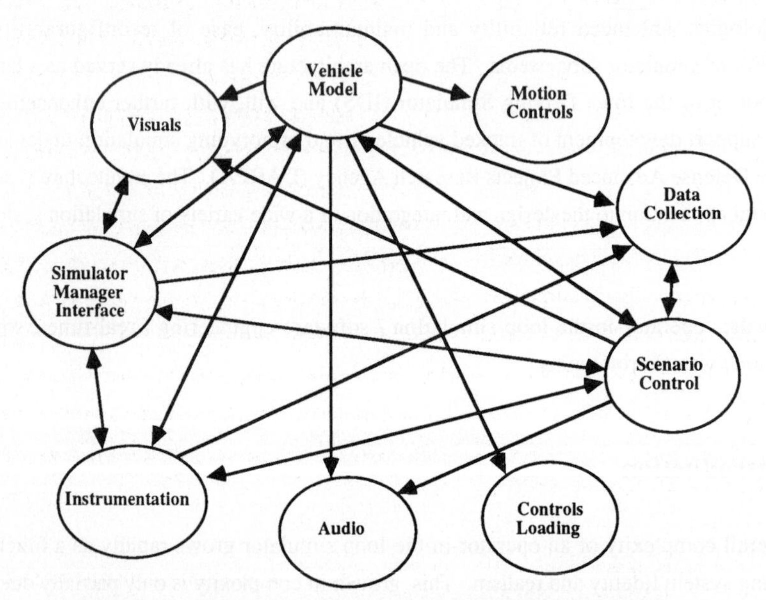

Figure 1. Software Architecture with Direct Interconnectivity

The resulting design challenges involve a complex set of closely tied functional, real-time, and communication requirements that must be mapped onto a distributed and parallel computing architecture. For example, the Iowa Driving Simulator (IDS) [1], which is a high fidelity ground vehicle driving simulator, consists of approximately a dozen software subsystems, distributed across four parallel processor computer systems comprising a total of 32 processors. Subsystem scheduling frequencies run as high as 1000 Hz, and there are several dozen distinct inter-subsystem data-flows that involve up to several kilobytes of data each, several of which cross computer system boundaries. Viewed as a whole, the overall complexity of such a system can be overwhelming.

A sound simulator design process must be prepared to deal with this high level of complexity in all phases of specification, design, implementation, and testing. This implies the need to employ sound software engineering practices throughout the process. For large simulator projects, such as the IDS, which have many concurrent and distributed components, numerous data-flows, and stringent real-time performance constraints, conventional software engineering methodologies may not be appropriate [2]. To support these large simulator projects, pragmatic software engineering methodologies must be employed that are specific to the needs and requirements of operator-in-the-loop simulator design.

In addition to rigorous software engineering paradigms, a simulator design methodology should enforce several additional properties. One of these is insuring that the resulting system provides for easy reconfiguration, upgrade, and extension; i.e., the ability to modify or replace individual subsystems, add new subsystems, or change the underlying computing environment, without modification of the overall simulator software architecture. To enhance modularity and reliability, subsystems should have standard interfaces and should be capable of being tested and validated independently. It is highly desirable to provide the potential for directly reusing subsystems developed for one simulator in the design of another.

To achieve these objectives, an open architectural framework for simulator design and implementation has been developed. This framework provides a set of high level abstractions that are general enough to allow specification of a broad range of simulator configurations and performance requirements. The framework also provides a set of associated design paradigms that support the independent specification, design, and implementation of simulator subsystems, with standard interfaces. Finally, the framework provides a standard simulator integration architecture. This structure allows modular subsystems to be "plugged into" an environment, or "simulator operating system," that provides the necessary integration services in accordance with system performance requirements. Such a framework is shown conceptually in Figures 2 and 3. Among the advantages of this open architectural approach are the ability to share and reuse simulator subsystems in different simulator designs, the potential for development of "off-the-shelf" subsystems, and a significant reduction in simulator design complexity and associated development and life-cycle costs.

The open software architecture framework consists of (1) a set of high level abstractions that are general enough to allow specification of a broad range of simulator configurations and performance requirements; (2) a set of associated, object-based, design paradigms; and (3) a suite of standard system support software that provides the integrating architecture and run-time interface for the simulator software subsystems. The architectural framework provides the following properties:

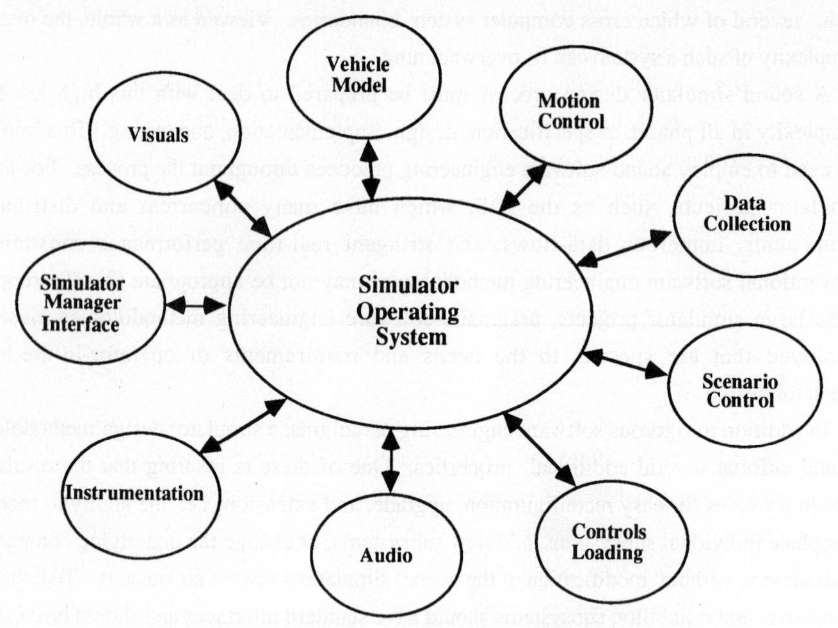

Figure 2. Modular Software Architecture with a Simulator Operating System

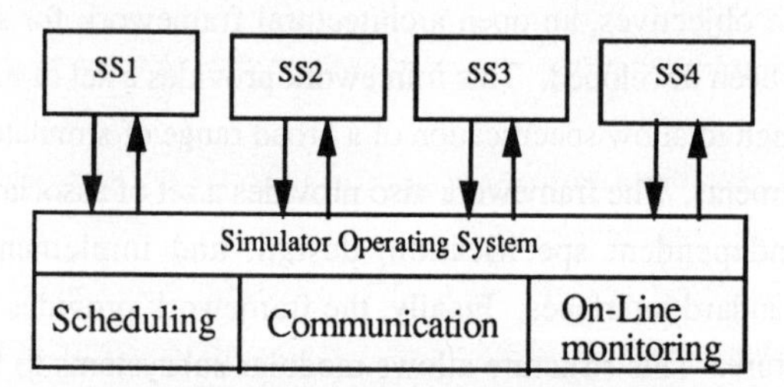

Figure 3. Simulator Operating System

(1) *Modularity*. The system functionality can be partitioned into completely modular subsystems that can be independently specified, designed, implemented, tested, maintained, and updated or replaced. These subsystem modules can also be shared by various simulator configurations.

(2) *Standard interfaces-and structure*. All subsystems are specifically designed and implemented with well-defined, standard interfaces through which all interactions of a subsystem with its external environment take place. These standard interfaces support the modularity requirement, and allow for the configuration of "off the shelf" subsystems into a simulator. Similarly, all subsystems are designed and implemented with a simple,

common, internal structure that completely insulates them from any knowledge of the overall simulator configuration or underlying hardware and communication architecture.

(3) *Separation of functionality and performance issues.* Communication and scheduling requirements are explicitly visible at all stages in the design process, to insure that real-time performance issues are not obscured and efficient mapping onto the target hardware system is not precluded.

(4) *Standard run-time environment.* The architecture provides a set of system services that are responsible for all real-time subsystem scheduling and inter-subsystem data communication. This run time environment insulates subsystems from one another in the sense that subsystems can interact only with the environment rather than directly with each other. The run-time environment integrates the three architectural properties described above. In particular, it provides the standard "software backplane" into which subsystems can be "plugged." The run-time environment (or simulation operating system) provides real-time scheduling of subsystems based upon information provided in a separate configuration database. In addition, the run-time environment transparently provides all underlying data movement and interprocess communication necessary to accomplish system data-flow requirements. The run-time system is implementable on a variety of hardware platforms and underlying vendor-supplied operating systems and allows subsystem designs to be independent of these factors. In this way, subsystems can be highly portable. The run-time environment also supports distributed computing architectures, and allows simple reconfiguration of a simulator without modification of subsystems.

The first two of the above properties are compatible with any object oriented design paradigm. However, the last two concern pragmatic real-time performance and communication issues that are not easily addressed within standard software design frameworks [3]. While specification and design methodologies have been proposed for real-time systems, based upon modeling constructs such as timed petri nets, temporal logic, and extended state machines [4-6], none of these are sufficiently developed at this time to deal with a system as complex as a high-fidelity virtual prototyping simulator. Furthermore, none provide a suitable basis for the standard run-time architecture identified in property (4) above.

The open simulator architecture is based upon the abstraction of all inter-subsystem communication into operations on logical, typed data structures, called cells. All external subsystem specifications are defined in terms of high level functional operations on cell data structures. To insure efficiency of implementation, particularly in a distributed environment that does not allow efficient access to typed data across system boundaries, cells are defined as a hybrid of shared data and message passing communication paradigms. Specifically, subsystems view cells as shared, typed data structures. However, subsystems acquire cells, for subsequent access, through the abstraction of an untyped read, similar to distributed message passing [7]. This approach, which balances abstraction purity and performance considerations, is in contrast to pure distributed shared memory approaches [8] that cannot be implemented efficiently enough to meet the real-time performance requirements of the simulator.

To complement the cell abstraction, a "simulator operating system" has been developed for the IDS. This distributed environment manages the cell abstraction transparently for all subsystems and provides completely deterministic subsystem scheduling. The simulator operating system provides a common "software back-plane," into which software subsystems can be inserted. The resulting structure is highly modular and easily reconfigurable. For instance, subsystems can be independently updated or changed, without affecting the remainder of the simulator. Subsystem scheduling and communication requirements, including those that cross computer system boundaries, are maintained separately from the subsystems themselves, in the form of a simple simulator configuration file. A subsystem can be moved from one computer to another, simply by updating several lines in the configuration file.

2 Architecture Overview

It should be noted that the notion of a "simulator operating system" embodied in Figure 3 differs substantially from the traditional notion of a "real time executive" that is often employed in the design of real-time systems. Typically, a real time executive provides a set of generic low-level services that can be utilized by application processes. These services may include a basic process-level scheduler (e.g., a frequency based scheduler), and a set of interprocess communication and synchronization primitives. These primitives provide a means for user processes to directly interact and communicate. As such, a simulator software architecture built directly on top of a real-time executive will look structurally like the architecture illustrated earlier in Figure 1. On the other hand, the simulator operating system approach, described herein provides a set of **high level** services that are directly oriented towards simulator integration. The simulator operating system can implement a set of abstractions that allow subsystems to see a highly regular and simple environment, enforce subsystem structure to uniform and rigorous standards, and limit the ability of subsystems to directly or indirectly impact the behavior of other subsystems except through these standard interfaces. The appropriate relationship between a real-time executive and a simulator operating system is shown in Figure 4.

While an architectural abstraction such as that shown in Figure 4 is clearly useful for specification and design purposes, high performance real-time systems often do not employ such a level in their actual implementation due to performance concerns. That is to say, the

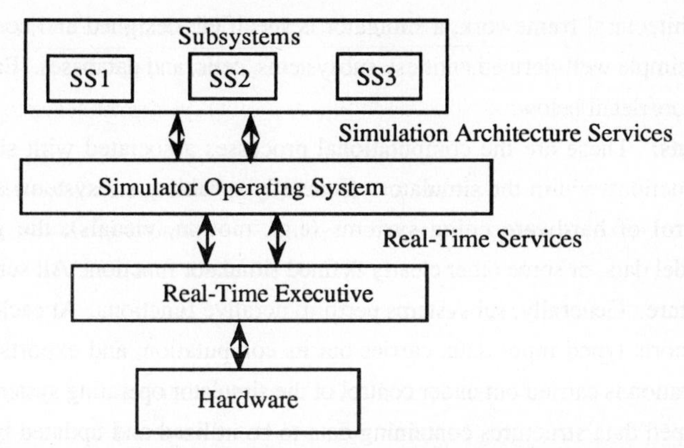

Figure 4. Relationship with Executive and Simulator Operating System

overhead associated with providing this layer in the run-time environment is deemed to be too high. However, if this simulator operating system layer is absent in the real-time environment the modularity and reconfigurability of the simulator will be severely restricted. The framework described in the paper is designed to be highly efficient and therefore incurs minimum run-time overhead. Furthermore, because the simulator operating system is layered on top of an underlying real-time executive, rather than directly implementing its own real time services, the architecture is easily ported to a variety of hardware platforms with different vendor-supplied operating systems and real-time executives. Portability is further enhanced by constraining the simulator architecture to use only a minimal set of real-time services, offered by virtually any real-time executive or operating system.

As noted earlier, the use of a standard simulator operating system offers the potential for code reusability at the subsystem level. An operational simulator can be integrated by simply combining an appropriate set of subsystems with a separate runtime specification. This specification includes the mapping of the subsystems to available resources, the location of the various global data structures, temporal scheduling constraints and other similar information. A specification can easily be modified to implement simulators with varying capabilities or in order to change to different resources due to hardware unavailability. An illustration of this integration process is shown in Figure 5.

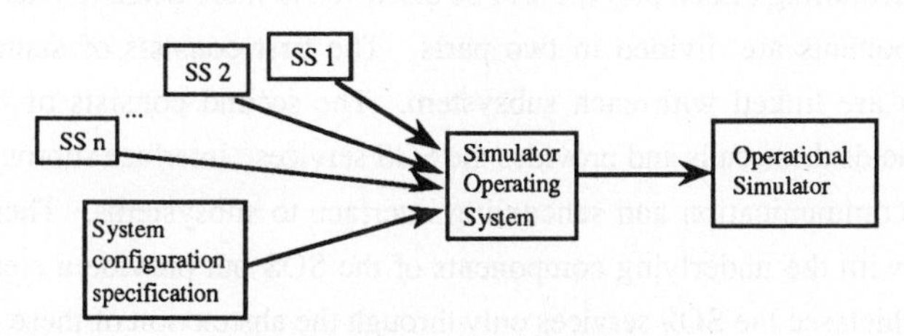

Figure 5. The Integration Process

In this architectural framework, a simulator is specified, designed and configured from three types of simple well-defined entities: subsystems, cells, and databases. Each of these is described in more detail below:

Subsystems: These are the computational processes associated with significant and identifiable functions within the simulator. Generally speaking, subsystems are associated with the control of hardware cuing systems (e.g., motion, visuals), the generation of simulation model data, or some other clearly defined simulator function. All subsystems have identical structure. Generally, subsystems perform iterative functions. At each iteration, the subsystem imports typed input data, carries out its computation, and exports typed output data. Each iteration is carried out under control of the simulator operating system.

Cells: Typed data structures containing data to be utilized and updated by subsystems are organized into larger logically related sets, called cells. The specification of the cells is done separately from subsystems. Each subsystem has a set of unique cells that it outputs its data into. All subsystems can import cells and thus gain access to all structured data contained there. Cells are the only form of communication among the subsystems.

Databases: Large collection of static data that must be accessed efficiently during the simulation is organized in entities termed databases. A database is an active entity, in that it has both code and data associated with it. The code provides uniform access methods so that all subsystems can access the data without consideration about where the data originates or how it is being accessed.

3 Simulator Operating System

The Simulator Operating System (SOS) is a collection of software that provides the integrating functionality to allow a collection of subsystems, cells and databases to be configured into a functioning simulator. Conceptually the SOS is a single entity. In practice, however, it consists of a collection of cooperating and potentially distributed software components, each implementing some part of the overall functionality.

As illustrated earlier in Figure 3, the SOS performs three primary functions: i) scheduling of subsystems, ii) management and movement of cell data among subsystems, and iii) on-line performance monitoring. Each of these will be discussed in more detail in later sections.

SOS components are divided in two parts. The first consists of standard SOS call interfaces that are linked with each subsystem. The second consists of the underlying software that fields these calls and provides the SOS services. Interface subroutines provide a uniform data communication and scheduling interface to subsystems. These subroutines communicate with the underlying components of the SOS but provide a clean interface to subsystems, which see the SOS services only through the abstraction of these call interfaces.

The remaining part of the SOS consists of a number of cooperating software components that implement the overall SOS functionality. The decomposition of SOS functionality is illustrated in Figure 6. Note that Figure 6 is the pragmatic implementation of the conceptual design shown in Figure 3.

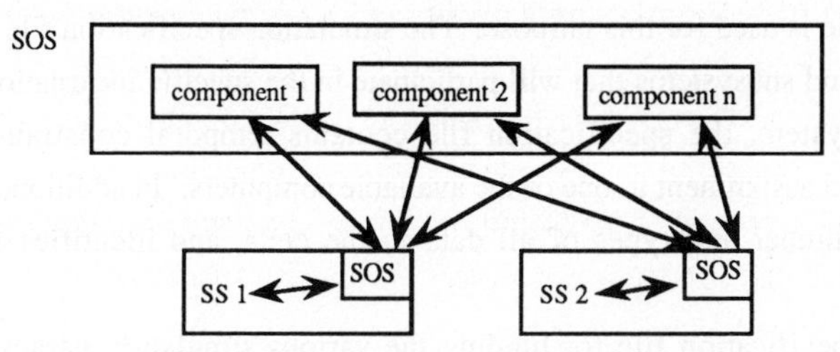

Figure 6. Implementation of the SOS

In addition to offering a clean subsystem interface, the encapsulation of simulator specific services in the SOS interface is an effective mechanism for integrating a simulator with subsystems executing on physically distributed computers. Subsystems running on different nodes of a distributed computer architecture have access to the same set of scheduling and communication facilities, and can be essentially unaware of the identity of their host computer. To achieve this hardware abstraction, the component of the SOS that implements the various services is implemented on each host in the simulator environment. Figure 7, illustrates the manner in which the architecture insulates subsystems from the underlying hardware architecture. Note that modifications to, or porting of, the SOS will not require modification of subsystem code so long as the interface specification is not changed.

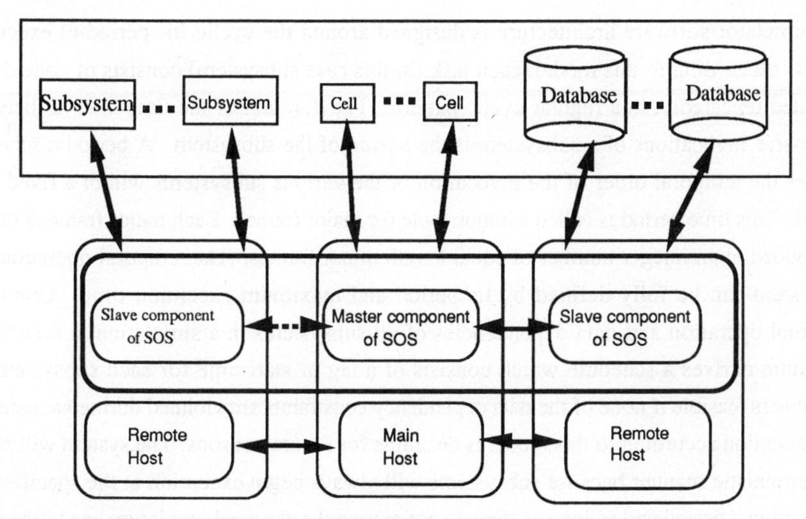

Figure 7. Illustration Hardware Insulation from Subsystems

As noted in the preceding discussion, virtually all details concerning the overall configuration and operation of a simulator have been abstracted and removed outside the subsystems. The SOS is designed to accept simulator configuration parameters as input at run-time. Specification of these parameters is then the only step required in order to integrate a working simulator given a set of subsystems, cells, and databases. A simulator specification file is used for this purpose. The simulation specification file contains a list of all computers and subsystems that will participate in the specific incarnation of a simulator. For each subsystem, the specification file contains temporal constraints, performance expectations and assignment to one of the available computers. In addition, the specification includes the number and types of all data in the cells, and identifies the databases to be utilized.

Use of a specification file for binding the various simulation parameters just before runtime has the advantage of allowing an integrator to configure a simulator tailored to the task at hand. Changes in configuration require no change in the code of either the subsystem or the SOS. In essence, subsystems cells and databases can be used as building blocks to create a variety of simulator configurations.

4 Details of S.O.S. Architecture

This section describes in more depth the various elements of the SOS.

4.1 Real Time Scheduling

The simulator software architecture is designed around the cyclic (or periodic) execution model. According to this model, each task (in this case subsystem) consists of code that is executed repeatedly, in a regular cyclic pattern. The duration of the time interval between successive invocations of a subsystem is the *period* of the subsystem. A periodic schedule defines the temporal order of the invocation of the various subsystems within a fixed time period. This time period is called a major cycle (or major frame). Each major frame is further subdivided in an integer number of equal length minor frames. The temporal operation of a subsystem can be fully defined by its period and maximum execution time. Given the temporal operation and data dependencies of all subsystems in a simulation, a scheduling algorithm derives a schedule which consists of a lag or start time for each subsystem. A schedule is feasible if none of the data dependency constraints are violated during execution.

Execution according to this model is desirable for several reasons: The system will run in a deterministic manner because subsystems will always begin execution at the specified lag time within a period and as long as they do not exceed the assumed maximum execution time, the system data flow will be intact. Furthermore, performance analysis of a single period can be extrapolated to any number of periods since, by definition, the periodic scheduler will repeat the major frame all through execution. A periodic scheduler can be easily implemented on any hardware platform, given an external timing source. This timing source will generate interrupts at the proper frequency and the interrupt service routine can look at the precomputed schedule and dispatch the appropriate subsystem.

Figure 8 illustrates a multi-processor, non-preemptive schedule. Even though this particular example is non-preemptive and has all processes starting at the boundaries of the minor frames, there is nothing in the fundamental design of this architecture that imposes these particular constraints on a periodic schedule.

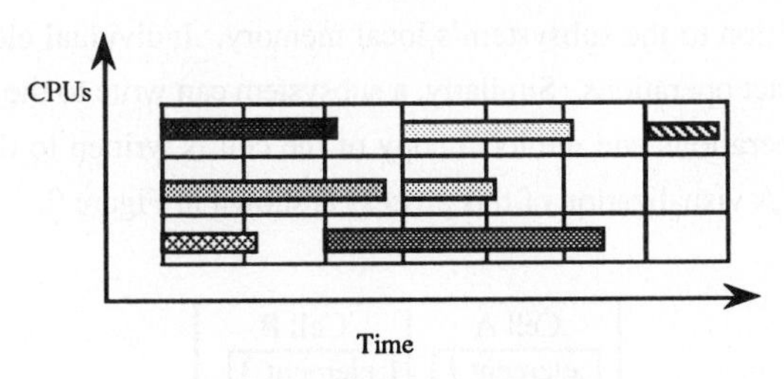

Figure 8. A Periodic Schedule

In order to interface to the scheduler, all subsystems are built around a fixed software structure that consists of a loop that executes one iteration per scheduling interval:

while (**scheduler_block()** == RUN)

do_work();

The scheduler_block function is part of the SOS interface and will block until the scheduling time of the subsystem within the major frame has arrived. After being released, the subsystem will perform its work and block again. This process will repeat until the simulation is terminated by the operator. Termination of the simulation or other simulation exceptions are communicated to the subsystems through the return value of the scheduler_block function.

4.2 Inter-Subsystem Communication

In order to avoid point-to-point communication and the associated dependencies among subsystems that would complicate their integration, all subsystems have a shared memory view of the global data in the system. This shared memory view is only a virtual view. The actual implementation depends on the capabilities of the particular hardware. In a shared memory multiprocessor, for example, shared memory will be used. But in a distributed system, the implementation may use some form of message passing. The only reflection of the implementation of the model will be in the timing characteristics of the communication primitives. The synthesis of the periodic schedule explicitly takes these timing characteristics into account.

As noted earlier, all data shared among subsystems in the simulation is partitioned into entities called *cells*. A cell is a collection of typed data items, called *elements*. All data in cells are system-wide global and responsibility for their handling lies with the SOS. The grouping of individual, but related items into one entity allows for efficient handling, especially when message-based interprocessor communication is used in the underlying implementation.

Each cell can be written by at most one subsystem which is called the *generator* of the cell. Gaining access to a data element is a two stage process. A cell_get operation copies a cell from its global location to the subsystem's local memory. Individual elements can then be read using cell_extract operations. Similarly, a subsystem can write to the local copy of a cell using cell_insert operations and the local copy of the cell is written to the global area by a cell_put operation. A visualization of this process is shown in Figure 9.

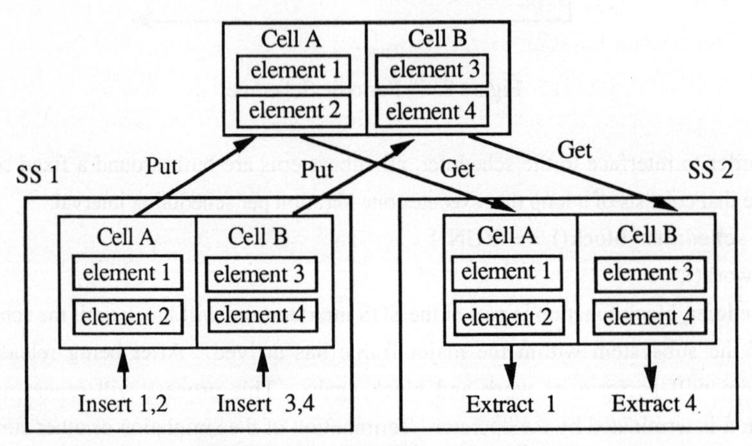

Figure 9. The Cell-Element Paradigm

Note that writing/reading cells is completely asynchronous; a cell_get will never block, and since there is only one cell generator there will never be a coherency problem for the data in the cells. Furthermore, cells do not buffer data. Thus successive writes overwrite the data previously in the cell. The responsibility for maintaining the correct dataflow relationships among subsystems lies with the creator of the scheduling specification, which is part of the simulation scheduling file. Even though there are automatic techniques for generating schedules that enforce dataflow relationships [9] for a small number of subsystems, a schedule can easily be created manually.

A cell can be designated as "double-buffered". A double-buffered cell contains two sets of data, the first one is used for reading and the second is used for writing, with these roles alternating at each scheduling frame. Double-buffered cells eliminate read/write conflicts but increase the delay between generation and consumption of data.

The actual location of the cells in the system is transparent to the subsystems, even in a distributed implementation. For example, a cell used by two subsystems running on the same computer may only reside on that computer; if a cell is used by subsystems running on different computers copies will reside on both computers. In the latter case, responsibility for the coherency of the two cell copies lies with the SOS and not with the subsystems. It is also the responsibility of the SOS to convert the format of the data when different representations are used in the hosts of the a distributed system. In this way, the portability of the subsystem is enhanced.

One of the advantages of the cell/element paradigm for inter-subsystem communication is the elimination of direct dependencies among subsystems. Since the cell specification is part of the system design, distinct from subsystem specifications, the functional specification of a subsystem depends only on the input cell(s) it requires, the output cell(s) it writes and the maximum execution duration of the subsystem between reading the input cell(s) and generating the output cell(s). This allows for efficient testing of individual subsystems in isolation by substituting the standard cell interface with an alternative interface that gathers its input from disk files and deposits the output into disk files. These data can then be compared with libraries of acceptable data and determine whether the subsystem performs according to its specification.

For more extensive testing, a collection of subsystems can be tested by feeding them data from disk files. In case of a nonfunctioning simulator, this semi-integrated mode can be used to narrow down the list of potentially erroneous subsystems by removing subsystems until the problem disappears.

Replaying a simulation is easily done by gathering the data that flows through the cells and feeding them back to the subsystems at a later time. Since the system runs in a completely deterministic manner due to the periodic scheduling, the state evolution can be exactly retraced using the captured input data.

4.3 Real Time Database Query

The previous section described the handling of read/write data. This data is small in size and can reside completely in memory. Often in simulators there is need to maintain large read-only databases. This data may represent the static environment in which a simulation takes place and is typically accessed by geographical location in the virtual world of the simulator.

An example is the terrain database for a ground vehicle simulator which provides the terrain height at given points in the simulator world.

Accessing a disk is necessary because, by definition, the database is too large to reside permanently in memory. Only a subset of its data can reside in main memory. When a query requires data that is not resident, the disk must be accessed. The latencies of current disk systems are in the 10-20 millisecond range and typical throughputs are 5 to 10 Mb/sec. If a query requires 100 bytes, it will take about 0.02 to 0.01 milliseconds to read this data but may involve 10-20 millisecond latency to reach it in the worst case. This latency may be far too high for the real-time performance requirements of subsystems. Coordinating database queries in a real time system can be a challenging problem due to the determinism and low latency required in such a system.

Traditionally, the solution to this problem has been to use dedicated high-speed disk arrays that utilize custom interface protocols that allow access of individual disk sectors and other similar shortcuts that significantly reduce the latency. Such system designs have low average latencies so that a deterministic upper bound can be can put on a read operation, but they are expensive and induce very specific hardware requirements. Furthermore, proliferation of modern disk interface protocols like SCSI that hide the actual sector mapping from the application software make this approach even less cost efficient.

The simulator architecture described here provides support for high performance real time database queries through a well defined paradigm that allows generic systems (systems without specialized disk I/O facilities) to be used even when latencies cannot be perfectly controlled, by minimizing the effects of the latency on the overall system design. This is achieved by using a prediction scheme that anticipates which data will be used in the future and prefetches this data into a memory resident query space. In this way, the only disk system performance requirement is that its throughput be high enough to support the worst case data rate required by the queries. This requirement is relatively independent of the disk latencies. The remainder of this section describes the design of the database entity within the framework of the simulator software architecture.

A *database* is an active entity that has both function and data associated with it. Databases have specific access methods provided by a set of interface subroutines. Only these methods/subroutines are available for querying the database. These access methods are available to all subsystems under all hosts in the simulation. An illustration of the database concept is shown in Figure 10.

There are two potentially conflicting design issues in the design of the database query mechanism: time efficiency and space efficiency. The time efficiency issue is addressed by allowing each database to be queried only for data within a region of interest (ROI). Space efficiency is addressed by utilizing variable resolution storage.

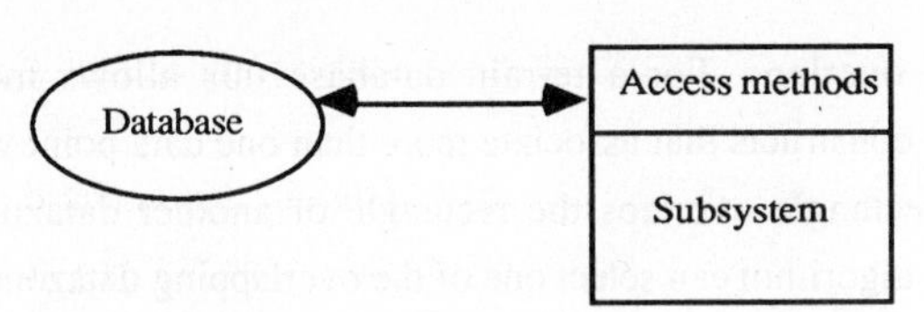

Figure 10. Conceptual View of a Database

The ROI is defined with respect to some well known point in the simulation--e.g., the location of an airplane or a vehicle. The ROI is not a fixed geometrical shape, it can shrink and expand during the simulation depending on the movement trend of the point that defines the ROI. The behavior and size of the ROI is part of specification of the database. The data required to query points within the ROI resides in main memory. As a result, queries for which the key lies within the ROI can be computed in real time by using data resident in memory while queries that lie outside the ROI will fail. Updating the memory resident data to reflect the ROI is done with a prediction scheme. Figure 11 illustrates the relationships between the various components that implement the database functionality. Noe that this process is transparent to the subsystems.

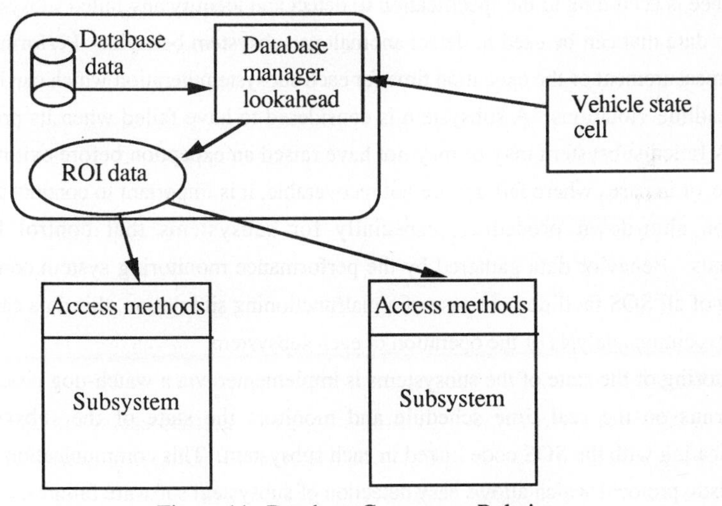

Figure 11. Database Component Relations

Variable density storage allows different portions of a database to have different resolutions. This reduces the storage requirements for large databases. The variable density storage is implemented by partitioning the xy plane into *datazones*. A datazone is an arbitrary rectangle that surrounds constant density data. The distance between adjacent data points is defined as the *resolution* of a datazone. By appropriately designing the datazone layout, a designer can find the optimum balance between resolution and storage requirements.

Datazones can overlap. For a terrain database this allows modeling of bridges, overpasses or other constructs that associate more than one data point with each query. All datazones whose rectangle overlaps the rectangle of another datazone will be properly flagged. The query algorithm can select one of the overlapping datazones by using an initial estimate of the query answer.

Each data structure associated with a datazone contains the coordinates of the rectangle that the datazone covers along with pointers that can be used to access the data in the datazone. To answer a query about a point X,Y, the interpolation algorithm must find the datazone(s) that contains the coordinates X,Y. Once a datazone is selected, the actual points can be accessed and used for the interpolation. When adjacent datazones are of different resolutions the assumption is the interpolation of points on the boundary will be coherent to both sides, and there should be no discontinuities when jumping across a resolution gradient.

4.4 On-Line Performance Monitoring

During a simulation all subsystems are monitored continuously in order to i) affirm that their performance is according to the specification ii) detect and identify any failed subsystems, and iii) gather data that can be used to detect anomalous subsystem behavior. Performance data includes measurement of the execution time for each subsystem iteration which can be used to derive deadline violations. A subsystem is considered to have failed when its process has exited. A failed subsystem may or may not have raised an exception before exiting. In the latter case, or in cases where failures are not recoverable, it is important to conduct an orderly simulation shut-down procedure, especially for subsystems that control hardware components. Behavior data gathered by the performance monitoring system consists of a usage log of all SOS facilities. In case of a malfunctioning simulation, this data can be used for post-execution analysis of the operation of each subsystem.

Monitoring of the state of the subsystems is implemented via a watch-dog process. This process runs on the real time schedule and monitors the state of the subsystems by communicating with the SOS code linked in each subsystem. This communication follows a deterministic protocol which allows easy detection of subsystem software failures.

Performance and behavior monitoring is implemented by the SOS component that is linked with each subsystem. Since each subsystem must use the SOS interface to interact with the simulation, it is a simple task to measure execution time and log the usage of the various resources by the subsystem. This approach to performance monitoring has the added advantage of not requiring special hardware or specific operating system features and as a result is very portable.

5 Distributed System Issues

Support for simulators that utilize a distributed computing environment is provided by splitting the SOS into a master and a slave component. The master SOS (MSOS) runs on a system that is designated as the **main host**. The slave SOS (SSOS) run on each of the remaining hosts which are designated as **remote**. Each of the SSOS components is responsible for monitoring the subsystems executing on the local host. Any out of the ordinary information is reported back to the MSOS. Communication between the master and slave part of the SOS is done though a deterministic communication link. Conceptually, there is a point to point connection between the main host and each of the remote hosts. During execution, each of the slave components of the SOS monitor the communication link for control and data messages from the MSOS. In response to control signals the SSOS will schedule subsystems running on its local host, terminate the simulation or perform other similar control functions. Untyped block messages are used to transfer cells and other data among the hosts. A diagram illustrating the remote execution model is shown in Figure 12.

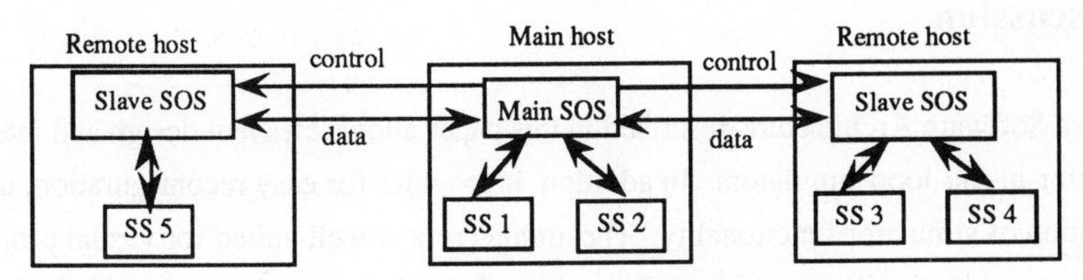

Figure 12. The Remote Execution Model

The implementation of the remote execution mechanism is geared towards efficient performance and conformance to real-time constraints, while maintaining the pure conceptual framework. Communication between the main host and remote subsystems is implemented by using *stub* processes similar to those typically used in remote procedure call facilities. When a subsystem is configured to run on a remote host, a stub process is scheduled on the main host using the temporal constraints that would apply to the subsystem. During run-time, the stub process sends commands to the slave SOS on the appropriate remote host in order to transfer cells (if needed) and schedule the remote subsystem. The execution time of the stub is a function of the latency and the bandwidth of the communication link between the main and remote host and is assumed to be known and deterministic. Once the intended simulator configuration is known, the stub's execution time can be easily measured for preparation of the final real-time schedule. Figure 13 illustrates a schedule that uses a stub for remotely executing a subsystem.

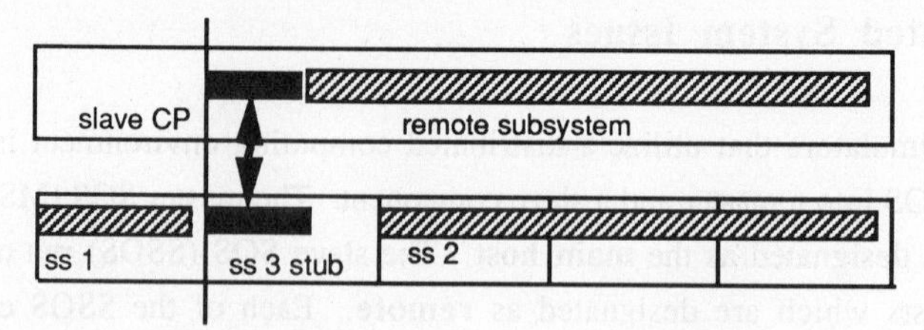

Figure 13. Remote Subsystem Scheduling

Use of a stub is an effective mechanism for factoring the latency and bandwidth of the communication hardware into the system at the real time scheduling level. The stub is treated as a subsystem whose execution time depends on the communication medium and amount of data being passed, while its temporal constraints are those of the subsystem scheduled on the remote host. Stubs are implemented as generic parametrized processes, so that they do not have to be written or tailored for a specific simulator configuration.

6 Discussion

The Open Software Architecture described in this paper allows efficient design and integration of operator-in-the-loop simulators. In addition, it provides for easy reconfiguration, upgrade, or extention of simulator functionality. The architecture is well-suited for virtual prototyping applications, since it allows rapid configuration of simulators, and the use of "off-the-shelf" subsystems as basic building blocks. This architecture has been used to design and develop the Iowa Driving Simulator (IDS). An IDS prototype system has been operational since August 1991. The software of the system consists of about a dozen subsystems using a configuration of 13 cells and a database that provides real time terrain height data. The hardware environment consists of four computers, including two multiprocessor systems, linked with various types of dedicated communication devices (DR-11, reflective memory, HSD).

The IDS prototype has illustrated the validity of this approach for the design and integration of simulators. The same subsystems have been used in a variety of different configurations consistent with the operational requirements of various experiments conducted on the simulator and the underlying computer and communication hardware has been changed without modification of subsystems. The largest configuration built thus far utilizes the entire hardware environment (4 computers with a total of 32 processors) and all the subsystems (visuals, vehicle model, motion, data collection, scenario control, control loading, audio instrumentation, manager interface). The smallest configuration consists of two hosts and only five subsystems and does not utilize the terrain database.

The experience gained by using this architecture for building an actual system has also shown the need for further refinement and enhancement of the architecture. Most notably better and more consistent mechanisms for error detection and recovery are needed. A system-level exception-handling mechanism is currently under development for this purpose. Improvements to the unit-test, integration test, and on-line performance monitoring facilities are also planned.

Acknowledgment: Research supported by NSF-Army-NASA Industry/University Cooperative Research Center for Simulation and Design Optimization of Mechanical Systems.

References

1. Stoner, J.W., et al., (1990), "Introduction to the Iowa Driving Simulator and Simulation Research Program," Technical Report R-86, Center for Simulation and Design Optimization, The University of Iowa, Iowa City, Iowa.
2. Harel, D, (1992), "Biting the Silver Bullet--Toward a Brighter Future for System Development," *COMPUTER*, Vol. 25, No. 1, pp. 8-20.
3. Stankovic, J.A., (1988), "Misconceptions About Real-Time Computing--A Serious Problem for Next-Generation Systems," *COMPUTER*, Vol. 21, No. 10, pp. 10-19.
4. Jahanian, F., and Mok, A.K., (1986), Safety Analysis of Timing Properties in Real-Time Systems," *IEEE Trans. on Software Eng.*, Vol. SE-12, No. 9, pp. 890-904.
5. Ostroff, J.S., (1989), *Temporal Logic for Real-Time Systems*, John Wiley & Sons, Inc., New York.
6. Reed, G.M., and Roscoe, A.W., (1986), "A Timed Model for Communicating Sequential Processes," *Proceedings, ICALP 86*, Springer LNCS 226, pp. 314-323.
7. Mullender, S., (ed.), (1989), *Distributed Systems*, Chapter 3, Interprocess Communication, ACM Press, New York, pp. 37-64.6.
8. Bisiani, R., and Rovisharkar, M., (1990), "Plus: A Distributed Shared-Memory System," Proceedings Seventeenth International Symposium on Computer Architecture," pp. 115-124, Seattle, Washington.
9. Chetto, H., Silly, M., and Bouchentouf, T., (1990) "Dynamic Scheduling of Real-Time Tasks under Precedence Constraints," *The Journal of Real-Time Systems*, 2, pp 181-194.

Man/Machine Interaction Dynamics and Performance Analysis

Harold P. Frisch

NASA/Goddard Space Flight Center, Greenbelt, MD 20771 USA

Abstract: The Man/Machine Interaction Dynamics and Performance (MMIDAP) analysis project, lead by NASA's Goddard Space Flight Center (GSFC), seeks to create an ability to study the consequences of machine design alternatives relative to the performance of both the machine and its operator. The MMIDAP problem highlights the conflicting needs and views of groups that focus on machine design and groups that focus on human performance, ergonomics, and cumulative injury potential. There is a critical need to integrate associated design and simulation tools and to establish multidisciplinary lines of communication. This will enable engineers to design mechanical systems and concurrently perform the system design trade studies needed to assess resultant machine-operator performance. Basic need within industries that design and manufacture human operated machinery, and the underlying complexity of cross disciplinary communication provide the motivation to closely coordinate evolving MMIDAP capability with the Concurrent Engineering community. This chapter outlines ongoing efforts by the GSFC and its university and small business collaborators to integrate both human performance and musculoskeletal databases with the host of analysis capabilities necessary for early design analysis and trade studies relative to the dynamic actions, reactions, and performance assessment of coupled machine-operator systems.

Keywords: human performance / human factors / concurrent engineering / multidisciplinary analysis / human machine interaction / biomechanics / biodynamics / ergonomics / biomechanical database / human performance database / musculoskeletal database / machine operator systems

1 Introduction

A confluence of diverse computer science, mechanical systems, and biosystem technologies is now forming. Advanced mechanical system dynamics analysis methodologies, biosystem measurement and modeling techniques, computer hardware configurations, Concurrent Engineering communications, database systems, anatomical, biomechanical, biodynamical, behavioral, and cognitive science research capabilities can, with reasonable effort and proper focus, be drawn together to create a Man/Machine Interaction Dynamics and Performance (MMIDAP) analysis capability. The envisioned capability is to build upon existing and readily extendable capabilities. Intelligent assistant interfaces are being envisioned to provide non-specialists with user friendly access to MMIDAP analysis capability. Boy provides an extensive development of the intelligent assistant system subject [1].

The final report of the 1985 Integrated Ergonomic Modeling Workshop [2] contains a detailed review of pre-1988 software capability along with a list of recommendations for future research. It specifically remarks that "there is a paucity of dynamic interface models" and that "an integrated ergonomic model is needed, feasible, and useful." The report's review of existing capability demonstrates that ergonomic modeling software has been primarily developed to support aerospace cockpit design, design for product maintainability, and whole body dynamics associated with automobile crash and pilot ejection. Some work exists under the general heading of optimization of sports motion; however, there is virtually nothing to support mechanical system designers that must evaluate machine operator interaction dynamics and performance with or without survival gear, in hostile environments, on-the-job, on earth, or in space.

There is no fundamental difference between machines that are used at the conventional workplace, machines that are used as aids for the disabled, and machines that are used as exercise equipment for athletic training or physical therapy. Within the context of Concurrent Engineering an analysis capability that can quantify machine operator performance vs. machine design alternatives during the early design phase is needed for both ground and space based applications. Within the context of physical therapy a capability is needed to measure and quantify progress, evaluate therapeutic procedures, and design patient specific treatment techniques.

The MMIDAP project supports the generic machine operator system design problem. It is directed toward machines that are controlled by a human operator's intelligent physical exertions. MMIDAP analysis tools will allow designers to introduce the physical and cognitive limitations of a specific operator or operator population class into the machine design process. The intent is to develop the MMIDAP analysis capability in as generic a manner as possible. This will enable its application within a broad range of aerospace, machine design, ergonomic, physical therapy and rehabilitation engineering problems.

This project intends to integrate dynamics, human performance, and biomechanical analysis tools with associated databases. It also intends to provide the nonspecialist user with intelligent assistant system interfaces. These will support both total system performance trade studies and detailed cause and effect analysis. There is no desire to duplicate the statics and kinematics based software systems that now support human factors investigations such as those identified in References 2 and 3. Our intent is to complement these with new techniques that support "what if?" studies that cannot ignore dynamics and human performance considerations.

2 Anthropometric and Biomechanical Databases

The MMIDAP project research group feels that it is time to take stock of present analysis methods and their associated data determination needs. A dedicated emphasis is needed to develop the infrastructure for the systematic, engineering approach to solving human system problems and to recognize current limitations. While it is true that this effort does not aim at the solution of a particular problem, the prospects for almost simultaneous progress across many diverse problem areas should not be underestimated as milestones in the infrastructure development process are achieved.

The National Library of Medicine (NLM) is currently undertaking a project that intends to build a digital image library of volumetric data representing a complete normal adult human male and female [4]. This "Visible Human Project" will include digital images derived from photographic images obtained from cryosectioning, computerized tomography, and magnetic resonance imaging, for example see Reference 5. NASA/GSFC is complementing this effort with a "whole body digital mapping project." The GSFC project seeks to develop a hierarchial tree of biomechanical and human performance analysis capability vs. data availability. The output of this project will be used by the MMIDAP project to both define objectives and to recognize state-of-the-art analysis limitations. This is a cooperative project involving biomechanics groups from The University of Iowa and Case Western Reserve University, human performance specialists from the University of Texas at Arlington, and anatomists from the University of Colorado at Denver. The objective is for the analysts to define data needs while anatomists define if it is feasible with modern technology to provide the requested data as a by-product of the "Visible Human Project."[1]

The musculoskeletal system is providing general focus to this effort while the human knee complex and the popliteus muscle in particular provide sharper focus. Methods to database and retrieve human structure and performance data are of primary concern. The sheer number of parameters and variety of combinations required to meet broad based application needs justify a formal organized approach. The approach proposed is a hierarchial relational data structure with embedded relations. A relational data structure with embedded relations provides structure to the data storage and retrieval problem.

Anatomical complexes may be viewed as objects that are decomposable into anatomical components. These too may be viewed as objects. Objects at every hierarchial level have

[1] The final report of this project is being prepared as an article for publication in the 1994 issue of "CRC Critical Reviews in Biotechnology.

attributes that can be referenced. The value of an attribute can be either numeric data, textural data, or another relation that points to finer detail information at the next lower hierarchial level. This data storage approach will allow both function and biostructural information to be integrated together within the same database system.

Special efforts will have to be directed toward the data normalization and scaling problem. This will enable users to appropriately transform digital anatomy information to suit a broad variety of patient specific and general biomechanical analysis needs. Additionally, computerized tomography and magnetic resonance imaging can be used to locate anatomical landmarks of particular patients. Methods are to be developed for using this information to scale and warp baseline digital anatomy information to a unique patient. This will provide the enabling technology needed to support both personalized biomechanical analysis and general anatomical studies.[2]

Kroemer [2] provides a review of currently supported anthropometric data bases and computer models used in the field of ergonomics. Winters [6] provides a source book for multiple muscle systems and movement organization along with a survey by Yamaguchi [7] listing human musculotendon actuator parameters from over 20 different published sources. Additionally Seireg [8] provides the anthropometric and musculoskeletal data used to support musculo load sharing research carried on at the University of Wisconsin at Madison.

One major problem with existing biomechanical data is that it comes from so many different sources with almost as many different measurement reference frames. A quick scan of data provided by Yamaguchi [7] reveals considerable numeric variation between reference sources for the same anatomical component. The data tables also reveal that there are considerable data gaps. The NLM's Visible Human Project is presenting the biomechanics community with a unique opportunity to fill these gaps and to obtain a consistent reference source of fundamental biomechanical data. GSFC's whole body digital mapping project seeks to precisely define what should be measured and how it should be databased, so as to be *useful* for follow-on analysis.

As a first step, it was recognized that computer based analyses beg for codification of terms and parameters. The MMIDAP project bases itself on an ability to integrate diverse models and analysis modules that were developed independently within narrow fields of research. The enabling of this integration of capabilities requires a standardized notational system. The current standards void not only reduces accessibility to available data but also serves to inhibit research progress. The development of a standardized systematic notation is

---

[2] It is interesting to speculate on the possibilities of an enhanced digital surgery capability. If large deformation and associated elastic properties of soft tissue components, such as muscle, veins, nerves, intestine, etc. can be defined, then it may be possible to develop the capability to both graphically cut and push these soft tissue components around in a virtual reality environment. Creation of an ability to cut, push, deform, and monitor large deformation strain in soft tissue should provide the basis for a new level of pre-surgical planning and training.

critical. Kondraske [9] provides a preliminary attempt to define the notational and design considerations associated with the development of such a coding process.

3 Human Performance Database

There is no lack of literature regarding the quantification of human function or performance. The literature as a whole can perhaps best be characterized by noting that it lacks a common conceptual framework upon which human performance quantification strategies can be based. As discussed by Kondraske [10], this has made it difficult to organize previous work and compare methods. The approach that is advocated herein for resolving this problem introduces the concept of a *functional unit*. This entity is defined in such a manner that it must possess a measurable resource level to accomplish a highly focused task. Considering all functional units collectively leads to the realization of a finite set of basic elements of performance (BEPs). In mathematical terms, the BEPs define a set of basis vectors while associated measured resource level defines vector magnitude. To specify a BEP one must delineate both the functional unit and its dimensions of performance.

Human BEPs may be organized into three primary domains:

1. Central processing
2. Physical: Environmental interface
3. Physical: Life-sustaining

The collective set of all BEPs forms a *performance pool*. This performance pool may be defined for an individual or for a population group. It defines levels of resources available relative to all dimensions of performance associated with all functional units. To accomplish any task (physical or mental), humans draw upon appropriate BEPs from the performance pool in the required amount. Successful task performance is determined by the availability of required BEPs. If insufficient BEP resources are available from the performance pool, the task cannot be accomplished. If just enough exist, task performance will be stressful. If more than enough exist, task performance will be comfortable. Unfortunately one cannot assume that all BEPs are functionally independent of each other. There are dependencies that must be recognized and accounted for in the performance analysis process. As stated by Fitts [11], we cannot study man's motor system at the behavioral level in isolation from its associated sensory mechanisms. We can only analyze the behavior of the entire receptor-neural-effector system. The implication here and the major challenge for MMIDAP is to develop and implement methods that automate the process of detecting and accounting for functional relationships between the sets of BEPs that must be simultaneously exercised during task performance.

3.1 Functional Units

A *functional unit*, by definition, serves some purpose or function. This unit may be composed of lower order units within a hierarchy, or it may be a component or subsystem within a higher order system. Functional unit simplicity enables identification of variables indicative of system performance.

From a mathematical perspective a functional unit possesses the capacity to operate along only a single dimension of function such as elbow flexion. The dimensions of performance are defined by unit vectors that are associated with performance such as range of motion, speed, strength, etc.; vector magnitude represents performance capacity. This measure represents *performance resources available* R_A along the relevant dimension. R_A is always positive and defined such that a larger number indicates greater performance resource available. Task performance can be judged as possible when *performance resources required* R_R are less than or equal to performance resources available R_A. Inadequate resources in any one dimension can prevent task execution.

Regardless of system complexity, the potential for success in a given task situation is determined by the comparison of R_R relative to R_A along each relevant dimension of performance. As previously mentioned, the realization of success must take into account functional relationships between dimensions of performance, e.g., speed and strength. The development of these functional relationships in a format compatible with codification in an intelligent assistant system interfaced to the human performance database represents many cutting edge research projects at the Human Performance Institute (HPI) at the University of Texas at Arlington. See Kondraske [10] for a review of ongoing work at the HPI.

3.2 Basic Elements of Human Performance

Applying the general concepts presented above, a human is viewed as an architectural structure composed of a finite set of interconnected functional units each of which possesses the capacity to operate along specific dimensions of performance. This gives rise to an *elemental resource model* for the human system.

A major decision is required to identify a level within the available hierarchy of subsystems which can be defined as the level of basic functional units. *Basic Functional Units* are those which are the first subsystems encountered for which a one-to-one mapping between a structure and its function can be identified. Considering all functional units collectively results in the realization of a finite set of distinct "Basic Elements of Performance" (BEPs). As previously noted, each BEP has an associated functional unit and one dimension of performance. For example, knee extensor (the functional unit) and speed (its dimension of performance) define a particular BEP.

The human performance task analysis approach advocated herein requires that the analyst, with appropriate intelligent assistant system aids, decompose a task into fundamental components. These fundamental components are the basic elements of performance. These can and are being databased by individual and population cross section. BEPs form the basis of the elemental resource model that is used for characterizing all aspects of the human system and its interface to the performance of tasks, e.g., musculoskeletal, central processing, and life sustaining.

Once task decomposition into BEPs is complete, human resource requirements can be estimated by both quasi-static biomechanical and inverse biodynamic analysis. Established norms for population cross sections of interest can be used to determine if tasks are within the available resource capabilities of target population groups. Individualized BEP data can be used to determine if a particular (e.g. disabled) individual can accomplish a defined task. Conversely, BEP requirements above the norm can be used to pinpoint exactly where the machine operator interface needs improvement. This approach to performance analysis forces consideration of those dimensions of performance that represent desirable or performance limiting qualities. Furthermore, it provides a consistent method for quantifying performance. Confusion resulting from dual concepts, such as strength vs. weakness and endurance vs. fatigue, is eliminated and a clear modeling framework emerges. Relative to Concurrent Engineering, the ability to estimate BEP requirements during the early phases of machine concept development provides the quantifiable information needed to perform rational trade studies and to guide engineering design toward an optimized machine operator interface.

3.3 Measurement and Databasing

Kondraske [12] provides a good overview of task decomposition via BEPs and the methods being used to database the BEP records of the 3000+ patients tested with systems developed at the Human Performance Institute (HPI). This work was originally focused within the field of Physical Therapy and Rehabilitation Engineering.

Measurement methods are employed which stress individual functional units relative to associated dimensions of performance. These provide quantifiable measures of performance resource availablity. Tests which stress the coordination of multiple functional units are also necessary and are therefore included. Procedures for recording scores, filing results, rating performance, etc., are implemented with computerized instrumentation. This is intended to maximize objectivity and eliminate subjective measures from the database.

A complete individualized data file is obtained by a mix of hard measures and data that can be confidently interpolated from the individual's general population group statistics. Hard measures are taken to characterize an individual's performance resources that are judged to be significantly above or below the norm for his/her population group.

3.4 Mapping BEPs to Tasks

The MMIDAP project is of the opinion that task decomposition into BEPs is the best state-of-the-art approach to providing a systematic basis for predicting human performance and for extrapolating human performance experimental test data to different population groups and task environments.

The quantification of performance capabilities for a task's functional components is an essential step in the process of establishing a system performance model. Mapping is the link between application independent and application dependent views. It represents the key to the useful application of the elemental resource model.

The mapping process involves task analysis in terms of task division, determination of dimensional requirements for each subtask, determination of resource availability, dependencies of biomechanical and biodynamical considerations on anthropometric parameters, gravity and inertial loading factors, and a resource utilization strategy. In complex systems such as humans, mapping is nontrivial. Specific mappings cannot be considered without a complete definition of the task (or subtask components). While this sounds complex it reflects the true complexity of the human to task interface.

Human anatomy defines the structural interconnections of functional units. This interconnection topology plays a significant role in developing the mapping from human performance space to task space. In external tasks, the connection between task and environmental interface functional units must first be identified. Once the task and the manner in which it is to be executed are defined, relatively simple physics is used to determine the mapping. Fortunately, biomechanics has, to a reasonable degree, derived most of the mathematical relationships required.

The development of an intelligent assistant system for developing the mappings that link tasks and BEPs is a critical element within the MMIDAP effort. It will provide the "user friendly" interface needed for product acceptance by the non-specialist user community. Conceptually it appears possible to create a support shell that would help users associate anatomy with function via the digital anatomy database developing via NLM's Visible Human Project. For human performance analysis, it would decompose a range of user defined high level tasks into associated basic elements of performance. Of course, this concept would be extendable to all biosystem functions that have anatomical association. It seems reasonable to expect that this capability could be developed as a hierarchial structure. BEPs combine to form simple tasks, simple tasks combine to form more complex tasks, etc.

3.5 Resource Allocation

When a human is presented with a task situation, he/she has some flexibility with regards to which functional units to employ and what fraction of available resources are to be used for task accomplishment. Redundancy results when the anatomic configuration allows for options. When two or more functional units with the same dimensions of performance are available, the problem of resource utilization or resource allocation must be resolved. In biosystems, this is usually determined automatically or subconsciously via a subconscious control strategy. It is reasonable to assume that some common strategy, i.e., *a resource utilization strategy*, is employed across the normal population which results in a more or less standard form. This is why individuals with normal amounts of performance resources appear to go about tasks in much the same manner. In a mathematics based model of a biosystem, a resource allocation control strategy must be defined as rules and constraints. The allocation rule being implemented within the MMIDAP analysis capability is to minimize stress on system resources while striving to accomplish the goal.

4 System Performance Analyses

A theory is presented by Kondraske [13] that develops a scientifically based conceptual framework for addressing many fields of concern relating to human performance. The theory involves the concepts of basic elements of performance and human resource economics.

Many questions regarding biomechanical behavior can and are being addressed without consideration of system performance; however, human performance questions associated with complex tasks cannot ignore biomechanics and biodynamics. Biomechanical models typically focus on the principles of materials and mechanical behavior, while a system performance model for a given subsystem recognizes dependency on components external to the biomechanical domain (vision, neuromotor control, etc.).

The basic difference between classic biomechanical analysis and performance analysis can be summarized as follows:

Biomechanical & Biodynamics Analysis - provide traditional static and dynamic analyses that depend upon the basic physical concepts of mass, inertia, geometry, stiffness, position, velocity, acceleration, musculotendon, and environmental loads.

Performance Analysis - uses biomechanical and biodynamics analysis information to quantify the qualitative parameters used to characterize a system's capacity to successfully accomplish a task. It focuses on providing analysts with an enabling capability to ask and quantify answers to the following three fundamental questions:

1. Can the task be accomplished? If not, why not.
2. How well can the task be accomplished?
3. What is the best way to accomplish the task?

These questions may be directed to either the machine operator, or the machine-operator system.

Performance analyses are simple in concept and yet powerful as total system design and evaluation tools. System performance can be modeled in terms of the *available performance resources* of the operator while quantitative task characterization can be expressed in terms of the *performance resource demands* required of the operator. For a detail development of these concepts see References 10 and 13.

By viewing the performance of a high level task as a hierarchy of elementary tasks, one can establish a foundation of fundamental resources. Since high level tasks are accomplished through the coordinated utilization of this performance resource pool, required resource deficiencies will place a limit on resultant high level task performance.

Non-trivial systems performance analysis rests on the ability to obtain numerical values for both performance resources availability and performance resource demands. Available performance resources can be determined from human performance measurements, databases thereof, and models that allow for their estimation from general population norms. The determination of performance resource demands starts with the decomposition of a high level task into a hierarchy of elemental level tasks. A knowledge of human structure and associated quasi-static and dynamics analysis capabilities are then used to define resource demands for each task within the hierarchy of elementary level tasks.

Adhering to the principle of resource economics (i.e., resource availabilities must meet or exceed resource demands for task completion to be realized), the highest level of high level task performance is limited by the most limiting resource.

This strategy can be adapted to result in:

- A simple yes or no regarding resource sufficiency in response to a given task requirement level
- A determination of a maximum performance level
- An analysis of a task's resources demands
- The stress on the resources involved
- The resources limiting better performance

Kondraske has asserted and argued subjectively that the general optimization rule to be applied to task decomposition is as follows: the "human is driven to accomplish the goal using that procedure which minimizes the stress across all performance resources." The problem of finding the optimal procedure can be formulated in terms of a trial and error resource utilization problem. From a machine-operator interface design, or a human task accomplishment perspective, it is normally sufficient to identify only the most limiting resources. These identify exactly where design improvement is needed or where modification of one or more components of the task scenario, goal, or operational procedure is required.

5 Integrated Musculoskeletal Machine Dynamics Equations of Motion

The creation of mathematical models for the characterization of system dynamics is a fundamental part of engineering analysis. Both mechanical and biomechanical groups frequently make use of lumped parameter models. These models consist of hinge connected rigid and flexible bodies, i.e., multibody systems. An excellent overview of existing automated methods for developing simulation models for complex mechanical systems via multibody dynamics analysis software systems is provided by Schiehlen [14]. In the late 1980s, several international groups discovered that equations of motion could be rederived in such a manner that computational speed could be greatly enhanced [15-17]. New implementations of these and analogous methods with improved speed and modeling capability are now in use [18-21].

Multibody simulation models have been successfully used to model certain classes of musculoskeletal systems. However, modeling weaknesses exist and these must be recognized before one attempts to use multibody tools for general biomechanical and biodynamical application. The following deficiencies associated with vertebrate biodynamics application have been recognized and plans are now underway to enhance the program NDISCOS [18,22], accordingly:

- Inverse dynamics for deterministic systems. This capability is necessary to predict resultant joint loads associated with the dynamic interaction between machine and operator.
- Intermittent loop closure and range of motion constraints. This capability is needed to model the interface between man and machine and to routinely include range of motion limits for anatomical joints.
- Biomechanical joints must now be approximated by conventional mechanical joints. To support more detailed analysis, an enhanced joint modeling capability that includes the full complement of human joints defined by Norkin [23] must be developed.
- Rolling/sliding contact of penetrating surfaces. This capability is needed to model the details of joint motion and loading and too adequately model the soft tissue interface contact between an operator's hand and the manipulandum used for machine control.
- Flexible body modeling. This capability is currently available within NDISCOS. It is required for stress distribution determination within the skeletal structure being stressed

by physical exertion. This is an important capability needed to support the joint prothesis design problem.

- Body clustering. This capability is needed to model joints such as the ankle and wrist. It is also needed to model the spine. In each case, relatively small bones are tied together by ligaments into a cluster that has limited range of motion. The desire is to develop a general cluster capability that will be applicable for generic mechanical system dynamics, biodynamic, and molecular dynamics research projects.

The dynamics analysis of mechanical systems is dominated by the need to solve the forward dynamics problem. That is, given a prescribed set of internal and external loads, predict system response. Attempts to perform forward dynamics analysis with neuro-musculo-skeletal systems are usually stopped by one's inability to mathematically characterize the human's cognitive processes that generate the neural activation signals that stimulate the body's musculo actuator system.

The MMIDAP project recognizes this fundamental limitation. It therefore concentrates on the inverse biodynamics problem. Graphical animation and laboratory testing techniques exist for obtaining an estimate of human dynamic response during a broad range of activities. If sufficient information can be obtained (displacement, rate, acceleration, and external loads) then inverse biodynamics methods can be used to predict what the resultant musculo actuator loads had to be to produce the defined response. These results can then be compared with performance resource availability information. The comparison is used to determine if the predicted musculo response, i.e., performance resource demands, required of a machine operator are within the performance resource limits of a particular operator or the average capability of the operator's population group.

6 Muscle Modeling and Load Sharing

Detailed neuro-musculo-skeletal modeling of the human system or any of its subsystems is an extremely complex problem that is beyond today's state-of-the-art capability. The First World Biomechanics Congress in August 1990 had over 80 oral presentations on the subjects of multiple muscle systems, biomechanics and movement organization. Formal reports on 46 of these presentations have been collected by Winters [6]. From these reports and others presented at the Congress, it is clear that muscle dynamics and neuro-musculo-skeletal organization and movement modeling is a subject that will occupy researchers for many more years.

There is great deal known about how nerve cells transmit signals, how these signals are put together, and how out of this integration higher functions emerge [24]. Nerve cells are connected through their synapses to form functional circuits; these are organized into the multineuronal circuits and assemblies that provide the basis for neural organization [25]. Muscles are controlled by nerves at neuromuscular junctions, and at these points activation signals are biochemically processed to initiate the muscle contraction process [26]. Any attempt to model the details of cognitive neuro-muscular-skeletal response would require the seemingly impossible development of a mathematical model of the biochemical, physiological, and psychological processes that lie between cognition and the innerviation of muscle fibers.

The details of muscle modeling have several more layers of complexity. For example, muscles are composed of muscle fibers that are differentiated by the biochemical properties that dictate their respective response speeds and resistance to fatigue. When the muscle is innerviated, select sets of muscle fibers called motor units contract while others remain in a rest or in an energetics recovery state. This motor unit apportionment issue further complicates the mathematical modeling of muscle contraction dynamics as can be seen from Hatze's complex mathematical formulation of the problem [27].

In spite of these outlined complexities in muscle dynamics modeling, progress is being made at a level compatible with real world application. Relatively simple mathematical approximations are appearing in the literature that are providing a basis for understanding how musculotendon systems produce force as a function of the associated reaction dynamics of the biochemical processes that produce muscle contraction, for examples see References 28-35. Also see Reference 36 for a rather detail review of the complexity associated with using system identification techniques to obtain the data needed to support studies associated with joint dynamics modeling. For example, it is stated therein in that despite the success of linear analysis, it is evident that the modeling of overall joint dynamics is not linear. The parameter values of linear second order models for joint dynamics are found to change dramatically with the operating point that is defined by such parameters as mean torque, perturbation amplitude, or mean position.

The incorporation of muscle dynamics into the framework of a multibody simulation capability is a rather straight forward process if the physics of muscle contraction can be assumed known. This has been demonstrated by Hatze [37] and Morris [38]. However, there is little debate that a mathematical model for the cognitive process is virtually impossible to define. Nevertheless, forward dynamics can still be used when known musculo innerviation is imposed, for example, by functional neuromuscular stimulation systems, as discussed by Chizeck [39]. It can also be used for well defined structured motion such as reflex response actions. The availability of measures for the biochemical dynamics of calcium ion concentration within the system of defining equations for muscle contraction dynamics, see References 27 and 31-33, provides an avenue to an understanding of the process of fatigue, discomfort, and pain. An extensive review of this connection is available from several papers published in a special issue on "Occupational Muscle Pain and Injury" by the European Journal of Applied Physiology [40,41].

Complexities associated with modeling muscle contraction dynamics are matched by the problem of resolving muscle load sharing and kinematic redundancy. The presence of redundant muscle actuators at virtually every anatomical joint implies that rules must exist for defining how muscles share the work load. Kinematic redundancy within the upper and lower extremety systems also presents mathematical modeling problems. Redundancy in the physical system to be modeled leads to a mathematical problem with an infinite set of solutions. This problem is resolved by optimization techniques that find the unique solution that minimizes a user-defined cost function. Zajac [42] provides an in-depth review of the complexity associated with modeling multijoint muscle systems.

Seireg [8] provides an extensive summary of cost functions relevant to ongoing research in muscle load sharing at the University of Wisconsin at Madison. This work models the entire musculoskeletal system as a collection of hinged rigid bodies. Each muscle is modeled as a linear actuator that may wrap around boney structure and act along a resultant line of action defined between the points of muscle insertion and origin. To support this effort Williams [43,44] has developed a computer program for determining muscle load sharing. The program uses Serieg's database of muscle characterization information. It accepts as input initial conditions of state, external loading, and parameters needed to define the cost function to be minimized. The cost functions are physically realizable minimization criteria such as minimizing all forces carried by muscles, work done by muscle, reaction forces and residual moments. The program then computes the musculo force required by each muscle to maintain the system in quasi-static equilibrium based upon user prescribed cost function assumptions.

Additionally, Zajac and students at Stanford University and the Rehabilitation R&D Center at the Palo Alto VA Medical Center are developing interactive graphics based modeling tools to study orthopedic surgical procedures [45]. In particular they have developed a model of the human lower extremity to study how surgical changes in musculoskeletal geometry and musculotendon parameters affect muscle force and its moment about the joints.

The message to the MMIDAP project is that underlying modeling constraints and limitations must be defined. The attempt to include biodynamics detail into a MMIDAP capability should not be viewed as the ultimate goal. For every general class of problems, it is critical to know when to put a stop to modeling fidelity. For MMIDAP, stopping at resultant joint loads appears appropriate; attempts to get into more modeling detail opens up some exceedingly complex and yet unresolved problems. For the envisioned MMIDAP application community, there is, without doubt, a point of diminishing returns relative to modeling capability and analysis complexity.

7 Prediction of Human Motion

One fundamental difference between repetitively testing human subjects and repetitively testing mathematical models is that the human's response is nonrepeatable [46]. The modeling goal for the prediction of human performance can therefore only be that the predicted motion be physically reasonable. Predictions and reasonable variations around them should be viewed as defining an envelop of possible human response. With this realization in mind, simplified motion prediction algorithms can justifiably be introduced into the motion prediction model. Physical realizability can be checked by viewing animated response, monitoring joint rates, acceleration, jerk, loading, and comparing these with norms in the BEP database.

The program JACK [47] has several unique features that make it ideal for MMIDAP application. Figure positioning by multiple constraints [48] is a capability that allows users to specify trajectories at several body fixed points (hand, feet, torso) and to then have motion trajectories for all other points predicted. Strength-guided motion [49] is a capability that allows for human strength and comfort data to be used in the motion prediction process. The creators of JACK make note of the fact that others such as Wilhelms [50] have used forward dynamics for human motion prediction. The JACK development team argues that utilization of a forward dynamics approach to human animation is difficult for the user to control because users must provide all joint torques. For a 3D system, this is a near impossible task. Kinematic and inverse kinematic approaches are easier to manipulate but suffer from the potential of unrealistic joint motion. JACK uses a blend of kinematic, dynamic, and biomechanical information when planning and executing a path. The task only needs to be described by a starting point, ending position, and external loads such as gravity and weights to be transported. An excellent review of the program JACK and computer graphics research as applied to the animation of human figures is provided by Badler [47,51,52].

8 Man-in-the-Loop Simulators for Complex Mechanical Systems

Major advances in formulating the multibody equations of motion needed to simulate complex mechanical equipment, along with the availability of low cost parallel processor computers, have provided a unique opportunity to create low cost real-time simulators for complex mechanical equipment with man in the control loop. Simulators accept real-time man-in-the-loop commands, graphically create a simulated visual environment, and drive other laboratory devices to create a simulated vibro-thermal-acoustic-shock environment. Stress and load information associated with machine components can be obtained directly from the predictive capabilities of the multibody dynamics software system that drives the simulator. Qualitative measures of system feel, cognition, and hand-foot-eye-ear coordination information can be obtained from operator comments. Human performance data can be obtained by monitoring operator response in the simulated environment. Machine performance can be obtained directly from the simulator. The ability to simulate in real-time all operator experienced loads at the operator machine interface and all other loads, such as sound and machine vibration that operator's subconsciously use for cueing purposes, sets this new effort apart from such systems as aircraft flight trainers. These loads are an integral part of the machine-operator feedback control cueing process.

The first step toward developing such a simulator capability for complex mechanical systems was taken at The University of Iowa with its development of a simulator for the J.I. Case backhoe [53]. The second step was the creation of the Iowa Driving Simulator [54]. The next step will be to develop the Department of Transportation's National Advanced Driving Simulator [55,56] at The University of Iowa in the mid 1990s.

9 Man/Machine Dynamic Interaction

Figure 1 provides a flow diagram of the proposed closed loop man/machine interaction dynamics and performance assessment process. The output of the program JACK is animated human system response. As for any engineering analysis study, the physical realizability of predicted response must always be checked. This is done by viewing animated response and resultant joint behavior. Performance parameters such as joint stiffness and comfort level within the JACK program allow users to tune predictions to bring them into the realm of physical realizability for the particular population group under study (old, young, normal, obese, handicapped, etc.) As a further check, JACK's predicted joint response information can be used as input to the program NDISCOS. This program offers a functionally complete capability for analyzing models of arbitrary complexity. NDISCOS can be used to create a detail dynamics model for the machine and the machine operator's musculoskeletal system. The associated equations of motion for the multibody model are exact, relative to the laws of Newtonian mechanics. The inverse dynamics capability of NDISCOS can be used to obtain a refined prediction for resultant joint loading. Differences between JACK and NDISCOS resultant load predictions stem from the simplifying assumptions within JACK's motion prediction algorithms and man/machine interaction dynamics effects.

The resultant joint load predictions made via NDISCOS's inverse dynamics capability can be used as input to a capability that addresses the nondeterministic muscle load sharing problem. If resultant joint loading and muscle load sharing predictions are acceptable, motion and load prediction information is ready to be used as input to the human performance

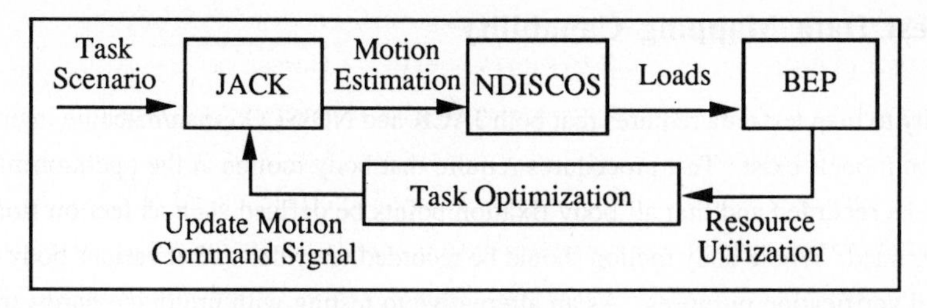

Figure 1. Closed-Loop Man/Machine Interaction Dynamics and Performance Analysis Assessment Process

BEP database at the HPI. The output of this step provides another assessment of physical realizability. If results violate physical realizability, JACK performance parameters can be adjusted and the process repeated.

If muscle allocation studies are required, the skeletal system model and associated computational theoretics will require non-trivial enhancement to include a detailed three dimensional characterization of critical joint complexes. An understanding of detail muscle load sharing is needed to explain, in a quantifiable sense, exactly why certain design options or operational scenarios have the potential of causing machine induced discomfort, fatigue, pain, or trauma.

In application, the assessment process will use an iterative refinement process that can be used until the successive approximations strategy converges to acceptable results. The predictions either confirm that man/machine interaction is acceptable or that some human performance parameters have exceeded database norms. If human performance requirements are excessive, machine design changes or operational scenarios can be refined until acceptable performance measures are achieved for the machine operator's population group. It is also possible to incrementally change population group by selecting different sets of anthropometric and BEP data from the database. Normally once an acceptable set of JACK performance parameters are obtained, they should be rather insensitive to modest changes in machine design, anthropometric, or BEP information.

The critical issue associated with the determination of an optimal scenario is the selection of a physically meaningful optimization criteria. This problem is compounded with the need and desire to minimize time, fatigue, and machine complexity while maximizing throughput and efficiency. The next key issue centers on how to develop a systematic means for determining the sensitivity of performance relevant motion parameters of interest to design variables. Design variables are the system variables that the engineer alters during the design optimization process.

10 Test Data Mapping Capability

The ability to map test data requires that both JACK and NDISCOS man/machine interaction simulation models exist. Test procedures require that body motion at the operator/machine interface be recorded and that all body fixation points be defined such as feet on floor and torso restrained. Whole body motion should be recorded via video and at various body points for model verification purposes. As an alternative to testing with prototype hardware, test data can also be obtained by using real-time man-in-the-loop simulators. During the early phase of product development testing via low cost simulators will be the sole source of MMIDAP design information.

Recorded motion at the machine-operator interface is used as input to JACK. Performance parameters are adjusted until whole body predicted motion agrees with the whole body data recorded during the test. To map test data to other population groups, simply change anthropometric data in the JACKand NDISCOS models. If protective/restrictive apparel and other environmental factors that effect body loading, such as reduced gravity, are important these are simply added or deleted from the mathematical model. Human response predictions are then evaluated relative to information in the BEP database. Some adjustment of the JACK performance parameters may be necessary. If adjustment is deemed excessive, this should be used as a signal that the data mapping capability is breaking down and that data extrapolation limits are being encountered.

11 Validation of MMIDAP Capability

The MMIDAP project intends to complement evolving capability with an extensive testing and validation program. Data mapping to different operator population groups can easily be validated by simply using different operators. Mapping to operators with restrictive apparel can also be easily tested in the laboratory. Data mapping to different environments and different machine design alternatives can be validated in the Iowa Driving Simulator and other facilities such as those at the Human Performance Institute at the University of Texas at Arlington.

12 Human Factors Based System Design

There is a need to accurately predict machine operation timelines. The step, from conceptualizations that meet basic kinematic reach, collision avoidance, and other quasi-static mission feasibility constraints, to the step used for accurate operational timeline predictions is exceeding difficult. Carefully constructed laboratory experiments are needed to ascertain operator variables and their impact upon operational timeline estimates. There is a need for a simulation-based capability that will both support essential laboratory testing efforts and provide a means for extrapolating test results to the actual work environment. Particular attention must be paid to modeling *all* loads that are feedback to the operator directly through the manipulandum used for machine control or indirectly, since they are all used by the operator for cueing purposes. If simulator load replication cannot be judged as realistic from the operator's perspective, both test data and associated extrapolations are worthless. The difficulties associated with simulator replication of loads in hostile environments, such as on-orbit, should not be trivialized. Environmental factors, such as zero gravity, frequently result in counter intuitive dynamic action and reaction effects. These must be throughly understood and carefully integrated into both the simulator's capability and all associated training or testing systems and procedures.

Simulator validation from the perspective of the operator must be accomplished. This must focus on validating the ability to create a simulator that portrays a realistic environment from the perspective of the operator. Once simulator realism is achieved, the simulator can be used for general timeline estimation work and system design tradeoff studies. Laboratory testing and support analysis will be needed to continuously validate realism as simulator capabilities expand to support the characterization of complex system components, in hostile environments, with complex feedback loading conditions.

13 Human-Machine-Task Computer-Aided-Design (HMTCAD)

HMTCAD is an integrated set of software tools and resources designed to work together to execute analyses that address such human-machine-task performance questions as margin for success or failure, performance limiting factors, operator/machine reserve capacities, maximal task performance, and optimal procedures. HMTCAD tools are based upon performance analyses of both the human-to-machine and the human-to-task interfaces as shown in Figure 2. Future versions will include the machine-to-task interface and the ultimate capability will enable analysis of the complete human-to-machine-to-task system chain.[3]

General systems performance theory and the Elementary Resource Model developed at the University of Texas at Arlington's Human Performance Institute (HPI) serve as the primary conceptual basis for the HMTCAD package [10,12]. These relatively new concepts are combined with the advanced multi-body dynamic analysis capabilities of NDISCOS, with several human performance/biostructure databases, and with many specialized human

[3] The text contained within this section, with the permission of Photon Research Associates, is from the introduction to their HMTCAD Systems Specifications & Architecture document.

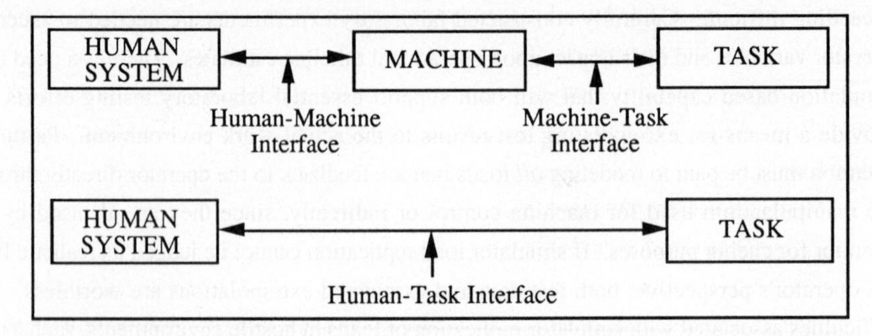

Figure 2. Key Components of the Generic Performance Analysis System that HMTCAD is being Designed to Analyze

performance analysis modules currently under development at the HPI. User access to this diverse range of capability is to be via a menu-driven point-click-and-drag user interface. The resultant HMTCAD software package consisting of an open ended architectural shell, associate computational infrastructure, interfaced databases, and analysis modules, is under development for the GSFC via a Small Business Research Initiative contract [57].

13.1 Purpose and Approach of HMTCAD

A graphical summary of the generic human-machine-task situation that arises during man-machine system modeling and simulation is shown above in Figure 2. The identified interfaces are important to recognize. Pilot studies have shown that failure to recognize these interfaces and to incorporate them into capability infrastructure sets the stage for terminology conflicts, confusion, and a weak theoretical foundation. The figure illustrates that analysis situations may include not only the complete "human-machine-task" situation, but also sub-situations. That is, situations that focus only on the "human-machine" interface, the "human-task" interface, or the "machine-task" interface. This figure represents the ultimate scope of HMTCAD capability. HMTCAD's initial product will focus on the development of an open-ended software tool for the human-task interface.

13.1.1 Performance Models/Analyses vs Biomechanical Models/Analyses

While many types of analyses are possible, HMTCAD focuses on *performance analyses* at each of the interfaces identified in Figure 2. Performance analyses are different from traditional kinematic or dynamic analyses commonly performed in biomechanics.

Nevertheless, an important class of performance analyses depend upon traditional methods to provide the intermediate numerical data needed to drive actual performance analysis.

Two primary classes of models, and corresponding analyses, can be found within the general field of biomechanics. For lack of established distinguishing terminology, these are termed here as:

1. biomechanical systems models
2. system performance models

In very general and perhaps obvious terms, a biomechanical system model is a simplified representation of a particular biological system whose *mechanical behavior* is of interest. A system performance model is a simplified, and different, representation of a system whose *performance* is of interest. Mechanical behavior typically entails static and dynamic analyses involving the basic physical concepts of mass, inertia, stiffness, position, velocity, and acceleration. In contrast, system performance pertains to all qualities that characterize the capacity of the system to execute its function(s). The quantification of these capacities is an essential component in the process of establishing system performance models. Some qualities, such as maximum speed and range-of-motion from position extremes, can be derived directly from primary biomechanical variables. Others, such as strength and smoothness of movement, tend to be more abstract but are still derived using parameterizations of basic variables. The description of mechanical behavior is often required as the first step in establishing quantifiable measures for many human performance capacities.

Considering the human biological system of interest as a whole, a systems performance model for a given subsystem recognizes dependencies on components external to the biomechanical domain such as vision and neuromotor control. In contrast, biomechanical models, almost by definition, typically focus on structural and mechanical properties, principles of materials, mechanical behavior, and dynamics. To address questions regarding performance, classical biomechanical models are frequently required to link the performance models of multiple subsystems together. Thus, while certain classes of questions regarding biomechanical behavior can be addressed without consideration of systems performance, the converse is not true.

Both types of models are necessary to investigate real physical systems. For example, consider models needed to address the following mechanical behavior vs. performance behavior questions:

1. Assess the *mechanical behavior* of a single force-transmitting anatomical structure, such as the popliteus muscle, as a function of knee configuration, motion, constraints, load, fiber type, fiber length, fiber pennation angle, subject, previous use, and external environment.

Assess the *performance capacity* of that muscle system, in concert with its supporting structures, to meet strength, speed, and endurance requirements in a specific athletic endeavor.

2. Assess the load-carrying capability and joint load distribution in all relevant anatomical structures in the neighborhood of a body joint such as the knee complex during a specific athletic endeavor such as running.

   Assess the ability of a given subject's subsystems, such as flexors and extensors, to meet worst-case strength, range, and speed demands of a specific athletic endeavor without the need to know--or at times the ability to describe--the biomechanical behavior of individual anatomical structures.

3. From a biomechanical linkage system model assess the force distribution in muscles, tendons, and ligaments in the arms and shoulder complex of a factory worker performing an assembly line task.

   From a performance model that includes a biomechanical linkage system model, assess the capacity of the factory worker's arm and shoulder system to achieve required speed, positioning, accuracy, and force at the hand based only on knowledge of performance capacities of shoulder and upper extremity flexors, extensors and similar subsystems associated with the involved joints.

While biomechanical models enjoy a strong theoretical rooting in disciplines with long histories, the concept of systems performance models and performance analysis are relatively new. Classical system theory [x] contains some relevant principles, but it does not directly address the distinct issue of *system performance*. The nature of problems addressed with these models, as well as the nature of their interrelationships, sometimes results in a blurring of the important distinctions raised herein.

13.2 Implementing Systems Performance Analysis at the Human-Task Interface

It is noted that the interface of two systems, such as a human and a workstation, is more appropriately modeled for the purpose at hand as the interface of the *system* (i.e., the human) to the *task* (i.e., the operation of the workstation). Performance analyses are simple in concept and yet powerful as tools. Using the general systems performance theoretical constructs presented in References 10 and 13, systems performance is modeled in terms of available performance resources. Quantitative task characterization is achieved in terms of performance resource demands. The ultimate step in the analysis involves comparison of resources available R_A to resource demands R_D for each and every unique performance resource. Direct comparison enables the analyst to obtain the quantifiable data needed to determine performance resource sufficiency and/or inadequacy. The data provides sufficient information for quantifiable "Y"or "N" answers, estimates of performance reserve capacity, and estimates of performance limiting factors.

The complexity of *executing* non-trivial systems performance analysis arises from the need to obtain numerical values for available performance resources and performance resource demands. Available performance resources can be determined from human performance measurements, databases thereof, and models that allow for their estimation. Resource availability for one or more performance resources associated with a given basic functional unit (e.g., a knee extensor) can frequently be estimated from measures of one or more other performance resources associated with the same functional unit. Note that resource demand values are also needed at the basic subsystem level, e.g., knee flexor, elbow extensior. This is the so-called "elemental level" of the Elementary Resource Model. To obtain numerical values for resource demands, the high level task must be defined in terms of function, goals(s), and possibly procedure. Then using knowledge of human structure, quasi-static and dynamic analyses can be developed that will support the mapping of generic task classes to performance resource requirements.

A particular class of dynamic analysis called "linked multi-body analysis" is required for human systems. Such analysis permits the influence of the size, shape, mass of body segments, and the physics of motion to be taken into account during the mapping process. As a consequence of this need for biomechanic support analysis, HMTCAD requires structural data in addition to human performance data. Data can be obtained for general population groups or by direct measurement of an individual. Estimations for so-called "missing data" are derived from models that utilize easily obtainable data such as height, sex, and weight. These are interfaced to relevant databases that have been compiled from data freely available in the open biomedical community literature.

Since the human system is highly redundant, there are different ways to accomplish a specific goal. This can be thought of as different ways to utilize the system's available performance resources. Whenever such flexibility exists, the thought arises regarding whether one task completion scenario is "better" than another. If so, this implies that a process of optimization must exist, i.e., *what is the optimal way to combine available resources to achieve a specific goal.* Resolution of the optimization issue introduces significant complexity into the problem of mapping a specified task goal into the estimation of resource demands at the elemental level. To complete the human performance analysis these estimates of resource requirements must be compared to measures and estimates of resources available. It is not sufficient to just obtain estimates for any resource demand profile that corresponds to "accomplishing the task goal." Rather, it is important to be able to obtain estimates of the resource demand profile (out of a family of possible alternatives) which corresponds to the optimal utilization of available performance resources.

Clearly, the full detail of performance analysis when applied to the human-task interface is complex. The goal of HMTCAD is to simplify this process. Intelligent assistant system aids are to be provided to guide non-specialist users who need quantifiable results from a human performance analysis. This system will prompt users to supply, select, and manage the vast amounts of data needed to characterize both the human system and the task. It will also attempt to unburden the user from the need to determine how to specify complex intermediate calculations associated with dynamic analysis as well as the interpolation of intermediate results.

13.3 Implementing Systems Performance Analysis at the System-Task Interface

The approach used here is similar to that employed for the Human-Task interface. General system performance theory is used to model "the system" in terms of its performance resources. The key difference is that, compared to the human system which has a fixed and stable architecture over time, artificial systems are quite varied and are likely to change quite rapidly during their useful life as part of either the basic design process or field utilization. Thus, HMTCAD must eventually evolve a capability for easily defining an architecture of subsystems and interconnections, a performance model for any generic mechanical system, and for importing the required performance model after it is created with another software tool. The development of this HMTCAD component is a long-term objective.

13.4 Integrating Systems Performance Analysis at "Human-Machine (Task)" and System-Task Interfaces

A complete performance analysis of many situations encountered involves a human operating a machine which in-turn is executing a task. A classic example is telerobotics. This integrated analysis capability can be achieved by combining performance models for the human-task interface and the system-task interface. Performance analyses, including those requiring optimization, in given task situations must be performed across the performance resource pools associated both the human and the artificial system. Task sharing between the two components can be modeled using resource substitution principles of general systems performance theory. While crude prototypes of such analyses are envisioned to be available over the next two years, the turn-key type of analysis with this capability is one of the longest term objectives of the HMTCAD project.

13.5 Further Evolution

Once analysis of the various interfaces discussed above is implemented, it would be natural to envision modeling situations which involve more than one human and/or artificial system working completely or partially together to accomplish tasks.

14 Summary

The Man/Machine Interaction Dynamics and Performance (MMIDAP) project, lead by NASA's Goddard Space Flight Center (GSFC), seeks to create an ability to study the consequences of machine design alternatives relative to the performance of both the machine and its operator. The envisioned MMIDAP capability is to be used for mechanical system design, human performance assessment, extrapolation of man/machine interaction test data, biomedical engineering, and soft prototyping within a Concurrent Engineering system. This chapter has reviewed the existing methodologies and techniques needed to create such capability. It has attempted to outline ongoing efforts to integrate both human performance and musculoskeletal databases with the host of analysis capabilities necessary for the early design analysis of dynamic actions, reactions, and performance assessment of coupled machine-operator systems. The multibody system dynamics software program NDISCOS of GSFC and Cambridge Research can be used for machine and musculoskeletal dynamics modeling. The program JACK from the University of Pennsylvania can be used for estimating and animating whole body human response to given loading situations and motion constraints. The basic elements of performance (BEP) task decomposition methodologies associated with the University of Texas at Arlington's Human Performance Institute's BEP database can be used for human performance assessment. Techniques for resolving the statically indeterminant muscular load sharing problems can be used for a detailed understanding of potential musculotendon or ligamentous fatigue, pain, discomfort, and trauma problems. Real time man in the loop simulators at The University of Iowa can be used to obtain real human performance data in a realistic, yet simulated, environment. The MMIDAP problem as defined herein highlights the conflicting needs and views of groups that focus on machine design and groups that focus on human performance and cumulative injury potential. An attempt has been made to show that there is a critical need to integrate design and simulation tools and to establish multidisciplinary lines of communication. Futhermore an outline is provided of planned integration efforts for human performance analyses, associated databases, and mechanical system design capabilities. This integration effort is expected to provide the Concurrent Engineering community with an ability to perform the early system trade studies needed to assess man/machine interaction dynamics and performance.

References

1. Boy, G., "Intelligent Assistant Systems" Knowledge-Based Systems Volume 6, Academic Press, 1991
2. Kroemer, K.H.E., et al. (Editors), "Ergonomic Models of Anthropometry, Human Biomechanics and Operator Equipment Interfaces," Proceedings, Workshop on Integrated Ergonomic Modeling, 1988
3. "Proceedings of the AFHRL Workshop on Human-Centered Design Technology for Maintainability," Air Force Human Resources Laboratory, Wright Patterson Air Force Base, Dayton Ohio, Sept 12-13, 1990
4. "Electronic Imaging," Report of the Board of Regents, National Library of Medicine Long Range Plan, NIH Publication Number 90-2197, U.S. Department of Health and Human Services, April 1990
5. Whitlock, D. and Spitzer, V., "Video 3D Atlas of the Human Knee in Cross Sections," From the Departments of Cellular and Structural Biology and Radiology at The University of Colorado School of Medicine, Denver CO., The C.V. Mosby Company, 1990
6. Winters, J.M. and Woo, S.L. (Editors), "Multiple Muscle Systems Biomechanics and Movement Organization," Springer Verlag, 1990
7. Yamaguchi, G.T., Sawa, A.G.U., Moram, D.W., Fessler, M.J., and Winters, J.M., "A Survey of Human Musculotendon Actuator Parameters," Appendix of [49]
8. Seireg, A. and Arvikar, R., "Biomechanical Analysis of the Musculoskeletal Structure for Medicine and Sports," Hemisphere Publishing Corporation, 1989
9. Kondraske, G.V., The HPI Shorthand Notation for Human System parameters," University of Texas at Arlington, Human Performance Institute, HPI Technical report TR 92-001R V1.0, 1992
10. Kondraske, G.V. "Quantitative Measurement and Assessment of Performance," Chapter 6 of "Rehabilitation Engineering", Smith R.V. and Leslie J.H. (Editors), CRC Press, 1990
11. Fitts, P.M., "The Information Capacity of the Human Motor System in Controlling the Amplitude of Movement," Journal of Experimental Psychology, Vol 47, No 6, pp 381-391, 1954
12. Kondraske, G.V., et al., "Measuring Human Performance: Concepts, Methods, and Applications," SOMA: Engineering for the Human Body (ASME), pp 6-13, January 1988
13. Kondraske, G.V., "Human Performance: Science or Art?," 13th Northeast Bioengineering Conference, Philadelphia PA Proceedings, pp 44-47, 1987
14. Schiehlen, W. (Editor), "Multibody Systems Handbook," Springer Verlag, 1990
15. Bae, D.S. and Haug, E.J., "A Recursive Formulation for Constrained Mechanical Systems, Part 1 - Open Loop," Mechanics of Structures and Machines, Vol 15, No 4, 1987
16. Bae, D.S. and Haug, E.J., "A Recursive Formulation for Constrained Mechanical Systems, Part 2 - Closed Loop," Mechanics of Structures and Machines, Vol 15, No 4, 1987
17. Bae, D.S., Haug, E.J., and Kuhl, J.G., "A Recursive Formulation for Constrained Mechanical Systems, Part 3 - Parallel Processor," Mechanics of Structures and Machines, Vol 16, No 2, 1988
18. Chun, H.M, Turner, J.D. and Frisch, H.P., "Order (N) DISCOS for Multibody Systems with Gear Reduction," Guidance and Control Conference, Portland Oregon, Aug 20-22, 1990
19. Hwang, R.S., and Haug, E.J., "A Recursive Multibody Dynamics Formulation for Parallel Computation," Technical Report R-13, The University of Iowa, Center for Simulation and Design Optimization of Mechanical Systems, 1988
20. Kim, S.S., et al., "New General Purpose Dynamics Simulation Code (NGDC)," The University of Iowa, Center for Simulation and Design Optimization of Mechanical Systems, 1990
21. Tsai, F.F. and Haug, E.J., "Automated Methods for High Speed Simulation of Multibody Dynamic Systems," Technical Report R-39, The University of Iowa, Center for Simulation and Design Optimization of Mechanical Systems, 1989
22. Chun, H.M, Turner, J.D. and Frisch, H.P., "A Recursive Order (N) Formulation for DISCOS with Topological Loops and Intermittent Surface Contact," AAS/AIAA Astrodynamics Specialists Conference, Durango, Colorado, Aug 19-22, 1991
23. Norkin, C.C and Levangie, P.A, "Joint Structure & Function, A Comprehensive Analysis," F.A. Davis Company, 1983
24. Kuffler, W.W., Nicholls, J.G., and Martin, A.R., "From Neuron to Brain: A Cellular Approach to the Function of the Nervous System," Second Edition, Sinauer Associates Inc., 1984
25. Shepherd, G.M., "Neurobiology," Oxford University Press, 1988
26. Bridgeman, B., "The Biology of Behavior and Mind," John Wiley \& Sons, 1988
27. Hatze, H., "Myocybernetic Control Models of Skeletal Muscle, Characteristics and Applications," University of South Africa, Muckleneuk, Pretoria, 1981
28. Hatze, H., "The Charge-Transfer Model of Myofilamentary Interaction: Prediction of Force Enhancement and Related Myodynamic Phenomena," pp 46-56 of \cite{kn:WINTERS-90}, 1990
29. Hoy, M.G., Zajac, F.E., and Gordon, M.E., "A Musculoskeletal Model of the Human Lower Extremety: The Effect of Muscle, Tendon, and Moment Arm on the Moment-Angle Relationship of Musculotendon Actuators at the Hip, Knee, and Ankle," Journal of Biomechanics, Vol 23, pp 157-169, 1990

30. Ma S.P. and Zahalak G.I., "A Distribution-Moment Model of Energetics in Skeletal Muscle," Journal of Biomechanics, Vol 24, pp 21-35, 1991
31. Zahalak, G.I., "A Distribution-moment Approximation for Kinetic Theories of Muscular Contraction," Math. Biosci. Vol 55, pp 89-114, 1981
32. Zahalak, G.I., "A Comparison of the Mechanical-Behavior of the Cat Soleus Muscle with a Distribution-Moment Model," Journal of Biomechanical Engineering, Vol 108, pp 131-140, 1986
33. Zahalak, G.I. and Ma, S.P., "Muscle Activation and Contraction: Constitutive Relations Based Directly on Cross-Bridge Kinetics," Journal of Biomechanical Engineering, Vol 112, pp 52-62, 1990
34. Zajac, F.E., Topp, E.L., and Stevenson P.J., "A dimensionless Musculotendon Model," IEEE/8th Annual Conference of the Engineering in Medicine and Biology Society, pp 601-604, 1986
35. Zajac, F.E., Topp, E.L., and Stevenson P.J., "Musculotendon Actuator Models for Use in Computer Studies and Design of Neuromuscular Stimulation Systems," RESNA 9th Annual Conference on Rehabilitation Engineering, Minneapolis Minnesota, pp 442-444, 1986
36. Kearney, R.E. and Hunter I.W., "System Identification of Human Joint Dynamics," Critical Reviews in Biomedical engineering, Vol 18, pp 55-87, 1990
37. Hatze, H., "A Comprehensive Model for Human Motion Simulation and its Application to the Take-off Phase of the Long Jump," Journal of Biomechanics, Vol 14, pp 135-142, 1981
38. Morris, G.R., Douglas, A.S. and Guoan, L. "A Comparison of Two Muscle Models for Simulating Human Saccadic Eye Motion," Final Report, 1990. Copy available from H.P. Frisch, Code 714.1 NASA/GSFC, Greenbelt, MD 20771
39. Chizeck, H.J., et al., "Control of Functional Neuromuscular Stimulation Systems for Standing and Locomotion in Paraplegics," Proceedings of the IEEE, Vol 79, pp 1155-1165, Sept 1988
40. Sejersted, O.M. and Westgaard, R.H., "Occupational Muscle Pain and Injury; Scientific Challenge," editorial, pp 271-274, special issue on Occupational Muscle Pain and Injury, European Journal of Applied Physiology, Vol 57, pp 271-372, 1988
41. Vollestad, N.K. and Sejersted, O.M. "Biochemical Correlates of Fatigue, A Brief Review," European Journal of Applied Physiology, Vol 57, pp 322-326, 1988
42. Zajac, F.E. and Gordon, M.E., "Determining Muscle's Force and Action in Multi-Articular Movement," Exercise and Sports Science Reviews, Vol 17, pp 187-230, 1989
43. Williams, R.J. and Seireg, A., "Interactive Computer Modeling of the Musculoskeletal System," IEEE Trans on Biomedical Engineering, BME-24, pp 213-219, 1977
44. Williams, R.J. and Seireg, A., "Interactive Modeling and Analysis of Open or Closed Loop Dynamic Systems with Redundant Actuators," Journal of Mechanical Design, Vol 101, pp 407- 416, 1979
45. Delp, S.L., Loan, J.P., Hoy, M.G., Zajac, F.E., Topp, E.L., and Rosen, J.M., "An Interactive Graphics-Based Model of the Lower Extremity to Study Orthopaedic Surgical Procedures," IEEE Trans. on Biomedical Engineering, Vol 37, No 8, pp 757-767, 1990
46. Hatze, H., "A Method for Describing the Motion of Biological Systems," Journal of Biomechanics, Vol 9, pp 101-104, 1976
47. Badler, N.L., "Human Factors Simulation Research at the University of Pennsylvania," Proceedings of the AFHRL Workshop on Human-Centered Design Technology for Maintainability," Air Force Human Resources Laboratory, Wright Patterson Air Force Base, Dayton Ohio, Sept 12-13, 1990
48. Badler, N.I., "Articulated Figure Positioning by Multiple Constraints," IEEE Journal on Computer Graphics and Application, pp 26-37, 1987
49. Lee, P., Wei, S., Zhao, J., and Badler, N.I., "Strength Guided Motion," Computer Graphics, Vol 24, pp 253-262, 1990
50. Wilhelms, J., "Using Dynamic Analysis for Realistic Animation of Articulated Bodies," IEEE Journal on Computer Graphics and Application, pp 12-27, 1987
51. Badler N.L., Lee, P., Phillips, C., and Otani, E.M., "The JACK Interactive Human Model," Proceedings of the First Annual Symposium on Mechanical System Design in a Concurrent Engineering Environment," The University of Iowa, Iowa City, Iowa, pp 179-198, 1989
52. Badler, N.L., Barsky, B.A., and Zeltzer, D., "Making Them Move, Mechanics, Control, Animation of Articulated Figures," Morgan Kaufmann Publishers, Inc. 1990
53. Chang, J.L., Kim, S.S., and Haug, E.J., "Real-Time Operator in the Loop Simulation of Multibody Systems," Technical Report R-72, The University of Iowa, Center for Simulation and Design Optimization of Mechanical Systems, 1990
54. Stoner, J.W., et al., "Introduction to the Iowa Driving Simulator and Simulation Research Program," University of Iowa, Center for Simulation and Design Optimization of Mechanical Systems, Technical Report R-86
55. Gunby, P., "Simulator Designed for Advanced Traffic Research," JAMA (Journal of the American Medical Association), Contempo Issue, Vol 268, No 3, pp 303, July 15, 1992

56. Haug, E.J., et al., "Feasibility Study and Conceptual Design of a National Advanced Driving Simulator," US Department of Transportation, DOT HS-807-596, 1990
57. Turner, J.D., "Integrated Ergonomic System Software Development," SBIR Final Report. Copy available from J.D. Turner, Photon Research Associates, Cambridge Research Division, 1033 Mass. Ave., Cambridge MA 02133, phone (617) 354-1522

Simulated Humans, Graphical Behaviors, and Animated Agents

Norman I. Badler

Department of Computer and Information Science, University of Pennsylvania, Philadelphia, PA 19104-6389, USA

Abstract: A variety of issues involved in visualizing human task behavior will be examined, focusing on the broad, yet vertically-integrated, effort at the University of Pennsylvania. Computer graphics visualization of the appearance, capabilities and performance of humans is a challenging task. From modeling reasonable body size and shape, through control of the highly redundant body linkage, to simulation of plausible motions, human figures offer numerous computational problems and constraints. Our research has produced a system, called *Jack*, for the definition, manipulation, animation, and human factors performance analysis of simulated human figures. Human motion can be visualized by interactive specification and simultaneous execution of multiple constraints. Enhanced control is provided by natural behaviors such as looking, reaching, balancing, lifting, stepping, walking, grasping, and so on. As an alternative to interactive specification, a simulation system allows a convenient temporal and spatial parallel "programming language" for behaviors. At an even higher level, we have been exploring the possibility of using Natural Language instructions, such as are found in assembly or maintenance manuals, to drive the behavior of our animated human agents.

Keywords: human figure models / human factors / animation / computer graphics / interactive systems

1 Introduction

Present technology lets us approach human appearance and motion through computer graphics modeling and three-dimensional animation. By properly delimiting the scope and application of human models, we can move forward, not to replace humans, but to substitute adequate computational surrogates in various situations otherwise unsafe, impossible, or too expensive for the real thing.

Our goals are to build computational models of human-like figures which, though they may not trick our senses into believing they are alive, nonetheless manifest animacy and convincing behavior. Toward this end, we:

- Create an interactive computer graphics human model.
- Endow it with reasonable biomechanical properties.
- Provide it with "human-like" behaviors.
- Use this simulated figure as an agent to effect changes in its world.
- Describe and guide its tasks through natural language instructions.

There are presently no perfect solutions to any of these problems, but significant advances have enabled the consideration of the suite of goals under uniform and consistent assumptions. Ultimately, we should be able to give our surrogate human directions that, in conjunction with suitable symbolic reasoning processes, make it appear to behave in a natural, appropriate, and intelligent fashion. Compromises will be essential, due to limits in computation, throughput of display hardware, and demands of real-time interaction, but our algorithms aim to balance the physical device constraints with carefully crafted models, general solutions, and thoughtful organization.

Our study will tend to focus on one particularly well-motivated application for human models: human factors analysis. Visualizing the appearance, capabilities and performance of humans is an important and demanding application. From modeling realistic, or at least reasonable, body size and shape through the control of the highly redundant body skeleton to the simulation of plausible motions, human figures offer numerous computational problems and constraints. Building software for human factors applications serves a widespread, non-animator user population. Our software design has tried to take into account a wide variety of engineering design-oriented tasks, rather than just offer a computer graphics and animation tool for the already skilled or computer-sophisticated animator.

The Computer Graphics Research Lab at the University of Pennsylvania has been involved in the research, design, and implementation of computer graphics human figure manipulation software since the late 1970s. The history of this effort is too lengthy to detail here; rather, we wish to describe the current state of our system, called *Jack*. The remainder of this paper discusses the major software features, organized around the topics of body and other geometric object structures, anthropometry, user interface, positioning, animation, analyses, rendering, virtual control, and animation from instructions.

2 Summary of *Jack* Features

The *Jack* software is built on Silicon Graphics workstations because those systems have the 3-D graphics features that greatly aid the process of interacting with highly articulated figures such as the human body. Of course, graphics capabilities themselves do not make a usable system. Our research has therefore focused on software to make the manipulation of a simulated human figure easy for a rather specific user population: human factors design engineers or ergonomics analysts involved in visualizing and assessing human motor performance, fit, reach, view, and other physical tasks in a workplace environment. Our program design has tried to take into account a wide variety of physical problem-oriented tasks, rather than just offer a computer graphics and animation tool for the already computer-sophisticated or skilled animator.

As we continue to interact with human factors specialists, particularly our research sponsors, we have come to appreciate the broad range of problems they must address. The challenge to embed a reasonable set of capabilities in an integrated system has provided dramatic incentives to study issues and solutions in 3-D interaction methodologies, multiple goal positioning, collision avoidance motion planning, locomotion, and strength guided motion (Section 2.5).

2.1 Body and Other Geometric Object Structure

Bodies as well as all other geometric objects, called *figures*, are represented externally to *Jack* in a language (*Peabody*) which describes their attributes and topological connections [1]. Figures consist of segments connected by joints, each with various degrees of freedom and joint limits. Important points are termed *sites* and are used, for example, to describe the attachment locations of joints or the positions of notable landmarks. Geometric *constraints* are used to position figures in the world coordinate reference frame.

The surface geometry associated with a segment has its own local coordinate system and is typically described as a network of polygons called *psurfs*. *Jack* is not intended to be, or substitute for, a "real" Computer-Aided Design system; we often obtain geometric data from other commercial CAD systems.

The default human figure in *Jack* consists of 39 segments, 38 joints, and 88 degrees of freedom, including a full 17 vertebra spine. Fully articulated hands add 30 more segments, 30 more joints, and 33 more degrees of freedom.

The body structure is not built into *Jack*; rather, the default body is there for user convenience. Any topological structure can be defined through *Peabody* and manipulated in *Jack*. In particular, this allows the use of figure models with greater or lesser articulation, as well as mechanisms, robots, insects, and so on.

The human shoulder has a complex joint structure. In *Jack* we model the shoulder accurately as a clavicle and shoulder set whose state is dependent on the position and orientation of the upper arm [2]. The shoulder joint center therefore moves in a biomechanically reasonable fashion. Spherical joint limits add to the motion realism.

Jack contains a hand model with fully articulated and joint-limited fingers and thumb. The more interesting feature, however, is an automatic grip. Given a geometric object that is to be grasped, the user can specify one of three types of grips -- power, precision, or disk [3] -- and *Jack* will move the hand to the object then move the fingers and hand into a reasonable grip position. The actual grip is completed by using real-time collision detection on the object's geometry to determine when finger motion should cease.

2.1.1 Independent Surface Geometry Per Segment

For interactive manipulation, detailed human figure surface geometry is usually unnecessary, however, the *psurfs* associated with each segment may be as simple or complex as desired. The default human model has a rather polyhedral appearance to keep the number of polygons low for graphical display update efficiency. The more accurate figures (the contour bodies, Section 2.1.2) may have hundreds of polygons per segment to give a smoother and more rounded appearance. The selections can be mixed from segment to segment: for example, a smoother head and face model with a default body.

Clothing a figure is important for ergonomic analyses since clothing often affects mobility and joint limits. *Jack* presently contains three types of clothing:

1. A rather simple kind which is simply a color differentiation for various segments (e.g. brown legs and lower torso yield "pants", blue upper torso and arms, a long-sleeved "shirt", etc.);
2. A more realistic "thick" clothing, which is the actual expansion of the segment geometry (hence its diameter) relative to the segment axis while still preserving the overall shape;
3. Additional equipment (such as helmets, tool belts, pockets, air supplies, etc.) attached or worn by simply adding appropriate geometric models to segments.

All three improve graphics appearance. The second and third approach are the most noteworthy since thick clothing and equipment should affect joint limits. The attachment of loose fitting or draped clothing is another matter entirely and is not addressed here.

2.1.2 Biostereometric Contour Bodies

One of the most interesting body databases in *Jack* is derived from biostereometric (photographically) scanned body surface data of 76 subjects. Originally supplied by Kathleen Robinette of the U.S. Air Force Armstrong Aerospace Medical Research Laboratory, the data consists of approximately 6000 data points for each subject, organized by body segment and arranged in parallel slices. We determined reasonable joint centers from the segment contours and surface landmark data, converted the segment topology into *Peabody* structures, and tiled the contours into polyhedral meshes [4]. The single torso segment in the original scanned body data was rigid. In the next section, we describe how we dramatically improved that situation.

2.1.3 Seventeen Segment Flexible Torso with Vertebral Limits

The lack of accurate flexibility in the torso is a notable weakness of most anthropometric models. Even the biostereometric bodies suffered from torso rigidity. We have constructed a 17 segment vertebral column (from lumbar to thoracic) whose movements are dictated by kinematic limits and some simple parameters [5]. The torso, in turn, is broken into 17 corresponding thick slices, one for each vertebra. With this arrangement, it is easy to have the contour body bend and "breathe" in a very realistic fashion. Movements of the torso are basically described by lateral, saggital, and axial rotations at the neck.

2.1.4 Facial Model

Humans have faces, and *Jack* provides a mechanism for presenting a face on a human figure. A photograph of a [real] face may be texture mapped onto a head *psurf*. The figure bears a close resemblance to a real person and the resulting image looks reasonable even when rotated. On the SGI VGX workstation, the texture appears on the surface in real-time, permitting interactive manipulation and animation of a recognizable figure. The only disadvantage is the rather delicate (for correct) positioning of the texture on the head. The facial features may also be *animated* [6]. While not important (perhaps) for human factors work, the expressions certainly enliven finished animations.

2.2 Anthropometry

Having a body model is one thing; being able to easily make it correspond to human size variation is another. Anthropometric scaling of body models is an important component of *Jack* [7,8]. *Jack* allows the manipulation and display of as many figures as desired up to the memory limits of the hardware. There are no restrictions whatsoever on the geometry, topology, or anthropometry used across the several figures.

2.2.1 Segment and Joint Attributes

The human figures used in *Jack* have various attributes associated with them that are used during manipulation and task analysis. The current set includes segment dimensions, joint limits, moment of inertia, mass, center of mass, and joint strength [7,8]. Raw anthropometric measurements (e.g. for specific landmarks or composite measurements such as "sitting height") can also be associated with an individual in the database.

The strength data may be based on tabular (empirical) data or strength prediction formulas [9]. Strength parameters may be either scaling (e.g. gender, handedness) or non-scaling (e.g. depending on the population). In any case, the user may alter the stored data or formulas to conform to whatever source or population model is desired.

2.2.2 Population Percentiles or Individuals

Either population statistics may be used to provide percentile data, or else an actual database of [real] individuals may be used. The former, e.g., is common in U.S. Army analyses, while the latter is often used by NASA for the specific individuals in the astronaut trainee pool.

The interface to the anthropometry database is through *SASS*: the Spreadsheet Anthropometric Scaling System [8]. As part of *Jack*, it offers flexible access to all the body attributes and a simple mechanism for changes. By making body segment *groups* such as the leg or the torso, dependencies between groups (such as *stature* = *head* + *neck* + *torso* + *leg*) can be defined and used in a spreadsheet-like fashion during body scaling. *Peabody* model files are created by *SASS* and made available to *Jack*. Alternatively, one can interactively manipulate the current body in *SASS* while displaying it in *Jack* to rapidly try out the effect of varying the individual, population percentile, gender, joint limits, etc.

2.2.3 Concurrent Display of Interactively Selected Dimensions

As noted above, a human figure may be modified by *SASS* while it is being displayed in *Jack*. In fact, the process is much more powerful than just updating a display. In Section 2.4, we will see that a figure may be subject to arbitrary goals for one or more of its joints. These goals are maintained (subject to joint limits and body integrity) during interactive manipulation. The process also applies to segment attribute changes done interactively in *SASS*: as the segment lengths change, e.g., the body will move to maintain the required position, orientation, or viewing constraints. It is therefore very easy to assess posture and viewing changes (as well as success or failure) across population percentiles or gender.

2.3 User Interface

One of the the most attractive features of *Jack* is the natural user interface into the three-dimensional world [1,10]. A significant part of the interface is offered by the hardware capabilities of the Silicon Graphics workstation upon which *Jack* is built. The software, however, makes this hardware power controllable.

Jack relies solely on the standard three button mouse and keyboard for interaction. The mouse is used to perform direct manipulation on the 3-D scene, e.g. selecting objects by picking their images, translating objects by holding down one or two mouse buttons corresponding to spatial coordinates, etc. The mouse is also actively used to negotiate through the pop-up command menus.

The keyboard is used for occasional command entry. The escape and control keys are used as meta-mouse buttons, e.g. to change the button interpretations from translation to rotation, or the affected coordinate frame from global to local.

2.3.1 Natural 3-D Interactive Interface

The naturalness of the interface arises from the coherence of user hand motions with the mouse and the correspondence between mouse cursor motion on the 2-D screen, a 3-D cursor (looking like a "jack") in the world, and 3-D objects displayed there. In particular, translations and rotations are selected with the mouse buttons (and perhaps a key), and the mouse motion is transformed into an appropriate 3-D cursor movement. Rotations display a wheel perpendicular to the selected axis; motion of the mouse cursor about the wheel display invokes a 3-D rotation about the actual axis. Any joint limits are respected.

Other motions that are easy to perform in *Jack* include real-time end-effector dragging (Section 2.4). The position and orientation of the end-effector is controlled by the same mouse and button interpretation method.

2.3.2 Multiple Windows

Jack supports multiple independent windows into the current environment. Thus one could be a global view, one could be a view from a figure's eye, another could be a view from a certain light source (to see what is being illuminated), etc. The camera and lights are represented as *psurfs* so that they may be positioned and observed just as any other object in the scene. Of course, as the camera is moved in one view, the corresponding camera view window shows the changing image. The same result obtains if a window view is attached to a figure's eye, hand, or any site for that matter.

2.4 Positioning

The manipulation power in *Jack* comes from novel real-time articulated figure positioning algorithms. These imbue the jointed figure with "behavioral fidelity"; that is, the ability to respond to varied positioning *goals* as well as to direct joint rotation.

Although joint angles may be manipulated directly to position a figure, spatial position or orientation goals are more convenient. These cartesian goals are transformed to body joint angles by a real-time inverse kinematics procedure based on nonlinear optimization (with linear constraints, such as joint limits) [11,12]. The solution moves the selected joint (end-effector) to the goal if it is reachable, otherwise it moves as close as feasible given the figure posture, the joint chain, and the joint limits. Any failure distance is reported numerically as well. This movement does not represent how a person would actually move, nor does it attempt to find the "best" or most "natural" position. It merely achieves goals. For better postures, additional goals can be created and maintained.

In practice, the constraints are organized into *behaviors* which may be applied to the figure [10,13]. Table 1 shows some of the behaviors currently implemented.

MANIPULATION PRIMITIVES	BEHAVIORAL PARAMETERS
Move Foot Move Center of Mass Bend Torso Rotate Pelvis Move Hand Move Head Move Eyes	Set Foot Behavior pivot hold global location hold local location keep heel on floor allow heel to rise Set Torso Behavior keep vertical hold global orientation Set Head Behavior fixate head fixate eyes Set Hand Behavior hands on hips hands on knees hold global location hold local location hand on site grab object
PASSIVE BEHAVIORS	
Balance Point Follows Feet Foot Orientation Follows Balance Line Pelvis Follows Feet Orientation Hands Maintain Consistent Orientation Root Through Center of Mass	
ACTIVE BEHAVIORS	
Take Step When Losing Balance Take Step When Pelvis is Twisted	

Table 1. Behavioral Controls

Since inverse kinematics is available, and since multiple goals may be active, *Jack* allows a joint to be moved interactively by attaching a position or orientation goal to the 3-D cursor controlled by the mouse [11]. The solution time is actually reduced because the current posture is likely to be close to the solution at the next input position, so the algorithm converges quickly. To avoid waiting for the solution, however, *Jack* updates the joint angles at every graphics window update by taking the solution obtained thus far. As the goal is moved or rotated, the posture changes as quickly as possible and "catches up" with the user whenever there is a significant pause in cursor motion.

One of the most interesting behaviors involves constraining the center of mass of a figure [10,13]. The center of mass is not a specific joint or point of the body, rather it is a computed quantity. Nonetheless, *Jack* permits it to be a participant in a goal. By constraining the center of mass to lie along a line goal above the figure's support polygon, a *balanced reach* may be effected. The motion is most dramatic when only one foot is constrained to the floor; moving a hand causes the other leg to lift off the floor for counterbalance when it is needed! This behavior allows us to interactively animate subtle weight shifting, single steps, and turning activities. In addition, since the computation of the center of mass will depend on any objects attached to the body (e.g. being held or worn), the motions automatically compensate for the new (static) distribution of weight.

2.5 Animation

The manipulations in *Jack* discussed so far are not really "animations" as we have already mentioned. For an animation, we expect some coherence and smoothness to the figure's motion; it is not enough to merely animate a numerical search no matter how clever or effective it is. *Jack* incorporates a number of mechanisms to produce human-like motion.

The ability to provide key animation parameters is extended in *Jack* to include behaviors, constraints and goals. Thus the beginning, duration, and end times of any of these may be specified. As the interpretation of the behaviors proceeds in temporal order, a body posture satisfying the current set of active constraints is produced for each frame time.

2.5.1 Strength Guided Motion

A rather more useful, but more restricted, animation method developed by Lee uses the inherent strength model stored for a human figure as the basis for computing certain types of motion. If the task involves moving a weight rather slowly to some goal position, then a *strength guided motion* algorithm computes a motion path based on the strength model and two additional parameters [14]. The parameters are the comfort level at which the motion should be performed and the allowed deviation from a straight-line path to the goal. Using a number of strategies based on the available torque at each joint in an arm (plus upper torso) joint chain, the algorithm computes an acceptable posture at every instant (e.g., 15 times a second) of the action. Strategies include moment reduction, pull back, adding joints, and recoil to bring comfort to acceptable levels. A useful range of lifting and reaching motions may be produced, including weight lifting, rising from a chair, pulling the body upward in a chin-up, and two-person coordinated lifting.

2.5.2 Collision Avoidance

Reaching goals with a highly articulated human figure is greatly complicated by the massive redundancy in the joint structure. To add collision avoidance is to confront the inherent exponential complexity of manipulating long and even branched joint chains. To manage this complexity, Ching [15] breaks the joint chains into groups of one, two or three degrees of freedom and then solves the goals sequentially. While a complete solution may theoretically require exponential time, in practice the search space heuristics find a solution much more efficiently. (In particular, most constrained reach goals are not so pathological that exact, worst-case solutions are in fact required.) This algorithm allows the figure to safely reach through small apertures, move feet to step over obstacles, or even climb a rock wall if the hand and foot holds are given in advance.

2.5.3 Posture Planning

Ching's collision avoidance algorithm still requires that the goal posture be known. This is often an untenable assumption if we are trying to assess whether or not some task can be performed at all. In this case we use a more qualitative approach called "posture planning" [16]. In this technique, two processes determine possible motions toward the desired goals while also avoiding (in a very conservative way) collisions with the workplace. The first process postulates motions of major body components (such as the torso elevation, the hip height, the palm orientation, etc.) that are helpful in achieving the goals. These motions will almost always be mutually dependent on one another (due to body joint space redundancy). So the second process uses "envisionment" to simulate these possible goal-directed actions to see that they advance the figure toward the goals. Moreover, this process also checks for potential collisions with the workplace and adjusts the postulated motions to avoid impacts. Once a successful set of motions is discovered, they can be assembled and re-played to show the figure's movement to the desired goals.

2.5.4 Locomotion

The figure in *Jack* can walk along a curved path [17]. The underlying gait is derived from biomechanics data. The curved path allows motion around obstacles and through doorways. The final position and orientation may be specified so that even a straight path will allow a different final facing direction. In addition, some of the primitive standing behaviors in *Jack* include taking a step or turning when the center of mass is moved or when the figure must regain balance. These minor stepping motions are an important part of the behavioral vocabulary of real people and have been ignored by animators.

2.6 Analyses

All the *Jack* features are available to compute certain aspects of some of the most commonly performed task analyses. We have already seen goal-directed reach. Other analyses include finding the reachable workspace, determining the eye view, showing static strength, torques, or force, showing the effect of population anthropometry on a particular situation, and demonstrating the results of an external task simulation.

2.6.1 Reachable Space

Jack can display a trace of any site; in particular, it can show the path of an end-effector as it is manipulated. The resulting trace gives a good idea of the reachable space as the end-effector is dragged about. Any joint chain can be used due to the general inverse kinematics solution. Other algorithms being studied by Alameldin can compute the reachable space boundary or volume off-line [18].

2.6.2 Eye View

We have already seen that *Jack* can show the view from any object, in particular, a figure's eye. Besides the normal perspective view, a simplified retinal projection window may be drawn. Objects in front of the eye are mapped into a (radius, angle) polar plot. When features such as foveal or peripheral areas are drawn in the retinal window, the relative visibility of scene features may be assessed. Much of the useful effort in this analysis mode was accomplished by a collaboration with The Lighthouse in New York and NASA Ames [19].

In addition to the retinal window, translucent view "cones" may be displayed from the eyes of a human figure. With the apex at the eye lens center, the shape of the cones follows any desired polygonal path, e.g. foveal area. By aiming the eyes with an interactive goal, the view cones follow the point of interest, converging or diverging as needed (subject to eye "joint" limits). Since the cones are translucent, workplace objects show through, giving the user a good impression of what can and cannot be seen by the subject.

2.6.3 Torque Load and Comfort

Interactive graphic displays of joint torque or end-effector forces may be shown in *Jack* as the user manipulates the figure [9]. Current as well as cumulative maximum forces or torques are displayed as moving bars in a *strength box* whose axes correspond to the joint's degrees of freedom. Individual, gender differentiated, and population percentile (e.g., 95th, 50th, and 5th) strengths may be compactly and comparatively displayed.

Torques along a joint chain may be shown, too. Given a force on an end-effector, *Jack* can compute and graphically display the reaction forces generated anywhere else in the body [9]. These torques may be graphically presented as color coded region on a contour body, instantly showing overloaded joints in red and safe loads in shades form white (no load) to blue (maximum load). In addition, a trace of the "safe" and "unsafe" regions (relative to the current strength model) is left in the display as the end-effector is moved about, producing a direct and real-time visualization of the accessible space.

Since the strength guided motion computes the instantaneous (static) joint torques in the current (changing) body posture, this information is available for display. During such actions, moving bar charts can show the level of comfort, physical work, or fatigue.

2.6.4 Interactive Body Sizing Under Active Constraints

We have already mentioned that changes to bodies made in *SASS* would maintain (as well as possible) any active constraints. Thus testing for the effectively reachable workplace over any population range is nearly trivial: e.g. constrain the feet or lower body, set the reach goal for the desired end-effector, and alter the percentile field of the appropriate *SASS* spreadsheet display. In another situation, suppose the eye is constrained to the design eye point of a cockpit, the hands and feet are positioned to appropriate goals, and the shoulders and hips are restrained by point goals representing a suitable restraint system. Then running through the percentiles with reach goals for the hands, feet, and hips will show how well or how poorly the population can carry out that task.

2.6.5 Simulation from an External System

The ultimate analysis tool is a simulation which executes some task and drives the human figure with a set of goals and timings. One of the principal issues involved in understanding and executing instructions is the form of the action planner. Classical planning strategies do not seem to suffice for human motion because people are highly redundant mechanisms and use flexible, incremental, and interruptible plan execution.

In one recent experiment we used a discrete event simulator running over a Knowledge Base (a system called *YAPS*: [20,21]). By building successively higher-level task actions from simple reach, attach, and release primitives, Levison was able to get *YAPS* to simulate the removal of a Fuel Control Valve from a hypothetical airplane [22]. A human performance rate predictor based on Fitts' Law (if appropriate) [23] was used to postulate reasonable task durations for reach and viewing actions [20].

The Fitts' Law formulation for task time performance is adequate for very simple reach and view tasks. For more generality, the strength model can be referenced to obtain estimates of minimum trajectory times. This approach, however, is limited to knowing the strength model and, moreover, does not adequately compute timings for more complex task units (e.g. inserting a bolt into a hole).

We are examining several task time databases to see how they might be incorporated into the simulator. These databases will be extremely useful for task analyses where nominal time-motion studies for such tasks have been extensively measured. In some cases, external task simulation software may provide the task sequence and timings, such as is done in the NASA Ames MIDAS helicopter simulation which uses *Jack* as the pilot mannequin. We have just begun another project to interface *Jack* with an Air Force Logistics planning and database system called DEPTH in order to study maintenance scenarios.

We have recently re-designed our in-house simulation system. It now behaves more like a continuous system with adaptive step sizes and, more importantly, it can interact with the larger set of *Jack* behaviors through an object-oriented, reactive paradigm. For example, the locomotion behavior may be controlled incrementally through the simulation so that the figure may pursue interactively changing targets while simultaneously avoiding obstacles in the direct path. The simulator is reactive in the sense that the collision-free paths do not need to be precomputed; rather, the next step is based on the local configuration of objects to be approached or avoided.

2.7 Rendering

Besides the hardware image synthesis available for polyhedral models on the Silicon Graphics workstation, the *Jack* system includes a sophisticated ray-tracer and a radiosity renderer. The ray-tracer does anti-aliasing, textures, specularity, translucency, reflections, shadows, multiple light sources, material properties, and chromatic aberrations. The radiosity renderer creates even more accurate diffuse lighting. The scene is automatically meshed to small *psurf* patches, then radiosity correctly portrays area light sources, soft shadows and color bleeding effects.

2.8 Virtual Humans

By using a number of 6 degree of freedom sensors from Ascension Technology Corp., motion of the entire human body may be controlled interactively. The sensors are placed on the wrists (or palms), the small of the back, and the forehead. The two hand sensors provide position and orientation information for the principal end effectors. The back sensor approximates the location of the center of mass and provides locomotion and stepping information as well as pelvis orientation. The forehead sensor supplies gaze direction. These inputs directly control the nearly full range of *Jack* behaviors. (Hand gestures are not sensed, but could be with readily available hand pose sensing gloves.) The result is a virtual human, controlled by a minimally encumbered operator.

2.9 Animation from Instructions

Controlling human motion tasks specified by language commands or instructions is a long-term goal of our research. Analysis of the form and content of instructions has begun in collaboration with Computer and Information Science department faculty members Bonnie Webber and Mark Steedman [24]. This part of the project is called *AnimNL*: Animation from Natural Language. By having our human figure understand Natural Language instructions, it becomes an agent able to carry out the tasks and intentions of the instructor.

In general, the idea is to process the language instructions into a plan graph of intended actions [25]. As the plan graph is incrementally elaborated to simulatable tasks, those actions are passed off to a context-dependent task--object motion manager. The manager determines the actual physical goals and approximate location for the agent. These are transferred to the incremental simulator which invokes necessary behaviors such as locomotion, posture planning, collision avoidance reach, balance, eye view directions, etc. Finally, *Jack* executes and animates the actions.

3 Conclusion

Even though *Jack* is under continual development, it has nonetheless already proved to be a substantial computational tool in analyzing human abilities in physical workplaces. It is being applied to actual problems involving space vehicle inhabitants, helicopter pilots, maintenance technicians, foot soldiers, and tractor drivers. This broad range of applications is precisely the target we intended to reach. The general capabilities embedded in *Jack* attempt to mirror certain aspects of human performance, rather than the specific requirements of the corresponding workplace.

We view the *Jack* system as the basis of a virtual animated agent that can carry out tasks and instructions in a simulated 3-D environment. While we have not yet fooled anyone into believing that the *Jack* figure is "real", its behaviors are becoming more reasonable and its repertoire of actions more extensive. When interactive control becomes more labor intensive than natural language instructional control, we will have reached a significant milestone toward an intelligent agent.

Acknowledgments: This research is or was partially supported by ARO Grant DAAL03-89-C-0031 including participation by the U.S. Army Human Engineering Laboratory, Natick Laboratory, TACOM, and NASA Ames Research Center; NASA Johnson Space Center through Lockheed Engineering and Sciences Company; U.S. Air Force DEPTH contract through General Dynamics Convair F33615-91-C-0001; MOCO Inc.; Siemens Research; Deere and Company; FMC Corp.; Martin-Marietta Denver Aerospace; National Science Foundation CISE Grant CDA88-22719 and their Instrumentation and Laboratory Improvement Program through Grant #USE-9152503; and the State of Pennsylvania Benjamin Franklin Partnership.

References

1. Phillips C, Badler N: Jack: A toolkit for manipulating articulated figures. ACM/SIGGRAPH Symposium on User Interface Software, Banff, Canada, pp. 221-229, 1988
2. Otani E: Software tools for dynamic and kinematic modeling of human motion. Master's thesis, Department of Mechanical Engineering, University of Pennsylvania, Technical Report, MS-CIS-89-43, Philadelphia, PA, 1989
3. Iberall T: The nature of human prehension: Three dextrous hands in one. IEEE Conference on Robotics and Automation, pp. 396-401, 1987
4. Fuchs H, Kedem Z, Uselton S: Optimal surface reconstruction from planar contours. Comm. of the ACM, Vol. 20, No. 10, pp. 693-702, 1977
5. Monheit G, Badler N: A kinematic model of the human spine and torso. IEEE Computer Graphics and Applications, Vol. 11, No. 2, 1991
6. Pelachaud C: Communication and coarticulation in facial animation. PhD Dissertation, Department of Computer and Information Science, University of Pennsylvania, 1991

7. Grosso M, Quach R, Otani E, Zhao J, Wei S, Ho P-H, Lu J, Badler N: Anthropometry for computer graphics human figures. Technical Report, MS-CIS-89-71, Dept. of Computer and Information Science, University of Pennsylvania, Philadelphia, PA, 1989
8. Grosso M, Quach R, Badler N: Anthropometry for computer animated human figures. In State-of-the Art in Computer Animation, N. Magnenat-Thalmann, D. Thalmann (eds.), Springer-Verlag, pp. 83-96, 1989
9. Wei S: Human strength database and multidimensional data display. PhD Dissertation, Department of Computer and Information Science, University of Pennsylvania, 1990
10. Phillips C: Interactive postural control of articulated geometric figures. PhD Dissertation, Department of Computer and Information Science, University of Pennsylvania, 1991
11. Phillips C, Zhao J, Badler N: Interactive real-time articulated figure manipulation using multiple kinematic constraints. Computer Graphics Vol. 24, No. 2, pp. 245-250, 1990
12. Zhao J, Badler N: Real time inverse kinematics with joint limits and spatial constraints. Technical Report MS-CIS-89-09, Department of Computer and Information Science, University of Pennsylvania, Philadelphia, PA, 1989
13. Phillips C, Badler, N: Interactive behaviors for bipedal articulated figures. Computer Graphics Vol. 25, No. 4, pp. 359-362, 1991
14. Lee P, Wei S, Zhao J, Badler N: Strength guided motion. Computer Graphics, Vol. 24, No. 4, pp. 253-262, 1990
15. Ching W: Motion planning for redundant branching articulated figures. PhD Dissertation, Department of Computer and Information Science, University of Pennsylvania, 1992
16. Jung M, Kalita J, Badler N, Ching W: Simulating human tasks using simple natural language instructions. Proc. Winter Simulation Conf., 1991
17. Ko H: Human locomotion on a curved path. Technical Report MS-CIS-92-02, Dept. of Computer and Information Science, Univ. of Pennsylvania, Philadelphia, PA, 1992
18. Alameldin T, Sobh T, Badler N: An adaptive and efficient system for computing the 3-D reachable workspace. IEEE International Conf. on Systems Engineering, Pittsburgh, PA, pp. 503-506, 1990
19. Stix G: Science and Business: Human Spec Sheet. Scientific American, pp. 132-133, November 1991
20. Esakov J, Badler N, Jung M: An investigation of language input and performance timing for task animation. Proc. Graphics Interface '89, Waterloo, Canada, pp. 86-93, 1989
21. Esakov J, Badler N: An Architecture for Human Task Animation Control. In Knowledge-Based Simulation: Methodology and Applications. P.A. Fishwick, R.S. Modjeski (eds.), Springer Verlag, New York, pp. 162-199, 1991
22. Levison L: Action composition for the animation of Natural Language instructions. Technical Report MS-CIS-91-28, Dept. of Computer and Information Science, Univ. of Pennsylvania, Philadelphia, PA, 1991
23. Fitts P: The information capacity of the human motor system in controlling the amplitude of movement. Journal of Experimental Psychology, Vol. 47, pp. 381-391, 1954
24. Badler N, Webber B, Kalita J, Esakov J: Animation from instructions. In Making Them Move: Mechanics, Control, and Animation of Articulated Figures. N. Badler, B. Barsky, D. Zeltzer (eds.), Morgan-Kaufmann, San Mateo, CA, pp. 51-93, 1990
25. Di Eugenio B: Action Representation for Natural Language Instructions. Proc. 1991 Annual Meeting of the Assoc. for Computational Linguistics, Berkeley CA, June 1991, pp. 333-334, 1991

Human Factors in Vehicle Driving Simulation

Arranger: P.A. Hancock, HFRL, University of Minnesota

Presenters: J. Flach, Department of Psychology, Wright State University
P.A. Hancock, HFRL, University of Minnesota
P. Green, UMTRI, University of Michigan
J.K. Caird, HFRL, University of Minnesota
A.D. Andre, NASA Ames Research Center

Overview

In considering operator-in-the-loop simulation much concern is rightly directed to the response of the human controller whose actions are vital with respect to performance of the overall system. Operator-in-the-loop simulation provides a cost-effective approach to understanding the performance envelope of any proposed system and is particularly relevant to the investigation of vehicles. However, the growth of full-fidelity manipulable simulation 'worlds' has reflexively begun to provide a new and exciting window from which to frame innovative and critical questions about how we can understand behavior itself. In the first of the present synopses, Flach argues that just such capabilities are central to a full evaluation of a control-theoretic approach to evaluating operator performance. In the second paper, Hancock argues that it is only through the use of such facilities that the design, test, and evaluation of prototype in-vehicle collision avoidance warning systems can be accomplished. As a critical component of the general IVHS effort, the safety gains potentially associated with an effective collision warning system clearly make the investment worthwhile. Hancock further argues that, exactly how such tests are to be conducted and such systems are to be designed relies heavily upon real-world application of perceptual field theories associated with ecological views of the coupling between driver perception and action.

However, full fidelity simulators are expensive systems and their cost/effectiveness has been questioned. Green explores ways in which many questions may be answered with reduced fidelity simulation. He develops a number of principals, which although simple in themself, have been largely ignored in previous interface developments. In citing a sequence of his own and others findings, Green indicates exactly how preferred design can be transferred from the realm of speculation to actual instantiation. In returning to the use of simulation, Caird points to a number of real and potential perceptual questions that might act to distort graphics worlds that the naive user may consider viable simply because they match in terms of metrical structure. He points to the lack of psychological knowledge and information that might resolve such issues. The general point that physical worlds are not simply equatable with perceptual worlds is one that has to be taken most seriously by graphic world designers and anyone hoping to extrapolate from high-fidelity simulation to real-world performance. In the final component, Andre also examines problem issues in simulation. He emphasizes the need to 'know' the simulation environment as well as the real-world performance environment and illustrates his argument with a number of traps ready to snare the unwary investigator. Finally, he points to a sequence of contemporary problems in design which could have been obviated if simulation had been used early in the design sequence.

The overall message from the collective presentations is clear. If concurrent engineering requires parallel consideration of multiple components of system design and manufacture, human factors issues should be given early prominence in any system that a human has to operate or maintain. Failure of early and adequate consideration of human factors will result in the spectacular failures that regale the news media. Although there are questions yet to be answered, the consensus of the present offering is that high-fidelity simulation is a vital tool in this process and is a central pillar of concurrent engineering approaches. While cost-effective in most environments, there are some systems that simply cannot be evaluated except using this form of assessment. The contemporary developments in computer capability to accomplish such simulation now make the widespread use of this approach a viable and advised strategy and is a facility especially welcome in the study of complex human behavior.

Acknowledgments: I would first like to thank Ed Haug for his kind invitation to participate in the most stimulating meeting. Further, I would like to thank the presenters in the session, Tony Andre, Jeff Caird, John Flach, and Paul Green for agreeing to participate and giving most interesting presentations. The comments of other participants in the Advanced Study Institute (ASI) were most welcome. Finally, I would like to thank the staff of the NATO ASI and The University of Iowa for all their help and assistance.

Evaluating In-Vehicle Collision Avoidance Warning Systems for IVHS

P.A. Hancock

Human Factors Research Laboratory, 164 Norris Hall, 172 Pillsbury Drive, S.E., University of Minnesota, Minneapolis, MN 55455

Abstract: Pursuant to the recent Intermodal Surface Transportation Efficiency Act (ISTEA), there has been a considerable apportionment of resources for developments in the area of Intelligent Vehicle Highway Systems (IVHS). IVHS promises, both here in the United States and in its various international forms, to address the questions of improved safe and efficient transportation. Such developments are mandated by the unacceptable levels of urban traffic congestion in most global conurbations and the epidemic level of traffic accidents which serve to rob society of human life and incur unsupportable burden on financial resources in the form of medical and insurance costs. IVHS can be divided into two major elements with respect to driver behavior. The first element, and one not dealt with in detail here, is assistance to navigation and congestion avoidance. The second, which is of central concern here, is collision avoidance. Collision avoidance systems seek to inform the driver of imminent or impending collision and to present assistance in conflict resolution. Just how such conflict resolution is to be enacted has yet to be determined. Various tactics have been suggested. They range from usurpation of control by some automatic system to messaging systems for preferred avoidance maneuvers. While the design and operational ramifications of these options are considered briefly here, the central theme is the critical use of simulation as a method for investigating and evaluating such alternatives. It is proposed that testing in high-fidelity simulation is currently the most, if not the only, viable option by which such technology can be safely instantiated. In examining this issue I contrast the situation specific approaches that appear to be favored in current research with an envelope approach based on the ecological analysis first posited by Gibson some fifty years ago. How simulation informs such design becomes a critical link if the safety aspects of IVHS are to reach fruition.

Keywords: collision avoidance / intelligent vehicle highway systems / driver behavior

Introduction

A recent estimate of the 1991 vehicle collision frequency suggested that there were 3.5 million vehicle accidents in the United States that were reported and potentially another 3.5 million that went unreported [1]. In 1990, the number of motor vehicle related deaths was 46,300, while the number of reported disabling injuries reach 1.7 million [2]. This fatality rate represents the equivalent death toll of approximately twice the number of fatalities in the Sioux City, Iowa air crash of a DC-10 every DAY. The societal burden is even greater when we add in the cost of care and the cost of suffering associated with permanent injury resulting from road traffic accidents. Such accidents are the leading cause of death in the United States up to the age of 38 and are the leading cause of all, accidental death up to the age of 78 [2]. Married to this grim picture is the realization that the number of vehicles on the roads of the developed and developing worlds is increasing at such a rate that unabated this number will exceed 1 billion soon after the turn of the century. At comparable accident rates this would result in at least one fatality per minute and one permanent injury per second on a global scale.

While these trends are daunting enough, they are not the whole picture. To these we can add the cost of the detriment associated with vehicle-related air pollution and the truly phenomenal costs that result from lost time as the workforce sits in traffic congestion waiting to get to and from the workplace. Little wonder that much serious consideration is given over to innovative solutions to these problems such as telecommuting. However, there is a more traditional approach to the question based upon the recognition that proliferation of freeways and arterials, even if it were possible, is not enough. In the United States such a program is generally recognized under the title Intelligent Vehicle-Highway Systems, or IVHS. IVHS mirrors a number of programs in the developed and developing world that seek to utilize the advantage of high technology in solving transportation questions. While the programs that have been initiated on differing continents are each designed to serve their own specific purposes, they are united in some general aims, foremost among which is the advancement of safety. It is on the central facet of this safety endeavor that the rest of the present work focuses.

In-Vehicle Collision Avoidance Warning Systems

It is perhaps the major selling point of IVHS that it promises improved road transportation safety. The principal way in which high technology promises to help reduce accident frequency is through the use of In-Vehicle Collision Avoidance Warning Systems (IVCAWS). These systems promise to provide the driver with timely information about the potential for imminent collision. Put in this way, the problem seems reasonable and somewhat straightforward. However, such a perspective is particularly misleading. One of the first steps on the road to a full understanding of the problems is the adoption of a systems perspective and to view collision-avoidance together with other common facets of IVHS development. In particular, the first efforts at in-vehicle information systems have been directed toward navigation aids that provide the driver with map information and more recently, dynamic, real-time information about the status of congestion in the immediate area. Predicated upon this assembly of information, intelligent aids can provide on-line recommendations about preferred route guidance and the appropriate ways in which to avoid stoppage and slow down. Yet in a fundamental way, collision-avoidance warning systems perform exactly the same function, except that the time constraint provides a quantitative different challenge. In many IVHS arenas, e.g., advanced traffic management systems, it is possible to import solutions that have been developed in other operational areas. However, this technology transfer does not appear to serve so well in collision-avoidance. For example, the comparable system in aviation are the various configurations of the TCAS system. Yet even here the time window does not fall into the millisecond range that appears to constrain road-vehicle operators. There are perhaps comparable technologies in the robotics realm in which robot vehicles have been asked to perform navigation tasks in which obstacle avoidance is a critical concern. But such systems represent totally automated functions and this raises perhaps the central design issue in IVCAWS development. That is how the component elements of the collision avoidance task are to be delegated.

One simple answer is to delegate all control to the automated systems upon detection of a collision situation. As one facet of the automated utopia (autopia), this solution has much appeal to the engineer who can then 'reduce' the problem, apparently, to one of algorithmic solution. In asking the rhetorical question, "would you want the first such system used in your car?", we can well imagine that the level of sophistication for flawless operation is unlikely to be forthcoming in the immediate future. One concern with such a system is the question of signal to noise ratio, such that we must consider the cost of a miss with respect to a false alarm. Missed signals (collisions) have a phenomenally high cost, yet their potential frequency is undoubtedly very low. Indeed some drivers may drive for decades without taking the sort of evasive action mandated of such a system. However, if great gain is focused on not missing signals, the problem of frequent false alarms is immediately encountered. It is well known from operation in other vehicles, e.g., single seat high-performance aircraft, that false alarms in warnings are highly distractive and disturbing to the operator. There is at present no simple solution to the trade-off that such a ratio implies other than the simple and simplistic affirmation of deriving a perfect detection algorithm.

Given an actual imminent collision situation, how is the automated vehicle expected to respond. It is probable that for the alert driver, a number of options open themselves up, such as running off the road onto a soft shoulder, and this may be considered a preferred action given the expected cost of collision. However, given the myriad of environmental contingencies, how is an automated decision expected to perform the task of detection, evaluation, and action initiation within the available time frame? The computational power presently required to accomplish such action, were it actually possible, would certainly cost many orders of magnitude more than the value of the car, which we cannot forget is manufactured in a competitive financial environment. Hopefully, such capabilities will become real within the coming years and the computational capability to support such action will drop in cost at rates comparable to today's logarithmic acceleration. However, in tthe absence of such availability, we need to keep the driver in the loop as we can assert that drivers currently are well able to avoid most obstacles set in their path.

Driver-in-the-Loop Collision Avoidance Systems

Despite the figures presented at the beginning of this paper, we cannot but wonder at the relative infrequency of collision given the number of opportunities on an everyday basis. Some may assert that this is especially true given their perceptions of the capabilities of other road users. It is difficult to speculate on the numerical potential for collisions in contemporary driving but from this perspective, it is clear that human controlled vehicles are able to navigate current road systems and avoid the vast majority of such events.[1] It is important not to remove this capability prior to the unequivocal establishment of a proven superior system. The combination of human response augmented by machine information or action falls under a heading of a type of system that I have previously referred to as hybrid systems. The use of hybrid systems allows the retention of flexible actions as compared to the highly constrained environment that has to be enacted in early stages of automation. In hybrid architecture, proximal warning systems have to act as information augmentation and decision support and this is the role explored below.

The idea of retaining driver-in-the-loop architecture, with the expectation of the final arbitration of avoidance action remaining with the human component is predicated upon a human-centered approach to IVHS system design [5,6]. While the human-centered approach presents numerous human factors related problems [5,7], the alternative systems architectures founded principally upon traffic engineering concerns hope to solve the problem via automation techniques with relatively little direct concern for the proximal user of the system, the driver. While such 'control' strategies have always had their appeal to the design engineer,

[1] At the recent IVHS America Meeting, Eugene Farber [3] speculated upon this frequency for one particular scenario which was rear-end collisions. By bounding the operational space to this one configuration he was able to derive the figure of approximately 0.5 million 'slow-downs' associated with the actions of a single driver for a lifetime of operation. Given the respective reliability of the "average' driver he further suggested some potential ways in which to consider the additive nature of warning aid reliability with this estimate of human reliability. Despite the inherent problems of integrating machine and human reliability [4], the approach advocated by Farber was one of shared actions which I put here under the title of a hybrid warning systems. It should also be noted that based on daylight driving, there is approximately one fatality per 77,519,380 miles of driving [2,5]. All of these observations attest to the current capability of the driving population for safe vehicle operation.

it has been shown with increasing frequency that such design approaches are almost always doomed to failure as a result of the intrinsic assumptions that form their basic premise and foundation. As indicated earlier, the state-of-the-art in automated collision detection and avoidance is certainly not well enough advanced at present to subsume such a function, hence the notion of a hybrid approach is probably the preferred if not the only viable contemporary strategy.

Having said this, we cannot adopt the design principles that have been enacted elsewhere in many complex systems of automating all that it is possible to automate and to leave the human as the "backup system of last resort." Indeed, the lessons from aviation, aerospace, process control and similar domains have taught the fallacies of such a strategy. This is particularly true of driving where the driver is liable to retain active control of steering, at least in the immediate future.[2] Thus, if task allocation is not to become a default approach [8,9] as outlined above, or based simply on descriptive, machine-oriented comparisons [10,11,12,13], we must seek new approaches that will stand up to the rigors of fast, real-time application that are the leitmotif of IVCAWS.

One potential solution lies in the application of adaptive systems through the medium of intelligent interfaces [14]. Adaptive human-machine systems [15], which represent a particular subset of hybrid systems, emphasize the use of mutual adaptive capability on behalf of both human and machine to promote flexible and rapid response to uncertain conditions and unusual task demands. As such, their major usage is focused on transient, unstable, emergency conditions as typified by the imminent collision scenario. Adaptive human-machine systems are facilitated by the use of intelligent interfaces which act as a translation intermediary between human and machine framing and managing respective queries and actions of both human and machine in the language and format which is appropriate to each element. The use of intelligent interfaces as envisaged by Hancock and Chignell [14], has been suggested for application to numerous human factors problems raised by IVHS development [16], and IVCAWS appears to be perhaps the prime example where such development could be immediately evaluated. As the full range of questions and problems concerning such application cannot be fully aired in the present brief space, it is perhaps best to consider a way in which the problem of collision avoidance warning can be approached, which contrasts with the more traditional thoughts on this issue.

---

[2] I should note here that there are a number of active research efforts that seek to provide automated steering for the driver such that active guidance is not required. If and when such systems are installed into IVHS environments, the evolving role of the driver will converge rapidly with many other operators to whom management and decision-making have replaced active control. How such a change in role and demand will be greeted and faced by the vast range of skills and capabilities in the driving public is essentially unknown. How drivers will respond in conditions where differing proportions of vehicles are equipped with such capabilities is also uncertain. Such human factors questions again emphasize the necessity for using a human-centered approach.

Traditional Versus Field Approaches to IVCAWS

While there are many practical questions as to the construction of collision-avoidance warning systems, behind such questions lie a more fundamental concern with how we can understand human interaction with complex technical systems. The time constraints involved in collision avoidance bring this discussion into particular prominence and such issues are aired briefly here. Traditional approaches to human-machine interaction are predicated upon an information-processing paradigm which has held sway for some forty years and had provided invaluable insights for a general understanding of human behavior. The quantitative nature of this endeavor has facilitated interaction with designers and engineers alike who seek to understand actions when humans are included in the control loop [17]. However, when we come to an understanding of the human response, primarily in a perceptual-motor domain, what is required is a much more thorough understanding of what constitutes the environment to which the individual must react. This is especially true for the dynamic and uncertain environments which connote a collision-avoidance situation. The notion advanced here is that we must draw upon more than the traditional information processing approach. Fortunately, such a complementary view is available and was first presented some fifty plus years ago by Gibson and Crooks [18]. Although this view blossomed into a full blown theoretical view of human and animal behavior [19], it is the initial work identified that is examined here in more detail.

In their original work, Gibson and Crooks [18] pointed to the commonalties between driving and forms of 'natural' locomotion such as walking and running. In essence, each of these tasks requires the individual to navigate toward a goal while avoiding manifest obstacles and areas that might also slow or inhibit transportation to the goal. They protested that driving, which is locomotion via a tool, is predominantly a perceptual-motor task. While this seems a reasonable assertion, it is the case that attention and failures of attention are intimately involved with normal driving and accident scenaria. Thus driving cannot be a task requiring only a perceptual-motor demand. One of the valuable constructs advanced was the notion of a field of safe travel which was represented as the field of possible paths which a vehicle may take unimpeded. This field is bounded by terrain, principally the roadway, and the moving objects. The task of steering is thus defined as keeping the vehicle headed into the center of this field. Allied to this notion of a field of safe travel is a minimum stopping zone. While acceleration is a function of goal achievement, deceleration also subserves the same function but principally by obstacle avoidance. Most frequently, the stopping zone is inside the field of safe travel and it is suggested that drivers often act to keep the ration of the field and zone close to constant in respect of the configuration and obstacles in the roadway. The field/zone ratio can be envisaged as an index of cautiousness. Gibson and Crooks observed that:

"A sudden frontal contraction or a shearing off of the field of safe travel which cuts it down or below the minimum stopping zone, produced in the driver a feeling of imminent collision, sometimes approaching panic, and an immediate and maximum braking reaction. There is an 'emergency.' Much more frequently, however, there occurs a gradual contraction of the field and, as it approaches the front boundary of the zone there follows a gradual slowing reaction of such strength as to keep the zone continually smaller than the field."

Intrinsic to this latter statement is a position which is strongly advocated here. That is that accident likely scenarios cannot be considered simply as extensions of normal driving. This implication is perhaps present in the notion of gap acceptance, in which collision can be inferred as a function of the extremes of the range of gap acceptance. I suggest here that accident conditions are qualitatively distinct from typical stable states of vehicle operation. Thus accidents and their prevention must be considered as an allied but independent investigation from that of the typical driver's task. One reason for positing the above assertion is again found in Gibson and Crook's work. They noted that while curbs, shoulders and shallow ditches are frequently excluded from the field of safe travel, in an emergency it is precisely these paths that may be sought by the driver and this can be seen in everyday avoidance on the freeway in potential 'shunt' or 'domino' accident situations. Thus field of safe travel is actually the field of safest travel given the current conditions. In one actual left-turn accident situation recorded on video-tape, we have determined that the complex interplay between drivers acts to rapidly change these fields and it is the case that in this collision, and we suspect in many others [20], that it is the iterative interactive and correction process between the drivers that eventually leads to impact [21]. With respect to fields and zones, Gibson and Crooks [18] noted that:

"These two (constructs) ... cannot of course be conceived as visible strictly speaking, in the sense that an object with a contour is visible. Nor are their boundaries sharply defined as are the lines and contours, although for convenience we may diagram them as if they were. They are fields within which certain behavior is possible. When they are perceived as such by the driver, the seeing and doing are merged in the same experience."

As a consequence of these observations we can see that driving is not simply locomotion in a free environment. First, the vehicles which act as locomotion prosthetics have velocity capabilities which exceed unaided human capability. Also most other moving obstacles are not insensate, but are purpose-directed, goal-oriented agents. Thus normal driving requires mutual cooperation and paradoxically some forms of collision require also require mutual 'interaction.' If driving requires the maintenance of mutual field exclusion, then collision-avoidance warning systems must inform as to these fields, and the progressive incursion into these fields by other entities. Thus:

"Safe and efficient driving is a matter of living up to the psychological travel and his minimum stopping zone must accord with the objective possibilities, and a ratio greater than unity must be maintained between them. This is the basic principle. High

speed, slippery roads, night driving, sharp curves, heavy traffic and thee like are 'dangerous' when they are, because they lower the field-zone ratio. Hidden obstacles are dangerous, when they are, because they tend to put the driver's field of safe travel out of correspondence with reality."

Implications for In-Vehicle Collision Avoidance Warning Systems

In this brief space, it is certainly not possible to explore all the nuances of the work reported by Gibson and Crooks [18], and further as part of an evolving view of animal behavior in general it is certainly insufficient to describe the complexities of Gibson's subsequent work which remained directly pertinent to vehicle control [19,22]. However, the observations that have been made are certainly of direct pertinence to collision-avoidance warnings and indeed complex systems control in general [23]. Before considering their direct implications, a number of problems need to be resolved. First, while the definition of the objective qualities of the identified fields and zones is well articulated, the connection with perception and the intimate linkage to action is less clear. This is critical, as the design of an effective warning is predicated on the development of interfaces that 'directly' display such fields and zones [24]. A further problem that has yet to be even addressed in an adequate manner, is how we start to structure a sensor array that can convey the information concerning the safest field of travel, given the constraints of a collision imminent environment. As we are uncertain how the human operator achieves this evaluation it is more than problematic as to how to construct a definitive algorithm to achieve the same objective. In fact, it is with respect to this problem that we must begin to identify the cost-effective nature of fully automated versus hybrid 'human-vehicle' collision avoidance control systems. Even given this information, how best to display it to the driver for maximum effectivity of response is still unclear. What is obvious is that the intrinsic time constraints defeat simple-minded alpha-numeric displays and suggest use of head-up multi-modal configurations. One manner with which to specify preferred zones of progress would be through color representations of fields directly projected onto the windscreen of the vehicle. How this aim is accomplished with the dual requirement of a 100% collision identification rate (unfortunately in signal detection terms often referred to as hit rate), together with a requirement for a zero false alarm rate, in other words flawless observations, is also as yet undetermined. While the theoretical take-over of all functions by automated control appears to offer the 'best of all possible worlds', we live in a society with rampant litigation; hence even marginal failure of such a system, or worse subsequent collision resulting from initial automated avoidance, appears to be mired in the questions of responsibility. Also, it is very clear that operators of any system are unhappy and reticent to have control literally lifted from their hands at any stage of operation. If these are the challenges, then what tools and methodologies can we use to get the work done.

On the Use of High-Fidelity Simulation

If the problems of IVHS safety developments, as pertinent to in-vehicle operations have been briefly outlined above, and some avenues of potential progress have been suggested, the question remains how to test and evaluate such fledgling systems. As can be readily inferred from the tenor of the present comments, and the general nature of his section of the whole argument on approaches to concurrent engineering, one obvious approach lies in simulation. Unfortunately, the word "simulation" now covers such a vast realm of capability that simply advocating simulation is insufficiently specific. Vehicle simulation ranges from hand-held video games and PC-based programs to multi-million dollar facilities which veridically recreate the dynamics of contemporary aircraft and spacecraft. Contemporary simulation is not confined to the recreation of existing vehicles, but more and more is being used to test the behavior and performance of proposed vehicles which can more readily, efficiently, and cheaply tested in electronic media. A strong rationale for such approaches lies in the engineering costs of prototype development, verification, and validation and proposed new systems; indeed, many would protest that such benefits are the major impetus that drives the increasing use of high-fidelity environments. Indeed, others in this volume have expounded on just this issue, so that I do not focus on such a rationale here. Rather, the question here focuses on driver-in-the-loop testing and, for this purpose, specification of simulation is a critical first step.

Some questions are $64 million questions and need many millions to derive a satisfactory answer. Some questions are $64 questions and the resources appropriated to such a question need also to be apportioned appropriately. In essence, simulation can be used to answer a wide realm of driver behavior questions and the central problem is the efficient choice of simulation level to provide useful and veridical answers to the questions posed. For example, we have conducted a number of head-up display evaluations for IVHS development in our fixed-base driving simulator [25,26]. It is an assumption, yet we believe a reasonable assumption, that the general principles that might be adduced from such experimental procedures would hold when applied to on-road driving. That is, the distortion and transfer effect, intrinsic in changing from fixed-base to on-road conditions, is insufficiently powerful to negate the main value of the findings given. Of course, it would be an advantage to perform all simulation functions in a full motion base, wrap-around graphics facility. However, in the present case, the nature and value of the question does not appear to dictate the use of such a facility, especially, as such findings have to undergo test-track and on-road evaluation anyway.

Indeed, as many simulation-based developments are destined for both test-track and on-road verification and validation anyway, some would question the utility of very high-fidelity simulators. The solution does not lie in a priori knowledge. That is, we cannot know before investigation which elements of performance are critically dependent upon veridical motion cues and which are not. [3] However, as with experimental procedures themselves, we have to begin with assumptions which are then tested against actual performance. The answer lies in a full range of simulation facilities that cover the spectrum of capability from the simple PC-based approaches to the on-going National Advanced Driving Simulator (NADS). Envisaging a suite of simulation facilities of ascendant capability is not a new insight, and indeed with simulation networks it is possible to provide collective interaction at remote sights with multiple capabilities. However, for some applications only high-fidelity is appropriate. It is asserted here that the initial prototype design test and evaluation and full verification of IVCAWS systems will have to proceed largely using such high-fidelity facilities. For this form of technology, which it should be remembered is a major rationale for IVHS itself, only simulation-based testing is immediately feasible. Test track and on-road evaluation, while they may come later in the testing program can only follow extensive evaluation in high-fidelity simulation.

Put simply, we do not as yet know enough about the way in which purely manual collision avoidance actions are initiated. Also, we have insufficient knowledge about how that process fails in actual collision events. We have only dim shadows of evidence provided by epidemiological data about accident frequency and obscure distortions that come from accident reconstructions. At present, our knowledge about the dynamics involved in collisions and near misses is such that the design of collision avoidance warnings is a hazardous endeavor and may potentially heighten rather than diminish collision risk. Also, with a user-centered approach, we may very well attempt technological solutions to vestigial problems which themselves might be more easily defeated by a careful analysis of what the driver is trying to achieve [5]. While each of these cautions may be valid, it is still the case that dynamic collision avoidance engenders rapid vehicle maneuvers in which the response of the driver is, most probably, critically dependent upon the interaction of visual and motion cues which they are receiving. That alone justifies the creation and use of high-fidelity simulation testing environments.

Summary and Conclusions

In the preceding sections I have tried to present the following train of argument. First, contemporary investment in the transportation infrastructure mandates not only improved

[3] On a somewhat arbitrary basis, I am using the presence of an integrated motion base, that is one that produces true dynamic response as compared to random ride motion, as the differentiate between medium and high-fidelity simulation. Thus while a fixed-base environment can provide strong visual cues as to motion, the full scale integration with a motion base, as typified by commercial flight simulators for pilot training, are those which begin to approach and surpass the Turing test for artificial realism.

efficiency in moving people and goods and a reduction in associated pollution, but also a critical focus on safety. Such a focus recognizes the unsupportable burden currently imposed by safety failures in road transportation. For IVHS, I have suggested that a major avenue through which safety improvements are expected to occur is through the use of proximity warning systems that can be generically referred to as In-Vehicle Collision-Avoidance Warning Systems (IVCAWS). I have further suggested that such systems are unlikely, at least in their early stages of development, to be purely automated systems which usurp control from the driver. Indeed, I have suggested that such a design approach is flawed and contains a number of inherent fallacies that have been put on view in other realms of technological development. I have suggested that intelligent interfaces should be of direct use in developing hybrid collision avoidance warning systems that seek to integrate the proven capabilities of human drivers with the nascent capabilities of proximal detection systems. However, the way in which humans currently accomplish such actions, and by extension how augmented information might be presented to them, might be better approached from an understanding of the tenets of ecological psychology, one approach to which, albeit from an older work, is given in some detail. I then argue what such a different approach might mean for IVCAWS. Finally, I suggest that design, test, evaluation, verification, and validation of proposed systems has to use simulation facilities and at least in most instances high-fidelity facilities such as those envisaged for the NADS. Absent such usage, we are not liable to reap the safety rewards hoped for and promised by IVHS.

Acknowledgement: Much of the IVHS and Road Safety Research conducted at the University of Minnesota, Human Factors Research Laboratory is supported through the Minnesota Department of Transportation (Mn/DOT) through the guidance of Commissioner Denn. In particular our IVHS efforts are especially supported through the GUIDESTAR program. I would particularly like to thank Mr. Jim Wright, Mr. Mike Robinson, Mr Mike Sobolewski and all GUIDESTAR staff for their support. Our work is also supported by the University of Minnesota, Center for Transport Studiés, directed by Mr. Richard Braun. I would like to thank members of the Center including Mr. Robert Johns and Ms. Lori McGuiness for all their help. I would also like to thank Paul Green for comments on an earlier version of this paper.

References

1. Burgett, A. (1992). *Contemporary accident frequency.* Paper presented in the Section Safety in IVHS: What is real? at the Annual Meeting of IVHS America, Newport Beach, CA, May.
2. National Safety Council (1991). *Accident facts.* National Safety Council: Chicago, IL.
3. Farber, E. (1992). Opportunities for accidents. Paper presented in the Section Safety in IVHS: What is real? at the Annual Meeting of IVHS America, Newport Beach, CA, May.
4. Adams, J.A. (1982). Issues in human reliability. *Human Factors*, **24**, 1-10.

5. Owens, D.A., Helmers, G., & Sivak, M. (1992). *Intelligent vehicle highway systems: A call for user-centered design*. Manuscript in press.
6. Hancock, P.A., Dewing, W., & Parasuraman, R. (1992). *Driver-centered system architecture for IVHS implementation*. Manuscript in Preparation.
7. Hancock, P.A., & Parasuraman, R. (1992). Human factors and safety in the design of intelligent vehicle-highway systems. *Journal of Safety Research*, in press.
8. Chapanis, A. (1965). On the allocation of functions between man and machines. *Occupational Psychology*, **39**, 1-10.
9. Chapanis, A. (1970). Human factors in systems engineering. In: K.B. DeGreene (Ed.). *Systems psychology*. New York: McGraw-Hill.
10. Bekey, G.A (1970). The human operator in control systems. In: K.B. DeGreene (Ed.). *Systems psychology*. New York: McGraw-Hill.
11. Fitts, P.M. (1951) (Ed.). *Human engineering for an effective air navigation and traffic control system*. Washington, D.C., National Research Council.
12. McCormick, E. J., & Sanders, M.S. (1982). *Human factors in engineering and design*. New York: McGraw-Hill.
13. Meister, D. (1971). *Methods of performing human factors analyses. Human factors theory and practice*. New York: Wiley.
14. Hancock, P.A., & Chignell, M.H. (1989). (Eds.). *Intelligent interfaces: Theory, research, and design*. North-Holland: Amsterdam.
15. Hancock, P.A., & Chignell, M.H. (1987). Adaptive control in human-machine systems. In: P.A. Hancock (Ed.). *Human factors psychology*. (pp. 305-345), Amsterdam: North-Holland.
16. Verwey, W. (1990). *Adaptable Driver-Car Interfacing and Mental Workload: A Review of the Literature*. Drive Project V1041. Traffic Research Center, University of Groningen, The Netherlands.
17. Flach, J.M. (1992). *Active psychophysics: A psychophysical program for closed-loop systems*. This volume.
18. Gibson, J.J., & Crooks, L.E. (1938). A theoretical field-analysis of automobile driving. *American Journal of Psychology*, **51**, 453-472.
19. Gibson, J.J. (1979/1986). *Ecological approach to visual perception*. Hillsdale, N.J. : Erlbaum.
20. Hancock, P.A., Caird, J.K., Shekhar, S., & Vercruyssen, M. (1991). Factors influencing drivers' left turn decisions. *Proceedings of the Human Factors Society*, **35**, 1139-1143.
21. Perrow (1984). *Normal accidents. Living with high-risk technologies*. New York: Basic Books.
22. Gibson, J.J. (1966). *The senses considered as perceptual systems*. Boston: Houghton-Mifflin.
23. Flach, J.M., Hancock, P.A., Caird, J.K., & Vicente, K. (1992). *The ecology of human-machine systems*. Erlbaum, in press.
24. Hansen, J.P. (1992). Representation of system invariants by optical invariants in configural displays. In: Flach, J.M., Hancock, P.A., Caird, J.K., & Vicente, K. (1992). *The ecology of human-machine systems*. Erlbaum, in press.
25. Coyle, M., Shargal, M., Shekhar, S., Yang, A., Caird, J.K., Hancock, P.A., & Johnson, S. (1991). Exploring headsup displays for driver workload management in intelligent vehicle highway systems. *Proceedings of the Second International Conference on Applications of Advanced Technologies in Transportation Engineering*, Minneapolis, MN, August.
26. Shekhar, S., Coyle, M.S., Shargal , M., Kozak, J.J., & Hancock, P.A. (1991). Design and validation of headsup displays for navigation in IVHS. *IEEE International Conference on Vehicle Navigation and Information Systems (VNIS/IVHS)*, Dearborn, Michigan.

Tools and Methods for Developing Easy to Use Driver Information Systems

Paul Green

University of Michigan Transportation Research Institute, Human Factors Division, 2901 Baxter Road, Ann Arbor, MI 48109-2150 USA

Abstract: This chapter describes approaches that can be used to determine if systems are safe and easy to use. Examples are given from the author's research for a variety of driver information systems, particularly warning systems, traffic information systems, and navigation displays, as well as conventional instrumentation such as speedometers. Methods are needed to provide for an early focus on users and their tasks, to provide for empirical measurement, and to support iterative design. Approaches examined include focus groups, paper and pencil studies at a driver licensing office, response time experiments, rapid prototypes using HyperCard and SuperCard, procedures involving driving simulators, and on-the-road experiments in instrumented vehicles. Simple driving simulators (task loaders) have proven to be particularly useful over the years. Application of these approaches varies with the design phase, and with the personnel, facilities, and funding available.

Keywords: human factors / ergonomics / driver interfaces / usability / navigation / design of controls / design of displays / vehicle design

Background

The purpose of this chapter is to

- convince you that ease of use is very important
- identify some of the issues of interest
- describe methods used to examine the issues (with examples)
- identify sources of additional information

Human factors or human factors engineering, ergonomics, human engineering, and engineering psychology have all been used to identify the work discussed in this chapter. They are all concerned with making systems, equipment, and facilities safe and easy to use and maintain. The differences among these terms are slight. Ergonomics and human factors engineering are the most widely used.

The author's experience has been that systems usually fail because the human element was ignored, not because of poor mechanical or electrical design. As an example, the author does not know of any cases where naval vessels failed to perform their mission where, say, the boiler foundation collapsed. He is aware of cases where boilers exploded because sailors were unable to monitor the equipment properly. That occurred because the compartment temperature was well above that at which sailors perform reliably. In other cases, the instrumentation was designed so errors were easy to make. While the human aspects of engineering had the greatest impact on system success, those aspects have proven to be the most challenging to solve and have received the least attention.

The emphasis of this section is on human factors as applied to the design of driver information systems, especially those related to IVHS (Intelligent Vehicle/Highway Systems). Of interest are Navigation/Route Guidance, Traffic Information/Hazard Warning, Motorist Information/Trip Services (Yellow Pages), Vehicle Monitoring, and Cellular Phones.

These systems can be made easy to use by applying human factors data from the literature, using analysis methods, and via experimentation. Devoted to the topic of human factors engineering are books [1,2], design standards [3], and journals (e.g., *Applied Ergonomics*, *Human Factors*, *Ergonomics*). Analysis methods such as the GOMS model [4] provide predictions of the time required to complete the task and insights into likely errors.

Key Design Principles

Good design will not result from mere knowledge of human factors data and calculation procedures, but from including human factors in the design process. The key, however, is getting the process right. In their 1985 paper, Gould and Lewis [5] identify three key design principles--early focus on users and tasks, empirical measurement, and iterative design. Early focus on users and tasks refers to identifying who will be using the equipment and what they will be doing with it. For example: How old are the users? How much do they know about the domain? How much experience have they had with similar systems? Which tasks are most important? What kinds of decisions will users make? Empirical measurement refers to collecting data on user performance and attitudes, especially early in development. Associated with this principle is the need for tools to prototype interfaces rapidly, as well as quantitative usability standards against which performance can be compared. Iterative design calls for repeated testing and re-designing of the interface until it passes the usability specifications. Application of these principles has been central to the research on driver interfaces conducted by the UMTRI Human Factors Division.

Paper and Pencil Methods

One technique used in a recent project involved Focus Groups [6]. Typical users of a product, here groups of 10-12 people who drove cars equipped with high technology instrumentation (HUDs, touch screen computers, etc.), met with a trained moderator. There were four groups, two in Los Angeles and two in New York. Participants responded to questions about driver information systems (e.g., How do you learn to use them?) that were recorded on videotape by a hidden camera. The comments provided insights as to how drivers would interact with future systems. For example, in response to the learning question drivers said,

The salesman showed me. It took more than an hour.

Very simple, you read the manual.

I like to play with gadgets. As a last resort I will read the manual. I like the challenge of solving problems and will only read the manual if I have failed.

I couldn't make heads or tails of the manual. My brother figured it out. It's in English but it sounds as if it is in Swedish. My mom can't understand anything at all.

It came with a cassette that went through everything. You just sit there. When you are all done it is in your head. You have heard it and seen it. The guy actually makes you work it while you sit in your car.

A second approach is to ask drivers directly what an interface is showing or how to use it. Paelke and Green [7] were interested in a gesture-based traffic information system interface. People working at UMTRI saw a skeleton map of Detroit (Figure 1). Participants held nontechnical positions (clerks, secretaries) and were not involved with IVHS-related projects. They demonstrated the motion they would use on a touch screen for zoom in, zoom out, and other actions. There was no consistent response, suggesting that a gesture-based interface requiring no training was infeasible, so the idea was abandoned. While there are limits to this experiment, it nonetheless shows how a simple human factors experiment conducted early on can influence design direction.

Another simple experiment concerned driver understanding of an in-vehicle warning display [8]. About 75 people were interviewed at a local driver licensing office. They were shown full size color printouts of various states of the display and asked "What is it telling

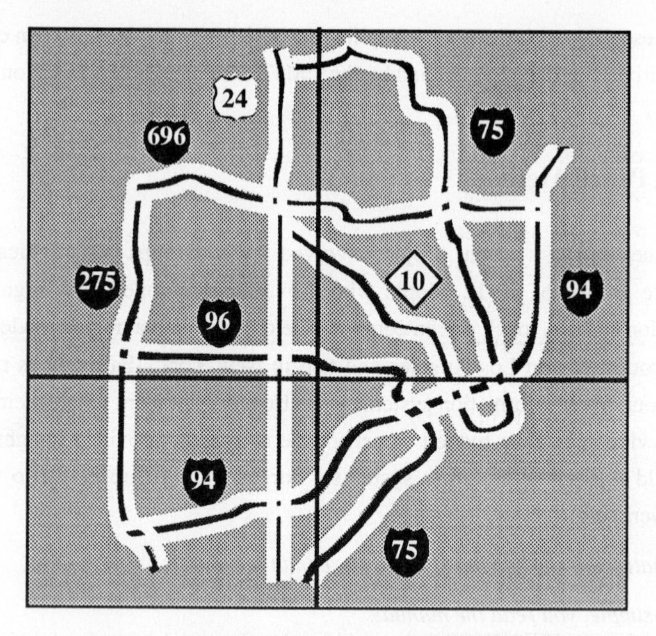

Figure 1. Skeleton Map Used in Gesture Display

you?" The display, Figure 2, consisted of printed legends with two LEDs to the left of each legend, green on the far left to indicate OK, and red next to it to indicate a malfunction. While the interface seemed simple, many drivers did not fully understand it. For example, for the air conditioner light, drivers would refer to problems with drive belts, even though the words "fluid level" were in plain view. To convince engineers that design modifications were necessary, a videotape was made showing the reactions of a subject along with a summary of the results. The videotape convinced engineers that the approach and results of the experiment were sensible.

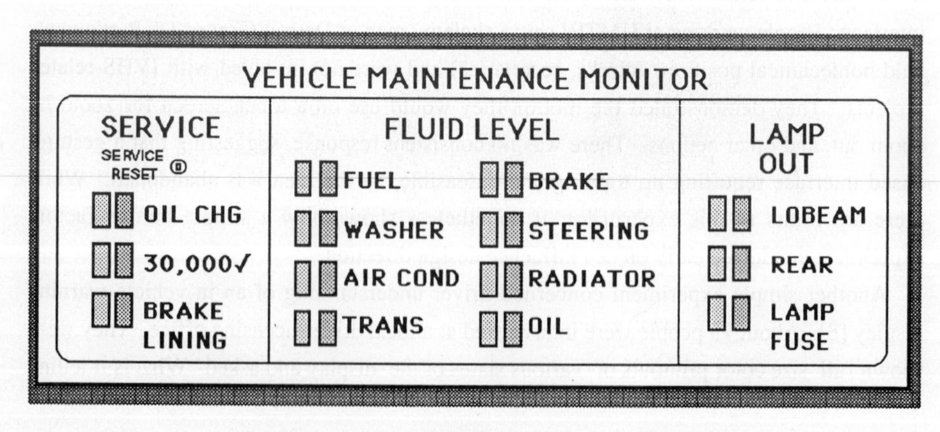

Figure 2. Vehicle Maintenance Monitor Display

Driving Simulation

To determine if user performance times meet system specifications, more sophisticated methods are required, typically involving either an instrumented vehicle or a driving simulator. For actions associated with the use of controls and displays, a high-fidelity simulator is often not necessary. For example, at UMTRI a simulator that runs on a Commodore-64 was developed [9]. The driver sits in a mockup of a 1985 Chrysler Laser and steers the vehicle down a single-lane road at night at a constant speed. A typical road scene is demonstrated in Figure 3. This scene has been shown on monitors of varying sizes and on projection video displays. The difficulty of the road (the sum of four nonharmonic sinusoids) can be varied and driver performance can be sampled at up to 10Hz. This system was originally developed for use in an experiment on driver fatigue [10], and has been used in about 20 experiments of brake lamp response times, the effect of highway sign materials on reading time, response times to brake lamps, the effects of mirror characteristics on driver performance, and so forth.

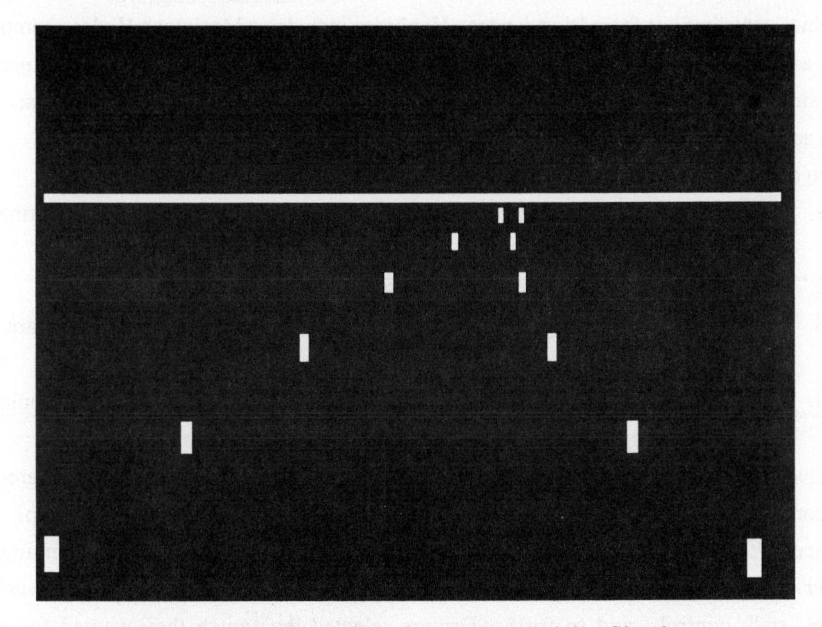

Figure 3. Road Scene from Commodore Driving Simulator

This system has proven extremely useful over the years and the hardware was inexpensive, less than $300 for the computer and a disk drive. On the other hand, over time as much as 100 times that amount has been spent on software development. In spite of that effort, the software has many drawbacks. Recently the software has been ported to a 286 computer and enhanced. The revised software has more comprehensive data logging and can communicate with other computers (so when the driver operates a control, the steering error is known). The enhanced simulator is now being used for studies on traffic information systems and car phones.

Similarly, the Biosciences Division at UMTRI developed a 386-based simulator that used a CRT as the display. Steering wheel and accelerator inputs from an instrumented vehicle buck controlled the simulator. The image presented was a curving two-lane road. This simulator provided a controlled environment for multiple hour studies of seating comfort.

As part of a demonstration, the Engineering Research Division developed a simulator that displayed a straight road on a Macintosh computer. The steering device was a mouse, which was used to counteract the effects of random side winds. The simulator allowed for representation of vehicles with a wide range of dynamic characteristics.

Currently underway at UMTRI is a combined effort of these three Divisions to develop a simulator for the Macintosh. It borrows baseline design requirements from the Human Factors Division simulator, dynamics from the Engineering Research simulator, and graphics contributions from Biosciences. Hardware includes a Macintosh II-class computer with a graphics accelerator card, the Human Factors Division buck, and a video projector. The simulator shows a two-lane curving road with a dashed centerline in color. It accepts both accelerator and steering wheel inputs.

In developing these simulators, several lessons have been learned.

1. Inexpensive simulators than can serve as task loaders and record driver performance are very useful.
2. There is pent-up demand for simulator use.
3. People are less willing to pay for simulator development than they are for simulator use.
4. For navigation and crash avoidance/warning studies more sophisticated simulators are needed, though at the current time interest in funding them is low.

One project using the Human Factors Division simulator concerned driver preferences for controls [11,12,13]. In two studies almost 200 drivers sat in a vehicle mockup. The surfaces where controls could be mounted were covered with Velcro®. Surrounding the driver on various mounting boards were over 250 types of push buttons, rocker switches, knobs, stalk controls, and so forth. Drivers selected the switch they wanted for each function (e.g., headlights on/off, hazard, etc.) and placed it at the location they preferred. Subsequently, drivers were asked to use each control while driving the simulator. About 10-15% of the time, drivers changed their preferences for switch types or location. This indicated that operation of the controls was important to examine, for which a simple simulator was needed. This switch selection task, nicknamed the "Mr. Potato-Head

Method," provided a simple and direct means to identify driver preferences for instrument panel controls.

Response Time Experiments

Another approach used at UMTRI involves a PC that controls two random-access slide projectors to obtain driver response times to displays [14,15]. Each response is examined in real time, a unique feature. If a response is too fast (representing a fast guess), too slow (when the subject stops paying attention), or in error, the test slide is repeated at the end of the test block. This results in an equal number of correct responses with reasonable times for each display, vastly simplifying analysis.

Three alternative methods of laboratory testing have been considered. In the control condition, drivers were shown slides of numeric speedometers on the instrument cluster. Their task was to press one of two buttons (speeding [over 55 mph] or not speeding). In a second condition drivers operated the Human Factors Driving Simulator. At random times a slide of a speedometer appeared on the instrument cluster and drivers responded to it. In a third condition drivers fixated ahead, looking for arrows pointing left or right instead of a road scene. On half of the trials those arrows appeared, in response to which the driver pressed a left or right key. On trials when an arrow did not appear, the driver looked down at the instrument panel and made a speeding/not speeding decision in response to the panel slide.

Performance in the control group (instrument panel alone condition) underpredicted performance of dual task conditions where legibility was poor because visual accommodation from the scene far ahead to the display just in front of them was absent. Variance in the arrows-panel task combination was less than that in the simulator condition. The arrows-panel combination was therefore used in a subsequent experiment.

That next experiment determined the time required to read a numeric speedometer. Five digit sizes, 3 locations, 3 contrast ratios, and 3 illumination levels were examined. From that research the following expression resulted.

$$RT\ (ms) = 1054 - 320(A) + 1050(1/H) + 202(L) + 89.6(1/\ln(C)) - 9.58(\ln(I)) + 4538(1/H^2)$$

where:

A = driver Age group (1=old, 2=young)
H = digit Height (5 to 19 mm)
L = Location (1=center, 2=sides)
C = Contrast ratio (1.5:1 to 20:1)
I = Illumination (1.08 to 915 lux)

Rapid Prototyping

Real driver interfaces are costly and time consuming to build, especially those related to IVHS. For human factors tests to impact design, testing must be completed before the actual hardware is available. As part of a DOT-funded project [16] and other research, the Human Factors Division has been using HyperCard and SuperCard on the Macintosh to prototype driver interfaces [17]. The use of Toolbook and Plus, object-oriented programming packages on the PC, has also been explored . These rapid prototyping tools have been used by others; for example, by Lockheed on the Space Station project, by GM on the TravTek project, by Motorola on the ADVANCE project, and by Ford for car phone studies.

One experiment in progress at UMTRI concerns the retrieval and display of traffic information [7]. Drivers are seated in a vehicle mockup. While operating a driving simulator questions are presented auditorally (for example, "Are there any traffic problems on I-96 west?"). Using one of several different touch screens (e.g., Figure 4), drivers retrieve the desired information (Figure 5), respond if there is a problem, and describe its location.

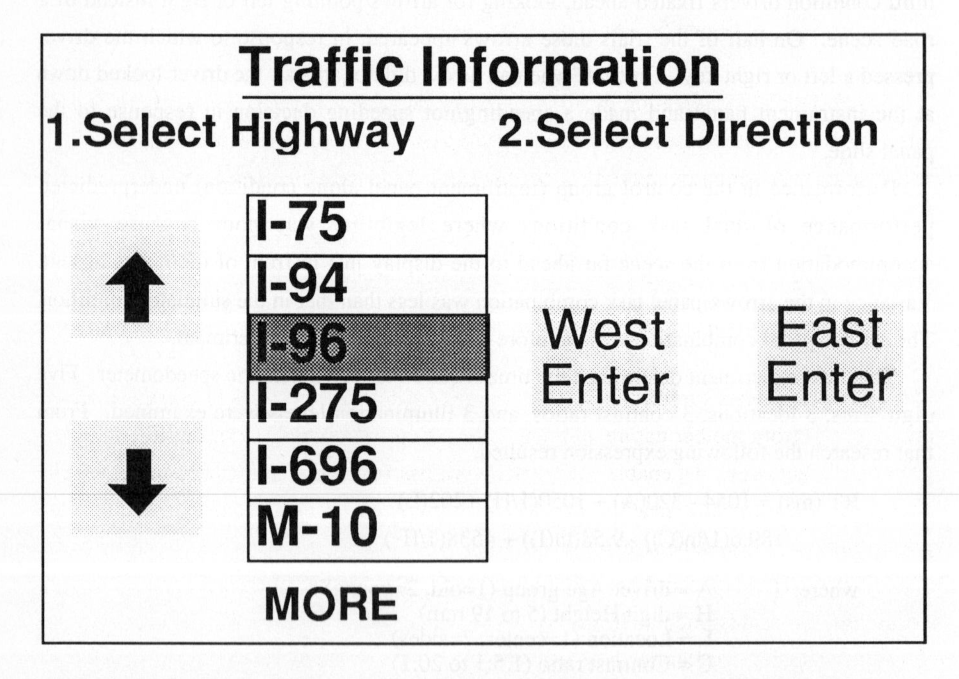

Figure 4. Traffic Information Retrieval Display, Bidirectional Scrolling Menu Design

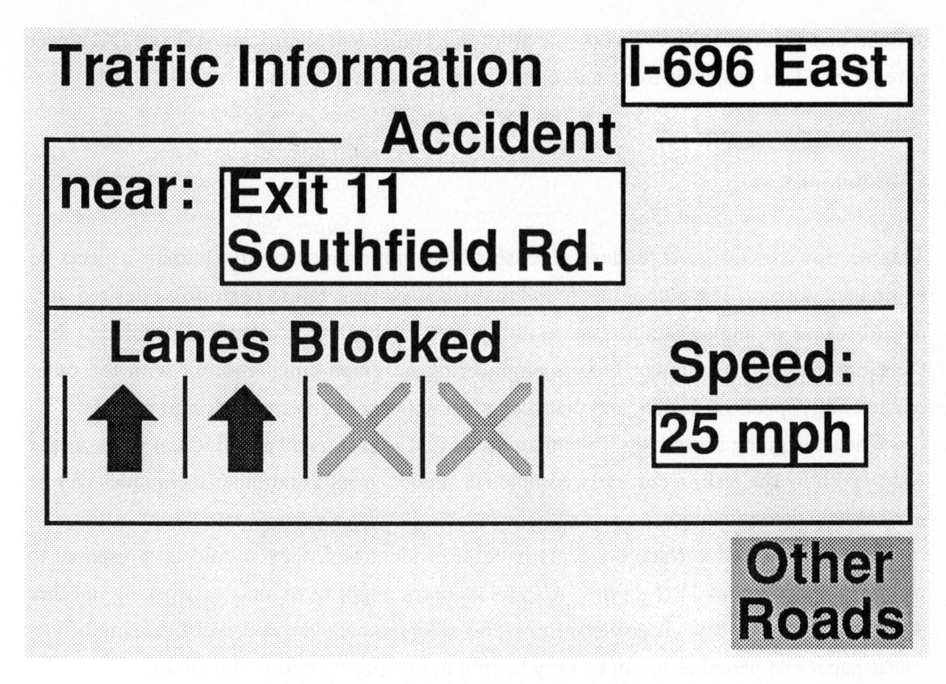

Figure 5. Text-Based Traffic Information Display

Despite their name, creation of initial versions of these interfaces has not been rapid, taking months in some cases. Modification of them, however, has been relatively quick. Furthermore, with complex systems it may be difficult to develop prototypes that react as quickly as real systems.

On-the-Road Tests

While laboratory tests provide for careful control of test conditions and are immune to complications from mother nature, systems intended for use by drivers must ultimately be tested on the highway. To enable such tests, work has begun on an instrumented research vehicle. Of interest is how steadily people drive (in terms of speed and lane position), when various foot controls and the steering wheel are used, and where drivers glance (and for how long). A 1991 Honda Accord has been fitted with a video-based lane tracker (giving lane deviation to the nearest inch at 10 Hz) and sensors for steering wheel position (accurate to the nearest degree), accelerator position, brake on/off, and speed (accurate to the nearest mph), all interfaced to a 486 personal computer. Also connected to it will be a NAC model V eye tracker, which gives x, y fixation coordinates in real time. To record the forward scene and the driver, two color video cameras are installed along with a VCR, screen splitter, and video monitor. To operate all of this equipment significant AC and DC power is required. This project is expected to be a large investment (costing about $100,000).

Conclusions

This section should give readers a sense of the importance of the human aspects of equipment design. The author has found that when systems fail to perform as desired, it is often because of human factors issues, not typical mechanical or electrical problems. It is therefore critical that human factors engineering be given equal footing with the other engineering aspects of design, and that such attention occur early in the design process.

When designing something for human use, a first step should be to find out what people want and how the task is currently completed. Focus groups and other techniques can be used to help determine this.

When the basic interface concepts have been identified, they should be presented to candidate users to obtain reactions. Candidate users might be associates working in other divisions, or, in the case of driver information systems, people at a driver licensing office. These paper and pencil tests can be very helpful in identifying design directions.

Subsequently, comprehensive laboratory tests are desired. These tests may use simulations of interfaces constructed using rapid prototyping software such as HyperCard, or, in later stages, may involve elements of real hardware. Such tests usually involve timing of user performance and, in the case of vehicle interfaces, use of a driving simulator. In many instances the simulator serves as a task loader, so a high fidelity simulator is often not required. On-the-road tests should be conducted as well.

Thus, for systems to be safe and easy for people to use, an understanding of who will be using those systems and how they will be used is critical. Further, once design approaches have been identified, user reactions and performance with them should be obtained. This should be done early in design. For the initial stages, paper and pencil mockups are sufficient, with rapid prototypes more appropriate for later use. To convince engineers of the significance of the kinds of problems users experience, videotapes of interactions should be made. Engineers often do not think of the human implications of what they design. Getting them to do so and to include human factors in the design process is a significant challenge.

References

1. Sanders, M.S. and McCormick, E.J. (1987). *Human Factors in Engineering and Design* (6th ed.), New York: McGraw-Hill.
2. Cushman, W.H. and Rosenberg, D.J. (1991). *Human Factors in Product Design*, Amsterdam, Netherlands: Elsevier.
3. United States Department of Defense. (1989). *Human Engineering Design Criteria for Military Systems, Equipment and Facilities* (Military Standard MIL-STD-1472D), Philadelphia, PA: Naval Forms and Publications Center.
4. Card, S.K., Moran, T.P., and Newell, A. (1983). *The Psychology of Human-Computer Interaction.* Hillsdale, NJ: Lawrence Erlbaum Associates.
5. Gould, J.D. and Lewis, C. (1985) Designing for Usability: Key Principles and What Designers Think, *Communications of the ACM*, March, **28**(3), 300-311.
6. Green, P. and, Brand, J. (1992). Future In-Car Information Systems: Input from Focus Groups (SAE paper 920614), Warrendale, PA: Society of Automotive Engineers.
7. Paelke, G. and Green, P. (1992). Development of a Traffic Information System Driver Interface, paper presented at the IVHS-America Annual Meeting.
8. Green, P. and Miller, D. (1983). *A Human Factors Evaluation of a Vehicle Maintenance Monitor.* Videotape report produced for Ford Motor Company (Electrical & Electronics Division and Automotive Safety Office). Ann Arbor, MI: The University of Michigan Transportation Research Institute.
9. Green, P. and Olson, A. (1989). *The Development and Use of the UMTRI Driving Simulator* (Technical Report No. UMTRI-89-25). Ann Arbor: The University of Michigan Transportation Research Institute. (NTIS No. PB 90 115940/AS, available from U.S. Department of Commerce, Washington, D.C.).
10. Green, P. (1985). *Human Factors Test of a Driver Alertness Device* (Technical Report No. UMTRI-85-49). Ann Arbor: The University of Michigan Transportation Research Institute. (NTIS No. PB 86 181229/AS, available from U.S. Department of Commerce, Washington, D.C.).
11. Green, P., Ottens, D., Kerst, J., Goldstein, S., and Adams, S. (1987). *Driver Preferences for Secondary Controls* (Technical Report No. UMTRI-87-47). Ann Arbor: The University of Michigan Transportation Research Institute. (NTIS No. PB 90 150541/AS, available from U.S. Department of Commerce, Washington, D.C.).
12. Green, P. and Goldstein, S. (1989). *Further Analysis of Driver Preferences for Secondary Controls* (Technical Report No. UMTRI-89-4). Ann Arbor, MI: The University of Michigan Transportation Research Institute. (NTIS No. PB 90 149782/AS, available from U.S. Department of Commerce, Washington, D.C.).
13. Green, P., Paelke, G. and Clack, K. (1989). *Instrument Panel Controls in Sedans: What Drivers Prefer and Why* (Technical Report UMTRI-89-15), Ann Arbor, MI: The University of Michigan Transportation Research Institute, July.
14. Bos, T., Green, P. and Kerst, J. (1988). *How Should Instrument Panel Legibility Be Tested?* (Technical Report UMTRI-88-35), Ann Arbor, MI: The University of Michigan Transportation Research Institute, November.
15. Boreczky, J., Green, P., Bos, T. and Kerst, J. (1988). *Effects of Size, Location, Contrast, Illumination, and Color on the Legibility of Numeric Speedometers* (Technical Report UMTRI-88-36), Ann Arbor, MI: The University of Michigan Transportation Research Institute, December.
16. Green, P., Williams, M., Serafin, C., and Paelke, G. (1991). Human Factors Research on Future Automotive Instrumentation: A Progress Report, *Proceedings of the 35th Annual Meeting of the Human Factors Society*, 1120-1124, Santa Monica, CA: The Human Factors Society.
17. Green, P., Boreczky, J., and Kim, Seung-Yun (Sylvia). (1990). Applications of Rapid Prototyping to Control and Display Design (SAE paper #900470, Special Publication SP-809), Warrendale, PA: Society of Automotive Engineers.

The United States Government does not endorse products or manufacturers. Trademarks or manufacturers' names appear herein only because they are considered essential to the object of the document.

The contents of this report reflect the views of the authors, who are responsible for the facts and accuracy of the data presented herein. The contents do not necessarily reflect the official policy of the Department of Transportation.

This report does not constitute a standard, specification, or regulation.

The Perception of Visually Simulated Environments

J. K. Caird

Human Factors Research Laboratory, 60 Norris Hall, 172 Pillsbury Drive, S.E., University of Minnesota, Minneapolis, MN 55455 USA

Abstract: The perception of simulated environments by the operator in-the-loop is examined. One objective of simulation is to provide realistic visual scenes for an operator to navigate through. While the physical re-creation of these synthetic environments has received considerable activity, less attention has been given to the perceived information conveyed by a simulated scene to an operator. In general, how perceptual differences between simulated and real scenes affect perception and subsequent action is not completely understood. To the extent it is understood, previous research done on the visual characteristics of flight and driving simulation is highlighted. The key issues of visual realism and fidelity, and distortion of spatial layout are elaborated.

Keywords: visual simulation / visual realism and fidelity / distortion of spatial layout / space perception

Introduction

Simulation is a useful and adaptable tool for Concurrent Engineering, but it must be constructed and used wisely. Those that build and use the hardware and software of simulation systems need to look beyond just the accurate re-creation of physical reality to the psychological reality of the operator. At this level of analysis, which seeks to characterize the operator in the simulation loop, inferences and generalizations are drawn about operator capabilities in the real world based upon their actions within simulated environments. If performance measures extracted during simulation are to have sufficient ecological validity and generalizability, then the accuracy of visual information contained within simulated scenes requires systematic verification. Since many visual simulation problems are common across types and uses, the present paper draws from previous work in flight and driving simulation and perception to frame a discussion of the critical issues of visual fidelity, realism, and distortion of layout.

The National Research Council [1-2] reviewed the important facets of simulation and recognized the following broad categories as important: fundamental behavioral processes, the fidelity of simulation, vehicle motion cues, performance assessment, modeling, visual simulation, training, and training methods. Visual simulation was further divided into visual display characteristics, and scene content categories. Visual display characteristics included such factors as field of view, resolution, luminance, color and contrast [3-10], whereas scene content and visual cues included studies of scene detail, the degree of object abstraction, terrain, object density, and textural qualities [1, 7-9, 11-14]. Two critical facets of simulation were highlighted. First, the technological substrate of computer graphics was (and is) advancing very rapidly. Second, at that time, a lexicon did not exist which enabled the identification of scene content. Of these points, the technical and perceptual are examined here.

Visual Fidelity and Realism

Throughout the visual simulation literature cited by the Committee on Human Factors [1], visual fidelity and realism received frequent mention. Physical visual fidelity was defined as, "The realistic degrees of freedom of spatial resolution, a correct rendering of luminance and color characteristics, the provision of field of view, as much depth of field in a flat plane presentation, and a continuous change in perspective to match the relative motion of the aircraft [or automobile] with respect to the outside world," [6, pg. 21]. Here, a simulated scene is taken to mean the reflected light from a projection surface which contains similar dynamic informational characteristics as the real world. In the most realistic case, which may not be achievable, the viewer should be unable to distinguish the pattern of information from a real or simulated scene. Considerable latitude, however, is possible in the parameters of visual fidelity. For example, although luminance levels far from replicate daylight conditions, they still exceed detection thresholds and therefore still afford active control and object recognition. Realism is not synonymous with fidelity, and is instead taken to mean the comprehensiveness or completeness of a simulated environment [7]. Overall, the various interpretations of visual fidelity and realism indicate that realism appears to be more related to scene content and visual cues, whereas, fidelity is more closely aligned with the accuracy of visual display characteristics. Over time, the use of the terms fidelity and realism have been used synonymously and without reference to the specific visual parameters which make a scene or display more or less realistic or accurate. As a consequence, the meaning of each term has become ambiguous and nominally useful.

The majority of papers reviewed did not attempt to specify visual fidelity parameters necessary for various task requirements. The Transportation Research Board [15] was one exception. They attempted to define high, medium, and low levels of visual simulation in terms of various display characteristics and scene cues (see Table 1). These levels are also categorized by research task (see 15, Tables A.1 to A.41). The difficult question of matching the level of visual fidelity or realism to specific tasks is another critical problem for simulation researchers to resolve, but is beyond the scope of the present review.

Visual System Fidelity	High		Medium	Low
Field of View	200 °		120°	60°
Brightness	High Movie		Medium Home TV	Low <Home TV
Contrast	≥ 16 S/G		12 S/G	≥ 8 S/G
Daytime scenes	Yes/No			
Nighttime Scenes	Yes/No			
Special Effects	Fog, Glare, Sun, Etc.			
Resolution	VH 4 arc min	High 6 arc min	Medium 13 arc min	Low 20 arc min
Moving Models	Number + Fidelity			
Scene Content	Dense/Urban;		Sparse; Urban	

Table 1: Characteristics of low, medium, and high levels of visual fidelity. Adapted from the Transportation Research Board [15, Appendix A, Table A.0].

Physical and psychological fidelity of simulation are further differentiable [16]. A number of psychological dimensions were suggested [16, pg. 5] and included: "the scope, extent, or segment of the environment represented in the simulation; the duration of the interaction between man and environment; the degree of effector interactions; the importance and degree of involvement with others; and the extent of perceived realism and related cognitive states." Psychological fidelity and realism are, undoubtedly, emergent properties of the repeated interaction between an operator and a simulated scene. While the physical dimensions of fidelity have received considerable attention, the identification of specific psychological dimensions of visual systems, given a set of task requirements, has been largely neglected. Typically, the perceptual impact of various fidelity and realism improvements has amounted to little more than confirmations of how good a particular scene looks to an "expert" viewer. For example, pilots and flight trainers, in their evaluations of visual systems, would judge a system and give improvement requirements such as; increased display resolution, additional texturing and shadowing, and reduced interaction lag [5, 16]. An interesting bias inherent in this type of evaluation was the skepticism of systems which were perceived to be unrealistic.

The by-product of this bias was an emphasis on more visual realism and fidelity as opposed to additional evaluation and perceptual experimentation to determine necessary perceptual requirements. Table 1 is important for providing fidelity and realism guidelines, however, it falls short of mapping physical dimensions to perceptual dimensions specific to driver behavior. Once the prerequisite experimentation has been executed, an extension to Table 1 would logically include physical system requirements for visual fidelity and realism mapped to important perceptual dimensions.

An argument was put forth by the National Research Council [1, pg. 39], namely: "It is possible (some say likely) that extremely high visual fidelity, in the engineering sense, will soon overtake the current need for perceptual research on visual displays for simulators." While advances in computer graphics architecture and algorithms will surely increase visual fidelity and realism, this argument fails to consider system cost, purpose or task requirements. Visual fidelity tends to covary with simulator cost [2, 8-9, 15-19]. Most simulation systems are constrained by cost and cannot include all the visual technologically available. Decisions on the side of providing greater visual realism and fidelity can be viewed as a conservative approach to capturing task requirements failing appropriate task decomposition and experimentation. Consequently, visual fidelity and realism, almost by default, have been advanced as the constructs necessary for "improving" simulation. Also, many system instantiations cannot wait until visual technology matures or become cost available. Failure to consider previous and future research on perception within visual simulation will surely result in high fidelity simulators which produce operator performance data of questionable validity and generalizability.

A comparison between flight and driving tasks also seems to indicate the need for greater scene accuracy and detail for driving. As a driver locomotes through a simulated traffic environment, she encounters a flow ground plane texture, fixed objects, and other moving vehicles. The rate of environmental flow is elevated in driving compared to flight since the driver is close to the ground, whereas, the pilot is far above the ground. Exceptions to this generalization are of course take-off, landing, and low-level flight, but in the general case the graphics power needed for driving appears to be greater than for flight. It is tempting to conclude that higher visual fidelity and realism are therefore necessary. However, for the task of driving, the minimal necessary perceptual information is not known. For example, is it sufficient to have only an accurately placed horizon line at infinity, a ground plane, and roadway edge-lines and center-stripes, all accurately coupled to driver inputs? How much scene detail is sufficient and how much is not needed for certain driving scenarios? While counter-intuitive, the removal or degradation of visual information helps establish minimal levels of task information. Many video driving games get by with minimal detail.

Distortion Of Perceived Layout

The placement of the viewpoint of the eye with respect to a simulated scene is important for accurate viewing. In computer graphics, the optimal placement of the eye with respect to a projection plane (the screen or CRT) is termed the viewport. Similarly, in perception the location of the eye as it views a picture plane is termed a viewpoint. If the eye is placed at the correct viewpoint, the spatial information of all objects on a flat (or curved) screen is the spatial layout. The perception of spatial layout is defined as the perception of distances, sizes and tilts of objects and surfaces within a simulated environment [20, pg. 1]. The perception of the layout of objects and planes represented in a scene or where objects are with respect to the operator is fundamental to active navigation through it. Therefore, the correct placement of objects in visual space by the underlying computational processes, the placement of the eye at the viewpoint, and any necessary corrections for distortions imposed by projection optics or screen curvature are paramount to layout fidelity. Further, perceived and actual layout are intimately dependent upon the field of view available to an observer. For example, in driving simulation, information for turning right or left is available by turning to that direction for information. Without a wrap-around presentation system or the coupling of a display to a driver's head movements, this information is simply not available. Similar field of view constraints are evidenced in flight simulation depending on needed task information.

As mentioned, a number of computational, material, and perceptual factors can produce distortions in the overall layout of a simulated scene. "Deviations from the correct viewing position result in distortions, blurring, and, ultimately, loss of view," [6, pg. 10]. Failure to place the eye at the correct viewpoint will result in the distortion of the scene [20-22]. Movement of the eye to the right and left, up or down, and in and out from the display surface produces a variety of layout distortions such as compression and shearing of images. What is not known, however, is the effect of these viewpoint changes on the overall perception of scene layout and subsequent actions to navigate through an environment. To experimentally test the various parameters affecting the perception of layout, requires the manipulation of viewpoint to determine the effect of distortion on a set of critical tasks. Assuming that the eye is at the correct viewport, distortions can also be subtly introduced by the accuracy of rendering geometry and display system properties [20, 24]. Curvature and material properties of wrap-around and CRT screens, screen joints, and projection optics also produce non-linear distortions of perceived spatial layout. In each case, the degree that these distortions in layout affect operator performance is not known. Finally, in addition to spatial layout other perceptual information has been identified as fundamental for movement through simulated environments and includes: textural distribution, vanishing limits and horizon ratio [25-28]. Manipulation of these perceptual invariants would serve to further validate the adequacy and accuracy of scene composition.

Summary and Conclusions

The utility of visual fidelity and realism as defining visual simulation characteristics and the effect of distortions on the perception of spatial layout was discussed. Two important conclusions can be drawn from this review. First, accepting the assumption that production of more realistic simulated scenes will eliminate the need to verify and experiment with the perceptual characteristics of simulation systems is particularly unwise. Failure to heed this conclusion will probably result in questionable generalizations about operator capabilities derived from such systems. Second, methodologies which directly and systematically compare real driving with simulated driving are absent and require the immediate attention of researchers. A comparative methodology which is capable identifying the minimal perceptual information for driving in real or synthetic environments should be developed. In combination, through further perceptual experimentation and comparison between real and simulated environments, visual simulation systems will become the tools of choice for many disciplines which seek to advance our understanding of safe human-machine operation.

Acknowledgments

The development of the University of Minnesota's Human Factors Research Laboratory simulation systems could not have been accomplished without the generous support of American Honda Motorvehicles Inc., The Minnesota Department of Transportation, The University of Minnesota's Center for Transportation Studies and Center for Research in Learning, Perception and Cognition. The author was supported by a pre-doctoral fellowship from NIH to the latter Center.

References

1. Jones, E.R., Hennessy, R.T., & Deutsch, S. (Eds.) (1985). Committee on Human Factors, National Research Council. *Human Factors Aspects of Simulation*. Washington, D.C.: National Academy Press.
2. Meister, D. (1990). Simulation and modeling. In J.R. Wilson & E.N. Corlett (Eds.) *Evaluation of human work: A practical ergonomics methodology*, (pp. 180-189), New York: Taylor and Francis.
3. Muckler, F.A., Nygaard, J.E., O'Kelly, L.L., & Williams, A.C. (1959). *Psychological variables in the design of flight simulation for training*. Technical Report 56-369, Wright Patterson Air Force Base, OH: Wright Air Development Center.
4. Smode, A.F., Hall, E.R., & Meyer, D.E. (1966). *An assessment of research relevant to pilot training*. Report AMRL-TR-66-196, Wright Patterson Air Force Base, OH: Aeromedical Research Laboratories.
5. Huff, E.M., & Nagel, D.C. (1975). Psychological aspects of aeronautical flight simulation. *American Psychologist*, 426-439.
6. National Research Council (1975). *Visual elements in flight simulation*. Assembly of Behavioral and Social Sciences, Washington, D.C.: National Academy of Sciences.
7. Hennessy, R.T., Sullivan, D.J., & Cooles, H.D. (1980). *Critical research issues and visual system requirements for a V/STOL training research simulator*. Report NAVTRAEQUIPCEN 78-C-0076-1, Orlando, Florida: Naval Equipment Training Center.

8. NATO-AGARD (1980). *Fidelity of simulation for pilot training.* Advisory Group for Aerospace Research and Development, Advisory Report 159, Neuilly sur Seine, France.
9. NATO-AGARD (1981). *Characteristics of flight simulator visual systems.* Advisory Group for Aerospace Research and Development, Advisory Report 159, Neuilly sur Seine, France.
10. Kraft, C.L., Anderson, C.D., & Elsworth, C.L. (1980). *Psychophysical criteria for visual simulation systems.* Report AFHRL-TR-79-30, Williams Air Force Base, AZ: Air Force Human Resources Laboratory.
11. Matheny, W.G. (1975). Investigations of the performance equivalence method for determining training simulator and training methods requirements, AIAA paper 75-108. *AIAA 13th Aerospace Science Meeting*, Pasadena, California, American Institute of Aeronautics and Astronautics.
12. National Research Council (1982). *Automation in combat aircraft.* Air Force Studies Board, Commission on Engineering and Technical Systems, Washington, D.C.: National Academy of Sciences.
13. Thorpe, J.A., Varney, N.C., McFadden, R.W., LeMaster, W.D., & Short, L.H. (1978). *Training effectiveness of three types of visual systems for KC-135 flight simulators.* Report AFHRL-TR-78-16, Williams Air Force Base, AZ: Human Resources Laboratory.
14. Prophet, W.W., Schelnutt, J.B., & Spears, W.D. (1981). *Simulator training requirements and effectiveness study (STRES): Future research needs.* Report AFHRL-TR-80-37, Brooks Air Force Base, TX: Air Force Human Resources Laboratory.
15. Transportation Research Board. (1992). Simulator technology: Analysis of applicability to motor vehicle travel. *Transportation Research Board Circular*, 388, National Research Council.
16. McCluskey, M.R. (1972). *Perspectives on simulation and miniaturization.* CONRAC Training Workshop, Fort Gordon, Georgia.
17. Flexman, R.E., & Stark, E.A. (1988). Training simulators. In G. Salvendy, (Ed.), *Handbook of Human Factors.* New York: John Wiley and Sons, 1012-1039.
18. Hancock, P.A., Caird, J.K., & White, H. (1990). *The use of driving simulation for the assessment, training, and testing of older drivers.* Report HFRL-NIA-90-01.
19. Husni, P. (1990). *Visual simulation white paper.* Mountain View, CA: Silicon Graphics.
20. Sedgewick, H.A. (1991). The effects of viewpoint on the virtual space of pictures. In S.R. Ellis, M.K. Kaiser & A.C. Grunwald (Eds.), *Pictorial communication in virtual and real environments.* (pp. 460-479), New York: Taylor and Francis.
21. Hagen, M.A. (1986). *Varieties of realism: Geometries of representational art.* New York: Cambridge University Press.
22. Hagen, M.A. (1991). How to make a visually realistic 3D display. *Computer Graphics*, **25** (2), 76-81.
23. Haber R.N. (1985). Toward a theory of the perceived spatial layout of scenes. *Computer Vision, Graphics and Image Processing*, **35**, 1-40.
24. Foley, J.D., van Dam, A., Feiner, S.K., & Hughes, J.F. (1990). *Computer Graphics.* Reading, MA: Addisson-Wesley Publishing Co.
25. Gibson, J.J. (1979). *The ecological approach to visual perception.* Boston: Houghton Mifflin.
26. Sedgewick, H.A (1980). The geometry of spatial layout in pictorial representation. In M. Hagen (Ed.), *The Perception of Pictures*, 1, 33-90. Academic Press.
27. Sedgewick, H.A. (1983). Environment-centered representation of spatial layout: Available information from texture and perspective. *Computer Vision, Graphics, and Image Processing.* **31**, 248-260.
28. Sedgewick, H.A. (1986). Space perception. In K. Boff, L. Kaufman, & J.P. Thomas (Eds.), *Handbook of perception and human performance* (pp.21.1 - 21.57). New York: Wiley.

Effective Vehicle Driving Simulation: Lessons from Aviation

Anthony D. Andre

Western Aerospace Labs, Inc., NASA Ames Research Center, Moffett Field, CA 94035-1000 USA

Abstract: Recent advances in computer and display systems technology have made vehicle driving simulators a viable training and evaluation resource for both civilian and military operations. Such computer-aided simulation has unlimited human factors applications for improving the safety, efficiency, and cost of a wide variety of mechanical vehicle systems. However, technological capabilities do not, by themselves, improve the design or training process. For simulators to be effective, users must understand the complexities of the simulation system, exercise control over the simulator's technological capabilities, and determine relevant criteria for simulation fidelity. Drawing on the experiences of the aviation community, this paper focuses on four important issues/applications for vehicle driving simulation: 1) modeling the simulation environment, 2) simulator fidelity and design, 3) automotive design evaluation (displays and controls), and 4) driver education and training.

Keywords: simulator fidelity / vehicle simulation / human factors / vehicle design

Introduction

Recent advances in computer processing and display systems technology have made vehicle driving simulators a viable research, training, and design resource for both civilian and military operations. Such computer-aided simulation has unlimited human factors applications for improving the safety, efficiency, and cost of a wide variety of mechanical vehicle systems. Nevertheless, our lack of knowledge of the complex simulator systems we use often inhibits us from conducting a proper study of the vehicle system or human-vehicle interface in question. Even in the aviation community, which has over 60 years of experience with aircraft simulators, there is no consensus pertaining to simulator effectiveness or design criteria [1]; a fact which underscores the complexity of seemingly straightforward simulation environments.

How can vehicle simulation be used to achieve its goal as an effective, low-cost, and safe tool for research, training, and design applications? It is argued here that simulator effectiveness, for the most part, lies in the hand of the user, who must understand the complexities of the simulation system, exercise control over the simulator's technological capabilities, and determine relevant criteria for simulation fidelity. In this context, and drawing on the experiences of the aviation community, the present paper focuses on four issues/applications for vehicle driving simulation: 1) modeling the simulation environment, 2) simulator fidelity and design criteria, 3) automotive design evaluation (displays and controls), and 4) driver education and testing.

Modeling the Simulation Environment

There is no doubt that good research starts with a thorough understanding of the actual vehicle, the driver, and the environment in which they interact. Understanding the complexities of the human-vehicle system allows the researcher to better specify and control those elements of the complex environment he or she is interested in. Of course, the careful study of complex problems almost always requires some form of reductionism. Accordingly, simulators, when used effectively, are powerful tools for studying complex human-machine behavior with a good deal of experimental control. Yet their effectiveness is often limited because the researcher lacks an understanding (or model) of the relationship between properties of the simulation system and those of the "real world" system they wish to optimize or study through simulation.

Modeling the simulation environment is difficult for two reasons: First, the sheer complexity of the simulation environment, which is often greater than that of the actual vehicle, is novel to the user and therefore difficult to assess off-hand. Second, demand characteristics of the task often interfere, or interact, with the experimenter's manipulations.

Simulators, much like the vehicles they emulate, are complex machines. Hundreds, if not thousands, of parameters related to displays, controls, and vehicle model characteristics can (and do) vary. But not all of these variables are controlled, let alone realized, by the researcher or designer, whose efforts are (at least) usually directed toward understanding and carefully manipulating task-relevant properties of the human-vehicle environment in question. Consider the following situations recently encountered by myself and Dr. Walt Johnson while conducting a full-mission simulation experiment in an advanced rotorcraft simulator with a helmet-mounted display [see Ref. 2].

One part of the study (unpublished) required the pilots to maintain a hover position at a fixed altitude while being disturbed by wind in the vertical axis. This maneuver was performed in the context of two viewing conditions (biocular and binocular) and various levels of scene detail (e.g., patterned and unpatterned ground texture) to assess the effects of visual cues on hover stability. So, what unexpected problems did we encounter in this part of the study? One thing that we didn't realize was that the simulator's collective, the device used for altitude control, was being implemented in a much more sensitive fashion than found in any operating helicopter. The over-sensitivity of the collective made it almost impossible for the pilots to effectively control the imposed disturbance. Thus, we failed to accurately simulate an important part of the real-world system.

Another example, from the same experiment, serves to illustrate how demand characteristics can interfere with experimental manipulations. Here, the task was to maintain a fixed hover position, in the absence of disturbances, near various ground textures and vertical structures (e.g., an open field versus a village). How was this task accomplished when good positional cues were absent? Pilots, in this case, used cues provided by burned-out fiber optic elements in their helmet-mounted display to effectively judge their own position and movement. Pilot comments during the post-mission interviews revealed that the relative motion between the small black holes left by the burned-out elements and features on the ground surface allowed the pilots to accurately detect very small positional variations of the craft.

The lesson here is "don't underestimate the will of the operator." Their goal is to succeed in the task; a goal that is made more desirable by the fun and excitement of operating a virtual vehicle. Accordingly, researchers must be aware that simulator subjects will often use whatever cues are available to perform the task well, even when they know such cues are unintentionally present, and/or are never available in the real world.

Simulator Fidelity and Design

Vehicle simulation, by definition, always represents some degree of abstraction from reality, for the simple reason that a simulator is not a vehicle. How large a degree of abstraction is tolerable depends mainly on the goals and purposes of the simulation. Simulator uses fall into three main categories--research, training, and testing--each having a unique, although not exclusive, set of fidelity criteria.

Research simulators must, by definition, be much more advanced and flexible than either training or testing simulators [3]. They must be capable of simulating a wide variety of visual scenes, displays, and controls, in addition to varying the computational models that connect them [3]. Further, some simulators must be capable of simulating more than one model or type of vehicle (e.g., automobile or aircraft). Given this technological flexibility, it is imperative that simulator fidelity be assessed according to the research questions being asked and the operational arena of generalization.

If, for example, one were to simulate and compare two different automotive speedometer formats (e.g., analog versus digital), it may be unnecessary to invest time, money, and computing resources generating a highly detailed visual scene or using a motion-base simulator. On the other hand, if one wanted to study how drivers maintain vehicle control and stability, then realistic representations of the dynamic visual and motion cues available while driving, as well as the vehicle controls, may be required. Still, other situations require some median level of simulator fidelity. For example, the study of car sickness in a simulator may require employing a realistic visual-motion response model but does not dictate using an actual steering wheel over a joystick (or similar device) for vehicle control. The lesson here is that simulator fidelity should not be measured in absolute terms but rather relative to the critical *information* components of the problem space; a notion echoed recently by Owen and Johnson [3].

Research simulators have unique and varying requirements for their own design as well. Because the researcher must be able to easily manipulate a wide variety of scenarios and collect performance data from each, the design of the simulator's experimenter interface can have a large impact on its effectiveness as a research tool. The ability to specify performance metrics and sample them at various rates, to communicate with subjects during and between trials, to manipulate experimental factors on-line, to record eye movements or other non-verbal behaviors, are but just a few of the research simulator's experimenter-interface requirements. Unfortunately, though, far less attention has been paid to such simulat**or** features as has been paid to some of the simulat**ion** features noted above (e.g., motion platforms, wraparound visual systems, etc.). Notwithstanding, the experimenter interface is often inundated with unnecessary, and often dysfunctional, high technology features much like the simulation testbed. For example, tightly-spaced touch panel controls and multi-function electronic displays do little to aid the experimenter's awareness, or control, of a complex and dynamic simulation.

Training simulators have been widely used and accepted in the aviation community for over 40 years because of their proven ability to enhance the crew's performance in a safe and costly manner [1]. Likewise, they have the potential to increase the safety and efficiency of automotive operations as well (discussed later in this paper). Certainly, fundamental engineering criteria, such as the extent to which the simulator's model reflects the actual vehicle's characteristics, is a more important consideration when a simulator is to be used for training. Yet, surprisingly, there is little evidence that increasing the *physical* similarity between the flight simulator and the actual aircraft produces corresponding increases in skill transfer [2]. For example, several simulation studies which have tested transfer to the aircraft have found little or no training benefit for the addition of a motion system [4]. In fact, some studies have indicated that a bad motion system may be worse than no motion system at all [1].

How can it be that departures from reality result in better transfer of training than high fidelity simulators? Here again, the answer lies in the concept of *information* [see Ref. 3]. Simply stated, if a training device, no matter how abstract, emphasizes the user's active manipulation of "the functionally useful properties in the structure of stimulation" [3]--that is, task-relevant information--high skill transfer to the actual vehicle environment is likely. Simply replicating the physical structure of the vehicle environment is not enough to guarantee high transfer of training. Thus, the former strategy, that of high "information fidelity," allows the researcher to focus on how the driver behaves in comparison to the driver of an actual vehicle. In contrast, the latter strategy, that of high "physical fidelity," limits the focus to what the simulator does in comparison to the actual vehicle.

Testing simulators for type ratings and recertification have become more common in the aviation industry as the technology to duplicate almost every aspect of modern aircraft and the aviation environment (air traffic control, weather, other aircraft) has become available. Here, the argument for exact replication is a defensible one, since we are interested solely in assessing how well the pilot (or driver) can operate a specific aircraft (or car) and only use simulation as a simple surrogate in order to save time and money. Moreover, as will be discussed later in this paper, simulators may provide an even better test instrument than actual vehicles.

Automotive Design Evaluation

Automotive displays and controls play an important role in the comfort, convenience, and safety of the driver. In the past, there was little variance between different makes and models as to the design and placement of automotive displays and controls; most displays were analog in format and located at the top center portion of the dashboard. Today, the same displays come in a variety of formats, including analog, digital, graphical, and auditory. Further, new displays and controls have been introduced that didn't even exist before, such as car phones, electronic navigation maps, and head-up displays (HUDs).

With no human factors standards dictating one format over another, automotive engineers and designers are often forced to make intuitive decisions and/or to conduct time-consuming research regarding display/control implementation. Unfortunately, neither intuitive reasoning nor part-task research can address the full array of contemporary display/control design issues. The reason is simple; many design issues can only be understood by studying the dynamic interactions of the person, vehicle, and environment. Naturally, vehicle driving simulators can provide the platform for such study early in the design process.

To examine this thesis, I refer the reader to a study conducted by Sojourner and Antin [5]. This study compared the effects of simulated head-up display (HUD) and dashboard-mounted digital speedometers on key perceptual driving tasks in a simulated driving environment [5]. Subjects viewed a videotape, taken from a driver's perspective, of a car traveling along a route previously memorized by the subject. In the HUD group, the speed of the car was indicated by a digital readout superimposed on the driving scene, thereby negating the need for the subject to take his/her eyes off the road. In the dashboard group, the speed of the car was indicated by a digital readout displayed on a small monitor located approximately 50-70 cm from the subjects. While viewing the test scene, subjects in both groups performed tasks related to navigation, speed monitoring, and salient cue detection [for more details, see Ref. 5].

Perhaps not surprisingly, the results indicated that the simulated HUD speedometer produced generally superior performance on the experimental tasks. So, what's wrong with this study? Several things, many, if not all, of which could have been eliminated via human-in-the-loop simulation. First, let's consider the task of speed monitoring as it exists in the actual driving environment. Although subjects in this study had only the speedometer displays to monitor speed, a number of other cues, such as engine pitch, optical flow of the visual scene, flow rate of passing ground texture, and vehicle kinematics, all help to specify our current speed while driving in the real world. Another missing cue, and perhaps the most important of them all, is the driver's active control of speed via the accelerator. Surely, there is no better cue to speed changes than the inputs of the driver. In fact, the speedometer is not a really speed cue--in the true sense--but rather a display that is attended to, on occasion, with the purpose of confirmation. Thus, it's no wonder that subjects, who had to rely solely on these displays for speed information, performed better when the display was projected onto the driving scene (i.e., with the HUD).

Clearly, then, the tasks of speed monitoring, navigation, and salient cue detection would take on new meaning in the context of human-in-the-loop vehicle simulation: natural speed cues would avail themselves, thereby reducing the subjects' over-reliance on the speedometer; attention would be directed to other components of the driving task (e.g., manual control, scanning other instruments); and various driving conditions (e.g., darkness, poor weather, etc.) would be compared.

How would these results have changed if the two display formats were evaluated using human-in-the-loop simulation? It's hard to say, for sure. But by preserving the ecological fidelity of human-vehicle-environment interaction (i.e., through simulation), we can better evaluate the effects of various display formats and be more confident that our findings will generalize to the actual vehicle system in question. Indeed, such research is needed before we can conclude that HUDs, and other modern vehicle amenities, are efficient and/or safe on the road.

Driver Education and Training

While simulators offer higher levels of information and physical fidelity for design evaluation, they may prove to be most beneficial for driver education and training. Driver education, at least in the United States, consists of two main components: 1) classroom instruction of the "rules of the road" and other safety and efficiency facts, and 2) behind-the-wheel experience driving an actual car. The former emphasizes declarative knowledge (knowing what to do), while the latter emphasizes procedural knowledge (knowing how to do it). In theory, what is learned in the classroom should transfer to actual driving conditions. But this is not always the case in reality, primarily because the amount of behind-the-wheel instruction time students receive is relatively small and is spent mainly on basic maneuvers (e.g., turning, parallel parking) and driving in "normal" conditions. The result is that students often do not have a good understanding of many (if not most) of the dynamic properties within the driving environment. In addition, they get little practice driving in challenging conditions for obvious reasons of safety and liability.

Simulators may provide an even better training instrument than the actual vehicles when they are used to allow aspiring drivers to develop and practice their skills in challenging conditions, such as at night, during poor weather, or when navigating an unfamiliar city. They can also be used to collect data on driver behaviors including eye scanning patterns, decision making, and risk taking; to train defensive driving skills in a safe, yet realistic, environment; and to assess the effects of age on driving skill and safety.

Imagine the effects on a student, learning what it's like to have to make a sudden stop on wet pavement with poor brakes; and then encountering the same situation with better brakes, or at a safer speed. These are lessons that cannot be learned through rote memorization--but can be experienced, as well as understood, in a vehicle driving simulator.

Summary

Vehicle simulators have many applications for improving the safety and efficiency of the driving environment. However, if we are to use simulators effectively, we must gain an understanding of the complexity of, and relationship between, vehicle simulation systems and their real-world counterparts. With this understanding, many important facets of the vehicle design and driver education/training process can, and will be, improved upon in the near future.

Acknowledgments: The author was supported by Cooperative Agreement No. NCC 2-486 from the NASA Ames Research Center. Sandra Hart was the technical monitor. Special thanks to Dr. Peter Hancock for his leadership on this project, and to Dr. Walt Johnson for his valuable comments and suggestions.

References

1. S. N. Roscoe, "Simulator qualification: Just as phony as it can be," *International Journal of Aviation Psychology*, Vol. 1(4), pp. 335-339, 1991.
2. A.D. Andre and W.W. Johnson, "Stereo effectiveness evaluation for precision hover tasks in a helmet-mounted display simulator," *In Proceedings of the 1992 IEEE International Symposium on Systems, Man and Cybernetics,* Chicago, IL: IEEE, in press.
3. D.H. Owen and W.W. Johnson, "An information-based approach to simulation research," *In R.A. Feik (Ed.), Proceedings of the Future Directions In Simulation Symposium Workshop*, Melbourne, Australia: DSTO, Aeronautical Research Laboratory, 1992.
4. W.L. Waag, "*Training effectiveness of visual and motion simulation*," (Tech. Rep. No. AFHRL-TR-79-72). Brooks Air Force Base, TX: Air Force Human Resources Laboratory, 1981.
5. R.J. Sojourner and J.F. Antin, "The effects of a simulated head-up display speedometer on perceptual task performance," *Human Factors*, Vol. 32(3), pp. 329-339, 1990.

Active Psychophysics: A Psychophysical Program for Closed-Loop Systems

John M. Flach

Wright State University, Psychology Department, Dayton, OH 45435, USA; Armstrong Laboratory, Wright-Patterson, AFB, Dayton, OH 45433, USA

Abstract: Concurrent Engineering requires a creative blend of basic research with design. This is particularly true for human-machine systems. Simulators provide both an opportunity and a challenge for basic research on human performance. Simulators provide an opportunity where researchers can manipulate goals, dynamics, and information in a way that will allow direct generalizations to operational environments. Simulators provide a challenge where human performance researchers must provide information with regard to critical design decisions --- How much realism is necessary? How can the simulator be used most effectively in training?

Keywords: human-machine systems / human-in-the-loop simulation / psychophysics / human performance

Introduction

A fundamental role of the human component in complex systems (e.g., vehicular control, air traffic control, or process control) is to "close-the-loop." That is, the human is included in the system because of the unique and adaptive abilities of perception, decision making, and motor control. Although there have been great advances in automated control systems, the adaptability and generality of the human have yet to be matched by automated sensing and control systems. This generality and adaptability of the human controller that makes her attractive (if not essential in many cases) as a component within complex systems poses a great challenge to basic researchers interested in modeling human performance, as well as for system designers who need to be able to integrate across human and electro-mechanical components to predict and evaluate system performance. In this chapter, I will briefly speculate on why basic research has failed to provide models of human performance that are adequate to the needs of system designers. I will then offer a framework for a basic research program that may prove more useful for characterizing the human's role in "closing-the-loop" in complex systems. An important component within this framework will be the use of simulation systems.

Traditionally, psychology has parsed the problem of human performance into problems of perception, cognition, and motor control. Research programs have evolved that focus on one or the other of these components in isolation from the others. For example, those who focus on perception often tightly control action (e.g., fixating the head with bite boards, using brief, tachistoscopic stimulus presentations, or using restricted response protocols such as key presses). Those who study motor control go to great lengths (e.g., deafferentation) to isolate motor control from perception. And those who study cognition select problems (e.g., tower of Hanoi, missionaries and cannibals, logic theorems) with minimal perceptual and motor demands.

Such strategies for studying perception and action have been successful in reducing complexity and allowing scientists to make inferences about the elementary cognitive processes that combine to control performance. However, these approaches miss the emergent properties that arise from the coupling of perception and action. Without an understanding of these emergent properties it may not be possible to integrate what we have learned about perception systems and action systems in isolation into a comprehensive and general theory of human performance. Thus, it is important not to ignore these emergent properties of the closed-loop, perception-action system. An *active psychophysics* [1,2,3] is needed to compliment the traditional work on passive psychophysics, perception, and motor performance.

Closed-Loop Systems Framework

One of the difficulties in moving from an approach to studying humans as isolated, open-loop systems to studying them as closed-loop systems is that in order to do so psychologists must relinquish control of the stimulus to their subjects. In traditional approaches, stimuli are under control of the experimenter. These stimuli are presented to the "subject" who makes an appropriate response. The response is scored, but it generally has no impact on the stimulus. If feedback is provided at all it's generally in terms of percent correct. An alternative, closed-loop systems approach is illustrated in Figure 1.

In a closed-loop system the input to the sensors (i.e., perceptual system) changes as a consequence of actions by the effector system. Thus, whenever the human is studied in a closed-loop context, the stimulus is under the subject's control. In these contexts, rather than stimulation, the experimenter manipulates and constrains goals, dynamics and information as independent variables. Experimenters will also provide disturbances (e.g., wind gusts).

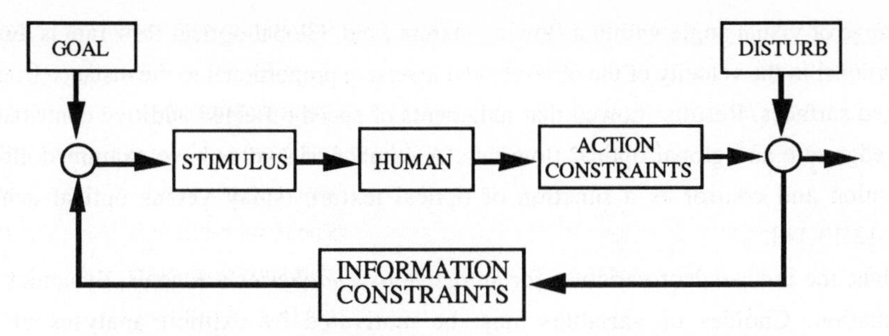

Figure 1. The experimental context seen as a closed-loop system.

However, although disturbances have interesting properties of their own, disturbances will primarily have the role of catalysts in this paradigm.

Independent Variables: Goals, Dynamics, and Information

Manipulation of goals defines functionality for a given task. Goals can be provided explicitly or implicitly. For example, the experimenter may ask the subject to maintain a particular track or speed (explicit). On the other hand, the experimenter may implicitly create goals by introducing consequences for actions within the task environment. For example, introducing obstacles to define a safe path of travel. Any aspects of the task environment that has consequences for the subject are part of the goal structure.

Manipulation of dynamics defines the action coupling between the actor and the environment—- for example, specification of the control law for the subject's "vehicle." The world is a very different place, in terms of safe fields of travel and potential landing sites, for a helicopter and a fixed wing aircraft. Flach, Hagen, and Larish [4] discuss how the dynamics of hovering versus straight and level flight influence the pick-up of information for the control of altitude. For laboratory tracking tasks, McRuer's Crossover Model is an excellent illustration of how performance adapts to changing constraints on dynamics [5,6]. See Flach [1] for a discussion of the Crossover Model in the context of active psychophysics.

Manipulation of information defines the perceptual coupling between the actor and the environment --- for example, manipulating the optical texture available. Denton [7] painted stripes on roadways so that they got progressively closer together as they neared traffic circles. This produced an increasing edge rate (specifying acceleration). The result was a reduction in approach speeds and a reduction in accidents. A most important source of information for the control of locomotion, whether walking, driving a car, or piloting an aircraft, is optical structure (i.e., invariants) within the dynamic optical flow field [8,9]. For example, Larish and Flach [10] compared edge rate with global optical flow rate as a source of information for the speed of self-motion. Global optical flow rate is an index of the global rate of change of visual angle within a flowing texture field. Global optical flow rate is directly proportional to the velocity of the observer and inversely proportional to the distance from the textured surfaces. Results showed that judgments of speed reflected additive contributions from edge rate and global optical flow rate. A number of studies have examined altitude perception and control as a function of optical texture (splay versus optical density) [4,11,12,13,14].

Thus, the independent variables for an *active psychophysics* are goals, dynamics, and information. Choices of variables must be motivated by explicit analyses of task environments. If research is to generalize to problems of driving or of aviation, then the goals, dynamics, and information variables studied in the laboratory must reflect levels found in the natural task environments. The problem of dynamics poses an important challenge. Simulating the dynamics of an actual vehicle requires precise engineering. As with any control system, time delays play a critical role. In the laboratory, computational and transport delays are formidable obstacles to providing realistic dynamics [15,16].

It is also difficult to replicate the information found in natural environments. First there is the problem of understanding the many sources of information. What are the structures within optical flow fields that are essential to the perception of self-motion? Caird [17] discusses some possibly critical aspects of visual information. How important is felt motion (e.g., vestibular and tactile sources of information)? Second, there is the problem of reproducing the information within each modality. Finally, there's the problem of coordinating across the various modalities of stimulation. Coordinating visual displays and motion platforms and maintaining realistic system response (e.g., time delays) poses a tremendous challenge to engineering science. To the extent that engineers create the technology to meet this challenge, a tremendous opportunity is created for basic research in human performance. Without such technology, it will be difficult for a behavior science to address the many human factors concerns posed by the problem of designing productive and safe human-machine systems [18].

Dependent Variables: Responses, Information Consequences, and Goals

Dependent measures, for active psychophysics, will not be exclusively responses. But to the extent possible we will want to measure the input and output for each of the boxes within Figure 1. We will want to know not simply what the actor does in terms of motor responses, but we will want to know the information consequences as well as the environmental (goal relevant) consequences of those actions. As Powers [19] has claimed, "behavior is the control of perception." Gibson [20] illustrates this nicely in the context of the visual control of locomotion. He writes, "an animal who is behaving in these ways is optically stimulated in the corresponding ways, or, equally, an animal *who so acts as to obtain these kinds of stimulation is behaving in the corresponding way*" [20]. Thus, stimulation becomes a dependent variable. For example, we might ask a pilot to maintain an aircraft at a specified altitude and then measure how he *makes the world look*. Does he *act so as to* maintain global optical flow rate constant? Does he maintain a constant splay angle; a constant optical density?

We are also interested in the converse relations of the consequences of information on action. For example, a driver may be instructed to drive at a particular speed. As experimenters, we can manipulate the information for speed by changing the ground texture (as Denton did) or by changing the distance to textured surfaces (this affects the global optical flow rate). Does the driver's control actions correlate with these manipulations of information?

Finally, the experimenter may be interested in goals as dependent variables. For example, the subject is given the task of driving at a safe speed or of following another vehicle at a safe distance. What speed or distance does the subject choose? Here we can ask questions about preferred levels of risk.

The relations across these measures (information, action, consequences) will be critical to understanding human performance. Because of the closed-loop structure, no single measure will be adequate. Behaviors will reflect trade-offs between information generated (exploratory activity) and the need to accomplish task goals (performatory activity). Because of this closed-loop structure, psychology will have to depend more on control theory for the logic of experiments and inferences. This logic has been used quite successfully in the area of vehicular control [21,22,23]. However, we are recommending it as a more general experimental paradigm for a wide range of perceptual and cognitive problems. It is important to keep in mind, however, that we are not recommending the servomechanism as a theoretical metaphor. Rather, we are recommending the tools of control theory for designing experiments and for guiding the logic of measurement and inference. For example, analysis in the frequency domain using the Bode space provides an important tool for relating input and output to infer a transfer function for a particular process. See Flach [1] for a more extended discussion of the use of control theory within active psychophysics.

Summary

Designers and engineers are often disappointed and frustrated when they seek answers from the behavioral science community to questions such as how realistic must the visual displays in our simulation be or do we need a motion base . The answers either don't exist or are so heavily qualified and laden with contingencies that they often raise more questions than they answer. Research has been tightly constrained in its abilities to manipulate and control variables that effect human performance. One response to these constraints has been the use of reductionistic strategies that decouple perception, action, and cognition as domains of study. This strategy has not been totally unsuccessful. However, there is some reason to doubt whether the data generated by such an approach can be integrated so as to address human factors concerns in human-machine systems [24,25].

Active psychophysics is an alternative framework that is directed at the emergent properties of perception-action systems. Research within this framework will also be reductionistic. However, the dimensions along which the problem is reduced will be goal, information, and action constraints, rather than perception, cognition, and motor control.

Fortunately, the promise for solutions is closely linked to a source of the concern -- simulations. Engineering advances in simulation technology (e.g., high resolution, real-time graphic displays) provide exactly the tools needed to conduct research within the active psychophysics framework. Simulation systems are ideal for controlling information, dynamics, and goals. Thus, there is the possibility for a concurrent and adaptive relationship between the engineering and human performance disciplines that will lead to the mutual enrichment of both fields. As a human performance researcher, I believe that the opportunities opened up by simulation technology may foreshadow an exciting paradigm shift within human experimental psychology. The result will be an experimental psychology whose implications for design of safe and productive human-machine systems will be direct an obvious, rather than post hoc an heavily qualified.

Acknowledgements: Thanks to Peter A. Hancock for organizing the human factors contribution to the Advanced Study Institute and inviting me to be a participant. Also, thanks to Ed Haug and his staff who created an ideal atmosphere for a productive exchange of ideas. During the preparation of this manuscript John Flach was partially supported by the Air Force Office of Scientific Research (AFOSR-91-0151).

References

1. Flach, J.M. (1990). Control with an eye for perception: Precursors to an active psychophysics. *Ecological Psychology*, **2**, 83-111.
2. Warren, R. (1988a). Active psychophysics: Theory and practice. In H.K. Ross (Ed.), *Fechner Day '88* (Proceedings of the 4th Annual Meeting of the International Society for Psychophysics) (pp. 47-52). Stirling, Scotland.
3. Warren, R. & McMillan, G. (1984). Altitude control using action-demanding interactive displays: Toward an active psychophysics. *Proceedings of the 1984 IMAGE III Conference*. (pp. 405-415). Phoenix, AZ: Air Force Human Resources Laboratory.
4. Flach, J.M., Hagen, B.A., & Larish, J.F. (In press). Active regulation of altitude as a function of optical texture. *Perception & Psychophysics*.
5. McRuer, D.T. & Jex, H.R. (1967). A review of quasi-linear pilot models. *IEEE Transactions on Human Factors in Electronics*, **HFE-8**(3), 231-249.
6. McRuer, D.T. & Weir, D.H. (1969). Theory of manual vehicular control. *Ergonomics*, **12**, 599-633.
7. Denton, G.G. (1980). The influence of visual pattern on perceived speed. *Perception*, **9**, 393-402.

8. Gibson, J.J. (1966). *The senses considered as perceptual systems.* Boston: Houghton-Mifflin.
9. Gibson, J.J. (1979/1986).. *The ecological approach to visual perception.* Hillsdale, NJ: Erlbaum.
10. Larish, J.F. & Flach, J.M. (1990). Sources of optical information useful for perception of speed of rectiinear self-motion. *Journal of Experimental Psychology: Human Perception and Performance,* **16**, 295-302.
11. Johnson, W.W., Tsang, P.S., Bennett, C.T. & Phatak, A.V. (1989). The visually guided control of simulated altitude. *Aviation, Space, and Environmental Medicine,* **60**, 152-156.
12. Warren, R. (1988b). Visual perception in high-speed low altitude flight. *Aviation, Space, and Environmental Medicine,* **59**, (11, Suppl.), A116-A124.
13. Wolpert, L. (1988). The active control of altitude over differing texture. *Proceedings of the 32nd Annual Meeting of the Human Factors Society* (pp. 15-19). Santa Monica, CA: Human Factors Society.
14. Wolpert, L. & Owen, D. (1985). *Proceedings of the 3rd Symposium on Aviation Psychology* (pp. 475-481). Columbus, OH: Ohio State University.
15. Middendorf, M.S., Fiorita, A.L. & McMillan, G.R. (1991). The effects of simulator transport delay on performance, workload, and control activity during low-level flight. *Proceedings of the AIAA Flight Simulation Technologies Conference.* Washington, DC: American Institute of Aeronautics and Astronautics.
16. Riccio, G., McMillan, G., Lusk, S., Gowran, B. & Bailey, R. (In preparation). *Temporal fidelity in flight simulators: Design guide.* Wright-Patterson AFB, OH: Armstrong Laboratory.
17. Caird, J. (This volume).
18. Hancock, P.A. (This volume). Evaluating in-vehicle collision avoidance warning systems for IVHS.
19. Powers, W.T. (1978). Quantitative analysis of purposive systems. Some spadework at the foundations of scientific psychology. *Psychological Review,* **85**, 417-435.
20. Gibson, J.J. (1958). Visually controlled locomotion and visual orientation in animals. *British Journal of Psychology,* **49**, 182-194. Also in E. Reed & R. Jones (Eds.) (1982). *Reasons for realism.* Hillsdale, NJ: Erlbaum. (148-163).
21. Wickens, C.D. (1984). *Engineering psychology and human performance.* Columbus, OH: Merrill.
22. Wickens, C.D. (1985). The effects of control dynamics on performance. In K.R. Boff, L. Kaufman, & J.P. Thomas (Eds.). *Handbook of perception and human performance,* Vol. II (pp. 39.1 - 39.60). New York: Wiley.
23. Sheridan, T.B. & Ferrell, W.R. (1974). *Man-machine systems.* Cambridge, MA: MIT Press.
24. Flach, J.M. (1989). An ecological alternative to eggsucking. *Human Factors Society Bulletin,* **32**(9), 4-6.
25. Vicente, K.J. (1991). A few implications of an ecological approach to human factors. *Human Factors Society Bulletin,* **33**(11), 1-4.

Keyword Index

Printing: Druckhaus Beltz, Hemsbach
Binding: Buchbinderei Schäffer, Grünstadt

NATO ASI Series F

Including Special Programmes on Sensory Systems for Robotic Control (ROB) and on Advanced Educational Technology (AET)

Vol. 40: Theoretical Foundations of Computer Graphics and CAD. Edited by R. A. Earnshaw. XX, 1246 pages. 1988.

Vol. 41: Neural Computers. Edited by R. Eckmiller and Ch. v. d. Malsburg. XIII, 566 pages. 1988.

Vol. 42: Real-Time Object Measurement and Classification. Edited by A. K. Jain. VIII, 407 pages. 1988. *(ROB)*

Vol. 43: Sensors and Sensory Systems for Advanced Robots. Edited by P. Dario. XI, 597 pages. 1988. *(ROB)*

Vol. 44: Signal Processing and Pattern Recognition in Nondestructive Evaluation of Materials. Edited by C. H. Chen. VIII, 344 pages. 1988. *(ROB)*

Vol. 45: Syntactic and Structural Pattern Recognition. Edited by G. Ferraté, T. Pavlidis, A. Sanfeliu and H. Bunke. XVI, 467 pages. 1988. *(ROB)*

Vol. 46: Recent Advances in Speech Understanding and Dialog Systems. Edited by H. Niemann, M. Lang and G. Sagerer. X, 521 pages. 1988.

Vol. 47: Advanced Computing Concepts and Techniques in Control Engineering. Edited by M. J. Denham and A. J. Laub. XI, 518 pages. 1988.

Vol. 48: Mathematical Models for Decision Support. Edited by G. Mitra. IX, 762 pages. 1988.

Vol. 49: Computer Integrated Manufacturing. Edited by I. B. Turksen. VIII, 568 pages. 1988.

Vol. 50: CAD Based Programming for Sensory Robots. Edited by B. Ravani. IX, 565 pages. 1988. *(ROB)*

Vol. 51: Algorithms and Model Formulations in Mathematical Programming. Edited by S. W. Wallace. IX, 190 pages. 1989.

Vol. 52: Sensor Devices and Systems for Robotics. Edited by A. Casals. IX, 362 pages. 1989. *(ROB)*

Vol. 53: Advanced Information Technologies for Industrial Material Flow Systems. Edited by S. Y. Nof and C. L. Moodie. IX, 710 pages. 1989.

Vol. 54: A Reappraisal of the Efficiency of Financial Markets. Edited by R. M. C. Guimarães, B. G. Kingsman and S. J. Taylor. X, 804 pages. 1989.

Vol. 55: Constructive Methods in Computing Science. Edited by M. Broy. VII, 478 pages. 1989.

Vol. 56: Multiple Criteria Decision Making and Risk Analysis Using Microcomputers. Edited by B. Karpak and S. Zionts. VII, 399 pages. 1989.

Vol. 57: Kinematics and Dynamic Issues in Sensor Based Control. Edited by G. E. Taylor. XI, 456 pages. 1990. *(ROB)*

Vol. 58: Highly Redundant Sensing in Robotic Systems. Edited by J. T. Tou and J. G. Balchen. X, 322 pages. 1990. *(ROB)*

Vol. 59: Superconducting Electronics. Edited by H. Weinstock and M. Nisenoff. X, 441 pages. 1989.

NATO ASI Series F

Including Special Programmes on Sensory Systems for Robotic Control (ROB) and on Advanced Educational Technology (AET)

Vol. 60: 3D Imaging in Medicine. Algorithms, Systems, Applications. Edited by K. H. Höhne, H. Fuchs and S. M. Pizer. IX, 460 pages. 1990.

Vol. 61: Knowledge, Data and Computer-Assisted Decisions. Edited by M. Schader and W. Gaul. VIII, 421 pages. 1990.

Vol. 62: Supercomputing. Edited by J. S. Kowalik. X, 425 pages. 1990.

Vol. 63: Traditional and Non-Traditional Robotic Sensors. Edited by T. C. Henderson. VIII, 468 pages. 1990. *(ROB)*Vol. 64: Sensory Robotics for the Handling of Limp Materials. Edited by P. M. Taylor. IX, 343 pages. 1990. *(ROB)*

Vol. 65: Mapping and Spatial Modelling for Navigation. Edited by L. F. Pau. VIII, 357 pages. 1990. *(ROB)*

Vol. 66: Sensor-Based Robots: Algorithms and Architectures. Edited by C. S. G. Lee. X, 285 pages. 1991. *(ROB)*

Vol. 67: Designing Hypermedia for Learning. Edited by D. H. Jonassen and H. Mandl. XXV, 457 pages. 1990. *(AET)*

Vol. 68: Neurocomputing. Algorithms, Architectures and Applications. Edited by F. Fogelman Soulié and J. Hérault. XI, 455 pages. 1990.

Vol. 69: Real-Time Integration Methods for Mechanical System Simulation. Edited by E. J. Haug and R. C. Deyo. VIII, 352 pages. 1991.

Vol. 70: Numerical Linear Algebra, Digital Signal Processing and Parallel Algorithms. Edited by G. H. Golub and P. Van Dooren. XIII, 729 pages. 1991.

Vol. 71: Expert Systems and Robotics. Edited by T. Jordanides and B. Torby. XII, 744 pages. 1991.

Vol. 72: High-Capacity Local and Metropolitan Area Networks. Architecture and Performance Issues. Edited by G. Pujolle. X, 536 pages. 1991.

Vol. 73: Automation and Systems Issues in Air Traffic Control. Edited by J. A. Wise, V. D. Hopkin and M. L. Smith. XIX, 594 pages. 1991.

Vol. 74: Picture Archiving and Communication Systems (PACS) in Medicine. Edited by H. K. Huang, O. Ratib, A. R. Bakker and G. Witte. XI, 438 pages. 1991.

Vol. 75: Speech Recognition and Understanding. Recent Advances, Trends and Applications. Edited by P. Laface and Renato De Mori. XI, 559 pages. 1991.

Vol. 76: Multimedia Interface Design in Education. Edited by A. D. N. Edwards and S. Holland. XIV, 216 pages. 1992. *(AET)*

Vol. 77: Computer Algorithms for Solving Linear Algebraic Equations. The State of the Art. Edited by E. Spedicato. VIII, 352 pages. 1991.

Vol. 78: Integrating Advanced Technology into Technology Education. Edited by M. Hacker, A. Gordon and M. de Vries. VIII, 185 pages. 1991. *(AET)*

Vol. 79: Logic, Algebra, and Computation. Edited by F. L. Bauer. VII, 485 pages. 1991.

Vol. 80: Intelligent Tutoring Systems for Foreign Language Learning. Edited by M. L. Swartz and M. Yazdani. IX, 347 pages. 1992. *(AET)*

NATO ASI Series F

Including Special Programmes on Sensory Systems for Robotic Control (ROB) and on Advanced Educational Technology (AET)

Vol. 81: Cognitive Tools for Learning. Edited by P. A. M. Kommers, D. H. Jonassen and J. T. Mayes. X, 278 pages. 1992. *(AET)*

Vol. 82: Combinatorial Optimization. New Frontiers in Theory and Practice. Edited by M. Akgül, H. W. Hamacher and S. Tüfekçi. XI, 334 pages. 1992.

Vol. 83: Active Perception and Robot Vision. Edited by A. K. Sood and H. Wechsler. IX, 756 pages. 1992.

Vol. 84: Computer-Based Learning Environments and Problem Solving. Edited by E. De Corte, M. C. Linn, H. Mandl and L. Verschaffel. XVI, 488 pages. 1992. *(AET)*

Vol. 85: Adaptive Learning Environments. Foundations and Frontiers. Edited by M. Jones and P. H. Winne. VIII, 408 pages. 1992. *(AET)*

Vol. 86: Intelligent Learning Environments and Knowledge Acquisition in Physics. Edited by A. Tiberghien and H. Mandl. VIII, 285 pages. 1992. *(AET)*

Vol. 87: Cognitive Modelling and Interactive Environments. With demo diskettes (Apple and IBM compatible). Edited by F. L. Engel, D. G. Bouwhuis, T. Bösser and G. d'Ydewalle. IX, 311 pages. 1992. *(AET)*

Vol. 88: Programming and Mathematical Method. Edited by M. Broy. VIII, 428 pages. 1992.

Vol. 89: Mathematical Problem Solving and New Information Technologies. Edited by J. P. Ponte, J. F. Matos, J. M. Matos and D. Fernandes. XV, 346 pages. 1992. *(AET)*

Vol. 90: Collaborative Learning Through Computer Conferencing. Edited by A. R. Kaye. X, 260 pages. 1992. *(AET)*

Vol. 91: New Directions for Intelligent Tutoring Systems. Edited by E. Costa. X, 296 pages. 1992. *(AET)*

Vol. 92: Hypermedia Courseware: Structures of Communication and Intelligent Help. Edited by A. Oliveira. X, 241 pages. 1992. *(AET)*

Vol. 93: Interactive Multimedia Learning Environments. Human Factors and Technical Considerations on Design Issues. Edited by M. Giardina. VIII, 254 pages. 1992. *(AET)*

Vol. 94: Logic and Algebra of Specification. Edited by F. L. Bauer, W. Brauer and H. Schwichtenberg. 1993.

Vol. 95: Comprehensive Systems Design: A New Educational Technology. Edited by C. M. Reigeluth, B. H. Banathy and J. R. Olson. IX, 437 pages. 1993. *(AET)*

Vol. 96: New Directions in Educational Technology. Edited by E. Scanlon and T. O'Shea. VIII, 251 pages. 1992. *(AET)*

Vol. 97: Advanced Models of Cognition for Medical Training and Practice. Edited by D. A. Evans and V. L. Patel. XI, 372 pages. 1992. *(AET)*

Vol. 98: Medical Images: Formation, Handling and Evaluation. Edited by A. E. Todd-Pokropek and M. A. Viergever. IX, 700 pages. 1992.

Vol. 99: Multisensor Fusion for Computer Vision. Edited by J. K. Aggarwal. XI, 456 pages. 1993. *(ROB)*

NATO ASI Series F

Including Special Programmes on Sensory Systems for Robotic Control (ROB) and on Advanced Educational Technology (AET)

Vol. 100: Communication from an Artificial Intelligence Perspective. Theoretical and Applied Issues. Edited by A. Ortony, J. Slack and O. Stock. XII, 260 pages. 1992.

Vol. 101: Recent Developments in Decision Support Systems. Edited by C. W. Holsapple and A. B. Whinston. XI, 618 pages. 1993.

Vol. 102: Robots and Biological Systems: Towards a New Bionics? Edited by P. Dario, G. Sandini and P. Aebischer. XII, 786 pages. 1993.

Vol. 103: Parallel Computing on Distributed Memory Multiprocessors. Edited by F. Özgüner, F. Erçal. VIII, 332 pages. 1993.

Vol. 104: Instructional Models in Computer-Based Learning Environments. Edited by S. Dijkstra, H. P. M. Krammer, and J. J. G. van Merriënboer. X, 510 pages. 1993. *(AET)*

Vol. 105: Designing Environments for Constructive Learning. Edited by T. M. Duffy, J. Lowyck, and D. H. Jonassen. VIII, 374 pages. 1993. *(AET)*

Vol. 106: Software for Parallel Computation. Edited by J. S. Kowalik, L. Grandinetti. IX, 363 pages. 1993.

Vol. 107: Advanced Educational Technologies for Mathematics and Science. Edited by D. L. Ferguson. XII, 749 pages. 1993. *(AET)*

Vol. 108: Concurrent Engineering: Tools and Technologies for Mechanical System Design. Edited by E. J. Haug. XIII, 998 pages. 1993.

Vol. 109: Advanced Educational Technology in Technology Education. Edited by A. Gordon, M. Hacker, and M. de Vries. VIII, 270 pages. 1993. *(AET)*

Vol. 110: Verification and Validation of Complex Systems: Human Factors Issues. Edited by J. A. Wise, V. D. Hopkin, and P. Stager. XV, 711 pages. 1993.

Vol. 111: Cognitive Models and Intelligent Environments for Learning Programming. Edited by E. Lemut, B. du Boulay, and G. Dettori. 1993. *(AET)*